FEATURES OF DISCOVERBIOLOGY.COM

THIRD EDITION

The *Discover Biology* Student Web Site and CD

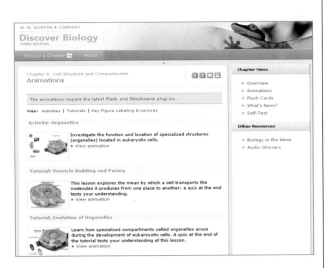

◉ Review Materials feature **chapter overviews, Key Term flash cards**, and **multiple-choice quizzes** to help students enhance their understanding.

◉ In-depth **animated tutorials** offer step-by-step explanations of key concepts.

◉ In addition to tutorials, **activities such as drag-and-drop labeling exercises, simulations of models and experiments, and What's News? exercises** are provided for each chapter.

◉ **Norton Gradebook-enabled quizzes** give students an opportunity to assess their understanding. Our new online Gradebook makes it easy for instructors to collect scores from Web site quiz assignments; a student version of the Gradebook lets students track their own quiz scores.

Additional highlights of DiscoverBiology.com

◉ **Focus On** sections offer helpful study advice for each chapter.

◉ **Biology in the News** weekly updates highlight relevant news stories and encourage students to apply what they learn in class to their daily lives.

◉ A **Glossary** provides definitions of all important terms in the text, with audio pronunciations for over 100 difficult words.

STUDENT WEB SITE AND CD

Icons throughout _Discover Biology_ indicate topics that are also covered as activities or tutorials on the student Web site/CD. Bold entries in the list below indicate an Animated Tutorial.

Media and Supplements Overview

Instructor Supplements

Norton Media Library

This multi-disc Instructor CD set features multimedia resources to enhance lectures and facilitate course planning and assessment. Contents include:

- All of the drawn art from the text, with and without labels, in JPEG and PowerPoint formats
- Many of the photographs from the text
- Supplemental photos not in the text
- Animations and activities from the student Web site, ready to use in lectures

Norton Resource Library

The Norton Resource Library provides comprehensive instructor resources in one centralized online location. In the library, instructors can download ready-to-use, one-stop solutions for online courses, such as WebCT e-Packs and BlackBoard Course Cartridges, or they can tailor these premade course packs to suit their own needs. This library's exceptional resources include:

- Lecture PowerPoints for every chapter, including in-class quizzes ideally suited for "clicker" polling questions
- Art PowerPoint slides, complete with artwork from the text
- Animations from the student Web site
- A Media Guide to the electronic resources available from W. W. Norton
- Chapter overviews, summaries, and outlines
- Test Bank in WebCT, BlackBoard, and ExamView formats
- Glossary

WebCT e-Pack

Available in the Norton Resource Library, the WebCT e-Pack includes summaries, PowerPoint lecture outlines, animations, Key Term flash cards, the Test Bank, and a glossary.

BlackBoard Course Cartridge

Available in the Norton Resource Library, the BlackBoard Course Cartridge includes summaries, PowerPoint lecture outlines, animations, Key Term flash cards, the Test Bank, and a glossary.

Norton Gradebook—wwnorton.com/web/gradebook

Norton Gradebook is an online resource that allows instructors and students to store and track their online quiz results. The results from each quiz students take on the student Web site are uploaded to the password-protected Gradebook, where instructors can access and sort them by section, book, chapter, student name, and date. Students can access the Norton Gradebook to review their personal results. Results can easily be downloaded to users' desktops. Registration for the Norton Gradebook is instant and no setup is required.

Test Bank/Computerized Test-Item File

by Richard Morel (Signal Mountain, TN) and Susan Weinstein (Marshall University)

Thoroughly revised and expanded for the third edition, the Test Bank contains multiple-choice, fill-in-the-blank, and true-false questions, including factual recall and conceptual questions. Available in both print and ExamView formats.

Overhead Transparencies

The *Discover Biology* transparency set includes all of the drawn art from the text in a relabeled and resized format that is optimal for projection and produces excellent image quality.

Biology Video Library

This extensive video library, available to qualified adopters of *Discover Biology*, Third Edition, includes VHS, DVD, and streaming video for each unit. Choose from short video clip collections to full-length films on topics linked to the text and your course.

Student Supplements

Student Web Site/CD

The student Web site (DiscoverBiology.com) offers a complete study plan that will improve students' biological literacy and help them get the maximum benefit from their reading.

- **Biology in the News** weekly updates highlight relevant news stories and encourage students to apply what they learn in class to their daily lives.

- **Animations and activities** include mini-lectures that offer step-by-step explanations of key concepts, drag-and-drop labeling exercises, simulated experiments, and What's News? exercises.

- **Chapter overviews** summarize key concepts and prompt students to explore animations and activities that will enhance their understanding.

- **Multiple-choice quizzes** provide students with an opportunity to assess their understanding. Connection of the quizzes with the **Norton Gradebook** allows instructors to collect scores easily and lets students track their own progress.

- **Key Term flash cards** test students' recall for new terminology.

- A **Glossary** provides definitions of all important terms in the text, with audio pronunciations for over 100 difficult words.

    A CD version of DiscoverBiology.com can be packaged with the text.

Study Guide

by David Demers (Sacred Heart University), Ed Dzialowski (University of North Texas), and Stephen Lebsack (Lynn-Benton Community College)

This resource offers a study plan and practice questions for each chapter to help students prepare for exams.

Art Notebook

The Art Notebook is an invaluable tool for taking effective notes during lectures. All of the drawn figures from the text are reproduced in color, with ample space for taking notes. This resource allows the student to focus on the lecture and not worry about trying to copy drawings from the blackboard or projector.

Discover Biology, Third Edition, eBook

An affordable and convenient alternative to the print textbook, the *Discover Biology* eBook retains the content of the print book and replicates actual book pages for a pleasant reading experience. In addition, a variety of features make the Norton eBook a powerful tool for study and review:

- Zoomable images allow students to get a closer look at the book's figures and photographs.

- Clear text, designed specifically for screen use, makes reading easy and comfortable.

- A search function facilitates study and review.

- A print function permits individual pages to be printed as needed.

- Sticky notes and highlighting allow students to notate the text.

- Links to DiscoverBiology.com direct students to the Web site for animations, activities, and review materials.

- Online and cross-platform: works on both Macs and PCs and allows students to access their eBook from home, school, or anywhere with an Internet connection.

Students can order either the full text of *Discover Biology*, Third Edition, or individual chapters. Visit NortonEbooks.com for more information.

THIRD EDITION

Discover Biology

Michael L. Cain
Bowdoin College

Hans Damman
Duncan, British Columbia

Robert A. Lue
Harvard University

Carol Kaesuk Yoon
Bellingham, Washington

Contributing Author
Richard Morel
Signal Mountain, Tennessee

W. W. NORTON & COMPANY
NEW YORK • LONDON

W. W. Norton & Company has been independent since its founding in 1923, when William Warder Norton and Mary D. Herter Norton first published lectures delivered at the People's Institute, the adult education division of New York City's Cooper Union. The Nortons soon expanded their program beyond the Institute, publishing books by celebrated academics from America and abroad. By mid-century, the two major pillars of Norton's publishing program—trade books and college texts—were firmly established. In the 1950s, the Norton family transferred control of the company to its employees and today—with a staff of four hundred and a comparable number of trade, college, and professional titles published each year—W. W. Norton & Company stands as the largest and oldest publishing house owned wholly by its employees.

PRINTED IN THE UNITED STATES OF AMERICA.
Third Edition.

Composition by TSI Graphics
Manufacturing by VonHoffmann, Jefferson City
Illustrations for the Third Edition by Dragonfly Media Group

Editor: Vanessa Drake-Johnson
Production Manager: JoAnn Simony
Cover and Book Design: Rubina Yeh
Development Editor: Susan Middleton
Project Editor: Christopher Miragliotta
Copy Editor: Norma Roche
Electronic Media Editor: April Lange
Associate Editor: Sarah England
Assistant Editor: Erin O'Brien
Photo Researchers: Neil Hoos, Motoko Oinuma, Ede Rothaus, Daniella Nilva

ISBN-13: 978-0-393-92867-9
ISBN-10: 0-393-92867-5

W. W. Norton & Company, Inc., 500 Fifth Avenue, New York, N.Y. 10110
www.wwnorton.com
W. W. Norton & Company Ltd., Castle House, 75/76 Wells Street, London WIT 3QT

2 3 4 5 6 7 8 9 0

About the Authors

Michael L. Cain received his Ph.D. in ecology and evolutionary biology from Cornell University and did postdoctoral research at Washington University in molecular genetics. He taught introductory biology and a broad range of other biology courses at New Mexico State University and Rose-Hulman Institute of Technology for thirteen years. He is now writing full time and is affiliated with Bowdoin College in Maine. Dr. Cain's interest in biology began early, stimulated by walks along the Maine coast and by the expert guidance of his high school biology teacher, Clayton Farraday. Dr. Cain has published dozens of scientific articles on such topics as genetic variation in plants, insect foraging behavior, long-distance seed dispersal, and factors that promote speciation in crickets. Dr. Cain is the recipient of numerous fellowships, grants, and awards, including the Pew Charitable Trust Teacher-Scholar Fellowship and research grants from the National Science Foundation.

Hans Damman received his Ph.D. in Entomology in 1986 from Cornell University. He was until 1999 an Associate Professor of Biology at Carleton University in Ottawa. At Carleton University he twice received the Faculty of Science Award for Excellence in Teaching. His teaching included courses in Animal Form and Function, Plant and Animal Interactions, and Ecology, and he supervised a large number of undergraduate Honors research projects. His research work began with a focus on the interactions between plants and the insects that eat them, and has recently expanded to include the population biology of plants. Dr. Damman currently works as a staff scientist at the Department of Fisheries and Oceans in British Columbia.

Robert A. Lue is Senior Lecturer on Molecular and Cellular Biology and the Director of Undergraduate Studies in the Biological Sciences at Harvard University, where he has received several teaching awards from the Faculty of Arts and Sciences and the Division of Continuing Education. He received his Ph.D. from Harvard and has taught undergraduate courses there since 1988. Dr. Lue has published award-winning educational multimedia on HIV and AIDS, and has chaired educator conferences on college biology, most recently for the National Science Foundation. Dr. Lue is a molecular cell biologist, and his own research interests include the role of cytoskeletal proteins in human cancer. His research has been funded by grants from the National Institutes of Health.

Carol Kaesuk Yoon received her Ph.D. from Cornell University and has been writing about biology for *The New York Times* since 1992. Her articles have also appeared in *The Washington Post, The Los Angeles Times,* and *Science* magazine. Dr. Yoon has taught writing as a Visiting Scholar with Cornell University's John S. Knight Writing Program, working with professors to help teach critical thinking in biology classes. She has also served as science consultant to Microsoft for their children's CD-ROM on tropical rainforests, part of the Magic School Bus series. Recent articles include stories on worldwide declines in penguins, a new date for the evolution of land plants, and the discovery of a new species of deep sea squid.

Richard Morel has worked as a senior developmental editor and acquisitions editor in the fields of chemistry, biochemistry, and biology. He also has held the positions of Curatorial Associate in Entomology at Harvard's Museum of Comparative Zoology and Research Associate in Chemistry at Smith College. He currently works as a freelance author and developmental editor.

Preface

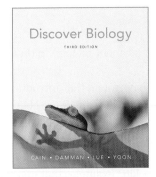

Seeking a cure for cancer. Exploring the wondrous diversity of life. Unraveling the mysteries of our development. These are but a few of the many reasons why biology is a gripping subject for us and, we hope, for students using this book. These topics are simultaneously intensely interesting and critically important. Because the scientific understanding of fundamental biological principles is growing by leaps and bounds, this is an exciting time to write, teach, and learn about all areas of the biological sciences.

Just consider that researchers are beginning to unlock the secrets of the human genome, allowing us to explore in unprecedented detail both the causes of and potential treatments for a host of human genetic disorders. And ecologists have begun to make real progress in understanding and predicting events that affect all of the world's ecosystems, such as the transport of pollutants and the contribution of human activities to global warming.

Discoveries such as these, along with a host of others, mean that biology is the subject of news stories on a daily basis. In recent months, for example, the news has focused on both the disturbing (the dangers posed by highly lethal varieties of the flu) and the encouraging (signs of progress in global efforts to restore the ozone layer that protects us from UV light). As these and other news reports show, biology carries important ramifications not only for individuals, but for all human societies, and it touches our lives every day as well as in long-term ways we are only beginning to comprehend.

This Book Is Written for Students

The very things that make biology so interesting—the rapid pace of new discoveries and the many applications of these discoveries by human societies—can make it a difficult subject to teach and to learn. The problem is only exacerbated by the wide variation in background and interests that nonmajors, ranging from nursing and other allied health students to business majors and liberal arts students, bring to the course. When we set out to write the third edition of *Discover Biology*, we asked ourselves, How can we convey the excitement, breadth, and relevance of biology to this varied group of students without burying them in facts and definitions?

We have answered this question in several ways. In considering which topics and details would be included in the book from among the vast group of possibilities, our goals were to

- highlight fundamental concepts
- provide human examples alongside comparative examples from other species
- discuss material of applied importance
- reveal something surprising or fascinating about the natural world

We also responded to this question by writing clear, streamlined chapters and by emphasizing the development of biological literacy to provide students with tools that will serve them well whether they continue on in a health-related field or simply become better consumers of information about biology in their everyday lives.

Over years of teaching, we have found that introductory, non-science students learn best from short chapters that provide a clear overview of key concepts, since these concepts serve as a "coat rack" on which to hang the details of the topic under consideration. We also think it is important to provide compelling examples that make the material come alive for students.

As described in detail below, to put our views into practice, we begin all our chapters with a single main message, a small set of key concepts, and an opening vignette to pique students' interest as they start reading. In the body of the chapter, we develop the main message to a depth that only slightly exceeds what can be covered in one class period. We illustrate key concepts with both applied topics and compelling examples from the natural world. We end the chapters by revisiting the key concepts and testing the student's ability to understand and apply those concepts.

Changes in the Third Edition: A Thorough Overhaul

As we began to revise the second edition of *Discover Biology*, we listened carefully to the comments of students and instructors who were using the book. They provided invaluable help, both by describing the book's strengths and by pointing out places where there was room for improvement. We also benefited greatly from the thoughtful responses to survey questions provided by dozens of instructors. These instructors took the time to examine the book closely and to provide us with extensive feedback on all of its features. During the long hours we spent carefully revising each chapter, the guidance and encouragement we received from these comments served us extremely well.

The message from all these comments was loud and clear: Stay with the existing model of chapters that emphasize key concepts, applied examples, and breaking news stories, but work to add more human examples and to more fully develop the presentation of critical topics in the cell biology and the animal and plant form and function units. Here, as always, our primary focus was on the student— we cut text that provided students with either too much or too little information, and we added text as needed to strengthen the explanation of fundamental concepts.

With this third edition, we have also strengthened our emphasis on developing biological literacy and illustrating the broad-based relevance of biology through the addition of new boxes called "Science Toolkit," "Biology on the Job," and "Biology Matters," new pedagogical elements such as marginal terminology notes and pronunciation guides, and the unique "Biology in the News" feature, all of which will be discussed in more detail below.

Keeping these basic principles in mind, we dove in and produced a major overhaul of the book. Foremost among the changes was a complete restructuring of the material on the form and function of organisms. In the previous two editions, we took an integrated approach to the coverage of form and function, discussing both plants and animals in each chapter. This allowed us to emphasize how plants and animals solve similar problems in different ways. But in a book such as this one—with short, focused chapters—this integrated approach meant that it was hard to cover the material in sufficient depth, especially given the varied needs of our audience.

To address this issue, we decided to separate the animal and plant material, creating two new units: "Animal Form and Function" (Unit 5) and "Plant Form and Function" (Unit 6). This change entailed a massive rewrite of previous chapters, with the end result that our coverage of how humans and other animals work was increased greatly. For example, we have expanded our discussion of the structure and function of human organs and organ systems, such as the kidney, brain, and digestive system (including the contributions of the pancreas, liver, and salivary glands). Other major changes to the animal form and function unit include the addition of two new chapters, one on nervous systems (Chapter 25, "The Nervous System") and the other on sensory systems (Chapter 26, "Sensing the Environment"). We also decided to increase our focus on humans in each chapter, especially on human health issues.

The material on plants has changed just as extensively. In two new chapters, Chapter 31 ("Plant Structure, Nutrition, and Transport") and Chapter 32 ("Plant Growth and Reproduction"), and one new Interlude ("Feeding a Hungry Planet"), we provide much greater detail on plant structure and function, and we explore fascinating topics such as how plants prevent their cells from freezing and hence ripping apart in extreme cold, how plants can live to be tens of thousands of years old, and how new developments in agriculture can help feed the growing human population. Care was also taken to relate the plant material to humans, as illustrated by a new "Science Toolkit" box ("Stirring Up Plant Clones") that describes how plant cloning can lead to increased crop yields, a new "Biology on the Job" box ("Growing Plants Without Soil") that describes what it is like to work in hydroponics, and a discussion of how chemicals that protect plants from heat stress, UV light, and other environmental factors may also improve human health.

Although the other five units have not changed as drastically, they, too, have undergone their most extensive revision since the first edition of *Discover Biology*, with the four goals mentioned above as the yardstick. For example, in Unit 2 ("Cells: The Basic Units of Life"), we have strengthened our presentation of fundamental concepts by writing a new chapter on membrane transport and cell communication (Chapter 6) and by adding new sections on isotopes, the four levels of protein structure, steroids, cilia and flagella, and the connection between metabolic rates and life span.

Similarly, our changes to Unit 4 ("Evolution") and Unit 7 ("Interactions with the Environment") aim to improve our coverage of fundamental concepts while simultaneously strengthening the human context. We have met these goals in Unit 4 by adding a unique new Interlude on human evolution (Interlude D, "Humans and Evolution," which describes how people affect—and are affected by—evolution); by expanding our coverage of the fossil record (for example, new sections on the origin of mammals and on exciting fossil discoveries that show that whales are most closely related to even-toed ungulates, such as pigs, camels, and hippos); and by increasing our coverage of the evolution of resistance to our best efforts to kill bacteria and other pathogen and pest species. In Unit 7, we have strengthened our coverage of how ecosystems work, and we have added a new Interlude (Interlude G, "Building a Sustainable Society," which provides a message of hope at the close of a unit that includes a lot of troubling information); new material on the connection between the delivery of water to New York City and ecosystem services; new material on science and public policy (the "Science Toolkit" box in Chapter 38 on repairing the ozone layer); and summaries of real news articles about topics such as the worldwide spread of Teflon and related chemicals throughout the environment (Chapter 37, "Marvel Chemicals Pop Up in Animals All Over World").

Features of the Third Edition: Approaches and Tools to Improve Student Learning

A primary goal of the third edition is to improve the biological literacy of our readers

Becoming biologically literate involves several challenges: understanding the fundamental concepts and becoming comfortable with the terminology of the field; acquiring some conceptual and critical skills to make sense of new information; and seeing the connections between the concepts covered in this text and the world around us. The third edition includes several features, some new, some honed from the previous two editions, that aim to prepare students to become biologically literate participants in society.

Several small additions to the new edition aim to help the varied audience of nonmajor students become competent and comfortable with biological terminology. **"Helpful to know" marginal notes** provide hints such as how to remember the difference between similar terms, or how prefixes and suffixes can be used as clues to a term's meaning. **In-text pronunciation guides** will increase students' confidence as they incorporate new terms into their vocabulary.

We think it is critical that students understand the biology behind the news; only in this way can they make informed choices when the issues involve biology—as more and more issues do. As part of our effort to help students become critical consumers of information about biology, each chapter in *Discover Biology* concludes with a unique feature called **"Biology in the News."** Each of these features includes a quote from a real news article, followed by a summary of the article and a discussion highlighting the issues raised by the news topic. "Evaluating the News" questions then encourage students to think about the social and ethical implications of the biological issues, providing a forum for students to develop critical thinking skills and apply what they learn in the textbook to real problems.

New **"Science Toolkit"** boxes provide insights into the methods, ideas, and tools that scientists employ, with the goal of helping students interpret the science and research presented not only in this text, but by the media and other sources.

We believe that biological literacy also involves seeing the connections between the science and our own lives. **"Biology on the Job"** boxes, unique to this text, feature interviews with people from a wide variety of fields—including a firefighter, a genetic counselor, a baker, a hostage negotiator, and many more. These boxes explore the connections between their day-to-day jobs and the biological topics in this text. With another new feature, **"Biology Matters,"** we present specific information of practical importance to students' lives. From making environmentally responsible choices to the importance of knowing your blood type to the common campus problems of depression and binge drinking, these boxes provide a down-to-earth connection to the science that we hope will engage and motivate students to make informed choices in their own lives.

Finally, biological literacy must involve engagement with the major biological issues of the day. Each of the book's seven units ends with an essay devoted to a significant issue confronting society: loss of biodiversity, cancer, genomics, humans and evolution, smoking, world hunger, and efforts to build a sustainable society. These **Interludes** draw on the biological concepts introduced earlier in the unit and give in-depth consideration to how biology is connected to the rest of our lives. By including seven different Interlude essays, we provide instructors with a menu of choices to adapt to the needs of their own courses.

The art and text work together to tell a story

In this and the previous two editions of *Discover Biology*, we worked hard to create beautiful, instructional art and to select superb photographs that integrate seamlessly with the text to guide students through even the most difficult material. Like the text, the figures in *Discover Biology* are streamlined to convey simple, direct messages that allow students to focus on essential concepts. Most figures in the book include balloons that highlight crucial information or explain the flow of an illustration. A large number of the figures are new; we also used feedback from students and reviewers to revise many figures from the second edition so as to further improve their ability to help "tell the story."

Chapters remain focused on a few key concepts

Each chapter's **main message** appears at the beginning of the chapter, along with a **chapter outline** and a list of **key concepts**. The chapter text also begins here with a vignette designed to capture students' interest and lead them into the material they will learn about in the chapter. The topics of these vignettes range from an examination of fetal alcohol syndrome to the story of the Iceman Ötzi (the mummy found in a melting glacier in Europe) to the use of forensic genomics to determine whether the Russian Grand Duchess Anastasia escaped execution in 1918 (she did not).

Following the vignette is an **overview** of the topics to be covered. Two levels of headings organize the material in a clear and straightforward fashion; second-level headings are usually in the form of declarative sentences that convey a key point or summarize an important message. Each chapter concludes by returning to explore the topic introduced in the vignette in more detail.

Chapters close with a **Chapter Review** section, featuring a summary of important points, a set of questions that review and apply key concepts, a list of key terms, a self-quiz, and the "Biology in the News" feature. Answers to all key concept and self-quiz questions are in the back of the text, along with a complete glossary and an index.

An expanded and enhanced set of instructor resource materials facilitates lecture preparation

For this edition we focused on improving and expanding the resources available to instructors. Highlights of this revision include new editable PowerPoint lecture outlines that feature text slides, selected images, and Concept Quiz "clicker" questions designed to encourage classroom participation. Additional figures and media elements are available on the multi-disc Norton Media Library Instructor's CD-ROM. For instructors who use WebCT or BlackBoard, our new course cartridge integrates selected elements from the media supplements that accompany *Discover Biology*. In response to requests from adopters of the second edition, we have also thoroughly overhauled and expanded our Test Bank, which now includes a broader selection of question types and difficulty levels. This item has been carefully examined by the authors and by outside reviewers to ensure that the questions therein offer a fair and accurate test of the content presented in each chapter of *Discover Biology*. New test-making software by ExamView makes it easy for instructors to adapt our bank of questions to suit their course needs.

For more detailed information about these and other resources available to instructors, please see p. iv.

DiscoverBiology.com helps students get the most from their reading assignments

No textbook about a subject as broad as biology can stand alone. Some topics require additional information or hands-on practice. In these cases, alternative media such as animation and interactive exercises can help students understand difficult material. For the third edition, *Discover Biology* offers an improved and comprehensive online study guide. A CD-ROM version of DiscoverBiology.com is also available.

For more information about this and other resources available to students, please see p. v.

We Thank the Many People Who Helped Us with This Revision

The process of getting from the kernel of an idea to a finished book involves a veritable army. Here we get to recognize and thank some of the major participants. First, we thank all of the manuscript reviewers of this and the first two editions, who are listed on pages xvi–xviii. Together, their criticism and encouragement contributed enormously to the crucial rewriting necessary to achieve a consistent level, clarity, and accuracy. Thanks also to Susan McGlew of Sinauer Associates and to Erin O'Brien and Chris Curcio of W. W. Norton for arranging and coordinating all of the reviews. We received valuable insights from several focus groups that took place near Chicago and in Portland, Oregon, and in particular we received useful advice from Tom Firak and Cecelia Hutchcraft of Oakton Community College; Deb Firak of McHenry County College; Barb Anderson, Lynn Fancher, Lynda Randa, Shyla Akkaraju, and other colleagues from the College of DuPage; Doug Ure, Chemeketa Community College; and April Fong, Mica Jordan, and other colleagues from Portland Community College.

As we revised *Discover Biology*, each of us contacted scientists from around the world to ask about their research or about breaking stories in their field of expertise. We are grateful for the time these scientists spent answering our many questions. We are also grateful for the scientific papers, illustrations, and photos they provided to help us as we worked on the third edition. We have communicated with many more scientists than we can thank individually, but collectively we thank them here. We are deeply in their debt.

It has been a great pleasure to work with Richard Morel, who joined us as a contributing author on this edition and made major contributions to Unit 2 ("Cells: The Basic Units of Life") and Unit 5 ("Animal Form and Function"). His formidable energy and creativity resulted in many improvements to the text and illustrations.

The talented artists at Dragonfly Media Group translated the ideas and sketches of the authors into beautiful art. Craig Durant, Mike Demaray, and Rob and Caitlin Duckwall deserve special thanks for their creativity, attention to detail, and speed. The wonderful photographs in the book were assembled by Motoko Oinuma, Ede Rothaus, Daniella Jo Nilva, and Neil Hoos, who endured many a round of rejects before they and we were satisfied. Susan Middleton, the manuscript editor, did a superb job of keeping the entire project on track, with good humor and amazing attention to detail. It is thanks to her that the level of writing is as clear and consistent across chapters as it is. Norma Roche did a great job in copyediting the manuscript. She has a knack for identifying troublesome sentences, and best of all, for fixing them.

There were many other people at W. W. Norton who played critical roles in bringing this third edition to fruition. Sarah England and Erin O'Brien worked miracles, keeping everything organized and on schedule and finding interesting photos on short notice. Rubina Yeh developed a striking new design for the book that holds to our original vision of an uncluttered book in which all of the elements have strong pedagogical value. Her clean design helps students focus on what is important. It is also extremely attractive. JoAnn Simony and Chris Miragliotta shepherded the book very effectively and creatively through the production process.

Getting the text right is only half the challenge in introductory biology. Many individuals have lent their talent and time to create a strong support package for the third edition.

We thank April Lange and Karen Misler for taking charge of the supplements to make sure we have the strongest possible support package. David Demers (Sacred Heart University), Ed Dzialowski (University of North Texas), and Stephen Lebsack (Lynn-Benton Community College) developed the study guide, approaching the task with enthusiasm and care.

In this edition we have expanded our offerings for instructors. We are grateful to Richard Morel for his careful work on the Test Bank, which builds on the foundation laid by Susan Weinstein in the second edition. It has now been significantly expanded and improved. We are also thankful to Donald Slish (State University of New York-Plattsburgh), Douglas Oba (University of Wisconsin/Marshfield), and Alana Synhoff (Florida Community College) for their work in developing the new instructor PowerPoint lecture outlines. We believe these will be a real asset for instructors. We appreciate the professionalism and expertise that Sumanas Inc. brought to bear on the development of the multimedia resources that accompany this text, especially the animations available to instructors and students.

We thank Andy Sinauer of Sinauer Associates and Vanessa Drake-Johnson of W. W. Norton for their editorial guidance as this text has evolved from an idea to a finished book, and now to a third edition.

This Book Is a Work in Progress

Like the field of biology itself, any textbook aiming to describe it is a work in progress. We hope you will contact us with your comments, questions, and ideas as you use this book.

Michael L. Cain
mcain@bowdoin.edu

Hans Damman
hdamman@shaw.ca

Robert A. Lue
robert_lue@harvard.edu

Carol Kaesuk Yoon
cky@cnw.com

Reviewers for the Third Edition

Marjay Anderson, Howard University
Caryn Babaian, Bucks County College
Sarah Barlow, Middle Tennessee State University
Gregory Beaulieu, University of Victoria
Neil Buckley, State University of New York/Plattsburgh
Heather Vance Chalcraft, East Carolina University
Van Christman, Ricks College
Jerry Cook, Sam Houston State University
Judith D'Aleo, Plymouth State University
Vern Damsteegt, Montgomery College
Gregg Dieringer, Northwest Missouri State University
Deborah Donovan, Western Washington University
William Ezell, University of North Carolina/Pemberton
Deborah Fahey, Wheaton College
April Fong, Portland Community College
Alexandros Georgakilas, East Carolina University
Glenn Gorelick, Citrus College
Robert Harms, St. Louis Community College/Meramec
Chris Haynes, Shelton State Community College
Thomas Hemmerly, Middle Tennessee State University
Tom Horvath, State University of New York/Oneonta
Paul Kasello, Virginia State University
Andrew Keth, Clarion University of Pennsylvania
Tasneem Khaleel, Montana State University
John Knesel, University of Louisiana/Monroe
Allen Landwer, Hardin-Simmons University
Harvey Liftin, Broward County Community College
Craig Longtine, North Hennepin Community College
Blasé Maffia, University of Miami
Patricia Mancini, Bridgewater State College
Roy Mason, Mount San Jacinto College
Daniela Monk, Washington State University
Brenda Moore, Truman State University
Jon Nickles, University of Alaska/Anchorage
Douglas Oba, University of Wisconsin/Marshfield

Donald Padgett, Bridgewater State College
Penelope Padgett, University of North Carolina/Chapel Hill
Brian Palestis, Wagner College
Snehlata Pandey, Hampton University
Robert Patterson, North Carolina State University
Nancy Pelaez, California State University/Fullerton
Pat Pendarvis, Southeastern Louisiana University
Patrick Pfaffle, Carthage College
Robert Pozos, San Diego State University
Jerry Purcell, Alamo Community College
Richard Ring, University of Victoria
Ron Ruppert, Cuesta College
Lynette Rushton, South Puget Sound Community College
Shamili Sandiford, College of DuPage
Harlan Scott, Howard Payne University
Erik Sculley, Towson University
Cara Shillington, Eastern Michigan University
Mark Shotwell, Slippery Rock University
Shaukat Siddiqi, Virginia State University
Donald Slish, State University of New York/Plattsburgh
Philip Snider, University of Houston
Ruth Sporer, Rutgers University/Camden
Neal Stewart, University of North Carolina/Greensboro
Bethany Stone, University of Missouri
Steven Strain, Slippery Rock University
Marshall Sundberg, Emporia State University
Alana Synhoff, Florida Community College
Joyce Tamashiro, University of Puget Sound
Steve Tanner, University of Missouri
William Velhagen, Longwood College
Mary Vetter, Luther College
Daniel Wang, University of Miami
Carol Weaver, Union University
Peter Wilkin, Purdue University North Central
Louise Wootton, Georgian Court University

Reviewers for the Second Edition

Gregory Beaulieu, University of Victoria
Robert Bernatzky, University of Massachusetts/Amherst
Nancy Berner, University of the South
Sarah Bruce, Towson University
Neil Buckley, SUNY/Plattsburgh
David Byres, Florida Community College/Jacksonville—
    South Campus
Van Christman, Ricks College
Jerry L. Cook, Sam Houston State University
Paul da Silva, College of Marin
Judith D'Aleo, Plymouth State College
Sandra Davis, University of Louisiana/Monroe
Pablo Delis, Hillsborough Community College
Deborah Fahey, Wheaton College
Wendy Garrison, University of Mississippi
Aiah A. Gbakima, Morgan State University
Kajal Ghoshroy, Museum of Natural History/Las Cruces
Jack Goldberg, University of California/Davis
Andrew Goliszek, North Carolina Agricultural and
    Technological State University
Glenn A. Gorelick, Citrus College
Bill Grant, North Carolina State University
Harry W. Greene, Cornell University
Barbara Hager, Cazenovia College
Robert Harms, Saint Louis Community College/Meramec
Chris Haynes, Shelton State Community College
Thomas E. Hemmerly, Middle Tennessee State University
Tasneem F. Khaleel, Montana State University/Billings
Hans Landel, North Seattle Community College
Katherine C. Larson, University of Central Arkansas
Ann S. Lumsden, Florida State University

Joyce Maxwell, California State University/Northridge
Bob McMaster, Holyoke Community College
Ali Mohamed, Virginia State University
Brenda Moore, Truman State University
Ruth S. Moseley, S.D. Bishop Community College
Benjamin Normark, University of Massachusetts/Amherst
Douglas Oba, Brigham Young University
Mary O'Connell, New Mexico State University
Marcy Osgood, University of Michigan
Donald Padgett, Bridgewater State College
Anthony Palombella, Longwood College
Snehlata Pandey, Hampton University
Robert P. Patterson, North Carolina State University
Jeffrey Podos, University of Massachusetts/Amherst
Richard A. Ring, University of Victoria
Kurt Schwenk, University of Connecticut
Erik P. Scully, Towson University
Cara Shillington, Eastern Michigan University
Barbara Shipes, Hampton University
Shaukat Siddiqi, Virginia State University
Julie Snyder, Hudson High School
Neal Stewart, University of North Carolina/Greensboro
Tim Stewart, Longwood College
Marshall D. Sundberg, Emporia State University
Cheryl Vaughan, Harvard University
John Vaughan, St. Petersburg College
Alain Viel, Harvard Medical School
William A. Velhagen, Jr., Longwood College
Mary Vetter, Luther College at the University of Regina
Carol H. Weaver, Union University
Jerry Waldvogel, Clemson University

Reviewers for the First Edition

Michael Abruzzo, California State University/Chico
James Agee, University of Washington
Craig Benkman, New Mexico State University
Elizabeth Bennett, Georgia College and State University
Stewart Berlocher, University of Illinois/Urbana
Juan Bouzat, University of Illinois/Urbana
Bryan Brendley, Gannon University
John Burk, Smith College
Kathleen Burt-Utley, University of New Orleans
Naomi Cappuccino, Carleton University
Alan de Queiroz, University of Colorado
Véronique Delesalle, Gettysburg College
Jean de Saix, University of North Carolina/Chapel Hill
Joseph Dickinson, University of Utah
Harold Dowse, University of Maine
John Edwards, University of Washington
Jonathon Evans, University of the South
Marion Fass, Beloit College
Richard Finnell, Texas A & M University
Dennis Gemmell, Kingsborough Community College
Laura Haas, New Mexico State University
Blanche Haning, University of North Carolina/
    Chapel Hill
Daniel J. Howard, New Mexico State University
Laura F. Huenneke, New Mexico State University
Laura Katz, Smith College
Katherine C. Larson, University of Central Arkansas

Kenneth Lopez, New Mexico State University
Phillip McClean, North Dakota State University
Amy McCune, Cornell University
Bruce McKee, University of Tennessee
Gretchen Meyer, Williams College
Brook Milligan, New Mexico State University
Kevin Padian, University of California/Berkeley
John Palka, University of Washington
Massimo Pigliucci, University of Tennessee
Ralph Preszler, New Mexico State University
Barbara Schaal, Washington University
David Secord, University of Washington
Nancy Stotz, New Mexico State University
Allan Strand, College of Charleston
John Trimble, Saint Francis College
Mary Tyler, University of Maine
Roy Van Driesche, University of Massachusetts/Amherst
Carol Wake, South Dakota State University
Jerry Waldvogel, Clemson University
Elsbeth Walker, University of Massachusetts/Amherst
Stephen Warburton, New Mexico State University
Paul Webb, University of Michigan
Peter Wimberger, University of Puget Sound
Allan Wolfe, Lebanon Valley College
David Woodruff, University of California/San Diego
Robin Wright, University of Washington

Guided Tour for Students

Features Designed to Help You Succeed in Your Study of Biology

Discover Biology's chapters are designed to be short and focused around key concepts that will guide you through the world of biology. Special features will help you understand as you read, master the terminology, and test what you've learned.

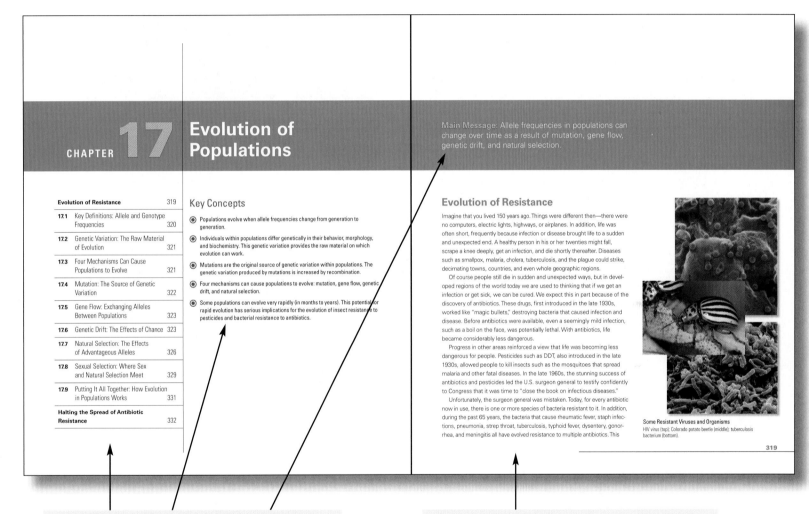

CHAPTER **17**

Evolution of Populations

Main Message: Allele frequencies in populations can change over time as a result of mutation, gene flow, genetic drift, and natural selection.

Key Concepts

- Populations evolve when allele frequencies change from generation to generation.

- Individuals within populations differ genetically in their behavior, morphology, and biochemistry. This genetic variation provides the raw material on which evolution can work.

- Mutations are the original source of genetic variation within populations. The genetic variation produced by mutations is increased by recombination.

- Four mechanisms can cause populations to evolve: mutation, gene flow, genetic drift, and natural selection.

- Some populations can evolve very rapidly (in months to years). This potential for rapid evolution has serious implications for the evolution of insect resistance to pesticides and bacterial resistance to antibiotics.

Evolution of Resistance

Imagine that you lived 150 years ago. Things were different then—there were no computers, electric lights, highways, or airplanes. In addition, life was often short, frequently because infection or disease brought life to a sudden and unexpected end. A healthy person in his or her twenties might fall, scrape a knee deeply, get an infection, and die shortly thereafter. Diseases such as smallpox, malaria, cholera, tuberculosis, and the plague could strike, decimating towns, countries, and even whole geographic regions.

Of course people still die in sudden and unexpected ways, but in developed regions of the world today we are used to thinking that if we get an infection or get sick, we can be cured. We expect this in part because of the discovery of antibiotics. These drugs, first introduced in the late 1930s, worked like "magic bullets," destroying bacteria that caused infection and disease. Before antibiotics were available, even a seemingly mild infection, such as a boil on the face, was potentially lethal. With antibiotics, life became considerably less dangerous.

Progress in other areas reinforced a view that life was becoming less dangerous for people. Pesticides such as DDT, also introduced in the late 1930s, allowed people to kill insects such as the mosquitoes that spread malaria and other fatal diseases. In the late 1960s, the stunning success of antibiotics and pesticides led the U.S. surgeon general to testify confidently to Congress that it was time to "close the book on infectious diseases."

Unfortunately, the surgeon general was mistaken. Today, for every antibiotic now in use, there is one or more species of bacteria resistant to it. In addition, during the past 65 years, the bacteria that cause rheumatic fever, staph infections, pneumonia, strep throat, tuberculosis, typhoid fever, dysentery, gonorrhea, and meningitis all have evolved resistance to multiple antibiotics. This

Some Resistant Viruses and Organisms
HIV virus (top); Colorado potato beetle (middle); tuberculosis bacterium (bottom).

319

Each chapter is structured around a **Main Message**, the central idea of the chapter, and the **Key Concepts** that support that message—the important information that we will be sure you master by the end of the chapter. The Main Message and Key Concepts, along with a **Chapter Outline**, are listed on the chapter opening pages for easy reference while you are reading.

Chapter Opening Vignettes begin each chapter. These vignettes present interesting stories or puzzles to arouse your curiosity as you start reading. At the conclusion of the chapter we revisit the opening vignette and explore it further in light of the information that you have picked up over the course of the chapter.

Following the vignette is a brief **Overview** of the topics that will be covered.

hormones. Ironically, this action also accounts for the negative side effects of aspirin. Armed with a growing knowledge of how aspirin affects these chemical reactions, researchers are now seeking to improve this old wonder drug by eliminating the negative aspects of its action.

All biological processes require energy, which living organisms must extract from their environment. They use this energy to manufacture and transform the various chemical compounds that make up living cells. The capture and use of energy by living organisms involves thousands of chemical reactions, which together are known as metabolism.

All the chemical reactions that occur in cells can be grouped into sequences called metabolic pathways. Just as chemical building blocks fall into a limited number of categories, the types of reactions that allow the cell to assemble and disassemble these building blocks are limited in number.

In this chapter we examine the role played by energy in the chemical reactions that maintain living systems. We also discuss the role of specialized proteins, called enzymes, used by the cell to speed up chemical reactions that would otherwise be too slow to sustain life. Finally, we explore the possible connections between life span and the overall rate of life's chemical reactions.

7.1 The Role of Energy in Living Systems

The discussion of any chemical process in the cell is a discussion about energy. The idea that energy is behind every activity in the cell seems natural and unsurprising, since all of us are accustomed to thinking of energy as a form of fuel. However, energy is more than just fuel, because its properties dictate which chemical reactions can occur and how molecules can be organized into living systems.

The laws of thermodynamics apply to living systems

The relationship between energy and the cell's activities is governed by the same physical laws that apply to everything else in the universe. These laws of thermodynamics define the ways cells transform chemical compounds and interact with the environment. The **first law of thermodynamics** states that energy cannot be either created or destroyed, only converted from one form to another.

Consider what happens when you use electrical hedge clippers, which have a small gas tank and generator attached to the clippers. The chemical energy in the covalent bonds of the gasoline molecules is converted into electrical energy by the generator. This electrical energy, in turn, is converted into the mechanical energy of motion in the hedge clippers themselves. Neither the generator nor the hedge clippers creates or destroys energy.

At a cellular level, the first law of thermodynamics is illustrated by mitochondria, which convert energy from food molecules such as sugars into the energy of covalent bonds in ATP (ADP + phosphate + energy → ATP, as we saw in Figure 4.9). Thus mitochondria do not create energy from nothing. They convert energy from one form (sugars) into another form (ATP), which can then be used by the cell.

The **second law of thermodynamics** describes how the cell relates to its environment. This law states that systems tend to become more disorderly. This statement may seem most appropriate to describe a household room or a toolshed, which, unless we spend energy tidying it up, tends toward disorder (Figure 7.1a). But it is true of all systems, including the internal organization of the cell, an organism, or even the whole universe.

As we saw in Chapters 4 and 5, a cell is made up of many chemical compounds assembled into complex ordered structures. Such a high level of organization may seem to fly in the face of disorder, but it has an explanation. The tremendous structural complexity of the cell and its organelles exists in a constant struggle against chaos. To counteract the natural tendency toward disorder, the cell must use energy to keep things orderly.

As order is created in living systems, those systems pass on or transfer disorder by releasing heat into the environment. Heat is a form of energy that causes rapid and random movement of molecules, a condition that is highly disordered. Thus, when cells release heat, they increase the degree of disorder in the molecules of the environment, which compensates for the increasing order inside the cell (Figure 7.1b). There is a direct connection between cellular organization and the transfer of heat energy because the chemical processes used to build well-ordered structures are the same ones that produce the

New order is at the expense of heat production.

There is a high degree of disorder in this cell, as shown by the random arrangement of its molecules.

The ordering of the molecules inside the cell is compensated for by the release of heat into the environment, which in turn becomes more disordered.

Figure 7.1 The Second Law of Thermodynamics
The disorder of a system tends to increase unless that tendency is countered by an input of energy. (a) Left unattended, all structures, such as this wooden toolshed, tend to lose their order and become disarrayed. An input of energy, here in the form of human effort, is needed to maintain the order of the structure. (b) Cells maintain their organization through a continuous input of energy from the environment. Thus they, too, obey the second law of thermodynamics.

heat. Hence the generation of order is directly coupled with the release of heat energy.

The flow of energy and the cycling of carbon connect living things with the environment

Where does the energy that creates order in the cell come from? We know from the first law of thermodynamics that the cell cannot create energy from nothing; thus it must come from outside of the cell. In other words, energy must be transferred into the cell in some fashion. In the case of photosynthetic organisms, the energy comes from sunlight. By using that energy to synthesize sugar molecules from carbon dioxide and water, those organisms convert it into the chemical bonds of sugars. For organisms that do not photosynthesize, energy comes from the chemical bonds in food molecules, such as sugars and fats.

The last two statements in the preceding paragraph reveal the chemistry of the relationship between producers and consumers. As we saw in Chapter 1, photosynthetic producers, such as plants, capture energy from sunlight, and nonphotosynthetic consumers, such as animals, obtain energy by consuming plants or other organisms that have consumed plants. This means that, thanks to photosynthesis, the sun is the primary energy source for living organisms.

However, plants are more than sources of consumable energy for nonphotosynthetic organisms. First, plants themselves use some of the sugars they make by photosynthesis; they do this especially at night, when there is no sunlight and no photosynthesis. Second, most organisms produce carbon dioxide (CO_2) as a by-product of the energy-harnessing process called respiration, and this CO_2, in turn, is a source of carbon for photosynthesis. In this way, carbon atoms are continually cycled from carbon dioxide in the atmosphere to sugars made by producers and back to carbon dioxide released by respiring producers and consumers (Figure 7.2). This kind of recycling occurs

Look for the numbered **main headings**, which highlight the major sections of the chapter, and the **subheadings**, which convey a key point or summarize an important message.

"Helpful to know" notes provide useful tips on how to understand and master biological terms.

life is a result of the repeated splitting of one species into two or more species, a process called **speciation**.

Speciation can result from a variety of processes. One of the most important is adaptation to different environments. Consider two populations of a species that are isolated from each other, as by a mountain or other barrier that prevents individuals from moving between the populations. Over time, natural selection may cause each population to become better adapted to its own particular environment on its own side of the barrier, leading to changes in the genetic makeup of both populations. Eventually, so many genetic changes may accumulate that if the barrier is removed, individuals from the two populations are no longer able to reproduce with each other. As we learned in Chapter 1, species are often defined in terms of reproduction: a species is a group of populations whose members can reproduce with each other but not with members of other such groups. Thus evolution by natural selection can lead to the formation of new species.

Organisms share characteristics due to common descent

The natural world is filled with puzzling examples of very different organisms that share certain characteristics. For example, the wing of a bat, the arm of a human, and the flipper of a whale all have five digits and contain the same kinds of bones (Figure 16.4a). Why do limbs that look so different and have such different functions have the same set of bones? Surely if the best possible wing, arm, and flipper were designed from scratch, their bones would not be so similar. Likewise, many organisms have **vestigial organs** (reduced or degenerate parts whose purpose is hard to discern). For example, why do we humans have a reduced tailbone and the remnants of muscles for moving a tail? And why do some snakes have rudimentary leg bones but no legs (Figure 16.4b)?

Evolution answers these and many other questions about shared characteristics of life. Many similarities among organisms are due to the fact that the organisms have a common ancestor. When one species splits into two, the two species that result share many features because they have evolved from a common ancestor. Features of organisms that are related to one another through common descent are said to be **homologous** [ho-*MOLL*-uh-guss]. For example, the wing of a bat, the arm of a human, and the flipper of a whale share the same set of bones because they are homologous (see Figure 16.4a). Similarly, some snakes have rudimentary leg bones because they evolved from reptiles with legs, and humans have rudimentary bones and muscles for a tail because our ancestors had tails.

Organisms can also share features as a result of **convergent evolution**, which occurs when natural selection causes distantly related organisms to evolve similar structures in response to similar environmental challenges.

Figure 16.4 Shared Characteristics
(a) The human arm, a whale's flipper, and a bat's wing are homologous structures, all of which have five digits and contain the same set of bones. (b) A python has rudimentary hind legs, as seen from the external surface and in the skeletal structure of the snake.

For example, the cacti found in North American deserts share many convergent features with distantly related plants found in African and Asian deserts (Figure 16.5). Similarly, although both sharks and dolphins have bodies streamlined for aquatic life, these species are very distantly related, and their overall similarities result from convergent evolution, not common descent. When species share characteristics because of convergent evolution, not common descent, those characteristics are said to be **analogous** [uh-*NAL*-uh-guss].

16.4 Strong Evidence Shows That Evolution Happens

Surveys taken over the past 10 years reveal that almost half of the adults in the United States do not believe that humans evolved from earlier species of animals. The results of these surveys are startling because evolution has been a settled issue in science for nearly 150 years. The vast majority of scientists of all nations, races, and creeds think that the evidence for evolution is very strong. In his landmark book, *The Origin of Species*, published in 1859, Charles Darwin argued convincingly that organisms are descended with modification from common ancestors. The scientific issue today is not whether evolution occurs, but how. To this question Darwin also offered an answer: he argued that the principal cause of evolutionary change is natural selection.

On this point Darwin was less successful in convincing other scientists, in part because at that time no one understood the underlying mechanisms of inheritance. For 60 years after the publication of *The Origin of Species*, many scientists thought Darwin was wrong to place so strong an emphasis on natural selection. However, the rediscovery of Gregor Mendel's work and the understanding of genetics that resulted made it clear that natural selection could cause significant evolutionary change, and hence that Darwin was at least partially correct.

Biologists still argue about the relative importance of natural selection and other mechanisms of evolution (such as genetic drift), but they do not dispute whether evolution occurs. Today's scientific debate about the causes of evolution can be compared to a dispute over what caused World War I to progress as it did: although we might argue over its causes, we all recognize that the war did indeed happen.

Why do scientists find the case for evolution so convincing? As we saw in Chapter 1, a scientific hypothesis must lead to predictions that can be tested, and hypotheses about evolution are no exception. Scientists have tested many predictions about evolution and have found them to be strongly supported by the evidence. Five lines of compelling evidence support evolution: fossils, traces of evolutionary history in existing organisms, continental drift, direct observations of genetic change in populations, and the present-day formation of new species.

Explore how a population evolves.

16.3

Figure 16.5 The Power of Natural Selection
Plants that grow in deserts often have fleshy stems (for water storage), protective spines, and reduced leaves. These three plants evolved from very different groups of leafy plants. They now resemble one another because of convergent evolution, driven by natural selection for life in a desert. Thus their shared structures (fleshy stems, spines, reduced leaves) are analogous, not homologous. (a) *Euphorbia* [you-*FOR*-bee-uh], a member of the spurge family. (b) *Echinocereus* [ee-*KY*-noh-*SEER*-ee-uss], a cactus. (c) *Hoodia*, a fleshy milkweed.

Each chapter closes by returning to explore the topic introduced in the vignette in more detail.

The **Chapter Review** section covers the main factual and conceptual points of the chapter. This section is designed to help you review and assess your comprehension of the chapter material.

long enough, such interbreeding could "reverse" the speciation process and cause species to go extinct. Pollution would not be likely to have this effect if the cichlid species had been separated by barriers to reproduction that occur *after* mating takes place, such as gametic isolation or zygote death (see Table 18.1).

Finally, remember that speciation in cichlids depends on the ability of females to recognize dif-

ferences in the colors of males; when cloudy water impairs that ability, new species cannot form. Thus pollution from human activities appears to have two profound effects: it halts the formation of new cichlid species while simultaneously causing existing species to go extinct. We must reduce that pollution if we are not to destroy one of nature's most amazing evolutionary experiments, the cichlids of Lake Victoria.

Chapter Review

Summary

18.1 Adaptation: Adjusting to Environmental Challenges

- Adaptations result in an apparent match between organisms and their environment, but are not caused by intentional design.
- Adaptive evolution is the process by which the fit between organisms and their environment is improved over time.
- Adaptations help organisms accomplish important functions, such as mate attraction and predator avoidance.
- Adaptations can be improved in short periods of time (months to years).

18.2 Adaptation Does Not Craft Perfect Organisms

- Adaptive evolution can be limited by genetic constraints: lack of genetic variation gives natural selection little or nothing on which to act.
- Adaptive evolution can be limited by developmental constraints: the multiple effects of developmental genes can prevent the organism from evolving in certain directions.
- Adaptive evolution can be limited by ecological trade-offs: conflicting demands faced by organisms can compromise their ability to perform important functions.

18.3 What Are Species?

- Species are often morphologically distinct, but morphology is not a reliable way to distinguish some species.
- A species is a group of interbreeding natural populations that is reproductively isolated from other such groups.
- The definition of species in terms of reproductive isolation has important limitations. It does not apply to fossil species (which must be identified by morphology), to organisms that reproduce mainly by asexual means, or to organisms that hybridize extensively in nature.

18.4 Speciation: Generating Biodiversity

- The crucial event in the formation of a new species is the evolution of reproductive isolation.

- Speciation usually occurs as a by-product of the genetic divergence of populations from one another caused by natural selection, genetic drift, or mutation.
- Speciation usually occurs when populations are geographically isolated from one another long enough for reproductive isolation to evolve. Most new species are thought to arise by this process, which is called allopatric speciation.
- Speciation can also occur without geographic isolation. This process, called sympatric speciation, acts when part of a population diverges genetically from the rest of the population.
- Polyploidy is one way that many plants evolve new species during a single generation.

18.5 Rates of Speciation

- Speciation occurs rapidly in some cases, but it requires hundreds of thousands to millions of years in other cases.

18.6 Implications of Adaptation and Speciation

- Adaptations are the means by which organisms adjust to challenges posed by new or changing environments.
- Speciation is the means by which the diversity of life has come into being.
- Adaptation and speciation influence such practical matters as how we fight diseases and develop domesticated species.

◉ Review and Application of Key Concepts

1. Select an organism (other than humans) that you are familiar with. List two adaptations of that organism. Explain carefully why each of these features is an adaptation.

2. What is adaptive evolution? Apply your understanding of adaptive evolution to organisms that cause infectious human diseases, such as bacterial species that cause plague or tuberculosis. How do our efforts to kill such organisms affect their evolution? Are the evolutionary changes we promote usually beneficial or harmful for us? Explain your answer.

3. Imagine that a species legally classified as rare and endangered is discovered to hybridize with a more common species. Since the two species interbreed in nature, should they be considered a single species? Since one of the two species is common, should the rare species no longer be legally classified as rare and endangered?

4. Should species that look different and are ecologically distinct, such as the oaks in Figure 18.7, be classified as one species or two? These oak species hybridize in nature. Should species that hybridize in nature be considered one species or two?

5. High winds during a tropical storm blow a small group of birds to an island previously uninhabited by that species. Assume that the island is located far from other populations of this species, and that environmental conditions on the island differ from those experienced by the birds' parent population. Is natural selection or genetic drift (or both) likely to influence whether the birds on the island form a new species? Explain your answer.

6. Hundreds of new species of cichlids evolved within the confines of Lake Victoria, but some of these species live in different habitats within the lake and rarely encounter one another. Would you consider such species to have evolved with or without geographic isolation?

7. How can new species form by sympatric speciation? Why is it harder for speciation to occur in sympatry than in allopatry?

Key Terms

adaptation (p. 340)
adaptive evolution (p. 340)
allopatric speciation (p. 348)
geographic isolation (p. 348)
hybrid (p. 346)
hybridize (p. 346)

polyploidy (p. 349)
reproductive isolation (p. 345)
ring species (p. 348)
speciation (p. 346)
species (p. 345)
sympatric speciation (p. 349)

Self-Quiz

1. Species whose geographic ranges overlap but which do not interbreed in nature are said to be
   a. geographically isolated.
   b. reproductively isolated.
   c. influenced by genetic drift.
   d. hybrids.

2. Which of the following evolutionary mechanisms acts to slow down or prevent the evolution of reproductive isolation?
   a. natural selection
   b. gene flow
   c. mutation
   d. genetic drift

3. The splitting of one species to form two or more species most commonly occurs
   a. by sympatric speciation.
   b. by genetic drift.
   c. by allopatric speciation.
   d. suddenly.

4. The time required for populations to diverge to form new species
   a. varies from a single generation to millions of years.
   b. is always greater in plants than in animals.
   c. is never less than 100,000 years.
   d. is rarely more than 1,000 years.

5. Adaptations
   a. match organisms closely to their environment.
   b. are often complex.
   c. help the organism accomplish important functions.
   d. all of the above

6. Prezygotic and postzygotic barriers to reproduction have the effect of
   a. reducing genetic differences between populations.
   b. increasing the chance of hybridization.
   c. preventing speciation.
   d. reducing or preventing gene flow between species.

7. Evidence suggests that sympatric speciation may have occurred or be in progress in three of the following four cases. Select the exception.
   a. apple maggot fly
   b. squirrels on opposite sides of the Grand Canyon
   c. cichlid fish
   d. polyploid plants (or their ancestors)

8. The diploid number of chromosomes in plant species A is 8; the diploid number in plant species B is 16. If plant species C originated when a hybrid between A and B spontaneously doubled its chromosome number, what is the most likely number of diploid chromosomes in C?
   a. 8   c. 24
   b. 12   d. 48

The **Summary** presents all of the major points introduced in the chapter, organized by chapter section.

Review and Application of Key Concepts and **Self-Quiz** questions help you determine whether you have developed a deep understanding of the key concepts that were listed at the beginning of the chapter. Answers to all of these questions are in the back of the text.

The **Key Terms** list is a helpful checkpoint as you review the material. Page numbers are included so that you can easily return to the page where each term was introduced.

Biology in the News

The **Biology in the News** feature relates the chapter topics to real news stories and gives you an opportunity to ponder the ethical and societal questions they raise. Learn how to analyze what you hear about in newspapers and magazines. *Only in Discover Biology*!

Biodiesel Boom Well-Timed

BY JOHN GARTNER

Biodiesel fueling stations are sprouting like weeds across America, where production of the alternative fuel rose 66 percent in 2003. Experts say the rapid growth of the renewable fuel will stretch the country's tenuous petroleum supply while helping people breathe a little easier.

Photosynthesis transforms the sun's energy into the usable chemical energy contained in plants. Fossil fuels, which consist of "aged" plant matter, are rich storehouses of energy that we received from the sun millions of years ago. But we are consuming fossil fuels rapidly, and the supply is limited. The search is on for new sources of energy, and new ways are emerging to connect photosynthesis to the immediate production of usable fuel. Some plants can produce large quantities of oils that can easily be converted to diesel fuel. The production of biodiesel means that the "millions of years" are no longer needed.

Biodiesel can be produced from many sources. Restaurants routinely discard huge quantities of cooking fats as waste. Plants such as mustard and soy produce large amounts of oils. And some types of algae that grow extremely rapidly in shallow salt ponds are composed of 50 percent oil. The oils from all of these sources can be used for production of biodiesel fuel. Consider the last source: salt pond algae on algae "farms" fed by wastewater from animal farms and sewage treatment plants could theoretically produce enough oil to supply 100 percent of the diesel fuel needs of the United States. Moreover, a valuable by-product of the conversion process is glycerine, which is used in soaps and many other products.

But there is more to the biodiesel story than renewability: biodiesel is also less polluting than petroleum-based diesel. It's true that biodiesel—like petroleum diesel—produces carbon dioxide when burned, so switching to biodiesel won't solve the problem of reducing the contribution that atmospheric carbon dioxide makes to global warming. However, use of biodiesel would lessen our impact on the environment in other ways. For one thing, biodiesel burns more cleanly than petro-

leum diesel, so it produces fewer emissions. It contains almost no sulfur, so it cannot contribute to the formation of acid rain. For another, biodiesel is as biodegradable as sugar. We have seen the effects of petroleum oil spills. In contrast, any biodiesel "spills" will quickly disappear as organisms break them down into the products of catabolism: water and carbon dioxide.

As public awareness grows, demand for biodiesel is likely to grow. As demand grows, so will production. Biodiesel is a breaking news story.

Evaluating the news

1. Investing in the biodiesel industry carries great risk. Oil producers could afford to cut their prices to drive biodiesel producers out of business. Should biodiesel development be carried out primarily by small producers, or by large, existing oil companies?
2. We are consuming fossil fuels at enormous and ever-increasing rates, and there is much controversy over whether we should stress conservation of existing reserves or drilling for more oil. How do you think the U.S. government should react to the emerging business of producing and distributing biodiesel fuel? Alone or with your classmates, write a letter that you might send to your representatives in Congress expressing your views.
3. Genetic engineering is in the news every day. Some algae and mustard plants are highly efficient producers of the oils that can be easily transformed into biodiesel fuel. Do you see genetic engineering playing a role in the "plant petroleum" business, and if so, how?

SOURCE: *Wired News*, June 1, 2004.

Science Toolkit

Science Toolkit boxes explore the tools that researchers use in their exploration of biology. These boxes are designed to illustrate the process of science, a crucial element for developing your understanding of biology as it is presented in this text, as well as the biology you will encounter in other places, such as the media. Did you know that there is more than one kind of cloning? (Check out chapter 15.) And do you know what the calorie listing on the back of that bag of chips actually means? (Check out chapter 7.)

Counting Calories with a Bomb

Practically all labels on food products include information on how many calories the food contains. The label on a bag of potato chips might read:

Nutrition Facts
Serving size: 1 oz. (28 g/15 chips)
Amount Per Serving: Calories—140, Calories from Fat—70

But just what are calories, and how can anybody determine how many calories are in "15 chips"?

Calories are units of heat energy. Nutritionists and chemists define 1 calorie as the amount of energy needed to raise the temperature of 1 liter of water 1°C. That definition hints at the way food manufacturers and nutritionists determine how many calories are in a sample of food, such as 1 ounce of potato chips: they burn the chips and compare the amount of heat given off to the amount needed to heat a known amount of water.

To accomplish this, the food sample is burned in a device called a bomb calorimeter. The calorimeter is composed of a sealed container (the "bomb") surrounded by another sealed container with a known amount of water in it. A thermometer shows the temperature of the water before and after the sample is burned. Once the researchers know how much the temperature of the water was increased by burning the food, they can calculate the number of calories that were in the sample. The value of 70 calories from fat can be obtained by extracting the fats from the sample and burning them separately.

Bomb calorimeters measure 100 percent of the energy in a food sample, but our bodies are not 100-percent efficient, so the number of calories our bodies actually use is less than the calorimeter's number. Beer, of course, doesn't burn, and many other foods and beverages have a high water content. In these instances the water is evaporated, and the calorimeter burns only the dry remains. So now you know all that goes into measuring the calories in a regular beer (146 calories for 12 fl oz or 99 for 12 fl oz), bananas (109 calories for 118 grams), or even iceberg lettuce (7 calories for 55 grams).

An ignition wire starts the burning process in the bomb.

The thermometer measures the change in the temperature of the water as the food burns.

The stirrer distributes the heat evenly in the water.

The outer chamber contains a known amount of water.

Water

The bomb contains the food sample and pure oxygen.

Inside a Bomb Calorimeter

of a single carbon atom bonded to four hydrogen atoms (Figure 7.3a). Because the bonds are covalent, the carbon and hydrogen atoms share their electrons. However, the electrons tend to be held more closely to the carbon atom than to the hydrogen atoms. The carbon has, in a sense, gained electrons and become

more "electron-rich"; we also speak of the carbon as being in a reduced state.

If we burn some of the methane gas, the products of the combustion reaction are carbon dioxide (CO_2) and water (H_2O). What has changed? The hydrogen atoms have left the carbon atom and joined oxygen to form water. And

Biology Matters boxes relate the chapter topics to practical issues and choices we make in our everyday lives. Plus these fast facts are fascinating: Is chocolate really addictive? (See chapter 8.) What is your metabolism, and how much more energy do you burn while running than while sitting here reading this book? (Find out in chapter 7.)

Biology Matters

You'll Just Have to Tough It Out

Being sick is never enjoyable, but the more you take antibiotics, the more likely they won't work for you—or others—in the future. In fact, antibiotics don't help with many of the most frequent illnesses that make us sick: cold, flu, sore throats (except strep), bronchitis, most runny noses, and most earaches. Although many doctors are prescribing antibiotics less than they did in the past, understanding when to use antibiotics, and when they *won't* help, will help you to protect yourself and others. And don't forget—when you do need antibiotics, take them exactly as your health-care provider prescribes, and continue to take them until they run out, even if you're feeling better.

The following information will help you care for yourself, and feel better, the next time you get sick:

Q: *If antibiotics will not help me, what will?*
A: There are many over-the-counter products available to treat the symptoms of your viral infection. These include cough suppressants which will help control coughing and decongestants to help relieve a stuffy nose. Read the label and ask your pharmacist or doctor if you have any questions about which will work best for you.

A cold usually lasts only a couple of days to a week. Tiredness from the flu may continue for several weeks. To feel better while you are sick:

- Drink plenty of fluids.
- Get plenty of rest.
- Use a humidifier—an electric device that puts water into the air.

Contact your doctor if:

- Your symptoms get worse.
- Your symptoms last a long time.
- After feeling a little better, you develop signs of a more serious problem. Some of these signs are a sick-to-your-stomach feeling, vomiting, high fever, shaking chills, and chest pain.

Source: U.S. Department of Health and Human Services.

- Insist on prudent use of antibiotics in human, plant, and animal health care. Medical doctors and agriculturists frequently use antibiotics inappropriately. For example, the U.S. government estimates that half of the 100 million antibiotic prescriptions written by doctors each year are not necessary—often the conditions for which they are prescribed (such as colds and flu) are caused not by bacteria, but by other disease agents, such as viruses, that are not affected by antibiotics. Similarly, antibiotics are commonly used to increase the growth rates of farm animals, a practice that encourages the development of antibiotic-resistant strains of bacteria, including strains that attack people. Such inappropriate use

of antibiotics encourages the evolution of antibiotic resistance in the many species of bacteria that are normally found in our bodies. As we have seen, these resistant, harmless bacteria can then transfer genes for antibiotic resistance to other, harmful species of bacteria.

- Improve sanitation, thus decreasing the spread of resistant bacteria from one person to another. This action is critically important in hospitals, where the abundant use of antibiotics has led to the emergence of highly resistant strains of bacteria that can cause a variety of "hospital diseases," some of which can be lethal.

What do you think we as a society should do?

Helpful to know
The word *antibiotic* has roots in Greek words meaning "against life." Because antibiotics work only on bacteria—and not against a wide range of disease agents, as many people think—we might do better to refer to them by the more accurate term *antibacterials*, to avoid confusing them with substances used to kill fungi (*antifungals*) and viruses (*antivirals*).

Biology on the Job boxes showcase interviews with people in the workplace and relate their experiences to the chapter topics. Learn what a dog show judge is actually looking for (chapter 18), what firefighters depend on (chapter 22), and what goes through a hostage negotiator's head (chapter 30). *Only in **Discover Biology**!*

Biology on the Job

Dog Show Judge Extraordinaire

Over the course of a single year in the United States, dogs of 150 breeds participate in over 15,000 dog shows, some of them local and state shows, others national. Judging all these contestants requires the services of many dog judges, who themselves are rated for their expertise. Mr. Edd Bivin was selected by a collection of dog judges, dog trainers, breeders, and kennel clubs as the "Judging Legend 2002." Below are excerpts from a 2002 interview with Mr. Bivin by TheDogPlace, a Web site devoted to the interests of dog fanciers.

When, and why, did you decide to become a judge? I probably decided to become a judge when I was a kid. I showed good dogs [but I would just] get patted on the head. I believed in the sport and knew if I ever were to become a judge, I wouldn't pay any attention to where a dog comes from or the age of the individual handler. I would evaluate what I consider to be the best dog.

What do you do in your "other" life? I've been Vice Chancellor at Texas Christian University for 16 years, but I have been at the University for 30 years. That's in Fort Worth, Texas, where I have lived all my life. I don't have a lot of time for other hobbies but [my wife and I] do enjoy travel. We are involved as much as possible in the community, certainly in the arts. . . . I'm very physical so I also work out and exercise a lot. Keeps me sane, or somewhat sane (laughing), or let's say more sane and less crazy.

What do you most enjoy about judging? Obviously the dogs, and the people. I also consider judging to be a personal competition of Edd Bivin with himself. Every time I go in the ring I compete with myself, to do the best job I can do, on that day, within the circumstances with which I find myself.

Let's talk about the sport today. Are most breeds better than 10 years ago? No. Breeds are cyclical; they progress and they fall back. So I can't accept the term "most." I will tell you that there are many breeds that are better today and many breeds not as good as they were 10 years ago. A big part depends on who is directing breeding programs and pockets of interest around the country.

When you first look down the line [of dogs for show], what draws your eye? Balance and proportion. Carriage and outline. [*smile*] Outline and character.

Judging a Dog Show

Should showmanship and presentation be considered? Certainly. One should never miss a good animal with proper type and character. I become concerned about individuals applauding dogs or saying it's not a great such and such but it's a great "show dog." Dog shows are a format for the evaluation of breeding stock. Generic dogs are not the strength of any breed.

The Job in Context

When he judges a show, Mr. Bivin evaluates the degree to which a dog achieves the ideal standard for its breed. Judges such as Mr. Bivin are approved for selected breeds only; this is done to ensure that each judge thoroughly understands the physical form and character of the dog breeds that he or she evaluates.

When judges compare a dog with the ideal standard for that dog's breed, they influence a process similar in result to adaptive evolution. Prize-winning dogs are sought after as breeding stock, so they leave more offspring than dogs that show poorly. One might think that this process—coupled with the efforts of breeders, who also strive to improve the quality of the dogs over time—would ensure that dogs today were "better" than dogs of 10 years ago. But Mr. Bivin begged to differ; he said that many breeds were better, but many others were worse. There are two underlying biological reasons why a breeding program might produce dogs of declining quality. One is excessive inbreeding (mating between close relatives), which can lead to the fixation of harmful alleles. The other is "genetic hitchhiking," a term that refers to the fact that while selecting for one characteristic (such as a certain physical feature of the dog), it is possible to inadvertently promote other, less desirable features that are linked genetically to that feature. It takes considerable care and skill for breeders to avoid such pitfalls and continue producing dogs that meet the high standards imposed by judges like Mr. Bivin.

Harnessing the Human Genome

Main Message: Knowledge of the human genome has revolutionized biology and medicine while raising difficult ethical questions.

A Crystal Ball for Your Health

In February of 2001, the world witnessed a scientific milestone, the fruit of the combined efforts of thousands of researchers over a period of 15 years. For the first time in history, a draft copy of the DNA sequence of the human genome was available for perusal by anyone with a personal computer and access to the Internet. To many scientists, this represented the crowning achievement of twentieth-century biology, and many press conferences and articles touted it as such.

The successful sequencing of complete genomes gave birth to a new field of biology. In earlier chapters of Unit 3 you learned that genetics is the study of how genes are expressed and transmitted in cells and organisms. The field of **genomics** builds on genetics by seeking to understand the structure and expression of entire genomes and how they change during evolution. Genomics can be further distinguished from genetics by the scale of the questions asked. Genetics has a smaller focus: it is concerned with how individual genes function—either alone or together with a limited set of other genes—to control a phenotype. Genomics, on the other hand, takes a far more comprehensive view, monitoring the coordinated activities of all the genes in the genome. The expanded scale of the issues addressed by genomics has already had a major effect on other fields in biology.

How will scientific achievements such as the sequencing of the human genome affect our lives? The simple answer is that knowing the entire sequence of the human genome amounts to knowing the blueprint that dictates every biological process in our bodies. Encoded in our 3.3 billion base pairs of DNA are variations in our individual genes that are likely to directly affect our future and our health. While all of us have genomes that are

99.9 percent identical to one another, the 0.1 percent of difference has great bearing on our susceptibility to certain diseases and genetic disorders—even on our overall life span. Our DNA sequences will reveal not just how our bodies work in the present, but also how they are likely to function in the future.

One of the great powers of knowing the roadmap of the human genome lies in its ability to predict an individual's predisposition for developing many diseases and genetic disorders. To use this information to predict the health of individuals, we must have a simple way to identify and compare relevant genetic variations. The DNA in two unrelated individuals differs, on average, by only one base in a thousand. Yet these single-base-pair differences, known as **single nucleotide polymorphisms (SNPs)** [*SNIPS*], are one important source of genomic variation (Figure C.1). If one could associate the presence of a specific SNP or group of SNPs with susceptibility to a disease such as breast cancer, individuals who have this SNP pattern could be forewarned of oncoming disease far in advance, allowing timely therapeutic intervention. Naturally, the detailed matching of SNP profiles with disease susceptibility depends on knowledge of the human genome sequence as a basis for comparison.

Even before the first draft of the human genome sequence was released in 2001, scientists in Great Britain began laying the foundation for a massive database of human SNP profiles. Today that database relies on SNP profiles from hundreds of thousands of blood samples donated by adult volunteers. Physicians refer these volunteers to the project, whose staff record each volunteer's current health status and

Figure C.1 This Technician Holds Ten Thousand Different SNPs in a Single DNA Microarray

C2

Interlude essays conclude each unit and explore a major issue confronting society, from the facts about cancer to the reality of world hunger, from human evolution to efforts to build a sustainable society. Even if your instructor doesn't have time to assign these essays, they are interesting and important enough to read on your own.

List of Boxes

Brief Table of Contents

Table of Contents

Unit 2—Cells: The Basic Units of Life

Chapter 4: Chemical Building Blocks 68

Looking for Life among the Stars 69

Proteins That Can Take the Heat 90

Chapter 19: The Evolutionary History of Life 356

Unit 5—Animal Form and Function

Chapter 20: Maintaining the Internal Environment 378

Unit 6—Plant Form and Function

Chapter 31: Plant Structure, Nutrition, and Transport 604

Chapter 32: Plant Growth and Reproduction 624

THIRD
EDITION

Discover Biology

CHAPTER 1

The Nature of Science and the Characteristics of Life

Key Concepts

◉ To investigate the natural world, scientists use the scientific method, which involves four basic steps: making an observation, forming a hypothesis to explain that observation, generating predictions from the hypothesis, and testing those predictions. When an experiment upholds the predictions, a hypothesis gains strength and support. When the predictions are not upheld, a hypothesis is discarded or modified.

◉ All living organisms are thought to have descended from a single common ancestor. Therefore they share certain characteristics: they are built of cells, reproduce using DNA, develop, capture energy from their environment, sense and respond to their environment, show a high level of organization, and evolve.

◉ Viruses test the limits of our definition of life because they exhibit some characteristics of living organisms but lack others.

◉ Living organisms are just one part of the biological hierarchy, which ranges in scale from molecules at the lowest level to cells, tissues, organs, organ systems, individuals (living organisms), populations, communities, ecosystems, biomes, and finally the biosphere.

◉ Energy maintains a characteristic flow through biological systems. In most ecosystems, energy goes from the sun to producers (such as plants) to consumers and decomposers (such as animals and fungi that consume plants and other organisms). Food webs depict the complex relationships between organisms that eat and are eaten.

Nanobes: A Tempest in a Teapot
Is this nanobe the smallest living organism ever discovered?
Or is it not alive at all?

They're Alive! Or Are They?

Minuscule and mysterious, they had remained hidden since time immemorial, buried deep within Earth, 3 miles beneath the ocean floor off the coast of Western Australia. Then several years ago, a team of scientists spotted the tiny oddities while using ultra-high-powered microscopes to study ancient rocks retrieved from an oil-drilling site. Researchers dubbed the miniature things "nanobes" because they are so small that they measure only billionths of a meter, or nanometers. The discovering scientists proclaimed them to be the world's smallest living organisms, thereby setting off a controversy.

Nanobes look much like other, larger living organisms, with shapes similar to small molds. Scientists have also discovered that nanobes grow when brought into the laboratory. In fact, the nanobes grew so quickly that within weeks they went from being visible only with the world's most powerful microscopes to forming clumps easily visible to the naked eye. When researchers inspected these nanobe colonies, what they saw were fast-expanding threadlike mats.

Some researchers have hailed this work as a fundamental and important discovery. Some have even suggested that nanobes are likely to be extremely widespread, found in many kinds of rock, and that they may be the most abundant form of life in and on Earth. These scientists suggest that nanobes could be critical in the decomposition of Earth's rocks and soils, fundamental chemical processes that shape Earth as we know it. These researchers also claim that the reason the mighty nanobe has escaped detection all these years is its incredibly small size.

More recently, certain scientists have reported finding these minuscule life forms in the blood of living organisms, suggesting that nanobes may be responsible for human diseases. For example, previously undetected nanobes may aid in the formation of kidney stones, an excruciatingly painful ailment.

But while all scientists agree that nanobes are indeed very small, they cannot seem to agree on whether nanobes are the world's smallest organisms or not. That is because there still is no agreement on whether nanobes are actually alive.

Those skeptical of the discovery contend that nanobes cannot be living because they are too small to contain all the materials and machinery basic to all forms of life. Some researchers suggest that nanobes are nonliving crystals that form and increase in size under the proper conditions, similar to the way drying salt water can create a growing crust of salt crystals. In this view, nanobes are not alive; they merely mimic the growth of a living organism.

But isn't it a simple matter to tell living from nonliving? Any schoolchild can distinguish between the inanimate stone and the living being who skips it across a pond. Why can't scientists agree on whether something is alive? How can scientists determine whether nanobes are living organisms that may cause painful diseases and alter the face of the planet, or just inconsequential inanimate objects? What, in fact, is life? In this chapter we will examine this question by considering what the characteristics of life are and how biologists study life.

What is life? This deceptively simple question is, in many ways, one of the most profound. It underlies medical controversies ranging from abortion and determining when life begins to the right to die and when life ends. The same question reaches deep into the sciences as researchers seek to understand when life on our own planet first took hold and whether life has ever existed on other planets. One of the sciences asking these questions is **biology**, which is the scientific study of life, and the subject of this book. The goal of biology is to improve our understanding of living organisms, from microscopic bacteria to giant redwood trees to human beings.

In this chapter we begin with an exploration of science and how scientists ask and answer questions about living organisms. Then we address the question, "What is life?" We will see that all living things, diverse though they are, share characteristics that unify them, and that all living organisms are part of a greater biological hierarchy of life. We will also see how organisms play different roles in biological systems, through which energy flows in different forms. We close by returning to the question we asked at the outset: whether nanobes are living are not.

1.1 Asking Questions, Testing Answers: The Work of Science

Science is a method of inquiry, a rational way of discovering truths about the natural world. Because it is such a powerful way of understanding nature, science holds a central place in modern society. For scientists and nonscientists alike, knowing how nature works can be exciting and fulfilling. In addition, applications of scientific knowledge influence all aspects of modern life. Every time we take medicine, instant-message a friend, or run on a treadmill, we are enjoying the benefits of science.

Yet few of us have a good picture of how science works, how it generates knowledge, and what its powers and limitations are. This lack of understanding is unfortunate, for several reasons. First, an understanding of science can be personally rewarding: it can add to our appreciation of day-to-day events, leading to a sense of awe about how nature works.

A second reason is that science plays an increasingly important role in decisions made by society as a whole, as well as in personal decisions made by individuals. As a society, for example, we must evaluate the discoveries made by scientists when making decisions about global warming, the courtroom use of DNA fingerprinting, and even whether teachers in our public schools are able to provide their students with the most current scientific knowledge about evolution. As individuals, we must evaluate daily the reports of scientific studies we see in the news. Should I avoid genetically engineered foods, or are they safe (Figure 1.1)? Is it okay to use my cell phone for hours at a time, or could it cause damage to my body? Is drinking red wine really good for my heart? To make good decisions on these and many other issues, everyone—not just scientists—benefits from understanding how the scientific process works.

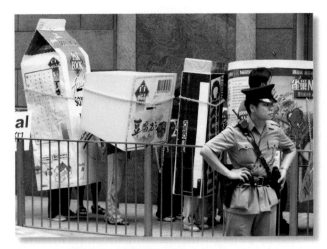

Figure 1.1 They Are What You Eat

More and more farmers around the world are growing genetically engineered crops. Soybeans and corn are among the most popular in this expanding list of biotech organisms. As scientists continue to debate the potential consequences of growing and eating such organisms, food products containing them have begun to appear on supermarket shelves—typically unlabeled. Here, Greenpeace activists dressed as genetically engineered food products protest in front of a dairy factory in Hong Kong, demanding that genetically engineered foods be labeled.

See this flowchart in action. 1.1

Figure 1.2 The Scientific Method

Scientists use the scientific method

To study the natural world, scientists follow a series of logical steps known as the **scientific method** (Figure 1.2). Many people assume that because it is logical, the scientific method must be a mechanical process done by rote, but as we shall see, serendipity and imagination have an important role to play as well.

The scientific method begins with **observations**. Scientists can make observations of the natural world in many different ways; some examples are looking through a microscope, diving to the ocean floor, and walking through a mountain meadow. The observation that started Dr. JoAnn M. Burkholder on her line of inquiry was one that many people made in North Carolina in the 1990s. Dr. Burkholder, a biologist at North Carolina State University, observed that huge numbers of fish were periodically being killed in mysterious die-offs; their bodies, covered with bleeding sores, were found floating by the millions in the region's estuaries, where the rivers meet the sea (Figure 1.3). While this observation is particularly dramatic, it shares a key feature with *all* observations that begin the scientific method: it strikes a biologist as curious or interesting, making him or her want to know, why is this happening?

After observation, the next step in the scientific method is the creative process of generating a **hypothe-**

sis, or an explanation for the observation. Some people describe a hypothesis as an educated guess. But generating a hypothesis to explain an observation is not always easy. For some time, researchers were thoroughly stumped by the North Carolina fish kills, unable to come up with explanations as to what might be causing them. As it turned out, several years earlier, colleagues of Dr. Burkholder had been dismayed to find that their laboratory fish were dying suddenly after exposure to local river water. When Dr. Burkholder looked into the problem, she found that a kind of microscopic organism called a protist greatly increased in numbers in the laboratory aquariums just before the fish died and decreased in numbers unless live fish were added. (Chapter 3 will describe protists in more detail.) When the fish die-offs began happening in the wild, Dr. Burkholder generated the hypothesis that the same tiny protist that appeared to have killed the laboratory fish, known by the name *Pfiesteria* [fih-STEER-ee-ah], was also causing the fish die-offs in local rivers.

But here is where the scientific method departs from other methods of inquiry. Dr. Burkholder used her hypothesis to generate **predictions** that she could then test. She asked herself, If my hypothesis is correct and *Pfiesteria* is causing fish die-offs in local rivers, what else can I expect to happen? Her first prediction was that *Pfiesteria* would be found in abundance in the river water during times

Dr. JoAnn Burkholder

(a)

Figure 1.3 A Fish Die-off Caused by *Pfiesteria*

The massive fish die-offs seen in many rivers appear to be caused by a protist called *Pfiesteria*. (*a*) Approximately 1 million fish were affected, as this photograph taken in 1991 of Blount Bay, in the Pamlico Estuary of North Carolina, illustrates. (*b*) Scientists observed bloody sores on the bodies of the dead fish.

(b)

when fish die-offs were happening and would not be found there when fish die-offs were not happening. Her second prediction was that the same *Pfiesteria* would be capable of killing healthy fish if introduced into the aquariums housing the fish in her laboratory.

The next step in the scientific process is testing the predictions of a hypothesis. One way scientists test their predictions is by devising and conducting **experiments**—controlled, repeated manipulations of nature. Another way to test predictions is by making further observations.

So what did Dr. Burkholder and her colleagues find when they began to test their predictions? First, they observed that *Pfiesteria* could indeed be found swarming in river regions where fish were dying, but not in those same regions when fish were not dying—upholding her first prediction. Then, in laboratory experiments, the researchers isolated *Pfiesteria* and exposed fish of many different kinds to the protist. The fish were quickly killed, upholding her second prediction.

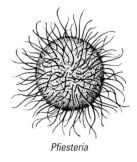

Pfiesteria

When scientists test a prediction of a hypothesis and find it upheld, the hypothesis is said to be supported. However, we cannot say that the hypothesis has been proved true. Proving a hypothesis true is not possible, because it could always fail when subjected to a different test. But when a prediction is not upheld, the hypothesis must be reexamined and changed, or else discarded. In both observation and experiments, Dr. Burkholder's predictions were upheld, providing strong support for the hypothesis that *Pfiesteria* is the culprit behind the massive fish kills.

Like all scientific studies, Dr. Burkholder's work raises as many questions as it answers, and her studies have generated much continuing research. Scientists still don't know exactly how *Pfiesteria* kills fish—by means of a toxin (a harmful or poisonous substance), by attacking the fish physically, or by some other method. Dr. Burkholder has hypothesized a toxin as the agent of destruction. Recent studies from other laboratories have found no evidence of a toxin coming from fish-killing strains of *Pfiesteria*. However, Dr. Burkholder maintains that the other labs tested the wrong strains of *Pfiesteria* and, in addition, handled them improperly. This kind of disagreement is a good example of how scientific progress is made.

So what will happen next? Researchers will continue their work using the scientific method—testing their hypotheses about toxins versus physical attacks, and keeping track as experiments bear out or fail to bear out their predictions. Along the way, their hypotheses will be supported, modified, or discarded. Slowly, in this trial-and-error fashion, scientists will come ever closer to a detailed understanding of this biological phenomenon.

In the meantime, debate about *Pfiesteria* is likely to remain heated. Some researchers hypothesize that *Pfiesteria*, and possibly *Pfiesteria* toxins, are also causing human health problems in the study area. And some biologists hypothesize that the fish die-offs are ultimately caused by big businesses—including factories and pig farms—dumping waste into waterways, which leads to outbreaks of killer strains of *Pfiesteria*.

As we have seen, scientists make observations, develop a hypothesis, make and test its predictions, then continue to test or else change or discard hypotheses. Together, these steps make up the scientific method. Clearly, Dr. Burkholder's studies are just the beginning of finding a thorough answer to why fish are dying, what the consequences are for humans, and what can be done about it.

The scientific method has limits

While the scientific method is a powerful way of studying the natural world, there are questions and areas of inquiry that science cannot address; this caution applies particularly to areas that do not lend themselves to testing. The scientific method cannot tell us, for example, what is morally right or morally wrong. Science can inform us, for example, about ways in which men and women differ physically, but it cannot tell us how we, as a society, should act on that information. In the realm of faith, science cannot prove or disprove the existence of a God or any other supernatural being. Nor can science tell us what is beautiful or lacking in beauty, which poems are most lyrical or which paintings most inspiring. So while science can exist comfortably alongside many different kinds of belief systems—religious, political, and personal—it cannot answer all their questions.

1.2 The Characteristics That All Living Organisms Share

Since living organisms range in size and shape from massive redwoods to microscopic bacteria, how can all the world's living organisms meet a single definition of life? In fact, the great diversity of body forms, habits, and sizes of organisms makes a simple, single-sentence definition of life impossible. But all living organisms are thought to be the descendants of a single common ancestor that arose billions of years ago (see the box on page 8). As a result, all forms of life, as diverse as they are, exhibit certain common features. Biologists define life by this set of shared characteristics (Table 1.1), which we describe in the sections that follow.

Table 1.1

The Shared Characteristics of Life

All living organisms

- are built of cells
- reproduce themselves using the hereditary material DNA
- develop
- capture energy from their environment
- sense their environment and respond to it
- show a high level of organization
- evolve

Living organisms are built of cells

The first organisms were single cells that existed billions of years ago. The **cell** remains the smallest and most basic unit of life. The simplest of organisms, such as bacteria, are still made up of just a single cell. Enclosed by a membrane, cells are tiny, self-contained units that make up every living organism.

Science Toolkit

Thinking Outside the Box: Thinking Inside the Soup

At the time Earth formed, 4.6 billion years ago, the planet was lifeless. The first sure signs of cellular life can be found in the fossil record as early as 3.5 billion years ago. But what happened during the intervening 1.1 billion years—how life arose from non-life—remains one of the most puzzling and hotly debated issues in science.

How do scientists devise hypotheses and experimental tests to explain an event—the origin of life—thought to have happened billions of years ago? This is one of the areas of biological research that not only benefits from thinking outside the box, but requires it. So what have scientists come up with? One of the many competing hypotheses suggests that life originated in hot springs deep on the ocean bottom. Another hypothesis proposes that life's birthplace lies near underwater volcanoes or in hot-water geysers on land, such as Yellowstone's Old Faithful. Still another suggests that life, or its building blocks, did not arise on Earth at all, but arrived here from another planet (such as Mars) by traveling through space on an asteroid or meteorite. All of these competing hypotheses have their strengths and weaknesses.

The best-known and longest-standing hypothesis, however, is the "soup theory" of life. In 1953, Stanley L. Miller, then a young graduate student, came up with a brilliant way to test this hypothesis, and his experiment is now considered a classic in modern biology for its creativity. Miller attempted to re-create the beginnings of life by simulating the conditions—hot seas and lightning-filled skies—of early Earth.

To make his "primordial soup," Miller began with water, which he kept boiling. The water vapor rose into the simulated atmosphere, which contained the gases methane, ammonia, and hydrogen—some of the gases that Earth's early atmosphere is thought to have harbored, and the very sort that could have belched forth from ancient volcanoes. Miller added an electrical spark to simulate lightning. He then cooled the sparking vapors until they turned to liquid, and the resulting liquid, like rain falling back to the seas, returned to the boiling water. Miller allowed his apparatus—a closed system in which the water continuously recycled between boiling seas and sparking skies—to "cook" for a week.

The whole situation was absurd—not least of all the idea that a person could simply cook up life's crucial components as Earth had done so long ago from such a seemingly simple recipe—except that it appeared to work. When Miller examined the contents of his primordial soup at the end of 7 days, he discovered that from water and simple gases alone, he had created an array of molecules critical to the origin of life, including two amino acids, the building blocks of proteins. Since then, other such "soup" experiments attempting to re-create Earth's early conditions have produced other key biological molecules, including all the common amino acids as well as sugars, lipids (fats), and the basic building blocks of the nucleic acids DNA and RNA.

Many questions remain: How closely do these primordial soup experiments actually mimic early Earth conditions? Once a soup of critical molecules developed, how did those molecules get organized into larger molecules and into the first cells able to gather energy and reproduce? Did the first cells float freely in such a soup, or did they not become cells until their parts fell out of the soup onto a surface, such as the ocean bottom?

Miller's simple experiment and others like it illustrate that even though life originated billions of years ago, scientists today can study those early processes in the laboratory. And researchers have continued to find fresh new ways to peer into the ancient origin of life. Some use DNA to study the family tree of all living organisms in an attempt to understand better the kind of organism most likely to have been the ancestor of us all. Others continue to scour meteorites and other extraterrestrial objects for hints of what might once have landed on Earth. How Earth made the crucial leap from barren stone to cradle of life remains one of science's most difficult and most fascinating questions—and the perfect subject on which scientists can continue to hone their creative talents.

The Primordial Soup of Life

Stanley Miller cooked up a brew containing many of the molecules necessary to life. Because the conditions inside his apparatus (shown here) imitated conditions on early Earth, his experiment supports the hypothesis that life could have originated in the early seas under lightning-filled skies.

Larger organisms, such as monkeys and oak trees, are made up of many different kinds of specialized cells and are known as **multicellular organisms**. In these cases, cells can be viewed as different kinds of building blocks that together make up an organism. For example, a monkey's body has skin cells, muscle cells, intestinal cells, brain cells, and so on (Figure 1.4).

Living organisms reproduce themselves via DNA

One of the key characteristics of living organisms is that they can reproduce. Single-celled organisms, such as bacteria, can reproduce without sex by dividing into two new genetically identical copies of themselves. In contrast, multicellular organisms reproduce in a variety of ways, both with sex and without. Humans and other mammals, for example, can reproduce by having sex, which allows a specialized reproductive cell in the male, called a sperm, to fertilize (join with) the female's specialized reproductive cell, the egg; a certain amount of time later—9 months in humans—the female gives birth

Figure 1.4 The Basic Building Block of Life: The Cell
Like all organisms, this Sykes' monkey is composed of cells. The intestinal cells shown in the inset are just one of the many different kinds of cells that make up the monkey's body.

to young. Some plants reproduce using a form of sex in which their flowers exchange pollen (the equivalent of sperm), which fertilizes cells equivalent to eggs inside the flowers. The plants then produce seeds, which develop into young plants. Multicellular organisms can also reproduce without sex—for example, by simply budding off new individuals, as sponges and some plants do. Familiar houseplants such as "mother of millions" and spider plants can produce tiny plantlets that drop off the parent plant and take root independently. In addition, some animals, such as the plant-sucking insects known as aphids, can reproduce clonally. A single female aphid can produce genetically identical copies of herself without ever mating with a male.

Whether organisms produce seeds, lay eggs, give birth, or just split in two, they all reproduce using a molecule known as **DNA** (**deoxyribonucleic** [dee-*OX*-ee-*RYE*-bo-noo-*CLAY*-ic] **acid**). DNA is the hereditary, or genetic, material that transfers information from parents to offspring. Briefly, the DNA molecule can be thought of as a blueprint or set of instructions for building an organism. A DNA molecule is shaped like a ladder that is twisted into a spiral along its length, a form known as the double helix (Figure 1.5).

This molecule contains a wealth of information—all the information necessary for an organism to create more cells or to grow from a fertilized egg into a complex multicellular organism that will eventually produce its own offspring. DNA is stored in every cell in every living organism. Life, no matter how simple or how complex, uses this inherited blueprint. We will discuss DNA in detail in Unit 3.

Learn more about the characteristics all life shares. 1.2

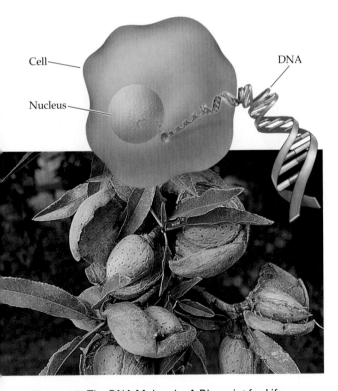

Cell——
Nucleus——
DNA

Figure 1.5 The DNA Molecule: A Blueprint for Life
DNA is a hereditary blueprint found in the cells of every living organism. DNA provides a set of instructions by which an individual organism can grow and develop, and which it can pass on to its own young so that they can grow and develop. This tree has produced young in the form of almonds, DNA-containing seeds that will eventually develop into new almond trees.

Living organisms develop

Using DNA as their blueprint, organisms come into being by building themselves anew every generation, a process known as **development**. A human sperm and a human egg fuse to form a single cell that grows and develops inside a woman's body. Nine months later, a fully developed baby is born as a living, breathing human being, who then develops further into an adult (Figure 1.6). All organisms arise through some process of development in which one organism arises from another organism's cell or cells, whether that organism completes its development by splitting off as a single cell or grows into something as complicated as a multicellular cactus or octopus.

Living organisms capture energy from their environment

To carry out their growth and development, and simply to persist, all organisms need energy. Organisms use a wide variety of methods to capture this energy from their environment.

Plants are among the organisms that can capture the energy of sunlight through a chemical process known as photosynthesis, by which they produce sugars and starches. (We will discuss photosynthesis in detail in Chapter 8.) Some bacteria can also harness energy from chemical sources such as iron or ammonia through an entirely different chemical reaction. Many organisms, including animals, fungi (mushroom-producing organisms and their kin), and certain one-celled organisms, can gather energy only by consuming other organisms. And some organisms acquire energy by multiple means; for example, certain plants can both photosynthesize and capture insects.

Animals exhibit many different ways of capturing energy (Figure 1.7). For example, some insects have mouthparts that they use to suck the nutritive juices from plants. Cheetahs run so quickly that they can chase and capture a fast-moving source of energy such as a gazelle. And many animals—including humans—get their energy by eating both plants and animals.

Living organisms sense their environment and respond to it

Living organisms are able to sense many aspects of their external environment, from the direction of sunlight (as

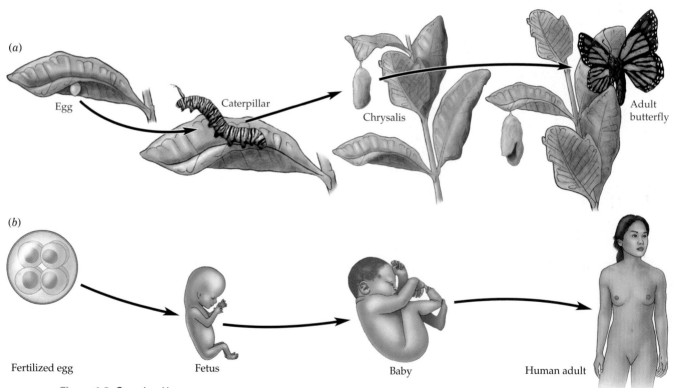

(a)

Egg   Caterpillar   Chrysalis   Adult butterfly

(b)

Fertilized egg   Fetus   Baby   Human adult

Figure 1.6 Growing Up

All living organisms develop. (*a*) A monarch butterfly develops through several stages, from egg to caterpillar to chrysalis to flying adult. (*b*) In humans, after a sperm fertilizes (fuses with) an egg, the resulting cell eventually develops into an adult.

Figure 1.7 Finding the Energy

While plants can capture energy from sunlight through photosynthesis, animals must get their energy by eating other organisms. This green tree python is ingesting a source of energy that it has captured.

Figure 1.8 Here Comes the Sun

All living organisms must be able to sense and respond to stimuli in their environment. These Maryland sunflowers have all detected rays of sunshine, turning toward their light and warmth.

many plants can do; Figure 1.8) to the presence of food and mates. Like humans, many animals can smell, hear, taste, touch, and see the environments around them. Some organisms can also sense things humans cannot, such as ultraviolet and infrared light, electrical fields, and ultrasonic sounds. Some bacteria can even act like a living compass, sensing which direction is north and which direction is up or down by means of magnetic particles within them.

Organisms have the ability to sense their internal environment as well. A plant, for example, reacts to a lack of internal water by taking action to conserve water, closing off openings in its body through which water could be lost and decreasing photosynthesis, an activity that uses water.

All organisms gather information about their internal and external environment by sensing it, then respond appropriately for their continued well-being.

Living organisms show a high level of organization

Living organisms are made up of parts that are spatially organized in a very specific way. Human bodies, for exam-

ple, exhibit highly organized internal organs and tissues. For the body to function properly, not only must most of those body parts be present; they must also have a particular spatial arrangement with respect to one another. If you look at a flower, you will see that the parts are far from randomly organized. A rock is still inanimate rock if broken into pieces, but a living organism must maintain a particular spatial organization to function properly (Figure 1.9).

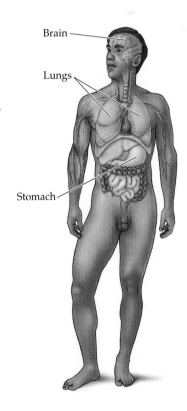

Brain

Lungs

Stomach

Figure 1.9 Staying Organized

Organization—having each part in its proper place—is crucial to an organism's functioning. The internal organs of a human being, such as the stomach, lungs, and brain, must be arranged in a specific way to function properly.

Biology Matters

How Much to Eat of What?

Humans require food for energy to survive, develop, and flourish. But as part of a disturbing recent trend, burger- and ice cream–loving Americans, along with increasing numbers of people throughout the developed world, are suffering from obesity, or an abnormally high proportion of body fat. This can lead to increased risk for conditions like high blood cholesterol, heart disease, and diabetes.

According to the National Center for Health Statistics, data collected in the United States from 1999–2002 indicates that:

64 percent of adults age 20 years and over are
    overweight or obese
30 percent of adults age 20 years and over
    are obese
15 percent of adolescents age 12–19 are
    overweight
15 percent of children age 6–11 years are
    overweight

GRAINS   VEGETABLES   FRUITS   MILK   MEAT & BEANS

To lose the extra weight, many people (college students among them), try diets of one sort or another. Recently, low-carbohydrate diets, such as the Atkins Diet and the South Beach Diet, have become very popular—and controversial. Such diets severely limit or entirely avoid carbohydrate-rich foods, from cakes and cookies to breads, rice, and even fruits. These diets do not count calories; instead, people are free to indulge in however much they'd like of high-fat foods (e.g., steak, fried eggs, and cheese), which were long considered dietary no-nos.

Many people are adamant advocates of these diets, saying they work when nothing else does. Yet some nutritionists say that low-carbohydrate diets are merely the latest fad. The startling popularity of the Atkins Diet, the South Beach Diet, and others like them has sparked a reassessment of long-held notions about nutrition. Even the government has reexamined its approach, replacing its traditional, one-size-fits-all food pyramid (developed in 1992 to represent the *Dietary Guidelines for Americans*, as published by the U.S. Department of Agriculture and the Department of Health and Human Services) with twelve different pyramids in 2005, tailored to varying lifestyles. This was done because "the American diet is not in balance. On average, Americans don't eat enough dark greens, orange vegetables, legumes, fruits, whole grains, and low-fat milk products. They eat more fats and added sugars."

New features of the current food pyramids include:

An emphasis on physical activity:

30 minutes most days to reduce the risk of chronic disease
60 minutes most days to help manage body weight
60 to 90 minutes daily to sustain weight loss

Encouragements to consume particular food groups:

2 cups of fruit and 2.5 cups of vegetables per day
3 or more ounce-equivalents per day of whole-grain products
3 cups per day of fat-free or low-fat milk or equivalent milk
    products

Limits on the consumption of other food groups:

Keep total fat intake between 20 and 35 percent of calories, with
    most fats coming from sources of polyunsaturated and mono-
    unsaturated fatty acids, such as fish, nuts, and vegetable oils.
Consume less than 2,300 mg of sodium per day (approximately
    one teaspoon of salt).
Limit alcoholic beverages to one drink per day for women and
    up to two drinks per day for men.

SOURCE: *Dietary Guidelines for Americans, 2005.*

Living organisms evolve

In the process of development, individual organisms change over short spans of time, developing from seeds into mature trees or from eggs into adult fish. Over longer time spans, whole groups of organisms change. A **species** is a group of organisms whose members can breed with one another to produce fertile offspring (that is, offspring that can themselves reproduce), but who do not, or cannot, breed with other organisms. For example, mountain lions, monarch butterflies, and Douglas fir trees are distinct species. When the characteristics of a species change over time or new species come to exist, the process is known as **evolution**. (See Unit 4 for more on evolution.)

Pronghorn antelope, for example, are the fastest-running creatures in North America. Over time, pronghorns as a species became more fleet because only those individual pronghorns that could outrace their predators survived to reproduce. The young of these survivors tended to be speedy themselves because they shared much of their DNA with their speedy parents.

A nineteenth-century Englishman named Charles Darwin, considered the father of evolutionary biology and one of the great thinkers of all time, made the first convincing argument that organisms evolve in his revolutionary book *The Origin of Species*. As Darwin explained, in the struggle to survive and the contest to reproduce, characteristics of species—such as the speed at which a pronghorn can run—tend to change over time. Any feature of a group of living organisms can change over time, or evolve (Figure 1.10). Features that are advantageous in the struggle to survive and reproduce, such as the pronghorn's ability to run quickly, are known as **adaptations**. Because evolutionary change can explain so many of the features of living organisms—from how fast pronghorns run to the sharpness of a shark's tooth—evolution is considered the central, unifying theme in biology.

1.3 Viruses: A Gray Zone Between Life and Nonlife?

Everyone eventually becomes all too familiar with viruses. Many people are laid low each winter by the influenza virus. If you come down with the flu, you cough and sneeze, your temperature rises, and your body aches all over. The reason you are suffering is that the influenza virus is infecting your body's cells, reproducing throughout your nose, throat, and lungs. In response, specialized defensive cells of your immune system fight back in several ways. One is

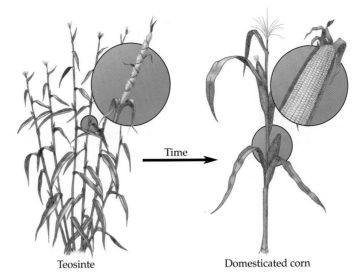

Figure 1.10 Living Organisms Evolve
Species of living organisms change over time, or evolve. Domesticated corn plants have many large kernels, or seeds, on fat cobs. These familiar plants evolved from the wild species known as teosinte, which has fewer, smaller seeds on skinnier cobs.

to attack and destroy cells infected by the virus. Another is to turn up your body temperature in order to prevent the virus from reproducing—hence your high fever.

Even as this microscopic virus proliferates throughout your cells—whether for days or weeks—it is evolving rapidly. Viruses evolve so quickly that they are able to evade many kinds of defenses, which makes them difficult for both medicines and your body to fight.

Many different viruses in addition to the influenza virus affect people, and still more kinds of viruses infect all the different forms of life. Viruses are such powerful foes that they certainly seem alive, just like the many organisms they attack. But, in fact, viruses are hard to characterize. Like living organisms, viruses reproduce, show a high level of organization, and evolve. Yet all viruses, including the influenza virus, lack some of the basic characteristics of life.

For one thing, viruses are not made up of cells. A virus is essentially just a hereditary blueprint, a piece of genetic material wrapped in a coat of proteins. Another difference is that viruses lack the structures necessary to perform nearly all the activities that living organisms do, including reproduction and energy collection. To accomplish these tasks, viruses must get the cells of other organisms to do their work for them, which they accomplish by invading those cells, just as the influenza virus invades the cells of your body. A third unusual feature of viruses is that, unlike living organisms, the genetic material they

Charles Darwin

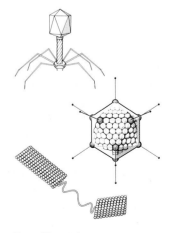

Three different viruses

pass from one generation to the next is not always DNA. Some viruses use a related molecule, known as RNA, or ribonucleic acid. (For more on RNA, see Chapter 13.)

Viruses defy easy definition, testing the limits of our definition of life, as they exhibit some characteristics of living organisms but lack others. Many scientists consider viruses to be nonliving. Other biologists place viruses in a gray zone between life and nonlife. Still others consider them a very simplified form of life.

Explore the biological hierarchy.

1.3

1.4 The Biological Hierarchy

Biologists find it useful to organize the great array of living organisms—individuals like each of us—into a **biological hierarchy** (Figure 1.11). But living organisms are just one part of that hierarchy; it also includes the many components that make up an individual organism as well as elements in the individual's environment. The hierarchy has many levels of organization, ranging from molecules at the lowest level up to the entire biosphere at the highest level. In scale, the hierarchy ranges from one-millionth of a meter (the approximate length of a molecule) to 12 million meters (the width of Earth).

Levels in the biological hierarchy provide biologists with a framework for studying living organisms

The biological hierarchy includes several levels below and including the individual. At its lowest level, it begins with the **molecules** found primarily in living organisms. An example is DNA, which carries the blueprint for building an organism. Many such specialized molecules are organized into the next level of the hierarchy; namely, cells, the basic unit of life. As mentioned earlier, some organisms, such as bacteria, consist of only a single cell. In other, multicellular organisms, cells of different kinds are organized into **tissues**, specialized, coordinated collections of cells that perform particular functions in the body, such as muscle or nerve tissues. Sometimes these tissues are organized into **organs**, body parts composed of different tissues that are organized to carry out specialized functions, such as hearts and brains. Groups of organs can function together in **organ systems**; for example, the stomach, liver, and intestines are all part of the organ system known as the digestive system. Groups of organ systems work together for the benefit of a single organism—an **individual organism**, often referred to simply as an **individual**.

There are also several levels in the hierarchy above the individual. Each individual is a member of a **population**, a group of similar organisms living and interacting in the same area. Examples include the population of field mice in one field and the population of blueberry bushes on a mountaintop. Groups of populations from different species that live and interact with one another in a particular area are known as a **community**—for example, the community of insect species living in a forest.

Communities, along with the physical environments they inhabit, are known as **ecosystems**; for example, a river ecosystem includes the river itself as well as the communities of organisms living in it. At the next level are **biomes**, large regions of the world that are defined on land by the plants that grow there (for example, the arctic tundra) and in water by the physical characteristics of the environment (for example, coral reefs). Finally, each biome is part of the one **biosphere**, which is defined as all the world's living organisms and the places where they live.

Knowing the biological hierarchy can be useful in everyday life

Few people are familiar with the concept of the biological hierarchy, so it might at first seem like a scientific idea important only to biologists. However, many of us interact with aspects of the biological hierarchy every day—whether at home or at work, shopping for food or getting health care, as the following examples show:

- When we take prescription drugs, we are using particular *molecules* that interact with very specific natural molecules in our bodies as well as with any other drugs and vitamins we may be taking at the same time.
- When we drink water from a household tap that is fed by a municipal water supply, we benefit from monitoring by water-quality engineers, who check for harmful *cells* such as toxic *E. coli* [ee KO-lye] bacteria and *Giardia* [jee-AR-dee-uh] (a protist), both of which can cause illness in people.
- If we get a professional massage, we trust that the massage therapist has been trained to know that the body has various kinds of *tissues*. With expert hands, the therapist uses techniques to relax certain tissues, particularly muscles and tendons.
- Some grocery shoppers are keenly aware that *organs* are made of specialized—and different-tasting—tissues, some days shopping for liver to cook up with onions while carefully avoiding the packages of brain or tongue. Other organs make their way onto the Thanksgiving dinner table: giblet gravy is made from

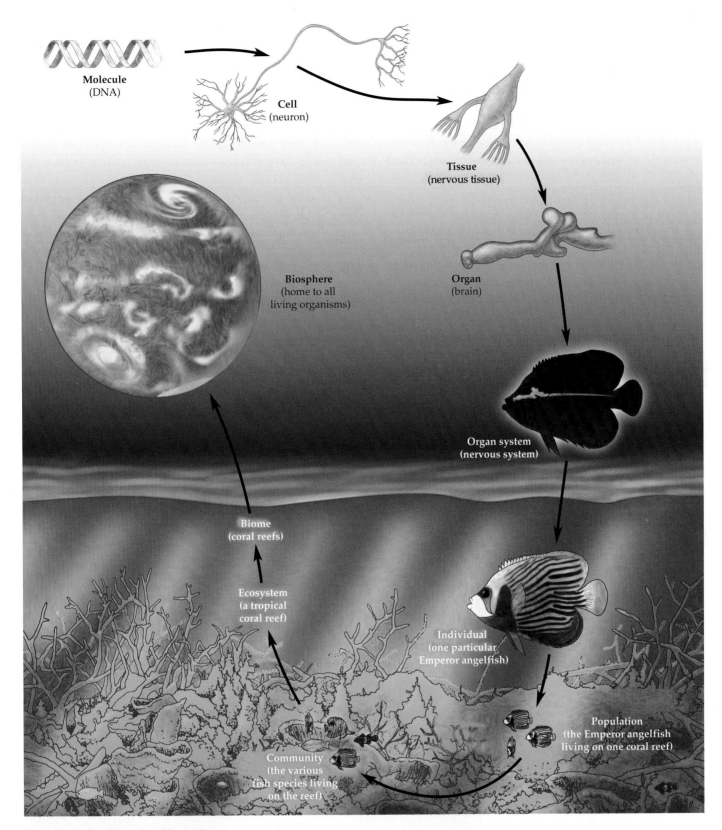

Molecule
(DNA)

Cell
(neuron)

Tissue
(nervous tissue)

Organ
(brain)

Biosphere
(home to all
living organisms)

Organ system
(nervous system)

Biome
(coral reefs)

Ecosystem
(a tropical
coral reef)

Individual
(one particular
Emperor angelfish)

Population
(the Emperor angelfish
living on one coral reef)

Community
(the various
fish species living
on the reef)

Figure 1.11 The Biological Hierarchy

Levels of biological organization can be traced from molecules found in organisms (such as DNA) all the
way up through levels of increasing organization to the biosphere, which includes all living organisms and
the places where they live. Aquarium enthusiasts may recognize the Emperor angelfish pictured in the
hierarchy.

some of the turkey's organs—such as heart and liver—which are wrapped in paper and packed inside most turkeys by supermarket butchers.

- Medical doctors are well versed in many levels of the biological hierarchy. When we visit a medical specialist, it is often to take advantage of his or her knowledge of the ailments of a particular *organ system*. For example, specialists dealing with the nervous organ system are called neurologists, while those dealing with women's reproductive organ systems are called gynecologists.
- *Individual organisms* are the most familiar level of the hierarchy because we deal with individuals every day. Each member of our household is an individual, and so is each pet and each separately potted houseplant. When the hostess in a restaurant asks us, "How many for dinner?" we know she's asking how many individuals are in our party.
- Many fish are declining in numbers around the world's oceans. As a result, a number of nations limit how many fish can be taken annually from certain *populations* (for example, Atlantic cod in the once-rich Georges Bank region off the coast of New England, as in Figure 1.12). Population size and regulation in turn affect which fish are available in the supermarket and how expensive they are.
- If you enjoy honey, you know that bees collecting pollen from different *communities* of flowers—say, a meadow of blooming clover and alfalfa or a field full of wild buckwheat—will produce very different-tasting honey.

- When you take a walk through the woods, you are encountering a forest *ecosystem*, though you may not be able to sense all the many organisms within that ecosystem.
- At any zoo or aquarium you are likely to find displays on *biomes* such as tundras and coral reefs. Natural history museums often illustrate biomes as dioramas.
- Only a small number of people—the men and women who fly in space—have so far been in a position to see the entire *biosphere*. Astronauts often say they have been forever changed by the awe-inspiring sight of that familiar blue and white orb, full of life, that is our biosphere. We can gain hints of their experience, however, by looking at color photographs and film footage taken aboard spacecraft.

1.5 Energy Flow Through Biological Systems

In most biological systems on Earth, the original source of energy is the sun. Energy flows first from the sun to plants and other photosynthesizing organisms such as green algae. These organisms are called **producers** because they use the sun's energy (in the form of light) to produce chemical energy in the form of sugars and starches. That energy can then be harvested by **consumers**, organisms that eat either producers or other organisms whose energy ultimately derives from producers or the remains of those organisms. Animals are a familiar example of consumers. There are

Figure 1.12
One Fish, Two Fish
In the icy waters off the New England coast, fishermen clean their haul of Atlantic cod. The capture of this tasty fish is regulated to protect its remaining populations. From the fishermen's point of view, it is the biological level of the individual that matters, since each fish, quite literally, counts.

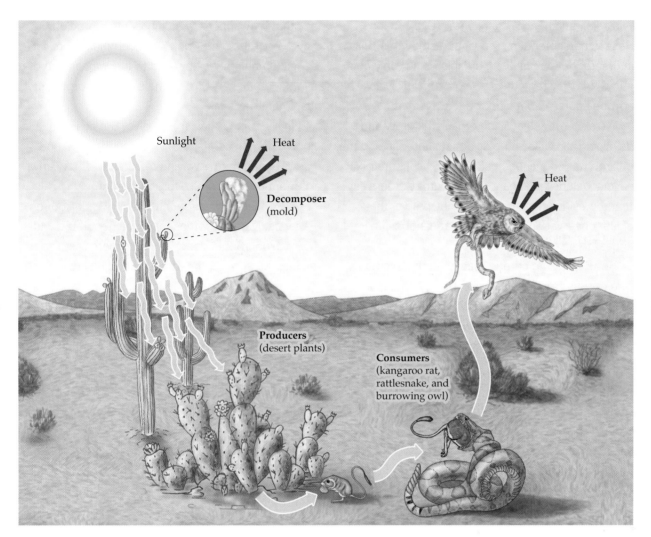

Figure 1.13
A Desert Food Web

Sunlight is captured by producers, in this case desert plants. Energy then flows to consumers, such as a fruit-eating kangaroo rat. These consumers are then eaten by other consumers, such as snakes, which can then be eaten by other consumers in turn. Decomposers, such as the mold on fruit, likewise get their energy either from producers or from organisms whose energy ultimately derives from producers. Decomposers and consumers give off energy in the form of heat. Energy flows from the sun to producers and then to consumers and decomposers throughout this food web.

Sunlight

Heat

Decomposer (mold)

Heat

Producers (desert plants)

Consumers (kangaroo rat, rattlesnake, and burrowing owl)

also **decomposers**, organisms that eat dead organisms or cast-off parts of living organisms, thereby returning nutrients to the biosphere. Many fungi are decomposers. As a result, energy flows almost entirely in one direction through the biosphere: from the sun to producers and then to consumers and decomposers, which give off energy as heat. A depiction of producers, consumers, and decom-posers that illustrates who eats whom is known as a **food web** (Figure 1.13).

Not all biological systems use the sun as the original source of energy, however. On the deep sea floor, far from the sun's light, there are entire ecosystems that depend instead on the energy that bacteria harness from nonliving materials, such as iron or ammonia.

Are Nanobes Indeed the Smallest Form of Life?

How do the newly discovered nanobes stack up against the characteristics of life? Do they meet all the criteria, or do we reject the hypothesis put forth by researchers that these are indeed the smallest form of life?

We know that all living organisms are built of cells. Scientists studying nanobes say that they appear to have a cell-like structure, with a cell-like outer surface, or membrane, encircling the nanobe. So far, so good.

Living organisms contain DNA. Do nanobes? Scientists have observed that when chemical stains known to produce a particular color in the presence of DNA are applied to nanobes, nanobes also stain this color, and hence must contain DNA. Researchers intend to recover and study that DNA in greater detail. But living organisms not only contain DNA, they reproduce using that DNA. Has anyone seen a nanobe reproduce? Not yet.

We also know that living organisms grow and develop. But while some scientists, after seeing the enlargement of nanobe colonies in the laboratory, have concluded that nanobes can grow, others say that the evidence is too weak. They predict that closer observations will show that nanobes only appear to be growing.

How about the criteria that living organisms capture energy from their environment, and sense and respond to their environment? Scientists still have not been able to determine whether nanobes eat or otherwise capture energy or respond to their environment.

Living organisms show a high level of organization. Scientists examining the structure of nanobes say that they appear to be highly structured and have very much the "look" of a complex microscopic organism, as opposed to the appearance of a simple, regular array of crystals or an amorphous, inanimate material.

Finally, living organisms evolve. But again, little is known here in connection with nanobes. No one has yet seen them evolving to adapt to a new environment or otherwise evolving over time.

Most scientists agree that the jury is still out on nanobes. More hypotheses will be generated and more experiments done, as much more remains to be learned about them, but, so far, like viruses, nanobes defy easy definition. However we choose to define such fascinating entities—as living or nonliving—viruses and nanobes show us the limits of our strict definitions while making it clear just how diverse life on Earth can be.

Chapter Review

Summary

1.1 Asking Questions, Testing Answers: The Work of Science

- Science is a rational way of discovering truths about the natural world.
- To answer questions about the natural world, scientists use the scientific method, which has four steps: make observations, devise a hypothesis to explain the observations, generate predictions from that hypothesis, and test those predictions.
- Scientists can test their hypotheses either by making further observations or by performing experiments (controlled, repeated manipulations of nature) that will either uphold or not uphold the predictions.
- A hypothesis cannot be proved true or false, only upheld or not upheld. If the predictions of a hypothesis are not upheld, the hypothesis is rejected or modified. If the predictions are upheld, then the hypothesis is supported. Even so, it is always possible that other predictions of the same hypothesis may not be upheld.
- The scientific method is not a useful way of addressing questions that cannot be tested, such as moral questions or the existence of a supernatural being.

1.2 The Characteristics That All Living Organisms Share

- The great diversity of life on Earth is unified by a set of shared characteristics.
- All living organisms are built of cells, the building blocks of living organisms. Some organisms are single-celled, composed of one building block; some are multicellular organisms, composed of many.
- All living organisms reproduce using DNA.
- All individual living organisms change during the course of their lives in a process called development.
- All living organisms capture energy from their environment.
- All living organisms sense and respond to their external and internal environment.
- The bodies of living organisms show a high level of spatial organization.
- Species of living organisms can change over generations, a process known as evolution. Adaptations are features that have evolved to allow organisms to survive and reproduce better. Evolution is the central theme in biology.

1.3 Viruses: A Gray Zone Between Life and Nonlife?

- Viruses share some of the characteristics of living organisms: they evolve, they reproduce, and they show a high level of spatial organization.
- Viruses lack some of the characteristics of living organisms: they are not made of cells, and they lack the structures necessary to perform activities essential to life, such as energy collection. Some viruses lack DNA.
- Because they exhibit only some of the characteristics of living organisms, biologists disagree on whether viruses should be considered life, nonlife, or something in between. Viruses also illustrate how difficult it is to define life precisely.

1.4 The Biological Hierarchy

- Living organisms are part of a biological hierarchy.
- Molecules are at the lowest level of the hierarchy and include such things as DNA. Specialized molecules can be organized into cells, the basic unit of life.
- Cells are organized into tissues, specialized tissues are organized into organs, and organs working together make up an organ system. Organ systems work in concert for the next level of the hierarchy, individuals.
- The individuals of a given species in a particular area constitute populations. Populations of different species in an area make up communities. Communities, along with the physical habitat they live in, constitute ecosystems.
- Ecosystems make up biomes, large regions of the world that are defined on land by the kinds of plants that grow there and in water by the physical characteristics of the environment. All the biomes on Earth make up our one single biosphere.
- The biological hierarchy can be seen and recognized by people in their everyday lives.

1.5 Energy Flow Through Biological Systems

- Biological systems are made up of three groups of organisms: (1) producers, such as plants, which create chemical energy in the form of sugars and starches from light; (2) consumers, such as animals, which eat producers or other organisms whose energy ultimately derives from producers; and (3) decomposers, such as fungi and certain bacteria, which eat dead organisms or cast-off parts of living organisms, releasing nutrients back to the biosphere.
- Most biological systems are fueled by energy from the sun. In these systems, energy flows in one direction: from sun to producers to consumers and decomposers. However, some deep-sea biological systems are fueled by energy that bacteria extract from nonliving materials such as iron or ammonia.
- Food webs diagram the relationships, in terms of who is eating whom, among producers, consumers, and decomposers in an ecosystem.

◉ Review and Application of Key Concepts

1. Describe one observation, one hypothesis, and one experiment from Dr. Burkholder's work.

2. Researchers are still unsure how *Pfiesteria* kills fish. Dr. Burkholder's studies suggest that a toxin released by *Pfiesteria* kills the fish. Other researchers say the *Pfiesteria* swarm the fish and attack them. Come up with another hypothesis or explanation for how *Pfiesteria* might be killing fish. Use your hypothesis to generate testable predictions and describe one experiment you might use to test one of your predictions.

3. Some people avoid eating beef because they fear contracting mad cow disease, which is thought to be caused by a class of molecules known as prions [pree-onz]. These still poorly understood molecules are a kind of protein that is able to infect the nervous system, eventually creating holes in the brain tissue, leading to loss of mental and physical function and eventually leading to death. Prions are thought to spread their effects through the body in a kind of "falling dominoes" effect—by causing other normal molecules to mimic the prion's shape and thus become infectious molecules themselves. Like a virus, a prion can cause disease. A prion can multiply or reproduce itself by making other molecules mimic its shape and behavior. Prions can spread from one part of the body to another. Are prions alive? Why or why not?

4. What are the levels of the biological hierarchy? Arrange them in their proper relationship with respect to one another, from smallest to largest. Give an example for each level.

5. Describe the flow of energy through the biological system that includes grasses, lions, sunshine, antelope, and ticks.

6. In the biological system described in the previous question, (a) identify which are producers, consumers, and decomposers; and (b) diagram the food web of this system.

7. Choose one of these two hypotheses:
   I. Nanobes are living organisms.
   II. Nanobes are not living organisms.

   Given what you know about the characteristics of life, (a) state some testable predictions about nanobes based on the hypothesis you chose; and (b) propose an experiment to test one of your predictions. (For example, if you chose hypothesis I, then one prediction might be, "Nanobes require energy." An experiment to test this prediction might be, "I will provide one group of nanobes with a lot of energy in the form of nutrients, heat, and light, and another group of nanobes with no energy. If the first group of nanobes grows and the second group does not, these observations will support my prediction and hypothesis.") In dreaming up your experiment, don't hold back. Remember—science thrives on creativity.

Key Terms

adaptation (p. 13)
biological hierarchy (p. 14)
biology (p. 4)
biome (p. 14)
biosphere (p. 14)
cell (p. 7)
community (p. 14)
consumer (p. 16)
decomposer (p. 17)
development (p. 10)
DNA (deoxyribonucleic acid)
   (p. 9)
ecosystem (p. 14)
evolution (p. 13)
experiment (p. 6)

food web (p. 17)
hypothesis (p. 5)
individual organism (p. 14)
molecule (p. 14)
multicellular organism (p. 9)
observation (p. 5)
organ (p. 14)
organ system (p. 14)
population (p. 14)
prediction (p. 5)
producer (p. 16)
science (p. 4)
scientific method (p. 5)
species (p. 13)
tissue (p. 14)

Self-Quiz

1. Which of the following is *not* an essential element of the scientific method?
   a. observations
   b. religious beliefs
   c. experiments
   d. hypotheses

2. A hypothesis is
   a. an educated guess explaining an observation.
   b. an observation.
   c. a test of a prediction.
   d. an experiment that works well.

3. Which of the following are both universal characteristics of life?
   a. the ability to move and the ability to reproduce
   b. the ability to reproduce using RNA and the ability to capture energy directly from the sun
   c. the ability to reproduce and the ability to sense the environment
   d. the ability to sense the environment and the ability to capture energy by eating

4. Viruses
   a. are able to photosynthesize.
   b. reproduce by budding.
   c. are inanimate objects.
   d. exist in a gray zone between life and nonlife.

5. Which of the following is the basic unit of life?
   a. plants
   b. DNA
   c. the cell
   d. the sun

6. Which of the following can reproduce without its own DNA?
   a. a human being
   b. a virus
   c. a single-celled organism
   d. none of the above

7. Which of the following is a multicellular organism?
   a. a beetle
   b. a brain
   c. a bacterium
   d. a forest

8. The energy in biological systems can originate from
   a. the sun and the moon.
   b. the sun only.
   c. plants and algae.
   d. the sun as well as nonliving materials such as iron.

Biology in the News

Baby's Fate in Summit Court

By Stephen Dyer

According to court records, Taylor [a 13-week-old baby] no longer can communicate, is on a ventilator at Children's Hospital Medical Center of Akron and doesn't respond to pain. She has almost no brain-stem function.

She got that way, police say, because she was shaken.

According to the social services worker in charge of Taylor's case, the baby had been at home with her father one November morning while Taylor's mother was at work. Then Taylor's father had called 911, saying that he had begun giving the baby her bath when she became "nonresponsive." According to her doctors, when the infant arrived at the hospital's pediatric intensive care unit, she was found to have suffered severe brain and spinal cord injuries.

The doctors agreed that the baby's condition was dire and soon predicted that this child, now hooked to a plethora of machines, had essentially no chance of recovery to any kind of normal or healthy life. As a result, the hospital clinicians suggested that it would be most appropriate to disconnect the baby from life support and let nature take its course. In her condition, the doctors said Taylor would not be expected to live much longer. Taylor's parents, however, would not consent to removing the infant from the machinery keeping her alive.

While it is understandable that making such a decision would be excruciatingly difficult for any parents, there was an added complication in the case of Taylor's parents. For by agreeing to discontinue life support, Taylor's parents might lose not only their child, but, some say, their freedom as well. Taylor's parents have come under investigation for causing her injuries. And as the baby's lawyer notes in the article, "If Taylor dies, whoever shook her would probably be charged with a homicide."

Evaluating the news

1. Baby Taylor is in what is called a vegetative state. Without life-support machines, she would probably die quickly. If only machines are keeping Baby Taylor's lungs breathing and her heart pumping, is she, in fact, alive? Do our definitions of life and nonlife provide any insights into Taylor's situation?

2. One of the most controversial issues in the United States is abortion. Some in the right-to-life movement say that abortion is murder of a person. Some defenders of the right to choose say that a fetus is not yet a living individual, but rather still tissue that is part of a woman's body. Do our definitions of life give more weight to one argument than another?

3. Some say that the right-to-life versus right-to-choose debate hinges on when life begins. Does life begin at conception, or at birth? How does our list of characteristics of life affect these arguments?

4. Does the definition of life given in this chapter cause you to alter or reaffirm your personal stance on whether "pulling the plug" on a baby like Taylor is murder or exercising the right to die? How so? Similarly, does the biological definition of life affect your personal stance on abortion?

Source: *The Beacon Journal,* December 1, 2002.

2 Organizing the Diversity of Life

Key Concepts

◉ Biologists examine all aspects of an organism's biology—its body structure, its behavior, its DNA—to look for inherited shared similarities with other organisms. The information is then summarized in an evolutionary tree.

◉ The most informative similarities for determining evolutionary relationships are shared derived features, which are shared by a group of organisms because the feature arose in their most recent common ancestor.

◉ Evolutionary trees can be used to predict behaviors and other attributes of organisms on the tree.

◉ The Linnaean hierarchy is a way of classifying all of life. It ranges from species at the lowest category level, up through genera, families, orders, classes, phyla, and kingdoms. Biologists also widely recognize the three-domain system as the most basic division of living organisms.

◉ Biologists are making dramatic progress in understanding the evolutionary relationships of the world's many organisms. In part, this progress is due to modern DNA studies, which are confirming some parts of the tree of life but also causing others to be restructured, resulting in some surprising relationships.

The Iceman Cometh

On a September day in 1991, hikers high in the Alps near the border between Austria and Italy came across a remarkable sight: the body of an ancient, mummified man in the melting glacial ice. Instantly dubbed the Iceman, the well-preserved body appeared to be a prehistoric hiker dating to the Stone Age, possibly a shepherd or traveler, who had died on the mountainside, his body captured by the ice, which had preserved it for some 5,000 years.

Extremely well preserved—right down to his underwear—this one-and-a-half-meter-tall visitor from another time promised a tantalizing peek into the past. Sporting tattoos on his body, he wore a loincloth, a leather belt and pouch, leggings and a jacket made of animal skin, a cape of woven grasses, and calfskin shoes lined with grass. He carried a bow and arrows, an axe, knives, and two pieces of birch fungus, perhaps a kind of prehistoric penicillin.

Perfectly suited for scientific examination, this first-ever corpse from the Stone Age found right in the path of hikers seemed too good to be true. Researchers began to fear that the Iceman might be an elaborate hoax. Speculation was rising that the body was a transplanted Egyptian or pre-Columbian American mummy. How could biologists determine the true identity of this long-lost wanderer? To know who the Iceman was, scientists realized they would have to figure out who his closest relatives had been. Was he most closely related to Columbian Indians? Or were his closest living relatives Egyptians or Northern Europeans? To solve the mystery, scientists would have to place him in the human family tree. How scientists identify organisms and place them on the family tree that includes all living things—the evolutionary tree of life—is the topic of this chapter.

A Stone-Age Mummy
Known as the Iceman, this mummy was discovered in melting ice by hikers in the Alps in 1991, some 5,000 years after he died.

Biologists often come across mysterious, previously unknown organisms. When they find a puzzling specimen, the first question they ask is, "What is this?" As with the Iceman, this question can be answered only by placing the organism into a kind of family tree of life, known as an evolutionary tree.

In this chapter we begin by examining how biologists build and use evolutionary trees. Next we examine the grand classification scheme known as the Linnaean hierarchy. Finally we look at some of the more interesting and important branches on the evolutionary tree of life that scientists are working to understand. We close with new information that scientists are uncovering about the Iceman and his place in the human family.

2.1   Building Evolutionary Trees

Genealogists (and increasingly, ordinary people with access to the Internet) gather information to show how members of a human extended family are related—both within a given generation (who is first, second, or third cousin to whom?) and from one generation to the next (who are my great-grandparents? my great-great-great-grandparents?). Once genealogists know how a group of people are related, they can organize that information into a family tree (Figure 2.1*a*). In the same way, **systematists** [*SISS*-tuh-muh-*TISTS*]—biologists who study the relationships among groups of different organisms—summarize that information in **evolutionary trees** (Figure 2.1*b*). (Systematists are also known as taxonomists.) Taken together, the many evolutionary trees of all the variety of living organisms form a single evolutionary tree of all life.

To determine where a new specimen belongs on the tree of life, biologists must ask themselves a series of questions. First, is this mystery organism a member of a species that is already named and placed on the tree of life? This is not a trivial question. Organisms that appear unique sometimes turn out to be a different form—even a different life stage or a different sex—of a species already known to exist. Conversely, organisms that look outwardly identical to known species can turn out to be quite distinct in terms of their DNA, thereby signaling to a biologist that the two specimens may be from different species.

If the mystery organism is not a member of a known species, then what are its closest relatives? If a plant, is it a close relative of a cycad or of a bluebell? If an animal, then is it more closely related to a guppy or to an eagle?

Once scientists know where an organism sits in the tree of life, they can name it. So, whether scientists are demystifying mummies or collecting new species from the rainforest, their first question always is, "Where does this organism fit in the evolutionary tree of life?"

Groups are related through their most recent common ancestor

Let's look more closely at the process of constructing an evolutionary tree. How *do* biologists determine which groups of organisms are more closely or more distantly related to one another?

As anyone who has been to a family reunion knows, the more closely related two people are, the more similar they tend to be to one another, in the way they look and sometimes even in the way they act—smiling or sneezing in just the same way. Close relatives tend to be similar even in how their bodies work, often exhibiting the same physical strengths or vulnerabilities to the same kinds of illnesses. In fact, the similarities extend right down to the level of a person's genetic material (DNA), and for good reason.

Recall that DNA is the molecule passed from one generation to the next, the genetic blueprint for the development of an individual. We inherited these blueprints for body structures and behaviors from our parents, who inherited their blueprints from their parents, and so on. As a result, we exhibit many of the characteristics that our relatives exhibit. In the same way, closely related groups of organisms (those that arose from the same ancestor) tend to resemble one another.

Unit 4 provides a detailed discussion of evolution and the origin of species. For now, it is enough to know that over time, populations within a species can evolve to become different enough to form new species. That is, an ancestor species can give rise to new species that are its descendants. On evolutionary trees, these descendants are depicted as the tips of branches (see Figure 2.1*b*). When tracing backward from the tips of branches, we are really tracing back through time, looking for the point at which one **lineage** hooks up with another lineage. This meeting point, depicted as a fork in the tree, represents the **most recent common ancestor** of the two lineages, the most recent ancestral group from which the descendants arose. The farther down the tree we go, the further back into history we are delving. Note that any two groups may have many common ancestors, but only one most recent common ancestor.

(a) Family tree of Britain's royalty

Prince William   Prince Harry

Prince Charles   Princess Diana

Queen Elizabeth   Prince Philip

(b) Swallowtail butterfly evolutionary tree

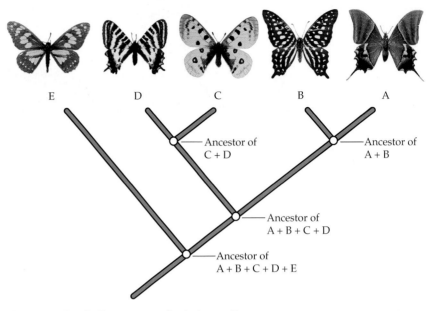

E   D   C   B   A

Ancestor of C + D

Ancestor of A + B

Ancestor of A + B + C + D

Ancestor of A + B + C + D + E

Figure 2.1 Family Trees versus Evolutionary Trees
(a) This family tree shows the relationships of a few familiar members of Britain's royal family, including Queen Elizabeth and Prince Philip, their son Charles, his first wife, the late Princess Diana, and Charles and Diana's two sons, William and Harry. (b) An evolutionary tree also diagrams relationships, in this case the evolutionary relationships among swallowtail butterfly species. This tree indicates that butterfly groups A and B are each other's closest relatives. They descended or evolved from a most recent common ancestor depicted on the tree at the point where their branches meet. Thus, from the most recent common ancestor, A and B branched off and evolved into the forms shown in the photographs. Groups A, B, C, and D all share the most recent common ancestor depicted at the next branching point down the tree, and so on.

Examining shared derived features helps determine the most recent common ancestor

Descendants often share key features because they share an ancestor. In human families, for example, a father's distinctive nose may be seen on the faces of all his children. Similarly, all the vertebrate animals—including humans, birds, snakes, and fish—can be placed together on an evolutionary tree because they all have the backbone that was a feature of their most recent common ancestor. Such key shared features are usually found in aspects of an organism's biology. Traditionally, biologists have compared species by looking at inherited structural characteristics of the body: the number of legs, the arrangement of petals in a flower, the anatomy of an animal's heart, and so on. In recent years, however, researchers have begun searching for similarities in other features, including behaviors.

How do biologists decide which structural or behavioral features to compare? It turns out that not all similarities between groups are equally useful for understanding relationships. The most useful are those unique features that evolved in a group's most recent common ancestor. Those features are then shared by the descendant species, having been passed down, or derived, from the most recent common ancestor. They are known as **shared derived features**, and they mark a group as a set of close relatives. Groups that are not as closely related—that is, not descended from the same most recent common ancestor—will not display those shared derived features. Instead, they display a different set of shared derived features that is unique to themselves and their close relatives.

Figure 2.2 depicts the relationships of some familiar animals, identifying some of the shared derived features that define these groups. For example, at the upper right-hand corner of the tree, the first shared feature is opposable thumbs (thumbs that are capable of being placed against one or more of the other fingers on the hand). This is one of the many features that set chimpanzees and humans apart from the other animals on the tree; no other organisms on the tree share this feature. Most likely we share it with chimpanzees because our most recent common ancestor had an opposable thumb, making it one piece of evidence that humans and chimpanzees share a more recent common

Test your knowledge
of these derived
characteristics.

2.1

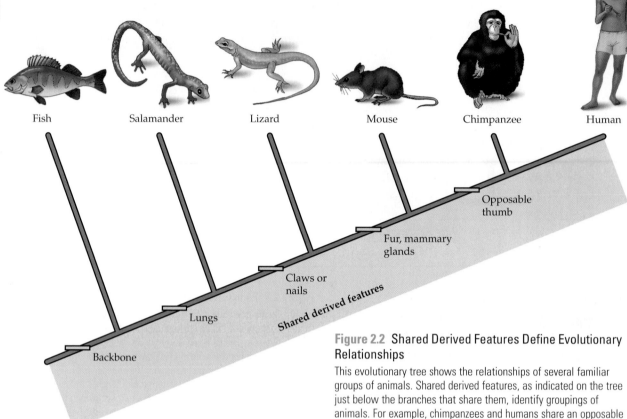

Figure 2.2 Shared Derived Features Define Evolutionary Relationships

This evolutionary tree shows the relationships of several familiar groups of animals. Shared derived features, as indicated on the tree just below the branches that share them, identify groupings of animals. For example, chimpanzees and humans share an opposable thumb, while chimpanzees, humans, mice, lizards, and salamanders all share lungs. All the animals depicted share a backbone.

Figure 2.3 Misleading Convergent Features

A panda's opposable thumb, which helps it grasp the bamboo shoots it loves so much, would seem to make it a very close relative of humans and chimpanzees. But in actuality its "thumb" is an enlarged wrist bone that evolved separately.

ancestry with each other than with the other animals on the tree.

Convergent features are similarities that do not indicate relatedness

Sometimes organisms share features not because they inherited them from their most recent common ancestor, but because they evolved the same feature independently. Such **convergent features** can be misleading. For example, the panda, like the human and the chimpanzee, has an opposable thumb (Figure 2.3). Should pandas be placed beside humans and chimpanzees on an evolutionary tree? The answer is a clear no, based on everything else scientists know about humans, chimpanzees, and pandas. Therefore we must share an opposable thumb with pandas—distant relations of ours—because pandas evolved this useful feature independently.

While it might seem simple to separate similarities that are convergent features from those that are true

shared derived features, in practice it can be quite difficult and remains one of the major challenges for systematists.

DNA comparisons are a powerful new way to study evolutionary relationships

In recent decades, advances in techniques for studying DNA have revolutionized all of biology, including the study of evolutionary relationships.

All living organisms use DNA as their hereditary material. As a result, by studying shared derived features of organisms' DNA, biologists have been able to study the relationships of many different groups that were difficult or impossible to study before. An example is the relationships among major groups such as bacteria, plants, and animals—groups whose bodies are so different that they are often impossible to compare with one another using structural features alone. Researchers are also learning more about groups that were difficult to study previously because their structures are so simple. For example, certain kinds of worms—in particular, those that live as parasites inside other organisms—have been difficult to compare because they are often tiny, smooth, and nearly featureless. Scientists can even unravel criminal mysteries using these techniques (see the box on page 28).

2.2 Using Evolutionary Trees to Predict the Biology of Organisms

Evolutionary trees are not merely a way to show which groups are related to which other groups. These trees have predictive powers as well, because researchers can expect close relatives to share many of the same features passed down by their most recent common ancestor.

As surprising as it might seem, there is now overwhelming evidence from living and fossil animals that the closest relatives of birds are the extinct creatures we know as dinosaurs. Of the animals shown in Figure 2.4, the

next-closest relatives of birds, after dinosaurs, are crocodiles and alligators, a group known as crocodilians. Knowing the relationships among these groups has made it possible for biologists to do the seemingly impossible: make predictions about the behavior of long-extinct dinosaurs.

Crocodilians and birds are known to be dutiful parents. They both build nests and defend their eggs and young. Scientists reasoned that if crocodilians and birds both exhibit extensive and complex parental care, then their most recent common ancestor probably exhibited this behavior as well. Because dinosaurs share this same common ancestor, scientists were able to predict that these long-gone and unobservable creatures tended their eggs and hatchlings, too—a shocking notion for creatures with a reputation for being big, vicious, and pea-brained.

Recently, researchers have confirmed that dinosaurs exhibited parental care. These researchers discovered the fossil of a dinosaur that had died 80 million years ago in what today is the Gobi Desert while sitting on a nest of eggs. At the time this dinosaur was originally

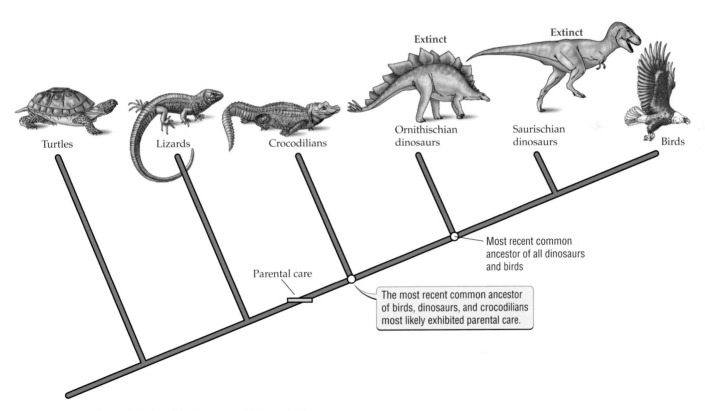

Figure 2.4 The Close Relationship Between Birds and Dinosaurs
On this evolutionary tree, birds and the two lineages of dinosaurs—ornithischians [or-nih-*THISS*-kee-uns] (plant eaters such as *Stegosaurus*) and saurischians [sore-*ISS*-kee-uns] (animals such as *Tyrannosaurus*)— share the most recent common ancestor. The next-closest relatives of this group are the crocodilians, including crocodiles and alligators.

Science Toolkit

The Guilty Dentist

While biologists use the study of evolutionary relationships mainly to build evolutionary trees of the world's organisms, scientists are finding wider and wider uses for evolutionary trees. Scientists have even used evolutionary trees to solve the mystery of whether a dentist infected his patients with HIV, the virus that causes acquired immunodeficiency syndrome (AIDS) in humans.

The dentist, who was himself infected with HIV, had been working, as usual, with his patients. Then some of his patients began turning up HIV-positive. Against a backdrop of increasing public controversy over whether health care workers with HIV pose a risk to their patients, many people wanted to know: had this dentist transmitted the deadly virus to his patients? Adding confusion to the situation was the fact that some of the infected patients had lifestyles or habits that put them at risk for HIV infection by other means. Those on both sides of the issue—some saying the dentist should not be blamed, others saying he was a threat—argued heatedly without resolution.

In order to answer the question, researchers took advantage of the observation that HIV's genetic material, like that of most viruses, evolves quickly, resulting in many different but closely related strains of HIV. The researchers created an evolutionary tree of the viral strains found in the dentist, in each of the infected patients, and in local people who were also infected with HIV but were not patients of the dentist. If the dentist had infected his patients, then the strains of HIV found in his patients should be more closely related to his strain and less closely related to the strains from other infected people.

The results showed that some of the dentist's patients probably were infected by him and that some probably were not. Two patients, X and Y, who could have been exposed to HIV as a result of their lifestyles or habits, had HIV strains that were not closely related to the dentist's strain. However, five other patients—A, B, C, D, and E—were carrying strains of the virus very closely related to the strain in the dentist. Moreover, those people were not at risk of contracting HIV by other means. So the scientists concluded that, at least in the case of these five patients, the dentist had infected them with his strain of the virus.

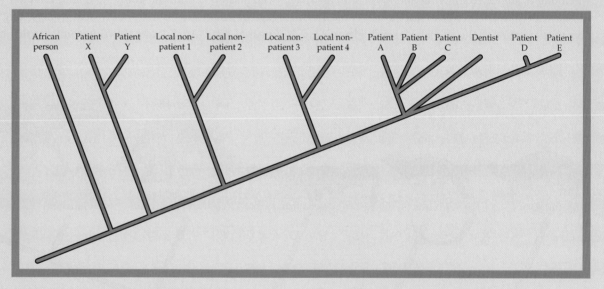

An Evolutionary Tree Solves the Whodunit

By examining the evolutionary relationships among the strains of HIV found in a number of infected people, including a dentist and his infected patients, researchers were able to solve the mystery of how his patients were likely to have been infected.

unearthed—in 1923—the idea that it might be sitting on a nest of its own eggs was so unthinkable that the assumption was that it was attacking and eating the eggs. Hence its discoverers gave it the name *Oviraptor* [*OH*-veeh-rap-ter], which means "egg seizer." Not knowing then that birds and dinosaurs were close relatives, biologists did not expect the two groups to show simi-lar behaviors. But in 1994, with a new evolutionary tree for dinosaurs, birds, and crocodilians in hand, the truth about *Oviraptor* became obvious. Rather than attacking the eggs, *Oviraptor* appears to have died in a sandstorm protecting its nest, its limbs encircling its unhatched young in a posture as protective as that of any bird (Figure 2.5).

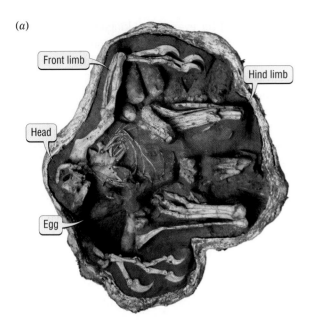

(a)

Front limb

Hind limb

Head

Egg

(b)

Figure 2.5 Parental Care by Dinosaurs

(*a*) The fossil of an *Oviraptor* dinosaur, which died sitting on its nest of eggs. (*b*) An artist's rendition of the dinosaur brooding its eggs shows how it might have looked shortly before the sandstorm began. Compare this dinosaur with (*c*) an ostrich as it would look on a nest of its eggs today.

(c)

2.3 A Classification System for Organizing Life: The Linnaean Hierarchy and Beyond

In addition to organizing the world's species into evolutionary trees, biologists have organized them into a classification system developed in the 1700s by the father of modern scientific naming, a Swedish biologist named Carolus Linnaeus [lih-*NEE*-us]. In the **Linnaean** [lih-*NEE*-un] **hierarchy** (Figure 2.6), the species is the smallest unit (lowest level) of classification. Closely related species are grouped together to form a **genus** (plural: genera). Using these two categories in the hierarchy, every species is given a unique, two-word Latin name called its **scientific name**. The first word tells what genus the organism is a part of, and the second defines the species. For example, humans are called *Homo sapiens*: *Homo* ("man") is the genus to which we belong, and *sapiens* [*SAY*-pee-enz] ("wise") is our species name. We are the only living species in our genus. Other species in our genus include *Homo erectus* ("upright man") and *Homo habilis* ("handy man"), both of which are extinct.

What about the other five levels in the Linnaean hierarchy, those above species and genus? All organisms belong to these levels as well, even though the scientific name does not mention them. Closely related genera are grouped together into a **family**. Closely related families are grouped into an **order**. Closely related orders are grouped into a **class**. Closely related

Homo habilis

classes are grouped into a **phylum** [*FYE*-lum] (plural: phyla). Finally, closely related phyla are grouped together into a **kingdom**.

Systematists often refer to groups of organisms at these various levels of classification as taxonomic groups, or more simply, as taxa (singular: **taxon**). Using Figure 2.6 as an example, we see that the species *Rosa gallica* is a taxon, but so are the higher levels of classification to which it belongs: Rosaceae, Rosales, Dicotyledonae, and so on up to Plantae. Each of these is a taxon, or taxonomic group, to which the moss rose belongs, along with all the other organisms grouped with it in that taxon.

Originally Linnaeus described just two kingdoms: plants and animals. Today biologists recognize anywhere

Review the Linnaean hierarchy. 2.2

from five to eight (or even more) kingdoms, depending on how finely it is useful for them to divide up the world's organisms. In this book we adopt the widely used six-kingdom system (Figure 2.7). Chapter 3 will give detailed descriptions of these six kingdoms.

In addition to the kingdoms designated by the Linnaean hierarchy, most biologists have also begun using an even higher level of organization called **domains** (see Figure 2.7). The three domains are **Bacteria** (which includes familiar disease-causing bacteria), **Archaea** [ar-*KEE*-uh] (bacteria-like organisms best known for living in extremely harsh environments), and **Eukarya** [you-*KARE*-ee-uh], which includes all the rest of the living organisms, from amoebas to plants to fungi to animals.

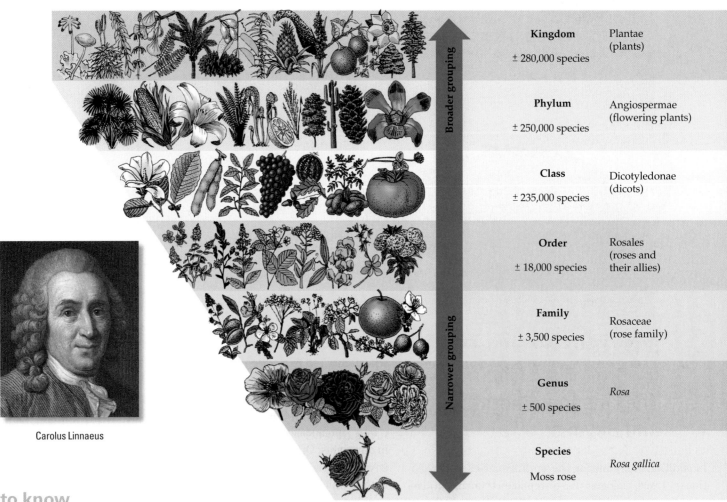

Carolus Linnaeus

Kingdom
± 280,000 species — Plantae (plants)

Phylum
± 250,000 species — Angiospermae (flowering plants)

Class
± 235,000 species — Dicotyledonae (dicots)

Order
± 18,000 species — Rosales (roses and their allies)

Family
± 3,500 species — Rosaceae (rose family)

Genus
± 500 species — *Rosa*

Species — *Rosa gallica*
Moss rose

Broader grouping → Narrower grouping

Helpful to know

An easy way to remember the levels of the Linnaean hierarchy is to memorize the following sentence, in which the first letter of each word stands for each descending level: *K*ing *P*hillip *C*leaned *O*ur *F*ilthy *G*ym *S*horts.

Figure 2.6 The Linnaean Hierarchy

The smallest unit of classification is the species (here, the moss rose, whose scientific name is *Rosa gallica*). This species belongs to the genus Rosa, which includes other roses. The genus *Rosa* lies within the family Rosaceae [roze-*ACE*-ee-ee], which lies within the order Rosales [roze-*AH*-leez], within the class Dicotyledonae [dye-kot-oh-*LEE*-duh-nee], within the phylum Angiospermae and the kingdom Plantae. All organisms can be classified using the same categories of "species" through "kingdom." This classification system was first devised by the Swedish naturalist Carolus Linnaeus (inset).

Figure 2.7 Kingdoms and Domains
This book uses both the three-domain system and the widely used six-kingdom system. The domain Bacteria is equivalent to the kingdom Bacteria, and the domain Archaea is equivalent to the kingdom Archaea. The domain Eukarya encompasses four kingdoms in this scheme: Protista (or protists, which include organisms such as amoebas and algae), Plantae (or plants), Fungi (including yeasts and mushroom-producing species), and Animalia (or animals).

The domains describe the most basic and ancient divisions among living organisms. DNA studies initially alerted biologists to the existence of the three domains and to the fact that the Archaea, which were once thought to be just more bacteria, belong to a separate and very ancient group. When the three-domain scheme was first proposed, it was highly controversial, as it set up Archaea as a major group on a par with the two huge, major groups that included all the other living organisms: Bacteria and Eukarya. Since then, researchers have discovered other evidence that corroborates the three-domain scheme, and it is now widely accepted.

2.4 Branches on the Tree of Life

The study of evolutionary relationships is one of the fastest-moving areas of biology, and new evolutionary trees are being constructed all the time. As mentioned earlier, these advances are due in part to new DNA studies. Some of these studies have confirmed certain branching relationships on the evolutionary tree of life, while others are revealing surprising connections.

Unexpected evolutionary relationships exist between the plant, fungus, and animal kingdoms

Very distantly related organisms—for example, a bacterium and a human—can be difficult to compare. What part of a bacterium would show any similarity to a person? As new studies show, an organism's DNA is an excellent place to look for shared derived features, as all living organisms carry DNA. Even extremely distantly related groups, such as domains and kingdoms, can be compared in order to reconstruct the arrangement of their branches on the tree of life.

Figures 2.8 and 2.9 show trees encompassing all of life, not just a portion of it: Figure 2.8 depicts the hypothesized relationships of the three domains, while Figure 2.9 provides the next level of detail by showing the hypothesized relationships among the kingdoms. Both of these trees are the subject of ongoing DNA comparison studies, and controversy remains, as not all studies are in agreement as to how the groups should be arranged on the trees.

Look at Figure 2.9: in general, the relationships among the kingdoms are not surprising. The most distant relatives of plants, animals, and fungi are the two kingdoms Bacteria and Archaea; bacteria and archaeans are single-celled organisms that at best seem like very distant cousins of ours.

But notice the relationships among plants, fungi, and animals: the two most closely related groups among the three are fungi and animals. For decades, fungi were thought to be more closely related to plants. Unable to move and very unlike animals, these faceless organisms, most familiar to us as mushrooms, seem more akin to trees, shrubs, and mosses. As a result, it came as a huge surprise to scientists and the general public alike when recent studies showed that fungi are actually more closely related to animals, including people, than to plants. That is, fungi share a more recent common ancestor with animals than with plants. Bringing this closer to home, the mushrooms on your pizza are more closely related to you than they are to the green peppers sitting next to them.

Could the slime in your bathroom shower really be more closely related to you than it is to your houseplants? How much do we really have in common with the likes of bread mold or yeast? A lot, it turns out. The finding that fungi and animals are more closely related to each other than either is to plants solved the long-standing mystery of why doctors often have such a difficult time treating fungal infections, particularly internal infections, in which a fungus has begun living inside the human body. Because fungi and animals are such close relatives, their cells work

Figure 2.8 Evolutionary Tree of Domains

This tree shows the relationships of the three domains. At the root of the tree is the universal ancestor, from which all living things descended. Of the three surviving lineages, the first split came between the Bacteria and the lineage that would give rise to the Archaea and Eukarya. The next split was between the Archaea and the Eukarya, making Archaea and Eukarya more closely related to each other than either group is to the Bacteria.

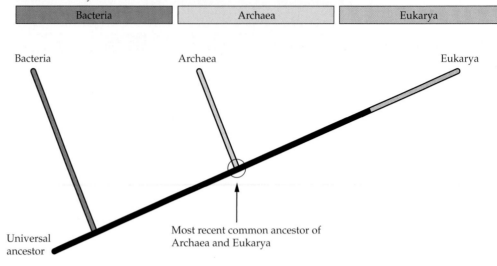

Three-domain system

| Bacteria | Archaea | Eukarya |

Bacteria

Archaea

Eukarya

Universal ancestor

Most recent common ancestor of Archaea and Eukarya

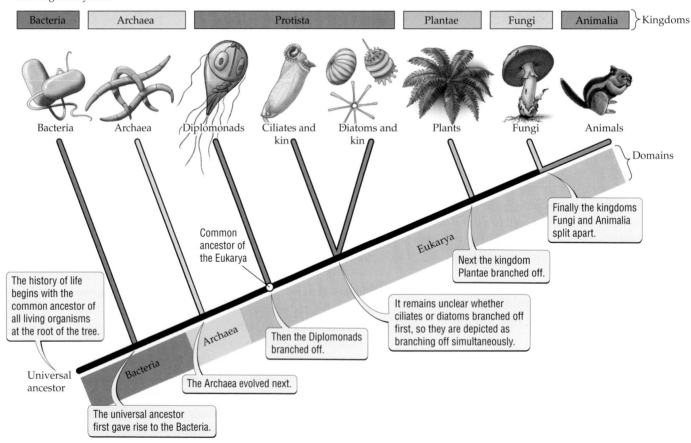

Six-kingdom system

| Bacteria | Archaea | Protista | Plantae | Fungi | Animalia | } Kingdoms

Bacteria

Archaea

Diplomonads

Ciliates and kin

Diatoms and kin

Plants

Fungi

Animals

} Domains

Common ancestor of the Eukarya

Eukarya

Finally the kingdoms Fungi and Animalia split apart.

Next the kingdom Plantae branched off.

It remains unclear whether ciliates or diatoms branched off first, so they are depicted as branching off simultaneously.

The history of life begins with the common ancestor of all living organisms at the root of the tree.

Archaea

Then the Diplomonads branched off.

Universal ancestor

Bacteria

The Archaea evolved next.

The universal ancestor first gave rise to the Bacteria.

Figure 2.9 The Tree of All Life

This evolutionary tree shows the relationships among the six kingdoms as well as the three domains. Each group branching off the tree can be thought of as a cluster of close relatives—a lineage, just like a lineage in a human family.

very similarly. Thus anything a doctor might use to kill off a fungus could kill, or nearly kill, the person as well.

So similar are humans and fungi that they even share a similar disease. Lou Gehrig's disease is a fatal disease of humans in which the nervous system degenerates quickly (Figure 2.10). Yeasts get a similar "disorder." Thus yeast is an excellent, if surprising, model for studying this dangerous human disease.

The primate evolutionary tree reveals the closest relative of humans

When we visit the ape house at a zoo, the striking similarities between the beings standing outside the cage looking in and those inside the cage looking out are obvious. These similarities are not surprising, since humans, apes, and monkeys all belong to the order Primates. But which of our primate relatives is our closest relation? Over the years this question has generated intense interest and heated debate. Researchers have studied everything from bone structure to behavior to DNA in attempts to determine which primate is humankind's closest kin.

While controversy still remains, the emerging consensus suggests that our closest relative is the chimpanzee, a fellow tool user with whom we share a remarkable degree of similarity in our DNA (Figure 2.11). More distantly related are gorillas, and beyond that orangutans, gibbons, monkeys (such as the spider monkey), and, most distantly, lemurs. In the essay at the close of Unit 4, we will take up this topic again in more detail.

A surprising tangle can be found at the roots of the tree of life

One of the oddest and most recent surprises to come out of DNA studies is the hypothesis that the tree of life might look a lot less like a tree than a highly interconnected web at its base. This idea arose when biologists began to find DNA in what seemed to be the "wrong" places. For example, scientists have found bacterial DNA in archaeans and in organisms in the Eukarya as well. How could this happen if the three lineages—Bacteria, Archaea, and Eukarya—diverged long ago? Each lineage should have evolved to be quite distinct over time. Yet scientists have found that some organisms' DNA looks like a grab bag of DNA collected from across the tree of life.

One hypothesis to explain the grab bag comes from Dr. W. Ford Doolittle, a biochemist at Dalhousie University in Canada. He suggests that throughout the

Figure 2.10 Astrophysicist Stephen Hawking

The brilliant scientist Stephen Hawking has suffered from amyotrophic lateral sclerosis, or Lou Gehrig's disease, for over 40 years. Despite its ravages to his body (he is paralyzed and cannot use his limbs or voice) he is responsible for breakthroughs in the fields of theoretical cosmology and quantum gravity. The study of yeasts with a similar disorder may provide hope to Hawking and the estimated 5,000 people in the United States who are diagnosed with Lou Gehrig's disease each year.

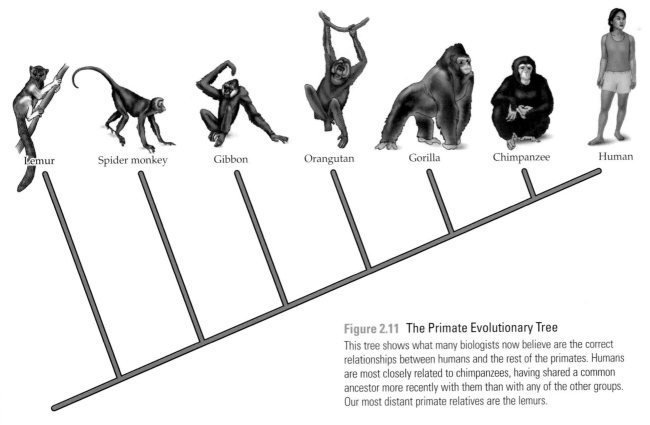

Lemur Spider monkey Gibbon Orangutan Gorilla Chimpanzee Human

Figure 2.11 The Primate Evolutionary Tree

This tree shows what many biologists now believe are the correct relationships between humans and the rest of the primates. Humans are most closely related to chimpanzees, having shared a common ancestor more recently with them than with any of the other groups. Our most distant primate relatives are the lemurs.

Figure 2.12 A Tangled Web at the Base of the Tree of Life?

On this representation of the evolutionary tree, green arrows represent multiple instances of horizontal gene transfer, in which genes moved between the lineages Bacteria, Archaea, and Eukarya, even after the three groups had become well established as separate lineages.

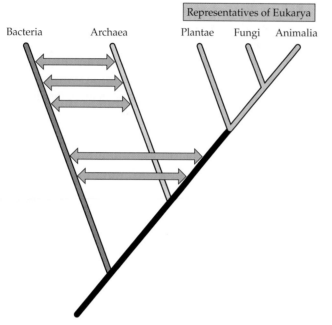

Representatives of Eukarya

Bacteria   Archaea   Plantae   Fungi   Animalia

moving across from one lineage to another—a process known as **horizontal gene transfer** (Figure 2.12). In this view, the tree of life stands as it is, but with a lot of extra, unexpected movement of genes in an unexpected direction—horizontally. According to Doolittle's hypothesis, the base of the tree of life might be most accurately pictured as a loosely knit community of very primitive cells that were freely exchanging genes, creating the grab-bag mixtures of genes that persist in organisms across the tree of life even today. Still controversial, this startling new hypothesis remains a matter of active debate.

The evolutionary tree of life is a work in progress

Although many of the major branching patterns on the evolutionary tree of life are well established and unlikely to change, scientists view evolutionary trees as hypotheses. They are the best approximations given what we know today. Biologists continue to study and reevaluate the relationships of all organisms using new information. In fact, the vast majority of relationships among the world's millions of species of plants, animals, fungi, bacteria, and protists remains to be worked out in detail.

early history of life, both before and after the three major lineages evolved, organisms in those three lineages were freely exchanging many different genes. So in addition to genes being passed "vertically" (from one generation to the next), genes also appear to have been

---

And the True Identity of the Iceman Is . . .

Hop hornbeam tree

Let's return to the mystery from the frozen Alps: who was the Iceman?

Scientists, hoping to uncover his true identity, looked for clues in the DNA still preserved in the Iceman's tissues. By comparing his DNA with the DNA of various groups of people around the world, scientists found key similarities—shared derived features in his DNA—showing that he was most closely related to modern peoples of northern Europe and much less closely related to Egyptian or South American peoples, from whom a mummy could have been stolen and moved. Researchers solved the mystery by placing the Iceman in the tree of life—specifically, in that part of the tree that depicts relationships among different groups of human beings. So the Iceman— dubbed Ötzi, since he was found in the Ötzal Alps

on the border of Austria and Italy—was no hoax after all.

Ongoing studies of Ötzi's remains continue to confirm that he lived in the region where he was found. They have also revealed details of what appear to have been the dramatic last hours of his life.

In Ötzi's gut, scientists discovered ancient pollen (microscopic spermlike cells released by flowers) from the hop hornbeam tree. Because the hop hornbeam blooms at a specific time of year and is found only in the Schnals Valley to the south of the Ötzal Alps, scientists know that Ötzi made his way out of the warm Schnals Valley and into the mountains on an early summer day. Scientists also know, from examining the minuscule remaining contents of his gut, that Ötzi had recently eaten a bit of tough, unleavened bread made from einkorn [INE-korn] wheat, one of the few grains known to have been domesticated by ancient Europeans at that time. And from a bit of bone and muscle fiber, researchers guess that the lucky Ötzi even had a bit of meat to chew on.

So what brought Ötzi to the mountains and his death, after his meal? Recent studies revealed the presence of blood from four other individuals—from two people on an arrow in his quiver, from a third person on his knife blade, and from the fourth on the back of his cloak. Scientists also found an arrow wound in Ötzi's back, a gash on his left hand, a slash on his right forearm, and three deep bruises on his side. Ötzi appears to have been in a fight for his life, attacked by several assailants and possibly carrying a wounded friend on his back—hence the stain on his cloak. Scientists hypothesize that in attempting to escape his attackers, Ötzi fled into the mountains, where he was caught at 300 meters altitude by a snowstorm, which killed him. Or perhaps the struggle took place nearby and as the glacier continues to melt, more mummies may be revealed—perhaps that of Ötzi's killer or his friend. Betting on such possibilities, scientists have begun scouring the mountains for more mummies.

If more mummies are found, researchers will again start the process of answering the question, "Who are these people?" by placing them on the tree of life. When it comes to solving biological problems in the real world, knowing where an organism fits in the tree of life is the first order of business.

Chapter Review

Summary

2.1 Building Evolutionary Trees

- Systematists study relationships among different organisms in order to create evolutionary trees. To understand an organism's place in the diversity of life, a systematist uses knowledge of its closest relatives to place it on the evolutionary tree of all life.
- Closely related groups of organisms share key features, such as distinctive anatomical structures or behaviors, inherited from their most recent common ancestor. Systematists use these shared derived features to recognize lineages of closely related organisms.
- Evolutionary trees depict the evolutionary relationships of groups of organisms. The tips of branches represent separate groups or species, and the forks represent the most recent common ancestors of those descendants.
- Convergent features are shared by distantly related organisms because the features evolved independently, not because they were inherited from the organisms' most recent common ancestor. Thus, they are not useful for determining relatedness.
- DNA comparisons are a useful new tool for studying evolutionary relationships.

2.2 Using Evolutionary Trees to Predict the Biology of Organisms

- Evolutionary trees can predict features of organisms and can lead to surprising discoveries.

- An example is the realization that long-extinct dinosaurs exhibited parental care, protecting nests of their eggs, just as crocodilians and birds do today. All three groups share a common ancestor that probably exhibited parental care.

2.3 A Classification System for Organizing Life: The Linnaean Hierarchy and Beyond

- The Linnaean hierarchy is a classification system for organizing all of life. In this scheme, every species of organism has a two-part scientific name indicating its genus and species.
- The lowest level of the Linnaean hierarchy is the species. From there the hierarchy moves upward to genera, families, orders, classes, phyla, and kingdoms.
- Different biologists divide up the living world into different numbers of kingdoms, depending on their purposes. In this book we use the common six-kingdom scheme.
- Biologists use a level of classification above kingdoms, known as domains. The domains are Bacteria, Archaea, and Eukarya.

2.4 Branches on the Tree of Life

- DNA studies have increased our understanding of the tree of life. In some cases, these studies confirm relationships suggested by similarities in structural and behavioral features; in other cases, DNA comparisons provide interesting surprises, often requiring that parts of the evolutionary tree of life be reorganized.

- From DNA studies, biologists now know that the kingdom Fungi shares a more recent common ancestor with the kingdom Animalia than it does with the kingdom Plantae. This finding overturns the long-held assumption that fungi were more closely related to plants.
- Scientists have greatly improved our understanding of the primate evolutionary tree. There is growing agreement that chimpanzees are humans' closest primate relatives.
- DNA studies indicate that the base of the tree of life may look more like a web than a simple root. The evidence for this web is the grab-bag nature of DNA in some organisms. Some biologists explain this pattern as a result of horizontal gene transfer between kingdoms early in the history of life.
- Evolutionary trees are the best hypotheses we have about evolutionary relationships; because most relationships have yet to be worked out in detail, evolutionary trees are works in progress.

◉ Review and Application of Key Concepts

1. How do biologists identify an unknown organism and place it on the tree of life?

2. How are family trees like evolutionary trees?

3. Define "shared derived features" and describe how they differ from other similarities between organisms. Why aren't the panda's thumb and the human thumb shared derived features?

4. How is an evolutionary tree like a hypothesis?

5. Why did biologists think that dinosaurs might exhibit parental behavior? Use the tree in Figure 2.4 to explain.

6. Look at the tree in Figure 2.4. It is well known that birds can sing, and that they sing often. Crocodilians are also known to make similar chirping vocalizations. What can you conclude about the existence of singing dinosaurs? And why?

7. What are the levels of the Linnaean hierarchy? Which is the smallest grouping? Which is the largest grouping?

8. Why has the study of DNA made such revolutionary changes in our understanding of the evolutionary tree of life? To defend your statement, give one example of a change made by a DNA study.

Key Terms

Archaea (p. 30)
Bacteria (p. 30)
class (p. 29)
convergent features (p. 26)
domain (p. 30)
Eukarya (p. 30)
evolutionary tree (p. 24)
family (p. 29)
genus (p. 29)
horizontal gene transfer (p. 34)
kingdom (p. 30)
lineage (p. 24)
Linnaean hierarchy (p. 29)
most recent common ancestor (p. 24)
order (p. 29)
phylum (p. 30)
scientific name (p. 29)
shared derived features (p. 25)
systematist (p. 24)
taxon (p. 30)

Self-Quiz

1. In Figure 2.1b, which two are most closely related?
   a. A and C
   b. D and E
   c. A and B
   d. B and D

2. The most powerful new tool being used by biologists to determine evolutionary relationships today is
   a. behavior.
   b. the cell.
   c. DNA.
   d. organs.

3. As depicted in Figure 2.11, the closest relative of humans is the
   a. chimpanzee.
   b. gorilla.
   c. orangutan.
   d. lemur.

4. Dinosaurs and crocodilians most likely exhibit similar parental behaviors because
   a. they are both scaly.
   b. they both lay eggs.
   c. they are both closely related to birds.
   d. they share a common ancestor that exhibited parental behaviors.

5. Which of the following can be concluded from Figure 2.9?
   a. Archaea and Bacteria are more closely related to each other than either is to a squirrel.
   b. Fungi, animals, and plants are equally closely related.
   c. Plants and fungi are more closely related to each other than either is to animals.
   d. Diplomonads are the most ancient group known.

6. Which of the following groupings list only domains?
   a. Eukarya, Bacteria, and Animalia
   b. Plantae, Protista, and Archaea
   c. Archaea, Bacteria, and Eukarya
   d. Bacteria, Archaea, and ciliates

Biology in the News

Irish Potato Famine Caused by Unidentified Fungus

The menacing strain of fungus long blamed for the Irish Potato Famine was not the culprit after all, according to genetic research.

During the 1840s, a blight raced through Ireland's potato crops, leading to a famine that killed at least 1 million people and spurred a wave of emigration to the United States.

Scientists have long known that the blight was caused by the fungus *Phytophthora infestans* [fye-*TOFF*-thuh-ruh in-*FESS*-tans]. And they have long assumed that the fungus was the same strain that is most widespread today, US-1.

But that assumption is wrong, said Jean Ristaino, a plant pathologist at North Carolina State University, who reported her findings in Thursday's issue of the journal *Nature*.

Many people are keenly interested in the disease that caused the Irish potato famine, not least of all historians, because the blight had huge social ramifications. One of the many changes that followed the ravages of the disease and the starvation it caused was the migration of many Irish families—perhaps some you know, or perhaps even your own ancestors—to America. So it was of particular interest when Ristaino and her colleagues were able to study diseased potato leaves from the actual Irish potato famine blight. The researchers studied leaves that had been collected in Ireland between 1845 and 1847. But after carefully searching the DNA found on more than two dozen such leaves, researchers were surprised to find no evidence of the presence of the so-called US-1 strain, the perpetrator that had long been blamed for the epidemic.

But this discovery is about the future as much as the past, making it of much more than mere academic interest. Scientists say that knowing which strain caused the blight could help researchers learn how to detect and prevent other such problems in the future.

(While *USA Today* refers to *Phytophthora infestans*, the organism that caused the blight, as a fungus, in fact it belongs to an entirely different kingdom: the protists. We will learn much more about both the fungus and protist kingdoms in Chapter 3. Many people mistakenly continue to call this well known disease-causing organism a fungus, however, because *Phytophthora infestans* is a protist that acts very much like a fungus.)

Evaluating the news

1. How would scientists determine whether the fungal strains on the potato leaves collected in the 1800s were the same as US-1 or not? That is, how would they determine what exactly these mystery fungi were?

2. In recent years, potato blight has begun to spread aggressively around the globe, in Russia, Mexico, Ecuador, and the United States. How might knowing that the killer fungus of the 1800s was not strain US-1, as originally thought, but in fact another strain, help scientists fend off epidemics?

3. What tangible benefits does the study of evolutionary trees have to offer society?

SOURCE: *USA Today*, June 6, 2001.

CHAPTER 3

Major Groups of Living Organisms

Key Concepts

⦿ Biologists use three systems to categorize living organisms: (a) the tree of life, showing evolutionary relationships; (b) three domains (Bacteria, Archaea, and Eukarya); and (c) the Linnaean hierarchy, which divides those domains into kingdoms (Bacteria, Archaea, Protista, Plantae, Fungi, and Animalia).

⦿ Organisms can also be categorized according to their cellular structure, as either prokaryotes (simple single-celled organisms) or eukaryotes (organisms with larger, more complex cells).

⦿ Bacteria and Archaea are prokaryotes. They are the most numerous organisms, the most widespread geographically, and the most diverse in methods of obtaining nutrition, acting as producers, consumers, and decomposers.

⦿ Protists are a diverse group of single-celled and multicellular organisms that can be animal-like, plantlike, or funguslike. Some represent early stages in the evolution of the eukaryotic cell and of multicellularity.

⦿ Plants pioneered living on land. After evolving vascular systems, plants evolved diverse shapes and sizes. Two other key innovations of plants were seeds and flowers. As essential components of land ecosystems, plants provide the food that almost all other organisms—the consumers—eventually use.

⦿ Fungi have a unique body plan, which allows them to penetrate other organisms, digest their tissues externally for food, and then absorb that material. Fungi are critical components of ecosystems, acting primarily as decomposers, but also as mutualists with algae and plants. Some fungi are parasites.

⦿ Animals are multicellular, often mobile, and range in complexity from sponges to mammals. Animals have evolved specialized tissues, organs and organ systems, body cavities, and a variety of shapes and sizes. They also show an astounding range of behaviors. Insects (an animal group) are the most species-rich group of all organisms. Animals are consumers, but also act as decomposers.

⦿ Viruses are not classified into any kingdom or domain.

Here, There, and Everywhere

In the spring of 2003, Michael Kovach, owner of a plant nursery in Virginia, was traveling in the Andes of South America looking for interesting new plants to buy and raise. That was when he came across it—the most spectacular orchid discovery in the last hundred years. The flower, a lush magenta and purple blossom that stretched 15 centimeters across, sat on a stem that was 30 centimeters tall. Huge, gorgeous, and extremely valuable, this species had gone undetected in the remote mountains despite an intense interest in orchids on the part of scientists and the public. When Kovach returned to the United States with samples of this orchid, scientists were so excited that, rather than taking the usual number of years to document, draw, and write about it, they did this work virtually overnight. The first scientific paper about the new species—*Phragmipedium kovachii* [frag-mih-*PEE*-dee-um koh-*VAH*-chee], named after its discoverer—was published in a record 8 days.

Elsewhere, far from the lush tropics, scientists drilling deep into Earth made a quite different, but similarly shocking, finding. Two miles beneath the surface, in what was thought to be barren rock, researchers discovered abundant life: bacteria that had been isolated from the rest of the world's organisms for millions of years. Among the species discovered was a bacterium for which the scientists immediately proposed the name *Bacillus infernus*, literally "bacterium from the inferno," or "bacterium from hell."

In the ocean—the same seas that humans have been traveling on, fishing in, and swimming through for many thousands of years—biologists have recently discovered scores of new microscopic species, mysterious new lives that have been washing ashore, splashing against boats, and swirling in the ocean depths for countless millennia.

Many Species Remain to Be Discovered by Scientists
Even large, showy new species continue to be discovered. This orchid, *Phragmipedium kovachii*, was found in South America in 2003 by Michael Kovach, after whom it is named.

The Rio Tinto
The waters of the Rio Tinto ("River of Fire"), once thought to be inhospitable to life, have been found to harbor a wide variety of microscopic organisms (inset).

Even in Europe, where scientists have been poking and prodding and collecting organisms for hundreds of years, scientists have found new life in an unexpected place. Spain's Rio Tinto, or River of Fire, was named for its deep red color and highly acidic waters, a combination producing an extremely hostile environment for living organisms. Scientists probing the river during a recent expedition, however, discovered a huge diversity of microscopic organisms there, including two unexpected groups: species of bacteria already known to survive similar harsh conditions elsewhere, and many other—some entirely new—microscopic organisms.

Such finds are becoming more common as biologists probe the last dark corners where life is hiding. Everywhere scientists are finding organisms of all sorts. As the world's species become better known, the lesson biologists continue to learn is that life is hardy and resourceful, thriving almost everywhere, even in the most unlikely places.

In the heat of the Sonoran Desert, cacti point their scorched green limbs to the sky, saved from drying out by the gallons of water stored safely inside them. High in the Himalayas, a woolly flying squirrel the size of a woodchuck glides through the skies. Wildflowers in the snow-covered Swiss Alps turn their blossoms to follow the movement of the sun and capture its precious warming rays. In the steaming waters of Yellowstone National Park's most famous geyser, Old Faithful, heat-loving bacteria comfortably persist, well suited to such extremes of temperature. And everywhere insects, the world's most abundant animals, fly and crawl, making meals out of everything from the nectar of flowers to poison-filled plant leaves to cow dung.

In this chapter we take a look at the diversity of life, the riot of living organisms that inhabits our planet.

As we saw in Chapter 1, all living organisms share a basic set of characteristics. Biologists believe that life shares this set of common properties because all living organisms descended from a single common ancestor, known as the universal ancestor. Since first appearing on this planet more than 3.5 billion years ago, life has evolved from single-celled organisms to a world full of wildly different organisms.

This great diversity is still far from being completely known, counted, or named. Most biologists estimate that there are between 3 million and 30 million species on Earth. This chapter is not a comprehensive examination of the world's many species, but instead an attempt to familiarize you with the diversity of life by introducing you to the major groups of organisms. First we review the primary systems used by biologists to classify and categorize the diversity of life, and we place the major groups of organisms in these systems. Then we explain key features characterizing each group—in particular, newly evolved features that allowed members of each group to thrive. In addition, we describe some of the more important and interesting members of each group. We close with a brief look at how biolo-gists are continuing the ongoing search for new forms of life.

3.1 The Major Groups in Context

The major groups of organisms are the Bacteria (including familiar disease-causing bacteria), the Archaea (bacteria-like organisms that are best known for living in extreme environments), the Protista (a diverse group that includes amoebas and algae), the Plantae [*PLAN*-tee] (plants), the Fungi [*FUNJ*-eye] (mushrooms, molds, and yeasts), and the Animalia (animals).

There are three major systems for classifying all of life

In order to provide a framework for thinking about the world's many organisms, we have included the key information from Chapter 2 (especially Figures 2.7 and 2.9) in a single figure. Figure 3.1 places the major groups of organisms in context in three different ways: (1) on the evolutionary tree of life, (2) in the Linnaean hierarchy—each of the major groups represents a kingdom, the Linnaean hierarchy's largest grouping—and (3) in the system of life's three domains. Recall that the Bacteria and Archaea are each recognized as both a kingdom and a domain, whereas the Eukarya is a single domain comprising four kingdoms.

The three systems shown in Figure 3.1 are widely used, making it important to learn all three. To help you keep track of where you are in each system, at the beginning of each section introducing a new group, a marginal diagram highlights that group on the tree of life and in the kingdom and domain systems. The description of each major group is accompanied by a figure that provides a broad overview of that group, an evolutionary tree of the subgroups within that group, and a photo gallery illustrating some of the prominent subgroups within that group. These figures provide an evolutionary framework for discussing each group's **evolutionary innovations**, new features that allowed its members to live and reproduce successfully. These figures can also serve as a ref-erence as you read later chapters and want to take a second look at groups you are learning about.

Organisms can also be identified as prokaryotes and eukaryotes

The Eukarya are distinguished from the Bacteria and the Archaea by the structure of their cells, which contain compartments with specialized functions. One compartment, the nucleus, holds the genetic material (DNA). The organisms in the domain Eukarya are also called **eukaryotes**. Both names come from the Greek *eu* ("true") and *karyote* ("kernel"), referring to the nucleus as a kind of kernel, or compartment, within the cell. In contrast, the single-celled bodies of the Bacteria and the Archaea do not have compartments. They are often referred to as **prokaryotes** (*pro*, "before") because they evolved before the evolution of cells with a nucleus and other compartments. Note that the terms *eukaryote* and *prokaryote* simply refer to the structure of an organism's cells and are not part of the domain, kingdom, or tree-of-life classifications.

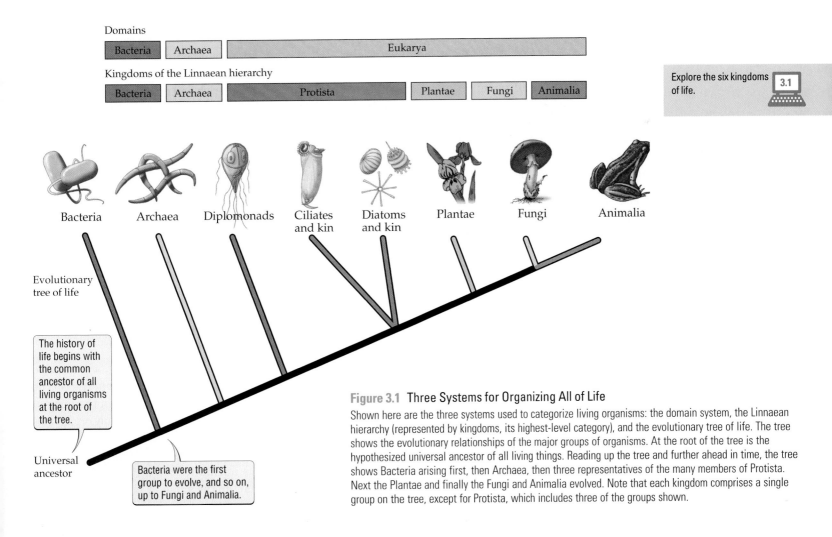

Explore the six kingdoms of life. 3.1

Figure 3.1 Three Systems for Organizing All of Life

Shown here are the three systems used to categorize living organisms: the domain system, the Linnaean hierarchy (represented by kingdoms, its highest-level category), and the evolutionary tree of life. The tree shows the evolutionary relationships of the major groups of organisms. At the root of the tree is the hypothesized universal ancestor of all living things. Reading up the tree and further ahead in time, the tree shows Bacteria arising first, then Archaea, then three representatives of the many members of Protista. Next the Plantae and finally the Fungi and Animalia evolved. Note that each kingdom comprises a single group on the tree, except for Protista, which includes three of the groups shown.

3.2 The Bacteria and Archaea: Tiny, Successful, and Abundant

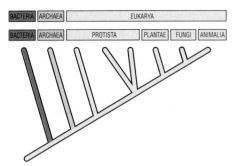

When life arose on Earth more than 3.5 billion years ago, the first living organisms were tiny, simple, single cells. The first modern lineage to arise—that is, the first to branch off the tree of life—was the Bacteria (Figure 3.2). Bacteria are probably most familiar as single-celled disease-causing organisms such as *Streptococcus pneumoniae* [noo-MO-nee-aye], which can cause pneumonia in humans. There is still uncertainty about this part of the tree of life, but one hypothesis suggests that the Archaea arose after the Bacteria, a scenario depicted by the evolutionary trees used in this book. Discovered in the 1970s, these single-celled, bacteria-like organisms are best known because some of them are **extremophiles** (literally, "lovers of the extreme"): they have been found thriving in boiling hot geysers, highly acidic waters, and the freezing cold seas off Antarctica. When the group Archaea arose, it split from the Eukarya lineage, which later gave rise to the rest of the world's organisms.

Although they are distinct groups, Bacteria and Archaea are similar in both size (microscopic) and structure (single-celled); they are similar in many other ways as well. So it is appropriate that we begin our introduction to the major groups of life by describing them and their ways of life together.

Bacteria and Archaea are not only the simplest organisms, but also the most successful at colonizing Earth. They are quite variable in shape, with some having shapes like spheres (called cocci [KOCK-eye]; singular: coccus), rods (called bacilli [ba-SILL-eye]; singular: bacillus), or corkscrews (called spirochetes [SPY-ro-keets]); however, they all share a basic structural plan (Figure 3.3). The picture of efficiency, these stripped-down organisms are nearly always single-celled and small. They typically have much less DNA than the cells of organisms in the Eukarya have. Eukaryotic genetic material is often full of what appears to be extra DNA that serves no known function. In contrast, prokaryotic genetic material contains only DNA that is actively in use for the survival and reproduction of the bacterial cell. Prokaryotic reproduction is similarly uncomplicated: prokaryotes typically reproduce by splitting in two, a process called fission.

Prokaryotes represent simplicity translated into success

While most people think of **biodiversity** in terms of butterflies, redwood trees, and other complex organisms built of many cells, the vast majority of life on Earth is single-celled and prokaryotic. Scientists estimate that the number of individual prokaryotes on Earth is about 5,000,000,000,000,000,000,000,000,000,000, or 5×10^{30}. Their success is due, in part, to how quickly they reproduce. Overnight, a single bacterium of the common species *Escherichia coli* (usually referred to by the abbreviation *E. coli*), which normally lives harmlessly in the human gut (see Figure 3.2), can divide to produce a population of 16 million bacteria.

Prokaryotes are also the most widespread of organisms, able to live nearly anywhere. They can persist even in places where most organisms would perish, such as the lightless ocean depths, the insides of boiling hot geysers, and miles below Earth's surface. Because of their small size, prokaryotes also live on and in other organisms. Scientists estimate that 1 square centimeter of healthy human skin is home to between 1,000 and 10,000 bacteria.

In addition, while many prokaryotes need the gas oxygen to survive (that is, they are aerobes, from *aero*, "air"; "bios," life), many others can survive without oxygen (they are anaerobes, from *an*, "without"). This ability to exist in both oxygen-rich and oxygen-free environments also increases the number of habitats in which prokaryotes can persist. But the real key to the success of these groups is the great diversity of ways in which they obtain and use nutrients.

Prokaryotes exhibit unmatched diversity in methods of obtaining nutrition

Every organism needs two things to grow and survive: a source of energy and a source of carbon. Carbon is the chemical building block used to make critical molecules for living, such as proteins and DNA. Prokaryotes are distinguished by having the most diverse methods of obtaining energy and carbon of any group of organisms on Earth.

When humans and other animals eat, we consume other species, from which we get both energy (in the form of chemical bonds) and carbon (in the form of carbon-containing molecules). In fact, many organisms, including all animals, all fungi, and some protists, get their energy and carbon by consuming other organisms—what we normally would think of as eating. Prokaryotes can do the same. A familiar example is

Figure 3.2 The Prokaryotes: Bacteria and Archaea

Prokaryotes are simple, microscopic, single-celled organisms that are the most ancient forms of life. Bacteria branched off first, then the Archaea split off from what would become the rest of the living organisms, the Eukarya.

- Number of species discovered to date: ~4,800
- Functions within ecosystems: Producers, consumers, decomposers
- Economic uses: Many, including producing antibiotics, cleaning up oil spills, treating sewage
- Did you know? The number of bacteria in your digestive tract outnumber all the humans that have lived on Earth since the beginning of time.

These bacteria, *Borrelia burgdorferi*, known as spirochetes because of their spiral-shaped cells, cause Lyme disease, which is transmitted to humans through a tick bite.

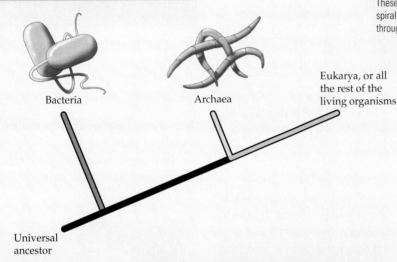

Bacteria

Archaea

Eukarya, or all the rest of the living organisms

Universal ancestor

This bacterium, known as *Escherichia coli*, is usually a harmless inhabitant of the human gut. However, toxic strains can contaminate and multiply on foods, such as raw hamburger, and cause illness or death in humans who eat them.

These archaeans, known as *Methanospirillum hungatii*, are shown in cross section (the two circular shapes) and as an elongated cell that is about to fission into two cells.

This bacterium is a member of the genus *Streptomyces*, which produces the antibiotics streptomycin, erythromycin, and tetracycline.

The bacterium *Chlamydia trachomatis* causes the most common sexually transmitted disease, chlamydia.

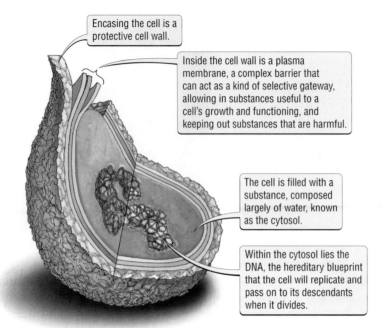

Encasing the cell is a protective cell wall.

Inside the cell wall is a plasma membrane, a complex barrier that can act as a kind of selective gateway, allowing in substances useful to a cell's growth and functioning, and keeping out substances that are harmful.

The cell is filled with a substance, composed largely of water, known as the cytosol.

Within the cytosol lies the DNA, the hereditary blueprint that the cell will replicate and pass on to its descendants when it divides.

Get to know the basic parts of a bacterial cell. 3.2

Figure 3.3 **The Basic Structure of the Prokaryotic Cell**
Prokaryotic cells tend to be about 10 times smaller than the cells of organisms in the Eukarya, and have much less DNA.

Clostridium botulinum [klaw-*STRID*-ee-um *BOTCH*-oo-*LINE*-um], a bacterium that causes food poisoning. It lives in and consumes food that humans have stored and produces a toxin that can make humans that eat that food sick. But while the rest of the world's eaters are restricted to consuming other organisms, or parts or products of other organisms, some prokaryotes can live by consuming carbon-containing compounds (such as pesticides) that are not parts of other organisms.

While animals, fungi, and many protists eat in order to live, plants and the remaining protists produce their own nutrition. These organisms get their energy from sunlight and their carbon from carbon dioxide, the gas that humans and other animals exhale, and use the two during photosynthesis to produce sugars. Some prokaryotes carry out photosynthesis as well. Cyanobacteria [sigh-*AN*-oh- . . .] (also called blue-green algae, although they are not algae), which make up the green slime more commonly known as pond scum, are a familiar example of prokaryotes that use sunlight and carbon dioxide to make their own nutrients (Figure 3.4).

But while the Eukarya can only eat or carry out photosynthesis, prokaryotes can also survive in two other ways. Some prokaryotes can use light as an energy source, the way plants and organisms such as cyanobacteria do, but instead of getting their carbon from carbon dioxide, they derive it from organic compounds. Finally,

Figure 3.4 Pond Scum: Bacteria That Photosynthesize
These photosynthetic cyanobacteria, sometimes called blue-green algae, can be found growing as slimy mats on freshwater ponds.

there are prokaryotes that use carbon dioxide as a carbon source, as plants and cyanobacteria do, but get their energy from such unlikely materials as iron and ammonia (Figure 3.5).

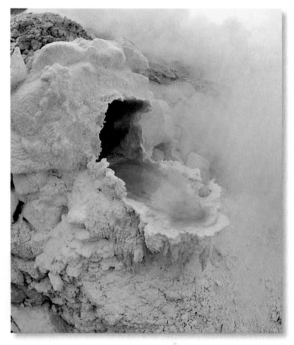

Figure 3.5 Curious Appetites
Sulfolobus [sul-*FALL*-uh-buss] gets its carbon from carbon dioxide, as plants do. However, this archaean gets its energy in an unusual way—not by harnessing sunlight (as plants do), nor even by eating other organisms (as animals do), but by chemically processing inorganic materials such as iron. This extremophile is living in a volcanic vent on the island of Kyushu in Japan.

Biology on the Job

A Passion for Life

Dr. Niles Eldredge grew up to get a job that is the stuff of dreams for many children—spending all day in one of the world's greatest natural history museums—and that is the stuff of nightmares for other people—stuck all day in the back hallways of a museum full of dead animals and pressed plants. And like many biologists, Dr. Eldredge is passionate not just about his job, but about the importance of the living world itself.

What is your official title? Curator in the Division of Paleontology at the American Museum of Natural History (AMNH) in New York City. I was also Curator-in-Chief of the Hall of Biodiversity—with overall responsibility for the scientific content—as we were developing the Hall. I am a research paleontologist and evolutionary theorist. I was lucky enough to be hired at the American Museum after I got my Ph.D. in 1969.

What are the everyday tasks of a museum curator? Our primary responsibility is to pursue original scientific research in our fields—including fieldwork, laboratory analysis, publication, and the raising of grant funds. We also are charged with the ultimate care of the collections—including adding to the collections, making sure they are properly maintained, and approving loan requests.

When did biodiversity become a separate field of scientific study? In a sense, biodiversity has always been a focus of biology—but only in the past 15 years or so have scientists joined others in seeing the rapidly accelerating loss of species as the world's ecosystems continue to be degraded at an alarming pace. Science can contribute a lot toward understanding what biodiversity is, why we should value it, what is happening to cause the "sixth extinction"—and what we can do to help slow the rate of species extinction on the planet right now.

What would you say is the goal of AMNH's biodiversity exhibit, or of biodiversity research in general? Our goal was to impress upon the visitor that life is beautiful, rich, vibrant, and still of great importance to humanity, and very much under threat—even though many of us live in cities and are not bombarded with life's beauties and richness every day (though there is still a lot of biodiversity in cities!).

What skills or personal characteristics do you think have been most beneficial to you in your job as a curator? Profound, unending curiosity and a passion for the history of life.

What do you enjoy most about your job? Thinking.

An exhibit at New York City's American Museum of Natural History.

The Job in Context

Dr. Eldredge is no doubt correct that unending curiosity and a passion for the entire history of living organisms are crucial for his work. How else could he handle the responsibility of managing the Hall of Biodiversity, which, by definition, covers all the living things we cover in this long chapter—as well as the many, many more organisms that once lived but have gone extinct? And while most people do not think of museums as dynamic scientific centers, they have become exactly that, with researchers like Dr. Eldredge seeking out and documenting the many fantastic organisms that live and have lived on Earth. The creation of the evolutionary trees shown in this chapter, the estimates of numbers of species in different groups, and the sorting of organisms into kingdoms and domains—all these activities are the work of people like Dr. Eldredge. And now, as he has pointed out, these researchers have become responsible not only for documenting what lives on Earth, but also for sending out a warning call that many of these organisms are in need of saving as well.

Prokaryotes can thrive in extreme environments

Prokaryotes are well known for living in nearly any kind of environment. While some bacteria thrive in unusual environments, Archaea is the group best known for the extreme lifestyles of some of its members. Some are extreme thermophiles (*thermo*, "heat"; *phile*, "lover") that live in geysers and hot springs. Others are extreme halophiles (*halo*, "salt"), thriving in high-sodium environments where nothing else can live—for example, in the Dead Sea and on fish and meat that have been heavily salted to keep most bacteria away (Figure 3.6).

Not all archaeans, however, are so remote from daily experience. Members of one group, the methanogens (*methano*, "methane"; *gen*, "producer"), inhabit animal guts and produce the methane gas in such things as human flatulence (intestinal gas) and cow burps.

Prokaryotes play important roles in the biosphere and in human society

Because they are able to use such a variety of food and energy sources and to live under such wide-ranging conditions, prokaryotes play numerous and important roles in ecosystems and in human society. For example, plants require the chemical nitrate, which they cannot make themselves. For this they depend on bacteria that can take nitrogen, a gas in the air, and convert it to nitrate. Without these bacteria there would be no plant life, and without plant life there would be no life on land.

Like plants, some bacteria, such as cyanobacteria, can photosynthesize. (Photosynthesis will be covered in detail in Chapter 8.) This ability makes them producers, the organisms at the base of food webs. Other bacteria are important decomposers. Oil-eating bacteria can be used to clean up ocean oil spills. Bacteria that can live on sewage are used to help decompose human waste so that it can be safely, usefully returned to the environment. Bacteria also live harmlessly in animal guts (including our own), aiding in food digestion.

Of course, not all prokaryotes are helpful. Many bacteria cause diseases; some are the source of nightmares, such as the flesh-eating bacteria able to destroy human flesh at frightening rates. With their ability to use almost anything as food, bacteria can also attack crops, stored foods, and domesticated livestock.

Bacteria and Archaea exhibit key differences

While similar in many ways, the Bacteria and Archaea are distinct lineages. In recent years, biologists have learned more about the Archaea, the more recently recognized of the two groups. One key distinction shows up in their DNA: much of archaean DNA is unique to archaeans. The Archaea and Bacteria also differ in how their metabolism (cellular machinery) is run. In addition, there are specific structural differences in the cells of the two groups: most prokaryotic cells have both a cell wall and a plasma membrane (see Figure 3.3), but some molecules in those structures differ between the two groups.

3.3 The Protista: A Window into the Early Evolution of the Eukarya

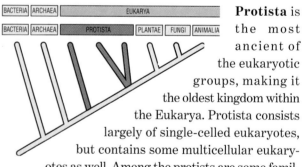

Protista is the most ancient of the eukaryotic groups, making it the oldest kingdom within the Eukarya. Protista consists largely of single-celled eukaryotes, but contains some multicellular eukaryotes as well. Among the protists are some familiar groups, such as single-celled amoebas and kelp (which are multicellular algae), as well as many unfamiliar groups. One of the few generalizations that can be made about this hard-to-define group is that its members are diverse in size, shape, and lifestyle.

Much remains unknown about the evolutionary relationships of the protists. Figure 3.7 presents one hypothesis for the evolutionary tree of some of the major

Figure 3.6 Better Than a Bag of Chips

For those who love salt, such as archaeans that are extreme halophiles, nothing beats a salt farm. Here in Thailand, seawater is evaporated, making it more and more salty and creating an environment that only an archaean could love.

Figure 3.7 The Protista

The protists form a diverse group of single-celled and multicellular organisms. The evolutionary relationships among protist groups are poorly understood. For example, dinoflagellates, apicomplexans, and ciliates are shown branching off this evolutionary tree simultaneously, to indicate that the order in which these groups evolved is still unknown.

- Number of species discovered to date: ~30,000
- Functions within ecosystems: Producers, consumers, decomposers
- Economic uses: Kelp, a multicellular alga, is raised for food. However, protists are best known for the damage they do, causing red tides and diseases such as amoebic dysentery.
- Did you know? Malaria, the second most deadly disease after AIDS, is caused by *Plasmodium*, a protist.

Seaweeds, such as this sea lettuce, are among the most familiar protists, green algae, seen along coastal shores.

Diplomonads   Dinoflagellates   Apicomplexans   Ciliates   Diatoms   Water molds   Green algae

Single-celled dinoflagellates, like this one, can multiply rapidly, forming deadly red tides.

Diatoms are important producers in freshwater and marine environments. Their glasslike outer shells are microscopic works of art.

Animal-like organisms such as this *Paramecium*, a ciliate, swim using tiny hairlike structures called cilia.

An apicomplexan surrounded by red blood cells. This protist parasite, *Plasmodium*, causes malaria.

Giardia

protist groups. Diplomonads are shown branching off first. Then three major groups—one including dinoflagellates, apicomplexans [ay-pee-kom-*PLEX*-uns], and ciliates; another including diatoms and water molds; and the last comprising the green algae—are shown splitting off at the same time. That is because scientists still do not know which of these three groups branched off first. Other protist groups are not shown on the tree because their placement is even more poorly understood.

Protists show great variety in their lifestyles. Some are plantlike; for example, green algae can photosynthesize, and plants are thought to have evolved from them. There are also animal-like protists—such as the ciliates—that move and hunt for food. Still others—such as the slime molds—are more like fungi. Most protists (such as *Paramecium* [pair-uh-*MEE*-see-um]; see Figure 3.7) are single-celled, but there are also many multicellular protists (such as kelp).

Protists represent early stages in the evolution of the eukaryotic cell

One of the most important events in evolutionary history was the evolution of eukaryotes. Unlike the cells of prokaryotes, the cells of the Eukarya contain internal compartments called **organelles**, which perform different functions. For example, the nucleus is the organelle that contains a eukaryotic cell's DNA, and mitochondria [my-to-*KON*-dree-uh] (singular: mitochondrion) are organelles that produce energy for the cell. Where did these organelles come from?

Biologists hypothesize that eukaryotes arose when free-living prokaryotic cells engulfed other free-living prokaryotic cells. Scientists now believe that this engulfing happened many times in the history of life. The engulfing cells evolved into the Eukarya as, over time, the captured cells evolved into organelles. Organelles developed specialized functions, on which the eukaryotic cells housing them came to depend for survival; organelles also lost the ability to survive as free-living organisms. For example, plant cells today contain chloroplasts, the organelles that carry out photosynthesis. The chloroplast is thought to have originated from a cyanobacterium captured by another prokaryotic cell. Thus a *combination of prokaryotes* appears to have resulted in primitive eukaryotes, the group from which complex multicellular organisms evolved.

Protists are of particular interest because different protist groups illustrate the variety of experimentation in engulfment that has gone on over time. For example, *Giardia lamblia* belongs to the group known as the diplomonads. This protist lives in streams and other

sources of fresh water. It results in a painful ailment of the digestive tract when consumed by humans, causing diarrhea and flatulence with a rotten-egg odor. Most single-celled eukaryotes contain a nucleus and one or more mitochondria, and chloroplasts if they photosynthesize. *Giardia lamblia*, however, appears to be a curious experiment in putting together a single-celled organism: it has two nuclei, no chloroplasts, and has lost the mitochondria it once had.

Protists provide insight into the early evolution of multicellularity

Some groups of protists are of interest to biologists because they have evolved from living as single-celled creatures to forming multicellular associations that function to varying degrees like more complex multicellular individuals. Among the more interesting of these experiments in the evolution of multicellularity are the slime molds, protists that were originally mistaken for molds (which are fungi). Commonly found on rotting vegetation, slime molds eat bacteria and live their lives in two phases: as independent, single-celled creatures and as members of a multicellular body. Like other protists that can live as either single-celled or multicellular organisms, slime molds are studied by biologists who hope to gain insight into the evolutionary transition from single-celled to multicellular living. The slime molds are among the protist groups excluded from the evolutionary tree in Figure 3.7 because their relationships to other protists remain poorly understood.

Protists had sex first

Prokaryotes reproduce simply by splitting in two, a form of asexual (nonsexual) reproduction. But eukaryotes typically reproduce sexually, by a process in which two individual organisms produce specialized sex cells known as gametes (for example, human gametes are eggs and sperm) that fuse together. This process combines the DNA contributions from two parents into one offspring. Protists were the group in which this form of reproduction first appeared, making sex one of its most noteworthy evolutionary innovations.

Protists are well known for their disease-causing abilities

Although most protists are harmless, many of the best-known protists are those that cause diseases. One example is the dinoflagellates (see Figure 3.7), a group of microscopic plantlike protists that live in the ocean and sometimes experience huge population explosions, known as blooms. Occasional blooms of toxic dinoflagellates cause dangerous "red tides." During red tides, any shellfish that have eaten these toxic dinoflagellates will in turn be poisonous to humans eating the shellfish. The animal-like protist *Plasmodium* causes malaria, which kills millions of people around the world each year—more than any other disease except AIDS. Finally, protists left their mark on human history forever when one of them (often mistakenly referred to as a fungus) attacked potato crops in Ireland in the 1800s, causing the disease known as potato blight. The resulting widespread loss of potato crops caused a devastating famine and a major emigration of Irish people to the United States in the 1840s.

3.4 The Plantae: Pioneers of Life on Land

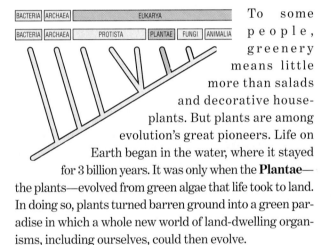

To some people, greenery means little more than salads and decorative houseplants. But plants are among evolution's great pioneers. Life on Earth began in the water, where it stayed for 3 billion years. It was only when the **Plantae**—the plants—evolved from green algae that life took to land. In doing so, plants turned barren ground into a green paradise in which a whole new world of land-dwelling organisms, including ourselves, could then evolve.

Today the diversity of the Plantae—a kingdom within the domain Eukarya—ranges from the most ancient lineages—mosses and their close relatives—to ferns, which evolved next, to gymnosperms, and finally to the most recently evolved plant lineage, the angiosperms (Figure 3.8).

Life on land requires special structures

Figure 3.9 shows the basic structure of a plant. The key evolutionary innovation of plants is their ability to harness chloroplasts in order to photosynthesize—to use light (energy from the sun) and carbon dioxide (a gas in the air) to produce food in the form of sugars. Most photosynthesis in plants takes place in their leaves, which typically grow in ways that maximize their ability to capture sunlight. A useful by-product of photosynthesis is the critical gas oxygen, which plants release into the air. Because plants are producers, they form the basis of essentially all terrestrial (land-based) food webs.

Organisms on land had to solve problems not faced by organisms living in the ocean. The most crucial of these was how to obtain and conserve water. One of the features allowing plants to do this is the **root system**, a collection of fingerlike growths that absorb water and nutrients from the soil. Another is the waxy covering over stem and leaves, known as the **cuticle**, which prevents plant tissues from drying out even when exposed to sun and air. A second challenge of life on land was gravity. While plants can float in water, they cannot "float" in the air. But plants cells have rigid cell walls. Composed of the organic compound cellulose, cell walls give the plant the rigidity it needs to grow up and into the air.

In addition to the features just mentioned, three major evolutionary innovations were critical to plants' highly successful colonization of land: vascular systems, seeds, and flowers. Each of these innovations marks the rise of a separate major plant group.

Vascular systems allowed ferns and their allies to grow to great heights

Early in their evolution, plants grew close to the ground. Mosses and their close relatives, which make up the most ancient plant lineage, represent those early days in the history of plants. These plants rely on each cell being able to absorb water directly. Thus the innermost cells of their bodies receive water only after it has managed to pass through every cell between them and the outermost layer. Because such movement of water from cell to cell, like the movement of water through a kitchen sponge, is relatively inefficient, these plants cannot grow to great heights or sizes.

Ferns and their close relatives, the next major plant group to arise, were able to grow taller because they evolved **vascular systems**—networks of specialized tissues that extend from the roots throughout the bodies of plants (see Figure 3.9). Vascular tissues can efficiently transport fluids and nutrients, much as the human circulatory system of veins and arteries transports blood. In addition, the presence of water and other fluids in the vessels helps make the plant firmer, in the same way that a balloon filled with water is firmer than an empty balloon. By providing sturdiness and efficient circulation of water and nutrients, the evolutionary innovation of vascular systems allowed plants to grow to new heights and sizes. All plants, except mosses and their close relatives, have vascular systems.

Figure 3.8 The Plantae

Plants are multicellular organisms that make their living by photosynthesis. They are a diverse group that pioneered life on land.

- Number of species discovered to date: ~250,000
- Functions within ecosystems: Producers
- Economic uses: Flowering plants provide all our crops: corn, tomatoes, rice, and so on. Fir trees and other conifers provide most of our wood and paper. Plants also produce important chemicals, such as morphine, caffeine, and menthol.
- Did you know? Of the 250,000 species of plants, at least 30,000 have edible parts. In spite of this abundance of potential foods, just three species—corn, wheat, and rice—provide most of the food the world's human populations eat.

Ferns and their close relatives evolved vascular systems that allowed them to grow to greater heights. This Ama'uma'u fern grows only in Hawaii.

Giant sequoia, like this huge tree, are conifers; the most familiar gymnosperms, they are important wood and paper producers.

Mosses   Ferns   Gymnosperms   Angiosperms

The orchids, the most species-rich family of angiosperms in the world, also produce some of the world's most beautiful flowers.

Mosses and their close relatives, the most ancient group of plants, do not have vascular systems and so cannot grow more than a few inches high.

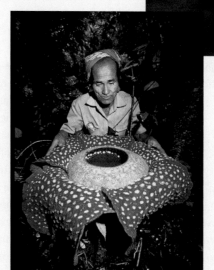

On Mt. Sago in Sumatra, the angiosperm *Rafflesia arnoldii* produces the world's largest blossoms, measuring as much as 1 meter across.

Gymnosperms evolved seeds as a way to protect their young

After mosses and ferns, the next group of plants to evolve was the **gymnosperms** [*JIM*-nuh-sperms]. This group includes pine trees and other conifers (cone-bearing plants; see Figure 3.8), as well as cycads and ginkgos.

Gymnosperms were the first plants to evolve **seeds**, structures in which plant embryos are encased in a protective covering and provided with a stored supply of nutrients (Figure 3.10). Gymnosperms (*gymno*, "naked"; *sperm*, "seed") have seeds that are relatively unprotected compared with those of angiosperms, the next major group of plants to arise. Gymnosperms were the dominant plants 250 million years ago, and the evolution of seeds was probably an important part of their success. Seeds provided nutrients that plant embryos could use to grow before they were able to produce their own food via photosynthesis. Seeds also provided embryos with protection from drying or rotting and from attack by predators.

Angiosperms produced the world's first flowers

Although typically we think of flowers when we think of plants, flowering plants are a relatively recent development in the history of life. Today the flowering plants, or **angiosperms** [*AN*-jee-oh-sperms], are the most dominant and diverse group of plants on the planet, including orchids, grasses, corn plants, and apple and maple trees. Angiosperms produce seeds that are well protected (*angio* means "vessel," referring to the tissues that encase the plant's embryo).

Highly diverse in size and shape, angiosperms live in a wide range of habitats—from mountaintops to deserts to salty marshes and fresh water. Almost any plant we can think of that is not a moss, a fern, or a cone-producing tree is an angiosperm. The key to the success of angiosperms, and their defining feature, is the **flower**. Flowers are specialized structures for sexual reproduction, or pollination, where the male and the female gametes meet (Figure 3.11).

Some flowers provide food, such as the sugary liquid known as nectar, to attract animals to visit them and in the process transport pollen (male gametes) from flower to flower. The transported pollen can fertilize another flower's ovules (structures containing female gametes). Thus animals can provide a means of sexual reproduction between even very distant plants. Plants also use wind to disperse their pollen from flower to flower. People with hay fever are reacting to this method of

Figure 3.9 The Basic Structures of a Plant

Shown here is a familiar garden vegetable, a pepper plant. Because it is a member of the angiosperms, the last of the major plant groups to evolve, all the evolutionary innovations that distinguish plants can be shown on this one plant.

plant sex, which sends nose-irritating pollen blowing through the air.

In addition to increasing the efficiency of fertilization through flowers, angiosperms have evolved a variety of ways to distribute their seeds to distant places in order to get their young off to a good start. One of these is the use of tasty fruits that attract animals. As the embryos of some angiosperms are developing, the surrounding ovary develops into a ripening fruit (see Figure 3.11). Animals eat the fruit and later excrete the seeds in their feces. These nutrient-rich wastes provide a good place for the seeds to begin life, often far from their parent plant where they will not compete with that parent for water,

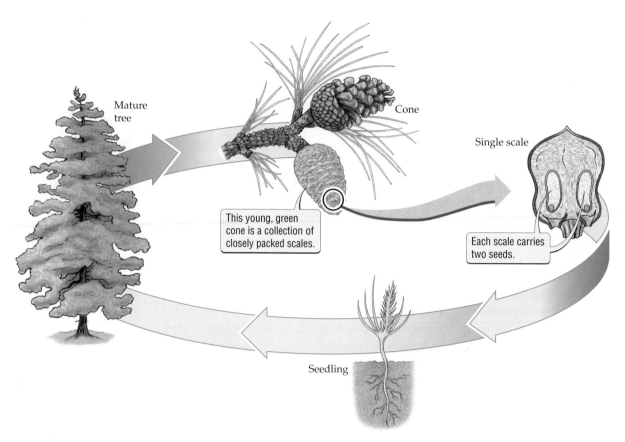

Figure 3.10 The Seed
The first plants with seeds were the gymnosperms, which include conifers such as pine trees.
Modern descendants of the first seeds can be found in pine cones.

Mature tree

Cone

Single scale

This young, green cone is a collection of closely packed scales.

Each scale carries two seeds.

Seedling

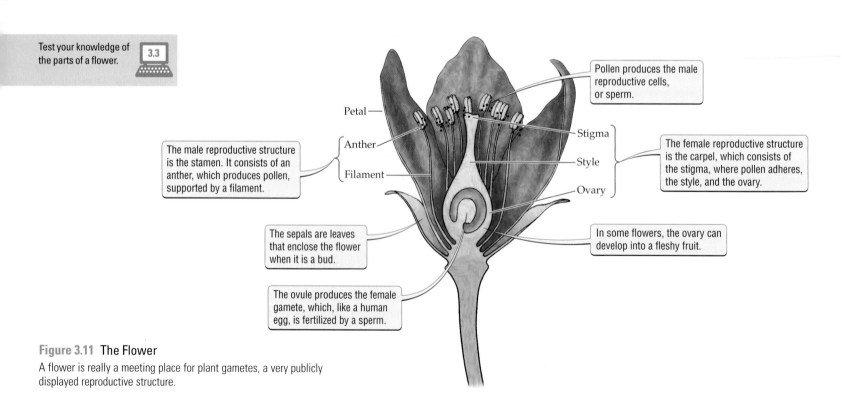

Test your knowledge of the parts of a flower. **3.3**

Pollen produces the male reproductive cells, or sperm.

Petal

Anther

Filament

The male reproductive structure is the stamen. It consists of an anther, which produces pollen, supported by a filament.

Stigma

Style

Ovary

The female reproductive structure is the carpel, which consists of the stigma, where pollen adheres, the style, and the ovary.

The sepals are leaves that enclose the flower when it is a bud.

In some flowers, the ovary can develop into a fleshy fruit.

The ovule produces the female gamete, which, like a human egg, is fertilized by a sperm.

Figure 3.11 The Flower
A flower is really a meeting place for plant gametes, a very publicly displayed reproductive structure.

(a)

(b)

Figure 3.12 Getting Around
Plants have evolved many ways of spreading to new areas. (*a*) A palm tree seed in a coconut can float for hundreds of miles until it reaches a new beach where it can take root and grow. (*b*) Some seeds have wings (for example, maple "keys") or other structures (such as dandelion fluff, shown here) that allow them to be carried by the wind, sometimes over great distances.

nutrients, or light. But hitching a ride in an animal's gut is not the only means plants have for overcoming their immobility; plant seeds have evolved many other ways to travel (Figure 3.12).

Plants are the basis of land ecosystems and provide many valuable products

It is difficult to overstate the significance of plants. As photosynthesizing organisms, plants use sunlight and carbon dioxide to make sugars, food that they and the organisms that eat them can use. Nearly all organisms on land ultimately depend on plants for food, either directly by eating plants or indirectly by eating other organisms (such as animals) that eat plants or that eat other organisms that eat plants, and so on. Many organisms live on or in plants, or on or in soils largely made up of decomposed plants.

Flowering plants provide humans with materials such as cotton for clothing and with pharmaceuticals such as morphine. Essentially all agricultural crops are flowering plants, and the entire floral industry rests on the reproductive structures of angiosperms. Gymnosperms such as pines, spruces, and firs are the basis of forestry industries, providing wood and paper.

As valuable as plants are when harvested, they are also valuable when left in nature. By soaking up rainwater in their roots and other tissues, for example, plants prevent runoff and erosion that can contaminate streams. Plants also produce the crucial gas oxygen.

3.5 The Fungi: A World of Decomposers

Most people are familiar with fungi as the mushrooms sliced on their pizza or sprouting from their lawns. However, the **Fungi**, a kingdom within the domain Eukarya, also includes yeasts (single-celled fungi) and molds. In fact, the familiar mushroom is just a small part of a fungus. Most fungal tissues typically are woven through whatever substance—often the tissues of another organism—the fungus is digesting and making its living from. Because fungal tissues are largely hidden from view, fungi are among the most poorly understood of the major groups of organisms.

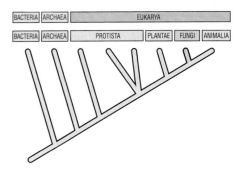

Fungi can be costly to human society. They can cause diseases, contaminate crops, rot food, and force us to clean our bathrooms more often than we might like. Other fungi are beneficial, providing us with pharmaceuticals, including antibiotics such as penicillin. Yeasts such as *Saccharomyces cerevisiae* [SAK-ah-roh-MICE-eez sair-uh-VEE-see-eye] can feed on sugars and produce two important products: alcohol and the gas carbon dioxide, both crucial to the rising of bread and the brewing of beer. Fungi also provide highly sought-after delicacies such as truffles, whose underground growing locations can be found only by specially trained dogs or pigs.

As Figure 3.13 shows, the fungi are divided into three distinct groups: zygomycetes, which evolved first, ascomycetes, and basidiomycetes. Each group differs in—and is named for—its unique reproductive structures.

Fungi play several roles in terrestrial ecosystems. Many fungi are decomposers. Playing the role of garbage processor and recycler, these fungi speed the return of the nutrients in dead and dying organisms to the ecosystem. Some fungi are **parasites** (organisms that live in or on other organisms and harm them), while others are **mutualists** (organisms that benefit from, and provide benefits to, the organisms they associate with). Sometimes the benefits and harms in these associations involve nutrition, sometimes not. Do not confuse these terms, however, with the categories of decomposer, consumer, and producer, even though they describe how organisms get their nutrition. *Mutualist* and *parasite* are a separate, unrelated pair of words to describe whether organisms are good or bad for the organisms they associate with. For example, consumers can be mutualists or parasites; likewise, mutualists can be consumers, producers, or decomposers.

Figure 3.13 The Fungi

Fungi are most familiar to us as mushrooms, but the main bodies of such fungi typically lie hidden underground or in another organism's tissues. Some fungi are decomposers, breaking down dead and dying organisms. Others are parasites, living on or in other organisms and harming them. Others are mutualists, living with other organisms to their mutual benefit. The three major groups of fungi are shown in the evolutionary tree.

- Number of species discovered to date: ~70,000
- Function within ecosystems: Decomposers and consumers
- Economic uses: Mushrooms are used for food, yeasts are used for producing alcoholic beverages and bread. Some fungi also produce antibiotics, drugs that help fight bacterial infections.
- Did you know? Highly sought-after mushrooms known as truffles can sell for $600 a pound.

Zygomycetes   Ascomycetes   Basidiomycetes

These foul-smelling basidiomycetes, known as stinkhorn mushrooms, attract flies, which get covered with their sticky spores and then scatter the spores as they fly to other locations.

Pilobolus, a zygomycete that lives on dung, can shoot its spores out at an initial speed of 50 kilometers per hour.

This ascomycete, *Penicillium*, is a relative of the original species that produced the antibiotic penicillin, a drug that fights bacterial infections and has saved the lives of countless people.

Fungi have evolved a structure that makes them highly efficient decomposers

One of the key evolutionary innovations of the Fungi is their body form. A typical fungus's body is a **mycelium** [my-*SEE*-lee-um] (plural: mycelia), a mat of threadlike projections called **hyphae** ([*HIGH*-fee]; singular: hypha). The mycelium typically grows hidden, either within the soil or within the tissues of the organism the fungus is decomposing (Figure 3.14).

Hyphae are composed of cell-like compartments encased in a cell wall. Unlike the cells in other multicellular organisms, the cell-like compartments of fungi are only incompletely separated by a partial divider known as a septum (plural: septa). Openings in the septum allow organelles—including nuclei—to pass from one compartment to another.

Like animals, fungi rely on other organisms for both energy and nutrients. Most animals, though, ingest food and then release digestive juices and proteins into a stomach to digest it. In contrast, fungi digest their food externally: after releasing special digestive proteins to break down the tissues or substance through which they grow, the hyphae then absorb the nutrients for the fungus to use.

The ability of fungi to grow through tissues makes them well suited to the roles of consumer and decomposer. In fact, fungi are among the most important groups of decomposers, recycling a large proportion of the dead and dying organisms on land.

Reproduction also sets fungi apart from other organisms

Characterized by complex mating systems, fungi come not in male and female sexes, but in a variety of mating types. Each type can mate successfully only with a different mating type. Another, more familiar aspect of fungal reproduction is **spores**, reproductive cells that typically are encased in a protective coating that shields them from drying or rotting. Known to most of us as the powdery dust on moldy food (Figure 3.15), fungal spores, like plant seeds, are scattered into the world by wind, water, and animals. Once carried to new locales, spores can begin growing as new, separate individuals.

The same characteristics that make fungi good decomposers make them dangerous parasites

Some fungi are parasites. Parasitic fungi grow their hyphae through the tissues of living organisms, causing diseases in animals (including humans) and plants (including crops).

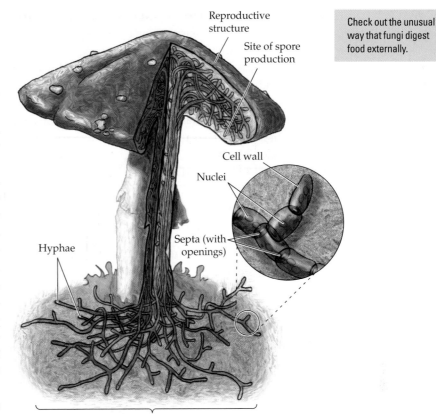

Check out the unusual way that fungi digest food externally. 3.4

Figure 3.14 The Basic Structures of a Fungus
Mats of hyphae, known collectively as its mycelium, form the main feeding body of a fungus. Hyphae are composed of cell-like compartments separated by septa. Openings in the septa allow organelles to move from one compartment to another. Unlike plant cell walls, the fungal cell walls encasing the hyphae are composed of chitin, the same material that makes up the hard outer skeleton of insects.

Test your knowledge of how hyphae help fungi grow. 3.5

Figure 3.15 Fungi Spread via Spores
This puffball fungus expels a cloud of spores into the air.

In humans, fungi can cause mild diseases such as athlete's foot. Fungal diseases can also be deadly, like the pneumonia caused by the fungus *Pneumocystis carinii* [*NOO*-moh-*SISS*-tiss kuh-*REE*-nee], the leading killer of people suffering from AIDS. Fungi attack plants, too. *Ceratocystis ulmi* [*SARE*-uh-toh-*SISS*-tiss *OOL*-mee] causes Dutch elm disease, which has nearly eliminated the elm trees that once formed arching canopies over streets all across the United States. Rusts and smuts are fungi that attack crops. Still other fungi are specialized for eating insects, and biologists are trying to use these fungi to kill off insects that are crop pests (Figure 3.16).

Some fungi live in beneficial associations with other species

Some fungi are mutualists, living in association with other organisms to their mutual benefit. One broad group of mutualists—found in all three groups of fungi (zygomycetes, ascomycetes, and basidiomycetes)—is known as **mycorrhizal** [*MY*-koh-*RYE*-zul] fungi. These species live in mutually beneficial associations with plants. The fungi form thick, spongy mats of mycelium on and in the plants' roots that help the plants absorb more water and nutrients. The fungi receive sugars and amino acids, the building blocks of proteins, from the plants. More than 95 percent of ferns (and their close relatives), gymnosperms, and angiosperms have mycorrhizal fungi living in association with their roots.

Morel

Figure 3.17 Mutualist Fungi
A lichen consists of an alga and a fungus intimately entwined in a mutually beneficial association. This lichen, known as British soldiers, is shown growing on an old log.

For example, morels—a group of mushrooms highly prized as food by some—are the reproductive structures of mycorrhizal fungi.

Another familiar fungal association is the **lichen** [*LIE*-kin], a lacy, orange or gray-green growth often seen on tree trunks or rocks. A lichen is an association of an alga (a photosynthetic protist, as we learned earlier) and a fungus (Figure 3.17). Both ascomycetes and basidiomycetes are known to form lichens. The alga and fungus in a lichen grow with their tissues intimately entwined, allowing the fungus to receive sugars and other carbon compounds from the alga. In return, the fungus produces lichen acids, a mixture of chemicals that scientists believe may function to protect both the fungus and the alga from being eaten by predators.

3.6 The Animalia: Complex, Diverse, and Mobile

The **Animalia**, or the animals, are a kingdom within the domain Eukarya. The Animalia is the most familiar major group, and the one to which humans belong. All animals are multicellular, and many of them are quite complex. The animals include flashy creatures such as Bengal tigers, peacocks, and you, as well as worms, sea

Figure 3.16 Fungal Parasites
Some fungi are parasites, making their living by attacking the tissues of other living organisms. This beetle, a weevil in Ecuador, has been killed by a *Cordyceps* [*KOR*-duh-seps] fungus, the stalks of which are growing out of its back.

stars, snails, insects, and other creatures that are less obviously animal-like, such as sponges and corals.

The sponges, the most ancient of animal lineages, were the first to branch off the evolutionary tree (Figure 3.18 on the next page). Next to evolve were the cnidarians [nye-*DARE*-ee-uns] (including jellyfish, sea anemones, and corals), and then the flatworms. The next group to evolve was the protostomes, a group that comprises more than 20 separate subgroups, including mollusks (such as snails and clams), annelids (segmented worms), and arthropods (including crustaceans, spiders, and insects), the three shown in Figure 3.18. These three protostome groups are depicted branching off together because it is unclear which of them evolved from the others first. What is known is that they are part of a single lineage descending from an ancestor that branched off the tree after flatworms, but before echinoderms [ee-*KYE*-noh-derms]. Next to evolve were the echinoderms (sea stars and the like) and the vertebrates (animals with backbones, such as fish, birds, and humans), both deuterostomes [*DOO*-ter-oh-stomes].

Like all fungi and some bacteria and protists, animals are consumers, making their living by eating the tissues of other organisms, from which they derive both carbon and energy. Animals differ from fungi and plants in that animal cells do not have cell walls surrounding their plasma membranes. Typically mobile and often in search of either food or mates, animals have evolved a huge diversity in their ways of life.

Animals evolved true tissues

Sponges are among the simplest of animals. They represent a time in the evolution of animals before tissues—specialized, coordinated collections of cells—had evolved. A sponge is a loose collection of cells (Figure 3.19). If it is put through a sieve and broken apart into individual cells, it will slowly reassemble as a whole sponge. Widespread and highly successful, sponges feed on amoebas and other tiny organisms in their aquatic environment, filtering a ton of water just to get enough food to grow an ounce.

One of the earliest animal groups to evolve true tissues was the cnidarians. Their name—Cnidaria—comes from the Greek word for "nettle," a stinging plant found on land. Cnidarians are characterized by stinging cells they use to immobilize prey and to protect themselves from predators. Like other cnidarians, jellyfish (Figure 3.20) have specialized nervous tissues, musclelike tissues, and digestive tissues. This specialization allows behavior—such as gracefully and rapidly swimming away from predators—that requires the coordination of many cells.

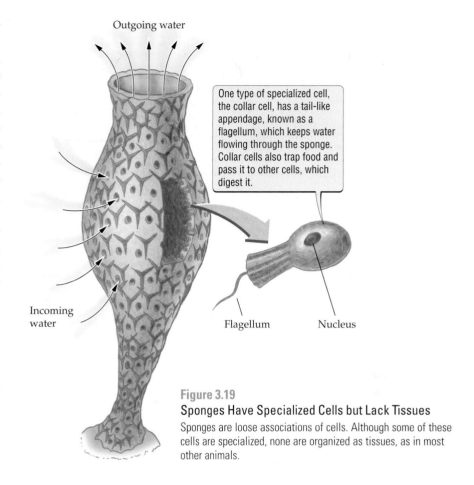

Outgoing water

One type of specialized cell, the collar cell, has a tail-like appendage, known as a flagellum, which keeps water flowing through the sponge. Collar cells also trap food and pass it to other cells, which digest it.

Incoming water

Flagellum   Nucleus

Figure 3.19

Sponges Have Specialized Cells but Lack Tissues

Sponges are loose associations of cells. Although some of these cells are specialized, none are organized as tissues, as in most other animals.

Animals evolved organs and organ systems

After tissues, the next level of complexity to evolve was organs and organ systems. Recall that organs are body parts composed of different tissues organized to carry out specialized functions. Usually organs have a defined boundary and a characteristic size and shape; an example is the kidney.

An organ system is a collection of organs functioning together to perform a specialized task. The human digestive system, for example, is an organ system that includes the stomach as well as other digestive organs, such as the pancreas, liver, and intestines. Flatworms, a group of fairly simple wormlike animals, were one of the earliest groups to evolve true organs and organ systems (Figure 3.21).

Animals evolved complete body cavities

Still later in the history of this group, animals evolved a complete body cavity—an interior space with a mouth at one end and an anal opening at the other. The two distinct evolutionary lineages that exhibit such cavities are the protostomes and the deuterostomes (see Figure 3.18). Protostomes include such animals as insects, worms, and

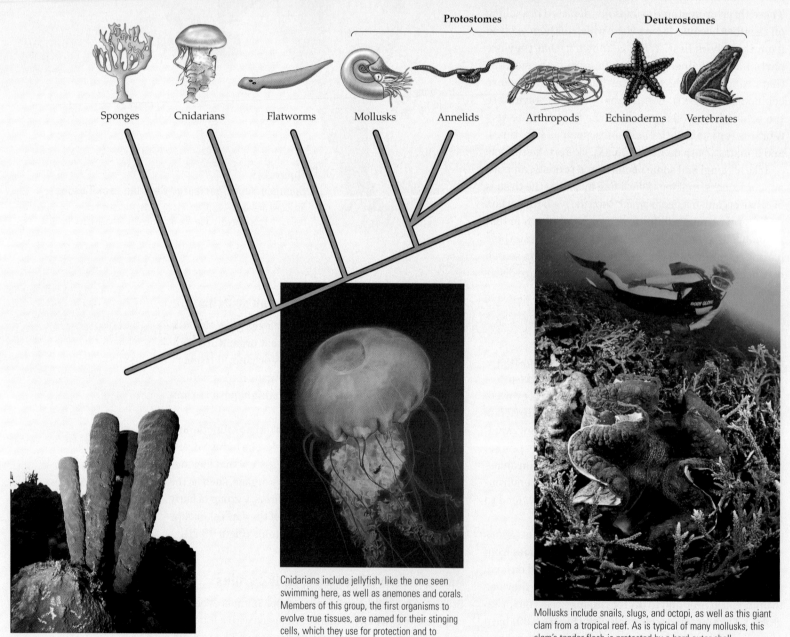

Figure 3.18 The Animalia

Animals are multicellular organisms that are typically mobile and display a wide variety of sizes and shapes. They range from sponges, which do not seem very animal-like, to more familiar forms such as elephants and whales.

- Number of species discovered to date: ~1,000,000
- Functions within ecosystems: Consumers and decomposers
- Economic uses: Humans use other animals as food, as workers, and as laboratory specimens.
- Did you know? Three-fourths of all known animal species are insects.

Flatworms, like this oceangoing flatworm from the West Coast of the United States, were among the earliest animals to evolve true organs and organ systems.

Protostomes

Deuterostomes

Sponges   Cnidarians   Flatworms   Mollusks   Annelids   Arthropods   Echinoderms   Vertebrates

Sponges are ancient aquatic animals. They have evolved some specialized cells, but no true tissues.

Cnidarians include jellyfish, like the one seen swimming here, as well as anemones and corals. Members of this group, the first organisms to evolve true tissues, are named for their stinging cells, which they use for protection and to disable prey.

Mollusks include snails, slugs, and octopi, as well as this giant clam from a tropical reef. As is typical of many mollusks, this clam's tender flesh is protected by a hard outer shell.

A key feature of annelids, also known as segmented worms, is segmentation. This body plan of repeating units can be seen as the series of distinct segments in this fire worm. The segmented body plan, which is also seen in arthropods and vertebrates, facilitated the evolution of many different body forms.

Echinoderms include sea stars, like the one from Indonesia shown here, and sea urchins. They are closely related to the vertebrates.

Arthropods include crustaceans, like lobsters and crabs, as well as millipedes, spiders, and the most species-rich of all groups, the insects. This *Morpho* butterfly, an inhabitant of the tropical rainforest, is one of the most spectacularly beautiful insects on Earth.

Amphibians, slimy creatures that include frogs, like this poison arrow frog from Costa Rica, and salamanders, typically spend part of their lives in water and part on land.

Vertebrates are the animals that have backbones, including fish, reptiles, amphibians, birds, and mammals. Shown here is a coral reef fish from Thailand. Fish were the earliest vertebrate animals.

Mammals are characterized by milk-producing mammary glands in females, as well as young that are born live (rather than being born in an egg that later hatches). These kangaroos are mammals, as are bears, dogs, lions, and humans.

Primates include monkeys, apes, and humans. In this group we find our closest relative, the chimpanzee (shown here), and the gorilla.

Review the tissue layers of a jellyfish.

3.6

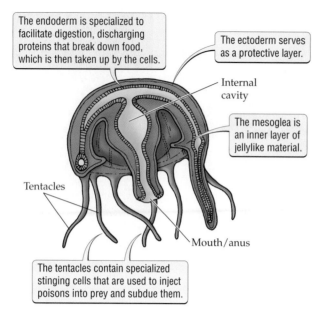

The endoderm is specialized to facilitate digestion, discharging proteins that break down food, which is then taken up by the cells.

The ectoderm serves as a protective layer.

Internal cavity

The mesoglea is an inner layer of jellylike material.

Tentacles

Mouth/anus

The tentacles contain specialized stinging cells that are used to inject poisons into prey and subdue them.

Figure 3.20 Jellyfish Have True Tissue Layers

Cnidarians (including jellyfish) were one of the earliest groups to evolve true tissues. These tissues include the ectoderm (*ecto*, "outer"; *derm*, "skin") and the endoderm (*endo*, "inner"). For clarity, these two layers are color-coded blue and yellow, respectively. Sandwiched between them is an inner (red) layer of secreted material known as the mesoglea (*meso*, "middle"; *glea*, "jelly"). In addition to serving as nervous tissue, the ectoderm coordinates with the endoderm to contract like muscle tissue. Tentacles bring food into the internal cavity through a single opening, which serves as both a mouth and an anus.

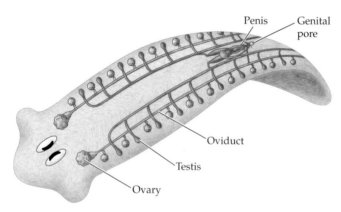

Penis

Genital pore

Oviduct

Testis

Ovary

Figure 3.21 Flatworms Evolved Organs and Organ Systems

One of several organ systems in the flatworm is the reproductive system. It contains both male and female structures, since every flatworm can function as both a male and a female. For clarity, we have color-coded the female structures pink (ovary, oviduct, and genital pore) and the male structures blue (penis and testis).

snails. Deuterostomes include animals such as sea stars (echinoderms) and all the animals with backbones (vertebrates), such as humans, fish, and birds.

The names for these two lineages refer to which of the two openings in the early embryo becomes the mouth. In **protostomes** (from *proto*, "first"; *stome*, "opening"), the mouth forms from the first opening to develop, and the anus forms elsewhere later. In **deuterostomes** (*deutero*, "second"), the first opening develops into the anus, while the second opening becomes the mouth. This developmental difference has led to radically different patterns of tissue organization in these two groups.

Animal body forms exhibit variations on a few themes

Animals exhibit a great variety of shapes and sizes, many of which are variations on a few basic body plans.

Arthropods (*arthro*, "jointed"; *pod*, "foot") have a hard outer skeleton called an **exoskeleton** (*exo*, "outer"), which is made of chitin [KYE-tin], the same material found in the cell walls of fungi.

One feature that has facilitated the evolution of arthropod bodies is their segmented body plan. Over time, individual body segments have evolved different combinations of legs, antennae, and other specialized appendages, resulting in a huge number of different types of animals, some of them extremely successful. Probably the best-known arthropod group is the **insects** (grasshoppers, beetles, butterflies, and ants, among others), which have six legs and live on land. Whereas prokaryotes dominate Earth in sheer numbers of individuals, insects dominate in number of species, having many more species than any other group of organisms.

Other arthropod groups include the arachnids [uh-RACK-nids] (spiders, scorpions, and ticks), which have eight legs and also live on land; the crustaceans (lobsters, shrimps, and crabs), which have ten or more legs and live primarily in water; and millipedes and centipedes, which live on land and have many more legs—but less specialization—than the previously mentioned groups. Arthropods are a wonderful illustration of how evolution can modify a basic body plan to produce many variations over time (Figure 3.22). Looking just at the evolution of the last segment (the rear ends) of these animals, one can see that the changes support a huge variety of shapes and lifestyles. The last segment has evolved into the delicate abdomen of a butterfly, the piercing abdomen of a wasp (which has a huge structure

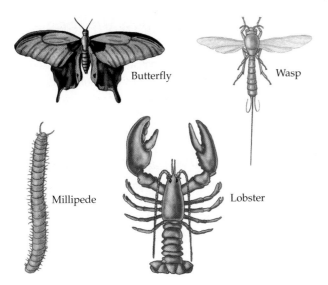

Butterfly

Wasp

Millipede

Lobster

Figure 3.22 Variations on a Theme

From a simple segmented body plan, arthropods have evolved a huge diversity of forms and sizes. The millipede can be viewed as the simplest form of these segmented animals, as all of its segments are similar. As segments have evolved and diversified, a variety of organisms have arisen, from lobster to swallowtail butterfly to parasitoid wasp.

for inserting and laying eggs deep in another animal's body), and the delicious tail of the lobster.

Such segmentation can also be seen in the annelids (segmented worms; see Figure 3.18). This group includes the familiar earthworm, whose body is made up of a repeated series of segments (Figure 3.23a).

Vertebrates—animals with an internal backbone—are also built on a (less obvious) segmented body plan (Figure 3.23b). The major vertebrate groups include fish, amphibians (frogs and salamanders), reptiles (snakes, lizards, turtles, and crocodiles), birds, and mammals (including humans and kangaroos). Like annelids and arthropods, vertebrates illustrate how a variety of very different forms can evolve from one basic body plan. The front appendage of vertebrates has evolved as an arm in humans, a wing in birds, a flipper in whales, an almost nonexistent nub in snakes, and a front leg in salamanders and lizards.

Animals exhibit an astounding variety of behaviors

Another fascinating characteristic of animals is their ability to move, which allows for a wide range of behaviors. Animals have evolved varied ways to capture prey,

eat prey, avoid being captured, attract mates and care for young, and migrate to new habitats. As we saw earlier, animals are quite useful to immobile organisms, such as plants, which have evolved ways to get animals to carry their pollen and seeds.

Animals play key roles in ecosystems and provide products for humans

Because they live by eating other organisms, and because most are mobile, animals play many roles in ecosystems. Most serve as consumers, preying on many species of plants and animals. Some animals, such as carrion beetles, serve as decomposers of dead animals.

Test your knowledge of the properties of multicellular eukaryotes.

3.7

(a)

(b)

Figure 3.23 Many Animals Are Segmented

Segmentation, a body plan in which segments repeat and often can evolve independently of one another, is shown here in (a) an earthworm (an annelid, or segmented worm), and (b) a vertebrate.

Animals also help spread plant seeds and fungal spores. And they can be pests, carrying diseases; ticks, for example, spread the bacterial parasite that causes Lyme disease. Animals, especially insects, can also be crop pests, such as the tomato hornworm, a caterpillar that attacks tomatoes.

Domesticated animals provide humans with food and material for clothing, including feathers and leather. They also provide transportation, as when we travel by horse or camel, communication, as when we send messages via carrier pigeon, and even good company, as when we spend time with our dogs or cats.

But we humans are the animal species having the greatest impact on life on Earth. Our rapidly growing population and our ability to drastically and rapidly modify Earth with cities, agriculture, and industries risk making the planet uninhabitable for ourselves and other species unless we take care (see "Biology Matters," below).

3.7 The Difficulty of Viruses

As we saw in Chapter 2, viruses are not represented on the tree of life or classified into any kingdom or domain. That is because they occupy a gray zone between living organisms and nonliving matter. A virus is simply some proteins wrapped around a fragment of DNA or RNA. Rather than branching off from any single point on the evolutionary tree of life, viruses may have arisen from the DNA or RNA of many different organisms. And this process may be ongoing, with new viruses appearing all the time.

As we saw in Chapter 1, viruses neither grow nor reproduce outside of the hosts they infect, nor do they exhibit many of the other characteristics of living organisms. For example, viruses do not gather energy, photosynthesize, or eat other organisms. Difficult to classify and lacking a clear evolutionary relationship to any one group, viruses cannot easily be placed into any of the existing classification schemes.

Biology Matters

What's on Your Plate?

The Audubon Society has a handy "Seafood Wallet Card" (available at http://seafood.audubon.org/seafood_wallet.pdf) which lists recommendations for seafood that can be eaten without concerns over damaging declining populations of particular species, as well as recommendations for seafood to eat sparingly or avoid altogether.

Using three color-coded categories, Audubon explains its recommendations as follows: "If a fish is in good shape—for example, if it's abundant, relatively well managed, or the fishing methods have little effect on habitat and catch few unintended creatures—it's ranked green. If there are some concerns about a species' status, fishing methods or management, it ranks yellow. Species with significant problems receive a red designation on our Fish Scale."

According to Audubon, "Consumer demand has driven some fish populations to their lowest levels ever. . . . You can help with everyday choices that make a difference."

SOURCE: http://seafood.audubon.org.

ENJOY	BE CAREFUL	AVOID
Anchovies	Cod (Pacific)	Caviar (imported/wild-caught)
Catfish (farmed)	Lobster (American)	Cod (Atlantic)
Crawfish	Mahi-Mahi	Chilean Sea Bass (Toothfish)
Dungeness crab	Oysters (wild-caught)	Flounder and Soles (Atlantic)
Halibut (Pacific)	Rainbow Trout (farmed)	Grouper
Mussels and Clams (farmed)	Scallops (bay and sea)	Halibut (Atlantic)
Oysters (Pacific farmed)	Shrimp (U.S. farmed or trawl-caught)	Monkfish
Sablefish (Alaska, British Columbia)	Squid (calamari)	Orange Roughy
Salmon (Wild Alaskan)	Swordfish (Atlantic)	Red Snapper
Sardines	Tuna (canned)	Salmon (farmed, including Atlantic)
Striped Bass (farmed)	Tuna: Ahi, Yellowfin, Bigeye, Albacore (longline caught)	Sharks
Tilapia (U.S. farmed)		Shrimp (imported)
Tuna: Ahi, Yellowfin, Bigeye, Albacore (pole/troll-caught)		Tuna: Bluefin

An Overflowing Cornucopia of Life

There is much to be seen even in a very brief sampling of the world's known organisms. But while much is already known of life on Earth, as we noted at the beginning of this chapter, scientists continue to discover life in unexpected places and situations. From glamorous orchids to the bacteria from hell, unknown species will continue to be discovered as biologists become more and more creative in where they look. Among the more imaginative places where biologists have begun to seek out undiscovered life are ancient, frozen lakes buried deep beneath Earth's surface in the Arctic, and on the ocean bottom, where biologists probe for life in submersible diving machines.

But while such rugged adventures will certainly lead to new discoveries, some of the greatest biological explorations involve venturing not out into the great beyond, but into uncharted regions within. New microscopic organisms are being found in the moist, warm mouths of many animals. Researchers say there are unexplored worlds of species in and on many animals—such as the mysteries of a toucan's gut or the world of plants and animals living in a slow-moving sloth's hair. It seems that wherever one can imagine that life could grab hold, it has.

We humans, too, are territory for exploration. The human body is normally host to millions of bacteria, both externally (on the skin) and internally (in the digestive system). In addition, we harbor hundreds of microscopic arthropods. The human body is also being continually invaded by viruses, making each person—as well as many other organisms—a densely populated community of living organisms. In fact, so overwhelmingly abundant is life in us and on us, that no matter how hard or often we try to ignore or scrub other species away, we will never walk alone.

Chapter Review

Summary

3.1 The Major Groups in Context

- The major groups are Bacteria, Archaea, Protista, Plantae, Fungi, and Animalia.
- All living organisms can be organized according to three systems based on evolutionary relationships: the evolutionary tree of life; the system of three domains (Bacteria, Archaea, and Eukarya); and the Linnaean hierarchy, in which the six major groups are kingdoms.
- Each group is characterized by evolutionary innovations, new features that allow the group to live and reproduce successfully.
- All organisms can also be classified according to cellular structure, as either prokaryotes (simple one-celled organisms without organelles) or eukaryotes (one-celled and multicellular organisms whose cells have organelles).

3.2 The Bacteria and Archaea: Tiny, Successful, and Abundant

- The two prokaryotic groups are Bacteria and Archaea. Both groups consist of microscopic, single-celled organisms. However, the two groups differ in significant ways: in their DNA, in key components of their cell walls and plasma membranes, and in their metabolism.
- Prokaryotes can reproduce extremely rapidly and are the most numerous life forms on Earth. They also have the most widespread distribution. Some prokaryotes, including many archaeans, are extremophiles, thriving in very hot (thermophiles), very salty (halophiles), or other unusual and extreme environments.
- Prokaryotes exhibit unmatched diversity in methods of getting and using energy and nutrients.
- Prokaryotes perform key roles in ecosystems, including photosynthesis, providing nitrate to plants, and decomposing dead organisms.

Prokaryotes are useful to humanity in many ways (for example, in cleaning up oil spills and helping us with our digestion), but they also cause deadly diseases.

3.3 The Protista: A Window into the Early Evolution of the Eukarya

- Protista, the most ancient eukaryote group, is highly diverse. It includes plantlike, animal-like, and funguslike organisms.
- The evolutionary relationships among members of the Protista remain poorly understood.
- Protists represent early stages in the evolution of the eukaryotic cell. In the past, some prokaryotic cells engulfed others. The engulfing cells evolved into primitive Eukarya, from which the first multicellular organisms evolved. The engulfed cells evolved into organelles, which perform essential, specialized functions for the cells that engulfed them, and have lost the ability to function on their own.
- Protists provide examples of the early evolution of multicellularity.
- Sex was a eukaryote innovation, and it evolved first in protists.
- Although protists include many harmless or helpful organisms, they also include many disease-causing organisms.

3.4 The Plantae: Pioneers of Life on Land

- The Plantae (plants) are multicellular, photosynthesizing organisms. They were the first group to live on land.
- Colonizing land presented two key problems: how to get and conserve water, and how to grow in the presence of gravity. Features that helped plants meet these challenges include a root system, the cuticle, and cell walls stiffened with cellulose. Other important evolutionary innovations that allowed plants to thrive and diversify include vascular systems, seeds, and flowers.
- Ferns and their allies were the first plants to evolve vascular systems, which in later plant groups allowed growth to great heights.
- Gymnosperms evolved the first seeds, which provide nutrients so that plant embryos can grow and develop before they are able to produce their own food via photosynthesis. Seeds also protect the plant embryo from drying, rotting, and attack by predators.
- Angiosperms evolved the first flowers, specialized reproductive structures that provide a place for male and female gametes to meet. Angiosperms have also evolved several means of pollen dispersal and seed dispersal.
- As producers, plants are the ultimate food source for nearly all land organisms and are therefore critical components of land-based food webs.
- Plants are harvested for food, clothing, pharmaceuticals, lumber, oxygen, and many other important products. Plants are also valuable when left in nature, where they provide oxygen and prevent erosion and water contamination, among other benefits.

3.5 The Fungi: A World of Decomposers

- Fungi have a unique body plan: a mat of hyphae called the mycelium, which grows as the hyphae penetrate, digest, and absorb food. This evolutionary innovation allows fungi to be successful as decomposers (breaking down dead and dying organisms), parasites (extracting nutrients from living organisms without providing benefit in return), and mutualists (living in close association with other organisms for their mutual benefit).
- Reproduction in fungi is complex and unique to this group. Instead of sexes, there are many mating systems and mating types. Encased reproductive cells called spores are protected from harsh conditions and able to disperse long distances via animals, wind, and water before beginning their development. Fungi are divided into three groups—zygomycetes, ascomycetes, and basidiomycetes—based on reproductive structures specific to each group.
- There are two groups of mutualist fungi. Mycorrhizal fungi live in and on the roots of most plants. In lichens, fungi and algae live in close association with each other.
- Fungi cause dangerous diseases, but they also produce valuable products such as foods and pharmaceuticals.

3.6 The Animalia: Complex, Diverse, and Mobile

- The Animalia make a living by eating other organisms and are typically mobile.
- The bodies of sponges, the most primitive animal group, have clusters of specialized cells, but cnidarians were one of the first groups to evolve true tissues (specialized cells that are also coordinated).
- Flatworms were one of the earliest groups to evolve true organs (different tissues organized for a specialized function, with a defined boundary and characteristic size and shape) and organ systems (groups of organs functioning together for specialized tasks).
- Complete body cavities (with both a mouth and an anus) did not show up in animal bodies until the evolution of protostomes (mollusks, annelids, and arthropods) and deuterostomes (echinoderms and vertebrates). In protostome embryos, the mouth develops from the first opening to form; in deuterostomes, the mouth develops from the second opening.
- Animals exhibit a great variety of forms and sizes, often variations on a single theme. One major theme is the segmented body plan, in which a given segment evolves different shapes in different animal groups in order to perform different functions. Annelids, arthropods, and vertebrates all exhibit variations on a basic segmented body plan.
- Insects are the most familiar of the arthropods, a group characterized by a hard outer skeleton known as an exoskeleton.
- Vertebrates (which include fish, amphibians, reptiles, birds, and mammals) are distinguished by having an internal backbone.
- Animals' ability to move allows them a great variety of behaviors, including ways to capture and eat prey, avoid being captured, attract mates and care for young, and migrate.
- Animals play a variety of roles in ecosystems, serving mainly as consumers and decomposers. Some spread diseases; others are crop pests. Animals also provide food, clothing, and other products to human society. As animals ourselves, we humans have had the greatest impact on ecosystems worldwide.

3.7 The Difficulty of Viruses

- Viruses are not placed on the tree of life or in any classification system involving kingdoms or domains.
- One reason for this is that viruses lack clear evolutionary relationships. They may have genetic material derived from the genetic material of many organisms all across the evolutionary tree.
- Another reason is that they lack many characteristics of living organisms, including the ability to grow and reproduce independently and to gather energy.

◉ Review and Application of Key Concepts

1. What are the three major systems used to categorize living organisms?

2. What two kingdoms make up the prokaryotes? Name three factors that contributed to the success of prokaryotes.

3. Why are *Giardia* and slime molds of particular interest to biologists interested in the early evolution of eukaryotes?

4. Describe the evolution of specialized cells, tissues, and organs in the Animalia. Include the name of the animal group for which each feature is an important evolutionary innovation.

5. To what kingdom and domain do viruses belong? Why?

6. Two of the major challenges facing plants when they colonized land were (a) obtaining and retaining water and (b) gravity. What were the evolutionary innovations that plants used to deal with these challenges?

7. Describe one way in which plant cells differ from animal cells. What are the consequences for plants?

8. Animals are typically mobile, whereas plants are not. In what ways have plants evolved to utilize the mobility of animals to aid in their survival and reproduction?

9. If viruses did indeed arise from many different organisms across the tree of life, how could we best accommodate them in the Linnaean hierarchy? In the domain system? Is it possible? Why or why not?

Key Terms

angiosperm (p. 51)
Animalia (p. 56)
arthropod (p. 60)
biodiversity (p. 42)
cuticle (p. 49)
deuterostome (p. 60)

eukaryote (p. 41)
evolutionary innovation (p. 41)
exoskeleton (p. 60)
extremophile (p. 42)
flower (p. 51)
Fungi (p. 53)

gymnosperm (p. 51)
hypha (p. 55)
insect (p. 60)
lichen (p. 56)
mutualist (p. 53)
mycelium (p. 55)
mycorrhizal (p. 56)
organelle (p. 48)
parasite (p. 53)

Plantae (p. 49)
prokaryote (p. 41)
Protista (p. 46)
protostome (p. 60)
root system (p. 49)
seed (p. 51)
spore (p. 55)
vascular system (p. 49)
vertebrate (p. 61)

Self-Quiz

1. Which group is the most abundant in numbers of individuals?
   - a. Animalia
   - b. Eukarya
   - c. Protista
   - d. Prokaryotes

2. Eukaryotes differ from prokaryotes in which of the following ways?
   - a. They do not have organelles in their cells, whereas prokaryotes do.
   - b. They exhibit a much greater diversity of nutritional modes than prokaryotes.
   - c. They have organelles in their cells, but prokaryotes do not.
   - d. They are more widespread than prokaryotes.

3. Which of the following groups contains organisms that represent early stages in the evolution of the eukaryotic cell?
   - a. Archaea
   - b. Protista
   - c. Fungi
   - d. Animalia

4. Which of the following groups was the first to succeed on land?
   - a. Plantae
   - b. Animalia
   - c. Bacteria
   - d. Fungi

5. What were key evolutionary innovations of the Plantae?
   - a. seeds, organelles, flowers
   - b. roots, cuticle, seeds, flowers
   - c. roots, hyphae, flowers
   - d. hyphae, cuticle, organelles

6. Fungi grow using
   - a. hyphae.
   - b. chloroplasts.
   - c. angiosperms.
   - d. prokaryotes.

7. Animals that are segmented include
   - a. vertebrates and annelids.
   - b. vertebrates and flatworms.
   - c. sponges and annelids.
   - d. sponges and flatworms.

8. Mycorrhizal fungi are
   - a. beneficial to plants because they help plants stay dry.
   - b. harmful to plants because they secrete acids.
   - c. beneficial to plants because they help plants take up water.
   - d. beneficial to plants because they help plants form lichens.

Bizarre Giant Virus Rewrites the Record Books

A bizarre new species of giant virus, found living inside an amoeba, has more genes than many bacteria and can be seen without an electron microscope, French researchers have discovered. . . . Similar in size to a small bacterium, the giant Mimivirus was named after its ability to "mimic a microbe" (hence "mimi"). Although large for a virus—one of the smallest forms of life known—it is still only 400 nanometers in diameter and 2,500 of them could fit into 1 millimeter.

This new and delightfully French discovery, whose name "Mimi" conjures a poodle more than a microscopic virus, has surprised researchers of the world's microscopic life forms in a variety of ways. According to scientist Jean Marie Claverie, this discovery reveals how ignorant we still are about the smaller members of the world. As he told *ABC Science Online*, "If such a big guy could be unnoticed for so long, what about much less conspicuous viruses?"

But, some say, more surprising still was where this strange new giant among viruses was found: inside a single-celled protist known as an amoeba, which scientists named *Acanthamoeba polyphaga* [AY-kan-tha-MEE-ba po-LIFF-uh-guh]. Researchers were examining water from a British cooling tower when they came across the amoeba and its strange stowaway. Like a Russian doll, the water tower, scientists say, has revealed ever smaller and more interesting finds, one hidden in another. So far, scientists don't know that the Mimivirus is responsible for any human disease, but early studies indicate that it might be capable of causing pneumonia.

The *ABC Science Online* report also quoted Professor Adrian Gibbs, a biologist at the Australian National University in Canberra, who surmised that finding even bigger viruses was a possibility: "You can't think of a more ordinary sort of place than a cooling tower in Bradford, England," he mused. "You can increase the probability of finding new things by looking in interesting places, like deep-sea vents or thermal pools, but you can also find them in your own backyard. We need more people out there having a look."

Evaluating the news

1. The Mimivirus was found in a water cooling tower. What other otherwise unexplored habitats—whether exotic, bizarre, or mundane—can you think of where scientists could search for new forms of life?

2. Some scientists say that the newly discovered virus should be considered its own new family of viruses—the Mimiviridae. Other scientists, who recently discovered a different microscopic organism that bears a resemblance to the Mimivirus, say that the two new organisms should be considered an entirely new kingdom. There are no set rules, however, on when a new group is different enough to be recognized as a new family or phylum, or even an entirely new kingdom. Given what you learned in Chapter 2 about the Linnaean hierarchy, how do you think biologists should decide when a group deserves to be a new kingdom or family or genus? Does it matter? Why or why not?

3. Scientists say that the Mimivirus blurs some of the distinctions between single-celled life—such as Archaea, Bacteria, and some protists—and viruses. The Mimivirus has genetic material more complex than that of a typical virus—more complex even than that in many bacteria. But the Mimivirus also lacks some genes known to be universal among bacteria. It also has a number of features that are typically viruslike. If viruses can be considered nonliving or on the edge of life, and bacteria are clearly living organisms, where does the Mimivirus fit in? And how does it affect our definition of viruses as living or nonliving?

SOURCE: *ABC Science Online*, March 31, 2003.

Biodiversity and People

Where Have All the Frogs Gone?

In 1987, a spectacularly beautiful creature known as the golden toad could be found in abundance in the only spot in the world where it was known to live: in a certain tropical forest high in the mountains of Costa Rica. That year, hundreds of the brightly colored animals were seen (Figure A.1). The next year, just a few were found. Within a few years the golden toad had disappeared entirely, never to be seen again.

While there is always a concern when a species plummets into extinction, most extinctions are easier to understand than the loss of the golden toad. When a forest-dwelling bird goes extinct because its forest is cut down, there is no lingering mystery. But the golden toads were living in a pristine area far from deforestation or development. There was no obvious reason for these frogs to go extinct.

Since the time when the golden toad was last seen, biologists around the world have documented population declines of numerous amphibians (a group of vertebrate animals that includes not only frogs, but also toads and salamanders). Many of these declines have occurred in preserved areas. In the United States, for example, in and around Yosemite National Park, where frogs and toads were once abundant, many species have declined or disappeared.

Figure A.1 Gone but Not Forgotten
Once abundant on a mountaintop in Costa Rica, the golden toad mysteriously went extinct in the late 1980s. Here the orange-colored male mates with the larger, and very differently colored, female.

Main Message: The diversity of life has risen and fallen naturally and dramatically in the past, but is now declining rapidly as a result of human activities.

Scientists currently believe that no single cause is responsible for these worldwide losses of amphibians. For some species living at high altitudes, increasing exposure to damaging ultraviolet light may be a problem as the protective ozone layer in the atmosphere thins. In many parts of the world, huge numbers of amphibians are being killed by fungal diseases. In other places, researchers report that pollution, in the form of artificial chemicals in the water, may be affecting processes of amphibian development and causing deformities. In some areas, parasites may be causing similar deformities in developing amphibians.

Whatever the problem in any particular region, amphibians are being lost around the world. Many scientists believe that these animals are more sensitive than other animals to environmental deterioration, and that—like canaries in a coal mine—the dying amphibians are warning signs of an ever more poisoned environment.

Meanwhile, as amphibian species mysteriously decline, biologists are finding that many other species are rapidly going extinct as well. Everywhere we hear warnings about the loss of species around the globe. How extensive is this loss of biodiversity, what is causing it, and do these losses really matter?

In this essay we look at how the numbers of species on Earth have changed over time, in mass extinctions in the past and in today's human-caused extinction. We also look at biodiversity and ask what value the totality of the world's species has for humanity.

How Many Species Are There on Earth?

Before we can understand what the current loss of species means, we must have some idea how many species there are on Earth. Surprisingly, in spite of intense worldwide interest, scientists do not know the exact number of living species. Estimates range widely, from 3 million to 100 million species. Most estimates, however, fall in the range of 3 million to 30 million species.

Scientists use indirect methods to estimate total species numbers

So far, the total number of species that have been collected, identified, named, and placed in the Linnaean hierarchy is about 1.5 million. Despite this massive cataloging, which has taken more than two centuries to complete, some researchers believe that biologists have barely scratched the surface. Some estimates suggest that 90 percent or more of all living organisms remain to be identified and named by biologists.

But how do we know? How do you count the number of organisms not yet discovered? Biologists use indirect methods to estimate how many species remain unknown. For example, in 1952, a researcher at the U.S. Department of Agriculture estimated, on the basis of the rate at which unknown insects were pouring into museums, that there were 10 million insect species in the world.

Using another method, Terry Erwin, an insect biologist at the Smithsonian Institution, shocked the world 30 years later with his estimate that the arthropods (the animal group that includes insects, spiders, and crustaceans) alone numbered more than 30 million species. That is 20 times the number of all the previously named species of all groups on Earth. Erwin believed that most of these arthropods were insects living in the tropical rainforest **canopy**—the nearly inaccessible habitat in the branches of rainforest trees.

Erwin based his estimate on actual species counts that he obtained by fogging (Figure A.2). This method starts by blowing a biodegradable insecticide high into the top of a single rainforest tree. Then the dead and dying insects that rain down to the ground are collected and counted, and the number of different species is tallied. Erwin found more than 1,100 species of beetles alone living in the top of one particular tropical tree. He estimated that 160 of these species were likely to be **specialists**; that is, able

Figure A.2 Fogging to Count Species
Tropical biologist Terry Erwin fogs the canopy of a tree to collect its many insects. On the basis of his studies of insects from tropical treetops, Erwin estimated that worldwide, there could be as many as 30 million species of tropical arthropods alone.

to live only in this species of tree and to be found nowhere else on Earth. From there, he was able to come up with a minimum estimate of the number of arthropod species in the world.

Here's how he did the calculation: There are an estimated 50,000 tropical tree species. If the tree species Erwin studied was typical, then the total number of beetle species living in tropical trees in the world should number 8 million (50,000 tree species × 160 specialist beetle species). Beetle species are thought to make up about 40 percent of all arthropod species. Eight million beetle species is 40 percent of 20 million. Therefore, the total number of arthropod species in the rainforest canopy should be 20 million. Many scientists believe that the total number of arthropod species in the canopy is double the number found in other parts of tropical forests, suggesting that there are another 10 million arthropod species in the rest of the tropical forests. Assuming these numbers, the total number of arthropod species in the tropics alone should be 30 million (20 million + 10 million). That means, of course, that there should be even more species of *all* kinds in the tropics, and more still when considering the entire world.

Like all such estimates, this one is based on numerous assumptions, some of which could be wrong. A change in even one of these assumptions could drastically alter the final number. While Erwin's calculation is among the most famous of these estimates, such indirect measures are typical of how the numbers of species yet to be discovered are determined, since it is impossible to count them directly. Scientists continue to argue over the exact figures and the assumptions on which they are based, but one thing is certain: the 1.5 million species discovered and named to date are far from the total number living in this world.

Some groups are well known and others are poorly studied

About half of the 1.5 million known species (about 750,000) are insects. All the remaining animals make up a mere 280,000 species or so (Figure A.3). The next-largest group is the plants, with about 250,000 known species. There are also approximately 69,000 named fungi, 30,000 animal-like protists known as protozoans, 27,000 algae, and some 4,800 prokaryotes, including both bacteria and archaeans.

Among these groups, some have been very well studied because they are large or easy to capture or popular with biologists. Others have been studied very poorly, often because they are microscopic or otherwise hard to collect and identify. The birds, for example, total 9,000 species and are among the best-studied organisms on Earth; relatively few new bird species are left to be discovered. Insects, on the other hand, remain poorly known; the majority, possibly the vast majority, are still undiscovered and unidentified.

Other groups, including whole kingdoms, such as Fungi, Bacteria, and Archaea, are also poorly known (Figure A.4). In a single gram of Maine soil, scientists have estimated that there may be as many as 10,000 species of bacteria. That's about 5,000 more species than have so far been named by biologists. The studies forming the basis of this estimate examined the numbers of different types of DNA contained in the soil. Scientists

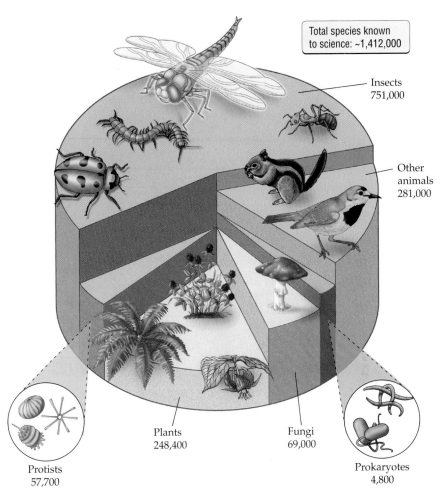

Total species known to science: ~1,412,000

Insects 751,000

Other animals 281,000

Plants 248,400

Fungi 69,000

Protists 57,700

Prokaryotes 4,800

Figure A.3 A Piece of the Pie

This pie chart breaks down all the known species on Earth into the major groups of organisms. Animals (particularly insects) and plants make up the vast majority of the known species, but many more species remain to be discovered.

estimated 10,000 different types, each of which probably represents a different species. There is an additional problem with poorly studied organisms: when little is known about a group of organisms, it can sometimes be difficult to determine whether two organisms from that group are members of the same species or two different species, making the tallying of total numbers of species even more complex.

Biologists continue to discover new species even in relatively well-known groups. For example, about 100 new fish species are discovered each year. And in 1992, although scientists were sure that all the large land mammals had already been accounted for, a large deerlike species was found in Vietnam. Then two years later, the barking deer, another large deer species, emerged from the mountains there as well.

Barking deer

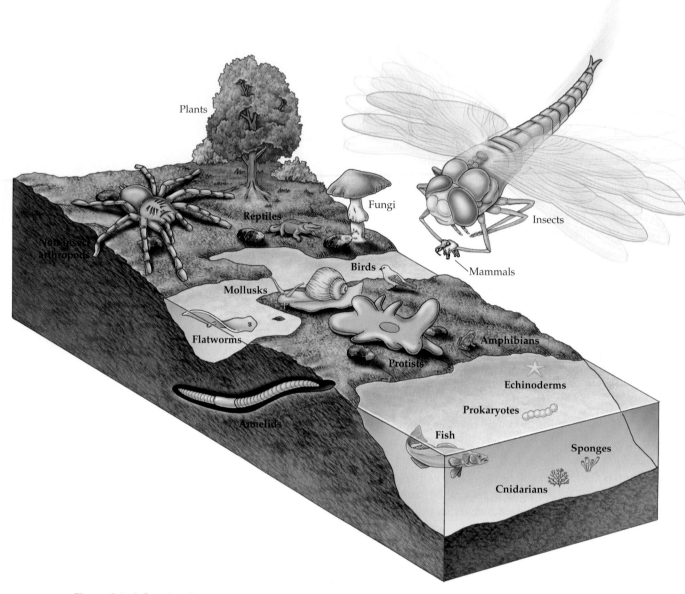

Figure A.4 A Species-Scape

In this strange world, each group of organisms is represented not by its true size, but by the number of known species in the group. As a representative of mammals, the elephant shown here is uncharacteristically tiny to reflect the relatively small number of mammalian species. The enormous size of the dragonfly holding the elephant reflects the vast number of insect species.

Explore the diversity of species on Earth. **A.1**

The Beginnings of a Present-Day Mass Extinction

The history of life on Earth includes a handful of drastic events, known as **mass extinctions**, during which huge numbers of species went extinct. Today, even as biologists struggle to get a total species count, many biologists assert that we are on our way toward a new mass extinction. In fact, many biologists say that the ongoing extinction—if it continues unabated—will lead to the most rapid mass extinction in the history of Earth. As with the total number of species on the planet, extinction rates are estimates. But even using conservative calculations, species are being lost at a staggering pace. As we shall see, the cause of this mass extinction is clear: the activities of the ever-increasing number of people living on Earth.

The current mass extinction probably began with early humans

Increasing evidence suggests that humans have been driving species to extinction around the world for a very long time. The fossil record shows that around the same time humans arrived in North America, Australia, Madagascar, and New Zealand, large animal species (including mammoths, giant ground sloths, camels, horses, and saber-toothed cats) began to disappear. Although some people suggest that these species extinctions may also have been due to climate changes, the coincidence of these species losses with the arrival of humans in three different parts of the world is striking and consistent. Of the genera of large mammals that roamed Earth 10,000 years ago, 73 percent are now extinct. (Remember from Chapter 2 that the genus is the next category above species in the Linnaean hierarchy.)

There are several ways that humans probably affected the animal life around them. Many large mammals would have made a hearty meal for prehistoric hunters. A similar number of birds, particularly those that would have been the easiest of prey for humans, disappeared. One species lost, for example, was a flightless duck—literally a sitting duck. Many large animals also may have suffered from having to compete with humans for the same prey. And in addition to predation and competition, humans had indirect effects. For example, numerous other animals that depended on large animals for their survival, such as vulture species that fed on their carcasses, went extinct as those animals declined in numbers.

Scientists have made estimates of species losses occurring today

The most devastating and obvious losses of species today are occurring in tropical **rainforests**—wet, lush forests in the tropical regions of the world (Figure A.5). Known to harbor huge numbers of species found nowhere else in the world, rainforests are being burned or cut at alarming rates. By 1989, less than half of the rainforest area that had existed in prehistoric times remained. And of the rest, an area equivalent to one football field was disappearing every second, which amounts to an area the size of Florida each year.

How do such large-scale losses of natural areas translate into numbers of species lost? According to Edward O. Wilson, a biologist at Harvard University, at the current rate of rainforest destruction, 27,000 additional species will be doomed to extinction each year, an average of 74 per day, or 3 every hour. And though rainforests

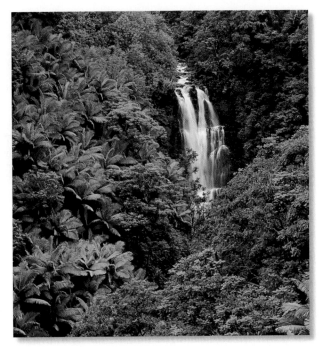

Figure A.5 The Rainforest
Tropical rainforests, like this one in Hawaii, are typically home to numerous species not found in other habitats.

are particularly rich in biodiversity, they are just one of many different **habitats**—characteristic places or types of environments in which different species can live.

While estimating extinction rates for the world as a whole can be difficult, scientists have definitively documented the extinction, caused by humans, of many hundreds of particular species in the last few thousand years. Twenty percent of the freshwater fish species known to be alive in recent history either have gone extinct or are nearly extinct. One large-scale study showed that 20 percent of the world's species of birds that existed 2000 years ago are no longer alive. Of the remaining bird species, 10 percent are estimated to be **endangered**—that is, in danger of extinction.

Although it may be tempting to assume that little of this extinction is happening close to home, evidence suggests otherwise (Figure A.6). As mentioned earlier, in the United States, frogs are disappearing in Yosemite National Park. In addition, some 200 plant species known to have existed in recent history have already gone extinct, and at least 600 more are predicted to disappear within the next few years. In North America overall, 29 percent of freshwater fish and 20 percent of freshwater mussels are endangered or extinct.

Altogether, biologists have amassed a wealth of data on species already gone or on their way out, and even

Giant ground sloth

Figure A.6 The First Not to Last

(a) Once fairly common on the sand dunes of the San Francisco peninsula, the Xerces Blue butterfly has earned a dubious fame as the first butterfly documented to have gone extinct in the United States because of human disturbance. Also known by the scientific name *Glaucopsyche xerces* [*GLOU*-ko-*SIGH*-kee *ZER*-seez], the Xerces Blue was last seen in the 1940s, and this photograph of a museum specimen is rare. Scientists say that the major cause of its extinction was probably destruction of its seaside habitat in the populous San Francisco area. (b) Researchers say that other factors may have contributed as well. For example, the Xerces Blue in its caterpillar phase was protected from predators by certain native ant species, which—also due to human disturbance—have declined in numbers or disappeared altogether. Here ants "milk" the caterpillar of another species of butterfly in a similar reciprocal arrangement. The ants drink a sugary substance released by the caterpillar and provide protection in return.

conservative analyses suggest that huge numbers of species have been, and continue to be, lost. Why are these species disappearing?

The Many Threats to Biodiversity

The remaining species of the world face continuing challenges to their survival. A number of human activities are threatening and destroying biodiversity around the globe and in our own backyards.

Habitat loss and deterioration are the biggest threats to biodiversity

Foremost among the direct threats to biodiversity is the destruction or deterioration of habitats. As human homes, farms, and industries spring up where natural areas once existed, habitats suited to nonhuman species continue to disappear or become radically altered. The term "habitat loss" for many people conjures up images of burning rainforest in the Amazon, but the problem is much more widespread.

Every time a suburban development of houses goes up where once there was a forest or field, habitat is destroyed (Figure A.7*a*). So widespread is the impact of growing human populations in urban and suburban areas that species are disappearing even from parks and

reserves in these areas. For example, Richard Primack and Brian Drayton, ecologists at Boston University, studied a large preserve in the midst of increasing suburban development outside Boston; there they found that 150 of the park's native plant species had disappeared. The immediate cause of the loss of species was most likely trampling and other disturbances as more and more people—including many likely nature lovers—used the park (Figure A.7*b*). But the increasing number of homes in the area and the decreasing number of nearby natural areas—from which seeds could have come to repopulate the park—also played an important role in the loss of species. In addition, pollution, erosion, and other effects of human activities and human population growth are altering natural habitats to the point where many species can no longer inhabit them.

Introduced foreign species can wipe out native species

Another devastating problem is the introduction of nonnative, or foreign, species—that is, species that do not naturally live in an area but are brought there on purpose or accidentally by humans. Researchers estimate that 50,000 such **introduced species** have entered the United States since Europeans arrived. Some of these introduced species, referred to as invasive species, are able to sweep through a landscape, wiping out native species as they go. The damage can happen directly, through their eating or parasitizing native species, or it

(a)

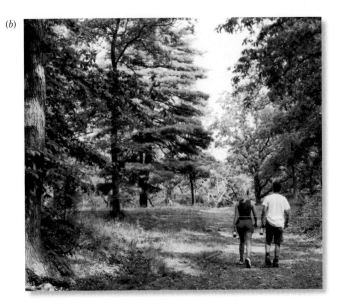

(b)

Figure A.7 The Threats of Habitat Loss and Deterioration
(*a*) Habitat loss doesn't have to mean the burning of tropical rainforest. Here a development of homes and swimming pools has replaced what was at one time a natural landscape. (*b*) Researchers discovered that 150 plant species have disappeared from the Middlesex Fells Reservation in Massachusetts, a 100-year-old preserve, despite the ban on development within the park. Many of the plants (which still exist in other locations) were suspected to have been lost as the result of increasing use by people from the growing suburban area surrounding the park.

can happen indirectly, by their outcompeting native species for food, soil, light, and other resources.

In Africa's Lake Victoria, roughly 500 species of native cichlid [*SICK*-lid] fish evolved in some 10,000 years, making this lake a trove of fish diversity. Fewer than half of those species now remain, however, and many of the surviving species are close to extinction. The Nile perch, introduced to the lake as a food fish for people, is responsible for the extinction of many of these species through predation. (Pollution and increased cloudiness of the water are other causes.)

In Guam, the brown tree snake (Figure A.8*a*), another invasive species, has drastically reduced the numbers of most of Guam's forest birds, leaving an eerie quiet where once the forest was noisy with tropical birdsong. The snake is thought to have been introduced accidentally, brought by U.S. military planes, from New Guinea, where it occurs naturally.

In Hawaii, introduced pigs that have escaped into and are living in the wild are devouring the native plant species. Domesticated cats and mongooses, also introduced species, have killed many of Hawaii's native birds, especially the ground-dwelling species whose nests are easy targets. Purple loosestrife, eucalyptus trees, and Scotch broom (Figure A.8*b*) are invasive plant species that are choking out native plants in various parts of the United States.

(a)

(b)

Figure A.8 The Threat of Introduced Species
Introduced species threaten native ecosystems everywhere. (*a*) The brown tree snake, which was inadvertently introduced to Guam, has not only decimated bird populations but also disturbed human populations, climbing into babies' cribs and swimming up sewer drains to appear in people's toilets. (*b*) Scotch broom, brought from Europe to the United States, threatens native plants across the country. This hardy plant has seedpods that explode to disperse its seeds. Spreading quickly, it can often be seen blooming along highways and other roadsides.

Climate changes also threaten species

Recent changes in climate, which many scientists believe are caused largely by human activities, constitute another threat that seems to be affecting many species. In Austria, for example, biologists have found whole communities of plants moving slowly up the Alps; apparently this movement is a response to global warming, as these plants are able to survive only at ever higher, cooler elevations. These plant communities are moving at an average rate of about a meter per decade. If the climate continues to warm, these alpine plants—which exist nowhere else in the world—will eventually run out of mountaintop and go extinct (Figure A.9). And in areas where organisms don't have cool mountains to climb, many species have begun moving to and living in ever more northern latitudes to escape the heat.

Some threats are difficult to identify or define

Learn more about the rapid explosion of the human population in the last few hundred years. A.2

Biologists now agree that many frogs, salamanders, and other amphibians—as well as many other kinds of organisms—are disappearing around the world. In some cases, like that of the amphibians, the reason for the disappearances remains unclear. For some amphibians, pollution appears to be the culprit. In other cases, increased ultraviolet radiation seems to be killing frogs off. In still other instances, diseases are wiping out amphibians.

According to one study, climate change may be at the root of the problem, making amphibians susceptible to both ultraviolet light and disease. Scientists studying Western toads found evidence that changing climate patterns were resulting in less rain falling in the mountains of the northwestern United States. Frog and toad eggs developing in what are now sometimes very shallow pools of water are subjected to stronger, damaging ultraviolet light. These weakened eggs become vulnerable to disease, including infection by a deadly fungus.

Despite what scientists have learned about many of the specific cases of amphibian demise, it remains unclear why so many amphibians all around the world are dying off at the same time. Is one force weakening them all? Lacking an obvious "smoking gun," biologists continue to confront a variety of potential threats without knowing which ones cause the biggest problems for the populations they study.

Human population growth underlies many, if not all, of the major threats to biodiversity

The biggest threat overall to nonhuman species is the growth of human populations. Many of the problems that we have mentioned here are the direct result of the increasing numbers of people on Earth. Our growth is what spurs continuing habitat deterioration as natural areas are converted to the farms, roads, and factories needed to support human life. As more people demand more cars, burn more gas and oil, and buy more paper, plastic, pesticides, herbicides, fertilizers, and food, more land must be devoted to making such items, and more pollution finds its way into the water and air. When people are finished using what they've bought, more waste crowds the landfills. And the effects of our growing population are further magnified by the fact that more resources are being used *per person* now than in the past. All these changes alter the environment and hasten the demise of other species. In the search for solutions to these problems, we need to consider not only direct destruction—for example, curbing deforestation itself—but also the indirect causes of biodiversity loss—in particular, the increasing use of resources by an ever-growing human population.

The factors we have discussed in this section are causing extinctions around the globe. Many more species have dwindled to dangerously low levels because of these same problems. While such species continue to hang on, their low numbers make them much more susceptible to future extinction.

Figure A.9 **Plants with Nowhere to Go**
The alpine androsace is part of a community of plant species that has been moving up the mountainsides of the Alps. On Mount Hohe Wilde in the Austrian Alps, this mountain wildflower has been migrating upward at a rate of about 2 meters per decade. Researchers say that if global warming continues, this species will soon run out of mountaintop to climb and go extinct.

Biology Matters

Does What I Do Matter?

The average college student produces 640 pounds of solid waste each year, including 500 disposable cups and 320 pounds of paper—think of all those notebooks, printer paper, and so on. A recent study suggests that more than 200 million tons of waste are generated yearly by college students alone. At the University of Colorado at Boulder, a once-weekly collection of recyclables has diverted about 40 percent of its disposable waste. Along similar lines, campuses offering reusable mugs and drink discounts have seen disposable waste decrease by as much as 30 percent.

Beyond such university-based efforts, more can be done. For instance, if all morning newspapers read around the country were recycled, 41,000 trees would be saved daily and 6 million tons of waste would never make their way to landfills. Americans throw away 25 *million* plastic beverage bottles every hour and discard enough aluminum to rebuild the nation's commercial airline fleet every three months.

By recycling a single ton of paper, we save:

- 17 trees
- 6,953 gallons of water
- 463 gallons of oil
- 587 pounds of air pollution
- 3.06 cubic yards of landfill space
- 4,077 kilowatt hours of energy

What has already been done is a good start. For example, 42 percent of all paper is now recycled. According to the U.S. Environmental Protection Agency, 64 million tons of material avoided ending up in landfills and incinerators because of recycling and composting activities in 1999. Today, the United States recycles 28 percent of its waste, a rate that has almost doubled over the past 15 years.

SOURCES: www.columbia.edu/cu/cssn/greens/waste.html; www.depts.drew.edu/admfrm/recycling.html; www.epa.gov.

Mass Extinctions of the Past

The current mass extinction is not, however, the first in the history of life on Earth. As described more fully in Chapter 19, since the time when life first took hold on this planet, more than 3.5 billion years ago, there have been five previous, well-documented mass extinctions. These mass extinctions took place around 440, 350, 250, 206, and 65 million years ago.

Unlike the current extinction, previous mass extinctions were not caused by humans (which had not yet evolved). Scientists have hypothesized other causes for the different extinctions, including climate change, increased volcanic activity, and reductions in sea level. One of the newest and best-supported explanations is that at least some mass extinctions were caused by the aftereffects of an extraterrestrial object—such as an asteroid—hitting Earth and filling the planet's atmosphere with a thick cloud of dust (Figure A.10).

Originally, this idea of an asteroid blasting the Earth and diminishing biodiversity was met with skepticism and ridicule. But no more. According to recent studies, extraterrestrial objects appear to have caused two of the five prior mass extinctions. One of these impacts occurred 65 million years ago, at the end of the Cretaceous period; it is the most famous of the five because it wiped out the last of the dinosaurs. But the largest of all prior extinctions occurred at the end of the Permian period, 250 million years ago; at that time, 80 to 90 percent of all marine species went extinct. With new evidence being revealed about the importance of extraterrestrial objects in mass extinctions, it's possible that other extinctions will prove to have been caused by extraterrestrial impacts as well.

During these mass extinctions, certain groups of organisms disappeared while others survived, apparently unscathed. And after each mass extinction, species numbers rebounded, as whole new groups of organisms evolved and colonized the planet. For example, when dinosaurs were dominant, only a few kinds of small mammals were around. But when the dinosaurs went extinct, mammals evolved into many new species. These species began living in new habitats and exhibiting new habits. This great diversification of the mammals eventually resulted in the evolution of our own species.

If Earth has recovered from mass extinctions in the past, why should we be concerned about the current mass extinction? Won't Earth recover as it always has? The number of species on Earth would probably increase again as new species evolved, but those that are being lost would be gone forever. Perhaps more importantly, recoveries from past mass extinctions have required many millions of years. Ocean reefs, for example, have been destroyed and have recovered from mass extinctions multiple times in the past 500 million years, but the recovery time was on the order of 5 to 10 million years (Figure A.11). Are we really willing to wait that long for species numbers to rebound?

The actions we take in the next 50 years will determine whether Earth's biodiversity is impoverished for millions of years. Will we doom our descendants to living in such a world?

The Importance of Biodiversity

Many people wonder whether the loss of one mouse species here or one beetle species there really makes a difference to humanity. To answer these questions, it helps to look at the situation from the perspective of biologists who have long wrestled with how to assess the value of diverse species within particular habitats. One question biologists are particularly interested in answering is how, if at all, biodiversity affects the forests, wetlands, oceans, rivers, and other wild ecosystems of the world. Since the 1990s, researchers have been studying how biodiversity can contribute to the health and stability of ecosystems.

Biodiversity can improve the function of ecosystems

From tiny experimental ecosystems in an English laboratory to experimental prairies in the midwestern United

Figure A.10
Fact or Fiction: The End of the World
Though humans are responsible for the current mass extinction, in the past various natural occurrences have been the culprit. Current research suggests that asteroids striking the planet destroyed biodiversity on at least two occasions, a "Chicken Little" idea that initially sparked ridicule, but is now widely accepted. This artist's conception depicts how such an event might look from space.

States (Figure A.12), researchers have found that the more species an ecosystem has, the healthier it appears to be. Biodiversity provides ecosystems with several benefits.

For one thing, researchers looking at a variety of ecosystems have found that the more diverse ecosystems are, the higher their **productivity**—that is, the higher the actual mass of plant matter (leaves, stems, fruits) they can produce. Why should biodiversity increase the productivity of an area? Different species are good at using differing resources; for example, some thrive in the sun-drenched portions of a habitat, whereas others can make the best use of the areas that lie mostly in the shade. The more different kinds of species there are, the more productively all parts and resources of a habitat can be utilized. The same goes for other resources, such as wet and soggy versus dry and parched soils. The more species there are in a habitat, the more likely it is that at least one species can use a given resource productively.

There is also evidence that the more species in an ecosystem, the greater the resilience of that ecosystem. For example, the more species are present in a patch of prairie, the more easily that area can return to a healthy state following a drought. Scientists have also found that an increased diversity of species in an area leads to a lower incidence of disease and lower rates of invasion by introduced species.

Productivity, resilience, resistance to disease, and resistance to invasion are all elements of good ecosystem health. Furthermore, diversity can even lead to more diversity: researchers have found that the more plant diversity there was in a plot of ground, the more species of insects were feeding on those plants and on one another. But why should people care whether ecosystems—such as forests, streams, marshes, and oceans—are healthy?

Figure A.11
Starting from Scratch
These coral beds are part of the Great Barrier Reef in Australia, which has rebounded from mass extinctions in the past, although it took millions of years to achieve the level of biodiversity that now flourishes there.

Biodiversity provides people with goods and services

Though we rarely think about it, the biosphere directly provides us with many goods and services. Even the most basic requirements of human life are provided by other organisms. Plants produce the oxygen that we breathe. Every bit of food we eat is provided by other species. A wide variety of crops and livestock, such as tomatoes and cows, have been domesticated from wild species, forming the basis of our agricultural system.

In addition, in many societies, wild species provide important foodstuffs. Insects—especially crunchy, tasty ones such as grasshoppers and ants—are an important source of protein for many peoples around the world. In Central America, many people dine on the green iguana, a huge lizard that likes to sunbathe in treetops. Known as the "chicken of the trees," this lizard has been a food source for 7,000 years. In some parts of South America, the guanaco (a relative of the llama) is an important source of

(a)

(b)

Figure A.12
Testing the Importance of Biodiversity
(a) The ecotron is a tiny experimental ecosystem created in England to test the importance of biodiversity under controlled laboratory conditions. These constructed ecosystems include grasses, wildflowers, snails, flies, and other organisms. (b) Much of the pioneering work on the importance of biodiversity has come from huge experiments in Minnesota prairies, in which scientists have created numerous plots containing different numbers of species.

Capybara

meat and hides. Also in South America, the capybara, the world's largest rodent, is a prized source of meat. In Japan, many different kinds of algae are used to make food, from sushi wrappings to seaweed soup. Many different fish are used to top off those pieces of sushi as well.

Biodiversity also comes to our assistance when we are ill. One-fourth of all prescription drugs dispensed by pharmacies are extracted from plants (Figure A.13). Quinine, used as an antimalarial drug, comes from a plant called yellow cinchona [sing-*KOH*-nah]. Taxol, an important drug for treating cancer, comes from the Pacific yew tree. Bromelain [*BRO*-muh-lin], a substance that controls tissue inflammation, comes from pineapples. Nearly as many other drugs derive from animals, fungi, or microscopic organisms such as bacteria.

In addition to food and medicine, wild species provide many other useful products. Among these are chemicals used as glues, fragrances, pesticides, and flavorings. Many species in the kingdom Plantae provide us with shelter and furniture; these include many of the world's trees, of course, but also various grasses (such as bamboo). Bacteria have turned out to be particularly useful to us, in part because of their great diversity and their ability to capture energy from so many different sources. Some bacteria, for example, can be used to "eat" oil spills and to clean up sewage. Bacteria also produce numerous extremely useful substances—including antibiotics, which they use to kill off competing bacteria, but which we use to rid our bodies of bacterial infections.

Whole ecosystems full of species can provide what are known as **ecosystem services**. For example, in coastal northern California, many hillsides were once covered with towering redwood trees, whose needles and branches gathered fog, mist, and rain, bringing it into the plants and into and onto the ground. Now that so many of these trees have been cut, much less water ends up making it into the ground, so less water is available to people living in the area. Without trees in the ground, hillsides can also easily erode and rivers become filled with sediments. Seemingly useless marshes full of reed and grass species can act as natural filters for the water moving through them, providing water-cleaning services to growing human populations.

Biodiversity also provides the world with aesthetic gifts. Scarlet macaws, parrot fish, sea anemones, tulip trees, and Shooting star wildflowers are among the species that make it clear that if there is a value to beauty, then biodiversity is worth a lot. For many people, the existence of such a rich, living world goes beyond mere beauty to providing spiritual refreshment and rejuvenation.

Finally, consider the argument that biodiversity has value in its own right, without having to fulfill human needs and desires. This viewpoint assumes that wildlife has the right to exist unperturbed and not be destroyed by other species, such as our own.

As great as the bounty of nature already is, with so many species yet undiscovered, the vast majority of nature's wealth remains untapped. If the sheer numbers of species yet to be discovered are any indication, much more awaits—beauty, food, shelter, medicine—if we can just find it, before it disappears along with the golden toad and so many other species.

Shooting star

(a) (b) (c) (d)

Figure A.13 Biodiversity and Your Health

Many important and common drugs originally were found in, and sometimes continue to be produced by and harvested from, other species, most often plants. A look at some of the drugs in your medicine cabinet will bring this point home. (*a*) The Madagascar periwinkle was originally found only on Madagascar, the large island off the southeastern coast of Africa. This flowering plant is the source of vinblastine, an important anti-cancer drug. (*b*) The opium poppy, which produces a beautiful garden flower, is the source of codeine and morphine, widely used painkillers that have prevented suffering in many ill and injured people. (*c*) Though we may not think of caffeine as a drug, this powerful stimulant derived from coffee beans has energized and sustained humanity through long days and nights of study and work. (*d*) The wild yam is the source of diosgenin [dye-*OSS*-juh-nin], the substance that is converted into progesterone for use in birth control pills.

◉ Review and Discussion

1. How did Terry Erwin estimate the number of insect species in the world, and why are such estimates so difficult to do? Try to think of one other way that biologists could try to estimate the remaining number of unknown species on Earth. Feel free to be creative—fogging trees seemed crazy at first, too.

2. Biologists don't know either the exact number of species on Earth or the exact rate at which species are going extinct. Given their lack of precise information, should we take seriously biologists' concerns about worldwide extinction of species? Why or why not?

3. How has biodiversity risen and fallen over time—when and in what way?

4. How has human population growth helped spur other threats to biodiversity?

5. Oftentimes people and endangered species come into conflict. If people continue to farm and build houses and shopping malls, many more habitats and species will disappear. But many people need and want such development. Do you think human beings have a right to such development? Do nonhuman species have a right to exist? If so, where do the rights of human beings end and the rights of other species begin?

6. In the past, when endangered species were threatened by a new building or other development, that development was often entirely forbidden by the courts and the Endangered Species Act. Now, more often, when development conflicts with the survival of an endangered species, developers, and conservationists compromise. Some land is used for building, and some is set aside for the endangered species. Why might such compromise be a good idea? Why not?

7. Every community has its conflicts between biodiversity and development. Consider the area where you live or study: Is there a habitat or environment that's endangered there because of human activity, or a way of life or profession endangered by environmental activism? Are loggers losing their jobs? Are the rivers becoming increasingly polluted with agricultural runoff? Get together with one or two classmates and write a short letter to the editor of your local newspaper, arguing either for or against the importance of preserving biodiversity or human lifestyles; back up your statements with evidence wherever possible. Every voice counts, so send your letter in to the paper.

Key Terms

canopy (p. A3)
ecosystem services (p. A13)
endangered species (p. A6)
habitat (p. A6)
introduced species (p. A7)

mass extinction (p. A5)
productivity (p. A12)
rainforest (p. A6)
specialist (p. A3)

4 Chemical Building Blocks

Key Concepts

◉ Living organisms are composed of atoms linked together by chemical bonds. Arrangements of linked atoms, called molecules, are essential for biological processes.

◉ Covalent bonds are the strongest chemical linkages that can form between two atoms. Most molecules found in living organisms are arrangements of atoms such as carbon, nitrogen, hydrogen, oxygen, phosphorus, and sulfur that are held together by covalent bonds.

◉ Weaker, noncovalent bonds form between two or more separate molecules and between parts of a single large molecule. They include hydrogen bonds and ionic bonds.

◉ The chemical characteristics of water are essential to the chemistry of life. Water is both the primary medium for and a key participant in life-supporting chemical reactions.

◉ Chemical reactions change molecules and associations of molecules. These changes are responsible for the many different processes observed in living organisms.

◉ The four major classes of chemical building blocks found in living organisms are carbohydrates (composed of sugars), nucleic acids (composed of nucleotides), proteins (composed of amino acids), and fats (composed of glycerol and fatty acids). Each class has several functions in living systems.

Looking for Life among the Stars

Throughout recorded history, human beings have gazed up at the stars and wondered what secrets those distant points of light might hold. The idea that the stars might tell us something about our origins has gained support over the past decade, thanks to several unlikely discoveries in the ice of Antarctica and Canada. These discoveries of meteorite fragments embedded in ice have yielded insights into the chemical composition of the universe and even the possibility of life on other planets.

An important meteorite discovery was made in a frozen lake close to the Yukon border of Canada. On a cold January night in 2000, a meteorite entered Earth's atmosphere, creating a shower of fireballs over Tagish Lake. The spectacle was witnessed by dozens of people on the ground and even by American military satellites. Given that roughly a thousand meteorites have been witnessed entering Earth's atmosphere over the past 200 years, what made this event so significant? This meteorite was an ancient C1 chondrite [*KON*-drite], which is not only very rare, but contains interstellar matter dating from the birth of our solar system.

One of the witnesses of the Tagish Lake event, Jim Brook, carefully gathered frozen samples a few days later and stored them in his freezer. By keeping the fragments frozen, Brook prevented them from losing chemical compounds that would have turned to gas upon thawing and avoided contaminating them with terrestrial compounds. The recovery of such pristine meteorite samples is invaluable for the insights they give us into the chemical compounds present at the birth of our solar system. The compounds in these samples, made up of carbon, nitrogen, and sulfur, may even be similar to those that fell on early Earth and contributed to the chemical components of life. Studying such meteorite

When Stars Fall

Meteorite and comet showers can be stunning spectacles for those who witness them. The meteorite shards that lodged in Tagish Lake (inset) also provide fascinating insight into the composition of the universe.

fragments amounts to studying the chemical building blocks that may have gone into the first living systems on Earth.

More recently, the search for chemical evidence of life on other planets has entered a new and exciting phase. Instead of simply studying the evidence available to us from random meteorite showers, researchers working with the National Aeronautics and Space Administration (NASA) have taken the search into space—specifically, to the planet Mars. NASA's Mars Exploration Rover Mission succeeded in landing two robotic explorers, or rovers, on the surface of the red planet in January of 2004. The two rovers, auspiciously named Spirit and Opportunity, were launched into space in the summer of 2003. Both were designed to act as robotic geologists seeking to determine whether there was ever water on Mars. Observations of the surface of Mars through telescopes have long revealed what appear be the dried-up beds of ancient Martian oceans. But without direct tests of the chemical composition of the Martian surface, these observations alone cannot prove that Mars once had water and thus might have supported life in the distant past. Consequently, scientists were thrilled when a combination of $410 million, the hard work of several hundred scientists and technicians, and a hefty dose of luck landed the rover Opportunity about 13 feet from just the sort of geologic formation that might answer the water question.

The stream of information sent back by Opportunity to eagerly waiting scientists on Earth included both detailed images of rock formations and information on their chemical composition. The images revealed tiny ball-shaped accumulations of minerals, dubbed "blueberries," on the surface of the weathered rock. More appropriately termed concretions, these balls are formed when mineral-laden water is present for an extended period of time. The chemical analysis also revealed the presence of jarosite, a mineral that is formed only in the presence of water. All of these observations strongly support the idea that a significant amount of water was present on the surface of Mars a few billion years ago.

For all its remarkable diversity, life as we know it is based on a rather limited number of chemical elements, which are found throughout the universe. The fact that such complexity can come from such simple components shows how we humans could have arisen from the chemical soup of a primitive Earth, using the same chemical elements that make up the very Earth we stand on. More than anything else, our origin in Earth's chemical elements reminds us of the fundamental unity between our bodies and our surroundings.

In this chapter we begin our exploration of the tremendous complexity of living organisms by first identifying their simplest chemical components. Then we examine how these simple components are linked together to form the many levels of organization in both the physical structures and the chemical processes of life. Many of the topics introduced in this chapter will be discussed in greater detail in later chapters of this unit.

4.1 Atoms Make Up the Physical World

All components of the physical world are made up of 92 natural chemical **elements**. Elements are the simplest building blocks of the physical world. Each element is identified by a one- or two-letter symbol; for example, oxygen is identified as O, calcium as Ca. Each element exists as units so small that more than a trillion of them could easily fit on the head of a pin. These tiny units are called **atoms**, and there are 92 different kinds, one for each natural element. An atom is therefore the smallest unit of an element that still has the characteristic chemical properties of that element.

It stands to reason that the chemical properties of an element such as oxygen must depend on the properties of oxygen atoms. So what makes the atoms of one element different from those of another? The answer lies in the different combinations of three atomic components. The first two components are electrically charged: **protons** have a positive charge (+), and **electrons** have a negative charge (–). As the name of the third component implies, **neutrons** lack an electrical charge and are therefore considered neutral. These three components, especially the electrons, determine how atoms behave in the physical world and interact with one another.

A single atom has a dense central core, called the nucleus, that contains one or more protons and is thus positively charged. With the exception of hydrogen atoms, the nucleus also contains one or more neutrons. The nucleus is surrounded by one or more negatively charged electrons (Figure 4.1), which move around it at relatively great distances. If a hydrogen nucleus were the size of a marble, the electron would move around it in a space as big as the Houston Astrodome. As a whole, the positively charged nucleus and the negatively charged electrons balance out such that atoms are electrically neutral. The number of electrons in a given atom is not fixed; the atom can lose, gain, and even share electrons with another atom.

When an atom loses one or more of its negatively charged electrons, it becomes positively charged. Likewise, when an atom gains one or more electrons, it becomes negatively charged. Atoms that become charged due to the loss or gain of electrons are called **ions**. Ions play critical roles in living systems.

The number of electrons associated with an atom determines the chemical properties of the element. These properties allow the atoms of one element to form linkages with one another or with atoms of other elements. Orderly associations of atoms of different elements form what are known as **chemical compounds**, and the smallest unit of a compound with the required arrangement of atoms is called a **molecule**. A molecule of the compound water, for example, is an arrangement of the elements hydrogen and oxygen.

Chemists have developed simple molecular formulas to represent compounds made up of molecules ranging in size from two atoms to many thousands of atoms. These formulas use the symbols for the elements. Table salt, for example, is a compound that has equal numbers of sodium (Na) and chlorine (Cl) atoms. The molecular formula for salt is therefore NaCl. If there is more than one atom of an element in a molecule, chemists use a subscript number after any atoms that are present more than once. For example, each molecule of water is made up of two hydrogen (H) atoms and one oxygen (O) atom, so the molecular formula for water is H_2O. The same notation is used for more complex compounds, such as

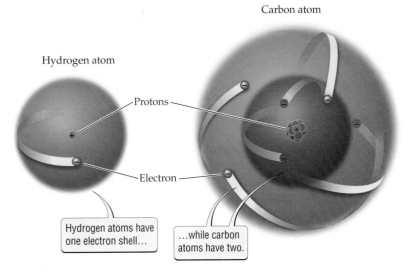

Figure 4.1 Atoms
The electrons, protons, and nuclei of these hydrogen and carbon atoms are shown greatly enlarged in relation to the size of the whole atom.

table sugar (sucrose), which has 12 carbons, 22 hydrogens, and 11 oxygens per molecule ($C_{12}H_{22}O_{11}$).

Atoms can be described by numbers

The letter symbols for different elements are not the only means of distinguishing one from another. The number of protons found in an atom's nucleus is the **atomic number** of that particular element. Hydrogen, with its single proton, has an atomic number of 1, while carbon, with its six protons, has an atomic number of 6.

Elements can also be distinguished by their **atomic mass number**, which is the sum of an atom's protons and neutrons. Since electrons have a negligible mass (about $1/2{,}000$ that of a proton), the atomic mass number is based on the number of protons and neutrons contained in an atom of a given element. Hydrogen has a single proton and no neutrons, so the atomic mass number for hydrogen is the same as its atomic number: namely, 1, written ^1H. In contrast, the nucleus of a carbon atom contains six protons and six neutrons, giving carbon an atomic mass number of 12 (^{12}C).

Some elements can exist as isotopes with different atomic mass numbers

So far we have discussed elements as if each is composed of only a single type of atom, but in the natural world some elements can contain different numbers of neutrons. These variants of elements are called **isotopes**.

All isotopes of an element have the same number of protons (the same atomic number), but they differ in their numbers of neutrons, and thus in their atomic mass numbers. For example, over 99 percent of the carbon atoms found in atmospheric carbon dioxide gas (CO_2) have an atomic mass number of 12 (^{12}C). However, a tiny fraction—about 1 percent—of those carbon atoms exist as an isotope containing eight instead of six neutrons, which, together with the six protons, result in an atomic mass number of 14 (^{14}C). The isotope is therefore referred to as carbon-14.

The carbon-14 isotope happens to be unstable, and as a result, it gives off energy in the form of radiation. For this reason, carbon-14 is described as a **radioisotope** [*RAY*-dee-oh-*EYE*-so-tope]. While only a fraction of known isotopes are radioactive, various radioisotopes such as carbon-14, phosphorus-32, and radioisotopes of hydrogen called deuterium (^{2}H) and tritium (^{3}H) have important uses in both research and medicine.

The radiation given off by radioisotopes can be physically detected by a variety of methods, ranging from simple film exposure to the use of sophisticated scanning machines. The detectability of radioisotopes means that their location and quantity can be determined fairly easily, a characteristic that makes them useful in medical diagnostics. For example, the thyroid gland takes up iodine for use in synthesizing hormones required by the body. When a low dose of an iodine radioisotope (iodine-131) is administered to patients with thyroid disease, the uptake of the radioisotope by the thyroid allows physicians to see the gland with an imaging device and determine how well it is functioning (Figure 4.2). When the patient is found to be suffering from cancer of the thyroid, repeated doses of iodine-131 can be continued as a therapy, since the accumulation of radioactivity in the thyroid tends to kill the cancer cells.

Figure 4.2 Isotopes in Medicine

Radioisotopes are useful in medical imaging. This gamma scan of the thyroid gland shows a normal left-hand lobe and an abnormal right-hand lobe swollen to twice its usual size because of a goiter. Note the higher metabolic activity (in red) in the normal lobe.

4.2 Covalent Bonds: The Strongest Linkages in Nature

All chemical linkages that hold atoms together to form molecules are called bonds. The strongest of these linkages are **covalent bonds**, which link atoms within a single molecule. In covalent bonds, atoms share electrons (Figure 4.3a). The electrons around the nucleus of every atom move in volumes of space that lie in layers, called shells, around the nucleus (see Figure 4.1). Each shell can contain a certain maximum number of electrons, and the atom is in its most stable state when all shells are filled to capacity. In the atoms of each element, shells are filled starting from the innermost shell. Atoms that have unfilled outer shells can react with one another. One way for an atom to fill its outermost shell is to share electrons with a neighboring atom. This sharing of electrons links the two atoms by forming a strong covalent bond.

Water and the natural gas called methane have the molecular formulas H_2O and CH_4, respectively. These formulas reveal the atomic components of each compound, but they say nothing about how the various atoms are bonded together and arranged in space. Another type of notation, known as a structural formula, is used to indicate both the atoms and the bonding arrangement of a compound. As Figure 4.3b shows, individual water molecules are held together by covalent bonds between the single oxygen atom and each of two hydrogen atoms. Likewise, methane has four covalent bonds that link the lone carbon atom to each of four hydrogen atoms.

The number of covalent bonds that an atom can form depends on the number of electrons needed to fill its outermost shell. Consider the electron sharing that occurs in the covalent bonds between hydrogen and oxygen in a water molecule. Hydrogen has one electron in its single shell, but that shell is filled only when it has two electrons. The inner shell of oxygen is filled, but its outer shell is not: it has six electrons, but is filled only when it has eight electrons. This situation can be resolved by mutual borrowing, on a "time-sharing" basis, between two hydrogen atoms and an oxygen atom: each hydrogen atom borrows one electron from oxygen, and the oxygen atom borrows two electrons, one from each of the hydrogen atoms. So the atoms contribute electrons in a way that makes the outer shells of all three atoms complete, at least on a shared basis. This kind of sharing requires the intimate association of a covalent bond.

(a)

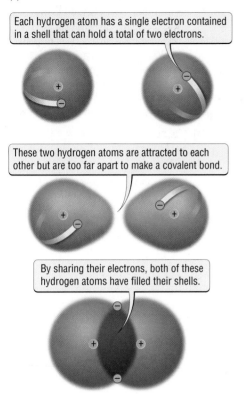

Each hydrogen atom has a single electron contained in a shell that can hold a total of two electrons.

These two hydrogen atoms are attracted to each other but are too far apart to make a covalent bond.

By sharing their electrons, both of these hydrogen atoms have filled their shells.

H—H (H₂)

(b)

Atom	Symbol	Number of possible bonds	Sample compounds	
Hydrogen	H	1 bond	H—H	Hydrogen gas (H₂)
Oxygen	O	2 bonds	(structure)	Water (H₂O)
Sulfur	S	2 bonds	(structure)	Hydrogen sulfide (H₂S)
Nitrogen	N	3 bonds	(structure)	Ammonia (NH₃)
Carbon	C	4 bonds	(structure)	Methane (CH₄)
Phosphorus	P	5 bonds	(structure)	Phosphoric acid (H₃PO₄)

Figure 4.3 Covalent Bonds
(a) If they are close enough to each other, two hydrogen atoms can share electrons, forming a covalent bond. The length of a covalent bond is such that the electrons can be shared, while the two positively charged nuclei are still kept far enough apart that they do not repel each other. *(b)* The number of covalent bonds that an atom can form depends on the number of electrons needed to fill its outermost shell. For example, an oxygen atom requires two more electrons to fill its shell and can form two covalent bonds. A carbon atom requires four more electrons and can form four covalent bonds.

Carbon, nitrogen, hydrogen, oxygen, phosphorus, and sulfur are the most common elements found in living organisms, and their atoms all form covalent bonds. Consequently, combinations of all or some of these atoms are found in the molecules that make up living organisms.

4.3 Noncovalent Bonds: Dynamic Linkages Between Molecules

Whereas covalent bonds link atoms to form molecules, **noncovalent bonds** are the most common linkages between separate molecules and between different parts of a single large molecule. By bringing molecules together in specific configurations, noncovalent bonds promote the complex organization of the physical structures and activities we associate with living organisms. Despite this important role, noncovalent bonds are far weaker than covalent bonds. Their strength lies in numbers. Noncovalent bonds do not involve the direct sharing of electrons between atoms. They are based on other chemical properties of two or more atoms.

Because of the relative weakness of noncovalent bonds, many such bonds are often required to hold two molecules together or to establish the configurations of large molecules. In this case, however, weakness is a virtue, because it also allows molecules to adjust to changes in their surroundings, a feature necessary for many biological processes. For example, when skin is pinched and then released, its ability to stretch and then spring back depends on the breaking and re-forming of a multitude of noncovalent bonds between many different classes of molecules.

See how temperature affects molecular movement. 4.1

Salt

Figure 4.4 Hydrogen Bonds Determine the Properties of Water and How Other Compounds Interact with Water
Each water molecule forms temporary hydrogen bonds with its neighbors, resulting in a network of molecules that account for water being liquid at room temperature. Water molecules are also attracted to electrically charged ions. Compounds that are held together by ionic bonds dissolve in water as each ion becomes surrounded by water molecules. Here the changes in a cube-shaped crystal of table salt, NaCl (yellow and green), are shown as it dissolves in water (blue and red).

Na$^+$ Cl$^-$

Explore the process of salt dissolving in water.   4.2

H$_2$O (Water)

H$_2$O (Water)

Slightly positive

Slightly negative

Hydrogen bond

Hydrogen bonds are constantly forming…

…and breaking.

Strong ionic bonds between sodium (Na$^+$) and chloride (Cl$^-$) ions…

…are broken by electrical attraction between water molecules and sodium and chloride ions.

Biology Matters

Elements 'R Us

Although there are traces of other atoms in our bodies, we are mostly made up of just eleven different elements. The below chart shows the percentage by weight of each element that is part of us. Much of the hydrogen and oxygen in our bodies is combined in the form of H_2O molecules, so that approximately 70 percent of our weight is water. Other trace elements make up about one-hundredth of one percent of our weight. We need to eat, in part, to replace the elements that we naturally lose each day. We also need food energy to keep us going.

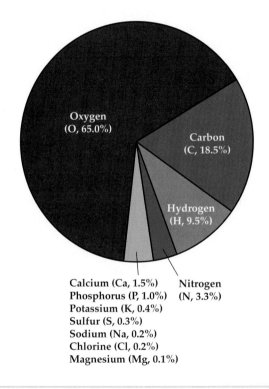

Oxygen (O, 65.0%)
Carbon (C, 18.5%)
Hydrogen (H, 9.5%)

Calcium (Ca, 1.5%)   Nitrogen
Phosphorus (P, 1.0%)   (N, 3.3%)
Potassium (K, 0.4%)
Sulfur (S, 0.3%)
Sodium (Na, 0.2%)
Chlorine (Cl, 0.2%)
Magnesium (Mg, 0.1%)

The nutrition label on a carton of whole milk specifies the amount of energy, expressed in calories, that is stored in the chemical bonds of one serving. Of the nutrient groups listed on the label, fats and carbohydrates contain atoms of the elements carbon (C), hydrogen (H), and oxygen (O). Cholesterol contains the same three elements. Protein contain nitrogen (N) and sulfur (S) in addition to C, H, and O. You can see that the elements sodium (Na), calcium (Ca), iron (Fe), and phosphorus (P) are included on the whole-milk label, as well. Vitamins also contain carbon, hydrogen, and nitrogen. So, whole milk is a source of at least nine of the eleven major elements that, combined in many different ways, make up our bodies.

Keeping an eye on food labels will help you better understand the amounts of energy and nutrients in your diet, and eating a variety of foods can help assure a healthy intake of elements. Potatoes and bananas are rich in potassium, an element important in nerve function. Dairy products are richer in calcium than most other foods. Calcium is a major component of bones and teeth. Yogurt is a good source of phosphorous, as are sunflower seeds. The monomers in DNA and RNA are bonded together by phosphate bonds. Processed foods often contain large amounts of sodium. Deficiencies in sodium are rare, and experts warn against ingesting too much of this element, because excess sodium is associated with high blood pressure.

Hydrogen bonds are important temporary bonds

The **hydrogen bond** is one of the most important kinds of noncovalent bonds in nature. Each hydrogen bond is about 20 times weaker than a covalent bond. During rapid biological processes such as muscle contraction, hydrogen bonds are broken and re-formed moment by moment.

The simplest example of a hydrogen bond can be found between water molecules (Figure 4.4). Hydrogen bonding is an important attribute of water and is the basis of many of water's chemical properties. Most organisms consist of more than 70 percent water by weight, and nearly every chemical process associated with life occurs in water.

As we have seen, each molecule of water is made up of two hydrogen atoms and one oxygen atom held together by covalent bonds. The positive nucleus of the oxygen atom tends to attract the negative electrons more powerfully than the nuclei of the two hydrogen atoms do, causing each water molecule to have an uneven distribution of electrical charge. That is, the oxygen atom carries a slightly negative charge, while the hydrogen atoms in turn carry a slightly positive charge. Molecules with an uneven distribution of charge are **polar**.

Because opposite charges attract, the slightly positive hydrogen atoms of one water molecule are attracted to the slightly negative oxygen atom of a neighboring water molecule. This attraction forms a hydrogen bond between the two water molecules, which in turn can form hydrogen bonds with other neighboring water molecules (see Figure 4.4). The resulting dynamic interplay of hydrogen-bonded water molecules is responsible for the properties of liquid water (in particular, for making it liquid at room temperature). The latticelike network of water molecules that forms in ice is also the result of hydrogen bonding.

Any polar compound that contains hydrogen atoms can form hydrogen bonds with a neighboring polar compound that contains a partially negative atom. Water molecules can form hydrogen bonds with other polar compounds, with the result that these compounds dissolve in water. Such compounds are said to be **soluble** in water, since each of the compound's molecules becomes surrounded by water molecules and can move freely throughout the liquid. For example, when sugar is added to water, the solid crystals dissolve; that is, they break apart as the sugar molecules in the crystal are surrounded by water molecules and scattered uniformly throughout the water. Since dissolved compounds abound in and around living cells, chemists and biologists use specific terms to describe these mixtures: a **solution** is any combination of a **solute** (a dissolved substance) and a **solvent** (the liquid, usually water, into which it is has dissolved).

The polarity of water molecules also means that they will not associate with other molecules that are not charged—that is, with **nonpolar** molecules. When they are added to water, nonpolar molecules—instead of becoming separated like polar molecules—are pushed into clusters. This is exactly what happens when olive oil is added to vinegar, a watery solution. In order to form a salad dressing, the two substances have to be shaken vigorously; otherwise, the oil separates from the vinegar. The distribution of electrons among carbon atoms and hydrogen atoms within oil molecules is nearly equal, making these molecules nonpolar and therefore insoluble. In the salad dressing, the water molecules in the vinegar do not mix with the nonpolar oil molecules. Instead, the oil molecules tend to cluster into tiny floating droplets when the dressing is shaken. Waxes are also nonpolar,

and automobile enthusiasts wax their cars not just to make them look shiny, but also to repel water and reduce the risk of rusting.

Molecules that interact freely with water (such as sugar) are called **hydrophilic** (*hydro*, "water"; *philic*, "loving"), while molecules that are repelled by water (such as oil) are called **hydrophobic** (*phobic*, "fearing").

Ionic bonds form between atoms of opposite charge

Ionic bonds, like hydrogen bonds, rely on the fact that opposite electrical charges attract. However, the attraction between two atoms of opposite charge has consequences different from those of hydrogen bonding: the electrons in the outer shell of one neutral atom are completely transferred to the outer shell of a neutral atom of a different element. This loss of electrons by one atom and the gain of electrons by another converts both neutral atoms into charged ions.

Ionic bonds between molecules dissolved in water are relatively weak. Like hydrogen bonds, however, they are essential for many temporary associations between molecules. For example, our ability to taste what we eat depends on the ionic bonds that form between food molecules dissolved in water and other specialized molecules in our taste buds. The rapid association and dissociation of multiple food molecules with our taste buds also allows us to discern several different tastes in a short time. If these associations were not brief, every meal would be dominated by the taste of the first bite.

Ionic bonds between molecules under dry conditions can be very strong. For example, a grain of dry table salt consists of countless sodium ions (Na^+) linked by ionic bonds to chloride ions (Cl^-). (Note that the charge of a given ion is indicated by a superscript plus or minus symbol.) In the absence of water, these ions pack tightly to form the hard, three-dimensional structures we know as salt crystals. When salt is added to water, the water molecules, because they are polar, are attracted to and surround both types of charged ions in NaCl. This interaction with water breaks up and dissolves the salt crystals, scattering both positive sodium ions and negative chloride ions throughout the liquid (see Figure 4.4).

Like sugar, then, salt is soluble in water. But there is one key difference: water molecules are attracted to and surround each ion in salt because of its charge, but hydrogen bonds do not form. Hydrogen bonds form only between polar compounds, not between water molecules and ions. Therefore, ions and polar molecules are two different classes of hydrophilic compounds that dissolve in water.

4.4 Chemical Reactions Rearrange Atoms Within Molecules

Molecules are not static arrangements of atoms. Biological processes require atoms to break existing connections and form new ones with other atoms. Consider the additional molecules that must be made to provide a growing plant with components for new leaves and stems. The plant does not produce these new molecules from a supply of raw atoms; rather, it acquires and rearranges preexisting molecules. The processes that break and form bonds between atoms are known as **chemical reactions**.

The standard notation for chemical reactions describes changes in the arrangement of atoms in molecules. Nitrogen and hydrogen, for example, can combine to produce ammonia gas (NH_3), which gives many window cleaners their sharp odor. This chemical reaction is written as follows:

$$3\,H_2 + N_2 \rightarrow 2\,NH_3.$$

The arrow indicates that the molecules on the left side of the equation, at the start of the reaction (hydrogen and nitrogen), are converted to the molecules on the right side, at the end of the reaction (ammonia). The numbers in front of the molecules define how many molecules participate in the reaction. In this case, three molecules of hydrogen gas (H_2) combine with one molecule of nitrogen gas (N_2) to produce two molecules of ammonia (note that "1" is usually not written). The atoms on the left side of the arrow are called reactants, and those on the right side are called products.

All chemical reactions rearrange atoms, but a chemical reaction can neither create nor destroy atoms. Therefore, the reaction must begin and end with the same number of atoms of each element. In the example here, each side of the equation has six hydrogen atoms and two nitrogen atoms.

Later in this chapter we'll see how chemical reactions promote more complex molecular organization. First, however, let's focus on the chemical reactions that determine how ions and molecules interact with water.

Acids and bases affect the pH of water

All chemical reactions that support life occur in water. Some of the most important are those that involve two classes of compounds: acids and bases. An **acid** is a polar compound that dissolves in water and loses one or more hydrogen ions (H^+). These hydrogen ions tend to bond with the surrounding water molecules, forming positively charged hydronium ions (H_3O^+). Because the formation of hydronium ions is easily reversed, hydrogen ions are constantly being exchanged between water molecules and other molecules dissolved in water.

Bases are also polar compounds, but unlike acids, they *accept* hydrogen ions from their surroundings. Thus acids and bases interact with water molecules in different ways and have opposite effects on the amount of free hydrogen ions in water. A base can accept one H^+ ion from water, leaving one hydroxide ion (OH^-) behind.

The concentration of free hydrogen ions in water influences the chemical reactions of many other molecules. This hydrogen ion concentration is expressed as a scale of numbers from 0 to 14, where 0 represents the highest concentration of free hydrogen ions and 14 represents the lowest. This scale is called the **pH** scale. Each unit of the scale represents a tenfold increase or decrease in the concentration of hydrogen ions in a sample of water.

When water contains no acids or bases, the concentrations of free hydrogen ions and hydroxide ions are equal, and the pH is said to be neutral, or in the middle of the scale, at pH 7. Below pH 7, the solution is said to be acidic because an acid has donated hydrogen ions, raising the concentration of free hydrogen ions in the solution. Above pH 7, the solution is said to be basic because a base has accepted hydrogen ions, lowering the concentration of free hydrogen ions in the solution (Figure 4.5).

We have all experienced acidic and basic substances. The tartness of lemon juice in a good homemade lemonade is due to the acidity of the juice (about pH 2). Our stomach juices are able to break down food because they are very acidic (about pH 2). At this low pH, many bonds between molecules are disrupted by the high concentration of free hydrogen ions, which associate with atoms that would otherwise be bonded to other atoms. At the other extreme is a substance such as oven cleaner, which is very basic (about pH 13) and can also cause molecules to be disrupted. This is why extremes of pH can be caustic, causing chemical burns on the skin. Figure 4.5 shows the approximate pH of some common solutions.

 Learn more about acids, bases, and pH. **4.3**

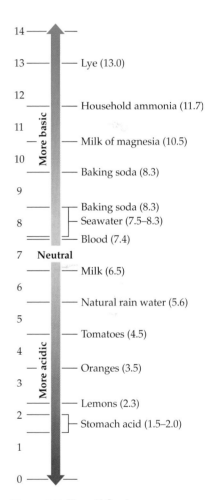

Figure 4.5 The pH Scale
A pH of 7 means that a solution is neutral, neither basic nor acidic. Values below 7 indicate acidic solutions; the lower the value, the more acidic the solution. A solution with a pH of 3 is ten times more acidic than one with a pH of 4. Values above 7 indicate basic solutions; the higher the value, the more basic the solution. A solution with a pH of 9 is ten times more basic than a solution with a pH of 8.

Buffers prevent large changes in pH

Most living systems function best at an internal pH that is close to neutral (see the box on page 80). Any change in pH to a value significantly below or above 7 adversely affects many biological processes. Because hydrogen ions can move so freely from one molecule to another during normal life processes, organisms need to prevent dramatic changes in the pH levels of their internal environments. Compounds called **buffers** meet this need by maintaining the concentration of hydrogen ions within narrow limits. They do so by releasing hydrogen ions when the surroundings become too basic (excessive OH^- ions, high pH) and accepting hydrogen ions when the surroundings become too acidic (excessive H^+ ions, low pH).

4.5 The Chemical Building Blocks of Living Systems

If we removed the water from any living organism, we would be left with four major classes of chemical compounds: carbohydrates, nucleic acids, proteins, and fats. These chemical compounds all consist of different arrangements of carbon atoms associated with hydrogen. Oxygen, nitrogen, phosphorus, and sulfur atoms may also be present.

Carbon is the predominant element in living systems partly because it can form large molecules that contain thousands of atoms. A single carbon atom can form strong covalent bonds with up to four other atoms. Even more importantly, carbon can bond to carbon, forming long chains, branched molecules, or even rings. Biological processes can therefore use the wide variety of large and complex structures that are possible with carbon-containing molecules. All carbon compounds found in nature are referred to as **organic compounds**.

Carbon compounds, most of which contain about 20 atoms, can either remain as individual molecules or bond with other molecules of about the same size to form larger structures called **macromolecules** (*macro*, "large"). Carbon compounds in living organisms often follow this principle of building very large structures from smaller units. Individual small molecules are called **monomers** (*mono*, "one"; *mer*, "part"). Macromolecules, which can contain hundreds of monomers bonded together (Figure 4.6), are also called **polymers** (*poly*, "many"). Polymers account for most of an organism's dry weight (its weight after all water is removed) and are essential for every structure and chemical process that we associate with life.

In living organisms, fewer than 70 different biological monomers are combined, in a nearly endless variety of ways, to produce polymers with many different properties. Polymers are therefore a step up from monomers in organizational complexity, and they have chemical properties that are not possible for a monomer. Furthermore, many organic polymers acquire chemical properties from attached groups of atoms called functional groups. These groups provide molecules with sites that can react with other molecules (Table 4.1).

As their name implies, **functional groups** are groups of covalently bonded atoms that have the same specific chemical properties on whatever molecule they are part of. Some functional groups help establish covalent linkages between monomers; others have more general effects on the chemical characteristics of a polymer.

The importance of these properties is illustrated in the monomers of each of the four groups of organic compounds found in all living systems: sugars, nucleotides, amino acids, and fatty acids. Each of these groups contains tremendously complex and changeable structures that are based on a limited range of organic compounds.

Table 4.1

Some Important Functional Groups Found in Organic Molecules

Functional group	Formula	Ball-and-stick model
Amino group	$-NH_2$	Bond to carbon atom
Carboxyl group	$-COOH$	
Hydroxyl group	$-OH$	
Phosphate group	$-PO_4$	

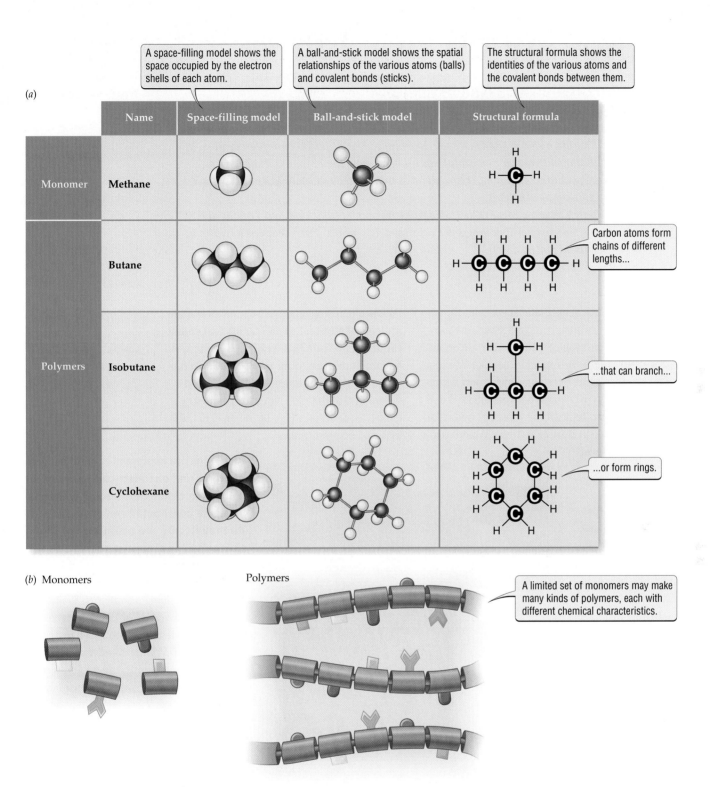

A space-filling model shows the space occupied by the electron shells of each atom.

A ball-and-stick model shows the spatial relationships of the various atoms (balls) and covalent bonds (sticks).

The structural formula shows the identities of the various atoms and the covalent bonds between them.

Name	Space-filling model	Ball-and-stick model	Structural formula
Methane (Monomer)			
Butane (Polymers)			
Isobutane (Polymers)			
Cyclohexane (Polymers)			

Carbon atoms form chains of different lengths...

...that can branch...

...or form rings.

(b) Monomers

Polymers

A limited set of monomers may make many kinds of polymers, each with different chemical characteristics.

Figure 4.6 Assembling Complex Structures from Smaller Components

In living organisms, the important principle of building large and complex structures from smaller components applies to both atoms and molecules. (a) A single carbon atom can form four covalent bonds with other atoms. When carbon atoms form bonds with other carbon atoms, a variety of chains can be formed, with different structures such as branches and rings. (b) Like single atoms, small chemical compounds can form covalent bonds with one another, giving rise to larger molecules with specific chemical properties. The small constituent compounds are called monomers, while the resulting large assemblage is called a polymer.

Biology on the Job

At the Chattanooga Nature Center

Tennessee's Chattanooga Nature Center (CNC) is a widely respected and popular destination for those interested in hands-on environmental education and habitat preservation. It has provided innovative environmental education for more than 250,000 students. And through its public programming, thousands of families and adults have learned to respect and care for their local habitat.

The center's director, Steve O'Neil, along with director of education Jean Lomino and school programs director Kyle Waggener, discussed the chemistry they teach as it relates to the environment and habitats.

Water is essential to life. What do you teach about water at the center? *Kyle responds:* We conduct a pond study every spring and fall. Students measure pH in our pond and compare it to the range of pHs that certain organisms can tolerate. Most algae and plants, for example, can survive in a pH range from 6.5 to 13.0. On the other hand, trout can survive only within a pH range of 6.5 to 7.5. In our pond the pH ranges between about 6 and 7. In the classroom we describe water molecules and hydrogen bonds that create the large-scale properties we see in natural settings. Water striders, for example, can walk on the surface, because hydrogen bonds among water molecules create a surface tension.

Jean, what other aspects of chemistry does the Chattanooga Nature center touch on? Sometimes—in our setting emphasizing habitats—chemistry as such does not take center stage. Overall, we focus on making connections between nature and the classroom. When we bring students out into the field, we look for what we call "teachable moments." These moments aren't pre-planned. They occur spontaneously, as we observe our outdoor surroundings. We try to seize upon these moments to demonstrate the connected dynamics of life. Some moments are related to the detection of odors and fragrances, and that certainly touches on chemistry. A teachable moment occurs, for instance, when we see pollinators such as bees and butterflies on flowers. We use that moment to discuss fragrance as one means employed by flowers to attract pollinators, and we follow up by describing the critical role of pollinators in plant reproduction.

In this chapter we describe an individual protein as being like a sentence that says only one thing, such as "detect the fragrance of a rose." So, when you see insect pollinators attracted to flowers, you are observing the chemistry between fragrances and the protein "sentences" in pollinators that detect them. *Jean and Kyle respond:* That will become an interesting point to add to our natural history teaching. It will help to demonstrate that the world of chemistry is closely connected to life at the level of habitats and ecosystems.

Water strider

Steve, you are interested in other aspects of biological chemistry that are part of something called constructed wetlands. What are constructed wetlands, and how do they fit into the chemistry of living things? Constructed wetlands are human-made wetland systems designed to treat wastewater using the complex chemistry of living organisms. We have a constructed wetland at the nature center that processes the wastewater from our visitors' center. Untreated water is distributed from a septic tank into an area with a waterproof liner that has been planted with bulrushes and other wetland grasses. The thick root mats of these plants provide a matrix for millions of bacteria that consume the suspended organic matter and ultimately break down toxic and harmful molecules. The water then flows to a gravel bed where other forms of bacteria and microorganisms continue to process the wastes and complete the job. Cleansed water and nutrients that are the products of this food chain then return to the environment. The plant life of a constructed wetland is appealing to the eye and attracts desired wildlife. All in all, constructed wetlands are an efficient way of treating wastewater in a controlled manner with no harmful by-products, just the way nature intended.

(continued)

Steve, Jean, and Kyle dedicate their efforts to increasing habitat awareness and promoting conservation. Their goal is to develop lasting connections—bonds—between people and nature, so that we become more acutely aware of the consequences of our actions, good and bad, through an understanding of how life works as a dynamic whole. In Chapter 8 we will see more of the Chattanooga Nature Center and learn of its involvement in saving an endangered species, the red wolf.

Carbohydrates provide energy and structural support for living organisms

Sugars are familiar to us as compounds that make foods taste sweet. Although only some sugars are perceived as sweet by human taste buds, most sugars are important food sources and serve as a major means of storing energy in living organisms. Sugars and their polymers are referred to as **carbohydrates**, since each carbon atom (*carbo*) is linked to two hydrogen atoms and an oxygen atom—the equivalent of a molecule of water (*hydrate*). The simplest sugar molecules are called **monosaccharides** [mon-oh-*SACK*-uh-rides] (*mono*, "one"; *sacchar*, "sugar").

Like all carbohydrates, monosaccharides are made up of units containing carbon, hydrogen, and oxygen atoms in the ratio 1:2:1. This ratio can also be expressed as the molecular formula $(CH_2O)_n$, with *n* ranging from 3 to 7. Monosaccharides are often referred to by the number of carbon atoms they contain; for example, a sugar with the molecular formula $(CH_2O)_5$ is called a five-carbon sugar. Note that the parentheses work the same in this notation as in multiplication. A more common way to express the molecular formula for this sugar is $C_5H_{10}O_5$.

The best-known example of a monosaccharide is **glucose** ($C_6H_{12}O_6$). Glucose has a key role in short-term energy storage, and nearly all of the chemical reactions that produce energy for living organisms involve the synthesis and breakdown of this sugar. Glucose is such a major player, particularly in the metabolism of animal cells, that elaborate control mechanisms regulate the concentration of glucose in response to changing energy needs.

Monosaccharides can combine to form larger, more complex molecules. For example, two covalently bonded molecules of glucose form maltose, a disaccharide (*di*, "two") (Figure 4.7*a*). Other common disaccharides are sucrose (our familiar table sugar) and lactose (found in dairy products). Similarly, up to thousands of monosaccharides can be linked together to form a polymer called a **polysaccharide**. Monosaccharides, disaccharides; and polysaccharides are all carbohydrates.

Carbohydrates perform several different functions in living organisms. One function is to provide structural support; cellulose, for example, is a polysaccharide that forms strong parallel fibers that help support the leaves and stems of plants (Figure 4.7*b*). Carbohydrates also provide fuel (energy), as we have already seen in the case of glucose. Starch—which is found in many plant-based foods and is familiar to many of us as the basis for "carbo loading" before a strenuous sporting activity such as a marathon—is a polysaccharide that provides energy (Figure 4.7*c*).

Nucleotides store information and energy

Nucleotides are profoundly important monomers that make up the genetic material in all organisms and act as energy "currency" in cells. Each nucleotide has three components (Figure 4.8): a **nitrogen base** is covalently bonded to a five-carbon sugar, which in turn is covalently bonded to a functional group called a phosphate group, which consists of a phosphate atom and four oxygen atoms (see Table 4.1).

This base–sugar–phosphate trio forms the building blocks of a class of polymers called **nucleic acids**, of which there are two kinds: deoxyribonucleic acid (DNA) and ribonucleic acid (RNA). They are distinguished chemically by the sugar in their nucleotides and by the nitrogen bases that bond with that sugar. The five different nucleotide bases are adenine, cytosine, guanine, thymine, and uracil. Deoxyribose, the sugar in DNA, can bond to adenine, cytosine, guanine, or thymine. Ribose, the sugar in RNA, differs from deoxyribose in that it has one more oxygen atom (see Figure 4.8). Ribose can bond to adenine, cytosine, guanine, or uracil (instead of thymine).

Figure 4.7 The Structure of Carbohydrates

Monosaccharides can bond with one another to form disaccharides and poly-saccharides. (*a*) Two molecules of glucose can be linked by a covalent bond to form the disaccharide maltose, releasing one water molecule. (*b*) Parallel strands of the polysaccharide cellulose are important components of plant cell walls. They help maintain the rigid structure of the cell walls, which is necessary for structural support of leaves and stems. (*c*) Highly branched polysaccharides, such as starch, are used to store energy in plants. That is why starch-rich foods such as potatoes are good sources of energy.

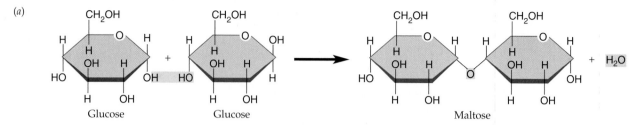

(*a*)

Glucose + Glucose → Maltose + H_2O

(*b*)

Cellulose fibers

Cellulose is a polymer of glucose, but its linkages are much stronger than in starch, giving plants structural support.

Cellulose

Potato plant

(*c*)

Starch granules

Starch

Starch is also a polymer of glucose. Its weaker bonds allow energy-rich monomers to be released.

Figure 4.8 Nucleotide Components

Each nucleotide consists of a five-carbon sugar linked to a nitrogen-containing base and one or more phosphate groups. The bases adenine, guanine, cytosine, and thymine, when linked to the sugar deoxyribose, form the building blocks of DNA. The bases adenine, guanine, cytosine, and uracil, when linked to the sugar ribose, form the building blocks of RNA.

Nucleotides have two essential functions in the cell: information storage and energy transfer—both of which highlight key characteristics of living systems. Every organism is defined by a nucleic acid "blueprint" that dictates what chemical building blocks are produced and how they are assembled to make up the organism. The order in which nucleotides are hooked together in the polymers of DNA or RNA determines all the physical attributes of, and chemical reactions that occur in, living organisms. We will learn a lot more about DNA and RNA in Unit 3, on genetics.

In addition to storing genetic information, nucleotides assist in energy transfer. A key player in this process is **adenosine triphosphate**, or **ATP**. Each ATP molecule consists of one nucleotide with an adenine base and three attached phosphate groups. The energy an organism consumes in food is transferred to and packaged in ATP. ATP is the universal fuel for living organisms, and many chemical reactions require energy from ATP in order to proceed. The energy of ATP is stored in the covalent bonds that link the three

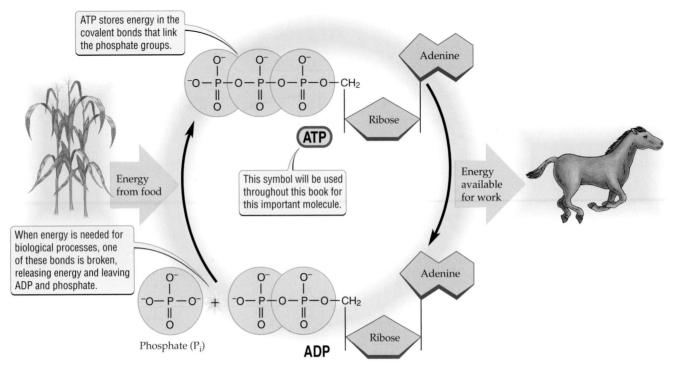

Figure 4.9 Production of ATP

ATP is the major form of short-term stored energy in living organisms. The energy obtained from digesting food is used to form ATP from ADP (adenosine diphosphate) and a phosphate group. Later, when energy is needed by the organism, a covalent bond in ATP is broken, forming ADP and phosphate again.

phosphate groups (Figure 4.9). The breaking of the bond between the second and third phosphates releases energy that is used to drive other chemical reactions.

Amino acids are the building blocks of proteins

Of the many different kinds of chemical compounds found in living organisms, **proteins** may be the most familiar. These polymers make up more than half the dry weight of living things. The steaks we throw on the grill in the summer consist mainly of proteins, and our ability to run to the finish line of a race depends on the coordinated actions of many proteins in our muscles. Our bodies are made up of thousands of different proteins. Some of these proteins form physical structures, such as keratin in hair and collagen in skin; others, called enzymes, help regulate the chemical reactions that drive biological processes.

Amino acids are the monomers used to build proteins. There are 20 different amino acids, all of which share some structural characteristics. All amino acids have a carbon atom called the alpha carbon, which forms a central attachment site for four other components. The alpha carbon is attached to a hydrogen atom, a chemical side chain called the R group, and two functional groups: an amino group ($-NH_2$), and a carboxyl group ($-COOH$) (Figure 4.10).

Figure 4.10 The Structure of Amino Acids

Amino acids are the building blocks that form polymers called proteins.

The different R groups give different amino acids different properties. R groups range from a single hydrogen atom to complex arrangements of carbon chains and ring structures (Figure 4.11). Thus organisms have a diverse pool of building blocks from which they can make proteins with many different properties.

To form proteins, linear chains of amino acids are linked together by covalent bonds between the amino group of one amino acid and the carboxyl group of another. These covalent bonds, among the most important bonds in biology, are called **peptide bonds** (Figure 4.12). Proteins can contain hundreds to thousands of

Figure 4.11 The Diversity of Amino Acids
Each of the 20 amino acids has a different R group bonded to the alpha carbon. The various R groups give each amino acid chemical properties that can be classified as hydrophobic, hydrophilic, or special.

Hydrophobic amino acids are nonpolar or uncharged and are repelled by water.

Hydrophilic amino acids are either charged or polar and can interact with water molecules.

Special amino acids have unique chemical properties or structures that influence how they fit into a folded protein chain.

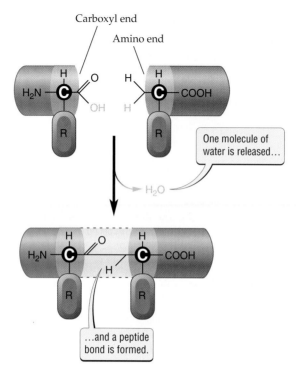

Carboxyl end

Amino end

One molecule of water is released…

H_2O

…and a peptide bond is formed.

Figure 4.12 Formation of a Peptide Bond

Peptide bonds form when the carboxyl group of one amino acid bonds covalently with the amino group of another amino acid. In the process, an OH group from the carboxyl end and a hydrogen atom from the amino end join to form one molecule of water.

peptide bonds, which bind the amino acids together like a string of letters. Each "letter" along the string can be any one of the 20 amino acids, and each protein has its own precise sequence of amino acids. A protein's sequence of amino acids—its string of letters, which is dictated by the sequence of nucleotides in DNA—is called its **primary structure** (Figure 4.13a).

In order for a language to have meaning, however, it must consist of more than just strings of letters; the letters must be grouped into phrases. Similarly, proteins need a level of organization beyond the primary strings of amino acids. In proteins, the next level of organization is sequences of amino acids that have been folded into spirals and sheets. The spirals are called alpha helixes, and the sheets are called beta pleated sheets. This "phrase" level of organization in a protein is called its **secondary structure** (Figure 4.13b).

But proteins work at a level beyond individual phrases. They operate as complete "sentences"—in particular, as commands for carrying out very specific functions. One protein's sentence might be, "Detect and pass along the fragrance of a rose." A different protein's might be, "Speed up the chemical reaction that breaks down starch in food."

Join amino acids to form proteins.

 4.4

In proteins, the secondary structure is folded into an overall three-dimensional form. This sentence-level organization is called the protein's **tertiary structure** (Figure 4.13c).

You may wonder how just 20 amino acids can generate the thousands of different proteins found in nature. But consider this: the number of different sentences that can be written using the 26 letters of the English alphabet seems to have no bounds. If you think of the protein alphabet as having 20 amino acid letters (only 6 fewer than in

(a) Primary structure (the "letters")

(b) Secondary structure (the "phrases")

Hydrogen bond

Hydrogen bond

(c) Tertiary structure (the "sentences")

(d) Quaternary structure

Figure 4.13 The Four Levels of Protein Structure

the English alphabet), you can see that the number of possible different protein sentences is enormous. The complexity of life is partly based on this diversity of proteins.

For some functions, single-protein sentences are not sufficient. Some proteins work at the level of whole "paragraphs"; that is, as groups of closely associated "sentences." For example, hemoglobin, the protein that carries oxygen in our blood, consists of four protein "sentences." This paragraph level of organization is called **quaternary structure** [KWAHT-er-nar-ee . . .] (Figure 4.13d).

The phrases, sentences, and paragraphs of a protein's structure result from the interactions of the R groups on individual amino acids with one another and with their immediate environment. Hydrogen bonds and covalent bonds play major roles in defining the tertiary structure and chemical properties of proteins. For most proteins, this specific three-dimensional structure is necessary for normal function, whether that function is a specific chemical activity or structural support.

In recent years our understanding of protein structure has grown tremendously, thanks to sophisticated new methods of visualizing a protein's three-dimensional shape. These breakthroughs have helped unravel the mysterious chemical forces that shape proteins, paving the way for the design of improved, synthetic proteins, as we'll see at the end of this chapter.

Fatty acids store energy and form membranes

Fatty acids are molecules composed primarily of long hydrocarbon (carbon and hydrogen) chains ending with a carboxyl group. They are the key components of fats and lipids. **Fats**, which are also known as triglycerides, are composed of three fatty acids combined with a simple three-carbon molecule called glycerol (a type of alcohol). They function in the long-term storage of energy for living organisms and are familiar to some of us as the prime targets of weight-loss programs. Fats are included in the broader category of **lipids**, a large group of compounds defined, in general, by the property of being insoluble in water. However, some very important lipids have one end that is soluble in water and one end that is not, as we shall see shortly. These lipids establish the boundaries of living cells, called *membranes*, and help control the exchange of molecules between cells and their environment.

The long hydrocarbon chains found in fatty acids usually contain 16 or 18 carbon atoms that can vary in the way they are covalently bonded together. Fatty acids in which all the carbon atoms are linked together by single covalent bonds are said to be **saturated** because each carbon is also bonded to a full complement of hydrogen atoms (Figure 4.14a). When one or more of the carbon atoms are linked together by double covalent bonds, the fatty acid is said to be **unsaturated**, because some of the

(a) Stearic acid is a saturated fatty acid. It contains no double bonds between its carbon atoms.

(b) Oleic acid is an unsaturated fatty acid. It has one double bond between two of its carbon atoms.

Stearic acid

Oleic acid

Figure 4.14
Saturated and Unsaturated Fatty Acids
(a) This space-filling model of stearic acid, with the chemical structure shown to the left of it, shows that the stearic acid molecule is straight; thus it can pack tightly to form a waxy solid at room temperature. (b) The double bond in oleic acid forms a kink in the molecule, preventing it from packing as tightly as stearic acid. Oleic acid therefore tends to be more liquid at room temperature and is commonly found in the storage fat of humans.

carbon atoms are not bonded to a full complement of hydrogen atoms (Figure 4.14b).

The significance of the double bonds in unsaturated fatty acids goes beyond simple differences in the number of hydrogen atoms linked to the carbon chain. Single bonds tend to adopt a straight configuration in space, while double bonds tend to introduce kinks into the hydrocarbon chain. Consequently, the straighter saturated fatty acids can pack very tightly together, forming solids such as fats and waxes at room temperature. The kinks in unsaturated fatty acids prevent such tight packing; thus they form oils, which are compounds that tend to be liquid at room temperature.

The role of fatty acids in energy storage and membrane structure depends on their covalent bonding to glycerol. Glycerol has three hydroxyl groups (OH), each of which can form a covalent bond with the carboxyl group (—COOH) at the end of a fatty acid chain. As the bond forms, the OH on the carboxyl group of a fatty acid combines with an H in one hydroxyl group on the glycerol, forming a molecule of water. When all three hydroxyl groups in glycerol are bonded to a fatty acid, the resulting compound is a triglyceride, or fat (Figure 4.15). Fats contain significantly more energy than does an equal amount of glucose, and they are the most common means of long-term energy storage in animals and plants.

The bonding of two fatty acid chains and a negatively charged phosphate group to glycerol produces molecules of a group of lipids called **phospholipids**. Phospholipids are the major component of all membranes in living cells. The negatively charged phosphate group on one end of the phospholipid is polar, which means that this region

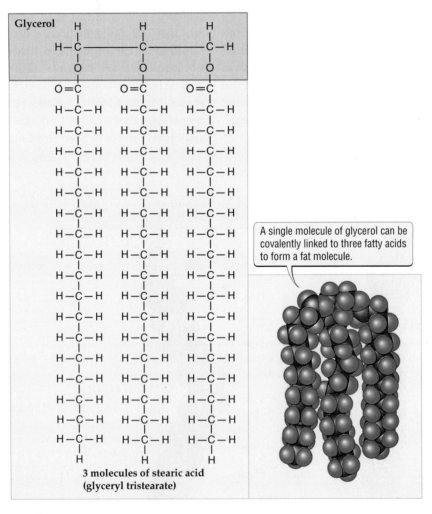

A single molecule of glycerol can be covalently linked to three fatty acids to form a fat molecule.

3 molecules of stearic acid
(glyceryl tristearate)

Figure 4.15 Fats
When the fatty acids that bond to glycerol are stearic acid, the resulting fat is glyceryl tristearate, the most common storage form of fat in animals and plants.

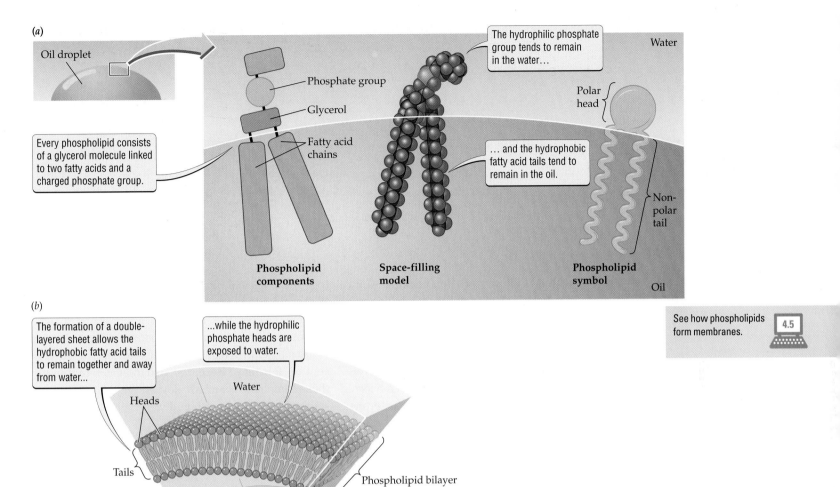

(a)

Oil droplet

Every phospholipid consists of a glycerol molecule linked to two fatty acids and a charged phosphate group.

Phosphate group

Glycerol

Fatty acid chains

Phospholipid components

The hydrophilic phosphate group tends to remain in the water...

... and the hydrophobic fatty acid tails tend to remain in the oil.

Space-filling model

Water

Polar head

Non-polar tail

Phospholipid symbol

Oil

(b)

The formation of a double-layered sheet allows the hydrophobic fatty acid tails to remain together and away from water...

...while the hydrophilic phosphate heads are exposed to water.

Water

Heads

Tails

Water

Phospholipid bilayer

See how phospholipids form membranes.

4.5

Figure 4.16 Phospholipids Form Membranes

(a) Three different representations show a phospholipid molecule at the boundary between an oil droplet and water. At an oil-water boundary, phospholipids are oriented in a specific way based on the different chemical properties of their component parts. (b) A double-layered sheet is the preferred orientation of phospholipids surrounded by water. Phospholipid bilayers form all the membranes of living organisms.

of the molecule (usually called its "head") can interact with polar water molecules or with positively charged ions and thus is hydrophilic. On the other hand, the fatty acid chains (the "tail" of the phospholipid molecule) are hydrophobic. They are entirely nonpolar and therefore are repelled by water molecules.

The resulting dual character of phospholipids allows them to form double-layered sheets in water that expose their hydrophilic heads to the water while keeping their hydrophobic tails isolated in the middle of the sheet. This double-layered sandwich of molecules is called a **phospholipid bilayer**, and it is the basis of all cell membranes (Figure 4.16). Membranes are a clear demonstration of how the chemical properties of molecules in water define a physical structure that is essential for living organisms.

Steroids are lipids that play vital roles in cell membranes and in development

We often hear cholesterol, testosterone, and estrogen mentioned in the news; these compounds all belong to a group called **steroids**. They are hydrophobic, and are classified as lipids. All steroids have the same fundamental structure of four carbon rings, but they differ in the groups of atoms that are attached to those rings (Figure 4.17).

Steroid ring structure   Progesterone   Cholesterol

Figure 4.17 Steroids
All steroids share the same basic four-ring structure, but have different groups of atoms attached.

See how heat changes a protein by frying an egg. 4.6

Cholesterol gets a lot of attention because of its effects on human health. Cholesterol plays a major role in cell membranes, and it is the "starting" molecule for other steroids and vitamin D. Estrogen and testosterone are called sex hormones because they are signaling molecules involved, among other things, in the development of features of our bodies, and even behaviors, that we recognize as "male" and "female." Testosterone is called an anabolic steroid (*anabolic*, "putting together") because one of its effects is the promotion of muscle and bone growth. Anabolic steroids have become the center of intense controversy and scandal in sports and have been widely banned. These steroids artificially enhance performance and destroy the concept of "fair play." Research also indicates that use of these compounds can have devastating, even fatal, consequences to the health and well-being of those who use them.

Proteins That Can Take the Heat

Understanding the chemistry of compounds from outer space and those found in living organisms on Earth today may allow us to predict the likelihood of life existing on other planets. In particular, exploring how the chemical characteristics of some proteins allow them to function under extreme conditions of heat and pressure will expand our understanding of what planetary environments can support life. On the other hand, this kind of information also has the potential to let us make improvements in the way some biological processes work here on Earth. Some of the reactions we are talking about are as common in the kitchen as they are in the body or the laboratory. For example, the texture of a hard-boiled egg is different from that of a raw egg because chemical changes during cooking cause the egg to change from a thick, sticky liquid, rich in the nutrients needed by a developing chick embryo, to the firm white and yellow mass that tastes so good to humans. What compounds are responsible for these changes? The answer is proteins, most of which simply cannot withstand heat.

Proteins are very particular about the temperatures at which they will remain properly folded and function normally. Most properly folded proteins tend to have hydrophilic R groups exposed on their surfaces and hydrophobic R groups isolated on the inside. This careful arrangement allows the protein to form hydrogen bonds with surrounding water molecules and remain dissolved.

When proteins are heated beyond a certain temperature, their weak noncovalent bonds break, and the proteins unfold, losing their regular three-dimensional structure. However, the peptide bonds that link one amino acid to another remain intact. In the ensuing chaos, hydrophobic R groups from several protein molecules randomly cluster together away from water, forming a featureless blob. In the case of an egg in boiling water, proteins such as albumin that were previously dissolved in the watery environment of the egg at room temperature are unfolded by the heat and randomly cluster together into the firm white of the boiled egg.

Until the 1970s it was believed that proteins could not be heated significantly above body temperature and be expected to function normally. Then came the discovery of bacteria that live and thrive in hot springs at temperatures just at the boiling point of water (100°C). Biologists realized that these bacteria must have proteins that remain functional at temperatures as high as 100°C. To improve our understanding of how proteins might be able to withstand such high temperatures, they set out to isolate and study the proteins from these heat-loving organisms. They discovered several proteins that promote chemical reactions in these bacteria and which are fully active at 100°C. One of these heat-stable proteins is even the basis of the forensic techniques that have figured so prominently in recent criminal trials in which matching of DNA samples is important evidence.

Biologists are now learning how these proteins are able to take the heat, and they are trying to alter proteins from other organisms that live at room temperature to be just as active at 100°C. This endeavor is not meant merely to satisfy scientific curiosity. Learning how to engineer a heat-resistant protein would mean discovering what kinds of

chemical bonds form a folded structure that is exceptionally stable. Thus proteins that are engineered to survive higher temperatures would also be far more stable at room temperature and could last longer on a supermarket shelf.

How do biologists make a protein that is more heat-resistant, or, in other words, stays folded? The answer lies in understanding how amino acids form the bonds that fold proteins into their three-dimensional shapes. One recent test case that proves the possibility of such an approach involves a protein called thermolysin. Found in bacteria that live at room temperature, thermolysin helps break down other proteins. Biologists have changed 8 of the 319 amino acids in thermolysin to produce a reengineered thermolysin that is fully active at 100°C. In making these amino acid changes, the researchers caused the formation of a new covalent bond that locked the three-dimensional structure of the protein in place, making it heat-resistant. When both the normal protein and the reengineered thermolysin were heated to 100°C, the reengineered thermolysin was 340 times more stable than the normal protein.

Although we may not want to engineer an egg that will not hard-boil, the ability to engineer proteins that remain active at high temperatures opens up exciting new possibilities. Not only could protein-rich food products be engineered to last far longer without chemical preservatives, but medically relevant proteins such as vaccines could be modified so that they no longer required cold-storage conditions, an important benefit for developing countries.

Chapter Review

Summary

4.1 Atoms Make Up the Physical World

- The physical world is made up of atoms. There are 92 different kinds of atoms, one for each natural chemical element.

- Individual atoms are made up of positively charged protons and neutrally charged neutrons in the nucleus and negatively charged electrons surrounding the nucleus. The particular combination of subatomic particles that makes up an atom determines its chemical properties.

- When atoms lose or gain electrons, they become either positively or negatively charged ions.
- The atomic number of an element is the number of protons in its nucleus, and its atomic mass number is the sum of protons and neutrons.
- Isotopes of an element have different numbers of neutrons, but the same number of protons. Radioisotopes are isotopes that give off radiation.
- Chemical compounds are formed by specific arrangements of atoms of different elements. These arrangements depend on the unique chemical characteristics of the atoms involved. The smallest unit of a chemical compound is called a molecule.

4.2 Covalent Bonds: The Strongest Linkages in Nature

- Chemical linkages that hold atoms together are called bonds.
- The nucleus of an atom is surrounded by a specific number of electrons that move in defined layers, or shells. Atoms share electrons with other atoms to fill their outermost electron shells to capacity. The bonding properties of an atom are determined by the number of electrons in its outermost shell.
- Covalent bonds, formed by the sharing of electrons between atoms, connect the atoms within a molecule. They are the strongest bonds in nature.

4.3 Noncovalent Bonds: Dynamic Linkages Between Molecules

- Noncovalent bonds link separate molecules as well as different parts of a single large molecule. They are weaker than covalent bonds and do not involve the sharing of electrons between atoms. Important types of noncovalent bonds include hydrogen bonds and ionic bonds.
- Hydrogen bonds form between polar compounds, which have an unequal distribution of charge. Partially positive hydrogen atoms in one compound form hydrogen bonds with partially negative atoms in another compound. Hydrogen bonding between water molecules, which accounts for the physical properties of water, is essential for life.
- Polar compounds, which interact freely with water and with one another, are hydrophilic; nonpolar compounds, which are repelled by water, are hydrophobic. Nonpolar compounds tend to group together in water. Polar compounds said to be soluble in water, while nonpolar compounds tend to be insoluble.
- Ionic bonds form between two atoms when electrons from one atom are transferred to another atom.

4.4 Chemical Reactions Rearrange Atoms Within Molecules

- Chemical reactions break and form bonds between atoms. They neither create nor destroy atoms, but merely alter the arrangement of atoms in molecules.
- All chemical reactions that support life occur in water. Polar compounds called acids and bases affect the amount of free hydrogen ions in water. Acids donate hydrogen ions to water molecules, form- ing hydronium ions (H_3O^+); bases accept hydrogen ions from water molecules, leaving hydroxide ions (OH^-).
- The concentration of free hydrogen ions in water is expressed by the pH scale, which ranges from 0 (most acidic, with the highest concentration of hydrogen ions) to 14 (most basic, with the lowest concentration). Most living systems function best near a neutral pH (about 7).
- Buffers are compounds that can both donate hydrogen ions to and accept them from water molecules. They help maintain the constant internal pH that is necessary for the chemical reactions of life.

4.5 The Chemical Building Blocks of Living Systems

- The ability of carbon atoms to form large and complex polymers has an important role in the generation of diverse organic building blocks. The four major classes of these building blocks contain carbon and hydrogen, and often atoms of other elements as well.
- Carbohydrates include simple sugars (monosaccharides) as well as more complex polymers (disaccharides and polysaccharides). Carbohydrates provide energy and physical support for living organisms.
- Each nucleotide consists of a five-carbon sugar, one of five nitrogen bases, and a phosphate group. Nucleotides are the building blocks of the nucleic acids DNA and RNA. DNA polymers made up of four types of nucleotides form the blueprint for life and dictate all the physical features and chemical reactions of a living organism. ATP functions in the storage of energy and its transfer from one chemical reaction to another.
- Amino acids are the building blocks of proteins. The chemical properties of the 20 different amino acids are determined by their different R groups. A series of amino acids are linked together by peptide bonds to give a protein its primary structure. The function and three-dimensional shape of a protein (its secondary and tertiary structure) are defined by the chemical properties of the amino acids it contains. Some proteins must be closely associated with other proteins (quaternary structure) in order to function.
- Fatty acids are the building blocks of fats and lipids. Fats are an important means of long-term energy storage; they can be saturated or unsaturated, depending on the nature of their covalent bonds. Phospholipids are the basic components of biological membranes. Steroids are lipids with specialized functions.

◉ Review and Application of Key Concepts

1. All the major chemical building blocks found in living organisms form polymers. Why are polymers especially useful in the organization of living systems?

2. A sample of pure water contains no acids or bases. Predict the pH of the water and explain your reasoning.

3. Describe how the chemical properties of water contribute to a type of noncovalent bond that is critical for life.

4. Describe the chemical properties of carbon atoms that make them especially suitable for forming biological polymers.

5. Polymers of amino acids have chemical characteristics that are important for life. How are these characteristics useful to a living organism?

6. Describe a single function for each of the following compounds that is relevant to biological processes: carbohydrates, nucleic acids, amino acids, fats.

Key Terms

Self-Quiz

1. The atoms of a single element
   a. have the same number of electrons.
   b. can form linkages only with atoms of the same element.
   c. can have different numbers of electrons.
   d. can never be part of a chemical compound.

2. Two atoms can form a covalent bond
   a. by sharing protons.
   b. by swapping nuclei.
   c. by sharing electrons.
   d. by sticking together on the basis of opposite electrical charges.

3. Which of the following statements about molecules is true?
   a. A single molecule contains atoms from only one element.
   b. The chemical bonds that link atoms into a molecule are arranged randomly.
   c. Molecules are found only in living organisms.
   d. Molecules can contain as few as two atoms.

4. Which of the following statements about ionic bonds is *not* true?
   a. They cannot exist without water molecules.
   b. They are not the same as hydrogen bonds.
   c. They require the loss of electrons.
   d. They are more temporary than covalent bonds.

5. Hydrogen bonds are especially important for living organisms because
   a. they occur only inside of organisms.
   b. they are very strong and maintain the physical stability of molecules.
   c. they allow biological molecules to dissolve in water, which is the universal medium for living processes.
   d. once formed, they never break.

6. Glucose is an important example of a
   a. protein.   c. fatty acid.
   b. carbohydrate.   d. nucleotide.

7. Peptide bonds in proteins
   a. connect amino acids to sugar monomers.
   b. bind phosphate groups to adenine.
   c. connect amino acids together.
   d. connect nitrogen bases to ribose monomers.

8. Alpha helixes are examples of
   a. primary protein structure.
   b. secondary protein structure.
   c. tertiary protein structure.
   d. quaternary protein structure.

9. Steroids are classified as
   a. sugars.   c. nucleotides.
   b. amino acids.   d. lipids.

Biology in the News

Canada Looks Likely to Ban Trans Fatty Acids in Foods

COMMENTARY BY STAFF WRITERS

- The government and opposition parties have declared war on trans fats, after the NDP [the New Democratic Party of Canada] introduced a motion that would make Canada the world's second country to ban heart-clogging compounds. . . .

- According to a spokesperson with the Food and Consumer Products Manufacturers of Canada, the introduction of trans fat-free snacks and take-out foods are proof the industry is heeding consumer concerns.

Trans fats are in the news as potentially dangerous dietary lipids. They are found throughout our modern diets. Trans fats are found in the oils used to cook French fries and other fast foods. They are used in commercial baked goods to protect against spoilage. Some vegetable oils and margarine contain trans fats. These fats are thought to play a role in increasing "bad" cholesterol and decreasing "good" cholesterol in the body. Studies have implicated some trans fats in contributing to coronary heart disease. As a result, the U.S. Food and Drug Administration has proposed including trans fatty acid content on food labels. As the article states, Canada may go even further.

Information about food and potential health hazards related to food is increasingly finding its way into the media. As consumers become better informed, they are becoming more selective about what they buy. Fast-food restaurants are adjusting the selection of food items they offer to match emerging public awareness of potential problems. Large chain supermarkets as well as smaller independent grocery stores are stocking their shelves with more "healthy" foods.

The use of steroid hormones in raising beef and pork is also in the news, and a matter of concern to many people. These hormones are similar to the hormones whose use in professional sports has been sparking controversy lately. Producers of beef, pork, and poultry also use antibiotics to promote the growth of their livestock and to cut costs. Resistant strains of bacteria that infect humans are becoming more widespread, and many scientists are concerned that this is due at least in part to the extensive use of these antibiotics in raising livestock.

The public has also become concerned about the use of artificial fertilizers and pesticides on fruits and vegetables. Organic farming allows none of these, and sales of organic fruits, vegetables, poultry, and meats are booming. Learning more about the potential health hazards related to food doesn't seem like a bad idea.

Evaluating the news

1. Fast-food and chain restaurants are part of our culture. In their advertising they make a point of responding to popular dietary concerns and trends. How would you like them to be more specific about providing nutritional information?

2. Supermarkets sell a broad variety of foods. Packaged foods have labels describing the contents in terms of basic nutrients. Fruits and vegetables in the produce section don't have similar labels. Do you think they should? Why or why not?

3. In addition to trans fats, many other compounds found in foods are spawning controversy. Organic foods are growing rapidly in popularity, even though they are more expensive. What information would you look for in the news or on the Internet that would help you decide whether the extra cost is worth it?

SOURCE: News Target Network, December 26, 2004.

CHAPTER 5

Cell Structure and Compartments

Key Concepts

- All living organisms are made up of one or more basic units called cells.

- The plasma membrane forms a boundary around every cell. It limits the movement of molecules into and out of the cell and determines how the cell communicates with the external environment.

- Prokaryotes are single-celled organisms that lack a nucleus and other internal compartments. Eukaryotes are single-celled or multicellular organisms whose cells have a nucleus and several other specialized internal compartments.

- The specialized internal compartments of the eukaryotic cell are called organelles. They concentrate and transport the molecules necessary for such processes as the production of energy and the breakdown of food material.

- The cytoskeleton consists of several distinct filament systems and their associated proteins. It plays an important role in cell shape and movement.

- Organelles such as mitochondria and chloroplasts are probably descendants of primitive prokaryotes that were engulfed by the ancestors of eukaryotic cells.

Traveling with Hijackers

Most of us don't realize it, but we are inhabited by a multitude of microscopic organisms that survive and reproduce thanks to the rich resources of our bodies. The majority of these microbes are harmless, and some, such as the bacteria in our digestive tracts, are actually required for our health. However, some microbes are just the opposite: their survival and reproduction occur at our expense to such a degree that they cause disease. In some cases the by-products of their biological processes can have serious toxic effects on the human body.

To combat undesirable microbes, we need to understand how their biological processes depend on and use our bodies' resources. Interestingly, many microbes don't just feed on the plentiful nutrients found in the human body; they have also evolved clever ways of hijacking specific processes that normally occur in our cells.

One such bacterium, *Listeria monocytogenes* [liss-*TEER*-ee-uh *MAW*-noh-sigh-*TOJ*-uh-neez], can cause severe inflammation of the stomach and intestines and damage to the nervous system. Nearly 2,000 people in the United States become seriously ill from *Listeria* each year, and more than 400 die from the infection. The bacterium is found in contaminated foods, and recent outbreaks have been traced to sources as seemingly harmless as chocolate milk.

Listeria moves from place to place inside an invaded body by hijacking several of that body's own proteins. Through a microscope, *Listeria* can be seen traveling from cell to cell in tissue samples taken from infected individuals. Under the microscope, these moving *Listeria* look very much like comets with rod-shaped bodies and trailing tails. Biologists now know that the "comet tail" of *Listeria* is made up of proteins that have been captured

Invaders in Inner Space
The bacteria *Listeria monocytogenes* can be seen inside an infected cell.

from the infected cell and made to form fibers that extend from the body of the bacterium. The growth of these fibers provides the propulsive force for the bacterium, literally pushing it through the cell. This force is so significant that when *Listeria* hits the inside of the cell's outer membrane, it pushes the membrane out to form a spike. The spike bursts, allowing the bacterium to leap to another cell and spread the infection.

Listeria's evolution of the ability to hijack a host cell's own proteins only hints at the many fascinating secrets harbored by the internal components of cells. In this chapter we explore some of these components and their normal functions in the life of a cell.

Every complex structure imaginable can be broken down into smaller and simpler parts. Even something as complicated as an airplane can be reduced to simpler components, such as engines, wiring, hydraulic systems, and a metal framework. Based on this assumption, we can begin to understand a complex system by first identifying and examining its elementary components.

The basic principle of building complex structures out of simple components applies to living systems as much as it applies to airplanes. In Chapter 4 we discussed the atoms and other chemical components that make up the building blocks of life. These components must be further arranged into living units before an organism such as a human being can exist. In other words, macromolecules such as proteins and DNA must be organized into more complex arrangements that promote the chemical reactions required for life.

This chapter explores that next level of organization in living systems, which defines the simplest unit of life: the cell. After a brief description of cells in general, we examine the structures and compartments that allow cells to support biological processes, beginning at the cell's physical boundary and working inward. We close with an exploration of how the complexity of the eukaryotic cell, with its internal compartments, may have evolved from prokaryotic cells.

such as a human being contains trillions of cells. Cells compose every organ in our bodies, and they determine how we look, move, and function as organisms.

Given the wide variety of different organs and specialized parts in each human body, it is not surprising that more than 200 different types of cells are required for those different parts. Most cell types in our bodies are specialists at something. For example, let's compare the cells that make up our muscles with those that form the clear lens of the eye. Muscle cells have the important task of generating the movements that we experience as muscle contractions and relaxations. They have specific protein components that allow them to change shape and generate physical force. But the cells in the lens of the eye do not need to generate physical force. Their task is to help focus light into the eye, allowing for vision. They therefore contain specific protein components that help focus light as it passes through the cells, in much the same way that a glass lens focuses light into a camera.

With millions of different species on Earth, the diversity of cell types found throughout the biosphere is enormous. None of the cell types found in our bodies, for example, is found in any plant. Yet even with so many different kinds of cells, certain basic components and structures are shared by all cells. Our ability to see these structures under the microscope led to the discovery of cells (see the box on page 99).

5.1 Cells: The Simplest Units of Life

The **cell** is the basic unit of life. In the same way that molecules such as proteins and fatty acids are made up of atoms, every living organism consists of one to many billions of membrane-enclosed units called cells. Bacteria are single cells, while a complex multicellular organism

5.2 The Plasma Membrane: Separating Cells from Their Environment

One of the key characteristics of living organisms is the existence of a boundary that separates the organism from its surrounding environment. If all the molecules within an organism were allowed to move freely in the environment,

Science Toolkit

Exploring Cells under the Microscope

To see something is to begin to know it. This simple statement is as true of biology as it is of fine art. Our awareness of cells as the basic units of life is based largely on our ability to see them. The instrument that opened the eyes of the scientific world to the existence of cells—the light microscope—was invented in the last quarter of the sixteenth century. The key components of early light microscopes were ground-glass lenses that bent incoming rays of light to produce magnified images of tiny specimens.

The study of magnified images began in the seventeenth century when Robert Hooke examined a piece of cork under a microscope and noticed that it was made up of little compartments. Hooke described these structures as small rooms, or cells, originating the term we use today. Ironically, the compartments Hooke saw under the early microscope were not living cells, since cork is dead plant tissue, but rather empty cell walls. However, the discovery of previously invisible living things proceeded rapidly, opening up a new world to scientific exploration.

While the light microscope has a place in the early history of biology, similar instruments are just as important in ongoing research today. The quality of lenses, however, has improved significantly since that time: the 200- to 300-fold magnification achieved in the seventeenth century has been improved to well over a thousand-fold achieved by today's standard light microscopes. This degree of magnification allows us to distinguish structures as small as $1/2{,}000{,}000$ of a meter, or 0.5 micrometer (μm). Light microscopes can therefore reveal not just animal and plant cells (5–100 μm), but also organelles such as mitochondria and chloroplasts (2–10 μm) as well as bacteria (1 μm).

Since the 1930s, an even more dramatic increase in magnification has been achieved by the replacement of visible light with streams of electrons that are focused by powerful magnets instead of glass lenses. Called electron microscopes, these instruments can magnify a specimen more than a hundred thousand times, revealing the internal structure of cells and even individual molecules such as proteins and nucleic acids. Both types of microscopy—electron and light—give us insights into how cells are organized and how each type of cell is physically adapted to a specific function in the body of a multicellular organism.

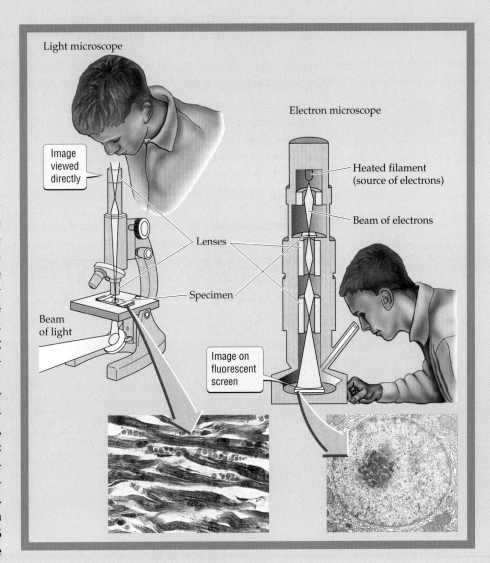

Light microscope

Electron microscope

Image viewed directly

Heated filament (source of electrons)

Beam of electrons

Lenses

Specimen

Beam of light

Image on fluorescent screen

they would not encounter one another frequently enough for life-sustaining chemical reactions to take place. So in nature we observe an enclosed compartment: the cell. Cells surround and concentrate the required compounds in a limited space and allow the chemical reactions required for life to occur.

The outer boundary that defines all cells is called the **plasma membrane**. As we saw in Chapter 4, biological membranes are composed mainly of a phospholipid bilayer. The phospholipids in each layer are oriented such that their hydrophilic heads are exposed to the watery environments inside and outside the cell and their hydrophobic fatty acid tails are grouped together in the interior of the membrane.

If the plasma membrane had no other function but to define the boundary of the cell and keep all its contents inside, a simple phospholipid bilayer would suffice. However, the plasma membrane must also allow the cell to communicate with the outside environment, capture essential molecules but keep out unwanted ones, and release waste products but prevent needed molecules from leaving the cell. In other words, the membrane must be selectively permeable.

The selective permeability of plasma membranes depends on various proteins that are embedded in the phospholipid bilayer. As Figure 5.1 shows, some of these proteins extend all the way through the phospholipid bilayer, forming gateways that allow the passage of selected ions and molecules into and out of the cell. Other proteins are used by the cell to recognize changes in the environment outside the cell, including signals from other cells.

Unless they are anchored to structures inside the cell, most proteins that are embedded in the plasma membrane are free to move about sideways, within the plane of the phospholipid bilayer. This freedom of movement supports what is known as the **fluid mosaic model**, which describes the plasma membrane as a highly mobile mixture of phospholipids and proteins. This mobility is essential for many cellular functions, including movement of the cell as a whole and the ability to detect external signals. Although the plasma membrane is a common feature of all cells, the set of proteins found in the membrane varies from one cell type to another. The specific combination of proteins determines how cells interact with the external environment and contributes to the unique properties of each cell type.

Test your knowledge of proteins in the plasma membrane. 5.1

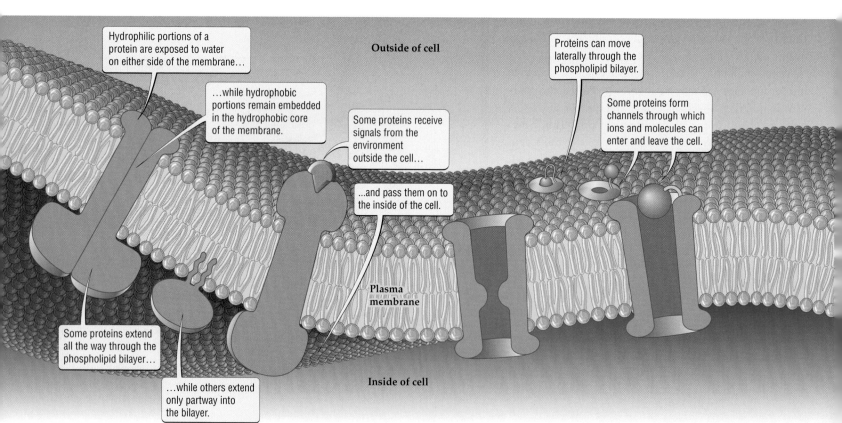

Figure 5.1 Proteins Are Embedded in the Plasma Membrane
Proteins can span the phospholipid bilayer of the plasma membrane in a variety of ways.

5.3 Comparing Prokaryotes and Eukaryotes

As we learned in Chapter 3, all living organisms can be classified into two groups based on whether their cells contain internal compartments. Organisms with cells that lack compartments are known as **prokaryotes**. Those that have internal membrane-enclosed compartments are known as **eukaryotes**.

Prokaryotic cells, which were probably the first cells to evolve, have little internal organization. Today, all prokaryotes are single-celled bacteria or archaeans. Most prokaryotes are spherical or rod-shaped and have a tough cell wall that forms outside the plasma membrane (Figure 5.2). The cell wall, which is made up of polysaccharides and proteins, helps maintain the shape and structural integrity of the organism.

The inside of a typical prokaryotic cell consists of a watery jellylike substance known as the **cytosol**. The cytosol contains a multitude of molecules, including DNA, RNA, proteins, and enzymes, all suspended in water. These components, together with a host of free ions, support the chemical reactions necessary for life. There are so many small and large molecules crowded into the cytosol that it behaves more like a thick jelly than a free-flowing liquid. In fact, many of the components found in the cytosol are probably similar to those found in the rich soup that first supported life billions of years ago.

Prokaryotic cells, on the whole, are smaller than eukaryotic cells. For example, the well-studied bacterium *Escherichia coli*, a common resident of the human intestine, is only 2 millionths of a meter (2 µm) long. About 125 *E. coli* would fit end to end across the period at the end of this sentence. The small sizes of bacteria and archaeans may account, in part, for their ability to get along without further internal organization. Their chemical components are contained in such a small volume of cytosol that further concentration into discrete regions within the cell is not needed.

Eukaryotes can exist in many forms, from single cells (such as yeasts) to large multicellular organisms (such as humans). All multicellular organisms are communities of eukaryotic cells that are specialized for different functions. A major distinction between prokaryotic and eukaryotic cells is that eukaryotic cells have an internal membrane-enclosed compartment called the **nucleus**, which prokaryotes lack. The nucleus houses most of the cell's DNA, effectively separating it from the remainder of the cell's components. The nucleus is therefore a cellular compartment with a specific function.

All eukaryotic cells are further organized by the presence of several other membrane-enclosed compartments (see Figure 5.2). Like the nucleus, each of these compartments is enclosed by a membrane and has a specific function. This internal compartmentalization is important because most eukaryotic cells are about a thousand times larger than bacteria by volume. Without this internal compartmentalization, they could not rely on their chemical components being close enough together for the necessary chemical reactions to occur.

5.4 The Specialized Internal Compartments of Eukaryotic Cells

Consider a large manufacturing plant with many rooms and different departments. Each department must have a specific function and an internal organization that contributes to the overall "life" of the manufacturing process. Specific departments represent an effective means of accomplishing particular tasks. For example, if all the members of the assembly line were scattered throughout the building, it would be difficult for the plant to assemble goods without wasting time or materials. Therefore, assembly workers and packers are all located in a centralized assembly department, effectively concentrating and coordinating their efforts.

Similarly, in the eukaryotic cell, specific processes can be carried out most efficiently by specialized "departments." The results of this enhanced efficiency include faster manufacturing of products, which for the cell could be energy-rich compounds such as ATP and structural components such as proteins. The specialized departments of the cell are the various membrane-enclosed compartments that divide its contents into smaller spaces. These smaller spaces contain, isolate, and concentrate the molecules necessary for different "manufacturing" processes. For example, ATP production and protein synthesis occur in two different kinds of compartments.

The cell's membrane-enclosed compartments are called **organelles**, a name that is especially appropriate because they are the "little organs" of the cell. Just as heart, lungs, and other organs have unique functions in the human body, each organelle has specific duties in the life of the cell.

In contrast to prokaryotes, in which all the contents of the cell inside the plasma membrane form the

Learn more about eukaryotic and prokaryotic cell structure: animal, plant, and bacterium.

5.2
5.3
5.4

EUKARYOTIC CELLS

Animal cell

Lysosome

Cytoskeleton

Golgi apparatus

Mitochondrion

Plant cell

Vesicle   DNA

Nucleus

Ribosomes

Rough endoplasmic reticulum (Rough ER)

Smooth endoplasmic reticulum (Smooth ER)

Chloroplast

Cytosol

Plasma membrane

Vacuole

Cell wall

PROKARYOTIC CELL

Bacterium

DNA

Ribosomes

Cytosol

Plasma membrane

Cell wall

Figure 5.2 Prokaryotic and Eukaryotic Cells Compared

The prokaryotic cell shown is a typical bacterium; the eukaryotic cells are a typical animal cell and plant cell. The bacterium is easily distinguished from the eukaryotic cells by its much smaller size and the absence of a nucleus and organelles. The plant cell is distinguished from the animal cell by the presence of chloroplasts, a large water-filled vacuole, and a rigid cell wall. Both types of eukaryotic cells contain nuclei and other organelles.

cytosol, the contents of a eukaryotic cell are divided between the cytosol and the organelles. In other words, the eukaryotic cytosol consists of all the cell contents inside the plasma membrane except the organelles. A related term, **cytoplasm**, describes a different set of contents of the eukaryotic cell: the cytosol plus all the organelles excluding only the nucleus. The internal space enclosed by the membrane of an organelle is called the **lumen**.

The nucleus is the repository for genetic material

The nucleus is the most distinctive organelle in eukaryotic cells because it represents a fundamental difference between eukaryotes and prokaryotes. Eukaryotic cells have a clearly delineated membrane-bound envelope—a nucleus—that contains most of their genetic material. Prokaryotic cells have their genetic material concentrated in one area, called the nucleoid region, but this region is not enclosed by a membrane. Returning to the comparison with a manufacturing plant, the nucleus of the cell is equivalent to the administrative offices of the plant. In other words, the nucleus is the specialized compartment that directs the activities of the cell in response to information it receives from other parts of the manufacturing plant and from communication with sources outside the plant. It fulfills this function by housing the cell's DNA, which contains the information necessary for all the structures and functions of the cell (Figure 5.3).

Inside the nucleus, long polymers of DNA are packaged with proteins into a remarkably small space. Since eukaryotic cells can have more than a thousand times the amount of DNA in prokaryotes, keeping the DNA tightly packed with proteins is necessary in order for it to fit inside the nucleus.

The arrangement of the membrane that surrounds the nucleus is different from that of the plasma membrane that surrounds the whole cell. The boundary of the nucleus, called the **nuclear envelope**, is a double membrane that contains small openings called **nuclear pores** (see Figure 5.3). These pores are gateways that allow molecules to move into and out of the nucleus, enabling it to communicate with the rest of the cell. The transfer of information encoded by the DNA depends on the movement of certain molecules out of the nucleus, while how and when this DNA information is transferred depend on the movement of specialized proteins into the nucleus.

The nucleus contains the cell's DNA...

...and is surrounded by a nuclear envelope consisting of two membranes.

Nucleus Nuclear envelope Nuclear pores

The nuclear envelope has openings called nuclear pores.

Nuclear pores Nuclear envelope

Figure 5.3 The Nucleus

The endoplasmic reticulum is the site of manufacture for lipids and many proteins

If the nucleus functions as the administrative offices of the cell, the endoplasmic reticulum is the factory floor where many of the cell's chemical building blocks are manufactured. The **endoplasmic reticulum** [. . . reh-*TICK*-you-lum] (**ER**) is surrounded by a single membrane that is connected to the outer membrane of the nuclear envelope. Unlike the nucleus, which is usually an irregular spherical structure, the ER is an extensive and complex network of interconnected tubes and flattened sacs stacked and connected to one another (Figure 5.4). The various lipids and proteins destined for other cellular compartments, for the plasma membrane, or for export from the cell are produced in these sacs. The lumen of the ER contains free-floating proteins. Proteins and lipids that are destined for other membranes of the cell are first inserted into the membrane of the ER.

When viewed under a microscope, ER membranes have two different appearances: rough and smooth (see Figure 5.4). In most cells, the majority of ER membranes

Smooth ER does not have ribosomes associated with it.

Rough endoplasmic reticulum

Ribosomes

Rough ER has ribosomes associated with it and is the major site of protein production.

Figure 5.4 The Endoplasmic Reticulum

The ER forms a series of flattened membrane sacs that are major sites for the synthesis of proteins and lipids.

Explore vesicle budding and fusing. 5.5

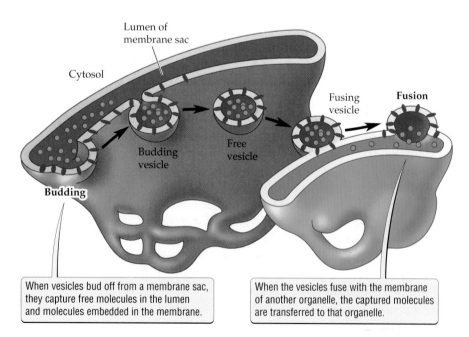

Lumen of membrane sac

Cytosol

Fusing vesicle

Fusion

Budding vesicle

Free vesicle

Budding

When vesicles bud off from a membrane sac, they capture free molecules in the lumen and molecules embedded in the membrane.

When the vesicles fuse with the membrane of another organelle, the captured molecules are transferred to that organelle.

Figure 5.5 How Vesicles Move Proteins and Lipids from One Compartment to Another

have small rounded particles associated with them that are exposed to the cytosol. Such ER is referred to as **rough ER**, and the particles are called **ribosomes**. Each ribosome can manufacture proteins from amino acid building blocks using instructions originating from the DNA in the nucleus. Ribosomes attached to the rough ER manufacture proteins that are destined for the ER lumen, for insertion into a membrane, or for export. But not all proteins are made by ribosomes in the rough ER; ribosomes that float freely in the cytosol manufacture proteins that remain in the cytosol.

In most cells, a small percentage of the ER membrane lacks ribosomes and is called **smooth ER** (see Figure 5.4). The smooth ER, which is connected to the rough ER, marks sites where portions of the ER membrane actively bud off to produce **vesicles**. Since each vesicle is formed from a patch of ER membrane and encloses a small portion of the ER lumen, it is an effective means of transporting proteins that either are embedded in the ER membrane or float free in the ER lumen (Figure 5.5). Vesicles are like the carts used to move goods between different departments of a factory, since they move proteins and lipids from the ER to other organelles.

The Golgi apparatus sends proteins and lipids to their final destinations in the cell

Another membranous organelle, the **Golgi apparatus**, directs proteins and lipids produced by the ER to their final destinations, either inside the cell or out to the external environment. The Golgi apparatus therefore functions as a sorting station, much like the shipping department in a manufacturing plant. In a shipping department, goods destined for different locations get address tags that indicate where they should be sent. Similarly, in the Golgi apparatus, the addition of specific chemical groups to proteins and lipids helps target them to other destinations in the cell. These cellular address tags include carbohydrate molecules and phosphate groups.

Under the electron microscope, the Golgi apparatus looks like a series of flattened membrane sacs stacked together and surrounded by many small vesicles (Figure 5.6). These vesicles transport proteins from the ER to the Golgi apparatus and between the various sacs of the Golgi. Vesicles are therefore the primary means by which proteins and lipids move from one sac to another in the Golgi apparatus and to their final destinations.

Lysosomes and vacuoles are specialized compartments for recycling, storage, and structural support

The proteins produced in the ER and sorted by the Golgi apparatus are destined either for the cell surface or for other organelles. In animal cells, distinct subsets of macromolecules, destined to be broken down, are addressed to organelles called lysosomes. **Lysosomes** contain enzymes used to break down macromolecules such as carbohydrates, proteins, and fats. They can adopt a variety of irregular shapes (Figure 5.7), but all are characterized by an acidic lumen with a pH of about 5. This acidic pH is the optimum environment for the lysosomal enzymes to do their work.

The macromolecules that are to be broken down are delivered to lysosomes in vesicles. The breakdown products, which include simple sugars, amino acids, and lipids, are then transported across the lysosomal membrane into the cytosol for use by the cell.

Plants and fungi have a different class of organelles, called **vacuoles**, that are related to lysosomes. A particular type of vacuole, commonly referred to as a **central vacuole**, is significantly larger than most lysosomes, usually occupying more than a third of a plant cell's total volume (Figure 5.8). Besides containing enzymes that break down substances, some central vacuoles store nutrients for later use by the plant cell. In seeds, central vacuoles in specialized cells store protein nutrients that are later broken down to provide amino acids for the

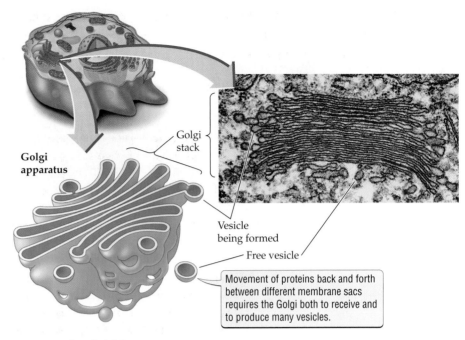

Movement of proteins back and forth between different membrane sacs requires the Golgi both to receive and to produce many vesicles.

Figure 5.6 The Golgi Apparatus
The Golgi apparatus consists of flattened membrane sacs, in which proteins are directed to their final destinations in the cell. Proteins move between the various membrane sacs of the ER and Golgi apparatus in vesicles.

growth of the plant embryo during germination. Seeds from the tombs of Egyptian kings have been successfully germinated thousands of years later, illustrating that central vacuoles are very good at preserving their contents. In many cases, large vacuoles filled with water also contribute to the overall rigidity of a plant by applying pressure against cell walls. Vacuoles therefore make many different contributions to the life of the plant cell, and a single cell can easily contain several vacuoles with different functions.

Mitochondria are the power stations of the eukaryotic cell

The previous discussion shows that various organelles are responsible for the production and storage of cellular components and their transport to the right places in the cell. The

Figure 5.7 Lysosomes in an Animal Cell
Lysosomes are vesicles full of enzymes that break down macromolecules. The cell shown here is from the stomach lining; it uses its lysosomes to break down food materials.

Vacuoles   Cell wall

Chloroplasts

Test your knowledge
of the parts of
mitochondria.

5.6

Figure 5.8 Vacuoles in a Plant Cell

Vacuoles are large organelles in plant cells that are used to store water, nutrients, or enzymes.

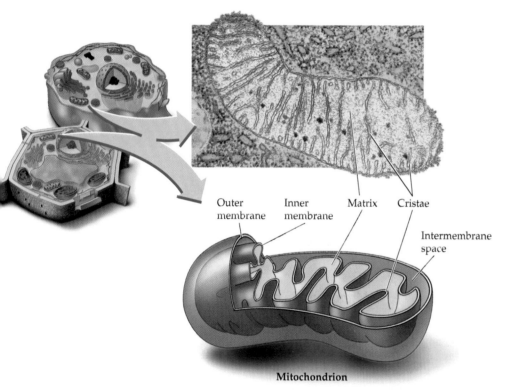

Outer   Inner   Matrix   Cristae
membrane   membrane

Intermembrane
space

Mitochondrion

Figure 5.9 Energy-Transforming Organelles: Mitochondria

Mitochondria are the major energy-converting organelles in all eukaryotic cells. Each mitochondrion has a double membrane, and the inner membrane is highly folded. The infoldings (cristae) of the inner membrane contain enzymes that participate in energy production. The inner lumen of the mitochondrion is called the matrix.

nucleus directs what proteins are to be produced, the ER is the site of production for most proteins, the Golgi apparatus modifies and sorts proteins for transport to their final destinations, and lysosomes and vacuoles break down and store cellular components, respectively. Thus, we have explored the administrative offices, assembly area, and shipping department of the cellular factory. However, none of these specialized departments could function without a source of energy to run the machines that produce the goods. The eukaryotic cell is no different: all the cellular processes discussed so far require a source of energy.

The producers of most of this energy are organelles called **mitochondria** (singular: mitochondrion), which are found in all eukaryotic cells—that is, in the cells of animals, plants, fungi, and protists. Mitochondria are like the factory's power plant: they use chemical reactions to transform the energy from many different molecules into ATP. The energy stored in the covalent bonds of ATP is used, in turn, to drive the many chemical reactions of the cell. Thus ATP is the universal cellular fuel.

Mitochondria are pod-shaped and are surrounded by a double membrane. In contrast to the nucleus, which also has a double membrane, the inner mitochondrial membrane has distinct folds, called **cristae** [*KRIS*-tee] (singular: crista). The space between the two membranes is called the intermembrane space, and the lumen interior to the cristae is called the matrix (Figure 5.9). The production of ATP by mitochondria depends both on the activities of proteins embedded in the cristae and on the fact that the mitochondrial matrix is completely separated from both the intermembrane space and the cytosol. With this unique setup, mitochondria are able to harness the energy released by the chemical breakdown of molecules produced from the partial breakdown of sugar to synthesize energy-rich ATP, as we will see in Chapter 7. In the process, oxygen is consumed and carbon dioxide and water are released.

Chloroplasts capture energy from sunlight

All eukaryotes have mitochondria that provide them with life-sustaining ATP, but the cells of plants and many protists have additional organelles called **chloroplasts** (Figure 5.10), which capture energy from sunlight and convert it to chemical energy. That chemical energy is stored as sugar molecules, which are assembled from carbon dioxide (CO_2) and water (H_2O) in a process called **photosynthesis**. The energy in these

Thylakoids

Granum

Thylakoid
space

Chloroplast

Thylakoid

Inner
membrane

Intermembrane
space

Outer
membrane

Test your knowledge
about chloroplasts.

5.7

sugars is used directly by plant cells and indirectly by
organisms that eat plants. At this very moment, as you
read this page, your brain and the muscles that move
your eyes are using chemical energy that was originally
produced in chloroplasts by photosynthesis.

In addition to chemical energy, photosynthesis produces a critical by-product: oxygen. The oxygen that we
breathe, and which many other organisms, including
plants, require for life, is produced during photosynthesis. Mitochondria depend on a continuous supply of that
oxygen to produce ATP in a process that is essentially
the reverse of photosynthesis.

Chloroplasts are enclosed by a double membrane,
within which lies a third, separate internal system of
membranes arranged like stacked pancakes (see Figure
5.10). The "stack" is called a **granum** [*GRAH*-num] (plural:
grana), and each "pancake" is called a **thylakoid** [*THIGH*-
luh-koid]. Thylakoids contain specialized pigments, such
as chlorophyll, that enable chloroplasts to capture energy
from sunlight. The green color of chlorophyll accounts

for the green coloration of most plants. Enzymes present in the space surrounding the thylakoids use the captured energy, water, and carbon dioxide to produce
carbohydrates.

5.5 The Cytoskeleton: Providing Shape and Movement

If a cell consisted only of a plasma membrane and
membrane-enclosed compartments, it would be a limp
bag of cytosol with organelles sloshing around inside.
Thanks to a system of protein filaments and tubules called
the cytoskeleton, such an unfortunate state of affairs is not
the case. As its name implies, the **cytoskeleton** is an
internal support system for the cell. In addition, it includes
cables that act as tracks along which vesicles move.

Helpful to know

Skeleton may seem an odd word to apply to cells, since cells don't have bones as we do. However, biologists give the word *skeleton* a larger meaning: any structure or framework that supports and/or protects a cell or an organism. They often add prefixes to describe the kind of skeleton they are referring to. Thus *cytoskeleton* refers to the structures within cells (*cyto*, "cell") that give them shape and structure. *Endoskeletons* (*endo*, "inner") are found in many animals, such as fish and mammals, and *exoskeletons* (*exo*, "outer") are found in insects and crustaceans.

Furthermore, the cytoskeleton is ever-changing and dynamic, allowing some cells to change their shape and move around on their own. Unlike the bony skeleton of an adult human, which has fixed connections between bones, the cytoskeleton of a cell has many noncovalent bonds between proteins that can break, re-form, and reshape the overall structure of the cell.

The cytoskeleton is composed of three basic types of protein filaments: microtubules, intermediate filaments, and microfilaments.

Microtubules support movement inside the cell

Microtubules are the thickest of the cytoskeleton filaments, with diameters of about 25 nanometers (nm). Each microtubule is a helical polymer of the protein monomer **tubulin** [*TOO*-byou-lin] (Figure 5.11a) and has two distinct ends. Microtubules can grow and shrink in length by adding or losing tubulin monomers at either end. This ability allows microtubules to form dynamic structures capable of rapidly changing a cell's shape and internal organization when necessary. The microtubules in most

(a) Microtubule

Monomers

Figure 5.11 Kinds of Filaments in the Cytoskeleton
(*a*) Microtubules are composed of tubulin monomers. (*b*) Microtubules form radial patterns in most cells, extending from the center of the cell to the plasma membrane. (*c*) Microtubules function as tracks along which vesicles are shuttled around the cell. (*d*) Intermediate filaments are multistranded, like a rope. (*e*) Microfilaments are composed of actin monomers.

Plasma membrane

(b)

Microtubules

Microtubules

Nucleus

Vesicles

Microtubules branch out from a central location near the nucleus to the plasma membrane.

These skin cells have been specially stained to make their microtubules fluorescent.

(c)

This high-magnification microscopic image shows vesicles moving along microtubules in a nerve cell.

(d) Intermediate filament

(e) Microfilament

eukaryotic cells radiate out from the center of the cell and end at the inner face of the plasma membrane (Figure 5.11b). This radial pattern of microtubules serves as an internal scaffold that helps position organelles such as the ER and the Golgi apparatus.

Microtubules also define the paths along which vesicles are guided in their travels from one organelle to another or from organelles to the cell surface. The ability of microtubules to act as "railroad tracks" for vesicles depends on **motor proteins** that attach to both vesicles and microtubules. These specialized proteins convert the energy of ATP into mechanical movement, which allows them to move along a microtubule in a specific direction, carrying an attached vesicle like cargo (Figure 5.11c).

Intermediate filaments strengthen the connections between cells

Intermediate filaments are a diverse class of ropelike filaments that are thinner than microtubules, about 8–12 nm in diameter (Figure 5.11d). Intermediate filaments serve as structural reinforcements. The layers of cells in your skin, for example, depend on intermediate filaments of a protein called keratin for mechanical strength and the ability to withstand physical stress. Intermediate filaments also hold some organelles in place. The nucleus, for example, is held in position by a surrounding network of intermediate filaments.

Microfilaments are involved in cell movement

Of the three filament types, **microfilaments** have the smallest diameter, 7 nm (Figure 5.11e), but they are the most important when it comes to cell movements. Each microfilament is a twisted polymer of monomers of the protein **actin**. Like microtubules, microfilaments can change their length frequently and rapidly.

Perhaps the best example of the rapid changes in microfilaments can be found in a cell moving across a flat surface. Under a microscope, certain skin cells, called fibroblasts, can be observed visibly crawling around. Their ability to move about in our bodies is an important part of wound healing, since fibroblasts migrate into the area of the wound to assist in closing up the damaged edges.

Let's examine the movement of fibroblasts in more detail. Fibroblasts move by extending flattened protrusions called **pseudopodia** (*pseudo*, "false"; *podia*, "feet") (singular: pseudopodium) while the opposite, trailing portion of the cell retracts (Figure 5.12). For these move-

ments, different arrangements of microfilaments are needed to alter the structure of the leading and trailing parts of the cell. In the protruding pseudopodia, the microfilaments tend to be aligned, pointing outward in a forward direction. As these filaments lengthen, they push against the plasma membrane, thereby extending the pseudopodia farther in the direction the cell is moving. At the same time, microfilaments in the trailing end of the cell, which are not well organized and angle in all directions, shorten. Here the breakdown of the actin network results in retraction of the plasma membrane at the rear end of the cell. The cell appears to be pulling up its rear end behind it.

Cilia and flagella act like oars and propellers

We have seen how some cells move by changing the organization of the cytoskeleton. Many protists and bacteria, and the sperm cells of some plants and all animals, are adept at propelling themselves rapidly through their environment using hairlike projections called **cilia** (singular: cilium) or whiplike flagella [fluh-*JELL*-uh] (singular: flagellum).

Figure 5.13a shows a protist covered with cilia, which it uses to move through its surroundings. Some cells use cilia not to move themselves around, but to move objects nearby. This is true of the cells that line parts of our respiratory system, which remain embedded in tissue and use their cilia to propel unwanted material out of our lungs

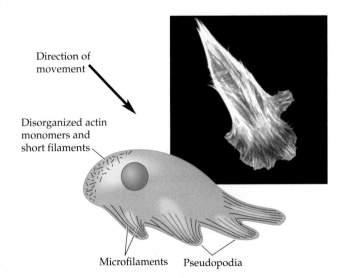

Figure 5.12 Microfilaments Allow Cell Movement
Microfilaments help cells crawl on flat surfaces by allowing them to extend flattened projections called pseudopodia.

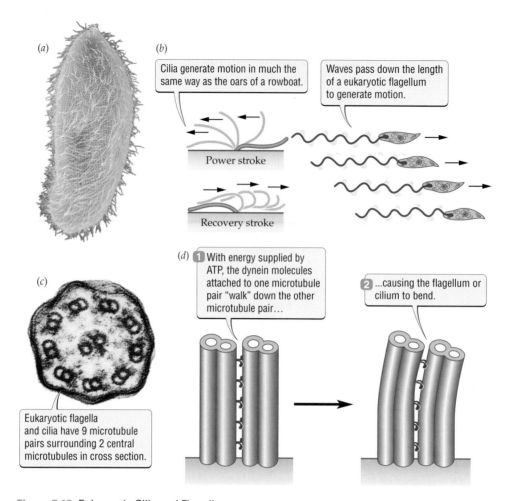

(a)

(b) Cilia generate motion in much the same way as the oars of a rowboat.

Waves pass down the length of a eukaryotic flagellum to generate motion.

Power stroke

Recovery stroke

(c) Eukaryotic flagella and cilia have 9 microtubule pairs surrounding 2 central microtubules in cross section.

(d) **1** With energy supplied by ATP, the dynein molecules attached to one microtubule pair "walk" down the other microtubule pair...

2 ...causing the flagellum or cilium to bend.

Figure 5.13 Eukaryotic Cilia and Flagella

Many eukaryotic organisms, especially single-celled ones, use cilia or flagella to generate movement. (a) This protist is covered with cilia. (b) Although cilia look like flagella, they generate movement by different kinds of bending motions. (c) The cilia and flagella of eukaryotes all share a characteristic arrangement of paired microtubules. (d) These microtubule pairs slide past each other as dynein proteins "walk" along them.

 Explore flagellar movement. 5.8

to our throats, where it can be eliminated. All cilia move much as the oars of a rowboat do. In contrast, **eukaryotic flagella** beat in a wavelike pattern (Figure 5.13*b*). Eukaryotic cilia and flagella have nine pairs of microtubules arranged around a central pair of microtubules (Figure 5.13*c*). The bending of cilia and flagella in eukaryotes depends on the movement of one pair of tiny tubules along another pair of tubules using a protein, called dynein [*DIE*-neen], that makes up arms projecting out from the tubules. These protein arms can "walk" up an adjacent microtubule pair using the energy derived from ATP (Figure 5.13*d*).

Although bacteria use a variety of interesting means of propelling themselves, the most widespread of these is **bacterial flagella**. A unique rotary "motor" rotates the stiff, corkscrew-like flagellum of the bacterium like a propeller (Figure 5.14*a*). This rotary motor is a truly remarkable machine that differs in both structure and action from anything found in eukaryotes. It rotates in response to hydrogen ions that diffuse into the bacterium from the environment (Figure 5.14*b*). To keep the flow going, the bacterium uses specialized proteins to pump hydrogen ions out of the cell. As these ions flow back into the cell from the external environment, they pass through the motor. It may take as many as 100 hydrogen ions diffusing back into the cell to rotate the flagellum just once, yet these bacterial propellers may rotate as rapidly as a thousand times a second. Some bacteria are very speedy and can move up to a hundred times their body length per second. That is equivalent to a 6-foot-tall athlete running the length of a football field in half a second!

The bacterial flagellum is made of the same actin protein that is found in muscle.

Bacterial flagella rotate like propellers...

...and lack the 9 + 2 arrangement of microtubules found in eukaryotic flagella.

Each time a hydrogen ion passes through the motor, the flagellum rotates another notch.

Figure 5.14 Bacterial Flagella
Bacteria have flagella that differ in form and action from eukaryotic flagella.

Explore the evolution of organelles. **5.9**

The Evolution of Organelles

The *Listeria* bacteria that we described at the beginning of this chapter disrupt a major component of a cell's cytoskeleton. *Listeria* is able to use the actin proteins of the host cell for its own purposes. In fact, biologists have found proteins on the surface of *Listeria* that capture actin monomers and start the process of polymerizing them to form filaments. Such a use of another cell's resources by an invader to further its own agenda might seem terribly unfair. However, we eukaryotes may not have the right to pass judgment on *Listeria*'s behavior. In the distant evolutionary past, the ancestors of eukaryotic cells probably did a similar thing by capturing and using their prokaryotic neighbors (Figure 5.15).

Mitochondria and chloroplasts in eukaryotic cells bear a striking physical resemblance to primitive prokaryotes. Both organelles have their own DNA and are able to make some proteins. They also reproduce independently of the cell by simply dividing in two. These striking characteristics imply that mitochondria and chloroplasts were once free-living prokaryotes that were engulfed by other cells, eventually forming a mutually beneficial relationship.

How did early single-celled eukaryotes benefit from capturing primitive prokaryotes? The answer may lie in the environment of Earth more than 3.5 billion years ago. At that time, when the first cells are thought to have arisen, there was virtually

Figure 5.15 How Primitive Eukaryotes May Have Acquired Organelles

Some organelles, such as mitochondria and chloroplasts, are likely descendants of engulfed prokaryotes. Other organelles, such as the endoplasmic reticulum, may have arisen from the gradual infolding of the plasma membrane.

no oxygen gas (O_2) in the atmosphere. Primitive prokaryotes were able to break down sugars and extract energy in the absence of oxygen; in fact, oxygen was toxic to these organisms.

However, as time passed, some prokaryotes evolved the ability to use the energy of sunlight to make energy-rich organic compounds from carbon dioxide. These photosynthetic reactions, which released oxygen as a waste product, slowly changed the atmosphere of Earth. As O_2 accumulated in the early atmosphere, some prokaryotes evolved the means to break down sugars and release energy using oxygen, which gave them a significant advantage over earlier prokaryotes. At this turning point in evolution, about 2 billion years ago, some primitive single-celled eukaryotes are thought to have captured some of these prokaryotes without destroying them. From these engulfed prokaryotes the host cells gained the ability to use atmospheric oxygen as part of an energy-extracting process.

The descendants of these captured prokaryotes are the mitochondria that we now find in every eukaryotic cell. All present-day eukaryotic cells depend on mitochondria for their ability to use oxygen to produce energy by breaking down sugars. Likewise, the ancestors of chloroplasts were probably primitive cyanobacteria that had evolved the ability to photosynthesize. Captured cyanobacteria, acting as primordial chloroplasts, would have conferred a great advantage on their host cells by providing them with their own food factories. Plant cells today depend on chloroplasts for photosynthesis.

So we are reminded that cells are made up of specialized compartments and chemical components that function together to support the processes of life. Since the division of labor seen in all eukaryotic cells may be based on the exploitation of the biological processes of other organisms in the evolutionary past, it is not surprising that the same principle can be used by current invaders such as *Listeria*. These phenomena confirm the old saying, "What goes around comes around."

Chapter Review

Summary

5.1 Cells: The Simplest Units of Life

- Cells are the basic units that make up all living organisms.
- Multicellular organisms are made up of different types of specialized cells.

5.2 The Plasma Membrane: Separating Cells from Their Environment

- Every cell is surrounded by a plasma membrane that separates the chemical reactions of life from the surrounding environment.
- According to the fluid mosaic model, the plasma membrane is a highly mobile assemblage of phospholipids and proteins that can move sideways within the membrane.
- Proteins in the plasma membrane allow the cell to communicate with and respond to the environment outside the cell, including signals from other cells.

5.3 Comparing Prokaryotes and Eukaryotes

- Living organisms can be classified as either prokaryotes or eukaryotes.
- Prokaryotes are single-celled organisms that lack internal compartments. Eukaryotes may be single-celled or multicellular, and their cells have internal membrane-enclosed compartments, such as the nucleus.
- The interior of a prokaryotic cell consists of the cytosol, made up of a multitude of molecules suspended in water.
- Eukaryotic cells can be a thousand times larger than prokaryotic cells in volume. They require internal compartments that concentrate and promote the chemical reactions of life.

5.4 The Specialized Internal Compartments of Eukaryotic Cells

- The specialized membrane-enclosed compartments inside a eukaryotic cell are known as organelles. Each type of organelle makes a specific contribution to the life of the cell. The cytosol plus all the organelles except the nucleus constitute the cytoplasm of the cell.
- The nucleus is the "administrator" organelle of the cell. It houses the DNA-encoded instructions that control every activity and structural feature of the cell. The nucleus is bounded by the nuclear envelope, a double membrane containing nuclear pores. Information in DNA is conveyed by molecules passing through nuclear pores.
- The endoplasmic reticulum (ER) is the site of manufacture for lipids and many proteins. The ER consists of rough ER and smooth ER, which serve different functions. Ribosomes in the rough ER manu-

facture proteins. The smooth ER buds off into vesicles, which carry proteins and lipids to other organelles.
- The Golgi apparatus receives proteins and lipids from the ER, sorts them, and directs them to their final destinations, either inside or outside the cell or in the plasma membrane.
- Lysosomes break down organic macromolecules such as proteins into simpler compounds that can be used by the cell. Vacuoles are similar to lysosomes but can also serve as storage structures and lend physical support to plant cells.
- Mitochondria produce energy for eukaryotic cells. Their unique structure includes a highly folded inner membrane (cristae) that isolates the matrix within it from both the intermembrane space and the cytosol.
- Chloroplasts harness the energy of sunlight and convert it to chemical energy in a process called photosynthesis. The specialized inner membrane system of chloroplasts contains a pigment called chlorophyll.

5.5 The Cytoskeleton: Providing Shape and Movement

- Eukaryotic cells depend on the cytoskeleton for structural support and the ability to move and change their shape.
- The cytoskeleton consists of three types of filaments: microtubules, intermediate filaments, and microfilaments.
- Microtubules (polymers of tubulin) and microfilaments (polymers of actin) can change their length frequently and rapidly. Microtubules are essential for the movement of organelles inside the cell, and microfilaments for the movement of the entire cell during locomotion using pseudopodia. Intermediate filaments strengthen connections between cells.
- Some protists, sperm cells, and bacteria move using cilia or flagella. Eukaryotic flagella are different in structure and action from bacterial flagella.

◉ Review and Application of Key Concepts

1. What fundamental features apply to both prokaryotic and eukaryotic cells, and why are these features advantageous for living organisms?

2. Describe the major components of a plasma membrane, and explain why we say that the membrane has a fluid dynamic nature.

3. All multicellular organisms are eukaryotic. Why do you think there are no multicellular prokaryotic organisms?

4. Describe the role of each major membrane-enclosed compartment in a eukaryotic cell.

5. Compare and describe the underlying mechanisms of movement of a fibroblast cell and of a bacterium with a flagellum.

6. What evidence is there to support the hypothesis that mitochondria and chloroplasts are descendants of primitive prokaryotes that were engulfed by ancestors of eukaryotic cells?

Key Terms

actin (p. 109)
bacterial flagellum (p. 110)
cell (p. 98)
central vacuole (p. 105)
chloroplast (p. 106)
cilium (p. 109)
crista (p. 106)
cytoplasm (p. 103)
cytoskeleton (p. 107)
cytosol (p. 101)
endoplasmic reticulum (ER)
   (p. 103)
eukaryote (p. 101)
eukaryotic flagellum (p. 110)
fluid mosaic model (p. 100)
Golgi apparatus (p. 104)
granum (p. 107)
intermediate filament (p. 109)
lumen (p. 103)
lysosome (p. 105)

microfilament (p. 109)
microtubule (p. 108)
mitochondrion (p. 106)
motor protein (p. 109)
nuclear envelope (p. 103)
nuclear pore (p. 103)
nucleus (p. 101)
organelle (p. 101)
photosynthesis (p. 106)
plasma membrane (p. 100)
prokaryote (p. 101)
pseudopodium (p. 109)
ribosome (p. 104)
rough ER (p. 104)
smooth ER (p. 104)
thylakoid (p. 107)
tubulin (p. 108)
vacuole (p. 105)
vesicle (p. 104)

Self-Quiz

1. Unlike prokaryotic cells, eukaryotic cells
   a. have no nucleus.
   b. have internal compartments.
   c. have ribosomes in their plasma membranes.
   d. lack a plasma membrane.

2. Which of the following would be found in a plasma membrane?
   a. proteins
   b. DNA
   c. mitochondria
   d. endoplasmic reticulum

3. Which of the following organelles are closely associated with ribosomes?
   a. the Golgi apparatus
   b. smooth endoplasmic reticulum
   c. rough endoplasmic reticulum
   d. mitochondria

4. Which organelle captures the energy from sunlight?
   a. mitochondria   c. the Golgi apparatus
   b. cell nuclei   d. chloroplasts

5. Which organelle is the "power plant" of eukaryotic cells?
   a. chloroplasts   c. the nucleus
   b. mitochondria   d. the plasma membrane

6. Which organelle contains both thylakoids and cristae?
   a. chloroplasts   c. nuclei
   b. mitochondria   d. none of the above

7. The internal system of protein fibers that contributes to a eukaryotic cell's structure and allows movement is called
   a. the endoplasmic reticulum.
   b. the cytoskeleton.
   c. the lysosomal system.
   d. the mitochondrial matrix.

8. Which of the following is *not* part of the cytoskeleton?
   a. pseudopodium   c. microtubule
   b. intermediate fiber   d. microfilament

9. Which of the following is powered by a hydrogen ion "motor"?
   a. pseudopodium   c. cilium
   b. bacterial flagellum   d. microtubule

10. Which of the following organelles are thought to have arisen from primitive prokaryotes?
   a. the endoplasmic reticulum and the nucleus
   b. the Golgi apparatus and lysosomes
   c. chloroplasts and mitochondria
   d. vacuoles and transport vesicles

High Doses of Vitamin E Supplements Do More Harm Than Good

Daily vitamin E doses of 400 international units (IU) or more can increase the risk of death and should be avoided, researchers reported at the American Heart Association's Scientific Sessions 2004.

In animal and observational studies, vitamin E supplementation was shown to prevent cardiovascular disease and cancer. However, other studies suggested that high doses could be harmful.

Chloroplasts make vitamin E, and green vegetables are especially rich in chloroplasts, making them good choices as a source of this vitamin. Yet supplementing our diets with vitamin E is a common practice. Supermarkets, health food stores, and pharmacies carry an enormous array of choices. However, the news about vitamin E tells us that too much of a good thing can be harmful. One common dose found in over-the-counter vitamin E supplements is 400 international units (IU) per tablet or gel capsule. The report says that any amount of vitamin E per day in excess of this dose is potentially harmful. However, the "percent daily value" information on vitamin labels can encourage us to think that "more is better." For example, bottles of vitamin E supplements say that a single pill containing 400 IU is giving us "1,333 percent" of the daily vitamin E we need! Knowing a little biology and paying attention to vitamin labeling can help us make better choices.

But what does "better choices" mean? Our perceptions of what vitamin supplements are necessary for our health are shaped not only by our knowledge of biology, but also by what we see in the news and in advertisements. Even for the best informed of us, what was considered a healthy choice at one time can turn out later to have negative consequences when additional studies are conducted. For example, initial research showed that vitamin E plays a role in preventing cardiovascular disease and cancer. Consumers concerned about these basic health issues might naturally assume that consuming more vitamin E per day would be healthier. However, as more recent research indicates, more may not be better—in fact, it could make worse the very conditions it was supposed to improve.

Taking supplementary iron further illustrates the "more is better" fallacy. It's well known that an iron deficiency is not a good thing. So taking supplementary iron or a multiple vitamin with iron may seem like a good idea—and it may well be for some people. But healthy people eating a balanced diet are probably getting all the iron they need; in addition, many food products are already "fortified" with iron. Nutritionists tell us that too much iron can cause health problems, including lethargy and liver damage. If we decide to take iron supplements—or any supplements, for that matter—it would be wise to find out how much is too much and to determine how much we are already getting in our diet.

Evaluating the news

1. It's easy to assume that more is better. However, the news about excess vitamin E shows that the opposite can also be true. If you decided to take a general-purpose multivitamin supplement, how would this news about vitamin E change the way you evaluated the list of doses of vitamins you'd find on the label?

2. More and more people are taking time to eat a healthier diet. Yet many of the same people also take vitamin supplements. What advice would you give a friend about combining a healthy diet with vitamin supplements?

3. Some vitamin and nutritional supplements carry labels that describe the product as "natural." Do you think "natural" vitamin E, or any vitamin supplement, would be safer to take in "overdose" quantities? Why or why not?

SOURCE: American Heart Association and *Annals of Internal Medicine*, November 10, 2004.

CHAPTER **6**

Cell Membranes, Transport, and Communication

Key Concepts

◉ The movement of materials into and out of a cell across the plasma membrane is highly selective. Cells have several means of moving substances across the plasma membrane.

◉ Cells must regulate the amount of water they contain. Water can move passively across plasma membranes. Water can also be moved across cell membranes actively, which requires an input of energy.

◉ Materials can move into and out of cells enclosed in vesicles created from membranes.

◉ Multicellular organisms require specialization of cell types and communication between cells.

◉ Neighboring cells in a multicellular organism are connected to one another by different types of cell junctions. These junctions help hold cells together and facilitate communication between neighboring cells.

◉ Cells can communicate with one another over short and long distances by means of signaling molecules. Cell signaling can be fast or slow.

Pollution Detectives

Lichens grow all around us. Although most lichens are easily overlooked, you may have noticed some of the more spectacular kinds, hanging like gray-green beards from tree branches or forming colorful crusts on grave-stones. Lichens are worth getting to know, both for their unique biology and for what they have to teach us about our environment.

Lichens do not belong to just one of the six kingdoms of life, as we saw in Chapter 3. Instead, each kind of lichen is an intimate association between a fungal species and a photosynthetic protist species. The fungus forms the visible surface of the lichen. When moistened, the fungal tissue becomes transparent, revealing a green layer of single-celled photosynthetic organisms living just beneath the surface.

Lichen survival depends entirely on the interplay between these two very different groups of cells. The fungus provides a protective environment for its photosynthetic partner by attaching to the surfaces of rocks or plants. It also provides mineral nutrients and water to its partner. The photosynthetic organism, in turn, uses the energy from sunlight to convert water and carbon dioxide into sugars, which it shares with the fungus.

A carefully controlled exchange of materials between the cell and its environment creates conditions inside each cell that can support the chemical reactions of life. These exchanges depend on the integrity of cellular membranes. Since lichens obtain nutrients and water directly from the surrounding atmosphere, many of their plasma membranes are directly exposed to airborne pollutants. For this reason, when the air becomes polluted, lichens begin to die before other organisms do. By monitoring the fates of lichens around factories and cities, we have access to an early warning system for developing environmental problems.

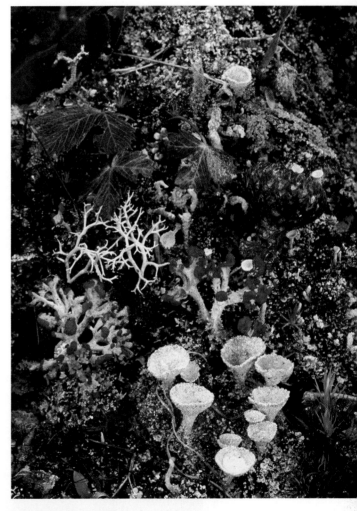

British Soldier and Pyxie Cup Lichens
Lichens arise from the interaction between single-celled photosynthesizers and fungi.

In the preceding chapters we have examined some of the chemical reactions that sustain life. Most of these reactions cannot take place outside of cells. Cells manage to maintain suitable conditions for the chemistry of life only by carefully controlling the uptake of materials from and loss of materials to their environment. All cells must have a way of moving materials into and out of themselves, as well as a way of controlling which materials can enter or leave them.

In this chapter we consider how cells manage their relationship with their surroundings. We begin by examining the role of the plasma membrane as gatekeeper and gate for substances entering and leaving the cell. Then we consider the cell's ability to regulate the amount of water entering and leaving it, a key component of cellular health. We show how cell membranes serve as transport "luggage," using a variety of "container" types. We discuss the ways cells are physically connected to one another, and how these connections facilitate communication between cells. We conclude with a look at the role of signaling molecules in cell communication.

6.1 The Plasma Membrane Is Both Gate and Gatekeeper

As we saw in Chapter 5, the plasma membrane separates the inside of a cell from the environment outside the cell. The structure of the plasma membrane is as universal a feature of life as the DNA-based genetic code. A double layer of lipids, the phospholipid bilayer, provides the framework for the plasma membrane (Figure 6.1). Although some biologically important materials can pass directly through the phospholipid bilayer, most cannot. Embedded in the phospholipid bilayer are many proteins,

Plasma membrane component	(a) Phospholipid bilayer	(b) Channel proteins	(c) Carrier proteins
Hydrophilic heads / Hydrophobic tails		Some channel proteins can open and close.	
Types of molecules that typically cross the membrane	Simple molecules, such as water, oxygen, and carbon dioxide Small hydrophobic molecules	Water and charged molecules, such as sodium and chloride ions Complex biochemicals, such as sugars, amino acids, and proteins	Charged molecules, such as sodium and chloride ions Simple hydrophilic molecules

Figure 6.1 The Plasma Membrane Controls What Enters and Leaves the Cell

(a) The basic structure of the plasma membrane consists of a double layer of phospholipid molecules with hydrophilic "heads" and hydrophobic "tails." This phospholipid bilayer is clearly visible in the electron micrograph at left. Proteins that span the plasma membrane (b, c) play an important role in moving materials into and out of cells. Different kinds of biologically important molecules cross the membrane by different routes.

which typically make up more than half the weight of the plasma membrane. Many of these proteins completely span the plasma membrane, providing pathways by which materials can enter or leave cells. The phospholipid bilayer and its associated proteins together act as a sophisticated filter that is **selectively permeable**, controlling which materials enter and leave the cell.

Cells can move materials across the plasma membrane with or without the expenditure of energy

Two general rules can help us understand how materials move into and out of cells:

1. Molecules move down **concentration gradients**—that is, from areas of abundance to areas of scarcity—and they do so *passively* (that is, with no input of energy). This movement is called **passive transport**.
2. Molecules can move up concentration gradients—that is, from areas of scarcity to areas of relative abundance—but they must do so *actively* (with an input of energy). This movement is called **active transport**.

These movements can be understood by using the physical example of a ball moving down or up a hill, in which the ball represents a chemical substance and the hill represents a concentration gradient (Figure 6.2). The ball rolls downhill on its own, but it cannot roll uphill unless we provide the energy needed to push it.

Imagine that you empty a packet of Kool-Aid powder into a pitcher of water. You can watch the coloring agents in the powder gradually spread from the area of high concentration—where you poured in the powder—throughout the water in the pitcher. This will happen even if you do not expend any energy to stir the mixture. The powder will dissolve and the color will continue to **diffuse**, or spread passively, until it is evenly distributed throughout the water. Once the color is distributed evenly, the concentration differences that make diffusion possible have disappeared, and diffusion stops.

In contrast, once it has diffused in a pitcher of water, the Kool-Aid will never re-form, all by itself, back into a concentrated dry powder, no matter how long you watch. The formation of concentrated powder requires that the molecules in Kool-Aid move up a concentration gradient from relatively low concentrations in the water to high concentrations in the powder. We can re-form the powder only by adding energy—for example, as heat—to evaporate the water.

Organisms rely heavily on both passive and active transport to take up and get rid of nutrients, gases, and wastes. Active transport, which requires energy, plays an essential role in allowing organisms to take up the raw materials for their biochemical reactions from the environment, where they exist at low concentrations, into cells, where they exist at high concentrations. Without active transport, organisms could not survive. As we shall see throughout this unit, however, the same diffusion process that allows Kool-Aid powder to spread through a pitcher of water also plays a key role in the transfer of some fundamental molecules, such as water, oxygen, and carbon dioxide, into and out of cells.

Test your knowledge of active versus passive movement of chemicals.

6.1

Small molecules can diffuse through the phospholipid bilayer

Materials that can cross the phospholipid bilayer of the plasma membrane do so strictly by moving passively down concentration gradients. Water, oxygen, and carbon dioxide usually enter and leave cells in this way (see Figure 6.1a). All these molecules are small and simple, consisting of just a few atoms each. In addition, some simple hydrophobic molecules can pass through the hydrophobic

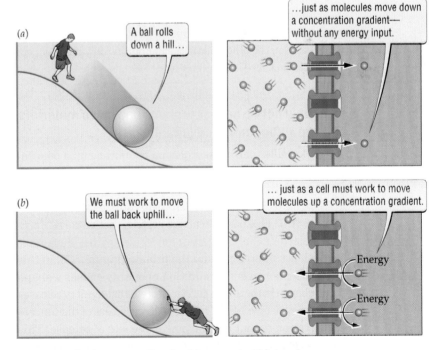

Figure 6.2 Active versus Passive Movement of Molecules
Materials can move into and out of organisms either passively (without an input of energy) or actively (with an input of energy). (a) Molecules can move passively through a membrane from areas of high concentration to areas of low concentration. (b) Energy is required to move materials from areas of low concentration to areas of high concentration.

core of the phospholipid bilayer. (Many of the early pesticides, such as DDT, were effective in killing insect pests precisely because they could easily get into their cells in this way.)

At the same time, the phospholipid bilayer acts as a barrier to the movement of most large, biologically important molecules. Even nutrients such as the simplest sugars and amino acids, for example, which consist of about 20 atoms each, are too large and too hydrophilic to diffuse through the hydrophobic core of the phospholipid bilayer. Various ions are also too hydrophilic to penetrate the phospholipid bilayer. Both groups of molecules require assistance to cross the plasma membrane.

Channel proteins and passive carrier proteins allow molecules to cross the plasma membrane passively

Two types of plasma membrane proteins allow large and hydrophilic molecules to cross the plasma membrane passively. **Channel proteins** provide a means for hydrophilic molecules of the right size and charge to move through the plasma membrane, as long as they are moving down a concentration gradient (see Figure 6.1*b*). **Passive carrier proteins** can bind to a particular molecule that fits into the folds of the protein. Once bound to its "cargo," the protein changes shape in such a way that it transfers the molecule from one side of the plasma membrane to the other (see Figure 6.1*c*). But the protein releases the molecule on the other side of the membrane only if its concentration there is relatively low.

Only active carrier proteins can move materials up a concentration gradient

Molecules can cross a plasma membrane up a concentration gradient only by active transport. **Active carrier proteins** can move molecules across the plasma membrane using energy from an energy storage molecule such as ATP. Like passive carrier proteins, active carrier proteins bind only to certain molecules with a shape that allows them to fit into the protein (Figure 6.3). In this case, however, the addition of energy causes a shape change in the active carrier protein that forcibly releases the molecule being transferred, regardless of the concentration of that molecule near the site of release. This mechanism allows active carrier proteins to move molecules from regions of low concentration to regions of high concentration.

Although active transport plays a critical role in the ability of cells to maintain appropriate internal condi-

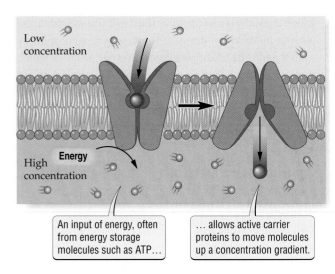

Figure 6.3 Active Carrier Proteins
Active carrier proteins use energy to move materials from areas of low concentration to areas of high concentration.

tions, its energy cost can be substantial. For example, 30 to 40 percent of the energy used by a resting human body fuels active transport across plasma membranes.

Active and passive transport often work together

To move molecules up concentration gradients, active carrier proteins often collaborate with passive carrier proteins. An active carrier protein pumps positively charged ions—typically sodium ions in animals and hydrogen ions in other organisms—against a concentration gradient so that the concentration of these ions builds up on one side of the plasma membrane. The ions then move passively back across the membrane, along with a second molecule that is moving against its concentration gradient.

Let's take a closer look at how this process works. The passive carrier proteins involved in this type of cooperative transfer differ from the simple ones described earlier, in that they bind two different molecules, not just one. Usually, such carrier proteins bind first to the positively charged ion, which changes the shape of the carrier protein so that it can bind to the second molecule. The act of binding to the second molecule transfers both molecules to the opposite side of the membrane. The carrier protein releases the ion readily on the side of the membrane where its concentration is low. Upon release of the ion, the carrier protein returns to its original shape, forcibly ejecting the second molecule on the side of the membrane where its concentration is high.

Isotonic solution	Hypertonic solution	Hypotonic solution
Neither gains nor loses water	**Loses water**	**Gains water**
When solute concentrations outside the cell equal concentrations inside the cell, the cell neither gains nor loses water.	When solute concentrations outside the cell exceed those inside the cell, the cell shrinks as it loses water to its environment.	When solute concentrations outside the cell are lower than those inside the cell, the cell swells as it gains water from its environment.

Plant cells

Animal cells

Normal red blood cell | Shrunken red blood cell | Bloated red blood cell

Figure 6.4 Water Moves Into and Out of Cells by Osmosis
Differences in the concentrations of dissolved materials outside and inside cells determine whether water moves into or out of cells.

Investigate the effects of water gain and loss in plant and animal cells.

6.2

In a **hypotonic solution**, the outside medium is more watery (has fewer solutes) than the cytosol of the cell, resulting in more water flowing into the cell than out of it; unchecked, this movement can cause a cell to swell until it bursts. In a **hypertonic solution**, the outside medium is less watery (has more solutes) than the cytosol, so more water flows out of the cell than into it. This movement causes the cell to shrink. For example, if a dog drinks seawater, which is hypertonic to its cells, the resulting cell shrinkage can cause life-threatening illness. Finally, in an **isotonic solution**, the concentration of the outside medium is "just right." The concentration of solutes is the same on both sides of the plasma membrane, and the amount of water flowing out equals the amount of water flowing in.

Cells, however, are not necessarily passive victims of the concentrations in their surroundings. Some organisms that live in hypertonic media, such as ocean-dwelling fish, have specialized cells that maintain the proper internal water concentration; these cells are able to do this despite the osmotic tendency of water to move out of the cells into the more concentrated salt water. Freshwater fish, on the other hand, have mechanisms that pump excess water out of their cells. The "balancing act" that cells must maintain in relation to their external environment is called **osmoregulation**, and it involves active transport and requires energy.

Helpful to know

The prefixes *hyper* ("more, too many, too much"), *hypo* (less, fewer, not enough), and *iso* ("equal") occur in a number of scientific terms. Here, for example, a hypertonic environment is one in which the surroundings have a higher concentration of solutes than the cytosol of the cell. Similarly, a hypotonic environment has a lower solute concentration than the cytosol, and in an isotonic environment, the solute concentrations inside and outside the cell are the same. Notice that these are all terms describing the relationship of one solution to another.

Notice that our carrier protein has moved the ion down a concentration gradient, which is typical passive transport. However, the carrier protein has moved the second molecule *against* a gradient, to an area of higher concentration. This indirect active transport mechanism allows a single power source—a concentration gradient in one kind of positively charged ion—to move many different molecules into or out of cells against concentration gradients.

6.2 Water Requires a Cellular Balancing Act

The environment inside and outside of a cell is watery. Water is constantly moving into and out of cells by a process called **osmosis**: the passive movement of water across a selectively permeable membrane. The water content of cells is continuously affected by osmosis, and too much or too little water in a cell can be disastrous. Cells can find themselves in external environments that are too watery, not watery enough, or just right (Figure 6.4).

6.3 Cell Membranes as Transport Luggage

We saw in Chapter 5 how molecules move around inside cells enclosed in vesicles. This kind of packaging of chemical substances into transport "luggage" also takes place at the plasma membrane. Many substances are exported from and imported into cells by becoming

Biology Matters

Water, Water Everywhere—But Not a Drop to Drink

After the tsunamis hit East Asia in December 2004, the availability of fresh water became a major problem in the stricken areas. Local water supplies became contaminated with seawater, threatening the spread of disease and dehydration. Seawater is not suitable for drinking, because it is hypertonic to our cells. When we drink seawater, our cells become dehydrated, because water flows from them into the hypertonic solution. Drinking too much seawater can quickly become fatal.

Certain ships in the U.S. Navy can extract fresh water from seawater through a process called reverse osmosis. Desalinization plants on these ships use high pressure and special membranes to force water molecules out of seawater and into storage tanks. For example, the USS *Bonhomme Richard* can produce over 30,000 gallons of fresh water a day from seawater by reverse osmosis. Such ships have the capacity to save many lives by supplying desperately needed fresh water, such as to victims of the tsunami disaster.

Why is fresh water so important? The body of an average person is about 70 percent water, meaning it contains about 10 gallons of life's most precious liquid. On an average day we lose around 10 cups of water, and if we don't replace what we lose, our bodies can't function properly. A child in a hot car can dehydrate and die within a few hours. In normal surroundings we would survive only about a week with little or no water. The importance of the mission of the *Bonhomme Richard* becomes apparent when we consider how vital it is to maintain a proper water balance within our bodies.

Experts recommend taking in the equivalent of about 11 cups of water a day for women and 16 cups a day for men. Beverages obviously consist largely of water, but alcoholic beverages are an exception. Drinking them can lead to excessive water loss, because alcohol reduces the level of a hormone that helps our bodies retain water. Urine volume increases and we end up losing water.

Many foods, such as lettuce and watermelon, also contain a lot of water, as shown in the table below.

Food	Percent water
Lettuce	95
Watermelon	92
Broccoli	91
Grapefruit	91
Milk	89
Orange juice	88
Carrot	87
Yogurt	85
Apple	84

SOURCES: American Dietetic Association (www.eatright.org/Public/Media/PublicMedia_21431.cfm, www.eatright.org/Public/NutritionInformation/index_19273.cfm) and *Scientific American* (www.sciam.com/askexpert_question.cfm?articleID=000AEAC0-93EC-1DEF-A838809EC588F2D7).

wrapped in, or unwrapped from, pieces of the plasma membrane.

Exocytosis [EX-oh-sigh-TOE-siss] (from *exo*, "outside"; *cyt*, "cell"; *osis*, "process") is the name given to the process by which cells release substances into their surroundings (Figure 6.5*a*). As a vesicle inside the cell approaches the plasma membrane, a portion of the vesicle's membrane fuses with the plasma membrane. It then creates an opening to the exterior of the cell, allowing the vesicle's contents to be discharged. The release of many chemical messages into the bloodstream depends on exocytosis. For example, after every meal and snack, specialized cells in your pancreas exocytose [EX-oh-sigh-TOZE] the hormone insulin, which moves through the bloodstream to other cells and signals them to take up glucose.

The reverse of exocytosis is **endocytosis** (*endo*, "inside"). In this process, a section of plasma membrane bulges inward around substances that are outside of the cell. The inward-budding portion of membrane eventually breaks free of the plasma membrane to become a closed vesicle, now wholly inside the cell and completely enclosing its contents. Endocytosis can be nonspecific (Figure 6.5b), in which case all of the material in the immediate area is surrounded and taken in. One form of nonspecific endocytosis is **pinocytosis** [*PINN*-oh . . .], often described as "cell drinking" because cells take in fluid in this way. The cell does not attempt to collect particular solutions; the vesicle budding into the cell contains whatever solutes were dissolved in the fluid the vesicle "drank."

Endocytosis can also be a very specific process, in which only one substance is enveloped and imported. How does a particular section of plasma membrane "know" what and when to endocytose? The answer depends on the kind of endocytosis. In **receptor-mediated endocytosis**, specialized receptor proteins embedded in some areas of the membrane determine what substances enter the cell (Figure 6.5c). The receptors do this by recognizing certain surface characteristics of the material they bring in. Cells in the liver, for example, take up cholesterol from our blood using receptor-mediated endocytosis. Receptors in the plasma membrane of these cells recognize proteins that tightly associate with cholesterol molecules.

Another form of endocytosis is "big-time" endocytosis from the cell's point of view (Figure 6.5d). In **phagocytosis**, or "cell eating," a cell ingests particles considerably larger than molecules, such as entire microorganisms. This remarkable process is restricted to specialized cells, such as the white blood cells that defend your body from infection. A single white blood cell can engulf a whole bacterium or yeast cell; this would be roughly like you swallowing a big Thanksgiving Day turkey whole! As is the case with cholesterol uptake, receptors in the membrane of the white blood cell allow it to recognize and phagocytose a harmful microorganism.

6.4 Cooperation among Cells in Multicellular Organisms

Sometimes it is tempting—even useful—to consider cells in isolation and to think of multicellular organisms as nothing more than collections of individual cells. However, this portrayal could not be further from the truth. Like any complex community, multicellular organisms benefit from specialization, but must coordinate the activities of all the specialized cells in the body. Over the course of

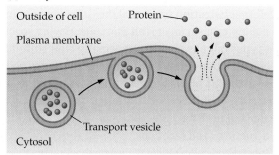

(a) Exocytosis

Outside of cell
Protein
Plasma membrane
Transport vesicle
Cytosol

(b) Endocytosis

Vesicle

(c) Receptor-mediated endocytosis

LDL particle
Phospholipid coat
Cholesterol
Protein
Receptor
Liver cell
Cholesterol processed

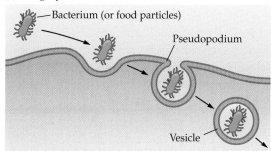

(d) Phagocytosis

Bacterium (or food particles)
Pseudopodium
Vesicle

Figure 6.5 Cells Are Importers and Exporters
(a) Exocytosis exports materials from the cell. (b) Endocytosis imports materials into the cell. (c) Receptor-mediated endocytosis is a specialized importing process. (d) Phagocytosis is endocytosis on a large cellular scale.

evolution, this simple need for coordination presented a tremendous challenge to living systems. It took almost 3 billion years for primitive single-celled life forms to evolve the systems of communication and cooperation that led to the first multicellular organisms. Clearly, cooperation is not an easy thing to achieve, especially when it involves billions upon billions of cells.

Two organizational principles apply to every large multicellular organism: the principle of cell specialization and the principle of cell communication. The principle of **cell specialization** states that the cells found in a multicellular organism are not all the same. Having specialized types of cells that are especially well suited for specific functions ensures that all the processes necessary for the life of the organism are carried out as efficiently as possible. Indeed, the larger the organism, the greater the need for different types of cells with different structures and functions.

Specialized cells fulfill a wide variety of needs in multicellular organisms (Figure 6.6). For example, just as every cell needs the structural support provided by the cytoskeleton, every multicellular organism requires specialized cells that support, maintain, and strengthen the physical form of the organism as a whole. In the case of humans, specialized cells deposit bone in the body to form a skeleton, which both supports the body and allows the movement of limbs. In plants, specialized hollow cells in the roots, stems, and leaves transport water between different parts of the plant as well as providing structural support.

The second organizational principle at work in all multicellular organisms is the principle of **cell communication**. Given their diversity of function and structure, cells must have some way to "talk" with one another. How they do this depends on how close they are to one another. Different kinds of physical connections help mediate communication between neighboring cells, while cells that are not in direct physical contact with one another rely on sending and receiving small signaling molecules to communicate with one another.

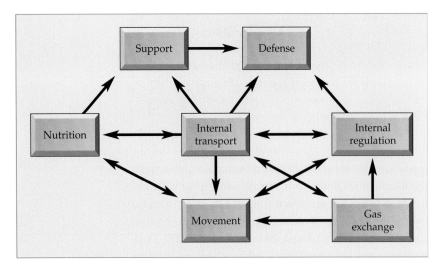

Figure 6.6 The Cellular Functions Required for Life
Specialized cells in a multicellular organism must carry out a number of interrelated functions if the organism is to survive.

6.5 Connections Between Neighboring Cells

In multicellular organisms, cells exist in close-knit communities. Neighboring cells are nestled against one another in the vastness of a whole organism. These cells must be interconnected if the organism is to survive. Microscopic studies have revealed that neighboring cells often share **cell junctions**. These connections not only contribute to the overall structural stability of cellular communities, but also allow neighboring cells to communicate by providing passages for critical substances that move from one cell to another.

Recall that plant cells are surrounded by cell walls. For plant cells to survive and communicate with one another, there must be cellular connections extending through these walls. **Plasmodesmata** [plaz-moe-*DEZ*-muh-tuh] (singular: plasmodesma) are tunnel-like channels that provide passageways for the flow of vital small molecules and water from cell to cell. Because of the plasmodesmata, the cytoplasm of plant cells is effectively continuous from one cell to the next (Figure 6.7*a*).

In animals, the situation is different. Most animal cells secrete a viscous coating, called an **extracellular matrix** (Figure 6.7*b*), that helps to hold cells together. Since animal cells have no cell walls, they have no plasmodesmata, but they do have cell junctions. These connections allow for cell-to-cell communication and complete the job of holding cells together—without them, our bodies would fall apart. There are three main types of cell junctions in animals.

Tight junctions (Figure 6.7*b*) hold cells together with strands of protein arranged in a belt beneath the plasma membrane of each cell. Cells held together with tight junctions form leak-proof sheets that prevent the passage of ions and small molecules between them. The sheets of cells that line the interiors of our intestines, for example, must control which products of digestion are absorbed, so they cannot be leaky. These cells are bound together by tight junctions.

Anchoring junctions act as protein "hooks" between cells or between a cell and the extracellular matrix. They allow materials to pass in-between cells while still holding them together. Lots of anchoring junctions are found among cells in tissues that experience heavy structural stress, such as heart muscle (see Figure 6.7*b*).

One of the most widespread types of cellular connections in animals is **gap junctions**. Without them, neighboring cells could not transmit the faint but vital chemical "whispers" that keep us alive. Gap junctions are direct channel protein connections between the

(a) Plasmodesmata

Walls of two adjacent plant cells

Vacuole

Plasmodesmata (channels between cells)

(b) Animal cell junctions

Extracellular matrix

Tight junctions

Anchoring junctions

Gap junctions

Plasma membranes of adjacent cells

Extracellular matrix

Figure 6.7 Cells in Multicellular Organisms Are Interconnected in Various Ways
(a) Plants cells are interconnected by plasmodesmata. (b) Many animal cells secrete an extracellular matrix and are joined by different types of junctions.

plasma membranes of two cells that allow the passage of ions and small molecules (see Figure 6.7b).

6.6 Signaling Molecules in Cell Communication

The use of small molecules to transmit signals between cells is widespread among multicellular organisms. In general, communication between cells uses small pro-

teins or other molecules that are released by one cell and received by another cell, which is called the **target cell**. These **signaling molecules** form the language of cellular communication.

The passing of signaling molecules from one cell to another is an effective way for two cells to communicate. However, the identities of these signaling molecules, and their effects on target cells, are highly variable. The particular signaling molecule used by the cells often depends on the speed required and the distance it must travel between cells to complete its communication. Some signaling molecules must work fast. They bring about a response in a target cell almost instantly and travel only between neighboring cells. These molecules are relatively fragile and short-lived. Other signaling molecules that must travel long distances through the bloodstream have a longer life span and are built to survive their journey. Once they reach their target cell, they work more slowly to bring about a response. If you've ever jumped in response to a sudden noise, you have experienced the almost instantaneous work of short-lived, fast-acting signaling molecules. If these signaling molecules were slow-acting, a "jump" would take days to occur. On the other hand, long-lived, slower-acting signals were at work as you grew in size throughout childhood. If those signals acted as fast as those between your nerves, "growth spurt" would have a whole new meaning!

When a signaling molecule reaches a target cell, that cell must have a specific means of receiving it and acting on its message. These tasks are the responsibility of a class of proteins called **receptors**. Some receptors are located on the surface of the target cell and encounter their matching signaling molecules there. Other receptors are located in the cytosol or inside the nucleus of the target cell. To reach these receptors, signaling molecules must cross the plasma membrane and perhaps the nuclear envelope (Figure 6.8). Locating where specific signaling-related receptor proteins reside in the cell is a necessary step in understanding how external signals affect the behavior of cells. Recent breakthroughs that allow scientists to tag and track various proteins in the cell have revealed many important features of the cell signaling process (see the box on page 127). These studies have confirmed that signaling molecules are important coordinators of a broad range of cellular processes (Table 6.1).

Review animal cell junctions. 6.3

Hormones are long-range signaling molecules

Depending on the size of the organism, cells that are anywhere from a few centimeters to several meters apart must be able to communicate with one another. Such

Cell surface receptor

Hydrophilic signaling molecule

Hydrophobic signaling molecule

Hydrophilic signaling molecules cannot pass through the plasma membrane and must bind receptors at the cell surface to indirectly affect processes inside the cell.

Plasma membrane

Intracellular receptor

Hydrophobic signaling molecules can pass through the plasma membrane and can directly affect processes inside the cell.

Figure 6.8 Receptors for Signaling Molecules
Two major classes of receptors bind to signaling molecules. Cell surface receptors are embedded in the plasma membrane and bind to signaling molecules that cannot cross the membrane. Intracellular receptors reside in the cytosol or the nucleus and bind to signaling molecules that can cross the plasma membrane.

Table 6.1

Examples of Signaling Molecules

Type of molecule	Name of molecule	Site(s) of synthesis	Function(s)
ANIMALS			
Amino acid derivatives	Adrenaline	Adrenal glands	Promotes release of stored fuels Promotes increased heart rate
	Thyroxine	Thyroid gland	Promotes increased metabolic rate
Choline derivative	Acetylcholine	Nerve cells	Assists signal transmission from nerves to muscles
Gas	Nitric oxide	Endothelial cells in blood vessel walls	Promotes relaxation of blood vessel walls
		Nerve cells	
Proteins	Insulin	Beta cells of the pancreas	Promotes the uptake of glucose by cells
	Nerve growth factor (NGF)	Tissues richly supplied with nerves	Promotes nerve growth and survival
	Platelet-derived growth factor (PDGF)	Many cell types	Promotes cell division
Steroids	Progesterone	Ovaries	Prepares the uterus for implantation Promotes mammary gland development
	Testosterone	Testes	Promotes the development of secondary male sex characteristics
PLANTS			
Acetic acid derivative	Auxin	Most plant cells	Promotes root formation Promotes stem elongation

Science Toolkit

Tagging Proteins

A living cell is a constant buzz of activity, of which proteins are a mainstay. Proteins are constantly interacting with one another and with other components of the cell. Many proteins move from place to place within the cell. Some proteins are being assembled or broken down. Enzymes (a type of protein) are speeding up reactions virtually everywhere. Clearly, the more researchers can learn about the movements, interplay, and fates of proteins, the closer they will get to solving the secrets of cellular life and developing treatments for diseases such as cancer.

It is no easy task keeping track of proteins in the heavy internal traffic through the intricate compartments and cytoskeleton of a cell. Yet the past decade has seen the emergence of a powerful tool that eases the difficulty considerably. It is a molecular tag that can be attached to proteins without disturbing their function much, if at all. The tag is also a protein, one that glows for a while after external light is shined on it (Figure A). Proteins that are "carrying lanterns" are a lot easier to track than those that are not.

This tagging tool provides great flexibility in studying proteins. Researchers can choose to tag and observe only particular proteins. They can then observe where these proteins are and where they go, measure how long they take to get to their destinations, and determine what other molecules they interact with along the way. This tool has revealed many properties of proteins in cells that were unknown before.

One technique for using this tool works as follows (Figure B): After particular proteins in the cell are tagged with the lantern protein molecule, light is shined on the proteins, causing the lantern tags to glow. Next, an area of the cell is "bleached" with intense light, depriving the

Figure A Molecular Model of the Lantern Protein

The original lantern protein came from a species of jellyfish. Genetically enhanced versions of this protein glow brighter and longer than the original.

lanterns in the bleached region of the capacity to glow. Now the movement of other lantern-tagged protein molecules into the bleached area can be tracked visually because their glow will show up against the bleached background.

Figure B Tracking the Movement of Proteins Tagged with the Lantern Protein

Proteins — Nucleus

1. Proteins in the cell have been exposed to light and are carrying glowing "lanterns."
2. Proteins in this area have been bleached with intense light, and their "lanterns" are no longer able to glow.
3. Glowing proteins can be observed as they move into the bleached space. Researchers can observe how fast they move.
4. Glowing proteins have completely reoccupied the bleached area.

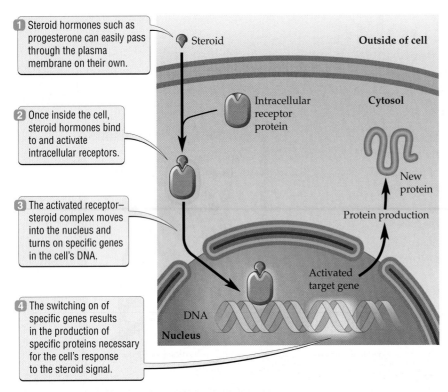

1 Steroid hormones such as progesterone can easily pass through the plasma membrane on their own.

2 Once inside the cell, steroid hormones bind to and activate intracellular receptors.

3 The activated receptor–steroid complex moves into the nucleus and turns on specific genes in the cell's DNA.

4 The switching on of specific genes results in the production of specific proteins necessary for the cell's response to the steroid signal.

Steroid

Outside of cell

Intracellular receptor protein

Cytosol

New protein

Protein production

Activated target gene

DNA

Nucleus

Figure 6.9 A Cell's Response to a Steroid Hormone

long-distance communication requires different kinds of signaling molecules from those used between neighboring cells. These long-distance signaling molecules are called **hormones**.

All multicellular organisms use hormones to coordinate the activities of different cells and tissues. Hormones are produced by cells in one part of the body and transported to target cells in another part of the body. In this chapter, we focus on how they transmit their messages to cells.

The transportation of hormones from their site of production to their target cells often depends on the circulation of fluids inside the organism. In plants, some hormones are dissolved in the sap, which moves between the roots and the rest of the plant. In animals, hormones are dissolved in the blood, which circulates throughout the body, ensuring rapid distribution.

Steroid hormones can cross cell membranes

Steroid hormones are an important class of signaling molecules that are essential for many growth processes, including the normal development of reproductive tissues in mammals. All steroid hormones are derived from cholesterol (see Figure 4.17). Because steroid hormones are hydrophobic, they can pass easily through the hydrophobic core of the target cell's plasma membrane and enter the cytosol. But being hydrophobic also means that they cannot move unaided through the bloodstream. They must be packaged with proteins that help them dissolve in the watery environment of the body. These associated proteins also extend the life span of steroids. The ability of steroids to remain in the bloodstream for up to several days improves their likelihood of reaching distant target cells.

When a steroid molecule arrives at its target cell, it crosses the plasma membrane and alters the production of specific proteins inside the cell (Figure 6.9). To do this, the hormone must bind to an intracellular receptor in the cytosol. Together, the steroid and the receptor protein form an active molecular complex that can enter the nucleus and interact with the target cell's DNA. Recall that the DNA in the nucleus carries the instructions for making all of the proteins needed by the cell. A gene is a segment of DNA that carries the instructions for making a particular protein, as we will see in Chapter 12. The steroid-receptor complex acts on specific genes, activating the production of the proteins they encode. In humans, for example, the genes activated by the steroid progesterone in target cells in the uterus (womb) encode proteins that are necessary for the proper growth of the uterine lining in preparation for pregnancy. In general, the action of steroid hormones initiates and coordinates the production of specific proteins necessary for changes in the cell.

Not all signaling molecules enter the cell as steroids do. Certain hormones send their signals into the cell via cell surface receptors. Some of the hormones that act in this way are small proteins; others are chemical derivatives of amino acids and fatty acids.

Why Are Lichens So Sensitive to Air Pollution?

The city of Sudbury, in the Canadian province of Ontario, is surrounded by some of the world's richest nickel deposits. Large smelter operations convert the ore into 2 million kilograms of nickel each year. Large amounts of sulfur dioxide (SO_2) are released into the air as a by-product of the smelting process. SO_2 reacts with water to produce sulfuric acid, which leads to acid rain. The high SO_2 concentrations in the air around Sudbury have given the city an unenviable reputation as a biological wasteland.

Lichens were one of the first groups of organisms to succumb to the SO_2 pollution around Sudbury (Figure 6.10). Unlike most plants, the fungal species in lichens are extremely efficient at absorbing nutrients from the atmosphere and from precipitation by both passive and active transport. In unpolluted environ-ments, this ability to take up nutrients from the air and rain allows the fungus to gather the chemical nutrients needed by the photosynthetic organisms. In the area surrounding Sudbury, however, it also ensured that the lichens would rapidly accumulate SO_2. Inside the lichen, the SO_2 combines with water to form sulfuric acid. The sulfuric acid breaks down the phospholipid bilayer of plasma membranes, disrupting the exchange of nutrients between the single-celled photosynthesizers and their associated fungus.

Since 1970, when air pollution around Sudbury was at its worst, a number of changes have reduced the amount of SO_2 deposited in the surrounding countryside. Improved SO_2 recovery at the smelters for the production of marketable sulfuric acid has reduced the amount of SO_2 emitted from the chimneys. The construction of the world's tallest chimney at one of Sudbury's three smelters has reduced SO_2 fallout by causing the chemical to travel farther from Sudbury through the air before settling to the ground. Finally, one of the other two smelters has closed. Together, these changes have reduced the SO_2 falling on the landscape around Sudbury from 2.5 million tons to less than 1 million tons each year. In response, the abundance and diversity of the lichens around Sudbury are increasing again.

(a) Before giant chimneys (1968)

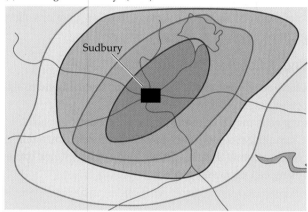

Sudbury

(c) After giant chimneys (1990)

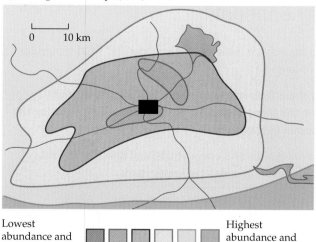

0   10 km

Lowest abundance and diversity of lichens   Highest abundance and diversity of lichens

(b)

Figure 6.10 The Effect of Sulfur Dioxide on Lichen Diversity

(a) Lichens growing on trees around Sudbury, Ontario, died when they took up SO_2 released into the air by nickel smelters. (b) The building of a 38.0-meter chimney in the 1970s reduced SO_2 pollution around Sudbury by ensuring that the SO_2 did not reach ground level near the city. (c) As local SO_2 levels decreased, the lichens around Sudbury recovered somewhat.

Chapter Review

Summary

6.1 The Plasma Membrane Is Both Gate and Gatekeeper

- Everything that enters or leaves a cell crosses a selectively permeable plasma membrane composed in part of a phospholipid bilayer.
- Cells can move materials across the plasma membrane either with the expenditure of energy (active transport) or without it (passive transport). In passive transport, substances are carried down concentration gradients, while in active transport, materials move up concentration gradients.
- Passive transport takes several forms. Certain small or hydrophobic molecules simply diffuse through the phospholipid bilayer unassisted. Proteins spanning the plasma membrane (channel proteins and passive carrier proteins) allow certain other large or hydrophilic molecules to cross the plasma membrane.
- Active transport requires the aid of active carrier proteins, which use energy to move materials across the membrane against a concentration gradient.
- Active carrier proteins can work together with special passive carrier proteins to move materials across the plasma membrane against a concentration gradient.

6.2 Water Requires a Cellular Balancing Act

- Osmosis is the passive movement of water across a selectively permeable membrane. This movement depends on the solute concentration of the medium surrounding the cell. In a hypotonic solution, water moves into cells. In a hypertonic solution, water moves out of cells. In an isotonic solution, water flows into and out of cells at the same rate.
- Cells can actively balance their water content by osmoregulation.

6.3 Cell Membranes as Transport Luggage

- Cells can export materials by exocytosis: a vesicle carrying a substance merges with the plasma membrane to release its contents outside the cell.
- Cells can import materials by endocytosis: part of the plasma membrane buds inward and encloses a substance, pinching off to form a transport vesicle moving inside the cell.
- Endocytosis can take several forms. In pinocytosis, cells ingest fluids. In receptor-mediated endocytosis, receptor proteins in the plasma membrane "recognize" the substance to be brought into the cell. In phagocytosis, the plasma membrane surrounds a particle larger than a molecule, such as a bacterium.

6.4 Cooperation among Cells in Multicellular Organisms

- There are two organizing principles in multicellular organisms.
- The principle of cell specialization states that cells are specialized for different tasks so that they can more efficiently serve the whole organism.
- The principle of cell communication states that the cells of an organism must be able to communicate with one another.

6.5 Connections Between Neighboring Cells

- Cell junctions hold cell communities together, increasing the stability of multicellular organisms and allowing neighboring cells to communicate.
- Cell junctions in plants, called plasmodesmata, are tunnel-like channels embedded in the cell walls.
- In animals, a viscous extracellular matrix coats most cells. There are several kinds of cell junctions in animal cells. Tight junctions bind cells together to form "leak-proof" sheets of cells. Anchoring junctions are protein "hooks" that hold cells together. Gap junctions allow the passage of small molecules between cells.

6.6 Signaling Molecules in Cell Communication

- Target cells have receptors, specialized proteins that respond to signaling molecules dispatched by other cells.
- Signal-sending cells may be close to or distant from target cells.
- Long-range signaling molecules are called hormones. Steroid hormones can cross cell membranes. Other kinds of hormones transmit signals using cell surface receptors.

◉ Review and Application of Key Concepts

1. Describe the ways in which a cell can "decide" what molecules to allow in or keep out.

2. Describe the effects of osmosis on a cell. Include in your discussion the three possible types of solutions constituting the cell's external environment.

3. Compare exocytosis and endocytosis, and describe the role of receptors in receptor-mediated endocytosis.

4. What organizing principles are evident in multicellular organisms? For each principle, consider what life might be like if

that principle were at work, but not the other. Would multicellular life be possible? And what might life be like if neither principle operated?

5. Describe the major types of cell junctions found in animals and plants.

6. Compare the pathways of a slow-acting and a fast-acting cell signal.

Key Terms

active carrier protein (p. 120)
active transport (p. 119)
anchoring junction (p. 124)
cell communication (p. 124)
cell junction (p. 124)
cell specialization (p. 124)
channel protein (p. 120)
concentration gradient (p. 119)
diffuse (p. 119)
endocytosis (p. 122)
exocytosis (p. 122)
extracellular matrix (p. 124)
gap junction (p. 124)
hormone (p. 128)
hypertonic solution (p. 121)
hypotonic solution (p. 121)

isotonic solution (p. 121)
osmoregulation (p. 121)
osmosis (p. 121)
passive carrier protein (p. 120)
passive transport (p. 119)
phagocytosis (p. 123)
pinocytosis (p. 123)
plasmodesma (p. 124)
receptor-mediated endocytosis
  (p. 123)
receptor (p. 125)
selectively permeable (p. 119)
signaling molecule (p. 125)
steroid hormone (p. 128)
target cell (p. 125)
tight junction (p. 124)

Self-Quiz

1. Which of the following is *not* a part of the plasma membrane?
   a. proteins
   b. phospholipids
   c. receptors
   d. genes

2. Energy is needed for
   a. diffusion.
   b. active transport.
   c. osmosis.
   d. passive transport.

3. Water would move out of a cell in
   a. a hypotonic solution.
   b. an isotonic solution.
   c. a hypertonic solution.
   d. none of the above

4. Which of the following describes movement of material out of a cell?
   a. pinocytosis
   b. phagocytosis
   c. endocytosis
   d. exocytosis

5. What cellular connection is "leak-proof?"
   a. anchoring junction
   b. tight junction
   c. plasmodesmata
   d. gap junction

6. Animal cells can directly exchange water and other small molecules through
   a. gap junctions.
   b. microfilaments.
   c. anchoring junctions.
   d. tight junctions.

7. Cell signaling involves
   a. receptors.
   b. signaling molecules.
   c. target cells.
   d. all of the above

8. Signaling molecules
   a. are all derived from cholesterol.
   b. are derived from active carrier proteins.
   c. affect only a cell's DNA.
   d. none of the above

9. A fast-acting cellular signal would probably
   a. travel through the bloodstream.
   b. affect nearby cells.
   c. be long-lived.
   d. affect distant cells.

10. Steroid hormones
    a. bind to receptors in the plasma membrane.
    b. must be actively transported.
    c. bind to receptors in the cytosol.
    d. change the shape of proteins in the nucleus.

Reclamation Techniques at Sudbury, Canada

BY DAN SHAW

The Sudbury region of Canada has been an example of how industry can have catastrophic effects on the environment. Fortunately, through the efforts of the multidisciplinary technical advisory committee, summer work crews, volunteer efforts, and industry itself, the region is now becoming an example of how degraded lands can be reclaimed.

Membranes and all, the lichens around Sudbury are returning to good health, as is the overall ecology of the formerly denuded area around the city. A massive effort to reduce air and water pollution has had award-winning results. Polluting emissions have been controlled. Volunteers have planted thousands of trees, a community has come together, and the land has been rejuvenated.

But in many other locations, the picture remains bleak. In the northeastern United States, for example, forests are being extensively damaged by acid precipitation. The acidity results from airborne pollutants, mainly sulfur dioxide and nitrogen oxides, that are generated by industries, automobiles, and power plants in the Midwest. Prevailing winds carry these pollutants eastward, dissolved in the water droplets of clouds and fog. When acid rain or snow falls from these clouds, or when acidified fog lingers, life suffers. In the northeastern United States, the water in lakes has become more acidic. The change in acidity affects aquatic life, since many aquatic life forms cannot survive the new conditions. Acid precipitation also affects soils. Calcium and magnesium ions, essential for plant growth, are washed away more rapidly by acid rain. Concentrations of aluminum can build up in the soil due to acid rain, and aluminum can prevent tree roots from absorbing nutrients. All these effects have a great deal to do with cell membranes, since in the final analysis, all exchanges between organisms and their surroundings take place where plasma membranes directly meet the environment outside the cell.

Although the environmental picture as a whole is bleak, there is some good news. Emissions of pollutants have been reduced by laws such as the U.S. Clean Air Act of 1970 and further amendments to it in 1990. Researchers are seeing gradual reductions in the acidity of the soil and water in some locations. They caution, however, that further reductions in emissions are needed, and that the recovery of damaged areas may take many years. Continued improvement in environmental conditions will require efforts on the part of many people and institutions. As in Sudbury, however, the actions of many people operating in a spirit of cooperation can work wonders.

Evaluating the news

1. Visible damage to lichens around Sudbury provided early warning signs of broader and deeper damage to the environment, and a community pulled together to solve the problem. How would you use this example to persuade lawmakers to deal with pollution problems you have seen described in the news?
2. Damage to the environment can take place at a much smaller scale than that of damage caused by large industries. "Do Not Walk on the Grass" signs, for example, are intended to protect and preserve lawns that we all find attractive. What small-scale biological damage, if any, can you observe in your neighborhood or on your campus?
3. Specific proteins in some of our cell membranes detect specific odors. Describe a situation in which those proteins could serve as our own pollution detectors.

SOURCE: *Restoration and Reclamation Review*, Spring 1999.

Key Concepts

⊙ Living organisms obey the universal laws of energy conversion and chemical change.

⊙ The sun is the primary source of energy for living organisms. Photosynthetic organisms capture energy from the sun and use it to synthesize sugars from carbon dioxide and water. Most cells break down sugars to release energy.

⊙ Metabolism consists of the chemical reactions that produce complex macromolecules such as sugars and proteins and which break down those macromolecules to yield smaller molecules and usable energy.

⊙ Enzymes greatly increase the rate of chemical reactions in cells. Metabolic pathways are sequences of enzyme-controlled chemical reactions.

"Take Two Aspirin and Call Me in the Morning"

If someone told you there was a wonder drug that reduced pain, fever, and inflammation and helped combat heart disease and cancer, would you believe them? As amazing as it may sound, such a drug does exist—in fact, it recently celebrated its hundredth birthday. It is aspirin. Even more remarkably, our understanding of how aspirin works and why it has so many remarkable effects on the human body is still in its infancy.

In the years following 1899, when the basic aspirin formula was developed, the drug became an important means of lowering pain and reducing fevers and inflammation. For decades these therapeutic benefits were more than enough to make aspirin a staple in every pharmacy and hospital around the world. They even spawned the well-known phrase "take two aspirin and call me in the morning," implying that aspirin could handle most medical problems overnight.

Today we know that aspirin can have even more amazing effects on human health than in its well-established role as a pain reliever. Well-publicized studies show that low doses of aspirin taken daily can reduce the risk of heart disease. In 1996, the importance of aspirin was further bolstered by studies showing that people who regularly take aspirin have lower rates of colon cancer. However, taking aspirin for prolonged periods does have negative side effects, including damage to the stomach and kidneys. Given all the possible health benefits of taking aspirin, researchers had some obvious questions: How does this wonder drug work, and can we make it better by reducing the side effects?

Today we know that the reason why aspirin affects so many processes in the body is that it blocks the synthesis of certain chemical messages, or

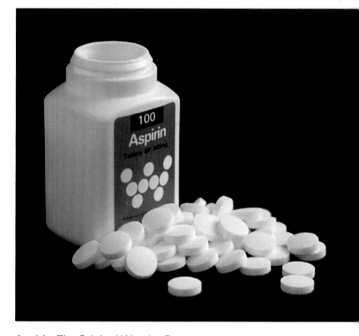

Aspirin: The Original Wonder Drug

This common painkiller turns out to have other medicinal uses as well, such as helping to prevent heart disease and certain cancers.

hormones. Ironically, this action also accounts for the negative side effects of aspirin. Armed with a growing knowledge of how aspirin affects these chemical reactions, researchers are now seeking to improve this old wonder drug by eliminating the negative aspects of its action.

All biological processes require energy, which living organisms must extract from their environment. They use this energy to manufacture and transform the various chemical compounds that make up living cells. The capture and use of energy by living organisms involves thousands of chemical reactions, which together are known as metabolism.

All the chemical reactions that occur in cells can be grouped into sequences called metabolic pathways. Just as chemical building blocks fall into a limited number of categories, the types of reactions that allow the cell to assemble and disassemble these building blocks are limited in number.

In this chapter we examine the role played by energy in the chemical reactions that maintain living systems. We also discuss the role of specialized proteins, called enzymes, used by the cell to speed up chemical reactions that would otherwise be too slow to sustain life. Finally, we explore the possible connections between life span and the overall rate of life's chemical reactions.

7.1 The Role of Energy in Living Systems

The discussion of any chemical process in the cell is a discussion about energy. The idea that energy is behind every activity in the cell seems natural and unsurprising, since all of us are accustomed to thinking of energy as a form of fuel. However, energy is more than just fuel, because its properties dictate which chemical reactions can occur and how molecules can be organized into living systems.

The laws of thermodynamics apply to living systems

The relationship between energy and the cell's activities is governed by the same physical laws that apply to everything else in the universe. These laws of thermodynamics define the ways cells transform chemical compounds and interact with the environment. The **first law of thermodynamics** states that energy cannot be either created or destroyed, only converted from one form to another.

Consider what happens when you use electrical hedge clippers, which have a small gas tank and generator attached to the clippers. The chemical energy in the covalent bonds of the gasoline molecules is converted into electrical energy by the generator. This electrical energy, in turn, is converted into the mechanical energy of motion in the hedge clippers themselves. Neither the generator nor the hedge clippers creates or destroys energy.

At a cellular level, the first law of thermodynamics is illustrated by mitochondria, which convert energy from food molecules such as sugars into the energy of covalent bonds in ATP (ADP + phosphate + energy → ATP, as we saw in Figure 4.9). Thus mitochondria do not create energy from nothing. They convert energy from one form (sugars) into another form (ATP), which can then be used by the cell.

The **second law of thermodynamics** describes how the cell relates to its environment. This law states that systems tend to become more disorderly. This statement may seem most appropriate to describe a household room or a toolshed, which, unless we spend energy tidying it up, tends toward disorder (Figure 7.1a). But it is true of all systems, including the internal organization of the cell, an organism, or even the whole universe.

As we saw in Chapters 4 and 5, a cell is made up of many chemical compounds assembled into complex ordered structures. Such a high level of organization may seem to fly in the face of disorder, but it has an explanation. The tremendous structural complexity of the cell and its organelles exists in a constant struggle against chaos. To counteract the natural tendency toward disorder, the cell must use energy to keep things orderly.

As order is created in living systems, those systems pass on or transfer disorder by releasing heat into the environment. Heat is a form of energy that causes rapid and random movement of molecules, a condition that is highly disordered. Thus, when cells release heat, they increase the degree of disorder in the molecules of the environment, which compensates for the increasing order inside the cell (Figure 7.1b). There is a direct connection between cellular organization and the transfer of heat energy because the chemical processes used to build well-ordered structures are the same ones that produce the

Helpful to know

The term *thermodynamics* comes from physics and derives from Greek words meaning the movement (*dynamics*) of heat (*thermo*). Heat always moves from warmer to cooler areas, a basic principle governing all biological processes.

Figure 7.1 The Second Law of Thermodynamics
The disorder of a system tends to increase unless that tendency is countered by an input of energy. (*a*) Left unattended, all structures, such as this wooden toolshed, tend to lose their order and become disarrayed. An input of energy, here in the form of human effort, is needed to maintain the order of the structure. (*b*) Cells maintain their organization through a continuous input of energy from the environment. Thus they, too, obey the second law of thermodynamics.

heat. Hence the generation of order is directly coupled with the release of heat energy.

The flow of energy and the cycling of carbon connect living things with the environment

Where does the energy that creates order in the cell come from? We know from the first law of thermodynamics that the cell cannot create energy from nothing; thus it must come from outside of the cell. In other words, energy must be transferred into the cell in some fashion. In the case of photosynthetic organisms, the energy comes from sunlight. By using that energy to synthesize sugar molecules from carbon dioxide and water, those organisms convert it into the chemical bonds of sugars. For organisms that do not photosynthesize, energy comes from the chemical bonds in food molecules, such as sugars and fats.

The last two statements in the preceding paragraph reveal the chemistry of the relationship between produc-

ers and consumers. As we saw in Chapter 1, photosynthetic producers, such as plants, capture energy from sunlight, and nonphotosynthetic consumers, such as animals, obtain energy by consuming plants or other organisms that have consumed plants. This means that, thanks to photosynthesis, the sun is the primary energy source for living organisms.

However, plants are more than sources of consumable energy for nonphotosynthetic organisms. First, plants themselves use some of the sugars they make by photosynthesis; they do this especially at night, when there is no sunlight and no photosynthesis. Second, most organisms produce carbon dioxide (CO_2) as a by-product of the energy-harnessing process called respiration, and this CO_2, in turn, is a source of carbon for photosynthesis. In this way, carbon atoms are continually cycled from carbon dioxide in the atmosphere to sugars made by producers and back to carbon dioxide released by respiring producers and consumers (Figure 7.2). This kind of recycling occurs

Figure 7.2 Carbon Cycling
Carbon atoms cycle among producers, consumers, and the environment. Carbon atoms become parts of different kinds of molecules as they cycle between living organisms and the environment.

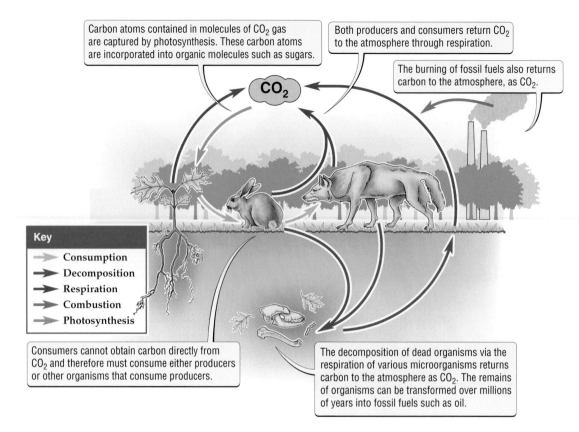

Carbon atoms contained in molecules of CO_2 gas are captured by photosynthesis. These carbon atoms are incorporated into organic molecules such as sugars.

Both producers and consumers return CO_2 to the atmosphere through respiration.

The burning of fossil fuels also returns carbon to the atmosphere, as CO_2.

Key
→ Consumption
→ Decomposition
→ Respiration
→ Combustion
→ Photosynthesis

Consumers cannot obtain carbon directly from CO_2 and therefore must consume either producers or other organisms that consume producers.

The decomposition of dead organisms via the respiration of various microorganisms returns carbon to the atmosphere as CO_2. The remains of organisms can be transformed over millions of years into fossil fuels such as oil.

not only for carbon, but also for other atomic building blocks of life, such as oxygen, nitrogen, and phosphorus.

7.2 Using Energy from the Controlled Combustion of Food

Living systems obtain energy from food by burning organic molecules such as sugars to form carbon dioxide and water. If our cells were to convert food into carbon dioxide and water in a single chemical reaction, however, we would burst into flame like a lit match. Here's the chemical equation that describes what happens when a wooden match burns:

$$Wood + O_2 \rightarrow CO_2 + H_2O + energy \text{ (as heat and light).}$$

This combustion reaction is similar to what happens when our cells burn food, but fortunately for us, there are some important differences.

The energy released from the burning match is dispersed into the environment as heat and cannot be regained by the match in any fashion. In contrast, living systems can capture and use energy that is gradually released through a series of reactions. The overall "combustion" reaction is broken down into a series of chemical reactions that release the energy in our food bit by bit. This multi-step process not only saves us from bursting into flame; it also gives our cells the opportunity to capture the small amounts of energy released at each step. On the other hand, the literal burning of food allows scientists to determine its nutritional value (see the box on page 139).

Capturing energy from foods requires the transfer of electrons

In the multiple chemical reactions that allow cells to capture energy from food, electrons are transferred from one molecule or atom to another. **Oxidation** is the loss of electrons from one molecule or atom to another, while **reduction** is just the opposite, the gain of electrons by one molecule or atom from another. Reactions such as these are called oxidation-reduction reactions, or simply **redox reactions**.

Let's consider the literal burning of an organic compound: the gas methane (CH_4). Its molecules consist

Science Toolkit

Counting Calories with a Bomb

Practically all labels on food products include information on how many Calories the food contains. The label on a bag of potato chips might read:

Nutrition Facts:
Serving Size: 1 oz. (28 g/15 chips)
Amount per Serving: Calories—140, Calories from Fat—70

But just what are Calories, and how can anybody determine how many Calories are in "15 chips"?

Calories are units of heat energy. Nutritionists and chemists define 1 Calorie as the amount of energy needed to raise the temperature of 1 liter of water 1°C. That definition hints at the way food manufacturers and nutritionists determine how many Calories are in a sample of food, such as 1 ounce of potato chips: they burn the chips and compare the amount of heat given off to the amount needed to heat a known amount of water.

To accomplish this, the food sample is burned in a device called a bomb calorimeter. The calorimeter is composed of a sealed container (the "bomb") surrounded by another sealed container with a known amount of water in it. A thermometer shows the temperature of the water before and after the sample is burned. Once the researchers know how much the temperature of the water was increased by burning the food, they can calculate the number of Calories that were in the sample. The value of 70 Calories from fat can be obtained by extracting the fats from the sample and burning them separately.

Bomb calorimeters measure 100 percent of the energy in a food sample, but our bodies are not 100-percent efficient, so the number of Calories our bodies actually use is less than the calorimeter's number. Beer, of course, doesn't burn, and many other foods and beverages have a high water content. In these instances the water is evaporated, and the calorimeter burns only the dry remains. So now you know all that goes into measuring the Calories in a regular beer (146 Calories for 12 fl oz or 99 for 12 fl oz), bananas (109 Calories for 118 grams), or even iceberg lettuce (7 Calories for 55 grams).

The thermometer measures the change in the temperature of the water as the food burns.

An ignition wire starts the burning process in the bomb.

The stirrer distributes the heat evenly in the water.

The outer chamber contains a known amount of water.

Water

The bomb contains the food sample and pure oxygen.

Inside a Bomb Calorimeter

of a single carbon atom bonded to four hydrogen atoms (Figure 7.3a). Because the bonds are covalent, the carbon and hydrogen atoms share their electrons. However, the electrons tend to be held more closely to the carbon atom than to the hydrogen atoms. The carbon has, in a sense, gained electrons and become more "electron-rich"; we also speak of the carbon as being in a reduced state.

If we burn some of the methane gas, the products of the combustion reaction are carbon dioxide (CO_2) and water (H_2O). What has changed? The hydrogen atoms have left the carbon atom and joined oxygen to form water. And

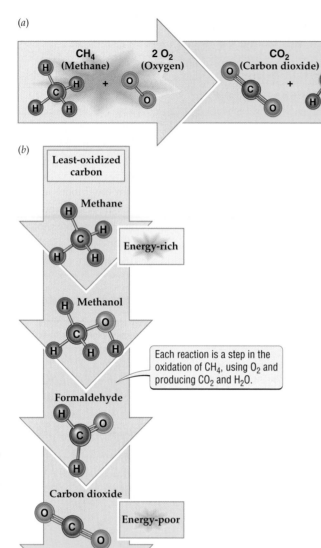

Figure 7.3
Oxidation of Methane
(a) When methane gas encounters a spark, it undergoes an explosive oxidation, a one-step combustion reaction. Energy is released to the environment as heat. (b) Methane can also be oxidized in a series of smaller steps. The single carbon atom becomes increasingly oxidized with each step. That is, the carbon atom is gradually surrounded by more oxygen atoms, which tend to attract its electrons, leaving it more oxidized.

Within the figure:
(a)
CH₄ (Methane) + 2 O₂ (Oxygen) → CO₂ (Carbon dioxide) + 2 H₂O (Water) + Energy (Heat)

(b)
Least-oxidized carbon
Methane — Energy-rich
Methanol
Each reaction is a step in the oxidation of CH₄, using O₂ and producing CO₂ and H₂O.
Formaldehyde
Carbon dioxide — Energy-poor
Most-oxidized carbon

once. Figure 7.3*b* shows such a process for methane. Instead of jumping from being part of the methane molecule to being part of a simple CO₂ molecule, the carbon atom passes through several chemical reactions and intermediate compounds. Each intermediate compound is a little more oxidized than the preceding one. In living systems, this kind of stepwise and controlled combustion allows the cell to couple each small energy-releasing oxidation reaction with other reactions, which store some of the released energy in newly formed chemical bonds. This transfer of energy from one compound to another is the basis of metabolic reactions.

Metabolism consists of all the chemical reactions within living organisms that produce complex macromolecules, such as sugars and proteins, and which break down those macromolecules to yield smaller molecules and usable energy. Metabolic reactions that create complex molecules out of smaller compounds are called **biosynthetic**, or **anabolic**, reactions; those that break down complex molecules to produce usable energy are called **catabolic** reactions.

Energy in a living system is transferred via the universal energy carrier, ATP (Figure 7.4). When ATP is produced from ADP and a phosphate group (see Figure 4.9), energy is stored in its chemical bonds. The energy released from ATP when it is broken down to ADP and phosphate is used to carry out the activities of the cell.

the carbon atom is now bound to two oxygen atoms. The oxygen atoms tend to pull the electrons *away* from the carbon atom. Compared with its situation in the methane molecule, the carbon atom in CO₂ has "lost" electrons, leaving it in an "electron-poor" state. The carbon atom has also gained oxygen atoms; it has become oxidized. Thus the combustion of an organic compound such as methane is an oxidation reaction.

Note that although oxidation is defined as the loss of electrons, generally speaking, it is also accompanied by the *addition* of oxygen to a compound—indeed, that is where the process originally got its name. Similarly, in reduction reactions, the carbon atom often *loses* oxygen atoms.

In contrast to the combustion of methane, biological oxidation takes place in a series of small steps, not all at

Figure 7.4 ATP Molecule
Adenosine triphosphate (ATP) is the universal energy carrier, with the chemical formula C10.H16.N5.O16.P3. It is made up of carbon (yellow), oxygen (red), nitrogen (blue), hydrogen (white), and phosphorus (green) atoms.

ATP production is an urgent priority for the human body; if it were halted, each cell would consume its entire supply of ATP in about a minute. In fact, in nearly every chemical reaction in the cell, ATP is consumed or synthesized.

The energy released from ATP when it is broken down to ADP and phosphate is used for a variety of activities in the cell—such as moving molecules and ions between various cellular compartments and generating mechanical force in a crawling cell—as well as for biosynthetic reactions. The catabolic reactions in the cell that release energy and harness it in the form of ATP are tightly coupled to the biosynthetic reactions that manufacture complex macromolecules from simpler chemical compounds. Thus the two kinds of reactions in metabolism—releasing energy by breaking things down, and using energy to build things up—are intimately related.

Chemical reactions are governed by simple transformations of energy

How does the cell control such a powerful event as combustion and break it down into smaller, more manageable and useful steps? The answer lies in the very nature of chemical reactions. Let's review some of the fundamental principles that govern chemical reactions, as represented by the following general example:

$$A + B \rightarrow C + D.$$

A and B are the starting materials, or reactants; C and D are the products formed by the reaction. Recall that a chemical reaction involves changing the arrangement of atoms in molecules. All chemical reactions tend to proceed in the direction that will result in products with greater stability and a lower energy state. Products whose energy state is lower than that of their reactants have less energy stored in their chemical bonds; these products are therefore in a less ordered state, as favored by the second law of thermodynamics. This tendency toward less order and a lower energy state encourages chemical reactions to proceed in a particular direction—namely, "downhill," energetically, from reactants to products.

Just because compounds A and B are present together, however, does not mean that they will react with each other. All compounds tend to be in a semi-stable state. They need to be destabilized, or activated ("jump-started"), before a chemical reaction can begin and proceed. The required jump start is the input of a small amount of energy, called the **activation energy** of the reaction. The activation energy alters the chemical configuration of the reactants so that the reaction can take

place. Once the reactants have overcome this activation energy barrier, the reaction proceeds, and a lower-energy product results (Figure 7.5a).

To make this concept clearer, imagine a basket full of puppies on a hilltop (Figure 7.5b). By themselves, even though they tumble about and jostle one another, the puppies don't have enough motivation (internal energy) to get out of the basket; the sides are too high. Then the puppies hear their mother barking, calling them from the bottom of the hill. This gives them just enough of an added impetus (an input of external energy) to try harder to get out of the basket (to jump-start the reaction). The

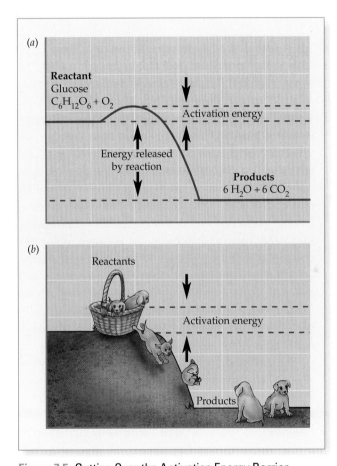

Figure 7.5 Getting Over the Activation Energy Barrier

(a) The oxidation of glucose must overcome an activation energy barrier to begin the series of "downhill" reactions that take place during respiration to produce water and carbon dioxide. The solid red line indicates the amount of energy associated with the reacting compounds. (b) To understand this, imagine that the reactants of a chemical reaction are a group of frisky puppies trying to get out of a basket sitting on a slope. The sides of the basket represent the activation energy barrier. The puppies' mother (not shown, but calling to them from somewhere at the bottom of the hill) represents the input of external energy that gets the puppies over the barrier, and the bottom of the slope represents the lower energy state of the products (the puppies that have made it down the hill).

puppies manage to scramble over the edge of the basket (overcome the activation energy barrier) and tumble downhill (the reaction going forward).

There they greet their mother, curl up against her, and go to sleep. Like the products in our hypothetical reaction, they are now in a lower energy state. Even if they tried, it would be more difficult for them to make it back up the slope and into the basket (return to being reactants); getting back up and in requires a lot more effort. Thus the preferred direction for the puppies is downhill, not back uphill. Similarly, the preferred direction for a chemical reaction is toward producing products, not backward from products to producing more of the reactants.

7.3 How Cells Speed Up Chemical Reactions

Where do chemical reactions in cells get the activation energy required for them to proceed? It comes from two sources. For the first, let's return to the example of a burning match: the activation energy required to light the match can come from the friction generated by moving the head of the match across a rough surface. Similarly, cells can acquire some of the activation energy they need from random collisions between molecules floating in the cytosol. These collisions become more frequent and energetic as the temperature increases and molecules in the cytosol move faster. At the normal body temperatures of most organisms, however, these collisions do not release enough energy to drive all the reactions required for life.

The second source of activation energy is a process also found in nonliving systems. In **catalysis**, one or more chemical substances (**catalysts**) participate in the chemical reaction in such a way as to lower the amount of activation energy required; this in turn greatly increases the rate at which the reaction proceeds. An important characteristic of catalysts is that, unlike reactants, they remain chemically unaltered after the reaction is over.

Enzymes speed up chemical reactions

In cells, the catalysts are a specialized class of proteins called **enzymes**. Nearly all chemical reactions that take place in living organisms require their assistance. Human cells, for example, contain several thousand different enzymes, each of which helps with a specific chemical reaction.

To increase the rate at which a chemical reaction proceeds, each enzyme binds to specific reactants, called its **substrates**. This binding action is what lowers the amount of activation energy the substrates need in order to react with each other. The enzyme brings them together in an orientation that favors the making or breaking of bonds required to form the products. For a particular enzyme to bind the correct reactants and alter them appropriately for the chemical reaction to proceed, it must have a high degree of specificity for those reactants. In other words, each enzyme must be specifically tailored to promote only one of the thousands of possible reactions that occur in the cell.

In the presence of many reactants, the presence of one enzyme rather than another can determine which chemical reaction takes place. However, an enzyme cannot make an impossible reaction happen. Nor can it promote a particular reaction by changing the amount of energy associated with the reactants or the products. It can only affect the rate at which a reaction occurs by lowering the activation energy barrier.

The hydration of carbon dioxide requires catalysis

A good example of how necessary enzymes are for life is the removal of carbon dioxide from cells in the human body. This process depends on an enzyme in red blood cells. Before we can exhale carbon dioxide, the gas must first be transferred from our cells into the bloodstream and then transported to the lungs. For this transfer to occur, carbon dioxide must react with water so that it can be transported in the blood as bicarbonate ions (HCO_3^-):

$$H_2O + CO_2 \rightarrow HCO_3^- + H^+.$$

This simple reaction, called the hydration of carbon dioxide (because it involves the addition of water), is necessary for both normal breathing and cellular respiration. The enzyme that catalyzes this reaction is called carbonic anhydrase; it speeds up the hydration of carbon dioxide by a factor of nearly 10 million. In fact, a single carbonic anhydrase molecule can hydrate more than 10,000 molecules of carbon dioxide in a single second. Needless to say, without carbonic anhydrase, the rate of carbon dioxide hydration would be so slow that we would not be able to rid our bodies of carbon dioxide fast enough to survive. When the circulating bicarbonate ions arrive at the lungs, they are converted back into carbon dioxide, which we then exhale as CO_2 gas.

Helpful to know

This chapter introduces several similar-sounding words beginning with *cata*. *Catabolism* refers to the "breaking down" reactions of metabolism. *Catalysis*, on the other hand, describes how reactions are facilitated, not whether molecules are broken down or built up. Most catabolic reactions are catalytic—they require enzymes (catalysts) to occur at the speed needed to sustain life—but the same is also true for most biosynthetic reactions.

The shape of an enzyme directly determines its activity

The specificity that enzymes have for their substrates depends on the three-dimensional shapes of both substrate and enzyme molecules. In the same way that a particular lock accepts only a key with just the right shape, each enzyme has an **active site** that fits only substrates with the correct three-dimensional shape and chemical characteristics (Figure 7.6a). Because of this specificity, enzymes are like molecular "matchmakers," bringing the right substrates together. The matching of an enzyme's active site to one or more substrates guarantees that a specific reaction will take place to yield the expected products.

Carbonic anhydrase, for example, is able to bind molecules of both carbon dioxide and water to its active site. By bringing these two substrates together in just the right positions, the active site of carbonic anhydrase promotes the hydration reaction (Figure 7.6b). If no enzyme were present, the two substrates would need to collide with each other in just the right way before the hydration reaction could take place. These sorts of molecular collisions do occur all the time, but not nearly as frequently as would be required for the continuous and rapid transfer of carbon dioxide from cells into the blood.

Sequences of reactions have energetic advantages

So far, we have discussed the activity of a single enzyme acting alone to promote a single chemical reaction, but this state of affairs is not common in the cell. Instead, groups of enzymes usually catalyze multiple steps in a sequence of chemical reactions known as a **metabolic pathway**. A metabolic pathway presents both advantages and challenges that illustrate important aspects of how enzymes usually behave in the cell.

Let's begin with the most noteworthy advantage of multi-step metabolic pathways: they permit the product whose production is catalyzed by one enzyme to immediately become a reactant in another reaction. In other words, the product of the first reaction is the immediate substrate for another enzyme and is rapidly consumed in a second catalyzed reaction. In general, a pathway of enzyme-catalyzed steps ensures a particular outcome from a sequence of chemical reactions. Such a sequence can be represented as follows:

$$A \xrightarrow{E1} B \xrightarrow{E2} C \xrightarrow{E3} D.$$

This equation says that enzyme E1 catalyzes the conversion of A to B, enzyme E2 catalyzes the conversion of B

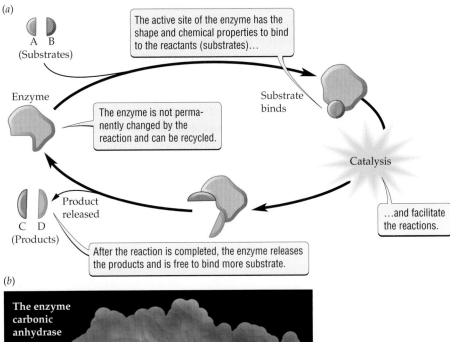

(a)

The active site of the enzyme has the shape and chemical properties to bind to the reactants (substrates)...

A B
(Substrates)

Enzyme

The enzyme is not permanently changed by the reaction and can be recycled.

Substrate binds

Catalysis

...and facilitate the reactions.

Product released

C D
(Products)

After the reaction is completed, the enzyme releases the products and is free to bind more substrate.

(b)

The enzyme carbonic anhydrase

Reactants

Active site

Figure 7.6 Enzymes as Molecular Matchmakers
(a) An enzyme brings together two reactants (A and B) such that a chemical reaction proceeds to form the products (C and D). (b) Carbonic anhydrase catalyzes the reaction of carbon dioxide and water to form bicarbonate.

to C, and so on, ensuring that D will be produced in the end. Metabolic pathways like this one produce most of the chemical building blocks of the cell, such as amino acids and nucleotides, and are necessary for the harnessing of energy from food or sunlight.

The challenge faced by all enzyme-catalyzed metabolic pathways is the need for the products of each reaction to find the enzyme that catalyzes the next reaction in a timely fashion. Enzymes and their substrates do not actively swim after each other like sharks looking for prey. Instead, they depend on random encounters or collisions inside the cell. Within the crowded environment of the cell, an enzyme collides with many molecules every second, but most of these molecules are not its substrates and do not fit into its active site, so nothing happens. Catalysis occurs only when the enzyme encounters appropriate substrates

Learn more about enzyme catalysis.

7.1

that fit its active site. Thus, although molecular collisions happen frequently in the cell, only some of these collisions result in enzyme-catalyzed reactions.

One way to increase the efficiency of a metabolic pathway is to increase the frequency of collisions between enzymes and their substrates. The enzymes involved in many of the multi-step pathways that are so common in metabolism are located close together. This arrangement means that the products of one reaction are physically close to the next enzyme that uses them as substrates, increasing the likelihood of their collision with it and thus promoting the next chemical reaction.

At the level of cellular organization, the enzymes necessary for a particular metabolic pathway can be contained inside a specific organelle (Figure 7.7). As we saw in Chapter 5, organelles concentrate the proteins and chemical compounds required for specific biological processes. Mitochondria, for example, are the sites where the breakdown products of food molecules are oxidized, generating most of the cell's ATP. The efficient production of ATP requires that the necessary enzymes and substrates be concentrated inside a small compartment. Several enzymes floating in the mitochondrial matrix participate in a series of reactions called the citric acid cycle, which forms the first part of the metabolic pathway that produces ATP. Other enzymes involved in the production of ATP are sequentially arranged in association with the inner mitochondrial membrane.

At the molecular level, several enzymes can be physically connected in a single giant multienzyme complex (see Figure 7.7). Many enzymes involved in the biosynthesis of cellular building blocks such as fatty acids and proteins function as large aggregates of multiple enzymes.

Figure 7.7 Grouping of Enzymes in the Cell
Enzymes are arranged in the cell so as to promote the multiple chemical reactions of metabolic pathways. These arrangements include the concentration of enzymes in organelles (in this case, a mitochondrion), their localization in membranes, and their clustering in multienzyme complexes.

7.4 Enzymes and Energy in Use: The Building of DNA

Enzyme-catalyzed metabolic pathways are involved in the synthesis and breakdown of most complex molecules in the cell. The biosynthesis of DNA is a good example. Before cells can divide, their DNA must be replicated so that each daughter cell can receive a complete set of the genetic material. This essential process also ensures that our genetic blueprint is passed from one generation to the next. Replicating a cell's DNA requires the synthesis of new molecules of DNA, which is a polymer of covalently bonded nucleotides. This process involves ATP and several enzymes, since individual nucleotides must be converted to an activated form before they can be added to a growing DNA chain.

Nucleotides can have one, two, or three phosphate groups bound to the sugar molecule, and are described as monophosphate, diphosphate, or triphosphate nucleotides, respectively (see Figure 4.8). Energy-rich triphosphate nucleotides are used for DNA synthesis. They are produced from monophosphate nucleotides with the help of two enzymes and ATP. Each of the enzymes catalyzes the transfer of a phosphate group from a molecule of ATP to the original monophosphate nucleotide (Figure 7.8*a*). Since the high-energy phosphate bonds of two ATPs are consumed in these reactions, the triphosphate nucleotide is now energy-rich and can form a covalent bond with another nucleotide. The chemical reaction that adds an "energized" triphosphate nucleotide to the end of a DNA chain is catalyzed by a third enzyme, resulting in the release of the two terminal phosphate groups (Figure 7.8b).

In addition to DNA, other complex molecules, such as sugars and fatty acids, are produced using ATP and specific enzymes. These and other biosynthetic pathways create and maintain the complex structures found in every living cell.

7.5 Metabolic Rates and Life Span

The idea of immortality has fascinated human beings throughout recorded history. Today, researchers are discovering which biological factors limit our life span and how those factors might be controlled to let us live longer. One key factor that has emerged is the overall metabolic rate of an organism. In general, small animals have faster metabolic rates and shorter life spans than large animals. Furthermore, slowing down the metabolism of an animal can alter its life span. For example, laboratory tests have shown that putting mice on a restricted diet slows down their metabolism (Figure 7.9), and that these mice live longer than mice that are allowed to eat as much as they like.

The idea that a higher metabolic rate can shorten one's life span may seem contradictory, since metabolism includes all of the chemical reactions that maintain living organisms. How might metabolism shorten the life span of an organism? The answer lies not in the reactions of metabolism themselves, but rather in the toxic chemical by-products that are sometimes accidentally produced when these reactions occur. These chemical compounds react with and damage cellular components such as DNA. The gradual accumulation of this cellular damage is an important contributing factor to aging and, ultimately, death.

The link between a slower metabolism and a longer life holds true for the nematode worm *Caenorhabditis elegans* [KIGH-no-rab-DIE-tiss EL-eh-ganz] as well as for mice. Worms that lack proteins responsible for maintaining a

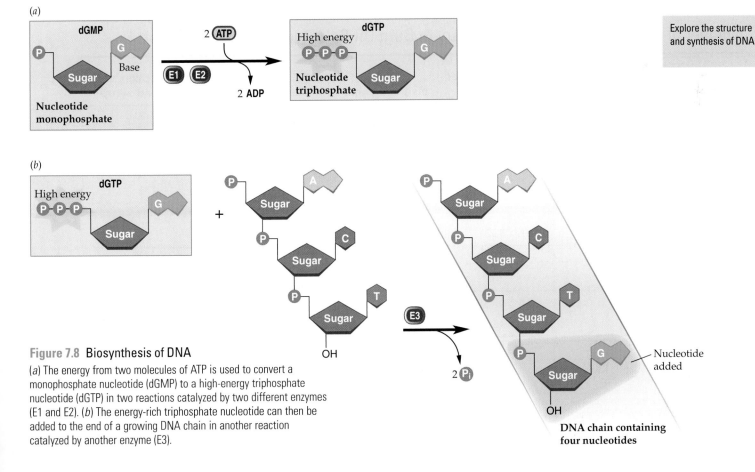

Figure 7.8 Biosynthesis of DNA
(a) The energy from two molecules of ATP is used to convert a monophosphate nucleotide (dGMP) to a high-energy triphosphate nucleotide (dGTP) in two reactions catalyzed by two different enzymes (E1 and E2). (b) The energy-rich triphosphate nucleotide can then be added to the end of a growing DNA chain in another reaction catalyzed by another enzyme (E3).

Explore the structure and synthesis of DNA. 7.2

Biology Matters

Energy Is a Weighty Subject

In biology, energy lies at the center of all activity, including simply staying alive. If you weigh about 140 pounds and are sitting still, you are using about as much energy per hour as a 75-watt light bulb. This energy allows you to breathe, keep your heart beating, and think. Nutritionists use the term *basal metabolic rate* (BMR) to describe how much energy we use each hour when we are at rest. The table at right shows average BMRs for males and females. Your metabolic rate varies with how active you are. During any activity, it will be higher than your resting rate, although how much higher depends on the activity.

For a 150-pound person, brisk walking consumes about 350 calories per hour. The same person bicycling at a speed of about 20 miles per hour burns around 720 calories per hour, and running at a moderate pace burns approximately 846 calories per hour.

If you take in more calories than your BMR consumes, you will gain weight, unless you increase your activity level to use up the extra calories. Energy also lies at the center of all weight gain. If you don't use up all of the energy contained in the food you eat, it is stored as fat. How, then, do we avoid taking in too much energy? One way to approach this question is to become more aware of the relationship between the number of calories in foods and how much energy particular activities consume. We all know that some foods contain more calories than others. Therefore, one way to avoid weight gain is to cut down on their intake and choose foods that contain less energy. Another option is to become more active. The below table lists the number of minutes it would take to consume the energy in some of our favorite foods by doing different activities. Once you become aware of the amount of walking necessary to burn off, for example, a cheeseburger and fries—over two and one-quarter hours!—you can develop a clearer picture of how you might balance dietary choices and physical activity.

	Weight in kilograms	Weight in pounds	Basal metabolic rate in watts per hour	Basal metabolic rate in calories per hour
Woman	60	132	68	58
Man	70	154	87	75

Number of Minutes of Exercise Necessary to Burn Off Selected Foods

Exercise	Cheeseburger and fries	Large hamburger and fries	Fish and chips	Chicken and fries	Three pancakes with butter and syrup
Aerobics					
Active	88	114	74	102	67
Golf					
With Trolley	224	290	188	260	171
Dancing					
Energetic	102	132	85	118	78
Jogging					
5 Miles per Hour	81	105	68	95	62
Swimming					
Steadily	81	105	68	95	62
Walking					
3 Miles per Hour	136	176	114	158	104

Note: All figures are approximate, and are based on a 150-pound woman. If you weigh more, you will burn more calories. If you weigh less, you will burn fewer.

normal metabolic rate, and which have an abnormally low metabolic rate as a result, live up to five times longer than "normal" worms. These mutant worms also take longer to mature and display a slowing of specific behaviors. Similar phenomena have been observed in fruit flies and in yeasts, implying that a broad range of species are subject to life-limiting metabolic accidents.

Given the supporting evidence gathered from several species, it is not surprising that the human life span may also be affected by metabolic rates. Women have slower metabolic rates than men, and they have a higher life expectancy (in the United States, 79 years versus 72 years for men). Even more striking is the fact that nine out of ten individuals 100 years old or older are women. Thus, although genetic factors that affect overall health also have a role in determining life expectancy, the accumulated mistakes from a lifetime of metabolism clearly take their toll. Scientists are considering ways to limit the metabolic activities that run the highest risk of producing toxic by-products, with the ultimate goal of extending our life span.

Figure 7.9 Metabolism in the Lab
To measure the metabolism of lab rats, the animals are put in glass containers that measure the volume of air exhaled while they are fed sugar.

Making a Wonder Drug Even Better

Enzymes control the rates of specific chemical reactions in the body, so altering enzyme activity sometimes causes illness and sometimes promotes healing. The effects of aspirin, it turns out, are a result of its blocking of the activity of two important enzymes, COX-1 and COX-2. (COX stands for cyclooxygenase [SIGH-kloh-OX-uh-jeh-nace].) COX-1 is continuously produced in the body and catalyzes the biosynthesis of hormones that help maintain the lining of the stomach. In contrast, COX-2 is produced only when an injury occurs, and it catalyzes the biosynthesis of different hormones that promote inflammation, fever, and the sensation of pain throughout the body. Although both enzymes are inhibited by aspirin, they participate in two different biosynthetic pathways and play different biological roles. The therapeutic benefits of aspirin (reduction of pain, inflammation, and fever) are due to its blocking of COX-2, while the negative side effects

(damage to the stomach lining) are due to its blocking of COX-1.

The blocking of COX-2 activity is probably also responsible for the effects of aspirin on colon cancer. Some cancerous cells have an abnormally high level of COX-2, which may encourage the growth of blood vessels into the tumor, thereby allowing the tumor to be nourished and thus to grow into a more serious cancer. By blocking COX-2 activity, aspirin may limit the blood supply to tumors and reduce the spread of the cancer.

How can we make aspirin better, with fewer negative side effects? The simple answer is to develop a drug that blocks only COX-2 activity and has little or no effect on COX-1. This challenge has been enthusiastically taken up by many research laboratories around the world, resulting in the recent development of a first generation of superaspirins.

The first successful step in developing super-aspirins was based on an understanding of the three-dimensional shape of the COX-2 enzyme. Recall that the shape of an enzyme defines its catalytic activity. Knowing the shape of the COX-2 enzyme allowed researchers to design inhibitor molecules that bind only to COX-2 and not to COX-1.

Over a dozen new compounds have been developed that bind to and block the activity of COX-2 while having no significant effect on COX-1. This first generation of superaspirins has been available for some time. While their benefits seem substantial, serious questions have arisen regarding the role of these superaspirins in causing heart disease.

Chapter Review

Summary

7.1 The Role of Energy in Living Systems

- Living organisms obey the same laws of thermodynamics that apply to the physical world.
- Consistent with the first law of thermodynamics, energy used by organisms is converted from one form to another, but is never created or destroyed.
- As predicted by the second law of thermodynamics, the creation of biological order is always accompanied by the transfer of disorder to the environment, most often in the form of heat.
- The sun is the primary energy source for living organisms.
- Atomic building blocks such as carbon are cycled between living organisms and the environment.

7.2 Using Energy from the Controlled Combustion of Food

- In oxidation, electrons are lost from one molecule or atom to another. In reduction, electrons are gained by one molecule or atom from another.
- Metabolism consists of two kinds of reactions that are tightly coupled with each other: catabolic reactions, which break down macromolecules and harness energy, and biosynthetic (anabolic) reactions, which build macromolecules and require energy.
- All chemical reactions require the input of a small amount of activation energy to proceed.

7.3 How Cells Speed Up Chemical Reactions

- Catalysis occurs when one or more chemical substances (catalysts) participate in a reaction but (unlike the reactants) remain chemically unaltered after the reaction is over. Catalysts speed up chemical reactions by lowering the amount of activation energy required.
- Most chemical reactions that support life require catalysis. In cells, the catalysts are proteins called enzymes.
- The activity of enzymes is highly specific. Each enzyme binds to a specific substrate or substrates and catalyzes a specific chemical reaction.

- The specificity of an enzyme depends on its three-dimensional shape and the chemical characteristics of its active site.
- Each of the multiple steps in a metabolic pathway is catalyzed by a different enzyme.

7.4 Enzymes and Energy in Use: The Building of DNA

- The replication of DNA during cell division involves building new DNA molecules by chaining nucleotides together with covalent bonds.
- Reactions catalyzed by two different enzymes use energy from ATP to add two phosphate groups to a monophosphate nucleotide, converting it into an energy-rich triphosphate nucleotide. A third reaction, catalyzed by a third enzyme, uses the energy in the two phosphate bonds to bind the nucleotide to the end of a growing DNA chain.

7.5 Metabolic Rates and Life Span

- The overall metabolic rate of an organism, in combination with genetic factors, may determine its life span.
- Higher metabolic rates are linked to higher production of toxic chemical by-products that can damage cells and shorten life span.

◉ Review and Application of Key Concepts

1. Describe the role of the second law of thermodynamics in living systems.

2. Describe carbon cycling in relation to photosynthesis and the catabolic synthesis of sugars.

3. Cells use several methods to increase the efficiency of enzyme catalysis in metabolic pathways. Describe two of these methods and how they apply to mitochondria.

4. Paying attention to the shapes of the enzymes involved, design a hypothetical metabolic pathway that changes a rectangle into two identical squares, then the squares into two identical triangles.

Key Terms

activation energy (p. 141)
active site (p. 143)
anabolic (p. 140)
biosynthetic (p. 140)
catabolic (p. 140)
catalysis (p. 142)
catalyst (p. 142)
enzyme (p. 142)
first law of thermodynamics
   (p. 136)

metabolic pathway (p. 143)
metabolism (p. 140)
oxidation (p. 138)
redox reaction (p. 138)
reduction (p. 138)
second law of thermodynamics
   (p. 136)
substrate (p. 142)

Self-Quiz

1. Which of the following statements is true?
   a. Cells are able to produce their own energy from nothing.
   b. Cells use energy only to generate heat and move molecules around.
   c. Cells obey the same physical laws of energy as the nonliving environment.
   d. Photosynthetic plants have no effect on the way animals obtain energy.

2. Living organisms use energy to
   a. organize chemical compounds into complex biological structures.
   b. decrease the disorder of the surrounding environment.
   c. cancel the laws of thermodynamics.
   d. prevent catabolic reactions.

3. The carbon atoms contained in organic compounds such as proteins
   a. are manufactured by cells for use in the organism.
   b. are recycled from the nonliving environment.
   c. differ from those found in CO_2 gas.
   d. cannot be oxidized under any circumstances.

4. Oxidation is
   a. the removal of oxygen atoms from a chemical compound.
   b. the gain of electrons by an atom.
   c. the loss of electrons by an atom.
   d. the synthesis of complex molecules.

5. Which molecule is most oxidized?
   a. CO_2 c. CH_3O
   b. CH_2O d. CH_4

6. The small input of energy required before a chemical reaction can proceed
   a. is called activation energy.
   b. is independent of the laws of thermodynamics.
   c. is provided by an enzyme.
   d. always takes the form of heat.

7. Activation energy is most like
   a. the energy released by a ball rolling down a hill.
   b. the energy required to push a ball from the bottom of a hill to the top.
   c. the energy required to get a ball over a hump and onto a downward slope.
   d. the energy that keeps a ball from moving.

8. Enzymes
   a. provide energy that speeds up reactions.
   b. are consumed during the reactions that they speed up.
   c. catalyze reactions that would otherwise never occur.
   d. catalyze reactions that would otherwise occur much more slowly.

9. The active site of an enzyme
   a. has the same shape for all known enzymes.
   b. can bind both its substrate and other kinds of molecules.
   c. does not play a direct role in catalyzing the reaction.
   d. can bring molecules together in a way that allows a chemical reaction to take place.

10. Metabolic pathways
    a. always break down large molecules.
    b. always bind smaller molecules together.
    c. are catalyzed by enzymes.
    d. occur only in mitochondria.

Biology in the News

===

More Heart Risks Found for COX-2 Inhibitors

Cautionary Information Was Released Last Month for Two More Drugs in This Class

By Susan J. Landers

More trouble was spotted in the family of medicines that includes the recently withdrawn Vioxx.

Pfizer Inc., which markets the COX-2 inhibitors Celebrex (celecoxib) and Bextra (valdecoxib), released new information on Dec. 17, 2004, indicating that Celebrex, like Merck & Co.'s Vioxx (rofecoxib), appears to pose an increased cardiovascular risk.

===

Sometimes an apparently exciting solution to a medical problem can give rise to serious side effects. This seems to be the case with the "superaspirins" that inhibit the action of the COX-2 enzyme. At first glance, they seemed to be a major step forward in pain relief, helping people suffering chronic pain from diseases such as arthritis without exposing them to the negative side effects of aspirin. But bad news has emerged. Further studies indicate that serious heart problems and even death could result from their use. The search for new medical solutions continues. As this research progresses, it may become clear that more extensive testing should be required of new superaspirins, so that a broader range of possible problems can be anticipated.

The COX-2 inhibitor story has broad implications for the use of medications, prescribed and otherwise, and for the assumptions we as consumers make about them. No consumer taking COX-2 inhibitors to relieve pain could have known about their possible dangers. Our physicians inform us to the best of their ability about the risks and benefits of the drugs they prescribe.

That informed advice is valuable, but the recent news about COX-2 inhibitors should lead us to be more careful. Not only should we ask our doctors more questions about potential risks as we consider whether or not to take their advice; we should also become more aware of the potential hazards of many over-the-counter medications and other health-related products. Many of these products contain explicit warnings on their labels or in leaflets inserted in the packages. But how many of us take the time to read these warnings carefully? Who would expect, for example, to see "KEEP OUT OF REACH OF CHILDREN" on the label of a bottle containing 1,000 mg tablets of vitamin C? Or a warning on a bottle of iron-containing multivitamins that reads, "Accidental overdose of iron-containing products is a leading cause of fatal poisoning in children under six"?

The COX-2 inhibitor story encourages us to take better notice of the warnings that are already available for prescription and nonprescription health products alike. It also warns us not to be surprised when new research on the effects and effectiveness of these products contradicts old news.

Evaluating the news

1. Advertisements for prescription drugs are flooding the airwaves. Almost all of them end with a "side effects may include" message. How does the COX-2 inhibitor news affect your reaction to these ads?
2. When we have common maladies such as colds and headaches, we routinely take over-the-counter medications without much thought about the possible harmful consequences. Yet even some brands of toothpaste have labels that caution against swallowing

the material. How will the implications of the COX-2 inhibitor case affect your purchasing behavior at the pharmacy?

3. Many "alternative" medicines are available without prescription. These products, most of which derive from long-established "folk" medicines, are claimed to cure or prevent many maladies and dis-eases. A growing number of respected physicians sell or recommend them in place of, or in addition to, prescription drugs. What general questions would you ask your doctor about these remedies?

SOURCE: *AMNews*, January 3/10, 2005.

CHAPTER **8** # Photosynthesis and Respiration

Key Concepts

⊙ The temporary storage and transfer of usable energy in the cell requires the production of energy carrier molecules, including ATP, NADH, and NADPH.

⊙ Photosynthesis is a series of chemical reactions that use sunlight, atmospheric carbon, and water to produce energy carrier molecules and to manufacture sugars. The process also releases oxygen gas into the environment.

⊙ There are two groups of reactions in photosynthesis: one set requires sunlight, while the other can occur in the dark.

⊙ Catabolism is a series of chemical reactions that break down food molecules and produce ATP.

⊙ In most eukaryotes, catabolism at the cellular level occurs primarily through aerobic respiration, which requires oxygen. Aerobic respiration has three stages: glycolysis, the citric acid cycle, and oxidative phosphorylation.

⊙ In some organisms, catabolism consists of glycolysis followed by other steps that do not require oxygen.

Food for Thought

The next time you feel hungry after skipping a meal, keep in mind that much of your body's demand for food is being made by your brain. A distinctive feature of the human brain is its size and need for energy. Although other animals, such as whales, certainly have larger brains by weight, the human brain is the largest when compared with the size of the human body. In other words, humans have the highest ratio of brain to body weight, which contributes to our status as the most intelligent of animals.

A daily challenge of having such a large brain is the urgent need to supply it with energy. Your brain consumes a large amount of energy while processing, sending, and receiving nerve impulses. Its energy demand is so high that more than half of the nourishment consumed by an infant is used by its brain.

Given the tremendous energy demands made by your brain, how does your body manage to keep it satisfied? The answer ultimately lies in how energy is captured from sunlight by photosynthetic organisms and used to manufacture sugars. These sugars are used in turn by all organisms, including humans, to provide the energy that supports biological processes.

A Scan of the Human Brain

iven all the complex structures and chemical reactions in every living system, obviously some kind of fuel must power it all. The capture of energy from the environment is one of the fundamental processes that support life. Ultimately, the sun is the primary source of energy for life on Earth. Solar energy is used by plants and other photosynthetic organisms to make sugars from carbon dioxide and water. Organisms that consume these producers acquire this chemical energy, which in turn is passed on to other organisms that eat those consumers.

As we saw in Chapter 7, the chemical reactions that transfer energy from one molecule to another and from one organism to another form the basis of metabolism. In this chapter we begin by examining how energy is transferred from one molecule to another. Then we explore the biosynthetic processes of metabolism by discussing how plants capture solar energy and use it to form the chemical bonds found in food molecules such as sugars. Finally we discuss the catabolic processes of metabolism, in which food molecules are oxidized and broken down to produce usable forms of energy.

8.1 Energy Carriers: Powering All Activities of the Cell

One method of energy transfer commonly found in the physical world depends on heat. When water in a kettle is boiled on a stove, the energy from the flame is transferred in the form of heat to the metal of the kettle and then to the water molecules. Living organisms generally cannot use such violent means of energy transfer. Instead, they use specialized molecules that transfer the chemical energy stored in covalent bonds in smaller, more manageable steps.

The temporary storage and transfer of energy in the cell depends on several **energy carrier** molecules. The most commonly used energy carrier is ATP, which stores energy in the form of covalent bonds between phosphate groups. The addition of a phosphate group to an organic molecule, as when a phosphate group is added to ADP to make ATP, is called **phosphorylation** [*FOSS*-for-uh-*LAY*-shun]. ATP can then contribute its stored energy to another molecule by transferring one of its phosphate groups to that molecule. This transfer energizes the recipient molecule, enabling it to change shape or react chemically with other molecules. An example of how ATP is used to add molecules to a complex macromolecule is described in Chapter 7.

The energy in chemical bonds is not the only form of energy that is transferred by energy carrier molecules. Two other important energy carriers—$NADP^+$ (nicotinamide adenine dinucleotide phosphate) and NAD^+ (nicotinamide adenine dinucleotide)—can pick up high-energy electrons and donate them to redox reactions. Each of these carriers can pick up two high-energy electrons along with a hydrogen ion (H^+), forming compounds known as **NADPH** and **NADH**, respectively. The ability of these compounds to donate these electrons to other molecules in turn means that NADPH and NADH can reduce other molecules. That is, they become oxidized by losing electrons, while the other compounds that accept the electrons are reduced. (For a review of oxidation and reduction, see Chapter 7.)

Although NADPH and NADH have equal abilities as reducing agents, there is a difference of one phosphate group between them. That small difference determines which target molecules they bind to and react with and which metabolic pathways they affect. NADPH is used in the biosynthetic reactions that manufacture sugar from carbon dioxide and water, and NADH is used in the catabolic reactions that produce ATP from the breakdown of sugars into water and carbon dioxide.

Two of the organelles discussed in previous chapters—chloroplasts and mitochondria—work together to capture and repackage the sun's energy into the energy carriers that power the activities of the cell. The activities of the two organelles are complementary: chloroplasts capture energy from sunlight and use it to synthesize sugars, while mitochondria extract energy from sugars and use it to synthesize energy carriers (Figure 8.1).

8.2 Photosynthesis: Capturing Energy from Sunlight

The next time you walk outside, look at the plants around you and consider the critical role they play in supporting the web of life that includes human beings. Plants and other photosynthetic organisms, which include some protists and bacteria, are able to capture energy from sunlight in the form of chemical bonds. The process of **photosynthesis** uses solar energy to synthesize energy carriers such as ATP and complex, energy-rich molecules such as sugars.

The chemical reactions of photosynthesis also result in the chemical splitting of water (H_2O) and the release of oxygen gas (O_2) into the environment (Figure 8.2). The O_2 by-product of photosynthesis is essential for all

PHOTOSYNTHESIS
(Chloroplasts)

AEROBIC RESPIRATION
(Mitochondria)

Figure 8.1 The Exchange of Molecules Between Chloroplasts and Mitochondria Produces Energy Carriers
This diagram summarizes the metabolic activities of chloroplasts and mitochondria. The activities of these two organelles, which we will examine in more detail as this chapter progresses, are complementary: chloroplasts capture energy from sunlight and use it to synthesize glucose, while mitochondria use energy from glucose to synthesize energy carriers.

oxygen-dependent life forms. In other words, plants and other photosynthesizers help support humans and virtually all other organisms, which either directly or indirectly depend on plants for both food and oxygen—another reason why plants are worthy of our respect.

Chloroplasts are the sites of photosynthesis

As mentioned in Chapter 5, photosynthesis in plants and protists takes place inside chloroplasts. These important organelles have a distinctive arrangement of membranes that makes photosynthesis possible (Figure 8.3). We can think of the chloroplast as a well-organized manufacturing plant, with its metabolic activities segregated into specialized departments. Like mitochondria, chloroplasts are surrounded by a double membrane, which divides the organelle into compartments. The outermost compartment, between the two membranes, is called the **intermembrane space**; the second compartment, enclosed by the inner membrane, is called the **stroma**. However, inside the stroma is a third, complex system of membrane-enclosed compartments that is not found in mitochondria. This system consists of groups of flattened, interconnected membrane sacs, called thylakoids, that lie one on top of another in stacks called grana. Each thylakoid is formed by a **thylakoid membrane** and in turn encloses a **thylakoid space**.

Photosynthesis consists of two sets of reactions. The first set of reactions takes place in the thylakoid membrane and directly captures energy from sunlight; because these reactions require light, they are collectively known as the **light reactions**. Both the thylakoid membrane and the thylakoid space house enzymes and other molecules needed to transfer solar energy to energy carriers such

as ATP and NADPH. The second set of reactions uses this captured energy to synthesize sugars from carbon dioxide and water. These reactions are known as the **dark reactions** because they do not require light to proceed (although they actually occur in both the light and the dark). The stroma contains enzymes that use energy carriers to manufacture sugars. Together the light and dark reactions form the foundation of life, not only for humans but also for virtually all other life forms on Earth.

See an overview of photosynthesis. 8.1

Test your knowledge of catabolism. 8.2

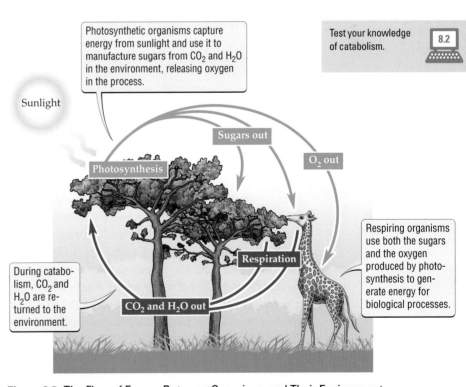

Photosynthetic organisms capture energy from sunlight and use it to manufacture sugars from CO_2 and H_2O in the environment, releasing oxygen in the process.

Sunlight

Sugars out

O_2 out

Photosynthesis

Respiration

During catabolism, CO_2 and H_2O are returned to the environment.

CO_2 and H_2O out

Respiring organisms use both the sugars and the oxygen produced by photosynthesis to generate energy for biological processes.

Figure 8.2 The Flow of Energy Between Organisms and Their Environment

Figure 8.3 Chloroplasts in a Leaf Cell

Leaf cells are rich in chloroplasts, the highly organized organelles in which photosynthesis occurs.

Leaf cross section

Leaf

Leaf cell

Vacuole

The dark reactions take place in the stroma.

Thylakoids

Thylakoid disk

Stroma

Chloroplast

The light reactions take place in the thylakoids.

Thylakoid membrane

Thylakoid space

Outer membrane

Inner membrane

Intermembrane space

Stroma

Granum (stack of thylakoids)

Each chloroplast has a distinctive triple membrane system.

The light reactions capture energy from sunlight

The capture of energy from sunlight is essential to photosynthesis and to life throughout the biosphere. In this process, solar energy is converted to the energy contained in the electrons and chemical bonds of organic compounds. Photosynthesis is carried out by specialized pigments, the most common being **chlorophyll** (from *chloro*, "greenish-yellow"; *phyll*, "leaf"), which accounts for the green color of most plant foliage. In fact, color plays a key role in photosynthesis. Rainbows show us that sunlight is composed of many colors. Plants absorb and use mostly red-orange and blue-violet light. For most plants, green light is "junk"—they cannot use it, so rather than absorbing it, they reflect it, which is why they look green to us.

Energy is captured by molecules of chlorophyll embedded in the thylakoid membrane. When exposed to light, the electrons associated with the covalent bonds of a chlorophyll molecule absorb energy from the light and become more energized. These energized electrons are often said to be "excited." To capture this energy, hundreds of chlorophyll molecules are arranged together in a formation called an **antenna complex**. Like the antenna on your car radio, which captures sound energy, the antenna complex captures solar energy (Figure 8.4). The energy from excited electrons is passed from one chlorophyll molecule to another in the antenna complex until it reaches the **reaction center**, which is composed of slightly different chlorophyll molecules.

At the reaction center, electrons accept the energy, become excited, and are passed to an **electron transport chain** (**ETC**), a series of electron-accepting proteins located next to one another, nearby in the thylakoid membrane. (ETCs are found in a number of metabolic processes that involve the transfer of energy, not just photosynthesis, as we shall see later in this chapter.) As electrons are passed down the ETC from one protein to another, small amounts of energy are released and used to generate the energy carriers ATP and NADPH. The combination of an antenna complex with a neighboring ETC is called a **photosystem**. Each thylakoid has many photosystem units; indeed, they make up more than half the thylakoid membrane by weight.

Photosystems come in two distinct types (see Figure 8.4); the transfer of the energy of each excited electron involves the activities of both kinds of photosystems. In **photosystem II**, electron flow along the ETC leads to the production of ATP, while **photosystem I** is primarily responsible for the production of the powerful reducing agent NADPH. (The reversed numbering of the two photosystems reflects the order in which each system was discovered by researchers, not the order of steps during photosynthesis.)

Let's consider how each photosystem contributes to photosynthesis as a single integrated process. The movements of electrons along the ETCs of the two photosystems enable the capture of energy from sunlight, but also have an important by-product; namely, the liberation of oxygen gas. As we shall see, oxygen is released by the splitting of water molecules, which donate electrons to the process of photosynthesis (Figure 8.5a). You could think of this as the splitting of water molecules by a light-dependent process—hence it is often referred to as *photolysis*.

The journey of excited electrons along the ETC in photosystem II includes their transfer to a channel protein that spans the thylakoid membrane. The channel protein uses the energy of the electrons to pump protons (H^+) from the stroma into the thylakoid space (Figure 8.5b). (Stripping a hydrogen atom of its single electron leaves just the nucleus of the H atom, which consists of a single proton. For this reason, protons are often denoted by H^+. Using this symbol also helps us keep track of ions and charged molecules that can interact in cellular situations.) As the protons accumulate inside the thylakoid space, they become relatively scarce in the stroma, causing a **proton gradient** (an imbalance in the proton concentration) across the thylakoid membrane.

Pumping of molecules to create a concentration gradient is a common means of harnessing energy for cellular processes. In this case, the gradient is used to manufacture ATP. Here's how it works. As we saw in Chapter 6, all dissolved substances, including protons, tend to move from a region of higher concentration to one of lower concentration. So the protons in the thylakoid space "want" to move back down the proton gra-

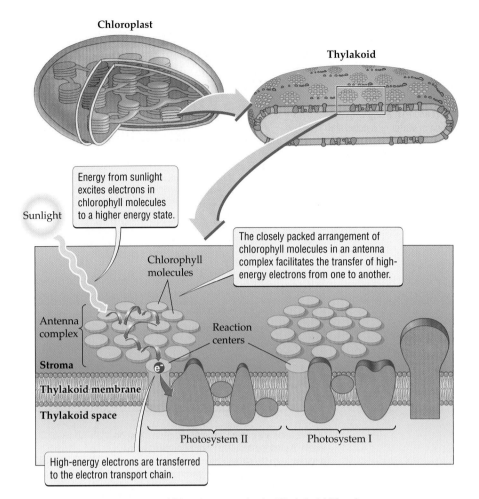

Figure 8.4 The Arrangement of Photosystems in the Thylakoid Membrane
The special arrangement of molecules in the thylakoid membrane facilitates the transfer of energy from the antenna complex (a collection of chlorophyll molecules; light green) to the proteins making up the electron transport chains (purple) of photosystems II and I.

dient to the stroma. Since the thylakoid membrane will not allow protons to pass through it, the only way for them to do so is through another channel protein that spans the thylakoid membrane, called ATP synthase. As the protons move through ATP synthase, they release energy, which in turn is used by ATP synthase to phosphorylate ADP to form ATP (Figure 8.5c).

Since there is more to the story of photosynthesis than the production of ATP, let's continue to follow the excited electrons as they journey down the ETC. Photosystem I receives electrons from the last protein in the ETC of photosystem II (see Figure 8.5b). These electrons, which have lost energy during their travel through photosystem II, get an additional boost of energy from sunlight again in the reaction center of photosystem I. Photosystem I eventually transfers these electrons to an ETC protein, which in turn transfers them to $NADP^+$. The transfer of electrons to $NADP^+$ gives it a negative charge, so it takes up

Figure 8.5 Production of Energy Carriers by the Light Reactions

The transfer of electrons between photosystems II and I results in the production of NADPH and ATP, along with the splitting of water molecules and the release of oxygen gas. Both the ATP and the NADPH are later used in the dark reactions.

(a)

(b) As electrons are passed down the electron transport chain of photosystem II, a channel protein pumps protons from the stroma into the thylakoid space.

(c) Because of the imbalance in the concentration of protons across the thylakoid membrane, protons move from the thylakoid space to the stroma via ATP synthase, thus driving ATP production.

When photosystems I and II are operating in series, the electrons ultimately are transferred to NADP⁺, which is reduced to NADPH.

H^+ from the stroma to form NADPH (see Figure 8.5b). In this manner, the two photosystems work together during the light reactions to produce both ATP and NADPH, which are then used in the dark reactions. Electrons flow out from photosystem I and are replaced at the "front end" of photosystem II. These replacement electrons were donated by water molecules, which split to produce hydrogen ions and oxygen gas (O_2) (see Figure 8.5a).

The dark reactions manufacture sugars

The energy carriers produced by the light reactions—ATP and NADPH—are used in the dark reactions to synthesize sugars from carbon dioxide and water. This process, known as **carbon fixation** (Figure 8.6), further emphasizes the interconnectedness between life and its environment. By capturing inorganic carbon atoms from CO_2 gas and fixing (incorporating) them into organic compounds such as sugars, the dark reactions make atmospheric carbon available to plants and, eventually, to other living organisms.

The dark reactions are catalyzed by enzymes that float freely in the stroma. The most abundant of these enzymes is called **rubisco**. Rubisco catalyzes the first reaction of carbon fixation, in which a molecule of the one-carbon compound CO_2 combines with a five-carbon compound called ribulose 1,5-bisphosphate [*RIB*-yu-loce . . . biss-*FOSS*-fate] to eventually produce two three-carbon compounds. This reaction can be expressed as an equation including just the carbon atoms in the compounds involved: 1C + 5C = 2 × 3C. This first reaction in carbon fixation is followed by a multi-step cycle of many reactions. These reactions manufacture sugars for use by the cell and regenerate more ribulose 1,5-bisphosphate, which keeps the dark reactions going. The entire process requires the input of energy from ATP and hydrogen ions from NADPH.

Three turns of the carbon fixation cycle bring in the three carbon atoms needed to produce one molecule of the three-carbon sugar glyceraldehyde 3-phosphate [gliss-er-*ALL*-duh-hide . . .]. We can follow this process by tracking the number of carbon atoms as they get rearranged into different compounds at each step of the cycle (see Figure 8.6). For every three molecules of CO_2 (3 × 1C = 3C) that combine with three molecules of ribulose 1,5-bisphosphate (3 × 5C = 15C), six molecules of the three-carbon compound are produced (6 × 3C = 18C). These molecules eventually produce three ribulose 1,5-bisphosphate molecules (3 × 5C = 15C) and one molecule of glyceraldehyde 3-phosphate (3C). As the arithmetic indicates, it takes three turns of the cycle to produce one three-carbon sugar molecule, with

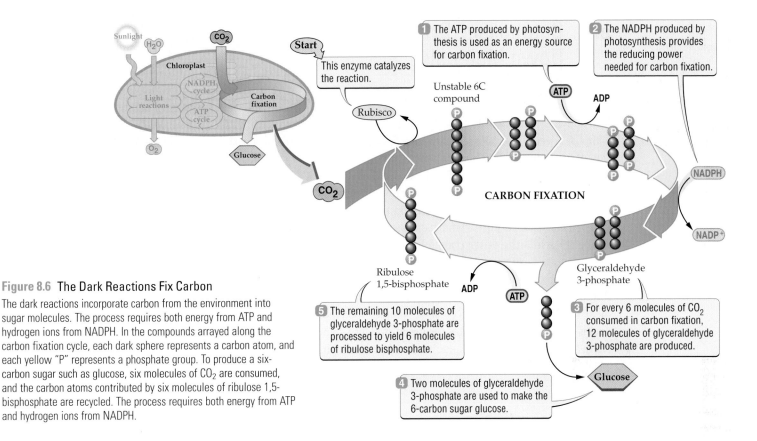

Figure 8.6 The Dark Reactions Fix Carbon
The dark reactions incorporate carbon from the environment into sugar molecules. The process requires both energy from ATP and hydrogen ions from NADPH. In the compounds arrayed along the carbon fixation cycle, each dark sphere represents a carbon atom, and each yellow "P" represents a phosphate group. To produce a six-carbon sugar such as glucose, six molecules of CO_2 are consumed, and the carbon atoms contributed by six molecules of ribulose 1,5-bisphosphate are recycled. The process requires both energy from ATP and hydrogen ions from NADPH.

Figure labels:
- Sunlight, H_2O, CO_2, Chloroplast, NADPH cycle, Light reactions, ATP cycle, Carbon fixation, O_2, Glucose
- Start
- This enzyme catalyzes the reaction.
- 1 The ATP produced by photosynthesis is used as an energy source for carbon fixation.
- 2 The NADPH produced by photosynthesis provides the reducing power needed for carbon fixation.
- Unstable 6C compound
- ATP, ADP
- Rubisco
- NADPH
- CARBON FIXATION
- NADP$^+$
- Ribulose 1,5-bisphosphate
- ADP, ATP
- Glyceraldehyde 3-phosphate
- 5 The remaining 10 molecules of glyceraldehyde 3-phosphate are processed to yield 6 molecules of ribulose bisphosphate.
- 3 For every 6 molecules of CO_2 consumed in carbon fixation, 12 molecules of glyceraldehyde 3-phosphate are produced.
- 4 Two molecules of glyceraldehyde 3-phosphate are used to make the 6-carbon sugar glucose.
- Glucose

the other carbon atoms being recycled to form ribulose 1,5-bisphosphate. The formation of one molecule of glyceraldehyde 3-phosphate also requires the input of nine molecules of ATP and six molecules of NADPH.

Glyceraldehyde 3-phosphate is the chemical building block used to manufacture glucose, from which other carbohydrates needed by the cell can be made. Most of the glyceraldehyde 3-phosphate made in the chloroplasts is exported from those organelles and eventually consumed in chemical reactions that produce ATP. Some of the exported molecules of glyceraldehyde 3-phosphate that are not immediately consumed in ATP synthesis are used to manufacture sucrose in the cytoplasm. Sucrose is an important food source for all the cells in a plant and is transported from the leaves, where photosynthesis takes place, to other parts of the plant. Significant amounts of sucrose are stored in the cells of sugarcane and sugar beets (Figure 8.7a), which is why these two plants are the major crops of the sugar industry worldwide.

Not all the glyceraldehyde 3-phosphate made in the chloroplasts is exported, however. Some of it is converted into starch by enzymes in the stroma. Starch, a polymer of glucose, is an important form of stored energy in plants. It accumulates in chloroplasts during the day and is then broken down into simple sugars at night. By setting aside food to generate energy at night, plants and other photo-synthesizing organisms are able to compensate for the lack of sunlight, and hence of photosynthesis, in the dark. Plant seeds, roots, and tubers are also rich in stored starch, which provides the energy required for germination and growth (Figure 8.7b). Indeed, the energy-rich nature of these plant components explains why they are such an important food source for animals.

(a)

(b)

Figure 8.7 Sugar and Starch
(a) This polarized light micrograph of sugarcane displays the crystalline structure of sucrose. (b) This scanning electron micrograph of a potato shows the starch granules stored in its cells.

8.3 Catabolism: Breaking Down Molecules for Energy

The conversion of food into useful energy, as we saw in Chapter 7, requires the breakdown and gradual oxidation of food molecules. The catabolic reactions that are constantly occurring in our bodies depend, either directly or indirectly, on the sugars produced by photosynthesis. In this respect, the two processes are complementary in that one harnesses energy from the environment (namely, sunlight), while the other transforms that harvested energy into a form that cells can use.

The first stage of catabolism is the digestive process that occurs in your stomach and intestines after every meal; namely, the breakdown of large, complex food molecules—such as carbohydrates, proteins, and fats—into their simpler components, such as amino acids and simple sugars. These compounds are then absorbed by the intestine and passed on via the bloodstream to other cells in the body. Each cell then converts the simple sugars supplied by digestion into fuel for its own use. This final, cellular stage of catabolism consists of three major processes in humans and most other eukaryotes: glycolysis, the citric acid cycle, and oxidative phosphorylation. These processes are referred to collectively as **aerobic respiration**.

Glycolysis is the first stage in the cellular breakdown of sugars

Glycolysis [gly-*KOLL*-uh-siss] literally means "sugar splitting." From an evolutionary standpoint, it was probably the earliest means of producing ATP from food molecules, and it is still the primary means of energy production in prokaryotes. In most eukaryotic organisms, which have more efficient ways of producing ATP

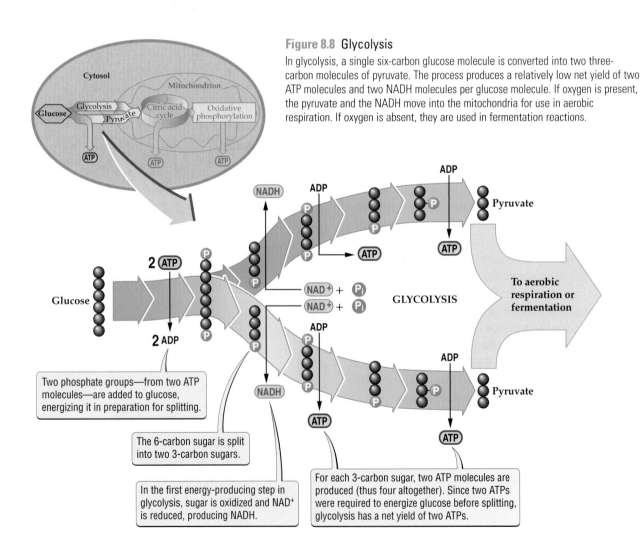

Figure 8.8 Glycolysis
In glycolysis, a single six-carbon glucose molecule is converted into two three-carbon molecules of pyruvate. The process produces a relatively low net yield of two ATP molecules and two NADH molecules per glucose molecule. If oxygen is present, the pyruvate and the NADH move into the mitochondria for use in aerobic respiration. If oxygen is absent, they are used in fermentation reactions.

Two phosphate groups—from two ATP molecules—are added to glucose, energizing it in preparation for splitting.

The 6-carbon sugar is split into two 3-carbon sugars.

In the first energy-producing step in glycolysis, sugar is oxidized and NAD+ is reduced, producing NADH.

For each 3-carbon sugar, two ATP molecules are produced (thus four altogether). Since two ATPs were required to energize glucose before splitting, glycolysis has a net yield of two ATPs.

using mitochondria, glycolysis is only a necessary preparation for later stages of catabolism.

Glycolysis consists of a series of chemical reactions that take place in the cytosol (Figure 8.8). These reactions break down the six-carbon sugar glucose and provide the mitochondria with the simpler substrate molecules they need in order to generate ATP. Through a series of enzyme-catalyzed reactions, glucose is converted to a six-carbon sugar intermediate, which is then split into two three-carbon sugars; these molecules are then converted into two three-carbon molecules of **pyruvate** [pie-*ROO*-vate], which is transported into the mitochondria for further processing.

So what does the cell gain from glycolysis? For each molecule of glucose consumed during glycolysis, four molecules of ADP are phosphorylated to produce four molecules of ATP, and electrons are donated to two molecules of NAD^+, generating two molecules of NADH. Since the early steps of glycolysis consume two molecules of ATP per glucose molecule, a single glucose molecule produces a net yield of two ATP molecules and two NADH molecules (see Figure 8.8). Glycolysis does not require oxygen gas (O_2), and it only partially oxidizes glucose to produce pyruvate. For complete oxidation using O_2, the pyruvate molecules enter the mitochondria for use in the eukaryotic cell's most productive ATP-generating system, as we shall see shortly.

Fermentation sidesteps the need for oxygen

Since glycolysis does not require O_2 from the atmosphere, its reactions are described as **anaerobic** [*ANN*-uh-ro-bick]. Glycolysis was probably essential for early life forms in the oxygen-poor atmosphere of primitive Earth. Some organisms today still use glycolysis as their only means of generating ATP. These organisms are anaerobic; that is, they can live without O_2. In these organisms, the pyruvate and the NADH produced by glycolysis remain in the cytosol. The pyruvate is converted into alcohol and CO_2, and the NADH is converted back to NAD^+, which can be used again in the reactions of glycolysis. When catabolism consists of glycolysis followed by these anaerobic reactions, it is known as **fermentation**.

Yeast is a familiar fungus that has an essential role in the production of beer. Fermentation by anaerobic yeasts converts pyruvate into the alcohol, called ethanol, and CO_2 gas that give beer its alcohol content and foamy effervescence (Figure 8.9*a*). This production of CO_2 also explains the role of baker's yeast in making bread rise, since the released gas expands the bread dough (see the box on page 162).

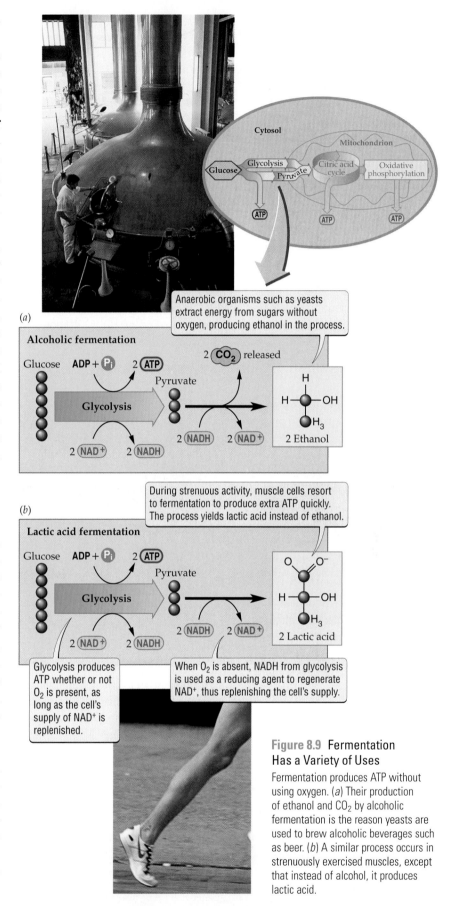

Figure 8.9 Fermentation Has a Variety of Uses

Fermentation produces ATP without using oxygen. (*a*) Their production of ethanol and CO_2 by alcoholic fermentation is the reason yeasts are used to brew alcoholic beverages such as beer. (*b*) A similar process occurs in strenuously exercised muscles, except that instead of alcohol, it produces lactic acid.

Biology on the Job

A Baker's Tale

Eric Morel worked for 16 years as head baker at the locally famous Black Sheep Deli, a bakery and food shop in the college town of Amherst, Massachusetts, while completing his degree in exercise science at the University of Massachusetts. Eric is now Dr. Morel and has moved on to a career as a chiropractor, having graduated from the University of Bridgeport, College of Chiropractic, all the while working at the Black Sheep part time.

Eric, as head baker, how did your day begin? I supervised a crew of four, and we began our day at three in the morning working with dough that had been developing from the night before.

What does "developing" mean? "Developing" means that different yeasts have been at work in different doughs during the night, producing the variety of flavors and textures that we need in order to provide our customers with the wide selection of bakery products they look for.

What do yeasts do when they are working in the dough? In bread dough one product of respiration, carbon dioxide, makes the dough rise and produces the holes we see inside baked loaves of bread. Lactic acid, a product of fermentation, contributes to the bread's flavor. Different combinations of time, yeasts, temperature, and other dough ingredients produce the different sizes of bubbles that characterize breads.

How does time play a role? In some breads we want the yeasts to work long enough to consume a good deal of the sugars in the dough to produce carbon dioxide–generated holes. In these breads we aim for less sweetness. But we don't want too much of the sugar consumed, because in baking, unconsumed sugars are caramelized on the outside of the loaf, giving the deep brown and tan tones of the crusts. Timing plays an important role. Usually doughs that rise longer are the sourdoughs. It's not that we want less sweetness, but rather that we want more sour flavor from lactate and other metabolites.

You mentioned sweetness. Do you want more sweetness in some doughs? For some yeasted pastries and breakfast rolls, like Danish and croissants, we do want a sweeter flavor. In that case we want more of the sugar to remain unconsumed.

So, what do you do? We use a highly aerobic yeast, and let it work for a short time, leaving more sugar intact and reducing the amount of carbon dioxide, lactate, and ethanol produced. Also, the recipe uses more sugar than other doughs.

It seems like you use a variety of yeasts. The Black Sheep is a European-style bakery; that is, we use primarily many types of wild yeasts originally captured from the air. Different yeasts yield different flavors and textures. One yeast, captured originally as a wild yeast, *Saccharomyces exiguous* [SACK-uh-ro-MY-seez ex-IJ-you-uss], works in the dough along with a bacterium, *Lactobacillus sanfrancisco*, to produce the unique aroma and taste of sourdough bread. For sweet doughs we use commercial yeast, called baker's yeast.

Does yeast have anything to do with your new profession as a chiropractor? In a way—the reliable biochemistry of yeast provided me with a job, as I worked my way through school.

The Job in Context

Yeasts provide bakers, brewers, and winemakers with the metabolic compounds they need to make their products. Winemakers and brewers are interested in anaerobic yeasts that produce ethanol. Bakers use both anaerobic and aerobic yeasts. Some of the anaerobic yeasts they use produce ethanol, while others produce lactic acid. Ethanol evaporates during baking, but lactic acid remains and is valued as a flavoring in breads.

Yeasts enrich our lives in many ways. They use the energy derived from fermentation and aerobic respiration to manufacture some of the molecules we perceive as favorite flavors and aromas. The biology of yeasts is very much a part of our everyday existence.

Macrophoto of Unrisen Dough (bottom) and Baked Dough (top)

Long molecules of proteins in wheat trap CO_2 gas generated by yeast fermenting sugar, causing the dough to rise when baked.

Fermentation is not limited to anaerobic organisms, however. It also takes place in the human body and the bodies of other **aerobic** [air-*OH*-bick] organisms (those that require O_2). When we exercise hard and push our muscles to the point of exhaustion, the pain we feel is largely a result of fermentation processes in the muscles. However, athletes' muscles do not produce alcohol and CO_2, as anaerobic yeasts do; instead, they convert pyruvate into lactic acid, which causes the burning sensation in aching muscles (Figure 8.9*b*). A rapid burst of strenuous exercise can exhaust the ATP stores in muscle in a matter of seconds; under these circumstances, the muscle cells resort to fermentation to produce extra ATP quickly. Both marathon runners and Olympic cyclists know from firsthand experience that to sustain strenuous physical activity, muscle cells must use both aerobic respiration and fermentation to generate enough ATP.

Most of the ATP in eukaryotes is produced by aerobic respiration

The bulk of ATP production in most eukaryotic organisms depends on the mitochondria. In contrast to glycolysis, which occurs in the cytosol and is anaerobic, the catabolic processes occurring in the mitochondria are aerobic. These organelles take the pyruvate and O_2 generated by glycolysis and use them in a series of oxidation reactions to release energy; this energy, in turn, is used to phosphorylate ADP to form ATP. Aerobic reactions dominate the ATP production of most organisms living under Earth's present, oxygen-rich atmosphere. Aerobic respiration is the term given to cellular catabolism that consists of glycolysis followed by a series of aerobic reactions in the mitochondria.

The important connections between oxygen, mitochondria, and ATP production turn up repeatedly in different cells in all living organisms. The muscle cells of the human heart, for example, have an exceptionally large number of mitochondria to produce the enormous amounts of ATP needed to keep the heart beating. Blind mole-rats, which live underground in oxygen-poor burrows and must dig continually every day, have also optimized their means of using oxygen to generate ATP in their muscles. Compared with the muscles of white laboratory rats, the muscles of blind mole-rats have 50 percent more mitochondria and 30 percent more blood capillaries. This means that their muscles are supplied with more blood, which maximizes the transfer of available oxygen, and have more mitochondria, which maximize the resulting production of ATP.

The Krebs cycle produces NADH and carbon dioxide

The second major stage in aerobic respiration, following glycolysis, is the **citric acid cycle**, a series of eight oxidation reactions that take place in the mitochondrial matrix (Figure 8.10). Before the cycle itself begins, however, there are several preliminary steps. After pyruvate is generated in the cytosol, it enters a mitochondrion, where it is converted into an acetyl group and bound to a large carrier molecule called coenzyme A. In the process, one of pyruvate's three carbon atoms is released in the form of CO_2, leaving acetyl CoA, a high-energy compound that serves as an important substrate for the citric acid cycle.

The Krebs cycle is sometimes referred to as the citric acid cycle because citric acid is the product of the first reaction involving acetyl CoA. The major consequence of this oxidation cycle is the production of high-energy electrons stored in NADH; CO_2 is released as a by-product. The molecules of NADH are then used in the next (third) stage of aerobic respiration.

Most of the oxidation reactions that take place in the cell are part of the citric acid cycle, emphasizing the importance of this process in energy production. In addition to sugars, stored fats can also enter the citric acid cycle. They are first broken down into fatty acids, which enter the mitochondria and are converted to acetyl CoA by a different set of reactions. The processes that produce energy from both sugars and fats come together at the beginning of the citric acid cycle.

Oxidative phosphorylation uses oxygen and NADH to produce ATP in quantity

The jackpot of ATP is generated in the third and last stage of aerobic respiration. This is also the stage at which the physical structure of the mitochondrion really comes into play. As we saw in Chapter 5, the mitochondrion has a double membrane that forms two separate compartments, the intermembrane space and the matrix. The NADH molecules produced in the matrix by the citric acid cycle donate their high-energy electrons to an electron transport chain (ETC) embedded in the inner membrane of the mitochondrion. A component of the

Blind mole-rat

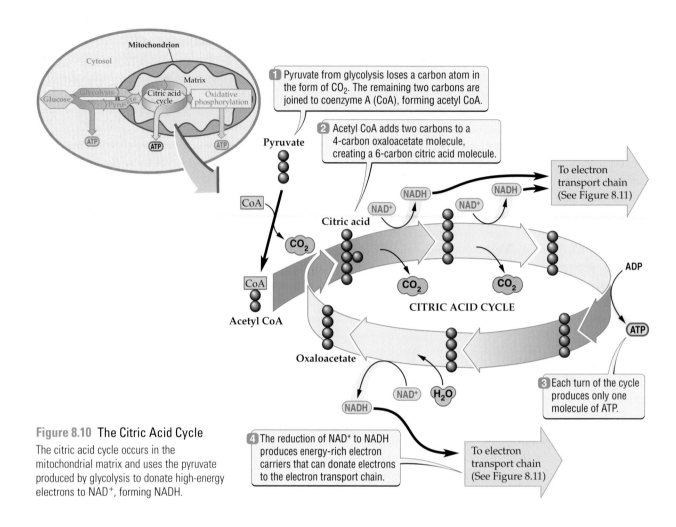

Figure 8.10 The Citric Acid Cycle
The citric acid cycle occurs in the mitochondrial matrix and uses the pyruvate produced by glycolysis to donate high-energy electrons to NAD$^+$, forming NADH.

1 Pyruvate from glycolysis loses a carbon atom in the form of CO_2. The remaining two carbons are joined to coenzyme A (CoA), forming acetyl CoA.

2 Acetyl CoA adds two carbons to a 4-carbon oxaloacetate molecule, creating a 6-carbon citric acid molecule.

3 Each turn of the cycle produces only one molecule of ATP.

4 The reduction of NAD$^+$ to NADH produces energy-rich electron carriers that can donate electrons to the electron transport chain.

ETC phosphorylates ADP to produce ATP. Since the electrons carried by the NADH molecules were gained by the oxidation of pyruvate in the citric acid cycle, and the components of the ETC are oxidized by the transfer of those electrons, the whole process is appropriately called **oxidative phosphorylation** (Figure 8.11).

If this process seems familiar, it is because the ETC in the inner mitochondrial membrane has a function similar to that of the ETC found in the thylakoid membrane of chloroplasts. In mitochondria, the electrons donated by NADH are passed along a series of ETC components, releasing energy that is used to pump protons through channel proteins from the matrix into the intermembrane space. The resulting buildup of protons in the intermembrane space of the mitochondrion has the same effect as the buildup of protons in the thylakoid space: it forms a proton gradient across the membrane. As in chloroplasts, the proton gradient causes protons to move through an ATP synthase in the inner mitochondrial membrane, driving the phosphorylation of ADP to form ATP (see Figure 8.11). Similarities in the way ATP is produced in chloroplasts and mitochondria demonstrate the evolutionary selection of a common biological mechanism.

In the final step of oxidative phosphorylation, electrons are donated to O_2 that has diffused into the mitochondrion. When O_2 accepts these electrons, it also combines with H$^+$ in the matrix to form water (H_2O). O_2 is the last electron acceptor in the series of electron transfers and is required for oxidative phosphorylation (see Figure 8.11).

Aerobic respiration (glycolysis, the citric acid cycle, and oxidative phosphorylation) has a net yield of well over thirty ATP molecules per molecule of glucose. It is clearly more productive than fermentation, which yields only two ATP molecules per molecule of glucose consumed. This marked contrast explains the need for aerobic respiration in large multicellular organisms with high energy demands. However, as we saw earlier, fermentation is still used by eukaryotic cells under certain circumstances when the availability of oxygen in the cellular environment is limited.

Figure 8.11 Oxidative Phosphorylation

Oxidative phosphorylation is the last stage of aerobic respiration and produces the most ATP of any process in the cell. Electrons from NADH produced in the citric acid cycle are donated to the electron transport chain in the inner mitochondrial membrane, where their energy is used in ATP production.

High-energy electrons in NADH are donated to the electron transport chain embedded in the inner mitochondrial membrane.

As electrons are passed down the electron transport chain, protons are pumped from the matrix into the intermembrane space, creating a proton imbalance across the inner mitochondrial membrane.

Oxygen is required as the final electron acceptor. In this final electron transfer, O_2 picks up protons and forms water.

The passage of accumulated protons from the intermembrane space to the matrix through ATP synthase drives ATP synthase to produce ATP.

Biology Matters

Our Fondness for Fermentation

"Fermented food" doesn't sound particularly appetizing, yet many of the most popular foods, drinks, and snacks depend on fermentation somewhere along the way to change unappealing starting ingredients into highly sought-after finished products. Consider that on average each year, Americans consume approximately 11+ pounds of chocolate, 9+ pounds of coffee, 31 pounds of cheese, $5\frac{1}{2}$ pounds of yogurt, 9 pounds of pickles, and 60 pounds of bread, all fermented foods.

Clearly, we love fermentation! Chocolate is a good example. Chocolate is a 60-billion-dollar-a-year industry. Worldwide, 5.5 billion pounds of chocolate are produced annually, mostly from Africa and South America. The Mayans and Aztecs of South America were making chocolate as long as 2,600 years ago, and the process is the same now as it was then. A natural fermentation process has always been central to the changes in cocoa beans that lead to chocolate.

To start the process of fermentation, beans are heaped into piles on plaintain leaves, where microorganisms naturally present in the environment begin to ferment the pulp surrounding the beans. Some of the products of fermentation penetrate the beans and disrupt cell membranes. Enzymes released from the disrupted cells produce the main flavors of chocolate by breaking down certain proteins. Without fermentation no flavor would arise. Aerobic bacteria then take over and, through respiration, consume the undesirable products of fermentation, such as acids and ethanol. Stirring and mixing the beans exposes them to more oxygen, which generates the brown color when it combines with compounds in the beans. And with more oxygen present, other flavors develop. Roasting

completes the process. The bitter but richly flavored product is more or less identical to its millennia-old counterpart. The Spanish originally brought chocolate to Europe, and Europeans contributed to our modern-day version by adding sugar to this enticing product of fermentation.

Nearly everyone enjoys chocolate, but is it addictive? Some chocolate consumers describe themselves jokingly as "chocoholics." But, among the products arising from the enzymes released from the beans' disrupted cell membranes, we find a stimulant similar to caffeine and another substance associated with pleasure responses in our nervous systems. Furthermore, some research suggests that substances in chocolate activate areas of the brain that are also activated by addictive drugs such as cocaine, and caffeine is known to have addictive properties.

Solving the Brain Drain

Given what we now know about obtaining energy from food, let's consider how we humans meet the high energy demands of our large brains. One way of doing so might be to increase our rate of metabolism so that we consume more food and generate more ATP. However, we do not generally have a higher rate of metabolism than do other mammals with far smaller brains, such as sheep. So the answer cannot lie in any special alteration of the catabolic processes described in this chapter.

Another way of keeping a large brain supplied with sufficient energy might be to evolve new ways of distributing energy to the various organ systems. Satisfying the energy needs of a living organism involves more than the mere production of ATP. The ATP that is available to a multicellular organism must be carefully distributed to satisfy the varying needs of different cell types. Approximately 70 percent of our resting metabolism (the processes involved in simply maintaining our bodies, without considering physical activity) supplies ATP to the heart, liver, kidneys, gastrointestinal tract, and brain. To supply our large brains, one or more of the other organs must either give up some of its precious energy resources or else become smaller. In fact, when compared with those of other primates, the human gastrointestinal tract is nearly 40 percent shorter. The larger energy demands of our brains are, in part, compensated for by our smaller gastrointestinal tracts. During their evolution, early humans switched from a strictly vegetarian diet to a more varied diet that included more easily digested foods such as meat. Thus there was no longer an advantage to having the kind of large gastrointestinal tract needed for slowly digesting vegetation.

A smaller gastrointestinal tract was probably not the only response to the evolution of our large brains. The amount of energy that a mother contributes to her offspring during pregnancy and early infancy is equally important to the development of the brain. By the time a human child reaches age 4, he or she has a brain that is 85 percent of its adult size. The human brain requires enormous amounts of energy during early development to support such rapid growth, and the only source for such metabolic resources is the mother. During pregnancy, the fetus pulls all of its nutrients from its mother to supply its growing energy needs; this transfer of energy continues after birth with the intake of breast milk. Thus we owe the development of our impressive brains in part to the evolutionary rearrangement of energy needs in our bodies and the generous investment of energy by our mothers. Thanks, Mom!

Chapter Review

Summary

8.1　Energy Carriers: Powering All Activities of the Cell

- Energy is transferred within living organisms via specialized compounds called energy carriers.
- ATP, the most commonly used energy carrier, donates energy stored in chemical bonds to chemical reactions. ATP is created through phosphorylation, the addition of a phosphate group to ADP.
- The energy carriers NADH and NADPH donate electrons to redox reactions. NADPH is used in photosynthesis, while NADH participates in catabolic reactions.

8.2　Photosynthesis: Capturing Energy from Sunlight

- Photosynthesis takes place in chloroplasts, which have a triple-membrane structure creating three compartments. The light reactions occur in the thylakoid membrane, and the dark reactions occur in the stroma.

- The light reactions capture energy from sunlight using chlorophyll molecules in the antenna complex. ATP and NADPH are produced as electrons flow in electron transport chains from photosystem II to photosystem I. The oxygen we breathe is a by-product of these reactions.
- The dark reactions use the ATP and NADPH produced by the light reactions to fix carbon from the atmosphere and synthesize sugars.

8.3　Catabolism: Breaking Down Molecules for Energy

- Catabolism is a series of oxidation reactions that break down food molecules and produce ATP.
- The form of catabolism most common in eukaryotes is aerobic respiration. It requires oxygen and has three stages: glycolysis, the citric acid cycle, and oxidative phosphorylation.
- Glycolysis occurs in the cytosol and splits glucose molecules to produce pyruvate and a small amount of ATP and NADH.

- Catabolism in the absence of oxygen is called fermentation. In fermentation, the pyruvate from glycolysis remains in the cytosol and is converted into either lactic acid (in aerobic organisms under certain circumstances) or alcohol and CO_2 (in anaerobic organisms such as some yeasts). Fermentation produces much less ATP than aerobic respiration.
- In the presence of oxygen, the pyruvate from glycolysis is used by mitochondria to generate many additional molecules of ATP.
- The citric acid cycle occurs inside the mitochondrial matrix and uses pyruvate to produce NADH and CO_2.
- Oxidative phosphorylation occurs in the intermembrane space and uses O_2 and NADH to produce most of the cell's ATP.

◉ Review and Application of Key Concepts

1. The transfer of electrons down an ETC produces a similar chemical event in both chloroplasts and mitochondria. Describe this chemical event and how it contributes to the production of ATP by both organelles.

2. Describe the sequence of events in the light and dark reactions of photosynthesis.

3. Certain drugs allow protons to pass through the inner mitochondrial membrane on their own, without the involvement of channel proteins. How do you think these drugs affect ATP synthesis?

4. Describe the flow of electrons from photosystem II to $NADP^+$ in the context of the chloroplast.

Key Terms

aerobic (p. 163)
aerobic respiration (p. 160)
anaerobic (p. 161)
antenna complex (p. 156)
carbon fixation (p. 158)
chlorophyll (p. 156)
citric acid cycle (p. 163)
dark reactions (p. 155)
electron transport chain (ETC) (p. 156)
energy carrier (p. 154)
fermentation (p. 161)
glycolysis (p. 160)
intermembrane space (p. 155)
light reactions (p. 155)
NADH (p. 154)

NADPH (p. 154)
oxidative phosphorylation (p. 164)
phosphorylation (p. 154)
photosynthesis (p. 154)
photosystem (p. 157)
photosystem I (p. 157)
photosystem II (p. 157)
proton gradient (p. 157)
pyruvate (p. 161)
reaction center (p. 156)
rubisco (p. 158)
stroma (p. 155)
thylakoid membrane (p. 155)
thylakoid space (p. 155)

Self-Quiz

1. The chemical compound most commonly used to transfer energy from one biological process to another is
   a. carbon dioxide.
   b. water.
   c. sugar.
   d. ATP.

2. The element carbon cycles
   a. between the sun and Earth.
   b. only from producers to consumers.
   c. only from consumers to producers.
   d. among producers, consumers, and the environment.

3. The oxygen produced in photosynthesis comes from
   a. carbon dioxide.
   b. sugars.
   c. pyruvate.
   d. water.

4. Photosynthesis occurs in
   a. chloroplasts.
   b. mitochondria.
   c. the citric acid cycle.
   d. glycolysis.

5. The light reactions in photosynthesis require
   a. oxygen.
   b. chlorophyll.
   c. rubisco.
   d. carbon fixation.

6. Glycolysis occurs in
   a. mitochondria.
   b. the cytosol.
   c. chloroplasts.
   d. thylakoids.

7. The electrons needed to replace those lost in the light reactions of photosynthesis come from
   a. sugars.
   b. channel proteins.
   c. water.
   d. electron transport chains.

8. Which of the following statements is *not* true?
   a. Glycolysis produces most of the ATP required by aerobic organisms.
   b. Glycolysis produces pyruvate, which is consumed by the citric acid cycle.
   c. Glycolysis occurs in the cytosol of the cell.
   d. Glycolysis is the first stage of aerobic respiration.

9. Which of the following is essential for oxidative phosphorylation?
   a. rubisco
   b. NADH
   c. plasma membrane
   d. chlorophyll

10. The citric acid cycle
    a. produces less ATP than glycolysis.
    b. produces simple sugars.
    c. produces more ATP than glycolysis.
    d. produces ETC proteins.

Biology in the News

Biodiesel Boom Well-Timed

By John Gartner

Biodiesel fueling stations are sprouting like weeds across America, where production of the alternative fuel rose 66 percent in 2003. Experts say the rapid growth of the renewable fuel will stretch the country's tenuous petroleum supply while helping people breathe a little easier.

Photosynthesis transforms the sun's energy into the usable chemical energy contained in plants. Fossil fuels, which consist of "aged" plant matter, are rich storehouses of energy that we received from the sun millions of years ago. But we are consuming fossil fuels rapidly, and the supply is limited. The search is on for new sources of energy, and new ways are emerging to connect photosynthesis to the immediate production of usable fuel. Some plants can produce large quantities of oils that can easily be converted to diesel fuel. The production of biodiesel means that the "millions of years" are no longer needed.

Biodiesel can be produced from many sources. Restaurants routinely discard huge quantities of cooking fats as waste. Plants such as mustard and soy produce large amounts of oils. And some types of algae that grow extremely rapidly in shallow salt ponds are composed of 50 percent oil. The oils from all of these sources can be used for production of biodiesel fuel. Consider the last source: salt pond algae on algae "farms" fed by wastewater from animal farms and sewage treatment plants could theoretically produce enough oil to supply 100 percent of the diesel fuel needs of the United States. Moreover, a valuable by-product of the conversion process is glycerine, which is used in soaps and many other products.

But there is more to the biodiesel story than renewability: biodiesel is also less polluting than petroleum-based diesel. It's true that biodiesel—like petroleum diesel—produces carbon dioxide when burned, so switching to biodiesel won't solve the problem of reducing the contribution that atmospheric carbon dioxide makes to global warming. However, use of biodiesel would lessen our impact on the environment in other ways. For one thing, biodiesel burns more cleanly than petroleum diesel, so it produces fewer emissions. It contains almost no sulfur, so it cannot contribute to the formation of acid rain. For another, biodiesel is as biodegradable as sugar. We have seen the effects of petroleum oil spills. In contrast, any biodiesel "spills" will quickly disappear as organisms break them down into the products of catabolism: water and carbon dioxide.

As public awareness grows, demand for biodiesel is likely to grow. As demand grows, so will production. Biodiesel is a breaking news story.

Evaluating the news

1. Investing in the biodiesel industry carries great risk. Oil producers could afford to cut their prices to drive biodiesel producers out of business. Should biodiesel development be carried out primarily by small producers, or by large, existing oil companies?

2. We are consuming fossil fuels at enormous and ever-increasing rates, and there is much controversy over whether we should stress conservation of existing reserves or drilling for more oil. How do you think the U.S. government should react to the emerging business of producing and distributing biodiesel fuel? Alone or with your classmates, write a letter that you might send to your representatives in Congress expressing your views.

3. Genetic engineering is in the news every day. Some algae and mustard plants are highly efficient producers of the oils that can be easily transformed into biodiesel fuel. Do you see genetic engineering playing a role in the "plant petroleum" business, and if so, how?

Source: *Wired News,* June 1, 2004.

CHAPTER **9**

Cell Division

Key Concepts

- Cell division is the process by which a cell divides to produce two daughter cells. It is necessary for an organism to grow and develop, to maintain and replace its tissues, and to pass on genetic information to the next generation.

- Cell division has several distinct stages, collectively known as the cell cycle. Each stage is marked by physical changes in the cell that either prepare the cell for division or that directly participate in the division process.

- Most of the events that prepare the cell for division occur during the interphase stage of the cell cycle. The physical events of cell division occur during mitosis or meiosis, and cytokinesis.

- Mitosis ensures that both daughter cells receive identical and complete sets of chromosomes. The number and shape of the chromosomes is uniquely characteristic of each species. Cells not involved in producing reproductive cells undergo mitosis.

- Meiosis occurs exclusively in reproductive cells and consists of two cycles of division. The end product is four daughter cells, each with half the number of chromosomes of the parent cell.

An Army Turned Against Itself

Your body is constantly under siege. With every breath you take, you bring in foreign invaders that must be repelled by your body's defenses. The continuous struggle against dangerous microorganisms in the environment is a normal part of your body's day-to-day existence.

The complex organization and functioning of your body require a great deal of effort to maintain and protect. Given the many kinds of specialized cells that make up the body, it is not surprising that a specific group of cells is specialized to defend it against foreign invaders. These cells are known collectively as the immune system. They serve as both the guards that protect exposed body surfaces from invasion and the soldiers that seek out and destroy any invaders that manage to get inside.

But the immune system, like any army, is not perfect. Many dangerous organisms have developed ways not only to evade the immune system's defenses, but to use them to damage the body. One such organism is the bacterium *Borrelia burgdorferi* [bore-AY-lee-uh berg-DOR-fer-ee], which is the primary cause of Lyme disease. Tiny ticks that live on animals such as deer and sheep carry these microscopic invaders. Each year more than 16,000 persons in the United States are bitten by infected ticks while camping or merely walking through tall grass. Just one bite from an infected tick can pass *Borrelia* to an unfortunate individual. The resulting Lyme disease usually begins with skin rashes and flulike symptoms; over time, it can lead to serious arthritic and neurological disorders.

The good news is that *Borrelia* is easily killed by powerful antibiotics. The bad news is that some patients experience severe pain in their joints long after ridding the body of the bacteria. This mysterious arthritic pain is due to an attack on the joints by cells of the immune system. Infection by *Borrelia*

Immune System Cells Attacking a Foreign Body, in This Case, Green Surgical Thread bacteria causes immune system cells that attack components of the body to divide rapidly, increase in number, and in effect turn the immune system against the body it is supposed to protect. This alarming event emphasizes the importance of cell division in diverse biological processes and the need the body has for ways of regulating it.

As you sit and read this book, you are probably not aware of the millions of cells in your body that are dividing, responding to growth signals, and replacing old cells with new ones. Cells proliferate both when tissue is damaged or lost, as when you fall and scrape your knee, and as a natural part of body repair and maintenance. The production of new cells is an essential activity for every multicellular organism.

In Chapter 6 we discussed how cells communicate with one another using signaling molecules. One of the most significant activities requiring cell-to-cell communication is cell division. Why, and when, is cell division important? The most obvious answer is during the growth and development of a fertilized egg into a mature multicellular organism. The increase in the size and complexity of the body during development requires not just more cells, but different kinds of cells. As a multicellular organism develops, a tremendous amount of cell division and proliferation takes place to expand existing tissues and create new tissue types.

Once an organism has achieved its mature size, cells must still divide to replace worn-out and damaged tissue. Skin, for example, must be continuously renewed as the dead cells on the surface are worn away. Simply put, the upper portion of the skin consists of multiple layers of cells that gradually move to the surface as old surface layers are lost to wear and tear. In the process, the cells undergo dramatic physical changes, such that by the time they reach the surface layer, they form dead, flattened scales of protein (Figure 9.1). As cells move up through the layers and are lost from the skin surface, they are replaced by new cells produced by the division of specialized cells in the deepest layer of the skin, called stem cells.

In this chapter we explore the mechanics of the two kinds of cell division and their different stages. First we consider the simpler process of mitosis, which all nonreproductive cells undergo when they divide. Then we examine the specialized division process used by reproductive cells.

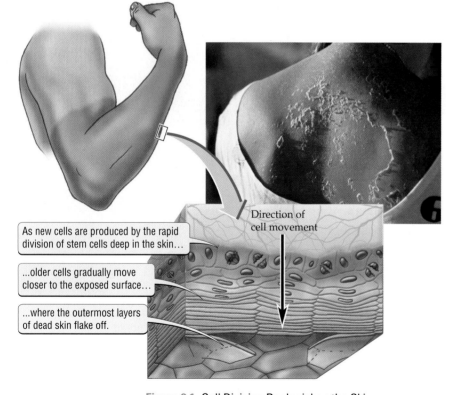

As new cells are produced by the rapid division of stem cells deep in the skin…

…older cells gradually move closer to the exposed surface…

…where the outermost layers of dead skin flake off.

Direction of cell movement

Figure 9.1 Cell Division Replenishes the Skin
Rapid cell division in the deepest layer of the skin is necessary to replace dead cells lost at the surface of the skin. This loss can be due to normal wear or to physical damage, such as the aftereffects of sunburn, as shown in the photo.

9.1 Stages of the Cell Cycle

What exactly does cell division mean? How do we get from one cell to two cells? The simple answer is that a single cell divides to form two so-called daughter cells. But although the outcome of cell division is simple, the process requires a great deal of cellular preparation. In order for the daughter cells to have the complete set of proteins they need to live and function normally, both must receive the genetic material that contains the blueprint for all of those proteins. In other words, both daughter cells must receive the full complement of DNA, in the form of chromosomes, that is characteristic of the organism. In addition, the parent cell must be large enough to divide in two and still contribute sufficient cytosol and organelles to each daughter cell.

Both requirements mean that before cell division takes place, key cellular components must be duplicated, including DNA, proteins, and lipids. These preparations are normally accomplished in a series of steps that make up part of the life cycle of every eukaryotic cell.

The life cycle of a single eukaryotic cell, which both begins and ends with cell division, is called the **cell cycle**. In the simplest terms, the cell cycle consists of two major stages that are very different from each other (Figure 9.2). The stage whose events are easily distinguished under the microscope is cell division, in which the parent cell splits into two daughter cells. As we shall see, this division stage has two parts: mitosis and cytokinesis. The stage between two successive divisions, called **interphase**, lasts considerably longer than cell division.

9.2 Interphase: Preparing the Cell for Division

It was once thought that the relatively long period of interphase was uneventful for the cell. Today we know that interphase is an active stage during which the cell prepares itself for division.

Interphase consists of three major stages: S, G_1, and G_2. These stages are defined by a key event that must occur for cell division to proceed: the entire DNA content of the nucleus must be replicated before it can be distributed to the daughter cells. Since replication requires the synthesis of new DNA, this stage is called **S phase** (S for "synthesis").

In most cells, S phase does not come immediately before or after division. Two other stages separate division and S phase in the cell cycle. The first, known as **G_1 phase** (G for "gap"), occurs after division but before S phase begins. The second, **G_2 phase**, occurs after S phase and before the start of division (see Figure 9.2). Although their names describe these stages as mere gaps between division and S phase, many essential processes occur during both G_1 and G_2.

G_1 and G_2 are important phases for two reasons. They are periods of growth, during which both the size of the cell and its protein content increase. Furthermore, each phase prepares the cell for the phase immediately following it. During G_1 phase, particular proteins must be made and activated for S phase to occur. Once the cell is large enough, these proteins promote the production of enzymes that catalyze the synthesis of DNA. Similarly, during G_2 phase, another set of proteins promotes the cellular events necessary for the cell to physically divide. These events include physical changes in the nucleus that

Learn more about the cell division cycle.

9.1

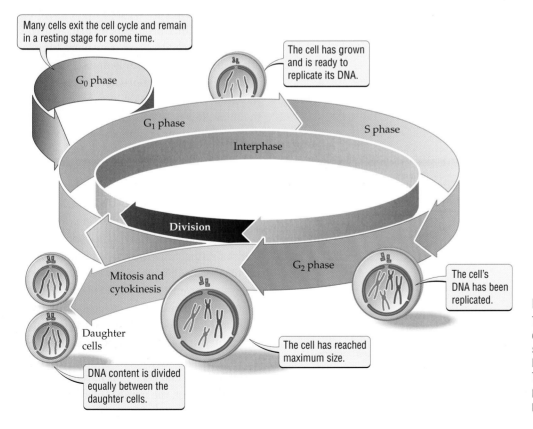

Figure 9.2 The Cell Cycle

The cell cycle consists of two major stages (shown by the central gray circle): cell division and interphase. In the division stage, the cell divides into two daughter cells. Interphase can be subdivided into three phases, as shown by the outer circle. The cell prepares itself for division by increasing in size and producing proteins needed for division during G_1 and G_2 phases and by replicating its DNA during S phase.

are required before the parent cell's DNA can be split equally between the two daughter cells.

The time it takes to complete the cell cycle depends on the organism, the type of cell, and the life stage of the organism. Dividing cells in tissues that require frequent replenishing, such as the skin or the lining of the intestine, require about 12 hours to complete the cell cycle. Most other actively dividing tissues in the human body require about 24 hours to complete the cycle. By contrast, a single-celled eukaryote such as yeast can complete the cell cycle in just 90 minutes.

Not all of the cells in your body go through the cell cycle. Many tissues do not require rapid replenishment of cells; if their cells divided continuously, it would be difficult to control the size of your body and its organs. Instead, the cells in most tissues pause in the cell cycle somewhere between division and S phase. This resting G_0 **phase**, which can last for periods ranging from days to years, is easily distinguished from G_1 by the absence of preparations for DNA synthesis (see Figure 9.2).

Cells in G_0 have exited the cell cycle. Liver cells remain in this resting state for up to a year before undergoing cell division. Other cells, such as those that form the lens of the eye, remain in G_0 for life, thus forming a nondividing tissue. Most of the nerve cells that make up the brain also exist in this nondividing state, which explains why brain cells lost as a result of physical trauma or chemical damage are usually not replaced.

9.3 Mitosis and Cytokinesis: From One Cell to Two Identical Cells

The climax of the cell cycle is cell division, which consists of two phases: mitosis and cytokinesis [sigh-toh-kih-NEE-sis]. These stages are not discrete in time; cytokinesis overlaps with the last phases of mitosis.

The central event of **mitosis** is the equal distribution of the parent cell's DNA to two daughter nuclei, each destined for a daughter cell. This process, called **DNA segregation**, is a distinctly physical process that requires the coordinated actions of several structural proteins. But before discussing the details of DNA segregation during mitosis, it is important to understand the preparation and packing of the DNA that occurs in the nucleus prior to mitosis.

The DNA of each species is organized into a distinctive karyotype

The DNA in the nucleus is not a random tangle of nucleotide polymers, but rather is tightly packed and highly organized into distinct individual structures. This packing is essential because about 2 meters of DNA must fit into the average nucleus, which has a diameter of less than 5 micrometers. DNA and its associated proteins are packed together to form thicker and more complex strands called **chromatin**. Chromatin, in turn, is further looped and packed to form even more complex structures, called **chromosomes** (Figure 9.3). We will learn much more about genes, chromosomes, and DNA packing and replication in Unit 3. For now, it is enough to know that each chromosome contains a single long molecule of DNA that carries a defined set of genes. At the beginning of mitosis, the chromatin is packed and condensed even more densely than usual, and it is at this stage that chromosomes can be seen under the microscope.

During mitosis, chromosomes adopt particular shapes that allow them to be identified under the microscope. The characteristic number and shapes of chromosomes found in each cell of a species is known as that species' **karyotype** (Figure 9.4a). For example, with the exception of eggs and sperm, the cells of the human body contain 46 chromosomes, horse cells have 64 chromosomes, and corn cells have 20 chromosomes. However, the number of chromosomes

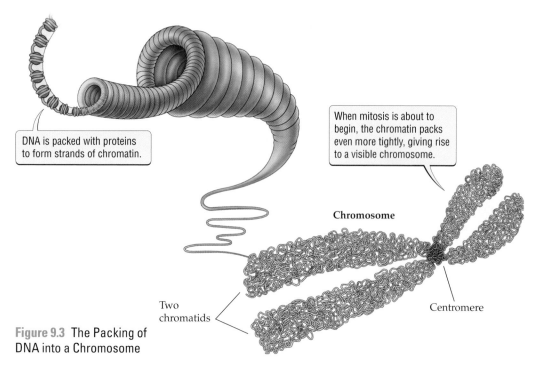

DNA is packed with proteins to form strands of chromatin.

When mitosis is about to begin, the chromatin packs even more tightly, giving rise to a visible chromosome.

Chromosome

Two chromatids

Centromere

Figure 9.3 The Packing of DNA into a Chromosome

(a)

(b)

Figure 9.4 **Human Chromosomes**
(*a*) The 46 chromosomes in this micrograph represent the karyotype of a human female. The chromosomes have been arranged in homologous pairs and numbered (or lettered, in the case of the sex chromosomes). One chromosome in each homologous pair has been artificially shaded pink and the other blue, as a reminder that one homologue comes from the individual's mother and the other from her father. (*b*) By the beginning of mitosis, the chromosomes have been duplicated. Each duplicated chromosome consists of two sister chromatids held together at a constriction point, the centromere. Note that these chromosomes have not been arranged into homologous pairs.

has no particular significance other than being an identifying characteristic of the species. It does not reflect the number of genes found in a particular species.

To understand how the DNA of a given species is replicated and segregated during the cell cycle, we must understand some similarities and differences within the set of chromosomes. Returning to the example of the human karyotype, our 46 chromosomes can be arranged in 23 pairs of **homologous chromosomes** [ho-*MOLL*-uh-guss]. In each homologous pair, one chromosome is inherited from the mother and the other from the father. In 22 of these homologous pairs (numbered 1 to 22), the two **homologues** (the paired chromosomes) are alike in length, shape, and the set of genes they carry. But the twenty-third pair is different. Its two homologues, individually identified as X or Y, can be similar to each other (XX) in length and gene set, or dissimilar (XY). They are called the **sex chromosomes** because they determine the sex of the organism; for example, in humans, individuals with XX chromosomes in their cells are female and those with one X and one Y chromosome are male.

Before cell division can proceed, the DNA of the parent cell must be replicated so that each daughter cell can receive a complete set of chromosomes. This replication occurs during S phase and produces chromosomes made up of two identical, side-by-side strands called **chromatids**. Thus, as mitosis begins, the nucleus of a human cell contains twice the usual amount of DNA, since each of the 46 chromosomes consists of a pair of identical sister chromatids, held together at a constriction point called the **centromere** (Figure 9.4*b*).

Chromosomes become visible during prophase

Mitosis is divided into five stages, each of which features easily identifiable events that are visible under the light microscope (Figure 9.5). The first stage of mitosis, called **prophase** (*pro*, "before"; *phase*, "appearance"), is characterized by the first appearance of visible chromosomes. In an interphase cell, the chromatin is dispersed throughout the nucleus, and specific chromosomes cannot be distinguished. As the cell moves from G_2 phase into prophase, the chromatin condenses, and the chromosomes become visible in the nucleus, looking like a tangled ball of spaghetti.

Important changes occur in the cytosol during prophase as well. Two protein structures called **centrosomes** (*centro*, "center"; *some*, "body") begin to move around the nucleus, finally halting at opposite sides in the cell. As we shall see, this arrangement of centrosomes defines the opposite ends, or poles, of the cell between which it will eventually separate to form two daughter cells.

At the same time that the centrosomes are moving toward the poles of the cell, microtubules are growing outward from each centrosome. These filaments are the beginnings of a structure called the **mitotic spindle**, which will later guide the movement of the chromosomes.

Chromosomes are attached to the spindle during prometaphase

In the next stage of mitosis, **prometaphase**, the nuclear envelope breaks down (see Figure 9.5). In the process,

Helpful to know

One way to avoid confusing the similar-sounding terms *centromere* and *centrosome* is to focus on the end of each word. Centromeres (*mere*, "part") are the part of the chromosome where two chromatids are joined at the beginning of mitosis. Centrosomes (*some*, "body") are not part of the chromosome; like most bodies, they can move—as they do during mitosis when they separate and move to different parts of the cell.

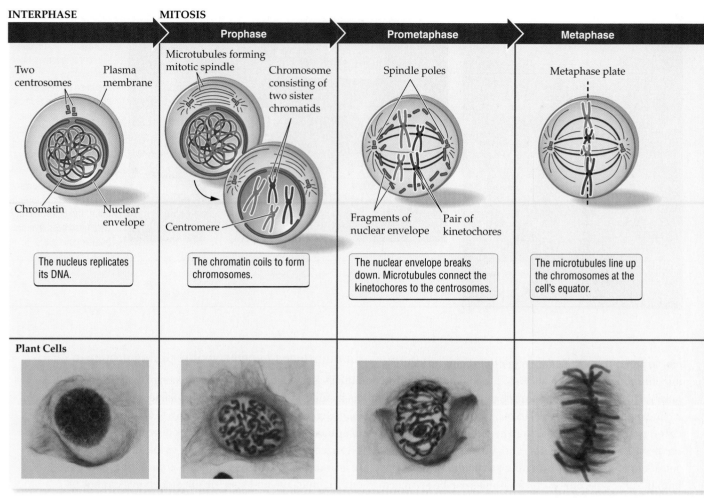

| | Prophase | Prometaphase | Metaphase |

Interphase

Two centrosomes — Plasma membrane

Chromatin — Nuclear envelope

The nucleus replicates its DNA.

Prophase

Microtubules forming mitotic spindle — Chromosome consisting of two sister chromatids

Centromere

The chromatin coils to form chromosomes.

Prometaphase

Spindle poles

Fragments of nuclear envelope — Pair of kinetochores

The nuclear envelope breaks down. Microtubules connect the kinetochores to the centrosomes.

Metaphase

Metaphase plate

The microtubules line up the chromosomes at the cell's equator.

Plant Cells

Figure 9.5 The Stages of Cell Division

The stages of mitosis and cytokinesis are shown by diagrams of a dividing animal cell (top) and microscopic images of a dividing plant cell (bottom).

See mitosis and cell division in action.

9.2

the mitotic spindle radiating from the centrosomes, which have now reached the poles, extends and enters the region of the cell that was once within the nucleus. The spindle microtubules then attach to the chromosomes at their centromeres, effectively linking each chromosome to both centrosomes.

The physical structure of the centromere dictates how each chromosome will be attached to the spindle microtubules. Each centromere has two plaques of protein, called **kinetochores** [kih-*NET*-oh-cores], that are oriented on opposite sides of the centromere. Each kinetochore forms a site of attachment for a single microtubule, so that the two chromatids that make up a chromosome end up being attached to opposite poles of the cell. This arrangement is essential for the later segregation of DNA.

Chromosomes line up in the middle of the cell during metaphase

Once each chromosome is attached to both poles of the spindle, its microtubule attachments lengthen and shorten, moving it toward the middle of the cell. There the chromosomes are eventually lined up in a single plane that is equally distant from both spindle poles. This stage of mitosis is called **metaphase** (*meta*, "after"), and the plane in which the chromosomes are arranged is called the *metaphase plate* (see Figure 9.5). Metaphase is so visually distinctive that its appearance is used as an indicator of actively dividing cells when tissues are examined under a microscope.

Chromatids separate during anaphase

During the next stage of mitosis, called **anaphase** (*ana*, "up"), DNA segregation takes place. At the beginning of anaphase, the sister chromatids separate (see Figure 9.5). Once they are separated, each chromatid is considered a separate daughter chromosome. The gradual

Anaphase	Telophase and Cytokinesis

Daughter chromosomes

Nuclear envelope forming

Chromosomes decondensing

Plane of constriction

The chromatids separate, and the new chromosomes (each derived from one chromatid) move toward the poles.

The separating chromosomes reach the poles. The nuclear envelopes re-form and the chromatin decondenses. Cytokinesis begins as the plane of constriction forms.

shortening of the microtubules pulls the newly separated daughter chromosomes to opposite poles of the cell. This remarkable event results in the equal segregation of chromosomes into the two daughter cells; each daughter cell will contain the same, complete genetic blueprint as the parent cell. In a human cell, for example, each of the 46 chromosomes is duplicated during S phase, yielding 46 pairs of identical sister chromatids. When the chromatids separate at anaphase, identical sets of 46 daughter chromosomes arrive at each spindle pole.

New nuclei form during telophase

The next phase of mitosis, **telophase** (*telo*, "end"), begins when a complete set of daughter chromosomes arrives at a spindle pole. Major changes also occur in the cytosol to prepare for division into two new cells. The spindle microtubules break down, and nuclear envelopes begin to form around each set of chromosomes (see Figure 9.5). As the two new nuclei become increasingly distinct in the cell, the chromosomes within them start to unfold, becoming less visible under the microscope. Telophase is the last stage of mitosis, and the cell is now ready for physical division into two daughter cells.

In most plants, additional changes during telophase prepare the plant cell for the physical process of dividing in two (Figure 9.6). Vesicles containing cell wall

Anaphase   Telophase   Cytokinesis

Cell wall

Plasma membrane

Vesicles filled with cell wall components

Vesicles fuse, forming a new cell wall.

Completed new cell wall

Figure 9.6 Cell Division in Plants
The tough cell wall that surrounds the cells of most plants makes cytokinesis more complex than in animal cells. Instead of pinching in two, as animal cells do, plant cells divide by building a new cell wall down the middle. Vesicles filled with cell wall components accumulate in the middle of the cell at the start of telophase. As the vesicles fuse, a new cell wall forms from the vesicles' contents, dividing the original cell into two new daughter cells at cytokinesis.

Figure 9.7 Sexual Reproduction Requires a Reduction in Chromosome Number

The fusion of a sperm and an egg at fertilization must produce a zygote with the complete diploid (2n) karyotype. This, in turn, requires that the two gametes each be haploid (n), with half the number of chromosomes found in other cells. Therefore, each gamete receives only one member of each homologous pair of chromosomes, so that when fertilization occurs, a full homologous pair re-forms in the zygote.

See meiosis in action.

9.3

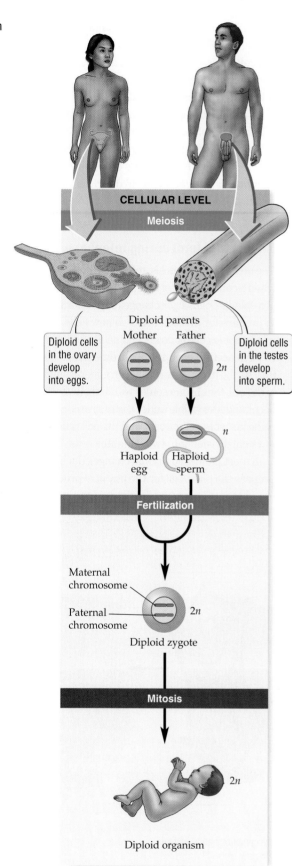

CELLULAR LEVEL

Meiosis

Diploid cells in the ovary develop into eggs.

Diploid parents
Mother   Father

2n

Diploid cells in the testes develop into sperm.

Haploid egg   Haploid sperm

n

Fertilization

Maternal chromosome

Paternal chromosome

2n

Diploid zygote

Mitosis

2n

Diploid organism

---

components accumulate in the region previously occupied by the metaphase plate. These vesicles fuse with one another, mingling their contents, and a new cell wall begins to form down the middle of the cell.

Cell division occurs during cytokinesis

Cytokinesis (*cyto*, "cell"; *kinesis*, "movement") features the division of the parent cell into two daughter cells (see Figure 9.5). In animal cells, the physical act of separation is performed by a ring of actin microfilaments that forms against the inner face of the plasma membrane like a belt at the equator of the cell. When the actin ring contracts, it pinches the cytoplasm of the cell and divides it in two. Since the plane of constriction by the actin ring lies in the plane of the metaphase plate, between the two newly formed nuclei, successful division results in two daughter cells, each with its own nucleus.

In plant cells, the new cell wall that began forming in telophase is completed, dividing the cell into two independent daughter cells (see Figure 9.6). Cytokinesis marks the end of the cell cycle, and once it is completed, the daughter cells are free to enter G_1 phase and start the process anew.

9.4 Meiosis: Halving the Chromosome Number

The remarkable process of mitosis occurs throughout your body during your entire lifetime. However, certain specialized cells in your body undergo a related but significantly different cell cycle, which produces daughter cells with half the number of chromosomes found in the parent cell. Before considering this cell cycle in detail, we need to know more about reproductive cells, or **gametes** (called *sperm* in males and *eggs* in females). Gametes are the only cells in the body produced by this specialized cell cycle. The reason for this lies in the role they play in sexual reproduction.

Gametes contain half the number of chromosomes as other cells

The creation of a new organism via sexual reproduction requires the fusion of two gametes in a process known as **fertilization.** A successful union of two gametes forms a single cell called a **zygote** (Figure 9.7). The zygote then undergoes multiple cell divisions to form an embryo, which develops into a new organism.

Biology on the Job

Wildlife Curator: Saving the Red Wolf

We were introduced to the Chattanooga Nature Center in Chapter 4. Tish Gailmard is another member of the center's team and carries the title Wildlife Curator. Tish and the center are part of a national program to save the endangered red wolf from extinction.

Tish, before we talk about breeding red wolves, could you describe your responsibilities at the Nature Center? At the center we provide a home for many mammals, birds, reptiles, and amphibians. I supervise and take an active part in caring for our animals. That means ordering correct foods for each type of animal and making sure they are fed regularly and on a schedule that matches their natural feeding cycles. Snakes, for instance, don't feed as often as wolves. We keep a detailed log on every animal that describes what foods they like, what they refuse. We even keep records of their temperament.

Besides providing adequate diets, in what other ways do you get involved in the lives of your animal residents? To us, a home means much more than simply providing an adequate diet. Just like us, animals need some fun, some distractions from the humdrum of daily life, and we develop enrichment programs for all of them. For the red wolves, we often hide treats in trees, about 5 feet above the ground. At that height they can expend a healthy effort to find them, and, once they do, they love to knock the treats down by stretching and jumping. At other times we spray a scent, such as deer scent, on old telephone books. The wolves love to tear them up. No one gets hurt, and the wolves have a chance to let loose some deeply rooted natural behaviors.

Speaking of the red wolves, tell us more about them and the program to save them. Red wolves are considered to be one of the most endangered animal species in North America. At the beginning of 2005 there were a little over 300 captive and wild red wolves alive. In the early 1980s there were only 17 true red wolves left in the world, and they were considered extinct in the wild. As a part of the U.S. Fish and Wildlife Species Survival Plan and a designated breeding site, we hope to increase those numbers. We have two breeding pairs at the center and hope they will produce pups, though they are getting on in wolf years at approximately 10 years old.

How does the overall program work? We are part of a captive breeding program established at the Point Defiance Zoo and Aquarium in Tacoma,

Red Wolf at the Chattanooga Nature Center

Washington. Over thirty zoos and nature centers cooperate in attempts to breed red wolves in an effort to restore them to the wild in sufficient numbers to survive as a species in the wild.

How does saving the red wolf fit into the overall biological scheme of things? As top predators in ecosystems, wolves keep other populations in check. Let me give you an example. One population of red wolves was kept on Bull's Island off the South Carolina shore as an intermediate step for release into the wild. Sea turtle populations were greatly diminished in that area, because the populations of raccoons and other animals that fed on their eggs had greatly increased. There were no natural predators to keep their numbers in check. The red wolves, natural predators of these animals, soon changed the balance, and the population of sea turtles rebounded. As wildlife curator I am concerned with contributing to nature's living balance through programs like saving the red wolf and through informing the public at our center.

The Job in Context

It may seem like Tish's job is far removed from the world of cell division. But, when you think about it, breeding is all about combining the products of meiosis. As you've learned from this chapter, gametes (sperm and eggs) are the only cells in our bodies, and in wolves' bodies, that are produced by meiosis. A sperm must fertilize an egg in order to produce a diploid zygote.

Getting the right sperm to the right egg is the key to the survival of this wolf species. For the remaining red wolf population, that means using carefully planned combinations of meiotic products—eggs and sperm—from those individuals that will give the best chance of producing strong, viable pups. Scientists at Point Defiance keep extensive records and pedigrees on red wolves and guide the destinies of specialized cells produced through cell division.

Test your knowledge of the meiotic stages.
9.4

Compare mitosis and meiosis.
9.5

If both the sperm and the egg contained a complete set of chromosomes (46 for humans), the resulting zygote would have double that number (92 chromosomes for humans), and this karyotype would be duplicated and passed on in all the later cell divisions. The resulting embryo would have double the normal number of chromosomes—an abnormality that would be lethal. Therefore, for offspring to have the same karyotype as their parents, fertilization must produce the normal number of chromosomes in the zygote.

The simple solution to this problem is for the gametes to contain half the number of chromosomes found in other cells. This arrangement is possible because the DNA is arranged in pairs of homologous chromosomes. In humans, for example, all body cells contain the same 23 homologous pairs of chromosomes. Each gamete a person produces, however, contains only one chromosome from each homologous pair. In the case of the sex chromosomes, this means that all gametes produced by women (eggs) contain a single X chromosome and all gametes produced by men (sperm) contain either an X or a Y chromosome. Gametes are thus **haploid** (*haploos*, "single"), and we assign the symbol *n* to the number of chromosomes in haploid cells. The cells that make up the rest of the body are **diploid** (*di*, "double") because they have 2*n* chromosomes—that is, twice the number of chromosomes that haploid cells have (see Figure 9.7).

Because each gamete contains the haploid number of chromosomes, the zygote formed by fertilization will

Figure 9.8 Similarities and Differences Between Meiosis and Mitosis

The major difference between meiosis and mitosis can be seen in meiosis I. The homologous chromosomes are paired during prophase I through metaphase I, resulting in a separation of the homologues at the end of meiosis I (the reduction division). In contrast, meiosis II is more similar to mitosis. For simplicity, not all the stages are shown.

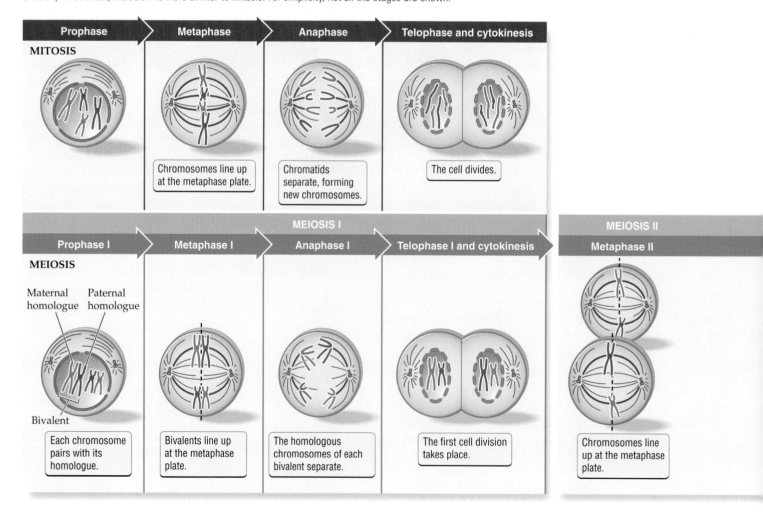

contain 2n chromosomes—that is, a complete human karyotype of 23 homologous pairs (including one pair of sex chromosomes). Furthermore, each pair of homologous chromosomes in the zygote will consist of one chromosome from the father and one from the mother (see Figure 9.7). This equal contribution of chromosomes by each parent is the basis for genetic inheritance. We will discuss the details of inheritance in Unit 3.

The day-to-day processes that maintain the life of an individual organism depend on the diploid state. On the other hand, since gametes are the only cells that can create a new organism, the continued propagation of a species depends on the haploid state. So there must be a way of generating haploid gametes from diploid cells. That process is **meiosis** [my-OH-siss] (*meio*, "less"), the specialized cell cycle in which a single diploid cell ultimately yields four haploid cells (Figure 9.8).

The stages of meiosis are very similar to those of mitosis. However, unlike mitosis, in which a single nuclear division is sufficient, meiosis involves two rounds of nuclear divisions. These successive divisions are called meiosis I and meiosis II, and each has a distinct role in producing haploid cells from diploid parent cells.

Meiosis I is the reduction division

Let's begin with the diploid cells of the reproductive tissue responsible for the production of gametes. **Meiosis I**, the first step in producing haploid gametes from these cells, reduces the chromosome number from 2n to n. It is achieved by a pairing of homologous chromosomes that

is not seen in mitosis. Otherwise, meiosis I has all the same stages as mitosis, and has a similar overall appearance (see Figure 9.8).

How does chromosome pairing during meiosis I reduce the number of chromosomes? The answer lies in both the preparation for and the mechanics of DNA segregation. During prophase I, each chromosome—now consisting of two chromatids—pairs with its homologue. In other words, the homologues of chromosome 6 pair up with each other, as do the copies of chromosome 3, and so on. These pairings, called **bivalents**, have a total of four chromatids each. The formation of bivalents provides an opportunity for the exchange of genetic sequences between homologous chromosomes (discussed in greater detail in Chapter 11).

An important consequence of the formation of bivalents becomes obvious during the later stages of prophase I, when the spindle microtubules extend to meet the chromosomes. In contrast to mitotic prophase, microtubules from only one pole attach to each chromosome of a homologous pair. When the bivalents move into position at the metaphase plate during metaphase I, the two chromosomes of the bivalent are attached to opposite spindle poles. When the microtubules shorten at anaphase I, the homologous *chromosomes* of each bivalent are pulled to opposite ends of the cell. This process is very different from what happens at anaphase of mitosis, in which the individual *chromatids* of each chromosome are separated to form new chromosomes (see Figure 9.8).

After anaphase I of meiosis, the events of telophase I follow the same patterns seen in mitosis, and cytokinesis produces two daughter cells. Unlike cells formed by mitosis, however, the daughter cells of meiosis I contain only half the number of chromosomes found in the parent cell. That is, each daughter cell has only one chromosome from each homologous pair, and is haploid. Meiosis I is therefore called a *reduction division* because it reduces the chromosome number by half: from diploid (2n) to haploid (n).

Meiosis II is similar to mitosis

The two haploid cells formed after meiosis I go through a second division, called **meiosis II**. This time the stages of the division cycle are more like those of mitosis. In particular, the chromatids separate at anaphase II, leading to an equal segregation of chromosomes into the daughter cells (see Figure 9.8). In this manner, the two haploid cells produced by meiosis I give rise to a total of

Anaphase II | Telophase II and cytokinesis

The second cell division takes place.

Four daughter cells form.

four haploid cells. These haploid cells are gametes, and they contain the appropriate number of chromosomes such that fertilization will produce a diploid zygote.

The reduction in chromosome number observed after meiosis I offsets the combining of chromosomes during fertilization and is nature's way of maintaining the constant chromosome number of a species during sexual reproduction.

Immune Cell Proliferation and Lyme Disease

Cell division is essential for the replacement of worn-out cells and tissues. Less obvious is the role it plays in defending the body from foreign invaders such as harmful bacteria, viruses, and fungi. These undesirable guests are collectively called **pathogens** (*patho*, "suffering"; *gen*, "producing"), a term that distinguishes them from the many microorganisms that are helpful to the body, such as beneficial intestinal bacteria.

The specialized cells collectively known as the immune system have the job of patrolling the body and defending it against pathogens. Just as human armies react to an attack on their country by recruiting new soldiers, immune system cells undergo cell division and proliferate when they are needed.

Cytotoxic T cells are crucial soldiers of the immune system, and when a pathogen attacks, they are prime candidates for proliferation. The T stands for thymus, a gland in the chest from which these cells originate, and they are called cytotoxic (*cyto*, "cell"; *toxic*, "poison") because their function is to kill other cells that have been infected or damaged by pathogens. Cytotoxic T cells are able to recognize infected cells because infection causes some of the pathogen's own proteins to appear on the surface of infected cells. Once the immune system identifies these proteins as "foreign," a T cell binds to them, killing the infected cell along with whatever pathogens are inside. This is how the body rids itself of infected cells and limits the illness caused by the pathogen. Usually, T cells and other cells of the immune system are able to discriminate between foreign proteins and the body's own proteins, and so avoid attacking the body's own healthy tissues.

However, certain pathogens, such as *Borrelia*, the bacteria that causes Lyme disease, are able to derail this normal process of self-tolerance.

As mentioned earlier, the painful arthritis associated with Lyme disease results from the fact that the immune system attacks the body's joints. Yet this attack continues in some patients long after the *Borrelia* bacteria have been killed by antibiotics—why? The arthritis may persist due to the similarity of a *Borrelia* protein to a normal cellular protein: a protein in the *Borrelia* cell wall contains amino acid sequences similar to those found in a human protein that is present on the surface of many normal human cells.

In brief, here is how the case of mistaken identity comes about. Certain cells of the immune system recognize the foreign *Borrelia* protein on infected cells, which induces them to divide in a normal immune response. But those same immune cells also recognize the similar human protein on healthy cells. When these cells move into the body's joints to clear the bacterial infection, they also attack and destroy normal cells with the normal protein. The death of cells in the joint is what causes inflammation and pain. Even after the bacteria have been eliminated, these T cells continue to proliferate and attack normal body cells, prolonging the arthritis. This painful consequence of inappropriate cell division reminds us that both the timing and the identities of the cells undergoing division must be carefully controlled in the body. Cancer, another dire consequence of inappropriate cell division, will be considered in the following Interlude.

Chapter Review

Summary

9.1 Stages of the Cell Cycle
- Cell proliferation requires repeated cell cycles, each of which divides a parent cell into two identical daughter cells.
- The cell cycle has two distinct stages: interphase and cell division.

9.2 Interphase: Preparing the Cell for Division
- Interphase consists of three phases: S, G_1, and G_2.
- During S phase, the cell's DNA is replicated. During phases G_1 and G_2, which come before and after S phase, the cell increases in size and produces specific proteins needed for division.
- Cells in the G_0 phase are in a resting state and do not divide.

9.3 Mitosis and Cytokinesis: From One Cell to Two Identical Cells
- During mitosis, the parent cell's replicated DNA is separated so that each daughter cell can receive a complete set of genes. This process is called DNA segregation.
- DNA in the nucleus is packed into strands called chromatin. The chromatin is further packed into chromosomes.
- The karyotype of a species is the specific number and shapes of its chromosomes.
- The chromosomes can be divided up into pairs of homologous chromosomes. One homologue in each pair is inherited from each parent. The sex chromosomes determine the sex of an organism.
- In the S phase (prior to mitosis), DNA replication produces chromosomes made up of two identical side-by-side strands, called chromatids, held together by a centromere.
- The five stages of mitosis are prophase, prometaphase, metaphase, anaphase, and telophase.
- In prophase, the chromosomes become visible. The two centrosomes begin to move to opposite parts of the cell, and the mitotic spindle forms between them.
- In prometaphase, the chromosomes are attached to the mitotic spindle at their kinetochores.
- In metaphase, the chromosomes align across the center of the cell at the metaphase plate.
- In anaphase, the two chromatids of each chromosome move to opposite ends of the cell. Once separated, they are considered daughter chromosomes.
- In telophase, new nuclei form around the two sets of daughter chromosomes on opposite sides of the cell.
- During cytokinesis, the cytoplasm of the parent cell is physically divided to create two daughter cells. In animal cells, a ring of actin microfilaments pinches the cell in two. In plant cells, a new cell wall forms to separate the two daughter cells.

9.4 Meiosis: Halving the Chromosome Number
- Meiosis is the special division process that produces gametes, which are haploid (n), containing only one chromosome from each homologous pair. Other cells in the body are produced by mitosis and are diploid ($2n$), containing a complete karyotype.
- Meiosis consists of two divisions.
- Meiosis I is called the reduction division because it produces haploid daughter cells from a diploid parent cell.
- During meiosis I, the chromosomes group into bivalents; each bivalent is a pair of duplicated homologous chromosomes, containing four chromatids in all.
- Meiosis II is similar to mitosis.
- Fusion of two gametes is called fertilization and results in a diploid zygote.

⊙ Review and Application of Key Concepts

1. Horses have a karyotype of 64 chromosomes. How many chromosomes would a horse cell undergoing mitosis have? How many would a horse cell undergoing meiosis II have?

2. Describe what happens in each stage of the cell cycle.

3. Compare the processes of mitosis and meiosis: how are they similar? How are they different?

4. What difference would you find in the karyotypes of male and female humans? How would this be reflected in the gametes of each sex?

5. How would the chromosome numbers of offspring of sexually reproducing organisms be affected if gametes were produced by mitosis instead of meiosis?

Key Terms

anaphase (p. 176)
bivalent (p. 181)
cell cycle (p. 173)
centromere (p. 175)

centrosome (p. 175)
chromatid (p. 175)
chromatin (p. 174)
chromosome (p. 174)

cytokinesis (p. 178)
diploid (p. 180)
DNA segregation (p. 174)
fertilization (p. 178)
G_0 phase (p. 174)
G_1 phase (p. 173)
G_2 phase (p. 173)
gamete (p. 178)
haploid (p. 180)
homologous chromosomes
    (p. 175)
homologues (p. 175)
interphase (p. 173)
karyotype (p. 174)

kinetochore (p. 176)
meiosis (p. 181)
meiosis I (p. 181)
meiosis II (p. 181)
metaphase (p. 176)
mitosis (p. 174)
mitotic spindle (p. 175)
pathogen (p. 182)
prometaphase (p. 175)
prophase (p. 175)
S phase (p. 173)
sex chromosome (p. 175)
telophase (p. 177)
zygote (p. 178)

Self-Quiz

1. In the cell cycle, duplication of DNA occurs in the
   a. G_1 phase.
   b. S phase.
   c. G_2 phase.
   d. division stage.

2. A karyotype is
   a. the number and shapes of a species' chromosomes.
   b. necessary for the physical separation of the daughter cells.
   c. a pair of identical chromosomes.
   d. the same in all species.

3. Which of the following statements is true?
   a. The cell lies dormant during interphase of the cell cycle.
   b. The key event of S phase is the synthesis of proteins required for mitosis.
   c. The cell increases in size during the G_0 phase.
   d. The cell increases in size during the G_1 and G_2 phases.

4. Which of the following statements is *not* true?
   a. DNA is packed into chromatin with the help of proteins.
   b. All chromosomes from a particular species adopt the same shape.
   c. Chromosomes are visible under the microscope only during mitosis or meiosis.
   d. Each species is characterized by a particular number of chromosomes.

5. Which of the following correctly represents the order of the phases in the cell cycle?
   a. mitosis, S phase, G_1 phase, G_2 phase
   b. G_0 phase, G_1 phase, mitosis, S phase
   c. S phase, mitosis, G_2 phase, G_1 phase
   d. G_1 phase, S phase, G_2 phase, mitosis

6. Cytokinesis occurs
   a. at the end of prophase.
   b. just before telophase.
   c. at the end of mitosis.
   d. at the end of G_1 phase.

7. In fertilization, gametes fuse to form
   a. a bivalent zygote.
   b. a haploid zygote.
   c. a diploid zygote.
   d. a triploid zygote.

8. Gametes contain
   a. twice the number of chromosomes as our skin cells.
   b. only sex chromosomes.
   c. half the number of chromosomes as our skin cells.
   d. only X chromosomes.

9. The reduction division is
   a. prophase of mitosis.
   b. anaphase II of meiosis.
   c. metaphase II of mitosis.
   d. meiosis I.

10. Meiosis results in
    a. four haploid cells.
    b. two diploid cells.
    c. four diploid cells.
    d. two haploid cells.

Biology in the News

***

It's Time to Play the Music, It's Time to Light the Lights

By Ashley Haley

We, of university age, grew up in the Muppet era—whether it was the lit stage of The Muppet Show, the friendly neighborhood of Sesame Street, the underground adventures of Fraggle Rock, or some wonderful mixture of them all. Sadly, the legacy of the Muppets has all but come to an end. The death of their creator and mastermind, Jim Henson, on May 16, 1990, resulted in the loss of the voice, personality, and talent of many of the Muppets we came to know and love. . . .

At age 53, Henson contracted streptococcus pneumonia and died a mere three days later, shocking the world.

***

We have focused on eukaryotic cell division in this chapter, but prokaryotes are good at cell division, too—sometimes with tragic results. The death of Jim Henson is a dramatic example. His lungs were literally consumed by the bacterium *Streptococcus pyogenes*. Once it gets a toehold in the body, this "flesh-eating" bacterium can reproduce extremely rapidly by a form of cell division called fission. Its capacity to reproduce so prolifically, combined with its ability to outflank the body's defense mechanisms, can render medical science helpless to stop the eating away of large amounts of tissue; death can result quickly.

Although Jim Henson died of pneumonia, these bacteria can also cause certain types of food poisoning. Food poisoning, whether caused by *S. pyogenes* or other bacteria, is a major health problem worldwide. Each year, in the United States alone, there are over 76 million cases of food poisoning, and five thousand deaths.

Contamination of food with bacteria can occur at any stage from harvest to kitchen. *Salmonella*, which reproduce in the digestive tracts of animals, are responsible for much of the food poisoning that occurs, though they seldom kill. The U.S. Department of Agriculture estimates that almost 40 percent of the poultry we buy is contaminated with *Salmonella*, as are about 12 percent of the pork and 5 percent of the beef. Ground beef is more likely to contain large numbers of these bacteria than whole cuts of beef, for two reasons. First, the hamburger may contain meat from several cows; if just one cow is contaminated, *Salmonella* can spread through the whole mix. Second, bacteria can grow more quickly in ground meat, which provides many more tiny surfaces on which to grow than do whole cuts of meat.

The widespread contamination of meat and poultry is why cooking them properly is so important. Experts recommend that meats be cooked to a temperature of at least 155°F. Consuming raw eggs, eggs fried sunny side up, or eggs boiled for less than 4 minutes can also be dangerous. Bacterial contamination isn't restricted to meat, poultry, and eggs. Several outbreaks of food poisoning have been traced to the presence of *Salmonella* on vegetables and fruits. Food safety professionals recommend thorough washing of any greens and vegetables to be served uncooked.

Evaluating the news

1. The cells of prokaryotes that cause food poisoning divide rapidly at room temperature. What should health inspectors look for in restaurants?
2. Given what you now know about bacterial reproductive power, what foods would you buy last and put away first? Why?
3. We routinely store foods in refrigerators. Why do you think some foods are considered more perishable than others? How do you think colder temperatures play a role in food preservation?

SOURCE: *The Muse,* November 23, 2001.

Cancer: Cell Division Out of Control

Enlisting a Virus to Fight Cancer

Cancer is the ultimate insult to the cooperative functioning of the cells in a multicellular organism. Cancerous tumors are like stubborn rebel colonies of cells ignoring the laws of cell-to-cell coordination that keep multicellular organisms alive. The cells that form a tumor divide with wild abandon, often failing to adopt the structures and activities required for the particular organ or tissue they are part of. At worst, these aggressive cells break out of the tumor and disperse to establish other similarly rebellious colonies in other parts of the body. This is the form of cancer that everyone fears—malignant cancer.

Cancer accounts for more than 500,000 deaths in the United States each year. While the past decade has seen improvements in treatment and prevention, in 2005 alone more than 1.3 million Americans were diagnosed with some form of cancer. The National Cancer Institute estimates that the collective price tag for the various forms of cancer is more than $100 billion per year—split almost evenly between costs due to individual deaths and the combination of direct medical costs and lost productivity.

Almost 30 years ago, President Richard Nixon declared a war on cancer in the United States by making anticancer research a high priority. Since then, some major victories have been won, thanks to improvements in radiation and drug therapies. Whereas in the early twentieth century very few individuals survived cancer, today roughly 40 percent of patients are alive 5 years after treatment is begun. Nevertheless, the war against cancer is far from over, and the need for powerful new treatments that can stop tumor growth and eliminate cancerous cells is as urgent as ever.

One of the most inventive potential methods of locating and destroying cancerous cells depends on the assistance of an unlikely ally: a virus. This adenovirus (Figure B.1), which infects mammals, takes over the biochemical machinery of specific cells to produce more viruses. In the process, the virus kills the infected cells, often causing respiratory tract disease.

To harness this destructive power for the benefit of individuals with cancer, researchers have mutated the adenovirus so that it successfully infects only cancerous cells. These mutant viruses are unable to multiply, but once they infect a cancerous cell, they are still able to destroy it. In the future, mutant adenoviruses could be administered to tumor sites, where they would infect and kill only the cancerous cells, leaving healthy cells untouched. Studies using cancerous cells grown outside of the body support the validity of this approach, as the mutant virus has been shown to selectively eliminate the cancerous cells.

This remarkable effort to tame a virus and turn it into an anticancer weapon depends on understanding how viruses take control of the cells they infect. New discoveries in this area have made it clear that many viruses have ways to bypass the normal controls that limit cell division and have revealed that some viruses play an important role in causing cancer.

In this essay we see how the "rules" that govern cell division are enforced inside each cell, and how the failure of these control systems results in cancer. Some controls are based on promoting cell division, others on preventing it. We explore the interplay of various factors—viruses, environment, and heredity—that can determine, or at least influence, whether cancer develops. We close with a look at one potential avenue for using a mutant adenovirus to combat cancer.

Figure B.1 Adenoviruses Swarming a Red Blood Cell: Are They Potential Weapons Against Cancer?

What Is Cancer?

To achieve a high level of organization, a multicellular organism must have a means of controlling and coordinating the behavior of its individual cells. Any large community that does not have rules quickly falls into chaos, and the same is true of a multicellular organism. Therefore, both the metabolic activities and the frequency of division of every cell are closely regulated. As we saw in Chapter 6, cells respond to a variety of signals, including chemical compounds and proteins secreted by other cells. Some of these signals work by activating cellular proteins required for the cell cycle. In effect, such positive growth signals enforce one set of rules that cells obey by promoting cell division when and where it is needed.

Until the late 1980s, cell division was thought to be controlled exclusively by "positive growth regulators"—positive signals that promote the cell cycle. Today we know that whether or not a cell divides is not determined solely by positive signals. The proper functioning of a multicellular organism also depends on "negative growth regulators"—signaling molecules that can counterbalance positive growth signals and halt the cell cycle. The life of every cell is therefore managed by a delicate interplay of positive and negative signals, both of which directly affect multiple proteins inside the cell.

Learn more about the development of cancer.

B.1

Because each multicellular organism is a cooperative community of cells, the failure of just one cell to maintain its balance of opposing positive and negative signals can have serious consequences. One of these consequences is **cancer**, a group of diseases caused by the rapid proliferation of cells due to inappropriate cell division (Figure B.2).

Positive Growth Regulators: Promoting Cell Division

Our understanding of how cells respond to the signals that promote cell division began with observations of cancer in animals. One of the best ways to study the effects of a particular control system is to discover what happens when it is no longer working. We can see how this principle operates by looking at a famous experiment conducted by the biologist Peyton Rous, who studied cancerous tumors called sarcomas in chickens in the first decade of the twentieth century.

Rous discovered that he could grind up sarcomas and extract an unidentified substance that, when injected into healthy chickens, caused cancer. He knew that the extract contained no bacteria because it had been carefully filtered, so the cause of the cancer had to be something much smaller—something that could pass through his filters.

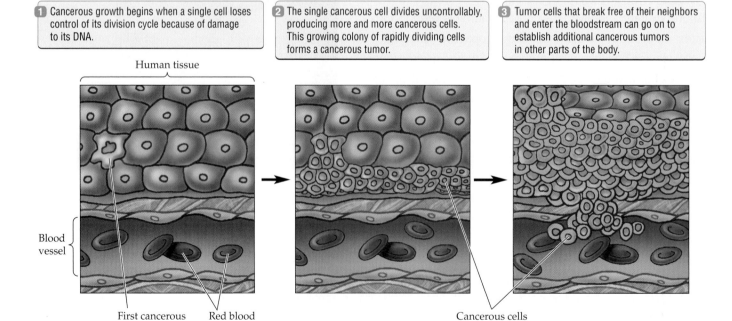

1 Cancerous growth begins when a single cell loses control of its division cycle because of damage to its DNA.

2 The single cancerous cell divides uncontrollably, producing more and more cancerous cells. This growing colony of rapidly dividing cells forms a cancerous tumor.

3 Tumor cells that break free of their neighbors and enter the bloodstream can go on to establish additional cancerous tumors in other parts of the body.

Human tissue

Blood vessel

First cancerous cell

Red blood cells

Cancerous cells

Figure B.2 Cancers Start with a Single Cell That Loses Control

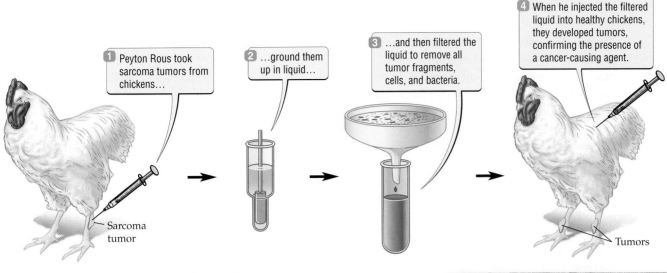

1. Peyton Rous took sarcoma tumors from chickens...

Sarcoma tumor

2. ...ground them up in liquid...

3. ...and then filtered the liquid to remove all tumor fragments, cells, and bacteria.

4. When he injected the filtered liquid into healthy chickens, they developed tumors, confirming the presence of a cancer-causing agent.

Tumors

Figure B.3 The Rous Sarcoma Virus Causes Cancer in Chickens

5. The cancer-causing agent isolated from the tumor is a virus called the Rous sarcoma virus, which is roughly 100 times smaller than the average animal cell.

His work led to the discovery of the first animal tumor virus, which was named the Rous sarcoma virus in honor of its discoverer and the type of tumor from which it was obtained (Figure B.3).

Some viruses can cause cancer

As we saw in Chapter 1, viruses are tiny assemblages of either RNA or DNA surrounded by protein. The nucleic acids found inside a virus contain genes that are necessary for the viral reproductive cycle. However, viruses are more than a hundred times smaller than the average animal cell. They are invaders that can multiply only by infecting the cells of other organisms and using the biochemical machinery of the infected cell for their own replication.

The discovery that a virus could cause cancer in animals was a major breakthrough in our understanding of this type of disease, but it took many more decades before scientists discovered how the Rous sarcoma virus derails the normal controls that regulate cell division. The solution to this mystery came with the discovery of a particular strain of Rous sarcoma virus that could multiply in cells without causing them to divide rapidly. Biologists could then compare the viral genes found in this virus with those found in cancer-causing strains. The virus that did not cause cancerous cell division was missing a single gene. An individual gene carries the code for a specific protein, so the absence of this gene meant that this strain of the Rous sarcoma virus was not able to produce one particular protein. Further research showed

that this protein, when present, is responsible for destroying the internal controls of the host cell.

What sort of protein might cause cancer? If you guessed it must be one that interferes with the cell's normal signaling process in some way, you would be right. The cancer-causing agent produced by the Rous sarcoma viral gene is a very active protein belonging to a class of enzymes called kinases.

Protein kinases activate their target proteins by adding phosphate groups to (phosphorylating) them. Under normal conditions, these activation events are counterbalanced by the action of other enzymes that remove the phosphates, effectively turning the target proteins off. When an overactive viral protein kinase acts on the same target proteins, however, there is no way to turn the signal cascade off. The avalanche of enzymatic reactions that drive the cell toward mitosis roars out of control, leading to a cell that just keeps dividing. Since the Rous sarcoma virus inserts all of its genes into the DNA of the infected cell, all the daughter cells also receive the cancer-causing gene. Eventually, the growing colony of rapidly dividing cells forms a tumor.

Oncogenes play an important role in cancer development

The protein kinase gene in the Rous sarcoma virus is named *Src* [SARK]. It is just one of several **oncogenes** (*onco*, "bulky mass"), or cancer-causing genes, found in viruses. However, the *Src* oncogene is not unique to the Rous sarcoma virus. Instead, it is an altered version of a gene normally found in the genetic material of the host organism's cells.

As we saw above, viruses like the Rous sarcoma virus multiply by becoming part of the infected cell's DNA. At some point in evolutionary history, a mutant protein kinase gene may have been picked up from an abnormal host cell and plugged into the genetic material of the virus. A **mutation**, as we shall see in Chapter 12, is a change in the DNA sequence of a gene. Mutations can alter the characteristics of the protein produced by the gene, either increasing or decreasing its activity or its ability to function. In the case of *Src*, the mutation results in the production of an overactive kinase that cannot be controlled like its normal counterpart in the cell.

The realization that the *Src* oncogene has a normal, controllable counterpart in host cells was an important step in identifying the cellular genes that regulate cell division. These normal cellular genes are called **proto-oncogenes** (*proto*, "first") because they are the predecessors of the viral oncogenes. Today scores of proto-oncogenes are known, most of which were first identified as oncogenic mutants in tumor viruses. Although most known tumor viruses cause cancer only in animals such as chickens, mice, and cats, all the proto-oncogenes identified in these animals are also found in human cells.

Oncogenes play a major role in human cancers. Human oncogenes have several sources. Some, as we have seen, are brought into the cell by an infecting virus. Most human oncogenes, however, come from mutations of proto-oncogenes caused by chemical pollutants and other environmental factors, such as exposure to too much sunlight. In rare cases, oncogenes can also be inherited.

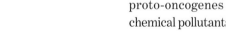

Negative Growth Regulators: Inhibiting Cell Division

As potentially dangerous as they may seem, oncogenes are not the sole villains responsible for the rampant cell growth that leads to cancer. Although this may be true in some cases, as with *Src* in chickens, usually something else must also go wrong before cancer can occur. That is because normal cells have internal safeguards that must be overcome before the controls on cell division are totally removed. These safeguards are a family of proteins that are called **tumor suppressors** because their normal activities were first discovered to stop tumor growth (Figure B.4). Tumor suppressors are therefore negative growth regulators that stop cells from dividing by opposing the action of the proteins encoded by proto-oncogenes.

Whether or not a normal cell divides depends on the activities of both proto-oncogenes and tumor suppressor genes. For a cell to divide, proto-oncogenes must be activated to promote the process, and tumor suppressor genes must be inactivated to allow the process to happen. Because controlling the timing and extent of cell division

Figure B.4 p53 Tumor Suppressor Protein
The p53 protein maintains normal cell division, stops tumor growth, and ensures the integrity of DNA replication. This graphic shows its DNA-binding regions.

is so important, cells have both of these counterbalancing control systems, which must be in agreement before a cell can divide.

How do tumor suppressor genes oppose the activity of proto-oncogenes under normal circumstances? The answer lies in the cascade of enzymatic events that lead to cell division. External signals that promote cell division often do so by triggering stepwise protein activations inside the cell that are collectively described as a **signal cascade**. **Growth factors** are a class of such signaling molecules that induce cell growth and division. The proliferation of cells in the human body is controlled largely by growth factors, which both initiate and maintain the processes needed for cell growth and division. Scores of proteins exported by cells function as growth factors, and in most cases their effects are confined to neighboring cells.

Tumor suppressors block specific steps of growth factor signal cascades

In the same way that proto-oncogene proteins induce cell division by activating the components of a growth factor signal cascade, tumor suppressors block cell division by inactivating some of the same components. A well-known example of tumor suppressor activity was discovered during a study of a rare childhood cancer known as retinoblastoma [*RET*-ih-noh-blass-*TOH*-ma]. As the name indicates, retinoblastoma (*retino*, "net"; *blastoma*, "bud") is a cancer that forms in the retina of the eye, and it often leads to blindness (Figure B.5). Retinoblastoma strikes one in every 15,000 children born in the United States and accounts for about 4 percent of childhood cancers.

What causes this kind of cancer to develop? As we saw in Chapter 9, each species has a characteristic set of chromosomes, termed the karyotype, that can be seen under the microscope. In cancerous cells from some children with retinoblastoma, a portion of chromosome 13 appeared to be missing, hinting that the cancer might be caused by the absence of a particular gene. Today we know that the gene in question normally produces a protein called Rb. The missing *Rb* gene, and the resulting lack of Rb protein, results in retinoblastoma.

This is not the effect we would expect an oncogene to have. An oncogene causes cancer by producing an overactive protein that pushes the cell to divide. The *Rb* gene must have the opposite effect, since its *absence* promotes cell division. A simple explanation would be that the Rb protein normally inhibits a process required for cell division; when it is missing, the brakes on cell division no longer work, and cells divide uncontrollably.

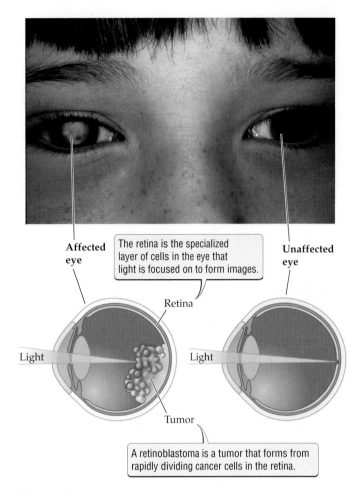

Affected eye

The retina is the specialized layer of cells in the eye that light is focused on to form images.

Unaffected eye

Retina

Light

Light

Tumor

A retinoblastoma is a tumor that forms from rapidly dividing cancer cells in the retina.

Figure B.5 Retinoblastoma
This child has a visible retinoblastoma in her right eye. A retinoblastoma both blocks the light and destroys the ability of retinal cells to respond to light, frequently resulting in blindness.

The Rb protein inhibits a key process in the cell's preparations for division. It binds to and inactivates a protein that is required for the cell's response to growth factor signals. When cells are stimulated to divide by growth factors under normal conditions, the resulting signal cascade involves not just the activation of proto-oncogenes, but also the inactivation of tumor suppressors such as Rb.

The Rb protein is inhibited by a protein kinase that is activated by the growth factor signal cascade. The activated kinase phosphorylates the Rb protein, causing it to change its shape and release its target protein, which can then activate the genes needed for cell division (Figure B.6). This example shows that the phosphorylation resulting from growth factor signals acts to turn on some proteins and turn off others, thus confirming the balance of positive and negative regulatory controls that must come into play before a cell can divide.

Helpful to know

By convention, the names of genes are always given in italic type, while the names of their protein products are in roman type. Often this font difference is the only thing distinguishing the protein name from the gene name; for example, the product of the *Rb* gene is the Rb protein.

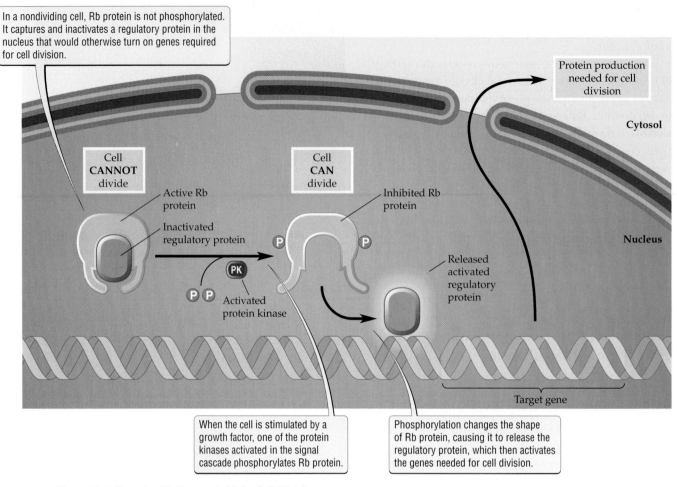

In a nondividing cell, Rb protein is not phosphorylated. It captures and inactivates a regulatory protein in the nucleus that would otherwise turn on genes required for cell division.

Protein production needed for cell division

Cytosol

Cell **CANNOT** divide

Active Rb protein

Inactivated regulatory protein

Cell **CAN** divide

Inhibited Rb protein

Nucleus

Released activated regulatory protein

Activated protein kinase

When the cell is stimulated by a growth factor, one of the protein kinases activated in the signal cascade phosphorylates Rb protein.

Phosphorylation changes the shape of Rb protein, causing it to release the regulatory protein, which then activates the genes needed for cell division.

Target gene

Figure B.6 How the Rb Protein Inhibits Cell Division

Both copies of a tumor suppressor gene must be mutated to cause cancer

The differences in how oncogenes and tumor suppressor genes function highlight differences in the kinds of genetic mutations that can lead to cancer. Because chromosomes exist in pairs, and because the two chromosomes in each pair have the same set of genes, there are two copies of each gene in the cell, one contributed by each parent. For an oncogene to promote cancer, only one copy of the proto-oncogene must be mutated to an oncogenic form. For example, one mutated copy of the gene might produce an overactive protein that can push the cell to divide.

In contrast, for a tumor suppressor gene to promote cancer, both copies of the gene must be mutated to an inactive form. In other words, if only one copy were inactivated, the other copy might still produce enough tumor suppressor protein to inhibit cell division. Therefore, complete loss of this negative control mechanism—meaning that no tumor suppressor protein is being made—

requires that both copies of a tumor suppressor gene be inactivated (Figure B.7).

An Interplay of Factors Can Cause Cancer

Most human cancers involve more than the mutation of one proto-oncogene or the complete inactivation of one tumor suppressor gene. Indeed, the complex series of events involved in cell division means that both inherited and environmental factors can come into play. As we shall see in Unit 3, a number of factors in the environment can cause changes in a cell's DNA. Only mutations found in the gametes or the gamete-producing cells, however, can be passed on to offspring. About 1 to 5 percent of all cancer cases can be traced exclusively to an inherited genetic defect. The remaining majority of cases involve either a combination of inherited and environmental factors or environmental factors alone.

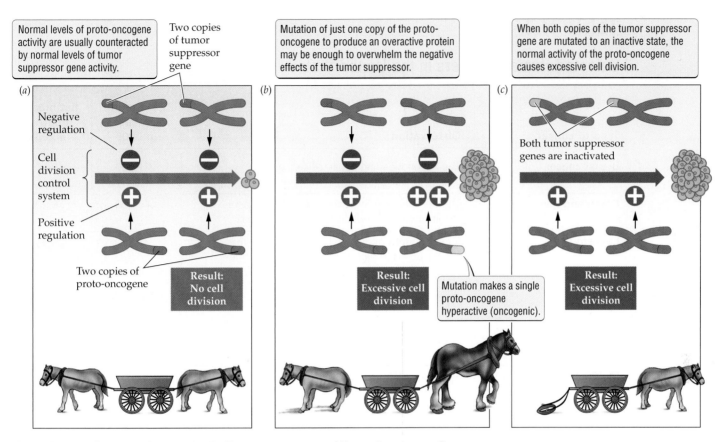

Figure B.7 The Control of Cell Division by Proto-oncogenes and Tumor Suppressor Genes

Whether or not a cell divides depends on the balance between proto-oncogene and tumor suppressor gene activity. (*a*) A normal cell that is not dividing can be compared to a cart attached to two ponies pulling in opposite directions. Because the ponies are of equal size and strength, the cart remains stationary. Likewise, the activities of proto-oncogenes, which promote cell division, are counterbalanced by the activities of tumor suppressor genes, which inhibit cell division, so no cell proliferation occurs. (*b*) The mutation of one copy of a proto-oncogene to an oncogene is like substituting a workhorse for one of the ponies. The workhorse is larger and stronger than the pony; hence it can pull the cart to one side. In a cell, the result is inappropriate cell division. (*c*) When both copies of a tumor suppressor gene are inactivated, the result is similar to completely eliminating one pony from the cart: the remaining pony can pull the cart to one side. Again, the result in a cell is inappropriate cell division.

In some ways this is good news, because many of the environmental factors that cause cancer are related to our lifestyles and behaviors, and we have the power to try to limit our exposure to those factors. To prevent cancer, we must try to reduce the likelihood of dangerous mutations accumulating in the DNA of our cells as well as to understand the genetic characteristics that may lead to cancer later in life.

Cancer is a multi-step process

Cancer is a group of diseases that is likely to affect many lives (Table B.1). Over the course of a lifetime, an American male has a 45 percent chance of developing an invasive cancer. American women fare slightly better, with a 38 percent chance of developing cancer. In the United States, one in four deaths is due to cancer, and more than 8 million Americans alive today have been diagnosed with cancer and are either cured or undergoing treatment. In 2005, more than 1,500 Americans died from cancer each day.

Given such a high incidence of cancer, you might think that only one or two mutations are sufficient to cause the disease. However, careful study of human cancers shows that several cellular safeguards have to fail before a cancerous tumor can form. The unlucky string of failures that produces a cancerous tumor includes both the mutation of proto-oncogenes and the loss of tumor suppressor activity.

Consider cancer of the colon (the large intestine), which is diagnosed in more than 100,000 individuals each

Table B.1

Selected Human Cancers in the United States

Type of cancer	Observation	Estimated new cases in 2005	Estimated deaths in 2005
Breast cancer	The second leading cause of cancer deaths in women	211,200	40,400
Colon and rectal cancer	The number of new cases is leveling off as a result of early detection and polyp removal	145,300	56,300
Leukemia	Often thought of as a childhood disease, this cancer of white blood cells affects more than 10 times as many adults as children every year	38,800	22,600
Lung cancer	Accounts for 28 percent of all cancer deaths and kills more women than breast cancer does	172,600	163,500
Ovarian cancer	Accounts for 3 percent of all cancers in women	22,200	16,200
Prostate cancer	The second leading cause of cancer deaths in men	232,100	30,400
Malignant melanoma	The most serious and rapidly increasing form of skin cancer in the United States	59,600	10,600

year in the United States. In many cases of colon cancer, the tumor cells contain at least one overactive oncogene and several completely inactive tumor suppressor genes. In fact, because the mutations in different genes that lead to colon cancer usually occur over a period of years, the gradual accumulation of these mutations can be linked with the stepwise progression toward cancer.

Let's look at the step-by-step sequence of chance mutations that might lead to colon cancer. In most cases, the first step is a relatively harmless, or **benign**, growth described as a polyp (Figure B.8). The cells that make up the polyp are undergoing division at an inappropriate rate. These cells are the descendants of a single cell in the lining of the colon that has suffered one or more mutations.

In many large polyps, the cells contain mutations that inactivate both copies of a tumor suppressor gene, together with a single mutation that transforms a proto-oncogene into an oncogene. The complete loss of one tumor suppressor's activity, combined with the presence of an overactive protein, is enough to allow inappropriate cell division. However, most such polyps do not spread to other tissues and can be safely removed surgically.

The progression from a benign polyp to a **malignant** tumor—that is, one that can spread throughout the body with life-threatening consequences—depends on the inac-

tivation of additional tumor suppressor genes. In many colon tumors, the start of true malignancy coincides with the loss of a part of chromosome 18 that contains at least two important tumor suppressor genes. This complete loss of two additional tumor suppressors results in a far more aggressive and rapid multiplication of the cancerous cells, greatly increasing the chance that they will spread to other tissues.

One of the last key events in the path to full malignancy is the complete inactivation of yet another tumor suppressor gene, named *p53*. For reasons that are not entirely clear, loss of the p53 protein seems to remove all controls on cell division, allowing the cancerous cells to break free of the original tumor and travel through the bloodstream to other parts of the body. At this point the cancerous cells are entirely resistant to signals from the body's regulatory and immune systems, and the worst possible scenario—a malignant tumor—has come true.

Cancer is often related to lifestyle choices

The relative contributions of inherited and environmental factors to an individual's cancer risk have been debated for decades. In recent years, large-scale studies have tried to settle this issue by tracking cancer incidence in thousands of pairs of identical twins, who share the same genetic makeup. If inherited genetic defects are more important than environmental factors in causing cancer, then one would expect to see a very similar incidence of cancer in both twins. On the other hand, if environmental factors play a greater role, one would expect to see significant differences in cancer incidence due to differences in the twins' adult environments or habits. In one Scandinavian study that tracked over 44,000 pairs of twins, the most important contributor to individual cancer risk by far was environmental factors, which included lifestyles and behaviors.

The contribution made by environmental factors to the vast majority of cancers confirms that changes in our personal behavior can reduce our risk. Let's consider one cause of cancer that is particularly amenable to behavioral change: tobacco use.

Since 1982, cigarette smoking has been recognized as the single leading cause of cancer mortality in the United States. This acquired behavior is a major cause of lung and oral cavity cancer, and it contributes to a wide range of other cancers, including those of the kidney, stomach, and bladder. There are close to 50 million American smokers, so it is not surprising that tobacco use accounts for one in five deaths in the United States. Among the thousands of chemical compounds that have been identified in tobacco smoke, over 40 have been confirmed to be carcinogens. A **carcinogen** [kar-*SIN*-uh-jin] is any physical, chemical, or biological agent that causes cancer.

Polycyclic [polly-*SIKE*-lik] aromatic hydrocarbons, or PAHs, are an important class of carcinogens found in tobacco smoke. These organic compounds can bind to DNA, forming a physical complex known as an *adduct*. Adducts cause mistakes in DNA synthesis (see Chapter 12), which introduce mutations into the DNA sequence. PAHs tend to form adducts at several sites on the *p53* gene in the lung cells of smokers. The resulting mutations in the *p53* gene prevent the production of functional p53 tumor suppressor protein in the affected cells. In the same way that the inactivation of *p53* contributes to colon cancer, its inactivation in lung cells allows them to divide uncontrollably, leading to lung cancer. Furthermore, the formation of adducts due to PAHs is not restricted to lung cells. The white blood cells of smokers also show PAH-related genetic damage, which can contribute to other forms of cancer.

The good news is that stopping smoking can dramatically reduce an individual's cancer risk. People who quit smoking before the age of 50 reduce their risk of dying in the subsequent 15 years by half. Regardless of age, people who quit smoking live longer than those who con-

tinue to smoke. While nicotine, which is the addictive drug in tobacco, makes quitting smoking difficult, all should find inspiration in the fact that one in five Americans is a former smoker.

Figure B.8 Colon Cancer Is a Multi-step Process
The sequential mutation of several genes that produce positive and negative growth regulators coincides with the progression from a benign polyp in the colon to a malignant tumor.

Section through colon

Complete loss of one tumor suppressor

Early polyp (benign)

Mutation of a proto-oncogene

Polyp

Complete loss of two more tumor suppressors

Late polyp

Complete loss of the p53 tumor suppressor

Malignant tumor

To bloodstream

Biology Matters

Before and After

Lung tissue from a non-smoker and a smoker.

The Truth about Cigarettes in School

About one-third of American college students smoke cigarettes regularly, although only about 13 percent admit to being daily smokers. And, in the 18- to 25-year-old age group, about half have tried marijuana at least once. Both tobacco and marijuana cigarettes contain a compound called benzopyrene, a powerful cancer-causing agent. This substance suppresses a gene that controls the cell cycle. If that gene is not working as it should, cells can begin to multiply without restraint. (As we have seen, cancer is unrestrained cell proliferation.) An average marijuana cigarette contains about 50 percent more benzopyrene than a cigarette made of tobacco. And, marijuana smoke is typically inhaled more deeply and held in the lungs longer than cigarette smoke. But smoking tobacco is much more addictive than smoking marijuana, and that means that people who start smoking cigarettes are far more likely to be trapped in the habit for a long time. And a regular marijuana smoker will consume far fewer cigarettes per day than even a moderate cigarette smoker. No matter what, however, both types of smoke are dangerous.

There's a lot that you can do to shape your own health destiny. Being informed and thinking ahead can be the road to a longer and healthier life.

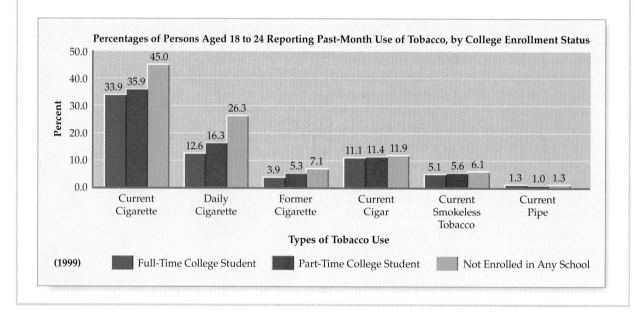

Percentages of Persons Aged 18 to 24 Reporting Past-Month Use of Tobacco, by College Enrollment Status

(1999) ■ Full-Time College Student ■ Part-Time College Student ■ Not Enrolled in Any School

Types of Tobacco Use

Making the Most of Losing p53

The connection between loss of the p53 tumor suppressor protein and cancers of the colon and lungs emphasizes the importance of this protein, but only hints at the broad range of its activities.

The p53 protein helps to maintain normal cell division

The p53 protein is perhaps most famous for its multiple roles in guarding the integrity of the cell (see Figure B.4).

It not only prevents the cell from dividing at inappropriate times, but also halts cell division when there is evidence of DNA damage that could result in harmful mutations. This protection gives cells the opportunity to repair the damage. If the repair process fails, p53 then goes so far as to induce a cascade of enzymatic reactions that kills the cell. In other words, if the cell's DNA is too badly damaged to repair, the cell commits suicide, rather than passing on mutations that could potentially harm the entire organism.

Given the important guardian functions of the p53 protein, it is not surprising that more than half of all

Explore making the most of losing p53.

B.3

cancers involve a complete loss of p53 activity in tumor cells. The number goes as high as 80 percent in some types of cancer, such as colon cancer.

The absence of p53 may be useful in cancer therapies involving adenoviruses

Researchers are exploring ways to use the absence of p53 activity as a means of identifying and destroying cancerous cells. Consider the proposed cancer therapy involving adenoviruses, introduced at the start of this essay. Researchers have discovered that in order to multiply, an adenovirus must inactivate the host cell's p53 protein. The same mechanism that enables the p53 protein to halt cell division also stops the DNA of the adenovirus in an infected cell from being used to make viral proteins. To avoid this defensive measure, an adenovirus gene produces a protein that binds to and disables the p53 protein, thereby allowing the virus to use the cell to make components for new viruses. In other words, the adenovirus can function effectively only in cells that have no active p53 protein. Since p53 also happens to be the tumor suppressor protein that is most often absent in cancerous cells, an important connection can be forged between the virus and cancerous cells that lack p53 activity.

In a clever turn of events, biologists have mutated the adenovirus such that it no longer produces the protein that disables p53. This mutant virus can multiply only in cells that already lack p53 activity, which results in the selective killing of these cells. Since the only cells in the body that are likely to lack p53 activity are cancerous cells, this virus works like a smart bomb that seeks out its target and destroys it. Experiments using cells isolated from cancers of the colon, cervix, and pancreas have shown that the mutant virus preferentially kills those cells. Clinical trials in patients with liver cancer have shown that injection of the mutant virus can reduce tumor size in approximately half of the patients. Larger-scale clinical trials are currently under way in patients with head and neck cancers. Only time will tell if this inventive application of our understanding of cancer will help people with the disease.

◉ Review and Discussion

1. Describe the possible consequences when a proto-oncogene becomes an oncogene.

2. Describe the interplay between cellular signals that promote cell division and those that inhibit cell division.

3. Colon cancer develops in a series of stages. Outline the stages and what happens in each stage.

4. Polycyclic aromatic hydrocarbons (PCHs) are found in tobacco. Discuss the possibility that commonly consumed products also contain carcinogens, some of which may not be identified.

5. In light of the clear link between tobacco usage and cancer, many have questioned the right of tobacco companies to continue selling such a deadly substance. Consider the issues of personal freedom versus public health policy and explain what restraints, if any, you think should be placed on the sale of tobacco.

6. As environmental causes of cancer receive increasing attention, the warning labels on food have become lengthier and more ominous in tone. Since many factors contribute to cancer, do you think that expanded food warning labels is an effective approach to reducing cancer risk? If so, how might one combat the public's tendency to ignore long and complex warning labels?

Key Terms

benign (p. B9)
cancer (p. B3)
carcinogen (p. B10)
growth factor (p. B6)
malignant (p. B9)

mutation (p. B5)
oncogene (p. B5)
proto-oncogene (p. B5)
signal cascade (p. B6)
tumor suppressor (p. B5)

10 Patterns of Inheritance

Key Concepts

◉ Genetics is the scientific study of genes, which are the basic units of inheritance. Genes contain instructions for building proteins.

◉ Organisms contain two copies of each gene, one inherited from each parent. When the two copies of a gene are identical, the individual is homozygous for that gene. When the two copies of a gene are different, the individual is heterozygous for the gene.

◉ Different versions, or alleles, of a gene produce different forms of a protein and cause hereditary differences among organisms. New alleles arise by mutation.

◉ During meiosis, the two inherited copies of each gene separate equally into gametes. With some exceptions, the separation of copies of one gene during meiosis is independent of the separation of copies of other genes.

◉ Some aspects of an organism's phenotype (observable characteristics) are controlled by single genes that have dominant alleles (alleles that determine the phenotype of the organism even when paired with nondominant alleles); in such cases, the phenotype of the offspring can be predicted accurately from the genotype (genetic makeup) of the parents.

◉ Many aspects of an organism's phenotype are determined by groups of genes that interact with one another and with the environment. As a result, offspring with identical genotypes can have very different phenotypes.

The Lost Princess

In the early hours of July 17, 1918, the Russian royal family was awakened and taken to the basement of a house in the industrial city of Ekaterinburg. Told they were to be photographed, the Tsarina, Alexandra, and her young son, Alexis, who suffered from hemophilia, were seated in chairs. The rest of the family—Tsar Nicholas II and his four daughters, Olga, Tatiana, Maria, and Anastasia—stood behind Alexandra and Alexis, as did the family physician, cook, maid, and valet. Suddenly, eleven men burst into the room, each with a different intended victim, and began firing with revolvers. In a brutal act that brought to an end the Romanov dynasty of pre-Communist Russia, all seven members of the royal family, and their four servants, were killed.

Or were they? In 1920, a woman was pulled freezing from a Berlin canal. At first known simply as "Fraulein Unbekannt," or "Miss Unknown," and later as Anna Anderson, she claimed that she was the Princess Anastasia. Her knowledge of minute details of life at the Russian imperial court convinced many that she was indeed Anastasia, the youngest daughter of Nicholas and Alexandra. Others, troubled by her inability to speak Russian and her bouts of erratic behavior, thought she was a pretender. Anna Anderson herself never doubted that she was Anastasia, a conviction she held to her death in 1984, at the age of 83.

Over the years, the legend of Princess Anastasia grew to become the subject of books, movies, and magazine articles. While the escape of a beautiful princess from execution made a wonderful story, was Anna Anderson really Anastasia? Ultimately, the mystery was solved with a combination of careful detective work and genetic analysis. Investigators used the basic

A Royal Mystery
(*a*) Princess Anastasia of Russia, fourth daughter of the Tsar.
(*b*) The actress Ingrid Bergman portrayed the princess in the 1956 movie *Anastasia*.

principles of genetics to determine whether Anna Anderson could have been the lost princess.

What are the rules that govern how characteristics are inherited? How could those rules be used to determine whether Anna Anderson was a member of the Russian royal family? To answer these and many other questions about inherited characteristics, we must understand the principles of genetics, the scientific study of genes.

Humans have used the principles of inheritance for thousands of years. Knowing that offspring resemble their parents, for example, people allowed only those individuals with desirable characteristics, such as large grains in wheat, to reproduce. Over time, people used this method to domesticate animals and to develop agricultural crops from wild plant species. As a field of science, however, genetics did not begin until 1866, the year that Gregor Mendel (Figure 10.1) published his landmark paper on inheritance in pea plants. Prior to Mendel's work, many facts about inheritance were known, but no one had organized those facts by describing and testing a hypothesis that could explain how inherited characteristics are passed from parent to offspring.

Mendel changed all that. His experiments led him to propose that the inherited characteristics of organisms are controlled by hereditary factors—now known as genes—and that one factor for each characteristic is inherited from each parent. Although he did not use the word "gene," Mendel was the first to propose the concept of the gene as the basic unit of inheritance. The emphasis that Mendel placed on genes continues today. In fact, we define **genetics** as the scientific study of genes.

In the more than 100 years since Mendel's work, we have learned a great deal about genes, especially about their physical and chemical properties. We now know that genes are located on chromosomes. Structurally, each gene is a segment of DNA within the long DNA molecule of the chromosome. As we shall see in this unit, most genes contain instructions for the synthesis of a single protein or protein subunit. Finally, most of the trillions of cells in our bodies contain exactly the same set of genes. In people and other organisms that reproduce sexually, each cell contains two copies of every gene, one inherited from each parent (Figure 10.2). There are two exceptions to this rule. One is our sperm and egg cells—our gametes—which, because they are haploid, have only one of the usual two copies of each gene (see Chapter 9). The other concerns sex-linked genes, in which only one copy is inherited in males (as we shall see in Chapter 11).

In this chapter, after defining some basic genetics terms, we outline Mendel's theory of inheritance, which provides the foundation for genetics even today. Then we explore various ways in which modern experiments have extended Mendel's laws to cover aspects of genetics not foreseen by Mendel.

Figure 10.1 Mendel and the Monastery Where He Performed His Experiments
Gregor Mendel (inset) was a monk at the monastery of St. Thomas, shown in this photograph. For many years it was believed that Mendel had performed his experiments behind the fence visible here immediately in front of the monastery. Staff members at a museum devoted to Mendel recently discovered that Mendel's garden, no longer evident, was located in the foreground of this photograph.

10.1 Essential Terms in Genetics

Organisms differ in many characteristics, or traits. A **trait** is a feature of an organism such as its height, flower color, or the chemical structure of one of its proteins. Many traits are determined at least in part by

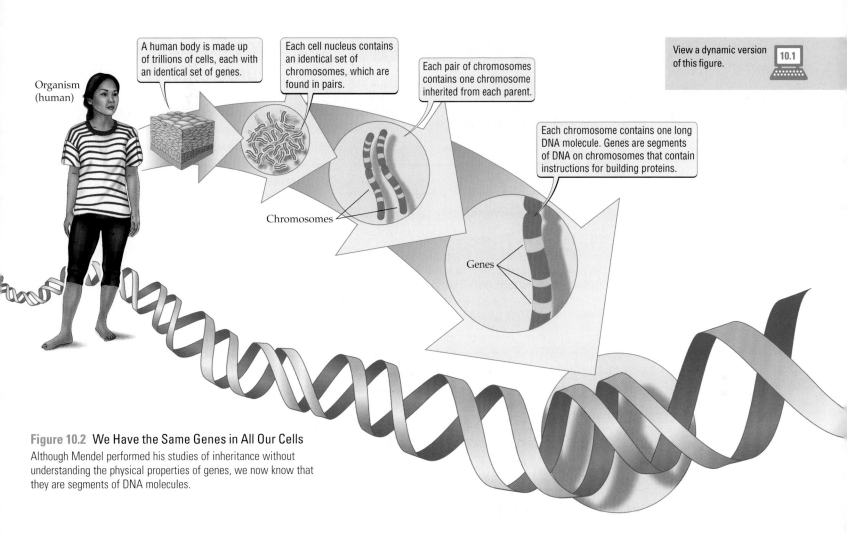

View a dynamic version of this figure. 10.1

A human body is made up of trillions of cells, each with an identical set of genes.

Each cell nucleus contains an identical set of chromosomes, which are found in pairs.

Each pair of chromosomes contains one chromosome inherited from each parent.

Each chromosome contains one long DNA molecule. Genes are segments of DNA on chromosomes that contain instructions for building proteins.

Organism (human)

Chromosomes

Genes

Figure 10.2 We Have the Same Genes in All Our Cells
Although Mendel performed his studies of inheritance without understanding the physical properties of genes, we now know that they are segments of DNA molecules.

genes, which are individual units of genetic information for specific traits. The gene is the basic unit of inheritance.

As we saw in Chapter 9, the body cells of most plants and animals are diploid; that is, they contain two of each chromosome type. (For example, people have 46 chromosomes—two of each of our 23 chromosome types.) Thus, since genes are located on chromosomes, cells in the body contain two copies of each gene, one inherited from each parent. These two copies are not necessarily identical, however. Such alternative versions of a gene are called **alleles**.

An allele that determines the physical characteristics of an organism even when it is paired with a different allele is referred to as a **dominant** allele. Dominant alleles are denoted by uppercase letters, such as A. An allele that has no physical effect when it is paired with a dominant allele is **recessive**. Recessive alleles are denoted by lowercase letters, such as a. An individual that carries two copies of the same allele, such as an AA or an aa individual, is referred to as a **homozygote**. An individ-

ual whose two gene copies differ, such as an Aa individual, is a **heterozygote**.

The genetic makeup of an organism is called its **genotype**; for example, the heterozygote referred to in the previous paragraph has genotype Aa. The **phenotype** of an organism is its observable physical characteristics. Aspects of an organism's phenotype include its appearance (for example, flower color), behavior (for example, the courtship display of a bird), and biochemistry (for example, the amount of a gene's protein product in the body). Two individuals with the same phenotype can have different genotypes (Figure 10.3).

A **genetic cross**, or cross for short, is a controlled mating experiment performed to examine the inheritance of a particular trait. "Cross" can also be used as a verb, as in "individuals of genotype AA were crossed with individuals of genotype aa." The parent generation of a genetic cross is called the **P generation**. The first generation of offspring in a genetic cross is called the **F_1 generation** ("F" is for "filial," a word that refers to a son or daughter). The second generation of a cross is called the **F_2 generation**.

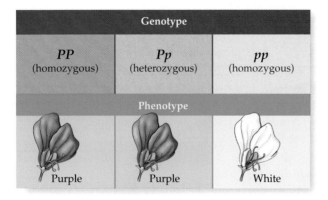

Figure 10.3 Genotype and Phenotype
Flower color in peas is controlled by a gene with two alleles (*P* and *p*). Although there are three genotypes (*PP*, *Pp*, and *pp*), there are only two phenotypes (purple flowers and white flowers). This happens because genotypes *PP* and *Pp* both produce purple flowers, while *pp* produces white flowers.

Definitions of these important genetics terms are collected in Table 10.1. Study these terms carefully, and refer to them as needed as you study the rest of this chapter.

10.2 Gene Mutations: The Source of New Alleles

Although a single individual has at most two different alleles for any given gene, when we examine the genotypes of a group of individuals, we may find that a particular gene has many different alleles. For example, in human populations, proteins often have three or more forms, each of which is produced by a different allele. Consider the ABO blood typing system. Three alleles determine a person's blood type: the I^A, I^B, and i alleles. As we describe on page 197, the first two are (equally) dominant, while the i allele is recessive.

The different alleles of a gene arise by **mutation**, which we can define briefly here as any change in the DNA that makes up a gene (see Chapter 13 for a more detailed discussion). When a mutation occurs, the new allele that results may contain instructions for a protein whose form differs from that of the version specified by the original allele (Figure 10.4). By specifying different versions of

Table 10.1

Basic Terms in Genetics

Term	Definition
Allele	One of two or more alternative versions of a gene.
Dominant allele	An allele that determines the phenotype of an organism even when paired with a different (recessive) allele.
F_1 generation	The first generation of offspring in a genetic cross.
F_2 generation	The second generation of offspring in a genetic cross.
Gene	An individual unit of genetic information for a specific trait. Genes are located on chromosomes and are the basic unit of inheritance.
Genetic cross	A controlled mating experiment, usually performed to examine the inheritance of a particular trait.
Genotype	The genetic makeup of an organism.
Heterozygote	An individual that carries one copy of each of two different alleles (for example, an *Aa* individual or a $C^W C^R$ individual).
Homozygote	An individual that carries two copies of the same allele (for example, an *AA*, *aa*, or $C^W C^W$ individual).
P generation	The parent generation of a genetic cross.
Phenotype	The observable characteristics of an organism.
Recessive allele	An allele that does not have a phenotypic effect when paired with a dominant allele.
Trait	A feature of an organism, such as height, flower color, or the chemical structure of a protein.

Review basic genetics terminology.

10.2

Figure 10.4 Not True to Its Name
This black rat snake has a mutation in a gene for color, causing it to be white, not black.

proteins, the different alleles of a gene cause hereditary differences among organisms.

Mutations are often harmful. Harm can occur, for example, if a mutation leads to the production of a protein that performs a vital function poorly. However, it is also common for mutations to have little effect, as when the new allele specifies a protein that is identical, or nearly identical, to the protein specified by the original allele. Occasionally, mutations produce alleles that improve on the original protein or carry out new, useful functions. Such mutations are beneficial.

Mutations have two other important characteristics. First, mutations occur at random with respect to their usefulness. There is no evidence, for example, that specific beneficial mutations occur because they are needed. Second, mutations can happen at any time and in any cell of the body. In multicellular organisms, however, only mutations that occur in gametes, or in the cells that ultimately produce gametes, can be passed on to offspring.

10.3 Basic Patterns of Inheritance

Now that we have defined some key genetic concepts and discussed how mutations produce new alleles, we are ready to explore how genes are transmitted from parents to offspring. Prior to Mendel, many people argued that the traits of both parents were blended in their offspring, much as paint colors blend when they are mixed together. According to this theory, which was known as the theory of blending inheritance, offspring should be intermediate in phenotype to their two parents, and it should not be possible to recover traits from previous generations. Thus, if a white-flowered plant were mated with a red-flowered plant, the offspring should have pink flowers, and the original flower colors of white and red should not be seen in later generations.

Many observations do not match these predictions, however. The features of offspring often are not intermediate to those of their parents, and it is common for traits to skip a generation (for example, a child may have blue eyes like one of its grandparents, but unlike its brown-eyed parents). How can such observations be explained? Gregor Mendel answered this question with a series of experiments on plants.

Mendel conducted genetic experiments on pea plants

During 8 years of investigation, Mendel conducted experiments on inheritance in pea plants. His results led him to reject the theory of blending inheritance. Mendel proposed instead that for each trait, offspring inherit two separate units of genetic information (genes), one from each parent.

Peas are an excellent organism for studying inheritance. Ordinarily, peas self-fertilize; that is, an individual pea plant contains both male and female reproductive organs, and it fertilizes itself. But because peas can also be mated experimentally, Mendel was able to perform carefully controlled genetic crosses. In addition, peas have true-breeding varieties, which means that when these plants self-fertilize, all of their offspring have the same phenotype as the parent. For example, one variety has yellow seeds and produces only offspring with yellow seeds; we now know that this happens because the parent plants are homozygous for the allele that causes seeds to be yellow. Mendel based all of his experiments on homozygous varieties that bred true for traits such as plant height, flower color, or the color or shape of the seeds.

In his experiments, Mendel observed inherited traits in each of three generations of plants, the parent and two offspring generations. For example, he crossed plants that bred true for purple flowers with plants that bred true for white flowers (Figure 10.5). Mendel then allowed the F_1 plants (the first generation of offspring) to self-fertilize, thereby producing the F_2 generation.

Inherited traits are determined by genes

According to the theory of blending inheritance, the cross shown in Figure 10.5 should have yielded F_1 generation

Figure 10.5 Three Generations in One of Mendel's Experiments

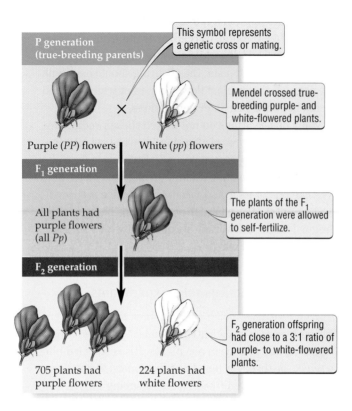

P generation (true-breeding parents)

This symbol represents a genetic cross or mating.

Mendel crossed true-breeding purple- and white-flowered plants.

Purple (*PP*) flowers   White (*pp*) flowers

F₁ generation

All plants had purple flowers (all *Pp*)

The plants of the F₁ generation were allowed to self-fertilize.

F₂ generation

F₂ generation offspring had close to a 3:1 ratio of purple- to white-flowered plants.

705 plants had purple flowers   224 plants had white flowers

plants bearing flowers of intermediate color. Instead, all the F₁ plants had purple flowers. Furthermore, when the F₁ plants self-fertilized, about 25 percent of the F₂ offspring had white flowers. Thus the occurrence of white flowers skipped a generation, something that should not happen under blending inheritance.

Mendel studied seven traits in peas, and his results for each of those traits were similar to those shown in Figure 10.5. These results led him to propose a new theory of inheritance, in which genes behave like separate units, or particles, not like colors of paints that blend together. Using modern terminology, Mendel's theory can be summarized as follows:

1. *Alternative versions of genes cause variation in inherited traits.* For example, peas have one version of a certain gene that causes flowers to be purple, and another version of the same gene that causes flowers to be white. These alternative versions of a gene are known as alleles.
2. *Offspring inherit one copy of a gene from each parent.* In his analysis of crosses like that in Figure 10.5, Mendel reasoned that for white flowers to reappear in the F₂ generation, the F₁ plants must have had two copies of the flower color gene (one copy that caused white flowers and one copy that caused purple flowers).

Mendel was right: with the exception of our gametes, all cells in the adult organism contain one maternal and one paternal copy of each of their many genes (see Figure 10.2).

3. *An allele is dominant if it determines the phenotype of an organism even when paired with a different allele.* For example, let's call the allele for purple flower color *P* and the allele for white flower color *p*. Plants that breed true for purple flowers must have two copies of the *P* allele (that is, they are of genotype *PP*), since otherwise they would occasionally produce white flowers. Similarly, plants that breed true for white flowers have two copies of the *p* allele (genotype *pp*). Thus the F₁ plants in Figure 10.5 must have genotype *Pp*; that is, they must have received a *P* allele from the *PP* parent with purple flowers and a *p* allele from the *pp* parent with white flowers. Since all the F₁ plants had purple flowers, the *P* allele is dominant and the *p* allele is recessive. A recessive allele like the *p* allele still produces a protein, but that protein has no effect on the phenotype.
4. *The two copies of a gene separate during meiosis and end up in different gametes.* Each gamete receives only one copy of each gene. If an organism has two copies of the same allele for a particular trait, as in the homozygous varieties used by Mendel, all of its gametes will contain that allele. However, if the organism has two different alleles, like an individual of genotype *Pp*, then 50 percent of the gametes will receive one of the alleles and 50 percent of the gametes will receive the other allele.
5. *Gametes fuse without regard to which alleles they carry.* When gametes fuse to form a zygote, they do so randomly with respect to the alleles they carry for a particular gene. As we'll see, this element of randomness allows us to use a simple method to determine the chance that offspring will have a particular genotype.

10.4 Mendel's Laws

Mendel summarized the results of his experiments in two laws: the law of segregation and the law of independent assortment. Let's take a look at each of Mendel's laws and how he developed them.

Mendel's first law: Segregation

The **law of segregation** states that the two copies of a gene separate during meiosis and end up in different gametes. This law can be used to predict how a single trait will be inherited. As an illustration, let's revisit the experiment shown in Figure 10.5. In that experiment, Mendel crossed plants that bred true for purple flowers (genotype *PP*) with individuals that bred true for white flowers (genotype *pp*). This cross produced an F_1 generation composed entirely of heterozygotes (individuals with genotype *Pp*). According to the law of segregation, when the F_1 plants reproduced, 50 percent of the pollen (sperm) should have contained the *P* allele, and the other 50 percent the *p* allele. The same is true for the eggs.

We can represent the separation of the two copies of a gene by a **Punnett square** (Figure 10.6), a method first used in 1905 by the British geneticist Reginald Punnett. In a Punnett square, all possible male gametes are listed on one side of the square, and all possible female gametes are listed on the perpendicular side of the square. Regardless of whether it has a *P* or a *p* allele, each sperm has an equal chance of fusing with an egg that has a *P* allele or an egg that has a *p* allele. Thus the four genotypes shown within the Punnett square are all equally likely.

Using the Punnett square method, we can predict that $1/4$ of the F_2 generation is likely to have genotype *PP*, $1/2$ to have genotype *Pp*, and $1/4$ to have genotype *pp*. Because the allele for purple flowers (*P*) is dominant, plants with *PP* or *Pp* genotypes have purple flowers, while *pp* genotypes have white flowers. Thus we predict that $3/4$ (75 percent) of the F_2 generation will have purple flowers and $1/4$ (25 percent) will have white flowers. This prediction is very close to Mendel's actual results for the F_2 generation: 705 (76 percent) had purple flowers and 224 (24 percent) had white flowers.

Mendel's second law: Independent assortment

Mendel also performed experiments in which he simultaneously tracked the inheritance of two traits. For example, pea seeds can have a round or wrinkled shape, and they can be yellow or green. Two different genes control these aspects of the plant's phenotype. With respect to seed shape, Mendel determined that the allele for round seeds (denoted *R*) was dominant to the allele for wrinkled seeds (*r*). With respect to seed color, he determined that the allele for yellow seeds (*Y*) was dominant to the allele for green seeds (*y*).

What would happen if round, yellow-seeded individuals of genotype *RRYY* were crossed with wrinkled, green-seeded individuals of genotype *rryy*? As might be expected, when Mendel performed this experiment, all of the resulting F_1 plants had genotype *RrYy* and hence round, yellow seeds.

Figure 10.6 The Punnett Square Method

The Punnett square method can be used to represent the separation of alleles into gametes and to predict the outcome of a genetic cross.

The F_1 offspring of *PP* × *pp* plants all have genotype *Pp*.

Each egg and each sperm produced by the F_1 plants has a 50% chance of receiving a *P* allele and a 50% chance of receiving a *p* allele.

Eggs and sperm combine randomly in four ways to form F_2 plants.

The Punnett square method predicts 3 purple-flowered offspring for every 1 white-flowered offspring, a 3:1 ratio.

Next, Mendel tested whether the inheritance of seed color was independent of the inheritance of seed shape. If this were true, then copies of one gene would be distributed into gametes independently from copies of the other gene, causing all possible combinations of the alleles to be found in the gametes (Figure 10.7). Mendel tested this prediction by crossing *RrYy* plants with each other. He obtained the following results in the F_2 generation: approximately $9/16$ of the seeds were round and yellow, $3/16$ were round and green, $3/16$ were wrinkled and yellow, and $1/16$ were wrinkled and green (a 9:3:3:1 ratio). As shown in Figure 10.8, Mendel's results were similar to what should have happened if the genes for these two traits were inherited independently of each other.

Mendel made similar crosses for various combinations of the seven traits he studied. His results led him to propose the **law of independent assortment**, which states that when gametes form, the separation of the two copies of one gene during meiosis is independent of the separation of the copies of other genes. (There are exceptions to this law, however; we will find out why in Chapter 11.)

Figure 10.7 Independent Assortment of Alleles

When alleles for two genes separate independently, all possible combinations of those alleles are found in the gametes. In the examples shown here, the dotted arrows show which parental copies of the two genes sort into the resulting gametes. (a) A parent of genotype *RrYy* produces four gametes, each with a different combination of alleles. (b) A parent of genotype *rrYy* also produces four gametes, but only two different combinations of alleles. If the parent genotype were *RRYY* or *rryy*, all four gametes would have the same combination of alleles (*RY* or *ry*) (not shown).

Investigate the law of independent assortment.

10.3

Figure 10.8
Are Genes Inherited Independently?

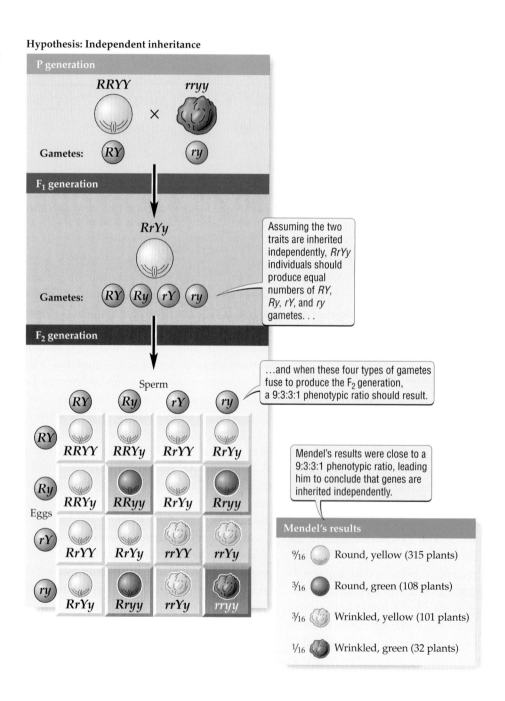

Test your understanding of linked inheritance versus independent assortment.

10.4

When developing both his first and second laws, it was important that Mendel observed the results of his genetic crosses in large numbers of offspring. We explore some general reasons for using a large number of "experimental units" in the box on page 196, but when performing genetic crosses in particular, it is a good idea to obtain as many offspring as possible because the chance of obtaining an offspring with a particular genotype is just that—a chance.

For example, if there is a ¼ chance that each offspring will be a homozygous recessive (*pp*) individual, that means that if many offspring are produced, it is likely that 25 percent of them will have genotype *pp*. But there is no guarantee that if there are four offspring, one of them will always have genotype *pp*. That may happen, but it is also possible for none of the offspring to have genotype *pp*, or for more than one to have genotype *pp*. Such outcomes can occur because the 25 percent probability that an offspring will have genotype *pp* applies not only to the first offspring, but to the second, third, and fourth offspring as well. Hence it is even possible (but not likely, since the chance is 0.25 × 0.25 × 0.25 × 0.25, or less than half of 1 percent) that all four offspring will have genotype *pp*. When many offspring are examined, the chance of obtaining unusual results (such as all of them having genotype *pp*) becomes very small.

10.5 Extensions of Mendel's Laws

Mendel's laws describe how genes are passed from parents to offspring. In some cases—such as the seven traits of pea plants that Mendel studied—these laws allow offspring phenotypes to be predicted accurately from parental genotypes. In particular, Mendel's laws allow us to make accurate predictions whenever an inherited trait is controlled by a single gene with two alleles, one dominant, the other recessive. But many traits are not under such simple genetic control. To account for such traits, extensions of Mendel's laws have been developed. These extensions supplement, rather than invalidate, Mendel's laws. Even when Mendel's laws do not accurately predict the characteristics of the offspring of a cross, the genes in question are inherited according to those laws. As we'll see, what differs is how the genes affect the phenotype of the organism, not how the genes are passed from parents to offspring.

Many alleles do not show complete dominance

For dominance to be complete, a single copy of the dominant allele must be enough to produce the maximum phenotypic effect; for example, one P allele ensures that even a Pp pea plant has purple flowers. But often dominance is not complete. In snapdragons, for example, when a homozygote with red flowers ($C^R C^R$) is crossed with a homozygote with white flowers ($C^W C^W$), the heterozygous offspring ($C^R C^W$) have pink flowers. Animals can also show a lack of complete dominance, as in the coat color of horses (Figure 10.9).

The colors of snapdragons and horses illustrate **incomplete dominance**: neither allele is dominant over the other, and the phenotype of heterozygotes is intermediate between the phenotypes of the two homozygotes. Although incomplete dominance superficially resembles the old

Snapdragon

Chestnut, genotype $H^C H^C$

Palomino, genotype $H^C H^W$

Cremello, genotype $H^W H^W$

  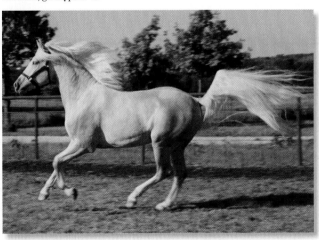

Figure 10.9 Incomplete Dominance in Horses
Palominos (genotype $H^C H^W$) are intermediate in color to chestnuts and cremellos.

Science Toolkit

Tossing Coins and Crossing Plants: Probability and the Design of Experiments

When a scientist conducts an experiment, he or she must think carefully about how many experimental units to use. An "experimental unit" is whatever the scientist applies a treatment or other test to, such as an individual in a genetic cross or a fish tank in an experiment designed to test the effect of a pollutant. Typically, scientists use as many experimental units as practical limitations (of cost, time, or space) will allow.

The basic principles of probability explain why scientists prefer to use many experimental units. The probability of an event is the chance that the event will occur. For example, there is a probability of 0.5 that a fair coin will turn up "heads" when it is tossed. A probability of 0.5 is the same thing as a 50% chance.

As an illustration, consider a hypothetical coin-tossing experiment. If a fair coin is tossed only a few times, the observed percentage of heads may differ greatly from 50%. For example, if you tossed a coin only 10 times, it would not be unusual to get 70% (7) heads. However, if you tossed a coin 10,000 times, it would be very unusual to get 70% (7,000) heads. If you got such a result, you would (and should) suspect that the coin was not fair after all.

Each toss of a coin is an independent event, in the sense that the outcome of one toss does not affect the outcome of another toss. For a series of independent events, we can estimate the probability of each of the possible outcomes from the results. Suppose we toss a coin 10,000 times and get 5,046 heads. From these results, it would be reasonable to estimate the chance of getting heads on the next toss as 50%, the percentage we expect from a fair coin. When only a small number of events are observed, our estimates of the underlying probabilities are less likely to be accurate, as when we toss a coin only a few times.

How does all this relate to the design of a scientific experiment? Consider Mendel's work on peas. When Mendel crossed heterozygous plants (for example, *Pp* individuals) with each other, the offspring always had a phenotypic ratio close to 3:1. The reason for this consistent ratio is that the chances of getting offspring with *PP*, *Pp*, and *pp* genotypes are 25%, 50%, and 25%, respectively (see Figure 10.5). Because the 25% *PP* individuals and the 50% *Pp* individuals have the same phenotype, 75% of the individuals should look alike, thus giving a 3:1 phenotypic ratio.

The Punnett square method predicts the percentages of *PP*, *Pp*, and *pp* offspring that Mendel should have observed. The method assumes that all sperm cells and all egg cells have an equal chance of achieving fertilization. When large numbers of offspring are considered, this assumption is not too far off, because the successes or failures of the different types of gametes tend to balance one another. But if Mendel had used only small numbers of offspring in his experiments, his results might have differed greatly from a 3:1 ratio. If that had been the case, he might not have discovered his two fundamental laws: the law of segregation and the law of independent assortment.

What was true in Mendel's experiments is true for scientific experiments in general: chance events not under the control of the scientist can affect the outcome of an experiment. This is true of any genetic cross (in which the researcher has no control over which sperm and which egg fuse to produce an offspring), but it is also true in many other settings.

For example, a scientist wanting to know whether agricultural pesticides affect the frequency of frogs with extra or missing legs might perform experiments comparing frogs in ponds that receive pesticides from nearby farm fields with frogs in other ponds that do not. The researcher would want to use as many ponds as possible to minimize the chance that some other factor, such as another pollutant or another organism found only in some ponds, skews the results one way or another. Biologists have indeed tested whether pesticides increase the chance of frog deformities. In one such experiment, although pesticides were not the direct cause of the deformities (a parasite was), the addition of pesticides weakened the frogs' immune systems, thus making it more likely that individuals would succumb to the parasite and develop deformities.

A Deformed Frog

Figure 10.10 Codominance in the Blood Typing System
People who have an AB blood type have codominant I^A and I^B alleles, whereas people with type A blood have two I^A alleles.

idea of blending inheritance, it is really just an extension of Mendelian inheritance. For example, if two heterozygous snapdragons ($C^R C^W$) are crossed, ¼ of the offspring are likely to have red flowers (genotype $C^R C^R$), ½ to have pink flowers (genotype $C^R C^W$), and ¼ to have white flowers (genotype $C^W C^W$). Work this out for yourself using the Punnett square method. You will see that Mendel's laws still apply; the main difference is that the heterozygotes ($C^R C^W$) look different from $C^R C^R$ individuals. Later generations can return to the original flower colors of red and white, which cannot occur under blending inheritance. Thus this example shows that incomplete dominance is very different from blending inheritance (a phenomenon that we now know does not occur).

A pair of alleles can also show **codominance**, in which the phenotype of the heterozygote is determined equally by each allele. The ABO blood typing system in people provides an example (Figure 10.10). Recall that three alleles determine a person's blood type: the I^A, I^B, and i alleles. The I^A and I^B alleles result in the production of A and B versions of a protein that is deposited on the surface of red blood cells. In $I^A I^B$ individuals, both versions are deposited, resulting in blood type AB. Since each allele has an equal effect on the phenotype of an $I^A I^B$ heterozygote, these alleles are codominant to each other. $I^A I^A$ and $I^A i$ individuals have blood type A, while $I^B I^B$ and $I^B i$ individuals have blood type B. Thus the i allele is recessive to the other two alleles. Individuals with genotype ii have blood type O.

Alleles for one gene can alter the effects of another gene

Try an interactive coin-flipping exercise. 10.5

A particular phenotype often depends on more than one gene. In such cases, the genes interact in the sense that the phenotypic effect of each gene depends partly on its own function and partly on the function of other genes. Such gene interactions are common in all types of organisms. For example, in yeast (a single-celled organism used to make bread and beer), each gene tested was found to interact with 34 other genes.

Gene interactions occur when the phenotypic effect of the alleles of one gene depends on which alleles are present for another, independently inherited gene; this phenomenon is known as **epistasis**. Coat color in mammals illustrates epistasis. In mice and many other mammals, a gene that controls production of the pigment melanin has a dominant allele (*B*) that produces black fur and a recessive allele (*b*) that produces brown fur. But the effects of the melanin alleles (*B* and *b*) can be eliminated completely, depending on which alleles are present at a second gene that interacts with the melanin gene. This second gene contains instructions for building a necessary biochemical precursor of melanin. Hence the second gene determines whether melanin can be produced at all. Individuals with at least one copy of the *C* allele at the second gene can produce melanin, while *cc* individuals cannot.

If a mouse, for example, has genotype *cc* at the second gene, it produces no pigment, regardless of which alleles it has for the melanin gene (Figure 10.11). A lack of pigment causes mice to have white fur. Thus, although we would expect *BB* and *Bb* mice to be black and *bb* mice to

However, mice with two *c* alleles at a gene that interacts with the melanin gene produce no pigment and are white in color, regardless of their genotype for the melanin gene (*BB*, *Bb*, or *bb*.)

Ordinarily, *BB* or *Bb* mice are black...

...and *bb* mice are brown.

Figure 10.11 Gene Interactions
Gene interactions are very common. One example is illustrated here by the effects in mice of the *c* allele of a gene that interacts with a melanin pigment gene.

be brown, *BBcc*, *Bbcc*, and *bbcc* mice all have white fur because they have the *cc* genotype at the gene that interacts with the melanin gene.

The environment can alter the effects of a gene

The effects of many genes depend on internal and external environmental conditions, including body temperature, carbon dioxide levels in the blood, external temperature, and amount of sunlight. An allele for coat color in Siamese cats (Figure 10.12), for example, is sensitive to temperature. This allele causes melanin to be produced only at low body temperatures. Because a cat's extremities tend to be colder than the rest of its body, melanin can be produced there, and hence the paws, nose, ears, and tail of a Siamese cat tend to be dark. If a patch of light fur is shaved from the body of a Siamese cat and covered with an ice pack, when the fur grows back, it will be dark. Similarly, if dark fur is shaved from the tail and allowed to grow back under warm conditions, it will be light-colored.

Chemicals, nutrition, sunlight, and many other environmental factors can also alter the effects of genes. In plants, genetically identical individuals (clones) grown in different environments often differ in many aspects of their phenotype, including height and the number of flowers they produce. Thus plants on a windswept mountainside may be short and have few flowers, while clones of the same plants grown in a warm, protected valley are tall with many flowers. Similar effects are found in people. For example, a person who was malnourished as a child will be shorter as an adult than if he or she had received plenty of food.

Most traits are determined by two or more genes

Mendel studied characteristics that were under simple genetic control: a single gene determined the phenotype for each of the traits he studied. Most traits, however, are **polygenic**—that is, they are determined by the action of more than one gene. Examples of polygenic traits include skin color, running speed, and body size in humans and height, flowering time, and seed number in plants. Let's look in more detail at one of these examples, the inheritance of skin color in humans.

The pigment melanin determines a person's skin color. Many of the differences among people in the amount of melanin in the skin are controlled by three genes. (There are probably more than three genes for skin color in people, but for simplicity, we'll consider only three of

Figure 10.12 The Environment Can Alter the Effects of Genes
Coat color in Siamese cats is controlled by an allele that produces dark pigment (as on the nose, tail, paws, and ears) only at low temperatures.

them here.) Each gene affects skin color equally. The skin colors that result from these three genes vary considerably (Figure 10.13). Differences between genotypes are smoothed over by suntans—exposure to the sun being another example of an environmental influence on the phenotype—causing human skin colors to vary nearly continuously from light to dark.

10.6 Putting It All Together: Genes and Inheritance

Patterns of inheritance are determined by genes that are passed from parent to offspring according to the simple rules summarized in Mendel's laws. Some traits are controlled by a single gene and are little affected by environmental conditions. For such traits, such as seed shape and flower color in pea plants, it is possible to predict the phenotypes of offspring just from knowing which alleles the parents have for a single gene.

Many other traits, however, are influenced by sets of genes that interact with one another and with the environment. For such traits, the relationship between genotype and phenotype is more complex. In such cases, a given gene does not act in isolation; rather, its effect depends not only on its own function, but also on the function of other genes with which it interacts and on the environment (Figure 10.14). When a gene does not act in

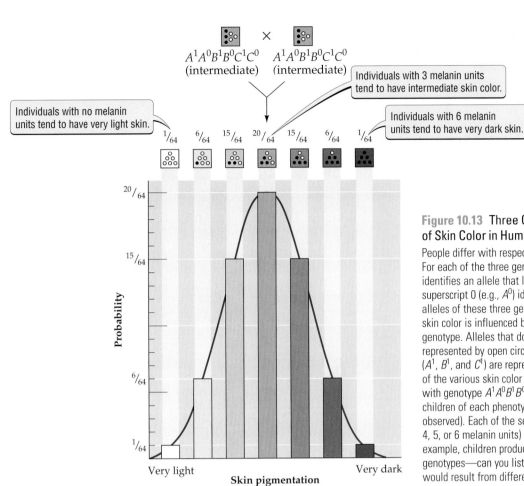

Figure 10.13 Three Genes Produce a Wide Range of Skin Color in Humans

People differ with respect to the amount of melanin, a dark pigment, in their skin. For each of the three genes shown here (A, B, and C), the superscript 1 (e.g., A^1) identifies an allele that leads to the production of one "unit" of melanin, while the superscript 0 (e.g., A^0) identifies an allele that produces no melanin. None of the alleles of these three genes is dominant. As illustrated in the figure, a person's skin color is influenced by the total number of melanin "units" specified by their genotype. Alleles that do not contribute to melanin production (A^0, B^0, and C^0) are represented by open circles, while alleles that contribute to melanin production (A^1, B^1, and C^1) are represented by solid circles. The bars represent the probability of the various skin color phenotypes that can occur in children of two parents both with genotype $A^1A^0B^1B^0C^1C^0$ (bar heights indicate the relative proportions of children of each phenotype when a large number of families of this cross are observed). Each of the seven different phenotypes (that is, children with 0, 1, 2, 3, 4, 5, or 6 melanin units) can be specified by one to several different genotypes. For example, children producing 2 melanin units can have any one of six different genotypes—can you list these six genotypes? Additional variation in skin color would result from different levels of sun exposure.

isolation, individuals of the same genotype may have very different phenotypes.

Many human diseases, for example, including heart disease, cancer, alcoholism, and diabetes, are strongly influenced by multiple genes and by many different environmental factors, such as smoking, diet, and overall mental and physical health. Predicting the phenotypes of offspring for such traits requires a detailed understanding of how genes and the environment influence the final product of the genes—the phenotype. Such prediction is a challenging and important task. We could reduce the death rate from heart disease and cancer, for example, if we knew how specific genes interacted with the environment to cause these diseases. Recent developments in genetics (as we shall see in Chapters 14 through 16) hold the promise of future improvements in our ability to predict these and other human disease phenotypes.

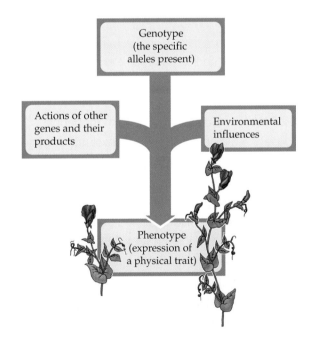

Figure 10.14 From Genotype to Phenotype: The Big Picture

The effect of a gene on an organism's phenotype depends on the gene's own function, the function of other genes with which it interacts, and the environment. As a result, two individuals with the same genotype for a gene may have very different phenotypes, as illustrated by the different fates of two pea plants.

Biology Matters

How to Avoid Life-Threatening Diseases

Whether or not a person becomes afflicted with degenerative disease can be strongly influenced by lifestyle—not just by the inheritance of alleles that cause genetic disorders. As shown here, 70 to 90 percent of cases of colon cancer, stroke, coronary heart disease, and type 2 diabetes can be avoided by following a "low-risk" lifestyle.

For colon cancer, the low-risk definition includes maintaining a proper body weight, engaging in physical activity equivalent to at least 30 minutes of brisk walking per day, taking a folic acid supplement of 100 µg/day or more, consuming fewer than three alcoholic beverages per day, not smoking, and eating fewer than three servings of red meat per week.

For stroke and coronary heart disease, the low-risk definition includes not smoking, eating a good diet (i.e., consuming low amounts of saturated and trans fats; low amounts of sugar or refined carbohydrates; and adequate amounts of polyunsaturated fats, n-3 fatty acids, cereal fiber, and folic acid), maintaining a proper body weight, engaging in physical activity equivalent to at least 30 minutes of brisk walking per day, and consuming fewer than three alcoholic beverages per day.

For type 2 diabetes, the low-risk definition is similar to that for stroke and coronary heart disease. However, n-3 fatty acids and folic acid are not necessary parts of the diet.

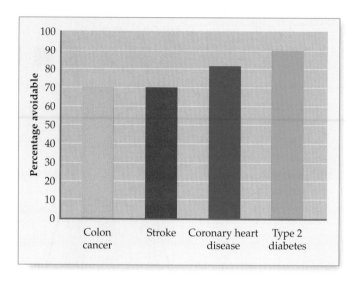

Solving the Mystery of the Lost Princess

Tsar Nicholas II abdicated in 1917, ending over three centuries of rule by the House of Romanov. While alive, Tsar Nicholas and his wife, the Tsarina Alexandra, served as a rallying point for people opposed to the new Communist revolutionary government. Even after death, the royal family was viewed as a threat, and so Communist officials spread misleading information about them. Tsar Nicholas, they said, had been shot, but his family had been moved to a place where they were safe from the turmoil of civil war. The resulting uncertainty over the fate and location of Alexandra and her daughters set the stage for the legend of Anastasia, the lost princess.

The mystery of Anna Anderson and the lost princess was finally solved in 1994. The secret grave of Tsar Nicholas and his family was unearthed in 1989, and the remains were carefully analyzed. Investigators electronically superimposed photographs of the skulls on archive photographs of the family, and they compared skeletal measurements with clothing known to have belonged to the Tsar and his daughters. They also matched the platinum dental work on one skull with the Tsarina's dental records. All these and other tests yielded a match: the skeletons seemed to be those of the Russian royal family. Finally, to make their case airtight, the investigators turned to the ultimate arbitrator, genetic analysis.

When DNA obtained from the skeletons was compared with DNA obtained from relatives of the Russian royal family (including the Tsar's brother, who died in 1899), the results showed conclusively that the skeletal remains were those of the Russian royal family. However, two sets of bones were missing, those of Prince Alexis and one of the two princesses, either Maria or Anastasia. Could it be that Anastasia had escaped and that Anna Anderson was who she claimed to be?

Here, too, DNA analysis provided an answer. First, DNA was obtained from small samples of Anna Anderson's intestinal tissue that had been preserved after a 1979 surgical operation. Next, her DNA was compared with DNA obtained from the skeletons of the Tsar and Tsarina. In one region of DNA, five codominant alleles (A^1, A^2, A^3, A^4, and A^5) were found. The Tsar had genotype A^1A^2, and the Tsarina had genotype A^2A^3. According to Mendel's laws, for Anna Anderson to have been the daughter of the Tsar and Tsarina, she should have had one of the following genotypes: A^1A^2, A^1A^3, A^2A^2, or A^2A^3. But her actual genotype was A^4A^5, which was not consistent with the Tsar and Tsarina being her parents. Three other regions of DNA yielded similar results, indicating that Anna Anderson was not the lost princess.

Chapter Review

Summary

10.1 Essential Terms in Genetics

- Genes—the basic units of inheritance—are segments of DNA that help determine an organism's physical characteristics, or traits.
- Many traits are controlled by genes. Individuals have two copies of each gene, one inherited from each parent.
- Alternative versions of a gene are called alleles.
- The genotype is an organism's genetic makeup, while the phenotype is its observable characteristics.
- A dominant allele is one that determines the phenotype of the organism even when it is paired with a different allele. A recessive allele has no phenotypic effect when it is paired with a dominant allele.

10.2 Gene Mutations: The Source of New Alleles

- In a population of many individuals, a particular gene may have one, a few, or many alleles.
- The different alleles of a gene arise by mutation.
- A new allele (produced by mutation) may cause the organism to build a protein that differs in form from the versions of the protein specified by other alleles of the gene.
- By specifying different proteins, the different alleles of a gene result in hereditary differences among organisms.
- Many mutations are harmful, many have little effect, and a few are beneficial. The effect of a mutation depends on the effect of the protein specified by the new allele.

- Mutations can happen at any time and in any cell in an organism, but only mutations in gametes or their precursor cells can be passed on to the next generation.
- Mutations occur at random with respect to their usefulness; thus the fact that a mutation may be useful does not make it more likely to occur.

10.3 Basic Patterns of Inheritance

- Over an 8-year period, Gregor Mendel performed a pioneering series of experiments on pea plants designed to elucidate how inheritance works.
- Mendel's results led him to reject the old theory of blending inheritance. Instead, they suggested that the inherited characteristics of organisms are controlled by specific units of inheritance, which we now know as genes.
- Mendel proposed a particulate theory of inheritance, summarized as follows: (1) Alleles of genes cause variation in inherited traits. (2) Offspring inherit one copy of a gene from each parent. (3) Alleles can be dominant or recessive. (4) The two copies of a gene separate into different gametes. (5) Gametes fuse without regard to which alleles they carry.

10.4 Mendel's Laws

- Mendel summarized the results of his experiments in two laws: the law of segregation and the law of independent assortment.
- The law of segregation states that the two copies of a gene end up in different gametes.

- The law of independent assortment states that when gametes form during meiosis, the separation of the copies of one gene is independent of the separation of the copies of other genes.
- The Punnett square method considers all possible combinations of gametes to predict the outcome of a genetic cross.

10.5   Extensions of Mendel's Laws

- For some traits, Mendel's laws do not predict the phenotype of the offspring. There are several reasons for this: (1) Many alleles do not show complete dominance; instead, they may show incomplete dominance (as in the coat color in horses) or codominance (as in the ABO blood typing in humans). (2) Alleles for one gene can alter the effects of another gene (epistasis). (3) The environment can alter the effect of a gene. (4) Most traits are polygenic; that is, they are determined by two or more genes.
- Even when Mendel's laws do not accurately predict offspring phenotype, the genes in question are inherited according to Mendel's laws. What differs is how the genes affect the phenotype, not how the genes are passed from parent to offspring.

10.6   Putting It All Together: Genes and Inheritance

- Some traits are controlled by a single gene that is little affected by other genes or by environmental conditions.
- Most traits, however, are influenced by sets of genes that interact with one another and with the environment.
- Thus the effect of a gene on an organism's phenotype can depend on the gene's own function, the function of other genes with which it interacts, and the environment. As a result, two individuals with the same genotype may have different phenotypes.

⊙ Review and Application of Key Concepts

1. Describe what genes are and how they work. Your explanation should include a description (using modern terminology) of Mendel's theory of inheritance, as well as a short summary of what we now know about (a) the chemical and physical structure of genes and (b) the information that they encode.

2. How many copies of each gene does a (sexually reproducing) organism have? Why? If all copies of a gene in an individual are identical (that is, the individual is homozygous for that gene), what can you infer about the genotypes of its parents?

3. Explain how new alleles arise, and how different alleles cause hereditary differences among organisms.

4. Draw a diagram that shows meiosis for an organism with four chromosomes (two pairs of chromosomes). Label two genes, one on each chromosome, and use the diagram to explain why the

separation of copies of one of these genes during meiosis is independent of the separation of copies of the other gene.

5. For flower color in peas, the allele for purple flowers (P) is dominant to the allele for white flowers (p). A purple-flowered plant, therefore, could be of genotype PP or Pp. What genetic cross could you make to determine the genotype of a purple-flowered plant? Explain how your cross enables you to do this.

6. Although we are accustomed to thinking of identical twins as looking exactly alike, they can have very different phenotypes. Why? Provide two specific reasons why identical twins might not look exactly alike.

7. Many lethal human genetic disorders are caused by a recessive allele, whereas relatively few are caused by a dominant allele. Why might dominant alleles for lethal human diseases be uncommon? (*Hint:* Solve problems 5 and 6 in the Sample Genetics Problems on page 203 and use the results to guide your answer to this question.)

Key Terms

allele (p. 189)
codominance (p. 197)
dominant (p. 189)
epistasis (p. 197)
F_1 generation (p. 189)
F_2 generation (p. 189)
gene (p. 189)
genetic cross (p. 189)
genetics (p. 188)
genotype (p. 189)
heterozygote (p. 189)
homozygote (p. 189)

incomplete dominance (p. 195)
law of independent assortment (p. 193)
law of segregation (p. 192)
mutation (p. 190)
P generation (p. 189)
phenotype (p. 189)
polygenic (p. 198)
Punnett square (p. 193)
recessive (p. 189)
trait (p. 188)

Self-Quiz

1. Alternative versions of a gene for a given trait are called
   a. alleles.   c. genotypes.
   b. heterozygotes.   d. copies of a gene.

2. If an allele for long hair (L) is dominant to an allele for short hair (l), then a cross of $Ll \times ll$ should yield
   a. ¼ short-haired offspring.
   b. ¾ short-haired offspring.
   c. ½ short-haired offspring.
   d. all offspring with intermediate hair length.

3. If A and a are two alleles of the same gene, then individuals of genotype Aa are

a. homozygous.   c. dominant.

b. heterozygous.   d. recessive.

4. Genes

    a. are the basic units of inheritance.

    b. are located on chromosomes and composed of DNA.

    c. usually contain instructions for building a single protein.

    d. all of the above

5. Coat color in horses shows incomplete dominance. H^CH^C individuals have a chestnut color, H^CH^W individuals have a palomino color, and H^WH^W individuals have a cremello (white) color (see Figure 10.9). What is the predicted phenotypic ratio of chestnut to palomino to cremello if H^CH^W individuals are crossed with other H^CH^W individuals?

    a. 3:1   c. 9:3:1

    b. 2:1:1   d. 1:2:1

6. Two alleles are said to show _____ when the phenotype of the heterozygote is determined equally by each allele.

    a. codominance   c. incomplete dominance

    b. complete dominance   d. epistasis

7. Traits that are determined by the action of more than one gene are

    a. recessive.

    b. not common.

    c. common in some organisms, but not people.

    d. polygenic.

8. Select the term indicating that the phenotypic effects of alleles for one gene depend on which alleles are present for another, independently inherited gene.

    a. phenotypic variation   c. gene-environment interaction

    b. codominance   d. epistasis

Sample Genetics Problems

1. One gene has alleles A and a, a second gene has alleles B and b, and a third gene has alleles C and c. List the possible gametes that can be formed from the following genotypes:

    a. Aa d. $AaBbCc$

    b. $BbCc$ e. $aaBBCc$

    c. $AAcc$

2. For the same three genes described in problem 1, what are the predicted genotype and phenotype ratios of the following genetic crosses? (Following our standard notation, alleles written in uppercase letters are dominant to alleles written in lowercase letters.)

    a. $Aa \times aa$ d. $BbCc \times BbCC$

    b. $BB \times bb$ e. $AaBbCc \times AAbbCc$

    c. $AABb \times aabb$

3. Sickle-cell anemia is inherited as a recessive genetic disorder in humans. That means that in terms of disease onset, the normal hemoglobin allele (S) is dominant to the sickle-cell allele (s). For two parents of genotype Ss, construct a Punnett square to predict the possible genotypes and phenotypes (does or does not have the disease) of their children. Also list the genotype and phenotype ratios. Each time two Ss individuals have a child together, what is the chance that the child will have sickle-cell anemia?

4. Alleles for a gene (C) that determines the color of Labrador retrievers show incomplete dominance. Black labs have genotype C^BC^B, chocolate labs have genotype C^BC^Y, and yellow labs have genotype C^YC^Y. If a black lab and yellow lab mate, what proportions of black, chocolate, and yellow coat colors would you expect to find in a litter of their puppies?

5. For any human genetic disorder caused by a recessive allele, let n be the allele that causes the disease and N be the normal allele (the capital "N" is for "normal" individuals).

    a. What are the phenotypes of NN, Nn, and nn individuals?

    b. Predict the outcome of a genetic cross between two Nn individuals. List the genotype and phenotype ratios that would result from such a cross.

    c. Predict the outcome of a genetic cross between an Nn and an NN individual. List the genotype and phenotype ratios that would result from such a cross.

6. For any human genetic disorder caused by a dominant allele, let D be the allele that causes the disorder and d be the normal allele (where the capital "D" stands for "disorder").

    a. What are the phenotypes of DD, Dd, and dd individuals?

    b. Predict the outcome of a genetic cross between two Dd individuals. List the genotype and phenotype ratios that would result from such a cross.

    c. Predict the outcome of a genetic cross between a Dd and a DD individual. List the genotype and phenotype ratios that would result from such a cross.

7. If blue flower color (B) is dominant to white flower color (b), what are the genotypes of the parents in the following genetic cross: blue flower × white flower yields only blue-flowered offspring?

8. The seed pods of peas can be yellow or green. In one of his experiments, Mendel crossed plants that were homozygous for the allele for yellow seed pods with plants that were homozygous for the allele for green seed pods. All seed pods in the F_1 generation were green. Which allele is dominant, the one for yellow or the one for green? Explain why.

DNA Tweak Turns Vole Mates into Soul Mates

Promiscuous mammals become stay-at-home dads in a study. There is no cure yet for humans.

By ALAN ZAREMBO

Scientists working with a rat-like animal called a vole have found that promiscuous males can be reprogrammed into monogamous partners by introducing a single gene into a specific part of their brains.

C an a single gene alter the social behavior of an animal species? This article describes a 2004 study that answers "Yes." The study sought to find out whether changing one gene could change the sexual behavior of male meadow voles. In captivity, male meadow voles are solitary and promiscuous—after they mate with a female, they do not spend extra time with or show preference for that female. Males of a closely related species, the prairie vole, are more monogamous than the meadow vole, in that they bond with the female they mate with, preferring to spend time with her and keeping other males away from her.

The news article summarizes how scientists inserted copies of a single gene into the brains of male meadow voles to find out whether that gene would change the way they behaved. The gene contains instructions for building the receptor of a hormone called vasopressin. Although male meadow voles already have such receptors in their brains, they do not have as many as the brains of monogamous prairie voles. By inserting copies of the vasopressin gene into the brains of male meadow voles, the researchers experimentally boosted the number of vasopressin receptors that these voles have in their brains.

The scientists suspected that vasopressin, which is released in the brain after sex, might influence the tendency for males to be monogamous. The results bore out this hypothesis: changing the number of receptors for vasopressin in the brain made the meadow vole males more sensitive to the hormone, and that, apparently, was enough to change their behavior from promiscuous to monogamous.

Evaluating the news

1. What, if anything, does the study on voles imply about human behavior? For example, do you think differences in one or a few genes could explain why some men have trouble being faithful to one partner? Or are humans so much more complex than voles that this study has little relevance to how people behave?

2. If we knew that a human behavior could be altered by a single gene, would it be ethical to insert that gene into the brain of a person in order to change his or her behavior?

3. Should research on the genetics of human behavior receive funding from national governments?

SOURCE: *Los Angeles Times*, Thursday, June 17, 2004.

CHAPTER 11

Chromosomes and Human Genetics

Key Concepts

◉ A gene is a region of the DNA molecule in a chromosome. Each gene has a specific location on the chromosome.

◉ Human males have one X and one Y chromosome, and human females have two X chromosomes. A specific gene on the Y chromosome is required for human embryos to develop as males.

◉ Genes that are located near one another on the same chromosome tend to be inherited together, or linked. Genes that are located far from one another on the same chromosome often are not linked; genes on different chromosomes also are not linked.

◉ The homologous chromosomes that pair during meiosis can exchange genes in a process called crossing-over.

◉ The genotypes of offspring can be different from that of either parent as a result of crossing-over, the independent assortment of maternal and paternal chromosomes into gametes, and fertilization.

◉ Many inherited genetic disorders in humans are caused by mutations of single genes. A far smaller number of human genetic disorders result from abnormalities in chromosome number or structure.

A Horrible Dance

As a child in the mid-1800s, George Huntington went with his father, a medical doctor, as he visited patients in rural Long Island, New York. On one such trip, the boy and his father saw two women by the roadside, a mother and daughter, who were bowing and twisting uncontrollably, their faces contorted in a series of strange grimaces. Huntington's father paused to speak with them, then left them, continuing on his rounds.

For any child, such an encounter would be unforgettable. For the young Huntington, it was that and more. Like his father and grandfather before him, George Huntington became a doctor, and in 1872 he described and named the disorder that had plagued the two women. He called it "hereditary chorea" (*chorea* [koh-*REE*-uh] comes from the Greek word for "dance").

Huntington described hereditary chorea as an inherited disorder, one that destroyed the nervous system and caused jerky, involuntary movements of the body and face. Hereditary chorea had no cure. Eventually, it killed its victims, but first it reduced them to a shell of their former selves: in addition to causing extreme motor impairments, it led to memory loss, severe depression, mood shifts, personality changes, and intellectual deterioration.

If a parent had hereditary chorea, it did not necessarily strike all the children of that parent. The children who did get it usually showed no symptoms until they were in their thirties, forties, or fifties. Thus hereditary chorea was like a genetic time bomb. In Huntington's words, the combination of the terrible symptoms of the disorder and its late and uncertain onset caused "those in whose veins the seeds of the disease are known to exist [to speak of it] with a kind of horror."

The Gene for Huntington Disease
Defects in a gene located at one end of chromosome 4 can cause Huntington disease, a devastating disorder of the nervous system. Here, chromosome 4 is highlighted in green, while the gene is highlighted in red and indicated by the white arrows.

Hereditary chorea is now known as Huntington disease, in honor of George Huntington. There still is no cure, but researchers have identified a mutation in a gene on chromosome 4 that causes the disorder, and they have isolated that gene. With the gene in hand, the quest continues for further understanding of, and ideally an effective treatment for, Huntington disease.

Humans are afflicted by many types of genetic disorders, with effects that range in severity from mild to deadly. For some of these conditions, such as some forms of breast cancer, an understanding of the genes that cause the disorder has contributed to effective means of treatment. For others, such as Huntington disease, the search for successful treatments is still under way.

In Chapter 10, we described Mendel's discovery that inherited traits are determined by genes. We begin this chapter with a second cornerstone of modern genetics, the chromosome theory of inheritance. We then explain how an individual's sex is determined in humans and other organisms, and how new combinations of alleles, different from those of either parent, can occur in offspring. This information about chromosomes, sex determination, and new allele combinations sets the stage for the discussion of human genetic disorders in the remainder of the chapter.

11.1 The Role of Chromosomes in Inheritance

When Mendel published his theory of inheritance in 1866, he had no idea what the physical properties of genes were. By 1882, studies using microscopes had revealed that threadlike structures—the chromosomes—exist inside of dividing cells. The German biologist August Weismann (Figure 11.1) hypothesized that the number of chromosomes was first reduced by half during the formation of sperm and egg cells, then restored to its full number during fertilization. In 1887, meiosis was discovered, thus confirming Weismann's hypothesis. Weismann also suggested that the hereditary material was located on chromosomes, but at that time there was no experimental evidence for or against that idea.

Genes are located on chromosomes

We now know that Weismann's hypothesis was correct: genes are located on chromosomes. Much experimental evidence supports this idea, which has come to be known as the **chromosome theory of inheritance**. As the photograph at the opening of this chapter illustrates, modern genetic techniques allow us to pinpoint which chromosome contains a particular gene and where on the chromosome that gene is located.

How are chromosomes, DNA, and genes related? Chromosomes that pair during meiosis and that have the same set of genes are called **homologous chromosomes**. In each homologous [ho-*MOLL*-uh-guss] pair, one member is inherited from the mother, the other from the father. Recall that each chromosome consists of a single long DNA molecule and many proteins; the proteins provide structural support for the DNA. Each gene is a small region of the DNA molecule, and there are many genes on each chromosome. For example, humans are estimated to have about 25,000 genes located on our 23 chromosomes. Thus, on average, we have 25,000/23, or 1,087, genes per chromosome.

The physical location of a gene on a chromosome is called a **locus** (plural: loci) (Figure 11.2). With some exceptions (to be discussed shortly), the gene at each locus has two copies, one on each homologous chromosome. The alleles we have at our genetic loci influence the inheritance of a broad range of phenotypic traits, one of which, blood types, is discussed in the box on page 210. Furthermore, the different genetic loci on a chromosome bear a physical relationship to one another; how close or far apart they are on the chromosome can affect the outcome of genetic crosses, as we shall also see shortly.

11.2 Autosomes and Sex Chromosomes

As we saw in Chapter 9, most pairs of homologous chromosomes are exactly alike in terms of length, shape, and the set of genes they carry. But in humans and many other organisms, this is not true of the chromosomes that determine the sex of the organism. To indicate these differences, these chromosomes are assigned different letter

Figure 11.1
August Weismann

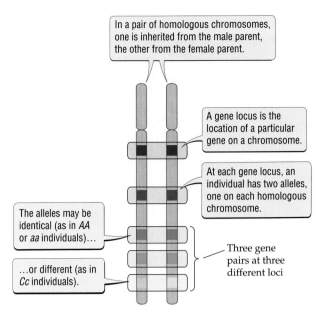

In a pair of homologous chromosomes, one is inherited from the male parent, the other from the female parent.

A gene locus is the location of a particular gene on a chromosome.

At each gene locus, an individual has two alleles, one on each homologous chromosome.

The alleles may be identical (as in *AA* or *aa* individuals)…

…or different (as in *Cc* individuals).

Three gene pairs at three different loci

Figure 11.2 Genes Are Located on Chromosomes
The genes shown here take up a larger portion of the chromosome than they would if they were drawn to scale.

names. In humans, for example, males have one X chromosome and one Y chromosome, whereas females have two X chromosomes (Figure 11.3). The Y chromosome in humans is much smaller than the X chromosome, and few of its genes have a counterpart on the X chromosome. Since human males have one X and one Y chromosome, they have only one copy (instead of the usual two) of each gene that is unique to either the X or the Y chromosome. In some other organisms, such as birds, butterflies, and some fish, males have two identical chromosomes, which we denote ZZ, whereas females have one Z chromosome and one W chromosome.

Chromosomes that determine sex are called **sex chromosomes**; all other chromosomes are called **autosomes**. Human autosomes are labeled not with letters, but with the numbers 1 through 22 (for example, chromosome 4).

In humans, sex is determined by males

Because human females have two copies of the X chromosome, all the gametes (eggs) they produce contain one X chromosome. Males, however, have one X chromosome and one Y chromosome, so, on average, half the gametes (sperm) they produce contain an X chromosome and half contain a Y chromosome. The sex chromosome carried by the sperm therefore determines the sex of a child. If a sperm carrying an X chromosome fertilizes an egg, the resulting child will be a girl; if a sperm carrying a Y chro-

mosome fertilizes the egg, the child will be a boy (see Figure 11.3).

Compared with the X chromosome, the Y chromosome has few genes. It does, however, carry one very important gene: the **SRY gene** (short for "sex-determining region Y"). The *SRY* gene functions as a master switch, committing the sex of the developing embryo to "male." In the absence of this gene, a human embryo develops as a female, but when this gene is present, the embryo develops as a male. The *SRY* gene does not act alone: in both males and females, other genes on the autosomes and sex chromosomes directly influence the development of sexual characteristics. However, the *SRY* gene plays a crucial role because when it is present, it causes the other genes to produce male sexual characteristics, whereas when it is absent, the other genes produce female sexual characteristics.

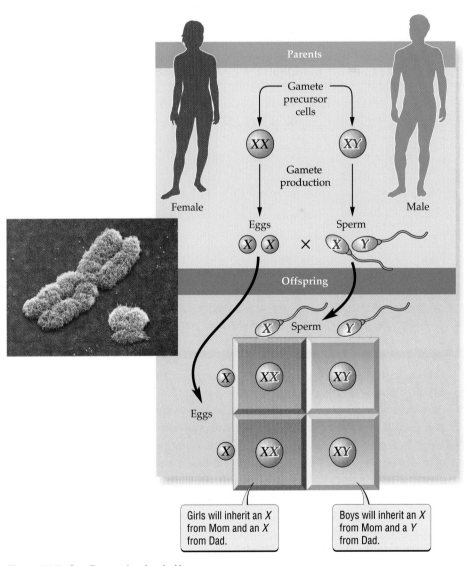

Parents

Gamete precursor cells

Gamete production

Female

Male

Eggs

Sperm

X X × X Y

Offspring

X Sperm Y

Eggs

X XX XY

X XX XY

Girls will inherit an *X* from Mom and an *X* from Dad.

Boys will inherit an *X* from Mom and a *Y* from Dad.

Figure 11.3 Sex Determination in Humans
Human females have two X chromosomes, while human males have one X and one Y chromosome (see inset).

Biology Matters

Know Your Type

One of the most important inherited characteristics to know about yourself and your family members is blood type, since the majority of Americans and Canadians will need donated blood at some point within their lives. Knowing your blood type and the blood types of your close relatives could save valuable time in an emergency, since the type of blood that you can receive is determined by the type of blood that you have. Type O– blood is the only type of blood that can be transfused to a person of any blood type, but only 7 percent of people have type O– blood. Blood centers often run short of type O and type B blood, but shortages of all types of blood occur during the summer and winter holidays.

The following table summarizes the ability of each blood type to either be given to or received from other blood types:

If your blood type is:	You can give blood to:	You can receive blood from:
A+	A+, AB+	A+, A–, O+, O–
O+	O+, A+, B+, AB+	O+, O–
B+	B+, AB+	B+, B–, O+, O–
AB+	AB+	Everyone
A–	A+, A–, AB+, AB–	A–, O–
O–	Everyone	O–
B–	B+, B–, AB+, AB–	B–, O–
AB–	AB+, AB–	AB–, A–, B–, O–

According to current data, 4.5 million Americans would die annually without life-saving blood transfusions. This statistic translates as approximately 32,000 pints of blood used each day in the United States, with someone receiving blood about every three seconds. Whole blood or its components are used for many surgical procedures as well as for ongoing treatment of chronic diseases. For example, people who have been in car accidents and suffered massive blood loss can need transfusions of 50 pints or more of red blood cells. As we saw in Chapter 10, sickle-cell disease is an inherited disease that affects more than 80,000 people in the United States. Some patients with complications from severe sickle-cell disease receive blood transfusions every month, up to 4 pints at a time.

We all expect blood to be there for us, but barely a fraction of those who can give, do. Yet sooner or later, virtually all of us will face a great vulnerability in which we will need blood. And that time is all too often unexpected. To find out where you can donate, visit www.givelife.org

or call 1-800-GIVE-LIFE (1-800-448-3543). To help you remember to schedule your blood donations and other regular screening tests, the College of American Pathologists encourages you to sign up for an e-mail reminder at www.myhealthtestreminder.com.

Some Facts about Blood

- An adult body contains 10 to 12 pints of blood.
- It takes about 6 to 10 minutes to donate a pint of blood and 24 hours for your body to replace the blood fluid volume. The red cells may take up to two months for full restoration.
- Just one unit of blood can be separated into components and used to treat up to three patients.
- Whole blood has a shelf life of 35 days. Red blood cells last 42 days, platelets only 5 days, and plasma up to one year.
- An average healthy person will be eligible to give blood every 56 days, or more than 330 times in his or her lifetime.
- Blood donation takes four steps: medical history, quick physical, donation, and snacks. The actual blood donation usually takes less than 10 minutes. The entire process, from when you sign in to the time you leave, takes about 45 minutes.
- Giving blood will not decrease your strength.
- You cannot get AIDS or any other infectious disease by donating blood.
- Fourteen tests, eleven of which are for infectious diseases, are performed on each unit of donated blood.
- 94 percent of all blood donors are registered voters.
- 60 percent of the U.S. population is eligible to donate, but only 5 percent do on a yearly basis. Seventeen percent of non-donors cite "never thought about it" as the main reason for not giving, while 15 percent say that they're "too busy." The number-one reason that donors say they give blood is that they "want to help others."
- Since a pint of blood weighs one pound, you lose a pound every time you donate blood.

Note: Facts and figures included here are courtesy of the American Red Cross, Northern Ohio Blood Services Region, and Blood Centers of the Pacific.

11.3 Linkage and Crossing-Over

As we saw in Chapter 10, the results of Mendel's experiments indicated that genes are inherited independently of one another. These results led him to propose his law of independent assortment, which states that the two copies of one gene are separated into gametes independently of the two copies of other genes. Early in the twentieth century, however, results from several laboratories indicated that certain genes were often inherited together, thus contradicting the law of independent assortment. Much of this work was done on fruit flies, a species that reproduces rapidly and is easy to study genetically.

Linked genes violate Mendel's law of independent assortment

Thomas Hunt Morgan discovered genes that were inherited together in his research on fruit flies, which began in 1909 at Columbia University in New York City. In one experiment, Morgan crossed a variety of fruit fly that was homozygous for both a gray body (*G*) and wings of normal length (*W*) with another variety that was homozygous for both a black body (*g*) and wings that were greatly reduced in length (*w*). That is, he crossed *GGWW* flies with *ggww* flies to obtain flies of genotype *GgWw* in the F_1 generation. He then mated those *GgWw* flies with *ggww* flies. As shown in Figure 11.4, Morgan's results were very different from the results he expected based on the law of independent assortment. What had happened?

Morgan concluded that the genes for body color and wing length must be located on the same chromosome. Because they were on the same chromosome, the genes were physically connected to each other; hence they were inherited together, and the law of independent assortment did not hold. Genes that are located near one another on the same chromosome and that are inherited together are said to be **genetically linked**. As we shall see shortly, some genes that are located far from one another on the same chromosome are not linked. Genes located on different chromosomes also are not genetically linked.

Crossing-over reduces genetic linkage

If the linkage between two genes on a chromosome were complete, a gamete could never have a chromosome type that was not originally present in one of the parents of the individual that produced that gamete. Consider, for example, the offspring of the *GGWW* × *ggww* cross shown in Figure 11.4. Recall that the two genes (one with alleles *G* and *g*, the other with alleles *W* and *w*) are on the same

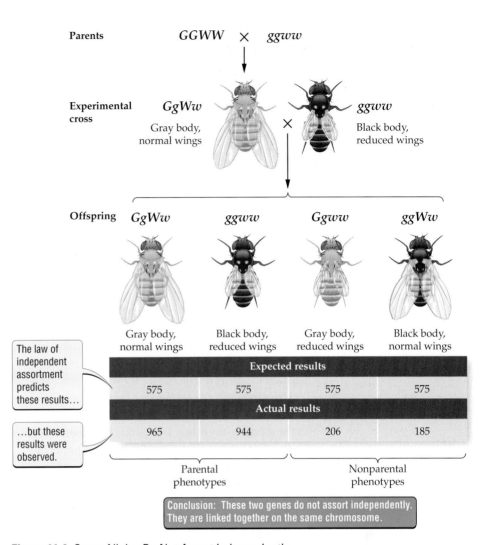

Conclusion: These two genes do not assort independently. They are linked together on the same chromosome.

Figure 11.4 Some Alleles Do Not Assort Independently
By crossing flies of genotype *GgWw* with flies of genotype *ggww*, Thomas Hunt Morgan found that the gene for body color (dominant allele *G* for gray, recessive allele *g* for black) was linked to the gene for wing length (dominant allele *W* for normal length, recessive allele *w* for reduced length). This linkage occurred because the two genes are located relatively close to each another on the same chromosome.

Explore some of Morgan's initial experiments.
11.1

chromosome. As a result, the *GgWw* flies would have inherited a *GW* chromosome from the *GGWW* parent and a *gw* chromosome from the *ggww* parent. Therefore, if linkage were complete, when the *GgWw* flies produced gametes, they would have been able to make only gametes with chromosomes like those in one of their parents; namely, *GW* gametes or *gw* gametes (Figure 11.5). In that case, half of the offspring from the *GgWw* × *ggww* cross shown in Figure 11.4 would have had genotype *GgWw*, and the other half would have had genotype *ggww*. Since the majority of the offspring did have those two genotypes, Morgan realized that the two genes were linked. But how can we explain the appearance of *Ggww*

Review the process of crossing-over.
11.2

and *ggWw* offspring, which have chromosomes (such as a *Gw* chromosome or a *gW* chromosome) that differ from those found in either parent?

To explain the appearance of these offspring genotypes, Morgan suggested that genes are physically exchanged between homologous chromosomes during meiosis. This exchange of genes is called **crossing-over**. To make this concept more concrete, imagine that the two chromosomes illustrated in Figure 11.6 come from one of your cells. You inherited one of these chromosomes from your father, the other from your mother. In crossing-over, part of the chromosome inherited from one parent is exchanged with the corresponding region of DNA inherited from the other parent. By physically exchanging pieces of homologous chromosomes, crossing-over combines alleles inherited from one parent with those inherited from the other. This exchange makes possible the formation of gametes with combinations of alleles that differ from those found in either parent, such as the gametes that resulted in the *Ggww* and *ggWw* offspring shown in Figure 11.4.

Crossing-over can be compared to the cutting of a string at a series of locations that stretch from one end of the string to the other. Two points that are far apart on the string will be separated from each other in most cuts, whereas points that are close to each other will rarely be separated. Similarly, genes that are far from each other on a chromosome are more likely to be separated by crossing-over than are genes that are close to each other. In fact, two genes on the same chromosome that are very far from each other may be separated by crossing-over so often that they are not linked. Such genes are inherited independently of one another even though they are located on the same chromosome. Among the traits that Mendel studied in pea plants, we now know that the genes for flower color and seed color

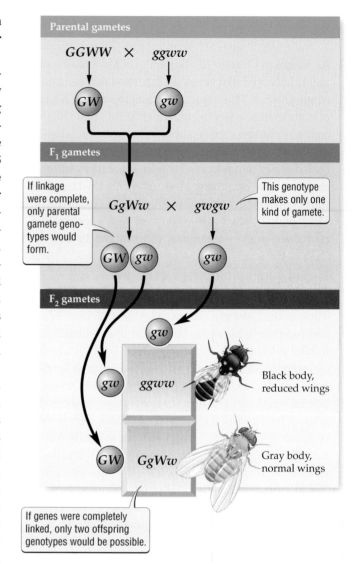

Parental gametes

$GGWW \times ggww$

GW   gw

F₁ gametes

If linkage were complete, only parental gamete genotypes would form.

$GgWw \times gwgw$

This genotype makes only one kind of gamete.

GW gw   gw

F₂ gametes

gw

gw   ggww   Black body, reduced wings

GW   GgWw   Gray body, normal wings

If genes were completely linked, only two offspring genotypes would be possible.

Figure 11.5 Linkage Is Not Complete
If genes were completely linked, the F₂ progeny in Morgan's experiment would all have had genotype *GgWw* or genotype *ggww*, as shown in this diagram. The results of Morgan's experiment (Figure 11.4) showed that this was not the case.

Figure 11.6 Crossing-Over Disrupts the Linkage Between Genes
In the case shown here, a crossing-over event occurs at a point on the chromosome between two linked genes, one gene with alleles *A* and *a*, the other with alleles *B* and *b*. As a result, half of the gametes have a parental genotype (*ABC* or *abc*), while the other half have a nonparental genotype (*Abc* or *aBC*). Although less likely, crossing-over also could have occurred between the gene with alleles *B* and *b* and the gene with alleles *C* and *c*, again producing nonparental genotypes (not shown). Finally, crossing-over could have occurred outside of the region bounded by the gene with alleles *A* and *a* and the gene with alleles *C* and *c*; in this case (also not shown), all of the gametes would have had a parental genotype (either *ABC* or *abc*).

Location of crossover

Before meiosis begins, the two chromosomes (one inherited from Mom, one from Dad) are each in an unduplicated state.

During meiosis, both of the chromosomes are in the duplicated state. Crossing-over between chromosome strands…

…results in two gametes with a parental genotype and two gametes with a nonparental genotype.

are both located on chromosome 1 (of the pea plant's seven pairs of chromosomes), but are so far apart that they are not linked. Thus the law of independent assortment holds for these genes.

11.4 Origins of Genetic Differences among Individuals

Inheritance is both a stable and a variable process. It is stable in that genetic information is transmitted accurately from one generation to the next. Despite this stability, offspring are not exact genetic copies, or clones, of their parents, and hence the individuals of a sexually reproducing species differ genetically. These genetic differences are important, for they provide the genetic variation on which evolution can act. They can also affect whether a person has a genetic disorder and, in some cases, how severe that disorder is. How do genetic differences among individuals arise? Our answer will focus primarily on people, but it also applies to other organisms that reproduce sexually.

First, as we saw in Chapter 10, new alleles arise by mutation. Once formed, those alleles are shuffled or arranged in new ways by crossing-over, independent assortment of chromosomes, and fertilization.

Let's examine how crossing-over typically causes a group of offspring of the same parents to have a wide range of different genotypes. Every time meiosis occurs, crossing-over produces some "new" chromosomes. These chromosomes are new in the sense that they contain some alleles inherited from one parent and other alleles inherited from the other parent. By exchanging alleles between chromosomes, crossing-over causes some offspring to have a genotype that differs from the genotype of either parent (see Figures 11.4 and 11.6).

New combinations of alleles can also be produced by the **independent assortment of chromosomes**, which is the random distribution of maternal and paternal chromosomes into gametes during meiosis. Independent assortment happens because the orientation of the maternal and paternal chromosomes varies at random when the chromosomes line up at the metaphase plate during meiosis (see Chapter 9); hence the chromosomes are shuffled into new combinations in the gametes.

The independent assortment of chromosomes has great potential for producing new combinations of alleles. In humans, for example, during meiosis, each pair of homologous chromosomes attaches to the spindle fibers at random in one of two ways: with the maternal chro-

mosome (red) oriented as shown here with respect to the paternal chromosome (blue):

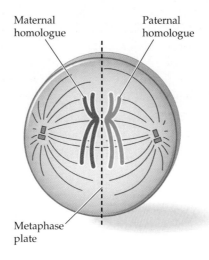

Maternal homologue   Paternal homologue

Metaphase plate

or the other way around:

Develop your understanding of how independent assortment produces gamete diversity through a dynamic animated tutorial.

11.3

Since each of the 23 pairs of homologous chromosomes lines up at random in one of two ways, there are 2^{23}, or 8,388,608, different ways that your chromosomes can be arranged whenever a sperm or egg cell is formed. Of these 8,388,608 ways of arranging the chromosomes, only two are the combination originally inherited from your parents. Thus, like crossing-over, the independent assortment of chromosomes can cause the formation of genotypes that differ from those of both parents.

Finally, fertilization has the potential to add a tremendous amount of genetic variation to that already produced by crossing-over and the independent assortment of chromosomes. In the previous paragraph we saw that each sperm cell represents one of over 8 million sperm cells that could have been formed in meiosis; similarly, each egg represents one of over 8 million possible egg cells. Even if we

Figure 11.7 Living with a Genetic Disorder

This boy with cystic fibrosis has to wear an inhaler to help him breathe, even while watching television. Cystic fibrosis is a lethal, recessive genetic disorder in humans in which mucus builds up in the lungs, digestive tract, and pancreas, causing chronic bronchitis, poor absorption of nutrients, and recurrent bacterial infections, often leading to death before the age of 35.

do not consider the variation caused by crossing-over, there are over 64 trillion (8 million possible sperm × 8 million possible eggs) genetically different offspring that could be formed each time a sperm fertilizes an egg. As a result, the chance that two siblings will be genetically identical is less than 1 in 64 trillion. (The exception, of course, is identical twins; because they form from a single fertilized egg, the chance that they will be genetically identical is 100 percent.) Furthermore, different parents have different alleles, so we are all unique: each time fertilization occurs, a new genetic individual with a unique set of alleles is formed.

11.5 Human Genetic Disorders

Many of us know someone who has suffered from a genetic disorder, such as a hereditary form of cancer, heart disease, or one of the many other disorders caused by gene mutations (Figure 11.7). Because human genetic disorders are so widespread, it is important to study them, since such studies can lead to the prevention or cure of much human suffering. But the study of human genetic disorders faces daunting problems. We humans have a long generation time, we select our own mates, and we decide when and whether to have children. As a result, geneticists cannot perform experiments designed to help them figure out how human genetic disorders are inherited. In addition, our families are much smaller than would be ideal in a scientific study (see the box in Chapter 10, page 196). How can we get around these problems?

Pedigrees are a useful way to study human genetic disorders

One way to study human genetic disorders is to analyze pedigrees. A **pedigree** is a chart similar to a family tree that shows genetic relationships among family members over two or more generations of a family's history. Pedigrees provide geneticists with a way to analyze information from many families in order to learn about the inheritance of a particular disorder. The pedigree shown in Figure 11.8, for example, shows the inheritance of cystic fibrosis, the most common lethal genetic disorder in the United States. Individuals 2 and 3 in generation III have cystic fibrosis, but their parents do not. The pedigree in Figure 11.8 indicates that the allele that causes cystic fibrosis is not dominant, for if it were, one of the parents of the affected individuals would have had the disorder.

Genetic disorders may or may not be inherited

As mentioned earlier, humans can be afflicted by a variety of genetic disorders. Some of these disorders—including most cancers (see Interlude B)—result from new mutations that occur in the cells of an individual

Test your understanding of human pedigree analysis with an interactive quiz. 11.4

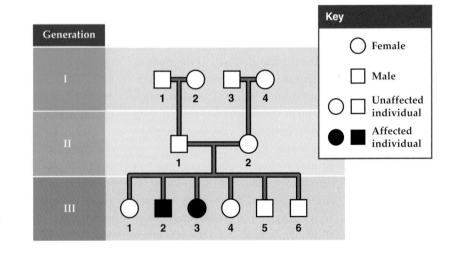

Figure 11.8 Human Pedigree Analysis

The pedigree shown here—in this case for cystic fibrosis—illustrates symbols commonly used by geneticists. The Roman numerals at left identify different generations. Numbers listed below the symbols identify individuals of a given generation. Individuals 1 and 2 in generation II each had genotype *Aa*, where the recessive *a* allele causes cystic fibrosis.

sometime during his or her life. Such mutations usually occur in cells other than gametes and hence are not passed down to offspring. Other genetic disorders, such as cystic fibrosis, are passed down from parent to child. These inherited genetic disorders can be caused by mutations in individual genes (Figure 11.9) or by abnormalities in chromosome number or structure. Clinical genetic tests can be used to determine whether a prospective parent carries an allele for one of these disorders, and variations on these methods can be used to test for genetic disorders long before a baby is born (see the box on page 217).

In the remainder of this chapter we focus on inherited genetic disorders that have relatively simple causes: those caused by mutations of a single gene and those caused by chromosomal abnormalities. As you read this

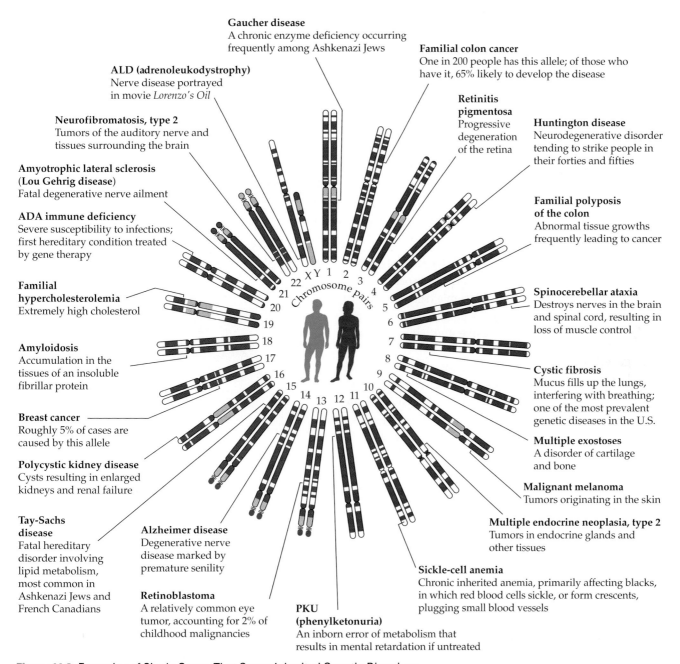

Gaucher disease
A chronic enzyme deficiency occurring frequently among Ashkenazi Jews

Familial colon cancer
One in 200 people has this allele; of those who have it, 65% likely to develop the disease

ALD (adrenoleukodystrophy)
Nerve disease portrayed in movie *Lorenzo's Oil*

Retinitis pigmentosa
Progressive degeneration of the retina

Huntington disease
Neurodegenerative disorder tending to strike people in their forties and fifties

Neurofibromatosis, type 2
Tumors of the auditory nerve and tissues surrounding the brain

Amyotrophic lateral sclerosis (Lou Gehrig disease)
Fatal degenerative nerve ailment

Familial polyposis of the colon
Abnormal tissue growths frequently leading to cancer

ADA immune deficiency
Severe susceptibility to infections; first hereditary condition treated by gene therapy

Familial hypercholesterolemia
Extremely high cholesterol

Spinocerebellar ataxia
Destroys nerves in the brain and spinal cord, resulting in loss of muscle control

Amyloidosis
Accumulation in the tissues of an insoluble fibrillar protein

Cystic fibrosis
Mucus fills up the lungs, interfering with breathing; one of the most prevalent genetic diseases in the U.S.

Breast cancer
Roughly 5% of cases are caused by this allele

Multiple exostoses
A disorder of cartilage and bone

Polycystic kidney disease
Cysts resulting in enlarged kidneys and renal failure

Malignant melanoma
Tumors originating in the skin

Tay-Sachs disease
Fatal hereditary disorder involving lipid metabolism, most common in Ashkenazi Jews and French Canadians

Alzheimer disease
Degenerative nerve disease marked by premature senility

Multiple endocrine neoplasia, type 2
Tumors in endocrine glands and other tissues

Retinoblastoma
A relatively common eye tumor, accounting for 2% of childhood malignancies

Sickle-cell anemia
Chronic inherited anemia, primarily affecting blacks, in which red blood cells sickle, or form crescents, plugging small blood vessels

PKU (phenylketonuria)
An inborn error of metabolism that results in mental retardation if untreated

X Y 1 2 3 4 5 6 7 8 9 10 11 12 13 14 15 16 17 18 19 20 21 22
Chromosome pairs

Figure 11.9 Examples of Single Genes That Cause Inherited Genetic Disorders
Mutations of single genes that cause genetic disorders are found on the X chromosome and on each of the 22 autosomes in humans. Thousands of single-gene genetic disorders are known; for clarity, we show only one such disorder per chromosome.

material, however, it is important to bear in mind that the tendency to develop some diseases, such as heart disease, diabetes, and some inherited forms of cancer, is caused by interactions among multiple genes and the environment. For most diseases caused by multiple genes, the identity of the genes involved and how they lead to disease is poorly understood.

11.6 Autosomal Inheritance of Single-Gene Mutations

We organize our discussion of single-gene genetic disorders by whether the gene is located on an autosome or a sex chromosome. We further subdivide our discussion of autosomal disorders by whether the disease-causing allele is recessive or dominant. As we shall see, recessive genetic disorders are much more common than dominant ones.

There are many autosomal recessive genetic disorders

Several thousand human genetic disorders are inherited as recessive traits. Most of these disorders, such as cystic fibrosis, sickle-cell anemia (see Figure 13.10 and Chapter 15), and Tay-Sachs disease (see Figure 11.9), are caused by recessive mutations of genes located on autosomes.

Recessive genetic disorders range in severity from those that are lethal to those with relatively mild effects. Tay-Sachs disease, for example, is a recessive genetic disorder in which the disease-causing allele encodes a version of a crucial enzyme that does not work properly, causing lipids to accumulate in brain cells. As a result, the brain begins to deteriorate during a child's first year of life, causing death within a few years. At the other end of the severity spectrum, albino skin color in humans can be caused by a variety of single-gene mutations, including one similar to that which produces a white coat color in mice and other mammals (see Figure 10.11).

The only individuals who actually get a disorder caused by an autosomal recessive allele (say, *a*) are those who have two copies of that allele (*aa*). Usually, when a child inherits a recessive genetic disorder, both parents are heterozygous; that is, they both have genotype *Aa*. (It is also possible for one or both parents to have genotype *aa*.) Because the *A* allele is dominant and does not cause the disorder, *Aa* individuals are said to be **carri-**

ers of the disorder: they carry the disorder allele (*a*), but do not get the disorder.

If two carriers of a recessive genetic disorder have children, the patterns of inheritance are the same as for any recessive trait: it is likely that ¼ of the children will have genotype *AA*, ½ will have genotype *Aa*, and ¼ will have genotype *aa*. Thus, as shown in Figure 11.10, each child has a 25 percent chance of not carrying the disorder allele (genotype *AA*), a 50 percent chance of being a carrier (genotype *Aa*), and a 25 percent chance of actually getting the disorder (genotype *aa*).

These percentages reveal one way in which lethal recessive disorders such as Tay-Sachs disease can persist in the human population: although homozygous recessive individuals (with genotype *aa*) die long before they are old enough to have children, carriers (with genotype *Aa*) are not harmed by the disorder. In a sense, the *a* alleles

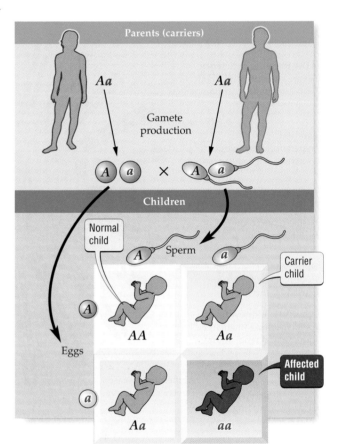

Figure 11.10

Inheritance of Autosomal Recessive Disorders

The patterns of inheritance for a human autosomal recessive genetic disorder are the same as for any recessive trait (compare this figure with the pattern shown by Mendel's pea plants in Figure 10.8). Recessive disorder alleles are colored red and denoted *a*. Dominant, normal alleles are black and denoted *A*. Here, the parents are a carrier male (genotype *Aa*) and a carrier female (genotype *Aa*).

Science Toolkit

Prenatal Genetic Screening

How is the baby? is one of the first questions we ask after a child is born. Usually everything is fine, but sometimes the answer can be devastating. Today, some parents choose to have one of several prenatal genetic screening methods performed to check their baby's health long before it is born.

This choice is not as new as you might think. In the 1870s, doctors occasionally withdrew some of the fluid in which the fetus is suspended to obtain information about its health. Modern versions of that practice have been standard medical procedure since the early 1960s. In **amniocentesis**, a needle is inserted through the abdomen into the uterus to extract a small amount of amniotic fluid from the pregnancy sac that surrounds the fetus. This fluid contains fetal cells (often sloughed-off skin cells) that can be tested for genetic disorders. Another method is **chorionic** [*KOR*-ee-ah-nik] **villus sampling (CVS)**, in which a physician uses ultrasound to guide a narrow, flexible tube through a woman's vagina and into her uterus, where the tip of the tube is placed next to the villi, a cluster of cells that attaches the pregnancy sac to the wall of the uterus. Cells are removed from the villi by gentle suction, then tested for genetic disorders.

Unfortunately, there are risks associated with amniocentesis and CVS, including vaginal cramping, miscarriage, and premature birth. One or two of every 400 women who undergo amniocentesis experience a miscarriage (0.25–0.5 percent). The risk of miscarriage is approximately twice that with CVS (0.5–1.0 percent). Because of such risks, these and other invasive tests are typically used only by parents who know they face an increased chance of giving birth to a baby with a genetic disorder. Older parents, for example, might want to test for Down syndrome, since the risk of that condition increases with the age of the mother. A couple in which one parent carries a dominant allele for a specific genetic disorder (such as Huntington disease) might also decide to have prenatal genetic screening done.

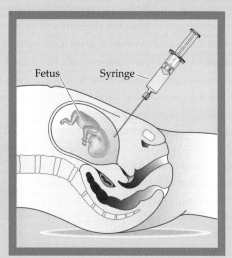

Amniocentesis

In amniocentesis, amniotic fluid, which contains fetal cells, is extracted from the uterus.

Until recently, couples who elected to have such tests performed had only two choices if their fears were confirmed: they could abort the baby, or they could give birth to a child who would have a genetic disorder. Since 1989, however, a third option has been available to couples who are willing, and can afford, to have a child by in vitro fertilization (in which fertilization occurs in a petri dish, after which one or more embryos are implanted into the mother's uterus). In **preimplantation genetic diagnosis (PGD)**, one or two cells are removed from the developing embryo, usually 3 days after fertilization occurs. (It is important to perform PGD at this time because then the embryo typically has 4–12 loosely connected cells; in another day or two, the cells will begin to fuse more tightly to one another, making PGD more difficult.) Next, the cell or cells removed from the embryo are tested for genetic disorders. Finally, one or more embryos that are free of disorders are implanted into the mother's uterus, and the rest of the embryos, including those with genetic disorders, are discarded.

PGD is typically used by parents who either have or carry alleles for a serious genetic disorder, such as cystic fibrosis or Huntington disease. Like all genetic screening methods, the use of PGD raises ethical issues. People who support the use of PGD think that amniocentesis and CVS provide parents with a bleak set of moral choices: if the fetus has a serious genetic disorder, the parents can either abort the baby or allow it to live a life that may be short and full of suffering. In their view, it is morally preferable to discard an embryo at the 4- to 12-cell stage than it is to abort a well-developed fetus, or to give birth to a child that will suffer the devastating effects of a serious genetic disorder. Those opposed to the use of PGD argue that the moral choices are the same: their view is that once fertilization has occurred, a new life has formed, and it is immoral to end that life, even at the 4- to 12-cell stage. What do you think?

can "hide" in heterozygous carriers, and those carriers are likely to pass the disorder allele to half of their children. Recessive genetic disorders also remain in the human population because new mutations produce new copies of the disorder alleles.

Dominant genetic disorders are less common

A dominant allele (*A*) that causes a genetic disorder cannot "hide" in the same way that a recessive allele can. In this case, *AA* and *Aa* individuals get the disorder; only *aa* individuals are symptom-free. When a dominant genetic disorder has serious negative effects, individuals that have the *A* allele may not live long enough to reproduce; hence few of them pass the allele on to their children. As a result, most dominant lethal alleles remain in the population primarily because in each generation, a few of these alleles are produced by mutation during gamete formation.

Huntington disease, which was described at the beginning of this chapter, illustrates another way in which dominant lethal alleles can remain in the population. The symptoms of Huntington disease begin relatively late in life, often after victims of the disorder have had children. Because the allele that causes the disorder can be passed on to the next generation before the victim dies, the disorder is more common than it would be if the symptoms began before childbearing age or if the disorder persisted by mutation alone.

11.7 Sex-Linked Inheritance of Single-Gene Mutations

Use an interactive Punnett square to explore the genetic mechanisms of several inherited diseases. 11.5

Roughly 1,200 of the estimated 25,000 human genes are found only on the X chromosome or only on the Y chromosome; such genes are said to be **sex-linked**. Because sex-linked genes are found on the X chromosome or the Y chromosome—but not both—males receive only one copy of each sex-linked gene (whether it is on the X chromosome or the Y chromosome), while females receive two copies of genes on the X chromosome and no copies of genes on the Y chromosome. About 15 genes are shared by the X and Y chromosomes. Hence males and females receive two copies of each of these genes, just as they do for all autosomal genes; as a result, these 15 genes are not sex-linked.

Approximately 1,100 of the 1,200 human sex-linked genes are located on the X chromosome, while 80 are located on the much smaller Y chromosome. Sex-linked genes on the

X chromosome are said to be **X-linked**; similarly, all sex-linked genes on the Y chromosome are said to be **Y-linked**. Although there are no well-documented cases of disease-causing Y-linked genes, X chromosomes do contain genes known to cause human genetic disorders (see Figure 11.9).

Sex-linked genes have different patterns of inheritance than do genes on autosomes. Consider how an X-linked recessive allele for a human genetic disorder is inherited (Figure 11.11). We label the recessive disorder allele *a*, and in the Punnett square we write this allele as X^a to emphasize the fact that it is on the X chromosome. Similarly, the dominant allele is labeled *A* and is written as X^A in the Punnett square. If a carrier female (with genotype X^AX^a) has children with a normal male (with genotype X^AY), each of their sons will have a 50 percent chance of getting the disorder (see Figure 11.11). This result differs greatly from what would happen if the same disorder allele (*a*) were on an autosome: in that case, none of the children, male or female, would get the disorder.

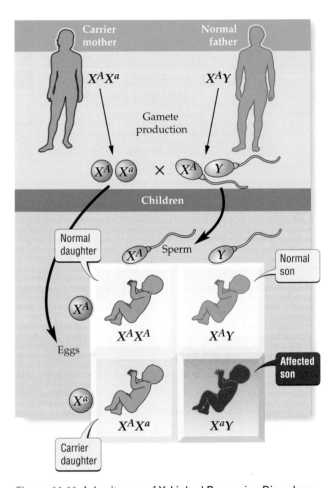

Figure 11.11 Inheritance of X-Linked Recessive Disorders
The recessive disorder allele (*a*) is located on the X chromosome and is denoted by X^a. The dominant normal allele (*A*) is also located on the X chromosome and is denoted by X^A.

For a recessive X-linked genetic disorder, males of genotype X^aY get the disorder because the Y chromosome does not have a copy of the gene, and hence a dominant A allele cannot mask the effects of the a allele. In general, males are more likely than females to get recessive X-linked disorders because they have to inherit only a single copy of the disorder allele, whereas females must inherit two copies to be affected. In contrast, both sexes are equally likely to be affected by autosomal recessive disorders.

X-linked genetic disorders in humans include hemophilia, a serious disorder in which minor cuts and bruises can cause a person to bleed to death, and Duchenne muscular dystrophy, a lethal disorder that causes the muscles to waste away, often leading to death in the early twenties. Both of these X-linked disorders are caused by recessive alleles. An example of a dominant X-linked disorder with a daunting name—congenital generalized hypertrichosis [*HY*-per-try-*KO*-siss], or CGH—is shown in Figure 11.12.

Changes in chromosome structure can have dramatic effects. If breakage occurs in the sex chromosomes, it can cause a change in the expected sex of the developing fetus. A deletion in which the portion of the Y chromosome that contains the *SRY* gene is lost produces an XY individual that develops as a female. A translocation that results in an X chromosome to which the Y chromosome's sex-determining region is attached produces an XX individual that develops as a male. XY females and XX males are always sterile (unable to produce offspring).

Changes to the structure of autosomes can have even more dramatic effects. Cri du chat [*KREE*-doo-*SHAH*] syndrome occurs when a child inherits a chromosome 5 that is missing a particular region (Figure 11.14). *Cri du chat* is French for "cry of the cat," which describes the characteristic mewing sound made by infants with this condition. Other characteristics of this condition are slow growth and a tendency toward severe mental retardation, a small head, and low-set ears.

Figure 11.12
Congenital Generalized Hypertrichosis (CGH)
A six-year-old boy with CGH, a rare genetic disorder that causes extreme hairiness of the face and upper body. CGH is caused by a dominant allele of a single gene on the X chromosome.

11.8 Inherited Chromosomal Abnormalities

Relatively few human genetic disorders are caused by inherited chromosomal abnormalities, probably because most large changes in the chromosomes kill the developing embryo. Two main types of chromosomal changes occur in humans and other organisms: changes in the structure (for example, the length) of an individual chromosome and changes in the overall number of chromosomes. (Bear in mind that either of these types of chromosomal change have to occur in the gametes or gamete-producing cells to be passed on to, and affect, offspring.)

The structure of chromosomes can change in several ways

When chromosomes are copied during cell division, breaks can occur that alter the length of one or more chromosomes. Sometimes a piece breaks off and is lost from the chromosome (**deletion**). Other times the broken piece is reattached, but in an erroneous way. In an **inversion**, the fragment returns to the correct place on the original chromosome, but with the genetic loci in reverse order. In a **translocation**, the broken piece is attached to a different, nonhomologous chromosome. Finally, a fragment from one chromosome can fuse to the homologous chromosome (a **duplication**), increasing the length of the chromosome that receives the fragment. Figure 11.13 diagrams these kinds of structural changes.

Changes in chromosome number are often fatal

Unusual numbers of chromosomes—such as one or three copies instead of the normal two—can be produced when

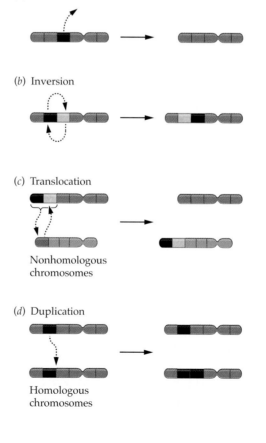

(*a*) Deletion

(*b*) Inversion

(*c*) Translocation

Nonhomologous chromosomes

(*d*) Duplication

Homologous chromosomes

Figure 11.13 Structural Changes to Chromosomes
(*a*) Deletion: A segment breaks off and is lost from the chromosome. (*b*) Inversion: A segment breaks off and is reattached, but in reverse order. (*c*) Translocation: A segment breaks off one chromosome and becomes attached to a different, nonhomologous chromosome. (*d*) Duplication: A fragment being copied from one chromosome breaks off and becomes attached to its homologue.

Cri du chat syndrome occurs when either the red-colored region or a larger portion of the top part of chromosome 5 is removed.

A young boy with cri du chat syndrome

Figure 11.14 Cri du Chat Syndrome
Cri du chat syndrome is caused by the removal of a portion of the top part of chromosome 5.

Normal chromosome 5

Chromosome 5 with deletion

Helpful to know

Earlier we noted the use of the suffix *some* in the names of cellular particles, such as chromosomes. Biologists use the related suffix *somy* to name conditions involving unusual numbers of chromosomes; two examples are *trisomy* ("three bodies") and *monosomy* ("one body").

chromosomes fail to separate properly during meiosis. In people, such changes in chromosome number often result in fetal death. At least 20 percent of human pregnancies abort spontaneously, largely as a result of such changes in chromosome number.

There is only one case in which a person who inherits the wrong number of autosomes commonly reaches adulthood: Down syndrome. Individuals with this condition usually have three copies of chromosome 21, the smallest autosome in humans; as a result, Down syndrome is also known as trisomy 21, where **trisomy** [*TRY-suh-mee*] refers to the condition of having three copies of a chromosome (instead of the usual two). A small minority (3–4 percent) of Down syndrome cases occur when an extra piece of chromosome 21 breaks off during cell division and attaches to another chromosome. People with Down syndrome tend to be short and mentally retarded, and they may have defects of the heart, kidneys, and digestive tract. With appropriate medical care, most people with this condition lead healthy lives, and many live to their sixties or seventies (their average life expectancy is 55). Live births can also result when an infant has three copies of chromosome 13, 15, or 18. However, such children have severe birth defects, and they rarely live beyond their first year.

Compared with having too few or too many autosomes, changes in the number of sex chromosomes can have relatively minor effects. Klinefelter syndrome, for example, is a condition found in males that have an extra X chromosome (XXY males). Such men have a normal life span and normal intelligence, and they tend to be tall. Many XXY males also have small testicles (about one-third normal size) and reduced fertility, and some have feminine characteristics, such as enlarged breasts. Females with a single X chromosome (Turner syndrome) have normal intelligence and tend to be short (with adult heights under 150 centimeters, or 4 feet 11 inches), to be sterile, and to have a broad, webbed neck. Other changes in the number of sex chromosomes, as in XYY males and XXX females, also produce relatively mild effects. However, when there are two or more extra sex chromosomes, as in XXXY males or XXXX females, a wide range of problems can result, including severe mental retardation.

Uncovering the Genetics of Huntington Disease

In 1872, George Huntington wrote the classic paper describing the disorder named for him. For over a century, there was little progress in treating the disorder. It was known that the gene for the disorder is on an autosome, and that the Huntington allele is dominant (*A*). But there was no cure, and little helpful information could be given to potential victims of Huntington disease (HD).

For example, by constructing a pedigree, a geneticist might learn that a person's father (who had HD) was of genotype *Aa*. If the mother did not have the disorder, and hence was of genotype *aa*, all the geneticist could say was that the person had a 50 percent chance of developing the disorder (see Problem 3 in the sample genetics problems at the end of this chapter).

In 1983, however, this situation changed dramatically when pedigree analyses that looked for patterns of genes inherited together indicated that the *HD* gene was linked to other genes known to be located on chromosome 4. This discovery set off an intense effort to isolate the *HD* gene. Over the next 10 years, researchers identified the genes on chromosome 4 that were linked most closely to the *HD* gene, a process that eventually allowed them first to pinpoint the gene's location on chromosome 4, then in 1993 to isolate the gene itself.

By isolating the *HD* gene, scientists were able to learn the identity of an abnormal protein produced by the Huntington allele. Portions of this protein form clumps in the brains of people with the disorder; these clumps are correlated with, and may cause, the symptoms of Huntington disease.

Dramatic new results within the past few years suggest that knowledge of the *HD* gene and its associated protein may help in the design of effective treatments. For example, in mice genetically engineered to have the human *HD* gene, scientists have developed ways to slow the progression of, or even reverse, the disorder symptoms. And human cells lacking the Huntington allele have been transplanted into an HD patient's brain, where they survived and remained free of HD protein clumps for 18 months. Collectively, these results offer hope that brain repair may eventually be possible.

In addition to providing clues to possible treatment, isolation of the *HD* gene had the immediate effect of allowing scientists to design a diagnostic genetic test for the disorder. With this test, a person at risk (because one parent had the disorder) can now learn with near certainty whether or not he or she also will get the disorder. In some cases, the genetic test offers hope. For example, people at risk who want to have a family without the fear of passing the disorder on to their children can take the test, and thus make an informed decision. But the test poses an agonizing choice for these and all other individuals at risk: If they take the test, they may experience tremendous relief if the results show that they will not get the disorder. Alternatively, they may experience a crushing loss of hope if they find out they have a lethal disease with horrible symptoms that, as of now, cannot be cured. Given these alternatives, would you take the test?

Chapter Review

Summary

11.1   The Role of Chromosomes in Inheritance

- The chromosome theory of inheritance states that genes are located on chromosomes.
- Each chromosome is composed of a single DNA molecule and many proteins.
- A gene is a region of DNA on a chromosome. The physical location of that region is called the gene's locus.
- Chromosomes that pair during meiosis and have the same set of genes are homologous chromosomes.

11.2   Autosomes and Sex Chromosomes

- Chromosomes that determine the sex of the organism are called sex chromosomes; all other chromosomes are called autosomes.

- Human males have one X and one Y chromosome. Human females have two X chromosomes.
- A specific gene on the Y chromosome (the *SRY* gene) is required for human embryos to develop as males.

11.3   Linkage and Crossing-Over

- Genes that are located near one another on the same chromosome are said to be genetically linked.
- Crossing-over, the exchange of genes between chromosomes, reduces the linkage between genes.
- Two genes that are far apart on a chromosome are more likely to be separated by crossing-over than are genes located near one another.

- Genes that are located very far apart on a chromosome may get separated by crossing-over so often that they are not genetically linked; Mendel's law of independent assortment holds for such genes.

11.4 Origins of Genetic Differences among Individuals

- Genetic differences among individuals underlie human genetic disorders and provide the genetic variation on which evolution can act.
- Maternal and paternal chromosomes are independently distributed into gametes during meiosis.
- Crossing-over, the independent assortment of chromosomes, and fertilization all cause offspring to differ genetically from one another and from their parents.

11.5 Human Genetic Disorders

- Pedigrees provide a useful way to study human genetic disorders.
- Humans suffer from a variety of genetic disorders, including those caused by mutations of a single gene and those caused by abnormalities in chromosome number or structure.
- Clinical genetic testing can be performed on both fetuses and adults. Three fetal testing procedures are amniocentesis, chorionic villus sampling (CVS), and preimplantation genetic diagnosis (PGD).

11.6 Autosomal Inheritance of Single-Gene Mutations

- Most genetic disorders are caused by a recessive allele (*a*) of a gene on an autosome. Only homozygous (*aa*) individuals get these disorders; heterozygous (*Aa*) individuals are merely carriers of the disorders.
- In dominant autosomal genetic disorders, which are much less common than recessive ones, both *AA* and *Aa* individuals are affected.
- Alleles that cause lethal dominant genetic disorders can remain in the population because the disorder symptoms begin late in life, as in Huntington disease, or because new disorder alleles are produced by mutation during each generation.

11.7 Sex-Linked Inheritance of Single-Gene Mutations

- Because males inherit only one X chromosome, genes on sex chromosomes have different patterns of inheritance than do genes on autosomes.
- Genes found only on one sex chromosome or the other, but not on both, are said to be sex-linked. Genes found only on the X chromosome are said to be X-linked; those found only on the Y chromosome are Y-linked.
- Males are more likely than females to have recessive X-linked genetic disorders because males need to inherit only one copy of the disorder allele to be affected, while females must inherit two copies to be affected. In contrast, both sexes are equally likely to be affected by autosomal genetic disorders.

11.8 Inherited Chromosomal Abnormalities

- Chromosomal abnormalities include changes in chromosome structure and chromosome number.

- Changes in the structure of an individual chromosome can happen if breakage occurs during cell division, resulting in deletion, inversion, duplication, or translocation of a chromosome fragment. Such structural changes can have profound effects.
- Changes in the number of autosomes in humans are usually lethal. Down syndrome, a form of trisomy in which individuals receive three copies of chromosome 21, is an exception to this rule.
- People that have one too many or one too few sex chromosomes can experience relatively minor effects; however, if there are four or more sex chromosomes (instead of the usual two), serious problems can result, including severe mental retardation.

⊙ Review and Application of Key Concepts

1. In terms of its physical structure, describe what a gene is and where it is located.

2. Consider the XY chromosome system by which sex is determined in humans. Do patterns of inheritance for genes located on the X chromosome differ between males and females? Why or why not?

3. Look carefully at Figure 11.4. Explain in your own words why the results shown in that figure convinced Morgan that genes located near one another on a chromosome tend to be inherited together, or linked. Why are genes located on different chromosomes not linked?

4. Explain how crossing-over occurs. Assume that genes *A*, *B*, and *C* are arranged in that order along a chromosome. Using your understanding of how crossing-over occurs, will it occur more often between genes *A* and *B* or between genes *A* and *C*?

5. Explain how nonparental genotypes are formed.

6. Are genetic disorders caused by single-gene mutations more common or less common than those caused by abnormalities in chromosome number or structure? Explain why.

Key Terms

amniocentesis (p. 217)
autosome (p. 209)
carrier (p. 216)
chorionic villus sampling (CVS) (p. 217)
chromosome theory of inheritance (p. 208)
crossing-over (p. 212)

deletion (p. 219)
duplication (p. 219)
genetic linkage (p. 211)
homologous chromosome (p. 208)
independent assortment of chromosomes (p. 213)
inversion (p. 219)
locus (p. 208)

Self-Quiz

1. Genes are
   a. located on chromosomes.
   b. composed of DNA.
   c. composed of both protein and DNA.
   d. both a and b

2. Which of the following is an autosomal dominant disorder in which the symptoms begin late in life and the nervous system is destroyed, resulting in death?
   a. Tay-Sachs disease
   b. Huntington disease
   c. Down syndrome
   d. cri du chat syndrome

3. Crossing-over is
   a. more likely between genes that are close together on a chromosome.
   b. more likely between genes that are on different chromosomes.
   c. more likely between genes that are far apart on a chromosome.
   d. not related to the distance between genes.

4. Comparatively few human genetic disorders are caused by chromosomal abnormalities. One reason is that
   a. most chromosomal abnormalities have little effect.
   b. it is difficult to detect changes in the number or length of chromosomes.
   c. most chromosomal abnormalities result in spontaneous abortion of the embryo.
   d. it is not possible to change the length or number of chromosomes.

5. Nonparental genotypes can be produced by
   a. crossing-over and the independent assortment of chromosomes.
   b. linkage.
   c. autosomes.
   d. sex chromosomes.

6. Sometimes a segment of DNA breaks off from a chromosome, then returns to the correct place on the original chromosome, but in reverse order. This type of chromosomal structural change is called
   a. crossing-over.   c. an inversion.
   b. a translocation.   d. a deletion.

7. The prenatal genetic screening method in which cells that attach the pregnancy sac to the wall of the uterus are tested for genetic disorders is called
   a. chorionic villus sampling (CVS).
   b. amniocentesis.
   c. preimplantation genetic diagnosis (PGD).
   d. in vitro fertilization.

8. Which of the following can most precisely be described as a master switch that commits the sex of the developing embryo to "male"?
   a. an X chromosome
   b. a Y chromosome
   c. an XY chromosome pair
   d. the *SRY* gene

Sample Genetics Problems

1. Recall that human females have two X chromosomes and human males have one X chromosome and one Y chromosome.
   a. Do males inherit their X chromosome from their mother or from their father?
   b. If a female has one copy of an X-linked recessive allele for a genetic disorder, does she have the disorder?
   c. If a male has one copy of an X-linked recessive allele for a genetic disorder, does he have the disorder?
   d. Assume that a female is a carrier of an X-linked recessive disorder. With respect to the disorder allele, how many types of gametes can she produce?
   e. Assume that a male with an X-linked recessive genetic disorder has children with a female who does not carry the disorder allele. Could any of their sons have the genetic disorder? How about their daughters? Could any of their children be carriers for the disorder? If so, which sex(es) could they be?

2. Cystic fibrosis is a recessive genetic disorder; the disorder allele, which we'll call *a*; is located on an autosome. What are the chances that parents with the following genotypes will have a child with the disorder?
   a. $aa \times Aa$ c. $Aa \times Aa$
   b. $Aa \times AA$ d. $aa \times AA$

3. Huntington disease (HD) is a genetic disorder caused by a dominant allele—we'll call it *A*—that is located on an autosome. What are the chances that parents with the following genotypes will have a child with HD?
   a. $aa \times Aa$ c. $Aa \times Aa$
   b. $Aa \times AA$ d. $aa \times AA$

4. Hemophilia is a recessive genetic disorder whose disorder allele, which we'll call *a*, is located on the X chromosome. What are the

chances that parents with the following genotypes will have a child with hemophilia?

a. $X^A X^A \times X^a Y$

b. $X^A X^a \times X^a Y$

c. $X^A X^a \times X^A Y$

d. $X^a X^a \times X^A Y$

e. Do male and female children have the same chance of getting the disorder?

5. Explain why the terms *homozygous* and *heterozygous* do not apply to X-linked traits in males.

6. Study the pedigree shown below. Is the disorder allele dominant or recessive? Is the disorder allele located on an autosome or on the X chromosome? What are the genotypes of individuals 1 and 2 in generation I?

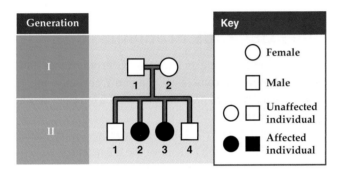

7. In the text, we state that males are more likely than females to inherit recessive X-linked genetic disorders. Are males also more likely than females to inherit dominant X-linked genetic disorders? Illustrate your answer by constructing Punnett squares in which

a. an affected female has children with a normal male.

b. an affected male has children with a normal female.

8. Study the pedigree shown below. Is the disorder allele dominant or recessive? Is the disorder allele located on an autosome or on the X chromosome? To answer this question, assume that individual 1 in generation I and individuals 1 and 6 in generation II do not carry the disorder allele.

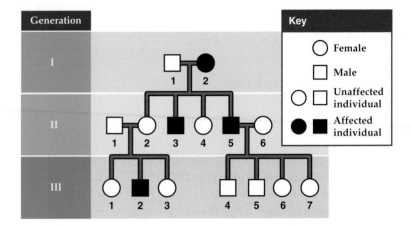

9. Imagine you are conducting an experiment on fruit flies, and you are tracking the inheritance of two genes, one with alleles *A* or *a*, the other with alleles *B* or *b*. *AABB* individuals are crossed with *aabb* individuals to produce F_1 offspring, all of which have genotype *AaBb*. These *AaBb* F_1 offspring are then crossed with *aaBB* individuals. Construct Punnett squares and list the possible offspring genotypes that you would expect in the F_2 generation

a. if the two genes were completely linked.

b. if the two genes were on different chromosomes.

Biology in the News

Heart Tumor Discovery

Researchers say they might be more common than previously thought, and they urge physicians to look more closely for them

Scientists studying a rare genetic disorder have made a surprising discovery that helps explain why certain heart tumors develop and suggests that they may be more common than had been believed.

The heart tumors described in this article are not cancerous, but they are dangerous: a portion of the tumor can break off and cause a stroke. In addition, if the tumor grows to a large size, it can block the flow of blood through the heart, causing breathing difficulties, dizziness, and chest pains. These heart tumors—called myxomas [mix-*OH*-muhs]—were thought to affect one out of every 100,000 people. However, a group of scientists who discovered a gene that causes myxomas think the condition may be more common than that.

The researchers who discovered the myxoma gene were studying a rare muscle disorder that causes tight or clenched hands or feet, an inability to open the jaw fully, and spots on the skin. No one had ever connected heart tumors to such muscle disorders, but the research group suspected there might be a link because sudden cardiac death in young people is correlated with these muscle problems. Genetic analyses showed they were right. They discovered a genetic defect that makes people prone to both muscle disorders and heart tumors.

Prior to this discovery, a person would know he or she had a heart tumor only by surviving a stroke or by a doctor finding the tumor while searching for the cause of symptoms such as difficulty breathing. If the results of this study are confirmed, this delay in diagnosis may soon end: people may be able to take a genetic test to see whether they have the allele for myxomas. Those with positive test results could then be checked regularly for signs of a tumor. (Doctors can detect heart tumors in a variety of ways, including listening to blood flow through the heart.) Any myxomas present could be removed surgically, thereby helping to reduce the risk of stroke or other serious medical problems.

Evaluating the news

1. Should doctors recommend that everyone be tested for the myxoma allele—even people with no known history of muscle disorders or heart tumors?

2. Tests like those mentioned in question 1 could serve as a form of preventative medicine and might also allow scientists to find out how common the myxoma allele really is. However, insurance companies or other organizations might want to know whether you have the myxoma allele. Given the potential benefits of the test and the possible invasion of your "genetic privacy," would you want to take the test?

3. Like the heart tumors described in this article, many human genetic disorders are caused by alleles that make a person prone to getting the disorder, but do not always produce the disorder. Would you want to know whether you have such alleles? Should anyone other than you have access to that information? Consider insurance companies again. As a society, we think it reasonable for them to know whether a person smokes or has a family history of cancer. Is more detailed information about whether a person has a broad range of disorder alleles any different? Why or why not?

SOURCE: *Associated Press*, Thursday, July 29, 2004.

CHAPTER 12 DNA

Key Concepts

◉ Genes are composed of DNA.

◉ Four nucleotides are the building blocks of DNA. Each of these nucleotides contains one of four nitrogen bases: adenine, cytosine, guanine, or thymine.

◉ DNA consists of two strands of nucleotides twisted together into a spiral. The two strands are held together by hydrogen bonds that form between adenine in one strand and thymine in the other, and similarly, between cytosine in one strand and guanine in the other.

◉ Each strand of DNA has an enormous number of bases arranged one after another. The sequence of bases in DNA differs among species and among individuals within a species. These differences are the basis of inherited variation.

◉ Because adenine pairs only with thymine and cytosine pairs only with guanine, each strand of DNA can serve as a template from which to duplicate the other strand.

◉ DNA in cells is damaged constantly by various physical, chemical, and biological agents. If this damage were not repaired, the organism might die.

Main Message: DNA is the genetic material. Its three-dimensional structure consists of two strands of nucleotides twisted into a double helix.

The Library of Life

The book you are reading has more than a million letters in it. How long would it take you to copy those letters by hand, one by one? How many mistakes do you think you would make? Would you check your work for mistakes, and if so, how long would that take?

Difficult as the job of copying all the letters in this book would be, it pales in comparison to the job your cells do each time they divide. Before a cell divides, it must make a copy of all its genetic information. That information is stored in deoxyribonucleic acid, or DNA, the substance of which genes are made. The amount of information stored in your DNA is mind-boggling: whereas this book contains roughly a million letters, the DNA in each of your cells has the equivalent of 6,600,000,000 letters. If it were printed with the same font in the same size as the letters on this page, the information that is in your DNA would fill thousands of books similar in length to this one.

Your cells copy the phenomenal amount of information in your DNA in a matter of hours. Yet despite the speed with which they work, on average, they make only one mistake for every billion "letters." How do they do this? What are the "letters" that make up the DNA molecule? More broadly, what is the structure of the DNA molecule, and what implications does this structure have for the processes of life? These are some of the questions we examine in this chapter.

A Model of the DNA Molecule

n the previous two chapters, we learned that genes control the inheritance of traits and are located on chromosomes. However, this knowledge leaves unanswered several fundamental questions: What are genes made of? When a cell divides to form two daughter cells, how is the information in the genes copied? How are errors in copying corrected, and how is damage to the cell's genetic material repaired?

To answer such questions, geneticists had to discover the substance of which genes are made, and they had to learn the physical structure of this substance. As they began their search, they were guided by three basic biological facts about the nature of the genetic material. First, the genetic material had to contain the information necessary for life. It had to contain, for example, the information needed to build the body of the organism and to control the complex metabolic reactions on which life depends. Second, the genetic material had to be composed of a substance that could be copied accurately. If this could not be done, reliable genetic information could not be passed from one generation to the next. Finally, the genetic material had to be variable; otherwise, there would be no genetic differences within or among species.

Parallel to the search for the chemical composition and physical structure of genes was a search for the function of genes: how exactly did genes produce their effects? In this chapter we describe how scientists discovered that genes are composed of DNA. We also describe the physical structure of genes, how the genetic material is copied and repaired, and how genetic disorders can arise. In Chapter 13 we will see how genes produce their effects.

12.1 The Search for the Genetic Material

By the early 1900s, geneticists knew that genes control the inheritance of traits, that genes are located on chromosomes, and that chromosomes are composed of DNA and proteins. With this knowledge in hand, the first step in the quest to understand the physical structure of genes was to determine whether the genetic material was DNA or protein.

Initially, most geneticists thought that protein was the more likely candidate. Proteins are large, complex molecules, and it was not hard to imagine that they could store the tremendous amount of information needed to govern the lives of cells. Proteins also vary considerably within and among species; hence it was reasonable to assume that they caused the inherited variation observed within and among species.

DNA, on the other hand, was initially judged a poor candidate for the genetic material, mainly because DNA was thought to be a small, simple molecule whose composition varied little among species. Over time, these ideas about DNA were shown to be wrong. In fact, DNA molecules are large and vary tremendously within and among species. Still, as we shall see, variations in the DNA molecule are more subtle than the variation in shape, electrical charge, and function shown by proteins, so it is not surprising that most researchers initially favored proteins as the genetic material.

Over a period of roughly 25 years (1928–1952), geneticists became convinced that DNA, not protein, was the genetic material. Let's consider three important studies that helped cause this shift of opinion.

Harmless bacteria can be transformed into deadly bacteria

In 1928, a British medical officer named Frederick Griffith published an important paper on *Streptococcus pneumoniae* [noo-*MO*-nee-ay], a bacterium that causes pneumonia in humans and other mammals. Griffith was studying two genetic varieties, or strains, of *Streptococcus* to find a cure for pneumonia, which was a common cause of death at that time. The two strains, called strain S and strain R, were named after differences in their appearance. When the bacteria were grown on a petri dish, strain S produced colonies that appeared smooth, while strain R produced colonies that appeared rough.

Griffith conducted four experiments on these bacteria (Figure 12.1). First, when he injected bacteria of strain R into mice, the mice survived and did not develop pneumonia. Second, when he injected bacteria of strain S into mice, the mice developed pneumonia and died. In the third experiment, he injected heat-killed strain S bacteria into mice, and once again the mice survived. His original plan was to test mice from the third experiment to see if they were resistant to later exposure to live strain S bacteria, but his 1928 paper did not include results from such a test.

The results from the first three experiments were not particularly unusual: Griffith had simply shown that there were two strains of bacteria, one of which (strain S) killed mice and was itself killed and rendered harmless by heat. In the fourth experiment, however, something unexpected happened. Griffith mixed heat-killed bacteria of strain S with live bacteria of strain R. On the basis of the results from the first three experiments, he expected the mice to survive. Instead, the mice died, and Griffith recovered large numbers of live strain S bacteria from the blood of the dead mice.

In Griffith's fourth experiment, something had caused harmless strain R bacteria to change into deadly strain S bacteria. Griffith showed that the change was genetic: when they reproduced, the altered strain R bacteria produced strain S bacteria. Overall, the results of Griffith's fourth experiment suggested that genetic material from heat-killed strain S bacteria had somehow changed living strain R bacteria into strain S bacteria.

This remarkable result stimulated an intensive hunt for the material that caused the change. We now know that the strain R bacteria had absorbed a small piece of DNA from the heat-killed strain S bacteria, causing the genotype of the strain R bacteria to change. Such a change in the genotype of a cell or organism after exposure to the DNA of another genotype is called **transformation**.

DNA can transform bacteria

For 10 years, Oswald Avery, Colin MacLeod, and Maclyn McCarty (all at Rockefeller University in New York) struggled to identify the genetic material that had transformed the bacteria in Griffith's experiments. They isolated and tested different compounds from the bacteria. Only DNA was able to transform harmless strain R bacteria into deadly strain S bacteria. In 1944, Avery and his colleagues published a landmark paper that summarized their results. The paper created quite a stir.

In addition to showing that DNA transforms bacteria, Avery, MacLeod, and McCarty's paper led many biologists to a broader conclusion: that DNA, not protein, is the genetic material. As a leading DNA researcher later remarked, the paper stimulated an "avalanche" of new research on DNA. Some biologists remained skeptical, arguing, for example, that DNA was too simple a molecule to be the genetic material. However, the tide was turning in favor of DNA.

The genetic instructions of a virus are contained in its DNA

In Griffith's experiments, heat killed the strain S bacteria, but did not destroy its genetic material. Since most proteins are destroyed by heat, this result suggested that protein was not the genetic material. Then the work by Avery and co-workers provided strong evidence that DNA was the genetic material. Additional proof came in 1952, when Alfred Hershey and Martha Chase published an elegant study on the genetic material of viruses.

Figure 12.1 Genetic Transformation of Bacteria
Griffith's four experiments showed that harmless strain R bacteria can be transformed into deadly strain S bacteria. We now know that this transformation occurs when a small piece of DNA from strain S bacteria enters and is incorporated into the DNA of strain R cells.

1 Injection
Strain R is harmless.
Living strain R

2 Strain S kills mice.
Living strain S

3 Heat renders strain S harmless.
Strain S being heated   Heat-killed strain S

4 Living cells of strain S isolated from dead mouse
Living strain R   Mixture of heat-killed strain S and living strain R
A substance from the heat-killed strain S changes strain R from harmless to deadly.

Hershey and Chase studied a virus that consists only of a DNA molecule surrounded by a coat of proteins (Figure 12.2). To reproduce, this virus attaches to the cell wall of a bacterium and injects its genetic material into the bacterium. The genetic material of the virus then takes over the bacterial cell, eventually killing it, but first causing it to produce many new viruses. Because the

Recreate Griffith's experiments. 12.1

virus is composed only of proteins and DNA, it provided an excellent experimental system in which to test which substance—DNA or protein—was the genetic material.

Hershey and Chase demonstrated that only the DNA portion of the virus was injected into the bacterium (see Figure 12.2). This result indicated that DNA was responsible for taking over the bacterial cell and for causing the production of new viruses. These experiments convinced most remaining skeptics that DNA, not protein, was the genetic material.

12.2 The Three-Dimensional Structure of DNA

By the early 1950s, genes were known to be composed of DNA. The next step was to determine the three-dimensional structure of DNA. This structure needed to be determined down to the level of atoms before gene function could be understood in molecular terms.

In 1951, Linus Pauling became the first person to figure out the three-dimensional structure of a protein. Pauling's success suggested that determining the three-dimensional structure of DNA should also be possible. Major research laboratories from around the world, including Pauling's, devoted great effort to reaching that goal.

The effort to discover the structure of DNA was a race to unlock some of the greatest mysteries of life: How is the cell's genetic material copied so that it can be passed from parent to offspring? How is genetic information stored in DNA?

DNA is a double helix

Working at Cambridge University in England, the American James Watson and the Englishman Francis Crick won the race to determine the physical structure of DNA. In a two-page paper published in 1953, they proposed that DNA was a **double helix**, a structure that can be thought of as a ladder twisted into a spiral coil

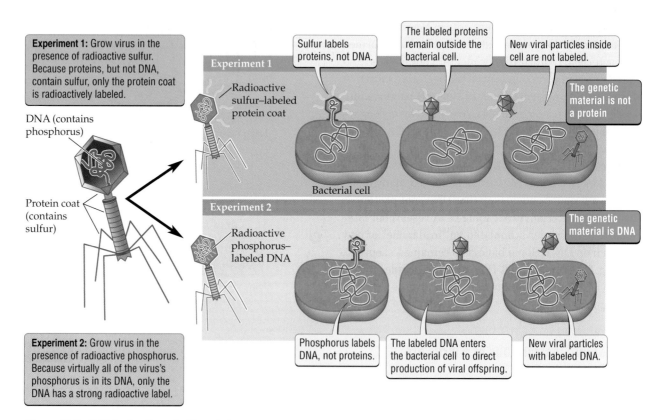

Experiment 1: Grow virus in the presence of radioactive sulfur. Because proteins, but not DNA, contain sulfur, only the protein coat is radioactively labeled.

DNA (contains phosphorus)

Protein coat (contains sulfur)

Radioactive sulfur–labeled protein coat

Sulfur labels proteins, not DNA.

The labeled proteins remain outside the bacterial cell.

New viral particles inside cell are not labeled.

The genetic material is not a protein

Bacterial cell

Experiment 1

Experiment 2

Radioactive phosphorus–labeled DNA

The genetic material is DNA

Phosphorus labels DNA, not proteins.

The labeled DNA enters the bacterial cell to direct production of viral offspring.

New viral particles with labeled DNA.

Experiment 2: Grow virus in the presence of radioactive phosphorus. Because virtually all of the virus's phosphorus is in its DNA, only the DNA has a strong radioactive label.

Figure 12.2 DNA Is the Genetic Material
The Hershey-Chase experiments on viruses that infect bacteria used a radioactive labeling technique. Hershey and Chase grew viruses in two different radioactive solutions that labeled either the viruses' DNA or their proteins. They then exposed bacteria to these viruses. They knew beforehand that the virus injects its genetic material into a bacterial cell, where it directs the production of new viruses. When Hershey and Chase found labeled DNA—but not labeled protein—inside bacterial cells and inside new viruses, they concluded that DNA was the genetic material.

See the Hershey-Chase experiment in motion. 12.2

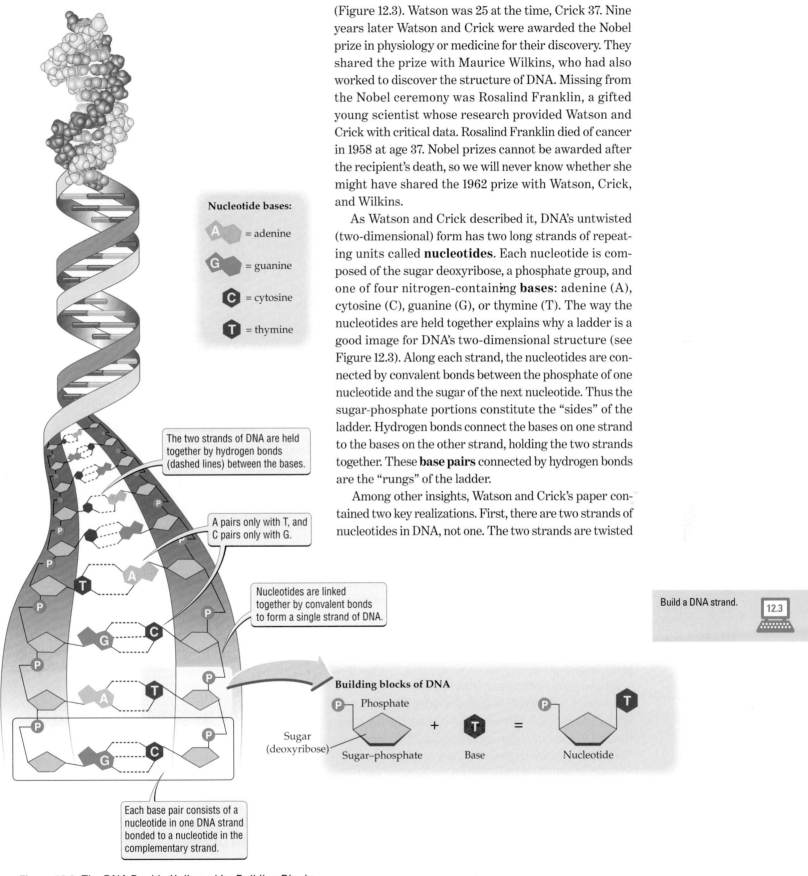

Nucleotide bases:

A = adenine

G = guanine

C = cytosine

T = thymine

The two strands of DNA are held together by hydrogen bonds (dashed lines) between the bases.

A pairs only with T, and C pairs only with G.

Nucleotides are linked together by convalent bonds to form a single strand of DNA.

Each base pair consists of a nucleotide in one DNA strand bonded to a nucleotide in the complementary strand.

Building blocks of DNA

P — Phosphate

Sugar (deoxyribose)

Sugar–phosphate

+ Base

= Nucleotide

Figure 12.3 The DNA Double Helix and Its Building Blocks

(Figure 12.3). Watson was 25 at the time, Crick 37. Nine years later Watson and Crick were awarded the Nobel prize in physiology or medicine for their discovery. They shared the prize with Maurice Wilkins, who had also worked to discover the structure of DNA. Missing from the Nobel ceremony was Rosalind Franklin, a gifted young scientist whose research provided Watson and Crick with critical data. Rosalind Franklin died of cancer in 1958 at age 37. Nobel prizes cannot be awarded after the recipient's death, so we will never know whether she might have shared the 1962 prize with Watson, Crick, and Wilkins.

As Watson and Crick described it, DNA's untwisted (two-dimensional) form has two long strands of repeating units called **nucleotides**. Each nucleotide is composed of the sugar deoxyribose, a phosphate group, and one of four nitrogen-containing **bases**: adenine (A), cytosine (C), guanine (G), or thymine (T). The way the nucleotides are held together explains why a ladder is a good image for DNA's two-dimensional structure (see Figure 12.3). Along each strand, the nucleotides are connected by convalent bonds between the phosphate of one nucleotide and the sugar of the next nucleotide. Thus the sugar-phosphate portions constitute the "sides" of the ladder. Hydrogen bonds connect the bases on one strand to the bases on the other strand, holding the two strands together. These **base pairs** connected by hydrogen bonds are the "rungs" of the ladder.

Among other insights, Watson and Crick's paper contained two key realizations. First, there are two strands of nucleotides in DNA, not one. The two strands are twisted

Build a DNA strand.

12.3

Helpful to know

The words *copy, replicate,* and *duplicate* have essentially the same meaning, but there is an important difference: only DNA gets *replicated* because the molecule serves as its own template in the copying process. In contrast, a sentence you copy from a book onto a piece of paper is not being replicated because no template is involved; similarly, molecules other than DNA may get *copied* or *duplicated,* but not *replicated.*

around each other, which is why DNA is called a double helix. Second, only certain combinations of bases are possible. Watson and Crick proposed a set of base-pairing rules, stating that adenine on one strand could pair only with thymine on the other strand; similarly, cytosine on one strand could pair only with guanine on the other strand. These base-pairing rules had an important consequence: when the sequence of bases on one strand of the DNA molecule was known, the sequence of bases on the other, **complementary strand** of the molecule was automatically known. For example, if one strand consisted of the sequence

ACCTAGGG,

then the complementary strand had to have the sequence

TGGATCCC.

Any other sequence would violate the base-pairing rules.

We now know that the physical structure of DNA proposed by Watson and Crick is correct in all its essential elements. This structure has great explanatory power. For example, as we shall see in the following section, the fact that adenine can pair only with thymine and that cytosine can pair only with guanine suggested a simple way in which the DNA molecule could be copied: the original strands could serve as templates on which new strands could be built. This suggestion turned out to be correct.

Knowledge of the three-dimensional structure of DNA also suggested that the information stored in DNA could be represented as a long string of the four bases: A, C, G, and T. Although A has to pair with T and C has to pair with G, the four bases can be arranged in any order along a strand of DNA. The fact that each strand of DNA is composed of millions of these bases suggested that a tremendous amount of information could be contained in the order of the bases along the DNA molecule, or **DNA sequence**; this suggestion has also proved to be correct (see Chapter 13).

The sequence of bases in DNA differs among species and among individuals within a species (Figure 12.4). We now know that different alleles of a gene have different DNA sequences, and hence that differences in DNA sequence are the basis of inherited variation. For example, people with a genetic disorder such as Huntington disease or cystic fibrosis inherit particular alleles that cause the disorder, as we saw in Chapter 11. At the molecular level, one allele causes a disease and another allele does not because the two alleles have a different sequence of bases. Sometimes a difference of only one base pair in hundreds or thousands within the

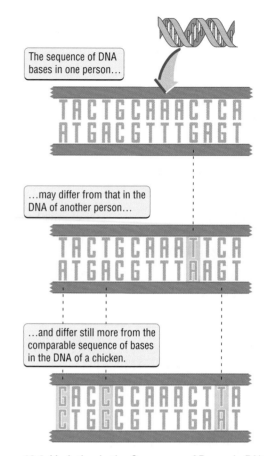

The sequence of DNA bases in one person…

TACTGCAAACTCA
ATGACGTTTGAGT

…may differ from that in the DNA of another person…

TACTGCAAATTCA
ATGACGTTTAAGT

…and differ still more from the comparable sequence of bases in the DNA of a chicken.

GACCGCAAACTTA
CTGGCGTTTGAAT

Figure 12.4 Variation in the Sequence of Bases in DNA
The sequence of bases in DNA differs among species and among individuals within a species. Here, the sequence of bases for a hypothetical region of DNA in two humans and a chicken is compared. Base pairs highlighted in yellow are different from the corresponding base pairs in the first human.

gene can make the difference between life and death. The severe consequences of some alleles that cause genetic disorders can lead people who are at risk of inheriting them to seek the guidance of a genetic counselor (see the box on page 234).

12.3 How DNA Is Replicated

As Watson and Crick noted in their historic 1953 paper, the structure of the DNA molecule suggested a simple way that the genetic material could be copied. They elaborated on this suggestion in a second paper, also published in 1953. Because A pairs only with T and C pairs only with G, each strand of DNA contains the information needed to duplicate the complementary strand. For this reason, Watson and Crick suggested that **DNA replication**—the

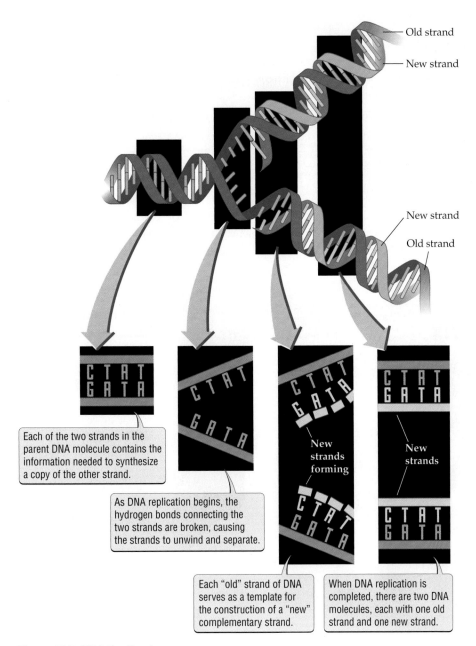

Figure 12.5 **DNA Replication**
In this overview of DNA replication, the parent DNA strands are dark blue, and the newly synthesized strands are light blue.

Image labels and callouts:

Old strand
New strand
New strand
Old strand

Each of the two strands in the parent DNA molecule contains the information needed to synthesize a copy of the other strand.

As DNA replication begins, the hydrogen bonds connecting the two strands are broken, causing the strands to unwind and separate.

New strands forming

New strands

Each "old" strand of DNA serves as a template for the construction of a "new" complementary strand.

When DNA replication is completed, there are two DNA molecules, each with one old strand and one new strand.

---

duplication of a DNA molecule—might work in the following way (Figure 12.5):

1. The hydrogen bonds connecting the two strands of the DNA molecule are broken.
2. Breaking of the hydrogen bonds causes the two strands to unwind and separate.
3. Each strand is then used as a template for the construction of a new strand of DNA.

4. When this process is completed, there are two identical copies of the original DNA molecule, each with the same sequence of bases. Each copy is composed of one "old" strand of DNA (from the original DNA molecule) and one newly synthesized strand of DNA.

Five years later, other researchers confirmed that DNA replication produces DNA molecules composed of one old strand and one new strand, as predicted by

Explore how DNA replicates.

12.4

Biology on the Job

Helping People Cope with Genetic Disorders

Each year, thousands of patients seek the guidance of genetic counselors like Robin L. Bennett, who is a senior genetic counselor and the associate director of the Medical Genetics Clinics at the University of Washington in Seattle.

What do you do during a typical day's work? I have an exciting job that makes a difference in people's lives every day. I work mostly in adult genetics—working with individuals who are affected with inherited conditions or who have concerns that they or their children may be at risk to inherit a condition (such as muscular dystrophy or cystic fibrosis). I usually see patients 2–3 days a week. The rest of the week I follow up with patients regarding their test results, how they are feeling after their clinic visit. I also teach a course for the School of Social Work, and I give lectures to the public, including students in all grade levels and many different types of health professionals.

How did you become interested in genetic counseling? Was there a key moment when you suddenly knew what you wanted to do? In the tenth grade in my introductory biology course, I learned about the new field of genetic counseling. I was interested in working in an area like this because my mother's best friend had a son with profound mental and physical handicaps. He finally had to be placed outside of the home because it was too hard for his family to take care of him. I saw how this devastated the family, and I wanted to make a difference for families like this. I can't imagine being anything but a genetic counselor.

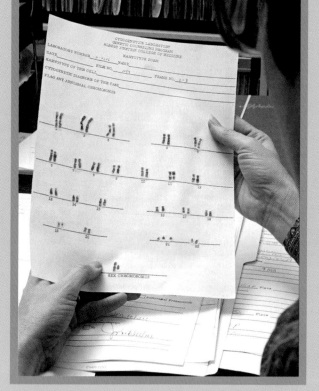

A genetic counselor reviewing the full set of chromosomes (the karotype) of a patient with Down syndrome

How does the type of genetic counseling that you do compare to the range of jobs that are open to a person trained as a genetic counselor? Genetic counselors have a wide range of career options. The majority work with either children or pregnant women with genetic conditions, or with adults and their families. There are more and more genetic counselors who are experts in a specific area such as cancer genetics, psychiatric genetics, or cardiac genetics. Most genetic counselors are involved in education at some level—they give talks or help develop educational materials (pamphlets, CDs, online programs). Many genetic counselors also do research related to genetic disorders and genetic counseling. Finally, more genetic counselors are becoming involved in setting policies related to genetic diseases, such as initiating newborn screening programs.

Is the certification program for genetic counseling difficult? All genetic counselors have a master's degree, which takes 18–24 months to complete. Genetic counselors take a certification exam from the American Board of Medical Genetics if they want to be clinical genetic counselors (most genetic counselors are certified). Most people who want to be genetic counselors don't think the exam is hard because it covers what they love to do.

What are some of the most exciting new discoveries or advances in your field? Individuals can now be tested for a predisposition to a genetic disease. This creates a new set of "healthy" people being evaluated for the potential for disease. There are many psychological factors to consider before pursuing this testing. Often there is no treatment for the disease, but there is a test to see if a person has a high chance of developing the condition—although the test won't predict exactly when the condition will occur, or even how bad the disease will be. Also, in the last few years, there are more treatments for genetic disorders (particularly enzyme replacement therapy). There are still no cures for genetic diseases (such as gene therapy).

You just mentioned that genetic tests raise psychological issues for patients. How often do genetic counselors face ethical and psychological issues? Genetic counselors face ethical issues almost on a daily basis. That is why most people are drawn to the field because it is never boring. Psychological issues are faced by each of our clients, that is why genetic counseling is so important before a person has a genetic test—not just after. In most cases, genetic counseling can help individuals make informed life choices. A person who has a fetus with multiple birth defects identified by ultrasound may choose to deliver the baby at a hospital that has a pediatric intensive care unit or doctors that specialize in the problems (for example, cardiac surgery). A person with a family history of colon cancer may be screened at age 25 instead of the usual age of 50, based on high risk for colon cancer identified by family history or a genetic test. A couple who are first cousins may have been afraid to have a child because of fears of having a child with multiple problems, but then be reassured that the risks of problems are lower than commonly believed.

What do you enjoy most about your job? Being a genetic counselor helps me to put my own life into perspective. I try to live life to the fullest because I can appreciate my blessings. Even when I have to give bad news, I feel like I'm giving people information in a supportive way that can bring meaning to their lives, even options and hope. I love the challenge of being a genetic counselor—every day is different—never boring.

Is there anything else you'd like to tell our readers? There are wonderful career resources through the National Society of Genetic Counselors Web site (www.nsgc.org), including spotlights on various genetic counselors and the NSGC Professional Status Survey. Also, the American Board of Genetic Counselors has information about genetic counseling programs (www.abgc.net).

The Job in Context

As we've seen in this and the preceding chapters, there have been rapid advances in our ability to test whether an adult, fetus, or embryo has specific alleles that influence human genetic disorders. Information from such tests can benefit patients greatly, relieving them of worry (if the test reveals they or their children don't have the disorder), or enabling them to begin treatment early or take other therapeutic measures (if the test reveals that they or their children have the disorder). But sometimes the results of these tests are hard to interpret. A genetic test may indicate that a person has a tendency to get a genetic disorder—not a clear "yes" they will get it, or "no" they will not. This uncertainty stems from a fundamental aspect of how genes work: often, the effect of a gene depends not only on its own function, but also on the function of other genes and on environmental factors. Although a person may have an allele that increases the chance that he or she will get a genetic disorder, he or she may also have other (often unknown) alleles that delay the onset of the disorder or reduce its severity. Even when the results of a genetic test are clear-cut, a person may feel overwhelmed by the news he or she receives. Genetic counselors are at the forefront of efforts to help patients understand and cope with the genetic conditions they face.

Watson and Crick. The main enzyme involved in the replication of DNA has now been identified and is called **DNA polymerase** [puh-*LIM*-er-ace].

The Watson-Crick model of DNA replication is elegant and simple, but the mechanics of actually copying DNA are far from simple. More than a dozen enzymes and proteins are needed to unwind the DNA, to stabilize the separated strands, to start the replication process, to attach nucleotides to the correct positions on the template strand, to "proofread" the results, and to join partly replicated fragments of DNA to one another.

Although DNA replication is such a complex task, cells can copy DNA molecules containing billions of nucleotides in a matter of hours—about 8 hours in people (over 100,000 nucleotides per second). This speed is achieved in part by starting the replication of the DNA molecule at thousands of different places at once. Despite their speed, cells make remarkably few mistakes when they copy their DNA.

12.4 Repairing Replication Errors and Damaged DNA

When DNA is copied, there are many opportunities for mistakes to be made. In humans, for example, more than 3 billion base pairs must be copied each time a cell divides, so there are over 6 billion opportunities for mistakes. In addition, the DNA in cells is constantly being damaged by various sources. Replication errors and damage to the DNA—especially to essential genes—disrupt normal cell functions. If not repaired, this damage would lead to the death of many cells and, ultimately, to the death of the organism.

Few mistakes are made in DNA replication

The enzymes that copy DNA sometimes insert an incorrect base in the newly synthesized strand. For example, if DNA polymerase were to insert a cytosine (C) across from an adenine (A) located on the template strand, an incorrect C–A pair bond would form instead of the correct T–A pair bond (Figure 12.6). Such mistakes are made about once in every 10,000 bases. However, nearly all of these mistakes are corrected immediately by enzymes (including DNA polymerase itself) that check, or "proofread," the pair bonds as they form. This form of error correction is similar to what happens as you type a paper, realize you made a mistake, and correct it right away with the "delete" key.

When an incorrect base is added but escapes the mechanism for immediate proofreading, a **mismatch error** has occurred. Mismatch errors occur about once in every 10 million bases. Cells contain repair proteins that specialize in fixing mismatch errors; these proteins play a role similar to the error checking you perform after you complete the first draft of a paper, print it, and carefully review it for mistakes. Proteins that fix mismatch errors correct 99 percent of those errors, reducing the overall chance of an error to the incredibly low rate of one mistake in every billion bases.

On the rare occasions when a mismatch error is not corrected, the DNA sequence is changed, and the new sequence is reproduced the next time the DNA is replicated. A change to the sequence of bases in an organism's DNA is called a **mutation**. Thus, when a mistake in the copying process is not corrected, a mutation has occurred. Mutations can also occur when cells are exposed to **mutagens**, substances or energy sources that alter DNA.

Mutations can result in the formation of new alleles. Some of the new alleles that result from mutation are beneficial, but most are either neutral or harmful. Among the harmful alleles are those that cause cancer and other human genetic disorders, such as sickle-cell anemia and Huntington disease. Note that our definition of mutation includes not only changes to the DNA sequence of a gene, but also changes to the number or structure of chromosomes (since such changes to chromosomes add bases to, delete bases from, or rearrange bases in the initial DNA sequence).

Normal gene function depends on DNA repair

Every day, the DNA in each of our cells is damaged thousands of times by chemical, physical, and biological agents. These agents include energy from radiation or heat, collisions with other molecules in the cell, attacks by viruses, and random chemical accidents (some of which are caused by environmental pollutants, but most of which result from normal metabolic processes). Our cells contain a complex set of repair proteins that fix the vast majority of this damage. Single-celled organisms such as yeasts have more than 50 different repair proteins, and humans probably have even more.

Although humans are very good at repairing damaged DNA, some organisms far exceed our abilities. A person exposed to 1,000 rads of radiation energy dies in a few weeks, in part because his or her DNA is damaged beyond repair. (We know this because this dose was received by some of the people who initially survived the atomic blasts at Hiroshima and Nagasaki.) Although 1,000 rads kills a person, such a dose would barely faze the bacterium *Deinococcus radiodurans* [DYE-no-KOK-us

Explore how DNA is repaired.

12.5

Figure 12.6 Mistakes Can Be Made in DNA Replication
Here, a cytosine (C) has been incorrectly inserted opposite an adenine (A). DNA repair enzymes almost always fix such mismatch errors before the cell's DNA is replicated again.

Biology Matters

Cancer in Children

Both children and adults are routinely exposed to mutagens that can cause cancer, including those outlined in the following table:

Some mutagens that damage DNA	Source of exposure
Arsenic	• Arsenic pesticides in wooden playsets and decks • Drinking water (contaminants from mining and power plants)
Mutagen *X* and other by-products of water chlorination	• Drinking water
Formaldehyde	• Indoor air (offgases from building materials) • Paper, dyes, paper coatings
Benzene	• Gasoline fumes, glue, paint, furniture wax, detergent
PAHs (*polycyclic aromatic hydrocarbons*, a group of chemicals released by burning fossil fuels)	• Food and drinking water (contaminants from gasoline and coal-fired power plants)

Children, however, are more vulnerable to carcinogenic mutagens than adults, in part because they are less able to detoxify chemicals, and their developing organ systems are more vulnerable to damage.

In addition, pound-for-pound, exposures are much higher in children, as the table below illustrates:

Children's average exposure relative to adults	
Air inhalation	3 to 1
Body surface area	2.25 to 1
Soil/dust consumption	3 to 1
Drinking water consumption	2.2 to 1
Dietary fat intake	3.4 to 1

As a result, slightly more than half of a person's cancer risk from many carcinogens accumulates by an early age.

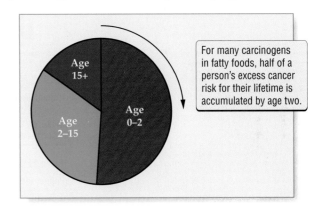

For many carcinogens in fatty foods, half of a person's excess cancer risk for their lifetime is accumulated by age two.

SOURCE: "EPA cancer policy revisions highlight risks to children," *Children's Health Policy Review*, March 3, 2003.

---

ray-di-o-*DER*-unz]. This species is so efficient at repairing damage to DNA that a dose of 1,000,000 rads merely slows its growth, but does not kill it. Even when the dose is raised to 3,000,000 rads—3,000 times greater than a lethal dose for a person—a small percentage of the bacteria survive.

In humans, *Deinococcus*, and all other organisms, there are three steps in **DNA repair**: the damaged DNA must be recognized, removed, and replaced (Figure 12.7). Different sets of repair proteins specialize in recognizing and removing different types of DNA damage. Once these first two steps have been accomplished, the final

Deinococcus radiodurans
(bacterium)

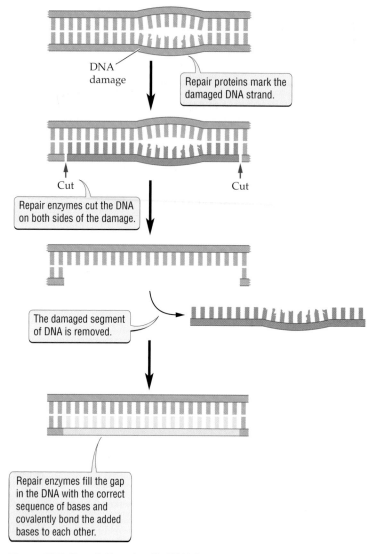

DNA
damage

Repair proteins mark the damaged DNA strand.

Cut   Cut

Repair enzymes cut the DNA on both sides of the damage.

The damaged segment of DNA is removed.

Repair enzymes fill the gap in the DNA with the correct sequence of bases and covalently bond the added bases to each other.

Figure 12.7 Repair Proteins Fix DNA Damage
Complex sets of DNA repair proteins work together to fix many types of damage to DNA. Some of these proteins bind to damaged DNA, thereby providing a molecular "tag" that indicates where the damage is located. Other proteins (enzymes) cut out the damaged DNA and replace it with newly synthesized DNA.

step is to add the correct sequence of bases to the damaged strand, replacing the nucleotides that were removed when the damaged section was cut out. This third step of the repair process is the same for most types of DNA repair (and, indeed, is similar to the correction of mismatch errors). When damage to DNA is not repaired, a mutation has occurred.

The importance of DNA repair mechanisms is highlighted by what happens when they fail to work. The child in Figure 12.8 has xeroderma pigmentosum (XP), a recessive genetic disorder in which even brief expo-

sure to sunlight causes painful blisters. The allele (a) that causes XP produces a nonfunctional version of one of the many human DNA repair proteins. The job of the normal form of this protein is to repair the kind of damage to DNA caused by ultraviolet (UV) light. The lack of this DNA repair protein makes individuals with XP highly susceptible to skin cancer. Several inherited tendencies to develop cancer, including some types of breast and colon cancer, also appear to be caused by defective versions of genes that control other kinds of DNA repair.

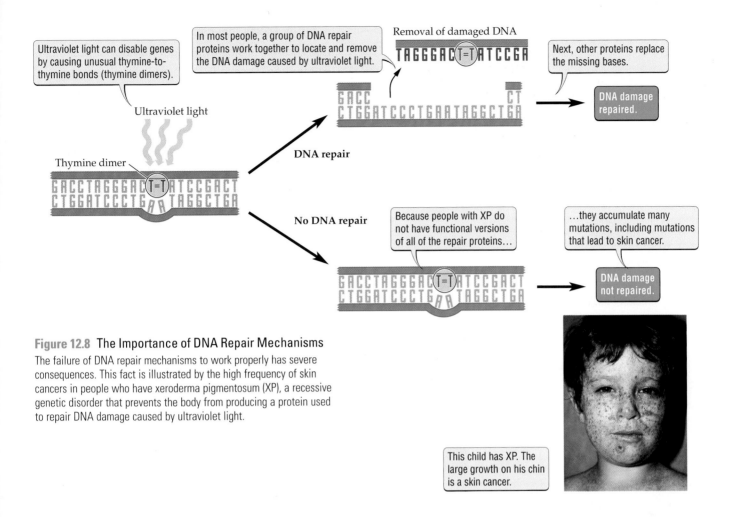

Ultraviolet light can disable genes by causing unusual thymine-to-thymine bonds (thymine dimers).

In most people, a group of DNA repair proteins work together to locate and remove the DNA damage caused by ultraviolet light.

Removal of damaged DNA

Next, other proteins replace the missing bases.

DNA damage repaired.

Ultraviolet light

Thymine dimer

DNA repair

No DNA repair

Because people with XP do not have functional versions of all of the repair proteins…

…they accumulate many mutations, including mutations that lead to skin cancer.

DNA damage not repaired.

Figure 12.8 The Importance of DNA Repair Mechanisms
The failure of DNA repair mechanisms to work properly has severe consequences. This fact is illustrated by the high frequency of skin cancers in people who have xeroderma pigmentosum (XP), a recessive genetic disorder that prevents the body from producing a protein used to repair DNA damage caused by ultraviolet light.

This child has XP. The large growth on his chin is a skin cancer.

Errors in the Library of Life

Usually, the DNA sequences of your genes help shape your phenotype, but do not cause problems. But all is not perfect in the library of life: errors in DNA replication can produce new, mutant alleles that cause genetic disorders, some of which are lethal. In Chapter 11 we saw that serious genetic disorders, including sickle-cell anemia, cystic fibrosis, and Huntington disease, can result from mutations in a single gene. In these and literally thousands of other cases, genetic disorders are caused by alleles that have a DNA sequence that differs from the sequence found in a normal allele.

An allele that causes a genetic disorder can differ by as little as one base from the normal allele, as the sickle-cell anemia allele does. In other genetic disor-

ders, the allele that causes the disorder differs by several bases from the sequence found in the normal allele. Cystic fibrosis, for example, is a fatal disorder caused by a recessive allele located on chromosome 7 (see Figures 11.8 and 11.9). In the critical portion of the normal allele, the sequence of bases is TAGTAGAAA, whereas in the cystic fibrosis allele, the sequence is TAGTAA. Thus cystic fibrosis is caused by an allele that is missing three bases (the sequence GAA) found in the normal allele (Figure 12.9a). The cystic fibrosis allele alters a protein that regulates the movement of salt (sodium chloride) into and out of cells; this change causes cells in the lungs, digestive tract, and other parts of the body to be covered with thick, sticky secretions.

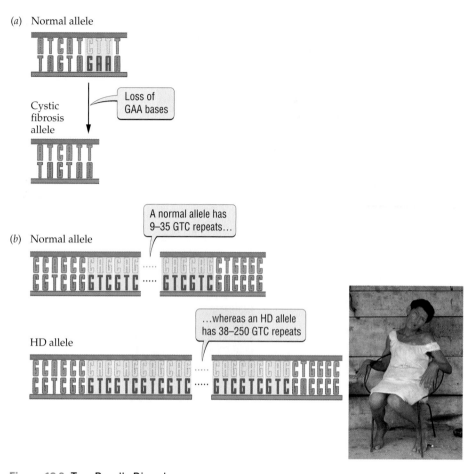

(a) Normal allele

Loss of
GAA bases

Cystic
fibrosis
allele

(b) Normal allele

A normal allele has
9–35 GTC repeats...

...whereas an HD allele
has 38–250 GTC repeats

HD allele

Figure 12.9 Two Deadly Disorders
(a) Cystic fibrosis is a deadly genetic disorder caused by a mutation in which three bases (the sequence GAA) are missing. (b) Huntington disease (HD) is a lethal neurological disorder caused by a family of mutant alleles in which 3 to 215 extra copies of the sequence GTC are inserted near the beginning of the *HD* gene.

Other alleles that cause genetic disorders differ from normal alleles by a larger number of bases. Huntington disease (HD) occurs when a person inherits a dominant allele that codes for an abnormal version of the protein usually produced by the *HD* gene. It turns out that many different alleles can cause Huntington disease, all sharing a common feature: each has extra copies of the sequence GTC inserted near the beginning of the *HD* gene (Figure 12.9*b*). Thus, instead of a single *HD* allele, there is a family of *HD* alleles that can cause the disorder.

Many other genetic disorders can be caused by more than one allele. In such cases, the particular allele a person inherits can be important because it can affect the severity of the disorder. For example, there are over 800 known alleles of the gene *BRCA1* that can cause inherited forms of breast cancer. If a woman with one of these alleles develops breast cancer, the severity of the disease—and hence the type of treatment a doctor might recommend—is influenced by which of the 800 *BRCA1* breast cancer alleles she possesses. Similarly, a person who inherits an *HD* allele with many extra copies of the sequence GTC is likely to get the disease at an earlier age than is someone whose *HD* allele has relatively few extra copies of the sequence GTC.

Chapter Review

Summary

12.1   The Search for the Genetic Material

- Geneticists initially thought that protein was the genetic material. Three landmark experiments showed that this initial view was wrong and that DNA, not protein, is the genetic material.
- The first experiment showed that harmless strain R bacteria could be transformed into deadly strain S bacteria when exposed to heat-killed strain S bacteria.
- The second experiment showed that only the DNA from heat-killed strain S bacteria was able to transform strain R bacteria into strain S bacteria.

- The third experiment showed that the DNA of a virus, not its proteins, was responsible for taking over a bacterial cell and producing the next generation of viruses.

12.2   The Three-Dimensional Structure of DNA

- In 1953, James Watson and Francis Crick determined that DNA is a double helix.
- The double helix is formed by two long strands of covalently bonded nucleotides. The two strands of nucleotides are held together by hydrogen bonds between the nucleotides' nitrogen-containing bases: adenine, cytosine, guanine, and thymine.

- Pairs of hydrogen-bonded nucleotides are called base pairs and follow specific base-pairing rules: An adenine on one strand pairs only with a thymine on the other strand. Similarly, a cytosine on one strand pairs only with a guanine on the other strand.
- These base-pairing rules allow each strand to serve as a template from which the other (complementary) strand can be duplicated.
- The sequence of bases in DNA, which differs among species and among individuals within a species, is the basis of inherited variation.

12.3 How DNA Is Replicated

- Because of the base-pairing rules (A pairs only with T, and C pairs only with G), each strand of DNA contains the information needed to duplicate the complementary strand.
- A complex set of enzymes and other proteins guides the replication of DNA; the primary enzyme involved is DNA polymerase. To replicate DNA, these enzymes must first break the hydrogen bonds connecting the two nucleotide strands.
- Breaking of the hydrogen bonds causes the two strands to unwind and separate. Each of these strands is then used as a template from which to build a new strand of DNA.
- DNA replication produces two copies of the DNA molecule, each composed of one old strand (from the parent DNA molecule) and one newly synthesized strand.

12.4 Repairing Replication Errors and Damaged DNA

- On rare occasions, mistakes occur during DNA replication. Most mistakes in the copying process are corrected, either immediately by "proofreading" or by later correction of mismatch errors.
- Mistakes in DNA replication that are not corrected are one source of mutations, which are changes in the DNA sequence.
- The DNA in each of our cells is altered thousands of times every day by mechanical, chemical, and radiation damage. If none of the DNA damage in an organism's cells were repaired, the cells, and ultimately the organism, would die.
- Replication errors and damage to DNA are fixed by a complex set of DNA repair proteins.
- Several inherited genetic disorders result from the failure of DNA repair proteins to work properly.

⊙ Review and Application of Key Concepts

1. Summarize the key findings of three experiments that helped to convince geneticists that genes are composed of DNA.

2. Draw a diagram that shows the three main components of a nucleotide from a DNA molecule. There are four types of nucleotides; what part of their structure makes each type different from the others? Finally, when covalent bonds link nucleotides to one another, what portion of the DNA molecule is produced?

3. Imagine that you could flatten a DNA molecule into a two-dimensional structure. What holds the two strands of the DNA molecule together?

4. A gene has two codominant alleles, A^1 and A^2. Each allele produces a unique version of a protein found on the surface of a type of white blood cell. In physical terms, each allele is a segment of DNA. What part of the structure of these DNA segments contains the genetic information of the alleles? In general terms, how does the DNA segment of one allele differ from the DNA segment of the other allele? Explain your answer.

5. Explain why the structure of DNA proposed by Watson and Crick suggested a way DNA could be replicated.

6. Explain how the following are related:
   a. the sequence of bases in DNA
   b. mutations
   c. alleles that cause human genetic disorders

7. Summarize how DNA repair works and why it is essential for cells to function normally.

Key Terms

base (p. 231)
base pair (p. 231)
complementary strand (p. 232)
DNA polymerase (p. 235)
DNA repair (p. 237)
DNA replication (p. 232)
DNA sequence (p. 232)

double helix (p. 230)
mismatch error (p. 236)
mutagen (p. 236)
mutation (p. 236)
nucleotide (p. 231)
transformation (p. 229)

Self-Quiz

1. The base-pairing rules for DNA state that
   a. any combination of bases is allowed.
   b. T pairs with C, A pairs with G.
   c. A pairs with T, C pairs with G.
   d. C pairs with A, T pairs with G.

2. DNA replication results in
   a. two DNA molecules, one with two old strands and one with two new strands.
   b. two DNA molecules, each of which has two new strands.
   c. two DNA molecules, each of which has one old strand and one new strand.
   d. none of the above

3. Experiments performed by Oswald Avery and colleagues showed that
   a. protein, not DNA, transformed bacteria.
   b. DNA, not protein, transformed bacteria.

c. carbohydrates, not protein, transformed bacteria.

d. either DNA by itself or protein by itself transformed bacteria.

4. The DNA of cells is damaged
   a. thousands of times per day.
   b. by collisions with other molecules, chemical accidents, and radiation.
   c. not very often and only by radiation.
   d. both a and b

5. The DNA of different species differs in
   a. the sequence of bases.
   b. the base-pairing rules.
   c. the number of nucleotide strands.
   d. the location of the sugar-phosphate portion of the DNA molecule.

6. If a strand of DNA has the sequence CGGTATATCC, then the complementary strand of DNA has the sequence
   a. ATTCGCGCAA.
   b. GCCCGCGCTT.
   c. GCCATATAGG.
   d. TAACGCGCTT.

7. Hershey and Chase conducted experiments with viruses that attack bacteria. They found that
   a. sulfur labeled DNA.
   b. phosphorus labeled protein.
   c. radioactively labeled protein entered bacterial cells.
   d. radioactively labeled DNA entered bacterial cells.

8. Mutation
   a. can produce new alleles.
   b. can be harmful, beneficial, or neutral.
   c. is a change in an organism's DNA sequence.
   d. all of the above

Biology in the News

Aral Catastrophe Recorded in DNA

By David Shukman

Fresh fears have been raised about the health of populations living near the shrinking Aral Sea in central Asia. A new study has now found high levels of DNA damage that could explain the region's abnormally high cancer rates.

This comes as the latest estimates say the Aral Sea is receding so rapidly it could vanish within the next 15 years.

In some places, people face two problems: they live near factories, farms, or other sources of pollution, and their community members seem to get sick from cancer or other diseases at an unusually high rate. Often they suspect that their families and friends get sick because of the pollutants in their area. For people living near the Aral Sea, such suspicions have just received some support from the results of a new scientific study.

The Aral Sea region is an ecological disaster zone. As described in the BBC article, the Aral Sea was once the world's fourth largest inland body of water. Now, however, starved of water by a poorly managed irrigation system that supplies water-intensive cotton crops, the sea is shrinking rapidly.

As its shoreline recedes, the sea leaves behind a dusty wasteland. The soil (once the bottom of the Aral Sea) is filled with herbicides, pesticides, and other toxins that flowed into the sea during decades of environmental abuse. Winds blow the soil into the air, exposing people to toxins with every breath they take.

The toxins appear to have caused widespread genetic damage. Dr. Spencer Wells, of the National Geographic Society and formerly of Oxford University's Welcome Trust Centre for Human Genetics, found that people living near the Aral Sea have 3.5 times more damage to their DNA than do people living in the United States. And among farmworkers, those closest to the source of agricultural chemicals, the damage rates skyrocket to 5 times the usual amount.

These unusually high levels of DNA damage seem to be taking a severe toll: cancer rates are also abnormally high. What's more, according to Dr. Wells, cancer rates may remain high for a long time. He reasons that the DNA of cells in the esophagus [eh-*SOFF*-uh-guss] (the tube that connects the mouth to the stomach, a common site of cancers in this population) and other parts of the body was damaged by toxic pollutants, and that this DNA damage caused the observed high rates of cancer. Dr. Wells and others fear that the pollutants may also have damaged the DNA in sperm and egg cells, in which case the genetic damage will be passed on to future generations. As Dr. Wells told the BBC, this would mean "not only that people are more likely to get cancer but also that their children and grandchildren are too."

Evaluating the news

1. Although the main result of this study (high levels of DNA damage) is striking, it does not provide direct proof (a) that people living in the area of the Aral Sea have high concentrations of toxins in their bodies, (b) that high concentrations of toxins in their bodies caused the DNA damage, or (c) that the DNA damage caused cancer. What is true for this study is true in general: it can be very hard to prove that a local source of pollution caused people in an area to get sick. In the face of such difficulties, what should we do? Should we demand more conclusive proof before taking action to change environmental policies, even though the delay may doom many to an early death or a debilitating disease? Or should we spend money and time protecting people from sources of pollution that may not actually be the cause of the observed high rates of illness?

2. The study described in this article was conducted in Uzbekistan. The cotton crops that divert water from the Aral Sea, causing the sea to shrink in size, are Uzbekistan's biggest export earner. One way to improve the health of people living in the Aral Sea region might be to allow more water to flow into the Aral Sea, thereby helping to restore the sea and reduce the amount of toxic chemicals in the soil and air. However, as one Uzbekistan official pointed out, such an action might leave people poor and hungry. Is there any way to protect jobs while simultaneously reducing threats from pollution?

3. Water, like the rivers that used to flow into the Aral Sea, is an example of a public resource whose use frequently leads to conflict. A business or farm may take water from a river, thereby depriving people and wildlife that live downstream of the use of that water. Similar conflicts occur where two nations share the use of bodies of water. How should such conflicts be resolved or prevented? Should people upstream be able to do whatever they want, or must they assume responsibility for how their actions affect those who live downstream? Do you think that the interests of species other than people should be taken into account, or should decisions regarding the use of resources such as water be based solely on human interests?

SOURCE: *BBC News*, Tuesday, July 29, 2004.

CHAPTER 13

From Gene to Protein

Key Concepts

◉ Most genes contain instructions for building proteins. The DNA sequence of a gene encodes the amino acid sequence of its protein product.

◉ A few genes encode ribonucleic acid (RNA) molecules as their final product.

◉ The flow of information from gene to protein requires two steps: transcription and translation.

◉ In eukaryotic cells, transcription occurs in the nucleus and produces a messenger RNA (mRNA) version of the information stored in the gene. The mRNA moves from the nucleus to the cytoplasm, where it is used to guide the construction of a protein.

◉ Translation occurs in the cytoplasm and converts the sequence of bases in an mRNA molecule to the sequence of amino acids in a protein.

◉ Gene mutations can alter the sequence of amino acids in a gene's protein product. Such changes, in turn, can alter the protein's function. Although changes in protein function are usually harmful, occasionally they benefit the organism.

Main Message: By controlling the production of proteins, genes play a key role in determining the organism's phenotype.

Finding the Messenger and Breaking the Code

We live in a global economy. The headquarters of a corporation may be located in one country—say, Germany—but the company's factories may be located elsewhere—say, the United States. Immediately a problem arises: decisions made in Germany need to be communicated to employees in the United States. This problem is easy to solve: A message must be sent from one location to the other. In addition, the message must be translated from German, the language in which the decision was made, to English, the language in which the decision must be implemented.

Eukaryotic cells face similar challenges. Genes work by controlling the production of proteins. Whereas genes are located in the nucleus of the cell, their protein products are made on ribosomes, which are located outside of the nucleus, in the cytoplasm. So a gene must control the construction of a protein from a distance. How is this accomplished? Like our imaginary corporate headquarters, the gene does this by sending a message. What is the chemical messenger that carries the gene's instructions from the nucleus to the ribosomes? And once the message reaches the ribosomes, how can the ribosomes "read" it?

This last question highlights another similarity between cells and our imaginary global corporation: to be effective, the information contained in genes must be translated from one "language" (that of DNA, which is based on a molecular alphabet of four nitrogen bases) to another (that of proteins, which is based on the 20 amino acids they contain). In the mid-1950s, biologists working on this problem realized that cells must have a

A Model of a tRNA Molecule
Transfer RNA molecules like this one read the genetic code.

"genetic code" that allows the information in the gene to be converted from the language of DNA to the language of proteins. The breaking of this genetic code was one of the crowning achievements of twentieth-century science.

Chapters 10 through 12 have described how genes are inherited, where they are located (on chromosomes), and what they are made of (DNA). But we have yet to describe how genes work. How do genes store the information needed to build their final products, proteins? How does the cell use that information?

Knowing how genes work can help us understand how mutations produce new phenotypes, including disease phenotypes. We begin this chapter by describing how genetic information is encoded in genes and how the cell uses that information to build proteins. We then describe how a change to a gene can change an organism's phenotype. At the end of the chapter, we apply what we've learned by focusing on the effects of alleles that cause two human genetic disorders, sickle-cell anemia and Huntington disease. Our discussion of how cells use the information stored in genes to build proteins focuses on eukaryotes, but except where noted, events are similar in prokaryotes.

13.1 How Genes Work

Learn about the flow of genetic information in a eukaryotic cell.

13.1

Proteins are essential to life. They are used by cells and organisms in many ways: some provide structural support, others transport materials through the body, still others defend against disease-causing organisms. In addition, the many chemical reactions on which life depends are controlled by a crucial group of proteins, the enzymes. Enzymes and other proteins influence so many features of the organism that they, along with the organism's internal and external environment, determine the organism's phenotype.

How do genes control the phenotype of an organism? Early clues came at the beginning of the twentieth century from the work of British physician Archibald Garrod, who studied several inherited human metabolic disorders. In 1902, he argued that these disorders were caused by an inability of the body to produce specific enzymes. Garrod was particularly interested in alkaptonuria, a condition in which the urine of otherwise healthy infants turns black when exposed to air. He proposed that infants with alkaptonuria have a defective version of an enzyme that in its normal form breaks down the substance that causes urine from these infants to turn black. But Garrod did not stop there: he and his collaborator, William Bateson, went on to suggest that in general, genes work by controlling the production of enzymes.

Genes contain information for the synthesis of RNA molecules

Garrod and Bateson were on the right track, but they were not entirely correct: genes control the production of all proteins, not just enzymes. Furthermore, although most genes contain instructions for building particular proteins, a few genes do not directly specify proteins. Rather, these genes contain instructions for building several kinds of ribonucleic acid (RNA) molecules that are used in the construction of proteins. Thus, directly and indirectly, genes control the production of proteins.

As we will see shortly, even genes that specify proteins make an RNA molecule as their initial product. Thus, modifying the definition in Chapter 10, we can redefine a **gene** as a DNA sequence that contains information for the synthesis of one of several types of RNA molecules used to make proteins.

RNA and DNA share a number of structural similarities as well as several important differences. Both are nucleic acids consisting of nucleotides covalently bonded to one another. But whereas DNA molecules are double-stranded, the various types of RNA molecules are all single-stranded; overall, the structure of an RNA molecule is similar to the structure of a *single* strand of DNA. The DNA molecule coils into a double helix, while RNA molecules assume various nonhelical shapes. As in DNA, each nucleotide in RNA is composed of a sugar, a phosphate group, and one of four nitrogen-containing bases (Figure 13.1). However, the nucleotides in RNA and DNA differ in two respects. First, RNA uses the sugar ribose, whereas DNA uses the sugar deoxyribose. Second, in RNA, the base uracil (U) replaces the base thymine (T), which is found only in DNA. The other three bases—adenine (A), cytosine (C), and guanine (G)—are the same in RNA and DNA.

Helpful to know

The names of proteins that are enzymes usually end with *ase*, while many names of proteins that are not enzymes end with *in*. For example, the protein RNA polymerase is an enzyme, but the protein huntingtin is not. (We'll encounter both of those molecules later in this chapter.)

Three types of RNA are involved in the production of proteins

The nucleic acids DNA and RNA play key roles in the construction of proteins. Several types of RNA, as well as many enzymes and other proteins, are required for the cell to make proteins. As already described, DNA controls the production of all these essential molecules, so DNA controls all aspects of protein production.

Cells use three main types of RNA molecules to construct proteins: **messenger RNA (mRNA)**, **ribosomal RNA (rRNA)**, and **transfer RNA (tRNA)**. The function of each of these three kinds of RNA is defined in Table 13.1 and discussed in more detail in the sections that follow. Cells also produce several other types of RNA that affect the production of proteins, but we will not discuss them in this chapter.

13.2 How Genes Control the Production of Proteins

In both prokaryotes and eukaryotes, the production of proteins happens in two steps: transcription and translation. Briefly, in **transcription**, an mRNA molecule is made using the information in the DNA sequence of a gene. The base sequence of that mRNA molecule specifies the amino acid sequence of a protein. In **translation**, the information in the mRNA molecule is used to synthesize the protein with the aid of several rRNA molecules, many tRNA molecules, and a number of proteins.

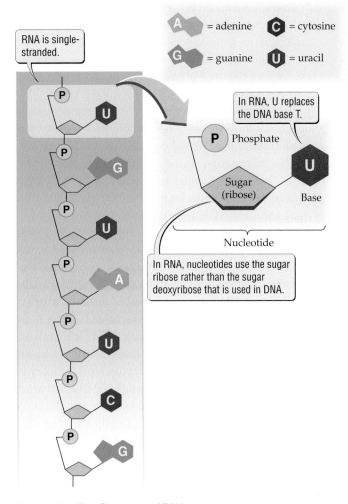

Figure 13.1 The Structure of RNA
To see the similarities and differences between DNA and RNA, compare this figure with Figure 12.3.

Table 13.1

RNA Molecules and Their Functions

Type of RNA	Function	Shape
Messenger RNA (mRNA)	Specifies the order of amino acids in a protein	
Ribosomal RNA (rRNA)	Major component of ribosomes, the molecular machines that make the covalent bonds that link amino acids together into a protein	
Transfer RNA (tRNA)	Transports the correct amino acid to the ribosome, based on the information encoded in the mRNA	

In the case of the few genes that encode rRNAs or tRNAs, transcription of the gene produces one of those molecules as its final product, and there is no second, translation step. For most of this chapter, however, we shall ignore this special case and focus on the transcription and translation of genes that encode proteins.

Before we discuss transcription and translation in detail, let's consider how genes work from the perspective of information flow. We will describe how eukaryotic cells use the information stored in genes to synthesize proteins. Events are similar in prokaryotes except that, because prokaryotes lack a nucleus, both genes and ribosomes are located in the cytoplasm.

For a protein to be made, the information in a gene must be sent from the gene, which is located in the nucleus, to the site of protein synthesis, a ribosome. This transfer of information requires an intermediary molecule because DNA does not leave the nucleus, whereas ribosomes are located in the cytoplasm (Figure 13.2). Messenger RNA is the intermediary molecule that transfers the information in the gene from the nucleus to the ribosomes. This transfer is made possible by transcription, in which the sequence of bases in mRNA is synthesized directly from the sequence of bases in one DNA strand of a gene. Because it is copied directly from the

Test your knowledge of protein synthesis. 13.2

gene's DNA sequence, mRNA provides the ribosome with all the information that is contained in the gene.

Once the mRNA molecule arrives at the ribosome, the information it contains must be translated from the language of DNA (nitrogen bases) to the language of proteins (amino acids). The information is translated at the ribosomes by tRNA molecules. To do this, a three-base sequence on each tRNA molecule binds to (by forming hydrogen bonds with) its complementary sequence on the mRNA, while another portion of the tRNA molecule binds to and carries the particular amino acid specified by the three-base "word" in the mRNA. We will examine the binding rules involved in this process shortly; for now, it is enough to know that their specificity allows the message in the mRNA molecule to be translated into the exact sequence of amino acids called for by the gene.

13.3 Transcription: Information Flow from DNA to RNA

Transcription is similar to DNA replication in that one strand of DNA is used as a template from which a new strand—in this case, a strand of mRNA—is formed. However, transcription differs from DNA replication in three important ways. First, a different enzyme guides the process: the key enzyme in DNA replication is DNA polymerase, while the key enzyme in transcription is **RNA polymerase**. Second, the entire DNA molecule is duplicated in DNA replication, but in transcription only the small portion of a DNA molecule that includes a particular gene is transcribed into mRNA. Finally, whereas DNA replication produces a double-stranded DNA molecule, transcription produces a single-stranded mRNA molecule.

Transcription of a gene begins when the enzyme RNA polymerase binds to a segment of DNA near the beginning of the gene, called a **promoter**. Although the promoters of different genes vary in size and sequence, all contain several specific sequences of 6–10 bases that help bind RNA polymerase. Once bound to the promoter, the RNA polymerase unwinds the DNA double helix at the beginning of the gene, thus separating a short portion of the two strands. Then the enzyme begins to construct an mRNA molecule (Figure 13.3). Only one of the two DNA strands (the one with the promoter), called the **template strand**, is used as a template; use of the complementary strand would result in a completely different sequence of amino acids, and hence a different protein.

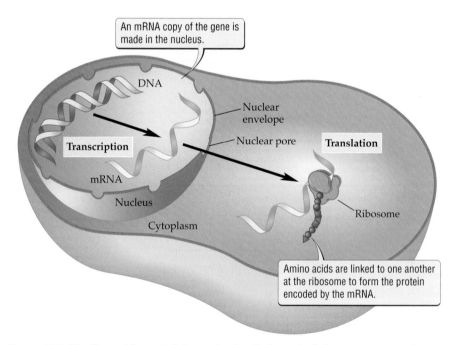

Figure 13.2 The Flow of Genetic Information in a Eukaryotic Cell

Genetic information flows from DNA to RNA to protein in two steps, transcription and translation. Transcription produces an mRNA molecule, which is then transported to the ribosome, where translation occurs and the protein is made. Different amino acids in the protein being constructed at the ribosome are represented by different purple shapes.

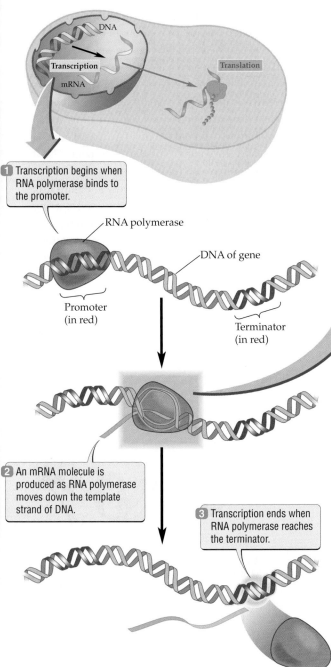

① Transcription begins when RNA polymerase binds to the promoter.

RNA polymerase

DNA of gene

Promoter (in red)

Terminator (in red)

② An mRNA molecule is produced as RNA polymerase moves down the template strand of DNA.

③ Transcription ends when RNA polymerase reaches the terminator.

Explore this overview of transcription. 13.3

Figure 13.3 An Overview of Transcription

RNA polymerase

New RNA strand

Template strand of DNA

RNA nucleotides

Direction of transcription →

Test your understanding of transcription. 13.4

then an mRNA molecule synthesized from this template will have the sequence

AAUACCGUGGC.

Notice that, by complementary base pairing with the DNA template strand, this mRNA sequence exactly duplicates (except that U substitutes for T) the sequence of the *other* DNA strand, the **coding strand**. Compare the sequence of the DNA coding strand—AATACCGTGGC—with the mRNA sequence above. This is why we can say that the information in DNA is directly copied to mRNA.

Synthesis of an mRNA molecule from the DNA template continues until the RNA polymerase reaches a sequence of bases called a **terminator**, at which point transcription ends and the newly formed mRNA molecule separates from its DNA template. The two strands of the gene's DNA then bond back to each other, ready to be used again when needed by the cell.

In eukaryotes, the newly formed mRNA molecule usually must be modified before it can be used to make a protein. Most eukaryotic genes contain internal sequences of bases, called **introns**, that do not specify part of the protein encoded by the gene (Figure 13.4); the base sequences within a gene that do encode parts of the protein are called **exons**. Base sequences copied from introns must be removed from the initial mRNA product if the protein encoded by the gene is to function properly.

The four kinds of bases in RNA pair with the four kinds of bases in DNA according to specific rules: A in RNA pairs with T in DNA, C in RNA pairs with G in DNA, G in RNA pairs with C in DNA, and U in RNA pairs with A in DNA. These base-pairing rules determine the sequence of bases in the mRNA molecule that is made from a DNA template. For example, if the sequence of the DNA template is

TTATGGCACCG,

Explore the removal of introns from RNA. 13.5

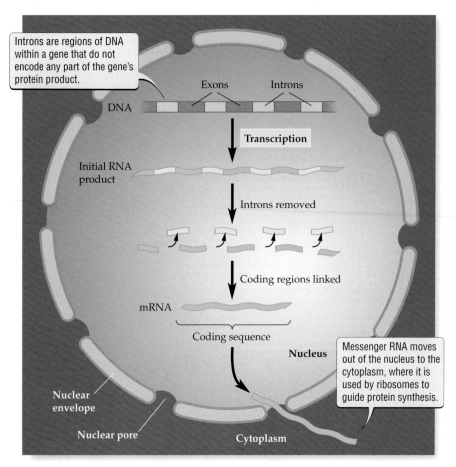

Exons   Introns

DNA

Transcription

Initial RNA product

Introns removed

Coding regions linked

mRNA

Coding sequence

Nucleus

Messenger RNA moves out of the nucleus to the cytoplasm, where it is used by ribosomes to guide protein synthesis.

Nuclear envelope

Nuclear pore

Cytoplasm

Figure 13.4 Removal of Introns by Eukaryotic Cells

Before some eukaryotic mRNA molecules can be used, enzymes in the nucleus must remove noncoding sequences (introns) and link the remaining coding sequences (exons) to one another.

After transcription and modification, the mRNA molecule is transported out of the nucleus, through a nuclear pore, to a ribosome in the cytoplasm, where the protein specified by the gene will be built. Thus the information for making a protein (encoded in the mRNA) will be carried from one part of the cell (the gene) to another (the ribosome).

13.4 The Genetic Code

Build a protein from mRNA.

13.6

The information in a gene is encoded in its sequence of bases. As we learned in the previous section, the gene's DNA sequence is used as a template to produce an mRNA molecule. Recall that the final products of most genes are proteins, and that proteins consist of one or several folded strings of amino acids. How does mRNA encode, or specify, the sequence of amino acids in a protein?

In the **genetic code**, each amino acid is specified by a sequence of three mRNA bases, called a **codon**, in an

mRNA molecule. Each mRNA base is part of only one codon, so the genetic code can be thought of as a molecular language of nonoverlapping words, in which the bases are the letters and the codons are the words. Let's consider the example shown in Figure 13.5, in which a portion of an mRNA molecule consists of the sequence UUCACUCAG. The first codon (UUC) specifies one amino acid (phenylalanine), the next codon (ACU) specifies a second amino acid (threonine), and the third codon (CAG) specifies a third amino acid (glutamine).

The genetic code is shown in its entirety in Figure 13.6. When reading the code, the cell begins at a fixed starting point on an mRNA molecule, called a **start codon** (usually the codon AUG), and ends at one of several **stop codons** (such as UGA or UAA). By beginning at a fixed point, the cell ensures that the message from the gene does not become scrambled. (To see why this is important, use Figure 13.6 to determine the amino acid sequence that would

Gene

In transcription, the DNA sequence of a gene is used to produce an mRNA molecule.

DNA strand

Transcription

mRNA

Translation

Protein

In translation, the genetic code is used to determine the amino acid that corresponds to each codon...

...A A G T G A G T C...

...U U C A C U C A G...

Codon Codon Codon

Phenyl-alanine Threo-nine Gluta-mine

...and the amino acids are linked to one another at the ribosome to form the protein specified by the mRNA molecule.

Figure 13.5 How Cells Use the Genetic Code

Figure 13.6 The Genetic Code

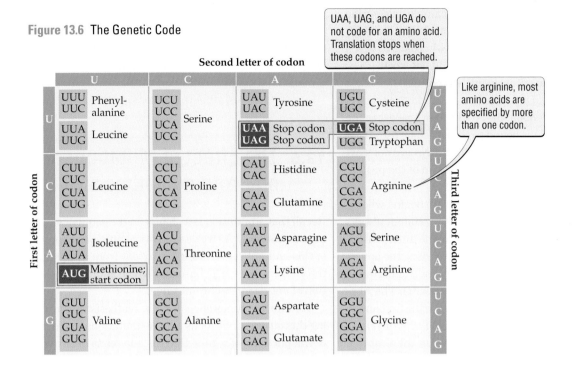

UAA, UAG, and UGA do not code for an amino acid. Translation stops when these codons are reached.

Like arginine, most amino acids are specified by more than one codon.

result if the sequence UUCACUCAG in Figure 13.5 were read in codons that began with the second U, not the first.)

The genetic code has several significant characteristics. First, the code is unambiguous: each codon specifies no more than one amino acid. Second, there are four possible bases at each of the three positions of a codon, so there are a total of 64 codons ($4 \times 4 \times 4 = 64$). Since there are only 20 amino acids, the code is redundant: several different codons may have the same "meaning" (that is, they call for the same amino acid). (We will explain why redundancy is important in the next paragraph.) Third, the code is virtually universal: nearly *all* organisms on Earth use the same code, a feature that illustrates the common descent of all organisms. The discovery of the genetic code and its near universality revolutionized our understanding of how genes work and helped pave the way for what is now a thriving biotech industry (see box on page 254).

There are a few exceptions to the code as shown in Figure 13.6. For example, in six species of yeasts, CUG codes for serine instead of leucine—but otherwise, the code these species use is identical. This and other changes indicate that the genetic code, like all other aspects of life, has evolved over time. As you might expect, the code evolves very slowly, because most changes produce proteins that do not work properly, thereby killing the organism. The reason some changes to the code are not lethal is that some organisms do not use one of the redundant codons at all. When mutations occur in these unused codons, the code can change in small ways without killing the organism.

13.5 Translation: Information Flow from mRNA to Protein

The genetic code provides the cell with the equivalent of a dictionary with which to transform the language of genes into the language of proteins. The conversion of a sequence of bases in mRNA to a sequence of amino acids in a protein, as mentioned earlier, is called translation.

Translation is the second major step in the process by which genes specify proteins (see Figure 13.2). It occurs at ribosomes, which are composed of several different sizes of rRNA molecules and more than 50 different proteins. The ribosomes are molecular machines that make the covalent bonds linking amino acids together into a particular protein. Because rRNA is a major component of ribosomes, it plays a central role in protein synthesis.

A crucial role in the synthesis of proteins at ribosomes is also played by tRNA molecules. There are many types of tRNA, but they all have a similar structure with two binding sites, as shown in Figure 13.7. At one end, each tRNA molecule has a site that binds to a particular amino acid. At the other end, each tRNA molecule has a sequence of three nitrogen bases, called an **anticodon**, that binds by complementary base pairing with a particular mRNA codon. Each tRNA molecule binds to and carries the amino acid that is specified by the mRNA codon to which it can bind; for example, the tRNA that can bind to the mRNA codon AGC carries the amino acid serine (Figure 13.7). If tRNAs carried amino acids that were different from those specified by the mRNA codons to which they could bind, the genetic code would not work.

For translation to occur, an mRNA molecule must first bind to a ribosome. Once this has occurred, translation begins at the start codon: the AUG codon nearest to the region where mRNA is bound to the ribosome (note that AUG codons located elsewhere in the mRNA are not start codons). Here's how the amino acid chain of the protein is built.

As Figure 13.8 illustrates, translation begins when a tRNA molecule binds to the AUG start codon, always bringing with it the amino acid methionine [meh-*THY*-oh-neen]. Next, another tRNA molecule, carrying the appropriate amino acid (in this example, glycine), binds to the second codon on the mRNA molecule (GGG). The ribosome then forms a covalent bond between the first amino acid (methionine) and the second amino acid (glycine). At the same time that the bond between the first two amino acids is formed, the bond between the first tRNA (the one bound to AUG) and the methionine it carries is broken. Next, the ribosome moves to the third mRNA codon, and the first tRNA is released from the mRNA.

Once the first tRNA is released, a tRNA molecule binds to the third codon, bringing with it the third amino acid of the growing amino acid chain (in this case, serine). The ribosome links the first two amino acids to the third one, then releases the second tRNA. This process continues until a stop codon is reached, at which point the mRNA molecule and the completed amino acid chain both separate from the ribosome. The new protein then folds into its compact, specific three-dimensional shape.

13.6 The Effect of Mutations on Protein Synthesis

Recall that a mutation is a change in the sequence of an organism's DNA. As we've seen in the previous chapters of this unit, mutations range in extent from a change in the identity of a single base pair to the addition or deletion of one or more chromosomes.

How do mutations affect protein synthesis? In answering this question, we focus here on mutations that occur in the portions of a gene that encode parts of proteins (exons), rather than on mutations that occur in introns or affect entire chromosomes.

Mutations can alter one or many bases in a gene's DNA sequence

There are three major types of mutations that can alter a gene's DNA sequence: substitutions, insertions, and deletions. For simplicity, we'll define these types for the case involving alteration of a single base, then extend the discussion to changes in multiple bases.

In a **substitution mutation**, one base is substituted for another at a single position in the DNA sequence of the gene. In the substitution mutation shown in Figure 13.9, for example, the sequence of the gene is changed when a thymine (T) is replaced by a cytosine (C). As the figure shows, this particular change causes the substitution of one amino acid for another because the mRNA codons made from the DNA sequences TAA and CAA encode different amino acids (isoleucine and valine, respectively; see Figure 13.6).

Insertion or **deletion mutations** occur when a single base is inserted into, or deleted from, a DNA sequence. Single-base insertions and deletions cause a genetic **frameshift**. Consider what happens in a multiple-choice test when you accidentally record the answer to a question twice on the answer sheet, which is equivalent to a single-base insertion: all your answers from that point forward are likely to be wrong, since each is an answer to the previous question. The same thing would happen if you forgot to answer a question but didn't leave space for

Learn more about translation of information from mRNA to protein.

13.7

Figure 13.7 Transfer RNA (tRNA)

Shown here are a computer model (left) and a diagrammatic version (right) of a tRNA molecule. Similar regions in the computer model and the diagram are drawn in matching colors. The genetic code works because each tRNA carries the amino acid specified by the mRNA codon to which it can bind.

it on the answer sheet (equivalent to a deletion). Similarly, a frameshift resulting from an insertion or a deletion shifts all the codons by one base, scrambling the message. As a result, during translation, the sequence of amino acids "downstream" from the site of the insertion or deletion is altered (see Figure 13.9).

In addition to single-base mutations, insertions and deletions involving more than one base—sometimes up to thousands of bases—are possible. Often, though not always, these events result in the synthesis of a protein that cannot function properly.

Mutations can cause a change in protein function

Mutations alter the DNA sequence of a gene, which in turn alters the sequence of bases in any mRNA molecule made from that gene. Such changes can have a wide range of effects on the resulting protein.

A mutation that produces a frameshift usually prevents the protein from functioning properly because it alters the identity of many of the amino acids in the protein. Frameshift mutations can also stop protein synthesis before it is complete: if a frameshift converts a codon specifying an amino acid into a stop codon, a full-length version of the protein will not be made. Regardless of whether it causes a frameshift, a mutation that alters an enzyme's binding site (the region of the enzyme that binds to its substrate) is usually harmful. Such mutations change the way the enzyme acts, decreasing or destroying its function. Finally, a mutation that inserts or deletes a series of bases causes the protein to have extra or missing amino acids, which can change the protein's shape and hence its function.

Sometimes changing a few bases in a gene's DNA sequence has little or no effect. For example, if a single-base substitution mutation does not alter the amino acids specified by the gene, then the structure and function of the protein will not be changed. Although a change in the DNA sequence from GGG to GGA would alter the mRNA sequence from CCC to CCU, both CCC and CCU code for the same amino acid, glycine (see Figure 13.6). In such cases, the substitution mutation is said to be "silent" because it produces no change in the structure (and hence the function) of the protein, and thus no change in the phenotype of the organism.

Figure 13.8 Translation

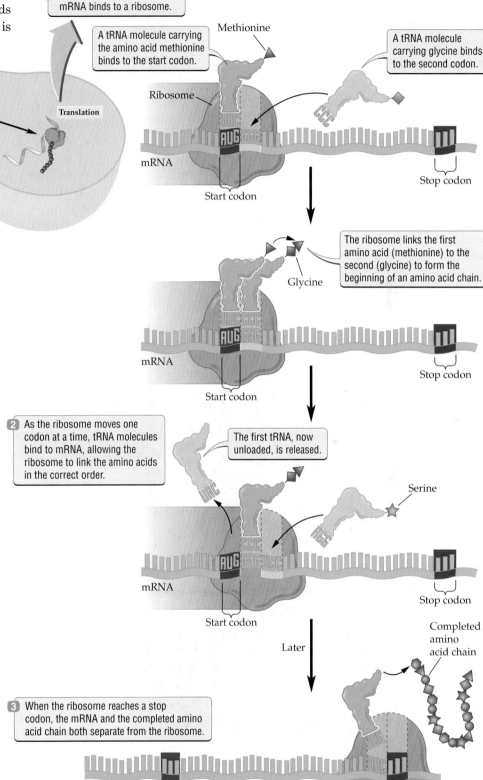

1. Translation begins when mRNA binds to a ribosome.

A tRNA molecule carrying the amino acid methionine binds to the start codon.

A tRNA molecule carrying glycine binds to the second codon.

Methionine

Ribosome

mRNA

Start codon

Stop codon

The ribosome links the first amino acid (methionine) to the second (glycine) to form the beginning of an amino acid chain.

Glycine

mRNA

Start codon

Stop codon

2. As the ribosome moves one codon at a time, tRNA molecules bind to mRNA, allowing the ribosome to link the amino acids in the correct order.

The first tRNA, now unloaded, is released.

Serine

mRNA

Start codon

Stop codon

Later

Completed amino acid chain

3. When the ribosome reaches a stop codon, the mRNA and the completed amino acid chain both separate from the ribosome.

mRNA

Start codon

Stop codon

Transcription

Translation

Cytoplasm

Biology on the Job

Investing in Biotech

Biotechnology or "biotech" refers to the use of biology to make many different types of products, including agricultural products, DNA fingerprinting kits, and medicinal drugs. As techniques for working with DNA and proteins improved in the 1960s and 1970s, new companies using these innovations sprang up around the world. Today, more than two billion dollars are invested annually in such start-up companies. Here we interview Russell T. Ray, a specialist in biotechnology companies and a managing partner for HLM Venture Partners, a venture capital firm.

What do you do during a typical day at work? On a typical day I arrive at the office around 7:30 A.M. and spend the next several hours responding to e-mail messages from coworkers and from companies we might invest in. I often meet with the managers of companies seeking to raise investment capital. I spend a few hours each day reviewing the scientific literature and reading health care news letters, many of which are published daily. I also call colleagues at other venture capital firms to compare notes on what they are seeing in the way of interesting investments, and to share information on projects that we might work on together.

How did you become interested in investing in biotechnology companies? My interest in biotechnology occurred almost by chance. In my first year in Merrill Lynch's investment banking department, I was selected to work on a financing project for a plant biotechnology company. Prior to joining Merrill Lynch, I had completed an MBA in finance, as well as a master of science in biology that focused on the territorial behavior in a species of Costa Rican hummingbird. My scientific training helped me to understand the newly emerging field of biotechnology. I have now been working in this area for twenty-three years and I still love it!

What do you look for in a successful business plan or idea? Our job is to find companies that offer our investors significant capital appreciation. In evaluating a biotechnology or medical device company, we focus on several areas:

Intellectual property—Does the company have patents that protect its technology and science from other companies? If not, this is usually a deal killer.

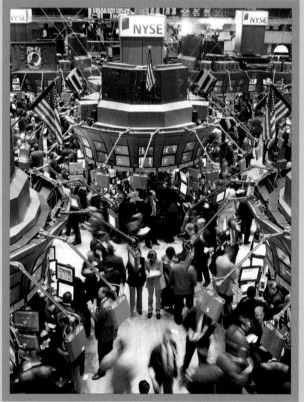

Stage of development—Does the company have drug candidates that have successfully passed through the early stages of development? If not, we are unlikely to invest due to the high risk of failure associated with products that have not progressed from animal testing to human clinical testing.

Market analysis—How big is the potential market? What the competing products?

Quality of management—Has management successfully done the job before? Who are the people behind the science?

Can you describe an example of an interesting company or a novel idea in the biotech industry? Take Cbr Systems, Inc. This company is the market leader in providing a service that collects, processes, and stores stem cells extracted from a newborn baby's umbilical cord. These cells are frozen and used in the future to provide the donor or a closely related family member with a stem cell infusion instead of a bone marrow transplant. The company has stored over 80,000 samples, and has used stored stem cells in 36 cases, with a success rate of 100 percent. The company receives a fee when the stem cells are collected from the umbilical cord, and an annual storage fee for 18 years. The company is profitable, and its revenues have been growing almost 120 percent per year for the past year or so.

How do you think the biotech industry is likely to change over the next five years? I would hope that we will see the arrival of targeted medicines that match the patient's unique genotype. We call this "personalized medicine" and believe that it represents the future in medicine.

What do you enjoy most about your job? I really enjoy working with bright entrepreneurs whose companies on are the cutting edge of science and medical technology. Each project I work on exposes me to new areas of science, so I'm constantly challenged to expand my horizons by learning more—something I love to do. Because the science is evolving so quickly, I never get bored and the people doing this work are very interesting and driven to succeed.

The Job in Context

Russell Ray and others like him straddle two worlds—they must understand both science and business. They invest in companies that take the information presented in this unit and apply it to solve real-world problems. As Mr. Ray suggests, we may be near a time when a doctor can determine your genotype and use that information to figure out how to best treat *you* ("personalized medicine"). The potential health benefits and business opportunities of personalized medicine are enormous.

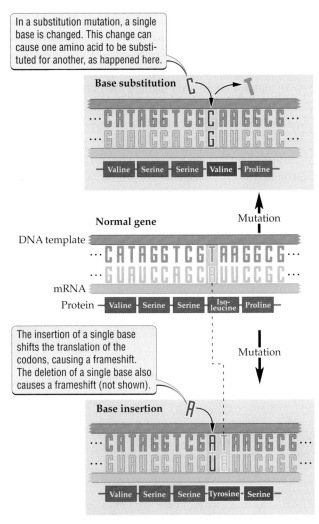

Figure 13.9 Effects of DNA Mutations on Protein Production

Two kinds of mutations are shown here: a substitution and an insertion. In each case, the mutation and its effects on transcription and translation are shown in red.

A few mutations may even be beneficial. Changes to the binding region of a protein, for example, might improve its efficiency or allow it to take on a new and useful function, such as reacting with a new substrate.

13.7 Putting It All Together: From Gene to Phenotype

Explore the impact of DNA mutations on protein synthesis.

Humans have approximately 25,000 genes, arranged linearly on 23 pairs of chromosomes. More than 99 percent of these genes code for proteins; the rest code for RNA molecules such as tRNA and rRNA. Here we review the major steps in how cells go from gene to protein to phenotype, focusing on genes that encode proteins. However, it is important to remember that transcription—the first step in the process that leads from gene to protein—is similar in all genes, including the small percentage of genes that specify tRNA and rRNA molecules. Translation does not occur for these genes because the tRNA and rRNA molecules are their final product.

Each gene is composed of a segment of DNA on a chromosome and consists of a sequence of the four bases adenine (A), cytosine (C), guanine (G), and thymine (T). The particular sequence of bases in the DNA of the gene specifies the amino acid sequence of the gene's protein product.

The two major steps in the synthesis of a protein from the information in its corresponding gene are transcription and translation (see Figure 13.2). In transcription, the sequence of bases in a gene is used as a template to produce an mRNA molecule. The cell then transports this mRNA molecule out of the nucleus to

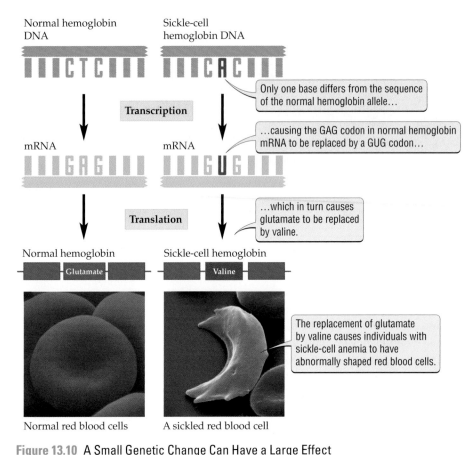

Normal hemoglobin DNA

Sickle-cell hemoglobin DNA

Only one base differs from the sequence of the normal hemoglobin allele...

Transcription

mRNA

mRNA

...causing the GAG codon in normal hemoglobin mRNA to be replaced by a GUG codon...

Translation

...which in turn causes glutamate to be replaced by valine.

Normal hemoglobin — Glutamate

Sickle-cell hemoglobin — Valine

The replacement of glutamate by valine causes individuals with sickle-cell anemia to have abnormally shaped red blood cells.

Normal red blood cells

A sickled red blood cell

Figure 13.10 A Small Genetic Change Can Have a Large Effect
Sickle-cell anemia is caused by a change in a single base.

a ribosome in the cytoplasm, where translation occurs. In translation, the sequence of bases in the mRNA molecule is used as a template to synthesize the gene's protein product by stringing together the correct sequence of amino acids.

The proteins encoded by genes are essential to life. A mutation in a gene can alter the sequence of amino acids in the gene's protein product, and this change can disable or otherwise alter the function of the protein. When a critical protein is disabled, the entire organism may be harmed. In people with the genetic disorder sickle-cell anemia, for example, a single base in the gene that encodes hemoglobin is altered (Figure 13.10). Hemoglobin is a protein involved in the transport of oxygen by red blood cells. The red blood cells of people with sickle-cell anemia become curved and distorted under low-oxygen conditions, such as those found in narrow blood vessels like our capillaries. The distorted red blood cells clog these narrow blood vessels, thereby leading to a wide range of serious effects, including heart and kidney failure. With little or no medical care, people with sickle-cell anemia usually die before they reach childbearing age (Figure 13.11). However, with intensive medical care, they can now live to their mid-forties or beyond.

In sickle-cell anemia, a gene mutation alters the gene's protein product, which in turn produces a change in the organism's phenotype. A similar chain of events occurs for other genes. Overall, the phenotype of an organism is determined by the organism's proteins and by its internal and external environment. Because genes control the production of proteins, they play a crucial role in determining the phenotype of the organism.

Figure 13.11 Children with Sickle-Cell Anemia
Originally people suffering from sickle-cell anemia would not live past young adulthood. In 1972 heavyweight boxing champion Joe Frazier (center, holding two children with the disease) participated in a fundraiser for the National Sickle Cell Anemia Foundation. With current medical developments people with the disease can now live longer.

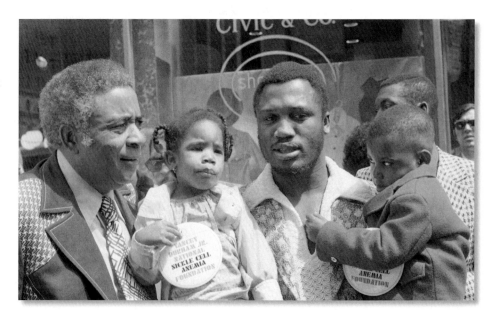

From Gene to Protein, to New Hope for Huntington Disease

Huntington disease (HD), as we saw in Chapters 11 and 12, is a dominant genetic disorder that strikes the brain, causing involuntary movements, personality changes, intellectual deterioration, and eventually, death. There is no cure, but since the isolation of the *HD* gene in 1993, new research has improved our understanding of the disease and provided hope that we may eventually be able to develop treatments that can control—or even reverse—its symptoms.

How did the isolation of the *HD* gene lead to these promising developments? In essence, scientists were able to use the genetic code to solve the communication problem described in the opening pages of this chapter. Specifically, they compared the DNA sequences of normal and *HD* alleles. This comparison found that *HD* alleles have 3 to 215 extra copies (called repeats) of the sequence GTC inserted near the beginning of the gene (see Figure 12.9); the rest of the sequence is normal. Thus, by using the genetic code to determine the amino acids specified by the *HD* gene, scientists identified both the normal and mutant versions of huntingtin, the protein produced by the *HD* gene (Figure 13.12).

After that, our understanding of HD improved rapidly. We learned that the number of GTC repeats in a mutant *HD* allele affects the age of onset and the severity of the disorder. Most people with Huntington disease have 5–15 extra GTC repeats. On average, their disease symptoms begin when they are in their fifties or sixties (if they have 5–6 extra repeats) or in their twenties or thirties (if they have 10–15 extra repeats). People with more than 15 extra GTC repeats usually develop symptoms before the age of 20.

Researchers have also learned that once mutant huntingtin enters a brain cell, enzymes in the cell cut the mutant protein in two, separating a long string of the amino acid glutamine from the rest of the molecule. (Examine Figures 13.6 and 13.12 carefully to be sure you understand why the mutant huntingtin protein contains a long string of glutamines.) The glutamine string that results can then enter the nucleus of the brain cell, where it forms clumps with other such pieces.

It remains an open issue whether these clumps of glutamine cause HD symptoms directly. Some researchers think the clumps are a harmless by-product, and that the disease symptoms are caused by the effects of mutant huntingtin on other genes. Using mice that had been genetically altered to have a human *HD* allele, scientists found that mutant huntingtin disrupted the normal activity of some of the genes in their brain cells. If this is how mutant huntingtin works in humans, then we may eventually be able to design treatments that block the protein's ability to produce these abnormal effects.

Other researchers think the clumps of glutamine directly cause brain cell death, which produces the symptoms of HD. If these researchers are right, we could search for drug therapies that stop the clumping process and thus possibly decrease, or even reverse, the symptoms of the disorder. Researchers were able to do this in mice. Here's what they did.

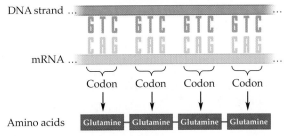

Figure 13.12 A Deadly Mutant
The mutant version of the protein huntingtin contains longer strings of the amino acid glutamine (shown in red), and has a different overall shape, than does the normal version of huntingtin (inset).

First, the researchers chose to work with the dye Congo red because they knew it could prevent the formation of glutamine clumps like those seen in people with HD. Next, they injected the dye into the brains of mice with the *HD* allele. Their findings were startling: the clumps were broken apart, and the mice lost less weight, walked in a more normal way, and survived significantly longer.

If these results are confirmed by other studies, they would suggest that the clumps are a direct cause of HD. Eight other neurological disorders, including Machado-Joseph disease and Haw River syndrome, are also caused by alleles that have extra GTC repeats. Furthermore, clumps of glutamine form in the nerve cells of people with these disorders. These observations provide a tantalizing prospect: if the glutamine clumps contribute directly to disease symptoms, the development of a therapy that prevents or reverses the formation of these clumps could offer hope for people with HD and other terrible neurological disorders.

Chapter Review

Summary

13.1 How Genes Work

- Genes work by controlling the production of proteins.
- Most genes contain instructions for building proteins. A few genes contain instructions for building several kinds of RNA that are used in the construction of proteins.
- An RNA molecule consists of a single strand of nucleotides. Each nucleotide is composed of the sugar ribose, a phosphate group, and one of four nitrogen-containing bases. The bases found in RNA—adenine (A), cytosine (C), guanine (G), and uracil (U)—are the same as those in DNA except that uracil replaces thymine (T).
- Three types of RNA (mRNA, rRNA, and tRNA) and many enzymes and other proteins are required for the cell to make proteins.

13.2 How Genes Control the Production of Proteins

- In both prokaryotes and eukaryotes, protein synthesis requires two steps: transcription and translation.
- In transcription, an RNA molecule is made using the DNA sequence of the gene.
- In translation, rRNA (plus some proteins), mRNA, and tRNA molecules together direct protein synthesis.
- In eukaryotes, information for the synthesis of a protein flows from the gene, located in the nucleus, to the site of protein synthesis, the ribosome, located in the cytoplasm.

13.3 Transcription: Information Flow from DNA to RNA

- During transcription, one strand of the gene's DNA serves as a template for synthesizing an mRNA molecule.
- The key enzyme in transcription is RNA polymerase.

- Each gene has a sequence (a promoter) at which RNA polymerase begins transcription, and another sequence (a terminator) at which the enzyme stops transcription.
- The mRNA molecule is constructed according to specific base-pairing rules: A, U, C, and G in mRNA pair with T, A, G, and C, respectively, in the template strand of DNA.
- In eukaryotes, most genes contain internal noncoding sequences of DNA (introns) that must be removed from the initial mRNA product while it is still in the nucleus. The remaining segments of mRNA correspond to exons, the DNA sequences of the gene that code for the protein.

13.4 The Genetic Code

- The information in a gene is encoded in its sequence of bases.
- In the genetic code, each amino acid is specified by a sequence of three bases, called a codon, in an mRNA molecule. During transcription, codons are "read" as a series of nonoverlapping molecular "words."
- When reading the genetic code, the cell begins at a fixed starting point on the mRNA (the start codon) and stops reading the code at one of several stop codons, thus ensuring that the message from the gene does not become scrambled.
- The genetic code is unambiguous (each codon specifies no more than one amino acid), redundant (several codons specify the same amino acid), and nearly universal (found in almost all organisms on Earth).

13.5 Translation: Information Flow from mRNA to Protein

- In translation, the information in the sequence of bases in mRNA determines the sequence of amino acids in a protein.

- Translation occurs at ribosomes, which are composed of rRNA and more than 50 different proteins.
- Transfer RNA molecules carry the amino acids specified by the mRNA to the ribosome. At the ribosome, the anticodon (a three-base sequence on the tRNA) binds by complementary base pairing with the appropriate codon on the mRNA. Each tRNA molecule carries the amino acid specified by the mRNA codon to which its anticodon can bind.
- The ribosome makes the covalent bonds that link the amino acids together into a protein.

13.6 The Effect of Mutations on Protein Synthesis

- Many mutations are caused by the substitution, insertion, or deletion of a single base in a gene's DNA sequence.
- Insertion or deletion of a single base causes a genetic frameshift, resulting in a different sequence of amino acids in the gene's protein product.
- Mutations involving insertions or deletions of a series of bases are also common.
- Mutations can change the function of a gene's protein product, especially when a frameshift is involved or when the mutation affects a protein's binding site. Mutations causing frameshifts usually destroy the protein's function.
- Other mutations have neutral effects, and a few even have beneficial effects.

13.7 Putting It All Together: From Gene to Phenotype

- More than 99 percent of genes specify proteins.
- Proteins are essential to life. Proteins, in conjunction with the environment, determine an organism's phenotype.
- Because genes control the production of proteins, they play a key role in determining an organism's phenotype.

◉ Review and Application of Key Concepts

1. What is a gene? In general terms, how does a gene store the information it contains?

2. Discuss the different products specified by genes. What are the function(s) of each of these products?

3. Describe the flow of genetic information from gene to phenotype.

4. How is the information contained in a gene transferred to another molecule? Why must the molecule that "carries" the information stored in the gene be transported out of the nucleus? In eukaryotes, why must the molecule that "carries" the information stored in the gene be modified before it is used to control the production of proteins?

5. Describe the roles played by rRNA, tRNA, and mRNA in translation.

6. Scientists have recently discovered a mutation in a gene that encodes a tRNA molecule that appears to be responsible for a series of human metabolic disorders. The mutation occurred at a base located immediately next to the anticodon of the tRNA, a change that destabilized the ability of the tRNA anticodon to bind to the correct mRNA codon. Why might a single-base mutation of this nature result in a series of metabolic disorders?

7. Write a paragraph explaining to someone with little background in biology what a mutation is and why mutations can affect protein function.

Key Terms

anticodon (p. 251)
coding strand (p. 249)
codon (p. 250)
deletion mutation (p. 252)
exon (p. 249)
frameshift (p. 252)
gene (p. 246)
genetic code (p. 250)
insertion mutation (p. 252)
intron (p. 249)
messenger RNA (mRNA) (p. 247)
promoter (p. 248)
ribosomal RNA (rRNA) (p. 247)
RNA polymerase (p. 248)
start codon (p. 250)
stop codon (p. 250)
substitution mutation (p. 252)
template strand (p. 248)
terminator (p. 249)
transcription (p. 247)
transfer RNA (tRNA) (p. 247)
translation (p. 247)

Self-Quiz

1. For genes that specify a protein, what molecule carries information from the gene to the ribosome?
   a. DNA
   b. mRNA
   c. tRNA
   d. rRNA

2. During translation, each amino acid in the growing protein chain is specified by how many nitrogen bases in mRNA?
   a. one
   b. two
   c. three
   d. four

3. Which molecule carries the amino acid specified by mRNA to the ribosome?
   a. rRNA
   b. tRNA
   c. a codon
   d. DNA

4. During transcription, which molecule or molecules are produced?
   a. mRNA
   b. rRNA
   c. tRNA
   d. all of the above

5. A portion of the template strand of a gene has the base sequence CGGATAGGGTAT. What is the sequence of amino acids specified by this DNA sequence? (Assume that the corresponding mRNA sequence will be read from left to right.)
   a. alanine–tyrosine–proline–isoleucine
   b. arginine–tyrosine–tryptophan–isoleucine
   c. arginine–isoleucine–glycine–tyrosine
   d. none of the above

6. What molecular machine makes the covalent bonds that link the amino acids of a protein together in the sequence specified by the gene that encodes that protein?
   a. tRNA
   b. mRNA
   c. rRNA
   d. ribosome

7. A mutation occurs in which the fourth, fifth, and sixth bases are deleted from a gene that encodes a protein with 57 amino acids. What will happen?
   a. The resulting frameshift will prevent protein production from being completed.
   b. A protein with 56 amino acids will be constructed.
   c. A protein that differs from the original one—but still has 57 amino acids—will be constructed.
   d. A protein with 54 amino acids will be constructed.

8. A protein-coding DNA sequence of a gene is called
   a. RNA polymerase.
   b. a protein polymerase.
   c. an intron.
   d. an exon.

Biology in the News

Injections Temporarily Turn Slacker Monkeys into Model Workers

By Alan Zarembo

Laboratory monkeys that started out as careless procrastinators became super-efficient workers after injections into their brains suppressed a gene linked to their ability to anticipate a reward.

The monkeys, which had been taught a computer game that rewarded them with drops of water and juice, lost their slacker ways and worked faster while making fewer errors.

Don't be surprised if a gleam comes into the eye of your boss if he or she reads Alan Zarembo's article in the *Los Angeles Times*. Based on a scientific study published this week in the *Proceedings of the National Academy of Sciences*, the news article describes an experiment that blocked the ability of a gene product to do its work in the brain, an action that transformed lazy monkeys into workaholics.

The monkeys had been trained to play a computer game in which they released a lever each time a computer screen turned from red to green. Only quick responses were counted as correct. As correct responses accumulated, a gray bar on the screen brightened, which the monkeys knew meant that a reward (water and juice) was imminent.

In the experiment, some monkeys received brain injections that blocked the action of a gene called *D2*. This gene normally produces receptors for dopamine, a chemical used in the brain for the perception of pleasure and satisfaction. Before the injections, the monkeys worked slowly when the gray bar was dim, speeding up and becoming more conscientious only when the bar became bright. It seems that the work ethic of these monkeys was similar to that in many people. "If the reward is not immediate, you procrastinate," says Dr. Barry Richmond, a neurologist who led the National Institute of Mental Health study.

The injections changed all that. Once the receptors encoded by the *D2* gene were no longer produced, the monkeys became model workers. They no longer understood the meaning of the gray bar, and as a result, they could not tell whether a reward was likely to come soon or some time later. Seeming to believe that a reward was always just around the corner, they worked rapidly and carefully at all times, stopping only after their thirst was quenched.

Evaluating the news

1. The scientists who conducted this study wanted to know how blocking the action of a specific gene affected the overall behavior of monkeys (which were used as a proxy for humans). It was hoped that results from the study would improve our understanding of human behavioral disorders, such as manic disorders (in which people work hard no matter how small the reward), depression (in which no reward seems worth the effort), and obsessive-compulsive disorders (in which rewards seem not to register). Thus a justification for this research is that it might someday benefit people with mental health problems. Does such a potential benefit make it ethical for people to perform experiments on the brains of primates—including species, such as the rhesus monkeys used in this experiment, that are similar enough to us that they can play computer games?

2. Imagine that research like that described in the *Los Angeles Times* article led to the development of a safe pill that enhanced your work productivity. Would you take the pill? What if the pill had harmful side effects? Would you take it anyway? (As you think about the not-so-safe version of this pill, consider the willingness some people show to take performance-enhancing drugs known to have harmful side effects, such as steroids.)

3. Should it be legal to develop drugs designed to enhance certain aspects of human behavior, such as our work ethic? If researchers identified a specific allele that enhanced a certain aspect of human behavior, should it be legal to permanently change a person's behavior by inserting the allele into his or her brain?

SOURCE: *Los Angeles Times*, Thursday, August 12, 2004.

Key Concepts

- Eukaryotic DNA is organized by a complex packing system that allows cells to store an enormous amount of information in a small space.

- Prokaryotes have relatively little DNA, most of which encodes proteins. Eukaryotes have more DNA, and more genes, than prokaryotes. Unlike prokaryotes, eukaryotes have large amounts of DNA that does not encode proteins.

- Organisms turn genes on and off in response to short-term changes in food availability or other features of the environment. Organisms also control genes over time, as when different genes are expressed at different times during embryonic development.

- In multicellular organisms, different cell types express different genes.

- The main way in which cells control gene expression is by regulating transcription. Cells can also control gene expression in other ways.

- Genetics has begun to move from the study of single genes to the study of interactions among large numbers of genes. This shift has the potential to revolutionize our understanding of genes and gene expression, leading to dramatic improvements in the practice of medicine.

Greek Myths and One-Eyed Lambs

Among his many adventures, the Greek mythological hero Odysseus encountered (and outwitted) a Cyclops, a gigantic humanlike creature with great strength and a single large eye. The faces of the Cyclopes of legend had characteristics that resembled those caused by some rare genetic and developmental disorders. Lambs, mice, and humans occasionally are born with a single large eye, along with other abnormalities of the brain and face. Such individuals die soon after birth.

What causes an animal to be born with only one eye? Two causes are known, both of which relate to the function of genes. Certain master-switch genes guide the development of an organism by activating, or "turning on," a series of other crucial genes. Some one-eyed individuals have a defect in one of these master-switch genes that prevents it from turning on other genes that control the normal development of the brain and face. Other one-eyed individuals were exposed as embryos to chemicals that prevented the protein product of the master-switch gene from having its normal effect. Whether it is due to a defective master-switch gene or exposure to chemicals, the formation of a single large eye results from the failure of cells to regulate (control) how a series of crucial genes are turned on.

Deciphering the causes of a single eye brings us to one of the most exciting areas of modern genetics: the control of gene expression. To develop and function normally, organisms must express the right genes at the right place and time, producing just the right amount of each gene's product. This is a task of bewildering complexity, but each of us does it, many times, every day. When the control of gene expression fails to work properly, disaster results: the organism may not develop properly (for example, it may form a single

Cyclops in Legend and Real Life

eye), or a group of cells may become cancerous, activating a very different set of genes than in a normal cell of the same type. As described at the close of this chapter, we now have the tools to study the expression of many genes at once. These tools have the potential to lead to major advances in our understanding of gene expression and our treatment of genetic disorders such as cancer.

Helpful to know

Biologists use several words to describe a gene whose product is made: the gene is "activated," "turned on," or is being "expressed."

Even though they all have the same genes, the various cells of a multicellular organism differ greatly in their structure and function (Figure 14.1). If genes are the blueprint for life, how can cells with the same genes be so different? The answer lies in how the genes are used: cells of different types differ in which of their genes are actually used to make proteins.

Recall from Chapter 13 that each gene contains instructions for the synthesis of a protein (or of an RNA molecule, such as tRNA or rRNA, used to make a protein). **Gene expression** is the synthesis of a gene's protein (or RNA) product, including any manipulation of the raw gene product needed to make it useful to the cell.

For different cells to use different genes, organisms must have ways to control which genes are expressed, and indeed they do. In addition to controlling which genes are expressed, organisms control where, when, and how much of a given gene's product is made. Factors that

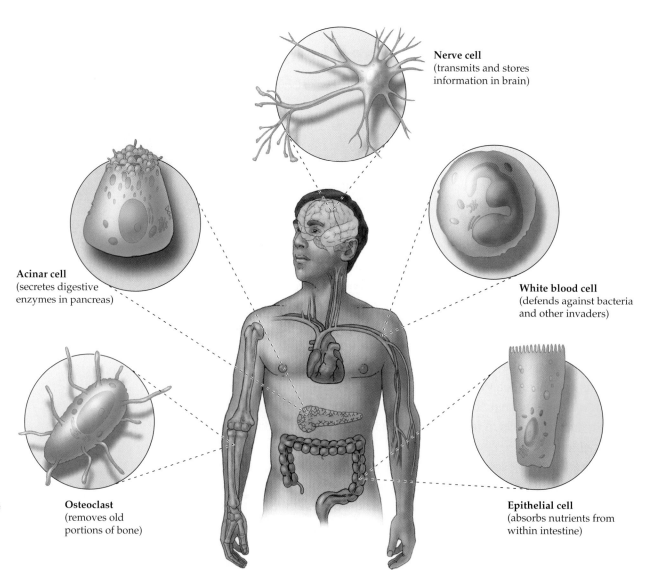

Figure 14.1 Different Cells Have the Same Genes

The nucleus of each cell type contains the same DNA. Although all cells have the same genes, they can differ greatly in structure and function because different genes are active in different types of cells.

Nerve cell
(transmits and stores information in brain)

White blood cell
(defends against bacteria and other invaders)

Acinar cell
(secretes digestive enzymes in pancreas)

Epithelial cell
(absorbs nutrients from within intestine)

Osteoclast
(removes old portions of bone)

influence gene expression include the cell type, the chemical environment of the cell, signals received from other cells in the organism, and signals received from the external environment.

In this chapter we describe how cells control the expression of their genes. We begin by examining the structural and functional organization of DNA—both of which are necessary for understanding patterns of gene expression in developing and adult organisms. Then we look at some of the ways in which cells control gene expression, and we consider the relationship between gene expression and an organism's phenotype. We close with a look at how one DNA technology—the DNA chip—is changing our views about gene expression and promising to improve many aspects of medical diagnosis and treatment.

14.1 The Structural and Functional Organization of DNA

How much DNA do different kinds of organisms have? What are the functions of different portions of an organism's DNA? Answering these questions calls for a comparison of prokaryotes (Bacteria and Archaea) and eukaryotes (all other organisms), as the DNA of these two major groups differs in several ways. First, a typical bacterium has several million base pairs of DNA, all in a single chromosome, whereas most eukaryotic cells have hundreds of millions to billions of base pairs distributed among several chromosomes (one DNA molecule per chromosome). Second, prokaryotic DNA is organized by function in a straightforward way: the different genes needed for a given metabolic pathway are usually grouped together along the DNA molecule. Although some eukaryotic genes with related functions are grouped near one another on a chromosome, many are not organized in this way; they may even be found on different chromosomes. Finally, most of the DNA in prokaryotes encodes proteins, and prokaryotic genes rarely contain noncoding segments of DNA. Eukaryotic genes, in contrast, contain noncoding DNA both within and between genes, as we will see shortly.

Overall, then, prokaryotic DNA is streamlined and organized by function in a simple, direct way, whereas eukaryotic DNA is more complex. Let's consider some of these differences in more detail.

Eukaryotes have much more DNA per cell than prokaryotes

The total component of DNA that an organism has in its cells is called its **genome**; in eukaryotes, this term refers to a haploid set of chromosomes, such as that found in a sperm or egg. The genome size of prokaryotes varies from 0.6 million to 30 million base pairs. Eukaryotic genomes show much greater variation, ranging in size from 12 million base pairs in yeast (a single-celled fungus) to over a trillion base pairs in one single-celled protist. Most vertebrates have genomes that contain hundreds of millions to billions of base pairs. For example, puffer fish have 400 million base pairs (Figure 14.2), the range among mammals is 1.5 to 6.3 billion base pairs (with humans at 3.3 billion base pairs), and some salamanders have 90 billion base pairs.

As these examples illustrate, eukaryotes usually have far more DNA than prokaryotes. Why is this so? In part, the reason is that eukaryotes are more complex organisms than prokaryotes and hence need more genes to run their metabolic machinery. A typical prokaryote has about 2,000 genes. Among the eukaryotes studied to date, the single-celled yeast *Saccharomyces cerevisiae* has 6,000 genes, the nematode worm *Caenorhabditis elegans* has 19,100 genes, several plant species each have an estimated 20,000 genes, and humans have about 25,000 genes.

Three *Caenorhabditis elegans*

Saccharomyces cerevisiae

Genes constitute only a small percentage of the DNA in eukaryotes

Although eukaryotes have roughly 3 to 15 times as many genes as a typical prokaryote, this difference in gene

Figure 14.2 Vertebrates with Millions to Billions of Base Pairs
This puffer fish has only 400 million base pairs compared to the diver's 3.3 billion.

Learn more about the structural organization of the chromosome.

14.1

Figure 14.3 The Composition of Eukaryotic DNA

Eukaryotic genes can be surrounded by spacer DNA (light blue) or by transposons (red and dark blue), as diagrammed here for five human genes (purple). Each of the two different types of transposons shown here is found in many copies throughout human DNA. Note that one transposon (dark blue) has inserted itself into one of the genes.

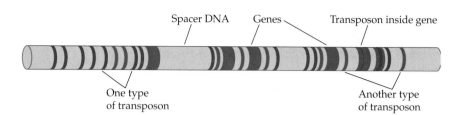

Spacer DNA   Genes   Transposon inside gene

One type of transposon   Another type of transposon

number does not fully explain why eukaryotes often have hundreds to thousands of times more DNA than prokaryotes. The main reason eukaryotes have more DNA is that only a small percentage of their DNA encodes proteins used by the organism. Most eukaryotic DNA is either "superfluous" or serves a function other than encoding a protein (or an rRNA or tRNA used to make a protein).

In humans, for example, scientists estimate that genes that encode a protein (or an rRNA or tRNA) make up less than 1.5 percent of the genome. Other genes encode a number of types of RNA molecules that are smaller than tRNA or rRNA; some of these small RNAs help control gene expression, as we shall see at the end of this chapter. In humans and other eukaryotes, the rest of the genome consists of various types of superfluous DNA, including noncoding DNA and transposons (Figure 14.3, Table 14.1).

Noncoding DNA, which does not encode proteins or RNA, includes introns and spacer DNA. Introns, as we saw in Chapter 13, are sequences of noncoding DNA within genes, while **spacer DNA** consists of stretches of DNA that separate genes. Eukaryotic DNA also contains many "jumping genes," or **transposons**: sequences that can move from one position on a chromosome to another, or even from one chromosome to another. Although many transposons encode proteins, most of those proteins are

used by the transposon, not by the organism. Transposons may surround genes or may even be inserted in the middle of a gene. Transposons can constitute a large proportion of a eukaryote's DNA; for example, they make up an estimated 36 percent of human DNA and more than 50 percent of the 5.4 billion base pairs in corn. Some transposons appear to be viral in origin, but for others, it is not known whether they are ancient viruses that became noninfective or former genes that acquired the ability to move throughout the genome.

14.2 DNA Packing in Eukaryotes

To be expressed, the information in a gene must first be transcribed into an RNA molecule; thus it is essential for the enzymes that guide transcription to be able to reach the gene. This task may sound simple, but it involves what may be the ultimate storage problem: how to store an enormous amount of information (the organism's DNA) in a small space (the nucleus), yet still be able to retrieve each piece of that information precisely when it is needed.

In humans and other eukaryotes, each chromosome in each cell contains one DNA molecule. These molecules

Table 14.1

Types of Eukaryotic DNA

Type	Subtypes	Description
Exons (of genes)		Most code for proteins used by the organism; others code for rRNAs, tRNAs, and various small RNAs that help regulate gene expression
Noncoding DNA	Introns	Sequences of noncoding DNA found within a gene that are removed from mRNA after transcription
	Spacer DNA	Sequences of noncoding DNA that separate genes
Transposons ("jumping genes")		Sequences of DNA that can move from one position on a chromosome to another, or from one chromosome to another

hold a vast amount of genetic information. As we learned in Chapter 9, the haploid number of chromosomes in humans is 23; these chromosomes together contain about 3.3 billion base pairs of DNA. If the DNA from all 46 chromosomes in a single human cell were held at full length, it would be more than 2 meters long (taller than most of us). That is a huge amount of DNA, especially considering that it is packed into a nucleus only 0.000006 meter (0.0002 inch) in diameter. The combined length of DNA in our bodies is staggering: the human body has about 10^{13} cells, each of which contains roughly 2 meters of DNA. Therefore, each of us has about 2×10^{13} meters of DNA in his or her body, a length that is more than 130 times the distance between Earth and the sun.

How can our cells pack such an enormous amount of DNA into such small spaces? They do so by organizing the very thin DNA molecule into a complex system with several levels of packing. At the highest level, a chromosome takes its characteristic shape because its DNA is condensed very tightly (Figure 14.4). As we begin to "unpack" the chromosome, we see that each portion of it consists of many tightly packed loops. Each loop, in turn, is composed of a fiber 30 nanometers (nm) wide. If we unpack that fiber, we see that it is made up of many **histone spools** packed together tightly. Histone spools are so named because each one consists of a segment of DNA wound around a "spool" composed of proteins called histones. Finally, if we unwind the DNA from the histones, we reach the level of the DNA double helix, which is 2 nm wide.

The packing scheme we've just described is found in eukaryotic cells during mitosis or meiosis, when the chromosomes are most condensed. Eukaryotic DNA is much less tightly packed during interphase, the part of the cell cycle when most gene expression occurs. During that time, much of the DNA is folded into 300 nm loops of the 30 nm fiber, but some remains more tightly packed. Genes in the tightly packed regions are not expressed because the proteins necessary for transcription cannot reach them.

Compared with eukaryotes, prokaryotic cells have much less DNA per cell, and their DNA packing system is less complex.

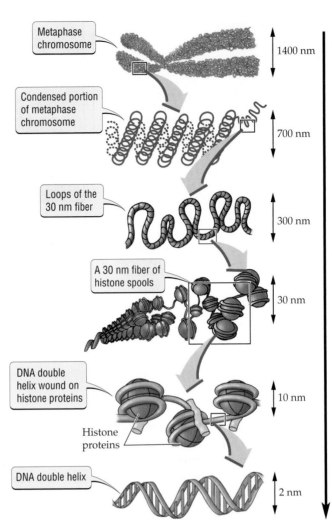

Figure 14.4
DNA Packing in Eukaryotes
The DNA of eukaryotes is highly organized by a complex packing system. This chromosome is shown at metaphase, the phase of the cell cycle at which DNA is most tightly packed.

Metaphase chromosome — 1400 nm

Condensed portion of metaphase chromosome — 700 nm

Loops of the 30 nm fiber — 300 nm

A 30 nm fiber of histone spools — 30 nm

DNA double helix wound on histone proteins — 10 nm

Histone proteins

DNA double helix — 2 nm

Unpacking

Test your knowledge of DNA packing. 14.2

14.3 Patterns of Gene Expression

It is through gene expression that genes influence the structure and function of a cell or organism. At any given time, roughly 5 percent of the genes in a typical human cell are being actively used (expressed). The rest of the genes are not in use. Different cells express different sets of genes, and within a given cell, the particular genes that are expressed can change over time. What determines which genes an organism expresses in a particular time, place, or type of cell?

Organisms turn genes on and off in response to short-term environmental changes

Single-celled organisms such as bacteria face a big challenge: they are directly exposed to their environment, and they have no specialized cells to help them deal with changes in that environment. One way they meet this challenge is to express different genes as conditions change.

Bacteria respond to changes in nutrient availability, for example, by turning genes on or off. If the nutrient lactose (a sugar found in milk) is added to a petri dish or beaker containing *E. coli* bacteria, within a matter of minutes the genes that encode the enzymes needed by the bacteria to break down lactose are activated (Figure 14.5).

Figure 14.5 Bacteria Express Different Genes as Food Availability Changes

The sugars lactose and arabinose are not always available, but both can serve as food for the bacterium *E. coli*.

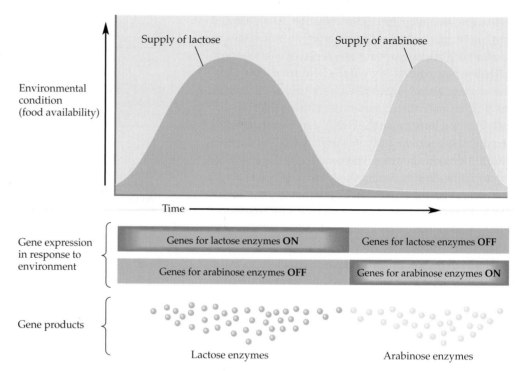

When the lactose is used up, the bacteria stop producing those enzymes. In effect, the bacteria specialize temporarily on an available resource. When that resource runs out, they switch to the next resource that becomes available (in Figure 14.5, the sugar arabinose). By producing the enzymes to process a particular food only when that food is available, bacteria do not waste energy and cellular resources making enzymes that are not needed.

Like single-celled organisms, multicellular organisms change which genes they express in response to short-term changes in the environment. For example, we humans change the genes we express when our blood sugar or blood pH levels change, allowing us to keep those levels from becoming too high or too low. Similarly, humans, plants, and many other organisms, when exposed to high temperatures, turn on certain genes to produce proteins that protect cells against heat damage.

Different cells in eukaryotes express different genes

In both developing embryos and adults, different types of cells express different genes. Whether a cell expresses a particular gene depends on the function of the gene's product. Not surprisingly, a gene that encodes a specialized protein is expressed only in cells that need to either use that protein within the cell or produce it for transport to other cells that will use it. For example, among the more than 200 types of cells in the human body, red blood cells are the only cells that use the oxygen transport protein hemoglobin; hence developing red blood cells are the only cells that express the gene for this protein (Figure 14.6). Similarly, the gene for crystallin, a protein that makes up the lens of the eye, is expressed only in cells of the developing eye lens. Finally, the gene for insulin, a hormone produced in the pancreas and used elsewhere in the body, is expressed only in certain cells of the pancreas.

Some genes, known as **housekeeping genes**, have an essential role in the maintenance of cellular activities and are expressed by most cells in the body. Genes for rRNA, for example, are expressed by most cells (see Figure 14.6). This is not surprising: virtually all cells need to make proteins, and rRNA is a key component of ribosomes, the sites at which proteins are built.

Development in eukaryotes relies on gene cascades and homeotic genes

Turning the correct genes on and off in response to changing environmental conditions is a challenging task. But multicellular organisms must also coordinate an even more difficult operation: developing from a single-celled zygote into a large, complex organism. The timing of gene activity during embryonic development is a task of great complexity, and errors in the control of this process can result in death or deformity (Figure 14.7).

During development, eukaryotic organisms control gene activity through cascades of gene expression. A **gene cascade**, in which a series of genes are turned on one after another, can be compared to a series of falling dominoes, in which one knocks over another, which in turn knocks over the next, and so on. When such a cascade occurs, the protein products of certain genes interact with one another and with signals from the environment to turn on different sets of genes in different cells. The proteins produced by those newly activated genes then interact with one another and with the environment to turn on still more genes, and so on. Eventually, genes are expressed whose protein products alter the structure and function of cells, allowing cells to become specialized for particular tasks.

The master-switch genes described at the beginning of this chapter, which biologists call **homeotic genes**, play a central role in the control of gene cascades. Each homeotic gene controls the expression of a series of other genes whose proteins direct the development of the organism. Given this crucial role, it is not surprising that defective versions of homeotic genes can have striking phenotypic effects, such as those described at the opening of this chapter or those shown in Figure 14.7.

At different times during an individual's development, different homeotic genes are active in the body's different cell types. For example, a homeotic gene that coordinates the development of the eye may be active in cells that will give rise to the eyes. In other parts of the body, although this gene is present, it is not in use; instead, other

homeotic genes are expressed. Finally, as the body changes during development, the homeotic genes expressed by cells also change.

Recently it was discovered that similar homeotic genes control development in similar ways in organisms as different as fruit flies, mice, and humans (Figure 14.8). This similarity indicates that these genes are ancient. Homeotic genes first evolved hundreds of millions of years ago, and since then they have been used in similar ways by a wide variety of organisms.

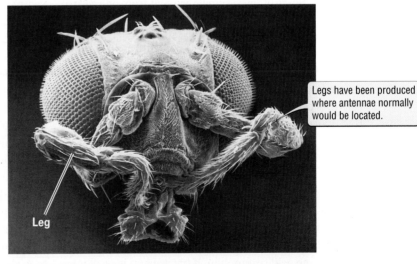

	Developing red blood cell	Eye lens cell (in embryo)	Pancreatic cell
Hemoglobin gene	ON	OFF	OFF
Crystallin gene	OFF	ON	OFF
Insulin gene	OFF	OFF	ON
rRNA gene	ON	ON	ON

Figure 14.6 Different Types of Cells Express Different Genes
Some genes, such as those encoding hemoglobin, crystallin, and insulin, are active ("on") only in the types of cells that use or produce the protein encoded by the gene. Other "housekeeping" genes, such as the rRNA gene, are active in most types of cells.

Head of a normal fruit fly

Eye

Antenna

Head of a developmental mutant

Leg

Legs have been produced where antennae normally would be located.

Figure 14.7 A Developmental Mutation with Bizarre Results
A mutation in a homeotic gene that controls development in fruit flies produces legs where antennae should be.

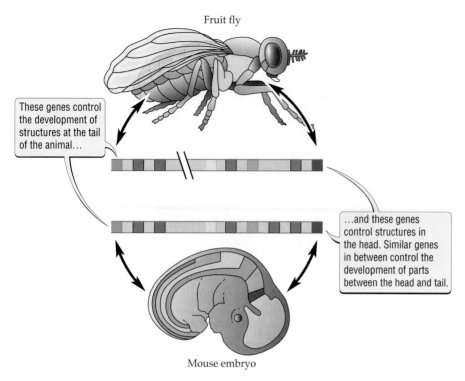

Fruit fly

These genes control the development of structures at the tail of the animal…

…and these genes control structures in the head. Similar genes in between control the development of parts between the head and tail.

Mouse embryo

Figure 14.8 Homeotic Genes in Different Organisms Are Similar

Development is controlled by similar homeotic genes in very different organisms. The homeotic genes that control development in fruit flies and mice are arranged in a similar order on fruit fly and mouse chromosomes. Moving from tail to head, the body parts affected by these genes are positioned in the same order in which the genes are arranged on the chromosomes. Similar homeotic genes and the structures whose development they control are matched here by color.

14.4 How Cells Control Gene Expression

Cells receive signals that influence which genes they turn on and off. Some of these signals are sent from one cell to another, as when one cell releases a signaling molecule that alters gene expression in another cell. Cells also receive signals from features of the organism's internal environment (for example, blood sugar level in humans) and external environment (for example, sunlight in plants). Overall, cells process information from a variety of signals and use that information to determine which genes to express.

Cells control the expression of most genes by controlling transcription

The most common way for cells to control gene expression is to turn the transcription of particular genes on or off. The bacterium *E. coli*, for example, requires a supply of the amino acid tryptophan. If tryptophan is available in the external environment, the bacterium absorbs it rather than wasting cellular resources making it. But if

tryptophan is not readily available, the bacterium expresses a series of five genes that together encode the enzymes used to make tryptophan.

E. coli controls these five genes in the following way. When tryptophan is present in the environment, it binds to a repressor protein in the bacterial cell (**repressor proteins** are so named because they stop the expression of one or more genes). This tryptophan-repressor protein complex can then bind to the tryptophan operator. An **operator** is a sequence of DNA that controls the transcription of a gene or group of genes—in this case, the five genes needed to make tryptophan. When bound to the operator, the tryptophan-repressor protein complex blocks access of RNA polymerase to the promoter of the tryptophan genes, thus blocking transcription of those genes (Figure 14.9*a*). In the absence of tryptophan, the repressor protein cannot bind to the operator, so RNA polymerase is free to bind to the promoter, and transcription of the genes occurs (Figure 14.9*b*). As a result, the cells do not make tryptophan when it is already present, but they do make it when tryptophan levels are low. This control of gene expression ensures two things: the cell always has an adequate supply of tryptophan, and the cell does not waste resources producing tryptophan when it is readily available in the environment.

A few genes, such as those in *E. coli* that encode the tryptophan-repressor protein, are always expressed at a low level; their transcription is not regulated. But most genes in prokaryotes and eukaryotes are regulated. In general, the control of transcription has two essential elements, both of which are illustrated by the control of tryptophan synthesis in *E. coli*. The first element is **regulatory DNA**, such as the tryptophan operator, that can switch a gene on and off. Second, to switch genes on and off, regulatory DNA must interact with **regulatory proteins** that signal whether a gene should be expressed. The repressor protein that binds to the tryptophan operator when tryptophan is present is an example of a regulatory protein. Together, regulatory DNA and regulatory proteins turn genes on and off in both prokaryotes and eukaryotes.

In eukaryotes, the situation is considerably more complex than in prokaryotes: there are many different types of regulatory proteins in eukaryotes, and dozens of regulatory DNA sequences may be involved in controlling gene expression. In part, this happens because in eukaryotes, unlike prokaryotes, regulatory DNA sequences and the genes they influence often are not located right next to each other. Despite this added complexity, the basic concepts are the same: regulatory proteins bind to regulatory DNA sequences, thereby turning genes on or off.

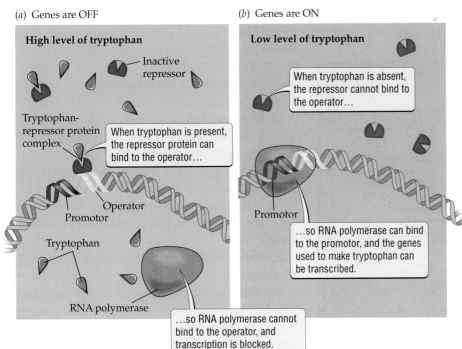

(a) Genes are OFF

High level of tryptophan

Inactive repressor

When tryptophan is present, the repressor protein can bind to the operator...

Tryptophan-repressor protein complex

Operator

Promotor

Tryptophan

RNA polymerase

...so RNA polymerase cannot bind to the operator, and transcription is blocked.

(b) Genes are ON

Low level of tryptophan

When tryptophan is absent, the repressor cannot bind to the operator...

Promotor

...so RNA polymerase can bind to the promotor, and the genes used to make tryptophan can be transcribed.

Figure 14.9 Repressor Proteins Turn Genes Off

In the bacterium *E. coli*, a repressor protein binds to an operator to control the transcription of a group of genes that encode the enzymes needed to make tryptophan. (*a*) When tryptophan is present, it binds to a repressor protein; this tryptophan-repressor protein complex then turns the genes off by binding to the operator. (*b*) The repressor protein by itself (that is, when tryptophan is absent) cannot bind to the operator; thus RNA polymerase can bind to the promoter, which turns on the genes used to make tryptophan.

Explore the tryptophan operon. 14.3

Cells also control gene expression in other ways

Cells can control gene expression at a number of key steps in the pathway from gene to protein (Figure 14.10):

1. *Tightly packed DNA is not expressed.* During interphase, when most gene expression occurs, the chromosomes are long and narrow and their DNA is relatively loosely packed. Even during interphase, however, some DNA remains tightly packed. Genes in this tightly packed DNA are not transcribed, in part because the proteins necessary for transcription cannot reach them.

2. *Transcription can be regulated.* Regulation of transcription, as described in the previous section, is the most common means of controlling gene expression.

3. *The breakdown of mRNA molecules can be regulated.* Recall from Chapter 13 that when cells express a gene to make a protein, an mRNA molecule is transcribed from the gene and then translated to make the gene's protein product. Most mRNA molecules are broken down by cells in amounts of time ranging from a few minutes to hours after they are made; a few persist for days or weeks. The longer the life of an mRNA molecule, the more protein molecules can be made from it. In some cases, cells determine how long mRNA molecules will persist by chemically modifying the mRNA.

4. *Translation can be inhibited.* Some proteins can bind to mRNA molecules and prevent their translation. This method of control is especially important for some long-lived mRNA molecules. It allows the cell to deactivate an mRNA molecule that otherwise might continue to produce a protein that is no longer being used.

5. *Proteins can be regulated after translation.* Cells must modify or transport many proteins before they can be used; both modification and transport can be used to regulate the availability of a protein. Completed proteins can also be rendered inactive by repressor molecules different from those used to control gene expression.

6. *Completed proteins can be destroyed.* Cells can target certain proteins for destruction, thus controlling gene expression at the final step in the pathway from gene to protein.

14.5 The Consequences of Controlling Gene Expression

Having described how cells and organisms control when, where, and how much of each gene's protein product is made, we turn now to a related question: how

Figure 14.10 Control of Gene Expression in Eukaryotes

Eukaryotes can control gene expression in many ways. Each point in the pathway from gene to protein represents a point at which cells can regulate the production of proteins.

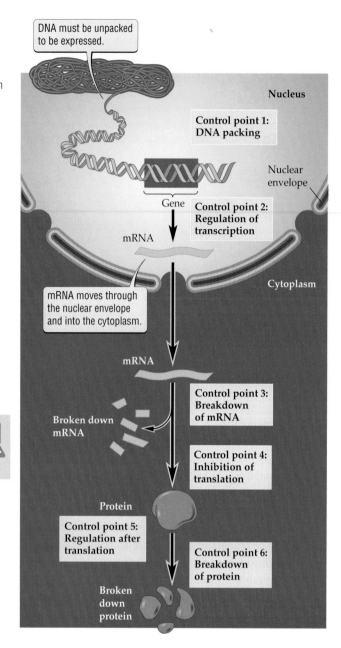

DNA must be unpacked to be expressed.

Nucleus

Control point 1: DNA packing

Nuclear envelope

Gene

Control point 2: Regulation of transcription

mRNA

Cytoplasm

mRNA moves through the nuclear envelope and into the cytoplasm.

mRNA

Control point 3: Breakdown of mRNA

Broken down mRNA

Control point 4: Inhibition of translation

Protein

Control point 5: Regulation after translation

Control point 6: Breakdown of protein

Broken down protein

Test your understanding of the control of gene expression in eukaryotes.

14.4

does the control of gene expression affect the phenotype of the organism? In many instances, the control of gene expression has a powerful and direct effect on the phenotype. Consider, for example, the effect of homeotic genes on a developing organism. As you will recall, a homeotic gene controls the expression of a series of other genes that influence development. If the protein specified by the homeotic gene does not function properly, the expression of those genes will not be controlled properly, and extreme phenotypic effects may result (see page 269 and Figure 14.7). For many other (nonhomeotic) genes, the same is true: if the control of gene expression is changed, the phenotype may change.

But the control of gene expression and the phenotype of the organism are not equivalent. Organisms turn genes on and off in response to signals from the environment, thus changing which proteins they produce. These proteins have the potential to influence the phenotype. However, as we saw in Chapter 10, the phenotype of an organism results from interactions between its genetic makeup and its environment. As a result, the environment not only can influence which genes are expressed, but also can alter the *effects* of a gene's protein product (see Figure 10.12). Thus, as summarized in Figure 10.14, the phenotype of an organism results from the combined effects of the organism's genotype (the specific alleles the organism has), the interactions among genes and their protein products (patterns of gene expression), and the environment.

From Gene Expression to Cancer Treatment

A revolution has begun in the field of genetics— one that may ultimately rival the computer revolution in its effect on society. Like the computer revolution, this genetic revolution is being pushed forward by technical advances. However, the revolution in genetics is not centered on a particular new type of machine or technology. Instead, it is due to a shift in perspective, from the study of single genes to the study of large numbers of genes simultaneously.

What can we learn by studying many genes at once? Scientists are using new technologies to discover what happens to patterns of gene expression when cells face changing environmental conditions. Do cells turn a small handful of genes on or off, or do they change the expression of large numbers of their genes? Early results suggest that environmental changes alter the expression of large numbers of genes.

For example, in 1996, scientists published the complete DNA sequence of *Saccharomyces cerevisiae*. They found that this yeast has about 6,000 genes. The researchers built multiple, identical copies of a DNA chip (see the box on page 274) that would hold all of these genes, and they used these chips to monitor the activity of all the yeast genes at once. Researchers learned, for example, that as the yeast adjusted to changes in food availability, 710 genes that had been inactive were turned on, while 1,030 genes that had been active were turned off.

Scientists have also begun to build DNA chips that have the potential to alter the practice of medicine. For example, by surveying how individuals with different genetic characteristics respond to different drugs, doctors may be able to use DNA chips to decide which of several possible drugs would be likely to work best in any given patient. In a family practice setting, mass-produced DNA chips could be used to identify the exact strain of bacteria causing a particular individual's sore throat, along with the antibiotics to which that bacterial strain is resistant (and which, therefore, should not be prescribed by the physician).

DNA chip results can even provide new hope for cancer patients. Consider breast cancer. In the United States, one in seven women will get breast cancer, and half of those women will die from it. When a breast cancer is detected and the tumor removed, doctors use characteristics of the tumor (such as its size and appearance) to help them decide what treatment the patient should receive after surgery. The goal is to use the most aggressive treatments—such as various forms of chemotherapy, which can have toxic side effects—only when needed; that is, only for those cancers that are most likely to spread to other parts of the body.

Unfortunately, existing methods of predicting whether the cancer will spread are not very accurate. As a result, many patients receive unnecessary, toxic treatments: roughly 80 percent of breast cancer patients who could be cured by surgery and radiotherapy alone are advised to have chemotherapy. To improve our ability to predict whether a patient's cancer will spread, a team of doctors in the Netherlands and the United States used DNA chips to study the expression of all 25,000 human genes in breast cancer tumors. Their findings revealed that different sets of genes are turned on or off in breast cancer tumors, depending on the severity of the cancer. Applying these results, the doctors correctly predicted whether the cancer would spread in 65 of 78 cases (83 percent), a big improvement over previous methods. In addition, the percentage of patients who could be cured by surgery and radiotherapy alone but who were advised incorrectly to have chemotherapy dropped from 80 percent to 25 percent.

In the examples we have just discussed, DNA chips could help doctors by removing the guesswork from the treatment of conditions ranging from sore throats to cancer. But it is important not to forget that many human diseases are heavily influenced by factors other than our genes—recall that 70 percent or more of cases of colon cancer, stroke, heart disease, and type 2 diabetes can be prevented simply by following a "low-risk" lifestyle (see page 200). For such disorders, results from DNA chips alone cannot provide doctors with the guidance they need. Instead, as we battle such disorders, the best approach may be to identify which of our many genes influence whether people with similar lifestyles will get particular diseases. Although this is a daunting task, it is well worth the effort: by understanding the effects of both lifestyle and genetics, we may be able to dramatically improve our ability to treat and even prevent disease.

Science Toolkit

Using DNA Chips to Monitor Gene Expression

Geneticists have long realized that the metabolism and phenotype of an organism are influenced by many genes. But organisms have thousands of genes, and until recently it was not possible to study how the expression of large numbers of genes changed during particular stages of development or under a particular set of environmental conditions. With the advent of DNA chips, first developed in the late 1990s, it is now possible to monitor the expression of many genes at once.

A **DNA chip** consists of thousands of samples of DNA placed in a regimented order on a small glass surface, or "chip," roughly the size of a dime. DNA chips are constructed in two main ways. In one method, a robotic arm that has multiple printing tips is used to deliver minute droplets of single-stranded DNA (500–5,000 bases long) to the chip's glass surface. The DNA is then treated and dried so that it will bind to the glass. The DNA sequence in each droplet is known (it may, for example, correspond to the exons of a gene), and each droplet has a known location on the chip.

In the second method, shorter (20–80 bases long) segments of DNA are used. These short pieces of single-stranded DNA can be synthesized directly on the glass chip, using a technique similar to the way computer chips are made (hence the name "DNA chip"). It is also possible to synthesize the fragments of DNA elsewhere (using a machine designed for that purpose); the fragments are then placed on the chip in a specific order and immobilized.

With either of these methods, DNA representing some or all of an organism's genes can be placed on a single DNA chip and used to screen the expression of many genes simultaneously. How is this done? Although a considerable number of technical steps must be performed, the basic idea is simple. When a gene is expressed, an mRNA copy of the information in that gene is produced. To monitor which genes are being expressed, mRNA is isolated from the organism or cells being tested, labeled (as with a dye that glows red or green), and then washed over the DNA chip. Both the mRNA and the DNA on the chip are single-stranded, so the labeled mRNA can bind to the DNA representing the gene from which it was originally produced.

Next, a scanner is used to detect the locations on the chip where DNA was able to bind to labeled mRNA. Since the gene that corresponds to each location on the chip is known, results from the scanner tell us which of the organism's genes produced mRNA—and hence which of the organism's many genes were expressed (and which were not). Finally, the genes expressed in different circumstances can be compared, providing valuable information about how organisms regulate development and cope with environmental variation.

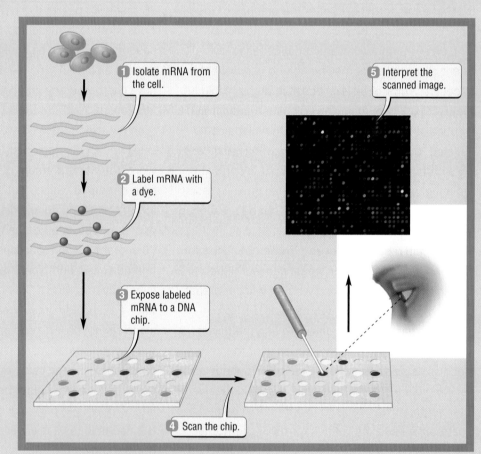

1 Isolate mRNA from the cell.

2 Label mRNA with a dye.

3 Expose labeled mRNA to a DNA chip.

4 Scan the chip.

5 Interpret the scanned image.

Using DNA Chips

Chapter Review

Summary

14.1 The Structural and Functional Organization of DNA

- Compared with eukaryotes, prokaryotes have a small amount of DNA, and it is all on a single chromosome. Most prokaryotic DNA encodes proteins, and functionally related genes in prokaryotes are grouped together in the DNA.
- With regard to DNA, eukaryotes differ from prokaryotes in several ways: (1) In eukaryotic cells, the DNA is distributed among several chromosomes. (2) Eukaryotes have more DNA per cell than prokaryotes do, in part because they have more genes than prokaryotes. In addition, genes constitute only a small portion of the genome in many eukaryotes; the rest consists of noncoding DNA (which includes introns and spacer DNA) and transposons. (3) Eukaryotic genes with related functions often are not located near one another.

14.2 DNA Packing in Eukaryotes

- Two features allow cells to pack an enormous amount of DNA into a very small space: the DNA molecule is very narrow, and in eukaryotes it is highly organized by a complex packing system. Segments of DNA are wound around histone spools, packed together tightly into a narrow fiber, and folded into a series of dense loops.
- The tightness of the DNA packing system is reduced during interphase, when most gene expression occurs.
- In DNA regions that are tightly packed, genes cannot be expressed because the proteins necessary for transcription cannot reach them.

14.3 Patterns of Gene Expression

- Both prokaryotes and eukaryotes selectively turn genes on and off in response to short-term changes in environmental conditions.
- Multicellular eukaryotes must also regulate gene expression over long periods of time during development.
- The different cell types of multicellular organisms express different genes.
- Housekeeping genes, which have an essential role in the maintenance of cellular activities, are expressed in most cells of the body.
- Eukaryotes control gene expression in development through gene cascades, which are regulated by homeotic (master-switch) genes.

14.4 How Cells Control Gene Expression

- Cells control most genes by controlling transcription.
- Transcription is controlled by regulatory DNA sequences that interact with regulatory proteins to switch genes on and off.
- An operator is a regulatory DNA sequence that controls the expression of a single gene or a group of genes in prokaryotes.

- One kind of regulatory protein—a repressor protein—binds to an operator to prevent transcription from taking place.
- Cells can also control gene expression at other points on the pathway from gene to protein.

14.5 The Consequences of Controlling Gene Expression

- The control of gene expression and the phenotype of the organism are not equivalent.
- The phenotype of the organism results from the combined effects of the organism's genotype, interactions among genes and their protein products, and the environment.

⊙ Review and Application of Key Concepts

1. The full length of the DNA in a eukaryotic cell is hundreds of thousands of times as long as the nucleus in which it is found. Describe how this is possible.

2. Summarize the major differences between DNA in prokaryotes and in eukaryotes, emphasizing differences in the amount and function of DNA.

3. Genes in eukaryotes are often separated on the DNA molecule by long segments of spacer DNA. Consider whether the presence of spacer DNA will increase, decrease, or not change how often crossing over occurs between a given pair of genes.

4. Imagine that you transferred a bacterium from an environment in which glucose was available as food to an environment in which the only source of food was the sugar arabinose. Special enzymes are required to digest arabinose, but not glucose. How do you think gene expression in the bacterium would change as a result of your action?

5. Cell types in multicellular organisms often differ considerably in structure and in the metabolic tasks they perform, yet each cell of a multicellular organism has the same set of genes. Explain how cells with the same genes can be so different (a) in structure and (b) in the metabolic tasks they perform.

6. Summarize how gene expression can be controlled at various steps in the process of producing a protein from the information stored in a gene.

7. Describe the function and method of operation of the tryptophan operator in *E. coli*.

8. What is a DNA chip? How can DNA chips be used to study many genes at once?

Key Terms

DNA chip (p. 274)
gene cascade (p. 269)
gene expression (p. 264)
genome (p. 265)
histone spool (p. 267)
homeotic gene (p. 269)
housekeeping gene (p. 268)

noncoding DNA (p. 266)
operator (p. 270)
regulatory DNA (p. 270)
regulatory protein (p. 270)
repressor protein (p. 270)
spacer DNA (p. 266)
transposon (p. 266)

Self-Quiz

1. A segment of DNA that separates two genes is called
   a. noncoding DNA.
   b. spacer DNA.
   c. a transposon.
   d. an exon.

2. In prokaryotes and eukaryotes, gene expression is most often controlled by regulation of which of the following?
   a. the destruction of a gene's protein product
   b. the length of time mRNA remains intact
   c. transcription
   d. translation

3. Which of the following is a regulatory DNA sequence?
   a. a repressor protein
   b. an operator
   c. an intron
   d. a housekeeping gene

4. During development, different cells express different genes, which results in
   a. the formation of different cell types.
   b. gene mutation.
   c. DNA packing.
   d. developmental abnormalities.

5. The DNA of eukaryotes is packed most loosely at which time?
   a. during mitosis
   b. when genes are expressed
   c. during interphase
   d. both b and c

6. Select the term that best describes all the DNA of an organism.
   a. genes
   b. spacer DNA
   c. gene expression
   d. genome

7. Genes that have an essential role in maintaining cellular activities and that are expressed in most cell types are called
   a. regulatory genes.
   b. housekeeping genes.
   c. homeotic genes.
   d. enzymes.

8. A DNA sequence that can move from one position on a chromosome to another is called
   a. an operator.
   b. a transposon.
   c. an exon.
   d. an intron.

9. Assume that an organism's genome has 20,000 genes and a large quantity of DNA, most of which is noncoding DNA. What kind of organism is it most likely to be?
   a. an insect
   b. a plant
   c. a bacterium
   d. either a or b

Biology in the News

RNA Treatment Lowers Cholesterol

Mouse study points way for targeting "interference" therapies

By Erika Check

Cholesterol levels have been lowered in mice by using tiny pieces of genetic material to block a particular gene. The demonstration shows that the experimental technique, called RNA interference (or RNAi), could be a practical way of curing a wide range of diseases.

RNAi works by using small molecules of RNA to trigger a cell to shut off a particular gene, for example, one that codes for a harmful protein.

For an increasing number of human genetic disorders, scientists know which gene or genes cause the disorder, but they still have no way to cure the condition. Researchers have attempted to cure such disorders by repairing the genes that cause the problem (an approach known as gene therapy), but this approach has proved more difficult than first envisioned. As described in this news article, another promising approach has entered the arena: RNAi.

Discovered in 1998, RNAi is the process of using specific sequences of RNA to reduce the expression of genes to which this RNA can bind. Mammals use RNAi as a natural defense mechanism against viruses such as HIV that have a genetic blueprint made from RNA, not DNA. The mammalian defense mechanism works by using an enzyme called Dicer to chop the viral genetic material into short pieces, each about 22 bases long. The mammalian cell then synthesizes a single-stranded RNA copy of the information in these short pieces; the resulting "short interfering" pieces of RNA (siRNAs) can bind to key viral genes, blocking their expression and helping the cell to overcome the infection.

In the laboratory, the natural RNAi process can be mimicked by designing a short RNA molecule that turns off a particular gene, such as one that produces a protein that causes health problems. In just 6 years, scientists and biotechnology companies have moved from describing the basic phenomenon to conducting clinical trials. The study described in this news article has been hailed as groundbreaking because, in mice at least, it solved a major problem: how to deliver the short RNA molecules to the part of the body where the problem genes are expressed (in this case, cells in the liver and small intestine that synthesize cholesterol). In previous studies, the short RNA molecules that blocked the effects of targeted genes tended to be damaged or destroyed before they reached the part of the body where they were needed.

Evaluating the news

1. Some people object to gene therapy on the grounds that it is not ethical to alter a person's DNA. Since RNA interference does not alter our genes, but rather blocks their effects, do you think RNAi constitutes a form of "genetic enhancement?"

2. Some mammals (including humans) have a rare mutation in which a protein that blocks muscle growth is produced in unusually low amounts; individuals with such mutant alleles have exceptionally well-developed muscles. In principle, short RNA molecules could be designed to shut down the gene that limits muscle growth. Thus RNAi has the potential to provide "normal" individuals (those that lack the rare mutant allele) with an easy way to increase muscle mass. Sports officials, who are aware of this potential, refer to this and other approaches that modify human genes or their effects as "gene doping." They are concerned that such treatments may offer athletes an undetectable way to gain an edge. Do you think it is likely that gene doping will be used by athletes and other people seeking to improve some aspect of their performance?

3. Some people are born with a genetic makeup that provides them an advantage in sports; the same is likely to be true of many human endeavors. Are such individuals "genetically enhanced?" If it proves possible to do so, should individuals that lack alleles providing an advantage in some aspect of human performance be allowed to use RNAi to level the playing field? How do you think it would affect human societies if people were allowed to use such an approach to influence any or all aspects of their phenotype?

Source: *Nature*, November 10, 2004.

CHAPTER **15**

DNA Technology

Key Concepts

◉ Recent innovations in the techniques used to manipulate DNA have greatly increased our ability to isolate and study genes and to alter the DNA of organisms.

◉ Restriction enzymes cut DNA molecules at specific target sequences. When used with gel electrophoresis, a technique that sorts the chopped pieces of DNA by size, restriction enzymes provide a powerful way to examine DNA sequence differences.

◉ A gene is said to be cloned when geneticists isolate it and produce many copies of it. Once a gene is cloned, automated sequencing machines can quickly determine its DNA sequence.

◉ Cloning and sequencing a gene can provide vital clues to its function, making these techniques critical to the study of genes that cause inherited genetic disorders as well as genes involved in normal cell function.

◉ In genetic engineering, a gene is isolated, modified, and inserted back into the same species or a different species. Expression of the transferred gene can change the phenotype of the genetically modified organism.

◉ DNA technology provides many benefits, but its use also raises ethical concerns and poses risks to human society.

Glowing Bunnies and Food for Millions

Some jellyfish produce flashes of light that may serve to ward off attacks from predators. In 2000, these same lights made the headlines: the gene that enables jellyfish to produce light was transferred to a rabbit named Alba, creating a piece of "living art" meant to confront people with a creature that was both lovable and alien. Alba was never shown in an art exhibit because the outcry over her creation caused her to be confined to the laboratory in which she was made.

While genes that cause organisms to glow in the dark are used by some artists to create controversial art exhibits, such genes are used primarily in scientific research and in medical applications. A light-producing gene from a firefly, for example, can help doctors treat the lung-destroying disease tuberculosis (TB), which is caused by a bacterium. To achieve this, the firefly gene is inserted into TB bacteria taken from a patient. The bacteria are then screened for resistance to different antibiotics. When exposed to an antibiotic, resistant bacteria are able to grow and express their genes, including the inserted firefly gene. As a result, resistant bacteria glow in the dark, whereas bacteria that are not resistant sicken and die. Doctors can use the results of this test to prescribe an antibiotic to which the bacteria are not resistant.

Techniques like those used to transfer light-producing genes are also being used to develop crop plant varieties that could improve the health and nutrition of millions of people. The basic principle is simple: new genes are transferred to crops such as rice or wheat, enabling them to produce essential nutrients that they otherwise could not make. Such genetically altered crops could save the lives of millions of undernourished children.

The creation of a glowing rabbit, the detection of resistant TB strains, and the development of crops with improved nutritional value were all made

Alba, a White Rabbit Genetically Modified to Glow in the Dark

possible by innovations in techniques for manipulating DNA. These new techniques have led to many medical and commercial applications, from the isolation of disease-causing genes to the production of industrial lubricants. Methods of manipulating DNA are also increasingly in the news, whether because of the millions of dollars made or lost in biotechnology stocks, the use of DNA fingerprinting in murder and rape trials, or the promise and controversy that surrounds gene therapy. In this chapter we describe the basic techniques for manipulating DNA, along with some of their applications and risks.

People have been indirectly manipulating the DNA of other organisms for thousands of years. This fact is well illustrated by the many differences between domesticated species and their wild ancestors. For example, because of genetic changes brought about through selective breeding, dog breeds differ greatly from one another and from their wild ancestor, the wolf. Similarly, due to selective breeding, food plants such as wheat and corn bear little resemblance to the wild species from which they arose.

Although we have a long history of altering the DNA of other organisms indirectly through traditional selective breeding methods, the past 35 years have witnessed a huge increase in the power, precision, and speed with which we can make such changes. For the first time, we can now select a particular gene, produce many copies of it, and then transfer it back into a living organism of our choice. In doing so, we can alter DNA directly and rapidly in ways that would never happen naturally, as when the light-producing gene from a jellyfish was inserted into a rabbit.

We begin this chapter by describing the methods by which scientists can isolate a gene and analyze its sequence. We then turn to some of the many practical applications of this technology, including DNA fingerprinting, therapeutic and reproductive cloning, and genetic engineering. We close by considering some of the ethical issues and social and environmental risks associated with DNA technology.

15.1 Working with DNA: Basic Techniques

The set of techniques that scientists use to manipulate DNA directly is referred to as **DNA technology**. As you know, DNA is the genetic material of all organisms. (Some viruses use RNA as their genetic material, but viruses are not truly living organisms.) Although the base sequence of DNA varies greatly among species, the general chemical structure of the DNA molecule (see Figure 12.3) is the same in all species. This consistency means that similar laboratory techniques can be used to analyze DNA from organisms that are as different as bacteria and people.

We introduced DNA chips, a recent innovation in DNA technology, in Chapter 14. In this section we describe several basic methods of DNA technology that made the invention of DNA chips possible. Most of these methods were invented and refined over the last 45 years.

To illustrate how these methods can be used, let's see how they can be applied to sickle-cell anemia. Recall that sickle-cell anemia is a lethal recessive genetic disorder in humans, caused by a mutation that alters a single amino acid in hemoglobin, a protein involved in the transport of oxygen by red blood cells (see Figure 13.10). Individuals who are homozygous for the disease-causing allele (genotype *ss*) suffer from many serious complications, including damage to the heart, lungs, kidneys, and brain.

People with sickle-cell anemia used to die at a young age and hence did not pass the sickle-cell allele on to the next generation. Now, however, with intensive medical care, these people can live into their forties or older, and hence can have children. Heterozygous individuals (genotype *Ss*), who carry the sickle-cell allele but usually have few or no symptoms, can also have children. As we shall see, DNA technology can be used to determine whether a person has no, one, or two copies of the sickle-cell allele. Such information can be very helpful to parents as they consider whether to have children.

Several key enzymes are used in **DNA technology**

We humans have 3.3 billion base pairs of DNA on 23 unique chromosomes. Each of these chromosomes contains a DNA molecule so large (140 million base pairs, on average) that it is difficult for scientists to work with. Thus, after DNA has been extracted from a person's cells, it must be broken into smaller pieces.

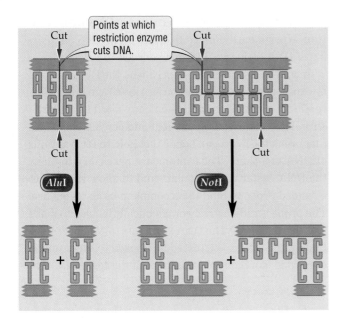

Figure 15.1 Restriction Enzymes Cut DNA at Specific Places

The restriction enzyme *Alu*I specifically binds to and then cuts the DNA molecule wherever the sequence AGCT (and no other sequence) occurs. Another restriction enzyme, *Not*I, specifically binds to and cuts the DNA sequence GCGGCCGC (and no other).

Gel electrophoresis sorts DNA fragments by size

Once a DNA sample has been cut into fragments by one or more restriction enzymes, researchers often use gel electrophoresis to help them see and analyze the fragments. In **gel electrophoresis** [uh-*LEK*-tro-fuh-*REE*-siss], DNA that has been chopped up by a restriction enzyme is placed into a depression (a "well") in a gelatin-like slab (a "gel") (Figure 15.2). An electrical current is then passed through the gel, causing the DNA (which has a negative electrical charge) to move toward the positive end of the gel. (Recall that opposite charges attract each other and like charges repel each other.) Longer pieces of DNA pass through the gel with more difficulty, and thus move more slowly, than shorter pieces. The distance the fragments travel through the gel is related to their speed of movement: after a fixed time period, the shorter, more rapidly moving fragments are found toward the positive end of the gel, while the longer, more slowly moving fragments are located closer to the negative end. Because the fragments are invisible to the human eye, they are stained or labeled by one of various methods.

By using restriction enzymes and gel electrophoresis together, we can examine differences in DNA sequences.

DNA can be broken into pieces by **restriction enzymes**, which cut DNA at highly specific sites. A restriction enzyme called *Alu*I, for example, cuts DNA everywhere the sequence AGCT occurs, but nowhere else (Figure 15.1). There are many different restriction enzymes, each of which recognizes and cuts its own unique target sequence of DNA. Restriction enzymes were discovered in bacteria in the late 1960s. In nature, restriction enzymes protect bacteria against foreign DNA, such as that of a virus (the enzymes chop up the viral DNA, thereby "restricting" viral growth).

When a restriction enzyme is used to chop up DNA in a test tube, the specificity of the enzyme ensures consistency: the same results are obtained each time DNA from one individual is used, regardless of what tissue (such as skin or hair) the DNA came from. Because restriction enzymes work on the DNA of all organisms, they are used in virtually all applications of DNA technology.

Two other important groups of enzymes are used in DNA technology. **Ligases** connect two DNA fragments to each other, making it possible to insert a gene from one organism into the DNA of another. DNA polymerase, the key enzyme involved in DNA replication, can also be used by geneticists to make many copies of a gene or other DNA sequence in a test tube.

Figure 15.2 Gel Electrophoresis

When subjected to an electrical current for a given period of time, DNA fragments move through a gel at different rates, depending on their size. Fragments found toward the positive end of the gel are shorter than fragments found toward the negative end of the gel. DNA cut by different restriction enzymes produces different patterns on the gel, an observation exploited in analyzing DNA for the presence of certain alleles or sequences.

For example, the restriction enzyme *Dde*I cuts the normal hemoglobin allele into two pieces, but it cannot cut the sickle-cell allele, thus providing a simple test for the disease allele (Figure 15.3).

DNA probes can test for the presence of an allele or gene

Another way of testing for the sickle-cell allele is to use a DNA probe. A **DNA probe** is a short, single-stranded segment of DNA with a known sequence, usually tens to hundreds of bases long. A probe can pair with another single-stranded segment of DNA if the sequence of bases in the probe (for example, CTGAGGA) is complementary to the sequence of bases in the other segment (for exam-

ple, GACTCCT). DNA probes are used in **DNA hybridization** experiments (Figure 15.4), which involve the pairing of DNA from two different sources.

In a DNA hybridization experiment, the DNA that will be exposed to a probe is first broken into fragments (using restriction enzymes) and then converted to a single-stranded form (by applying heat and certain chemicals that break the hydrogen bonds that hold the two strands of DNA together). These steps are necessary to make the DNA easier to manipulate and to allow the single-stranded probe to bind to its complementary sequence. The probe is radioactively or chemically labeled to make it easier to identify the DNA segments to which it binds. Finally, the DNA to be tested is exposed to the probe. If the probe can bind to a fragment of the test DNA, that

Learn more about genetic testing.

15.1

Figure 15.3 Identifying the Sickle-Cell Allele with Restriction Enzymes and Gel Electrophoresis

The restriction enzyme *Dde*I cuts DNA wherever it encounters the sequence GACTC (shown in yellow in the top panels). Only one such sequence occurs in the normal allele of the hemoglobin gene. A single base-pair mutation (in which T–A becomes A–T) causes sickle-cell anemia; this mutation occurs in the GACTC sequence. Since *Dde*I cannot recognize the mutant sequence GACAC, it does not cut the DNA at this location. As a result, the sickle-cell allele produces only one band on the gel, whereas the normal hemoglobin allele shows up as two bands on the gel, representing two fragments of different sizes.

Figure 15.4 DNA Hybridization
When exposed to several single-stranded segments of test DNA (shown in blue), a DNA probe (shown in orange) can bind only to the segment that contains the base sequence that is complementary to the probe.

Radioactively labeled DNA probe

Single-stranded test DNA

The probe is mixed with single-stranded test DNA.

The probe pairs with the test DNA if its complementary base sequence is present.

shows that the sequence complementary to the probe is present on that fragment.

Scientists can use DNA hybridization experiments to test whether a person has no, one, or two copies of the mutant allele that causes sickle-cell anemia. In such experiments, two 21-base-long DNA probes are used: one that binds only to the sickle-cell allele (*s*), and another that binds only to the normal hemoglobin allele (*S*). The probes that bind to a person's DNA indicate whether that person is normal (only the *S* probe binds), heterozygous (both probes bind), or homozygous for the disease (only the *s* probe binds).

DNA sequencing and DNA synthesis are used in basic research and practical applications of modern genetics

DNA sequencing allows researchers to determine the sequence of bases in a DNA fragment, a gene, or even the entire genome of an organism. Sequences can be determined by several methods, the most efficient of which rely on automated sequencing machines (Figure 15.5). One of these machines can identify over a million bases per day, thus making it possible to determine the sequence of a single gene quickly. DNA can also be sequenced manually by relatively slow, but still highly effective, methods.

The synthesis of probes and other DNA fragments can also be automated. DNA synthesis machines can rapidly produce segments of a specified base sequence up to hundreds of bases long. For example, in less than an hour, a DNA synthesis machine can produce the two 21-base probes used to test for sickle-cell anemia.

15.2 Working with DNA: DNA Cloning

A single copy of a gene is difficult to study, so geneticists often **clone** it by isolating and making many identical copies of it. "Cloning" can also refer to making many copies of any DNA fragment, not just a gene. A gene that is cloned can be sequenced, transferred to other cells or organisms, or used as a probe in DNA hybridization experiments. In addition, the cloning and sequencing of a gene enables researchers to use the genetic code to determine the amino acid sequence of the gene's protein product. Knowledge of a gene's product can provide vital clues to the gene's

Figure 15.5 A DNA Sequencing Machine
Automated DNA sequencing machines can determine DNA sequences rapidly.

function, as it did in the case of the *HD* gene involved in Huntington disease (see Chapter 13). For this reason, DNA cloning is a key step in the study of genes that cause inherited genetic disorders and cancers. This section introduces two of the most common methods of cloning DNA.

Cloning can be done by constructing a DNA library

A **DNA library** is a collection of DNA fragments from one organism, each of which is stored in a host organism, such as a bacterium. For humans, a complete DNA library would contain tens to hundreds of thousands of DNA fragments, which collectively would include all the 3.3 billion bases in the human genome.

The concepts behind the formation of a DNA library are simple. First, the DNA is broken into pieces by a restriction enzyme. The fragments are then inserted into **vectors**, which are pieces of DNA that are used to transfer genes or other DNA fragments from one species to another. **Plasmids**, which are small circular molecules of DNA that are found

Watch how a DNA library is created. 15.2

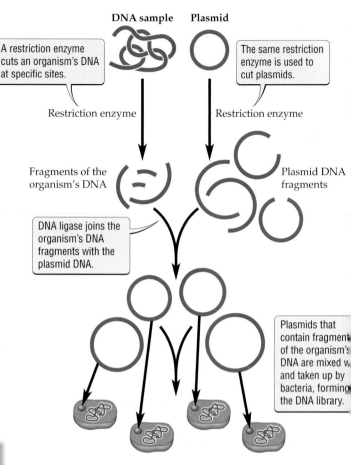

Figure 15.7 Construction of a DNA Library
This DNA library consists of a population of host bacteria, each of which contains a plasmid that has been manipulated to contain a different fragment of DNA from another organism.

Figure 15.6 Plasmids
Plasmids are small, circular molecules of DNA found naturally in bacteria. Plasmids are separate from, and much smaller than, the single main chromosome of the bacterium. Here, a ruptured *E. coli* bacterium spills out its chromosome and several plasmids, two of which are indicated with arrows.

naturally in bacteria (Figure 15.6), are commonly used as vectors; DNA from viruses can also be used for this purpose. Vectors can move DNA fragments into a host organism such as a bacterium; as we'll see below, this is done so that multiple copies of each DNA fragment can be made.

To use plasmids to construct a DNA library of the human genome, for example, fragments of human DNA are inserted into plasmid vectors and then mixed with bacteria under conditions that cause some of the bacteria to take up a plasmid (Figure 15.7). Once the bacteria have taken up the plasmids, the library is complete: each bacterium now contains a different fragment of human DNA, and collectively the population of bacteria contains many fragments of the human genome.

How is a DNA library used? Bacteria reproduce rapidly, and as they do so, they make new copies of the DNA fragments in the plasmids. So one use of a DNA library is for cloning DNA. In the first step of this process, a few bacterial cells from the library are grown on nutrient gel in a petri dish. Each bacterial cell gives rise to a mass of cells, called a colony, on the surface of the gel. Every cell in the colony contains a copy of the DNA fragment inserted into the bacterium that started the colony.

To clone a particular gene, we first must locate bacterial colonies that carry that gene. Therefore, the colonies are screened (tested) by DNA hybridization to see if their DNA can pair with a probe for the gene of interest. Colonies whose DNA can pair with the probe contain all or part of the desired gene. It is often necessary to screen the colonies on many petri dishes before such a colony is found. Bacteria from a colony that contains all or part of the gene are then grown in a liquid broth, producing billions of bacterial cells. Each of these cells contains a plasmid with the gene of interest embedded in it. Thus, by screening a DNA library, we can isolate a particular gene from a large number of different human DNA fragments, then produce many copies of that gene.

The polymerase chain reaction can clone DNA very rapidly

In some cases, DNA can be cloned by the **polymerase chain reaction (PCR)**, a method that uses DNA polymerase to make billions of copies of a targeted sequence of DNA in just a few hours (Figure 15.8).

To clone a particular gene by PCR, two short segments of synthetic DNA, called **DNA primers**, must be used. Each primer is designed to bind to one of the two ends of the gene of interest by complementary base pairing. By the series of steps shown in Figure 15.8, DNA polymerase then produces many copies of the sequence of DNA that is between the primers; that is, it produces many copies of the gene. To use PCR in gene cloning, scientists must know the DNA sequence of both ends of the gene; without this knowledge, they cannot synthesize the specific primers that will pair with the ends of the gene.

See the polymerase chain reaction in motion. **15.3**

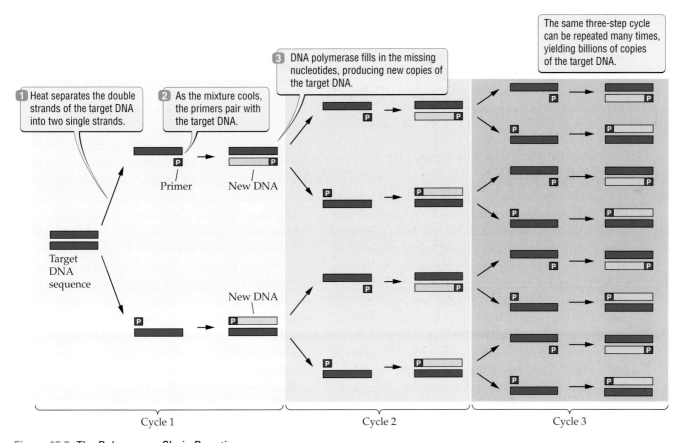

Figure 15.8 The Polymerase Chain Reaction
Short primers that can pair with the two ends of a gene of interest are mixed in a test tube with a sample of an organism's fragmented DNA, the enzyme DNA polymerase, and nucleotides (containing the bases A, C, G, or T). A machine then processes the mixture through the three steps shown here, in which the temperature is first raised and then lowered, to double the number of double-stranded versions of the desired sequence. The doubling process can be repeated many times (only three cycles are shown here). The targeted DNA sequence for a given cycle (dark blue), the DNA added in a given cycle (light blue), and the primers (red) are color-coded here for clarity only—the colors do not represent biological differences in the DNA of these sequences.

PCR technology offers two key advantages over other methods of DNA cloning: it can be used with very small amounts of DNA, and it clones genes very rapidly. As a result, PCR has come to be used widely in both research and commercial applications. The technique became so successful so quickly that in 1991, only six years after the first paper on PCR was published, the PCR patent was sold for $300 million. Two years later, in 1993, Kary Mullis won a Nobel prize for his discovery of PCR.

15.3 Applications of DNA Technology

There are many important applications of DNA technology, recent examples of which include prenatal screening for genetic disorders and the use of genetic profiling to determine which of several drugs should work best in a given patient (see page 254). Other examples include gene therapy (addressed later in this chapter) and the use of DNA technology in therapeutic and reproductive cloning (see the box on page 287). In this section we focus on two common uses of DNA technology: DNA fingerprinting and genetic engineering.

DNA fingerprinting can distinguish one individual from another

The use of DNA analyses to identify individuals is called **DNA fingerprinting**. Scientists use DNA fingerprinting in much the same way that traditional fingerprints are used. A laboratory can take a biological sample, such as blood, tissue, or semen, from a crime scene and develop a DNA fingerprint, or profile, of the person from whom the sample came. That profile can then be compared with another profile—for example, that of a crime victim or suspect—to see if the two profiles match (Figure 15.9).

When two DNA profiles don't match, as when DNA from a rape suspect differs from DNA obtained from semen found in the victim, the results are definitive (in this case, proving that the suspect is innocent). A match, like that shown in Figure 15.9, provides evidence that a crime scene sample (for example, a drop of blood on a shirt) could have come from the tested victim or suspect. However, a match does not provide definitive proof, since it is theoretically possible for two people to have the same DNA profile. In most legal cases, however, the probability that two people will have the same DNA profile is between one chance in 100,000 and one chance in a billion. (The range depends in part on the methods used to create the DNA profile, as

Figure 15.9 Catching the Bad Guys Using DNA
The DNA profile on the far left is that of the defendant (D) in a murder trial, while the profile on the far right is that of the victim (V). The defendant's jeans and shirt were splattered with blood—the DNA from which matches the victim's.

we shall see shortly.) Widely used in criminal cases, DNA fingerprints have been used to convict criminals of murder and rape and to prove suspects' innocence, sometimes freeing convicts after years of wrongful imprisonment.

DNA fingerprinting takes advantage of the fact that all individuals (except identical twins and other multiples) are genetically unique. In order to distinguish between different people, researchers examine regions of DNA that vary greatly from one person to the next. Such highly variable regions include noncoding portions of our DNA, such as introns and spacer DNA, which tend to vary greatly in size and base sequence from one person to the next.

DNA fingerprinting can be done by various methods, including RFLP analysis and PCR amplification. In **RFLP analysis**, restriction enzymes are used to cut a DNA sample into small pieces. Differences (between individuals) in the lengths of these fragments are called restriction fragment length polymorphisms, or RFLPs. Next, the fragments are sorted by size using gel electrophoresis. Finally, DNA probes are used to identify the number and size of the fragments on the gel, enabling DNA profiles like those in Figure 15.9 to be formed. Usually several probes are used, each of which binds to a DNA region known to vary greatly between individuals. The more probes are used, the lower the chance that two individuals will have the same DNA fingerprint.

Science Toolkit

Human Cloning

On December 27, 2002, members of a Quebec-based sect announced that they had produced the first human clone, "Eve," a baby girl born the previous day. For several months, a flurry of news coverage focused on this story, which, if true, would mean that a mere 6 years after the birth of the first **clone** of a mammal (Dolly the sheep), people would be added to a growing list of mammals that have been cloned. Most scientists reacted with skepticism to the idea that this sect, which believes that life on Earth was created by aliens, had actually cloned a person, and after several weeks during which the sect failed to produce proof that the baby was indeed a clone, the story was dropped abruptly.

But what exactly is cloning, and why would anyone want to clone a sheep, a pig, or a cow, let alone a person? When referring to tissues, organs, or whole organisms, *cloning* can refer to two technologies: therapeutic cloning and reproductive cloning. Although related to each another, these technologies differ in both process and intent.

In **therapeutic cloning**, the goal is to produce stem cells, which have potential uses in a wide range of medical applications. Recall from Chapter 9 that stem cells are a special class of undifferentiated cells that can be used to produce any of the body's many different cell types. There are three key steps in therapeutic cloning. First, the nucleus of an unfertilized egg cell is replaced with the nucleus of a nonreproductive donor cell, such as a skin cell. Next, chemicals are used to activate the egg—that is, to trick it into dividing so that it begins to form an embryo. Finally, stem cells are removed from the developing embryo and stimulated to grow into a wide range of human cell types.

In the distant future, it may be possible to use these stem cells to grow entire human organs. A more immediate goal is to transplant the cells to damaged organs, such as a heart after a heart attack, where they can divide, thereby replacing injured or dead cells. Progress toward this second, more limited goal has been made: stem cells have been produced and transplanted successfully to mouse organs, and results from a 2003 study suggest that a similar approach can improve heart function in human heart attack victims. In addition, cloned skin is already being used for victims of severe burns.

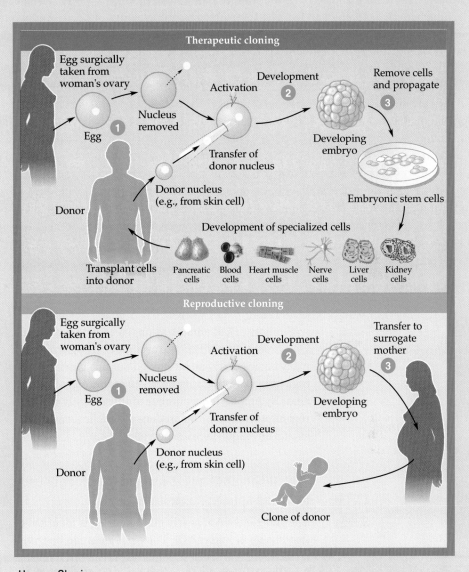

Human Cloning

In therapeutic cloning, the goal is to produce stem cells with the same genotype as the donor; these stem cells could then be used to repair damaged organs without causing the donor's immune system to reject them as foreign. In reproductive cloning, the goal is to produce a genetically identical copy, or clone, of the donor individual.

The goal of **reproductive cloning** is to produce an offspring that is an exact genetic copy of another individual. The first two steps in reproductive cloning are the same as those in therapeutic cloning, but stem cells are not removed from the embryo. Instead, the embryo is left intact and transferred to the uterus of a surrogate mother, where, if all goes well, it continues to develop, ultimately resulting in the birth of a healthy baby. Such a baby—whether a sheep, pig, or person—is referred to as a

(continued)

clone, since it is genetically identical to the individual who provided the donor nucleus.

To date, clones have been developed in a variety of mammals, including sheep, pigs, mice, cows, and cats. In each instance, the goal was to produce offspring that were genetically identical to the individual that provided the donor nucleus. Why would anyone want to do this?

Reproductive cloning can be used to produce multiple copies of an organism with useful characteristics. For example, a company in South Dakota has produced cloned calves that have been genetically engineered to produce human disease-fighting proteins called immunoglobins. The ultimate aim is to create herds of genetically identical cows, each of which would serve as a "biological factory," producing large quantities of commercially valuable immunoglobin proteins.

Reproductive cloning is also being used to produce pigs that could save the lives of people in need of organ transplants. Each year, thousands of people die while waiting for an organ transplant. Pig organs are roughly the same size as human organs, and could work well in people except for one major problem: the human immune system rejects them as foreign. In a series of recent studies, scientists have used reproductive cloning to produce pigs whose organs lack a key protein that stimulates the human immune system to attack. Thus this work represents an important step toward the production of pig clones whose organs could be used to save the lives of people who would otherwise die for lack of a suitable transplant organ.

And what about people? Should human reproductive cloning be allowed so that people could clone themselves or attempt to "rebuild" a loved one lost to disease or a tragic accident? Recall that an individual's phenotype is determined not only by its genotype, but also by the interaction of its genes with one another and with the environment. Thus a clone would be a different person, with a different personality, from the individual whose genes they shared (this is true even for identical twins who grow up together). Would it be fair to subject a human clone to the expectation that they be just like the person they were intended to replicate? In addition, reproductive cloning of animals can produce individuals that suffer from multiple genetic and physical defects and may die at an early age. Given this fact, do you think it would be ethical to clone a person?

Explore whether human cloning is possible. 15.4

Similarly, PCR amplification (so called because the PCR process "amplifies" the original DNA by producing many copies of it) can be carried out using primers for regions of DNA that vary greatly from one person to the next. By amplifying a number of these highly variable regions, technicians can produce a DNA fingerprint that should be unique for each person.

Learn more about genetic engineering. 15.5

Genetic engineering is used to transfer genes from one species to another

The genetic code of nearly all organisms is identical to that shown in Figure 13.6. Thus, if a gene can be transferred from one species to another, it can often make a functional protein product in the new species. The light-producing gene from a firefly mentioned at the beginning of this chapter, for example, has been transferred to and expressed in organisms as different as plants, mice, and bacteria. The deliberate transfer of a gene from one species to another is an example of genetic engineering. **Genetic engineering** is a three-step process in which a DNA sequence (often a gene) is isolated, modified, and inserted back into the same species or into a different species. An individual receiving such a DNA sequence is called a **genetically modified organism (GMO)**.

Several techniques can be used to insert genes into organisms. We have already seen how plasmids can be used to transfer a gene from humans or other organisms to bacteria. Plasmids can also be used to transfer genes to plant or animal cells (Figure 15.10). In some species, including many plants and some mammals, genetically modified (GM) adults can be generated, or cloned, from these altered cells (see the box on page 287). Other techniques for gene transfer include the use of viruses to "infect" cells with genes from other species and gene guns that fire microscopic pellets coated with the gene of interest into target cells.

Genetic engineering is commonly used to alter the phenotype (often some aspect of the performance or productivity) of the GMO. Atlantic salmon, for example, have been given genes that cause them to grow up to six times faster than normal (Figure 15.11). Crop plants have been genetically engineered for a wide variety of traits, including insect resistance, disease resistance, frost tolerance, and herbicide resistance (to allow the crops to survive the application of weed-killing chemicals). In 2002, such GM crops were planted in 59 million hectares (145 million acres) worldwide, a coverage that included over 20 percent of the world's soybeans, corn, cotton, and canola.

Genetic engineering is also commonly used to produce many copies of a DNA sequence, a gene, or a gene's protein product (Table 15.1). One example is the human hormone insulin, which is mass-produced by *E. coli* bacteria that have been engineered to contain the human insulin gene. Since 1978, the insulin produced by these bacteria has made it possible to treat millions of people suffering from

Firefly

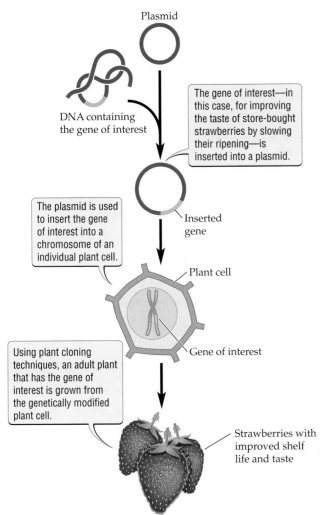

Plasmid

DNA containing the gene of interest

The gene of interest—in this case, for improving the taste of store-bought strawberries by slowing their ripening—is inserted into a plasmid.

Inserted gene

The plasmid is used to insert the gene of interest into a chromosome of an individual plant cell.

Plant cell

Using plant cloning techniques, an adult plant that has the gene of interest is grown from the genetically modified plant cell.

Gene of interest

Strawberries with improved shelf life and taste

Figure 15.10 Genetic Engineering of Plants via Plasmids
Plasmids can be used to insert a gene of interest into plant cells. Adult plants can then be grown from the genetically modified cells.

15.4 Ethical Issues and Risks of DNA Technology

As we have seen, DNA technology provides many benefits to human society. At the same time, the immense power and scope of genetic engineering and other aspects of DNA technology raise ethical concerns and pose a variety of risks. At the most basic level, what gives us the right to alter the DNA of other species? We typically do so for our own advantage, but is this an ethical thing to do? As you think about this question, remember that although humans can make striking changes to other species using genetic engineering, we can do the same using conventional plant or animal breeding (Figure 15.12). Many people might agree that there was no ethical conflict in altering the DNA of a bacterium or a virus, but does that mean there is also no conflict associated with altering the DNA of a plant, a dog, a chimpanzee, or a person?

With respect to altering human DNA, how do we distinguish between acceptable and unacceptable uses of genetic engineering? If it is ethical to genetically engineer a human being to cure a horrible disease, is it also acceptable to make less critical changes? For example, if it were possible to do so, would it be ethical to alter the future intelligence, personality, looks, or sexual orientation of our children before birth? According to a 1990s

diabetes more safely and less expensively than in the past, when insulin derived from pigs and cows was sometimes in short supply and could cause allergic reactions.

Another application of genetic engineering is in the manufacture of certain vaccines. Bacteria have been genetically engineered to produce large amounts of the surface proteins of a particular disease agent (that is, a virus, bacterium, or protist that causes an infectious disease). When injected into the human body, these proteins stimulate the immune system to recognize and destroy the disease agent that normally carries them, thus protecting us against future attack. This approach has been used to develop effective vaccines for flu and malaria, among other diseases. Similarly, despite a recent setback in which a genetically engineered AIDS vaccine was only moderately successful, there are ongoing attempts to develop an AIDS vaccine using genetically engineered surface proteins from the AIDS virus.

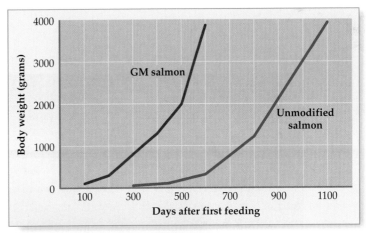

Figure 15.11 Feed Me!
In fish farms, genetically modified salmon eat more food and grow more rapidly than their unmodified relatives.

Table 15.1

Methods of Production and Uses for Some Products of Genetic Engineering

For each product, the gene or DNA sequence that codes for the product is either inserted into host cells, such as *E. coli* or mammalian cells, or used in one of several automated procedures, such as a DNA synthesis machine or PCR. These cells or automated procedures are then used to make many copies of the product.

Product	Method of production	Use
PROTEINS		
Human insulin	*E. coli*	Treatment of diabetes
Human growth hormone	*E. coli*	Treatment of growth disorders
Taxol biosynthesis enzyme	*E. coli*	Treatment of ovarian cancer
Luciferase (from firefly)	Bacterial cells	Testing for antibiotic resistance
Human clotting factor VIII	Mammalian cells	Treatment of hemophilia
ADA	Human cells	Treatment of ADA deficiency
DNA SEQUENCES		
Sickle-cell probe	DNA synthesis machine	Testing for sickle-cell anemia
BRCA1 probe	DNA synthesis machine	Testing for breast cancer mutations
HD probe	*E. coli*	Testing for Huntington disease
M13 probe	*E. coli*, PCR	DNA fingerprinting in plants
33.6 and other probes	*E. coli*, PCR	DNA fingerprinting in humans

(a)

(b)

Figure 15.12 Genetic Engineering vs. Breeding
When it comes to strange appearances, genetic engineering and more conventional breeding methods compete. (a) These "naked" chickens were genetically engineered as part of a research project to develop low-fat poultry that is environmentally friendly and cheap. (b) This British Belgian Blue bull developed due to selective breeding practices promoted by the meat industry.

Biology Matters

Think Before You Drink

The use of genetically modified organisms (GMOs) in agriculture and food production has expanded dramatically since biotech crops were first commercially grown in 1996. So has controversy over its risks and benefits. Global biotech crop acreage grew to 200 million acres in 2004, according to the International Service for the Acquisition of Agri-biotech Applications (ISAAA). In 2003, the United States grew 63 percent of the transgenic crops grown worldwide; Canada grew 6 percent.

Because there are no labeling requirements for GMO foods in the United States, many consumers are unaware that between 70 and 75 percent of all processed foods available in U.S. grocery stores may contain ingredients from genetically engineered plants, according to estimates from the Grocery Manufacturers of America. Breads, cereal, frozen pizzas, hot dogs, and soda are just a few of them. Soybean oil, cottonseed oil, and corn syrup are ingredients used extensively in processed foods. Soybeans, cotton, and corn dominate the 100 million acres of genetically engineered crops that were planted in the United States in 2003, according to the U.S. Department of Agriculture (USDA). Through genetic engineering, these plants have been made to ward off pests and to tolerate herbicides used to kill weeds. Other crops, such as squash, potatoes, and papaya, have been engineered to resist plant diseases.

Corn syrup, derived from corn, is a common ingredient in many juices and sodas, and genetically modified corn syrup is used in most

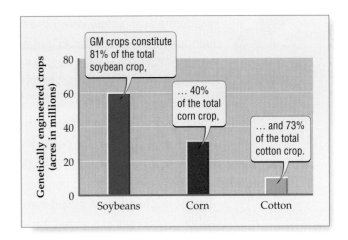

brand-name sodas (from Coke to 7-Up and Dr. Pepper) and juices (from Gatorade to Ocean Spray and V-8). If this concerns you, True Food Now!, an organization concerned about GMO use in the food supply, has created a shoppers guide. For more information visit their Web site at www.truefoodnow.org. While the battle rages between supporters and opponents of GMOs in food, it will be up to consumers—like you—to decide whether to purchase or avoid these products.

March of Dimes survey, more than 40 percent of Americans would make such modifications if given the chance. Is it fair for parents to make such decisions on behalf of their children? Often there are no easy answers to questions like these.

The use of DNA technology also involves risks. For example, 12 of the world's 13 most important crop plants can mate and produce offspring with a wild plant species in some region where they are grown. If a crop plant is genetically engineered to be resistant to an herbicide, the potential exists for the resistance gene to be transferred (by mating) from the crop to the wild species (one instance in which this has already occurred is described in the box above). Thus there is a risk that by engineering our crops to resist herbicides, we will unintentionally create "superweeds" that are resistant to the same herbicides. Similar risks exist for most efforts to alter the performance of an organism. In general, a gene that is good for humans when it is in one species (or one set of

circumstances, such as an agricultural field) may be bad for us if it is in another species (or another set of circumstances, such as a more natural field environment).

There may also be environmental or social costs associated with genetic engineering. Engineering crops to be resistant to herbicides, for example, might promote increased use of herbicides, many of which are harmful to the environment. Alternatively, a product might be environmentally safe yet still have social costs. Consider bovine growth hormone (BGH), which is mass-produced by GM bacteria. Among other effects, BGH increases milk production in cattle. Before the introduction of genetically engineered BGH in the 1980s, milk surpluses already were common. The use of BGH by large milk producers has created even larger milk surpluses, driving down the price of milk and forcing small producers of milk—the traditional family farms—out of business. Are lower milk prices for consumers worth the social cost of driving small dairy farms into bankruptcy?

Human Gene Therapy

On September 14, 1990, 4-year-old Ashanthi DeSilva made medical history when she received intravenous fluid that contained genetically modified versions of her own white blood cells. She suffered from adenosine deaminase (ADA) deficiency, a genetic disorder that severely limits the ability of the body to fight disease and can make ordinary infections, such as colds or flu, lethal. This disorder is caused by a mutation to a single gene that affects the ability of white blood cells to combat disease. Earlier, doctors had removed some of Ashanthi's white blood cells and added the normal ADA gene to them. Thus her cells were genetically engineered in an attempt to fix a lethal genetic defect. Ashanthi responded very well to the treatment and now leads an essentially normal life (Figure 15.13).

The treatment Ashanthi received was the first clinical gene therapy experiment ever performed. Human **gene therapy** seeks to correct genetic disorders by modifying the genes that cause them. The possibility of curing even the worst of genetic diseases by reaching into our cells and repairing the mutations that cause them is a bold and captivating prospect.

As such, gene therapy has attracted much media attention—some of it, unfortunately, bordering on hype. Take Ashanthi's case: in addition to gene therapy, she received other treatments for

Figure 15.13 Gene Therapy for ADA Deficiency
ADA deficiency, a lethal disorder caused by a genetic defect in white blood cells, was the first genetic disorder to be treated by gene therapy.

Ashanthi DeSilva has ADA. Gene therapy for ADA may have contributed to her remarkable good health.

Test your knowledge of how gene therapy can help cure ADA deficiency.

15.6

1 A blood sample is taken.

2 White blood cells are isolated from the sample.

3 A corrected ADA gene is inserted into a harmless virus.

4 The patient's white blood cells are infected with genetically modified viruses.

5 White blood cells containing the corrected ADA gene are returned to the patient.

White blood cells containing corrected ADA gene

Corrected ADA gene

ADA deficiency. Hence, contrary to some reports in the media, her remarkable good health cannot be attributed to gene therapy alone. Overall, although there are now more than 600 gene therapy experiments in progress worldwide, there are few success stories. Why? Has gene therapy been oversold?

Until recently, proponents of gene therapy could point to a pioneering French study and answer with a confident "No." That study, which was published in 2000, described how researchers cured X-SCID, a disease similar to ADA deficiency that cripples the immune system, by inserting a corrected gene into bone marrow cells. Eleven children with X-SCID were treated by gene therapy alone, and nine of them were cured. Thus, for the first time, scientists had achieved the holy grail of gene therapy: a human genetic disorder was cured solely by fixing the genes that caused the disorder.

Unfortunately, two of the children who were cured of X-SCID went on to develop leukemia, a form of cancer that strikes white blood cells (one of these children died in 2004). What went wrong boils down to a delivery problem: by mistake, the corrected gene was inserted near, and promoted the expression of, a gene that can cause leukemia. In early 2005, a third child (who received gene therapy for X-SCID in 2002) also developed leukemia; at the time of this writing, the cause of the cancer was not yet known.

The three cases of leukemia followed close on the heels of another terrible event: in 1999, a young man participating in a gene therapy experiment died from an allergic reaction to the virus used to deliver an engineered gene. The combined effect of these tragedies sent shock waves through the gene therapy field. Worldwide, many gene therapy trials were placed on hold, and gene therapy advocates no longer had a single success story in which gene therapy alone cured a genetic disorder without causing serious side effects. Some have suggested that gene therapy efforts should be abandoned. Should they?

Most scientists think not. Major setbacks are expected in innovative fields like gene therapy—medical history is filled with cases in which tragedies occurred during the development of techniques that we now take for granted (such as vaccines and organ transplants). In addition, despite the setbacks, promising results continue to pour in. For example, in the past few years researchers have used gene therapy to cure mice of sickle-cell anemia, causing an astonishing 99 percent of mouse red blood cells to express the corrected gene. In other experiments, also in mice, the devastating effects of muscular dystrophy were partially corrected by insertion of a gene that produces a protein lacking in people with that fatal disease. In both of these cases, the work with mice offers hope that similar progress can be made with people.

Gene therapy offers hope for curing terrible human diseases, but before it can reach its full potential, formidable hurdles remain. One of the most important is finding a way to deliver the engineered gene safely and effectively to the cells where it is needed. Harmless viruses are often used for this purpose, but to date viral delivery methods have had limited success. In part, the reason for this low rate of success is that the human body defends itself so well against viruses that the viruses often are destroyed before they deliver the corrected gene to enough of the target cells where it is needed.

Researchers are constantly striving to improve the effectiveness with which viruses deliver engineered genes, and they are also working hard to ensure that the viruses do not insert the modified genes into the wrong tissues or the wrong location within the DNA of the target cells, thereby causing cancer. With respect to X-SCID, for example, researchers are currently modifying the delivery virus so that it targets fewer types of cells, and hence will be less likely to cause cancer. Overall, although many challenges lie ahead, there is cause for excitement and hope.

Chapter Review

Summary

15.1 Working with DNA: Basic Techniques

- Scientists can manipulate DNA using a variety of laboratory techniques. Because the structure of DNA is the same in all organisms, these techniques work on all species.
- Restriction enzymes are used to break DNA into small pieces. Gel electrophoresis separates the resulting DNA fragments by size.
- Two other groups of enzymes—ligases and DNA polymerase—are also commonly used in DNA technology.
- DNA probes are used in DNA hybridization experiments to test for the presence of a particular allele or gene.
- Automated machines greatly speed up the processes of sequencing and synthesizing DNA.

15.2 Working with DNA: DNA Cloning

- A gene is said to be cloned if it has been isolated and many copies of it have been made.
- After being cloned, a gene can be sequenced, transferred to other organisms, or used in DNA hybridization experiments.
- Two of the most common methods of cloning are by constructing a DNA library and by using the polymerase chain reaction (PCR).
- To build a DNA library, a vector such as a plasmid is used to transfer DNA fragments from the organism whose gene is to be cloned to a host organism, such as a bacterium.
- To clone a gene by PCR, primers (short segments of DNA that can bind to the beginning and end of the gene) are synthesized and used to produce billions of copies of the gene in a few hours.

15.3 Applications of DNA Technology

- Reproductive cloning has been used to produce clones of whole organisms, including nonhuman mammals. A related technology, therapeutic cloning, is used to produce stem cells that can be used to repair damaged organs and may, in the distant future, be used to grow entire organs in need of replacement.
- DNA fingerprinting, which results in a unique genetic profile for each individual, is widely used in criminal cases. DNA fingerprinting can be done by RFLP analysis or PCR amplification.
- In genetic engineering, a DNA sequence (often a gene) is isolated, modified, and inserted back into the same species or into a different species. Genetic engineering works because, with rare exceptions, all organisms share an identical genetic code.
- Genetic engineering is used to alter the phenotype (especially the performance or productivity) of the genetically modified organism (GMO) or to produce many copies of a DNA sequence, a gene, or a gene's protein product.

15.4 Ethical Issues and Risks of DNA Technology

- The use of DNA technology provides potential benefits, but also raises ethical questions and poses potential environmental and social risks. The fine line between acceptable and unacceptable changes made by us to our own DNA and the DNA of other organisms is not always clear.
- Benefits and risks must be considered carefully in evaluating the potential effect of any particular use of DNA technology on human society.

◉ Review and Application of Key Concepts

1. Discuss the extent to which our current ability to manipulate DNA differs from what people have done for thousands of years to produce a wide range of domesticated species, such as dogs, corn, and cows.

2. A couple is considering whether to have children. The woman knows that her grandmother had sickle-cell anemia, and hence that one of her parents may have passed the sickle-cell allele on to her; the same is true of the man. Describe in detail a series of DNA technology methods that would allow the man and the woman to determine whether they carry the sickle-cell allele or not.

3. Define DNA cloning and describe how it is done.

4. Discuss the advantages of cloning a gene.

5. When a DNA library is made, tens to hundreds of thousands of fragments of an organism's DNA are stored in a large number of host organisms (one fragment per host organism). How is the library screened so that scientists can find a particular gene of interest?

6. What is genetic engineering? How is it accomplished? Select one example of genetic engineering and describe its potential advantages and disadvantages.

7. Is it ethical to modify the DNA of a bacterium? A single-celled yeast? A worm? A plant? A cat? A human? Give reasons for your answers.

8. Are there some modifications to the DNA of humans that are not acceptable? Assuming you think so, what criteria would you use to draw the line between acceptable and unacceptable changes?

Key Terms

clone (of a gene) (p. 283)
clone (of an organism) (p. 287)
DNA fingerprinting (p. 286)
DNA hybridization (p. 282)
DNA library (p. 284)
DNA primer (p. 285)
DNA probe (p. 282)
DNA sequencing (p. 283)
DNA technology (p. 280)
gel electrophoresis (p. 281)
gene therapy (p. 292)
genetic engineering (p. 288)

genetically modified organism
   (GMO) (p. 288)
ligase (p. 281)
plasmid (p. 284)
polymerase chain reaction
   (PCR) (p. 285)
reproductive cloning (p. 287)
restriction enzyme (p. 281)
RFLP analysis (p. 286)
therapeutic cloning (p. 287)
vector (p. 284)

Self-Quiz

1. Which of the following cuts DNA at highly specific target sequences?
   a. ligases
   b. DNA polymerase
   c. restriction enzymes
   d. RNA polymerase

2. A collection of an organism's DNA fragments stored in a host organism is called a
   a. DNA library.
   b. DNA restriction site.
   c. plasmid.
   d. DNA clone.

3. The pairing of complementary DNA strands from two different sources is called
   a. DNA replication.
   b. DNA hybridization.
   c. genetic engineering.
   d. DNA cloning.

4. Genetic engineering
   a. can be used to make copies of a DNA sequence, a gene, or a gene's protein product.
   b. can be used to alter the phenotype of the genetically modified organism.
   c. raises ethical questions and poses risks to society.
   d. all of the above

5. When DNA fragments are placed on an electrophoresis gel and subjected to an electrical current, the _____ fragments move the farthest in a given time.
   a. smallest
   b. largest
   c. PCR
   d. DNA library

6. A short, single-stranded sequence of DNA whose bases are complementary to a portion of the DNA on another DNA strand is called a
   a. DNA hybrid.
   b. clone.
   c. DNA probe.
   d. mRNA.

7. Small circular segments of DNA that are found naturally in bacteria are called
   a. plasmids.
   b. primers.
   c. vectors.
   d. clones.

8. If the DNA sequence of the beginning and end of a gene is known, which of the following methods could be used to produce billions of copies of the gene in a few hours?
   a. construction of a DNA library
   b. reproductive cloning
   c. therapeutic cloning
   d. PCR

Biotech Grass Found Far Afield

BY MIKE LEE

Genes from a genetically engineered grass travel much farther than previously measured and can spread biotech traits to related plants at least 13 miles away, according to a study made public Monday.

"That is just huge," said Norm Ellstrand, genetics professor at the University of California, Riverside, and an expert in biotech gene movement. "How could you possibly contain it?"

Critics of genetic engineering have long argued that genes from genetically modified (GM) plants or animals could spread to wild species, potentially wreaking environmental havoc. Prior to now, however, their concerns tended to be dismissed as hypothetical and alarmist. Such glib dismissals will no longer be possible.

Some previous studies had shown that genes from GM crops do not spread far. In most cases, such studies were conducted with small test plots (for example, 0.1 acre) that contained few GM plants. The chance that genes from a GM plant will spread to a wild species is much lower in a small test plot than it would be in a large farm field with hundreds to thousands of acres.

In a study published in *Proceedings of the National Academy of Sciences*, scientists used a commercial-scale (400-acre) plot to examine the spread of genes for resistance to the herbicide Roundup. These resistance genes had been inserted into a variety of GM bentgrass developed by Monsanto Company and The Scotts Company; they hope to use the engineered bentgrass as turf in golf courses.

Grasses are wind-pollinated; thus the concern was that resistance genes could spread long distances if pollen blown by wind from GM bentgrass were to fertilize any one of a dozen other grass species that can breed with bentgrass. That appears to be what happened: DNA tests discovered the modified genes in non-GM bentgrass located up to 13 miles away from the test plot; the genes were also found in other grass species growing wild up to 9 miles from the test plot.

Representatives from Monsanto and Scotts suggested that these results are not all that worrisome, since any plants that contained the Roundup-resistance genes could be controlled by other herbicides. But even some advocates of the use of GM crops are not buying this argument. Bill Rose, a leading turfgrass marketer who favors the use of GM grasses, says that releasing wind-pollinated GM grasses is "like turning a stallion loose with no fences." He suggests short-circuiting this problem by using GM grasses engineered to have sterile pollen, hence making it much less likely that genes from such grasses could escape to wild species.

Evaluating the news

1. Based on the results of the bentgrass study described here, the United States Department of Agriculture (USDA) has ordered that an environmental impact study be conducted on GM bentgrass before the agency decides whether Scotts can sell it. This is the first time USDA has demanded such a review for a GM crop. Many other GM crops are wind-pollinated; corn, rice, wheat, rye, and barley are all derived from plants in the grass family. Should environmental impact studies be conducted for other crops as well?

2. If modified genes are transferred to a wild species and, as a result, the wild species is transformed into a "superweed," who should pay for the control of this new pest—the company that produced the GM crop that spawned the superweed, or the public (via taxes)?

3. Pollen from a GM plant engineered to produce an insecticide may contain that insecticide. Such pollen can be spread by wind, then consumed by animals (either directly, or indirectly when they eat leaves on which the pollen lands). Some studies (for example, on monarch butterflies) have shown that wild species can be harmed by eating such pollen, while other studies (some of which also were conducted on monarch butterflies) have found no such ill effects.

Given the conflicting scientific results, is it better to err on the side of caution (by implementing controls that may prove expensive), or should companies be allowed to continue planting large fields of GM crops until conclusive evidence is found that such practices cause serious environmental harm?

SOURCE: *Sacramento Bee*, Tuesday, September 21, 2004.

Harnessing the Human Genome

A Crystal Ball for Your Health

In February of 2001, the world witnessed a scientific milestone, the fruit of the combined efforts of thousands of researchers over a period of 15 years. For the first time in history, a draft copy of the DNA sequence of the human genome was available for perusal by anyone with a personal computer and access to the Internet. To many scientists, this represented the crowning achievement of twentieth-century biology, and many press conferences and articles touted it as such.

The successful sequencing of complete genomes gave birth to a new field of biology. In earlier chapters of Unit 3 you learned that genetics is the study of how genes are expressed and transmitted in cells and organisms. The field of **genomics** builds on genetics by seeking to understand the structure and expression of entire genomes and how they change during evolution. Genomics can be further distinguished from genetics by the scale of the questions asked. Genetics has a smaller focus: it is concerned with how individual genes function—either alone or together with a limited set of other genes—to control a phenotype. Genomics, on the other hand, takes a far more comprehensive view, monitoring the coordinated activities of all the genes in the genome. The expanded scale of the issues addressed by genomics has already had a major effect on other fields in biology.

How will scientific achievements such as the sequencing of the human genome affect our lives? The simple answer is that knowing the entire sequence of the human genome amounts to knowing the blueprint that dictates every biological process in our bodies. Encoded in our 3.3 billion base pairs of DNA are variations in our individual genes that are likely to directly affect our future and our health. While all of us have genomes that are

Main Message: Knowledge of the human genome has revolutionized biology and medicine while raising difficult ethical questions.

Helpful to know

Polymorphism (Greek for "many forms") can have different meanings in different contexts. In SNPs, a particular human DNA sequence is polymorphic due to differences in a single nucleotide. In restriction fragment length polymorphisms, or RFLPs (see Chapter 15), *polymorphism* refers to the different fragment lengths into which a person's DNA is cut by a given restriction enzyme.

99.9 percent identical to one another, the 0.1 percent of difference has great bearing on our susceptibility to certain diseases and genetic disorders—even on our overall life span. Our DNA sequences will reveal not just how our bodies work in the present, but also how they are likely to function in the future.

One of the great powers of knowing the roadmap of the human genome lies in its ability to predict an individual's predisposition for developing many diseases and genetic disorders. To use this information to predict the health of individuals, we must have a simple way to identify and compare relevant genetic variations. The DNA in two unrelated individuals differs, on average, by only one base in a thousand. Yet these single-base-pair differences, known as **single nucleotide polymorphisms** (**SNPs**) [*SNIPS*], are one important source of genomic variation (Figure C.1). If one could associate the presence of a specific SNP or group of SNPs with susceptibility to a disease such as breast cancer, individuals who have this SNP pattern could be forewarned of oncoming disease far in advance, allowing timely therapeutic intervention. Naturally, the detailed matching of SNP profiles with disease susceptibility depends on knowledge of the human genome sequence as a basis for comparison.

Even before the first draft of the human genome sequence was released in 2001, scientists in Great Britain began laying the foundation for a massive database of human SNP profiles. Today that database relies on SNP profiles from hundreds of thousands of blood samples donated by adult volunteers. Physicians refer these volunteers to the project, whose staff record each volunteer's current health status and

Figure C.1 This Technician Holds Ten Thousand Different SNPs in a Single DNA Microarray

match it with their SNP profile. In addition, the volunteers are tracked over time so that changes in their health can also be matched with their SNP profile. Using this database and others like it, specific SNPs will eventually be identified as indicators of disease susceptibility and even particular patterns of aging. The continued proliferation of such databases across the globe is redefining the concept of health profiling, and with it creating new concerns about the possible effects on individual privacy and even health insurance coverage.

SNP databases are just one example of the tremendous promise and accompanying ethical problems that result from having access to our complete genetic blueprint. In this essay we explore some of the medical benefits and ethical challenges this new knowledge is likely to bring us.

The Quest for the Human Genome

Efforts to understand how inheritance works have been proceeding for more than two millennia. There have been numerous intellectual and technical breakthroughs over the last 140 years, several of which we have highlighted in earlier chapters of this unit. All of these discoveries paved the way for the focused drive to sequence the human genome, which took a mere 15 years (Figure C.2). By 1990, the National Institutes for Health (NIH) and the U.S. Department of Energy (DOE) had created an international consortium of sequencing laboratories or centers, collectively known as the **Human Genome Project** (**HGP**). The HGP was originally scheduled to complete the first draft of the human genome sequence by 2005. The DNA to be sequenced was obtained from several anonymous donors from diverse backgrounds. To ensure privacy, the identities of the donors were never associated with the DNA samples. Although human genomes are generally 99.9 percent identical, using DNA from several individuals would avoid the possibility of sequencing a single genome with significant mutations and mistakenly using it as the genomic benchmark.

In the early stages of the HGP, the decision was made to sequence the smaller genomes of several model organisms as a rehearsal for tackling the human genome. These smaller sequencing projects would allow scientists to improve on existing methods and develop the computer programs necessary for knitting together and analyzing huge quantities of sequence data. The selected genomes came from a bacterium (*Escherichia coli*) with 4.6 million base pairs, budding yeast (*Saccharomyces cerevisiae*), with 12 million base pairs, and a nematode worm (*Caenorhabditis elegans*) with 97 million base pairs.

Interestingly, however, the first complete genome sequence of a free-living organism was published in 1995 by a group that was not involved in the HGP. The 1.8 million-base-pair genome of the *Haemophilus influenzae* bacterium was completely sequenced in a collaborative effort spearheaded by Craig Venter, a former NIH scientist. Herein lay the seed of a controversy that continues to swirl around large-scale research efforts like the Human Genome Project even today. The issue that lies at the heart of the controversy is whether competition among researchers, and possibly between commercial and federally funded research projects, is in the best interests of science and the public. In the case of Craig Venter, few would dispute that by eventually forming a commercial company to sequence the human genome, he spurred a competitive race with the HGP that ultimately accelerated the entire effort. On the other hand, might such competitive situations impede the sharing of scientific data that is an essential part of research?

A Preview of Our Blueprint

Now that we can browse our genetic blueprint, what do all those sequences of base pairs tell us? The sequence produced in 2001 was only a first draft; scientists are still filling in gaps of missing sequence, checking for accuracy, and confirming their predictions of which sequences truly represent genes and which are stretches of noncoding DNA (see Chapter 14). Analysis of the human genome sequence is therefore still in the early stages. However, many significant observations have already emerged. For one thing, the human genome sequence reveals the number of genes required to encode the physical complexities of the human organism.

Explore more about how the human genome was sequenced. C.1

Figure C.2 Milestones in the Quest for the Human Genome

Oswald Avery

Frederick Sanger

Aristotle believes that parents pass biological information to their offspring.

Oswald Avery, Colin MacLeod, and Maclyn McCarty show that DNA is the transforming factor and the possible basis for heredity.

Frederick Sanger and Allan Maxam and Walter Gilbert develop two different methods for sequencing DNA. Sanger's approach eventually becomes the preferred technique.

Fourth century BC   1865   1944   1953   1977

Gregor Mendel performs his groundbreaking studies on inheritance in peas, defining the basic laws of heredity: equal segregation and independent assortment.

James Watson and Francis Crick discover the double helical structure of DNA.

Purple flowers

White flowers

Round, yellow peas

Round, green peas

Wrinkled, yellow peas

Wrinkled, green peas

Complexity in organisms is determined by more than just the number of genes

Prior to the initial publication of the human genome sequence, scientists had predicted that humans would have at least 100,000 different genes. The first big surprise from reviewing the sequence was a revised estimate of 30,000 to 40,000 human genes. Today the estimate is lower still, with a range of 20,000 to 30,000 human genes. Estimates of gene numbers are largely based on computer analyses of genome sequences, using programs that predict the beginnings and ends of previously unknown genes. Such analyses can be difficult, and they will have to continue for several more years before the estimate can be considered definitive. The estimated number of human genes is expected to rise modestly due to improvements in both the quality of the sequence data and the computer programs used to make the predictions. Nevertheless, it appears that constructing and maintaining a human requires only six to ten times the number of genes needed to construct a bacterium such as *E. coli* (Table C.1).

Before taking offense at the implication that humans are not much more complex than bacteria, let's consider several factors that help determine the complexity of organisms. First, let's assume that every gene can independently adopt two states, either on or off. The number of different possible combinations of messenger RNA molecules that could exist in an organism at any given time is therefore determined by the formula 2^n, where n represents the number of genes. Because the formula is exponential, the number of different mRNA combinations (and hence the number of different populations of proteins or other gene products) increases extremely rapidly as the number of genes increases. For example, let's imagine that an organism has only two genes. As Figure C.3 shows, this tiny genome could produce four ($2^2 = 2 \times 2$) different combinations of mRNAs. Similarly, a four-gene genome could have 16 ($2 \times 2 \times 2 \times 2 = 2^4$) different combinations of mRNAs. And so on.

Figure C.2 Milestones in the Quest for the Human Genome (*continued*)

Craig Venter and collaborators publish the complete genome sequence of the *Haemophilus influenzae* bacterium.

Venter announces the formation of a commercial company, Celera Genomics, that will sequence the entire human genome by 2001. HGP scientists publish the complete genome sequence of the nematode worm *Caenorhabditis elegans*.

1990	1995	1996	1998	1999

NIH and the DOE establish the publicly funded Human Genome Project. An international consortium of sequencing centers is scheduled to completely sequence the human genome by 2005, but finishes in 2001.

The HGP publishes the complete genome sequence of budding yeast, *Saccaromyces cerevisiae*.

Both the public and commercial sequencing efforts pass the 1 billion-base-pair milestone. HGP scientists publish the first complete human chromosome sequence, that of chromosome 22.

Table C.1

A Sample of the Genomes Sequenced to Date

Organism description	Scientific name	Date	Estimated genome size (millions of base pairs)	Predicted number of genes
Bacterium	*Haemophilus influenzae*	1995	1.8	1,740
Toxic shock bacterium	*Staphylococcus aureus*	2005	2.8	2,600
Bacterium	*E. coli*	1997	4.6	3,240
Anthrax bacterium	*Bacillus anthracis* [. . . an-THRASS-iss]	2003	5.2	5,000
Budding yeast	*Saccaromyces cerevisiae*	1996	12	6,000
Fruit fly	*Drosophila melanogaster*	2000	180	13,600
Nematode worm	*Caenorhabditis elegans*	1998	97	19,100

Anthrax bacterium

Celera and collaborators publish the genome sequence of the fruit fly, *Drosophila melanogaster*. Venter and Collins agree to collaborate. Disagreement over the data-release policy ends the short-lived collaboration between the HGP and Celera.

Since publication of the first-draft sequence, hundreds of laboratories around the world continue to fill in gaps in the sequence, proofread and correct the existing sequence, and confirm the identities of possible genes.

2000   **2001**   **2005**

The Human Genome Project and Celera both publish complete draft sequences of the human genome in separate journals: HGP in *Nature*; Celera in *Science*.

The new field of comparative genomics depends on continuing efforts to sequence the genomes of other organisms such as the human, lab rat, *C. elegans* worm, and mustard plant. By June 2005, 266 complete genome sequences had been published, 80% from bacterial sources.

Table C.1

A Sample of the Genomes Sequenced to Date (*continued*)

Organism description	Scientific name	Date	Estimated genome size (millions of base pairs)	Predicted number of genes
Laboratory rat	*Rattus norvegicus* [. . . nor-*VAY*-juh-kuss]	2004	2,750	25,000
Flowering plant	*Arabidopsis thaliana* [uh-*RAB*-ih-*DOP*-siss *THAH*-lee-*AH*-nuh]	2000	125	25,500
Human	*Homo sapiens*	2001	3,200	20,000–30,000
Puffer fish	*Takifugu rubripes* [*TAHK*-ih-*FOO*-goo roo-*BRIPE*-eez]	2002	400	31,000

Puffer fish

Figure C.3 How Genes Can Produce Multiple RNAs

The number of combinations of mRNA transcripts an organism can produce rises exponentially as the number of genes increases. Two genes (shown here) could produce $2^2 = 2 \times 2 = 4$ different combinations of RNAs. Similarly, four genes would yield $2^4 = 16$ different mRNA combinations, eight genes would produce $2^8 = 264$ combinations, and so on.

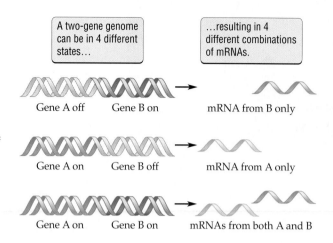

A two-gene genome can be in 4 different states...

...resulting in 4 different combinations of mRNAs.

Gene A off   Gene B on   mRNA from B only

Gene A on   Gene B off   mRNA from A only

Gene A on   Gene B on   mRNAs from both A and B

Figure C.4
Gene Expression in a Fruit Fly Embryo

Many human proteins are similar to fruit fly proteins, indicating that certain protein activities and interactions were possibly established early in evolution. This micrograph of a *Drosophila* embryo identifies various gene products, including the hairy protein in red.

What about our two real-life organisms? The *E. coli* genome contains 3,240 genes (see Table C.1), which would yield $2^{3,240}$ different possible combinations of mRNAs. This number is huge. But for contrast, consider the human genome: the number of possible mRNA combinations with 30,000 genes is $2^{30,000}$. This means that, even based on such a simplified model of gene expression, the human genome can produce a vastly larger number of different combinations of mRNA molecules than can the genome of *E. coli*. A tenfold increase in gene number produces far more than a tenfold increase in complexity.

But this calculation is still somewhat crude. A more accurate comparison of human and bacterial mRNA combinations would have to take into account a second factor; namely, the selective removal of introns from human mRNA products. As we saw in Chapter 13, introns must be removed from most eukaryotic mRNA transcripts before they can be used to make proteins. If different combinations of introns are removed, a single gene can produce many different mRNA transcripts. This variable removal of introns from eukaryotic mRNAs greatly increases the number of different mRNAs produced by eukaryotic genomes such as ours. Bacterial genes do not have introns, which means that each gene produces only one mRNA product.

A third factor to consider is the different mechanisms humans and bacteria use for regulating gene expression (see Chapter 14). Since all organisms can alter which genes are expressed in response to changes in the environment or important life events, the complexity of the mRNA population produced by the genome can vary dramatically within individual cells. Many years down the road, when total numbers of genes and mRNA populations have been characterized for each genome, truly accurate comparisons of complexity among different organisms will become possible.

Comparing the genomes of different organisms yields valuable insights

Important observations stemming from the human genome sequence are often based on comparisons with genome sequences from other organisms. These comparisons shed light on the common sets of genes that have been conserved throughout evolution. Perhaps the most striking patterns emerge when we compare the protein products of these genes from one species to the next. Most proteins can be categorized according to their functional domains. A *domain* is a segment of a given protein that has a specific function; for example, enzymatic activity or the ability to bind to another protein. Interestingly, nine out of ten protein domains found in humans have counterparts in fruit flies and nematode worms. While the arrangements of protein domains are more complex in humans, their very existence in fruit flies (Figure C.4) and worms implies that certain protein activities and interactions were established early in evolution.

On a broader scale, roughly 60 percent of human proteins are similar to proteins found in fruit flies and worms. These proteins participate in the core processes required for life, such as glycolysis and putting together the monomers that constitute the building blocks of life. Unsurprisingly, the protein sets that have been conserved over millions of years include DNA polymerases, proteins that regulate DNA transcription, enzymes involved in the processes of metabolism, and many of the receptors and kinases involved in signal cascades.

Evolutionary conservation across species also means that well-understood processes in other organisms may yield insight into similar but less understood processes in humans. The regulation of biological clocks is just one essential process that is likely to involve similar proteins in different species. If you have ever flown on an airplane across several time zones, you have probably experienced a disturbance in your sleep patterns. This phenomenon, known as jet lag, is due to a disruption of your biological clock. In mammals, biological clocks dictate the daily cycling of many physiological events, including sleep. Earlier studies in fruit flies had uncovered many so-called clock genes that control the timing of the fly's activities. Comparisons between fruit fly clock gene sequences and the human genome have already yielded several previously unknown candidate genes that may play a role in the daily activity cycles of humans. Furthermore, the human genome sequence has revealed the chromosomal locations of the known human clock genes, at least one of which is responsible for an inherited sleep disorder.

Studies like these that analyze and compare genomes from multiple organisms are part of a new field in biology called **comparative genomics**. The valuable insights to be gained from comparative genomics have motivated an explosion of efforts to sequence the genomes of a broad range of species. In March of 2005, the NIH announced a new round of genome sequencing projects aimed at 12 more species, including a marmoset, a sea slug, a pea aphid, and three fungi (Figure C.5). Each selected organism represents a potential model for human disease or has significant economic importance. The sea slug *Aplysia californica* [uh-*PLEEZE*-ya . . .], for example, has very large nerve cells and has been used as a model to study memory and its loss due to disease. Likewise, the marmoset—*Callithrix jacchus*, a Central and South American monkey—is an important model for human diseases such as multiple sclerosis. The pea aphid, *Acyrthosiphon pisum* [*AYE*-ser-thoh-*SIGH*-fun *PEE*-zum], is responsible for hundreds of millions of dollars of crop losses each year and is exceptionally resistant to pesticides. By June 2005, more than 260 complete genome sequences were published and available for analysis, and the gold rush to sequence still more species shows no signs of abating.

Health Care for You Alone

Long before the attempt to sequence the human genome began, scores of genes were already known as culprits in human disease. The link between mutations in a given gene and the likelihood of coming down with a particular disease formed the basis for **genetic screening**. The early days of genetic screening involved looking for gene mutations in order to assess a person's future health risks. By the end of the twentieth century, people had the option of testing themselves, and often their unborn fetuses, for mutations linked to diseases such as breast cancer and genetic disorders such cystic fibrosis and Huntington disease (see the box on page 217). With hundreds of such tests available, people were able to learn more about their children, their families, and themselves than they had ever thought possible.

However, as we emphasized in earlier chapters, many diseases, such as breast cancer, are not caused by mutations in just one or two genes. Many malignant cancers,

(a)

(b)

(c)

(d)

Figure C. 5 More Organisms Will Be Sequenced
The NIH has scheduled projects to sequence the genomes of 12 more organisms, including the four shown here: (a) marmoset (*Callithrix jacchus*); (b) sea slug (*Aplysia californica*); (c) pea aphid (*Acyrthosiphon pisum*); (d) yeast (*Schizosaccharomyces octosporus*).

for example, require mutations in several different genes, resulting in a far more complex genetic profile than can be revealed by traditional tests of a few genes (see Interlude B). Knowledge of the human genome sequence and new genomic techniques are sure to exponentially increase the number of such variations in an individual's genome that can be analyzed, radically expanding the scope of genetic screening.

SNPs are a powerful means of characterizing individual genomes

By the end of the twentieth century, only a few thousand SNPs were known—a number that was simply not adequate to cover the entire human genome. The completion of the draft sequence of the human genome, followed by a concerted effort by laboratories across the world, had uncovered the identities and locations of more than 10 million SNPs as of 2005. Such a huge number of SNPs brings the variation in individual genomes into sharper focus by allowing genomes to be compared with a vastly more detailed SNP map of the human genome.

Of course, for this breakthrough to be medically relevant, there has to be a rapid and affordable way to detect thousands of SNPs in an individual DNA sample. Using traditional polymerase chain reaction (PCR) technology (see Chapter 15), it would take weeks and cost tens of thousands of dollars to create a detailed SNP profile for a single patient. However, the DNA chip technology described in Chapter 14 has recently been modified to allow the screening of 10,000 SNPs in less than an hour (Figure C.6). Within a few years, this technology is likely to be further refined to allow the screening of 100,000 SNPs in a matter of hours. People will then have the option of viewing their genomic fingerprint and subjecting it to analysis.

SNP profiling will help identify the risk of getting certain diseases

Only a minority of SNPs fall within gene coding sequences (exons), and only some of them will be found to affect the activity of the proteins produced by those genes. Nevertheless, a significant number of SNPs with dramatic effects on human health have already been identified.

The presence of a particular SNP in the apolipoprotein E [*APE*-oh-*LIP*-poh . . .] (*apoE*) gene, for example, is associated with a risk of developing Alzheimer disease (AD). The leading cause of cognitive degeneration in the elderly, AD is characterized by the formation of plaques of protein in the brain, leading to extensive cell death. The

Figure C.6 From Tissue to Test Results
Using cells from tissues or body fluids, scientists can isolate a person's DNA and use a variety of methods to screen it for single nucleotide polymorphisms (SNPs).

INTERLUDE C

normal ApoE protein plays a role in delivering choles-terol to cells and is thought to help remodel the plasma membranes of brain cells. The exact role played by *apoE* in AD plaque formation is unclear, but clinical studies have shown that a particular *apoE* SNP correlates with an increased risk of AD. On the other hand, the presence of a different *apoE* SNP correlates with a reduced likeli-hood of AD. Thus determining a person's SNP profile for the *apoE* gene can reveal whether or not he or she is at high risk for AD later in life.

Alterations in structural proteins due to the presence of a particular SNP in the gene sequence can alter a person's susceptibility to a range of diseases. For exam-ple, the presence of a particular SNP in the gene encod-ing the cartilage intermediate layer protein (CILP) is associated with enhanced risk for lumbar disc disease (LDD). LDD is characterized by the gradual degenera-tion of the collagen-rich discs that lie between the verte-brae of the lumbar region of the backbone (Figure C.7). About 75 percent of all adults will experience prolonged lower back pain at some point in their lives, and the major-ity of these cases will be due to LDD. Recent research in Japan has revealed that the presence of a particular SNP in the *CILP* gene significantly enhances CILP's ability to block the activity of a growth factor in the lumbar disc; this blocking may contribute to the degeneration of lumbar discs. Individuals found to have this *CILP* SNP could therefore take early steps to avoid physical activi-ties that might contribute to lower back pain.

SNP profiling may help provide individuals with tailored drug therapies

SNP profiles can also give us insight into how an individ-ual will respond to drug therapy. The wide range of pos-sible responses a patient may have to a particular drug often complicates the prescribing of safe and appropri-ate therapies. These variations in response may be due to preexisting disease, the presence of other drugs in the body, or nutritional status. Genetic variations in drug-metabolizing enzymes can also have a significant effect on patient response.

Several genes encoding enzymes that process drugs in the body have been identified. Some of these genes con-tain SNPs that give an indication of their likely activity. For example, an enzyme called CYP2D6 is required for the activation of painkillers such as codeine. Nearly 10 percent of the population is homozygous for a SNP that renders this enzyme inactive. These individuals fail to respond to codeine and do not get adequate pain relief from the drug. Similarly, certain active carrier proteins

tend to pump anticancer drugs out of cells, resulting in the need for higher drug doses. SNP profiling of the genes that encode these proteins may identify variations that either decrease or increase their pumping activity, allow-ing physicians to prescribe drug doses tailored to the individual.

As the number of SNPs with known effects on gene functions increases, the possibility of using a SNP profile to optimize an individual's drug treatment plan will move closer to reality. In the near future, patients will sit down with their physicians, review their SNP profile, and find out what it reveals about their susceptibility to various diseases and how well a particular drug therapy might work for them. These methods will result in a personalized health program designed to head off future problems while simultaneously treating current illnesses as effectively as possible. In fact, submitting a blood sample for DNA chip analysis of hundreds of thousands of SNPs could become the first health-related activity of childhood.

Genetic Testing Raises Ethical Issues

The analysis of the human genome is already transform-ing the way we think about the many diseases that are caused by a combination of genetic and environmental fac-tors. Understanding the genetic component of a disease can allow physicians to better predict the likelihood of that disease occurring in a particular individual, giving them the option to either prescribe preventative measures or improve the course of treatment. This is a brave new world of opportunity for improving human health and quality of life and is likely to affect all of us within the next decade. At the same time, we will all face a number of ethical ques-tions, some of which have already arisen from the appli-cation of traditional methods of genetic testing.

New prenatal genetic screening methods provide added choices—and dilemmas—for parents

Genetic screening for potential health problems does not have to wait until early childhood. While still pregnant, a woman can carry out numerous tests on the genetic char-acter of her fetus. Depending on the outcome of a SNP profile (or a more traditional genetic test, as described in the box on page 217), she and her partner may or may not choose to terminate the pregnancy.

Figure C.7 Structural Proteins Foreshadowing Lumbar Disk Degeneration This X-ray of a lumbar spine with advanced degenerative disc disease shows narrowing of the intervertebral disc spaces that can cause back pain, a common affliction among adults. This may be associated with a particular SNP in the gene that encodes the protein for the cartilage in the back.

Explore how to get from tissue to test results.

C.2

Learn more about prenatal screening.

C.3

Such decisions are already altering which babies are being born. In the United States, the number of Down syndrome babies is decreasing as a result of genetic screening (Figure C.8). According to some reports, the abortion rates for Down syndrome fetuses range from 50 up to 90 percent. One study reported a 99 percent abortion rate of fetuses found to suffer from thalassemia [tha-luh-*SEE*-mee-uh], a genetic disorder of the blood. In Asia, where boys are more highly valued than girls, female fetuses are regularly being aborted. Demographers are predicting a highly skewed sex ratio as these babies grow up over the next few decades.

What should prospective parents do if they discover that their fetus has a genetic disorder that will destine them and their child for a difficult life? A person with Down syndrome, for example, experiences mild to severe mental retardation along with problems, such as heart disease, that can require repeated and difficult surgeries. In addition, caring for a child with Down syndrome can be difficult for a family—financially, emotionally, and physically. For many parents, the answer is to terminate the pregnancy. Others argue that children with Down syndrome are among the most loving of people, with characteristically pleasant dispositions, able to enjoy life despite their impairments. Some individuals with Down syndrome hold part-time or full-time jobs and are able to live independently from their parents.

And what about cystic fibrosis, another genetic disorder for which testing is available? Children with cystic fibrosis often have great difficulty breathing, requiring painful daily therapies to loosen the thick secretions in their lungs. In addition, these children often suffer from infections and pneumonias, experience digestive problems, and have trouble growing. Yet they have the potential to lead productive lives as adults: though babies with cystic fibrosis once routinely died as children, today many live into their forties or even their fifties.

When prenatal genetic screening reveals disorders such as these, the prospect of deciding whether to continue or terminate a pregnancy can be daunting. Many parents feel ill equipped to decide whether the quality of life that a child with a genetic disorder is likely to have will be worthwhile. Their choice is often complicated further by feelings of guilt and the question of how caring for such a child might affect their lives and the lives of their other children.

Prenatal genetic screening combined with other technologies may encourage parents to choose embryos with the right genetics

In an even more extreme application of prenatal genetic screening, couples becoming pregnant via in vitro fertilization already have their genetic choice of fetuses before a pregnancy has even begun, as we saw in Chapter 11 (see the discussion of preimplantation genetic diagnosis in the box on page 217). Even before

(a)

(b)

Figure C.8 A Disappearing Condition?
(*a*) This child shows the typical facial symptoms of Down syndrome, or trisomy 21. The major effect of the extra chromosome is mental retardation, but there are often other health problems as well. The number of Down syndrome babies being born is on the decline, a trend attributed to prenatal screening for the condition. (*b*) Individuals with Down syndrome carry two copies of all chromosomes except chromosome 21, for which they carry three copies, as shown.

the completion of the human genome sequence, couples were able to assess the risk of passing on certain defective alleles to their fetus.

Let's return to the example of cystic fibrosis. Recall from Chapter 11 that cystic fibrosis is an autosomal recessive disorder: Individuals who have two copies of the defective allele develop cystic fibrosis. Those who have only a single defective allele are known as carriers; they are able to pass the allele for the disorder on to their children while remaining healthy themselves. When both parents are carriers, each of their children has a 1 in 4 chance of being homozygous for the defective allele and developing the disorder.

In 1994, one couple, both cystic fibrosis carriers, made reproductive history by undergoing in vitro fertilization, having the resulting embryos tested for the cystic fibrosis allele, and then choosing which embryos to keep. Two homozygous embryos were rejected, while three embryos that were found not to be homozygous were implanted. From these embryos, a single baby boy grew to term and was born. He is a carrier—that is, he is heterozygous for the cystic fibrosis allele, so he could pass it on to some of his children, some of whom might get the disease if his future partner had cystic fibrosis or was a carrier—but he himself will not get the disease. While these parents were only interested in the cystic fibrosis gene, there are no legal restrictions in place to stop parents from testing embryos before implantation for any number of genetic disorders or predispositions to disease using the growing diagnostic power of SNP profiling.

SNP profiling may affect individual privacy and freedom in new ways

Despite the exciting promise that detailed genetic profiling may hold, it raises a number of ethical problems. We have touched on some of these problems already, but there are others. For example, will such detailed biological profiles be used to discriminate against individuals? Health insurance companies will be tempted to raise premiums for people whose profiles show a high susceptibility to disease later in life. Individuals with a high risk of serious illness might even find themselves denied health and life insurance.

Furthermore, as the new field of behavioral genomics enters the stage, many believe that SNP profiling will reveal a person's susceptibility to syndromes such as alcoholism, schizophrenia, and clinical depression. The revelation that certain people have a heightened susceptibility to drug addiction, for example, might cost them their jobs. By understanding our genetic blueprint too well, we run the risk of losing sight of our humanity in an avalanche of predictions and genetic susceptibilities.

To prevent the potential misuse of SNP profiles, new guidelines for personal privacy must be established. To this end, about 5 percent of the budget for the Human Genome Project is devoted to studying the ethical, legal, and social issues surrounding the availability of genetic information. If we are truly to reap the benefits of this scientific achievement, addressing these issues will be as important as filling the gaps in the human genome sequence.

◉ Review and Discussion

1. Based on what you have learned about cells, organelles, and metabolism, describe some of the genes you probably share with a tomato plant. What sets of genes do you think you do *not* share with plants?

2. What are SNPs, and what role do they play in genetic evaluations related to health issues?

3. How do you think knowing the entire genome of the mosquito that carries malaria might help eradicate that deadly disease?

4. Do you think patients should be genetically screened before doctors issue them drug treatments? Why or why not? What ethical issues, if any, need to be considered?

5. Why might a pharmaceutical trade association be against widespread genetic screening?

6. As researchers discover more and more genes that predict certain tendencies—for example, toward obesity, high blood pressure, or depression—do you think doctors should screen their patients for some or all of these genes to help them adopt the healthiest lifestyles possible? Why or why not?

Key Terms

comparative genomics (p. C8)
genetic screening (p. C8)
genomics (p. C1)
Human Genome Project (HGP) (p. C3)
single nucleotide polymorphism (SNP) (p. C2)

Key Concepts

- Evolution is change in the genetic characteristics of populations of organisms over time. For evolution to occur, there must be inherited differences among the individuals in a population. Populations evolve, whereas individuals do not.

- Individuals with certain inherited characteristics may survive and reproduce at a higher rate than other individuals do, a process known as natural selection. The inherited characteristics of individuals that leave more offspring become more common in the following generation.

- Adaptations are features of an organism that improve its performance in its environment. Adaptations are products of natural selection.

- The great diversity of life on Earth has resulted from the repeated splitting of species into two or more species.

- When one species splits into two, the two species that result share many features because they have evolved from a common ancestor.

- An enormous amount of evidence shows that evolution has occurred. One strong line of evidence comes from the fossil record, which allows biologists to reconstruct the history of life on Earth and shows how new species arose from previous species. Features of existing organisms, patterns of continental drift, and direct observations of genetic change also provide strong evidence for evolution.

A Journey Begins

The Galápagos [guh-*LOP*-uh-goce] Islands, located off the west coast of South America, are isolated, encrusted with lava, and home to bizarre creatures found nowhere else on Earth. Life here offers odd twists on the usual: tortoises that reach giant size, land-dwelling lizards that take to the sea, and vampire finches that suck the blood of other birds.

The unusual species of the Galápagos Islands have long fascinated biologists. One such biologist was the 22-year-old Charles Darwin, who visited the Galápagos as part of a remarkable 5-year journey on the ship *Beagle*. The *Beagle* left England on a surveying voyage in 1831 and sailed to South America, from there to the Galápagos Islands, and eventually back to England in 1836. Throughout this long voyage, Darwin made careful observations of the regions he visited and collected specimens of the organisms he found there.

Upon his return to England, Darwin consulted other scholars and thought deeply about what he had seen on his journey. Early in 1837, he learned from taxonomists that many of the specimens he had collected in the Galápagos were new species, and that many of these species were confined to a single island. Where the organisms he had collected from different islands turned out to be groups of new species, those species often shared unique characteristics, and thus appeared to be closely related. Darwin also noticed that the species of the Galápagos Islands differed considerably from the species of another group of islands he had visited, the Cape Verde Islands off the west coast of Africa. He was struck by this observation because, in terms of their physical environment, the two groups of islands were very similar—yet the Galápagos species were more similar to species that lived in very different environments in South America than to the Cape Verdean species.

Charles Darwin's Voyage

(*a*) The course sailed by the *Beagle*. (*b*) The Galápagos Islands, located 1,000 kilometers to the west of Ecuador. (*c*) Bartolomé, one of the Galápagos Islands.

These and many other observations led Darwin to a bold conclusion: Species were not, as virtually everyone in his time thought, the unchanging result of separate acts of creation by God. Instead, species had descended with modification from ancestor species; that is, they had changed over time, or evolved—in the case of the species of the Galápagos Islands, most likely from South American species. Darwin's recognition in 1837 that species had changed over time was a key step in an intellectual journey that would blossom, 22 years later, into the publication of a book that shook the world: *The Origin of Species*.

E arth teems with organisms, many of which are exquisitely matched to their environments. The soaring flight of a hawk, the beauty and practicality of a flower, and the stunning camouflage of a caterpillar each provide a glimpse into the remarkable designs of organisms. How did organisms come to be as they are, seemingly engineered to match their surroundings (Figure 16.1)? What has caused the amazing diversity of life? And within that diversity, why do organisms share so many characteristics? Scientists who study evolution seek to answer questions such as these.

Evolution is biological change over time. In this context, "biological change" can refer to changes in the genetic characteristics of populations or to changes in the kinds of species living on Earth. We open this chapter by defining the term *evolution* more fully. We then provide an overview of the mechanisms that cause evolution, its consequences for life on Earth, the evidence that shows it is happening, and the impact of evolutionary thought on human society. We close the chapter by describing an example of evolution in action: the short- and long-term changes that have occurred in finches that live on the Galápagos Islands.

16.1 Biological Evolution: The Sum of Genetic Changes

Defined broadly, evolution is descent with modification, often with an increase in the variety of the descendant forms. The term can be applied to organisms, cars, computers, or hats. In each of these cases, new items represent modified versions of previous items, and often several varieties arise where only one existed before. But there is an important and fundamental difference between biological evolution and, say, the evolution of hats: hats change over time because of deliberate decisions made by their designers. As we will see throughout this chapter, biological evolution is not guided by a "designer" in nature (though humans can, and do, direct the course of evolution in some species).

In the context of biology, evolution can be defined in several ways. In this book, we define **evolution** as change in the genetic characteristics of a population of organisms over time. We take this approach so we can use our knowledge of genetics to help us understand how evolution works. Note that evolution is defined at the level of the population, not the individual. Populations evolve, but individuals do not. (That is because, overall, the genotypes of individuals do not change over the course of their lives. Mutations can and do occur within the cells of an individual, but most of the genes in any given cell do not mutate, and the mutations that do occur differ from cell to cell.) In the next section we take a brief look at two mechanisms that cause evolution; we will examine these and other such mechanisms in greater detail in

Figure 16.1 **The Match Between an Organism and Its Environment**
Dolphins (members of the Odontoceti [oh-*DAHNT*-uh-set-ee], or toothed whales, a group that also includes killer whales) are species whose form beautifully matches their marine environment. Genetic, anatomical, and fossil evidence indicates that dolphins and other whales evolved from a terrestrial ancestor, thus illustrating the extent to which evolution has altered their body form to fit life in the ocean.

Chapter 17, where we discuss small-scale evolutionary changes, or **microevolution**.

Evolution can also be defined as the pattern of large-scale changes in life on Earth over long time periods. From this perspective, evolution is history; specifically, it is the history of the formation and extinction of species and higher taxonomic groups over time. Focusing on the rise and fall of different groups of organisms over time provides us with a grand view of the history of life on Earth, a view to which we will return in Chapter 19 when we discuss large-scale evolutionary changes, or **macroevolution**.

16.2 Mechanisms of Evolution

Think of the students in your biology class: if they were lined up in front of you, their many differences would make it easy to tell one person from the next. What is true for people is true for all organisms. Individuals of all species vary in many of their characteristics, including aspects of their **morphology** (form and structure), biochemistry, and behavior.

What causes organisms to be different from one another? As we learned in Unit 3, individuals may differ if they have different alleles for genes that influence their phenotype. Such differences in the genetic makeup of individuals are essential for evolution. To see why this is so, consider what would happen if (1) some individuals in a population are larger than others; (2) body size is an inherited characteristic; and (3) large individuals produce more offspring than small individuals. How might such a population change over time?

Here is one way that evolution could change this population: Body size is an inherited trait, so the offspring of large parents will themselves grow up to be large. Furthermore, since large individuals produce more offspring than small individuals, from one generation to the next, more and more of the offspring will be large. Thus, over time, the average size of individuals in the population will increase, and the alleles that cause large size will become more common. In other words, since we define evolution as change in the genetic characteristics of a population over time, the population will evolve.

As we have seen, evolution can occur if some individuals within a population survive and reproduce at a higher rate than other individuals. In nature, there are two main reasons why some individuals survive and reproduce at a higher rate than others: natural selection and genetic drift.

In many cases, consistent differences in the survival and reproduction of individuals are based on inherited traits. For example, running speed in pronghorn antelope is at least partially under genetic control, and pronghorns that are fast may escape predators more easily than pronghorns that are slower. Thus, as we saw in Chapter 1, more of the fast pronghorns are likely to survive and produce offspring, and those offspring will tend to be fast, too. A similar process occurs in garter snakes that eat highly poisonous rough-skinned newts (Figure 16.2). These newts produce one of the most potent neurotoxins known, tetrodotoxin [teh-*TROH*-doh-tox-in], or TTX. Newt toxicity varies from one geographic region to another, and the amount of TTX that a snake can tolerate is under genetic control. In regions where newts are highly toxic, snakes that tolerate more of the poison have an advantage over other snakes. As a result, snakes in these regions have evolved to tolerate TTX concentrations that would kill 25,000 mice.

The pronghorn and garter snake examples illustrate **natural selection**, the process by which individuals, because of particular inherited characteristics, survive and reproduce at a higher rate than do other individuals. Because the characteristics that enhance survival and reproduction are inherited, they are passed on from one generation to the next. Thus, over time, natural selection can cause a population to evolve so that more and more individuals have the beneficial characteristics.

Chance events can also cause some individuals to leave more offspring than others. For example, when a windstorm causes trees to fall amid a population of forest wildflowers, some of the plants are crushed (and leave no offspring), while others survive (and leave offspring). In this case, whether a plant dies or survives is

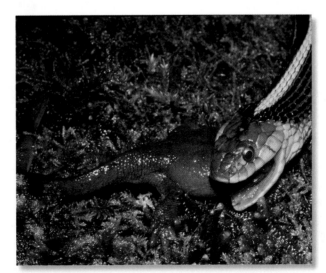

Figure 16.2 Natural Selection at Work
Some garter snakes, like this one, have the genetic advantage of being able to survive eating toxic rough-skinned newts.

Biology Matters

You Can't Live with 'Em, You Can't Live Without 'Em

Evolution is a natural process, but it also occurs in response to human interventions like the increasing development of antibacterial (and, to a lesser extent, antiviral) consumer products. Once largely limited to hospitals and other places at high risk for infections, antibacterial agents are now routinely added to soaps, lotions, dishwashing liquids, and other cleaning products. A recent study published in *Annals of Internal Medicine* estimated that 75 percent of liquid and 29 percent of bar soaps available in the United States contain antibacterial ingredients. Furthermore, many additional products are now coated or impregnated with antibacterial agents, including tissues, cutting boards, toothbrushes, bedding, and children's toys.

So what's the problem? Wouldn't anyone prefer to avoid disease-causing bacteria? If it were that simple, the answer would be yes, but these antibacterial products also kill the beneficial bacteria that surround us in our environment and, according to the World Health Organization, contribute directly to the rise in antibiotic-resistant bacteria. In 2000, the American Medical Association (AMA) advised consumers to avoid extensive use of "antibacterial soaps, lotions, and other household products" and also called for greater regulation of antibacterial products.

Perhaps more obviously relevant to each of us on a day-to-day basis is the conclusion emerging from numerous studies that the use of antibacterial consumer products doesn't decrease the frequency of illness in people that use them. The following guidelines (based on information from the American Society for Microbiology) will help you stay healthy and keep your home clean, while avoiding the risks associated with antibacterial products:

- Plain old soap and hot water remain the best for washing your hands, body, and dishes.
- Wash your hands thoroughly and often before you:
    Prepare or eat food
    Treat a cut or wound or tend to someone who is sick
    Insert or remove contact lenses
and after you:
    Use the bathroom
    Handle uncooked foods, particularly raw meat, poultry, or fish
    Change a diaper
    Blow your nose, cough, or sneeze
    Touch a pet, especially reptiles and exotic animals
    Handle garbage
    Tend to someone who is sick or injured
- Limit your use of antibacterial products to situations when you are most at risk, such as when you are unable to wash your hands.
- Use bleach to clean your bathroom.
- Use separate cutting boards for raw meat and foods that may not be cooked before eating (e.g., fruits and vegetables).
- Wash all fruits and vegetables either in soapy water (rinse thoroughly, of course) or in one of the new fruit and vegetable washes.
- Wash all kitchen surfaces, dishes, and utensils in hot, soapy water. Make sure you rinse thoroughly. If possible, put everything (including cutting boards) in the dishwasher.
- Every time you run your dishwasher, throw in your kitchen sponge.
- Don't wipe your counters with a sponge that's been sitting on your sink. This can deposit even more bacteria on your countertops. Use paper towels or replace your dish rag every day with a clean one.

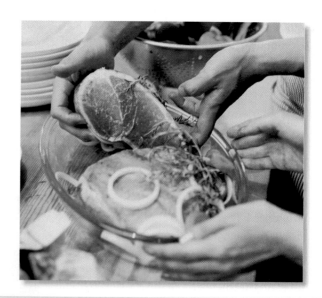

a matter of chance alone. As shown in Figure 16.3, such chance differences in survival can alter the genetic makeup of a population. Differences in reproduction or survival caused by chance events can lead to **genetic drift**, a process in which the genetic makeup of a population fluctuates randomly over time, rather than being shaped in a nonrandom way by natural selection. Like natural selection, genetic drift can cause populations to evolve.

Although it is a simplification, we can summarize evolution as a two-step process: (1) individuals differ genetically for many characteristics, and (2) natural selection

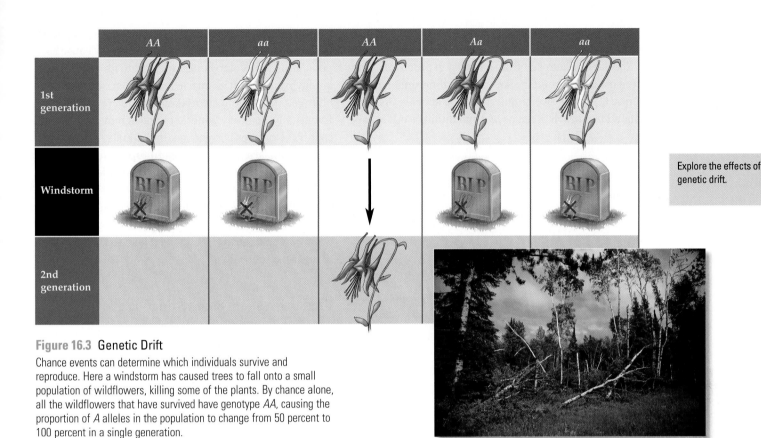

	AA	aa	AA	Aa	aa
1st generation					
Windstorm	RIP	RIP		RIP	RIP
2nd generation					

Explore the effects of genetic drift. 16.1

Figure 16.3 Genetic Drift
Chance events can determine which individuals survive and reproduce. Here a windstorm has caused trees to fall onto a small population of wildflowers, killing some of the plants. By chance alone, all the wildflowers that have survived have genotype *AA*, causing the proportion of *A* alleles in the population to change from 50 percent to 100 percent in a single generation.

or genetic drift can cause individuals of one genotype to survive or reproduce better than individuals of other genotypes, thus producing evolutionary change.

16.3 Consequences of Evolution for Life on Earth

Life on Earth is distinguished by matches between organisms and their environments, by a great diversity of species, and by many puzzling examples of organisms that differ greatly in many respects yet share certain key characteristics. Evolutionary biology seeks to explain all these features of life on Earth. This attempt to understand why the living world is the way it is motivates three basic themes of evolutionary biology: adaptation, the diversity of life, and the shared characteristics of life.

Adaptations result from natural selection

Some of the most striking features of the natural world are the complex characteristics of organisms and the often remarkable ways in which they are suited to their environments (see Figure 16.1). Such striking features of life result from **adaptations**, which are defined as characteristics of an organism that improve the performance (involving the growth, reproduction, or survival) of that organism in its environment.

Adaptations are the product of natural selection. Here's the reasoning behind this statement: If their reproduction were not restricted in some way, all organisms would reproduce so much that their populations would outstrip the limited resources available to them. Because organisms produce more offspring than can survive, the individuals in a population must struggle for existence. In this struggle, the individuals whose inherited characteristics provide the best match to their environments tend to leave more offspring than other individuals. This is natural selection in action. Over time, natural selection leads to the accumulation of favorable features—adaptations—within the population.

The diversity of life results from the splitting of one species into two or more species

A major focus of evolutionary biology is to understand how the great diversity of life on Earth arose. Here, too, evolution provides a simple, clear explanation: the diversity of life is a result of the repeated splitting of one species into two or more species, a process called **speciation**.

Speciation can result from a variety of processes. One of the most important is adaptation to different environments. Consider two populations of a species that are isolated from each other, as by a mountain or other bar-rier that prevents individuals from moving between the populations. Over time, natural selection may cause each population to become better adapted to its own particular environment on its own side of the barrier, leading to changes in the genetic makeup of both populations. Eventually, so many genetic changes may accumulate that if the barrier is removed, individuals from the two populations are no longer able to reproduce with each other. As we learned in Chapter 1, species are often defined in terms of reproduction: a species is a group of populations whose members can reproduce with each other but not with members of other such groups. Thus evolution by natural selection can lead to the formation of new species.

Organisms share characteristics due to common descent

The natural world is filled with puzzling examples of very different organisms that share certain characteristics. For example, the wing of a bat, the arm of a human, and the flipper of a whale all have five digits and contain the same kinds of bones (Figure 16.4a). Why do limbs that look so different and have such different functions have the same set of bones? Surely if the best possible wing, arm, and flipper were designed from scratch, their bones would not be so similar. Likewise, many organisms have **vestigial organs** (reduced or degenerate parts whose purpose is hard to discern). For example, why do we humans have a reduced tailbone and the remnants of muscles for moving a tail? And why do some snakes have rudimentary leg bones but no legs (Figure 16.4b)?

Evolution answers these and many other questions about shared characteristics of life. Many similarities among organisms are due to the fact that the organisms have a common ancestor. When one species splits into two, the two species that result share many features because they have evolved from a common ancestor. Features of organisms that are related to one another through common descent are said to be **homologous** [ho-*MOLL*-uh-guss]. For example, the wing of a bat, the arm of a human, and the flipper of a whale share the same set of bones because they are homologous (see Figure 16.4a). Similarly, some snakes have rudimentary leg bones because they evolved from reptiles with legs, and humans have rudimentary bones and muscles for a tail because our (distant) ancestors had tails.

Organisms can also share features as a result of **convergent evolution**, which occurs when natural selection causes distantly related organisms to evolve similar structures in response to similar environmental challenges.

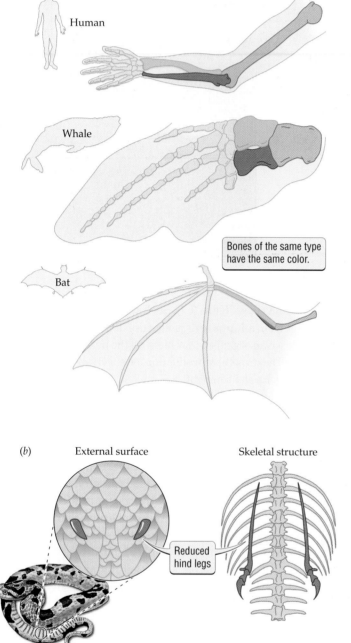

Bones of the same type have the same color.

Figure 16.4 Shared Characteristics
(a) The human arm, a whale's flipper, and a bat's wing are homologous structures, all of which have five digits and contain the same set of bones. (b) A python has rudimentary hind legs, as seen from the external surface and in the skeletal structure of the snake.

For example, the cacti found in North American deserts share many convergent features with distantly related plants found in African and Asian deserts (Figure 16.5). Similarly, although both sharks and dolphins have bodies streamlined for aquatic life, these species are very distantly related, and their overall similarities result from convergent evolution, not common descent. When species share characteristics because of convergent evolution, not common descent, those characteristics are said to be **analogous** [uh-*NAL*-uh-guss].

16.4 Strong Evidence Shows That Evolution Happens

Surveys taken over the past 10 years reveal that almost half of the adults in the United States do not believe that humans evolved from earlier species of animals. The results of these surveys are startling because evolution has been a settled issue in science for nearly 150 years. The vast majority of scientists of all nations, races, and creeds think that the evidence for evolution is very strong. In his landmark book, *The Origin of Species*, published in 1859, Charles Darwin argued convincingly that organisms are descended with modification from common ancestors. The scientific issue today is not whether evolution occurs, but how. To this question Darwin also offered an answer: he argued that the principal cause of evolutionary change is natural selection.

On this point Darwin was less successful in convincing other scientists, in part because at that time no one understood the underlying mechanisms of inheritance. For 60 years after the publication of *The Origin of Species*, many scientists thought Darwin was wrong to place so strong an emphasis on natural selection. However, the rediscovery of Gregor Mendel's work and the understanding of genetics that resulted made it clear that natural selection could cause significant evolutionary change, and hence that Darwin was at least partially correct.

Biologists still argue about the relative importance of natural selection and other mechanisms of evolution (such as genetic drift), but they do not dispute whether evolution occurs. Today's scientific debate about the causes of evolution can be compared to a dispute over what caused World War I to progress as it did: although we might argue over its causes, we all recognize that the war did indeed happen.

Why do scientists find the case for evolution so convincing? As we saw in Chapter 1, a scientific hypothesis must lead to predictions that can be tested, and hypotheses about evolution are no exception. Scientists have tested many predictions about evolution and have found them to be strongly supported by the evidence. Five lines of compelling evidence support evolution: fossils, traces of evolutionary history in existing organisms, continental drift, direct observations of genetic change in populations, and the present-day formation of new species.

(a)

(b)

(c)

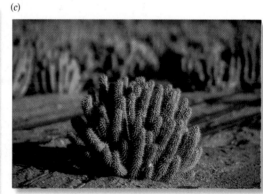

Figure 16.5 The Power of Natural Selection
Plants that grow in deserts often have fleshy stems (for water storage), protective spines, and reduced leaves. These three plants evolved from very different groups of leafy plants. They now resemble one another because of convergent evolution, driven by natural selection for life in a desert. Thus their shared structures (fleshy stems, spines, reduced leaves) are analogous, not homologous. (*a*) *Euphorbia* [you-*FOR*-bee-uh], a member of the spurge family. (*b*) *Echinocereus* [ee-*KY*-noh-*SEER*-ee-uss], a cactus. (*c*) *Hoodia*, a fleshy milkweed.

Evolution is strongly supported by the fossil record

Fossils are the preserved remains (or their impressions) of formerly living organisms. The fossil record allows biologists to reconstruct the history of life on Earth, and it provides some of the strongest evidence that species have evolved over time. For example, fossils document the fact of extinction and illustrate how the descendants of previously living species changed over time (Figure 16.6). The fossil record also reveals how environments have changed over time: fossils of whales have been found in the Sahara desert, and fossils of trees, dinosaurs, and tropical marine organisms have been found in Antarctica.

As we saw in Chapter 2, the evolutionary relationships among organisms—their pattern of descent from a common ancestor—can often be determined by comparing their anatomical characteristics. When this technique is applied to fossils, we find that the fossil record contains excellent examples of how major new groups of organisms arose from previously existing organisms. We will discuss one of these examples, the evolution of mammals from reptiles, in Chapter 19. Fossils showing how new organisms evolved from ancestral organisms also exist for many other groups, including fish, amphibians, reptiles, birds, and humans.

Finally, the times at which organisms appear in the fossil record match predictions based on evolutionary patterns of descent. For example, the evolutionary relationships among modern horses and their ancestors can be determined by comparing anatomical features of fossils. On the basis of those evolutionary relationships and the ages of fossils of horses thought to be ancestors to the modern horse, we can conclude that the modern horse (genus *Equus*) evolved relatively recently, about 5 million

Fossils of marine organisms: marine ammonite (top), and Ginkgo (bottom).

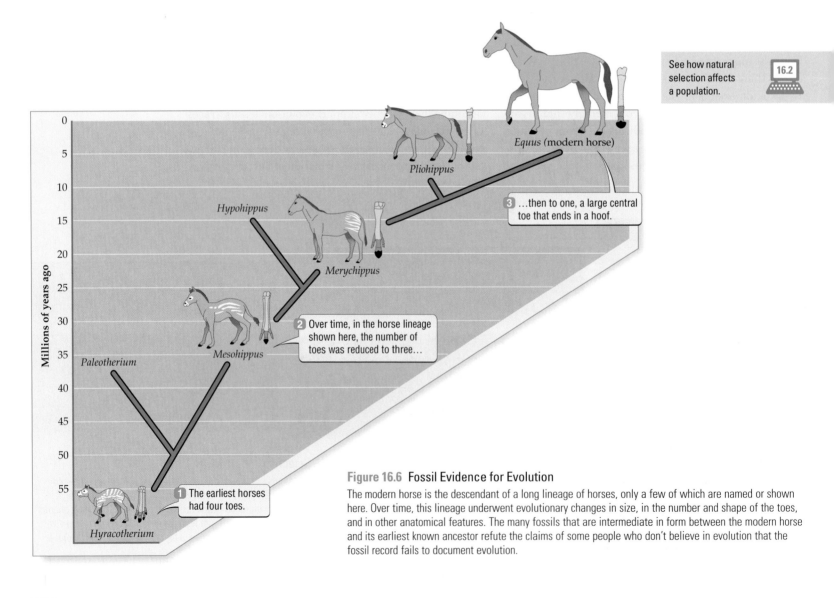

See how natural selection affects a population. 16.2

3 ...then to one, a large central toe that ends in a hoof.

2 Over time, in the horse lineage shown here, the number of toes was reduced to three...

1 The earliest horses had four toes.

Figure 16.6 Fossil Evidence for Evolution
The modern horse is the descendant of a long lineage of horses, only a few of which are named or shown here. Over time, this lineage underwent evolutionary changes in size, in the number and shape of the toes, and in other anatomical features. The many fossils that are intermediate in form between the modern horse and its earliest known ancestor refute the claims of some people who don't believe in evolution that the fossil record fails to document evolution.

years ago (see Figure 16.6). Thus we can predict that *Equus* fossils should not be found in very old rocks, and so far they have not.

Organisms contain evidence of their evolutionary history

A major prediction of evolution is that organisms should carry within themselves evidence of their evolutionary past—and they do. We described some examples of such evidence earlier in this chapter; namely, the vestigial organs found in some organisms (for example, the "legs" of a snake and the "tail" of a human) and the remarkable similarity in design of limbs that differ greatly in function (for example, the bat wing and the human arm).

Patterns of embryonic development provide further evidence of organisms' evolutionary past. For example, anteaters and some whales do not have teeth as adults, but as embryos they do. Why should the embryos of these organisms produce teeth and then reabsorb them? Or consider the embryos of fish, amphibians, reptiles, birds, and mammals (including humans), all of which have gill pouches. In fish, the gill pouches develop into gills that the adults use to breathe under water. But why should the embryos of organisms that breathe air develop gill pouches?

Our understanding of evolution provides an answer to these puzzles: similarities in patterns of development are caused by descent from a common ancestor. Fossil evidence suggests that anteater and whale embryos have teeth because anteaters and whales evolved from organisms with teeth. Similarly, fossil evidence indicates that the first mammals and the first birds each evolved from a (different) group of reptiles, that the first reptiles evolved from a group of amphibians, and that the first amphibians evolved from a group of fish. Thus it is likely that the embryos of air-breathing organisms such as humans, birds, lizards, and tree frogs all have gill pouches because all of these organisms share a common (fish) ancestor. In general, unless there is strong natural selection to remove anatomical features from the embryos of descendant groups (to remove teeth from whale embryos, or gill pouches from the embryos of air-breathing animals, for example), these features tend to remain by default. In the adults, however, they may be modified to serve other purposes (as gill pouches develop into gills in fish, but into parts of the ear and throat in humans), or they may disappear (as with the teeth of whales and anteaters).

Finally, as we saw in Chapter 2, the evolutionary relationships among organisms—their patterns of descent from a common ancestor—can often be determined from anatomical features. These patterns of descent can be used to make predictions about the similarity among organisms of molecules such as DNA and proteins. In such studies, scientists examine DNA or proteins whose function is not related to the anatomical characteristics originally used to determine the pattern of evolutionary relationships. Maintaining this separation keeps the approach from becoming circular: it ensures that the DNA and proteins provide a source of information that is independent from that provided by the original anatomical features. Biologists have predicted that the DNA sequences and proteins of organisms that share a more recent common ancestor should be more similar—and they are (Figure 16.7). If organisms were not related to one another by common descent, there would be no reason to expect that the degree of similarity found in DNA and proteins could be predicted from evolutionary relationships based on anatomical features. Thus the finding that the patterns of descent based on these two different sources of information are usually the same provides strong evidence for evolution.

Continental drift and evolution explain the geographic locations of fossils

Earth's continents move over time, by a process called **continental drift**. Each year, for example, the distance between South America and Africa increases by about 3 centimeters. Although they are separating from one another now, about 250 million years ago South America, Africa, and all of the other landmasses of Earth had drifted together to form one giant continent, called **Pangaea** [pan-*JEE*-ah]. Beginning about 200 million years ago, Pangaea slowly split up to form the continents we know today (see Figure 19.7).

We can use knowledge about evolution and continental drift to make predictions about the geographic locations where fossils will be found. For example, organisms that evolved when Pangaea was intact could have moved relatively easily between what later became widely separated regions, such as Antarctica and India. For that reason, we can predict that their fossils should be found on most or all continents. In contrast, the fossils of species that evolved after the breakup of Pangaea should be found on only one or a few continents, such as the continent on which they originated and any connected or nearby landmasses.

Predictions about the geographic distributions of fossils have proved correct, and they provide another important line of evidence for evolution. For example, today the lungfish *Neoceratodus fosteri* [nee-oh-sair-*AH*-to-dus *FAW*-ster-ee] is found only in northeastern Australia, but

Gill pouch

Human embryo with gill pouches

Figure 16.7 Independent Lines of Evidence Yield the Same Result

The pattern of evolutionary relationships among the animals shown here is based on the number of mutations required to change the amino acid sequence of cytochrome *c* from that found in humans to that found in the other organisms. Cytochrome *c* is an enzyme that functions in aerobic respiration and is found in all eukaryotes. The pattern of evolutionary relationships shown here, in which humans are most closely related to rhesus monkeys and least closely related to moths, matches the pattern derived independently from anatomical features.

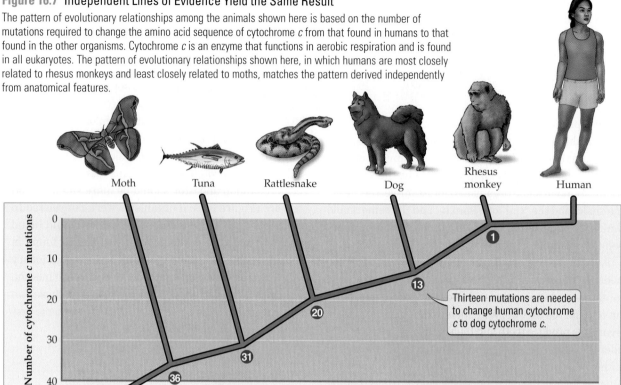

Thirteen mutations are needed to change human cytochrome *c* to dog cytochrome *c*.

Common ancestor

Moth   Tuna   Rattlesnake   Dog   Rhesus monkey   Human

its ancestors lived during the time of Pangaea, and fossils of those ancestors are found on all continents except Antarctica (Figure 16.8). At the other extreme, modern horses in the genus *Equus* first evolved in North America about 5 million years ago, long after the breakup of Pangaea. The oldest *Equus* fossils are found only in North America, as we would predict. The land bridge that today connects North and South America was formed roughly 3 million years ago. As a result, we would predict that *Equus* fossils found in South America should be less than 3 million years old; to date, all such discoveries have been less than 3 million years old.

Direct observation reveals genetic changes within species

In thousands of studies, researchers have observed populations in the wild, in agricultural settings, and in the laboratory changing genetically over time. Such observations provide direct, concrete evidence for evolution. Consider how farmers have altered the wild mustard, *Brassica oleracea* [BRASS-i-ka oh-ler-ACE-ee-uh], to produce several distinct crops, all part of the same species (Figure 16.9). By

allowing only individuals with certain characteristics to breed—a process called **artificial selection**—they have crafted enormous evolutionary changes within this species. The tremendous variation that we have produced within dogs, ornamental flowers, and many other species illustrates the power of artificial selection to produce evolutionary change. Natural selection can produce similar evolutionary changes, as shown by the often striking match between organisms and their environment (see Figures 16.1 and 16.5) and by case studies such as that of the medium ground finch, which we will examine shortly.

New species can be produced experimentally and their formation can be observed in nature

Biologists have directly observed the formation of new species from previously existing species. The first experiment in which a new species was formed took place in the early 1900s, when the primrose *Primula kewensis* [PRIM-you-lah kee-WEN-siss] was produced. Scientists have also observed the formation of new species in nature. For example, two new species of salsify [SAL-suh-fee] plants were discovered in Idaho and eastern Washington in 1950.

Salsify

Figure 16.8 Once It Lived Throughout the Earth
Ancestors of the freshwater lungfish *Neoceratodus fosteri* (inset) lived during the time of Pangaea. Fossils of these fish have been found on all continents except Antarctica. *N. fosteri*, which is currently found only in the orange-shaded region of northeastern Australia, is the only surviving member of its family. Places where fossils of *N. fosteri*'s ancestors have been found are indicated by the red circles.

Neither of the new species had been found in those regions or anywhere else in the world in 1920. Genetic data reveal that both of the new species evolved from previously existing species, and field surveys indicate that this event occurred between 1920 and 1950. The two new species continue to thrive, and one of them has become common since its discovery in 1950.

16.5 The Impact of Evolutionary Thought

Before the concept of evolution was developed, adaptations often were taken as logical evidence for the existence of a creator. Organisms that seemed well designed for their environments implied to many people that there must be a supernatural designer—a God—much as the existence of a watch implies the existence of a watchmaker. In addition, from long tradition—dating from Plato, Aristotle, and other Greek philosophers more than two thousand years ago—species were viewed as unchanging over time. Darwin's work shook these ideas to their very foundation. No longer could species be viewed as unchanging, nor could examples of apparent design in nature be offered as logical proof that God existed, since evolution by natural selection provided an alternative, scientific explanation for the morphology of organisms.

The evolution of species was a radical idea in the midnineteenth century, and the argument that the morphology of organisms could be explained by natural selection

Figure 16.9 Artificial Selection Produces Genetic Change
Humans have directed the evolution of the wild mustard *Brassica oleracea* (bottom center) to produce several different crop plants. Despite their obvious differences, all the plants shown here, both wild and domesticated, are members of the same species, *B. oleracea*.

Biology on the Job

Evolution's Champion

Dr. Eugenie Scott is executive director of the National Center for Science Education (NCSE), a not-for-profit membership organization that provides information and resources for schools, parents, and concerned citizens working to keep evolution in public school science education. The NCSE educates the press and the public about the scientific, educational, and legal aspects of the evolution/creationism controversy. Dr. Scott has writ-ten a book about this controversy, entitled Evolution vs. Creationism.

What do you do on a typical day at work? At work, I'm the "Dear Abby" of people who want to be sure that evolution continues to be taught in our public schools. People call me on the phone if they have questions about evolution, or when they feel the wrong people have been elected to their school board. I advise parents, teachers, legislators, and school boards about the difference between science and religion, and why reli-gious views should not be taught as science. I also do a lot of public speaking at science teacher conferences, universities, and fund-raising events.

It sounds like you are mainly an educational activist. How did you become involved in that line of work? It was almost an accident. I started out as a professor of physical anthropology at the University of Kentucky. While I was there, I became involved in a local controversy over whether creationism should be taught in the biology classrooms of public schools. I was hooked. Working as an activist was challenging, in part because many people view the issue of whether to teach creation-ism as a battle between science and religion. I think of the issue much more narrowly—as a conflict between science and those opposed to teaching scientific princi-ples that differ from a literal interpretation of the Bible.

What are some of the major political and social issues at stake in the debate about evolution? There's a wide range of political and social issues at stake, including separation of church and state, public understanding of science, and the quality of science education. Evolution is central to the teaching of biology: without evolution as a central organizing principle, biology becomes a set of facts to memorize. Evolution is the periodic table of biology—it makes biology make sense, makes it hang together.

What do you like best about your job? I love my job because I think the issues are so important, and I really enjoy edu-cating people about the evolution/creationism contro-versy. Another great thing is that I get to read a broad range of scientific material—I love science and I have to read broadly in order to counter claims that creationists make, not only about evolution, but also about physics, geology, and gen-eral aspects of biology.

The Job in Context

Creationism states that all species were created by God roughly 10,000 years ago and that they have not evolved since then. As scientific statements, these assertions are not correct. The scientific evidence indicates that (1) Earth is more than 4 billion years old, (2) life on Earth began about 3.5 billion years ago, and (3) evolution has occurred, continues to occur, and is responsible for the great diversity of life on Earth.

A surrogate for creationism, known as the intelligent design move-ment, doesn't insist that life began on Earth as recently as 10,000 years ago. Nevertheless, both intelligent design and creationism explain the

Keeping Evolution in the Classroom
This teacher explains how this skeleton evolved, and where it came from.

diversity of life in terms of the actions of a supernatural creator. Such an explanation rings true to some people because their religious beliefs suggest that a supernatural being was directly or indirectly responsible for life on Earth. Science, however, is limited to natural, not supernatural, explanations. Furthermore, as we saw in Chapter 1, the scientific method is central to how science works: experiments are designed to test predictions made from a hypothesis, and if the results fail to support the hypothesis, it is modified or discarded. In contrast, creationism is based on supernatural, not natural, explanations, and creationists do not accept data that fail to support their ideas.

Like Dr. Scott, we think it is important for students—the future leaders of society—to understand evolution. If medical doctors, for example, had no understanding of evolution, they would not realize that overuse of antibiotics has the disastrous effect of causing bacteria to evolve resistance to those antibiotics (a topic we will explore further in Chapter 17). If creationists have their way and students are prevented from learning what science has to offer, we run the risk of having students unable to compete effectively in college classrooms or in today's global economy. When scientific understanding and nonscientific beliefs come into conflict, as illustrated by the conflict between evolutionary biology and creationism, the resulting debates can be enlightening. Such debates, however, should not be used as an excuse to keep students from being taught according to our best and most current scientific understanding of how the world works.

was even more so. These ideas not only revolutionized biology, but also had profound effects on other fields, ranging from literature to philosophy to economics.

The idea of Darwinian evolution had a profound effect on religion as well. Evolution was viewed initially as a direct attack on Judeo-Christian religion, and this presumed attack prompted a spirited counterattack by many prominent members of the clergy. Today, however, most religious leaders and most scientists view evolution and religion as compatible but distinct fields of inquiry (see the box on page 310). The Catholic Church, for example, accepts that evolution explains the physical characteristics of humans, but maintains that religion is required to explain our spiritual characteristics. Similarly, although the vast majority of scientists accept the scientific evidence for evolution, many of those same scientists have religious beliefs. Overall, most scientists recognize that religious beliefs are up to the individual and that science cannot answer questions regarding the existence of God or other matters of religious import.

The emergence of evolutionary thought has also had an effect on human technology and industry. For example, an understanding of evolution has proved essential as farmers and researchers have sought to prevent or slow the evolution of resistance to pesticides by insects. Information about the evolutionary relationships among organisms can also be used to increase the efficiency of the search for new antibiotics and other pharmaceuticals, food additives, pigments, and many other valuable products.

Finally, in what may be a sign of changes that lie ahead, engineers have begun to use evolutionary principles to solve a variety of design problems in nonliving systems. In 2003, for example, a patent was granted for a new type of controller. A controller is a device like the cruise control on a car that can regulate processes such as car speed, heat output from a home furnace, or the access of information from a computer disk. Usually, such a discovery would be the direct brainchild of a human inventor. However, the controller patented in 2003 was "invented" by a computer program that mimicked the biological processes of mutation, mating, and natural selection. The program functioned much as natural selection would in the wild: from one generation to the next, the program selected the "digital individual" that functioned as the best controller, allowing that individual to produce the most "offspring." The end result of this process was a digital individual whose features outperformed the best controllers currently on the market.

Evolution in Action

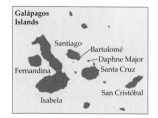

Galápagos Islands

Since Darwin's time, the Galápagos Islands have provided a natural biological laboratory in which scientists have studied evolution. The climate of these islands is usually hot and relatively wet from January to May, and cooler and drier for the rest of the year. But in 1977 the wet season never arrived; very little rain fell in the entire year. On Daphne Major, a small island near the center of the Galápagos Islands, the lack of rain withered the plants that lived there (Figure 16.10). Soon the effects were also felt by a seed-eating bird, the medium ground finch. During the drought, the number of these birds on Daphne Major plummeted from 1,200 to 180.

The lack of rain not only caused many birds to die, but also induced evolutionary change in the medium ground finch population. One effect of the drought,

Before drought

After drought

Figure 16.10 A Drought Results in Rapid Evolutionary Change
The 1977 drought on Daphne Major in the Galápagos Islands had a dramatic effect on the plant life there, thus setting the stage for natural selection to cause rapid evolutionary change in birds that depended on the plants for food.

which prevailed until the wet season of 1978, was that the seeds available to the finches were larger than normal. This happened because the finches had already eaten most of the smaller seeds by the time the drought began, and plants on the island produced few seeds during the drought. Large seeds can be difficult or impossible for birds with small beaks to crack open and eat. Thus finches with larger beaks had an edge, and many more large-beaked than small-beaked finches survived the drought to contribute offspring to future generations. As a result, the beak size of the medium ground finch population evolved toward a larger size.

Medium ground finches on Daphne Major have been studied continuously since 1972. As Figure 16.11 shows, the drought of 1977 had a strong but temporary effect on beak size: the average beak size of these birds shot up in the late 1970s, but then declined over time to end up at values similar to those observed at the beginning of the study period. Beak size was not the only feature of the birds that was subject to bouts of natural selection. After nearly 30 years of study, the body size of the birds had decreased, and their beak shape had changed from a more blunt to a more pointed condition.

As the research on the medium ground finch shows, we live in an ever-changing world in which species are constantly being shaped by evolutionary forces. In particular, this research shows how natural selection can cause traits to vary over time within a species. Natural selection and other evolutionary mechanisms can have even greater effects, such as causing entirely new species to form.

New species have evolved many times on the Galápagos Islands. These islands are home to many unique and unusual plant and animal species, including 13 species of finches (Figure 16.12). The Galápagos finches are closely related to one another, they are found nowhere else on Earth, and they exhibit many behaviors that are unusual for finches. For example, whereas most finches around the world are specialized to eat seeds, different Galápagos finches are specialized to eat a variety of foods, including seeds, insects, green leaves—even blood.

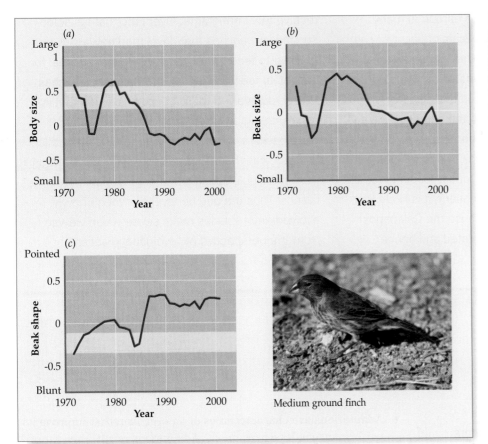

Figure 16.11 Evolutionary Change Within a Species
Several morphological features of the medium ground finch population on Daphne Major fluctuated over a 30-year period. Although body size (a), beak size (b), and beak shape (c) all fluctuated over time, only body size and beak shape changed significantly from the beginning to the end of the study period. The light-blue bands indicate the range of values that would be expected in the absence of evolutionary change.

Medium ground finch

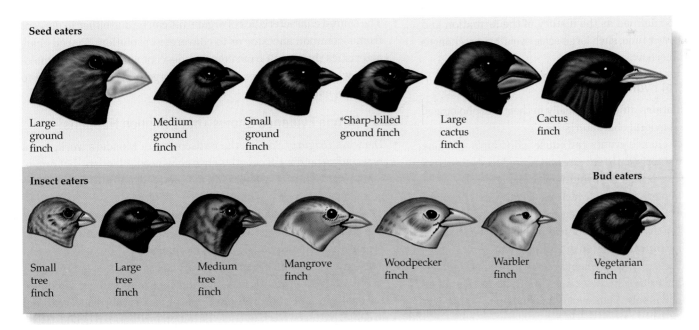

Figure 16.12 The Galápagos Finches
Thirteen unique species of finches evolved on the Galápagos Islands in isolation from other finches. Recent DNA evidence suggests that the warbler finch may actually comprise two different (but morphologically similar) species. The sharp-billed ground finch—the "vampire finch" mentioned at the beginning of this chapter—feeds on the blood of other birds in addition to eating seeds.

Why do these small islands harbor so many unique but closely related species of finches? The answer seems to be that all the finches on the Galápagos Islands descended from a single species, most likely a finch that reached the islands (perhaps blown there by a storm) from the nearest mainland, South America, about 3 million years ago. Upon its arrival, this species found itself in a place where many kinds of birds, such as insect-feeding woodpeckers and warblers, were absent; it is likely that these birds were absent because the Galápagos Islands are geographically isolated and hence receive few immigrants. Over time, natural selection favored finches that developed new ways (for finches) to feed themselves. While it is unusual for finches to feed on insects, the Galápagos finches that do so evolved to fill an ecological role that is usually taken by other birds, such as the absent woodpeckers and warblers. The end result of this process was that new species of finches evolved from the single species that originally colonized the islands. Thus the odd behaviors of these newly evolved finch species make sense when viewed from the perspective of "evolution in action."

Chapter Review

Summary

16.1   Biological Evolution: The Sum of Genetic Changes

- Evolution can be defined as change in the genetic characteristics of populations of organisms over time; such small-scale evolutionary changes are referred to as microevolution.
- Evolution can also be defined as the history of the formation and extinction of species over time; such large-scale evolutionary changes are referred to as macroevolution.

16.2   Mechanisms of Evolution

- Individuals in populations differ genetically in their morphological, biochemical, and behavioral characteristics.
- Natural selection or chance events can cause individuals with one set of inherited characteristics to survive and reproduce better than other individuals, thus producing evolutionary change.
- Natural selection is a mechanism by which some individuals, because they have particular inherited characteristics, consistently survive and reproduce better than do other individuals. Thus evolutionary changes wrought by natural selection are not random.
- Genetic drift is a process by which chance alone may cause individuals with one set of inherited characteristics to survive or reproduce better than other individuals. Thus evolutionary changes caused by genetic drift are random.

16.3   Consequences of Evolution for Life on Earth

- Evolution can explain three aspects of life on Earth: adaptations, the great diversity of species, and many puzzling examples in which otherwise dissimilar organisms share certain characteristics.

- Adaptations are characteristics of an organism that improve its performance in its environment. Adaptations result from natural selection.
- The diversity of life is a result of speciation, the repeated splitting of one species into two or more species.
- The shared characteristics of organisms can be due either to descent from a common ancestor or to convergent evolution. Shared characteristics that result from common descent are said to be homologous, while those that result from convergent evolution are said to be analogous.

16.4   Strong Evidence Shows That Evolution Happens

- Darwin's *Origin of Species*, the rediscovery of Mendel's work, and a new understanding of genetics have convinced scientists that organisms have evolved—that is, descended with modification from common ancestors.
- The fossil record provides clear evidence for the evolution of species over time. It also documents the evolution of major new groups of organisms from previously existing organisms.
- The extent to which organisms share characteristics is consistent with patterns of evolutionary relationships among them. Evidence for evolution includes remnant anatomical structures (vestigial organs), patterns of embryonic development, and molecular (DNA and protein) similarities.
- As predicted by our understanding of evolution and continental drift, fossils of organisms that evolved when the present-day continents were all part of the supercontinent Pangaea have a wider geographic distribution than do fossils of more recently evolved organisms.

- In thousands of studies, researchers have observed genetic changes in populations over time, providing direct evidence of small evolutionary changes. Some of these studies involve artificial selection, in which people alter a population by choosing which individuals to breed; others involve observations of natural selection.
- Biologists have observed the evolution of new species from previously existing species.

16.5 The Impact of Evolutionary Thought

- Darwin's ideas on evolution and natural selection revolutionized biology, overturning the views that adaptations must be proof of God's existence and that species do not change over time. His ideas also had a profound effect on many other fields, including literature, economics, and religion.
- Evolutionary biology has many practical applications in agriculture, industry, and medicine.

◉ Review and Application of Key Concepts

1. Explain what evolution is and why we state that "populations evolve, but individuals do not."

2. A population of lizards lives on an island and eats insects found in shrubs. Because the shrubs have narrow branches, the lizards tend to be small (so they can move effectively among the branches). A group of lizards from this population migrates to a nearby island where the vegetation consists mostly of trees; the branches of those trees are thicker than the shrubs on which the lizards used to feed. A few of the lizards that migrate to the new island are slightly larger than the others; large size is not a disadvantage for moving on the trees (because the branches are thicker) and is advantageous when males compete with other males to mate with females. Assuming that large size is an inherited characteristic, explain what is likely to happen to the average size of the lizards in their new home, and why.

3. How does evolution explain (a) adaptations, (b) the great diversity of species, and (c) the many examples in which otherwise dissimilar organisms share certain characteristics?

4. Why are scientists throughout the world convinced that evolution happened? Consider the five lines of evidence discussed in this chapter.

5. Although biologists agree that evolution occurs, they debate which mechanisms are most important in causing evolutionary change. Does this mean that the theory of evolution is wrong?

6. Genetic drift occurs when chance events cause some individuals in a population to contribute more offspring to the next generation than other individuals. Are such chance events likely to have a greater effect in small or in large populations? [*Hint:* Examine Figure 16.3. Consider whether the proportion of the *A* allele in the population would be likely to change from 50 percent (5 of the 10 alleles originally present were *A* alleles) to 100 percent if there were 1,000 plants instead of 5 plants in the population.]

Key Terms

adaptation (p. 303)
analogous (p. 305)
artificial selection (p. 308)
continental drift (p. 307)
convergent evolution (p. 304)
evolution (p. 300)
fossil (p. 306)
genetic drift (p. 302)

homologous (p. 304)
macroevolution (p. 301)
microevolution (p. 301)
morphology (p. 301)
natural selection (p. 301)
Pangaea (p. 307)
speciation (p. 304)
vestigial organs (p. 304)

Self-Quiz

1. Which of the following provides evidence for evolution?
   a. direct observations of genetic changes in populations
   b. shared characteristics of organisms
   c. the fossil record
   d. all of the above

2. In natural selection,
   a. the genetic composition of the population changes at random over time.
   b. new mutations are generated over time.
   c. all individuals in a population are equally likely to contribute offspring to the next generation.
   d. individuals that possess particular inherited characteristics consistently survive and reproduce at a higher rate than other individuals.

3. Adaptations
   a. are features of an organism that hinder its performance in its environment.
   b. are not common.
   c. consistently result from natural selection.
   d. consistently result from genetic drift.

4. The fossil record shows that the first mammals evolved 220 million years ago. The supercontinent Pangaea began to break apart 200 million years ago. Therefore, fossils of the first mammals should be found
   a. on most or all of the current continents.
   b. only in Antarctica.
   c. on only one or a few continents.
   d. none of the above

5. The fact that the flipper of a whale and the arm of a human both have five digits and the same kinds of bones can be used to illustrate that
   a. genetic drift can cause the evolution of populations.
   b. organisms can share characteristics simply because they share a common ancestor.
   c. whales evolved from humans.
   d. humans evolved from whales.

6. The Galápagos Islands provide examples of
   a. microevolution only.
   b. macroevolution only.
   c. both micro- and macroevolutionary change.
   d. none of the above

7. Differences in survival and reproduction caused by chance events can cause the genetic makeup of a population to change at random over time. This process is called
   a. mutation.   c. macroevolution.
   b. natural selection.   d. genetic drift.

8. The splitting of one species into two or more species is called
   a. speciation.   c. common descent.
   b. macroevolution.   d. adaptation.

9. Features of organisms that are related to one another through common descent are
   a. convergent.   c. divergent.
   b. homologous.   d. analogous.

Biology in the News

What a Story Lice Can Tell

By Nicholas Wade

A spectator with an especially intimate view of human evolution is beginning to tell its story and has so far divulged [some] quite unexpected findings.

The human louse finds people so delicious that it will accept no substitutes and cannot live more than a few hours away from the sustenance and warmth of the human body. This devotion to the human cause means that the evolutionary history of human lice dovetails with that of their hosts and reflects several pivotal events that affected both species.

One of the mysteries that surrounds our evolution is how and when modern humans (*Homo sapiens*) and our humanlike ancestors spread across the globe. A study published in the October 5, 2004, issue of the scientific journal *PLoS Biology* uses species that travel with us—lice—to shed new light on when and where we came into contact with our humanlike ancestors.

Many scientists think that modern humans originated in Africa, then migrated from there to other parts of the world during the last 100,000 years. As people moved out of Africa, they would have brought their lice with them. Dr. David Reed, a curator at the Florida Museum of Natural History, has analyzed the DNA of human lice and found that it clusters into two groups: a group from North America and Central America, and a group that includes lice from all over the world. Dr. Reed's calculations suggest that these two groups of lice evolved in isolation from each other for nearly 1.2 million years—long before modern humans existed. So how did modern humans come to have these two different strains of lice?

The answer may lie with *Homo erectus*, a humanlike species that migrated from Africa to Asia more than a million years ago. Some scientists think that *H. erectus* may have survived in Asia until as recently as 30,000 years ago. Dr. Reed's results lend some support to this idea.

As modern humans traveled from Africa to Asia, they may have come into contact with surviving *H. erectus* individuals, picking up their lice. Then, when modern humans migrated to the Americas (about 13,000 years ago), they would have brought with them the two strains of lice—the one they picked up from *H. erectus* and the African strain they already had. As a result, the DNA of lice on modern humans clusters into two groups, one of which was picked up from *H. erectus* and now is found only in the Americas. As interesting as these findings are, prepare to learn more as more research on these lice is done. As Dr. Reed notes, "I really think we are only scratching the surface of what lice can tell us."

Evaluating the news

1. An underlying assumption of this article is that humans, like other species, can evolve. Do you agree with this assumption? Why or why not?

2. This news article reports that DNA sequences were used to reconstruct the evolutionary history of lice, an organism that transmits several lethal human diseases (including typhus). It is also possible to use DNA sequences to reconstruct the evolutionary history of a disease-causing bacterium or virus. From that history, it may then be possible to estimate when and where the disease originated. This has been done for a variety of diseases, such as AIDS and the disease caused by the hantavirus. Why might information about when and where a disease originated be of use to medical researchers? (*Hint:* Species can become resistant to a disease over time. In what region of the world would you be most likely to find species that are resistant to the disease? How might those species be of use to researchers?)

3. Given that some members of society are opposed to evolution (say, for religious reasons), should tax dollars be used to fund evolutionary studies, some of which can provide social benefit (for example, by developing the methods used to track down the origin of a lethal disease)? Why or why not?

4. Does using tax dollars to fund evolutionary studies differ from using tax dollars to fund military operations? Both are opposed by some members of society for religious reasons, but also may provide some social good. Explain your answer.

SOURCE: *The New York Times*, Tuesday, October 5, 2004.

CHAPTER 17

Evolution of Populations

Key Concepts

- Populations evolve when allele frequencies change from generation to generation.

- Individuals within populations differ genetically in their behavior, morphology, and biochemistry. This genetic variation provides the raw material on which evolution can work.

- Mutations are the original source of genetic variation within populations. The genetic variation produced by mutations is increased by recombination.

- Four mechanisms can cause populations to evolve: mutation, gene flow, genetic drift, and natural selection.

- Some populations can evolve very rapidly (in months to years). This potential for rapid evolution has serious implications for the evolution of insect resistance to pesticides and bacterial resistance to antibiotics.

Evolution of Resistance

Imagine that you lived 150 years ago. Things were different then—there were no computers, electric lights, highways, or airplanes. In addition, life was often short, frequently because infection or disease brought life to a sudden and unexpected end. A healthy person in his or her twenties might fall, scrape a knee deeply, get an infection, and die shortly thereafter. Diseases such as smallpox, malaria, cholera, tuberculosis, and the plague could strike, decimating towns, countries, and even whole geographic regions.

Of course people still die in sudden and unexpected ways, but in developed regions of the world today we are used to thinking that if we get an infection or get sick, we can be cured. We expect this in part because of the discovery of antibiotics. These drugs, first introduced in the late 1930s, worked like "magic bullets," destroying bacteria that caused infection and disease. Before antibiotics were available, even a seemingly mild infection, such as a boil on the face, was potentially lethal. With antibiotics, life became considerably less dangerous.

Progress in other areas reinforced a view that life was becoming less dangerous for people. Pesticides such as DDT, also introduced in the late 1930s, allowed people to kill insects such as the mosquitoes that spread malaria and other fatal diseases. In the late 1960s, the stunning success of antibiotics and pesticides led the U.S. surgeon general to testify confidently to Congress that it was time to "close the book on infectious diseases."

Unfortunately, the surgeon general was mistaken. Today, for every antibiotic now in use, there is one or more species of bacteria resistant to it. In addition, during the past 65 years, the bacteria that cause rheumatic fever, staph infections, pneumonia, strep throat, tuberculosis, typhoid fever, dysentery, gonorrhea, and meningitis all have evolved resistance to multiple antibiotics. This

Some Resistant Viruses and Organisms
HIV virus (top); Colorado potato beetle (middle); tuberculosis bacterium (bottom).

319

widespread resistance of bacteria to antibiotics has serious implications. Some biologists worry that we may enter a "postantibiotic" era, in which bacterial diseases ravage human populations on a scale not seen since the discovery of antibiotics.

And bacteria are not the only disease agents affected by this trend. Most viruses, fungi, and parasites that cause disease have evolved resistance to our best efforts to destroy them. Similar problems have arisen in our efforts to control insects that spread disease and damage crops. At first, just like antibiotics, DDT and other newly discovered pesticides were remarkably effective: insects are notori-ously hard to kill, but the new poisons did an excellent job. Over time, however, pesticides have worked less and less well. Over 500 species of insects are now resistant to one or more pesticides. The speed with which resistance to pesticides evolves in insects is worrisome. It used to take decades, but now resis-tance often appears in a few years or less.

Overall, the evolution of resistance sends a sobering message: the pests and killers we seek to destroy evolve resistance rapidly in response to our best efforts to defeat them. How does this happen? What forces have led to the evolution of resistance in so many different organisms?

In the nineteenth century, Charles Darwin wrote that evolution occurs too slowly to observe, but over the last 80 years, thousands of studies have observed and documented the evolution of populations. Collectively, these studies indicate that Darwin had the basic ideas about evolution right, but that he was wrong about the time required for evolution to occur. Modern studies show that populations evolve slowly in some cases and very rapidly in others (within a few generations, covering a time span of months to years). Similarly, new species form slowly in some cases but rapidly in others (in a year to a few thousand years).

In this chapter we describe evolutionary changes that take place in populations, sometimes over short time spans. In particular, we focus on how the frequencies (proportions) of alleles in populations can change from generation to generation. These changes in allele frequen-cies over time are referred to as **microevolution**, so called because they represent the smallest scale at which evo-lution occurs. An example of this type of evolutionary change would be if the percentage of a particular allele (such as *a*) significantly increased or decreased in a par-ticular population over several generations.

We begin our discussion of microevolution with defini-tions of two essential terms, genotype frequency and allele frequency, and with a brief look at what constitutes the "raw material" of evolution. With that information in hand, we then discuss four mechanisms that can cause allele fre-quencies to change over time: mutation, gene flow, genetic drift, and natural selection. We end the chapter by consid-ering some possible ways to combat infectious diseases more effectively by slowing the evolution of resistance.

17.1 Key Definitions: Allele and Genotype Frequencies

If we want to know whether a population is evolving with regard to a particular trait, we need to determine the allele frequencies for the genes involved, since (as men-tioned above) microevolution is defined as change in allele frequencies. **Allele frequency** refers to the proportion, or percentage, of an allele (such as the *A* or *a* allele) in a population. Similarly, **genotype frequency** refers to the proportion, or percentage, of a genotype (such as the *AA*, *Aa*, or *aa* genotype) in a population.

To see how these proportions are calculated, let's imag-ine that flower color in a population of 1,000 plants is determined by a single gene, which has two alleles: a dom-inant allele, *R*, for red flower color, and a recessive allele, *r*, for white flower color. If the population contains 160 *RR* individuals, 480 *Rr* individuals, and 360 *rr* individuals, then we can obtain the frequencies for the three geno-types (*RR*, *Rr*, and *rr*) by dividing their numbers by the total number of individuals in the population (1,000). Thus the genotype frequencies for *RR*, *Rr*, and *rr* are 0.16, 0.48, and 0.36, respectively. These genotype frequencies add up to 1.0—as they should, because *RR*, *Rr*, and *rr* are the only three genotypes possible and a frequency of 1.0 is equivalent to 100 percent. Hence one way to check our genotype calculations is to make sure they add up to 1.0.

Allele frequencies can be computed by the following method, which we illustrate for the *R* allele (though we could have chosen the *r* allele instead). There are 1,000 individu-als in our plant population, each of which has two alleles of the flower color gene. Thus the total number of alleles in the population equals 2,000. There are 160 *RR* individuals, each of which carries two *R* alleles, for a total of 320 *R* alleles. *Rr* individuals have one *R* allele each, for a total of 480 *R* alleles, and *rr* individuals have no *R* alleles. Thus there are 800 (320 + 480 + 0) *R* alleles in the population. Finally, we calculate the frequency of the *R* allele by dividing the number of *R* alleles by the total number of alleles in the population: (800 *R* alleles)/(2,000 total alleles) = 0.4. Since the gene for flower color has only two alleles, *R* and *r*, the sum of their frequencies must equal 1.0. Hence the frequency of the *r* allele is 1.0 − 0.4 = 0.6. We can check this value by performing a calculation for the *r* allele similar to the one we made for the *R* allele.

17.2 Genetic Variation: The Raw Material of Evolution

Within a population, individuals differ in their behavior and morphology (Figure 17.1). As described in Unit 3, much of this variation is under genetic control. Organisms also vary greatly in biochemical traits that are under direct genetic control, such as the amino acid sequences of their proteins. The underlying cause of all inherited differences among individuals is variation in their DNA sequences. Such genetic differences among the individuals of a population are collectively referred to as **genetic variation**; this variation is important because it provides the raw material on which evolution can work.

Mutation is one source of genetic variation

Recall from Chapter 12 that mutations are changes in the sequence of an organism's DNA, and that mutations give rise to new alleles. Thus mutations are the original source of all genetic variation. Gene mutations are caused by various accidents, such as mistakes in DNA replication, collisions of the DNA molecule with other molecules, or damage from heat or chemical agents. As a result, mutations and the new alleles they produce do not appear when an organism needs them; instead, mutations occur at random and are not directed toward any goal.

Despite the efficiency with which repair proteins fix damage to DNA and correct errors in DNA replication, mutations occur regularly in all organisms. Humans, for example, have two copies each (one copy from each parent) of their approximately 25,000 genes. On average, two or three of these 50,000 gene copies have mutations that make them different from those of either parent.

Recombination provides another source of genetic variation

In sexually reproducing organisms, as we saw in Chapter 11, the different alleles produced by mutation are grouped in new arrangements by crossing-over, the independent assortment of chromosomes, and fertilization; collectively, these three processes are known as **recombination**. Recombination causes offspring to have many new combinations of alleles that differ from those found in either parent. Thus recombination greatly increases the genetic variation produced originally by mutation.

17.3 Four Mechanisms Can Cause Populations to Evolve

Evolution occurs when allele frequencies in a population change over time. There are four primary ways by which such changes can occur: mutation, gene flow, genetic drift, and natural selection. Here we provide a brief overview of how these mechanisms work, saving more detailed discussion of each mechanism for the sections that follow.

Figure 17.1 Morphological Variation
The individuals within a population often vary greatly in many morphological traits, such as the color patterns of the sea stars shown here.

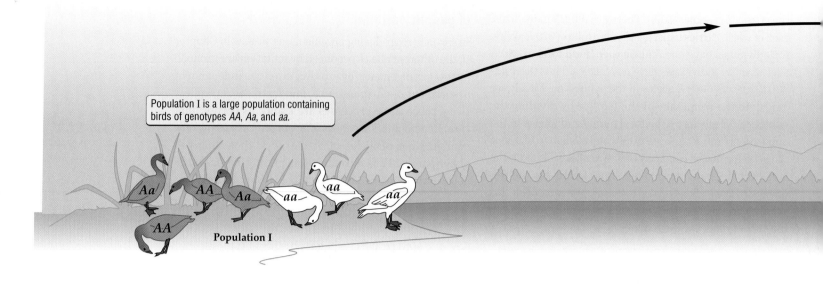

Population I is a large population containing birds of genotypes *AA*, *Aa*, and *aa*.

Population I

Mutation and gene flow (the movement or exchange of alleles between populations) can introduce new alleles into a population, thus changing allele frequencies and causing evolution to occur. Although genetic drift is driven by chance events, it too can cause allele frequencies to change and hence evolution to occur, as we saw in Chapter 16. Finally, individuals with certain alleles may have an advantage over individuals that do not have those alleles. In this case, natural selection causes the frequency of the favored alleles to increase, thus causing evolution to occur.

If we know the genotype frequencies in a population, then the Hardy-Weinberg equation allows us to test whether evolution is occurring (see the box on page 324). This test works the same regardless of the mechanisms that are causing allele frequencies to change.

17.4 Mutation: The Source of Genetic Variation

As we have seen, mutation creates new alleles at random, thereby providing the raw material for evolution. In this sense, all evolutionary change depends ultimately on mutation.

We can also view mutation in another way: as a mechanism for changing allele frequencies in a population, thus causing evolution to occur. However, mutations occur so infrequently in any particular gene that they cause little direct change in the allele frequencies of populations. In contrast, allele frequencies in populations often change

rapidly, indicating that mutation, acting alone, cannot be directly responsible for most evolutionary change.

Although mutations have little direct effect on allele frequencies, in some cases new mutations play a critical role in the evolution of populations. For example, HIV (the human immunodeficiency virus), which causes AIDS, has a high mutation rate and thus produces many new mutations, even within a single patient's body. Some of these mutations may allow the virus to resist new clinical treatments, which makes combating the virus difficult because it is a "moving target."

In another example, genetic evidence indicates that the resistance of the mosquito *Culex pipiens* [*KYOO*-leks *PIP*-yens] to organophosphate pesticides was caused by a single mutation that occurred in the 1960s. Since that time, mosquitoes blown by storms or moved accidentally by people have carried the initially rare mutant allele from its place of origin in Africa or Asia to both North America and Europe. This mutant allele is highly advantageous to the mosquito: individuals that have the non-mutant allele die when exposed to organophosphate pesticides. Thus, when the mutant allele is introduced (by gene flow) into a population that is exposed to organophosphate pesticides, natural selection causes the mutant allele to increase rapidly in frequency, leading to the evolution of resistance within the new population.

Mutations like those that allow disease agents or pests to resist our best efforts to kill them are obviously beneficial to the organisms in which they occur. Most mutations, however, are either harmful to their bearers or have little effect. In general, the effect of a mutation often depends on the environment in which the organism lives. For example, certain mutations that provide houseflies with resistance

Explore the process of gene flow.

17.1

Figure 17.2 Gene Flow

A bird of genotype *aa* migrates from population I to population II, thereby introducing the *a* allele to population II.

Population II is located far from population I and initially has only birds of genotype *AA*.

Population II

to the pesticide DDT also reduce their rate of growth. Flies that grow more slowly take longer to mature, and thus do not produce as many offspring in their lifetimes as flies that grow at the normal rate. Thus, in the absence of DDT, such mutations are harmful. When DDT is sprayed, however, these mutations provide an advantage great enough to offset the disadvantage of slow growth. As a result, the mutant alleles have spread throughout housefly populations and can now be found globally.

17.5 Gene Flow: Exchanging Alleles Between Populations

When individuals move from one population to another, an exchange of alleles between populations, or **gene flow**, occurs (Figure 17.2). Gene flow can also occur when only gametes move from one population to another, as happens with windblown pollen from plants.

Gene flow can play a role similar to that of mutation by introducing new alleles into a population. Such introductions of new alleles can have dramatic effects. For example, in the case of the mosquito *Culex pipiens*, discussed earlier, a new allele that made the mosquito resistant to organophosphate pesticides spread by gene flow across three continents. This spread of a new mutant allele allowed billions of mosquitoes to survive the application of pesticides that otherwise would have killed them.

Because it consists of an exchange of alleles between one population and another, gene flow makes the genetic composition of different populations more similar. In this way, gene flow can counteract the effects of mutation, genetic drift, and natural selection, all of which can cause populations to become more different from one another. In some plant species, for example, neighboring populations live in very different environments, yet remain genetically similar. Natural selection, by favoring different alleles in the different environments, would tend to make the populations differ genetically. The lack of genetic difference between such populations appears to be due to gene flow, which occurs at a rate high enough to cause the populations to remain genetically similar despite the effects of natural selection.

Explore the Hardy-Weinberg equation.

17.2

17.6 Genetic Drift: The Effects of Chance

As we learned in Chapter 16, chance events may determine which individuals contribute offspring to the next generation (see Figure 16.3). As a result, chance events can cause alleles from the parent generation to be selected at random for inclusion in the next generation. The process by which alleles are selected at random over time is called **genetic drift**. This process leads to random changes in allele frequencies from generation to generation.

Genetic drift affects small populations

The chance events that cause genetic drift are much more important in small populations than in large populations.

Science Toolkit

Testing Whether Evolution Is Occurring in Natural Populations

There are four mechanisms that can cause evolutionary change in natural populations: mutation, gene flow, genetic drift, and natural selection. This sounds simple enough, but when a biologist begins to study a population in nature, he or she may have little idea whether one or more of these four mechanisms are actually important in the study population. Fortunately, a quick genetic calculation can be performed, the results of which provide an initial indication of whether the population is evolving.

Let's assume we are dealing with a gene with just two alleles, A and a. And let's assign new letters to represent the allele frequencies: p for the frequency of A, and q for the frequency of a. (Adding two more letters—p and q—to the mix is not strictly necessary when calculating genotype frequencies, but it will prove very helpful shortly when we calculate allele frequencies.) Because this gene has exactly two alleles, we know that the sum of the two allele frequencies is $p + q = 1$.

The quick genetic calculation to determine whether evolution is occurring in a population relies on the **Hardy-Weinberg equation**, which has the general form

Frequency of genotype AA

Frequency of genotype aa

$$p^2 + 2pq + q^2 = 1$$

Frequency of genotype Aa

As described in Appendix 1, this equation is derived by assuming that the population is *not* evolving—in other words, that mutation, gene flow, genetic drift, and natural selection *do not* cause allele frequencies to change. Thus the Hardy-Weinberg equation gives us the genotype frequencies for a population that is not evolving. Because there are exactly three genotypes in this population, the three frequencies must sum to 1 (equivalent to 100 percent), as the Hardy-Weinberg equation shows.

The value of the Hardy-Weinberg approach is that we can use it to test whether the genotype frequencies in a real population match those pre-

dicted by the equation. If the actual frequencies differ considerably from the frequencies predicted by the Hardy-Weinberg equation, then the assumptions used to derive the equation must not be correct. Hence we can conclude that one or more of the four evolutionary mechanisms (mutation, gene flow, genetic drift, or natural selection) is at work. As a result, the population will evolve, and allele frequencies will change over time.

To find out whether genotype frequencies in a real population differ from those predicted by the Hardy-Weinberg equation, we must first determine the genotypes of individuals in the population; often this determination is made using the techniques described in Chapter 15. Let's assume that a population of 1,000 individuals has 460 individuals of genotype AA, 280 individuals of genotype Aa, and 260 individuals of genotype aa. Using this information, we calculate that the observed frequency of the A allele is $p = (2 \times 460 + 280)/2{,}000 = 0.6$, from which we also know that q, the observed frequency of the a allele, is 0.4.

Now we can put all of this information together to determine whether the population is evolving or not. The observed frequency of the A allele is 0.6. If the population is not evolving, the Hardy-Weinberg equation will hold, and the frequency of the AA genotype should be p^2, or 0.36. Similarly, the frequency of Aa should be 0.48 ($2pq$), and the frequency of aa should be 0.16 (q^2). Therefore, since there are 1,000 individuals in the population, if the population is not evolving, we would expect to find 360 ($0.36 \times 1{,}000$) AA individuals, 480 ($0.48 \times 1{,}000$) Aa individuals, and 160 ($0.16 \times 1{,}000$) aa individuals. In fact, there were 460 AA individuals, 280 Aa individuals, and 260 aa individuals. Thus there are more AA and aa individuals, and fewer Aa individuals, than expected.

The differences between the actual and expected genotype frequencies that we have just described are large. A biologist who obtained real data like those in our example would conclude that the actual genotype frequencies differed significantly from those in the Hardy-Weinberg equation, and thus that the population was evolving. Next, he or she would begin to wonder what evolutionary mechanisms might be driving the population away from the predictions of the Hardy-Weinberg equation. Once you've finished reading this chapter, look again at the differences between the observed and the expected genotype frequencies. Can you suggest one or more evolutionary mechanisms that might explain these differences?

To understand why, consider what happens when you toss a fair coin. It would not be all that unusual to get four heads in five tosses—there is in fact about a 15 percent chance you will get that result. But it would be astonishing to get 4,000 heads in 5,000 tosses. Even though the frequency of heads would be the same in both cases (80 percent), the chance of getting many more than the expected 50 percent heads is much greater if the coin is tossed a few times than if it is tossed thousands of times.

In natural populations, the number of individuals in a population has an effect similar to the number of times a coin is tossed. Consider the small population of wildflowers shown in Figure 17.3. By chance alone, some individuals leave offspring and others do not. In this example, such chance events alter the allele frequencies of a gene with two alleles (R and r). The changes are so rapid that one of the alleles (r) is lost from the population in just two generations. The other allele (R) reaches **fixation**, a frequency of 100 percent. When there are many individuals in a population, the likelihood that each allele will be passed on to the next generation greatly increases. Thus, if there had been many more individuals in the population shown in Figure 17.3, it is unlikely that chance events could have caused such dramatic changes in so short a time.

Genetic drift also occurs in large populations, but in these cases its effects are more easily overcome by natural selection and other evolutionary mechanisms. Thus, in large populations, genetic drift causes little change in allele frequencies over time.

What types of chance events cause genetic drift? One important source of genetic drift is the random alignment of alleles during gamete formation, which causes (by chance alone) some alleles, but not others, to be passed on to offspring. Another source of genetic drift is chance events associated with the survival and reproduction of individuals. In this case, even though a particular genotype may increase in frequency from one generation to the next, it is important to remember that the increase is due to chance events (as in Figure 17.3), not to that genotype having an advantage over other genotypes because of the alleles it carries (as would occur in natural selection).

Figure 17.3 Genetic Drift

In this small population of wildflowers, chance events determine which plants leave offspring without regard to which individuals are better equipped for survival or reproduction. Here, chance events cause the frequency of the R allele to increase from 60 percent to 100 percent in two generations. Note that this particular outcome is just one of many ways that genetic drift could have caused allele frequencies to change at random over time.

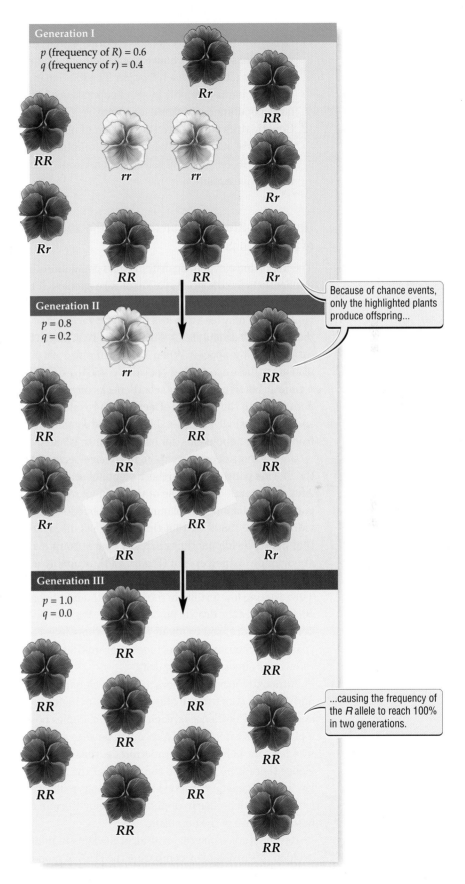

Generation I

p (frequency of R) = 0.6
q (frequency of r) = 0.4

Because of chance events, only the highlighted plants produce offspring…

Generation II

p = 0.8
q = 0.2

Generation III

p = 1.0
q = 0.0

…causing the frequency of the R allele to reach 100% in two generations.

Overall, genetic drift can affect the evolution of small populations in two ways:

1. Genetic drift can reduce genetic variation within small populations because chance alone eventually causes one of the alleles to reach fixation. The fixation of alleles can happen rapidly in small populations, but in large populations it takes a long time.
2. Genetic drift can lead to the fixation of alleles that are neutral, harmful, or beneficial. As emphasized in Chapter 16, only natural selection consistently leads to adaptive evolution.

Genetic bottlenecks can threaten the survival of populations

The importance of genetic drift in small populations has implications for the preservation of rare species. If the number of individuals in a population falls to very low levels, genetic drift may lead to a loss of genetic variation or to the fixation of harmful alleles, either of which can hasten the extinction of the species. When a drop in the size of a population has these effects, the population is said to experience a **genetic bottleneck**. Genetic bottlenecks often occur in nature due to the **founder effect**, which results when a small group of individuals establishes a new population far from existing populations (for example, on an island).

Genetic bottlenecks are thought to have occurred in the Florida panther, the northern elephant seal, and the African cheetah. In the case of the endangered Florida panther, population sizes plummeted to about 30–50 individuals in the 1980s. At that time, biologists discovered that male Florida panthers had low sperm counts and abnormally shaped sperm (Figure 17.4), probably as a result of the fixation of harmful alleles by genetic drift. The resulting low fertility in the males is thought to have contributed to the drop in population size. Panther numbers have increased to about 80 individuals in recent years, in part because of breeding programs designed to reduce the effects of genetic drift.

Although Florida panthers, northern elephant seals, and African cheetahs all show signs of having experienced genetic bottlenecks, each of these examples poses a problem: We don't know how much genetic variation was present before the population decreased in size. Hence there is no way to be sure whether the observed low levels of genetic variation in these animals really were caused by a decrease in population size or were just a natural feature of the organism.

Recent studies on greater prairie chickens in Illinois avoided this problem by comparing the DNA of modern

Elephant seal

birds with the DNA of their prebottleneck ancestors, obtained from (nonliving) museum specimens. There were millions of greater prairie chickens in Illinois in the nineteenth century, but by 1993 the conversion of their prairie habitat to farmland had caused their numbers to drop to only 50 birds in two isolated populations (Figure 17.5). This drop in numbers caused a genetic bottleneck: the modern birds lacked 30 percent of the alleles found in the museum specimens, and they suffered poor reproductive success compared with other prairie chicken populations that had not experienced a genetic bottleneck (only 56 percent of their eggs hatched, versus 85–99 percent in other populations). From 1992 to 1996, 271 birds were introduced to Illinois from large populations in Minnesota, Kansas, and Nebraska; this was done to increase both the size and the genetic variation of the Illinois populations. By 1997, the number of males in one of the two remaining Illinois populations had increased from a low of 7 to over 60 birds, and the hatching success of eggs had risen to 94 percent.

17.7 Natural Selection: The Effects of Advantageous Alleles

Natural selection is a process by which individuals with particular inherited characteristics survive and reproduce at a higher rate than other individuals in a population. Natural selection acts by favoring some phenotypes over others, as when mosquitoes that are resistant to a pesticide survive at a higher rate and produce more offspring than other mosquitoes. Although natural selection acts directly on the phenotype, not on the genotype, the alleles that code for forms of a trait that are favored by natural selection tend to become more common in the offspring generation than in the parent generation. For example, if large size is an inherited characteristic, and if natural selection consistently favors large individuals, then alleles that cause large size will tend to become more common in the population over time.

Even natural selection does not always lead to evolutionary change

Of the four mechanisms of evolution, natural selection is the only one that consistently improves the reproductive success of the organism in its environment. As shown by the study of the medium ground finch

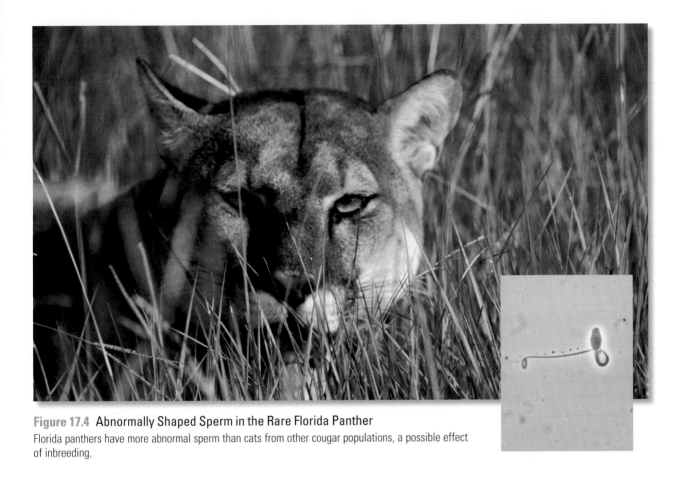

Figure 17.4 Abnormally Shaped Sperm in the Rare Florida Panther
Florida panthers have more abnormal sperm than cats from other cougar populations, a possible effect of inbreeding.

Illinois

By 1993, only 50 greater prairie chickens remained in Illinois, causing both the number of alleles and the percentage of eggs that hatched to decrease.

	Illinois		Kansas	Minnesota	Nebraska
	Prebottleneck (1933)	Postbottleneck	No bottleneck		
Population size	25,000	50	750,000	4,000	75,000 – 200,000
No. of alleles at 6 genetic loci	31	22	35	32	35
Percentage of eggs that hatch	93	56	99	85	96

Prebottleneck (1820)   Postbottleneck (1993)

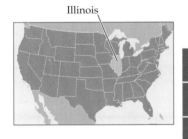

In 1820, the grasslands in which greater prairie chickens live covered most of Illinois.

In 1993, less than 1% of the grassland remained, and the birds could be found only in these two locations.

Figure 17.5 A Genetic Bottleneck
The Illinois population of greater prairie chickens dropped from 25,000 birds in 1933 to only 50 birds in 1993. This drop in population size caused a loss of genetic variation and a drop in the percentage of eggs that hatched. Here, the modern, postbottleneck Illinois population is compared with the 1933 prebottleneck Illinois population, as well as with populations in Kansas, Minnesota, and Nebraska that never experienced a bottleneck.

described in Chapter 16, natural selection can sometimes cause traits to evolve rapidly in response to changes in the environment.

However, even when natural selection favors one allele over another, it does not necessarily lead to evolutionary change. The other evolutionary mechanisms—genetic drift, gene flow, and mutation—may oppose its effects and prevent allele frequencies from changing. It also bears repeating that unless individuals within a population differ genetically, and unless some of them have beneficial mutations on which selection can act, natural selection is powerless. If none of the mosquitoes in a population, for example, carry alleles for resistance to a pesticide, then natural selection cannot promote the evolution of resistance to that pesticide.

There are three types of natural selection

Natural selection can be divided into three types: directional selection, stabilizing selection, and disruptive selection. Despite these categories, it is important to remember that all types of natural selection operate by the same principle: individuals with certain forms of an inherited phenotypic trait tend to survive better and produce more offspring than individuals with other forms of that trait.

In **directional selection**, individuals with one extreme of an inherited phenotypic trait have an advantage over other individuals in the population. For example, if large individuals produce more offspring than small individuals, then there will be directional selection for large body size (Figure 17.6a).

Directional selection can be illustrated by the rise and fall of dark-colored forms in various moth species. For example, dark-colored forms of the peppered moth were favored by natural selection when industrial pollution blackened the bark of trees in Europe and North America. The color of these moths is a genetically determined trait, and the allele for dark color is dominant to alleles for light color. Peppered moths are active at night and rest on trees during the day. The proportion of the dark-colored moths increased over time, apparently because they were harder for predators such as birds to find against the blackened bark of trees (Figure 17.7). The rise in the proportion of dark-colored forms took less than 50 years. The first dark-colored moth was found near Manchester, England, in 1848. By 1895, about 98 percent of the moths near Manchester were dark-colored, and proportions of over 90 percent were common in other heavily industrialized areas of England.

Today, however, light-colored moths outnumber dark-colored moths. The reason, once again, is directional selection, but this time it was the light-colored moths that were favored, not the dark-colored moths. This turnaround was due to the Clean Air Act passed in England in 1956. As the air quality improved, the bark of trees became lighter, and light-colored moths became harder for predators to find than dark-colored moths. As a result, the proportion of dark-colored moths plummeted (see Figure 17.7).

Learn more about the three models of natural selection.

Figure 17.6 The Three Types of Natural Selection
Directional selection (a), stabilizing selection (b), and disruptive selection (c) affect phenotypic traits such as body size differently. The graphs in the top row show the relative numbers of individuals with different body sizes in a population before selection. The phenotypes favored by selection are shown in yellow. The graphs in the bottom row show how each type of natural selection affects the distribution of body size in the population.

Dark-colored moth

Light-colored moth

Figure 17.7 Directional Selection in the Peppered Moth
The frequency of dark-colored peppered moths declined dramatically from 1959 to 1995 in regions near Liverpool, England, and Detroit, Michigan. Before 1959, dark-colored moths had risen in frequency in both England and the United States after industrial pollution had blackened the bark of trees, causing dark-colored moths to be harder for bird predators to find than light-colored moths. A reduction in air pollution following clean air legislation enacted in 1956 in England and in 1963 in the United States apparently removed this advantage, leading the dark-colored moths to decline in a similar way in the two regions. (No data were collected in Detroit for the 30-year period 1963–1993.)

In **stabilizing selection**, individuals with intermediate values of an inherited phenotypic trait have an advantage over other individuals in the population (Figure 17.6b). Birth weight in humans provides a classic example. Historically, light or heavy babies did not survive as well as babies of average weight, and as a result there was stabilizing selection for intermediate birth weights (Figure 17.8). By the late 1980s, however, selection against small and large babies had decreased considerably in some countries with advanced medical care, such as Italy, Japan, and the United States. This reduction in the strength of stabilizing selection was caused by advances in the care of very light premature babies and by increases in the use of cesarean deliveries for babies that are large relative to their mothers (and hence pose a risk of injury to mother and child at birth).

In **disruptive selection**, individuals with either extreme of an inherited phenotypic trait have an advantage over individuals with an intermediate phenotype (Figure 17.6c). This type of selection is probably not common, but it appears to affect beak size within a population of African seed crackers, in which birds with large or small beaks survive better than birds with intermediate-sized beaks (Figure 17.9).

17.8 Sexual Selection: Where Sex and Natural Selection Meet

The males and females of many species differ greatly in size, appearance, or behavior. Many of these differences seem to be related to the ability of individuals to obtain mates. In lions, for example, males are considerably larger than females, and males fight, sometimes violently, for the privilege of mating with females. Since larger males are stronger and tend to be more successful in fights for females, natural selection may have favored large size in males, but not in females, thus causing the difference between the size of male and female lions.

When individuals differ in inherited characteristics that cause them to differ in their ability to get mates, a

> Babies that weigh 8 pounds at birth survive at a higher rate than do lighter or heavier babies.

Figure 17.8 Stabilizing Selection for Human Birth Weight

This graph is based on data for 13,700 babies born from 1935 to 1946 in a hospital in London. In countries that can afford intensive medical care for newborns, the strength of stabilizing selection has been reduced in recent years; a graph of such data collected today would be flatter (less rounded) at its peak than the graph shown here.

Figure 17.9 Disruptive Selection for Beak Size

In African seed crackers, differences in feeding efficiencies may cause differences in survival. Among a group of young birds hatched in one year, only those with a small or large beak size survived the dry season, when seeds were scarce; all the birds with intermediate beak sizes died. Thus natural selection favored both large-beaked and small-beaked birds over birds with intermediate beak sizes. Red bars indicate the beak sizes of young birds that survived the dry season, while yellow and blue bars indicate the beak sizes of young birds that died.

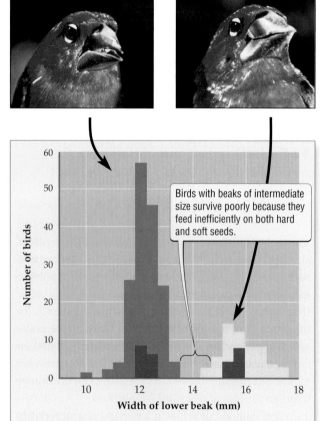

> Small-beaked birds feed more efficiently on soft seeds.

> Large-beaked birds can crack hard seeds.

> Birds with beaks of intermediate size survive poorly because they feed inefficiently on both hard and soft seeds.

special form of natural selection, called **sexual selection**, is at work. Sexual selection favors individuals that are good at getting mates, and, as with the lions we just described, it often helps to explain differences between males and females in size, courtship behavior, and other features. It is important to realize, however, that characteristics that increase the chance of mating sometimes decrease the chance of survival. For example, male túngara frogs perform a complex mating call that may or may not end in one or more "chucks." Females prefer to mate with males that emit "chucks," but frog-eating bats use that same sound to help them locate their prey (see Figure 18.5). Thus an attempt to locate a mate can end in disaster.

In many species, the members of one sex—often females—are choosy about whether or not to mate. In such species, the choosy partner acts as the "brake," while the suitor, who tries to convince its reluctant partner to mate, acts as the "gas." In birds, for example, bright-colored males may perform elaborate displays in their attempts to woo a mate (Figure 17.10a). In other species, males may attract attention by other means, such as calling vigorously; females then select as their mates the males with the loudest calls.

For the process we've just described to make sense, we would expect that if the choosy partner bases her (or, occasionally, his) choice of a mate on a trait such as

(a)

(b)

Figure 17.10 Sexual Selection at Work
(a) A male peacock displays his magnificent tail for a peahen. (b) A male gray tree frog issues a call for mates.

color or calling vigor, then that trait should serve as a good indicator of the quality of the mate. A series of recent studies have provided support for this hypothesis. In blackbirds, females choose males with orange beaks more often than males with yellow beaks. It turns out that orange-beaked males have had fewer infections than yellow-beaked males; thus, by selecting males with orange beaks, the females are selecting males in good health. Similarly, in mice, females can tell from the odor of a male's urine how many parasites he has, and they use this information to choose their mates.

In the two examples we have just mentioned, females use a trait such as beak color or odor to tell how healthy the males currently are. In other cases, females seem to test whether males have "good genes." Female gray tree frogs, for example, prefer to mate with males that give long mating calls (Figure 17.10b). In an elegant experiment, scientists collected unfertilized eggs from wild females, sperm from males with long calls, and sperm from males with short calls. The eggs of a given female were then divided into two batches, and one was fertilized with sperm from long-calling males, the other with sperm from short-calling males. The offspring of long-calling male frogs grew faster, were bigger, and survived better than the offspring of short-calling male frogs. Thus

the long-calling males seem to have been genetically superior to the short-calling males.

17.9 Putting It All Together: How Evolution in Populations Works

In this chapter we've described how four mechanisms—mutation, gene flow, genetic drift, and natural selection—can cause allele frequencies to change in populations. With this information in hand, let's revisit our earlier description of how evolution works. On pages 301–302 of Chapter 16, we described how natural selection and genetic drift, operating on the genetic variation found in a population, can cause allele frequencies to change over time.

We can now describe how evolution works more fully, summarizing it as a three-step process. First, mutations and the genetic rearrangements caused by recombination occur at random. Second, these random events generate inherited differences in the characteristics of individuals in populations. And third, gene flow, genetic drift, and natural selection acting on that genetic variation cause allele frequencies in the population to change over time. By changing allele frequencies directly, mutation can also cause evolution to occur, although, as we have seen, this typically happens at a very slow rate.

It is important to remember the role that chance can play in evolution, since three of the four mechanisms that cause allele frequencies to change in populations are influenced by chance events: (1) genes can mutate at random to produce new alleles; (2) alleles can be transferred by gene flow from one population to another (and the alleles a migrating individual carries usually can be viewed as a random selection of the alleles in its population); and (3) alleles can increase in frequency in a given population due to genetic drift.

But evolution in populations is by no means restricted to chance events. The fourth mechanism, natural selec-

tion, is not a random process. The directional selection exerted on peppered moths illustrates that point beautifully. Once pollution controls were imposed in the mid-twentieth century, light-colored moths were favored consistently—not by chance, but by the changing environmental conditions. This consistent advantage caused the proportion of dark-colored moths in the population to drop rapidly (see Figure 17.7). As this example shows, natural selection favors some individuals over others because they have particular characteristics that increase their ability to survive and reproduce in their environment. As a result, over time, natural selection improves the match between organisms and their environment, a topic we will discuss in more detail in the next chapter.

Learn about the crisis in antibiotic resistance.

17.4

Halting the Spread of Antibiotic Resistance

The footprint of evolutionary change can be found throughout nature. We see it in battles between predator and prey, parasite and host, herbivore and plant. In these three cases, natural selection favors individuals whose characteristics either improve their ability to consume others or improve their ability to avoid being consumed. Similar battles occur between people and organisms that we want to kill, such as fungi and insects that destroy crops and bacteria that cause disease. Unfortunately, our foes have proved hard to vanquish. In fact, our attempts to kill them are having the unintended effect of actually promoting the evolution of resistance. Individual bacteria, fungi, and insects that are not resistant to our antibiotics or pesticides die, leaving behind ever larger proportions of individuals that are resistant—hence our efforts to kill them impose selection for resistance to the very weapons we try to kill them with.

No group of organisms illustrates the problem of resistance better than bacteria. Bacterial populations rapidly evolved resistance to antibiotics after these drugs were introduced in the 1930s. In 1941, for example, pneumonia could be cured in several days if patients took 40,000 units of penicillin per day. Today, however, even with excellent medical

care and the administration of 24 million units of penicillin per day, a patient can still die from complications of this disease.

What is true of pneumonia is also true of other bacterial diseases. As a group, bacteria have many mechanisms for coping with once-lethal antibiotics. Some of these mechanisms provide resistance only to specific antibiotics, such as a cellular pump that removes a specific antibiotic, an enzyme that destroys a certain type of antibiotic, or a structural change that prevents an antibiotic from affecting its usual target (for example, a bacterium whose cell wall structure differs from that of other bacteria may be impervious to penicillin, which normally kills bacteria by attacking their cell walls). Other bacterial defense mechanisms provide resistance to multiple antibiotics. For example, some bacteria have developed pumps that remove many types of antibiotics, while other bacteria grow in layers that prevent the antibiotics from reaching cells that are not near the surface.

How do bacteria develop these and other resistance mechanisms so rapidly? In part, this happens because bacteria have such a short time between generations. As we saw in Chapter 3, bacteria reproduce by fission, so only two daughter cells are produced when each bacterium divides. However, because the time between divisions can be as little as 20 to 30 minutes, one bacterium can produce enormous numbers of offspring in just a few days. Some of these offspring may carry new mutations that provide resistance to one or more antibiotics. Once such mutations occur, our widespread use of antibiotics causes alleles that provide resistance to increase rapidly in frequency.

This problem is then made much worse by the fact that bacteria can transfer resistance genes within and among species with ease, often very quickly (Figure 17.11). The movement of resistance genes from one species to another is especially troubling: genes for resistance that evolve in a relatively harmless species of bacteria can be transferred to a highly dangerous species, creating the potential for a public health disaster.

Halting the spread of antibiotic resistance, and thereby preventing this sort of doomsday scenario, will require concerted efforts on our part; this problem will not go away by itself. Consider the costs and benefits of the following actions:

- Devote greater resources to the study of the biology of bacteria and other disease agents, thereby improving our ability to design drugs or management strategies that attack weak points in their life cycles.
- Learn to live with bacteria, rather than seeking to annihilate them. For example, instead of using disinfectants designed to kill bacteria, we could try to limit bacterial numbers by following safe cooking practices and by keeping our homes clean. A decrease in the use of "antibacterial" disinfectants would be helpful because some genes that confer resistance to these products can also lead to antibiotic resistance (see Figure 17.11). Similarly, in combating bacterial diseases, we could search for drugs that make bacteria harmless but do not destroy them; because this approach is not lethal for the bacteria, resistance should evolve more slowly.

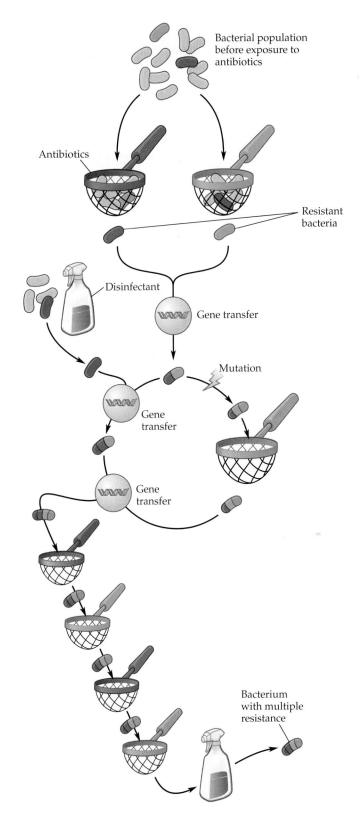

Figure 17.11 The Rise of Antibiotic Resistance
Before antibiotic use was common, most bacteria could be killed by antibiotics. Early use of antibiotics selected those few individuals that were resistant. Over time, new mutations and repeated episodes of selection—either by antibiotics (represented by sieves) or by disinfectants—caused the frequency of resistant individuals to increase. In addition, transfer of genes within and among bacterial species allowed some bacteria to develop resistance to multiple antibiotics.

Biology Matters

You'll Just Have to Tough It Out

Being sick is never enjoyable, but the more you take antibiotics, the more likely they won't work for you—or others—in the future. In fact, antibiotics don't help with many of the most frequent illnesses that make us sick: cold, flu, sore throats (except strep), bronchitis, most runny noses, and most earaches. Although many doctors are prescribing antibiotics less than they did in the past, understanding when to use antibiotics, and when they *won't* help, will help you to protect yourself and others. And don't forget—when you do need antibiotics, take them exactly as your health-care provider prescribes, and continue to take them until they run out, even if you're feeling better.

The following information will help you care for yourself, and feel better, the next time you get sick:

Q: *If antibiotics will not help me, what will?*

A: There are many over-the-counter products available to treat the symptoms of your viral infection. These include cough suppressants which will help control coughing and decongestants to help relieve a stuffy nose. Read the label and ask your pharmacist or doctor if you have any questions about which will work best for you.

A cold usually lasts only a couple of days to a week. Tiredness from the flu may continue for several weeks. To feel better while you are sick:

- Drink plenty of fluids.
- Get plenty of rest.
- Use a humidifier—an electric device that puts water into the air.

Contact your doctor if:

- Your symptoms get worse.
- Your symptoms last a long time.
- After feeling a little better, you develop signs of a more serious problem. Some of these signs are a sick-to-your-stomach feeling, vomiting, high fever, shaking chills, and chest pain.

SOURCE: U.S. Department of Health and Human Services.

- Insist on prudent use of antibiotics in human, plant, and animal health care. Medical doctors and agriculturists frequently use antibiotics inappropriately. For example, the U.S. government estimates that half of the 100 million antibiotic prescriptions written by doctors each year are not necessary—often the conditions for which they are prescribed (such as colds and flu) are caused not by bacteria, but by other disease agents, such as viruses, that are not affected by antibiotics. Similarly, antibiotics are commonly used to increase the growth rates of farm animals, a practice that encourages the development of antibiotic-resistant strains of bacteria, including strains that attack people. Such inappropriate use of antibiotics encourages the evolution of antibiotic resistance in the many species of bacteria that are normally found in our bodies. As we have seen, these resistant, harmless bacteria can then transfer genes for antibiotic resistance to other, harmful species of bacteria.

- Improve sanitation, thus decreasing the spread of resistant bacteria from one person to another. This action is critically important in hospitals, where the abundant use of antibiotics has led to the emergence of highly resistant strains of bacteria that can cause a variety of "hospital diseases," some of which can be lethal.

What do you think we as a society should do?

Chapter Review

Summary

17.1 Key Definitions: Allele and Genotype Frequencies

- An allele's frequency is the proportion of that allele in a population. Calculating allele frequencies is an important part of determining whether a population is evolving.
- A genotype's frequency is the proportion of that genotype in a population.

17.2 Genetic Variation: The Raw Material of Evolution

- Individuals within populations differ in morphological, behavioral, and biochemical traits, many of which are under genetic control.
- Genetic variation provides the raw material on which evolution can work. The two sources of genetic variation are mutation and recombination.
- Mutations (changes in the sequence of an organism's DNA) give rise to new alleles.
- Recombination (crossing-over, independent assortment of chromosomes, and fertilization) causes the offspring of sexually reproducing organisms to have new combinations of alleles that differ from those found in either parent.

17.3 Four Mechanisms Can Cause Populations to Evolve

- Allele frequencies in populations can change over time as a result of mutation, gene flow, genetic drift, or natural selection.
- The Hardy-Weinberg equation can be used to test whether one or more of these four mechanisms is causing a population to evolve.

17.4 Mutation: The Source of Genetic Variation

- All evolutionary change depends ultimately on the production of new alleles by mutation.
- Mutations cause little direct change in allele frequencies over time.
- Mutations can, however, stimulate the rapid evolution of populations by providing new genetic variation on which natural selection, genetic drift, or gene flow can act.

17.5 Gene Flow: Exchanging Alleles Between Populations

- Gene flow can introduce new alleles into a population, providing new genetic variation on which evolution can work.
- Gene flow makes the genetic composition of populations more similar.

17.6 Genetic Drift: The Effects of Chance

- Genetic drift causes random changes in allele frequencies over time.
- Genetic drift can cause small populations to lose genetic variation.

- Genetic drift can cause the fixation of harmful, neutral, or beneficial alleles.
- In a genetic bottleneck, a drop in population size causes genetic drift to have pronounced effects, including reduced genetic variation or the fixation of alleles. A genetic bottleneck can threaten the survival of a population.

17.7 Natural Selection: The Effects of Advantageous Alleles

- In natural selection, individuals that possess certain forms of an inherited phenotypic trait tend to survive better and produce more offspring than individuals that possess other forms of that trait.
- Natural selection is the only evolutionary mechanism that consistently favors alleles that improve the reproductive success of the organism in its environment.
- There are three types of natural selection. In directional selection, individuals at one extreme of a trait have the advantage. In stabilizing selection, the advantage goes to individuals with intermediate values of a trait. In disruptive selection, individuals at both extremes have an advantage over individuals with an intermediate phenotype.

17.8 Sexual Selection: Where Sex and Natural Selection Meet

- Sexual selection occurs when individuals with different inherited characteristics differ in their ability to get mates.
- Sexual selection underlies many differences between males and females, such as differences in size and courtship behavior.
- Forms of a trait favored by sexual selection can lead to decreased survival.
- Sexual selection can occur when members of one sex—often females—are choosy about whether or not to mate.

17.9 Putting It All Together: How Evolution in Populations Works

- Evolution can be summarized as a three-step process: (1) Mutations and the genetic rearrangements caused by recombination occur at random. (2) These random events then generate inherited differences in the characteristics of individuals in populations. (3) Finally, mutation, gene flow, genetic drift, and natural selection can cause allele frequencies to change over time.
- Of the four mechanisms of evolutionary change, three—mutation, gene flow, and genetic drift—are influenced by chance events. Natural selection, on the other hand, is a nonrandom process that favors individuals with particular characteristics that increase their ability to survive and reproduce in their environment.

Review and Application of Key Concepts

1. Select one of the four evolutionary mechanisms discussed in this chapter (mutation, gene flow, genetic drift, and natural selection). Describe how it can cause allele frequencies to change from generation to generation.

2. Summarize the role of genetic variation in evolution.

3. Describe how recombination increases the genetic variation originally produced by mutation.

4. Using your own words, define the following terms: gene flow, genetic drift, natural selection, sexual selection.

5. One way to prevent a small population of a plant or animal species from going extinct is to deliberately introduce some individuals from a large population of the same species into the smaller population. In terms of the evolutionary mechanisms discussed in this chapter, what are the potential benefits and drawbacks of transferring individuals from one population to another? Do you think biologists and concerned citizens should take such actions?

6. See the toads in question 2 of the Self-Quiz. How do the numbers of toads with genotypes *AA*, *Aa*, and *aa* compare with the numbers you would expect based on the Hardy-Weinberg equation (see the box on page 324)? Discuss the factors that could cause any differences you find.

7. Explain the reasoning behind the statement in the text (on page 332) that "in fact, our attempts to kill [bacteria] are having the unintended effect of actually promoting the evolution of resistance." If we shifted our efforts away from killing bacteria and toward reducing our exposure to them or slowing their growth, why might such a change in our approach slow the evolution of antibiotic resistance?

Key Terms

allele frequency (p. 320)
directional selection (p. 328)
disruptive selection (p. 329)
fixation (p. 325)
founder effect (p. 326)
gene flow (p. 323)
genetic bottleneck (p. 326)
genetic drift (p. 323)
genetic variation (p. 321)

genotype frequency (p. 320)
Hardy-Weinberg equation (p. 324)
microevolution (p. 320)
natural selection (p. 326)
recombination (p. 321)
sexual selection (p. 330)
stabilizing selection (p. 329)

Self-Quiz

1. A population of 1,500 individuals has 375 individuals of genotype *AA*, 750 individuals of genotype *Aa*, and 375 individuals of genotype *aa*. The genotype frequencies for genotypes *AA*, *Aa*, and *aa* are
   a. 0.33, 0.33, 0.33   c. 0.375, 0.75, 0.375
   b. 0.25, 0.50, 0.25   d. 0.125, 0.25, 0.125

2. A population of toads has 280 individuals of genotype *AA*, 80 individuals of genotype *Aa*, and 60 individuals of genotype *aa*. What is the frequency of the *a* allele?
   a. 0.24   b. 0.33   c. 0.14   d. 0.07

3. A study of a population the goldenrod *Solidago altissima* [soll-uh-DAY-go al-TISS-ih-ma] finds that large individuals consistently survive at a higher rate than small individuals. Assuming that size is an inherited trait, the most likely evolutionary mechanism at work here is
   a. disruptive selection.
   b. directional selection.
   c. stabilizing selection.
   d. natural selection, but it is not possible to tell whether it is disruptive, directional, or stabilizing.

4. Based on the material in the box on page 324, if the frequency of the *A* allele is 0.7 and the frequency of the *a* allele is 0.3, what is the expected frequency of individuals of genotype *Aa* in a population that is not evolving?
   a. 0.21   b. 0.09   c. 0.49   d. 0.42

5. Over time, a population of birds ranges in size from 10 to 20 individuals. If allele frequencies were observed to change in a random way from year to year, which of the following would be the most likely cause of the observed changes in gene frequency?
   a. stabilizing selection   c. genetic drift
   b. disruptive selection   d. mutation

6. Two large populations of a species that are found in neighboring locations with different environments are observed to become genetically more similar over time. Which evolutionary mechanism is the most likely cause of this trend?
   a. gene flow   c. natural selection
   b. mutation   d. genetic drift

7. Assume that individuals of genotype *Aa* are intermediate in size and that they leave more offspring than either *AA* or *aa* individuals. This situation is an example of
   a. directional selection.   c. stabilizing selection.
   b. disruptive selection.   d. sexual selection.

8. The process by which differences in the inherited characteristics of individuals cause them to differ in their ability to get mates is most accurately called
   a. natural selection.   c. mate choice.
   b. reproductive success.   d. sexual selection.

Biology in the News

WHO Urges End to Use of Antibiotics for Animal Growth

By Marc Kaufman

The World Heath Organization will recommend today that nations phase out the widespread and controversial use of antibiotic growth promoters in animal feed, saying that the move will help preserve the effectiveness of antibiotics for medicine and can be done without significant expense or health consequences to farm animals.

Based on a study of Denmark's experience following a 1998 voluntary ban on antibiotic growth promoters, WHO concluded that under similar conditions the use of low-dosage antibiotics "for the sole purpose of growth promotion can be discontinued."

The WHO report is sure to stir up a hornet's nest of differing opinions. On the one hand, it supports calls from the scientific community and public action groups to stop the routine use of antibiotics—primarily because such use speeds up the evolution of resistance to antibiotics in bacteria. As described in the *Washington Post* article, such appeals have had an effect, as indicated by the McDonald's Corporation's recent decision to tell their suppliers to cut back on antibiotic use.

But according to representatives for meat producers and makers of animal drugs in the United States, the WHO report makes little sense because not all animal drugs pose the same risk, and because the most important cause of bacterial resistance is the overuse of antibiotics in hospitals and doctors' offices.

The WHO is sticking to its guns. WHO officials cite a scientific study of the voluntary ban in Denmark, which found that the ban drove up the cost of raising pigs by about 1 percent and that the use of antibiotics to treat sick animals increased. Even so, the *overall* use of antibiotics on farms decreased by about 50 percent. As a result, the occurrence of resistant bacteria dropped considerably. For example, before the ban, 60 to 80 percent of chickens harbored bacteria resistant to three widely used antibiotics. After the ban, 5 to 35 percent of chickens had resistant bacteria. According to the WHO, the results in Denmark are likely to apply to the United States as well, since farm practices in the two countries are similar.

As the WHO report also noted, there were indications that after the ban, the levels of antibiotic-resistant bacteria in people also dropped, although the data from Denmark were too limited to draw firm conclusions. Previous studies, however, had suggested that when antibiotics are banned in livestock feed, the percentage of bacteria in people that are resistant to those antibiotics drops. One such drop occurred after the 1996 European Union ban of an animal antibiotic that promotes resistance to the human antibiotic vancomycin [*VANG*-ko-*MY*-sin]. Before the ban, 5.7 percent of hospital patients harbored vancomycin-resistant bacteria; by 1 year after the ban, that percentage had dropped to just 0.6 percent.

Evaluating the news

1. Representatives of firms that make animal antibiotics argue that if antibiotic use were reduced in the United States, the cost of raising chicken, beef, and pork would increase by considerably more than the 1 percent observed after the Danish ban. If these firms are right, would you be willing to pay more—say, 5 to 10 percent more—for meat products that contained fewer antibiotic-resistant bacteria?

2. Animals raised for meat often live under very crowded conditions, which make it easy for diseases to spread. Should animals produced for meat be raised under less crowded conditions in order to decrease the need for antibiotics? Or do you think we should continue to raise animals under crowded conditions in order to produce more meat per unit of farmland to feed Earth's growing human population? Explain your answer.

Source: *The Washington Post*, Wednesday, August 13, 2003.

Key Concepts

◉ An adaptation is a feature of an organism that improves the performance of the organism in its environment. Adaptations result from natural selection.

◉ The process by which natural selection improves the match between an organism and its environment over time is called adaptive evolution. Adaptive evolution causes organisms to adjust to environmental change, sometimes over short periods of time (months to years).

◉ A species is a group of interbreeding natural populations that is reproductively isolated from other such groups.

◉ Speciation, the process by which one species splits to form two or more species, is usually a by-product of genetic differences between populations that are caused by factors such as natural selection or genetic drift.

◉ Speciation often occurs when populations of a species become geographically isolated. Such isolation limits gene flow between the populations, which makes the evolution of reproductive isolation more likely.

◉ New species can also form in the absence of geographic isolation.

Main Message: Adaptive evolution causes organisms to adjust to environmental change and helps to generate the great diversity of life.

Cichlid Mysteries

The surface of Earth changes slowly but dramatically over time. Islands rise from the sea, new lakes form and old ones disappear, and mountains are thrust up to divide once-continuous landmasses. Such changes alter the environments in which species live, setting the stage for grand natural experiments in evolution.

No evolutionary experiment has been more wondrous than that of the cichlid fish of Lake Victoria, the largest of the Great Lakes of East Africa. Lake Victoria first formed about 750,000 years ago, but geologic evidence indicates that the lake was nearly or completely dry as recently as 15,000 years ago. Whenever a lake forms or refills with water, it may be colonized by one or more species of fish, some of which may then evolve to form new species unique to the new lake. Such a sequence of events has happened many times in lakes around the world, but nowhere more spectacularly than in Lake Victoria.

Until recently, Lake Victoria harbored over 500 species of cichlids, a greater number of fish species than is found in all the lakes and rivers of Europe combined. The cichlids are a diverse and colorful group of fish, well known to the aquarium trade. Genetic evidence suggests that the cichlids of Lake Victoria originated about 100,000 years ago; thus they did not die out during the dry period 15,000 years ago (some may have survived in pools within the lake or in nearby rivers). Genetic data also indicate that the cichlids in Lake Victoria descended from two ancestor species from Lake Kivu, a smaller and older lake located about 275 kilometers from Lake Victoria. The finding that 500 species of fish descended from two ancestor species in 100,000 years brings us to the first of our "cichlid mysteries": how did so many Lake Victoria cichlid species form in a relatively short period of time?

A Lake Victoria Cichlid

The mystery deepens when we realize that many of the Lake Victoria cichlid species differ considerably from one another in color, jaw structure, and feeding specialization, yet genetically, all the species in the lake are extremely closely related to one another. What evolutionary forces have caused some of these closely related species to differ so much? Furthermore, in the past 30 years, roughly 200 Lake Victoria cichlid species have disappeared. Some, we know, were driven to extinction by an introduced predatory fish species, the Nile perch. But many species rarely eaten by Nile perch have also vanished. What drove these species to extinction? We'll return to these cichlid mysteries at the close of this chapter.

We discussed three great themes of evolutionary biology in Chapter 16: adaptation, the diversity of life, and the shared characteristics of life. In this chapter we return to two of these themes, adaptation and biodiversity. We examine the characteristics of adaptations and discuss how they are shaped by natural selection. Then we reconsider the concept of the species and how to define species effectively. We focus the remainder of the chapter on speciation, the process that generates the diversity of life.

18.1 Adaptation: Adjusting to Environmental Challenges

An **adaptation** is a feature of an organism that improves aspects of the performance (such as growth, reproduction, or survival) of that organism in its environment. Many adaptations result in what appears to be a remarkably well-designed match between the organism and its environment. As we saw in Chapter 16, adaptations do not result from any intentional "design," but rather from natural selection: individuals with inherited characteristics that allow them to survive and reproduce better than other individuals replace those with less favorable characteristics. This process, which improves the match between organisms and their environment over time, is called **adaptive evolution**. A similar process can occur when people influence the way domesticated species such as dogs change over time (see the box on page 342).

There are many different kinds of adaptations

There are many striking examples of adaptations in the natural world. Consider how weaver ants construct nests of living leaves. Nest building requires the concerted actions of many individual ants, some of which draw the edges of leaves together while others weave them in place by moving silk-spinning larvae (immature ants) back and forth over the seam of the two leaves. These actions are not the result of conscious planning on the part of the ants; rather, they illustrate how a simple evolutionary mechanism—natural selection—can produce a complex behavioral adaptation (cooperative nest building, which benefits the ants by providing them with shelter).

In another example of an adaptation, the caterpillars (the larval or immature stage) of a certain moth species develop different shapes depending on which part of their food plant (an oak tree) they feed on—the flowers or the leaves. Those that feed on flowers resemble oak flowers; those that feed on leaves resemble oak twigs (Figure 18.1). Thus the larvae develop in a way that causes them to match whatever background they feed on, making them more difficult for predators to locate.

Natural selection has also shaped some astonishing adaptations that facilitate reproduction. The flowers of some orchid species, for example, use chemical attractants and appearance to mimic female wasps, thereby attracting male wasps and fooling them into attempting to mate with the flowers. In the course of these attempts, the insects become coated with pollen, which they then transfer from one plant to another.

All adaptations share certain key characteristics

Look carefully at the photograph and diagram of the eye of the four-eyed fish, *Anableps anableps* (Figure 18.2). Although this fish really has only two eyes, they function as four, enabling the fish to see clearly through both air and water. The four-eyed fish is a surface feeder, so the ability to see above water helps it locate prey such as insects. Its unique eyes also allow it to scan

Weaver ants at work

(a)

(b)

Figure 18.1 Caterpillars That Match Their Environments
Caterpillars of the moth *Nemoria arizonaria* [nee-*MORE*-ee-uh ar-ih-zoh-*NARE*-ee-uh] differ in shape depending on their diet. (*a*) Caterpillars that hatch in the spring resemble the oak flowers on which they feed. (*b*) Caterpillars that hatch in the summer eat leaves and resemble oak twigs. Experiments have demonstrated that chemicals in the leaves control the switch that determines whether the caterpillars will mimic flowers or twigs.

simultaneously for predators attacking from above (such as birds or people) or below (such as other fish). The four-eyed fish can also walk on land, and it often escapes trouble by jumping out of the water. Thus, while it would be interesting to watch, this fish would make a poor choice for a home aquarium.

Although there are literally millions of examples of adaptations, the four-eyed fish and the other examples we've discussed so far illustrate their most important characteristics:

1. Adaptations have the appearance of having been designed to match the organism to its environment (consider the caterpillars in Figure 18.1).
2. Adaptations are often complex (consider the eye of *Anableps anableps* and the nest-building behavior of weaver ants).
3. Adaptations help the organism accomplish important functions, such as feeding, regulation of body chemistry, defense against predators, and reproduction. (This point is illustrated by all of the examples we have discussed.)

Populations can adjust rapidly to environmental change

Male guppies in the mountain streams of Trinidad and Venezuela have bright and variable colors that serve to attract females. But the bright colors that help the males succeed in attracting mates also make them easier for predators to find. How do guppy populations evolve in response to such conflicting pressures?

Field observations show that guppies from streams where few predators lurk have bright colors, but guppies

(a)

(b) Light from above
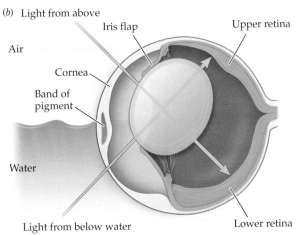

Figure 18.2 The Four-Eyed Fish, *Anableps anableps*
The four-eyed fish really only has two eyes, but each eye has a special design that lets it see clearly in both air and water. (*a*) The fish usually swims so that the band of pigment in the eye is level with the waterline, thus dividing the eye into above- and below-water halves. (*b*) The iris flap shields the upper pupil from glare off the surface of the water. When the fish looks above the water, light passes through a flattened lens and then strikes the lower retina (the retina is composed of photoreceptor cells that receive the image and nerve cells that send the image to the brain). Humans also have a flattened lens, which provides the best image when looking through air. When the fish looks below the water, light passes through a rounded lens before striking the upper retina. Ordinary fish also have a rounded lens, which provides the best image when looking through water.

Biology on the Job

Dog Show Judge Extraordinaire

Over the course of a single year in the United States, dogs of 150 breeds participate in over 15,000 dog shows, some of them local and state shows, others national. Judging all these contestants requires the services of many dog judges, who themselves are rated for their expertise. Mr. Edd Bivin was selected by a collection of dog judges, dog trainers, breeders, and kennel clubs as the "Judging Legend 2002." Below are excerpts from a 2002 interview with Mr. Bivin by TheDogPlace, a Web site devoted to the interests of dog fanciers.

When, and why, did you decide to become a judge? I probably decided to become a judge when I was a kid. I showed good dogs [but I would just] get patted on the head. I believed in the sport and knew if I ever were to become a judge, I wouldn't pay any attention to where a dog comes from or the age of the individual handler. I would evaluate what I consider to be the best dog.

What do you do in your "other" life? I've been Vice Chancellor at Texas Christian University for 16 years, but I have been at the University for 30 years. That's in Fort Worth, Texas, where I have lived all my life. I don't have a lot of time for other hobbies but [my wife and I] do enjoy travel. We are involved as much as possible in the community, certainly in the arts. . . . I'm very physical so I also work out and exercise a lot. Keeps me sane, or somewhat sane (laughing), or let's say more sane and less crazy.

Judging a Dog Show

What do you most enjoy about judging? Obviously the dogs, and the people. I also consider judging to be a personal competition of Edd Bivin with himself. Every time I go in the ring I compete with myself, to do the best job I can do, on that day, within the circumstances with which I find myself.

Let's talk about the sport today. Are most breeds better than 10 years ago? No. Breeds are cyclical; they progress and they fall back. So I can't accept the term "most." I will tell you that there are many breeds that are better today and many breeds not as good as they were 10 years ago. A big part depends on who is directing breeding programs and pockets of interest around the country.

When you first look down the line [of dogs for show], what draws your eye? Balance and proportion. Carriage and outline. [*smile*] Outline and character.

Should showmanship and presentation be considered? Certainly. One should never miss a good animal with proper type and character. I become concerned about individuals applauding dogs or saying it's not a great such and such but it's a great "show dog." Dog shows are a format for the evaluation of breeding stock. Generic dogs are not the strength of any breed.

The Job in Context

When he judges a show, Mr. Bivin evaluates the degree to which a dog achieves the ideal standard for its breed. Judges such as Mr. Bivin are approved for selected breeds only; this is done to ensure that each judge thoroughly understands the physical form and character of the dog breeds that he or she evaluates.

When judges compare a dog with the ideal standard for that dog's breed, they influence a process similar in result to adaptive evolution. Prize-winning dogs are sought after as breeding stock, so they leave more offspring than dogs that show poorly. One might think that this process—coupled with the efforts of breeders, who also strive to improve the quality of the dogs over time—would ensure that dogs today were "better" than dogs of 10 years ago. But Mr. Bivin begged to differ; he said that many breeds were better, but many others were worse. There are two underlying biological reasons why a breeding program might produce dogs of declining quality. One is excessive inbreeding (mating between close relatives), which can lead to the fixation of harmful alleles. The other is "genetic hitchhiking," a term that refers to the fact that while selecting for one characteristic (such as a certain physical feature of the dog), it is possible to inadvertently promote other, less desirable features that are linked genetically to that feature. It takes considerable care and skill for breeders to avoid such pitfalls and continue producing dogs that meet the high standards imposed by judges like Mr. Bivin.

from streams with more predators are drab in comparison. Predators can influence guppy coloration because, with more predators present, a higher proportion of the most brightly colored guppies are likely to be eaten each generation, and so do not pass their genes on to the next generation. The match between the color of guppies and the number of predators evolves very rapidly: when guppies are experimentally transferred to a different environment (from an area with few predators to an area with many predators, or vice versa), their color patterns evolve to match the new conditions within 10 to 15 generations (14 to 23 months).

The ability to evolve rapidly in response to changing environmental conditions is not limited to guppies. In soapberry bug populations in Florida, for example, beak length has evolved rapidly to match the size of the fruits on which these insects feed (Figure 18.3). Similarly, as we saw in Chapter 16, beak sizes in the medium ground finch evolve from year to year to match the size of the seeds that are available. In addition, as we saw in Chapter 17, viruses, bacteria, and insects can evolve resistance to our best efforts to kill them in only a few months or years. Collectively, these examples illustrate an important point: evolution by natural selection can improve the adaptations of organisms in short periods of time.

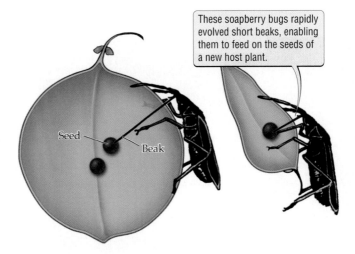

Figure 18.3 **Rapid Evolution in an Insect**
In Florida, soapberry bugs traditionally fed on seeds within the large, round fruit of a native plant species, the balloon vine (left). Over the past 30–50 years, some populations of soapberry bugs have evolved short beaks, enabling them to feed on seeds within the fruit of an introduced species, the golden rain tree (right).

18.2 Adaptation Does Not Craft Perfect Organisms

As impressive as the adaptations we see in nature may be, we do not mean to suggest that natural selection results in a perfect match between an organism and its environment. In many cases, genetic constraints, developmental constraints, or ecological trade-offs prevent further improvements in an organism's adaptation. In this section we look at these barriers to perfection.

Lack of genetic variation can limit adaptation

For the quality of an adaptation to increase over time, there must be genetic variation for traits that can enhance the match between the organism and its environment. In some cases, the absence of such genetic variation places a direct limit on the ability of natural selection to cause adaptive evolution. For example, the mosquito *Culex pipiens* is now resistant to organophosphate pesticides, but this resistance is based on a single mutation that occurred in the 1960s, as we saw in Chapter 17. Before this mutation occurred, adaptation to these pesticides was not possible,

and billions of mosquitoes died because their populations lacked genetic variation for resistance to the pesticides.

The multiple effects of developmental genes can limit adaptation

As we saw in Chapter 14, changes in genes that control development can have dramatic effects on the phenotype. Such a change in the developmental program of an organism often influences more than one part of its phenotype. Thus changes in genes that control development can have many effects, some of which may be advantageous while others may harm or kill the organism.

The multiple effects of developmental genes can limit the ability of the organism to evolve in certain directions, which in turn may limit what adaptive evolution can achieve. For example, the larval stages of some insects, such as beetles and moths, lack wings or well-developed eyes, two important adaptations that the adult forms of these insects have (Figure 18.4). Beetle and moth larvae have a wide range of lifestyles, so wings or well-developed eyes probably would benefit many of these larvae. Since the adult and larval forms of these insects carry exactly the same genetic instructions, it is likely that turning on or expressing the genes that control wing or eye production in adults would have other, extremely harmful effects in larvae (in other words, it would kill them). Thus the lack of wings and eyes in beetle and moth larvae probably results from developmental limitations.

Figure 18.4 An Apparent Developmental Limitation
The larval form of this moth (*a*) does not have wings or eyes, two
important adaptations that are found in the adult form (*b*).

(a)

> Wings or eyes might benefit some moth larvae.

> Since larvae and adults have the same genetic instructions, developmental limitations probably prevent the larvae from having wings or eyes.

(b)

Ecological trade-offs can limit adaptation

To survive and reproduce, organisms must perform
many functions, such as finding food and mates, avoiding predators, and surviving the challenges posed by
the physical environment. Within the realm of what is
genetically and developmentally possible, natural selection increases the overall ability of the organism to sur-

Figure 18.5 Does Love or Death Await?
Male túngara frogs face an ecological trade-off: the same type of call
that is most successful at attracting females also makes it easier for
predatory bats to locate calling males.

vive and reproduce. However, the many and often conflicting demands that organisms face result in trade-
offs or compromises in their ability to perform important
functions.

High levels of reproduction, for example, are often
associated with decreased longevity. In some cases, such
a trade-off is due to relatively subtle costs of reproduction: resources directed toward reproduction are not
available for other uses, such as storing energy to help
the organism survive a cold winter. In red deer, for example, females that reproduced the previous spring have a
higher rate of death during winter than do females that
did not reproduce. But costs associated with reproduction can sometimes be immediate and dramatic, as illustrated by the mating calls of the túngara frog described
in Chapter 17 (Figure 18.5). In general, the widespread
existence of trade-offs between reproduction and other
important functions ensures that organisms are not perfect, for the simple reason that it is not possible to be
the best at all things at once.

18.3 What Are Species?

Before we discuss the processes that have generated the
great diversity of species on Earth, we must first define
what a species is.

Species are often morphologically distinct

In a practical sense, species are usually defined in terms
of morphology; that is, two organisms are classified as

members of different species if they look sufficiently different. All of us use such a definition, as when we distinguish bald eagles (Figure 18.6) from other birds by how they look.

Most species can be identified by morphological characteristics. Indeed, morphology is the only way we have to identify and distinguish fossil species. However, for living species, a morphological definition does not always work well: there are pairs of species whose members look exactly alike, but which cannot breed with each other. Conversely, there are populations whose members vary phenotypically (sometimes dramatically), yet remain part of the same species because they can interbreed. So what "holds a species together" and distinguishes it from other species?

Species are reproductively isolated from one another

In most cases, members of different species cannot reproduce with each other. When barriers to reproduction exist between species, the species are said to be **reproductively isolated** from one another. Barriers to reproduction can act before (prezygotic barriers) or after the formation of a zygote (postzygotic barriers) (Table 18.1). While there are a wide range of barriers to reproduction, they all have the same net effect: few or no alleles are exchanged between species. This restriction ensures that the members of a species share a common set of genes and alleles. Because members of a species exchange alleles with one another, but not with members of other species, they usually remain phenotypically similar to one another and different from members of other species.

Species, then, can be defined in terms of reproductive isolation: a **species** is a group of populations that can interbreed with one another but which are reproductively isolated from other such groups. Note that reproductive isolation is not necessarily the same as geographic isolation: our definition of species includes populations that could interbreed if they were in contact with one another, but do not because they have no opportunity to do so (for example, because they are too far away from one another).

A definition of species based on reproductive isolation has important limitations. For example, it is of no use when defining fossil species, since no information can be obtained about whether or not two fossil forms were reproductively isolated from each other. (So, as mentioned earlier, fossil species are defined on the basis of morphology.) Nor does our definition apply to organisms, such as bacteria and dandelions, that reproduce mainly by asexual means.

Our definition of species also fails to work well for the many plant and animal species that mate in nature to

(a)

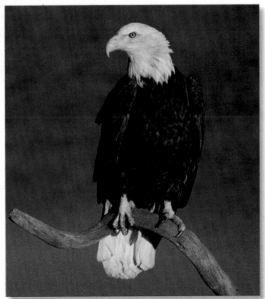

(b)

Figure 18.6 In Most Species, Members Look Alike
Bald eagles that live in Alaska (a) look the same as bald eagles that live in Colorado (b). Although these birds live far apart, they remain phenotypically similar.

Table 18.1

Barriers That Can Reproductively Isolate Two Species in the Same Geographic Region

Type of barrier	Description	Effect
PREZYGOTIC BARRIERS		
Ecological isolation	The two species breed in different portions of their habitat, at different seasons, or at different times of the day	Mating is prevented
Behavioral isolation	The two species respond poorly to each other's courtship displays or other mating behaviors	Mating is prevented
Mechanical isolation	The two species are physically unable to mate	Mating is prevented
Gametic isolation	The gametes of the two species cannot fuse, or they survive poorly in the reproductive tract of the other species	Fertilization is prevented
POSTZYGOTIC BARRIERS		
Zygote death	Zygotes fail to develop properly and die before birth	No offspring are produced
Hybrid performance	Hybrids survive poorly or reproduce poorly	Hybrids are not successful

produce fertile offspring. Species that interbreed in nature are said to **hybridize**, and their offspring are called **hybrids**. Although they can reproduce with each other, species that hybridize often look different from each other (Figure 18.7) or are distinct ecologically (for example, they are usually found in different environments, or they differ in how they perform important biological functions, such as obtaining food).

Despite the limitations of a definition of species based on reproductive isolation, most biologists define species this way, and this is the definition we will use in this book. Many alternative species definitions exist, but those definitions have problems as well. It is perhaps best to recognize that we have defined species based on a simple concept (reproductive isolation), but to keep in mind that the reality of a species in nature can be considerably more complicated.

18.4 Speciation: Generating Biodiversity

The tremendous diversity of life on Earth is caused by **speciation**, the process in which one species splits to form two or more species that are reproductively isolated from one another. The study of speciation is fundamental to understanding the diversity of life on Earth.

How do new species form? The crucial event in the formation of new species is the evolution of reproductive isolation, which requires that populations that once could interbreed diverge enough from one another so that they are no longer able to do so. However, as we saw in Chapter 17, populations within a species can be connected by gene flow, which tends to keep them genetically similar to one another. How does reproductive isolation develop within a species, whose members interbreed and therefore share a common set of genes and alleles?

Speciation can be explained by the same mechanisms that cause the evolution of populations

Speciation is usually considered a secondary consequence of the evolution of populations. In essence, populations evolve genetic differences from one another—due to mutation, genetic drift, or natural selection—and some of these genetic differences result in partial or total reproductive isolation.

(a) Gray oak

Hybrids

Gambel oak

(b)

Figure 18.7 Some Species Interbreed Yet Remain Distinct
(a) The gray oak and the gambel oak can mate to produce fertile hybrids. However, the two species remain phenotypically different, as is evident from their leaves. (b) Hybrid individuals, with varying degrees of fall color, are found throughout the hillside shown in this photograph. The two oak species are found nearby, one at lower elevations (gray oak), the other at higher elevations (gambel oak).

This idea can be illustrated by the results of an experiment with fruit flies. A population of flies was separated into several smaller populations, all of which were placed in similar environments, except that some flies were fed maltose (a simple sugar) and others were fed starch. Over time, the flies raised on these two different foods started to become reproductively isolated from each other (Figure 18.8). This happened even though the experimenter was not actively selecting for reproductive isolation. Instead, the fly populations changed genetically over time (most likely in response to natural selection for the ability to use different food sources), and those genetic changes had the side effect of causing partial reproductive isolation. The experiment was not continued long enough for speciation to occur, but it does illustrate how reproductive isolation can evolve as a by-product of other evolutionary changes.

As suggested by Figure 18.8, natural selection can cause populations to diverge genetically when populations located in different environments face different selection pressures. Populations can also diverge from one another as a result of mutation and genetic drift. In contrast, gene flow always operates to prevent the genetic divergence of populations. Thus, for populations to accumulate enough genetic differences to cause speciation, the factors that promote divergence must have a greater effect than does the amount of ongoing gene flow.

See speciation in action.

18.1

Figure 18.8
Selection Can Cause Reproductive Isolation

An initial sample of fruit flies were separated into four populations and raised on two different kinds of food (starch or maltose) for several generations. In the experimental group, flies from the populations that had become adapted to feed on starch were then placed with flies adapted to feed on maltose, and mating frequencies were observed. In the nonexperimental group, flies from one population adapted to feed on starch were placed with flies from the other population adapted to feed on starch; a similar procedure was followed for flies adapted to feed on maltose.

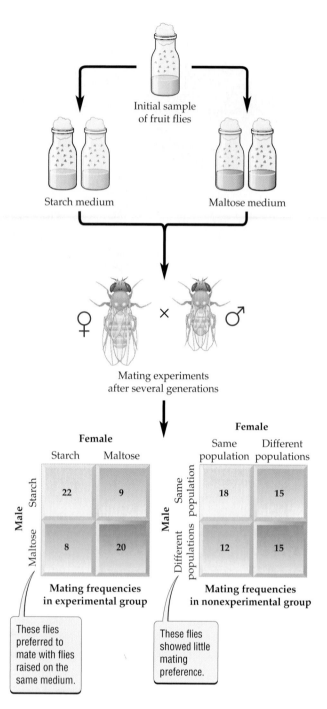

Initial sample of fruit flies

Starch medium   Maltose medium

♀ × ♂

Mating experiments after several generations

Mating frequencies in experimental group

	Female	
Male	**Starch**	**Maltose**
Starch	22	9
Maltose	8	20

These flies preferred to mate with flies raised on the same medium.

Mating frequencies in nonexperimental group

	Female	
Male	**Same population**	**Different populations**
Same population	18	15
Different populations	12	15

These flies showed little mating preference.

Speciation often results from geographic isolation

A new species can form when populations of a single species become separated, or geographically isolated, from one another. This process can begin when a newly formed geographic barrier, such as a river or a mountain chain, isolates two populations of a single species. Such **geographic isolation** can also occur when a few members of a species colonize a region that is difficult to reach, such as an island located far outside the usual geographic range of the species.

The distance required for geographic isolation to occur varies tremendously from species to species. Populations of squirrels and other rodents that live on opposite sides of the Grand Canyon have diverged considerably, whereas populations of birds—whose members can cross the canyon relatively easily—have not. In general, geographic isolation is said to occur whenever populations are separated by a distance that is great enough to limit gene flow.

However they arise, geographically isolated populations are connected to other populations by little or no gene flow. For this reason, mutation, genetic drift, and natural selection can more easily cause the populations to diverge genetically from one another. If the populations remain isolated for a long enough time, they can evolve into new species (Figure 18.9). The formation of new species from populations that are geographically isolated from one another is called **allopatric speciation**.

Much evidence indicates that geographic isolation can lead to speciation. One line of evidence is the observation that in many groups of organisms, the number of species is greatest in regions where strong geographic barriers increase the potential for geographic isolation. Examples include species that live in mountainous regions (such as birds in New Guinea), on island chains (such as plants in the Hawaiian Islands), or in small, isolated lakes (for example, the cichlid species found in isolated African lakes differ from one another).

A second line of evidence comes from cases in which individuals from a population found at one extreme end of a species' geographic range reproduce poorly with individuals from a population at the other end, even though individuals from both ends reproduce well with individuals from intermediate portions of the species' range. A special case of this phenomenon occurs when the populations loop around a geographic barrier; in this case, **ring species** form, in which populations at the two ends of the loop are in contact with one another, yet individuals from these populations cannot interbreed. Ring species have been found in salamander populations that loop around the mountains of the Sierra Nevada in California.

A third line of evidence comes from the results of laboratory experiments in which populations that are separated from one another develop reproductive isolation over time (see Figure 18.8).

Speciation can occur without geographic isolation

There is a greater potential for gene flow between populations whose geographic ranges overlap or are adjacent to one another than there is between populations that are geographically isolated from one another. Thus the majority of speciation events are thought to occur by allopatric speciation. Nevertheless, it has long been known that plants can form new species in the absence of geographic isolation, and recent work has provided convincing evidence that animals can as well. The formation of new species in the absence of geographic isolation is called **sympatric speciation**.

In plants, rapid chromosomal changes can cause sympatric speciation. New plant species can form in a single generation as a result of **polyploidy**, a condition in which an individual has more than two sets of chromosomes, usually due to the failure of chromosomes to separate during meiosis (see Chapter 9). Polyploidy can also occur when a hybrid spontaneously doubles its chromosome number. Polyploidy is invariably fatal in people, but it is not lethal in many plant species. While it does not kill them, a doubling of the chromosomes can lead to reproductive isolation because the chromosome number in the gametes of the new polyploid no longer matches the numbers in the gametes of either of its parents. Although relatively few plant species originate directly in this way, polyploidy has had a large effect on life on Earth: more than half of all plant species alive today are descended from species that originated by polyploidy. A few animal species also appear to have originated by polyploidy, including several species of lizards and fish and one mammal (an Argentine rat).

Evidence is mounting that sympatric speciation can occur in animals by means other than polyploidy. For example, there is compelling evidence that new fish species can form in the absence of geographic isolation. In one such case, genetic data indicate that 9 and 11 cichlid species, respectively, have originated within the confines of Lake Bermin and Lake Barombi Mbo, two small lakes in West Africa. The formation of new species within these lakes provides strong evidence for sympatric speciation because these lakes are small and simple in structure. As a result, fish populations within these lakes cannot evolve into new species while living apart from one another in different environments within the lake. In contrast, large lakes such as Lake Victoria, which we described at the opening of this chapter, usually provide a wide range of different environments, and hence species can be geographically isolated from one another even within the confines of a single lake.

Strong evidence in support of sympatric speciation has also been found in other animals. Researchers believe that North American populations of the apple maggot fly, *Rhagoletis pomonella* [rag-oh-*LEE*-tiss po-mo-*NELL*-uh], are in the process of diverging into new species even

A single plant species is distributed over a broad geographic range.

The sea level rises and isolates plant populations from each other. The populations may adapt to different environments on opposite sides of the barrier, indirectly causing genetic changes that reduce their ability to interbreed with one another.

When the barrier is removed, the plants recolonize the intervening area and mingle, but do not interbreed.

Range of overlap

Figure 18.9 Allopatric Speciation
New species can form when populations are separated by a geographic barrier, such as a rising sea.

though their geographic ranges overlap. Historically, *Rhagoletis* usually ate native hawthorn fruits, but in the mid-nineteenth century these flies were first recorded as pests on apples, an introduced nonnative species. *Rhagoletis* populations that feed on apples are now genetically distinct from populations that feed on hawthorns. Members of the apple and hawthorn populations mate at different times of the year and usually lay their eggs only on the fruit of their particular food plant. As a result,

Rhagoletis pomonella

there is little gene flow between the apple and hawthorn populations. In addition, researchers have identified alleles that benefit flies that feed on one host plant, but are detrimental to flies that feed on the other host plant. Thus natural selection operating on these alleles acts to limit whatever gene flow does occur. Over time, the ongoing research on *Rhagoletis* may well provide a dramatic case history of sympatric speciation.

18.5 Rates of Speciation

When speciation is caused by polyploidy or other types of rapid chromosomal change, new species form in a single generation. As we have seen, new species also appear to have formed relatively rapidly in the case of some cichlid fish. As we saw at the opening of this chapter, genetic analyses indicate that the 500 species of cichlids in Lake Victoria descended from just two ancestor species. Genetic evidence also indicates that these 500 species evolved over the past 100,000 years. Thus, in roughly 100,000 years, 500 fish species evolved from two species.

In many—perhaps most—cases, speciation occurs more slowly. Among freshwater fish, the time required for speciation has been estimated to range from a minimum of 3,000 years (in pupfish) to over 9 million years (in characins [*KAIR*-uh-sins], a group that includes carp, piranha, and many aquarium fish). In other groups of organisms, including fruit flies, snapping shrimp, and birds, the time required for speciation has been estimated to range from 600,000 to 3 million years. Furthermore, some populations can be geographically isolated for a long time without evolving reproductive isolation. North

American and European sycamore trees, for example, have been separated for more than 20 million years, yet the two populations remain morphologically similar and can breed with each other.

18.6 Implications of Adaptation and Speciation

Adaptation and speciation are, respectively, the means by which organisms adjust to the challenges posed by new or changing environments and the means by which the diversity of life has come into being. Both, therefore, are critical to understanding how evolution works.

Adaptation and speciation are also very important from an applied perspective. For example, to combat rapidly evolving disease agents—such as HIV, the virus that causes AIDS—we must have a detailed understanding of the new adaptations that enable them to overcome our best efforts to kill them. Speciation has long been of practical importance to humans, as our development of domesticated crop and animal species readily attests.

In addition, understanding the often slow pace of speciation gives us a strong incentive to stop the ongoing extinctions of species. Speciation can require hundreds of thousands to millions of years, yet humans are driving species extinct in decades to hundreds of years. If we continue to drive species extinct at the present rate, it will take millions of years before the speciation process can replace the large number of species that are currently being lost.

Rapid Speciation in Lake Victoria Cichlids

As we saw earlier in this chapter, the Lake Victoria cichlids formed new species at a rapid rate, resulting in an evolutionary expansion unmatched by any other group of vertebrates. The rapid speciation of these fish brings us back to one of the cichlid mysteries introduced at the opening of this chapter: how did so many species form in just 100,000 years? The answer hinges on two key aspects of cichlid biology.

First, cichlids in Lake Victoria appear to use color as a basis for mate choice: females prefer to mate with males of a particular color (an example of sexual selection; see pages 329–331). Within Lake Victoria, there are many pairs of ecologically similar cichlid species that differ from each other in color, but little else. In these pairs of species, the males of one species tend to be blue, while the males of the other species are red or yellow. Females of a given species show a strong preference for mating with males of their own species. This preference appears to be based on the color of the males: when researchers experimentally changed the quality of the light so that the females could not see the difference between blue and red/yellow males, the female preference for mating with males of their own species broke down. This research suggests that if males in two populations of a cichlid species happened to have different colors, female preferences for mating with males of a certain color could cause these populations to become reproductively isolated, thus setting the stage for the formation of new species.

In addition, cichlids have unusual jaws that can be modified relatively easily over the course of evolution to specialize on new food items. This feature of their biology causes the cichlids in the lake to vary greatly in form and feeding behavior (Figure 18.10). How do their easily modified jaws relate to speciation? If female mate choice caused two populations to begin to be reproductively isolated from each other, the resulting lack of gene flow could allow the populations to specialize on different sources of food, which would make it increasingly likely that they would continue to diverge and form new species.

Another cichlid mystery remains: we saw that predation by the introduced Nile perch has driven many species to extinction, but what has caused species not eaten by the Nile perch to vanish? Here, too, the answer may be related to female mate choice. As the experiments described above showed, female cichlids need to be able to see the colors of males to distinguish males of their own species from males of closely related species. However, pollution due to human activities has caused the water of Lake Victoria to become cloudy. If females cannot distinguish the colors of potential mates in the murky water, reproductive barriers between species might break down. As reproductive barriers lost their effectiveness, species that once were distinct might interbreed freely and become more similar to each other. If continued

Figure 18.10 Lake Victoria Cichlids
The four species shown here illustrate some of the differences in morphology and feeding behavior found among Lake Victoria cichlids.

Haplochromis chilotes
(feeds on insects)

Haplochromis macrognathus
(feeds on other fish)

Astatotilapia elegans
(generalized bottom feeder)

Macropleurodus bicolor
(feeds on snails and other molluscs)

long enough, such interbreeding could "reverse" the speciation process and cause species to go extinct. Pollution would not be likely to have this effect if the cichlid species had been separated by barriers to reproduction that occur *after* mating takes place, such as gametic isolation or zygote death (see Table 18.1).

Finally, remember that speciation in cichlids depends on the ability of females to recognize dif-ferences in the colors of males; when cloudy water impairs that ability, new species cannot form. Thus pollution from human activities appears to have two profound effects: it halts the formation of new cich-lid species while simultaneously causing existing species to go extinct. We must reduce that pollution if we are not to destroy one of nature's most amaz-ing evolutionary experiments, the cichlids of Lake Victoria.

Chapter Review

Summary

18.1 Adaptation: Adjusting to Environmental Challenges

- Adaptations result in an apparent match between organisms and their environment, but are not caused by intentional design.
- Adaptive evolution is the process by which the fit between organ-isms and their environment is improved over time.
- Adaptations help organisms accomplish important functions, such as mate attraction and predator avoidance.
- Adaptations can be improved in short periods of time (months to years).

18.2 Adaptation Does Not Craft Perfect Organisms

- Adaptive evolution can be limited by genetic constraints: lack of genetic variation gives natural selection little or nothing on which to act.
- Adaptive evolution can be limited by developmental constraints: the multiple effects of developmental genes can prevent the organism from evolving in certain directions.
- Adaptive evolution can be limited by ecological trade-offs: conflict-ing demands faced by organisms can compromise their ability to per-form important functions.

18.3 What Are Species?

- Species are often morphologically distinct, but morphology is not a reliable way to distinguish some species.
- A species is a group of interbreeding natural populations that is reproductively isolated from other such groups.
- The definition of species in terms of reproductive isolation has impor-tant limitations. It does not apply to fossil species (which must be identified by morphology), to organisms that reproduce mainly by asexual means, or to organisms that hybridize extensively in nature.

18.4 Speciation: Generating Biodiversity

- The crucial event in the formation of a new species is the evolution of reproductive isolation.

- Speciation usually occurs as a by-product of the genetic divergence of populations from one another caused by natural selection, genetic drift, or mutation.
- Speciation usually occurs when populations are geographically iso-lated from one another long enough for reproductive isolation to evolve. Most new species are thought to arise by this process, which is called allopatric speciation.
- Speciation can also occur without geographic isolation. This process, called sympatric speciation, acts when part of a population diverges genetically from the rest of the population.
- Polyploidy is one way that many plants evolve new species during a single generation.

18.5 Rates of Speciation

- Speciation occurs rapidly in some cases, but it requires hundreds of thousands to millions of years in other cases.

18.6 Implications of Adaptation and Speciation

- Adaptations are the means by which organisms adjust to challenges posed by new or changing environments.
- Speciation is the means by which the diversity of life has come into being.
- Adaptation and speciation influence such practical matters as how we fight diseases and develop domesticated species.

⊙ Review and Application of Key Concepts

1. Select an organism (other than humans) that you are familiar with. List two adaptations of that organism. Explain carefully why each of these features is an adaptation.

2. What is adaptive evolution? Apply your understanding of adaptive evolution to organisms that cause infectious human diseases, such as bacterial species that cause plague or tuberculosis. How do our efforts to kill such organisms affect their evolution? Are the evolutionary changes we promote usually beneficial or harmful for us? Explain your answer.

3. Imagine that a species legally classified as rare and endangered is discovered to hybridize with a more common species. Since the two species interbreed in nature, should they be considered a single species? Since one of the two species is common, should the rare species no longer be legally classified as rare and endangered?

4. Should species that look different and are ecologically distinct, such as the oaks in Figure 18.7, be classified as one species or two? These oak species hybridize in nature. Should species that hybridize in nature be considered one species or two?

5. High winds during a tropical storm blow a small group of birds to an island previously uninhabited by that species. Assume that the island is located far from other populations of this species, and that environmental conditions on the island differ from those experienced by the birds' parent population. Is natural selection or genetic drift (or both) likely to influence whether the birds on the island form a new species? Explain your answer.

6. Hundreds of new species of cichlids evolved within the confines of Lake Victoria, but some of these species live in different habitats within the lake and rarely encounter one another. Would you consider such species to have evolved with or without geographic isolation?

7. How can new species form by sympatric speciation? Why is it harder for speciation to occur in sympatry than in allopatry?

Key Terms

adaptation (p. 340)
adaptive evolution (p. 340)
allopatric speciation (p. 348)
geographic isolation (p. 348)
hybrid (p. 346)
hybridize (p. 346)

polyploidy (p. 349)
reproductive isolation (p. 345)
ring species (p. 348)
speciation (p. 346)
species (p. 345)
sympatric speciation (p. 349)

Self-Quiz

1. Species whose geographic ranges overlap but which do not interbreed in nature are said to be
   a. geographically isolated.
   b. reproductively isolated.
   c. influenced by genetic drift.
   d. hybrids.

2. Which of the following evolutionary mechanisms acts to slow down or prevent the evolution of reproductive isolation?
   a. natural selection
   b. gene flow
   c. mutation
   d. genetic drift

3. The splitting of one species to form two or more species most commonly occurs
   a. by sympatric speciation.
   b. by genetic drift.
   c. by allopatric speciation.
   d. suddenly.

4. The time required for populations to diverge to form new species
   a. varies from a single generation to millions of years.
   b. is always greater in plants than in animals.
   c. is never less than 100,000 years.
   d. is rarely more than 1,000 years.

5. Adaptations
   a. match organisms closely to their environment.
   b. are often complex.
   c. help the organism accomplish important functions.
   d. all of the above

6. Prezygotic and postzygotic barriers to reproduction have the effect of
   a. reducing genetic differences between populations.
   b. increasing the chance of hybridization.
   c. preventing speciation.
   d. reducing or preventing gene flow between species.

7. Evidence suggests that sympatric speciation may have occurred or be in progress in three of the following four cases. Select the exception.
   a. apple maggot fly
   b. squirrels on opposite sides of the Grand Canyon
   c. cichlid fish
   d. polyploid plants (or their ancestors)

8. The diploid number of chromosomes in plant species A is 8; the diploid number in plant species B is 16. If plant species C originated when a hybrid between A and B spontaneously doubled its chromosome number, what is the most likely number of diploid chromosomes in C?
   a. 8   c. 24
   b. 12   d. 48

Humans' Ancestral Tree Adds a Twig

By Robert Lee Hotz

On an isolated Indonesian island, scientists have discovered skeletons of a previously unknown human species—tiny, Hobbit-sized figures who lived among dwarf elephants and giant lizards as recently as 12,000 years ago.

Experts in human origins called the discovery, made public Wednesday, of an extinct human species barely three feet tall the most important—and surprising—human find in the last 50 years.

It was the discovery of a lifetime: a skeleton of an adult female that stood about 3 feet tall—shorter than the average height of a 3-year-old girl in the United States—and lived within the last 20,000 years. The skeleton was found on Flores, a small island west of Java. The Flores female would hardly be imposing if you met her face-to-face, but she is causing quite a stir in scientific circles, based on a report in the journal *Nature*.

With a brain roughly the size of a grapefruit, at first glance she is reminiscent of the small-brained human ancestors that lived about 3 million years ago. But *Homo floresiensis*, as her species has been named, made stone tools, cooked with fire, stood upright, and, most astonishingly of all, lived as recently as 12,000 years ago. While the best-preserved skeleton was from 18,000 years ago, fragments from six others ranged in age from 95,000 to 12,000 years ago. As the news article stated, "This is remarkably recent and overlaps by tens of thousands of years with modern humans in the region." Some of the more recent bones were not even completely fossilized.

Our diminutive relative may have been wiped out by a massive volcanic explosion that covered the island with ash 12,000 years ago. There is no evidence of modern humans living on Flores until after the eruption, so there is no reason to think that *H. sapiens* and *H. floresiensis* lived on the island at the same time. But for some, the discovery of *H. floresiensis* shakes our notion of who and what we are as a species. As Marta Lahr and Robert Foley, members of Cambridge University's Department of Biological Anthropology, said in their commentary in the same issue of *Nature*, "It is breathtaking to think that such a different species of [humans] existed so recently. . . . For most of its 160,000-year history, *Homo sapiens* seems to have shared the planet with other bipedal and cultural beings—our global dominance may be far more recent than we thought."

Evaluating the news

1. If *Homo floresiensis* had persisted into modern times and members of that species had been discovered within the last 200 years by our species, how do you think we would have treated them? What makes you think our species would act in the manner you describe?

2. Many people take it for granted that we humans have an inherent right to manipulate the environment and other species in ways that benefit us. Imagine that *H. floresiensis* was still alive. If we shared the planet with another human species, would that affect our assumption that we have the right to alter nature as we please? As you think about this question, consider how the existence of *H. floresiensis* would affect both your personal views and the views of society at large.

3. Human history and current events contain examples in which people of one culture slaughter those of another, as well as examples in which humans deliberately drive other species to extinction or the brink of extinction. Few other animal species behave in this way, and none take the systematic, deliberate approach that humans employ in our efforts to destroy other cultures or species. Why do we behave in this way?

SOURCE: *Los Angeles Times*, Thursday, October 28, 2004.

19

The Evolutionary History of Life

Key Concepts

- The fossil record documents the history of life on Earth and provides clear evidence of evolution.

- Early photosynthetic organisms released oxygen to the atmosphere as a waste product, thereby setting the stage for the evolution of the first eukaryotes, and later, the first multicellular organisms.

- An astonishing increase in animal diversity occurred 530 million years ago, when large forms of most of the major living animal phyla appeared suddenly in the fossil record.

- The colonization of land by the first plants and animals marked the beginning of another major increase in the diversity of life.

- The history of life can be summarized by the rise and fall of major groups of protists, plants, and animals. This history has been greatly influenced by continental drift, mass extinctions, and adaptive radiations.

Puzzling Fossils in a Frozen Wasteland

Antarctica is a crystal desert, an ice-covered land in which heat and liquid water are very scarce. Few organisms can survive the extreme cold and lack of available water, and most of those that can are small and live near the sea. The entire continent has only two species of flowering plants (Antarctic hair grass and Antarctic pearlwort), and its largest terrestrial animal is a fly 5 millimeters long.

In the interior of the continent the organisms are even smaller: in most places the only living things are microscopic bacteria and protists, including some that survive in a state of suspended animation (frozen but alive in the ice). Some of the interior valleys have little ice, and hence seem a little less forbidding. But these valleys are so dry and cold that they support no visible life. There, the only organisms found on land are photosynthetic bacteria and lichens that spend their entire lives in a narrow zone just under the translucent surface of certain types of rocks.

Despite the nearly lifeless appearance of this continent, fossils reveal that Antarctica used to be very different from today's frozen landscape. Where life now maintains an uncertain foothold, the land was once bordered by tropical reefs and later covered with forests. At different times, ferns, freshwater fish, large amphibians, aquatic beetles, and trees as tall as 22 meters thrived in what is now among the harshest of environments. Dinosaurs once roamed these lands, and millions of years later mammals and reptiles were pursued by the terrorbird, a fast-running flightless bird that stood 3.5 meters tall.

Early explorers and scientists were amazed when they discovered these fossils of ancient life forms in Antarctica. The fossils showed that life used to be rich and abundant where now it barely exists. These scientists and explorers were left with a simple question: What happened?

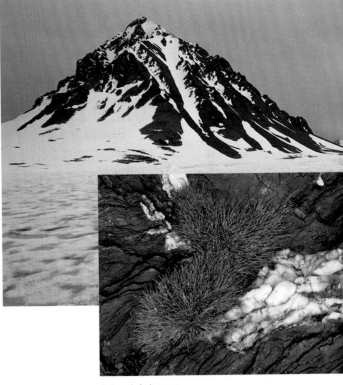

Antarctic hair grass

The Antarctic Landscape
Only two species of flowering plants are found on the continent, one of which is Antarctic hair grass.

Earth abounds with life. About 1.5 million species have been described, millions of species have been collected and await formal description, and millions more await discovery (most estimates for the total number of species on Earth range from 3 to 30 million, as we saw in Interlude A). While these numbers are large, the species alive today are thought to represent far less than 1 percent of all the species that have ever lived.

In previous chapters of this unit we have discussed how the diversity of life arose, focusing on the mechanisms that drive the evolution of populations (microevolution) and lead to the formation of new species (speciation). In this chapter we broaden our scope to discuss large-scale evolutionary changes, or macroevolution.

Macroevolution is evolution above the species level. Examples of macroevolutionary change include the origin of new genera or higher taxonomic groups as well as increases or decreases in the number of members in such groups (for example, the number of species in a genus). The study of macroevolution addresses evolutionary expansions that bring new groups to prominence and large-scale extinctions that greatly alter the diversity of life on Earth. Thus macroevolution emphasizes how life on Earth has changed over time. Such an emphasis leads us to think of evolution as a historical process—specifically, as the history of the formation and extinction of species and higher taxonomic groups over time.

We begin this chapter with a look at how the history of life on Earth is documented in the fossil record, followed by a summary of major events in that history. We then consider some of the forces that increase and decrease biodiversity over the long term—continental drift, mass extinctions, and adaptive radiations—and examine their effects with regard to the rise of our own group, the mammals.

19.1 The Fossil Record: A Guide to the Past

Fossils are the preserved remains or impressions of individual organisms that lived in the past (Figure 19.1). In many fossils, portions of the bodies of dead organisms are replaced with minerals; thus the original body form is maintained, but the fossil contains new material. Fossils are often found in sedimentary rock (rock that consists of layers of sediments that have hardened), but can also form in a few other situations. Insects, for example, have been found in amber, the fossilized resin of a tree (see Figure 19.1e), and many mammals, including mammoths and a 5,000-year-old man (see the photograph on page 23), have been found in melting glaciers.

As we saw in Chapter 16, the fossil record documents the history of life and is central to the study of evolution. Fossils provided the first compelling evidence that past organisms are unlike living forms, that many forms have disappeared from Earth completely, and that life has evolved through time.

The relative depth or distance from the surface of Earth at which a fossil is found is referred to as its order in the fossil record. The age of fossils corresponds to their order: usually, older fossils are found in deeper, older rock layers. The order in which organisms appear in the fossil record agrees with our understanding of evolution based on other evidence, thus providing strong support for evolution. For example, analyses of the morphology, DNA sequences, and other characteristics of living organisms indicate that bony fish gave rise to amphibians, which later gave rise to reptiles, which still later gave rise to mammals. That is exactly the order in which fossils from these groups appear in the fossil record. The fossil record also provides excellent examples of the evolution of major new groups of organisms, such as the evolution of mammals from reptiles (see page 371).

While knowing the order of various organisms and groups of organisms in the fossil record is very helpful, it can provide only *relative* ages of fossils. In some cases, a better approximation of a fossil's age can be made using **radioisotopes**, which are unstable, radioactive forms of elements that decay to more stable forms at a constant rate over time. For example, for a given amount of the radioisotope carbon-14 (^{14}C), half of it decays to the stable element carbon-12 every 5,730 years. By measuring the amount of ^{14}C that remains in a fossil, scientists can estimate the age of the fossil. Carbon-14 can be used to date only relatively recent fossils (too little ^{14}C remains to date fossils formed more than 70,000 years ago), but elements such as uranium-235, which has a half-life of 700 million years, can be used to date much older materials. If, as commonly occurs, a fossil does not contain any radioisotopes, an approximate date for the fossil is determined by dating rocks found above and below the fossil.

The fossil record is not complete

Although many fossils have been found, the fossil record still has many gaps. Most organisms decompose rapidly after death; hence very few form fossils. Even if an organism is preserved initially as a fossil, a variety of common

Figure 19.1 Fossils Through the Ages

(a) Soft-bodied animals such as the ones that left these fossils dominated life on Earth 600 million years ago (mya). (b) A fossil of a trilobite [TRY-lo-bite] that lived in the Devonian period (410–355 mya). Note the rows of lenses on each eye. (c) This leaf of a 300-million-year-old seed fern was found near Washington, DC. The fossil formed during the Carboniferous [kar-buh-NIFF-er-us] period (355 to 290 mya). The great forests of this period led to the formation of the fossil fuels (oil, coal, and natural gas) that we use today as sources of energy. (d) A fossil of a Velociraptor entangled with a Protoceratops, which bit down of the predator's claw, locking both in a death grip. (e) This 20-million-year-old termite is preserved in amber, the fossilized resin of a tree.

geologic processes (such as erosion and extreme heat or pressure) can destroy the rock in which it is embedded. Finally, fossils can be difficult to find. Given the unusual circumstances that must occur for a fossil to form, remain intact, and be discovered by scientists, a species could evolve, thrive for millions of years, and become extinct without our ever finding evidence of its existence in the fossil record.

Nevertheless, the fossil record shows clearly that there have been great changes in the groups of organisms that have dominated life on Earth over time. As we will see throughout this chapter, these changes have been caused by the extinction of some groups and the expansion of other groups. Furthermore, while gaps in the fossil record remain, each year new discoveries fill in some of those gaps.

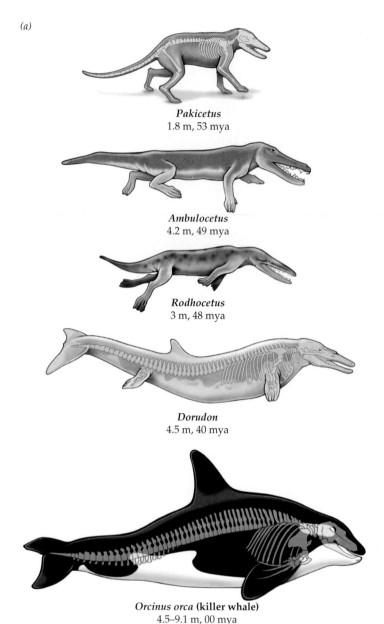

(a)

Pakicetus
1.8 m, 53 mya

Ambulocetus
4.2 m, 49 mya

Rodhocetus
3 m, 48 mya

Dorudon
4.5 m, 40 mya

Orcinus orca (killer whale)
4.5–9.1 m, 00 mya

Fossils reveal that whales are closely related to a group of hoofed mammals

Let's take a closer look at what recent fossil discoveries have revealed about one group of mammals, the whales. The origin of whales has long puzzled biologists. Most mammals live on land, and it is hard to imagine how a land mammal could be transformed into something as different as a whale. But recently discovered fossils provide a glimpse of how that transformation occurred (Figure 19.2*a*).

The bone structure of an early whale ancestor, *Pakicetus* [*PACK*-uh-*SEET*-us], suggests that it probably spent most of its time on land. However, *Pakicetus* shared features (such as unusual bones in its inner ear) with modern whales and with the more whalelike creatures shown in Figure 19.2*a*. Over many generations, the ancestors of whales became increasingly similar to modern whales: their legs became smaller, and their overall shape took on the streamlined form of a fully aquatic mammal.

The recently discovered fossils also confirm the results of genetic analyses; namely, that whales are most closely related to the artiodactyls [are-tee-oh-*DACK*-tuls]—eventoed, hoofed mammals such as camels, cows, pigs, deer, and hippopotamuses. In all artiodactyls, an ankle bone called the astragalus [uh-*STRAG*-uh-luss] has an unusual shape, in which both the top and bottom surfaces of the bone resemble those of a pulley (Figure 19.2*b*). In 2001, the ankle bones of several whale ancestors, including *Pakicetus* and *Rodhocetus* [*ROD*-oh-*SEET*-us], were discovered. These early whales had ankle bones with the same unusual shape as those of artiodactyls. Since this shape is an adaptation for running on land, it is highly unlikely that whale ancestors developed such bones as a result of convergent evolution. Instead, these new fossils strongly suggest that whale ancestors had such bones because they shared a (recent) common ancestor with the artiodactyls.

(b)

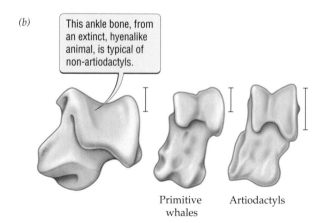

This ankle bone, from an extinct, hyenalike animal, is typical of non-artiodactyls.

Primitive whales

Artiodactyls

Figure 19.2 Shape-Shifters

(*a*) It took roughly 15 million years for whale ancestors to make the transition from life on land to life in water. The oldest whale ancestor, *Pakicetus*, lived on land 53 million years ago. *Ambulocetus* [am-byu-lo-*SEE*-tus] had strong, well-developed legs and probably was semiaquatic, living at the water's edge and hunting much as a crocodile does today. In *Rodhocetus*, the body is more streamlined, and the front legs are shaped more like flippers. By 40 mya, *Dorudon* was fully aquatic. The drawings are based on reconstructed fossil skeletons, which have been superimposed on two of the whale ancestors. Compare this sequence of whale ancestors with *Orcinus*, a modern toothed whale. (*b*) The astragalus (ankle bone) of two whale ancestors, *Pakicetus* (shown here) and *Rodhocetus* (not shown), is similar in shape to that of artiodactyls (hoofed mammals), but very different from that found in most other mammals. Scale bar is 1 cm.

19.2 The History of Life on Earth

Figure 19.3 provides a sweeping overview of the history of life on Earth; study it carefully. In this section we focus on three of the main events in the history of life: the origin of cellular organisms, the beginning of multicellular life, and the colonization of the land.

The first single-celled organisms arose at least 3.5 billion years ago

Our solar system and Earth formed 4.6 billion years ago. The oldest known rocks on Earth (3.8 billion years old) contain carbon deposits that hint at life. The first solid evidence for life, however, comes from 3.5-billion-year-old fossilized mats called stromatolites [stro-MATT-uh-lights], which resemble similar mats formed by present-day bacteria. Although the earliest forms of life may have arisen between 4 billion and 3.5 billion years ago, fossils of that age have yet to be found. (See Chapter 3 for more discussion of the origin of life.)

Eukaryotes first appear in the fossil record at about 2.1 billion years ago. Thus, after the origin of prokaryotes 3.5 billion years ago, it took well over a billion years for the first eukaryotes to evolve. During this long period, the evolution of eukaryotes may have been limited in part by low levels of oxygen in the atmosphere. Chemical analyses of very old rocks indicate that Earth's atmosphere initially contained almost no oxygen. Roughly 2.75 billion years ago, however, some groups of bacteria evolved the ability to conduct photosynthesis, which releases oxygen as a waste product. As a result, the oxygen concentration in the atmosphere increased over time (Figure 19.4).

Eukaryotes are larger than most prokaryotes. Because of their size, eukaryotes depend on aerobic respiration, which requires oxygen and provides more energy per unit of food than do metabolic reactions (such as fermentation) that do not use oxygen (see Chapter 7). In addition, oxygen and other materials spread more slowly through a large cell than through a small cell. Overall, because of their relatively large size, eukaryotic cells would not have been able to get enough oxygen to meet their needs until the atmospheric concentration of oxygen reached at least 2–3 percent of present-day levels. Once those levels were reached, about 2.1 billion years ago, the first single-celled eukaryotes—organisms that resembled some modern algae—evolved (see Figure 19.4). As oxygen levels continued to increase, the evolution of larger and more complex multicellular organisms became possible.

Oxygen was toxic to many early forms of life. Thus, as the oxygen concentration in the atmosphere increased, many early prokaryotes went extinct or became restricted to environments that lack oxygen. Because the biologically driven increase in the oxygen concentration of the atmosphere drove many early organisms extinct while simultaneously setting the stage for the origin of multicellular eukaryotes, it was one of the most important events in the history of life on Earth.

Multicellular life evolved about 650 million years ago

All early forms of life evolved in water. About 650 million years ago (mya), there was an increase in the number of organisms appearing in the fossil record. At that time, much of Earth was covered by shallow seas, which were filled with plankton (protists, small multicellular animals, and algae that float freely in the water).

By 600 mya, larger, soft-bodied multicellular animals had evolved (see Figure 19.1a). These animals were flat and appear to have crawled or stood upright on the seafloor, probably feeding on living plankton or their remains. No evidence indicates that any of these animals preyed on the others. Many of these early multicellular animals may have belonged to groups of organisms that are no longer found on Earth.

The early to middle Cambrian period (530 mya) witnessed an astonishing burst of evolutionary activity. In a dramatic increase in the diversity of life that is known as the **Cambrian explosion**, large forms of most of the major living animal phyla, as well as other phyla that have since become extinct, appeared suddenly in the fossil record. (The word "explosion" here refers to a rapid increase in the number and diversity of species, not to a physical explosion.) The Cambrian explosion lasted only 5 to 10 million years—a blink of an eye in geologic terms (compare this time span with the 1.4 billion years it took for eukaryotes to evolve from prokaryotes).

The Cambrian explosion was one of the most spectacular events in the evolutionary history of life. It changed the face of life on Earth: from a world of relatively simple, slow-moving, soft-bodied scavengers and herbivores, suddenly there emerged a world filled with large, mobile predators in pursuit of herbivores with hard body coverings that defended them against those predators (Figure 19.5).

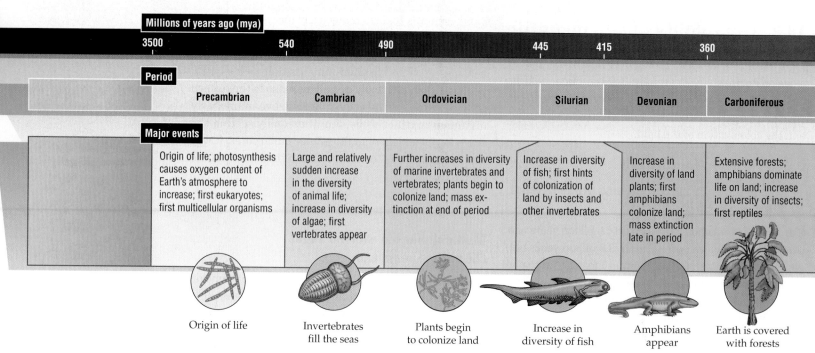

Millions of years ago (mya)						
3500	540	490	445	415	360	

Period

Precambrian	Cambrian	Ordovician	Silurian	Devonian	Carboniferous

Major events

Origin of life; photosynthesis causes oxygen content of Earth's atmosphere to increase; first eukaryotes; first multicellular organisms	Large and relatively sudden increase in the diversity of animal life; increase in diversity of algae; first vertebrates appear	Further increases in diversity of marine invertebrates and vertebrates; plants begin to colonize land; mass extinction at end of period	Increase in diversity of fish; first hints of colonization of land by insects and other invertebrates	Increase in diversity of land plants; first amphibians colonize land; mass extinction late in period	Extensive forests; amphibians dominate life on land; increase in diversity of insects; first reptiles
Origin of life	Invertebrates fill the seas	Plants begin to colonize land	Increase in diversity of fish	Amphibians appear	Earth is covered with forests

Figure 19.3 The History of Life on Earth: The Geologic Timescale

The history of life can be divided into 12 major geologic time periods, beginning with the Precambrian (3,500 to 540 mya) and extending to the Quaternary (1.8 mya to the present). This timescale is not drawn to scale; to do so and to include the Precambrian would require extending the diagram off the book page to the left by more than 5 *feet*.

 Test your knowledge of the major geologic time periods. **19.1**

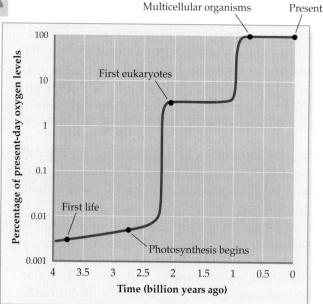

Figure 19.4 Oxygen on the Rise

The release of oxygen as a waste product by photosynthetic organisms caused its concentration in Earth's atmosphere to increase greatly over the last 3 to 4 billion years, facilitating the evolution of eukaryotes and multicellular organisms.

Colonization of land followed the Cambrian explosion

Because life first evolved in water, the colonization of land posed enormous challenges. Indeed, many of the functions basic to life, including support, movement, reproduction, and the regulation of ions, water, and heat, must be handled very differently on land than in water. Descendants of green algae were the first organisms to meet these challenges about 500 mya. These early terrestrial colonists had few cells and a simple body plan, but from them, land plants evolved and diversified greatly. By the end of the Devonian [deh-*VO*-nee-un] period (360 mya), Earth was covered with plants. Like plants today, the plants of the Devonian included low-lying spreading species, short upright species, shrubs, and trees.

As new groups of land plants arose, they evolved a series of key innovations, including a waterproof cuticle, vascular systems, structural support tissues (wood), leaves and roots of various kinds, seeds, the tree growth form, and specialized reproductive structures. These and other important changes allowed plants to cope with life on land. Waterproofing, stems with efficient transport mechanisms, and roots, for example, were important features that helped plants acquire and conserve water while living on dry land.

Although there are hints of land animals as early as 490 mya, the first definite fossils of terrestrial animals are of spiders and millipedes that date from about 410 mya. Many of the early animal colonists on land were predators; others, such as millipedes, fed on living plants

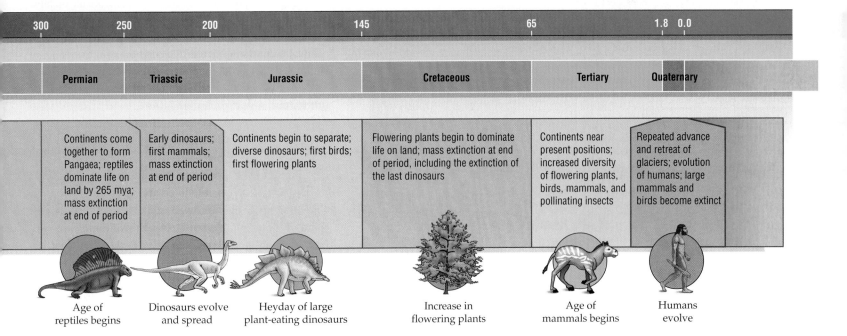

| 300 | 250 | 200 | 145 | 65 | 1.8 0.0 |

| Permian | Triassic | Jurassic | Cretaceous | Tertiary | Quaternary |

| Continents come together to form Pangaea; reptiles dominate life on land by 265 mya; mass extinction at end of period | Early dinosaurs; first mammals; mass extinction at end of period | Continents begin to separate; diverse dinosaurs; first birds; first flowering plants | Flowering plants begin to dominate life on land; mass extinction at end of period, including the extinction of the last dinosaurs | Continents near present positions; increased diversity of flowering plants, birds, mammals, and pollinating insects | Repeated advance and retreat of glaciers; evolution of humans; large mammals and birds become extinct |

| Age of reptiles begins | Dinosaurs evolve and spread | Heyday of large plant-eating dinosaurs | Increase in flowering plants | Age of mammals begins | Humans evolve |

or decaying plant material. Insects, which are currently the most diverse group of terrestrial animals, first appeared roughly 400 mya, and they played a major role on land by 350 mya.

The first vertebrates to colonize land were amphibians, the earliest fossils of which date to about 365 mya. Early amphibians resembled, and probably descended from, lobe-finned fish (Figure 19.6). Amphibians were the most abundant large organisms on land for about 100 million years. In

Millipede

mya
2500   **545**   **500**   440

The Cambrian explosion

Precambrian   Cambrian   Ordovician

Figure 19.5 Before and After the Cambrian Explosion
The Cambrian explosion greatly altered the history of life on Earth.

Before the Cambrian explosion, Earth was dominated by soft-bodied scavengers and herbivores.

After the Cambrian explosion, life was dominated by more complex animals, including predators and well-defended herbivores.

Millions of years ago (mya)

2500	545	500	440	**410**	**355**	290	250
Precambrian	Cambrian	Ordovician	Silurian	Devonian	Carboniferous	Permian	Triassic

The fins of this fish, which had bones and were muscular, could have provided support on land.

(a)

Although early amphibians probably spent considerable time in water, the muscles and bones in their legs allowed movement on land.

(b)

Figure 19.6 The First Amphibians
(*a*) Amphibians probably descended from a lobe-finned fish like the one shown here. (*b*) This early amphibian was reconstructed from a 365-million-year-old (late Devonian) fossil.

Learn more about the major events of each geologic time period.

19.2

19.3

the late Permian period, the reptiles, which had evolved from a group of reptile-like amphibians, rose to become the most common vertebrate group. Reptiles were the first group of vertebrates that could reproduce without returning to open water (for example, to lay eggs). As a result, reptiles were the first vertebrates that could fully exploit the available opportunities for terrestrial life. Reptiles,

including the dinosaurs, dominated vertebrate life on land for 200 million years (265 mya to 65 mya), and they remain important today. Mammals, the vertebrate group that currently dominates life on land, evolved from reptiles roughly 220 mya (see page 363).

19.3 The Effects of Continental Drift

The enormous size of the continents may cause us to think of them as immovable. But this notion is not correct. The continents move slowly relative to one another, and over hundreds of millions of years they travel considerable distances (Figure 19.7). This movement of the continents over time is called **continental drift**. The continents can be thought of as plates of solid matter that "float" on the surface of Earth's mantle, a hot layer of semisolid rock.

How can something as big as a continent move from place to place? Two forces cause the continental plates to move. First, hot plumes of liquid rock rise from Earth's mantle to the surface and push the continents away from one another. This process can cause the seafloor to spread, as it is doing between North America and Europe, which are separating at a rate of 2.5 centimeters per year.

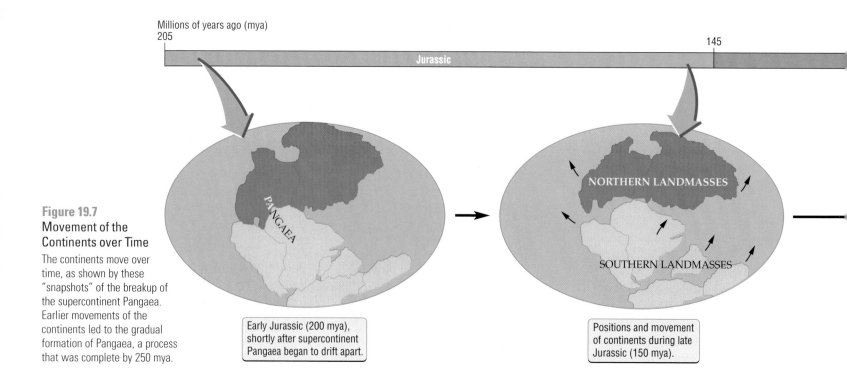

Figure 19.7 Movement of the Continents over Time

The continents move over time, as shown by these "snapshots" of the breakup of the supercontinent Pangaea. Earlier movements of the continents led to the gradual formation of Pangaea, a process that was complete by 250 mya.

Millions of years ago (mya)
205 — Jurassic — 145

PANGAEA

Early Jurassic (200 mya), shortly after supercontinent Pangaea began to drift apart.

NORTHERN LANDMASSES

SOUTHERN LANDMASSES

Positions and movement of continents during late Jurassic (150 mya).

This process can also cause bodies of land to break apart, as is currently happening in Iceland and East Africa. Second, where two plates collide, one can sink into the mantle below the other. This sinking action gradually pulls the rest of the plate along with it; the sinking plate slowly melts as it slides beneath the other plate.

Patterns of continental drift—most notably the breakup of the ancient supercontinent Pangaea—have had dramatic effects on the history of life. Pangaea began to break apart early in the Jurassic period (about 200 mya), ultimately separating into the continents we know today (see Figure 19.7). As the continents drifted apart, populations of organisms that once were connected by land became isolated from one another.

As we learned in Chapter 18, geographic isolation reduces or eliminates gene flow, thereby promoting speciation. The separation of the continents was geographic isolation on a grand scale, and it led to the formation of many new species. Among mammals, for example, kangaroos, koalas, and other marsupials that are unique to Australia evolved in geographic isolation on that continent, which broke apart from Antarctica and South America about 40 mya.

Continental drift also affects climate, which has a profound effect on organisms. Shifts in the positions of the continents alter ocean currents, and these currents have a major influence on the global climate. At various times, changes in the global climate caused by the movements of the continents have led to the extinctions of many species.

19.4 Mass Extinctions: Worldwide Losses of Species

As the fossil record shows, species have gone extinct throughout the long history of life. The rate at which this has happened—that is, the number of species that have gone extinct during a given period—has varied over time, from low to very high. At the upper end of this scale, the fossil record shows that there have been five **mass extinctions**, periods of time during which great numbers of species went extinct throughout most of Earth. Each of these upheavals left a permanent mark on the history of life, driving more than 50 percent of Earth's species to extinction (Figure 19.8). The causes of the five mass extinctions are difficult to determine, but are thought to include such factors as climate change, massive volcanic eruptions, asteroid impacts, changes in the composition of marine and atmospheric gases, and changes in sea levels. In addition to the five mass extinctions revealed by the fossil record, we may be entering a sixth, human-caused mass extinction today (see page 367 and Interlude A).

Of the five previous mass extinctions, the largest occurred at the end of the Permian period, 250 mya. The **Permian extinction** radically altered life in the oceans

See continental drift in action.

19.4

Explore the causes of continental drift.

19.5

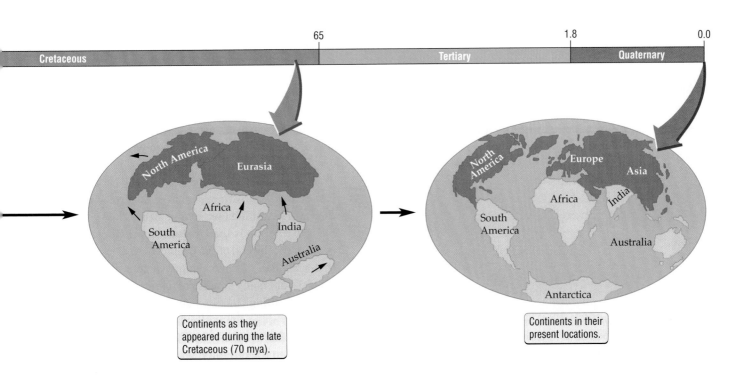

Continents as they appeared during the late Cretaceous (70 mya).

Continents in their present locations.

Figure 19.8 The Five Mass Extinctions

Each of these mass extinctions drastically reduced the diversity of marine and terrestrial animals, as shown here. Plant groups (not shown) were also severely affected.

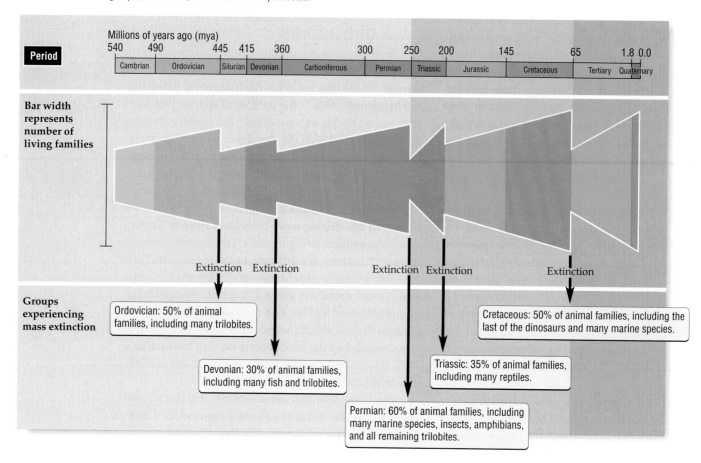

Ordovician: 50% of animal families, including many trilobites.

Devonian: 30% of animal families, including many fish and trilobites.

Permian: 60% of animal families, including many marine species, insects, amphibians, and all remaining trilobites.

Triassic: 35% of animal families, including many reptiles.

Cretaceous: 50% of animal families, including the last of the dinosaurs and many marine species.

Before   **After**

Figure 19.9 Before and After the Permian Mass Extinction

The Permian mass extinction wiped out nearly all of the rich variety of burrowing, bottom-dwelling, and swimming organisms that lived in the oceans before it occurred.

(Figure 19.9). Among marine invertebrates (animals without backbones), an estimated 50 to 63 percent of the existing families, 82 percent of the genera, and 95 percent of the species went extinct. The Permian mass extinction was also highly destructive on land. It removed 62 percent of the existing terrestrial families, brought the reign of the amphibians to a close, and caused the only major extinction of insects in their 400-million-year history (8 of 27 orders of insects went extinct).

Although not as severe as the Permian extinction, each of the other mass extinctions also had a profound effect on the diversity of life (see Figure 19.8). The best-studied mass extinction is the **Cretaceous extinction**, which occurred at the end of the Cretaceous period, 65 mya. At that time, half of the marine invertebrate species perished, as did many families of terrestrial plants and animals, including the dinosaurs. The Cretaceous mass extinction was probably caused at least in part by the collision of an asteroid with Earth. A 65-million-year-old, 180-kilometer-wide crater lies buried in sediments off the

Biology Matters

Is a Mass Extinction Underway?

The International Union for the Conservation of Nature (IUCN) maintains what it calls its Red List, which identifies the world's threatened species. To be defined as such, a species must face a high to extremely high risk of extinction in the wild. The 2004 Red List contains 15,589 species threatened with extinction, of a total of approximately 57,000 species assessed. Because this assessment accounts for less than 3 percent of the world's 1.9 million described species, the total number of species threatened with extinction worldwide would actually be much larger.

Among the major species groups that were assessed for the 2004 Red List, the number of threatened species ranges between 12 and 52 percent, with, for example, 12 percent of birds listed as threatened, 23 percent of mammals, 32 percent of amphibians, and 52 percent of cycads.

The Red List is based on an easy-to-understand system for categorizing extinction risk; it is also objective, yielding consistent results when used by different people. Because of these two attributes, the Red List is internationally recognized as an effective method to assess extinction risk. Table 1 shows some worldwide data for a few species groups that have been heavily assessed. In all four cases, the number of threatened species has risen since 1996–1998, and in the case of amphibians and gymnosperms, the increase has been dramatic.

For types of organisms other than those listed in Table 1, extinction risk has been evaluated for only a few known species; for example, it has been evaluated for only 771 out of 950,000 described insect species. If the threatened species listed in the table do go extinct, and the percentage of species under threat in other taxonomic groups turns out to be similar to those listed, then the percentage of species that will go extinct will approach those in some of the previous mass extinctions.

Table 2 contains data on the numbers of species that the 2004 Red List identified as threatened in North America.

Table 1

Group	Number of known species	Number of species evaluated for extinction risk	Number of threatened species in 1996–1998	Number of threatened species in 2004	Threatened species in 2004 (as a percentage of species evaluated)
VERTEBRATES					
Mammals	5,416	4,853	1,096	1,101	23 percent
Birds	9,917	9,917	1,107	1,213	12 percent
Amphibians	5,743	5,743	124	1,770	31 percent
PLANTS					
Gymnosperms	9,80	907	142	305	34 percent

Table 2

North America	Mammals	Birds	Reptiles	Amphibians	Fish	Mollusks	Other invertebrates	Plants	Total
Canada	16	19	2	1	24	1	10	1	**74**
Saint Pierre and Miquelon	0	1	0	0	1	0	0	0	**2**
United States	40	71	27	50	154	261	300	240	**1,143**

For more information about the 2004 Red List, visit www.redlist.org.

Yucatán coast of Mexico; this crater is thought to have formed when an asteroid 10 kilometers wide struck Earth. An asteroid of this size would have caused great clouds of dust to hurtle into the atmosphere; this dust would have blocked sunlight around the globe for months to years, causing temperatures to drop drastically and driving many species extinct.

The effects of mass extinctions on the diversity of life are twofold. First, as noted earlier, entire groups of organisms perish, changing the history of life forever. Second, the extinction of one or more dominant groups of organisms can provide new ecological and evolutionary opportunities for groups of organisms that previously were of relatively minor importance, thus dramatically altering the course of evolution.

19.5 Adaptive Radiations: Increases in the Diversity of Life

After each of the five mass extinctions, some of the surviving groups of organisms diversified to replace those that had become extinct. These bursts of evolution, which lasted roughly 10 million years each, were just as important to the future course of evolution as the extinctions themselves. For example, when the dinosaurs went extinct 65 mya, the mammals diversified greatly in size and in ecological role (Figure 19.10). If mammals had not diversified to replace the dinosaurs, humans probably would not exist, and the history of life over the past 65 million years would have been very different from what actually happened.

When a group of organisms expands to take on new ecological roles and to form new species and higher taxonomic groups, that group is said to have undergone an **adaptive radiation**. Some of the great adaptive radiations in the history of life occurred after mass extinctions, as when the mammals diversified to replace the dinosaurs. In such cases, the adaptive radiations may have been caused by the release from competition that occurs after a dominant group of organisms (such as the dinosaurs) goes extinct.

In other cases, adaptive radiations have occurred after a group of organisms has acquired a new adaptation that allows it to use its environment in new ways. The first terrestrial plants, for example, possessed adaptations that allowed them to thrive on land, a new and highly challenging environment. The descendants of those early colonists radiated greatly, forming many new species and higher taxonomic groups that were able to live in a broad range of new environments (from desert to arctic to tropical regions).

The term *adaptive radiation* can refer to relatively small evolutionary expansions, as seen in the radiation of finches in the Galápagos Islands (see Figure 16.12). It can also refer to much larger expansions, such as those that followed the movement of vertebrates onto land or the origin of flowering plants.

19.6 The Origin and Adaptive Radiation of Mammals

The fossil record shows that mammals evolved from reptiles. Living mammals differ from living reptiles in many respects, including the way they move (Figure 19.11), the nature of their teeth, and the structure of their jaws. However, it is difficult to draw the line between mammals and reptiles in the fossil record. Some fossil species have features that are intermediate between the two groups; such fossils provide a beautiful illustration of an evolutionary shift from one major group of organisms to another.

The mammalian jaw and teeth evolved from reptilian forms in three stages

Let's examine the evolution of mammals from reptiles from the perspective of two traits that fossilize well: the nature of their teeth and the structure of their jaws.

Compared with reptiles, mammals have complex teeth and jaws. For example, mammalian teeth differ considerably from one portion of the jaw to another: some teeth are specialized for tearing (incisors), others for hunting or defense (canines), still others for grinding (molars). In contrast, reptilian teeth change little in form or function from one position along the jaw to another (Figure 19.12a).

With respect to jaws, the reptilian jaw has a hinge at the back for the attachment of muscles that simply snap the top and bottom of the jaw together. In mammals, the hinge has moved forward and is controlled by strong cheek muscles, including some positioned in front of the hinge. As a result, mammalian jaws are more powerful and more accurate than reptilian jaws. To see why this is so, think of how well you could close a door by pulling it shut with a handle located near its hinges (analogous to the reptilian method) versus

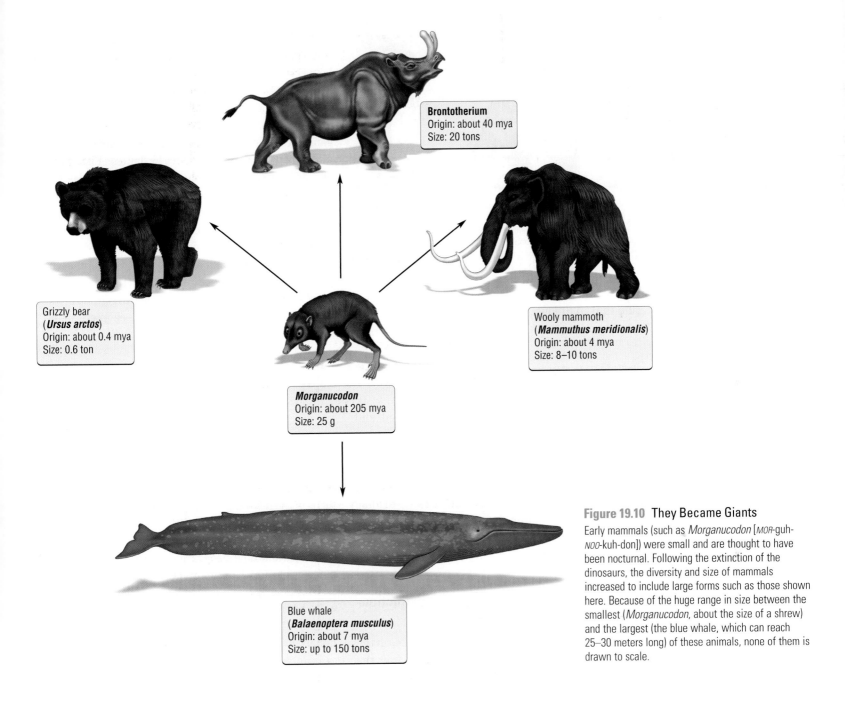

Figure 19.10 They Became Giants
Early mammals (such as *Morganucodon* [MOR-guh-NOO-kuh-don]) were small and are thought to have been nocturnal. Following the extinction of the dinosaurs, the diversity and size of mammals increased to include large forms such as those shown here. Because of the huge range in size between the smallest (*Morganucodon*, about the size of a shrew) and the largest (the blue whale, which can reach 25–30 meters long) of these animals, none of them is drawn to scale.

Brontotherium
Origin: about 40 mya
Size: 20 tons

Grizzly bear
(**Ursus arctos**)
Origin: about 0.4 mya
Size: 0.6 ton

Wooly mammoth
(**Mammuthus meridionalis**)
Origin: about 4 mya
Size: 8–10 tons

Morganucodon
Origin: about 205 mya
Size: 25 g

Blue whale
(**Balaenoptera musculus**)
Origin: about 7 mya
Size: up to 150 tons

pulling it shut with a handle located away from the hinges, closer to the actual position of a doorknob (analogous to the mammalian method).

How did these differences in the teeth and jaws of mammals and reptiles arise? The fossil record shows that these changes arose gradually, over the course of about 80 million years (from about 300 mya to 220 mya). During this time, there were three key steps in the transition from reptile to mammal, as shown in Figure 19.12. First, a group of reptiles evolved to have an opening in the bones behind the eye; this opening is called the tem-

poral fenestra. In the living animal, a muscle passed through this opening and served to increase the power with which the jaw could be closed. Second, a group of reptiles known as the therapsids [thuh-RAP-sidz] evolved a larger temporal fenestra (and hence more powerful jaw muscles), and their teeth showed the first signs of specialization.

The third step occurred when jaws very similar to mammalian jaws arose in one subgroup of the therapsids, the **cynodonts** [SIGH-no-donts] (see Figure 19.11). In these animals—the last in a long line of mammal-like

Helpful to know

Cynodont comes from Greek roots meaning "dog-toothed." In cynodonts, the canine teeth (from *canis*, Latin for "dog") were prominent, though not as large as in some of their predecessors.

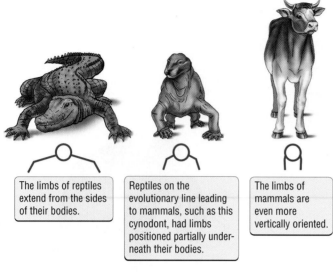

| The limbs of reptiles extend from the sides of their bodies. | Reptiles on the evolutionary line leading to mammals, such as this cynodont, had limbs positioned partially underneath their bodies. | The limbs of mammals are even more vertically oriented. |

Figure 19.11 A Gradual Change in Gait
The legs of most living reptiles stick out to the sides of their bodies, giving them a sprawling gait. Over time, the legs of mammal-like reptiles became positioned under the body, leading eventually to the vertical orientation of the legs and the upright gait of living mammals.

reptiles—the teeth became still more specialized, and the hinge of the jaws moved forward, using a different set of bones than in other reptiles (Figure 19.12c–e). Changes in the jaw hinge are particularly clear in some cynodont species (Figure 19.12e) that had jaws with both a reptilian hinge (in reduced form) and a mammal-like hinge. Over time, mammals lost the reptilian hinge (Figure 19.12g), the bones of which evolved to become bones in the inner ear.

Mammals increased in size after the extinction of the dinosaurs

There were many species of mammal-like reptiles in the early Triassic period, 245 mya. By 200 mya, however, the mammal-like reptiles had declined as other reptiles, most notably the dinosaurs, came to dominate Earth. Although the mammal-like reptiles became extinct, they left behind the first mammals as their descendants. The earliest mammals, which were small, rodent-sized organisms, evolved about 220 mya, at roughly the same time as the first dinosaurs.

Throughout the long reign of the dinosaurs, most mammals remained small. Many appear to have been nocturnal (active at night) because they had large eye sockets, as do many living nocturnal organisms. By being nocturnal and small, early mammals may have been to dinosaurs what a mouse is to a lion: hard to notice and too small to eat.

Fossil and genetic evidence suggests that several of the orders of living mammals diverged from one another between 100 and 85 mya, well before the extinction of the dinosaurs. But most of the major radiations within these and other groups of mammals did not occur until after the dinosaurs went extinct (65 mya). After the dinosaurs were gone, the mammals radiated greatly to include many new forms that were large and active by day (see Figure 19.10). Some land mammals reached enormous sizes. An example is the extinct Beast of Baluchistan, which was over three times as large as an elephant. Other mammals (the whales; see Figure 19.2) became specialized for life in water, while still others (the bats) became specialized for flight and hunting at night. Finally, one group of mammals (the primates) became specialized for life in trees and evolved especially large brains. We will examine this group in more detail in Interlude D, which follows this chapter.

19.7 An Overview of the Evolutionary History of Life

As we have seen, the history of life on Earth can be summarized by the rise and fall of major groups of protists, plants, and animals (see Figure 19.3). These broad patterns in the history of life are caused by the extinction or decline of some groups and the origin or expansion of other groups. Taken together, mass extinctions and adaptive radiations have been largely responsible for shaping macroevolution. Continental drift also plays an important role, in that the movement of the continents can promote both mass extinctions and adaptive radiations.

In addition to offering this broad view of the evolutionary history of life, we want to emphasize two important and related concepts: how major new groups of organisms evolve, and the difference between the evolution of populations (microevolution) and the evolution of higher taxonomic groups (macroevolution).

Major new groups of organisms evolve from existing groups

The Nobel prize-winning geneticist François Jacob [*YAH-kub*] described evolution as similar to "tinkering"; that is, as a process in which existing forms are modified and adjusted slightly. This view of evolution is especially relevant when considering the origin of major new groups of organisms. New groups do not arise from scratch, but rather arise as modifications of organisms that already exist. The origin of mammals from reptiles provides an excellent example of this process. Mammals have novel features, such as specialized teeth and powerful jaws, but the fossil record reveals that these features arose as gradual modifications of the teeth and jaws found in reptiles (see Figure 19.12). These modifications occurred in the absence of a plan or design; instead, over time, bones and teeth that served one purpose in reptiles were gradually put to new uses in mammals.

Macroevolution differs from the evolution of populations

As we saw in Chapter 18, the often exquisite adaptations of organisms result from natural selection. Evolution by natural selection is a short-term process: adaptations are shaped by natural selection to match the organism's current environment. Can natural selection provide a complete explanation of macroevolution, the rise and fall of higher taxonomic groups? The answer is no, in part because natural selection does not operate in a vacuum. Large-scale processes, such as the movements of continents, influence which features of organisms are adaptive and which are not; thus an understanding of natural selection alone is not sufficient to understand patterns of evolution above the species level.

Furthermore, mass extinctions can remove entire groups of organisms, seemingly at random—even those that possess unique and highly advantageous adaptations. A group of predatory gastropods (snails and their relatives), for example, went extinct in the Triassic mass extinction, shortly after they had evolved the ability to drill through the shells of other gastropods. The ability to drill through shells had opened up a major new way of life for these organisms (see the box on page 372). If these

Figure 19.12 From Reptile to Mammal

Over an 80-million-year period, the jaws and teeth of the reptilian ancestors of mammals gradually changed to resemble those of living mammals. In addition to those shown here, there are dozens of other fossil species of mammal-like reptiles—species with features that are intermediate to those of reptiles and mammals. When fossils from all of these species are lined up next to one another, the transition from reptile to mammal appears very smooth. Here [red] shows the size and position of the dentary bone, which ultimately formed the entire lower jaw in mammals. The muscles that close the jaw pass through an opening called the temporal fenestra (tf); a larger temporal fenestra allows for larger and more powerful jaw muscles. In reptiles (a), the hinge of the jaw is formed by the articular/quadrate (art/q) bones; in mammals (g), the hinge is formed by two entirely different bones, the dentary/squamosal (d/sq). Advanced cynodonts and early mammals had two hinges, the reptilian (art/q) hinge and the mammalian (d/sq) hinge.

(a) Ancestral reptile (*Haptodus*)

These reptiles had large jaw muscles, multiple bones in the lower jaw, and single-point teeth.

(b) Therapsid (*Biarmosuchus*)

Therapsids had large canine teeth, long faces, and a single hinge (art/q) at the back of the jaw.

(c) Early cynodont (*Procynosuchus*)

In early cynodonts, the temporal fenestra was further enlarged, allowing for very powerful jaw muscles.

(d) Cynodont (*Thrinaxodon*)

The dentary (the major jaw bone, colored in red) became enlarged, and the back teeth had multiple cusps.

(e) Advanced cynodont (*Probainognathus*)

In advanced cynodonts, complex, multi-cusped teeth enhanced chewing. The jaw had two hinges, art/q and d/sq.

(f) *Morganucodon* (early mammal)

Morganucondon had typical mammalian teeth. The jaw had two hinges, but the reptilian art/q hinge was reduced (not visible in this diagram).

(g) Tree shrew (*Tupaia*)

In tree shrews and other mammals, the teeth are highly specialized. The lower jaw is composed of a single bone, and the jaw has one hinge (d/sq).

Biology on the Job

Geerat Vermeij: A Hands-On Approach to Evolution

Geerat J. Vermeij [GHEER-aht ver-MAY] is Distinguished Professor of Geology at the University of California, Davis, where he studies the history of life on Earth. Dr. Vermeij has wide-ranging interests in evolutionary biology. He has written over a hundred scientific papers and five books, including Evolution and Escalation: An Ecological History of Life *and an autobiography entitled* Privileged Hands: A Scientific Life. *Dr. Vermeij has been blind since he was 3 years old, when his eyes were removed to prevent a rare disease from causing neural damage.*

What do you do during a typical day's work? On a typical day I get up, have breakfast, and then write for an hour or so before going to the office, where I spend my time reading scientific publications (10 to 30 per week), including manuscripts I am reviewing for journals; talking to students; giving an hour's lecture; doing research in and working on my research collection; and corresponding with colleagues and often members of the public.

How did you become interested in studying the history of life? What was your first job as an evolutionary biologist? My interests in science and natural history date back to my earliest childhood. I began collecting shells seriously at age 10, at which point I knew in general what I wanted to do with my life. My scientific interests were greatly encouraged throughout, but especially at Princeton during my undergraduate years and at Yale as a Ph.D. student. My first job was as instructor—later as assistant professor to full professor—in the Department of Zoology at the University of Maryland, College Park.

Your research often takes you to remote locations in the field. What sort of places have you visited? I have been fortunate to have visited every continent except Antarctica, and numerous islands ranging from Iceland and the Aleutians to many Caribbean islands, Hawaii, the Society Islands, Guam, and Palau. In all these places I have carried out extensive fieldwork. I have collected fossils in places ranging from Panama and Florida to the Aleutians, and I have worked in marine laboratories in many places, including Panama, Jamaica, Hawaii, Japan, and aboard research ships.

What are some of the more unusual or hair-raising situations that you have experienced in the field? While aboard the R.V. *Alpha Helix* in Indonesia and the Philippines, we were stopped three times by military patrols, one of which came close to arresting my wife and me in the belief we were Taiwanese poachers. In the northern Mariana Islands, I jumped from a small boat onto a slippery lava shore in substantial surf and ended up scarring various parts of my body. I have also been stung by stingrays in Panama and bitten by moray eels at Moorea in the Society Islands.

You have a legendary ability to learn about the ecology of a place and the history of life by touching, smelling, and listening to nature. Most scientists rely primarily on visual observations. To what extent does your use of senses other than sight allow you to discover new things about the natural world? While in the field, I use every sense available to me to observe and to keep track of my surroundings. My hands are essential in quickly identifying desirable specimens and potentially dangerous animals; I carefully listen to patterns of water movement for oncoming waves or other sudden changes. My hands, fingers, fingernails, and tactile extensions such as pins are essential for making detailed observations on shells or other natural objects. For the kinds of comparative research I do, it is essential to have a good working memory, so that I can compare one place with another I may have visited long before. Observation and comparison are at the heart of my science, done while I think about what I am observing in relation to previous experiences.

What do you think is the most exciting and important discovery or advance in understanding about the history of life that has occurred in the last 10 or 20 years? With advances in our understanding of patterns of evolutionary descent, gained in part from fossils and in part from analyses of DNA sequences in living species, we have come to grips with major evolutionary trends and patterns through time, and we are unraveling the events surrounding the origins and early diversification of animals during and just before the Cambrian period.

Shell Collection of Geerat Vermeij

Is the history of life important for the general public? Why do you feel this way? The history of life should interest the public at large for several reasons. First, it provides the most direct evidence of evolution and reveals how organisms over the past 3.5 billion years have reacted to and created major environmental changes. This history provides the only long-term record of the consequences of extinction and diversification and enables us to study species invasion over the long run. Moreover, it places human actions and capacities in the larger perspective of the earlier history of life. For the imagination, the history of life transports us back in time to a very unfamiliar world, yet one in which all the principles and processes that govern life today were also in force.

What do you enjoy most about your job? Discovering new things, either through my own original research or by reading others' work, provides novelty every single day. I enjoy imparting information and insight, both through writing and by lecturing, and I thoroughly enjoy working with students as they become independent scientists and scholars.

Is there anything else you'd like to tell our readers? A burning curiosity about what is in the world and how the world works is what drives my life, greatly and importantly augmented by a strong sense of aesthetics. I want to know, to understand, to impart, to marvel. My interests range widely from ancient organisms to human affairs, even if the shells of fossil and living molluscs remain my first scientific love. Science (and scholarly work of all kinds) is a communal effort; I therefore have a great interest in preserving and improving scientific ethics and standards, and in involving professionals and amateurs alike in the pursuit of knowledge and understanding.

The Job in Context

In over 30 years of studying fossil shells, Dr. Vermeij has used his hands to examine how the history of life in the oceans was influenced by ongoing battles between predator and prey. His results show that over long periods of time, predators such as crabs and shell-drilling snails evolved increasingly sophisticated ways to crush, pry open, or drill through the shells of their prey. And the prey responded in kind, developing tougher, better-defended shells. Dr. Vermeij realized that this evolutionary "arms race" had proceeded in a single direction for hundreds of millions of years, shaping the history of life in the oceans. His hands revealed a fundamental story about life in the oceans, a story that no sighted biologist had discovered before him.

---

predatory gastropods had not gone extinct, they probably would have thrived and formed many new species that had the ability to drill through shells (as did another group of shell-drilling gastropods that evolved 120 million years later). As this example shows, even species that have highly beneficial adaptations don't always survive over the long term.

Overall, broad patterns in the history of life cannot be predicted solely from an understanding of how populations evolve (microevolution). For a full comprehension of the history of life on Earth, we must consider factors such as mass extinctions, adaptive radiations, and continental drift, all of which can have a tremendous effect on evolution above the species level (macroevolution).

Oyster drill

When Antarctica Was Green

Antarctic fossils of dinosaurs, forest trees, and tropical marine organisms are vivid testimony to the fact that we live in a dynamic world. These fossils reveal great changes over time, ranging as they do from Cambrian marine organisms to early land plants to birds and mammals. The very different organisms that have lived in Antarctica at different times illustrate the broad changes in the history of life described in this chapter, such as the Cambrian explosion, the colonization of land, and the periods of domination by amphibians, reptiles, and mammals.

The Antarctic fossils also show the striking contrast between the diverse life forms that once lived in Antarctica and the few that live there today. The small number of present-day organisms in Antarctica is due in part to continental drift. As Pangaea broke apart, ocean currents were rerouted, causing Earth's climate to grow colder and ice caps to form at the poles. This process was especially pronounced in Antarctica, which experienced an ever-colder climate as it moved toward its present position over the South Pole. Once Antarctica separated from

Australia and South America, about 40 mya, the organisms on Antarctica were trapped there. Thus, as the climate of Antarctica became increasingly cold, most Antarctic species perished.

However, the same continental movements that brought destruction to Antarctica sowed the seeds of creation elsewhere. The rerouting of ocean currents that contributed to the formation of the Antarctic ice cap also produced the largest differences in temperature between the poles and the tropics that Earth has ever known. The wide range of new habitats that resulted from these temperature differences helped set the stage for adaptive radiations in many organisms, including humans.

Chapter Review

Summary

19.1   The Fossil Record: A Guide to the Past

- The fossil record documents the history of life on Earth.
- Fossils reveal that past organisms were unlike living organisms, that many species have gone extinct, and that there have been great changes in the dominant groups of organisms over time.
- The order in which organisms appear in the fossil record is consistent with our understanding of evolution gained from other kinds of evidence, including morphology and DNA sequences. Sometimes the approximate age of a fossil can be determined through analysis of radioisotopes.
- Although the fossil record is not complete, it provides excellent examples of the evolution of major new groups of organisms.
- Recently discovered fossils show that whales are most closely related to artiodactyls (a group of hoofed mammals), supporting the results of genetic analyses.

19.2   The History of Life on Earth

- The first single-celled organisms resembled bacteria and evolved at least 3.5 billion years ago.
- About 2.75 billion years ago, some groups of bacteria evolved the ability to conduct photosynthesis, which releases oxygen as a waste product.
- The release of oxygen by photosynthetic bacteria caused oxygen concentrations in the atmosphere to increase. Rising oxygen concentrations made possible the evolution of single-celled eukaryotes about 2.1 billion years ago. Multicellular eukaryotes followed about 650 mya.
- Life on Earth changed dramatically during the Cambrian explosion (530 mya), when large predators and well-defended herbivores suddenly appear in the fossil record.

- The land was first colonized by plants and invertebrates (about 500 and 410 mya, respectively), which were followed later by vertebrates (about 365 mya).

19.3   The Effects of Continental Drift

- Continental drift has had profound effects on the history of life on Earth.
- The separation of the continents over the past 200 million years has led to geographic isolation on a grand scale, promoting the evolution of many new species.
- At different times, climate changes caused by the movements of the continents have led to the extinctions of many species.

19.4   Mass Extinctions: Worldwide Losses of Species

- There have been five mass extinctions during the history of life on Earth.
- The extinction of some groups and the survival of others greatly alters the subsequent course of evolution.
- The extinction of a dominant group of organisms can provide new opportunities for other groups.

19.5   Adaptive Radiations: Increases in the Diversity of Life

- The history of life has been heavily influenced by adaptive radiations, in which a group of organisms diversifies greatly and takes on new ecological roles.
- Adaptive radiations can be caused by the release from competition that follows a mass extinction.
- Adaptive radiations can also occur when a group of organisms evolves a new adaptation that allows them to fill new ecological roles.

19.6   The Origin and Adaptive Radiation of Mammals

- Mammals evolved from reptiles over the course of about 80 million years (300 to 220 mya). During this time, a group of reptiles gradually evolved mammalian features such as vertically oriented legs, specialized teeth, and powerful jaws.

- The first mammals evolved from cynodonts, the last of a line of mammal-like reptiles. Mammals and dinosaurs originated at roughly the same time (220 mya), about 20 million years before dinosaurs became the dominant land vertebrates.
- Throughout the long reign of the dinosaurs (200 mya to 65 mya), most mammals remained small and nocturnal. Following the extinction of the dinosaurs, the mammals radiated to include many species that were large and active by day.

19.7 An Overview of the Evolutionary History of Life

- Major new groups of organisms do not evolve from scratch, but rather arise through a series of modifications of existing organisms.
- Macroevolution cannot be predicted solely from an understanding of the mechanisms that cause microevolution.

⦿ Review and Application of Key Concepts

1. The fossil record provides clear examples of the evolution of new groups of organisms from previously existing organisms. Describe the major steps of one such example.

2. How did the evolution of photosynthesis affect the history of life on Earth?

3. What is the Cambrian explosion, and why was it important?

4. Life arose in water. Explain why the colonization of land represented a major evolutionary step in the history of life. What challenges—and opportunities—awaited early colonists of land?

5. Mass extinctions can remove entire groups of organisms, seemingly at random—even groups that possess highly advantageous adaptations. How can this be?

6. Evidence from the fossil record indicates that it usually takes 10 million years for adaptive radiation to replace the species lost during a mass extinction. Discuss this observation in light of your understanding of the speciation process (see Chapter 18). What does it suggest about the consequences of the losses of species that are occurring today?

7. Is macroevolution fundamentally different from microevolution? Can macroevolutionary patterns be explained solely in terms of microevolutionary processes? Does any evolutionary mechanism that we have studied link macroevolution and microevolution?

Key Terms

adaptive radiation (p. 368)
Cambrian explosion (p. 361)
continental drift (p. 364)
Cretaceous extinction (p. 366)
cynodont (p. 369)
macroevolution (p. 358)
mass extinction (p. 365)
Permian extinction (p. 365)
radioisotope (p. 358)

Self-Quiz

1. Continental drift
   a. can occur when liquid rock rises to the surface and pushes the continents away from one another.
   b. no longer occurs today.
   c. has led to the geographic isolation of many populations, thus promoting speciation.
   d. both a and c

2. The fossil record
   a. documents the history of life.
   b. provides examples of the evolution of major new groups of organisms.
   c. is not complete.
   d. all of the above

3. Mass extinctions
   a. are always caused by asteroid impacts.
   b. are periods of time in which many species go extinct worldwide.
   c. have little lasting effect on the history of life.
   d. affect only terrestrial organisms.

4. The Cambrian explosion
   a. caused a spectacular increase in the size and complexity of animal life.
   b. caused a mass extinction.
   c. was the time during which all living animal phyla suddenly appeared.
   d. had few consequences for the later evolution of life.

5. The history of life shows that
   a. biodiversity has remained constant for about 400 million years.
   b. extinctions have little effect on biodiversity.
   c. macroevolution is greatly influenced by mass extinctions and adaptive radiations.
   d. macroevolution can be understand solely in terms of the evolution of populations.

6. _____ are radioactive forms of elements that decay to more stable forms over time.
   a. X-rays
   b. Carbon-12 and carbon-14
   c. Radioisotopes
   d. Adaptive radiations

7. Evolution above the species level, characterized by the rise and fall of major groups of organisms, is called
   a. macroevolution.
   b. microevolution.
   c. mass extinction.
   d. adaptive radiation.

8. Which of the following terms most specifically describes what occurs when a group of organisms expands to take on new ecological roles, forming new species and higher taxonomic groups in the process?
   a. speciation
   b. evolution
   c. mass extinction
   d. adaptive radiation

Biology in the News

When Bats and Humans Were One and the Same

BY CARL ZIMMER

Scientists have used computer analysis to read evolution backward and reconstruct a large part of the genome of an 80-million-year-old mammal. This tiny shrewlike creature was the common ancestor of humans and other living mammals as diverse as horses, bats, tigers, and whales.

Actual DNA molecules cannot survive such lengths of time. Mammal fossils from this period are extremely rare. But by tracking the course of mammalian evolution, scientists can pinpoint when a common ancestor existed and what, in general terms, it was like.

In a scientific study that may conjure visions of the movie *Jurassic Park*, researchers have used a computer program to reconstruct part of the genetic blueprint of an early mammal that was an ancestor to many living mammals. Previous studies on the fossils and DNA of mammals indicated that this early mammal lived 80 million years ago and was a small, nocturnal organism that resembled a shrew.

The new study, published in the December 2004 issue of *Genome Research*, used a computer program to estimate the sequence of a 1.1-million-base-long region of this early mammal's DNA. Dr. David Haussler, a Howard Hughes Medical Institute investigator at the University of California, Santa Cruz, and his colleagues used an approach that was similar to comparing medieval manuscripts, each of which has its own set of copying errors, and trying to reconstruct the original text. By comparing the analogous 1.1 million bases from 19 living mammal species, the researchers found all of the places where a mutation had occurred in this segment of DNA. By working backward from the resulting 19 different sets of mutations—one set for each species studied—Dr. Haussler and his colleagues determined what they think is the original DNA sequence of our shrewlike ancestor.

A nagging problem in a study like this one is accuracy—how could the researchers possibly know if the computer program they used did a good job of estimating ancestral DNA? To find out, the scientists created mock ancestral DNA sequences and then allowed those sequences to mutate along several simulated evolutionary branches. Next, they fed the mutated sequences into their computer program, challenging it to estimate the original (mock) ancestral sequence. In a repeated series of such tests, they found that the program had a 98.5 percent accuracy rate—much higher than they had hoped.

Will scientists soon build an ancient mammal from its DNA sequence? Not likely, says Dr. Haussler. He notes that making an organism from its DNA would be a "fascinating challenge, but it's far beyond our capability at this point." Instead, scientists plan to reconstruct the sequences of an entire series of extinct genomes. Dr. Haussler's approach could be used, for example, to predict the sequence of the genome of the common ancestor of primates. From that sequence, the key genetic changes at different branch points in our own evolutionary history could be determined. Scientists also hope to insert the DNA sequences of ancient mammalian genes into mice. The products of those genes may provide important clues to the brain chemistry, color vision, and overall metabolism of early mammals.

Evaluating the news

1. The mistaken idea that humans once were the same as bats (as stated in the title of the *New York Times* article) can give the impression that we descended from bats. That is not true—the evolutionary path leading to bats has been separate from the path leading to humans for tens of millions of years. You may have seen similar statements that imply (also incorrectly) that humans descended from chimpanzees. Do you think that such statements contribute to public misunderstanding of how evolution works, or cause some people to dismiss evolution as absurd?

2. Let's assume that scientists succeed in reconstructing the genome of the extinct common ancestor of living primates. If the result makes it possible to identify how genes in the evolutionary lineage leading to humans differ from those in the evolutionary lineages leading to other primates, how should we use that information? Do you think people would be tempted to improve our own species, or to modify other primate species (for example, to make them more humanlike)? Would such modifications of our own or other species be ethical?

3. Several of the possible future research projects described in this article would be expensive. Should tax dollars be used to pay for such research? If so, why? If not, why not?

SOURCE: *The New York Times,* Tuesday, December 7, 2004.

Humans and Evolution

Who Are We? Where Do We Come From?

Imagine the commotion that would result if a group of Neandertals were to stroll down the street today. Neandertals had large, arching ridges above their eyes, a low, sloping forehead, no chin, and a face that, compared with ours, looked as if it had been pulled forward. Without shirts, they would be even more striking. Slightly shorter than humans of today, Neandertals had thick necks, were heavily boned, and, as indicated by markings on their skeletons where their muscles were attached, were very strong (Figure D.1).

Neandertals take their name from the Neander Valley in eastern Germany (*tal* comes from the German for "valley"). A fossilized skeleton discovered there in 1856 came to be known as the Neandertal Man. At first, these bones were the subject of great debate: some people thought them to be those of a bear, while others argued that they were the remains of an ancient human. Finally, after similar fossils were found in other places, scientists became convinced that the bones were indeed of human origin.

The Neandertal fossils shook our understanding of ourselves, for they provided dramatic proof that different forms of humans once existed. The fossils of our human and humanlike ancestors provide an eerie sense of recognition, for they reveal creatures that were like us, yet not like us. What do such fossils tell us about who we are and where we came from? Do findings from the fossil record match results from analyses of the human genome? In this essay we examine questions like these as we explore our evolutionary past and speculate about our evolutionary future. We also consider how human actions affect evolution, both in ourselves and in other species.

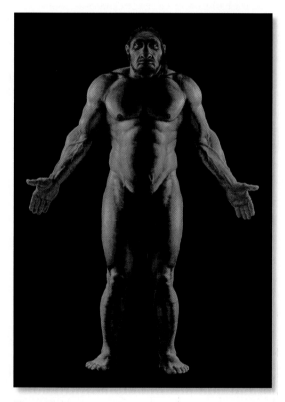

Figure D.1
An Artist's Interpretation of a Neandertal's Appearance

We Are Apes

Human beings—*Homo sapiens*—are animals, classified as members of the chordate phylum, the mammal class, the primate order, and the hominid family. We share with all other mammals certain unique features, including body hair (which provides insulation in many mammals) and milk produced by mammary glands. Like all **primates**, we also have flexible shoulder and elbow joints, five functional fingers and toes, thumbs that are **opposable** (that is, they can be placed opposite other fingers), flat nails (instead of claws), and brains that are large in relation to our body size.

As we think about the type of animal we are, we begin to address the questions posed at the beginning of this essay: Who are we? Where do we come from? For several hundred years, we viewed ourselves as very distinct from other animals. Although we classified ourselves as primates, we placed ourselves in one family, the hominid family, and the species most similar to us—chimpanzees, gorillas, and orangutans—in a separate family, the pongid family. The decision not to place any other living species in our family both reflected and reinforced our view of ourselves as different from other animals.

After the publication of Darwin's *The Origin of Species*, people realized that as animals, we descended from earlier animals. But even so, we continued to view ourselves as distinct from other animals. For example, before the early 1960s, when genetic analyses were first used to determine evolutionary relationships among primates, it was widely believed that the evolutionary lineage leading to humans split from the lineage leading to chimpanzees about 30 million years ago. Thus, although scientists recognized that we shared a common ancestor with chimpanzees and other apes, they believed that humans had evolved separately from the apes for tens of millions of years.

As we shall see, in the past 40 years, genetic analyses and a series of spectacular fossil discoveries have changed that view. Scientists now believe that the human–chimpanzee divergence occurred much more recently, about 5–7 million years ago (Figure D.2). Similarly, a combination of genetic analyses and fossil discoveries suggest that the evolutionary lineage leading to humans diverged from that leading to gorillas about 7–8 mya, and from the lineage leading to orangutans about 12–16 mya.

In fact, genetic analyses published in the past few years suggest that our relationship to the apes is so close as to be entangled. Our DNA is very similar to the DNA of apes, especially chimpanzees. On average, the DNA sequence of a human differs from the DNA sequence of a chimpanzee at about 1 percent of the nucleotide bases. In addition, our chromosomes contain a complex mix of DNA. Most of our DNA is closer in sequence to that of chimpanzees than to that of other apes, but in some regions, our DNA is more like that of gorillas. In still other chromosomal regions, the chimpanzees and gorillas are more closely related to each other than either is to us.

The broad conclusion that emerges from studies published over the last 40 years is that we are not just closely

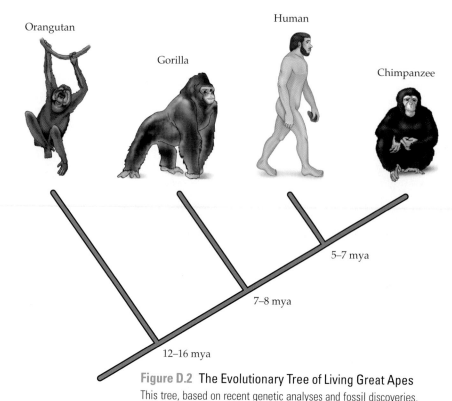

Figure D.2 The Evolutionary Tree of Living Great Apes
This tree, based on recent genetic analyses and fossil discoveries, shows that humans are most closely related to chimpanzees.

Orangutan
Gorilla
Human
Chimpanzee

5–7 mya
7–8 mya
12–16 mya

related to apes, we *are* apes. Some commentators have suggested that if space aliens existed and observed life on Earth, they would classify humans as the "fourth chimpanzee" (there are three living species of chimpanzees). Be that as it may, we share many characteristics with apes, especially chimpanzees, including the use of tools, a capacity for symbolic language, the performance of deliberate acts of deception, and, it seems, a sense of self-awareness (Figure D.3). These similarities—as well as our many differences—make the story of how humans evolved all the more interesting. We begin that story by describing the origin of the first **hominids**, the family that contains humans and our now extinct, human-like ancestors.

Hominid Evolution: From Climbing Trees to Walking Upright

Primates are thought to have originated 80–65 mya from small nocturnal mammals, similar to tree shrews, that ate insects and lived in trees. The fossil evidence of primate origins is sketchy, however, and the first definite

Tree shrew

primate fossils are 56 million years old. These early primates resembled modern lemurs (see Figure 2.11). Over time, the primates diversified greatly, eventually giving rise to the first hominids, roughly 5–7 mya.

Hominids have large brains, an upright walking posture, and complex toolmaking behaviors. Of these traits, our intelligence and toolmaking abilities, and the cultures associated with them, are central to what it means to be human. In an evolutionary sense, however, the increases in our intelligence and toolmaking abilities were secondary changes: they occurred relatively late in our evolutionary history and were part of a general trend toward large brain size in primates. The first big step in human evolution was the shift from being quadrupedal (moving on four legs) to being **bipedal** (walking upright on two legs), a change that occurred long before hominids evolved large brains. Many skeletal changes accompanied the switch to walking upright, including the loss of opposable toes (the big toe is opposable in all primates except humans).

It is not necessary—or even feasible—to walk upright in a tree, and for an organism that lived primarily in trees, the loss of an opposable big toe would be a handicap. On the ground, however, walking upright would have provided several advantages, including freeing the hands for carrying objects or using tools and improving the line of sight (that is, being able to see over nearby objects). Thus it is likely that the evolution of an upright posture was

Figure D.3 Our Closest Living Relative
Actress Dorothy Lamour and her co-star, Jiggs, on the set of the 1938 movie *Jungle Love*. Note the similarity in posture, especially the positioning of the legs and hands.

linked to a switch from life in the trees to life on the ground, a change that probably occurred between 8 and 5 mya.

The shift to life on the ground was probably not sudden or complete. The skeletal structure of some of the oldest fossil hominids (dating from 4.4 mya) indicates that they walked upright. However, foot bones and fossilized footprints from 3 to 3.5 mya show that the hominids living at that time still had partially opposable big toes (Figure D.4), perhaps because they continued to use trees some of the time.

The earliest known hominid is *Sahelanthropus tchadensis* [sah-heel-*AN*-thro-puss chuh-*DEN*-siss]. A 6- to 7-million-year-old skull of this species was discovered by Michel Brunet in 2002 (Figure D.5). Other early hominids include *Ardipithecus ramidus* [ar-dih-*PITH*-eh-kuss *RAM*-uh-duss] (5.8–4.4 mya) and several *Australopithecus* [*AW*-strah-loh-*PITH*-uh-kuss] species that lived between 4.2 and 3.0 mya. Not much is known about *Sahelanthropus*, but *Ardipithecus* and *Australopithecus* are thought to have walked upright. However, the size of their brains (as measured by the volume of the braincase) was relatively small, and their skulls and teeth were more similar to those of other apes than to those of humans. (Compare the skull of *A. afarensis*—one of the early *Australopithecus* species—with the skull of *Homo sapiens* in Figure D.6.)

Evolution in the Genus *Homo*

The oldest *Homo* fossil fragments were found in Africa and date from 2.4 mya, suggesting that the earliest members of the genus *Homo* originated in Africa 2 to 3 million years ago. More complete early *Homo* fossils exist from the period 1.9–1.6 mya; these fossils have been given the species name *Homo habilis* [*HAB*-uh-liss]. The oldest *H. habilis* fossils resemble those of *Australopithecus africanus*, a slightly more recent hominid than *A. afarensis* (see Figure D.6). In more recent *H. habilis* fossils, the face is not pulled forward as much, and the skull is more rounded. In these and other ways, more recent *H. habilis* specimens have features that are intermediate between those of *A. africanus* and *Homo erectus*, a species that evolved after *H. habilis*. Thus *H. habilis* fossils provide an excellent record of the evolutionary shift from ancestral hominid characteristics (in *Australopithecus*) to more recent ones (in *H. erectus*).

Taller and more robust than *H. habilis*, *H. erectus* also had a larger brain and a skull more like that of modern humans (see Figure D.6). It is likely that by 500,000 years ago *H. erectus* could use, but not necessarily make, fire. In addition, they probably hunted large species of game animals. The evidence to support the latter conclusion includes the remarkable discovery in Germany of three 400,000-year-old spears, each about 2 meters long and

See the hominid species timeline.

Figure D.4 Early Hominids Had Partially Opposable Big Toes

(*a*) Foot bones of early hominid

Partially opposable big toe

These hominid foot bones, 3 to 3.5 million years old, were discovered in 1995.

These bones (shown in tan) are based on fossils of a similar age.

(*a*) Fossilized hominid foot bones showed that some hominids living 3.5–3 mya walked upright, but had partially opposable big toes. (*b*) These footprints of two early hominids walking upright, side by side, were found in Africa. (*c*) Chimpanzees, the closest living relatives of humans, have fully opposable big toes.

Figure D.5 The Million Dollar Skull

Paleontologist Michel Brunet is holding what has been called the find of the century: a nearly complete skull of the oldest known hominid, *Sahelanthropus tchadensis*. (The species' name refers to the fact that the fossil was found in the Sahel region of Chad.) For this discovery, Dr. Brunet received a $1 million Dan David prize; three of these prizes are awarded each year for achievements that have a major scientific, technological, cultural, or social effect on our world.

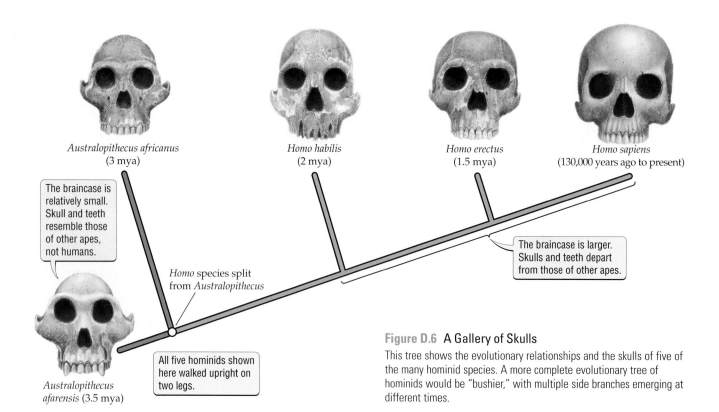

Australopithecus africanus (3 mya)

Homo habilis (2 mya)

Homo erectus (1.5 mya)

Homo sapiens (130,000 years ago to present)

The braincase is relatively small. Skull and teeth resemble those of other apes, not humans.

Homo species split from *Australopithecus*

The braincase is larger. Skulls and teeth depart from those of other apes.

All five hominids shown here walked upright on two legs.

Australopithecus afarensis (3.5 mya)

Figure D.6 A Gallery of Skulls

This tree shows the evolutionary relationships and the skulls of five of the many hominid species. A more complete evolutionary tree of hominids would be "bushier," with multiple side branches emerging at different times.

designed for throwing with a forward center of gravity (like a modern javelin).

It was long thought that *H. habilis* gave rise to *H. erectus*, which then spread from Africa about 1 mya and later evolved into *Homo sapiens*. This simple picture has become more complicated with recent fossil discoveries. Some evidence now suggests that the fossils labeled *H. habilis* are from two different species, and there is debate over which of these species gave rise to *H. erectus*. In addition, it now appears that *H. erectus* or an earlier form of *Homo* migrated from Africa much longer ago than previously thought. *Homo* fossils dating from 1.9 to 1.7 mya have been found in Java, the Central Asian republic of Georgia, and China. Finally, in 2004, a new miniature species of *Homo* was discovered on an Indonesian island (see page 354). This species, known as *H. floresiensis*, appears to have lived on that island 95,000 to 12,000 years ago. Other fossil evidence indicates that *H. erectus* lived on nearby islands from 1,000,000 to 25,000 years ago, while *H. sapiens* lived in the same general region from 60,000 years ago to the present time.

Overall, current research on *H. habilis*, *H. erectus*, and other early *Homo* species indicates that there were more species of *Homo* than once thought, and that several of these species existed in the same places and times. A complete hominid evolutionary tree would therefore be much "bushier" than the version shown in Figure D.6. More research and evidence will be necessary before general agreement is reached regarding the number of early *Homo* species and their evolutionary relationships.

The Origin and Spread of Modern Humans

The fossil record indicates that a number of species with features that were intermediate between those of *Homo erectus* and those of *Homo sapiens* originated between 400,000 and 130,000 years ago; some of these species persisted as recently as 30,000 years ago. Known as "archaic *H. sapiens*," these fossils have been found in Africa, China, Java, and Europe. These ancestors of modern humans developed new tools and new ways of making tools, used new foods, built complex shelters, and controlled the use of fire. Toolmaking and other technologies continued to improve with the origin of anatomically modern humans (Figure D.7).

Early populations of archaic *H. sapiens* gave rise to both Neandertals (an advanced type of archaic *H. sapiens* that lived from 230,000 to 30,000 years ago) and anatomically modern humans. The oldest fossils of anatomically modern humans, dating from 130,000 years ago, have been found in Africa (Figure D.8). More recent fossils of anatomically modern humans have been found in such places as Israel (115,000 years ago), China (60,000

(a)

(b)

Figure D.7 Advanced Stone Age Tools and Art
(*a*) Over the past 300,000 years, *H. sapiens* developed a rich set of new tools and technologies, first in Africa, then in other parts of the world. The archaeological record shows that these tools and technologies did not appear suddenly together, but rather developed over a long period of time and a broad geographic area. (*b*) This spearhead and the figurine of a bison date from 40,000 to 10,000 years ago.

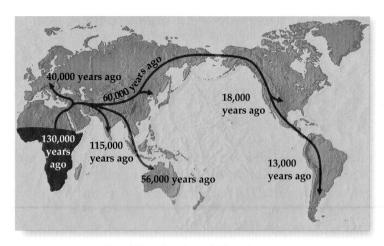

Figure D.8 Migration from Africa

The earliest known fossil and archaeological specimens found in various parts of the world suggest that anatomically modern humans originated in Africa and dispersed from there. The dates provided give the age of the earliest evidence that humans lived in different regions of the world.

unknown populations of archaic *H. sapiens*, which evolved, also in Africa, from *H. erectus*). They then spread from Africa to the rest of the world, completely replacing (and not breeding with) all other *Homo* populations, including *H. erectus*, the Neandertals, and *H. floresiensis*. In contrast, the **multiregional hypothesis** (Figure D.9*b*) proposes that modern humans evolved over time from *H. erectus* populations located throughout the world. According to this hypothesis, regional differences among human populations developed early, but worldwide gene flow caused these different populations to evolve modern characteristics simultaneously and to remain a single species.

Which of these hypotheses is correct? Let's consider some of the evidence. According to the multiregional hypothesis, when different populations of early humans came into contact, extensive gene flow should have caused them to become more similar to one another. Thus we would not expect different types of early humans to coexist in the same geographic region, yet remain distinct for long periods of time. But in fact, Neandertals and more modern humans coexisted in western Asia for about 80,000 years. Even as recently as 12,000 to 25,000 years ago, *H. sapiens* may have shared some parts of their range with *H. floresiensis* and *H. erectus*. Thus the fossil record indicates that modern humans overlapped in time, yet remained distinct from, *H. erectus*, *H. floresiensis*, and Neandertal populations. This finding calls into question

years ago), Australia (56,000 years ago), and the Americas (18,000 to 13,000 years ago).

There has been considerable controversy over the origin of anatomically modern humans. Two conflicting hypotheses have been proposed. According to the **out-of-Africa hypothesis** (Figure D.9*a*), modern humans first evolved in Africa sometime within the past 200,000 years (from

Figure D.9
The Origin of Anatomically Modern Humans

(*a*) According to the out-of-Africa hypothesis, anatomically modern humans (red) originated in Africa within the past 200,000 years, then migrated to Europe and Asia, replacing *Homo erectus* (blue) and archaic *H. sapiens* populations. (*b*) In contrast, the multiregional hypothesis proposes that populations of *H. erectus* in Africa, Asia, and Europe evolved simultaneously into anatomically modern humans.

the extensive gene flow assumed by the multiregional hypothesis.

The best fossil evidence for the shift from archaic to modern *H. sapiens* comes from Africa, providing some support for the out-of-Africa hypothesis. Recent analyses of human DNA sequences are also consistent with the out-of-Africa hypothesis. However, the "complete replacement" part of the hypothesis may not be correct. Fossils have been found that some scientists interpret as showing a mix of Neandertal and modern human characteristics. Similarly, some genetic studies (in which evolutionary trees were constructed based on DNA sequences from living humans) indicate that genes from archaic *H. sapiens* populations outside Africa may have contributed to the genetic makeup of modern humans, suggesting that limited interbreeding did take place.

In summary, many scientists think that anatomically modern humans arose in Africa and spread from there to other parts of the world. However, the origin of modern humans continues to be an issue; this debate is especially active concerning the extent to which early *Homo sapiens* interbred with, and hence did not completely replace, more ancient *Homo* populations.

The Evolutionary Future of Humans

Now that we have examined our evolutionary past, what can it tell us about our evolutionary future? Will we go the way of science fiction stories and evolve into beings with huge brains? Such a direction appears unlikely, because for the last 75,000 years human brain size appears to have decreased slightly, not increased. The average brain size of Neandertal fossils (75,000 to 35,000 years ago) and anatomically modern human fossils (35,000 to 10,000 years ago) was about 1,500 cubic centimeters, while the brain size of an average human today is about 1,350 cubic centimeters. Most of this difference is due to body size: both Neandertals and early modern humans were larger than living humans, and large organisms have bigger brains than small organisms. Although our brains may not be shrinking much (relative to body size), there is no evidence that they are getting larger. But what changes can we expect as a result of genetic drift, gene flow, and natural selection? (To review these concepts, see Chapter 17.)

To answer this question, let's consider several recent and current aspects of human societies. Before the development of agriculture about 10,500 years ago, human populations were small and widely scattered. On a large geographic scale, these populations were isolated from one another by geographic barriers. For example, 30,000 years ago, the chance of an African meeting an Australian was virtually nil. Early human populations were probably isolated on a much smaller geographic scale as well, as illustrated by the fact that before the twentieth century, people from nearby valleys separated by mountains in New Guinea had little contact with one another and spoke different languages (even today, 823 different indigenous languages are spoken in New Guinea).

Translated into evolutionary terms, the conditions of early human populations were exactly those under which genetic drift should be important: population sizes were small, and there was probably little gene flow among populations. Thus we would predict that genetic drift has played a major role in causing genetic differences among human populations.

Some evidence supports this claim: analyses of the rate of evolution of the skulls and teeth of modern humans indicate that genetic drift has been a more important factor in their evolution than natural selection. Because the human population is now large and mobile, however, genetic drift is less likely to play a major role in future human evolution. Instead, high rates of gene flow among human populations could substantially reduce their differences over time (Figure D.10), even for traits such as skin color that appear to have resulted from natural selection rather than genetic drift.

Figure D.10
Gene Flow in Our Future
Gene flow among human populations could reduce some of the features that distinguish different groups of people. Future humans might look something like this computer composite image, which was formed from photographs of eight Afro-Caribbean models, eight Caucasian models, and eight Japanese models.

What about the role of natural selection? Due to our reliance on tools and technology, there is now relatively little selection pressure on many characteristics (such as poor vision) that might have been greatly disadvantageous at previous times in our evolutionary history. This does not mean that natural selection will have no effect on us in the future. Infectious diseases take a terrible toll each year in human death and suffering, so there remains strong selection pressure for the evolution of increased disease resistance in human populations. For example, in a recent genetic analysis, scientists found that alleles that provide resistance to malaria increased in frequency so rapidly in populations exposed to malaria that their rise is best explained by natural selection—not by genetic drift, gene flow, or mutation.

Finally, at various times and places in recent history, humans have been tempted to direct our own evolution, a topic we shall return to later in this essay.

The Effects of Humans on Evolution

Reports in the news media make it clear that humans have a profound effect on their environment, but rarely do we hear mention of our effect on evolution. In fact, it is precisely because we have such a large effect on the environment that we also have a large effect on evolution.

People have profound effects on microevolution in other species

Humans greatly affect how allele frequencies in populations of other species change over time. In many cases, we exert strong natural selection on other species, and they change genetically in response to our actions. For example, when we apply pesticides to kill insects that spread disease or eat our crops, those insects rapidly evolve resistance to our best efforts to exterminate them (Figure D.11). Similarly, disease-causing bacteria and viruses rapidly become resistant to new drug therapies. The evolution of resistance has large social costs in terms of human suffering and lost food production. It also costs us a great deal of money: at least $33 billion to $50 billion per year in the United States alone (see page D11).

Rapid evolution caused by human actions is not limited to the evolution of resistance in disease or pest organisms. Intense fishing efforts, for example, have caused fish to evolve slower growth rates and thinner bodies,

In one year, the percentage of mosquitoes that were resistant to a standard dose of DDT rose from 5% to 80%.

Figure D.11
Directional Selection for Pesticide Resistance
Over a 2-year period, mosquitoes were captured from a population at different times and sprayed with the pesticide DDT. Use of DDT in the area where the mosquitoes were captured caused the population to evolve resistance rapidly, thus limiting the effectiveness of the pesticide. Mosquitoes were considered resistant if they were not killed by a standard dose of DDT (4 percent DDT) in 1 hour.

allowing them to slip through nets more easily. Similarly, in hatchery populations of salmon, natural selection is producing males that are smaller and return from the sea earlier than in the wild (since such males survive better at the hatcheries).

In addition to exerting strong selection on other species, our actions can also change patterns of genetic drift and gene flow in natural populations, again affecting microevolution. For example, when we modify or destroy the habitat in which a species lives, the number of individuals in a population is often reduced, sometimes dramatically. When this occurs, genetic drift can reduce genetic variation and cause the fixation of harmful alleles, as in the greater prairie chicken (see Figure 17.5). Similarly, when we destroy portions of what once were large regions of continuous habitat (as when we convert native grasslands to farmlands, Figure D.12), distances between remnants of the original habitat are increased, thereby reducing gene flow between natural populations.

In general, human actions alter large regions of Earth: we cut down forests; drain wetlands; plow grasslands; add chemicals to the air, water, and soil; transport species to new environments; kill large predators; and harvest species for food, pets, and clothing. These activities cause

(a)

(b)

Figure D.12 Human Effect on the Landscape
When habitats are changed from native grasslands and wetlands to farmlands, the impact on natural populations can be severe. This impact can be seen by comparing the species found in the protected wetlands of a Michigan state park (*a*) with those found in Michigan farmland (*b*).

profound changes in the environments where species live. Because species evolve in response to their environments, the changes we make to those environments alter the course of evolution in many species.

While our effects on evolution are pervasive and have large costs for human societies, all is not hopeless. By understanding how we affect evolution, we can take steps to limit our negative effects on ourselves and on other species. We can use our knowledge of resistance, for example, to implement strategies designed to slow the rate at which it evolves (see Chapter 17). Similarly, we can prevent negative effects of genetic drift by taking actions to increase population sizes in rare or endangered species. As we shall see in the next section, it is also within our power to halt what may become a sixth mass extinction, in which large numbers of species go extinct worldwide.

People also alter macroevolutionary patterns

In addition to affecting microevolution, humans have profound effects on the rise and fall of major taxonomic groups of organisms. Our effects on macroevolution are most easily seen in human-caused extinctions of species. Large numbers of species are in danger of extinction as a result of habitat destruction, introductions of invasive species, and overharvesting. For many different types of organisms, threats such as these are causing a dramatic increase in extinction rates. For example, based on the extinctions of birds and mammals known to have occurred over the last 400 years, human actions seem to have increased extinction rates in these groups by 100

to 1,000 times the usual (or "background") rates found in the fossil record.

Are we currently in the midst of another mass extinction? If not, are we at risk of experiencing one in the near future? A comparison of current extinction rates with those of prior mass extinctions indicates that we probably have not entered a sixth mass extinction yet (since we have currently driven far fewer than 50 percent of the known species on Earth extinct, whereas more than 50 percent of species perished in each previous mass extinction). Scientists estimate, however, that if the current trends in our behavior continue, extinction rates over the next 100 years will be at least 10,000 times the background rate for mammals, birds, plants, and many other organisms. For example, based on current rates of deforestation, scientists estimate that virtually all of the world's tropical forests will be cut down in the next 50 years. If this happens, many of the species that live in tropical forests will go extinct. Since over 50 percent of the world's species live in tropical forests, the single action of removing tropical forests may cause extinctions of species on a scale that does rival the "Big Five" mass extinctions of the past.

When we drive species extinct, we alter the course of evolution and change the history of life on Earth forever. We don't know yet whether people will cause a sixth mass extinction. However, if we do, it will be unique: it will be caused by one species, not many, and not by an environmental event. And the species causing it—humans—will be one that could understand the consequences of its actions and could have prevented the mass extinction from occurring. It takes millions of years to recover from mass extinctions, as we saw in Chapter 19. Therefore, if

Biology Matters

Humans: The World's Dominant Evolutionary Force?

Human impact on world ecology has "increased to the point where humans may be the world's dominant evolutionary force," according to a study reported in *Science* magazine in 2001. A wide range of organisms has evolved resistance to human efforts to kill or control them, and this evolution is costing vast sums of money. For example, as the HIV (the virus that causes AIDS) has evolved, AZT, a relatively inexpensive drug used to treat it, has become less and less effective, leading to higher costs both for treating patients with newer and more powerful drugs, and for developing new drugs to keep up with the evolving virus. It is estimated that a pharmaceutical company must spend roughly $150 million to develop one new drug.

The following table gives estimates of the annual costs of human-induced evolution in the United States, in several areas of activity. These areas include resistance in insects and in two disease agents: *Staphylococcus aureus* (a bacterium that can cause food poisoning, toxic shock syndrome, and infections of the skin and other soft tissues) and HIV. The total annual cost, including items not cited below, may exceed $100 billion.

Scientists are devising ways to slow human-induced evolution. For instance, one method used to thwart microbes' resistance to medicinal drugs is combination therapy, in which a patient takes several drugs in combination so that almost all of the microbes are killed, leaving few to reproduce and evolve. Another is direct observation therapy, in which health-care providers administer all drug doses to patients to ensure that they take the drugs as frequently and as long as necessary to succeed in controlling the disease. Although some of these methods are labor-intensive, they are likely to cost society far less in the long run.

Some areas of increased spending due to resistance	Annual cost (in billions of dollars)
Additional pesticide use (to combat resistant insects)	1.2
Loss of crops	2–7
Treatment of patients infected with resistant *S. aureus*:	
Patients infected outside hospitals	14–21
Hospital infections: penicillin-resistant *S. aureus*	2–7
Hospital infections: methicillin-resistant *S. aureus*	8
HIV drug resistance	6.3
Total cost	**33–50**

SOURCE: S. R. Palumbi, *Science 293*, 2001: pp. 1786–1790.

we cause a mass extinction, our descendants will live in a biologically impoverished world for a long time to come.

People sometimes attempt to control human evolution

In addition to the effects humans have on other species, we have also sought to control our own evolution. In the United States, for example, the early part of the twentieth century saw the creation of a **eugenics** [you-*JENN*-iks] **movement**, an effort to breed better humans by encouraging the reproduction of people with "good" genes and discouraging the reproduction of people with "bad" genes (Figure D.13). The eugenics movement tried to achieve its goals by passing sterilization laws, marriage laws, and immigration laws; we shall focus here only on sterilization laws.

In 1907, the state of Indiana passed the first sterilization law in the United States, allowing state officials to sterilize people deemed unfit to breed. Two years later California passed an even stricter law, making it legal for officials there to castrate males or to remove the ovaries of women who were "feeble-minded." This vague phrase, popular at the time, was sometimes used to refer to people with mental handicaps, but also to people with social behaviors held to be deviant, such as drunkenness or promiscuity. Similar fates could await prisoners considered to have sexual or moral perversions or anyone who had more than three criminal convictions. By 1940, 33 states had passed forced sterilization laws (few as severe as the one in California), and by 1960, over 60,000 people in the United States had been sterilized without their consent.

Sterilization laws gradually passed out of favor, in part because of public revulsion against similar (and more extreme) practices conducted by the Nazis in Germany. Members of the public also realized that it was difficult to establish who really was feeble-minded. For example, government officials argued that Carrie Buck, the first woman forcibly sterilized in Virginia, should be prevented from having more children because she and her daughter were feeble-minded—but later examination of school records showed that her daughter had been on the honor roll in first grade. But even if society had found eugenics morally acceptable, is it likely that the movement could have been effective in its goal of breeding better humans?

The answer is probably no. Eugenics laws aimed to reduce the occurrence of conditions such as so-called feeble-mindedness. It was assumed that these conditions were inherited, and hence that if people with a condition held to be undesirable were prevented from having children, the frequency of the condition would decrease over time. However, it can take many years to remove an allele from a population, which could make it difficult for sterilization laws to have their intended effect. "Feeble-mindedness," for example, was assumed to be a recessive condition, and calculations made in the 1920s by opponents of the eugenics movement showed that it would take about 250 years of forced sterilization to reduce the frequency of feeble-minded individuals in the population from 1 percent to ¼ percent. Thus, for sterilization laws to work as intended, society would have had to tolerate forced sterilizations for hundreds of years.

In addition, we now know that many of the traits of interest to the eugenics movement are not under simple genetic control; instead, they are strongly influenced by the environment or controlled by many genes. It takes even longer to reduce the frequency of traits under complex control than single-gene traits. Therefore, to make rapid changes in the human population, we would have to

Figure D.13 A 1913 News Article Promoting Eugenics
The scientists and government officials involved with the eugenics movement believed that they were benefiting mankind by halting the spread of "bad" genes and lifting "the burden of the feeble-minded," as described in this pamphlet, which reads "it is but humanity and economy to segregate them." The heredity chart at the bottom illustrates how the "mental defects" were passed from generation to generation.

take drastic actions, such as preventing all carriers of any "bad" allele from reproducing, even if the carrier did not have any "undesirable" characteristics. Furthermore, because mutation would continually generate new copies of "bad" alleles, we would have to continually test the entire population for undesirable alleles and prevent anyone who carried even one of these alleles from reproducing. What do you think it would be like to live under such a system?

⊙ Review and Discussion

1. As discussed in this essay, most biologists classify humans as apes. How do you react to being considered an ape?

2. Fossil evidence indicates that in the relatively recent past (about 30,000 years ago), our species shared the planet with at least three other species of hominids: *Homo erectus*, the Neandertals, and *H. floresiensis*. These species differed from us considerably, yet they also shared many similarities with us, and some may even have had the capacity for language. If one or more of these species were alive today, what effect would that have on the world as we know it?

3. Imagine what the world would be like if half of the species you are familiar with were no longer alive. Since it is our actions that may cause a sixth mass extinction, resulting in the loss of 50 percent of the species on Earth, do we have an ethical responsibility to prevent that extinction from occurring?

4. There are instances in which human society limits the rights of an individual to protect the greater social good (for example, by placing a person who has committed a crime in prison). Do you think an ethical case can be made for limiting the right of people who carry genes with harmful effects to reproduce? Ethics aside, what do you think it would be like to live in a society in which everyone was tested for the presence of harmful alleles, and if you were found to carry one, you would not be allowed to have children?

Key Terms

bipedal (p. D3)
eugenics movement (p. D11)
hominid (p. D3)
multiregional hypothesis (p. D7)

opposable (p. D2)
out-of-Africa hypothesis (p. D7)
primate (p. D2)

CHAPTER 20 Maintaining the Internal Environment

Key Concepts

⦿ To survive, animals must maintain a consistent set of internal conditions regardless of their external environment. This regulatory process is called homeostasis.

⦿ Animals can maintain their internal environment only within a certain range of external conditions.

⦿ Homeostasis follows a few general rules: (1) gains must equal losses; (2) close regulation of internal conditions requires feedback from the conditions themselves; (3) large animals can maintain homeostasis more easily than small ones; and (4) the greater the difference between internal and external conditions, the more energy homeostasis requires.

⦿ In large multicellular animals, homeostasis requires maintaining consistent conditions in the fluid around the cells and in the blood.

⦿ The kidney plays a central role in regulating water and solute concentrations in the bodies of many animals, including humans.

⦿ Animals gain heat in two ways: externally from the sun, and internally from their own metabolism. To maintain a consistent body temperature, animals exchange heat with their external environment by three routes: conduction, radiation, and evaporation.

⦿ Heatstroke represents a failure of homeostasis.

Surviving in the Desert

Imagine what it would be like to live in one of the world's hot deserts without refrigerators or air conditioners. Desert-dwelling organisms survive despite scarce water, intense sunlight, and high temperatures. The highest air temperature ever recorded—58°C (about 136°F) in the shade—was reported in the Sahara desert of northern Africa. Yet the body temperature of a person living in the desert heat would be about the same as yours, and on a hot desert day, much lower than that of the surroundings. On the other hand, if you are sitting in a room at 20°C (about 68°F), your internal temperature will be much higher than that of your surroundings. Our bodies have ways of keeping our internal temperature steady despite external variation.

In addition to being hot, deserts are dry, being defined as areas receiving very little precipitation—less than 25 centimeters (about 10 inches) annually. Yuma, Arizona, in the Sonoran Desert, for example, averages only 6.5 centimeters (about 2.5 inches) of rainfall annually. The chemical reactions of life, however, take place in the internal environment of cells, and require water. For this reason, desert-dwelling organisms have evolved ways of adjusting to the dry conditions by conserving and retaining water. In spite of the harsh conditions, many organisms not only survive in the desert, but thrive.

No animal symbolizes life in the desert better than the camel. In one day, an Arabian (one-humped) camel can carry a rider 150 kilometers through the desert without rest or water. The Arabian camel was domesticated over 4,000 years ago in the Arabian desert and in Egypt. From there, it spread throughout the deserts of North Africa and the Middle East. Camels proved invaluable as beasts of burden because they could tolerate conditions too harsh for most other domesticated animals. Their abilities made two major groups of trade routes possible: the Silk Road, which passed through the

Humans and Camels Can Survive in the Inhospitable Desert

Both humans and camels can survive in the desert. Unlike camels, however, people cannot remain in the desert for more than a day without resorting to technology not available to camels: these Bedouins in the Sinai desert depend on water piped from 300 miles away.

Middle East and connected medieval Europe with China, and the north-south trade routes across the Sahara, which made the city of Timbuktu, in what is now Mali in West Africa, the center of a powerful African empire at about the same time.

Although the Arabian camel is a consummate desert dweller, it was desert-dwelling humans who domesticated and relied on this animal. In other words, humans can survive the same desert environment in which the camel thrives. The Bedouin of the Arabian Peninsula and the Touareg of the Sahara still travel through deserts in the company of their camels. But there are limits: although camels can survive the heat of the Sahara in summer without water or shade for well over 2 weeks, humans under the same conditions will die after just 2 days. Whereas camels can lose more than a third of their body water and still survive, humans losing a much smaller percentage under desert conditions will die. How do camels manage these extreme water losses when humans cannot? And how do humans compensate for their greater vulnerability in order to live and travel in the desert?

Like all organisms, animals must maintain an overall internal environment that supports the vital chemical reactions occurring inside their cells. The delicate chemistry of life, introduced in Unit 2, takes place only in the presence of the right temperature, the right amount of water, and the right concentrations of inorganic and organic chemicals. Rarely does the external environment continuously match these needed internal conditions. **Homeostasis** [HO-mee-oh-STAY-siss] (from *homeo*, "same"; *stasis*, "standing") is the name given to the process of regulating internal conditions so as to keep them suited to biochemical reactions when the physical environment does not match those conditions.

Animals have specialized tissues that serve as a means of exchange between the body surface and the external physical environment. Like opened screened windows, which allow air to move in and out but filter out other things (such as mosquitoes), these tissues allow certain substances to cross but not others. Many aspects of homeostasis involve controlling under what circumstances water, solutes, and heat can cross these tissues.

We begin our discussion of homeostasis by considering the internal conditions animals must maintain to support biological processes. Homeostasis follows a few general rules, which we examine next. Then we discuss how animals in general, and humans in particular, regulate their water content and internal body temperature. We finish with a look at what happens when homeostasis fails.

Brine shrimp

20.1 Conditions That Limit Life

Animals live in a wide range of environments, from parched hot deserts to the frigid Arctic, from salty seas to freshwater lakes. Many mammal and bird species survive the bitter −70°C cold of Siberian winters, and some ants can survive on desert sands heated to 70°C. Although animals can survive an amazing spectrum of external environmental conditions, the cells within their bodies require a much narrower range of conditions to remain viable.

Animals need the right amount of internal water

Water accounts for more than half the weight of most animals. Water makes up 60 percent of our own body weight, for example. The loss of just 5 percent of our body water reduces our ability to do physical work by one-third, and the loss of as little as 12 percent can kill us. Although some animals can survive much more dramatic losses of water, they cannot remain active when badly dehydrated. For example, brine shrimp, which are sold as pets under the intriguing name "sea monkeys," can survive the loss of almost all water from their cells. If their water content drops below 40 percent, however, brine shrimp become inactive. Pet shops sell living, but dormant, brine shrimp eggs in this dehydrated state. The eggs hatch into active brine shrimp only when enough water is added to rehydrate their bodies to 45–60 percent water, the proportion typical of most active animals.

Liquid water is the solvent essential to life. It dissolves the materials needed for life, and it is the medium in which the chemistry of life occurs. As we learned in Chapter 4, materials dissolved in water are referred to as solutes. Solutes inside cells must encounter one another to react chemically. The higher the percentage of cellular water, the more spaced out the solutes are within the cell. If solutes are separated too widely, they will not encounter one another often enough for chemical reactions to occur.

Excess water can also change the three-dimensional shape, or tertiary structure, of proteins. Recall from Chapter 4 that the proper functioning of proteins such as enzymes depends on their tertiary structure, so too much water can interfere with protein-mediated functions inside the cell. In addition, too much water can cause cells to burst because the delicate plasma membrane cannot hold up under the increased internal pressure (see Chapter 6).

Too little water can also spell trouble. Just as too much water changes the ability of proteins to function, too little water can make it impossible for enzymes to do their work. Too little water also causes the fluid inside cells to become thicker and flow less easily, which interferes with the movement of solutes within cells.

Animal cells must stay within certain temperatures

At 0°C, water freezes, which has important consequences for living things. As water freezes, it expands, and it may rip apart the plasma membrane and other cell structures (Figure 20.1). This is what happens to cells in our tissues when we suffer frostbite. At the other end of the scale, as temperatures rise above 40°C, the enzymes that mediate chemical reactions in cells begin to lose their terti-ary structure. Only a handful of animals can survive temperatures well above 40°C. These animals have enzymes that hold their shape at these high temperatures. The larvae of one fly species, for example, can survive body temperatures that reach 49°C in the hot springs of Yellowstone National Park.

20.2 Basic Rules Governing Homeostasis

The processes that regulate an animal's internal environment follow a few general rules:

Gains must equal losses

In order for an animal to maintain relatively constant internal conditions, increases in water, solutes, or heat must equal losses of water, solutes, or heat. When gains do not equal losses, internal conditions change. The body then responds to bring internal conditions back to within a particular range.

On a cold winter's day, for example, we may not be able to produce metabolic heat as rapidly as we are

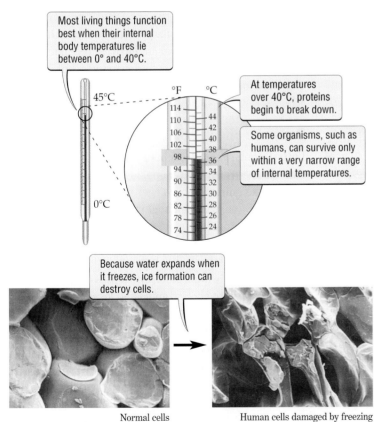

Frost-bitten human hand

Figure 20.1 The Temperature of Life
Most living organisms require a body temperature that falls between 0°C and 40°C. Although most animals can survive a range of temperatures, they usually perform best within a relatively narrow range.

Normal cells   Human cells damaged by freezing

losing heat to the environment. In that case, heat losses start to exceed heat gains, and the body temperature begins to drop. Cells in the brain that can sense this drop initiate processes that produce more heat. We might respond by shivering and getting "goose bumps." Shivering and goose bumps are caused by muscle contractions, and those contractions use energy to produce heat. Blood flow to the skin decreases, so less heat from the blood escapes at the body surfaces. If the external temperature isn't extremely cold, these measures may produce enough heat to balance the loss. Unlike other animals, we can help our bodies' homeostatic efforts with hats, heavy coats, and mittens that reduce the amount of heat leaving our bodies (Figure 20.2).

Homeostasis requires feedback

As we have already indicated, homeostasis depends on regulation. Regulation requires monitoring so that the regulatory process being monitored can either continue or stop at an appropriate time. A process called **negative feedback** provides this monitoring. We experience negative feedback every day. If we are thirsty, for example, we drink water until feedback signals in the body tell us our thirst has been quenched, at which point we stop drinking. **Feedback** simply means that the results of a process in turn affect, or "feed back" into, the process itself. The word "negative" emphasizes that this feedback slows down or stops a process.

A nonliving example from everyday life—the thermostat of a centrally heated and air-conditioned home—illustrates how negative feedback works. We set the thermostat at a certain temperature, called the set point. The thermostat has a mechanism that allows it to "sense" when the temperature inside the house moves above or below the set point. When the temperature rises above the set point, the thermostat turns on the air-conditioning system, and the cooling process begins. But the temperature doesn't start falling immediately; it continues to rise for a brief period (though more slowly) before leveling off and beginning to fall. When it drops to the set point again, the thermostat shuts off the air-conditioning process. Again, because there is a delay in the response, the temperature continues to fall for a while before rising again.

As with the mechanical example of the thermostat, negative feedback provides a "steadying hand" in the homeostasis of many animals. Negative feedback keeps the internal environment in a constant state within a range of external conditions. Such closely controlled homeostasis is dynamic, toning down the extent of internal change so that large shifts in the external environment produce only small, acceptable shifts internally. For this reason, negative feedback is common in homeostatic processes. Humans, for example, have a set point for body temperature. When we get too hot, we perspire, or sweat. When our temperature drops to the set point, negative feedback stops the cooling process (perspiration). Having our thirst quenched is also an example of negative feedback: when we are no longer thirsty, we stop drinking.

Positive feedback is the opposite of negative feedback in that the effects of a process increase the intensity or speed of that process. Positive feedback is not generally involved in animal homeostasis, but it is nevertheless found in a number of biological processes.

Body size affects homeostasis

Large animals tend to exchange water, solutes, and heat with their environment more slowly than small animals do. A large animal has a larger volume compared with its surface area than a smaller animal. The ratio between these two quantities—surface area and volume—determines how quickly or slowly an animal can gain or lose water, solutes, or heat. When the surface area-to-volume ratio is relatively high, gains and losses are rapid; when the ratio is relatively low, gains and losses happen more slowly.

Let's see how this works by comparing two imaginary animals at the cellular level (Figure 20.3a). For simplicity, we've made the two animals cubical, though the same principles apply to noncubical cells and multicellular animals with more than eight cells. For each animal, we assume

Figure 20.2 Human Resources Against the Cold
Just as in the Sinai desert, humans have technology not available to animals. In Siberia this Nenet woman can bundle up against the chill to help her body's homeostatic efforts, unlike her reindeer.

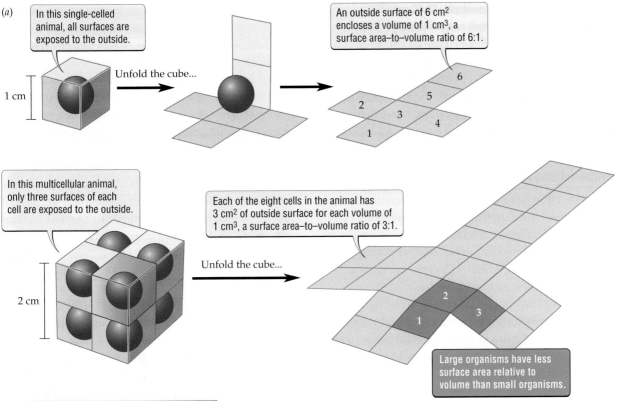

(a)

In this single-celled animal, all surfaces are exposed to the outside.

Unfold the cube...

An outside surface of 6 cm² encloses a volume of 1 cm³, a surface area–to–volume ratio of 6:1.

1 cm

6
5
2
3
4
1

In this multicellular animal, only three surfaces of each cell are exposed to the outside.

Each of the eight cells in the animal has 3 cm² of outside surface for each volume of 1 cm³, a surface area–to–volume ratio of 3:1.

2 cm

Unfold the cube...

2
1
3

Large organisms have less surface area relative to volume than small organisms.

(b)

Figure 20.3 Size Affects Homeostasis

(a) Increasing an animal's size decreases the amount of surface area available for each of its cells (and thus for the animal as a whole) to exchange water, solutes, and heat with its external environment. For simplicity, the two organisms being compared here are depicted as cubical. *(b)* Newborn babies have a larger surface area-to-volume ratio than adults do, so they lose heat and moisture to the environment faster. When babies are born prematurely, they spend time in an incubator until they gain enough weight (volume) to lower their disproportionately high surface area-to-volume ratio.

Investigate the relationship between surface area and volume.

20.1

that water, solutes, and heat are being exchanged across all cell surfaces exposed to the external surroundings at the same rate. The single-celled animal can exchange water, solutes, and heat across six surfaces, since all six surfaces of its one cell are exposed to the surroundings. However, in the eight-celled animal, each cell can exchange water, solutes, and heat with the external environment across only three surfaces. Yet in both animals, each cell, regardless of how many surfaces it has exposed, is busy generating heat at the same rate and can hold the same amount of water and solutes. The single-celled animal has all six cell surfaces available for exchanges, but each cell in the eight-celled animal has only half as many surfaces available. So the whole eight-celled animal has only half as much surface area, compared with its volume, as the single-celled animal. These numbers tell us that the cell in the single-celled animal can exchange water, solutes, and heat with its environment more quickly than the cells in the multi-

cellular animal can because it has a higher surface area-to-volume ratio.

If we calculate each animal's surface area-to-volume ratio and compare the two, we get a picture of how each animal might respond to its environment. The surface area of our single-celled cubical animal is 1 square centimeter (cm²) per side × 6 sides = 6 cm², and its volume is 1 cm × 1 cm × 1 cm = 1 cm³, giving it a surface area-to-volume ratio of 6:1. Compare this with the ratio for the multicellular cubical animal: its eight cells give it a surface area of 24 cm² (2 cm × 2 cm × 6 sides) and a volume of 8 cm³ (2 cm × 2 cm × 2 cm). So the eight-celled animal has a surface area-to-volume ratio of 3:1 (24/8 = 3).

This geometry has far-reaching implications for whole animals of any size. It explains, for example, why newborn babies are more susceptible to changes in temperature and to dehydration than adults. A newborn has a large surface area compared with its body volume and so loses heat and water more rapidly than an older child or an adult. That's why we are careful to protect babies from temperature extremes (Figure 20.3*b*). More generally, we would expect all small animals (whether newborns, juveniles, or adults) to exchange water, solutes, and heat with their environment more rapidly, and to have a harder time maintaining homeostasis, than larger animals.

Homeostasis requires energy

The more effectively an animal can regulate its internal environment, the closer it can keep its internal conditions to the ideal for its cellular activities—in particular, for its enzymes. The ability of humans to maintain a high internal temperature within a 1°C range creates the kind of internal environment in which our cells' enzymes are specialized to function at their best. Animals such as butterflies, whose body temperatures change dramatically over the course of a day as environmental temperatures fluctuate, tend to have enzymes that work less efficiently than those of animals that closely regulate their internal temperature. But their enzymes work over a much wider range of temperatures.

If maintaining near-constant internal conditions is best for cellular activities, why do only some animals closely regulate them, while many others allow their internal conditions, including temperature, to change as their surroundings change? The answer is that homeostasis requires energy. The narrower the range of conditions that the body tries to maintain, the more energy the body must expend. Put another way, the greater the difference between external conditions and ideal internal conditions, the greater the energy cost for the animal.

We can see how this works in humans by observing how homicide investigators determine a victim's time of death (Figure 20.4). Imagine that investigators arrive at a murder scene on a 25°C summer night. From roughly the time of death, the victim's body, still clothed, has slowly cooled as heat was lost and no longer replaced from within by respiration. Suppose the temperature of the victim is 34°C, about 3.5°C below the normal body temperature of a healthy person, which is 37.5°C. Investigators can estimate the time of death by using a formula or graph that shows how long it would take the body to cool to a temperature of 34°C from 37.5°C in 25°C surroundings. In this case they would estimate the time of death at between 3 and 8 hours ago. From the point of view of homeostasis, we can think of this loss of heat as the amount of energy the victim would have to have generated for 3 to 8 hours were he still alive and clothed. Because of the rule that gains must equal losses, to maintain a constant internal temperature, his body would have been replacing the heat he lost to the environment. The victim would have had to produce enough metabolic heat to prevent a 3.5°C drop in temperature (from 37.5°C to 34°C) in surroundings of 25°C for 3 to 8 hours.

If this were a crisp fall evening, the time of death calculation would change because the difference between normal body temperature and the air temperature would be greater, and heat would have left the body more rapidly. Let's assume an air temperature of only 5°C. Now the environmental temperature is 32.5°C below the normal body temperature. The victim, were

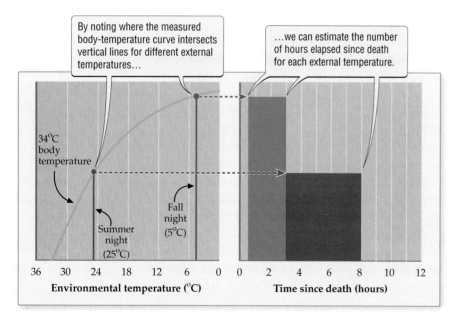

By noting where the measured body-temperature curve intersects vertical lines for different external temperatures…

…we can estimate the number of hours elapsed since death for each external temperature.

34°C body temperature

Summer night (25°C)

Fall night (5°C)

Environmental temperature (°C)
36 30 24 18 12 6 0

Time since death (hours)
0 2 4 6 8 10 12

Figure 20.4 Crime Scene Investigators Can Use Loss of Body Heat to Estimate Time of Death
Complex curves relating body temperature, environmental temperature, and time of death have been created to help investigators estimate the amount of time elapsed since death. Only one curve (for a measured body temperature of 34°C) is shown here, for estimating the time of death for a clothed human body at two different air temperatures: 25°C and 5°C. In the right-hand graph, time of death is indicated not by bar height, but as a range of times determined by where the left and right sides of each bar intersect the horizontal axis.

he still alive, would have to have produced the same amount of heat over a much shorter time period (a half hour to 3 hours) in order for gains to equal losses. Only by using more energy to produce heat more rapidly could our victim have maintained his body temperature on that cool evening.

Now let's return to our earlier question about humans and butterflies: granted that close regulation of internal conditions has high energy costs, if it facilitates cellular activities, why don't all animals simply "pay the price"? Why do butterflies (and other animals) put up with less efficient enzymes? The answer is that for some animals, especially small ones, the costs of close regulation of internal conditions outweigh the benefits. A small butterfly, with its relatively large surface area, loses heat to the environment much more quickly than we do. As a result, the butterfly would have to expend much more energy relative to its size to replace lost heat. Close regulation of body temperature would, therefore, work for the butterfly only if the resulting benefits were also disproportionately large.

20.3 Regulation of Water and Solute Concentrations

Why do animals need to regulate their water and solute concentrations? To answer this question, let's consider the world of a deeply embedded single cell in a multicellular animal and how it is connected to the animal's external environment through the movement of these materials.

Internal cells exchange substances directly with the interstitial fluid

Recall that all eight cells in our imaginary multicellular animal in Figure 20.2*a* had some direct contact with the external environment. However, animals with even a few more cells (in our cubical example, more than 27 cells) will include some internal cells that have no direct contact with the animal's external environment. And in multicellular animals with thousands to many billions of cells—such as most animals visible to the naked eye, including humans—*most* cells have no direct connection with the outside.

The distance between a deeply embedded cell and the external environment can be huge, especially when measured in cell widths. If all water and solutes entering and leaving an animal's body could reach deeply embedded cells only by traveling from one adjoining cell to the next, passing through cell junctions or diffusing across plasma membranes, their movement would be far too slow for the animal to survive. Obviously, internal cells must rely on some other means of connecting with, and exchanging materials with, that faraway external universe. They do so through a series of exchanges across surfaces and tiny spaces between them and the outside world.

If we begin at the level of the cell, the first exchange is with the only surrounding medium directly available to most cells: the **interstitial fluid** [in-ter-*STISH*-ul . . .] (*inter*, "between"; *stitial*, "standing"), a watery substance that surrounds and bathes the cells (Figure 20.5). From the cell's viewpoint, the interstitial fluid that comes into contact with its plasma membrane is its only "window on the world." Normally, the interstitial fluid delivers all the substances cells need, such as nutrients and oxygen, which cells constantly consume; this fluid also carries away all the waste products cells are constantly producing. Some cells produce molecules needed elsewhere in the body, such as hormones and other signals. These molecules move into the interstitial fluid for further distribution. The plasma membrane forms the selectively permeable surface that regulates this two-way traffic between the interstitial fluid and the cytosol inside the cell.

Recall from Chapter 6 that energy must often be expended to transport molecules into or out of a cell against a concentration gradient. Water, in contrast, crosses the plasma membrane passively by diffusion. Water is the key to many exchanges that take place across the plasma membrane: interstitial fluid is a watery solution, and cells are largely water. Recall also from Chapter 6 that the concentrations of the solutions inside of and outside of a cell have a lot to do with the cell's survival. When the external solution is hypertonic (more concentrated than the cytosol), more water molecules leave the cell than enter, and the cell shrinks. When the external solution is hypotonic (less concentrated than the cytosol), the cell takes in water and swells. In an isotonic solution, the concentrations are equal on each side of the membrane. In healthy animals, homeostasis works to keep the interstitial fluid as close to isotonic as possible, creating an environment that is "just right" for cells.

So cells have a connection with the interstitial fluid, but what connections does the fluid have that allow it to acquire and discharge materials? That is where the circulatory system comes into the picture.

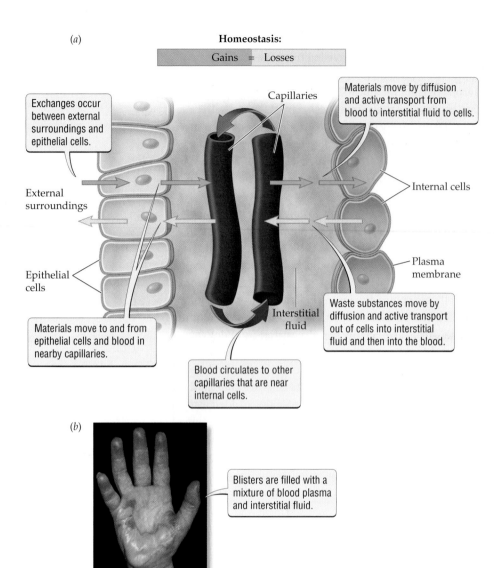

(a)

Homeostasis:

Gains = Losses

Exchanges occur between external surroundings and epithelial cells.

Capillaries

Materials move by diffusion and active transport from blood to interstitial fluid to cells.

External surroundings

Internal cells

Epithelial cells

Plasma membrane

Materials move to and from epithelial cells and blood in nearby capillaries.

Interstitial fluid

Waste substances move by diffusion and active transport out of cells into interstitial fluid and then into the blood.

Blood circulates to other capillaries that are near internal cells.

(b)

Blisters are filled with a mixture of blood plasma and interstitial fluid.

Figure 20.5 A Series of Regulated Exchanges

(a) All movements of water and solutes between internal cells and an animal's external environment involve exchanges across surfaces such as the plasma membranes of cells, the walls of capillaries, and epithelial tissues. Materials move from regions of higher concentration to regions of lower concentration. (b) A blister, such as on the lower part of this 4-year-old's hand with second-degree burns, is a raised portion of skin that is filled with a mixture of blood plasma and interstitial fluid.

The interstitial fluid exchanges substances with the body fluid

The next step away from the cell and toward the external world is the exchange between the interstitial fluid and the body fluid, which in humans we call the **blood plasma**. In our bodies, the interstitial fluid gets its supplies and disposes of wastes as it exchanges water and solutes with blood plasma, the liquid component of our blood (Figure 20.5a). Materials in the circulatory system, which distributes blood in vessels of different sizes throughout the body, can move close to cells through highly branched networks of tiny blood vessels called **capillaries**.

Exchanges between interstitial fluid and blood plasma take place across the capillaries' very thin, highly permeable walls. As a cell takes up the dissolved substances it needs, they become depleted in the surrounding interstitial fluid. Those substances are then more concentrated in the blood plasma than in the interstitial fluid. So those substances easily diffuse from the blood plasma across capillary walls into the interstitial fluid, then into the cell. Waste solutes, on the other hand, are more concentrated in the cell than in the interstitial fluid surrounding them. Thus they diffuse into the fluid, and from there across the capillaries into the blood plasma. where the concentration of waste solutes is less than in the fluid. The plasma carries them away. The concentrations of solutes in blood plasma must be regulated to keep them within the bounds that allow these exchanges to take place as needed by the cells.

The distances between capillaries and cell surfaces are so small that diffusion in the interstitial fluid to and from cells and capillaries takes very little time. You can see a mix of blood plasma and interstitial fluid inside blisters that form on damaged skin (Figure 20.5b).

The body fluid exchanges substances with the environment across epithelial tissues

The last steps away from internal cells are the interfaces between the blood and epithelial tissues and, finally, between epithelial tissues and the external environment (see Figure 20.5a). **Epithelial tissues** are specialized to regulate how the animal's internal environment exchanges water, solutes, and heat directly with the external environment. By this means, needed materials are imported, wastes dissolved in the blood plasma are disposed of, and the contents and solute concentrations of the blood plasma are adjusted as needed.

Epithelial tissues form the outermost living layer of the skin, the lining of the digestive system, the gas exchange surfaces of the lungs, and parts of the kidneys. We shall see in later chapters how the epithelial tissues that line our digestive system and lungs facilitate the exchange of materials.

The environment affects the regulation of body fluid composition

The challenges that an animal faces in regulating its water and solute concentrations depend on its environment (Figure 20.6). As noted earlier, animal cells constitute a watery environment that supports the chemistry of life. The proportions of water and solutes inside animals

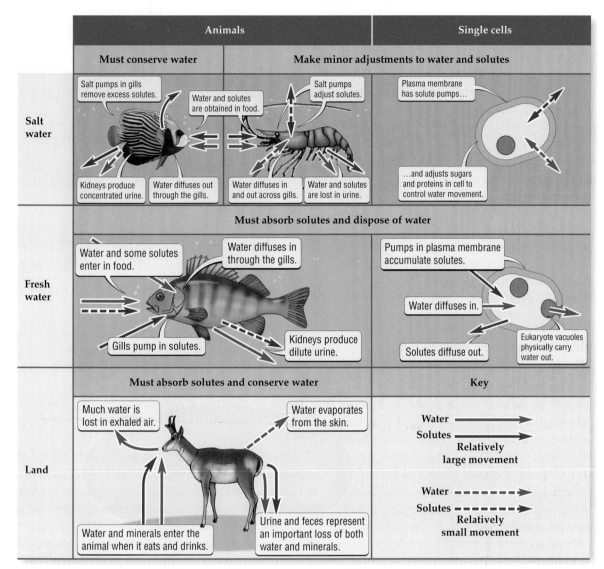

Figure 20.6 How an Animal Regulates Its Water and Solute Concentrations Depends on Its Environment

remain remarkably constant among different animal species. The effort a marine animal must put into maintaining its internal environment depends mostly on the characteristics of the external environment.

The body fluids of some marine animals (such as the lobster) closely resemble ocean water (Table 20.1), probably because life first evolved in that environment. These marine animals need to make only minor adjustments to their body fluids to maintain homeostasis.

Marine fish, on the other hand, spend about 5 percent of their metabolic energy maintaining the composition of their body fluids. Saltwater fish have body fluids that contain lower concentrations of solutes than seawater does. As a consequence, marine fish tend to

Table 20.1

A Comparison of the Solute Compositions of Marine Animal Body Fluid with That of Seawater and Fresh Water

Solute	Marine animal (lobster)	Seawater	Fresh water
Sodium	541	470	0.17
Potassium	8	10	Not detectable
Chloride	552	548	0.03

gain solutes from, and lose water to, the ocean. They must expend energy to hold onto water and to get rid of excess solutes. Freshwater fish have the opposite problem: their body fluids contain solutes in much higher concentrations than found in their external environment. As a result, they tend to lose solutes to, while constantly gaining water from, their environment. Freshwater animals must expend energy to retain the solutes they need and to keep too much water from accumulating in their bodies.

In terms of maintaining the composition of their body fluids, land animals live in perhaps the most challenging environment of all: they tend to lose both water and solutes to their environment. Water constantly escapes from a terrestrial animal's body. Animals must offset these constant losses by drinking water and getting solutes from their food.

Metabolism changes the composition of body fluids

The solute composition of an animal is inevitably altered by its metabolism. As its cells metabolize substances, they use up some of the chemicals dissolved in its body fluids and produce some new ones. Often these changes require only minor adjustments, but some waste products can pose serious problems. The metabolism of proteins, for example, produces ammonia, the same chemical found in many household cleaners. A buildup of ammonia can become poisonous to cells. To protect against this buildup, cells dump ammonia into the surrounding interstitial fluid, from which it diffuses into the blood plasma. Ammonia in the blood continues to pose a potential danger, however, so it must be removed from the body altogether.

Removal of ammonia is accomplished in several ways. Some animals remove ammonia directly from their bodies, usually with a large amount of water, while others use energy to convert ammonia into less dangerous molecules. Animals living in fresh water do not expend energy to convert ammonia; instead, they expel it in a dilute water solution. They can afford the water loss because they need to get rid of water anyway (see Figure 20.6). Some terrestrial animals, including humans, that can get plenty of water to drink convert ammonia into a much less toxic compound called urea, which they excrete along with water in liquid urine. Other terrestrial animals, such as birds, reptiles, and insects, conserve precious water by converting ammonia into white crystals of uric acid, which they then excrete using relatively little water (Figure 20.7).

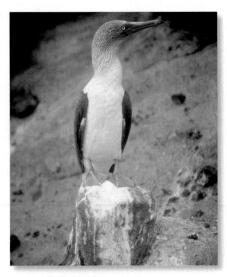

Figure 20.7 Birds Convert Toxic Ammonia into Harmless Uric Acid
The white color associated with bird droppings, such as these deposits made by seabirds, comes from uric acid. Birds and other animals, including reptiles and insects, use energy to convert the toxic ammonia produced by protein metabolism into white crystals of uric acid.

Human kidneys regulate body water and solutes

All animals must remove from their cells and body fluids certain metabolic products, such as ammonia, that, if left in the body too long, would become poisonous to their cells. All animals must also regulate the concentrations of solutes, such as sodium and calcium, in their body fluids. Terrestrial animals face the additional challenge of conserving water and retaining vital solutes. All these functions are performed by the **kidneys**.

Kidneys filter and regulate the composition of the blood as it moves through them—and they do so in surprisingly large volumes. Though our paired, fist-sized kidneys make up only a tiny fraction (0.5 percent) of our total body weight, they process over 20 percent of the blood each time it circulates through the body. That means that about 1 liter of blood enters and leaves our kidneys each minute. When blood leaves the kidneys and returns to the circulatory system, it is largely cleansed of metabolic wastes and is carrying water and solutes in normal amounts. The volume of blood leaving the kidneys is slightly smaller than the entering volume because some water is lost to make **urine**, the watery solution of wastes we expel from our bodies. We lose about 1.5 liters (less than a half gallon) of water a day in urine and replace it when we drink water and other fluids. That's one reason we get thirsty: thirst is a signal that we need to replenish lost water.

The blood-cleansing work of the kidneys—**filtration**—is performed by the kidney's basic functional unit, the **nephron** [*NEF*-ron] (Figure 20.8). Each human kidney has about a million of these tiny units. Blood enters and leaves each nephron through a bed of capillaries called a **glomerulus** [glo-*MARE*-you-luss] (plural: glomeruli) (see Figure 23.10). The glomerulus lies in a cup-shaped capsule that opens at its far end into a U-shaped **tubule**. Plasma from the blood flowing through the glomerulus is filtered—forced by blood pressure to pass through tiny pores in the capillary walls into the tubule. All ions and molecules small enough to pass through these pores travel with the plasma into the tubule, while substances too large to pass through (such as large proteins, lipids, and blood cells) remain in the blood. The **filtrate** (the filtered plasma in the tubule) includes many of the small solutes, such as urea, that the body needs to get rid of. As the filtrate passes down the tubule, urine gradually forms. Each tubule connects to a larger tube called a collecting duct, which leads to a series of ever larger tubes. All these tubes eventually join the main tube, which drains urine from each kidney and carries it to the bladder. Here urine is stored until it can be excreted into the external environment.

Removing wastes from the blood plasma is only part of the kidney's job. In addition to urea and other waste substances, the filtrate just starting its journey down the tubule contains materials that the body needs to take back; these materials include water, certain dissolved ions, and dissolved nutrients such as minerals, sugars, and small proteins. A second important function of the nephron is **reabsorption** of valuable solutes and water

(a)

Nephron

Glomerulus

Tubule

Collecting duct

Renal artery (blood in)

Renal vein (blood out)

Blood in

Glomerulus

Blood out

Capillary

1 Under pressure generated by the heart, blood plasma from the capillaries of the glomerulus filters into the tubule of the nephron.

Collecting duct

2 Sodium is actively pumped out of the tubule into the surrounding fluid.

3 The high sodium concentration in the surrounding fluid pulls water out of the tubule and collecting duct.

5 Water is reabsorbed by the capillaries.

Tubule

4 Removal of water leads to urine four times as concentrated as blood.

Urine

See the human kidney in action.

20.2

(b)

Figure 20.8 The Human Kidney Regulates Internal Water Content and Removes Toxic Wastes

(a) Animals rely on specialized kidneys of many different forms, all of which, like the human kidney shown here, remove the toxic by-products of protein metabolism and assist in regulating internal water and solute concentrations. (b) The human kidney has a dense network of blood vessels and capillaries (red). Urine flow is shown in yellow.

from inside the tubule before they can drain out of the kidney with the rest of the urine. All along the length of the tubule within the nephron are sites where water and needed solutes are moved back across the tubule walls into close contact with capillaries, into which they pass, rejoining the blood plasma.

The reabsorption of water causes urine to become increasingly concentrated in waste solutes as it moves down the tubule. The magnitude of this concentration is truly remarkable. Each day about 180 liters of water—enough to fill a large bathtub—are filtered out through the capillaries in the glomeruli. Of course, we cannot afford to lose that amount of water each day. And even if we could, it wouldn't be easy to replace that much lost water by drinking it. So we are indebted to our kidneys for selectively reabsorbing almost all of the valuable nutrients and 99 percent of the water from the urine: only about 1.5 liters of urine actually leave our bodies each day.

The tubules manage this impressive efficiency almost entirely through active transport: cells in the tubule walls pump sodium ions out of the tubules into the interstitial fluid outside their walls. This causes water to diffuse from inside the tubules to these zones of high sodium concentration in the interstitial fluid (see Figure 20.8). As Table 20.2 shows, water and valuable nutrients are very effectively reabsorbed from the tubule, whereas wastes are reabsorbed inefficiently or not at all. It may be surprising to see from the table that 50 percent of the urea is reabsorbed. This reabsorbed urea also becomes a solute in the interstitial fluid that helps draw water out of the filtrate in the tubule.

Table 20.2

Substances Processed by the Human Kidney

	Substance	Percentage reabsorbed
NUTRIENTS	Glucose	100
	Sodium	99.4
	Chloride	99.1
WASTES	Urea (a waste product of the breakdown of proteins)	50
	Creatinine (a waste product produced by muscle tissue)	0

A third function of the kidney is **secretion**: certain substances, such as potassium ions, hydrogen ions, and some medications and toxins, are actively transported from the capillaries into the filtrate in the tubule. Secretion is not part of the initial filtration process in the glomerulus, but occurs farther along the tubule, before it joins the collecting duct.

20.4 How Animals Gain and Lose Heat

In addition to regulating their internal concentrations of water and solutes, many animals, including humans, maintain a relatively stable body temperature. To do so requires animals to respond to heat from two major sources. **Metabolic heat** is an inevitable by-product of an animal's metabolism; as we saw in Chapter 7, heat is released as the breaking and forming of bonds in biochemical reactions produces and consumes energy. Metabolic heat keeps the human body temperature at 37°C, even though a comfortable room may be at 20°C. The other major source of heat is the sun. As we shall see shortly, energy from the sun can reach us directly, when we are exposed to sunlight, or indirectly, when we are exposed to air and the surfaces of objects that have been heated by the sun.

The surface area that an animal presents to a heat source affects the rate at which it exchanges heat with the external environment. If we want to warm ourselves in the heat of the sun, we face the sun fully to place more surface area of skin in full sunlight. On the other hand, if we have to sleep in a cold room, we tend to curl up into a ball; this minimizes our surface area, thus slowing down the rate at which we lose metabolic heat to our surroundings.

Animals have two basic ways of regulating their internal temperature

Humans rely mostly on metabolic heat to maintain their body temperature. Only a few groups of animals generate enough metabolic heat to contribute significantly to their heat gain. These animals, known as **endotherms** (*endo*, "inside"; *therm*, "heat"), maintain a high and nearly constant body temperature even in an environment with variable temperatures. Birds and mammals are endotherms, and most scientists believe that the ancestors of birds—the dinosaurs—were endotherms as well.

Maintaining a constant high body temperature regardless of environmental temperatures is not possible for most animals. These **ectotherms** (*ecto*, "outside") depend on environmental heat sources to regulate their body temperature. Ectotherms include most fish, invertebrates, amphibians, and reptiles.

As endotherms, mammals and birds can remain active even when the lack of environmental heat sources forces ectotherms into inactivity. This advantage comes at a huge energetic cost, however. Endotherms use energy about ten times as quickly as ectotherms of a similar size. As a result, endothermic animals must eat much more food in a given period of time than ectotherms of the same size. For example, if your house were guarded by a 50-kilogram endothermic dog, you would have to feed it roughly 500 kilograms of dog chow each year, whereas if you relied on a 50-kilogram ectotherm, such as an alligator, for the same service, you would have to provide it annually with a mere 50 kilograms of alligator chow (Figure 20.9).

Regardless of whether an animal is an endotherm or an ectotherm, heat can enter and leave its body by three different routes: conduction, radiation, and evaporation (Figure 20.10). We now consider each of these routes in more detail.

Animals can gain or lose heat by conduction

Heat **conduction** can occur by direct contact between objects or substances. For example, if you put a heavy metal frying pan on a stove, the handle soon becomes too hot to hold. The heat moves by conduction from the pan along the handle and to your hand when it comes into direct contact with the pan. Animals often gain heat conductively by lying in contact with a rock or other surface that is warmer than the body (Figure 20.11a), or get rid of heat conductively by digging down into cooler desert sands during the hot desert daytime. Conduction can also happen within an object, as heat is transferred from molecules in one part of the object to molecules elsewhere in the object (as when it moves to the handle of the frying pan). Metabolic heat reaches an animal's surface partly by conduction of this kind as heat is passed from cell to cell.

Whether an animal gains or loses heat conductively depends on which has the higher temperature—the animal or the substance it comes into contact with. That is because all heat exchanges in nature follow a universal rule: heat always flows from warmer areas to cooler areas. If an animal is in contact with an object (such as a rock) that is warmer than itself, the animal will gain heat. If the surface temperature of the animal is warmer than

(a)

Endotherms require a lot of food to fuel their high metabolism.

Endotherms can maintain a constant internal temperature.

Endotherms minimize heat exchange with the external environment through insulating fur.

DAY	NIGHT
Air temperature = 30°C	Air temperature = 20°C
Body temperature = 37°C	Body temperature = 37°C

(b)

Ectotherms require relatively little food.

The internal temperature of ectotherms changes when the environmental temperature changes.

Ectotherms depend on heat exchange with the external environment.

DAY	NIGHT
Air temperature = 30°C	Air temperature = 20°C
Body temperature = 37°C	Body temperature = 20°C

Figure 20.9

Temperature Regulation in Endotherms versus Ectotherms
(*a*) Endothermic dogs gain independence from environmental heat sources by using a lot of energy to generate metabolic heat.
(*b*) Ectothermic alligators use much less energy to regulate their internal temperature by relying almost completely on environmental heat sources.

that of the object, heat moves in the opposite direction, from the animal to the object. When you hold a glass of ice water, for example, it feels cold because heat is being conducted out of your hand.

How fast heat is gained or lost conductively depends on several factors. The materials involved are one factor: solids and liquids conduct heat much more quickly than air does. We can remain comfortable indefinitely in 20°C (68°F) air, but start to shiver with cold if we stay in a

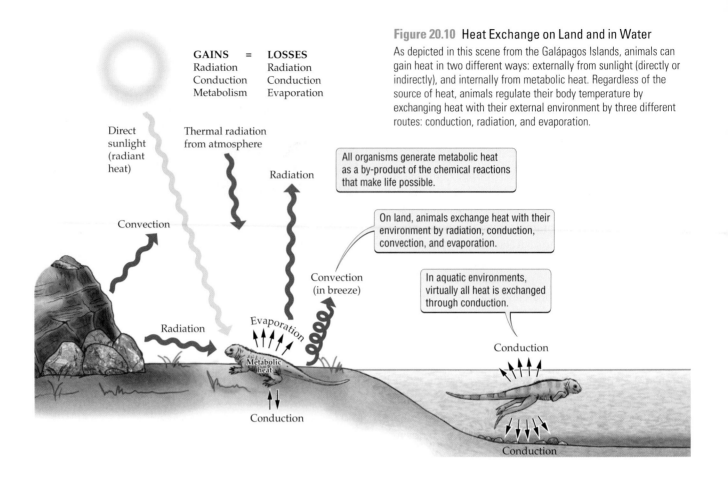

GAINS = LOSSES
Radiation   Radiation
Conduction   Conduction
Metabolism   Evaporation

Direct sunlight (radiant heat)

Thermal radiation from atmosphere

Radiation

Convection

Radiation

Evaporation

Metabolic heat

Conduction

Convection (in breeze)

Conduction

Conduction

All organisms generate metabolic heat as a by-product of the chemical reactions that make life possible.

On land, animals exchange heat with their environment by radiation, conduction, convection, and evaporation.

In aquatic environments, virtually all heat is exchanged through conduction.

Figure 20.10 Heat Exchange on Land and in Water
As depicted in this scene from the Galápagos Islands, animals can gain heat in two different ways: externally from sunlight (directly or indirectly), and internally from metabolic heat. Regardless of the source of heat, animals regulate their body temperature by exchanging heat with their external environment by three different routes: conduction, radiation, and evaporation.

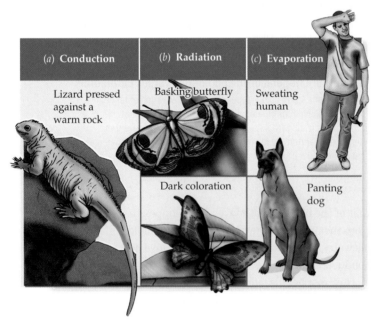

(a) Conduction	(b) Radiation	(c) Evaporation
Lizard pressed against a warm rock	Basking butterfly	Sweating human
	Dark coloration	Panting dog

Figure 20.11 Some Strategies for Dealing with Different Kinds of Heat Exchanges

20°C swimming pool for too long, because water conducts heat from the body 25 times faster than air does. This difference in conduction rates means that aquatic animals lose or gain heat conductively much more quickly than land animals do. As a result, terrestrial animals can maintain a constant body temperature that differs from that of their environment relatively easily, whereas aquatic animals can do so only with great effort. For all but a few aquatic animals, the energy cost of raising and maintaining the body temperature above the temperature of the surrounding water is too high. The bodies of most aquatic animals, therefore, remain at about the same temperature as their environment.

The rate of conductive heat exchange also depends on how wide the temperature difference is between the two objects that are exchanging heat. As an animal exchanges heat with its environment by conduction, the temperature difference where the two surfaces meet decreases, so the rate of heat exchange slows down. As we saw earlier, a murder victim's body will cool faster on a cool autumn day than on a warm summer night. In fact, some calculations used by medical examiners for estimating time of death include a much larger drop in temperature in the first hour after death, when the temperature difference between the victim and the surroundings is greatest.

Animals can exchange heat with their environment through radiation

Radiation is a means of exchanging heat in the form of light or infrared (heat) waves. The heat we feel from sunlight is radiant heat. All objects, including animals, can both absorb and emit radiant heat. During the day, radiant heat from the sun provides the major heat source for animals that rely heavily on an external source of heat. At night, without the sun's light, the environment and the animals that depend on environmental heat sources slowly cool as they radiate heat as infrared light.

As with conduction, the rate of radiant heat exchange depends on the materials involved. Objects with dark surfaces absorb sunlight better than objects with light-colored surfaces, which reflect sunlight. A light-colored butterfly basking in the sun will take longer to absorb heat than a dark-colored butterfly (Figure 20.11*b*). Similarly, on hot summer days we wear light-colored clothing to stay cool because it absorbs less radiant heat than dark clothing.

Convection increases the rate of heat conduction and radiation

Convection is the physical movement of heat in a moving gas (such as wind) or liquid (such as water currents). Convection can speed the exchange of heat by conduction or radiation. For example, a breeze cools us by carrying away conductively heated air sitting next to our skin, replacing it with cooler air. In this way, convection maintains a large temperature difference between the skin surface and the air immediately next to it. The large temperature difference speeds the conduction of heat from the skin to the air it touches.

The nature of an animal's outer surface can affect convection. Anything that prevents the physical movement of air or water near the surface reduces convection. For example, fur, feathers, and the fill in sleeping bags trap a layer of conductively heated air next to the body surface (Figure 20.12*b*), reducing the temperature difference between skin and adjacent air to slow heat loss in a cold environment.

Convection can occur within an animal's body as well as at its surface. The circulation of blood, for example, transfers heat convectively from one part of the human body to another, especially between the body core and

(*a*)   (*b*)

Figure 20.12 Convection In and Out
(*a*) This runner's flush comes from the convection of heat from inside the body out, to cool off. (*b*) On the other hand, this wolf's fur reduces convection and keeps heat inside its body, to stay warm.

the outer surfaces. When we begin to overheat, we flush bright red as the capillaries in the skin expand to carry heated blood from the body core to the skin surface (Figure 20.12*a*). Warming of the skin in this way leads to a bigger temperature difference between the skin and the environment, increasing the rate of heat loss through conduction and radiation.

Evaporation is a cooling process

Evaporation allows animals to lose heat to their environment from liquid (mostly water) on a body surface that is exposed to air. The transition from a liquid (water) to a gas (water vapor) requires a lot of energy; water evaporating from an animal draws most of this energy from body heat. Thus the animal loses body heat to the external environment.

Many animals living in hot terrestrial environments have ways of exposing wet surfaces to the external environment. When we sweat, we feel cooler because the sweat evaporates from our skin (Figure 20.11*c*). Human skin makes a good evaporative surface because humans have relatively little hair. Most other mammals have evolved other forms of evaporation because their outer skin is covered by fur. Dogs, for example, rely on the tongue, whose wet surface loses moisture to the air when it is exposed by panting. Evaporation is enhanced by convective cooling as air moves back and forth over the panting dog's tongue.

The most important feature of evaporation is that it provides a way of losing heat that does not depend on the temperature of the animal relative to its environment. An animal can lose heat by conduction or radiation only if the temperature of the environment is lower than its body temperature. However, the same animal can lose heat evaporatively even in extreme heat. In deserts, where environmental temperatures often exceed internal temperatures, the value of cooling through evaporation often outweighs the accompanying loss of precious water. On the other hand, in cold terrestrial environments, animals tend to minimize evaporation to preserve heat; even so, the unavoidable evaporation of water during gas exchange (breathing) often represents an important heat loss.

Warmer is generally better for body temperatures

When animals have the ability to raise their body temperature above the temperature of their environment, they generally do so. Thus most land animals, whether endotherms or ectotherms, and whether large or small, function in ways that allow them to have internal temperatures greater than the air temperature for as much of the day as possible. Animals do this because high temperatures speed the chemical reactions on which life depends, allowing them to move more quickly.

We experience the effects of temperature on body function when our bare hands cool down on a cold winter day: our hands become too stiff and too weak to pick up and handle things. Similarly, many insects and lizards—both ectotherms—can barely move early in the morning before the sun provides sufficient radiant energy to warm them up. When ectotherms can warm up, they often maintain their body temperature in the 35°C–40°C range typical of endotherms. The major advantage we, as endotherms, get from our metabolically expensive lifestyle is that we can remain active no matter what the weather or the time of day.

Marine iguanas gain heat on land to sustain them while underwater

Marine iguanas, which are ectotherms, feed in the cold ocean waters surrounding the Galápagos Islands (see Figure 20.10). The challenges they face illustrate how ectotherms can take advantage of environmental heat sources to maintain relatively constant internal temperatures.

Marine iguanas function best at a body temperature of about 37°C. Because they quickly lose heat by conduction to the relatively cool (20°C) ocean water in which they feed, they must warm up before entering the water. On land, they exchange heat with their environment con-

ductively and radiatively to raise their internal temperature. They maximize their radiant heat gain by having black skin and by basking, a behavior that involves presenting as much of their surface area to the sun's rays as possible. In addition, they gain heat conductively by pressing their bellies against the islands' black volcanic rock, which the sun can heat to temperatures of over 40°C. Basking also increases the iguanas' convective heat gain: as they get warmer, their heart rate increases, in turn increasing the flow of blood through the skin to help carry heat from the warmed skin to the tissues in the body core, and bringing cool blood to the skin surface where it can warm up.

After about an hour of basking, when their internal temperature has reached 37°C, the iguanas dive into the water. Negative feedback ends their warming behaviors when they have gained enough heat. While in the water, their heart rate slows, slowing the flow of blood to the skin. This strategy reduces convective heat transfer from the body core to the skin surface. Nonetheless, heat is conducted rapidly from the body into the cool water, allowing the iguanas only about 30 minutes to feed before their falling body temperature forces them to return to land.

Chickadees must reduce heat loss to survive cold winters

Winter represents an especially stressful period in the lives of endotherms. Consider the chickadees that commonly visit North American bird feeders in winter. These small birds must survive winter temperatures of −40°C (almost 80°C lower than their core body temperature) at a time of year when food is scarce. Because chickadees are small, they can lose heat over a large surface area relative to their body volume.

How do chickadees survive the bitter winter cold? They develop a greater feather mass in winter, which allows them to trap a thicker layer of still air next to their bodies. As temperatures drop, chickadees fluff out their feathers, further reducing convection. When not feeding, chickadees nestle among the branches of conifers, which (unlike trees that lose their leaves every winter) can protect them from the wind. While roosting, they decrease their surface area by tucking their unfeathered legs against their bodies and huddling into a rounder shape (thus lowering their surface area-to-volume ratio). Tucking the beak under the wing reduces the amount of evaporative heat (and water) lost with each breath.

When all these actions are not enough, chickadees, like many animals, begin to shiver by contracting their

Chickadee

Biology on the Job

Balancing Your Body on a Bike: More Than Just Staying Upright

John Hepburn is a mechanical engineer who lives in Greensboro, North Carolina. He is the president of the 200-member Greensboro Velo Club (velo is French for "bicycle"). The club regularly participates in many events, including the Tour de Cure sponsored by the American Diabetes Association. Some members of the club treat events like these as competitive races; for them, riding is a serious hobby, almost a job. Every member, however, shares a good time with friends who have an intense interest in bicycling and bicycle racing. We spoke with Mr. Hepburn about the challenges of maintaining homeostasis during a race.

What do you need to think about in terms of keeping your body "up to speed," so to speak, during an event? Three things are basic: we need to take in enough water, enough nutrients, and to try keep our bodies from overheating or getting too cold.

How about water? How do you deal with the water loss that occurs? We combine a need for water with a need to replenish certain nutrients that we are losing. We carry water bottles filled with what you could call "sports drinks." These contain water with sugar and salts. We can also carry the sports drink in a reservoir around our shoulders, which has a tube a rider can drink from. The manufacturer's name, appropriately enough, is CamelBak.

Do these drinks contain lots of sugars and nutrients? Not really. It's important to keep these drinks dilute enough so that they don't draw water out of your body. If they are too concentrated, they can have the same effect as drinking salt water from the ocean. On the other end of the scale, riders have become ill from drinking too much water.

And what do you do about food? There are aid stops along the way, usually three or four in a 60-mile event. We can get food there as well as liquids. Fruits like bananas and apples are common. Other foods tend to be high-calorie and heavy on the carbohydrates, like cookies and peanut butter and jelly sandwiches.

How do you deal with temperature? Planning is the key, and not always easy. For example, a forecast is some help, but if the route takes you to different elevations, the temperature can vary, sometimes a lot. If the event starts early in the day and finishes late in the day, the temperatures you experience can cover a pretty wide range. There are several ways to deal with these challenges. If you are starting off in cool temperatures, you can wear two layers of clothing. The clothing is designed to allow you to take off one layer easily, if temperatures rise as the day progresses. Arm and leg warmers are available for use in cold weather. They wrap around limbs from ankle to thigh or shoulder to wrist.

How about hot weather? You can wear thinner garments. And the material is made from polypropylene or polyethylene that "breathes" and acts as a wick, so cooling evaporation can occur at a high rate. It may surprise you to know that these materials are much cooler and allow much more cooling than, say, a white T-shirt.

Technology Helps Maintain Bike Racers' Homeostasis

The Job in Context

To maintain homeostasis during a bike race, cyclists need to assist their body's regulatory processes in three ways. They must replace water lost through evaporation (sweating) and elimination (urine). They must take in nutrients to maintain solute balances. And they must aid their internal temperature regulation mechanisms by dressing, or undressing, appropriately. John Hepburn and other organizers of these events need to plan ahead for the stresses that will be placed on participants. For instance, they need to space aid stations at intervals that match the demands of the route. A route with long inclines places added stress on riders, so aid stations must be placed at shorter intervals along inclines to assure that the materials needed to maintain homeostasis under these conditions are available to riders.

muscles rapidly. These muscle contractions do no useful work in terms of movement, but they do generate much-needed metabolic heat. Shivering takes a lot of energy, however, so chickadees cannot sustain this activity for an entire long winter night. When their energy reserves begin to run out, chickadees may have to lower their body temperature by up to 12°C, which saves about 20 percent of the metabolic cost of maintaining a normal body temperature. This strategy may make the difference between surviving to morning and dying in their sleep.

Figure 20.13 Heatstroke

Heat waves happen all over the world and can be deadly. During 1995, more Americans died of heatstroke than from any other climate-related cause, and in Chicago alone 465 people died of heat-related causes during the heat wave that hit the city in July of that year. In China in 2004, a heat wave sent many suffering from heatstroke to hospitals, such as this one in Shaanxi Province, for intravenous treatments, and killed at least 39 people in one two-day period.

20.5 Heatstroke: What Happens When Homeostasis Can't Keep Up?

Heatstroke in humans illustrates the limits of homeostasis. Normally, we keep our core body temperature within the narrow range of 36.5°C to 37.5°C, even though the temperature of the air around us usually differs from that temperature. Under some circumstances, such as during strenuous exercise, we must pay close attention to our internal balance to compensate for the extra heat generated (see the box on page 395). Heatstroke occurs when the body temperature rises above 41°C (Figure 20.13). Symptoms include dizziness, nausea, confusion, and loss of muscle control; in severe cases, death can result.

What causes the body temperature to rise to this dangerous level? Normally, when our core body temperature begins to rise above normal, we start to sweat. As the water in the sweat evaporates, it takes with it a surprising amount of heat. A heavily sweating person can get rid of 900 kilocalories in an hour—enough heat to raise the body temperature by over 12°C.

Heatstroke can occur if the sweating response fails, which can happen in two situations. If the body has already lost a large amount of water, the sweating mechanism shuts down to prevent further water loss; this is the main cause of heatstroke in young people. Or the sweating response may be impaired by age or certain prescription drugs, the main cause of heatstroke in the elderly.

How Camels and People Beat the Desert Heat

Now that we have looked at many aspects of homeostasis in more detail, let's return to our comparison of how camels and humans survive the harsh desert climate. The major challenge facing both species is how to keep the body temperature from climbing to dangerous levels without losing too much precious water to evaporative cooling. Whereas the camel survives in the desert through adaptation, humans survive through technology.

Camels can tolerate much greater fluctuations in both body fluid composition and internal temperature than humans. A camel can lose up to 30 percent of its body water and can allow its body temperature to vary by as much as 6°C. In contrast, we can survive the loss of only 12 to 15 percent of our body water, and we must keep our body temperature within a very narrow 1°C range.

Letting their body temperature fluctuate over such a wide range provides camels with several benefits. For one thing, it reduces the need for evaporative cooling. As a camel's internal temperature rises from 34.5°C in the cool desert morning to 40.5°C in the afternoon heat, the camel effectively stores heat. It takes roughly 2,900 kilocalories of heat to raise the body temperature of a typical 500-kilogram camel by 6°C. To lose this much heat evaporatively would cost the camel about 5 liters of water. By simply allowing itself to heat up, the camel conserves that amount of water in its body. The camel then loses the heat by conduction and radiation at night, when air temperatures in the desert drop below the camel's body temperature.

Allowing its body temperature to increase as the day heats up gives the camel another benefit as well: at a typical daily high temperature of 40°C, the temperature difference between the environment and a hot camel is less than that for a cool camel. As its internal body temperature climbs, the rate at which a camel gains heat from its environment slows down.

In contrast to camels, humans do not store much heat. Instead, as the air temperature increases above 30°C, humans sweat freely to offset the increased heat gain. In this way we can keep our body temperature constant. The invention of light-colored clothing provided us with another way to reduce heat gain by both conduction and radiation.

Nonetheless, maintaining a constant body temperature comes at a cost of about 7 liters of water lost through evaporation each day. We must replace the water lost as sweat by drinking regularly.

In spite of allowing their body temperature to fluctuate, camels do use water to regulate it. When their body temperature reaches 40.5°C, camels begin to sweat freely. This happens for sufficiently brief periods that the amount of water lost is small in relation to their body weight—about 5.5 liters of water in a day. Camels will continue, slowly, to lose up to 30 percent of their body water. They can tolerate such severe water losses because the water level in their blood is maintained even when they are badly dehydrated. In humans, the proportion of water in the blood drops as they dehydrate, causing the blood to thicken and flow through the circulatory system too slowly to effectively bring nutrients to and carry wastes from cells and tissues.

Because of their economical use of water, camels can travel for over 2 weeks through waterless stretches of desert. Humans in the same environment must drink several times a day to prevent fatal water loss. It is only because we have developed water-storage technologies that allow us to carry water with us that humans have been able to follow the camel into the desert.

Chapter Review

Summary

20.1 Conditions That Limit Life

- Animals must have the right amount of water in their cells (typically 40 to 65 percent of their body weight) to survive and remain active. Too much water can interfere with chemical reactions, alter protein function, and burst cell membranes. Too little water can also alter protein function.
- Animals must also keep their cells within a certain temperature range. Most cells cannot survive at temperatures below 0°C, at which water freezes, destroying cell membranes, or at temperatures above 40°C, which deform proteins, destroying their function.

20.2 Basic Rules Governing Homeostasis

- Gains of heat, water, and solutes must equal their losses in order to maintain consistent internal conditions.
- Most homeostasis is controlled by negative feedback, in which the results of a process cause that process to slow down or stop.
- Body size affects homeostasis. Large animals, which have a smaller surface area relative to their volume than small animals, do not lose or gain heat as rapidly as small animals. Thus large animals can maintain internal conditions different from those of their environment more easily than small animals.

- Homeostasis requires energy. The greater the difference between external conditions and the desired internal conditions, the greater the energy cost to the animal.

20.3 Regulation of Water and Solute Concentrations

- The cytosol of a cell exchanges water and solutes with the interstitial fluid surrounding the cell across the cell's selectively permeable plasma membrane. The content of each fluid must be regulated to facilitate these exchanges.
- Interstitial fluid exchanges water and solutes with the blood plasma across very thin capillary walls. The circulatory system distributes blood throughout the body, bringing it close to the cells and the interstitial fluid that surrounds them.
- Epithelial tissues provide the surfaces for exchange between the blood plasma and the external environment. They line the skin, the digestive system, the gas exchange surfaces of the lungs, and parts of the kidneys.
- The external environment affects the regulation of body fluid composition. Marine fish, freshwater fish, and terrestrial animals each face different challenges in maintaining the composition of their body fluids.
- Metabolism changes the composition of body fluids. Some waste products, including ammonia, are toxic to cells, requiring continuous removal.
- In many animals, including humans, kidneys regulate body water and solute concentrations. The kidney's basic unit, the nephron, performs three functions: filtration, reabsorption, and secretion. The resulting concentrated solution (urine) is carried by ducts to the bladder and excreted from the body.
- In filtration, blood plasma is filtered across capillary walls in the glomerulus, sending water and small solutes into the tubule. In reabsorption, water and valuable solutes in the filtrate pass back across the tubule walls and return to the capillaries. In secretion, larger waste substances are actively secreted into the tubule.

20.4 How Animals Gain and Lose Heat

- Animals gain heat from two sources: externally from the sun (directly or indirectly) and internally from metabolic heat.
- Endotherms (including birds and mammals) depend on metabolic heat to maintain high, nearly constant body temperatures regardless of environmental temperatures. This close regulation requires a lot of energy to maintain, but allows animals to be active in a wide range of external temperatures.
- Ectotherms (fish, invertebrates, amphibians, and reptiles) depend more on environmental heat than on metabolic heat, and their internal temperatures vary more. This form of regulation requires less energy but forces animals to be inactive in cold environments.
- Animals can exchange heat with their environment by three means: conduction (contact with a substance, usually a solid or liquid), radiation (absorbing or giving off light or infrared energy), and evaporation (conversion of water to water vapor on a body surface).

- Heat (energy) flows from warmer to cooler places. An animal gains heat by conduction when its surroundings are warmer than it is, and loses heat when it is in contact with colder substances.
- Convective movement of air or water can speed radiant or conductive heat exchange.
- An animal can lose heat evaporatively even when the environmental temperature is greater than its internal temperature.

20.5 Heatstroke: What Happens When Homeostasis Can't Keep Up?

- Homeostasis in humans keeps the core body temperature within a 1°C range. At high environmental temperatures, we sweat, losing heat evaporatively.
- Heatstroke occurs when the sweating mechanism shuts down, allowing the body temperature to rise. Sweating shutdown may be caused by the loss of large amounts of water, old age, and certain prescription drugs.

◉ Review and Application of Key Concepts

1. Describe the limits within which life can exist and how these limits affect the cells in an organism.

2. What are the general rules of homeostasis?

3. Imagine that you were getting ready for an expedition to the North Pole. What tactics might you use to maintain homeostasis in this cold, dry environment? Discuss some of the ways humans use technology to assist homeostasis.

4. Describe negative feedback in terms of eating your favorite meal.

5. Describe the process of filtration and reabsorption in a nephron.

6. What are the three routes by which animals exchange heat with the external environment? Describe each route in terms of an animal's body and its surroundings.

7. Describe heatstroke from the point of view of an internal cell and its immediate surroundings. How might heatstroke affect the cell's ability to function?

Key Terms

blood plasma (p. 386)	epithelial tissue (p. 386)
capillary (p. 386)	evaporation (p. 393)
conduction (p. 391)	feedback (p. 382)
convection (p. 393)	filtrate (p. 389)
ectotherm (p. 391)	filtration (p. 389)
endotherm (p. 390)	glomerulus (p. 389)

homeostasis (p. 380)
interstitial fluid (p. 385)
kidney (p. 388)
metabolic heat (p. 390)
negative feedback (p. 382)
nephron (p. 389)

positive feedback (p. 382)
radiation (of heat) (p. 393)
reabsorption (p. 389)
secretion (p. 390)
tubule (p. 389)
urine (p. 388)

Self-Quiz

1. Which of the following typically happens when water inside cells freezes?
   a. The cells function at full capacity.
   b. Animal cells begin to produce ammonia.
   c. The plasma membranes are ripped apart.
   d. Animals encourage evaporative heat loss to restore homeostasis.

2. Interstitial fluid is
   a. found only in kidney tubules.
   b. the fluid portion of blood.
   c. the fluid surrounding cells.
   d. found in kidney capillaries.

3. Blood plasma transports primarily
   a. heat.
   b. water.
   c. solutes.
   d. all of the above

4. In the nephron, water is reabsorbed in the
   a. glomerulus.
   b. urine.
   c. tubule.
   d. plasma membrane.

5. An important function of the human kidney is to
   a. produce a concentrated urine.
   b. produce a dilute urine.
   c. convert toxic wastes into harmless ones.
   d. all of the above

6. Endothermic animals
   a. are usually cooler than their surroundings.
   b. cannot gain heat by conduction.
   c. are usually warmer than their surroundings.
   d. generate relatively small amounts of metabolic heat.

7. Ectothermic animals
   a. produce large amounts of metabolic heat.
   b. rely primarily on external sources of heat.
   c. cannot radiate heat.
   d. cannot lose heat by convection.

8. The primary heat source for endothermic animals is
   a. metabolic heat.
   b. conduction.
   c. radiation.
   d. evaporation.

9. Fur and feathers
   a. reduce heat loss in endothermic animals.
   b. reduce convective heat transfer.
   c. reduce heat loss by trapping air, which conducts heat poorly.
   d. all of the above

10. Chickadees can survive cold winters by
    a. developing a greater feather mass.
    b. fluffing their feathers.
    c. tucking their legs against their bodies.
    d. all of the above

France Heat Wave
Death Toll at 15,000

Paris—An estimated 15,000 people died in France's scorching heat wave last month, the country's largest undertaker said Tuesday, surpassing the official government estimate of 11,435.

Isabelle Dubois-Costes, a spokeswoman from General Funeral Services, said the updated estimate included deaths in the second half of August, after record-breaking temperatures had abated.

During the summer of 2003, Europe was hit by an unprecedented heat wave. Record-setting temperatures topped 40°C (104°F) in European countries used to much milder temperatures. These record high temperatures persisted throughout much of July and August. The effects proved disastrous.

As the article indicates, in France alone, 15,000 deaths, most of them among the elderly, were attributed to the heat. Although the French population suffered most, the officially acknowledged number of heat-related deaths approached 20,000 across Europe. Patients with heatstroke and dehydration swamped hospitals, and the bodies of the deceased piled up more quickly than morgues could handle them.

Europeans have access to medical care as good as any in the world. Why, then, did temperatures that most North Americans have experienced take such a grim toll? In the wake of the disaster, fingers have been pointed at a number of potential culprits.

Some of the suggested causes are cultural. In countries whose populations are used to peak summer temperatures 15°C lower than those recorded in the summer of 2003, air conditioning is much less common than in North America. Thus people had fewer options for escaping the heat. In addition, the heat wave peaked during August, a time when many Europeans traditionally take their summer holidays. As a result, hospitals struggling to deal with the influx of patients were understaffed. And in many cases, families left elderly relatives at home alone when they headed off on vacation.

Other causes were biological. Most of the victims were elderly. With increasing age, humans have a harder time regulating their body temperature. To make matters worse, many older people are afflicted with illnesses or take medications that interfere with their ability to regulate their body temperature.

Evaluating the news

1. In July of 1995, a week-long heat wave in Chicago is reliably estimated to have contributed to the deaths of between 700 and 800 residents of the city (see Figure 20.11). Can you think of why this might have happened in a North American city?

2. Many of us have elderly relatives. What would you do to ensure their well-being during periods of hot weather, which most places in North America now experience at some point during the summer?

3. A disproportionate number of heat-related deaths occur in cities rather than in the countryside. What are some of the physical and social aspects of cities that make their residents particularly vulnerable to extreme summer heat?

SOURCE: *Boston Globe*, September 9, 2003.

CHAPTER 21

Nutrition and Digestion

Key Concepts

◉ Nutrients provide animals with energy and with the molecular building blocks that make up their bodies.

◉ Animals depend on a relatively small number of chemical elements, which they get by eating other organisms.

◉ The human digestive system is a continuous passageway between mouth and anus. It has specialized regions and organs that break food down into a form that can be swallowed, then process it into small molecules that can be absorbed.

◉ Despite great differences among the diets of animals, their digestive systems share the same general features. Most animal digestive systems process food mechanically and chemically as it moves from mouth to anus.

◉ The digestive systems of herbivores are specialized for extracting nutrients from their nutritionally poor diet of plant tissue.

Body Image: The Mirror and the Index

We have to live in our bodies, but we certainly don't have to like them. Only 1 percent of women responding to a recent survey in England admitted that they were "completely happy" with the shape of their bodies. At any one time, 45 percent of women and 30 percent of men in the United States are trying to lose weight. Apparently following their parents' lead, even children as young as 5 years old worry about their weight. Are we right to worry so much about our weight?

On the "No" side, much of the pressure that drives people to want to change their bodies arises from cosmetic, rather than health, reasons. We do not have to look far to find magazines, films, and television programming filled with young and unrealistically slender, well-toned bodies. Research shows that our society discriminates against people seen as being over-weight, both in social life and in the workplace. These pervasive social pressures create a distorted self-image that leads people to want to lose weight even when it is neither necessary nor healthy to do so. For example, a survey of Australian adults with a healthy weight revealed that almost half of the women and a quarter of the men felt that they should lose weight. More alarmingly, a number of surveys have shown that over 20 percent of underweight women see themselves as overweight. In an attempt to improve their self-image, many people go on inappropriate or even danger-ous diets or turn to cosmetic surgery. At the extreme, some people (prima-rily young girls) develop eating disorders such as anorexia (excessive dieting and exercise) and bulimia (purging just-eaten meals with laxatives and vom-iting), which over time can cause them literally to starve to death.

On the "Yes" side of the question, both health professionals and govern-ment officials warn that many people weigh too much. Their concern lies

Variation in Body Mass Index
Pedestrians on any city sidewalk are likely to show a range of BMI values.

Biology Matters

The Body Mass Index

	Normal						Overweight					Obese										Extreme obesity							
BMI	19	20	21	22	23	24	25	26	27	28	29	30	31	32	33	34	35	36	37	38	39	40	41	42	43	44	45	46
Height (inches)												**Body weight (pounds)**																
58	91	96	100	105	110	115	119	124	129	134	138	143	148	153	158	162	167	172	177	181	186	191	196	201	205	210	215	220
59	94	99	104	109	114	119	124	128	133	138	143	148	153	158	163	168	173	178	183	188	193	198	203	208	212	217	222	227
60	97	102	107	112	118	123	128	133	138	143	148	153	158	163	168	174	179	184	189	194	199	204	209	215	220	225	230	235
61	100	106	111	116	122	127	132	137	143	148	153	158	164	169	174	180	185	190	195	201	206	211	217	222	227	232	238	243
62	104	109	115	120	126	131	136	142	147	153	158	164	169	175	180	186	191	196	202	207	213	218	224	229	235	240	246	251
63	107	113	118	124	130	135	141	146	152	158	163	169	175	180	186	191	197	203	208	214	220	225	231	237	242	248	254	259
64	110	116	122	128	134	140	145	151	157	163	169	174	180	186	192	197	204	209	215	221	227	232	238	244	250	256	262	267
65	114	120	126	132	138	144	150	156	162	168	174	180	186	192	198	204	210	216	222	228	234	240	246	252	258	264	270	276
66	118	124	130	136	142	148	155	161	167	173	179	186	192	198	204	210	216	223	229	235	241	247	253	260	266	272	278	284
67	121	127	134	140	146	153	159	166	172	178	185	191	198	204	211	217	223	230	236	242	249	255	261	268	274	280	287	293
68	125	131	138	144	151	158	164	171	177	184	190	197	203	210	216	223	230	236	243	249	256	262	269	276	282	289	295	302
69	128	135	142	149	155	162	169	176	182	189	196	203	209	216	223	230	236	243	250	257	263	270	277	284	291	297	304	311

SOURCE: Adapted from *Clinical Guidelines on the Identification, Evaluation, and Treatment of Overweight and Obesity in Adults: The Evidence Report.*

You can find your BMI in the Body Mass Index table above by finding your height in inches at the left, then moving your finger to the right until you find your weight. Keep in mind that a high BMI doesn't necessarily indicate a weight problem. Muscle weighs more than fat, so a highly conditioned athlete with a lot of muscle mass would have a high BMI. Abnormally low BMIs, on the other hand, often indicate an eating disorder. An active person who is 5 feet 6 inches tall and weighs 85 pounds has a BMI of 13.7, a strong indicator of the eating disorder anorexia nervosa.

At the opposite end of the scale, obesity is fast becoming a major threat to good health in the United States. About 127 million American adults are overweight (BMI 25–29), 60 million are obese (BMI 30–39), and 9 million are severely obese (BMI over 40). As the table at right shows, there have been large increases in the percentage of adults who are overweight, obese, or severely obese in the past 30 years.

Even more worrisome is the fact that obesity in American children is reaching epidemic proportions. Today, one child in five is overweight. Obese children often have high cholesterol and blood glucose levels, increasing their risk of heart disease and diabetes. Obese children have a seven in ten chance of being obese as adults.

	Percent of adults over 20 who are overweight	Percent of adults over 20 who are obese	Percent of adults over 20 who are severely obese
1999–2000	64.5	30.5	4.7
1988–1994	56.0	23.0	2.9
1976–1980	46.0	14.4	No data

not so much with body image as with body mass index (BMI), a measure that takes into account both height and weight (see the box on page 404). A person 5 feet 8 inches tall who weighs over 200 pounds would be classified as obese using their BMI of 30.4. However, a muscular athlete may have a high BMI and still look "trim" because muscle is denser than fat. So a high BMI doesn't necessarily indicate a weight problem.

Nevertheless, for the average person, the BMI provides a good indication of whether the person is at risk for various weight-related health problems, including heart disease, diabetes, and reduced mobility. Anyone who is obese (defined as having a BMI of 30 or more) has an increased risk of suffer-ing from weight-related illness. In both developed and developing countries, and for males and females of all ages, the BMI has increased over the past decade. In the early 1990s, for example, only 14 percent of American men between 20 and 34 years of age were considered obese. Today that percentage has almost doubled, to 24 percent. The potential future cost to health care systems has governments worried about this trend.

The BMI can provide individuals with a more sensible guide to what weight is reasonable for them than the images seen in the media or messages coming from peers. Having a high BMI is a good reason to lose weight. But having a lower than normal BMI also carries many risks. How to gain that happy medium?

Some people may live to eat, but all people must eat to live. The nutrients that we get from food play two critical roles in our survival (Figure 21.1). Nutrients provide us with the energy we need in order to live; they also provide our bodies with the raw materials for constructing the molecules on which our survival depends.

Like all animals, we are consumers. This means that we cannot get either the energy or the chemical building blocks we need directly from our nonliving physical surroundings, as plants do. Instead, we must get both by eating other organisms. Thus the **nutrients** required by animals include biologically produced chemical compounds, such as proteins, as well as inorganic chemicals, such as sodium. Eating is the first stage of **digestion**, the chemical breakdown of food that takes place in the bodies of all animals, including humans. Through digestion, our bodies obtain the essential nutrients they need to function properly.

In this chapter we look at how animals convert the organisms they eat into forms that allow the nutrients in those organisms to be used for their own life processes. We begin with an overview of the nutrients animals need and the forms in which animals can use them. Next we take a tour of the human digestive system and then compare it with those of other animals. We conclude by considering some of the ways animals convert different types of food into absorbable nutrients.

Figure 21.1 The Fate of Food: A Brief Overview
Food contains materials from which we obtain energy and the chemical building blocks needed for building our own molecules. Digestion converts food into usable nutrients that either supply energy or become incorporated into body molecules.

21.1 Nutrients That Animals Need

Although nutrients are contained in the food ingested by animals, animals usually cannot use nutrients in the form in which they are found in food. The process of digestion converts food into forms that allow nutrients to enter the body and move into cells.

Food supplies animal cells with chemical elements

Animal cells depend on relatively few of the known chemical elements (Figure 21.2). Some elements are needed in relatively large amounts, making up over 99 percent of an animal's body weight. Of these eleven elements, carbon, hydrogen, and oxygen alone make up about 93 percent of the body. Carbon, sulfur, phosphorous, oxygen, nitrogen, and hydrogen are contained in the large organic molecules that characterize life, including carbohydrates, amino acids, fatty acids, and nucleotides. These organic compounds not only make up the molecules that give animal bodies their structure, but also store energy in the chemical bonds that hold those molecules together. By breaking these chemical bonds during glycolysis and respiration, animals release the energy that they need to survive. Animals use these elements for structures or functions that occur throughout their bodies. For example, without nitrogen we could not make the proteins that form the cytoskeletons of cells and control all of the chemical reactions that keep us alive. Nor could we make DNA, the genetic material that encodes all of the information that makes us what we are.

The remaining elements needed in relatively large amounts (magnesium, calcium, potassium, sodium, and chlorine) together make up only 6 percent of our body weight. These elements are often used in their ionic form to regulate cell functions.

Ten other elements together make up less than 1 percent of an animal's body weight. Although these elements occur in tiny amounts, they are essential: animals cannot survive without them. The human body, for example, contains an amount of iodine that would come to less than a quarter teaspoon. Without this minute quantity, however, the body cannot make the chemicals that determine how we burn energy and how the body grows.

Animals get their nutrients by eating other organisms

Animals must eat other organisms, either living or dead. We call animals that eat plants **herbivores**. **Carnivores** eat other living animals. Some animals (including humans), called **omnivores**, eat both plants and animals.

Plants and animals provide the animals that eat them with the three main classes of nutrients produced only by living organisms: carbohydrates, fats, and proteins. But animals cannot absorb these nutrients in the form in which they are ingested. Most carbohydrates, and all fats and proteins, occur in food as very large molecules that cannot cross the plasma membranes of the epithelial cells that form the absorbing surface of an animal's digestive system. Animals must break down these molecules into the simpler subunits of which they are built before they can absorb them. Thus animals must break

Figure 21.2 The Elements Essential to Life
Here the periodic table used by chemists has been adapted to show the chemical elements that play an essential role for animals. Some elements (shown in red) are essential to all animals in relatively large amounts, but others (shown in pink) are needed in only small amounts.

carbohydrates into monosaccharides (simple sugars), fats into fatty acids and **monoglycerides** [*MAH*-noh-*GLISS*-er-ides], and proteins into amino acids (Figure 21.3). Animals then reassemble these subunits into new carbohydrates, fats, and proteins inside their cells.

Once absorbed, the monosaccharides, fatty acids, monoglycerides, and amino acids serve a second important function in animals: they provide the energy for respiration (Table 21.1). Carbohydrates and fats provide us with most of the energy we need each day, which for inactive people ranges from about 1,600 kilocalories for women to 2,200 kilocalories for men. Animals use **carbohydrates** primarily as a convenient source of quick energy. **Fats**, because they pack more calories into each gram than carbohydrates do, provide a particularly good way of storing energy. By packing on fat before they hibernate or migrate long distances, animals can store a lot of energy while putting on relatively little weight.

Table 21.1

Nutrients Provide Animals with Energy

Note: One calorie (abbreviated cal) = the energy needed to raise 1 gram of water 1°C. One kilocalorie (abbreviated kcal) = 1,000 calories, or the amount of energy needed to raise 1 kilogram of water 1°C. An 180-pound human can walk about a mile on 70 kcal.

Nutrient	Absorbable units	Energy content of 1 gram (kcal)	Major use
Carbohydrates	Monosaccharides	4.8	Energy
Fats	Fatty acids, monoglycerides	9.6	Energy storage
Proteins	Amino acids	4.8	Building other proteins

Only certain forms of nutrients can pass from the outside of the plasma membrane to the inside.

A few di- and tri-peptides can cross plasma membranes.

Figure 21.3 The Absorbable Forms of Nutrients
Most animal cells can absorb only the simple breakdown products of the carbohydrates, proteins, and fats contained in the food they eat. Many minerals can cross membranes only in their ionic forms.

Learn more about the digestion and absorption of food.

Proteins provide as much energy per unit of weight as carbohydrates do when broken down, but proteins play a much more important role as a source of amino acids, which animal bodies use to build their own proteins, than they do as a source of energy. During periods of extreme starvation, however, the body begins to digest the proteins in its own tissues, although only after depleting its internal stores of carbohydrates and fats. Death can result when the body is forced to use its own protein in this way for too long.

Certain nutrients can be obtained only from food

The nutritional labels on our foods give us a good overview of the kinds of nutrients we need. For example, the label on the cereal box shown in Figure 21.4 first lists the nutrients we need in relatively large amounts—fats, proteins, and carbohydrates—and then lists those nutrients needed in smaller amounts (including vitamins and minerals). All of these nutrients come from the grains used to make this particular cereal. Let's consider these nutrient groups in a little more detail.

Animals get energy from organic compounds in their food.

The diet of animals consists mostly of proteins, fats, and carbohydrates.

Vitamins are organic micro-nutrients that are essential to animals in small amounts.

Figure 21.4
Nutritional Labels Can Tell Us a Great Deal
The label from a cereal box indicates the important sources of energy and nutrients for humans. We obtain energy from carbohydrates, fats, and proteins.

Although humans can synthesize some of the 20 amino acids needed to make proteins, we must get eight of them, called **essential amino acids**, from our food (Figure 21.5). Whereas our bodies can get all eight essential amino acids by breaking down the proteins found in any animal food, people who eat only plant foods must include a variety of those foods to do the same.

Vitamins are organic nutrients needed by our bodies, but only in tiny amounts. Vitamins have two other characteristics that distinguish them from the groups of nutrients we have already discussed: unlike proteins, vitamins are not used to generate the physical building blocks of the body. And unlike carbohydrates and fats, vitamins do not provide us with energy. Rather, they help our bodies regulate metabolic processes.

Our bodies either cannot manufacture vitamins at all or can make them only in amounts that are insufficient to meet our needs. Thus we have to get vitamins from our food. Humans require two classes of vitamins: water-soluble vitamins, which help speed essential chemical reactions in the body, and fat-soluble vitamins, which play a variety of roles (Table 21.2).

Often spoken of in the same breath with vitamins are **minerals**, small inorganic compounds that, like vitamins, are needed by the body in small amounts. A few of the nutrients in cereal, or almost any food, come in mineral form, such as sodium, potassium, and zinc. An example is table salt, which is composed of sodium chloride (NaCl) crystals. We often get minerals in their elemental form.

Insufficient amounts of vitamins and minerals in the diet can lead to deficiency diseases such as pellagra (Figure 21.6). Although vitamin and mineral supplements are widely available commercially, there is little scientific evidence of any benefit from taking more than the recommended daily dose of a vitamin. The body quickly rids itself of excess water-soluble vitamins, and fat-soluble vitamins may build up in the body to harmful levels (see Table 21.2).

21.2 The Human Digestive System

After we eat a meal, our bodies still have a lot of work to do before absorbing the nutrients that the food contains. We know that a slice of pizza can't just be swallowed whole and absorbed into our bodies. Even after a thorough chewing, pizza contains carbohydrates, fats, and proteins that are too big and complex to absorb.

The **digestive system** (Figure 21.7) converts the food animals eat into nutrients in a form their bodies can absorb and use. The human digestive system consists of a tubular passageway through which food flows from mouth to anus. Associated with the digestive system are various glands and organs that secrete substances into this tube. (A *gland* is an organ that secretes one or more substances.) The tube consists of several regions, each of which plays a specific role in digestion. Epithelial tissues lining the digestive system are specialized to carry out the particular functions associated with each section of the tube. As food passes through the tube, it is broken down both physically and chemically, until it is eventually converted into simple molecules that can cross the lining of the digestive system and move into the body. Keep in mind that food that is still in the digestive system (that has not yet been absorbed across the lining of the digestive system into other tissues) remains outside the body as far as the cells needing the nutrients are concerned.

During **ingestion** (eating), food enters the **oral cavity** (mouth), where mechanical breakdown and some chemical breakdown begins. Our teeth are shaped to cut, crush, or grind ingested food into smaller pieces, which provide a greater surface area on which digestive enzymes can work. The muscular tongue mixes these food particles with **saliva**, a fluid substance produced and delivered by six **salivary glands** associated with the mouth. Saliva contains enzymes that start to break down starches in the food chemically; it also turns the food into a moist mass that can slip easily down the throat. (Try swallowing dry crackers!)

Figure 21.5 Essential Amino Acids in the Human Diet
Humans cannot produce eight of the twenty amino acids that we use to build proteins, so we must get them from our food. Meat contains all eight essential amino acids, but vegetarians must combine different grains and vegetables to get all eight. The pink box lists the amino acids that both grains and legumes provide in adequate quantities.

Eight essential amino acids must be obtained from foods:

Tryptophan
Methionine

Grains

Valine
Threonine
Phenylalanine
Leucine

Isoleucine
Lysine

Legumes

Test your knowledge of the human digestive system.

21.2

Figure 21.6 Insufficient Vitamin Intake Can Produce a Deficiency Disease
Pellagra, which results from a deficiency in the B vitamin niacin, often occurs among people whose diets depend heavily on corn. Symptoms of pellagra include diarrhea, mental confusion, and, as shown here, a shiny scaly rash that develops on skin exposed to the sun.

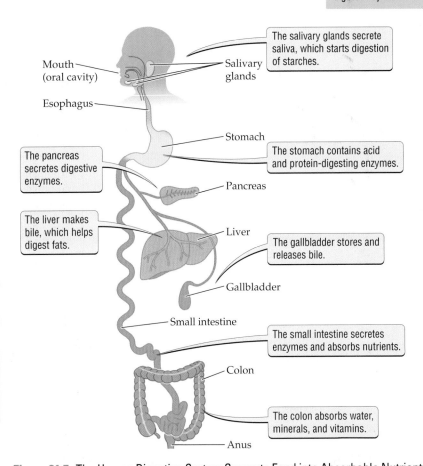

Mouth (oral cavity)

Esophagus

The salivary glands secrete saliva, which starts digestion of starches.

Salivary glands

Stomach

The stomach contains acid and protein-digesting enzymes.

The pancreas secretes digestive enzymes.

Pancreas

The liver makes bile, which helps digest fats.

Liver

The gallbladder stores and releases bile.

Gallbladder

Small intestine

The small intestine secretes enzymes and absorbs nutrients.

Colon

The colon absorbs water, minerals, and vitamins.

Anus

Figure 21.7 The Human Digestive System Converts Food into Absorbable Nutrients
As it moves through the digestive system, food is broken down into individual small molecules that can be absorbed across the lining of the intestine. Different organs and regions along the way have different functions.

Table 21.2

Vitamins Needed in the Human Diet

The human body cannot make these essential vitamins, or else makes them in insufficient amounts, and so must get what it needs from food. Vitamin C is a vitamin only for primates (including humans) and a few other animals (such as guinea pigs). All other animals can make vitamin C as needed.

Class	Vitamin	Main functions	Possible symptoms of deficiency and *excess*
Water-soluble	B vitamins: thiamine (B_1), riboflavin (B_2), niacin (nicotinamide), pyridoxine (B_6), pantothenic acid, folate (folic acid), cyanocobalamin (B_{12}), biotin	Act with enzymes to speed metabolic reactions, or act as raw materials for chemicals that do so. Work with enzymes to promote necessary biochemical reactions.	Deficiency: B vitamins act in concert; deficiency in one can cause symptoms related to deficiency in others. Deficiency diseases include pellagra and beriberi (damage to heart and muscles). *Excess: B_6 in excess can cause neurological damage.*
	Vitamin C (ascorbic acid)	Assists in maintenance of teeth, bones, and other tissues.	Deficiency: scurvy (teeth and bones degenerate), increased susceptibility to infection. *Excess: diarrhea, kidney stones with chronic overuse.*
Fat-soluble	Vitamin A (carotene)	Produces visual pigment needed for good eyesight; also used in making bone.	Deficiency: poor night vision; dry skin and hair. *Excess: nausea, vomiting, fragile bones.*
	Vitamin D	Promotes calcium absorption and bone formation.	Deficiency: poor formation of bones and teeth, irritability. *Excess: diarrhea and fatigue.*
	Vitamin E	Protects lipids in cell membranes and other cell components.	Deficiency: very rare. *Excess: heart problems.*
	Vitamin K	Produces clotting agent in the blood.	Deficiency: prolonged bleeding, slow wound healing. *Excess: liver damage.*

Swallowing brings the food into the tubular **esophagus** [eh-SOF-uh-guss], where waves of muscular contraction carry it from the mouth to the **stomach**. Digestion begins in earnest in the stomach, which contains acid and pepsin, an enzyme that begins to break proteins into shorter chains of amino acids. Muscles in the wall of the stomach alternately contract and relax to mix the food particles with the acid and pepsin. The mixture is stored in the stomach until it can move into the small intestine.

The **small intestine** is a thin tube (about 1 inch, or 2.5 centimeters, in diameter) that is highly coiled; if straightened, it would extend about 20 feet (6 meters). The upper and lower parts of the small intestine serve different functions. The upper part of the small intestine, which lies nearest the stomach, digests the large molecules coming from the stomach into simpler forms that the body can absorb. Here a number of enzymes that specifically digest proteins, carbohydrates, and lipids complete the digestion of the food. The small intestine itself produces some of these enzymes, but many are produced in organs closely associated with the small intestine. The **pancreas** produces a broad range of enzymes that reach the upper small intestine through a duct. **Bile**, which contains the chemicals needed to convert fats, in particular, into an absorbable form, is produced by the **liver**, an organ closely associated with the upper small intestine and serving a multitude of functions. The liver secretes bile to the **gallbladder**, which stores it and dispenses it to the upper small intestine as needed.

The lower portion of the small intestine provides a large surface area for the absorption of nutrients and water. This absorption takes place across huge numbers of tiny fingerlike projections called **villi** (singular: villus) (Figure 21.8). Each villus is about 1 millimeter long, with a surface consisting of specialized cells that have microscopic extensions of their plasma membranes. Because these extensions give the outer surfaces of the villi a fuzzy, brushlike appearance under the microscope, this feature of the intestinal lining is called the **brush border**. Each projection in the brush border is less than one-thousandth of a millimeter long. This complexly folded surface provides the lower small intestine with its impressive surface area for absorption—almost 300 square meters, roughly equivalent to the area of a tennis court!

For absorption to occur, the digested nutrients must move across the epithelial lining of the small intestine and into the body. The inside of each villus has two ways of carrying the absorbed nutrients off to other cells: blood capillaries (part of the circulatory system) carry amino acids and sugars, and a thin-walled lymph vessel (part of the lymphatic system) carries monoglycerides and fatty acids.

Figure 21.8
The Lining of the Small Intestine Is Complexly Folded
Small projections called villi increase the absorptive area of the small intestine. The cells of the villi have microscopic extensions of their plasma membranes that increase the absorptive area still further.

By the time the material remaining in the digestive system leaves the small intestine, it contains relatively few nutrients. In the **colon** (also called the large intestine), this digested matter is prepared for removal from the body as waste, a process that involves three different functions. First, the colon absorbs almost all remaining minerals from the waste stream as ions. Second, the colon absorbs almost all of the remaining water from the waste stream. Finally, large numbers of bacteria living in the colon break down some of the remaining waste to release additional nutrients that the body can absorb. These bacteria also produce some vitamins that are absorbed from the colon. The end result is a relatively solid waste product, politely referred to as **feces** [*FEE*-seez], that consists mostly of indigestible material and the remains of bacteria. The feces leave the body through the **anus**.

21.3 Comparing Animal Digestive Systems

As it turns out, the human digestive system makes a good model for animal digestive systems in general. Using our understanding of what happens in the human digestive system, let's highlight some of the general patterns in how animals break down and absorb nutrients from the food they eat.

Test your knowledge of the typical digestive system in animals.

Animal digestive systems share a similar overall design

Considering the diversity of form and diet among animals, their digestive systems follow a strikingly similar pattern (Figure 21.9*a*). In humans and most other animals, the digestive system moves food through the body from mouth to anus by means of muscular contractions. This one-way flow sets up a sort of food-processing assembly line, along which the different regions of the digestive system perform specialized functions.

The functions of the regions of the human digestive system follow a consistent sequence. The first step is mechanical and physical breakdown of food into ever smaller and simpler pieces. This is followed by chemical breakdown, nutrient absorption, and waste disposal. Other animals have digestive functions laid out in much the same sequence (Figure 21.9*b*). Parts of the animal that share a function with part of the human digestive system are often given the same name, even if their evolutionary origins differ greatly.

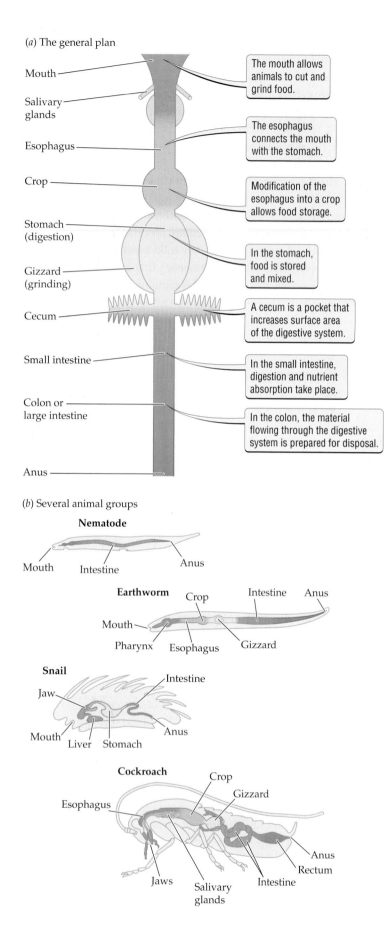

(a) The general plan

Mouth — The mouth allows animals to cut and grind food.

Salivary glands

Esophagus — The esophagus connects the mouth with the stomach.

Crop — Modification of the esophagus into a crop allows food storage.

Stomach (digestion)

Gizzard (grinding) — In the stomach, food is stored and mixed.

Cecum — A cecum is a pocket that increases surface area of the digestive system.

Small intestine — In the small intestine, digestion and nutrient absorption take place.

Colon or large intestine — In the colon, the material flowing through the digestive system is prepared for disposal.

Anus

(b) Several animal groups

Nematode
Mouth   Intestine   Anus

Earthworm   Crop   Intestine   Anus
Mouth
Pharynx   Esophagus   Gizzard

Snail   Intestine
Jaw
Mouth   Liver   Stomach   Anus

Cockroach   Crop
Esophagus   Gizzard
Anus
Rectum
Jaws   Salivary glands   Intestine

Figure 21.9
Animal Digestive Systems Share a Similar Organization
(a) The digestive systems of all animals are organized in a remarkably consistent way, as shown in this generalized diagram. (b) Even the digestive systems of animals from very different groups show variations on the same theme.

Many animals break food into increasingly smaller pieces to speed digestion

As they ingest food, many animals use muscles in combination with hard surfaces to mechanically break large food particles into smaller ones. These smaller pieces are then moistened for easier swallowing. Smaller food particles expose more of their surface area, relative to their volume, to acid and enzymes, which greatly speeds their digestion.

Humans and many other animals use hard teeth embedded in jaws to crush, rip, tear, or grind large food particles into smaller ones (see Figure 21.9b). In other animals, including earthworms and birds, part of the digestive system is (or was, in the case of some extinct dinosaurs) modified into a muscular **gizzard**, which grinds food together with rocks or sand grains that the animal has swallowed. Certain other consumers, including animals that lack teeth or even jaws, depend completely on chemical digestion to break down food. Their food must be ingested in small particles or as a liquid. For example, orb-weaving spiders cannot mechanically break up their food because they lack chewing mouthparts; however, their piercing jaws can inject venom into their prey. After the venom has liquefied the tissues of its prey, the spider sucks out the fluid and discards all of the hard parts (Figure 21.10).

Animals convert complex molecules into simpler ones for absorption

All animals must convert complex nutrients into simpler forms that can cross the epithelial cells lining the small intestine. Like those of humans, their digestive systems can absorb carbohydrates only as monosaccharides, but in most animal diets carbohydrates are ingested largely as starches and cellulose. Recall from Chapter 4 that these large molecules consist of many monosaccharide monomers bonded together. Similarly, other animals' digestive systems can absorb the building blocks needed to build proteins only as individual amino acids, but the amino acids in food come "packaged" in proteins that often consist of hundreds of amino acids strung together.

To digest complex carbohydrates, fats, and proteins, animals, including humans, use a variety of enzymes. (For a review of how enzymes work, see Chapter 7.) Each digestive enzyme acts on a specific type of chemical bond. Rennin, for example, is an enzyme that helps newborn mammals digest milk proteins. It works by specifically cutting the bonds joining hydrophobic amino acids and leaving intact the bonds joining other types of amino acids. Humans learned thousands of years ago how to exploit rennin to separate milk solids (also called curds, such as those found in cottage cheese) from the liquid (the whey) as a first step in cheese production.

Most animals produce enzymes that can digest most of the nutrients in their diet. For example, the saliva of mammals contains amylase, which breaks starches into sugars. The stomach produces acid and pepsin, which together start the process of breaking proteins into shorter strings of amino acids. Pepsin is specialized to work well in the acidic (low pH) environment of the stomach. The partially digested food leaving the stomach is also very acidic, creating a potential problem as it enters the small intestine: the enzymes that carry out most of the digestion of our food in the small intestine do not work well in a highly acidic environment. To adjust the pH, the pancreas secretes sodium bicarbonate—perhaps more familiar in its powdery form as baking soda—which neutralizes the acid in the upper small intestine.

The mammalian pancreas also secretes several enzymes into the small intestine. Pancreatic amylase breaks long carbohydrate molecules into simpler sugars. Trypsin and chymotrypsin [KIGH-moh-TRIP-sin] break down proteins into amino acids. Pancreatic lipases break fats into absorbable fatty acids and monoglycerides. The digestion of carbohydrates and proteins started by the pancreatic enzymes is completed by epidermal cells lining the small intestine. These cells secrete a variety of enzymes that split short chains of sugars into absorbable monosaccharides and short chains of amino acids into absorbable single amino acids.

The digestion of fats poses a particular problem because fats are not very soluble in water. Fats form globules that are too big for the epithelial cells lining the small intestine to absorb. These globules must be broken down and made to mix more evenly with the watery contents of the digestive system. Like the human liver, the livers of many animals produce bile. This substance acts the same way dishwashing detergent does on greasy dishes: it puts a coating on fat globules that lets them partially dissolve in water. Large globules break into tiny droplets, on which lipases can work more easily.

Figure 21.10 Orb-Weaving Spiders Rely on Chemical Digestion
Araneus pima, like all orb-weaving spiders, injects its prey with venom, which contains toxins that quickly liquefy the tissues of the prey, allowing the spider to drink its meal.

Missing enzymes can prevent the uptake of nutrients

When an animal lacks the enzymes needed to break nutrients into absorbable units, it cannot use those nutrients. Consider the enzyme lactase, which breaks the disaccharide lactose into monosaccharides (Figure 21.11*a*). Human infants, like all other young mammals, need lactase when nursing because lactose is the sugar in their mothers' milk. Until humans began dairy farming, however, adults rarely encountered lactose in their diets. Accordingly, most humans stop producing lactase as they mature. Only adult individuals of European, northern Indian, Arabian, or central African descent frequently retain the ability to produce lactase as adults (Figure 21.11*b*).

The ability of adults in these populations to digest lactase in milk probably arose by chance mutation. Adults who come from populations that do not produce lactase and who drink milk containing lactose experience digestive problems: bacteria living in the digestive system break down the lactose, and the by-products of this digestion are painful gas and diarrhea. Some adults who do not produce lactase come from populations that normally produce the enzyme; these adults are said to have a condition known as lactosis. For all of these people, bacterially altered milk products, such as cheese and yogurt, are generally digestible because the bacteria used in their production convert much of the lactose to a digestible form.

A second example of a missing enzyme involves cellulose. This complex carbohydrate is a major component

(a)

| Lactose is a disaccharide that our small intestines cannot absorb. | The enzyme lactase breaks lactose into… | …two easily absorbable monosaccharides. |

Lactose + H_2O $\xrightarrow{\text{Lactase}}$ Galactose + Glucose

(b)

Human populations native to only a few places (•) produce lactase as adults.

Figure 21.11 Lactase and Milk in the Human Diet
(a) Lactase is the enzyme that digests lactose, the major sugar in milk, breaking it down into the monosaccharides galactose and glucose. Most people do not produce lactase as adults. (b) Adults from populations native to only a few places (red) produce lactase. In areas where virtually no one produces lactase as an adult (green), most milk products—for example, cheese and yogurt—are made with bacteria or fungi that break down lactose.

of the cell walls that surround all plant cells; thus most plant parts—wood, leaves, and stems—contain a lot of cellulose. Cellulose is composed of long chains of glucose molecules bound together, and the bonds holding those glucose molecules together represent a lot of chemical energy. However, the glucose molecules in cellulose are bound in such a way that animals cannot break them apart. Animals do not produce any enzymes that can digest cellulose, so they cannot release these bound glucose molecules into the energy-releasing pathways of their cells. However, as we shall see shortly, some animals have found a way to get around this problem.

The capacity for absorbing nutrients depends on surface area

To stay alive, an animal must absorb nutrients from its digestive system within a certain time frame. If it took us a year, for example, to absorb a slice or two of pepperoni pizza, we wouldn't be around very long. The amounts of nutrients an animal can absorb over a given time period depend on the surface area available for absorption—the greater the surface area, the faster the absorption. Like those of humans, the digestive systems of many animals have specialized absorptive surfaces that provide areas large enough to take in an adequate supply of nutrients for distributing in a timely manner to all the cells in the body.

The specialized surfaces of each digestive system maximize the surface area available for absorbing nutrients. As we have seen, the villi and the brush borders in our own lower small intestine provide a very large surface area packed into a very small volume. In other animals, the anatomical solutions to the problem of increasing the surface area of the small intestine show interesting variations. For example, the small intestine of sharks contains a spiral valve, a spiraling sheet of tissue that greatly increases the area for absorption of nutrients. Earthworms have a ridge of tissue sticking into the intestine to increase its surface area for absorption.

Herbivore digestive systems are modified for digesting plants

Herbivores have digestive systems that are specialized to allow them to extract nutrients from plant tissues. Herbivorous animals face a different set of digestive challenges from carnivorous animals. Plant tissues contain much less protein and much more indigestible cellulose than animal tissues do. In terms of bulk, herbivores must eat more food than carnivores, and they must invest more time in breaking down their food. To highlight the evolutionary modifications the animal digestive system has made for herbivory, let's compare two similarly sized mammals: a sheep (an herbivore) and a dog (a carnivore) (Figure 21.12).

Several parts of the digestive systems of dogs and sheep reflect the differences in their diets. These differences begin in the mouth. Dogs have spiked and blade-like front teeth suited for slicing and tearing meat. Their flattened back teeth come into contact with their food only briefly, just long enough to grind the torn pieces of meat into chunks small enough to swallow. In contrast, sheep have broad, flattened back teeth that can grind tough plant tissues into small pieces.

There are also noticeable differences in the stomachs of herbivores and carnivores. Pound for pound, a carnivore's diet has a much higher protein content than an herbivore's diet. The single-chambered stomach of a dog produces enzymes that partially break down meat. The sheep's stomach has evolved into a more complex, four-chambered structure that harbors thriving populations of bacteria, fungi, and single-celled eukaryotes. These

Dogs have spiked, sharp front teeth for cutting and tearing meat.

Sheep have broad, flattened teeth for grinding plant material.

Flattened back teeth

Spiked and bladelike teeth

Flattened teeth

Dogs have a simple stomach…

…whereas sheep have a complex one.

Dogs have a relatively short small intestine…

…whereas sheep have an extremely long one.

Figure 21.12 Herbivores and Carnivores Compared
Although carnivorous dogs and herbivorous sheep are similar in size, they have remarkably different digestive systems that reflect their different diets.

microorganisms break down the cellulose in plant matter, making the nutrients and energy contained in the cellulose available to the sheep. The sheep provides the microbes with a well-maintained home and abundant food, so both parties benefit.

Dogs have a small intestine that, relative to the sheep's, provides a small surface area for absorption. The differ-ence is primarily due to length: sheep have a small intestine that is six times as long as that of a similarly sized dog. This design provides a huge surface area over which the sheep can absorb the scarce protein in its diet. The elaborate digestive system of the sheep is what allows it to survive on a diet that could not support a dog.

Dieting: The Long and Short of It

Even for those who would genuinely benefit from losing weight, dieting isn't easy, especially when it comes to keeping the weight off over the long term. And, as some studies show, questions can arise as to what changes in the diet are best (see the box on page 416). The vast majority of people who diet have tried unsuccessfully to lose weight many times in the past. Those people who do manage to lose weight in the short term almost always gain the weight back.

Many diets promise a dramatic weight loss in a short time. Although this promise is appealing, the chances of long-term success are not good. If a diet involves a dramatic decrease in calories consumed, then your body may quickly shift into "starvation mode": your metabolism slows (in order to conserve

Science Toolkit

Why Scientific Research Can Lead to Ambiguous Results

Most of us get our scientific information from the popular media. These sources provide us with easily understandable versions of complex research by focusing on just a few aspects of the research, which they distill into one or two clear conclusions. If you have the time or resources to delve into the original research, however, the messages often become much less simple.

Low-carbohydrate diets, such as the Atkins diet, have been publicized as the best way to lose weight. Diets seem to come and go, but the "low-carb" diet has had remarkable staying power. Does the low-carb diet work better than other diets? In 2003, the *New England Journal of Medicine* published the results of two studies that tried to answer this question.

Upon publication of the two papers, the media quickly spread the news: the low-carb diet works better. But does it? Whereas we get our information in one-line sound bites, the results of medical research come in harder-to-digest formats. If we look more closely, we often find that the answer is the ever-frustrating "It depends."

Study 1 compared a low-carbohydrate diet with a traditional low-fat diet under conditions that reflect how we typically diet—at home, with little professional guidance—over a period of 12 months. Slightly to severely overweight participants were randomly assigned to one of the two diets. The participants met with a dietitian and were given a popular book that described their diet, but otherwise managed the diet on their own. Initially, the people on the low-carb diet lost more weight than those on the traditional low-fat diet, but by the end of the year both groups had lost similarly small percentages (3–4 percent) of their initial body weight. The low-carb diet did manage to decrease the type of cholesterol most strongly linked to heart disease. Forty percent of the participants in both groups failed to follow their diet for the full year.

Study 2 compared how well a low-carbohydrate diet, as compared with a low-fat diet, helped severely overweight people to lose weight, also under real-life conditions, over a 6-month period. As in study 1, the participants were assigned to one of the two diets, received some instruction and written materials describing their diet, and then managed the diet themselves. At the end of the study, the low-carb group had done

Low-Fat (left) versus Low-Carb (right) Diets

better, losing 6 percent of their initial body weight, as compared with 2 percent for the low-fat group. Though some aspects of health improved more with the low-carb diet, cholesterol levels did not. As in study 1, 40 percent of the participants were no longer following their diets by the end of the study.

The results of the two studies are much less clear-cut than might be expected from the way they were reported in the news. Only study 2 suggested that the low-carbohydrate diet is more effective than the low-fat diet as a way to lose weight. Only study 1 indicated that the low-carb diet more effectively reduced harmful cholesterol. Both studies agreed that weight loss on either diet was small, and that a major problem with both diets was the high dropout rate.

Why didn't these two studies provide a clear answer to what appears to be a relatively straightforward question?

1. The two studies addressed slightly different questions. Study 1 asked if the low-carb and low-fat diets worked equally well for people who were at least slightly overweight. Study 2 focused on how well the diets worked for people who were severely overweight. We could reasonably expect the two groups of participants to respond to the diets differently.

2. The two studies had different designs. Whereas study 1 ran for 12 months, study 2 ran for only 6 months. This difference may have been important because it was during the final 6 months of study 1 that differences between the two diets disappeared.

3. Because the effects of both diets were relatively small, small differences in how the studies were run could have had a large effect on their conclusions.

4. Media reports warp our expectations. News stories play up the exciting conclusions and often ignore the less exciting details. People want to hear that the Diet-of-the-Day works; they do not want to hear that it works only some of the time. The consistent result that people have a hard time sticking to any diet also does not make as good a story.

energy), so that, paradoxically, it becomes harder and harder to lose weight. (This metabolic slowdown is probably a vestige of our evolutionary history, resulting from times in the past when starvation presented a real threat and it made sense for the body to conserve energy whenever it lost weight.) Furthermore, diets that promise quick weight loss often focus on just one or a few kinds of foods. We cannot sustain such unbalanced nutrition indefinitely without running the risk of deficiencies. Doctors and nutritionists recommend that any weight loss program include a change in lifestyle that supports long-term maintenance of a balanced diet.

A simple reality governs all weight loss: you can lose weight only if you use more calories than you eat each day. To accomplish this, professionals recommend not only reducing the daily number of calories you take in, but also increasing the number of calories you use—in other words, "diet and exercise."

The more drastically you change your eating habits and lifestyle, the less likely you are to stick with your diet. The diet that leads to the greatest long-term success, therefore, usually involves a moderate reduction in calories eaten while maintaining a healthy balance of proteins, carbohydrates, and fats, and a moderate increase in activity. Such a diet-and-exercise regimen should lead to a moderate weight loss of a few pounds per month when maintained over many months. Once the goal is achieved, the chances of keeping the weight off are much greater.

Chapter Review

Summary

21.1 Nutrients That Animals Need

- Life depends on only a few chemical elements. Some of these elements are needed in relatively large amounts, others in only tiny quantities, although they are still essential to survival.
- Plant-eating animals are herbivores, animals that eat other animals are carnivores, and animals (including humans) that eat both plant and animal foods are omnivores.
- Animals must break down carbohydrates into monosaccharides, fats into fatty acids and monoglycerides, and proteins into amino acids before these nutrients can be absorbed into the body.
- Animals rely on nutrients for both chemical building blocks and energy. Respiration releases the energy in carbohydrates and fats. Proteins are used mostly as a source of amino acids.
- Human bodies cannot make, or make enough of, certain amino acids. These essential amino acids must be obtained from our food.
- Vitamins are organic compounds obtained from food that our bodies use to help regulate metabolic processes. Minerals are small inorganic molecules needed by the body in small amounts. Vitamin and mineral intake should be neither too low nor too high for optimum functioning.

21.2 The Human Digestive System

- The digestive system is a tubular passageway lined with epithelial tissue that runs through the body, and through which food normally flows from mouth to anus.

- During ingestion, food enters the oral cavity, where teeth break it into smaller pieces for easier chemical breakdown. Salivary glands secrete saliva, which moistens the food and begins the chemical breakdown of carbohydrates.
- Food passes from the oral cavity down the esophagus and into the acid environment of the stomach, where protein digestion begins.
- Partially digested food moves from the stomach into the upper small intestine. Here enzymes secreted by the pancreas and the intestine itself complete the digestion of the food. The liver produces bile (stored in the gallbladder), which helps digest fats.
- The lower small intestine is where digested nutrients are absorbed, via capillaries and lymph vessels, into the body. The relatively huge surface area of the lower small intestine results from its highly folded structure (villi and their brush borders).
- From the small intestine, the remaining unabsorbed material moves into the colon, which reabsorbs water and minerals from this waste material. Here bacteria break down some of the waste and release nutrients that the body can absorb. The remaining waste (feces) leaves the body through the anus.

21.3 Comparing Animal Digestive Systems

- The digestive systems of most animals share a remarkably similar design, based on their similar function.
- Animals use muscles and hard surfaces to break food into increasingly smaller pieces to speed digestion. Some animals process their food for ingestion as a liquid. In some animals, gizzards help grind food.

- Digestion breaks large molecules into smaller, simpler ones that the small intestine can absorb, using enzymes to speed chemical reactions.
- When individuals or groups of animals lack particular enzymes, digestion of certain nutrients is impeded. Some humans lack lactase, the enzyme that digests the milk sugar lactose. All animals lack enzymes for digesting cellulose.
- The absorptive surfaces of intestines are extensively modified to increase their surface area and, therefore, their rate of nutrient absorption.
- Herbivores extract nutrients from plants through adaptations to their digestive systems. Their teeth are shaped for thorough grinding of tough plant tissues; their multi-chambered stomachs house microorganisms that digest cellulose, and a relatively longer small intestine increases the surface area for absorbing the scarce protein in their diets.

◉ Review and Application of Key Concepts

1. Name the chemical elements that are necessary for animals to obtain in their diets and describe a few of the body's molecules in which they would be found.

2. A typical apple, including its core, contains less than 1 gram of protein and fat, 32 grams of carbohydrates, and 6 grams of indigestible materials collectively called fiber. Describe the journey of an apple through your digestive system as it is eaten, digested, and absorbed. Identify the different regions of the digestive system and what happens in each region. What do you think an apple contains that would be indigestible to humans?

3. Describe the general features common to most animals' digestive systems. Why do you think animals share these features?

4. Describe the features of an herbivore's digestive system that are specialized for a plant diet.

Key Terms

anus (p. 411)
bile (p. 410)
brush border (p. 411)
carbohydrate (p. 407)
carnivore (p. 406)
colon (p. 411)
digestion (p. 405)
digestive system (p. 409)
esophagus (p. 410)
essential amino acid (p. 408)

fat (p. 407)
feces (p. 411)
gallbladder (p. 411)
gizzard (p. 412)
herbivore (p. 406)
ingestion (p. 409)
liver (p. 411)
mineral (p. 408)
monoglyceride (p. 407)
nutrients (p. 405)

omnivore (p. 406)
oral cavity (p. 409)
pancreas (p. 410)
protein (p. 408)
saliva (p. 409)

salivary gland (p. 409)
small intestine (p. 410)
stomach (p. 410)
villus (p. 411)
vitamin (p. 408)

Self-Quiz

1. Which of the following is *not* an essential nutrient for animals?
   a. hydrogen
   b. nitrogen
   c. cellulose
   d. iodine

2. During digestion, proteins are broken down into
   a. monosaccharides.
   b. fats.
   c. sugars.
   d. amino acids.

3. Which of the following would contain all eight essential amino acids?
   a. all-wheat cereal
   b. corn chips
   c. beans and rice
   d. roasted peanuts

4. Digestion
   a. reassembles nutrients into molecules for use by the body.
   b. breaks nutrients into a form that can be absorbed.
   c. forms proteins from amino acids.
   d. forms fats from carbohydrates.

5. Which of the following is a strongly acidic environment?
   a. mouth
   b. small intestine
   c. stomach
   d. large intestine

6. Digestive enzymes are secreted by the
   a. salivary glands.
   b. pancreas.
   c. small intestine.
   d. all of the above

7. Bacteria that assist in human digestion are found in
   a. the stomach.
   b. the mouth.
   c. the gallbladder.
   d. the colon.

8. Which of the following *cannot* be digested by animals?
   a. proteins
   b. cellulose
   c. fats
   d. carbohydrates

9. In the digestive systems of some animals, gizzards
   a. secrete digestive enzymes.
   b. absorb nutrients.
   c. grind food.
   d. store bile.

10. What adaptation allows herbivores to digest cellulose?
   a. a long intestine
   b. intestinal villi
   c. the pancreas
   d. resident bacteria

Biology in the News

Nuts to Atkins, Try the Amish Diet

Despite a high-calorie diet, an Old Order community's intensely physical lifestyle produces a strikingly low rate of obesity

By Jeannine Stein

Forget the standard-issue health and fitness resolutions that include joining a gym, going to yoga and trading meatball subs for white-meat turkey. It may just be that the best way to get in shape is to plow the back 40, toss a few bales of hay and wash buckets of wet clothes by hand.

Call it the Amish paradox. An exercise science professor has discovered that a pocket of Old Order Amish folks in Ontario, Canada, has stunningly low obesity levels, despite a diet high in fat, calories and refined sugar—exactly the stuff doctors tell us not to eat.

They're at a paltry 4% obesity rate, compared to a whopping 31% in the general U.S. population, which, as we all know, is getting fatter by the minute. This group of Amish manages to keep its overweight levels low despite a diet that includes meat, potatoes, gravy, cakes, pies and eggs. So what's their secret? Exercise, people. Exercise.

A study contrasting the lifestyles of the Old Order Amish with "normal" Americans found that the proportion of obese Amish was a mere 4 percent, compared with 30 percent for the rest of us. In our technologically advanced Western society, we tend to battle bulging waists with diets that cut our caloric intake, such as the low-carbohydrate Atkins diet. The Amish, in contrast, live a lifestyle more typical of the mid-1800s. They shun technology, depending on manual labor instead.

The Amish eat an average of about 3,600 calories, one and a half times more than a typical American. The Amish way of life is not a low-calorie one.

The secret to staying slim in the face of a diet that makes a calorie-conscious consumer cringe? Activity. The Amish have about six times the level of physical activity of the general population. Most American adults tend to walk 2,000 to 3,000 steps in a day. Amish men, in contrast, typically walk over 18,000 steps in a day, and Amish women walk an average of over 14,000 steps in a day. Because they rely much less on machines to ease their workload, the Amish typically do more than six times the physical work of a typical American.

The Amish diet works because the calories taken in balance the calories burned. If you eat more calories, the only way to maintain your weight—and your health—is to burn more calories by doing more work. Conversely, if you burn fewer calories by living an inactive life, then you can maintain your weight only by cutting back on your caloric intake accordingly.

The Amish maintain their weight not by following some secret diet regime, but by following the simple and un-faddish advice long given by family doctors: maintaining your weight depends on a combination of healthy eating and an active lifestyle.

Evaluating the news

1. How does the Amish lifestyle compare to your own?
2. What do you think of the chances of success of an Amish diet in the broader North American culture? Why might it appeal to overweight people? What barriers to its success do you see?
3. How have work-saving devices affected our lifestyle and our health? Can you think of some specific examples?

Source: *Los Angeles Times*, Friday, January 12, 2004.

Key Concepts

- Humans have an elaborate respiratory system with extensively folded lungs that provide a huge surface area for gas exchange.

- Oxygen and carbon dioxide move across animals' gas exchange surfaces exclusively by diffusion. Several principles of diffusion govern the biology of gas exchange systems in animals.

- Aquatic animals exchange gases across the surfaces of gills that project into the environment, which allow for effective gas exchange in oxygen-poor water. Terrestrial animals exchange gases across the surfaces of internal lungs to minimize water loss.

- In many animals, circulatory systems transport oxygen and carbon dioxide between gas exchange surfaces and respiring cells.

- Many animals use oxygen-binding pigments in the blood to carry oxygen more efficiently. The pigment used by humans and many other animals is hemoglobin, which is contained in red blood cells.

In High Places

When hiking high in the mountains, you may find yourself gasping for air. Shortness of breath is one of the first signs of oxygen deprivation, which we suffer at high altitudes. The amount of oxygen inhaled with each breath drops by half for every 5,500 meters (18,000 feet) of vertical altitude that we climb. Symptoms of "mountain sickness" start to appear at about 3,000 meters (10,000 feet), and these symptoms become more severe as a climber moves higher. From shortness of breath, these symptoms progress to headaches, tiredness, and loss of appetite, and, in severe cases, to nausea, heart palpitations, and hallucinations. At 7,100 meters (23,300 feet), where each breath of air provides only one-third as much oxygen as at sea level, most people pass out. Altitude sickness is a real health risk: each year two to three people die while on high-altitude treks in the Himalayas of Nepal.

Although the consequences of reduced oxygen availability at high altitudes can be severe, humans can adapt to these conditions. The highest permanent human settlement lies at 5,700 meters (19,000 feet) in the Andes of South America. Several climbers have reached the top of Mount Everest (8,848 meters, or 30,000 feet) without using oxygen tanks.

In an attempt to demonstrate that humans can function in the low-oxygen environment found at high elevations, a group of Bolivian soccer enthusiasts organized a game in the snow-covered crater of Mount Sajama, a 6,500-meter (21,500-foot) volcano near Bolivia's border with Chile. Seventeen players, all acclimatized to the alpine environment, set out, equipped with orange soccer balls and black goal posts that would stand out against the snow. Two players had to turn back after suffering from altitude sickness, but the remaining fifteen made it to the crater and completed a short 40-minute game on a half-size soccer field.

High Altitude Record Holders
Residents of the Andes have adapted to the low oxygen availability of their high-altitude homes. The griffon vulture of Africa holds the altitude record for animals.

Other animals can also function at high altitudes. The record for the highest known altitude reached by an animal is held by an unfortunate griffon vulture that met its end in an encounter with a plane flying over Africa at 11,300 meters (37,000 feet), where the amount of oxygen is only one-fourth that at sea level.

So certain people and other animals not only tolerate high altitudes, but thrive "up there." How do they do it?

Life requires energy. All animals, including humans, supply their cells with energy through respiration, a process that both consumes and produces gases. As we learned in Chapter 8, respiration converts the nutrient glucose (a monosaccharide) in several steps into molecules of carbon dioxide and water. Oxygen is consumed in the process, and the energy that was stored in the chemical bonds of glucose is released along the way (Figure 22.1). To support respiration, animals need a way of taking in oxygen, and getting rid of carbon dioxide.

In humans and other large multicellular animals, the distances these two gases must travel between the outside of the body and the mitochondria in all the body's cells would be too great for survival if they could travel only by passive movement (diffusion) from cell to cell. In such animals, therefore, respiration requires not only a specialized body surface where these gases can be exchanged with the outside environment, but also a system to transport them. Oxygen must be extracted from the air and transported to the mitochondria inside the cells. Simultaneously, the carbon dioxide that leaves the mitochondria must be carried to where it can be expelled into the animal's external environment.

In this chapter we look at how humans and other animals exchange oxygen and carbon dioxide between the outside and inside of their bodies. As a starting point, we look at breathing and gas exchange in the human respiratory system. Then we expand our view to include other animals, noting the general principles that apply to gas exchange in all animals as well as examining the problems certain animal groups face and how those animals solve them. Next we trace the routes by which gases are transported between the body surface and the mitochondria, where respiration takes place. We close with a look at how humans and other animals survive in environments where oxygen is in short supply.

Figure 22.1 Respiration Occurs in Mitochondria

During respiration, oxygen is consumed as energy is extracted from glucose for use by the cell. Heat, water, and carbon dioxide are also produced.

22.1 Breathing in Humans

The process of taking oxygen into the lungs and expelling carbon dioxide from them is called **breathing**. In active animals such as humans, breathing must move a lot of air into and out of the lungs in order to exchange enough gases to survive. At rest, when our oxygen needs are lowest, humans move about 360 liters of air into their lungs each hour—an amount that would fill 360 half-liter pop bottles. Oxygen makes up 21 percent, or about one-fifth, of the air we breathe. So out of the 360 liters of air we breathe every hour, the human body takes in and transports about 76 liters of oxygen. During strenuous exercise, our breathing rate increases so that we can

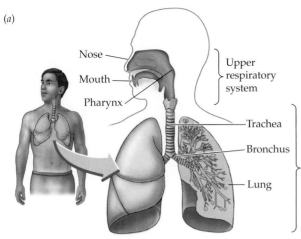

(a)

Nose

Mouth

Pharynx

Upper respiratory system

Trachea

Bronchus

Lung

Lower respiratory system

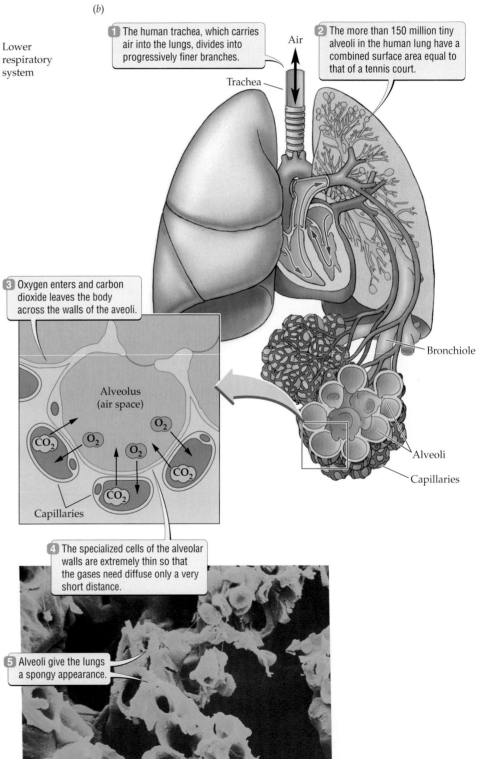

Figure 22.2 Gas Exchange in Humans
(a) The upper respiratory system leads to the trachea, the first part of the lower respiratory system. The trachea branches into two bronchi, which branch into bronchioles in the lungs. (b) The structure of the lungs speeds the diffusion of oxygen and carbon dioxide into and out of the body by providing a large surface area for gas exchange.

(b)

1 The human trachea, which carries air into the lungs, divides into progressively finer branches.

Air

2 The more than 150 million tiny alveoli in the human lung have a combined surface area equal to that of a tennis court.

Trachea

3 Oxygen enters and carbon dioxide leaves the body across the walls of the aveoli.

Alveolus (air space)

O_2

CO_2 O_2

O_2

CO_2

CO_2

Capillaries

Bronchiole

Alveoli

Capillaries

4 The specialized cells of the alveolar walls are extremely thin so that the gases need diffuse only a very short distance.

5 Alveoli give the lungs a spongy appearance.

move a much greater volume—up to 6,000 liters per hour in trained athletes.

What do we do when the unthinkable happens, and a person stops breathing? Any standard first aid or CPR (cardiopulmonary resuscitation) course teaches what to do under those circumstances. Humans must breathe to exchange gases with their environment. Although we can survive for more than a week without food and several days without water, a mere 4 minutes without oxygen can lead to irreversible brain damage. The box on page 424 describes some emergency situations in which special measures are needed to aid or restore normal breathing. A key part of giving first aid to someone who has stopped breathing is to "check their ABCs": *A*irway, *B*reathing, and *C*irculation. By checking the ABCs, we are systematically checking the path that oxygen takes during breathing and respiration.

The airway allows air to move between the environment and the gas exchange surfaces in the lungs that lie inside the body. When we breathe in, air passes through a series of passageways in the head, neck, and chest before moving into the lungs, where the actual exchange of gases takes place. The circulation of blood then carries the gases to and from cells throughout the body. Let's look at this pathway in more detail.

With each breath, our lungs alternately draw in (inhale) and force out (exhale) 0.4 to 0.5 liters of air, depending on whether we are female or male. When we inhale, air enters the nose, moves through the nasal cavity, and enters the **pharynx** [*FAIR*-inks] (throat), the area where the back of the mouth and the nasal cavity join together into a single passageway. (If we are breathing through the mouth, then the air passes directly from the mouth into the pharynx.) Together the nose, mouth, and pharynx make up the **upper respiratory system** (Figure 22.2a).

Biology on the Job

Where There's Fire, There's Smoke

George Middleton is a fireman and emergency medical technician (EMT) for the town of Carlisle, Massachusetts. As a firefighter, he must protect himself against hot and hostile gaseous environments. As an EMT, he helps others survive in catastrophic and often life-threatening circumstances. In both roles, he employs expertise and technology that aid him in maintaining the internal environments of himself and other human beings.

Let's begin by talking about your role as a firefighter. What dangers do you face at a fire scene? There is the heat and smoke, of course, but we also face invisible toxic fumes given off by burning materials. And the dangers change as we fight a fire.

How so? The first phase of fighting a fire is called the "initial fire attack," where the goal is to quickly gain access to the main body or seat of the fire, and knock down the "heavy" obvious fire with a controlled but powerful application of water from a high-pressure hose. The heat, flames, and highly combustible smoke are the primary hazards.

What happens after knockdown? After the obvious fire has been knocked down, we begin the "overhaul" process. During overhaul the dangers become less obvious. At this time we've usually begun aggressively ventilating the structure to clear the smoke. We also sometimes need to "open up" and check concealed spaces behind walls and ceilings to be sure the fire is out in these less accessible areas. If either of these actions are done improperly, the fire can flare back up. Typically after all visible flame is put out, a lot of smoldering material remains, and must be systematically and completely extinguished. Usually far more water is used in the overhaul process than in the initial fire attack. During the overhaul process, the dangers are many: fire and water damage can weaken structural components and present collapse hazards. The remaining smoldering materials give off large quantities of invisible poisonous gases such as carbon monoxide.

What do you do to protect yourself? We wear self-contained breathing apparatus called "SCBA." This consists of a special face mask and a cylinder that holds compressed air. They're rated for 30 minutes of use in hazardous environments, but under the stressful conditions of firefighting, we're lucky to get 15 or 20 minutes out of them before the low-air

Fires Produce Many Toxic Gases That Require Special Breathing Apparatus

The photo shows a firefighter using the self-contained breathing apparatus called "SCBA," which consists of a special face mask and a cylinder that holds compressed air.

alarm goes off, indicating that only a quarter of a tank remains. During overhaul, when the smoke has mostly been cleared, it's sometimes tempting to remove the face mask, because the dangerous gases and fumes aren't really visible. But carbon monoxide, for example, is still present in high concentrations and is very toxic, and you can't see it or smell it. This is one reason why we have strict safety regulations and standard operating procedures that require wearing and using SCBA during the "overhaul" phase of firefighting operations.

What about your job as an EMT? How do gases come into play in that role? As EMTs, we administer oxygen to victims of smoke inhalation, which can cause swelling in the upper respiratory tract, making it hard for victims to breathe. Providing large quantities of O_2 in these cases relieves the stress on the patient's circulatory system resulting from the diminished gas exchange. We transport these patients to the emergency room quickly because of the imminent danger of the swelling causing a partial or complete obstruction of the patient's airway.

In what other situations do you administer oxygen? We respond to many different situations that require the use of supplemental oxygen. Patients with chest pain, victims of cardiac arrest, and victims with traumatic injuries are a few examples. In addition to the pain and discomfort of broken bones, for instance, there is often associated internal bleeding, which can cause shock. Administering oxygen helps the patient's body compensate for the blood loss and can prevent shock—a dangerous and life-threatening clinical condition.

How do you know when a victim needs oxygen? Sometimes it's obvious. If there is a question, we have an instrument, called a pulse oximeter, that we can clamp onto a finger. It gives a reading telling the percentage of oxygen in the blood. Normal readings range from 99 percent to 95 percent. If we get readings below that, we administer oxygen.

The Job in Context

Firefighters face gas exchange problems that most of us never encounter. They need to be expert and disciplined in the use of equipment that makes breathable air available to them as they enter environments that would otherwise be deadly. Even when the air looks clear, toxic fumes can be present. Emergency medical technicians must become expert not only in administering oxygen, but in determining when and under what conditions to do so. Firefighters also respond to toxic chemical spills and leaks, where protective breathing equipment is necessary to prevent the entry of poisonous fumes into their respiratory systems.

From the pharynx, air moves into the **trachea** (windpipe), the largest breathing tube in our respiratory system. By touching your throat, you can feel ridges under the skin. These ridges, like the ridges in some vacuum cleaner hoses, reinforce the trachea, giving it the strength it needs to maintain its shape during breathing. The trachea branches into two smaller tubes, called **bronchi** [*BRONG-kee*] (singular: bronchus). Each bronchus leads to one of the paired **lungs**, the organs where the exchange of gases occurs. Together, the trachea, bronchi, and lungs make up the **lower respiratory system**.

Inside the lungs, the bronchi divide into **bronchioles**, a series of branching, ever smaller tubes. The tiniest bronchiole extensions open into the **alveoli** [al-*VEE*-oh-lie] (singular: alveolus), minute sacs where the actual exchange of gases takes place (Figure 22.2b). The alveoli are organized into groups resembling clusters of grapes, each cluster joined to a tiny bronchiole. In each cluster, oxygen and carbon dioxide are exchanged across the alveolar surfaces, which are directly exposed to air brought in via the bronchiole.

After crossing the thin cells that form the walls of the alveoli, oxygen moves into tiny blood capillaries that surround the alveoli. From there it is carried by the blood to respiring cells throughout the body. With each breath, a new supply of oxygen arrives at the surface of each alveolus to replace the oxygen of the previous breath that has already moved across the alveolar cells and into the blood. Damage to these delicate gas exchange surfaces by cigarette smoking can have serious health consequences (see the box on page 426).

22.2 Principles of Gas Exchange

Breathing in many animals, including humans, involves actively pumping air into and out of the lungs. However, at the surfaces of the alveoli and body cells, where the exchange of oxygen and carbon dioxide actually takes place, these gases move solely by the passive process of diffusion. Thus animals must be able to transport sufficient quantities of gases to the very close proximity of their cells. Three basic rules governing diffusion should help us understand what animal bodies can do to facilitate gas exchange.

Rule 1: Gases diffuse from areas of high concentration to areas of low concentration

Two important points follow from the fact that oxygen and carbon dioxide diffuse from areas of high concentration to areas of low concentration. First, the concentration of a gas at its source must be higher than that at its destination in order for it to diffuse toward its destination. Second, greater differences in concentration result in more abundant and rapid flow of gases from source to destination (Figure 22.3).

Helpful to know

Some of the Latin and Greek words on which biological terminology is based have endings for singular and plural that differ from those in English. Words ending in *us* in the singular (such as *villus* from Chapter 21, *bronchus*, and *alveolus*) substitute *i* for the plural to get *villi*, *bronchi*, and *alveoli*. Some words ending in *a* in the singular add an *e* to make the plural, so *trachea*, for example, becomes *tracheae*.

Helpful to know

In Chapter 20 we mentioned *ule* and *ulus* as endings indicating a small version of something. The ending *ole* has the same effect. For example, *tracheole* and *bronchiole* literally mean "little trachea" and "little bronchus."

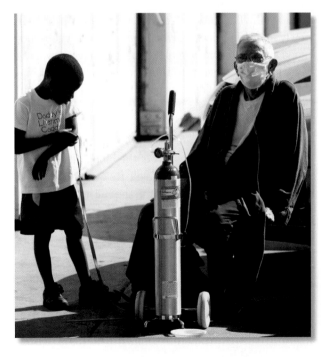

Figure 22.3 Oxygen from a Tank Can Increase the Oxygen Gradient Between Source and Destination
Many of the diseases of the respiratory system that accompany aging decrease our ability to take up oxygen from the air. By breathing supplemental oxygen, we can compensate for our failing respiratory system by increasing the gradient of oxygen between the lungs and the blood.

Biology Matters

Emphysema, Bronchitis, and Your Lungs

Emphysema is a serious and often fatal disease that is closely related to smoking. It develops gradually over many years. More than 3 million people in the United States were suffering from this condition in 2002. In the lungs of people with emphysema, the alveoli and the tiny blood capillaries lying close to them are damaged, so oxygen cannot get into the blood as easily as it should. The body tissues don't get enough oxygen, and muscles can waste away.

X-Ray of Emphysema, with the Infected Lung on the Right Filled with Fluid

Cigarette smoking is a leading cause of emphysema. Quitting can slow down the progression of the condition, but it cannot reverse damage already done. Emphysema can also have a genetic basis: people born with a deficiency of a protein that protects lung tissue almost inevitably develop this inherited form of the disease. The personal tragedy and the economic costs generated by emphysema are staggering.

Chronic bronchitis is another lung condition attributable to cigarette smoking. Normally, cells lining the airway produce mucus that traps unwanted particles, and cilia propel these particles out of the airway. The mucus-producing cells become overactive in people with chronic bronchitis. The excess mucus impedes the cilia and obstructs the airway. Less oxygen gets into the blood, and the blood cannot get rid of carbon dioxide as easily. The condition also puts added stress on the heart. In 2000 over 2.5 million people were suffering from chronic bronchitis attributable to smoking.

It is worth protecting our lungs. As the sufferers of emphysema and chronic bronchitis know, without healthy lungs we cannot expect to lead a long and active life.

First, we can consider the concentration gradient in the alveoli. The concentration of oxygen is higher in newly inhaled air in the alveoli than in the oxygen-poor blood arriving at the alveoli, so oxygen moves from the alveolar air (the source) into the blood (the destination). Blood that is now carrying oxygen is then circulated to respiring cells. The concentration of oxygen is higher in this blood (now the source) than in the respiring cells (the destination), where oxygen is being consumed. So oxygen moves out of the blood and into the cells. As oxygen moves into a cell, its cytosol becomes richer in oxygen. Mitochondria in the cell are constantly using oxygen. When the concentration of oxygen in a mitochondrion becomes lower than the concentration in the surrounding cytosol, oxygen diffuses into the mitochondrion, and respiration can continue.

The same rule holds for the journey of carbon dioxide, but the direction is reversed. At the same time they are using oxygen, the mitochondria are producing carbon dioxide. When the concentration of carbon dioxide in a

mitochondrion becomes greater than that in the cytosol, carbon dioxide diffuses out of the mitochondrion into the cytosol and then out of the cell. The concentration of carbon dioxide near respiring cells is higher than that in the blood, so CO_2 moves into the blood.

The concentration of carbon dioxide in the air in the alveoli is low relative to that in the blood arriving at the thin alveolar walls. Thus CO_2 in the blood diffuses across the capillary walls, across the plasma membranes of the alveolar cells, through the cytosol of those cells, and into the air in the alveolar sacs. Even at similar concentrations, CO_2 diffuses 20 times faster than oxygen in both cytosol and blood plasma, so moving CO_2 out of the body presents much less of a challenge than getting O_2 into the body. Between breaths, CO_2 concentrations in the alveoli build up rapidly, to as high as 5.3 percent, whereas fresh air (outside the body) contains only 0.04 percent CO_2. So the air we inhale can accept a great deal of CO_2 when it reaches the alveoli and mixes with the air that has acquired CO_2 from the blood.

The second important corollary to rule 1 is that the higher the concentration of a gas in the environment compared with that inside an animal, the faster the gas will diffuse across exchange surfaces. In the human lungs, for example, the greater the difference between the O_2 concentration in the air entering an alveolus and the O_2 concentration in the blood in the capillaries surrounding it, the more abundantly and rapidly O_2 can move across the alveolar cells and into the blood, and, in turn, the more oxygen will become available to supply the mitochondria in respiring cells.

This second corollary has important biological implications. Animals that live in habitats with a high O_2 concentration can take up O_2 more quickly than can animals living in habitats low in O_2. For terrestrial animals, the concentration of O_2 differs greatly depending on where an animal lives (Figure 22.4). As we saw at the opening of this chapter, the air at 5,500 meters up a mountain contains only half as much O_2 as the air at sea level. This means that only half as much O_2 is available for diffusion across the alveolar surfaces in the lungs of mountain climbers—one reason they have difficulty breathing atop the world's tallest peaks.

Aquatic animals live in habitats that contain much less oxygen than air does. On a typical summer day, a liter of lake water contains only $\frac{1}{20}$ as much O_2 as the adjacent air. We can estimate, then, that over equal surface areas available for absorption, O_2 can diffuse into our blood about 20 times more quickly than into the blood of a fish. Moreover, warmer water holds less dissolved gas than cooler water. (You can demonstrate this for yourself: let a glass of cold tap water stand at room temperature; small bubbles of gas will form in the glass because, as the water warms up, it cannot hold as much gas.) It is for this reason that goldfish in a warm pond often come to the surface and gulp air in the summertime.

Rule 2: The number of molecules exchanged per unit of time depends on the diffusion surface area

The surface area that an organism has available for gas exchange determines the capacity of that organism to absorb oxygen. Twice as much gas can diffuse across an area of 4 square feet than an area of 2 square feet. Because respiring cells need a lot of oxygen, most animals cannot supply enough oxygen to their cells by simply allowing diffusion across their outer surface. In humans, for example, even if the skin were specialized for gas exchange, it would not have enough surface area to supply the oxygen needs of all our cells. It turns out that, although each alve-

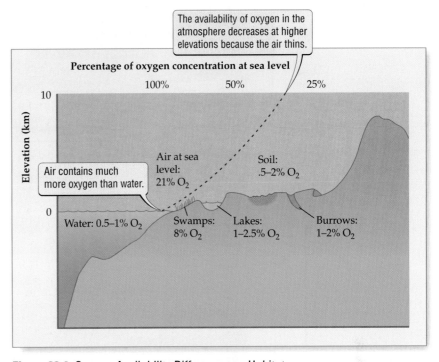

Figure 22.4 Oxygen Availability Differs among Habitats
A liter of air contains more oxygen (O_2) than a liter of water. In air, the amount of O_2 per liter of air decreases with increasing altitude. Measurements for aquatic habitats and habitats at sea level are given as the percentage of oxygen contained in those habitats; measurements for terrestrial habitats at high altitudes are given as a percentage of the oxygen concentration in air at sea level.

olus in our lungs is extremely small, together the 150 million alveoli in a single lung provide a combined surface area for gas exchange that is 90 times the surface area of the skin. Put another way, if the surface area of just one lung could be spread out flat, it would cover about the same area as a tennis court. Without this specialized and very large gas exchange surface, we could never manage to meet the body's oxygen demand. Like the alveoli in the human lung, specialized gas exchange structures in other animals are characterized by highly folded sheets of specialized epithelial tissue that pack an immense surface area into a small space.

The oxygen demand of an animal depends not only on its size, but also on its level of activity. The more active an animal, the more oxygen it needs, and consequently, the greater the surface area it needs for gas exchange. For example, the surface area of the gills in fish and other aquatic animals matches the level of activity of those animals very closely (Figure 22.5). As we saw in Chapter 20, endothermic animals (such as humans), which use metabolic heat to maintain high body temperatures, respire ten times as quickly as

Learn more about simple diffusion.

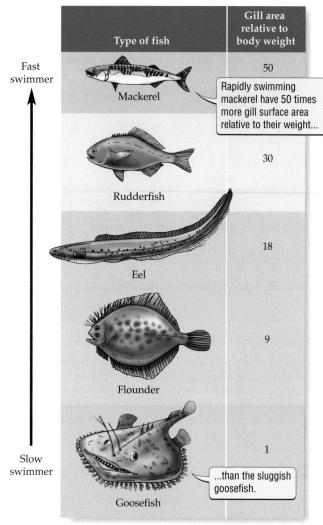

Figure 22.5

The Surface Area of Fish Gills Varies with Activity Level

Actively swimming fish have a greater gill surface area relative to body weight than do slower-swimming fish. This difference suggests that natural selection favors gas exchange surfaces large enough to meet the oxygen demand of the organism.

Flatworm

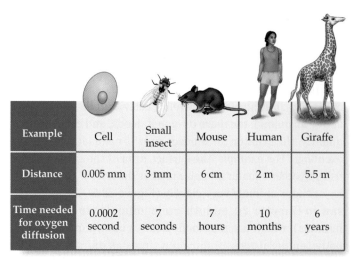

Example	Cell	Small insect	Mouse	Human	Giraffe
Distance	0.005 mm	3 mm	6 cm	2 m	5.5 m
Time needed for oxygen diffusion	0.0002 second	7 seconds	7 hours	10 months	6 years

Figure 22.6

The Relationship Between Body Size and Diffusion Time

As body size increases, the time it would take to transport gases from one end of an organism to the other by diffusion alone increases greatly. The times reported here are based on the speed of oxygen diffusion through water, the main component of living things.

ectothermic animals (such as lizards, frogs, and insects), which rely primarily on environmental heat. As a result, endothermic animals have a much higher oxygen demand, and so must have a larger lung surface area, than ectothermic animals of similar size.

Rule 3: The shorter the distance over which a gas must diffuse, the faster it can reach its destination

Oxygen and carbon dioxide cannot diffuse over long distances fast enough to meet the metabolic needs of large animals. Over short distances (about 0.001 millimeter), such as those across plasma membranes or within a cell, gases and other dissolved materials can diffuse within a fraction of a second (Figure 22.6). However, even for animals that we would consider small—such as an insect a few millimeters in diameter—it can take more than a few seconds for gases to diffuse from one end of the body to the other. In biological terms, that's a long time. At the scale of a human, it would take months for oxygen to diffuse from our lungs to the tips of our toes.

For this reason, most animals must have either a shape that brings all their cells within a few millimeters of the air or water around them or a way of transporting gases within their bodies. Very small or flat, sheet-shaped animals, such as flatworms, can allow O_2 and CO_2 to diffuse all the way from their gas exchange surfaces to their respiring cells. Because of their size and shape, most other animals must have a way of transporting gases to their respiring cells. Many animals transfer gases from their gas exchange surfaces to the blood, which is pumped to respiring cells throughout the body.

22.3 Gas Exchange on Land and in Water

Having covered some of the general patterns of gas exchange in animals, let's look more closely at how animals exchange gases in the very different environments of water and air.

The gas exchange structures of terrestrial and aquatic animals differ

If we look at the gas exchange surfaces of aquatic animals, including fish and a variety of invertebrates, we find that they differ in some important ways from the lungs of land-dwelling humans. Whereas our lungs, like the gas exchange surfaces of most land animals, consist of an infolding of the body surface that brings the gas exchange surface inside the animal (Figure 22.7a), the **gills** used by most aquatic animals represent an outfolding of the body surface (Figure 22.7b). Gills, therefore, are exposed directly to the environment. Because land animals have internal gas exchange surfaces, they must breathe by actively pumping air into and out of the internal organs that contain those gas exchange surfaces. This pumping brings air with a high oxygen content into close contact with the exchange surfaces. If we did not move air into and out of our lungs, the oxygen concentration of the air in our lungs would drop to the point at which the concentration gradient would be too small to allow enough oxygen to diffuse into our blood. That is why people pass out if they hold their breath too long. Fish and many other aquatic animals can keep a supply of fresh water flowing over their gills with relative ease; it takes a lot less work to move water over external gills because the one-way flow requires no in-and-out pumping.

Compare diffusion in air and in water.

Lungs do not work well under water

When people breathe water, they drown. Why do lungs that work so well on land do so poorly under water?

We have already seen that water contains only a fraction of the oxygen found in air at sea level. But the answer also lies in some of the physical differences between water and air. Oxygen diffuses about 300,000 times more slowly in water than in air. If our lungs were filled with water, the oxygen in each alveolus could not diffuse into the blood fast enough to supply our needs. In addition, water is much denser than air. A liter of water weighs over a thousand times as much as a liter of air. Even if oxygen were as plentiful in water as in air and diffused as rapidly, we would have to breathe the same volume of water as we do air, about 360 liters of water an hour. That would mean inhaling and exhaling a thousand times more weight in water than we do in air to fulfill our need for oxygen, which translates into a thousand times more work—an impossible amount, given the design of the human respiratory system.

With these facts in mind, we can compare how humans and trout exchange gases. We use about 2,500 times more oxygen per hour than the trout, both because we are much larger and because, as endotherms, we use energy much more quickly. The smaller, ectothermic trout must meet its oxygen demand by passing 150 kilograms of heavy, oxygen-poor water over its gills every hour. Because the water need only flow over its gills, rather than into and out of an internal chamber (as with our lungs), it is possible for a 1-kilogram trout to expose its gills to this much water. But, in order to meet our much higher oxygen demand, we have to move only about 3 kilograms of air into and out of our lungs each hour. If we breathed water instead of air, our lungs would have to move an impossible 350,000 kilograms of water in an hour. Even if reduced to the size and metabolism of a trout, we would still have to pump 150 kilograms of water into and out of our lungs. People drown not because water contains no oxygen, but because their lungs cannot possibly pump enough water in and out of their lungs fast enough to meet their oxygen demand.

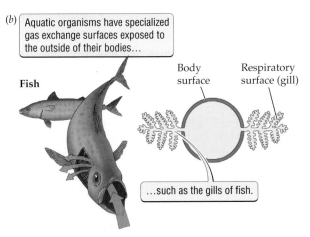

Figure 22.7 Lungs and Gills Compared
(a) The lungs of terrestrial animals represent an infolding of the body surface. (b) The gills of aquatic animals represent an outfolding of the body surface.

Trout

Why don't land animals have gills?

We can also ask the reverse question: why don't terrestrial animals have gills? The answer lies in a problem that goes hand in hand with gas exchange on land. Any surface that allows oxygen and carbon dioxide to diffuse into and out of an animal also allows water to diffuse out of that animal. For terrestrial animals, gas exchange means water loss. Indeed, the bodies of land animals represent an evolutionary compromise between the conflicting demands of water conservation and gas exchange. Most of the outer surfaces of terrestrial animals are waterproof, making those surfaces useless for gas exchange, but well suited for retaining water. By having internal lungs as gas exchange surfaces, terrestrial animals gain more control over what crosses those surfaces. Having the gas exchange surfaces inside the body ensures that the environment next to the lung's inner surface is moist. In this moist environment, the rate at which water diffuses out of the animal decreases, just as clothes dry more slowly on a humid day.

Animals that live in particularly dry environments often have even more elaborate strategies for avoiding water loss in gas exchange. Kangaroo rats, for example, live in a dry desert environment and must be particularly careful to conserve water. These animals take advantage of the fact that cool air cannot hold as much moisture as warm air. When a kangaroo rat inhales dry desert air, that air picks up moisture from the tissues that line its complexly folded nasal passage (Figure 22.8). The inhaled air warms to body temperature and picks up even more water in the lungs. The kangaroo rat recovers much of this water, however, when it exhales the air through its nasal passages. Much as moisture condenses on the outside of a glass of ice water on a hot, humid day, moisture in the warm exhaled air condenses onto the cooler nasal passages and is reabsorbed.

22.4 How Animals Transport Gases to Respiring Cells

After oxygen enters the body of an animal, it must still reach the respiring cells. As we have seen, diffusion allows for the rapid movement of materials only over very short distances. Diffusion works too slowly to adequately supply most respiring cells at distances much more than a fraction of a millimeter from the gas exchange surface. Animals that are too large to rely on diffusion alone must have a way of transporting oxygen and carbon dioxide between their gas exchange surface and their respiring cells. Many groups of animals transfer these gases to the **blood**, a liquid tissue that, in animals such as humans, consists primarily of blood plasma and blood cells. Blood moves rapidly from the gas exchange surfaces to cells throughout the body and then back to the gas exchange surfaces through a **circulatory system**, which consists of a series of interconnected tubular vessels through which blood is pumped. (In this chapter, we limit our consideration of this internal transport system to the movement of gases into and out of the blood, saving a discussion of how the circulatory system pumps the blood throughout the body for Chapter 23.)

Blood is a sophisticated tissue capable of carrying oxygen and carbon dioxide between an animal's gas exchange surfaces and the respiring cells throughout its body. However, blood plasma—the liquid component of blood—by itself is poorly suited to the transport of

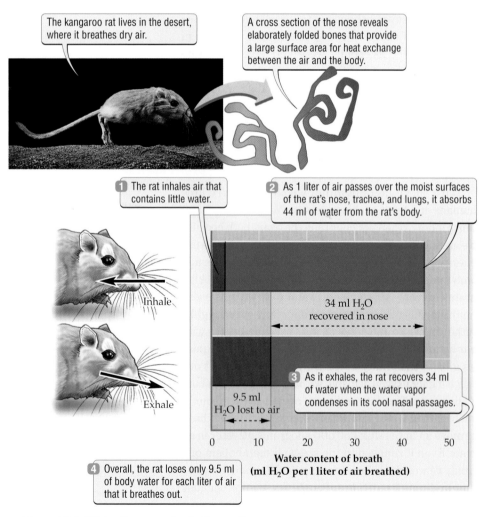

Figure 22.8 Water Conservation in Desert Rats
Desert-dwelling kangaroo rats recover most of the body water that is taken up by the dry air they inhale.

oxygen. Water, the main component of plasma, cannot hold much dissolved oxygen; this is especially true at the high body temperatures of endotherms. Human plasma at body temperature, for example, can carry only about 2 milliliters of oxygen per liter, or roughly one-third the amount of oxygen that fresh water at 20°C can hold—a capacity much too low to meet an endotherm's high oxygen demand. Animals have solved the problem of blood plasma's poor oxygen-carrying capacity in two ways.

Pigment molecules increase the efficiency of oxygen transport

One way animals have overcome their blood plasma's poor capacity for transporting oxygen is by employing a class of large proteins called **oxygen-binding pigments**. As their name suggests, these molecules can bind to oxygen molecules. However, what makes these pigments useful is not just their ability to bind oxygen, but their ability to let go of it. In addition to this capacity for **reversible binding** with oxygen, under specific conditions, they are very good at picking up oxygen in regions of high oxygen concentration and at letting oxygen go in regions of relatively low oxygen concentration. All three of these attributes are necessary for the successful transport of oxygen from outside the body to respiring cells.

The oxygen-binding pigment used by many animals, including humans, is **hemoglobin**. Each hemoglobin molecule can carry up to four oxygen molecules (Figure 22.9). Hemoglobin greatly increases the oxygen-carrying capacity of blood, from the 2 milliliters per liter that can be dissolved in the plasma to 190 milliliters per liter. The oxygen atoms are bound to iron atoms in the hemoglobin molecule. When we do not get enough iron in our diet, the body cannot produce enough hemoglobin. The oxygen-carrying capacity of the blood is reduced, sometimes leading to constant fatigue. This condition is known as anemia.

Oxygen-binding pigments appear to have evolved independently in many different groups of animals, but they all share a similar basic design (a complex protein that usually contains iron) and exhibit a similar feature (the ability to bind reversibly to oxygen depending on the oxygen concentration in the plasma). We call these molecules *pigments* because they have distinctive colors that change depending on whether or not they are bound to oxygen. Hemoglobin, for example, changes color from purple-red when it is not bound to oxygen to bright red when it is.

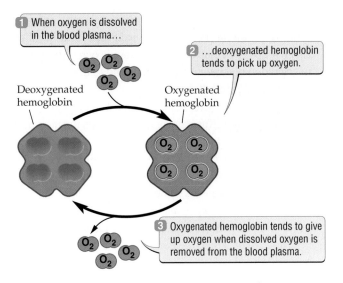

1 When oxygen is dissolved in the blood plasma…

2 …deoxygenated hemoglobin tends to pick up oxygen.

Deoxygenated hemoglobin

Oxygenated hemoglobin

3 Oxygenated hemoglobin tends to give up oxygen when dissolved oxygen is removed from the blood plasma.

Figure 22.9 Hemoglobin Can Bind and Release Oxygen
Hemoglobin is a complex molecule that allows the blood to carry much more oxygen than could be dissolved in blood plasma. Hemoglobin picks up or releases oxygen molecules depending on the amount of oxygen dissolved in the blood plasma around it.

Hemoglobin is carried in the blood by red blood cells

The second way some animals, including humans, have overcome the oxygen-carrying limitations of their blood plasma is by packaging their oxygen-binding pigment into cells. In animals that use hemoglobin, these cells are called **red blood cells**. Each human red blood cell contains about 250,000 hemoglobin molecules. That adds up to a million molecules of oxygen that can bind to the hemoglobin molecules in one red blood cell!

For O_2 in the alveoli to reach the hemoglobin molecules within red blood cells requires some additional diffusion steps that are not required for oxygen dissolved in the blood plasma. After O_2 diffuses into the capillaries surrounding the alveoli, it must then diffuse across the plasma membrane of a red blood cell, and then into the red blood cell until it encounters a hemoglobin molecule, where it is quickly bound. Despite these extra diffusion steps, having oxygen carried by hemoglobin in red blood cells vastly increases the efficiency with which oxygen can be transported in the bloodstream to respiring cells.

Hemoglobin

Explore how hemoglobin picks up and delivers oxygen.

22.3

Oxygen dissolved in blood plasma plays an important role in gas exchange

Despite the blood plasma's limited ability to transport oxygen, it still has a role to play in the process. The

Helpful to know

Don't confuse the single windpipe in humans and other mammals (the *trachea*) with insect *tracheae,* tubes that branch into tracheoles the way the bronchi in human lungs branch into bronchioles.

transport of oxygen in the blood of animals with oxygen-binding pigments involves several steps, and in all of these steps the relationship between oxygen dissolved in the blood plasma and oxygen bound to hemoglobin is important (Figure 22.10).

When oxygen enters the blood at the gas exchange surface, it first diffuses into and dissolves in the blood plasma. Once it diffuses into a red blood cell and binds with hemoglobin, the oxygen is no longer dissolved in the plasma. In this way, as hemoglobin picks up oxygen, it keeps the plasma oxygen concentration low. This, in turn, ensures that the concentration gradient of oxygen between the plasma and the alveoli remains large, speeding the rate at which oxygen diffuses out of the alveoli and into the blood.

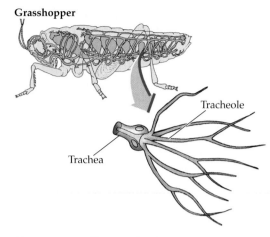

Grasshopper

Tracheole

Trachea

Figure 22.11 Insects "Breathe" Differently
Once the tracheae deliver oxygen to the tracheoles, it can diffuse relatively rapidly through the gas contained in the tracheoles. The outside openings of the tracheae open and close to regulate oxygen uptake. All of the cells in an insect must lie close to a tracheole.

Once the oxygen-carrying blood gets near the respiring cells, this sequence is reversed. Respiring cells are using oxygen, so the concentration of oxygen in the blood plasma near a respiring cell is constantly being lowered. When the concentration is low enough, oxygen is released from the hemoglobin into the plasma.

The carbon dioxide produced by respiring cells also helps to trigger the release of the oxygen that those cells need. The carbon dioxide diffuses into the plasma. As it builds up in the plasma, it causes the plasma and the red blood cells in it to become slightly more acidic. When a hemoglobin molecule encounters a more acidic environment, the shape of the molecule changes in such a way that it tends to release any oxygen molecules it is carrying.

Insects have a different way of moving gases to and from respiring cells

Instead of using their circulatory system to carry gases, insects move air in and out through a series of tubes that branch inward from the surface of the body (Figure 22.11). The largest tubes, called **tracheae** [tray-kee-ee] (singular: trachea), branch repeatedly; the smallest of these, called **tracheoles** [*TRAY*-kee-olz], allow gases to diffuse directly to and from respiring cells. But insects don't rely on diffusion alone. They actively pump air in and out of their tracheal system. If an insect lands on a surface near you, don't panic. See if you can see its abdomen (the body section farthest from the head) making pumping movements.

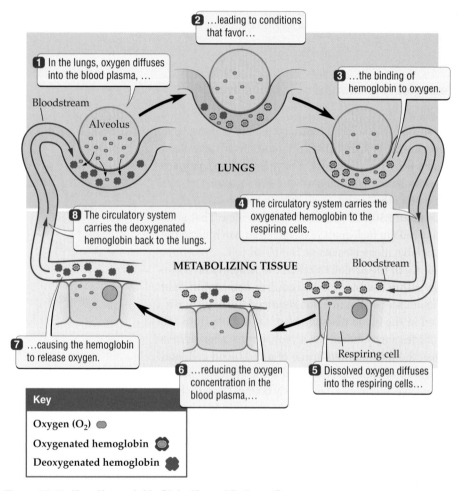

2 …leading to conditions that favor…

1 In the lungs, oxygen diffuses into the blood plasma, …

3 …the binding of hemoglobin to oxygen.

Bloodstream

Alveolus

LUNGS

8 The circulatory system carries the deoxygenated hemoglobin back to the lungs.

4 The circulatory system carries the oxygenated hemoglobin to the respiring cells.

METABOLIZING TISSUE

Bloodstream

7 …causing the hemoglobin to release oxygen.

Respiring cell

6 …reducing the oxygen concentration in the blood plasma,…

5 Dissolved oxygen diffuses into the respiring cells…

Key

Oxygen (O$_2$)

Oxygenated hemoglobin

Deoxygenated hemoglobin

Figure 22.10 How Hemoglobin Picks Up and Delivers Oxygen
In humans, hemoglobin picks up oxygen in the lungs, where the concentration of oxygen dissolved in the blood plasma is relatively high. Hemoglobin releases oxygen as it passes through respiring tissue, where the concentration of oxygen in the blood plasma is relatively low. Most of the oxygen in our blood at any time is bound to hemoglobin rather than dissolved in the blood plasma itself.

Breathing in Thin Air

People who live at high altitudes, where each breath of air provides relatively little oxygen, have characteristics that allow them to survive in such environments. First, they tend to have large lungs. In addition, the hemoglobin concentration in their blood is higher than that in people living near sea level, often giving their skin a red color. The hemoglobin itself picks up oxygen more readily from the blood plasma, which allows their blood to leave the lungs 100 percent saturated with oxygen, despite the smaller difference in oxygen concentration between air and blood. When individuals who are native to lowland areas visit high altitudes, the hemoglobin in their blood typically leaves their lungs only partially oxygenated.

Without previous exposure to high-altitude environments, our initial response to thin mountain air is similar to an emergency response. Our breathing rates and heart rates increase. In essence, our bodies act as if we were exercising at a lower altitude. The increased breathing rate maintains as great a difference in oxygen concentration as possible between the air and the body, and the rapidly pumping heart carries oxygenated blood to respiring cells more quickly.

As athletes and mountaineers know, the human body can adjust to low-oxygen environments if given time. One recommendation for would-be mountain climbers is to allow one day of adjustment time for each 600 meters they wish to ascend above 2,000 meters. With more time spent at high altitudes, the body adjusts by increasing the hemoglobin concentration in the blood and by increasing the ability of the hemoglobin to unload oxygen. The first response increases the oxygen-carrying capacity of the blood, and the second allows the blood to unload oxygen more quickly near respiring cells. Note, however, that an increased ability to unload oxygen to cells means a decrease in the ability to pick up oxygen in the lungs. The decreased attraction of hemoglobin to oxygen found in short-term adaptation to high altitudes is just the opposite of the condition found in human populations evolutionarily adapted to high altitudes: their hemoglobin readily picks up oxygen from the thin air. This difference illustrates an interesting contrast between a short-term physiological response and a long-term evolutionary response.

Among bird species, the bar-headed goose has evolved a remarkable ability to fly in extremely thin air. These birds migrate from the lowlands of India over the Himalayas to Tibet and back again. Sometimes they fly right over Mount Everest! Like all birds, bar-headed geese have lungs that work quite differently from our own. Instead of passing back and forth over saclike alveoli, the air in bird lungs passes in one direction through tiny gas exchange tubes. This structure allows birds, as a group, to extract oxygen from air more efficiently than mammals. Thus bar-headed geese can pull more scarce oxygen out of each breath of high-altitude air.

In addition, bar-headed geese have a few unique features that allow them to fly in a low-oxygen environment. Their hemoglobin is especially good at picking up oxygen. At the same time, their flight muscles are especially good at drawing oxygen from the blood. In addition, they can breathe rapidly without causing changes in their blood chemistry that lead to a reduced blood flow to the brain. When we breathe in and out rapidly, or hyperventilate, blood flow to the brain can decrease to the point that our ability to think effectively diminishes. In contrast, hyperventilating bar-headed geese continue to function normally. This combination of a highly efficient bird lung and a few special features allows bar-headed geese to fly great distances at altitudes where we could not function for even a short time without supplemental oxygen.

Chapter Review

Summary

22.1 Breathing in Humans

- The human respiratory system carries air from the nose (or mouth) to the lungs through a series of tubular passageways: from nose to pharynx (throat) to trachea (windpipe) to the two bronchi, which branch into smaller and smaller bronchioles, eventually reaching clusters of tiny sacs in the lungs called alveoli.
- Actual gas exchange takes place in the alveoli, where oxygen diffuses into the blood and carbon dioxide diffuses out.

22.2 Principles of Gas Exchange

- Oxygen and carbon dioxide enter and leave organisms exclusively by diffusion.
- Gases diffuse from areas of high concentration to areas of low concentration. This principle has two biological consequences: First, the concentration of a gas at its source must be lower than that at its destination. Second, the higher the concentration of a gas in the environment compared with that inside an organism, the faster the gas will diffuse into the organism.
- The greater the surface area available for diffusion, the more gas can diffuse across it per unit of time.
- The shorter the distance over which a gas must diffuse, the faster it can reach its destination.

22.3 Gas Exchange on Land and in Water

- Aquatic animals rely on gills, which provide a large gas exchange surface on the outside of the body. Water can flow over external gills without the in-and-out pumping required by internal gas exchange systems.
- Terrestrial animals rely on internal gas exchange structures, such as lungs, which require active pumping of air in and out.
- The weight of water, combined with the low level of oxygen in water, would make it impossible for lungs to pump enough water in and out to support an animal's oxygen demand. Because gills are external, they can extract sufficient oxygen from the water flowing over them to support an animal without requiring too much expenditure of energy.
- Internal lungs allow terrestrial animals to reduce the water loss that inevitably goes along with gas exchange on land. External gills on land would lead to the loss of far too much body water during gas exchange to support an animal's survival.

22.4 How Animals Transport Gases to Respiring Cells

- Most groups of animals rely on a circulatory system that pumps liquid blood through the body to carry gases between gas exchange surfaces and respiring cells more quickly than is possible by diffusion alone.
- Blood plasma has a low capacity for transporting dissolved oxygen. To compensate, animals use oxygen-binding pigments, such as hemoglobin, which greatly increase the oxygen-carrying capacity of blood. These proteins bind reversibly to O_2, picking O_2 up at the gas exchange surfaces, where the concentration of oxygen is high, and releasing O_2 near respiring cells, where the concentration of oxygen is low.
- Hemoglobin molecules are housed in red blood cells, which further increases the oxygen-carrying capacity of the blood.
- The level of dissolved oxygen in the blood plasma plays an important role in triggering the uptake of oxygen by hemoglobin at gas exchange surfaces and its release at respiring cells.
- Insects' unique form of gas transport allows O_2 and CO_2 to diffuse between the external environment and respiring cells through air-filled tracheae and tracheoles.

⊙ Review and Application of Key Concepts

1. Describe the human respiratory system from nose and mouth to alveoli.

2. Describe the transport of oxygen from an alveolus to a respiring cell and the movement of carbon dioxide from a cell to the air in an alveolus. Include the directions of diffusion for oxygen and carbon dioxide in an alveolus and in the blood plasma near a respiring cell.

3. What are some of the advantages of gills in aquatic environments? What would their disadvantages be in terrestrial environments? Similarly, describe the advantages of lungs in a terrestrial environment and the disadvantages of lungs in an aquatic environment.

4. Hemoglobin greatly increases the amount of oxygen our blood can carry. Describe why this is so, as well as the conditions under which oxygen is loaded and unloaded from the hemoglobin molecule.

5. The only aquatic organisms that maintain their body temperatures permanently above that of the water in which they live are endotherms: mammals (such as whales, seals, and sea otters), and birds (such as penguins). Even though they live in

the water, these animals all breathe air using lungs. What advantage might air-breathing aquatic endotherms have over a hypothetical aquatic endotherm that depended on gills for gas exchange?

Key Terms

alveolus (p. 425)
blood (p. 430)
breathing (p. 422)
bronchus (p. 425)
bronchiole (p. 425)
circulatory system (p. 430)
gill (p. 429)
hemoglobin (p. 431)
lower respiratory system
    (p. 425)

lung (p. 425)
oxygen-binding pigment (p. 431)
pharynx (p. 423)
red blood cell (p. 431)
reversible binding (p. 431)
trachea (in mammals) (p. 425)
trachea (in insects) (p. 432)
tracheole (p. 432)
upper respiratory system
    (p. 423)

Self-Quiz

1. The largest tube leading to our lungs is the
   a. bronchiole.
   b. alveolus.
   c. trachea.
   d. bronchus.

2. In humans, gas exchange takes place in the
   a. bronchioles.
   b. trachea.
   c. alveoli.
   d. bronchi.

3. The rate at which oxygen moves from the alveoli of our lungs into our blood depends on
   a. the difference in oxygen concentration between the alveoli and the blood.
   b. the distance between the alveoli and the respiring cells.
   c. the availability of energy to transport gases across the alveoli.
   d. none of the above

4. In which of the following ways does water differ from air as a source of oxygen for animals?
   a. Water contains a lower concentration of oxygen than does air.
   b. Oxygen diffuses more rapidly in air than in water.
   c. As water warms up, the amount of dissolved oxygen it can hold decreases.
   d. all of the above

5. Carbon dioxide diffuses from
   a. hemoglobin to cells.
   b. blood to mitochondria.
   c. alveoli to blood.
   d. blood to air in the alveoli.

6. When added together, the surface area of the tiny alveoli in a single human lung have a combined surface area that is
   a. about half that of the skin.
   b. about 90 times that of the skin.
   c. about equal to that of the skin.
   d. none of the above

7. Oxygen is transported throughout the human body primarily
   a. as a dissolved gas in the blood plasma.
   b. bound to hemoglobin molecules in red blood cells.
   c. through tracheoles.
   d. by diffusion.

8. Which of the following represents a feature that would allow humans to get enough oxygen at high altitudes?
   a. an increase in the amount of hemoglobin in the blood
   b. a decrease in the area of the gas exchange surface inside the lung
   c. a decrease in the breathing rate, combined with a decrease in lung size
   d. a reduction in the amount of hemoglobin in the blood

9. Hemoglobin
   a. binds to carbon dioxide.
   b. binds to oxygen.
   c. retains water.
   d. contains red blood cells.

10. Insects breathe through
    a. lungs.
    b. alveoli.
    c. tracheae.
    d. their skin.

Biology in the News

Study "Proves" Asthma Cause

Evidence has been found of a link between air pollution and asthma in children.

Although air pollution has been linked with respiratory problems for some time, a proof of a consistent connection has always proved elusive.

But a study published on Friday lends support to a direct association.

The researchers are calling for improved public health strategies to target children at risk.

A study led by a Dutch researcher demonstrated that as the levels of air pollution increase, so does the chance that children will suffer from a variety of problems associated with the lower respiratory system. The researchers found that children who suffered from both allergies and asthma were more likely to suffer breathing problems on days when the air was more polluted with either particulates (tiny dust particles suspended in the air) or the chemicals common in car exhaust.

The tubes of the respiratory system, especially the bronchi, play an important role in keeping dust and other contaminants (including live ones such as bacteria and fungal spores) out of the lungs. The surfaces of these passageways are lined with epithelial cells. Some of these cells are specialized to release mucus, which traps the dust, microorganisms, and other contaminants that inevitably enter the upper respiratory system when we inhale. Other epithelial cells bear bunches of cilia; these tiny, hairlike appendages move in waves to transport the mucus, laden with contaminants, back up the bronchioles, bronchi, trachea, and into the throat. Normally, we help the movement of mucus up from the lungs by coughing and clearing the throat. Then we eliminate the mucus from the respiratory system either by spitting it out or by swallowing it (thereby eliminating it through the digestive system).

In people with asthma, however, the response to these contaminants goes beyond coughing. Several studies have shown that about 5 percent of the human population responds to contaminants in the lower respiratory system with a hypersensitive reaction in the bronchioles. This reaction involves swelling of the bronchiole walls, which makes it difficult for air to enter or leave the lungs. Other environmental factors can also trigger an asthmatic response in some people; these factors include a sudden air temperature change and too much exercise in too short a time ("sports asthma"). Regardless of the cause, until the bronchiole swelling goes down, asthma sufferers can have serious difficulty breathing.

Evaluating the news

1. Do you know anyone who suffers from asthma? How does it affect their life?

2. One comment made by the authors of this study was that there seemed to be *no* level of air pollution that produced no effect on asthma and allergy sufferers. How reasonable does it seem to you to try to eliminate air pollution altogether? Explain your point of view.

3. A scientist was quoted in the *BBC News* article as asserting that the most effective way of dealing with asthma might be to try reducing the symptoms caused by the contaminants rather than to try eliminating the contaminants themselves. This researcher believed that so many asthma-provoking factors are present in the air in addition to those associated with air pollution that it might prove impossible to prevent asthma. What do you think of this approach to the issue? How do you think the costs of eliminating the symptoms of asthma would compare with the costs of eliminating its causes?

SOURCE: *BBC News*, March 12, 1999.

CHAPTER 23 Blood and Circulation

Key Concepts

- The human circulatory system, like that of many other animals, depends on a muscular heart that contracts and relaxes as it pumps blood through one-way valves and vessels.

- Some blood vessels transport blood to all parts of the body, while other vessels allow the exchange of materials between blood and cells. The sizes and structures of blood vessels are suited to their different functions.

- Humans and other vertebrates have a closed circulatory system, with vessels that form a complete closed loop and a muscular heart with one-way valves. Other animals, including arthropods, have an open circulatory system, with open-ended vessels.

- High blood pressure and constricted blood vessels in humans can have a number of serious health consequences, including heart attack, stroke, and kidney failure.

High Blood Pressure on the African Savanna

The human heart generates pressure that drives blood through the circulatory system. As part of a routine checkup, doctors measure blood pressure to assess the health of the circulatory system. A resting blood pressure of slightly less than 120/80 millimeters of mercury (mm Hg) usually indicates a heart and circulatory system that are working normally. The high and low numbers refer to the maximum and minimum pressures, respectively, that our blood vessels experience as blood pulses through them. The unit of measurement used, millimeters of mercury, refers to the height to which the pressure can push a column of the heavy liquid mercury.

Blood pressures higher than 120/80 indicate potential health problems. In resting adults, blood pressures of 140/90 or greater characterize high blood pressure. About half of North American adults suffer from mild high blood pressure, and about a fifth of that population suffers from severe high blood pressure, with blood pressures over 160/95. High blood pressure represents one of the major causes of death in aging North Americans. If blood pressures remain over normal levels for an extended time, the risk of serious health problems such as heart attack, stroke, and kidney damage increases.

Scientists have turned to other animals in their efforts to understand more about the control of blood pressure. On the African savanna, giraffes move gracefully across sweeping grasslands, munching on high, flat-topped acacia trees. A giraffe's heart must generate a lot of pressure to get blood up that long neck to a head 6 meters above ground level. These animals live with blood pressures of 260/160. How do giraffes manage to thrive while

Rising to the Top
The giraffe's heart must produce high pressures to pump blood all the way up to its head when it is feeding.

maintaining blood pressures that would threaten human lives? And how do they prevent the flow of too much blood to the head and brain when they bend down to drink?

Explore the human circulatory system. 23.1

As we have seen repeatedly in earlier chapters, animals need to move materials from one part of the body to another. Respiration requires the movement of nutrients and gases from their sources to their destinations. Coordinated activities such as growth and homeostasis also depend on the transfer of information and materials within the body.

In previous chapters we have discussed the importance of diffusion as a means of transporting materials in animal bodies. Over short distances (less than a fraction of a millimeter), such as across plasma membranes, within a single cell, or across the tiny distances between blood capillaries and interstitial fluid, gases and dissolved materials can diffuse within a fraction of a second. However, the distances between sources and destinations in most animals are far too great for diffusion to work effectively. As multicellular animals evolved from their single-celled ancestors, they developed internal transport systems based on muscular **hearts**, which pump blood carrying nutrients and wastes over long distances. Blood circulates through their bodies, either by bathing tissues in a loose flow or by moving through **blood vessels** to particular destinations.

In this chapter we consider how animals transport vital materials throughout their bodies. We start by introducing the structure and function of the human circulatory system. Next we examine how the structure and size of the blood vessels affect blood flow. We then look briefly at patterns of circulatory systems in other animals to note some basic similarities and differences. We close with a discussion of some of the health problems that beset the human circulatory system.

23.1 The Human Cardiovascular System

In vertebrates, the circulatory system is called the **cardiovascular system** (from *kardia*, "heart"; *vasculum*, "small vessel"). It consists of a muscular heart that pumps blood, which circulates through a complex series of loops formed by blood vessels. The human cardiovascular system includes many of the basic elements common to such systems in a variety of animals. To answer the question of how animals move materials around inside the body, therefore, we begin with a look at the human cardiovascular system.

The human heart pumps blood to the body through two circuits

In the human cardiovascular system, tubular blood vessels form a series of loops, or circuits, through which blood moves from the heart to respiring cells and then back to the heart (Figure 23.1). There are two main

Figure 23.1 The Human Cardiovascular System
The human cardiovascular system, like those of all other mammals, depends on a heart to push blood through loops formed by blood vessels. There are two main circuits, the pulmonary circuit and the systemic circuit.

Exchange of materials between the blood and the surrounding cells occurs in tiny capillaries.

Site of gas exchange in the lungs

In the pulmonary circuit, blood that has delivered oxygen to cells and picked up carbon dioxide from them is pumped to the lungs, where gas exchange takes place.

Red blood cell

Extremely thin walls formed by epithelial cells are all that separate the blood in the capillaries from the surrounding cells.

The muscular heart contracts to generate the pressure that pushes blood through the circulatory system.

In the systemic circuit, blood that has picked up oxygen from the lungs is pumped into body cells.

Respiring tissues

Valves in the veins ensure that the blood flows in only one direction.

Capillaries

Vein walls and artery walls contain muscle.

Vein

Artery

circuits within this system. In the **pulmonary circuit** (from *pulmon*, "lung"), the heart pumps blood to the lungs. This blood, which has delivered oxygen to distant cells and picked up carbon dioxide from them, eventually flows through the capillaries surrounding the alveoli in the lungs (see Figure 22.2). Oxygen diffuses from air in the alveoli across the thin capillary walls and into the blood. At the same time, the CO_2 in the blood diffuses across the capillaries into the alveolar air. The blood, now rich in oxygen and low in carbon dioxide, flows back to the heart, completing the pulmonary circuit. The heart then pumps this blood into the **systemic circuit**, which delivers oxygen to cells throughout the body and picks up carbon dioxide from them. These exchanges take place across the capillary walls of the systemic circuit.

In a typical adult, 5 to 6 liters of blood—carrying blood cells and a rich mixture of dissolved minerals, gases, nutrients, and products of metabolism—circulate continuously through the blood vessels. Nearly all of our cells lie no more than 0.03 millimeter (less than one-third the thickness of this page) from a blood vessel with which they can exchange materials. Over these very short distances, diffusion works well.

Carrying blood so close to all the trillions of cells in our bodies requires an extensive system of blood vessels. If laid end to end, the approximately 17 billion blood vessels in the human body would stretch an amazing 20,000 kilometers, or from Seattle to London and back again with about 1,000 kilometers to spare.

There are three major kinds of blood vessels in the human body (see Figure 23.1):

1. **Arteries** carry blood from the heart for distribution to the body. These vessels branch many times into smaller and smaller vessels. The finest branchings (called arterioles) are only 0.3 millimeter in diameter, about the same diameter as a fine thread used for sewing. Arterioles control the flow of blood to the smallest vessels, the capillaries.

2. **Capillaries**, less than 0.01 millimeter in diameter, allow the exchange of materials between the blood and the surrounding interstitial fluid and cells. A piece of muscle tissue the size of a pencil tip may contain more than 1,000 capillaries. Taken together, the capillaries in our bodies provide a combined surface area equal to the floor space in three large houses.

3. **Veins** carry blood from the body back to the heart. The return journey begins as blood leaves the capillaries in tiny veins called venules, which join together into ever larger veins, eventually collecting all the blood from the body for return to the heart.

The human cardiovascular system is considered a **closed circulatory system** for two reasons: blood is enclosed in the vessels and the heart, and each of the two circuits is a closed loop—from heart to arteries to capillaries to veins and back to the heart again.

The human heart has two pumps and four chambers

It is the contractions of the muscular heart that pump blood through our blood vessels. The human heart—like the hearts of all mammals, and of birds as well—has two sides, which pump in a coordinated way, but which are physically independent of each other, separated by a thick wall of tissue. This arrangement allows each side to act as a separate pump (Figure 23.2*a*). The larger left side receives oxygenated blood returning from the lungs and pumps it through the systemic circuit to respiring cells. The smaller right side receives blood returning from the systemic circuit that is low in oxygen and laden with CO_2, and pumps it through the pulmonary circuit for gas exchange in the lungs. The left side is smaller because it doesn't need to pump blood as far as the right side does; it pumps blood to the lungs, which lie near the heart, whereas the right side pumps blood to the rest of the body.

Each side of the heart is further divided into two chambers: an upper chamber, the **atrium** (plural: atria), and a lower chamber, the **ventricle**. Contraction of the smaller, thinner-walled atrium pumps blood into the corresponding ventricle. Then contraction of the ventricle, which has a thicker, more muscular wall, pumps blood out of the heart and into the blood vessels heading for either the lungs or the body. Note that even though the two sides of the heart serve separate functions, they pump together at the same time. So the left and right atria contract at the same time, followed by the left and right ventricles contracting at the same time.

One-way valves, separating the atrium from the ventricle and the ventricle from the artery leading out of it, ensure that blood flows through each side of the heart in one direction only—from the atrium to the ventricle to the body, and not the reverse (Figure 23.2*b*). Heard through a stethoscope, the *lub-dub* sound of the heartbeat is actually the sound of these valves snapping shut (Figure 23.3). The *lub* sound is made when the valve separating the atrium from the ventricle closes; the pressure that develops in the ventricle as it contracts snaps this valve closed. The *dub* sound occurs when the valve separating the ventricle from the artery closes; this valve closes when the pressure in the artery becomes greater than the pressure in the relaxing ventricle.

Test your knowledge of the human circulatory system. 23.2

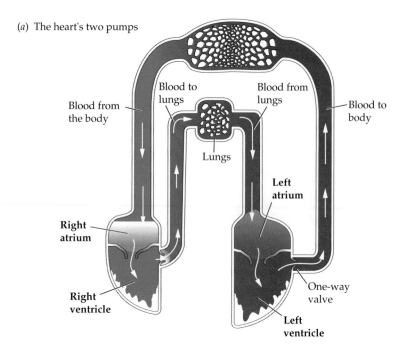

(a) The heart's two pumps

Blood from the body

Blood to lungs

Blood from lungs

Blood to body

Left atrium

Right atrium

Right ventricle

Lungs

One-way valve

Left ventricle

(b) How a one-way heart valve works

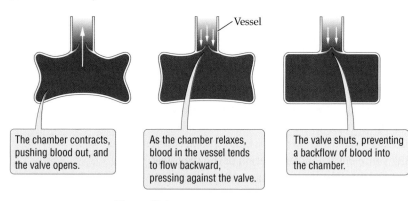

Vessel

| The chamber contracts, pushing blood out, and the valve opens. | As the chamber relaxes, blood in the vessel tends to flow backward, pressing against the valve. | The valve shuts, preventing a backflow of blood into the chamber. |

Figure 23.2

A Generalized View of Blood Flow in the Heart Chambers

(a) The pattern of flow in the heart is from atria to ventricles. Each side of the heart functions as a separate pump, although the two sides pump at the same time. (b) A one-way valve allows blood flow out, but not back in.

Examine the beat and sound of the heart.

23.3

What's in a heartbeat?

Our hearts beat, we hope, without fail about 60 to 75 times each minute, which amounts to about 3 billion beats in a 70-year lifetime. The number of beats per minute is referred to as the **heart rate.** When we are at rest, our hearts pump all of our blood (5 to 6 liters) each minute—in other words, the equivalent of 7,000 liters of blood passes through the cardiovascular system each day. Each contraction of the heart generates a sharp increase (spike) in pressure that propels the blood (Figure 23.4).

The blood pressure measured in a doctor's office reflects the pressures in the arteries leading to the body from the left ventricle. A **blood pressure** reading of 120/80, for example, means that contraction of the left ventricle generates 120 mm Hg of pressure in the arteries, followed by a drop to 80 mm Hg when the ventricle relaxes and refills. Similar measurements made in the arteries leading from the smaller right ventricle into the shorter pulmonary circuit would reveal lower pressures, changing between only 8 and 25 mm Hg.

By the time the blood enters the capillaries, the pressure has dropped to 35 mm Hg. When the blood leaves the capillaries, the pressure is only 10 mm Hg. The drop in pressure as the blood progresses through the circulatory system reflects the cumulative effect of friction between the blood and the blood vessel walls, as we shall see in the following section.

Each **heartbeat** we experience lasts (for most people) a little less than 1 second and consists of a series of events called the **cardiac cycle** (Figure 23.5). This cycle has two phases: relaxation and contraction. The relaxation phase is called **diastole** [die-ASS-toh-lee] (from a Greek word for "expansion"). The lower number in a blood pressure reading represents the pressure at diastole. Diastolic pressure is the number doctors worry about more if it is high because it represents the "background" pressure

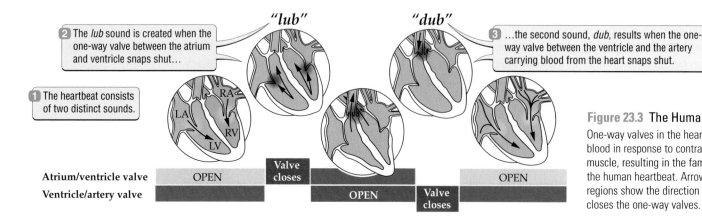

"lub"

"dub"

2 The *lub* sound is created when the one-way valve between the atrium and ventricle snaps shut...

3 ...the second sound, *dub*, results when the one-way valve between the ventricle and the artery carrying blood from the heart snaps shut.

1 The heartbeat consists of two distinct sounds.

RA

LA

RV

LV

Figure 23.3 The Human Heartbeat

One-way valves in the heart direct the flow of blood in response to contraction of the heart muscle, resulting in the familiar *lub-dub* sound of the human heartbeat. Arrows pointing to the red regions show the direction of blood flow that closes the one-way valves.

| Atrium/ventricle valve | OPEN | Valve closes | | | OPEN |
| Ventricle/artery valve | | | OPEN | Valve closes | |

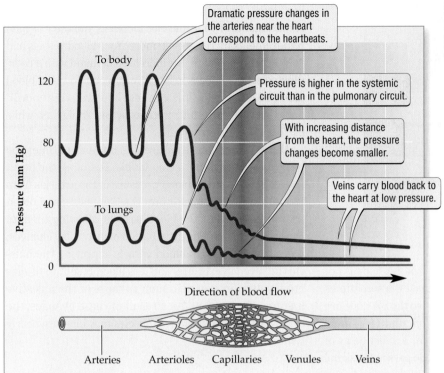

Figure 23.4

Pressure Changes in the Human Circulatory System

Both the pressures and the pressure differences in human blood vessels decrease as the blood moves through the circulatory system.

Dramatic pressure changes in the arteries near the heart correspond to the heartbeats.

Pressure is higher in the systemic circuit than in the pulmonary circuit.

With increasing distance from the heart, the pressure changes become smaller.

Veins carry blood back to the heart at low pressure.

To body

To lungs

Pressure (mm Hg)

120

80

40

0

Direction of blood flow

Arteries   Arterioles   Capillaries   Venules   Veins

---

that is constantly working on blood vessels. During diastole, blood flows into the relaxed heart. **Systole** [*SISS*-toh-lee] (Greek for "contraction") is the pumping phase: blood is pumped first from the atria to the ventricles, then from the ventricles to the lungs and body.

The regularity of the normal heartbeat comes from signals the heart receives from a group of specialized cells in a region within the heart itself, called the **pacemaker** (Figure 23.6). This signaling center is also called the **sinoatrial (SA) node** [*SIGH*-noh-*AY*-tree-ul . . .]. Signals from the SA node travel across the atria, causing them to contract

Learn more about the cardiac cycle.

DIASTOLE | SYSTOLE

1 Blood flows into the relaxed heart.

2 Blood is forced into the ventricles as the atria contract.

0.1 sec

0.4 sec | 0.3 sec

3 Blood is pumped to the body and lungs as the ventricles contract.

Figure 23.5 The Cardiac Cycle

A heartbeat lasts less than a second and has two phases, diastole and systole.

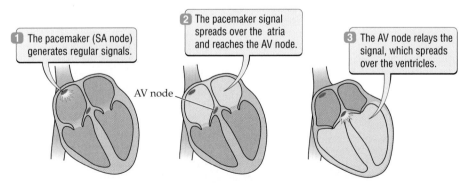

1 The pacemaker (SA node) generates regular signals.

2 The pacemaker signal spreads over the atria and reaches the AV node.

3 The AV node relays the signal, which spreads over the ventricles.

AV node

Figure 23.6 The Heart's Signaling System

A heartbeat is initiated by a signal from the pacemaker that spreads over the heart in an orderly way.

Figure 23.7
Artificial Pacemakers
Can Save Lives
When the heart's natural pacemaker fails to function, an artificial pacemaker can be implanted surgically in a patient's chest, where it generates regular electrical signals that have the same effect as those from the natural pacemaker.

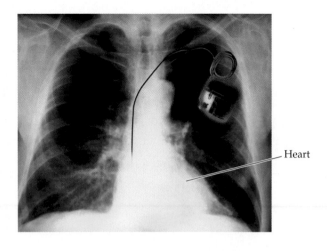

Heart

and pump blood into the ventricles. If the SA node is not working correctly, an artificial pacemaker is sometimes implanted (Figure 23.7). The signals from the SA node are relayed to another node, the **atrioventricular (AV) node** [*AY*-tree-oh-ven-*TRICK*-you-ler . . .]. After a short delay of about one-tenth of a second, the AV node passes the signal on to the ventricles, causing them to contract and send blood to the lungs and body. The short delay allows the atria to empty more completely. It is the signals from these two nodes that are measured in an **electrocardiogram (EKG)**.

The cardiovascular system responds to the body's changing needs

If you are sitting still and reading this book, the demands from your muscles for oxygen and nutrients are at a "rest-

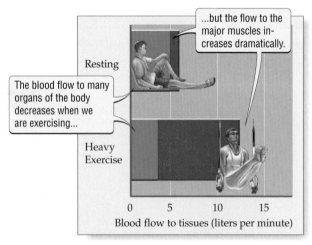

...but the flow to the major muscles increases dramatically.

Resting

The blood flow to many organs of the body decreases when we are exercising...

Heavy Exercise

0   5   10   15
Blood flow to tissues (liters per minute)

Figure 23.8 The Response of the Human Cardiovascular System to Exercise

The human cardiovascular system responds to increased activity by increasing the flow of blood to the active muscles (measured in liters per minute).

ing" level. But when you are exercising, the nutrient requirements of your muscles increase, and their oxygen requirements increase even more. Your blood vessels and your heart respond to this increased demand with a coordinated set of changes in heart rate, blood flow, and blood distribution. Physicians can test how well this coordinated system works under stress by recording data while patients run on a treadmill.

During strenuous exercise, the number of times the heart beats per minute can double. Doubling the heart rate can increase the blood pressure in the arteries from 120 mm Hg to about 180 mm Hg. The increased blood pressure stretches the blood vessels, allowing them to carry more blood. The distribution of blood also changes: more blood is sent to blood vessels supplying the muscles involved in exercise, and less blood is sent to tissues not involved in exercise, such as those of the digestive system (Figure 23.8). As a result of these changes, the supply of blood to the active muscles may increase nearly tenfold.

23.2 Blood Vessels and Factors Affecting Blood Flow

Blood vessels, as we have seen, serve different purposes, depending on whether they are arteries, veins, or capillaries, large or small. In both size and structure, blood vessel form matches function.

Arteries and veins are specialized for transporting blood in quantity

Large vessels—arteries and veins—carry blood in large volumes. Their purpose is not to exchange significant amounts of materials directly with the interstitial fluid around the cells, but rather to transport blood rapidly throughout the body. Vessels specialized for bulk transport of blood have relatively large diameters and muscular, elastic walls (see Figure 23.1).

Decreasing the diameter of a blood vessel greatly increases the surface area of the vessel relative to the volume of blood it carries, which greatly increases the frictional drag on the blood. To picture this, slide your hand along a tabletop; the resistance you feel is due to friction between the palm of your hand and the table's surface, which slows down the movement of you hand. Similarly, drag in blood vessels results from friction between the blood vessel walls and the layer of blood in contact with them. In large vessels, only a small portion of the blood experiences frictional drag because most of the blood is not in contact with the vessel walls. The smaller the diameter of the blood vessel, however, the greater the proportion of blood close enough to the wall to experience drag. Decreasing the diameter of a blood vessel by half, for example, increases the drag that the blood experiences by 16 times; in other words, a heart would have to work 16 times harder to maintain the same blood flow in the smaller vessel (Figure 23.9). Thus, to reduce the amount of work the heart must do to transport blood, the size of blood vessels is matched to the relative distance the blood must travel in them. The vessels that carry blood the farthest have the largest diameters, and thus the blood experiences relatively little drag.

The complex walls of the largest arteries equip them to regulate blood distribution. Layers of muscle tissue in the walls contract and relax to alter the amount of blood flowing through the arteries. This ability allows us to control the flow of blood to different parts of the body during rest or strenuous exercise. The elasticity of the artery walls also allows them to stretch when a heart contraction increases the blood pressure. When you feel your pulse in an artery, you are actually feeling the artery walls bulge and relax in response to the surge of pressure generated by the contraction of the heart. The ability of the artery walls to stretch and relax reduces these pressure

changes as distance from the heart increases, helping to protect the most delicate blood vessels, the capillaries, from damaging pressure changes.

Veins are not as muscular as arteries (see the box on page 446). In addition, because venous blood pressure is relatively low (see Figure 23.4) and blood returning to the heart often must rise against the force of gravity, veins have an added feature that arteries don't need: one-way valves that keep blood flowing back to the heart. These valves have flaps that work like the one-way valves in the heart: Blood actively flowing toward the valve forces the flaps open. When the pressure declines below a certain level, some blood starts to flow in reverse due to gravity, forcing the flaps closed; this prevents any appreciable backflow from occurring. When these valves fail to close properly, as can happen in older or overweight people, some backflow does occur, causing varicose (abnormally swollen) veins, which are common in the legs.

Capillaries are specialized for the exchange of materials

Human circulatory systems, like those of vertebrates generally, exchange materials across the thin walls of the tiny capillaries that complete the circulatory loop by connecting arteries to veins. Closed circulatory systems provide a distribution system that can attend to the finest details at the cellular level, such as the complex exchange of water and solutes that occurs in the nephrons of the kidneys (Figure 23.10; see Figure 20.8 for a review of this process).

Compared with arteries and veins, capillaries have extremely thin, porous walls, formed exclusively of

Figure 23.9 Blood Vessel Diameter and Frictional Drag
As the diameter of a blood vessel increases, the amount of blood that it can transport in a given time increases dramatically—more so than simple size differences alone would explain. With the same pressure from the heart, about 256 times as much blood can flow through a 4-millimeter vessel as through a 1-millimeter vessel. Based on size alone—that is, if drag were not a consideration—the larger vessel would only be able to carry 16 times what a smaller vessel could carry.

Figure 23.10
Capillaries Are Specialized for Exchanging Materials
Each one of the millions of glomeruli in the nephrons of a kidney is a network of capillaries.

Science Toolkit

What Was Old Is New Again

You probably would not expect to find a leech in a surgeon's toolkit. Increasingly often, however, the humble, and rather disgusting, leech plays a critical role in the success of surgeries. Like a number of other modern medical treatments, leeches represent a good example of ancient medical tools that still play a role in contemporary medicine.

Leeches are related to earthworms. Although most are predators or scavengers, the leech of medical interest is the aptly named *Hirudo medicinalis*, which feeds on human blood. The medicinal leech attaches to the human body using a specialized suckerlike mouth. With this mouth it pierces the skin and sucks up the blood that begins to flow.

In feeding, the leech faces three important challenges: how to avoid drawing the attention of its victim, how to get as much blood as quickly as possible, and how to make sure that the natural clotting reaction of blood doesn't stop the blood flow. The solution to all three challenges, and the reason for its usefulness in surgery, is its spit. The saliva of the medicinal leech contains a mild anesthetic that dulls the pain, another chemical that causes blood vessel diameter to increase, allowing more blood flow, and an anticoagulant that interferes with the clotting reaction.

The last two properties, increasing blood flow and preventing clot formation, have made the medicinal leech the tool of choice in ensuring the success of many delicate surgeries involving skin grafts (in which damaged skin is replaced by skin from elsewhere on the body) and the reattachment of body parts such as fingers, ears, and nipples (during breast reconstruction). In each of these operations, surgeons must reestablish blood flow to the newly attached skin or body part. They can relatively easily connect up the relatively thick-walled arterioles, but it is much more difficult to reattach the more delicate venules. As a result, blood can often flow into the reattached tissue, but not out. Thus blood may pool in the new tissue rather than flowing through it. Unless the blood has a way out, the skin graft or body part may die.

By increasing blood flow and preventing clotting, medicinal leeches create a small wound that bleeds profusely for several hours. By attaching leeches every few hours for 3 to 5 days, surgeons can give the blood that enters the reattached tissue a way out long enough to give the venules a chance to reattach themselves.

All of this brings us to the question of how a surgeon came up with the idea of using a blood-sucking parasite to aid in surgery to begin with. The idea is far from new. As long as 2,500 years ago, Egyptians used leeches to bleed their patients. In medieval times, leeches were commonly used to remove blood from people suffering from various illnesses. These early physicians recognized that the large amount of blood that flowed from each leech bite provided an effective way of removing from their patients the bad blood (or "humors") that was, at the time, understood as the cause of illness.

Thus modern medicine has borrowed an ancient medical tool. We no longer bleed patients to remove bad blood, but the unique ability of a leech's bite to bleed freely and painlessly remains useful.

Medicinal Leeches (*Hirudo medicinalis*) Being Used to Drain Blood from an Injured Thumb

---

epithelial cells, across which materials diffuse easily (see Figure 23.1). In some places, the entire circumference of the capillary wall is composed of just one epithelial cell! In addition, some capillaries are so small that red blood cells must pass through them in single file.

Their tiny diameters give capillaries a relatively large surface area relative to the volume of blood passing through them. You can understand the relationship between large vessels and capillaries by picturing a telephone cable containing hundreds of small wires. If you removed all the wires from the cable and compared their volumes and surfaces, you would find that the volume of the many small wires taken together is essentially equal to the volume of the empty cable sheath, but that the combined surface area of the wires is far greater than the surface area of the sheath.

Their large surface area suits the purpose of the capillaries: exchange of gases, nutrients, and other materials with the interstitial fluid surrounding the capillaries, and thus with respiring cells. Blood flows through capillaries slowly enough to ensure adequate time for the exchange of substances with nearby cells. The dramatic increase in drag that accompanies a decrease in capillary diameter strongly resists blood flow and forces the heart to work harder.

23.3 Comparing Circulatory Systems among Animals

So far in this chapter we have looked only at the human cardiovascular system. How does that system compare with the circulatory systems of other animals?

Some animals have open circulatory systems

As mentioned earlier, the human cardiovascular system is a closed circulatory system; so are the systems of other vertebrates (fish, amphibians, reptiles, birds, and other mammals). In contrast, the circulatory systems of insects and some molluscs (such as snails and clams) are structured quite differently. These creatures have **open circulatory systems**, in which the blood vessels leading from the heart are open-ended. The insect heart is little more than a muscled vessel in which waves of contraction push blood out the open ends of the short vessels leading from the heart into the general body cavity.

Animals with open circulatory systems have much less control over how their blood circulates than do animals with closed systems. Blood leaves the open-ended vessels and flows into the spaces surrounding the organs (Figure 23.11). Blood makes its way back to the heart by flowing through these spaces and back in through openings in the heart. Thus the blood contents are distributed less precisely than in closed circulatory systems. Many of the organs of animals with open circulatory systems have large surface areas. As the surfaces of these organs are bathed in blood, the large surface area provides for adequate exchange of materials.

Open circulatory systems may seem inefficient, but they work well enough to have allowed insects and other arthropods to diversify enormously, far more than any other group of organisms (experts estimate that there may be over 10 million species of insects alone). Consider also that in these animals, blood is not burdened with the task of carrying and distributing oxygen. Recall that in insects gases are distributed and exchanged through a separate network of tracheae that is not coupled with the circulatory system (see Figure 22.11).

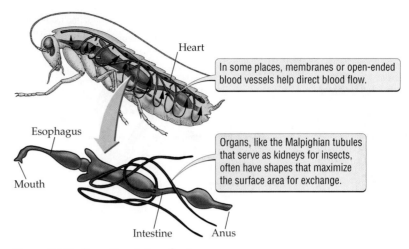

Figure 23.11 Open Circulatory Systems
In open circulatory systems like those of insects, blood flows through the body cavity and bathes whole organs.

Most animal hearts share a number of features

Like human hearts, the hearts of other animals rely on muscles to pump blood through the circulatory system, whether open or closed. The heart not only has to overcome the frictional drag generated by blood flowing past the blood vessel walls, but also must lift the blood against the pull of gravity (though in small animals such as insects, the effect of gravity is negligible).

Hearts come in several forms in the animal world, but they all use a muscle tissue unique to animals to generate pressure, and their vessels all have one-way valves to direct blood flow. In addition, all animals with closed circulatory systems have hearts that consist of at least two chambers—an atrium and a ventricle—separated from each other and from the blood vessels by one-way valves (Figure 23.12*a*). Human hearts are an example of one widespread design, the four-chambered heart (see Figure 23.2).

Some animals use auxiliary means to pump blood. A simple type of pump found in some animals (such as humans and sharks) that helps to power blood flow consists of a blood vessel that contains one-way valves and is surrounded by muscle tissue (Figure 23.12*b*). Contractions of the surrounding muscle tissue during normal activity squeeze the vessel, and one-way valves ensure that the blood will move in only one direction as a result of those contractions. In some animals, this sort of pump supplements a chambered heart. For example, the veins in our legs work in this way to ease the task of lifting the blood back up to the heart.

As we have seen, hearts of animals with open circulatory systems, such as insects, may consist of a muscular vessel. In these tubular hearts, the body fluid is pushed ahead of a wave of contraction from back to front (Figure 23.12*c*).

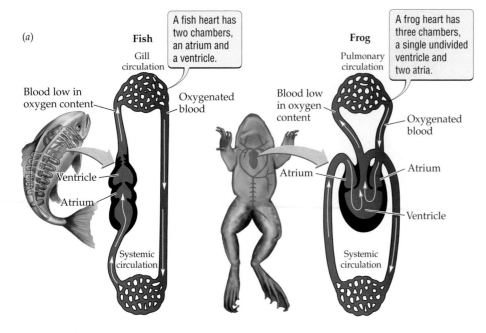

Fish

Gill circulation

A fish heart has two chambers, an atrium and a ventricle.

Blood low in oxygen content

Oxygenated blood

Ventricle

Atrium

Systemic circulation

Frog

Pulmonary circulation

A frog heart has three chambers, a single undivided ventricle and two atria.

Blood low in oxygen content

Oxygenated blood

Atrium

Atrium

Ventricle

Systemic circulation

(*b*) Blood vessels surrounded by muscle tissue

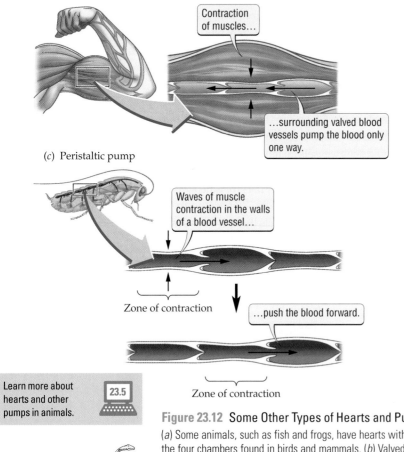

Contraction of muscles…

…surrounding valved blood vessels pump the blood only one way.

(*c*) Peristaltic pump

Waves of muscle contraction in the walls of a blood vessel…

Zone of contraction

…push the blood forward.

Zone of contraction

Zone of contraction

Learn more about hearts and other pumps in animals.

23.5

Figure 23.12 Some Other Types of Hearts and Pumps
(*a*) Some animals, such as fish and frogs, have hearts with fewer than the four chambers found in birds and mammals. (*b*) Valved blood vessels surrounded by muscle tissue provide an auxiliary means of pumping blood. (*c*) Insect hearts have no chambers; they are simply large, muscular vessels with one-way valves.

Barosaurus

Regardless of their structural variations, all animal hearts must overcome drag and lift blood. To perform adequately, a heart must generate enough pressure to overcome the drag on the blood as it flows past the walls of the vessels. The heart's pumping action must also be strong enough to overcome the downward pull of gravity and lift blood to the highest part of the body. For example, of the 7,000 liters of blood that the human heart pumps through the circulatory system each day, roughly 1,000 liters go to the brain. Because a liter of blood weighs about 1 kilogram, each day we must do the work necessary to lift 1,000 kilograms of blood the 30- to 40-centimeter distance from our hearts to our heads. The extinct dinosaur *Barosaurus*, which had to lift blood the length of its extraordinary 10-meter-long neck, would have needed a heart that could generate a pressure of 646 mm Hg simply to support the weight of this column of blood.

23.4 Health Conditions Affecting the Human Cardiovascular System

A number of serious health problems can affect the human cardiovascular system. Two conditions with adverse and overlapping health effects are high blood pressure and atherosclerosis.

Hypertension can have serious consequences

Hearts need to generate pressure in order to keep blood circulating, but sometimes that pressure can be abnormally high. The technical term for abnormally high blood pressure is **hypertension**. High blood pressure in humans can have several causes. One is narrowed arteries due to plaque formation (explained later in this section).

Hypertension can also result from an upset of the water and solute balance in the body. In response to the concentrations of solutes in our body fluids, water naturally diffuses between our cells, the interstitial fluid, and the blood plasma, as we saw in Chapter 20. If the buildup of water in the interstitial fluid becomes too great, this "extra" water diffuses into the capillaries, increasing the volume of the blood plasma beyond its normal 5 to 6 liters. Increasing the volume of any closed system means increasing the pressure within that system.

The kidneys play a central role in hypertension because they retain or dispose of water in response to changes in the body's blood volume. Normally, the kidneys respond to an increase in blood volume by producing lots of dilute urine, which has the effect of removing water from the body. However, in many cases of hypertension, the kidneys respond as if a high blood volume were normal, regulating the blood volume at levels that maintain high blood pressure. Unless there are visible symptoms, however, hypertension can go unnoticed for years. In the meantime, the constant high pressure is silently taking a toll on the cardiovascular system as well as on other organs designed to work at lower (normal) blood pressures, such as the kidneys themselves.

For one thing, high pressure means that the heart has to work harder. The heart responds as any muscle would to increased work—by bulking up. The wall of the left ventricle, which pumps blood throughout the body, becomes thicker. Rather than helping the situation, however, this thickening eventually interferes with the ventricle's ability to pump blood. It also increases the ventricle's demand for oxygen, further adding stress on the heart.

Blood vessels, too, are adversely affected by hypertension. The constant high pressure makes vessel walls weaker. Vessel walls compensate for the high pressure they are under by getting thicker; this thickening reduces blood vessel diameters, which in turn increases drag.

Vessels under abnormally high pressure can also develop aneurysms [AN-yuh-RIZMS], weakened places where the vessel walls bulge out and can burst (Figure 23.13). Aneurisms are always in danger of rupturing. When an aneurysm bursts in the brain, a stroke results. A stroke occurs when a patch of brain tissue is deprived of its blood supply, either by blockage (from a narrowed blood vessel)

or by the rupture of a blood vessel (from a burst aneurysm). Either way, localized brain damage occurs as brain cells die. Their loss can disrupt organ functioning, speech, or body movement, depending on the part of the brain affected.

A third significant risk of hypertension is kidney damage. The high pressures exerted on the many delicate capillaries in the glomeruli of the kidneys can damage them and inhibit their ability to function, sometimes leading to total kidney failure and death.

Fat in our diet affects cholesterol levels in our blood

The health of the cardiovascular system is strongly influenced by hereditary factors and by levels of activity (see the box on page 451). But the diet—especially the intake of fats—is an important factor as well. Not all fats are equal, however. Monounsaturated fats (found in olive oil) and unsaturated fats (in cooking oils such as canola, corn, peanut, or sesame) not only do less harm than saturated fats; small amounts of these oils in the diet are actually beneficial. **Saturated fats** (such as butter, lard, cheeses, and the fat in meats), on the other hand, have a negative effect on the body. That is because the body converts saturated fats into **cholesterol**. Generally, the more saturated fats we eat, the more cholesterol we produce, although the body chemistry of some people also has an inherited tendency to maintain high cholesterol levels regardless of diet.

At certain levels, cholesterol is essential to the body; it is important in cell membranes and is the precursor molecule for making steroid hormones and vitamin D, among other things. But the body must get rid of amounts in excess of what it needs for maintenance. The blood carries cholesterol in two forms, each with very different effects on our health. As a fatty material, cholesterol does not dissolve easily in blood, which is water-based, so it is packaged with proteins into molecules called lipoproteins.

There are two kinds of cholesterol-carrying lipoproteins. **Low-density lipoproteins (LDLs)** carry cholesterol to the cells that need it, while **high-density lipoproteins (HDLs)** carry excess cholesterol back to the liver, where it is removed from circulation (Figure 23.14a). The process is similar to delivery and trash removal in a city or town: LDLs are like delivery trucks, bringing items to individual households (cells), and HDLs are like garbage trucks, removing trash and taking it to the dump (the liver).

The proportions of these two kinds of lipoproteins circulating in the blood provide a measure of your health in relation to cholesterol. A high percentage of HDLs means

Figure 23.13 Cerebral Aneurysm
Aneurysms, such as this particularly large one of the internal carotid artery, are abnormal enlargements of blood vessels. This giant aneurysm is greater than 2.5 cm (about 1 inch); when aneurysms burst they can cause strokes or death.

(a)

LIVER

HDL returns excess cholesterol to the liver.

Production of cholesterol is increased by a diet high in saturated fat and by inherited factors.

HDL cholesterol

LDL cholesterol

TISSUES

The amount of cholesterol removed from the blood circulation is affected by the cells' need for cholesterol.

ARTERY

If there is a high level of cholesterol in the blood, the artery wall takes up too much LDL.

(b)

Figure 23.14 Arteries and Cholesterol: A Fatal Attraction

(a) Cholesterol, when not in use by the cells or sequestered by the liver, circulates in the bloodstream, transported by two different kinds of lipoproteins. (b) Plaques that build up on the insides of artery walls can narrow arteries to dangerous levels. The artery on the left is normal, while the one on the right has become badly clogged (the plaque is yellow). Heart attack or stroke can result.

that waste cholesterol is being ferried from tissues back to the liver at an efficient rate. But a low percentage of HDLs in the blood indicates that cholesterol is being delivered to cells faster than it can be taken away to the liver. Waste cholesterol builds up in the bloodstream, much as trash collects on sidewalks during a garbage collectors' strike. What does the body do with this excess cholesterol circulating in the blood?

Because cholesterol does not dissolve easily in blood, the body's solution is to store excess cholesterol in the inner lining of blood vessel walls, where it eventually builds up to form deadly plaques; this condition is called atherosclerosis [*ATH*-uh-roh-skler-*OH*-siss] (Figure 23.14b). Recall that decreasing the diameter of a blood vessel dra-

matically increases the pressure the heart must generate to maintain an adequate flow of blood; thus the heart must work harder to force blood through the narrowed arteries, increasing the blood pressure.

Over many years, continued hypertension can reduce the heart's ability to pump blood, leading eventually to congestive heart failure, an often fatal condition in which blood backs up in the heart and lungs. More dramatically, plaques in the major blood vessels supplying the tissues of the heart itself can reduce the flow of blood carrying oxygen and nutrients to the heart muscle. If the restriction is severe enough to kill significant numbers of cells in the heart muscle, the result is a heart attack. Narrowing or blockage of an artery leading to the brain can cause a stroke.

Biology Matters

Check Your Physical Activity and Heart Disease I.Q.

Test how much you know about how physical activity affects your heart by taking this quiz prepared by the National Heart, Lung, and Blood Institute of the National Institutes of Health. Mark each statement true or false, then read the answers that follow.

1. Regular physical activity can reduce your chances of getting heart disease. __ T __ F

2. Most people get enough physical activity from their normal daily routine. __ T __ F

3. You don't have to train like a marathon runner to become more physically fit. __ T __ F

4. Exercise programs do not require a lot of time to be very effective. __ T __ F

5. People who need to lose some weight are the only ones who will benefit from regular physical activity. __ T __ F

6. All exercises give you the same benefits. __ T __ F

7. It doesn't take a lot of money or expensive equipment to become physically fit. __ T __ F

8. There are many risks and injuries that can occur with exercise. __ T __ F

9. You should consult a doctor before starting a physical activity program. __ T __ F

10. To help stay physically active, include a variety of activities. __ T __ F

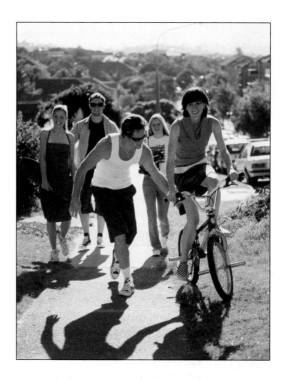

Answers

1. True. Heart disease is almost twice as likely to develop in inactive people. Being physically inactive is a risk factor for heart disease along with cigarette smoking, high blood pressure, high blood cholesterol, and being overweight. Regular physical activity (even mild to moderate exercise) can reduce this risk.

2. False. Most Americans are very busy but not very active. Every American adult should make a habit of getting 30 minutes of low to moderate levels of physical activity daily. This includes walking and climbing stairs.

3. True. Low- to moderate-intensity activities, such as pleasure walking, stair climbing, and dancing, can have both short- and long-term benefits. At least 30 minutes of physical activity every day can help improve your heart health.

4. True. If you don't have 30 minutes in your schedule for an exercise break, try to find two 15-minute periods or even three 10-minute periods. These exercise breaks will soon become a habit you can't live without.

5. False. People who are physically active experience many positive benefits. Regular physical activity gives you more energy, reduces stress, and helps you to sleep better.

6. False. Low-intensity activities—if performed daily—can have some long-term health benefits and can lower your risk of heart disease.

7. True. Many activities require little or no equipment. For example, brisk walking requires only a comfortable pair of walking shoes. Your college or university may offer free or inexpensive recreation facilities and physical activity classes.

8. False. The most common risk in exercising is injury to the muscles and joints. Such injuries are usually caused by exercising too hard for too long, particularly if a person has been inactive.

9. True. You should ask your doctor before you start (or greatly increase) your physical activity *if* you have a medical condition such as high blood pressure, have pains or pressure in the chest and shoulder, feel dizzy or faint, get breathless after mild exertion, or plan a vigorous activity program. If none of these apply, start slow and get moving.

10. True. Pick several different activities that you like doing. You will be more likely to stay with it.

How well did you do? As you can see from the quiz, you don't need to get involved in team sports or expensive exercise programs to keep fit. You will reap big dividends both now and in later years if you form healthy patterns of activity now.

High Blood Pressure on the Savanna Revisited

Drinking giraffe

A blood pressure of 260/160 would be cause for grave concern in humans, but it is not in giraffes, for which such a reading is considered normal. To lift a column of blood extending 2 meters from heart to brain, giraffes have large, powerful hearts. A giraffe's heart weighs about 24 pounds and can generate high blood pressures without coming under undue stress. As a result, it functions normally at these high pressures and does not thicken. But the brain gets preferential treatment: because of the high pressures needed to provide the brain with a continuous blood supply, capillaries elsewhere in the giraffe's body experience pressures so great that they would destroy delicate human capillaries. To compensate, giraffe capillaries have thicker walls than their human equivalents.

When a giraffe bends its neck down to drink water, the systolic pressure in the brain could increase from its normal 100 mm Hg to the 260 mm Hg measured in the animal's feet. The capillaries in the giraffe's brain lack the thicker walls of capillaries elsewhere in the body and could easily be destroyed by such high pressures. One-way valves in the long neck veins, however, prevent a backflow of blood from the neck veins to the brain when the giraffe drinks, thus preventing excessive pressures in the brain's capillaries. The combination of a strong heart and specialized blood vessels makes it possible for a giraffe to get a drink of water without passing out.

But protection against fainting is not the only challenge the giraffe faces. Think for a moment about the high pressures in the lower regions of its legs, as well as in its neck when it is drinking. There is a lot of blood in a giraffe's body above its legs, and all that blood is being pulled downward by gravity. This results in blood pressures in the lower parts of the legs that would rupture human vessels. How do giraffes keep their blood vessels from bursting under these great pressures?

The form and function of the giraffe's skin provides the answer. In the lower regions of the body, the skin forms a tight, thick sheath that applies counterpressure, just as the pressurized flight suits that pilots of high-performance aircraft wear prevent blood from plunging to their lower bodies during accelerated maneuvers. Without the suits, the forces of gravity that are generated during rolls and dives would pull blood away from the pilots' brains and lead to loss of consciousness. In a way, giraffes wear flight suit leggings. For giraffes, living a high-pressure life on the savanna requires many adaptations in the circulatory system.

Chapter Review

Summary

23.1 The Human Cardiovascular System

- Humans and other vertebrates have a cardiovascular system, a closed circulatory system with a chambered heart that pumps blood through closed loops (circuits) of enclosed vessels.
- The human cardiovascular system has two main circuits. In the pulmonary circuit, blood is pumped to the lungs, where it disposes of CO_2 and picks up O_2, before returning to the heart. In the systemic circuit, oxygenated blood returning from the lungs is pumped out to body tissues, where it delivers O_2 and nutrients and picks up CO_2 and other wastes, before returning to the heart.
- Arteries are large vessels that transport blood away from the heart. Veins are large vessels that that carry blood back to the heart. Capillaries are the smallest vessels and exchange materials by diffusion with nearby cells.

- The vertebrate heart is composed of four chambers that make up two separate muscular pumps, each composed of an atrium and a ventricle. The left atrium and ventricle pump blood to the body, while the right atrium and ventricle pump blood to the lungs.
- The heart rate is the number of times a heart beats per minute. A heartbeat consists of one cardiac cycle, which has two phases. During diastole, the heart rests briefly, while during systole, the heart muscle contracts and pumps blood.
- The heartbeat's regularity is maintained by electrical signals from the heart's natural pacemaker, the sinoatrial (SA) node.
- The human circulatory system adjusts the heart rate and patterns of blood distribution according to the body's needs.

23.2 Blood Vessels and Factors Affecting Blood Flow

- Large vessels (arteries and veins) are built for mass transport of blood, whereas capillaries are built for exchange of materials with surrounding cells.
- In large vessels, the walls are thick and strong. Muscle tissue in artery walls allows arteries to contract and relax to control the flow of blood to different parts of the body. Veins have valves to keep blood flowing in one direction, back to the heart.
- Large diameters in arteries and veins yield low surface area-to-volume ratios, which reduce frictional drag between blood and the vessels' inner walls, thus promoting easy long-distance flow.
- In capillaries, thin, porous walls allow for easy diffusion. The small diameters of capillaries produce high surface area-to-volume ratios, providing a large area across which diffusion can occur, as well as slowing the flow of blood, thus ensuring adequate time for exchanges to take place.

23.3 Comparing Circulatory Systems among Animals

- Some animals have open circulatory systems, in which blood leaves open-ended vessels and flows freely into spaces surrounding the organs. Hearts in these systems are weakly contracting regions of certain vessels.
- Open circulatory systems allow much less control over how blood circulates than closed systems do. To compensate, the organs of many animals with closed systems have large surface areas for exchange of materials. In arthropods, blood is not burdened with the task of carrying oxygen or CO_2.
- Chambered hearts include atria that forcefully fill ventricles, which then pump blood through vessels. In insect hearts, muscles associated with the tubular heart push blood from back to front.
- All muscular hearts must create sufficient pressure to overcome the frictional drag associated with blood flowing past inner blood vessel walls and to lift blood against the pull of gravity.

23.4 Health Conditions Affecting the Human Cardiovascular System

- Hypertension and atherosclerosis are serious long-term diseases.

- Hypertension can be caused by an upset in the water and solute balance in the body, in which the kidneys treat higher-than-normal blood pressures as normal. Hypertension can also be caused by atherosclerosis.
- Long-term, untreated hypertension has many adverse effects on the body. It puts strain on the heart, which has to pump harder to bring the same amount of blood to tissues.
- Hypertension weakens vessel walls, which can produce aneurysms. Burst aneurysms in the brain are one cause of strokes.
- Hypertension can damage the capillaries in the glomeruli, leading to kidney failure and death.
- Atherosclerosis is the buildup of fatty plaques on inner blood vessel walls. It can be caused by a diet high in saturated fats, by hereditary factors, or a combination of both.
- Saturated fats are converted by the body to cholesterol, which is transported to cells by low-density lipoproteins (LDLs) and removed as waste to the liver by high-density lipoproteins (HDLs). When the ratio of HDLs to LDLs is low, excess cholesterol is deposited as plaques in blood vessels.
- Vessels narrowed by plaques put a greater demand on the heart. Over time, this can cause congestive heart failure. Narrowing or blockage in one of the vessels serving the heart muscle itself can cause a heart attack; in a vessel serving the brain, it can cause a stroke.

⊙ Review and Application of Key Concepts

1. Diffusion is required for exchanges of the materials needed to sustain life. Transport of large volumes of blood is also a necessity. Describe the relationship between diffusion and the bulk transportation of blood in your body.

2. Humans have a closed circulatory system. Describe the voyage of a drop of blood from the left atrium of your heart and back again.

3. Red blood cells transport oxygen. Our hearts and large blood vessels transport red blood cells. Describe how the meaning of "transport" changes from the viewpoint of a red blood cell as it is moved from the heart to a capillary near a respiring cell.

4. The amount of blood that can flow through a blood vessel depends in great part on the diameter of the vessel. Explain how this works.

5. Open circulatory systems work just fine for many organisms. Describe why an open circulatory system would not work in a human.

6. We learned about kidney nephrons in Chapter 20. Describe how high blood pressure could damage a nephron and its function.

Key Terms

artery (p. 441)

atrioventricular (AV) node
    (p. 444)

atrium (p. 441)

blood pressure (p. 442)

blood vessel (p. 440)

capillary (p. 441)

cardiac cycle (p. 442)

cardiovascular system (p. 440)

cholesterol (p. 449)

closed circulatory system
    (p. 441)

diastole (p. 442)

electrocardiogram (EKG)
    (p. 444)

heart (p. 440)

heart rate (p. 442)

heartbeat (p. 442)

high-density lipoprotein (HDL)
    (p. 449)

hypertension (p. 448)

low-density lipoprotein (LDL)
    (p. 449)

open circulatory system (p. 447)

pacemaker (p. 443)

pulmonary circuit (p. 441)

saturated fat (p. 449)

sinoatrial (SA) node (p. 443)

systemic circuit (p. 441)

systole (p. 443)

vein (p. 441)

ventricle (p. 441)

Self-Quiz

1. Which blood vessels carry blood back toward the heart?
   a. ventricles
   b. arteries
   c. veins
   d. capillaries

2. The pulmonary circuit of the human cardiovascular system moves blood to and from
   a. the kidneys.
   b. the digestive system.
   c. the systemic circuit.
   d. the lungs.

3. In the human cardiovascular system, blood flows from
   a. artery to atrium to ventricle to vein.
   b. vein to atrium to ventricle to artery.
   c. vein to ventricle to atrium to artery.
   d. artery to ventricle to atrium to vein.

4. During diastole
   a. only the atria are pumping blood.
   b. only the ventricles are pumping blood.
   c. neither the atria nor the ventricles are pumping blood.
   d. both the atria and the ventricles are pumping blood.

5. The signals that establish a regular heartbeat come from the
   a. ventricles.
   b. atrium.
   c. SA node.
   d. AV node.

6. As the diameter of a blood vessel increases, the frictional drag that must be overcome by the heart
   a. is not affected.
   b. increases.
   c. decreases.
   d. increases and then decreases.

7. Which of the following statements is correct?
   a. The circulation to the lungs works at a lower pressure than the circulation to the body.
   b. Both atria pump blood to the lungs, and both ventricles pump blood to the body.
   c. Blood returning to the heart from the lungs carries little oxygen, while the blood returning to the heart from the body carries a lot of oxygen.
   d. Both atria pump blood to the body, and both ventricles pump blood from the lungs.

8. Which of the following correctly describes valves in closed circulatory systems?
   a. They are all one-way.
   b. They are all two-way.
   c. Valves in veins are two-way.
   d. Valves in arteries are two-way.

9. Plaques can cause
   a. atherosclerosis.
   b. congestive heart failure.
   c. heart attacks.
   d. all of the above

10. Which of the following is true of hypertension in humans?
    a. It is a common health problem.
    b. It can increase the risk of severe kidney problems.
    c. It can lead to a thickening of the heart muscle.
    d. all of the above

Biology in the News

Tour Rider Fails Drugs Test

An unnamed rider in the Tour de France has tested positive for the banned substance EPO. . . .

"This positive test for EPO was carried out in mid-Tour and up to now he is the only rider to test positive on the Tour," said deputy Tour director Daniel Baal.

"We now must wait for the follow-up test that the rider asked for, but I can tell you that it is not one of the race leaders."

In the highly competitive world of sports, a factor affecting blood transport of oxygen sometimes becomes important. Erythropoietin (EPO) [eh-*RITH*-roh-*POY*-uh-tin] is a hormone the liver produces to stimulate the production of red blood cells. Such stimulation is necessary because each red blood cell lasts for only about 17 weeks before it needs to be replaced. Red blood cells transport oxygen, so the more red blood cells the blood has, the more oxygen it can carry. Knowing this, some athletes participating in endurance sports take EPO supplements with the intent of increasing their stamina. By increasing red blood cell production (from a normal 40 percent to as much as 60 percent of blood volume), EPO increases the oxygen-carrying capacity of blood.

EPO comes with clear health risks. That is because an abnormally high number of red blood cells can make the blood too thick to move around efficiently. This increased thickness slows blood flow and increases the likelihood of vessel blockage. A number of otherwise healthy endurance athletes have died of cardiovascular problems, including heart attacks, during competitions over the past decade, apparently because of doping with EPO.

For this reason, as well as to "level the playing field," EPO supplements are banned for participants in a number of endurance competitions, including long-distance cycling, long-distance running, swimming, and cross-country skiing. When presented with the chance of increas-

ing their stamina and improving their chances of winning a competition, however, many athletes cannot resist the temptation to use EPO. Identifying athletes who abuse EPO is difficult because the body produces this chemical naturally. Instead, many tests focus on whether an athlete has an exceptionally high red blood cell count.

Evaluating the news

1. Based on your experience, and on your understanding through the media, what pressures do you think athletes face that lead them to risk their reputations, and perhaps their lives, by using performance-enhancing drugs?

2. Some people would argue that if sport is viewed as entertainment, rather than as competition, then whatever people do to be most entertaining (that is, to produce the most spectacular performances) is acceptable. How would you defend or argue against this point of view?

3. Like so many abused drugs, EPO has an important and legitimate use in helping people overcome medical conditions, such as cancer, that lead to low red blood cell counts. How do we balance the legitimate and illegitimate uses of such drugs?

Source: *BBCsport*, July 27, 2003.

CHAPTER 24

Animal Hormones

Key Concepts

- Hormones allow widely separated cells within an organism to communicate and coordinate their activities. The circulatory system carries hormones from sites of production to sites of action.

- A single hormone may have several different effects depending on the target tissues on which it acts. The effects of a small amount of a hormone are amplified throughout the body.

- Several hormones often work together to coordinate a biological process or activity. Often this coordination involves pairs of hormones with opposite effects.

- Hormones control growth, reproduction, and development.

From Caterpillar to Butterfly

Butterflies live a double life. They hatch from eggs as wingless caterpillars devoted completely to eating. Armed with strong cutting and chewing mouthparts, caterpillars munch their way through plant tissues as they accumulate the nutrients that they need to grow. Caterpillars have short, stubby legs better suited to holding onto a plant than to moving between plants. Indeed, they often complete their life as caterpillars on a single plant.

As adults, butterflies devote themselves to reproduction. Their brilliantly colored wings allow them to search widely for and identify potential mates. Once mated, females search their surroundings for good places to lay eggs, while males look for more females to mate with. Finding mates and laying eggs both require that butterflies fly long distances. Butterflies get the energy needed for flight by using strawlike mouthparts to suck up the sugary nectar produced by flowers. Unlike the plant tissues on which caterpillars feed, nectar contains few nutrients other than those needed to supply energy.

If you compared a caterpillar with an adult butterfly without knowing their relationship, it would be hard to imagine that the two forms bear any relationship to each other. From most perspectives, a caterpillar differs more from the butterfly into which it develops than one species does from another. By having two very different body forms during their development, butterflies can essentially pack two specialized "species" into one individual: (1) a larval stage (the caterpillar) that is well suited for eating and growing, and (2) an adult stage (the butterfly) that is well suited for mating and egg laying. Compare this life cycle with that of humans, who must deal with the dual demands of feeding and reproduction using a single, all-purpose body form. How does the butterfly produce such radically different forms?

Metamorphosis
A black swallowtail changes its form from a caterpillar to an adult butterfly.

Multicellular animals must coordinate the functioning of their many specialized cells, tissues, and organs. Just as a group of people cannot effectively complete a task without frequent communication, the many cells within an animal need to communicate to allow the whole organism to develop and function effectively. Just as we send memos, letters, or e-mail messages, cells send signaling molecules to let cells in other parts of the organism know what to do.

Hormones are signaling molecules that are produced by specialized cells or glands and circulate within the animal body. All animals rely on hormones. Animals also rely on electrical and chemical signals in the nervous system, a much faster means of communication than hormones. However, the actions of hormones and of the nervous system are often coordinated.

In this chapter we explore the role hormones play in coordinating the biology of multicellular organisms. We begin by providing an overview of how hormones work. Then we build on this general perspective by introducing examples of how animals use hormones to coordinate essential functions over different lengths of time, from short-term to long-term. We close by returning to consider how several hormones work together to coordinate butterfly metamorphosis.

24.1 How Hormones Work

Before discussing specific ways in which hormones allow animals to coordinate their many functions, let's briefly review some of what we learned in Chapter 6 about how signaling molecules function in communication between cells.

Hormones trigger a response in target cells

A hormone released by one cell causes a response in another cell or group of cells called the **target** (Figure 24.1). The target cells for a particular hormone may lie in several different kinds of tissues and may respond to that hormone in different ways. **Testosterone**, for example, is a hormone that influences the development of many features associated with maleness in humans and has a number of different effects on different target tissues at different times: it controls the development of the male sex organs in male fetuses, stimulates growth in cells

throughout the body, stimulates the production of sperm by cells in the testes, stimulates the production of facial hair by cells in the skin of the face, and causes behavioral changes through interactions with cells in the brain. In general, animals can get by with fewer distinct hormones by having most hormones serve multiple functions.

Hormones can act on target cells in two different ways (see Figure 6.8):

1. Some hormones, particularly steroid hormones, can pass through the plasma membrane of a target cell to act inside that cell, often by interacting directly with the genes inside the nucleus (see Figure 6.9).
2. Some hormones cannot pass through the plasma membrane of a target cell; they can affect the target cell only by binding to cell surface receptors embedded in the target cell's plasma membrane.

Most hormones travel through the circulatory system to act on distant cells

Hormones often act on target cells located far from the cells that produced them. In animals, most hormones reduce their travel time by using the circulatory system. Hormones can travel only as fast as the blood moves, which means that they take several seconds or more to arrive at their target cells. A few seconds may not seem

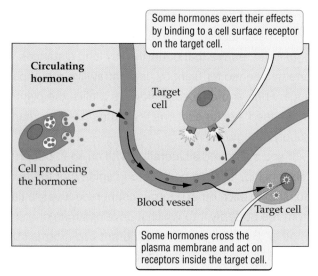

Figure 24.1

Hormones Allow Cells to Communicate with One Another
Hormones released by one cell travel through the circulatory system to cause a reaction in target cells in other parts of the organism.

See hormone action through the circulatory system.   24.1

like a long time, but it does mean that hormones cannot coordinate activities that require very quick responses. Only the nervous system can coordinate activities, such as muscle contraction and movement, that require virtually instantaneous reactions. Hormones coordinate functions that take place over timescales of seconds to months (Figure 24.2).

To survive their journey through the circulatory system, hormones must resist chemical breakdown. On the other hand, their breakdown by enzymes after they have performed their function at the target prevents them from continuing to affect target cells for longer than intended.

Target cells can amplify the hormonal signal

Animals usually produce hormones in tiny amounts measured in micrograms. For example, an adult human female may produce between 100 and 200 micrograms a day of estrogens, an important type of hormone that has many effects associated with femaleness throughout the body. (A single tiny grain of common table salt weighs about 10 micrograms.) How can such minuscule doses of a hormone so dramatically regulate the function and appearance of an animal? When a single hormone molecule binds to its receptor in the target cell, it sets in motion a chain of events that may activate thousands of protein molecules in the target cell. This signal **amplification** means that a single hormone molecule can have a dramatic effect on the target cell.

24.2 Regulating Short-Term Processes: Glucose and Calcium Homeostasis

Hormones play a central role in coordinating the biological processes of animals, and that role is reflected in the structural organization of the cells and organs that produce them. Animal hormones are produced in groups of specialized cells within tissues as well as in specialized organs called **endocrine glands**. Unlike many glands, endocrine glands do not have ducts (special tubes that deliver secretions from the gland directly to the site of action); instead, they release their products directly into the bloodstream, where they circulate throughout the body until they reach their target cells. Some hormone-secreting cells are also found within tissues that have other functions, such as the pancreas (which is a ducted gland) and the lining of the intestine. Endocrine glands and specialized hormone-secreting cells together make up the **endocrine system** (Figure 24.3). The **hypothalamus**, a small region at the base of the vertebrate brain, coordinates the endocrine system and integrates it with the nervous system.

Blood glucose levels are regulated by hormones from the pancreas and the adrenal glands

Glucose is one of the most important sugars in animal nutrition. It is the starting point for glycolysis and respiration, which produce the energy storage molecule ATP (see Chapter 8). Here we describe the role of human hormones in coordinating the availability of glucose to show how hormones can affect different target tissues differently and how several hormones can interact to coordinate a biological process.

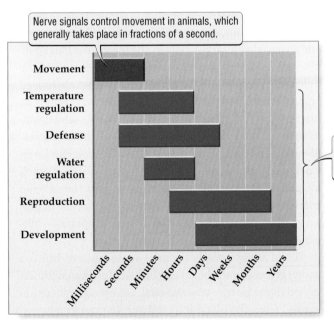

Figure 24.2

Hormones Act More Slowly Than the Nervous System

The activities animals must coordinate occur over timescales that range from fractions of a second to many years. In animals, hormones, which travel relatively slowly through the bloodstream, tend to coordinate longer-term activities (blue bars), while the nervous system coordinates activities that must occur rapidly, such as muscular contractions (red bar).

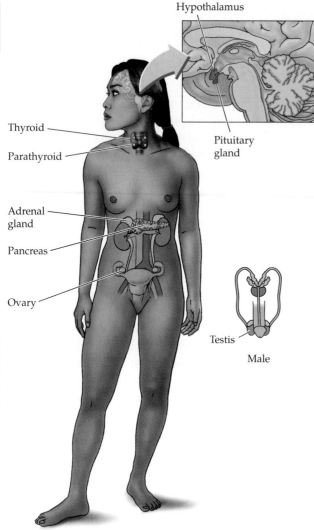

Figure 24.3 The Human Endocrine System
The endocrine system consists of glands that secrete hormones directly into the circulatory system, as well as specialized cells within other kinds of tissues (such as the pancreas). The hypothalamus is the coordinator of the endocrine system.

Helpful to know

The prefixes *glyco* ("sweet") and *gluco* ("glucose") show up in many terms relating to sugars in general or glucose in particular. Examples in this chapter include *glucagon*, a hormone involved in regulating glucose; *glycogen*, the storage form of glucose in the liver; and *glycolysis*, the first stage in the chemical breakdown of glucose in the cytosol.

Most animals, including humans, obtain glucose from their food. Glucose is absorbed from digested food in the small intestine, from which it travels through the circulatory system to respiring cells. Animals, including humans, can store excess glucose, but they must be able to use it quickly when it is needed. Four hormones—insulin, glucagon, epinephrine, and norepinephrine—coordinate the storage and release of glucose in humans.

The pancreas, in addition to producing digestive enzymes (see Chapter 21), contains groups of specialized cells called **islet cells** [*EYE*-let . . .] that produce and release insulin and glucagon. You will recall that the pancreas sends its digestive enzymes to the small intestine through

a duct. The islet cells, however, secrete their hormones directly into the bloodstream without using the duct. The pancreas releases **insulin** after a meal, when the blood has absorbed more glucose than the body can use. Insulin acts on target cells in the liver, which respond by absorbing increased amounts of glucose from the blood. These liver cells assemble individual glucose molecules into long chains to form a storage carbohydrate called **glycogen** [*GLY*-kuh-jin]. Insulin also acts on other target cells to promote the storage of fatty acids from digested fats and amino acids from digested proteins.

Insulin helps ensure that we store excess nutrients rather than lose them by eliminating them from the body. When levels of glucose in the blood drop, the pancreas produces less insulin. The blood glucose level has a homeostatic set point that shuts off production of insulin when it is reached—an example of negative feedback at work (see Chapter 20 for a review of feedback).

The pancreas also produces and releases **glucagon**, a hormone whose effect is opposite to that of insulin. When blood glucose falls to low levels, in addition to secreting less insulin, the pancreas also releases more glucagon, which stimulates the storage cells in the liver to convert stored glycogen into glucose and release it into the bloodstream. Figure 24.4 illustrates the insulin-glucose-glucagon relationship. When two hormones have such opposite effects, they are said to be **antagonistic**.

The **adrenal glands** are a pair of endocrine glands that sit on top of the kidneys. They release **epinephrine** [*EP*-ih-*NEF*-rin] (also called *adrenaline*) and **norepinephrine** [*NOR*-ep-ih-*NEF*-rin], a pair of hormones involved in regulating the availability of glucose. They are released in response to nerve signals from the brain that report on the body's stress levels or warn of danger (Figure 24.5). These hormones, too, travel through the bloodstream until they reach target cells in the liver. Like glucagon, they stimulate glycogen breakdown in the liver and the release of glucose into the bloodstream. They also speed up the heartbeat so that glucose is delivered throughout the body more rapidly. In this way, glucose becomes available to fuel a rapid response to danger. If you see a rattlesnake in front of you about to strike, you can jump back instantly because of your nervous system's quick response. The nervous system also signals the adrenal glands to kick in right away, causing the release of glucose, which speeds up the rest of your response—running away quickly.

Blood always needs to carry some glucose, but too much glucose in the blood can be harmful. In a condition called diabetes [die-uh-*BEET*-eez], not enough glucose is moved from the blood into storage. In type 1 diabetes,

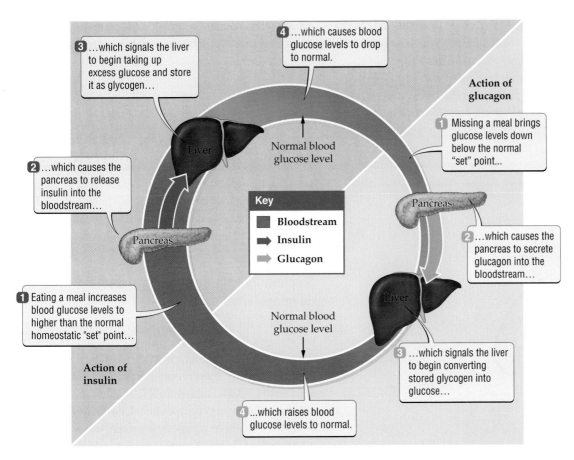

Figure 24.4 Balancing Levels of Glucose in the Blood
Two hormones, insulin and glucagon, work in opposite ways to regulate blood levels of glucose. The width of the red circuit indicates the relative amounts of glucose in the bloodstream at different times.

See more about balancing glucose levels in the blood.

damage to the islet cells either prevents insulin production altogether, or the pancreas produces a defective form of insulin that cannot bind to receptor proteins on the surfaces of target cells. In type 2 diabetes, the pancreas either produces too little insulin, or the receptors on target cells no longer respond to insulin. Over time, the resulting high levels of glucose in the blood damage capillaries, causing poor circulation and slowing the healing of injured tissues. Damage to blood vessels in the eyes can cause vision problems (in severe cases, blindness). Thanks to our understanding of the biology of insulin and our ability to produce the hormone in commercial quantities through genetic engineering techniques, as described in Chapter 15, most people with diabetes can now monitor and manage their own glucose levels by modifying their diet and injecting additional insulin as needed (Figure 24.6). In addition, by eating well and getting exercise, people can reduce their chances of developing type 2 diabetes, which is closely linked with accumulating excess body fat during adulthood.

Calcium is regulated by hormones from the thyroid and parathyroid glands

In addition to regulating energy-supplying nutrients such as glucose, hormones coordinate the uptake and release of mineral nutrients, such as calcium. Calcium in a mineral form is a major component of human bones. (In our blood it exists as calcium ions, Ca^{2+}.) Two antagonistic hormones, calcitonin and parathyroid hormone, work together to regulate the amount of calcium circulating in the human bloodstream.

An excess of calcium in the blood triggers the **thyroid gland**, an endocrine gland in the neck, to release calcitonin. **Calcitonin** removes calcium from the blood by promoting its storage in bones and by stimulating the kidneys to reabsorb less calcium, allowing more calcium to be removed from the body in the urine. When there is too little calcium in the blood, the **parathyroid gland**, a gland lying beside the thyroid gland, releases **parathyroid hormone** (**PTH**), which stimulates the release of calcium from bones and increases the reabsorption of calcium in the kidneys, so that less is excreted from the body in the urine. PTH also activates vitamin D, which increases the uptake of calcium from food in the intestines.

Figure 24.7 illustrates how calcitonin and PTH maintain proper levels of calcium in the blood while storing the excess and strengthening bones. Negative feedback comes into play with each hormone: when either hormone causes blood calcium to reach a normal level, the gland producing that hormone slows down or stops.

24.3 Regulating Medium- and Long-Term Processes: Human Reproduction

Animals rely on hormones to coordinate their reproduction. In humans, hormones influence both long-term and short-term aspects of sexual development and reproduction.

Sex hormones play a role in sexual development before birth

Hormones influence human sexual development even before birth. The genetic program for developing into one sex or the other in humans is determined by the presence or absence of a Y chromosome, as we saw in Chapter 11: females are XX and males are XY. It is through the action of hormones circulating in its blood, however, that a fetus develops into either a male or a female. If the fetus has a

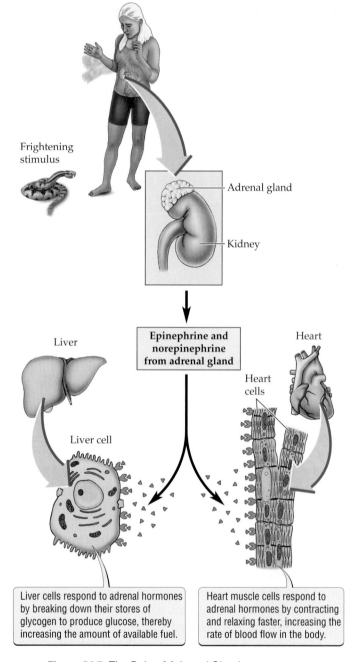

Frightening stimulus

Adrenal gland

Kidney

Liver

Epinephrine and norepinephrine from adrenal gland

Heart

Heart cells

Liver cell

Liver cells respond to adrenal hormones by breaking down their stores of glycogen to produce glucose, thereby increasing the amount of available fuel.

Heart muscle cells respond to adrenal hormones by contracting and relaxing faster, increasing the rate of blood flow in the body.

Figure 24.5 The Role of Adrenal Glands

The adrenal glands produce epinephrine (adrenaline) and norepinephrine (noradrenaline) that allow quick access to stored energy.

(a)

(b)

Figure 24.6 Compensating for a Defective Pancreas

(a) People with diabetes must monitor their blood glucose levels, such as with this glucometer. (b) Diabetics inject measured quantities of insulin to prevent abnormally high levels of blood glucose.

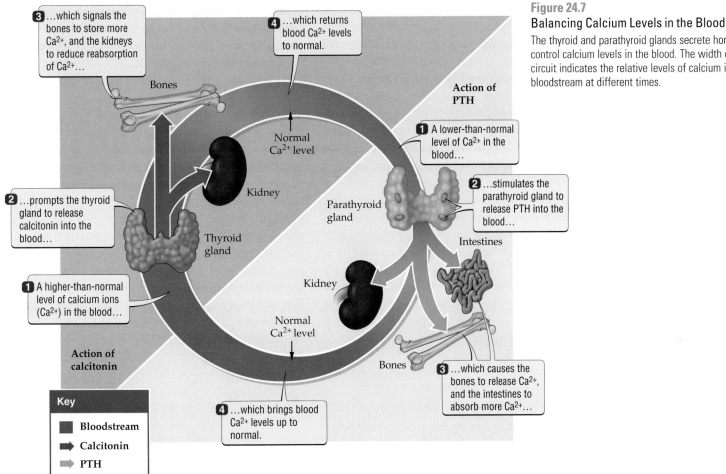

Figure 24.7

Balancing Calcium Levels in the Blood

The thyroid and parathyroid glands secrete hormones that control calcium levels in the blood. The width of the red circuit indicates the relative levels of calcium in the bloodstream at different times.

3 ...which signals the bones to store more Ca²⁺, and the kidneys to reduce reabsorption of Ca²⁺...

4 ...which returns blood Ca²⁺ levels to normal.

Bones

Action of PTH

1 A lower-than-normal level of Ca²⁺ in the blood...

Normal Ca²⁺ level

Kidney

2 ...stimulates the parathyroid gland to release PTH into the blood...

2 ...prompts the thyroid gland to release calcitonin into the blood...

Parathyroid gland

Thyroid gland

Intestines

Kidney

1 A higher-than-normal level of calcium ions (Ca²⁺) in the blood...

Normal Ca²⁺ level

Bones

3 ...which causes the bones to release Ca²⁺, and the intestines to absorb more Ca²⁺...

Action of calcitonin

4 ...which brings blood Ca²⁺ levels up to normal.

Key

- Bloodstream
- Calcitonin
- PTH

Y chromosome, testes (the male reproductive organs) begin to form just 4 to 6 weeks after fertilization of the egg. In the absence of a Y chromosome, ovaries (the female reproductive organs) develop instead. These sex glands—testes and ovaries—are more generally called **gonads**. By the seventh week of the fetus's development, its gonads are producing steroid hormones specific to its sex. These **sex hormones** signal genes in their target cells to begin the process of sexual development (Figure 24.8).

The gonads produce three major types of hormones: estrogens, progesterone, and androgens. Both males and females produce all three, but in different ratios: males have more androgens than estrogens, and females have more estrogens than androgens. **Estrogens** play a role in determining female characteristics such as wide hips, a voice that is pitched higher than that of males, and breasts developed for producing milk. **Progesterone** prepares the wall of the uterus to provide an environment for a developing fetus. **Androgens** (*andro*, "male") stimulate cells to develop the characteristics of maleness discussed earlier; the primary androgen is testosterone.

The testes secrete three hormones that together coordinate the development of male reproductive structures. Testosterone and a second, closely similar androgen direct the development of internal reproductive structures such as the sperm ducts and prostate gland; the third androgen directs the development of external structures such as the penis. Similarly, the development of all female reproductive structures, both internal and external, falls under the control of three estrogens.

Sex hormones coordinate sexual maturation at puberty

At an age of about 10 to 12 years, young humans begin to make the transition to sexually mature adults capable of reproduction. During this transition, called **puberty**, the levels of sex hormones produced by the gonads rise markedly. Moreover, the hypothalamus activates the production of two other hormones by the **pituitary gland**, a double-lobed gland that lies at the base of the brain. These two hormones are **luteinizing hormone** (LH) and

Helpful to know

The suffix *gen* means "produces" or "is produced by or from." For example, *estrogens* and *androgens* produce the characteristics of femaleness and maleness, respectively, and *glycogen* is produced from glucose.

follicle-stimulating hormone (FSH); they are referred to collectively as the **gonadotropins** [goh-*NAD*-uh-*TROPE*-ins]. Together they coordinate the development of sperm in males and play a role in regulating the menstrual cycle in females. In combination with growth hormone, which we shall discuss below, the sex hormones also control the development of other characteristics associated with sexual maturity: a deepening voice, development of facial and pubic hair, and growth of the penis and scrotum in males; and development of breasts and pubic hair and changes in body shape in females.

Gonadotropins also help maintain the functioning of the reproductive organs and glands throughout an individual's lifetime. In males, gonadotropins further stimulate the production of sperm in the presence of testosterone. At birth, the ovaries of a female already contain her entire lifetime supply of immature egg cells. These cells remain in a state of suspended development until the release of gonadotropins at puberty, when they resume their development.

Sex hormones coordinate the menstrual cycle of human females

In mature human males, sperm are continuously being produced. Human females, however, do not produce mature eggs continuously. Instead, individual eggs mature and are released in a 28-day **menstrual cycle**. Figure 24.9 summarizes the fluctuations in hormone levels that drive this cycle.

During the first 2 weeks of the menstrual cycle, estrogen levels rise steadily as a single egg completes its devel-

opment in the ovary. The egg develops inside an organized group of cells that surround it, called a **follicle**. The developing follicle secretes increasing amounts of estrogens. Estrogen levels peak in about 12 days, causing a massive and sudden release of gonadotropins by the pituitary gland. The gonadotropins trigger **ovulation**, the release of the egg from the follicle. The cells of the ruptured follicle, which remain behind in the ovary after ovulation, now develop into a **corpus luteum** (Latin for "yellow body"). The same spike in gonadotropin levels that triggers ovulation also stimulates the corpus luteum to begin secreting progesterone.

Once ovulation has occurred, the gonadotropin and estrogen levels drop quickly, and progesterone becomes the dominant hormone. Progesterone and estrogens prepare the lining of the uterus to receive a fertilized egg: the lining thickens steadily, and new blood vessels develop that can supply a growing embryo. If the egg is fertilized, the corpus luteum continues to produce progesterone throughout the pregnancy to support the lining of the

Human embryo, approximately 12 weeks old

Figure 24.8 Sex-Specific Steroid Hormones
Male and female sex hormones coordinate the development of reproductive structures in the human fetus.

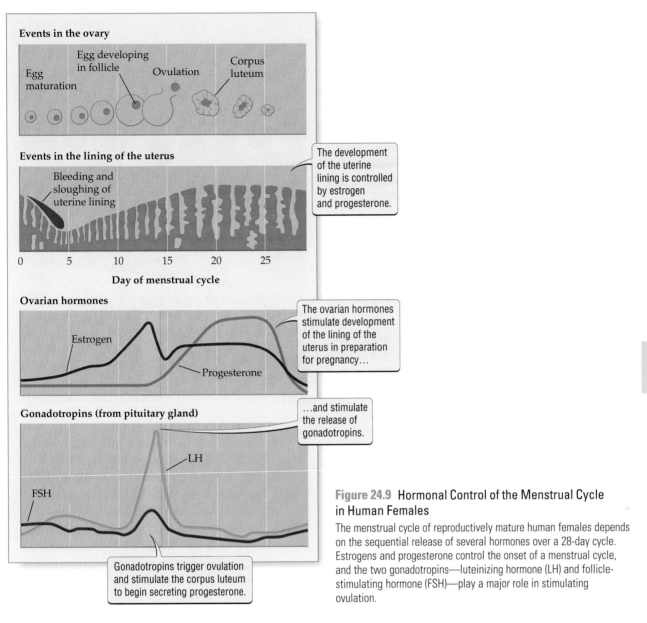

Events in the ovary

Egg maturation
Egg developing in follicle
Ovulation
Corpus luteum

Events in the lining of the uterus

Bleeding and sloughing of uterine lining

The development of the uterine lining is controlled by estrogen and progesterone.

Day of menstrual cycle

Ovarian hormones

Estrogen
Progesterone

The ovarian hormones stimulate development of the lining of the uterus in preparation for pregnancy...

Gonadotropins (from pituitary gland)

...and stimulate the release of gonadotropins.

LH
FSH

Gonadotropins trigger ovulation and stimulate the corpus luteum to begin secreting progesterone.

Learn more about hormone control of the menstrual cycle.

24.4

Figure 24.9 Hormonal Control of the Menstrual Cycle in Human Females

The menstrual cycle of reproductively mature human females depends on the sequential release of several hormones over a 28-day cycle. Estrogens and progesterone control the onset of a menstrual cycle, and the two gonadotropins—luteinizing hormone (LH) and follicle-stimulating hormone (FSH)—play a major role in stimulating ovulation.

uterus. (One reason the birth control pill usually contains synthetic hormones that mimic progesterone is to trick the body into thinking it is already pregnant; see the box on page 466.) Other hormonal interactions take place during pregnancy, a story we shall continue in Chapter 29.

If the egg is not fertilized, no embryo develops, and progesterone levels drop about 12 days after ovulation. Without high levels of progesterone to support them, the new blood vessels and thickened lining of the uterus do not last long. They separate from the uterus and are expelled from the body during menstruation. The onset of menstruation marks the beginning of a new menstrual cycle as estrogen levels begin to rise again and another new egg begins to mature.

After about 500 cycles of menstruation, a new sequence of hormonal events occurs that is, in some ways, the reverse of puberty. Typically when a woman reaches between 40 and 50 years of age, her ovaries begin to produce less estrogens and progesterone. Eventually the drop in the levels of these two hormones results in **menopause**, in which the menstrual cycle ceases permanently.

24.4 Regulating Long-Term Processes: Human Growth

In addition to sexual development and reproduction, hormones regulate the long-term process of the overall growth of animals: how much and when we grow. The

Biology Matters

"The Pill": Be Safe, Not Sorry

Oral contraception, or "the Pill," has been around since the 1960s and has proved to be highly reliable. The most commonly used type of oral contraceptive contains a combination of synthetic versions of two hormones, progesterone and an estrogen. These "combination" pills prevent the release of mature eggs each month. But, if, perchance, an egg does become fertilized, they provide a second line of defense by making the lining of the uterus inhospitable to a fertilized egg. Without a receptive uterine wall, a pregnancy cannot become established.

Among women who take the Pill exactly according to directions, its failure rate is much less than 1 percent. But "exactly according to directions" is an important caveat. Success or failure depends more on the person than on the Pill. Only about 1 in 1,000 women who take the Pill as directed become pregnant. But among a population of less careful "typical" women, the failure rate is 50 times greater.

As far as contraception goes, the Pill can provide reliable protection, but there are other sides to the story. For example, smoking while taking an oral contraceptive greatly increases the risk of heart attack, stroke, and blood clots, especially in women over 35 who smoke 15 or more cigarettes a day. And there are drugs and medications, including some antibiotics and painkillers, that can decrease the level of protection the Pill provides. Experts advise consulting with your doctor about any medication you are taking.

Oral contraceptives also have a "hidden" side effect: feeling "safe" can mean an increase in sexual activity, and that can mean increased exposure to sexually transmitted diseases. By the age of 24, one in three sexually active Americans will have contracted an STD; moreover, at least one in four—perhaps as many as one in two—will contract an STD at some point in their lives. Oral contraceptives offer no protection against these diseases, so the need to practice safe sex remains very important.

---

pituitary gland produces a hormone appropriately called **growth hormone (GH)**. Its main target cells lie in bones and in muscles associated with bones. GH promotes the growth of bones and stimulates increases in muscle mass. Normally, the human pituitary gland produces GH in amounts that promote adequate growth during childhood and puberty, and then stops producing it when we become adults.

Too much or too little GH can have lifelong consequences. If too much growth hormone is present before a person reaches adulthood, the result is gigantism (Figure 24.10a). People with this condition can grow to be 8 feet tall or taller. On the other hand, if the pituitary begins producing excess growth hormone after a person

has reached adulthood, the person will not become a giant because the growth centers in the bones that contribute the most to giving us our height close during puberty. Instead, bones in other parts of the body (in particular, the hands, feet, and parts of the face and head) will begin to grow again, becoming enlarged. This condition is called acromegaly (Figure 24.10b).

Other conditions can occur if the body receives too little GH. Dwarfism results when too little GH is available as growth progresses during childhood and adolescence. (Figure 24.10c). Adult dwarfs are usually under 4 feet tall. Growth hormone supplements are now being manufactured by processes like those described in Chapter 15, using genetically engineered bacteria. Under careful medical

(a) A pituitary giant

(c) Pituitary dwarves

(b) Face and hands of acromegaly

Figure 24.10 When GH Production Goes Awry

(a) Too much GH before puberty produces gigantism; Robert Wadlow, in a photograph from 1938 with normal-sized women, was reportedly the world's tallest man, reaching a height of 8 feet 11 inches.
(b) GH production restarting after maturity results in acromegaly, and the enlargement of bones in the face, arms, and legs. (c) Too little GH produces dwarfism. Dwarfism that results from too little growth hormone results in small stature, but relatively normal proportions, as is the case with these pituitary dwarves photographed in 1934 en route to the World's Fair in New York.

guidance, young people producing too little GH—whether simply shorter than normal or potential dwarfs—can be helped to increase their height somewhat.

Although the use of growth hormone in the production of beef, pork, and lamb has been in the news recently, no connections have been found linking these uses with human health problems. But there is an increasing demand for organic meats from livestock that have been raised without the use of added hormones (see the box on page 468).

Because it promotes the growth of muscle mass, some athletes have used, and abused, growth hormone in an effort to improve their performance levels. Used in this way, growth hormone can have serious side effects, including cancer and heart failure.

Biology on the Job

Selling Food You Care About

Chuck Pruett is a third-generation grocer and owner of two very successful markets in the southeastern United States, one in Asheville, North Carolina, and one in Chattanooga, Tennessee. Both markets go by the name of Greenlife Grocery and feature organic foods. Tammy Grafe manages the Chattanooga market. Both stores have grown rapidly in popularity, especially as people have become increasingly concerned about the quality of what they eat.

Chuck, how did your interest in organic foods develop? I have been working in the traditional grocery business since I was eleven. As I grew up, I began looking at the ingredients listed on the labels of different food products. I didn't like what I saw. So I did a lot of investigating and developed a natural foods section in my father's store.

Tammy, are you a third-generation grocer too? No, my background is in management, but I have a serious interest in food quality and safety. I was raised on a vegan [*VEE*-gun] diet. I am still a vegan, meaning I don't eat animal products or food derived from animals.

What about hormones that may affect humans in fruits and vegetables? *Tammy:* Hormones are present in plant tissue. One group is called phytoestrogens [*FITE*-oh-*ESS*-troh-jins]; these are similar to human estrogens but have much weaker effects and are not stored in the body. There is some evidence that these "mild" hormones can help to maintain bone strength in postmenopausal women. There's still a lot to learn about them. Soybeans contain large amounts of phytoestrogens.

What food products might contain hormones that could have an effect on people? *Chuck:* The U.S. Department of Agriculture allows hormones to be used in raising beef cattle, hogs, and lamb. These hormones, which must be approved by the Food and Drug Administration, help to promote growth, and they are similar to human hormones.

If these substances are approved for use by the USDA and the FDA, what's the concern? *Chuck:* There are no proven links between the use of these substances and health problems, but there is growing public concern about what is in the food that people eat. Our position as organic grocers is that we offer an alternative that avoids use of these hormones.

What, in your view, is a major advantage of selecting organic meat products? *Tammy:* The rules for certifying meat as organic are strict. They state that no growth hormones or antibiotics are to be used in raising the animals, and that they must be fed 100 percent organic feed. So we feel our customers can rely on getting meats that are free of substances that are, for some people, a matter of concern.

What about chickens—does "free range" mean "no hormones"? *Chuck:* "Free range" simply means that the chickens have free access to a space outside of the building in which they are housed. The USDA does not allow hormones to be used in raising chickens, so hormones aren't a concern.

Tammy: However, organic chickens are raised without the use of antibiotics. And the use of antibiotics has become an issue due to the possibility of resistant strains of certain bacteria developing.

What organic food issues come up with milk and dairy products? *Tammy:* Milk and milk products contain a hormone called bovine somatotropin [su-*MAT*-oh-*TROH*-pin], BST for short. Even organic milk contains BST. This hormone helps young cows grow, and in adult cows it promotes milk production. Some milk producers provide extra BST in the diets of their cows.

Chuck: I should add that studies have shown that the amounts of this hormone are about the same in organic milk and in milk produced from cows that have been given added hormone. We are an organic store, and, simply as a matter of philosophy, we buy only from producers who do not use added hormone.

But in this chapter we talk about how growth hormone can adversely affect human growth; what effect does BST have on humans? *Chuck:* BST doesn't get into our bloodstream from consuming dairy products. It's a protein hormone that is broken down into amino acids during digestion. So, it's really just another nutrient.

Organic Produce and Meats at a Greenlife Grocery

What food concerns are there with seafood? *Tammy:* The major factor we are careful about in seafood and fish is the growth of bacteria. A trustworthy supplier and careful handling are our major concerns. Hormones that could affect people are not found in fish.

The Job in Context

The general public is becoming more aware of the importance of being well informed about the foods they eat. People are reading more about healthy eating, and organic foods are growing in popularity. The Greenlife Grocery markets are booming, as is the organic foods industry in general. Chuck and Tammy spend a great deal of time keeping tabs on new developments and carefully screening new suppliers. Knowing more about the hormones we may be taking in with our food can only increase our understanding of healthier diets.

The National Institutes of Health (NIH) says the following about phytoestrogens: "Clearly, additional research is needed to further evaluate the effects of phytoestrogens before judgments regarding their safety and usefulness can be made. Current research sponsored by the National Institutes of Health should provide important evidence concerning the safety of phytoestrogens and their potential role in the skeletal health of women after menopause."

Metamorphosis: A Hormonal Symphony

We are commonly familiar with only the two active forms of a butterfly's life cycle—caterpillar and adult—but butterflies (and moths) experience four distinct stages over the course of their lives: egg, larva, pupa, and adult. Butterflies undergo the amazing transformation from caterpillar to adult during a sometimes inconspicuous stage called the pupa. During this stage the tissues of the caterpillar are broken down and new tissues and body parts develop to form the adult. This dramatic change, known as **metamorphosis** [met-uh-*MOR*-fuh-sis], depends on the coordination of a complex process of breaking down old tissues and building up new ones.

The development of a human from a fertilized egg to maturity involves the precisely timed turning off and on of specific genes in specific parts of the body. Imagine, then, the complex orchestration needed to control the development of not just one, but several physically different body forms during the lifetime of a butterfly. Butterflies, like other animals, rely on hormones to orchestrate their development, providing another clear example of how a few different hormones can act together to control even tremendously complex events.

In a very real sense, adult butterflies are different animals than the caterpillars from which they arise. Adults develop from clusters of cells, called imaginal discs, that reside within the growing caterpillar. During the pupal stage, the structures of the caterpillar body are disassembled. The cells of the imaginal discs begin to divide and grow to form adult structures, scavenging materials from the broken-down tissues and cells of the caterpillar. Clearly, without careful coordination of the disassembly of the caterpillar and subsequent assembly of the adult, metamorphosis could lead to a hopeless mess rather than a functioning butterfly.

Considering the complexity of metamorphosis, its control by a mere three hormones is astonishingly simple and elegant (Figure 24.11). Molting hormone, released by the prothoracic glands, has two different effects. Throughout the life of the caterpillar, it stimulates molting, or shedding of the exoskeleton. Molting hormone also promotes imaginal disc development. A second hormone, called juvenile hormone, released by the corpora allata, a pair of glands closely associated with the insect brain, controls the outcome of the molt.

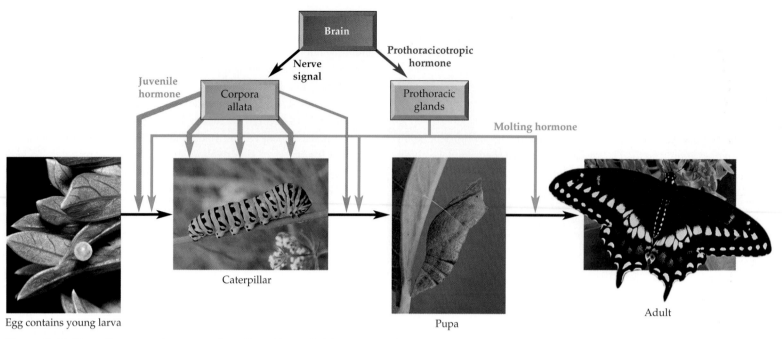

Figure 24.11 Three Hormones Interact to Coordinate Butterfly Metamorphosis
Hormone-producing cells link the nervous and endocrine systems to control the timing of events in metamorphosis. The release of molting hormone in response to the release of prothoracicotropic hormone determines when the insect sheds its exoskeleton. The amount of juvenile hormone released by the corpora allata determines whether the individual takes the form of a caterpillar, pupa, or adult following the molt.

High concentrations of juvenile hormone in the body fluid before a molt lead to a larger caterpillar, intermediate concentrations lead to a pupa, and an absence of juvenile hormone produces an adult.

The timing of events during metamorphosis depends both on signals from the nervous system and on hormones. When signals from the nervous system indicate that a caterpillar has reached the proper stage of development, the brain halts the release of juvenile hormone, thus initiating metamorphosis. The timing of the release of molting hormone depends on a more complex series of events. The brain produces a third hormone, called prothoracicotropic hormone, which stimulates the prothoracic glands to release molting hormone—but only if the following two conditions are met:

1. The level of juvenile hormone indicates to target cells in the brain that the caterpillar has reached a suitable stage in development to begin metamorphosis.
2. Information reaching the brain indicates that the release of prothoracicotropic hormone will cause the molt to the pupal stage to take place at a relatively safe time of day.

Thus three factors allow butterflies to efficiently coordinate the complex process of metamorphosis. First, not one chemical, but a combination of interacting chemical messages, is used; second, different target tissues respond differently to each hormone; and third, as is true of most animals, the functioning of hormones is integrated with that of the nervous system.

Chapter Review

Summary

24.1 How Hormones Work

- The release of a hormone by one cell triggers a response in a target cell. A single hormone may affect many different kinds of target cells, causing a different response in each.
- Hormones may act on target cells by moving through the plasma membrane to the cell's interior or by acting on receptors embedded in the plasma membrane.
- Hormones typically move through the circulatory system to reach distant target cells. Because they can move only as quickly as the blood moves, they tend to coordinate animals' slower, longer-term functions.
- The hormonal signal is usually amplified in target cells; the binding of a single hormone molecule to its receptor may activate thousands of proteins in the target cell.

24.2 Regulating Short-Term Processes: Glucose and Calcium Homeostasis

- Animal hormones are produced by groups of specialized hormone-secreting cells within other kinds of tissues as well as in endocrine glands, which release their products directly into the bloodstream. The glands and specialized cells together make up the endocrine system.
- The hypothalamus coordinates the endocrine system and integrates it with the nervous system.
- Hormones typically act together to coordinate biological processes. A pair of hormones with opposite effects are said to be antagonistic.
- A pair of antagonistic hormones produced by islet cells in the pancreas—insulin and glucagon—regulate glucose levels in human blood. Insulin increases the storage of glucose in liver cells, while glucagon causes the liver to release glucose into the bloodstream.
- Under stressful conditions, the adrenal glands produce epinephrine and norepinephrine, which cause the liver to release glucose, providing quick energy for escape from danger.
- In the disease diabetes, the pancreas produces either too little insulin or insulin that cannot bind to its target cells. The high levels of blood glucose that result can damage tissues.
- Calcitonin and parathyroid hormone (PTH) are antagonistic hormones that regulate blood levels of calcium. Calcitonin (produced by the thyroid gland) lowers calcium levels by promoting its storage in bones and by stimulating the kidneys to reabsorb less calcium. Parathyroid hormone (PTH, released by the parathyroid gland) raises calcium levels by stimulating bones to release calcium, by prompting the kidneys to reabsorb more calcium, and by causing the intestines to take up more calcium during digestion.

24.3 Regulating Medium- and Long-Term Processes: Human Reproduction

- In humans, hormones influence both long-term and short-term aspects of sexual development, pregnancy, and birth.
- Sex hormones guide the development of a fetus's sexual characteristics based on its genotype (XX for a female or XY for a male). By the seventh week of fetal development, the fetus's developing sex organs are producing steroid hormones specific to its genotype.
- The gonads (testes in males, ovaries in females) produce three major types of hormones: estrogens stimulate the development of female characteristics, progesterone prepares the uterine wall to support a developing fetus during pregnancy, and androgens stimulate the development of male characteristics. Both men and women produce all three types of sex hormones, but in different proportions.
- At puberty, humans make the transition to reproductive maturity. Puberty is caused by rising levels of sex hormones and by the production of gonadotropins, two hormones that coordinate the development of sperm in males and eggs in females.
- Gonadotropins, which are produced by the pituitary gland, also help maintain the functioning of the reproductive organs and glands throughout an individual's lifetime.
- Hormones coordinate the menstrual cycle of human females. Individual eggs mature and are released in a 28-day menstrual cycle. During the first 2 weeks of the menstrual cycle, the immature egg completes its development in a follicle in the ovary. Estrogen levels rise; progesterone levels remain low.
- When the egg has completed its development, a massive, sudden release of gonadotropins triggers ovulation, releasing the egg from the ovary. This spike in gonadotropin levels also signals the ruptured follicle, whose cells remain in the ovary as the corpus luteum, to begin releasing progesterone.
- If the egg is fertilized, progesterone levels remain high throughout the pregnancy, keeping the uterine lining well developed to protect the developing fetus.
- If the egg is not fertilized, no embryo develops, and about 12 days after the egg's release, progesterone levels drop, initiating menstruation, in which the additional blood vessels and thickened lining of the uterus are expelled from the body.

24.4 Regulating Long-Term Processes: Human Growth

- The pituitary gland produces growth hormone (GH), which promotes the growth of bones and stimulates increases in muscle mass.
- Too much GH present in the body during childhood and puberty results in gigantism. If the pituitary gland resumes release of GH

after puberty, acromegaly results, a condition in which bones in the face, hands, and feet begin to grow again.

- Too little GH during childhood and puberty results in dwarfism. Young people producing too little GH naturally can now receive GH supplements derived from genetically engineered bacteria.

◉ Review and Application of Key Concepts

1. Describe the journey insulin takes from where it is produced in the body to its target cells, and explain what it does when it gets to its destination.

2. Testosterone has many effects. Describe the effects it has on human males.

3. Hormones can work quickly, but not as quickly as the nervous system. Describe how your adrenal glands and nervous system would work together if you were suddenly to come face to face with an angry, snarling dog.

4. The menstrual cycle in human females involves fluctuating levels of several hormones working together to create a series of events that repeats about every 28 days. If you were "inventing" an animal like a female white-tail deer, which breeds just once a year, how would you design a hormonal program for it? Describe the hormones involved and the timing of the shifts in their levels.

5. Growth hormone has lifelong effects. What are three possible consequences of abnormalities in the production of growth hormone in humans?

Key Terms

adrenal gland (p. 460)
amplification (p. 459)
androgen (p. 463)
antagonistic (p. 460)
calcitonin (p. 461)
corpus luteum (p. 464)
endocrine gland (p. 459)
endocrine system (p. 459)
epinephrine (p. 460)
estrogen (p. 463)
follicle (p. 464)
follicle-stimulating hormone
    (FSH) (p. 464)
glucagon (p. 460)
glycogen (p. 460)
gonadotropin (p. 464)
gonad (p. 463)
growth hormone (GH) (p. 466)
hormone (p. 458)

hypothalamus (p. 459)
insulin (p. 460)
islet cell (p. 460)
luteinizing hormone (p. 463)
menopause (p. 465)
menstrual cycle (p. 464)
metamorphosis (p. 469)
norepinephrine (p. 460)
ovulation (p. 464)
parathyroid gland (p. 461)
parathyroid hormone (PTH)
    (p. 461)
pituitary gland (p. 463)
progesterone (p. 463)
puberty (p. 463)
sex hormone (p. 463)
target (p. 458)
testosterone (p. 458)
thyroid gland (p. 461)

Self-Quiz

1. Which of the following is true of target cells?
   a. Target cells occur only in the endocrine system of animals.
   b. Target cells can amplify the chemical message carried by hormones.
   c. Some target cells have cell surface receptor proteins embedded in the plasma membrane that bind a specific hormone.
   d. both b and c

2. Endocrine glands secrete hormones directly into
   a. the kidneys.   c. the bloodstream.
   b. the pancreas.   d. the hypothalamus.

3. Which of the following is an endocrine gland?
   a. the thyroid gland   c. the pituitary gland
   b. the adrenal gland   d. all of the above

4. Which one of the following increases the uptake of glucose by the liver?
   a. glucagon   c. calcitonin
   b. insulin   d. parathyroid hormone

5. Islet cells in the pancreas produce
   a. epinephrine.   c. insulin.
   b. norepinephrine.   d. growth hormone.

6. Adrenal glands secrete
   a. epinephrine.   c. glucagon.
   b. calcitonin.   d. androgens.

7. Estrogens play a role in
   a. developing female characteristics.
   b. developing male characteristics.
   c. regulating butterfly metamorphosis.
   d. developing male gonads.

8. The pituitary gland produces
   a. glucagon.   c. gonadotropins.
   b. insulin.   d. calcitonin.

9. Progesterone prepares the uterus for
   a. producing gonadotropins.
   b. receiving a fertilized egg.
   c. initiating puberty.
   d. producing more testosterone.

10. Acromegaly results from
    a. not enough growth hormone after puberty is complete.
    b. not enough growth hormone during puberty.
    c. too much growth hormone before and during puberty.
    d. too much growth hormone after puberty is complete.

Biology in the News

Warning over Long-Term HRT Use

Hormone Replacement Therapy should only be used for the short-term relief of menopausal symptoms, say experts.

The new guidance suggests that if a woman has taken HRT for around five years, she should consider if it is in her best interest to continue.

The guidance, in a Royal College of Obstetricians and Gynaecologists book, follows a review of HRT research.

Hormone replacement therapy (HRT) was until recently a widely prescribed treatment for the symptoms of menopause. Menopause is by no means an illness; as mentioned earlier in the chapter, it represents a normal stage in female development. Nonetheless, the reproductive changes that define menopause go hand in hand with several symptoms not directly related to reproduction: hot flashes, sleeplessness, headaches, irritability, and an increased tendency toward osteoporosis [OSS-tee-oh-pu-ROH-siss], a long-term net loss of calcium from the bones, making them porous and weak. For the 75 percent of women who suffer these symptoms, menopause can decrease the quality of life. The weakened bones that accompany osteoporosis can lead to disabling hip fractures.

An apparently simple way to make menopause more bearable was to add as supplements the hormones that the female body stops producing at menopause. Thus doctors began prescribing HRT, in the form of pills containing estrogens or a combination of estrogens and progestins (synthetic hormones that act like progesterone), as a way of allowing the lives of menopausal women to return to normal.

Recent studies, however, have shown that long-term exposure to all forms of HRT significantly increases the risk of heart attack, stroke, and possibly breast cancer. The U.S. National Heart, Lung, and Blood Institute stopped a large clinical study of estrogen-plus-progestin HRT because the researchers conducting the study found that the risks of long-term HRT outweighed its benefits. Specifically, they found "increased risks of heart attacks, stroke, invasive breast cancer, and blood clots in study participants taking estrogen plus progestin compared with women taking placebo pills. There were noteworthy benefits of estrogen plus progestin, including fewer cases of colon cancer and hip fractures, but on balance, the harm was greater than the benefit."

Current guidelines now recommend that doctors prescribe HRT only for women suffering severely from the side effects of menopause, and then only after providing their patients with a clear discussion of the associated risks and benefits.

Evaluating the news

1. Whereas the symptoms associated with menopause are not life-threatening, heart attacks, strokes, and breast cancer are potentially fatal diseases. Given this, why do you think that regulatory agencies have not stopped the use of HRT altogether?
2. In contrast to many medical treatments that rely on the introduction of completely foreign chemicals into the body, HRT seeks only to replace chemical compounds that the body can no longer produce. This would seem to be a relatively innocuous sort of treatment. Why then, does the long-term use of HRT cause such serious health problems?
3. Can you think of other treatments that involve significant benefits as well as significant risks? What would you want to know from your doctor before following such a treatment?

SOURCE: *BBC News*, November 30, 2004.

CHAPTER 25

The Nervous System

Key Concepts

◉ Neurons transmit information in the form of electrical signals called action potentials. They pass that information on to other cells by means of chemical signals.

◉ Connections among neurons in the central nervous system (the brain and spinal cord) allow animals to integrate sensory information coming from, and motor commands going out to, nerves in the peripheral nervous system.

◉ Humans have large brains relative to our body size. During our evolution from primate ancestors, some parts of the human brain increased to a much greater extent than others.

◉ The nervous system interacts closely with the endocrine system.

Coffee, Tea, and Chocolate: Stimulants in Human Society

If you need coffee to get going in the morning, you are not alone. One of the constants of human societies is the use of mild stimulants of some sort. Most North Americans know coffee, tea, and chocolate. Coffee originated as a drink made from the ground seeds of the coffee tree on the Arabian Peninsula about AD 1000. From there, it spread throughout the Middle East with the spread of Islam. Today, coffee represents the most popular and economically valuable stimulant worldwide, bringing in over $8 billion in trade.

Chocolate originated about 2,000 years ago among the Maya of Central America as a drink derived from the seeds of the cacao tree. In contrast to the sweet hot chocolate that you've probably had, the Mayan version did without the sugar, but added hot pepper to give it a little zing. The Spanish introduced chocolate to Europe shortly after they began their exploration of Central America; by mixing sugar and spices with the ground cacao (cocoa) beans, they produced a sweet drink more like the hot chocolate we know. Not until 1847 did a British company produce the first chocolate bar.

Tea has an even longer history than coffee or chocolate. The tea that the Chinese learned to make was first brewed from the leaves of *Camellia sinensis*, a small tree native to China, around 2700 BC. From China, the use of tea spread to Japan in AD 500. With the arrival of European explorers in the Far East around 1600, tea spread rapidly, first to Europe and then to North America. The tea trade today is worth almost $3 billion worldwide.

Coffee, chocolate, and tea, along with two other plant-based drinks— cola, made with the African cola nut, and the Argentine drink yerba maté, made from the leaves of *Ilex paraguariensis*, a small tree—supply most of

Coffee

Chocolate

Tea

Most Human Cultures Use Drinks That Contain Stimulants

the world's demand for stimulants. All of these drinks contain the chemical compound caffeine. Caffeine acts on the nervous system to produce the stimulating effects we associate with these drinks: greater alertness, a faster heart rate, more energy, and wakefulness. Why does caffeine have these effects on us?

A ll animals must coordinate the functions of their cells and tissues. As we saw in Chapter 24, animals use hormones to send information between different parts of the body through the bloodstream. In addition to this slower endocrine system, animals have a unique high-speed internal communication system. The **nervous system** consists of specialized cells called **neurons** (from *neuro*, "nerve, sinew") that can transmit information throughout the body in a fraction of a second. Animals need this almost instantaneous form of communication to coordinate the rapid contractions of their muscles, which play such an essential role in their ability to run, swim, and fly.

In this chapter we begin with the basic anatomy of neurons and the nervous system before turning to consider how the nervous system transmits information so quickly. Then we discuss how the arrangement of individual neurons into a nervous system affects the ability of humans and other animals to process information. Because of the human brain's importance in initiating and coordinating the enormous quantity of information transmitted through the nervous system, we spend some time looking at its organization, and at the significance of its size for human evolution. We close by examining the close connection between the nervous and endocrine systems.

25.1 Neurons, Nerves, and More: Basic Features

A neuron is a cell that can receive information, transmit that information rapidly over a distance, and then pass that information on to another cell. Neurons interact with other neurons, with muscle cells, and with endocrine cells.

The form of a neuron reflects its unique function (Figure 25.1*a*). The cell body of a neuron contains a nucleus and the same organelles found in any animal cell. In addition, a neuron has multiple extensions from the central cell body, which are identified by different terms depending on whether they receive signals from, or send signals to, other cells. A neuron's many, highly branched **dendrites** (from *den*, "tree") receive signals from adjacent cells. Signals from the dendrites travel to the cell body, where they are collected and transmitted down the neuron's single **axon**, a thin extension of the cell body that transmits the signals to other cells, sometimes over a long distance. Although we think of cells as small, human neurons can have axons extending up to a meter in length.

Cells closely associated with neurons produce an insulating sheath made of a fatty material called **myelin**, which encases the axons of most neurons in vertebrates, including humans. Neurons with a myelin sheath are said to be *myelinated*; those without one are *unmyelinated*. As we shall see, myelinated axons can carry signals more rapidly than unmyelinated ones.

At its end, the single axon divides up into many axon terminals, which transmit the signal to other cells. The axons of many individual neurons may be bundled together (along with supporting cells, blood vessels, and connective tissue) to form a major communication pathway called a **nerve** (Figure 25.1*b*).

Just as axons are bundled together to form nerves, neuronal cell bodies combine in various ways to create larger structures. A **ganglion** (plural: ganglia) is a cluster of neuronal cell bodies that serves as a local or regional integrating center. Vertebrates have a number of ganglia throughout their bodies. In animals that have a head end and a tail end, the head end usually houses a dominant aggregation of neurons that controls and coordinates the signals moving throughout the entire body. In relatively simple animals, this control center amounts to nothing more than a ganglion; even the lowly earthworm has a head end with a tiny ganglionic control center.

Vertebrates have a much more complex structure, called a **brain**, which serves as the major clearinghouse for information. Notice, too, that the "brainy" end of a vertebrate contains most of its sensory structures, such as eyes to see with, ears to hear with, and a nose to smell with. The evolutionary advantage to having the brain near these sensory structures is that it reduces the time

(a)

Dendrites collect information from other neurons.

The cell body contains all of the structures basic to eukaryotic animal cells.

1 A small gap called the synaptic cleft separates the transmitting and receiving neurons.

2 Neurotransmitters are stored in the axon terminal.

The signal travels from the dendrites to the axon.

The axon conducts the signal away from the cell body.

A SYNAPSE

Axon terminals transmit information to other cells.

3 The transmitting neuron releases neurotransmitters into the synaptic cleft.

4 Neurotransmitters bind to the receptors in the plasma membrane of the receiving cell.

(b)

Connective tissue

Veins

Nerve

Arteries

Individual neurons

Axons

Myelin sheath

Figure 25.1

Neurons Carry Information

(*a*) Neurons rapidly transmit signals from one cell to another. Neurons meet at structures called synapses; here electrical signals in the axon terminals are converted into chemical signals in the form of neurotransmitters, which diffuse across a small gap (the synaptic cleft) between the neurons. (*b*) Neurons are bundled together with supporting tissues and blood vessels into cable-like nerves.

it takes for the brain to receive sensory information. For humans and many other vertebrates, this arrangement has increased our ability to avoid danger and to detect and capture food more quickly.

Another major component of most nervous systems is a collection of nerves that carry information between the head-end processing center and the tail end of the animal. Vertebrates, including humans, have a specialized **spinal cord** that collects and transmits information coming from the body; it then either fashions appropriate responses itself or sends the information on to the brain for further processing. The brain, in turn, sends instructions back to the spinal cord. In contrast to the small numbers of large nerves used by some other animals, such as insects, the vertebrate spinal cord contains large concentrations of dendrites and axon terminals that allow information exchange among huge numbers of neurons along largely myelinated pathways.

25.2 Neurons Transmit Rapid Electrical Signals

The electrical signals transmitted by neurons are generated as a result of a stimulus in the environment or within the body. A **stimulus** is anything that can cause cells receiving it to generate a signal. You are experiencing stimuli as you read this page. Millions of neurons in your eyes are generating almost countless signals as light coming from the page into your eyes produces a constantly changing pattern of visual stimuli. These signals result from the flow of charged ions through channel proteins in the plasma membranes of the neurons. The channels open and close, and the electrical charges on the inside and the outside of the neurons' plasma membranes change rapidly. Let's look at this process more closely.

Axons transmit electrical signals called action potentials

Axons transmit information to other cells using a type of self-sustaining electrical signal called an **action potential** (Figure 25.2a). These signals depend on the rapid movement of positively charged sodium ions (Na^+) across the plasma membrane of the axon. To understand how this works, it helps to begin when the axon is not transmitting an action potential. Since the axon is essentially at rest, the electrical state of the axon at this time is called its **resting potential**. In this state, active carrier proteins in the plasma membrane are pumping Na^+ from the axon into the surrounding interstitial fluid. This continuous removal of positively charged ions from the axon interior maintains a relatively negative charge inside the axon. Thus the constant pumping keeps the axon in a **polarized** state—that is, it maintains a difference in electrical charge between the inside and the outside of the axon. This difference in charge contains the potential energy needed to generate an action potential.

Events that may cause an axon to generate an action potential begin when electrochemical signals traveling along the dendrites and across the neuronal cell body reach a patch of the axon's plasma membrane near where it joins the cell body. In response to an arriving signal, some of the channels in this portion of the axon open, allowing Na^+ from the surrounding interstitial fluid to enter the axon. This causes the axon to become locally **depolarized**: the charge inside the axon here is still negative, but not as negative as before—in other words, the charge difference across the plasma membrane of this portion of the axon has lessened.

If the depolarization is slight, not much will happen. But if the electrochemical signals arriving at the axon are strong enough, they will cause more channels to open, so that the depolarization reaches a threshold level. Once this threshold is reached, an action potential automatically begins. Immediately, a very large depolarization occurs: even more channels open in the depolarized part of the plasma membrane, allowing a large number of sodium ions to flow rapidly into the axon. So many sodium ions enter so quickly that the local charge difference across the plasma membrane actually reverses; for a short time, the inside of the affected portion of the axon takes on a relatively positive charge.

This reversal of charge depolarizes the plasma membrane of the adjoining portion of the axon, which in turn depolarizes the next portion, and so on down the axon away from the cell body. The signal cannot reverse, heading back toward the cell body, because the channels that just opened in each depolarized segment of the axon then close for a brief time (see Figure 25.2a). The only channels that can open in response to a local depolarization are the channels "downstream" from the previous region. Once the action potential has moved on, that portion of the axon returns to its resting potential. Thus, once an action potential has begun, an inevitable wave of charge reversals travels down the axon, much like a lit fuse or a line of falling dominoes.

Because action potentials in unmyelinated neurons involve the movement of ions across plasma membranes, the signal moves more slowly than an electrical current moving down a wire. Whereas we notice no delays in a telephone conversation carried by an electrical current between two cities 1,000 kilometers apart, it would take an action potential traveling at a typical speed of 5 meters per second more than 2 days to cover the same distance. For most animals, however, an action potential moving at this clip travels the length of the body in a fraction of a second, which is sufficient for most of their internal communication needs.

Myelinated axons transmit action potentials especially rapidly

Among vertebrates and in a few invertebrates, a myelin sheath surrounds the axons of most neurons, which greatly speeds signal transmission (Figure 25.3). The myelin sheath covers segments of the axon, leaving unmyelinated gaps between the myelinated zones, giving the axon the appearance of a chain of sausage links (Figure 25.2b). The gaps between myelinated segments are called **nodes of Ranvier**. Action potentials can occur only at these nodes,

(a) Neuron with unmyelinated axon

Cell body

A ⟶ 1 m : ~0.1 s ⟶ B

Axon

Plasma membranes

Na⁺ entry locally depolarizes the axon…

Na⁺

Channels closed

Na⁺

…which sufficiently depolarizes the adjacent region of the axon to open more Na⁺ channels, re-creating the action potential there.

(b) Neuron with myelinated axon

A ⟶ 1 m : ~0.007 s ⟶ B

Myelin sheath

Na⁺ channels open, generating an action potential.

The signal spreads many times faster than in unmyelinated nerves.

Myelin

Axon

Na⁺

Na⁺

Na⁺

Nodes of Ranvier

Na⁺

Charge moves to next node.

The action potential is triggered at the new node and continues from node to node as fast as 150 m/s.

Na⁺

Axon

Na⁺

Sodium channels closed

Na⁺

Figure 25.2 How Axons Transmit Signals
Action potentials (red) move down axons differently, depending on whether the axon is myelinated or not. (*a*) Unmyelinated axons carry self-sustaining action potentials that rely on changes in the charge difference across their plasma membranes. (*b*) Myelinated axons transmit action potentials very rapidly as the signal leaps from node to node.

because only there can Na⁺ move across the plasma membrane. There are no sodium channels along the portions of an axon's plasma membrane that are surrounded by myelin. Action potentials move faster in myelinated fibers because they leap from node to node.

When an action potential occurs at one node, the region of the axon's interior adjacent to that node becomes positively charged, and the charge, in the form of an electrical current, moves very quickly to the next node. Notice that Na⁺ cannot flow across the axon's plasma membrane in the region between nodes because there are no sodium channels there. Instead, the electrical charge flows rapidly *inside* the axon, from the depolarized node to the next node along the axon (see Figure 25.2*b*). As a result, the axon depolarizes at the next node,

Figure 25.3 Myelinated and Unmyelinated Nerve Fibers
Colored transmission electron micrograph of a cross-section through nerve fibers sheathed in myelin (brown).

and a new action potential begins there. In this way, the signal jumps rapidly from one node of Ranvier to the next node down the axon.

The electrical charge can move down the interior of a myelinated axon much more quickly than can a wave of charge reversals along the plasma membrane of an unmyelinated neuron. Unmyelinated neurons transmit action potentials down axons at a maximum speed of about 30 meters per second (and an average speed of about 5 meters per second), while in myelinated axons, action potentials average about 120 meters per second and can jump along at up to 150 meters per second.

Disruption of the myelin sheath can lead to serious problems in humans. Diseases that affect the myelin sheath, such as multiple sclerosis and Guillain-Barré [ghee-*LAN*-bar-*AY*] syndrome, can disrupt vision, speech, balance, and muscular coordination, and may eventually kill their victims.

Action potentials have several important features

Action potentials—whether in myelinated or unmyelinated axons—have several important characteristics that make them well suited for sending information rapidly throughout the body. First, as we have just seen, they move along the axon in only one direction.

Second, an action potential remains consistently strong as it moves from one end of the axon to the other; it does not weaken with distance. If these signals weakened as they moved along, they might lose their capacity to initiate events in other neurons or cells, or disappear altogether before they reached their target cells.

Third, action potentials occur either "all the way" or not at all; that is, they all have the same strength. A strong stimulus initiates more action potentials, not stronger action potentials. This allows information to be "digitized" in much the same way that computers digitize information as series of zeros and ones. A 0 is an absence of an action potential, and a 1 is an action potential; there is no potentially confusing in-between. The sound of a whisper generates action potentials that have the same strength as those generated by a shout, but there are fewer of them. A whisper could be described as 100010001, a shout as 101010101, and an explosion as 111111111.

Explore how neurotransmitters are released. 25.1

Neurotransmitters transmit signals between adjacent cells

If neurons are to be part of a whole-body communication network, they must be able to pass their signals along to other cells. Once a signal reaches the end of an axon, it is passed on to the next cell at a junction called a **synapse** (see Figure 25.1). Neurons form synapses with other neurons, with muscle cells (Figure 25.4), and with endocrine cells. In a synapse between two neurons, the axon terminal of the sending cell may interact with a dendrite or with the cell body of the receiving cell.

The transmission of information at a synapse takes place across a minute gap between the two cells, called the **synaptic cleft**. In the axon terminal, the electrical signal is transformed into a chemical signal in the form of molecules called **neurotransmitters**. The arrival of an action potential at the axon terminal causes it to release these molecules into the synaptic cleft by exocytosis. The neurotransmitters diffuse virtually instantaneously across the narrow cleft and bind to receptor proteins in the plasma membrane of the target cell.

If enough molecules of the appropriate neurotransmitter bind to enough target cell receptors, they will affect the electrical charge of the target cell. In some cases, the plasma membrane of the target cell depolarizes, triggering an action potential that begins its rapid journey along the dendrite or across the cell body. In other cases, the neurotransmitter inhibits the target cell, making it *less* likely that it will produce an action potential. In target cells that are not neurons, the transfer of signals occurs in the same way.

Given the major advantage in speed the nervous system has over the endocrine system for transmitting information, it may seem odd that neurons depend on

Figure 25.4 Synapses Between Neurons and Muscle Cells
These nerve cells (green) are connected to muscle fibers (red) via synapses. The neurotransmitter released by the neuron across the synapse is the chemical acetylcholine, which makes the muscle contract.

the process of diffusion at synapses between cells, which is slower than the transfer of electrical charges. However, the journey neurotransmitters make across the synaptic cleft is extremely short—less than a millionth of a millimeter—which allows them to reach their destinations in tiny fractions of a second. Moreover, the very long axons of some neurons ensure that action potentials need to cross only a few synapses to reach their destinations.

The connections between neurons form an essential component of nervous systems because it is through these networks that neurons exchange information. The subtlety and specificity of this communication is possible due to the huge number of unique connections that exist. Several factors contribute to this variety. Every action potential is a small unit of information, just as a byte is in computer "talk." A single neuron can generate up to 4,800 action potentials per minute, and thus 4,800 "bytes" of signal information. Like the bodies of most vertebrates, the human body has many billions of neurons. Some neurons in humans may interact with as many as 10,000 other neurons in a dense tangle of dendrites and axon terminals. The scale of these numbers allows us to put the operation of nervous systems into a clearer perspective.

In addition, biologists point out that some neurons both produce and respond to more than one neurotransmitter (Table 25.1). The types of receptors on the target cell plasma membrane determine what events result from neurotransmitter binding. One neurotransmitter may bind several different receptor types and trigger different responses in different target cells. Acetylcholine, for example, is the major neurotransmitter responsible for signaling skeletal muscles to contract, while in the heart the same neurotransmitter signals muscles to slow down—to contract less often and with less strength. Yet things can go wrong with these complex neurotransmitter systems (see the box on page 482).

25.3 Making Sense of Action Potentials in the Body as a Whole

The nervous system as a whole converts stimuli into **sensory input**—action potentials that convey information about stimuli—which it sends to processing centers, where the information is processed and integrated with

Table 25.1

Functions of Common Human Neurotransmitters

Neurotransmitter	Major functions
Acetylcholine [ASS-uh-teel-KOH-leen]	Controls muscle contractions
Dopamine	Affects muscle activity; stimulates neurons in the pleasure center of the brain
GABA (gamma-aminobutyric acid) [. . . uh-MEEN-oh-byu-TEER-ik . . .]	Assists muscle coordination by inhibiting counterproductive or unneeded neurons
Serotonin [SAIR-uh-TOH-nin]	Regulates temperature, sleep, mood
Melatonin [MEL-uh-TOH-nin]	Helps regulate day-night cycles, including sleep
Enkephalins and endorphins	Inhibit transmission of signals from pain sensors, pain perception
Substance P	Regulates transmission of signals from pain sensors

other information. The nervous system then fashions a response (consisting of more action potentials) appropriate to the initial stimuli. You can easily demonstrate this for yourself by taking a moment to perform this simple experiment: bring the tips of your two index fingers together in front of you until they touch. You have just converted visual stimuli (printed words) gathered through your eyes (sensory input) into a response (bringing your index fingers together) that "makes sense" in terms of those stimuli. How did your nervous system act as a unified whole to convert billions of signals every fraction of a second from your eyes into an appropriate and coordinated response?

Sensory information can move through the nervous system by paths that range from simple circuits involving only a few neurons to complicated pathways involving vast numbers of neurons in different areas of the brain and body. As a rule, the more complicated the required response, the more processing the information receives.

Learn more about the propagation of nerve impulses. 25.2

Test your knowledge of the function of neurotransmitters. 25.3

Signals travel between the peripheral and the central nervous systems

In vertebrates, the nervous system is generally divided into two main parts. The **central nervous system (CNS)** consists of all the neurons (and their supporting cells) in the brain and the spinal cord and is devoted to the exchange and integration of information among neurons

Biology Matters

Increases in Numbers of Students Seeking Counseling

As our brains process vast amounts of information from second to second and day to day, our moods and sense of psychological well-being, or not-so-well-being, are shaped and shipped to our conscious minds. If the chemistry of neurons in our brain isn't just right, or if incoming information is abnormally stressful, emotional trouble can arise.

In recent years, stress and the demands of everyday living seem to have become more severe. The effect on college students in particular has been dramatic. One study covering a total of 13 years (1988–2001) demonstrated that there were large increases in the numbers of students seeking counseling and psychological help at Kansas State University. The study included over 13,000 students. It showed that the number of students suffering from depression doubled, and the number of students with suicidal tendencies tripled, from 1988 to 2001. The number of students who sought help related to sexual assault quadrupled!

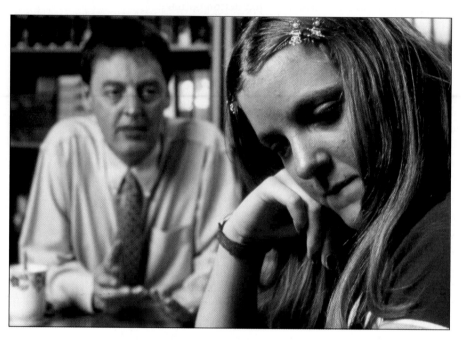

Anecdotal evidence suggests that this trend is not limited to Kansas State—many colleges and universities report similar dramatic increases in the number of students seeking counseling for depression and anxiety, and college suicides have made headlines repeatedly.

The growing problem has gained some political attention. One bill introduced in the U.S. House of Representatives in 2003 called for $10,000,000 in grants for counseling centers on college campuses. Part of the argument for allocating the money pointed out how serious the problem is. The sponsors of the bill cited evidence that, for many students needing help, alcohol consumption and drug use go up and academic performance goes down. On a positive note, the sponsors pointed out that students who get counseling are more likely to stay in school—14 percent more likely than students who don't get help.

Counseling seems to help, and in addition, research on the brain chemistry underlying these problems is producing results. At the purely biological level, medical researchers refer to psychological difficulties as "chemical imbalances" because chemicals such as neurotransmitters lie at the very foundation of brain function. Brain chemicals are generated according to instructions from the genetic blueprint in a person's DNA. Researchers point out, for example, that if depressed people have blood relatives who suffer from depression, the tendency to become depressed may be inherited. But increasing evidence indicates that environmental factors, and even a person's own behavior, may affect patterns of gene expression. Research is discovering many new types of neurotransmitters, and those discoveries are leading to the availability of more drugs that can treat psychological disorders. Combined with counseling, these new drugs provide more hope for students suffering from psychological problems.

(Figure 25.5). The **peripheral nervous system (PNS)** acts as a kind of messenger service, ferrying signals to and from the CNS. The PNS consists of the sense organs (which we will cover in Chapter 26) plus all the nerves that are not part of the CNS. We will spare you from having to learn the many categories and subcategories of neurons devised by biologists to describe different parts of the nervous system—except for the three main classes of neurons.

Sensory neurons in the PNS convey sensory input to the CNS. Most signals first pass to **interneurons**, which lie wholly within the CNS, primarily in the spinal cord. Interneurons process the sensory input from sensory neurons, then pass the signals on to other neurons, in one of

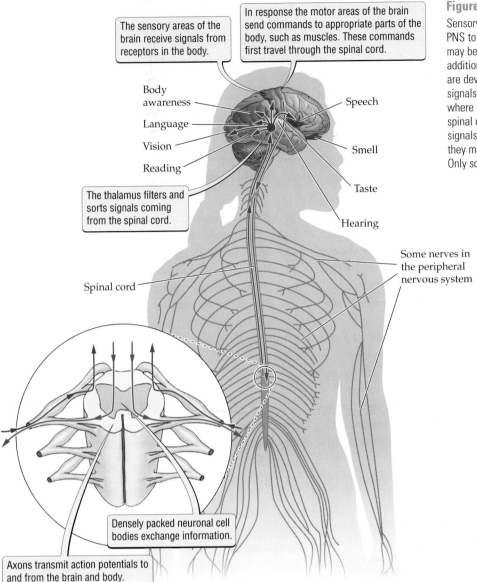

The sensory areas of the brain receive signals from receptors in the body.

In response the motor areas of the brain send commands to appropriate parts of the body, such as muscles. These commands first travel through the spinal cord.

Body awareness

Language

Vision

Reading

The thalamus filters and sorts signals coming from the spinal cord.

Speech

Smell

Taste

Hearing

Spinal cord

Some nerves in the peripheral nervous system

Densely packed neuronal cell bodies exchange information.

Axons transmit action potentials to and from the brain and body.

Figure 25.5 Organization of the Human Nervous System

Sensory information (red arrows) travels along sensory neurons in the PNS to the spinal cord, which is part of the CNS. Here the information may be processed and acted on or, more often, sent on to the brain for additional integration and interpretation. Different parts of the brain are devoted to processing different kinds of information. Command signals from the brain (blue arrows) travel back down the spinal cord, where they are passed on to motor neurons in the PNS, leaving the spinal cord by a slightly different route. In the body, these motor signals may stimulate or inhibit muscles, organs, or endocrine cells; they may also override the actions triggered by spinal cord processing. Only some of the PNS nerves from the neck down are shown here.

Learn more about the organization of the human nervous system.

25.4

several ways. First, they may send signals directly to **motor neurons**. These neurons in the PNS carry signals back out to the body, thereby causing changes in muscles, organs, or endocrine glands in response to the initial stimuli. Second, interneurons may pass the signals up the spinal cord to the brain for further processing; the brain may (or may not) respond by sending new signals down the spinal cord for relay out through motor neurons to the body (see Figure 25.5). Finally, interneurons may send signals *both* up to the brain and out to the body at the same time.

Sensory information can be processed with or without the brain

The human spinal cord—a collection of nerve tissue about as thick as a finger—is an organized association of neu-

rons. Dense collections of neuronal cell bodies, dendrites, and axon terminals exchange information there, and large bundles of axons (nerves) carry this information to and from the brain (see Figure 25.5). The vertebrate spinal cord forms an intermediate site of information processing that acts as a kind of filter between the brain and the sensory neurons bringing input from the far reaches of the body. Although much information needs to be passed on to the more sophisticated processing centers in the brain, some simpler processing can be taken care of by the spinal cord neurons themselves.

You know from experience that if you touch something hot, such as a candle flame, you jerk back your hand without having to think about it. A pain message is sent by sensory neurons to your spinal cord, and a "get your hand out of there" message reaches the muscles in your arm via motor neurons (Figure 25.6). You have just experienced the action of a **reflex arc** (also called a **spinal reflex**). This reflex arc consists of just three neurons: a sensory neuron that sends the message to the spinal cord, an interneuron, and a motor neuron that creates a response in the body by sending a message to, in this case, a muscle. All of the information processing takes place in the spinal cord—the brain is not required.

There are many reflex arcs in the human body. One of the best known is the "knee-jerk response," which you may have encountered at the doctor's office: the physician taps just below your kneecap, which automatically forces your lower leg to twitch forward. We also have a number of reflex arcs that respond to internal stimuli, such as those involved in fine-tuning the operation of some of our organs. A few reflexes are even simpler than the one shown in Figure 25.6, consisting of only two neurons: a sensory neuron links directly with a motor neuron in the spinal cord.

Even though this reflexive motor response to pain doesn't require the brain's involvement, many messages do get sent to the brain. We are conscious of the pain from the candle flame, and the brain may process that information into additional responses, including a stream of words, many unfit to print here. However, because the brain requires more time to process information than a

reflex arc does, these brain-directed responses come later—perhaps only a matter of milliseconds later, but that time lag is significant. If our bodies relied on the brain to process the motor part of the reflex arc—jerking the hand away from a painful stimulus—we might get burned more severely.

25.4 Organization of the Human Brain

In contrast to the exchange of information between a few neurons in a reflex arc, the brain receives information in the form of action potentials from millions of sensory neurons, and it can direct a response to that information to millions of motor neurons. The human brain consists of huge numbers of neurons organized and interconnected to allow the efficient exchange of information. Its vast numbers of neurons can sort and interpret a bewildering array of incoming information.

The brain has three major regions

Far from being a homogeneous mass of neurons, the brain has distinct regions: the forebrain, the midbrain, and the

Explore how reflex arcs work.

25.5

Figure 25.6

Reflex Arcs Do Not Require Processing in the Brain

A simple circuit of neurons allows us to take quick action without having to "think" about it.

Cross section of spinal cord

2 Information is processed in the spinal cord by interneurons.

Interneuron

Sensory neuron

Motor neuron

Muscles

3 Signals to the arm muscles travel down motor nerves, causing the arm to jerk away from the stimulus.

Painful stimulus

1 A pain message goes up a sensory nerve to the spinal cord.

hindbrain (Figure 25.7). Each region has a different set of functions.

The **forebrain** might be described as the supercomputer region. It has three important components: the thalamus, the cerebrum, and the hypothalamus. The **thalamus** is the central switching area of the brain. It acts as a critical filter, saving the brain from processing all the stimuli our bodies are constantly receiving from both within and without. The thalamus determines which of the sensory signals coming from the spinal cord should go to the conscious perception centers in the cerebrum and which should go to the other portions of the brain that act on sensory inputs without our being aware of it. For example, when you touch a candle flame, the thalamus passes along the pain message to one part of your brain, and you are very much aware of that. But it also passes along information to other parts of the brain that you probably won't be aware of until the next time you see a lit candle. Then you'll remember what *not* to do.

The **cerebrum** forms the most conspicuous portion of the brains of humans and many mammals (though not those of many other vertebrates). The cerebrum has four major lobes, as shown in Figure 25.8. The frontal lobe does most of the thinking. The parietal lobe handles aspects of speech as well as taste, reading, and sensation.

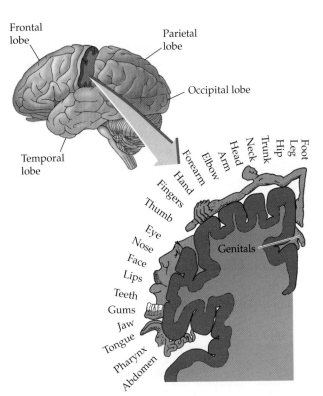

Figure 25.8 How the Brain "Sees" the Body

Different parts of the cerebral cortex process sensory information received from different parts of the body. The four lobes of the neocortex represent physically distinct areas of the brain. The cross section of the brain "maps" body parts to different parts of the brain. The size distortion of the body parts reflects the relative numbers of neurons in the brain devoted to sensory information received from those parts. For example, more neurons process sensory information from the foot, a relatively small body part, than from the leg, a much larger body part.

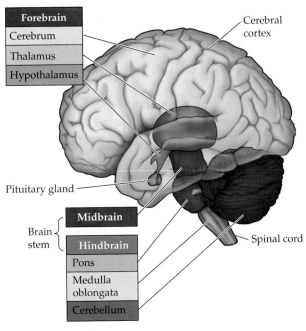

Figure 25.7

The Human Brain Has Three Specialized Regions

Shown here are the major parts of the brain viewed from the left side (with the back of the brain to the right). Note that the cerebrum surrounds much of the brain stem, which extends up into the center of the brain.

The occipital lobe is the region that deals with vision. And the temporal lobe processes sound and smell. Bundles of axons carry information between the left and right halves of the cerebrum. The outer layer of the cerebrum, called the **cerebral cortex**, is only about 5 millimeters thick, but because it is highly and intricately folded, it accounts for approximately 80 percent of our brain mass. In its billions of neurons and even more billions of connections rest our personalities, our capacity to talk, calculate, create art, and consciously sense the world around us. If the thalamus sends signals to the cerebrum, they usually arrive in a specific part of the cerebral cortex that corresponds to the area of the body that received the original stimulus (see Figure 25.8). These sensory signals then become part of our conscious awareness. Coordinated responses to these signals arise in the motor areas of the cerebrum.

We met the third component of the forebrain—the hypothalamus—in Chapter 24, where some of its functions as an integrator of the nervous system and the endocrine system were described. It is the center for regulating homeostasis in blood pressure, body temperature, sex drive, hunger, and thirst. The emotions of rage and pleasure also arise from the hypothalamus. Most of us have regular patterns of sleep and wakefulness; the timing of these patterns rests in the hypothalamus's biological clocks.

Learn more about the cerebral cortex. 25.6

The **midbrain** helps to maintain muscle tone and sends some sensory data to higher brain centers. Higher brain centers are called "higher" because they process information in much more sophisticated ways than "lower" brain centers; they are also literally higher up when we are standing.

The **hindbrain**—which includes the medulla oblongata, the pons, and the cerebellum—coordinates information dealing with breathing rhythms and balance, among many other functions. The midbrain and hindbrain together make up the **brain stem**.

Imaging technologies help us see inside the brain

With the development of sophisticated medical imaging technologies, such as **magnetic resonance imaging**

Figure 25.9 Different Activities Stimulate Different Parts of the Brain

Magnetic resonance imaging (MRI) can show active portions of the brain. In this three-dimensional MRI, the parts of the brain that are active when the left hand is moving are highlighted in red.

(MRI), it has become possible to map out in great detail how the structure of the brain relates to its function. MRI can detect which parts of the brain are active at a particular time (Figure 25.9). By exposing people to a certain stimulus or having them do a particular task while monitoring the brain with MRI, we can identify the parts of the brain that are activated.

When we listen to conversation, for example, the left temporal lobe of the brain becomes active. This portion of the brain is involved in language comprehension. When speakers of Mandarin Chinese listen to a conversation in that language, however, both the right and left temporal lobes of the brain become active. Typically, the right temporal lobe is involved in understanding music. In Mandarin Chinese, unlike most languages, the meaning of words depends not just on the sounds of vowels and consonants, but also on intonation, or how the tone changes as a word is spoken: whereas the word *ba* spoken like a question (with rising intonation) means "to pull out," the same word spoken like a command (with descending intonation) means "tyrant." The activity on the right sides of the brains of Mandarin Chinese speakers reflects their need to understand this musical component of their language. The box on page 488 describes positron emission tomography (PET), another imaging tool used to understand the workings of the brain.

25.5 Why Do Humans Have Big Brains?

Although the human brain is not the largest in the animal kingdom—an honor that belongs to the 6-kilogram brain of the 130,000-kilogram blue whale—it is the largest relative to body size. An average animal of our size would be expected to have a brain about one-sixth of the size of a human brain.

If we look at the brain sizes of our primate ancestors as they evolved to become modern humans (*Homo sapiens*), we can see that not all parts of the brain increased at the same rate. While the brain stem and the areas surrounding it have remained about the same size, a part of the cerebral cortex found only in mammals, called the **neocortex** (*neo*, "new"), has increased in size to a far greater extent than other parts (Figure 25.10). The neocortex is where the ability to think complex thoughts, use sophisticated language, and create new devices such as spearheads and plows resides. Similarly, our ability to develop tactics for hunting prey and for organizing villages arises from the powers of neurons in the cerebral cortex and its "neo" dimensions. The enlarged neocortex also gave us the ability to calculate mathematically and anticipate outcomes of future scenarios—an enormous advantage in a world shared with other organisms that must rely mostly on sheer speed, strength, and endurance.

Rapid increases in brain size occurred after bipedalism

The large size that characterizes the modern human brain has arisen very recently in our evolutionary history. Four million years ago, *Australopithecus* had a brain volume of about 600 milliliters; 2.5 million to 2 million years later, *Homo erectus*, the first (oldest) member of the genus to which we belong, had a brain that was only 100 milliliters bigger—an increase of only about 17 percent. Over the past 2 million to 1.5 million years, brain size in humans doubled, eventually averaging about 1,400 milliliters.

As we saw in Interlude D, the rapid increase in human brain size occurred roughly 2 million years *after* fossils of early humanlike species gave clear evidence of bipedalism (literally, "walking on two legs," from *bi*, "two," and *ped*, "foot"), so it was not directly associated with this change. Instead, the increase occurred more or less in parallel with the increasing use of stone tools. In other words, increased brain size appears to be more closely related to advances in thinking and technology than to changes in the way we walk.

Why human brain size evolved so rapidly over our recent evolutionary history remains a mystery. Among other mammals, only two biological features seem consistently to go hand in hand with larger-than-expected brain size: living in trees and feeding on foods that are both scarce and distributed in patches. Both of these characteristics require the handling of complex spatial information. By themselves, these two explanations seem unlikely to account for the dramatic increase in the size of the human brain. Early bipedal primates neither spent much time climbing trees nor fed on food types that were any more scarce or patchily distributed than those of other organisms.

Instead, it may be that once brains in ancestral human species reached the size of those in *Homo erectus*, a small increase in overall brain size meant a relatively large increase in the size of the neocortex, which plays such an important role in how we perceive and respond to our world. The neocortex may have proved especially useful to humans who lived and hunted in social groups and who were beginning to develop language and technology.

Australopithecus africanus *Australopithecus robustus* *Homo erectus* Modern *Homo sapiens*

The neocortex has increased in size greatly.

This area of the brain is about the same as in earlier species.

Figure 25.10 The Rise of the Neocortex
During human evolution, the neocortex increased in size more rapidly than other brain regions, so that it is now proportionately much larger than it was in our primate ancestors. (The "old" cortex has not disappeared; rather, it has become enveloped by the neocortex.) The shapes and contours of the brains shown here (seen from the right side) were determined by making casts of the insides of fossil skulls.

In what would appear to be a logical extension of this trend in human evolution, society has come to equate bigger brains with greater intelligence. Our culture is filled with images of highly intelligent Earthlings and aliens with big heads that presumably contain large brains (Figure 25.11). Recent studies relating brain size to scores

Figure 25.11 Intelligent Aliens
Our science fiction stories and films give aliens big heads in order to indicate that they have big brains and are therefore much more intelligent than we are. Shown here is an alien from the planet Metalune, from the 1955 movie *This Island Earth*.

Science Toolkit

Watching People Think

As we think or respond to our surroundings, the pattern of activity in the brain changes. If we listen to music, for example, certain parts of the brain become more active. These patterns of brain activity can be seen using scanning devices and computers. Figure 25.7 showed images of brain activity created using magnetic resonance imaging (MRI). Another method used for brain imaging is **positron emission tomography (PET)**.

Researchers or clinicians who use PET attach small amounts of radioactivity to molecules used naturally by the brain, such as glucose, then inject those molecules into a volunteer's or patient's bloodstream. A PET scanner has sensors that detect where these radiolabeled molecules become concentrated in the person's brain. Depending on the molecule used, a PET scanner's sensors can monitor blood flow, oxygen use, or glucose metabolism in the brain. Computers convert information from the sensors into color images of the brain that show where increased activity is taking place. In this way, researchers can see what areas are active in different situations.

PET scans are allowing researchers to understand a great deal about how our brains function, both under normal circumstances and in response to various abnormal situations, such as the use of illegal drugs (including cocaine and heroin). In medicine, PET scans are aiding doctors in the diagnosis of difficult diseases and conditions. For example, PET scans can detect brain cancers early, when they are much more treatable. PET scans can also differentiate between benign and cancerous tumors, and they can help in diagnosing Alzheimer's disease and Parkinson's disease. If a person suffers from seizures, PET can locate the area of the brain that is not functioning properly.

The PET scans in the accompanying figure show the activity in different parts of a person's brain while resting and while performing different tasks. In this case, the scanner measured the relative amounts of glucose being used by different parts of the brain, which indicates the intensity of neuronal activity in those areas. (Recall that the more active a cell of any type is, the more glucose it requires for respiration.)

(a) Resting

(b) Listening to music

(c) Colors and patterns

(d) Thinking

PET Scans

These scans show the intensity of activity in different parts of the brain (as measured by glucose uptake) in people doing different tasks, ranging from greatest (red-orange) to medium (yellow-green) to least (blue-black). In series *b*, *c*, and *d*, the red arrow indicates the region showing the most significant changes in activity from series *a* (the brain at rest). (*b*) In a person listening to music, activity occurs primarily on the right side of the brain. (*c*) Looking at colors and visual patterns activates regions at the back of the cerebral cortex. (*d*) When a person is asked to carry out a "thinking" task, portions at the front of the cerebral cortex become active.

As with MRI, each image can show only what is going on in one horizontal "slice" of the brain, so a series of images are created to represent different levels in the brain. In the accompanying figure, only two images from each series, from the top of the brain (left) and from near the bottom (right), are shown. Series *a* shows a scan of a brain "at rest." As you can see, even a resting brain requires differing amounts of glucose in different areas, so the images in series *a* serve as a kind of "control," or baseline, for comparison with scans of specific activities. As you compare the patterns in series *b*, *c*, and *d* with those in series *a*, notice what *doesn't* change as well as what does: regions of the brain not involved in the specific activities remain remarkably similar in each set of images.

As you go through your day, try to imagine what your brain patterns may look like and how they may be changing as you take an exam, listen to music, or study an illustration in a textbook.

on intelligence tests support this stereotype, although these studies also show that test scores can vary widely for a given brain size.

It seems unlikely that the human race will evolve even larger brains, however. In fact, human brain size has actually been shrinking, not increasing, as we saw in Interlude D. It may be that the costs of a larger brain would outweigh the benefits. At rest, the human brain, which constitutes 2 percent of our body weight, consumes 15 percent of our oxygen intake. In other words, it is energetically expensive to maintain and use our large brains. In addition, our heads are already large enough to complicate the process of giving birth.

Large parts of the human brain are devoted to hands and to speech

In comparing the organization of the modern human brain with that of our closest living relatives, the chimpanzees, we find a number of notable differences related to our extensive use of our hands and to language. A greater proportion of the human brain is devoted to processing sensory input from the hands and providing detailed control of their movements. This feature of the human brain almost certainly followed from the evolution of a precision grip in our first bipedal ancestors. In combination with the depth perception skills that we inherited from our primate ancestors, our ability to make fine movements with our fingers may have contributed to our unique tool-making abilities.

We also devote a lot of our large neocortex to understanding and producing speech, whereas chimpanzees devote only a small part of their neocortex to this function. The left and right sides of our brains have become specialized to handle different aspects of using hands and language. In contrast, chimpanzees process information related to hands and to sound production and detection in a similar way on both sides of the brain.

25.6 Integration of the Nervous and Endocrine Systems

Although we have thus far considered the endocrine system and the nervous system as two alternative means of internal communication, it is important to recognize that they are closely connected. The hypothalamus, located at the base of the human brain, links the nervous system to the endocrine system. The hypothalamus sits atop the pituitary gland, which releases hormones that directly or indirectly control almost all aspects of the endocrine system (Figure 25.12).

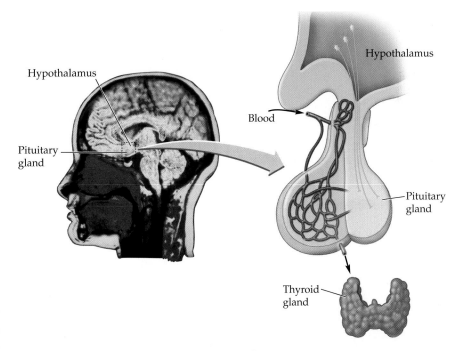

Figure 25.12 **The Hypothalamus and Pituitary Gland Coordinate the Nervous and Endocrine Systems**
In this side view of the human head, note the proximity of the hypothalamus and the pituitary gland at the base of the brain. The enlargement shows how hormones get from the hypothalamus to the pituitary and then into the bloodstream.

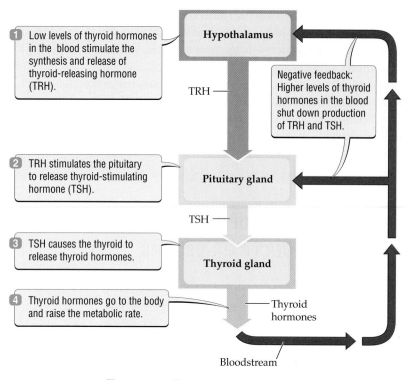

1. Low levels of thyroid hormones in the blood stimulate the synthesis and release of thyroid-releasing hormone (TRH).

Hypothalamus

TRH

Negative feedback: Higher levels of thyroid hormones in the blood shut down production of TRH and TSH.

2. TRH stimulates the pituitary to release thyroid-stimulating hormone (TSH).

Pituitary gland

TSH

3. TSH causes the thyroid to release thyroid hormones.

Thyroid gland

4. Thyroid hormones go to the body and raise the metabolic rate.

Thyroid hormones

Bloodstream

Figure 25.13 The Endocrine and Nervous Systems Work Together to Control the Rate at which Energy Is Released from Food

The interaction between the hypothalamus, pituitary gland, and thyroid glands illustrates how the endocrine and nervous systems can work together to control a biological process.

The hypothalamus consists of specialized neurons that release hormones in response to action potentials. These hormones enter blood vessels that connect the hypothalamus directly to the pituitary gland. The hypothalamic hormones may either inhibit or stimulate specific cells in the pituitary, which then stop producing hormones or start producing hormones that enter the general circulation. Different types of hormone-producing cells in the pituitary each produce a specific hormone in response to the equally specific hormones released by neurons in the hypothalamus. The eight different hormones produced by the pituitary broadly control how the body functions.

The thyroid gland, for example, controls the rate at which the body releases energy from food through the action of thyroid hormones (Figure 25.13). If these hormones are present in the blood at low levels—meaning that release of energy from food will go slowly—the appropriate neurons in the hypothalamus detect those low levels. In response, the hypothalamus releases a hormone called **thyrotropin-releasing hormone (TRH)** into the blood vessels connecting it with the pituitary. In response to TRH, the pituitary gland releases **thyroid-stimulating hormone (TSH)** into the general circulation. The thyroid gland responds to TSH by releasing its two thyroid hormones, which then increase the rate at which cells release energy. Higher levels of thyroid hormones in the blood shut down the production of TRH and TSH by negative feedback. Overall, the nervous system interacts closely with the endocrine system, primarily through the hypothalamus, which integrates the functions of the two systems.

Caffeine and Addiction

Why are caffeinated drinks so popular around the world? A typical American gets over 300 milligrams of caffeine from a variety of sources each day. Table 25.2 shows the amount of caffeine in selected beverages and different kinds of chocolates. It isn't hard to dose yourself with 300 milligrams of caffeine.

Most of us understand that caffeine increases our alertness: a cup of coffee or a can of cola helps us stay awake and seems to give us more energy. A close look at how the body responds to caffeine reveals several additional effects: the heart rate increases, blood pressure rises, more blood flows to the muscles and less to the digestive system, the airway opens up to make it easier to breathe, and the pleasure center in the brain is stimulated to make us feel good. It should be clear that whatever caffeine does, it has a powerful effect on how the body works.

Caffeine works by affecting the levels of three substances in the body: adenosine, epinephrine, and norepinephrine. Normally, the release of adenosine into a brain synapse inhibits neuronal activity, causing drowsiness. Adenosine also causes muscle tissue in the blood vessels supplying the brain to relax, increasing their blood-carrying capacity. Caffeine interferes with adenosine's action by binding to (and thus blocking) the receptors for this neurotransmitter at synapses in the brain. The neurons with the blocked receptors thus continue to produce action potentials, causing us to feel alert. Blocking the adenosine receptors also reduces the blood supply to the brain.

The increased activity of neurons in the brain has a second, indirect effect: it sends a signal to the adrenal glands. In response, the adrenal glands release epinephrine and norepinephrine, each of which act can act as a neurotransmitter as well as a hormone. These neurotransmitters/hormones travel through the circulatory system to various target tissues, triggering a series of responses collectively known as the **fight-or-flight response**: they increase heart rate, blood pressure, blood flow to the brain, and energy availability—all of which contribute to increased activity throughout the body.

Caffeine also increases levels of dopamine, a neurotransmitter that excites brain "pleasure centers." This effect may be related to the observation that regularly ingesting caffeine (whether through drinks or through solid food such as chocolate bars) can lead to addiction. In laboratory studies, rats that could push a lever to inject dopamine into the region of the brain associated with the perception of pleasure would push the lever until the dosage killed them. Human addiction to drugs such as methamphetamine and cocaine may involve a similar mechanism.

In addition to caffeine's relationship to addiction, our bodies may come to need caffeine to achieve a level of alertness to which we have become accustomed. Skipping a morning coffee can, in these cases, lead to irritability and headaches. The

Table 25.2

Caffeine Content of Some Selected Items

Item	Typical content (mg)	Range (mg)
Coffee (8 oz. cup)		
Brewed, drip method	100	60–180
Instant	65	30–120
Decaffeinated	3	1–5
Tea (8 oz. cup)		
Brewed, major U.S. brands	40	20–90
Brewed, imported brands	60	25–110
Instant	28	24–31
Iced	25	9–50
Some soft drinks (8 oz.)	24	20–40
Cocoa (8 oz.)	6	3–32
Chocolate milk (8 oz.)	5	2–7
Milk chocolate (1 oz.)	6	1–15
Dark chocolate, semi-sweet (1 oz.)	20	5–35
Baker's chocolate (1 oz.)	26	26
Chocolate-flavored syrup (1 oz.)	4	4

headaches arise because regular caffeine intake forces the body to become more sensitive to adenosine in order to maintain the blood supply to the brain. In other words, the blood supply increases in response to smaller-than-normal amounts of adenosine. When we do not get our daily "dose" of caffeine, the blood vessels in the brain relax more than they should because there is no caffeine to balance the increased sensitivity to adenosine. The result is a pounding headache—and the cure is another fix of caffeine.

Chapter Review

Summary

25.1 Neurons, Nerves, and More: Basic Features

- Animals have unique cells called neurons that rapidly carry electrical signals along specialized extensions called dendrites and axons.
- Neurons interact with other neurons, with muscle cells, and with endocrine cells.
- Most axons are sheathed in myelin. Axons are bundled together with supporting tissues into nerves.
- Ganglia are clusters of neuronal cell bodies that act as integrating centers. At the head end of some simple animals is a ganglion for integrating information for the whole body. Vertebrates have a more complex structure, called a brain. Important sensory organs are near the brain, which saves valuable time getting electrical signals to it.
- Vertebrates have a specialized spinal cord that runs from the brain to the "tail" end of the body.

25.2 Neurons Transmit Rapid Electrical Signals

- Electrical signals are transmitted by the nervous system in response to stimuli, either from the environment or from within the body. The signals carried by neurons are called action potentials.
- When not generating an action potential, an axon has a resting potential because it is polarized (with a more negative charge on the inside of the plasma membrane than the outside).
- If an electrical signal arriving from other parts of the neuron is strong enough to depolarize the axon beyond a threshold level, it can initiate a brief reversal of charge—an action potential. Once the action potential starts, it continues automatically down the length of the axon: a brief charge reversal in one part of the axon depolarizes the adjacent "downstream" section of the axon in turn.
- In myelinated axons, action potentials leap from one node of Ranvier (a gap in the myelin sheath) to the next, allowing signals to move more rapidly down the axon than they can along the plasma membranes of unmyelinated axons.
- Action potentials have several important features: (1) they move down the axon in only one direction; (2) the strength of charge remains consistent from one end of the axon to the other; and (3) the strength of the original stimulus is communicated through the nervous system by variations in the frequency, not the strength, of action potentials.
- Neurons transmit information to other cells at synapses. An action potential arriving at an axon terminal causes the release of neurotransmitter molecules, which diffuse across the narrow synaptic cleft almost instantly to trigger a response in the target cell.

25.3 Making Sense of Action Potentials in the Body as a Whole

- The vertebrate nervous system is divided into the central nervous system (CNS), consisting of the brain and spinal cord, and the peripheral nervous system (PNS), consisting of the sense organs and all the remaining nerves.
- Sensory neurons carry sensory input to the CNS, interneurons in the CNS process sensory input, and motor neurons carry command signals from the CNS back to the PNS for an appropriate response. Interneurons also direct information to the brain and receive information from it.
- Reflex arcs process sensory input from the PNS without input from the brain. Reflex arcs give the body a much quicker response to danger than if processing by the brain were required.

25.4 Organization of the Human Brain

- The human brain has three major regions: forebrain, midbrain, and hindbrain. The midbrain and hindbrain together are called the brain stem.
- The major components of the forebrain are the thalamus, the cerebrum, and the hypothalamus.
- The thalamus filters incoming signals and sends those that need further processing to the appropriate part of the brain.
- The most conspicuous portion of the human brain is the cerebrum. The cerebral cortex, containing billions of neurons with many billions of connections, accounts for 80 percent of human brain mass.
- The hypothalamus coordinates many endocrine functions; it is the center for regulating homeostasis, for emotions of rage and pleasure, and for regulating cycles of sleep and wakefulness.
- The midbrain helps to maintain muscle tone and relays some sensory input to higher brain centers.
- The hindbrain coordinates information dealing with breathing rhythms and balance, among many other functions.
- Imaging technologies, including magnetic resonance imaging (MRI) and positron emission tomography (PET), make it possible to "see" what brains are doing while a person is performing various activities.

25.5 Why Do Humans Have Big Brains?

- Modern humans have large brains relative to our body size.
- During human evolution, the neocortex increased disproportionately in size. This increase occurred long after our humanlike ancestors became bipedal.

- Our large neocortex gives us the ability to think complex thoughts, use sophisticated language, and invent tools and weapons. A large neocortex may have proved especially useful to humans who lived and hunted in social groups and who were beginning to develop language and technology.
- The human brain is unlikely to become larger. Maintaining and using our large brains is energetically expensive, and an increased head size would make giving birth even more difficult.
- In humans, a larger proportion of the brain is devoted to the hands (and their precision grip) and to language than in chimpanzees, our nearest primate relatives. The left and right sides of the human brain are specialized for different tasks, whereas chimpanzee brains do not show this specialization.

25.6 Integration of the Nervous and Endocrine Systems

- The nervous system works closely with the endocrine system, primarily through the hypothalamus, to control how the body functions. Hormones from the hypothalamus inhibit or stimulate specific cells in the pituitary gland, which in turn produce hormones that enter the general circulation. These hormones control most aspects of the endocrine system.
- The thyroid gland produces hormones that control the speed at which the body releases energy from food. Low levels of these hormones trigger the hypothalamus to produce thyroid-releasing hormone (TRH). TRH causes the pituitary to release thyroid-stimulating hormone (TSH), which causes the thyroid to produce more thyroid hormones.

◉ Review and Application of Key Concepts

1. Describe the movement of action potentials down the axon of one neuron and across a synapse to the next neuron.

2. A simple reflex arc can coordinate an appropriate response to a stimulus such as heat with just two or three neurons. If the incoming information says "too hot," then the instruction "move hand" is sent out. If no "too hot" message is sent, then a "move hand" instruction does not get sent out. Using four neurons, including one interneuron, design a nervous system that could decide to send the message "move" or "stay still" in response to two different stimuli that do not necessarily occur at the same time.

3. Our endocrine and nervous systems work together. Compare the roles of the endocrine and nervous systems in internal communication.

4. Why has human brain size increased rapidly over the past 2 million years? How has this increase in brain size affected what humans can do?

Key Terms

action potential (p. 478)
axon (p. 476)
brain (p. 476)
brain stem (p. 486)
central nervous system (CNS) (p. 481)
cerebral cortex (p. 485)
cerebrum (p. 485)
dendrite (p. 476)
depolarized (p. 478)
fight-or-flight response (p. 491)
forebrain (p. 485)
ganglion (p. 476)
hindbrain (p. 486)
interneuron (p. 482)
magnetic resonance imaging (MRI) (p. 486)
midbrain (p. 486)
motor neuron (p. 483)
myelin (p. 476)
neocortex (p. 486)
nerve (p. 476)
nervous system (p. 476)

neuron (p. 476)
neurotransmitter (p. 480)
node of Ranvier (p. 478)
peripheral nervous system (PNS) (p. 482)
polarized (p. 478)
positron emission tomography (PET) (p. 488)
reflex arc (p. 484)
resting potential (p. 478)
sensory input (p. 481)
sensory neuron (p. 482)
spinal cord (p. 477)
spinal reflex (p. 484)
stimulus (p. 478)
synapse (p. 480)
synaptic cleft (p. 480)
thalamus (p. 485)
thyroid-stimulating hormone (TSH) (p. 490)
thyrotropin-releasing hormone (TRH) (p. 490)

Self-Quiz

1. In neurons, dendrites
   a. send signals to other neurons.
   b. secrete neurotransmitters.
   c. receive signals from other neurons.
   d. secrete hormones.

2. Action potentials are relayed from one neuron to the next at
   a. motor neurons.   c. a node of Ranvier.
   b. an axon.   d. a synapse.

3. Neurotransmitters
   a. move across synaptic clefts by diffusion.
   b. can only diffuse out of axons that are myelinated.
   c. are found only in interneurons.
   d. move across synaptic clefts only by active transport.

4. In the human brain, the thalamus
   a. is a center for language use.
   b. is the central switching area.
   c. is part of the cerebral cortex.
   d. is part of the brain stem.

5. The part of the human brain that is proportionally much larger than in other primates is
   a. the thalamus.
   b. the pituitary gland.
   c. the neocortex.
   d. the midbrain.

6. The brain stem
   a. contains the cerebral cortex.
   b. contains the pons.
   c. contains the cerebrum.
   d. contains the thalamus.

7. Compared with that of chimpanzees, a larger part of the human brain is devoted to
   a. walking.
   b. hand use.
   c. language.
   d. both b and c

8. The function of the hypothalamus
   a. is to release hormones that control the function of the pituitary gland.
   b. is to release hormones that directly control the thyroid gland.
   c. was largely responsible for the increase in human brain size over evolutionary time.
   d. is to serve as a neurotransmitter.

9. In response to low levels of thyroid hormones in the blood,
   a. the hypothalamus produces thyroid-releasing hormone.
   b. the pituitary gland produces thyroid-releasing hormone.
   c. the hypothalamus produces thyroid-stimulating hormone.
   d. the thalamus produces thyroid-releasing hormone.

10. In the thyroid hormone control system, the source of negative feedback is the
    a. hypothalamus.   c. thyroid gland.
    b. pituitary gland.   d. cerebrum.

Biology in the News

Nail Biting "Damages IQ"

Children who bite their fingernails may be damaging their IQ, a study suggests.

Researchers in Russia say children who chew their nails are at greater risk of lead poisoning.

This is because lead can gather under their nails simply by playing in dusty conditions, both indoors and outdoors.

It has long been known that exposure to lead may contribute to developmental problems in some children. Previous studies have suggested it may also damage the nervous system.

The U.S. Environmental Protection Agency lists lead among the most important and widespread environmental pollutants. Lead is a soft metal that has found many uses. Until the 1980s, it was a common additive to gasoline and was an important component of paint pigments. The lead from these two sources, along with lead still being discharged into the environment from industrial sources, eventually makes its way into the soil and dust that surrounds us. In cities around the world, including those in North America, the levels of lead in the soil can be high enough to pose a real risk to our health.

Young children are particularly vulnerable to the effects of lead because they live a life that brings them into close contact with dirt. Think of typical toddlers: they spend much of the day playing on the ground, where soil and dust get on their skin and under their fingernails. In addition, they rarely wash their hands and often place their fingers in their mouths. Thus lead can easily make its way into a child's body.

Children are also particularly vulnerable to lead because their brains and nervous systems are still developing. Lead can affect the developing nervous system at very low doses. When lead gets into the blood of children under school age at levels as low as 50 micrograms (about the weight of a single grain of table salt) of lead per liter of blood (a 5-year-old child has about 2 liters of blood), it can lead to reduced scores on intelligence tests. Children with negligible lead levels in their blood score 7 points higher on standard IQ tests than children with 100 micrograms of lead per liter of blood. Moreover, the evidence suggests that these effects on intelligence persist after brain development stops.

Regulations introduced since the 1970s restrict the use of lead in gasoline and in indoor paints in North American and European countries, including Russia. These regulations have dramatically reduced the levels of lead in the environment, resulting in a steady decrease in the amount of lead contaminating our bodies. However, many countries have yet to control the lead content of either gas or paint, and lead products continue to be sold there even though alternatives have been found. Because lead does not break down in the environment, even people in countries where it is regulated continue to be exposed to lead from lead-based paints applied to houses built before 1970 and to lead from gasoline that has accumulated in roadside and urban soils.

Evaluating the news

1. Why might children, especially young children, be at particular risk from lead in their environment? Think about how their size and behavior (other than biting their nails) may play a role.

2. What do you think about the continued use of leaded products by nations less affluent than those of North America and Europe? Why would companies there continue to sell such products in spite of known risks and proven alternatives? Are any of the reasons you come up with defensible ones?

3. Mercury is another dangerous metal that has effects similar to lead. Imagine you are a reporter; spend some time on the Internet or at your local library investigating the dangers mercury may pose to people in the United States. What kinds of situations pose a threat from mercury, and what do you think should be done about them?

SOURCE: *BBC News*, March 3, 2003.

Key Concepts

- Three basic types of receptors respond to stimuli in the environment to allow us to hear, see, taste, smell, and feel. All receptors respond to stimuli by generating action potentials that carry sensory information back to the brain.

- Individual receptors for smell and taste are specialized to respond to one chemical (smell) or one class of chemicals (taste).

- Mechanoreceptors respond to physical changes in the environment. They underlie touch and hearing as well as other kinds of perception, such as balance, body position, temperature, and pain.

- Photoreceptors respond to light and underlie vision. Lenses and retinas in our eyes allow us to perceive detailed and focused images.

- The brain plays a critical role in sensory perception. It integrates information from the different senses and, as much as possible, resolves conflicts between the different senses to provide a consistent sensory experience.

Eyes as Cameras

Vision provides our quickest access to the world. Other senses, such as taste, smell, touch, and hearing, all inform us about our surroundings, but vision is the only sense that receives information traveling at light speed. For example, we can see a flash of distant lightening immediately, but the sound of thunder takes a few seconds to reach us. When someone brings a rose into a room, it is visible to us before the diffusing molecules of its fragrance reach our noses. When broadcasting a description, the police tell us what the suspect looks like, not how the person smells or what the person's voice sounds like. Product designers make sure their packaging emphasizes color and shape, even when we buy the products inside for their taste or smell. We make this assumption about vision ourselves: we choose to remember our travels mostly by looking at images we have created and saved, such as photos and "home" videos.

The human eye does a remarkable job of gathering light. In a general way, our eyes work like cameras to record images. A photograph represents a record of the pattern of light that hits light-sensitive chemicals that coat a piece of film. Digital cameras use an electronic sensor in place of film to record the pattern of light. Simply holding a piece of film up to an object would not result in a picture. Instead we would simply get a piece of film that was uniformly overexposed. The function of a camera is to carefully control how the light hits the film so that a recognizable image forms.

The camera accomplishes this control in two ways. First, the light that enters the camera is focused on the film by a lens. The lens bends light in such a way that all of the light coming from a single point of the object being photographed hits a single point on the film. Without a way of focusing, light reflecting from a particular point on the object would spread over the entire

We See Lightning Before We Hear the Sound of Thunder
This photograph was taken with a camera that has many features in common with the human eye.

497

area of the film. In a camera, the lens can move closer to and farther from the film to bring into focus the light reflecting from objects close to and far from the camera. Second, the camera controls the amount of light reaching the film. It accomplishes this in two ways: First, we can limit the light entering the camera by controlling the amount of time the shutter is open. Second, we can vary the size of the opening, or aperture, through which light enters the camera; the larger the aperture, the more light reaches the film in a given amount of time.

All the important features of a camera are also found in the human eye. The "film" in the human eye is the retina, a sheet of light-sensitive cells that covers the back of the eye. The eye also has a lens that bends light to bring it into focus on the retina. Like a camera lens, the human lens can adjust its focus to accommodate both nearby and distant objects. Unlike the camera lens, however, the eye adjusts its focus by changing the shape of the lens, rather than by changing the distance between the lens and the retina. The pupil of the eye, like the aperture in a camera, increases or decreases in diameter to let in more or less light. You can see the pupil change size if you shine a flashlight on a friend's eye.

Inventing the camera was no mean feat. How could such remarkable structures as eyes have evolved?

Animals must have ways of gathering information about their environment. Perceiving such things as a potential mate, the time of day, the temperature, or the presence of enemies or food makes it possible for animals to coordinate their behavior effectively. We are able to sense our world in great detail. Simply standing at a street corner, for example, brings most of our senses into play. An attractive person or an unusual scene might "catch our eye." We can feel a breeze, smell exhaust from cars and buses, and hear traffic passing, horns blowing, radios blaring. We know whether or not it's a hot day. If we're eating a snack at the same time, we know how it tastes. And we "know" how to keep our balance while standing there.

The vast numbers of cells devoted to sensing our environment continually generate an equally vast amount of information, which is encoded as action potentials passed by sensory neurons to the brain. Additional processing in the brain allows us to make sense out of this amazing volume of information so that we know where we are, what we are doing, and, we hope, what to do next.

In the previous chapter we explored the nervous system, which carries all of this information. Now we explore what happens at the "receiving ends" of the nervous system: how sensory input is generated in response to different kinds of environmental stimuli (in this chapter) and how the body makes its responses (in the next chapter). In this chapter we introduce the three most widespread types of sensory receptors found in animals—chemical, mechanical, and visual—with an emphasis on how these receptors work in humans to produce the familiar "five senses" and some other kinds of perception that are critical, but not usually part of our consciousness. We don't consciously "think" about our balance, for example, but it, too, depends on sensory structures. After considering the role of the brain in processing sensory information, we return to the question raised at the opening of this chapter: how has the sophisticated human eye evolved?

26.1 Sensory Structures: Making Sense of the Environment

Today, computers are everywhere, and we may think of ourselves as living in the Information Age. But animals have always lived in an information age: the surroundings of any animal contain a great diversity of information, and animals are well equipped to process it. We, like all animals, are constantly bombarded with information in the form of stimuli that can be converted into images, sounds, smells, tastes, and various perceptions of touch. And, like all animals, we need to sense our "inner environment" as well, to check on the status of various functions in our bodies.

There are three basic types of sensory receptors

We are taught early on that humans have five basic senses: smell, taste, vision, hearing, and touch (see the box on page 500 for an example of our senses in action in the workplace). Yet only three distinct classes of sensory cells provide the sensory input for all our senses (including some additional senses less familiar to us, which we shall touch on later in this chapter):

1. **Chemoreceptors** [*KEE*-moh-reh-*SEP*-ters] respond to chemicals. Chemoreceptors are found in two senses: receptors on the tongue give us our sense of taste, while those in the nasal passages give us our sense of smell. Note that taste involves sensing chemicals in direct contact with our bodies (food in the mouth), while smell involves sensing chemicals from a distant source.

2. **Mechanoreceptors** [*MEK*-uh-noh-re-*SEP*-ters] respond to various kinds of physical stimuli. Mechanoreceptors are responsible not only for sensing physical changes directly affecting the surface and interior of the body (our senses of touch, pain, temperature, body position, and balance), but also for responding to physical changes happening at a distance (hearing).

3. **Photoreceptors** respond to light. Only one sense (vision) uses photoreceptors.

In addition to the senses humans have, many animals respond to stimuli that we cannot sense. Some animals, for example, can detect Earth's magnetic field. (Though humans cannot detect magnetic fields directly, our large neocortex has allowed us to detect them by inventing instruments such as the compass.) Animals with magnetoreceptors, such as certain migratory birds, use their geomagnetic sense to navigate, much as we use a compass.

Many aquatic animals can detect electrical fields. Although we can certainly detect and respond to large jolts of electricity, animals with electroreceptors are able to respond to tiny electrical currents generated by the muscular contractions of other animals. Sharks, rays, and salmon are among the organisms that have the ability to sense electrical fields. Sharks may use this sense to detect prey that are not visible, such as flounder buried just under the sand on the ocean bottom. Table 26.1 summarizes the various senses and their receptors.

Sensory structures share certain features

As we examine the details of how the different senses work, it will be helpful to keep in mind three general observations:

1. *All sensory organs and other structures convert some property of the environment into action potentials.* This observation applies primarily to the external environment, but some senses also respond to the "inner environment" of the body's interior. Receptor cells have synapses with the dendrites of sensory neurons (or, in some cases, are sensory neurons themselves). The sensory neurons, in turn, carry sensory input as action potentials back to the central nervous system for processing—in the spinal cord, the brain, or both. If appropriate, responses are carried by motor neurons from the CNS to the peripheral nervous system to cause reactions. Most simple receptors consist of a specialized receptor cell attached to a sensory neuron whose axon leads back to the CNS. In some senses there is no specialized receptor cell, only sensory nerve endings. More complex sensory structures contain groups of receptor cells and supporting cells, all interconnected by intermediate neurons that pass their signals to sensory neurons.

2. *The more receptor cells an animal has that are sensitive to a particular stimulus, the greater the amount of detail*

Table 26.1

Different Ways to Sense the World

The sensory receptors listed here are found in many animals, including most vertebrates. Those marked with an asterisk are not found in humans.

Receptor type	Stimulus	Sense
Photoreceptors	Light	Vision
Chemoreceptors	Chemicals	Taste
		Smell
Mechanoreceptors	Physical changes	Touch
		Hearing
		Proprioception (body position)
		Balance
		Thermoreception (gradations of heat and cold)
		Pain
*Electroreceptors	Electrical fields (especially those generated by muscle contractions of other animals)	*Electrical sense
*Magnetoreceptors	Magnetic fields (especially of Earth)	*Magnetic sense

A ray

Biology on the Job

Sensitive Dentistry

Dr. Jenna Bagley began her dental practice in Greenfield, Massachusetts, in 1997 after graduating from Tufts University's School of Dental Medicine. She joined the practice established in 1973 by her father, Dr. Peter Bagley.

Dentistry must involve all five senses, but do you use one sense more than the others? I'd say that vision is the most important, for a couple of reasons. For one thing, I have to be able to see a tooth before I can work on it. And I have to be tuned in to looking for fine detail.

Since your work requires seeing things in great detail, do you use any special equipment to enhance your vision? Yes. Because I need to be able see small details, I wear something called loupes. These are essentially eyeglasses that magnify what I'm looking at by two and one-half times. The loupes make a tooth look about as big as a thumb, and when I'm using them, I can see much more detail.

How about lighting? Lighting is very important. I have a lamp on an extension arm that I can move into position to illuminate the patient's mouth. I have another headlamp attached to my loupes, which allows me to focus even more light on the small area that I'm working on.

Vision Is Important in Dentistry
Dentists often wear loupes, which allow them to see a patient's teeth in much greater detail, and have lamps for further illumination.

Do you use vision in other ways? I need to evaluate X-rays, though I don't need color vision to do that. In X-rays, I look for subtle shades of black, gray, and white. These can tell me a lot, especially about places I can't see when looking in someone's mouth with my own eyes, such as between teeth.

Does your sense of smell contribute to evaluating problems? Bad breath isn't a pleasant subject for most people, but for dentists, the odors in breath can provide valuable diagnostic clues. Certain types of infection have their own recognizable odor. For example, periodontal disease—disease of the gums—is associated with a distinct odor.

How does hearing play a role in your practice? The most important role of hearing is pretty obvious: I listen carefully to patients and make sure I attend to their descriptions of what is bothering them. I need to ask questions and hear answers. But there is one sound that I listen to that isn't a patient's voice. If I'm going to use a high-speed drill, I run it for a moment before I use it. If the sound's not right, I can tell it's not running at the proper speed.

How do you use your sense of touch in your work? In several ways. I palpate the soft tissues of a patient's mouth, cheeks, and tongue to check for possible problems. I use touch to determine whether a tooth is abnormally mobile in its socket. Healthy teeth can be moved only slightly. In addition, when working on a patient, I need to position my fingers very precisely in order to do a procedure properly. These "finger rest" positions require a lot of practice in dental school.

In the dentist's office, people are often asked if their teeth are overly sensitive to hot or cold. What do these sensations mean to you? Sensitivity to cold can indicate a cavity or gum recession that leaves a root exposed. On the other hand, sensitivity to heat may mean that there is an inflammation inside a tooth or impending nerve death in a tooth.

A concern of patients is to avoid experiencing one of the senses, pain. How did you learn to administer local anesthetics? In dental school, students practice on each other after extensive training. I had the advantage of having a father who, as a dentist, had years of experience. He helped me hone my skills by having me practice on him. We also had to learn in school about how nerves work and how local anesthetics function as sensory blockers.

Some dentists use gas as an anesthetic. Do you use nitrous oxide in your practice? I don't, but many dentists do. The gas acts to reduce patient anxiety. Children are often very anxious when they visit the dentist, so pediatric dentists commonly use it. Those who do are keenly aware that they, and the assistants in the room with them, need to monitor its administration at all times. A patient is breathing a mixture of oxygen and nitrous oxide that is very different from the air around us, so the balance of gases must be very carefully maintained. In Massachusetts we have to have special training and pass an examination before we can be licensed to use it.

Overall, then, you use your senses in many highly focused and specific ways in your practice. Yes. I believe dental school and experience teach us to apply our senses to make good judgments and refine skills that end up helping people lead healthier lives.

Dr. Bagley spent 4 years of intense study in dental school, a good part of which included building new and refined relationships between her senses and the skills required by her profession. Her vision and sense of touch play a large role in the precise manipulations she must perform while working on teeth. When she examines X-rays, her eyes send information to her brain, which interprets that information and guides her diagnosis and treatment. Her sense of smell also has diagnostic value, and without hearing she couldn't listen to her patients' descriptions of problems and concerns. Good senses are essential to the quality of the care she provides.

the animal is capable of detecting in that feature of the environment. Similarly, the more densely packed receptors are in a particular sensory structure, the more information can be sent to the CNS for processing. Bloodhounds, for example, have about 200 million chemoreceptors in their nasal passages, at least 40 times more than humans, giving them an exquisitely sensitive sense of smell.

3. *Animals use their senses to tell them about the condition of their surroundings and to locate the sources of distant stimuli.* Some senses rely on paired sensory organs that make it possible for an animal to locate the source of a distant stimulus. The two sensory organs, one on each side of the body, receive slightly different sets of sensory input, which the brain can integrate and analyze. Humans, like all vertebrates, have paired organs for vision and hearing. (Though smell also involves distant stimuli, we do not locate the sources of smells with paired organs.)

26.2 Chemoreceptors: Smell and Taste

The ability to sense the presence of chemicals allows animals to find and identify important features of their environment. Animals use chemical stimuli to identify food, detect predators, and find mates. Humans and other terrestrial animals have distinct senses of smell and taste. Smell involves detecting airborne chemicals that often reach the receptor cells at lower concentrations than do the chemicals perceived in tasting. Taste involves detecting chemicals at relatively high concentrations that are dissolved in water or in the moist mixture of food being chewed. For aquatic creatures, smell and taste are virtually the same, since all chemicals reach aquatic animals dissolved in water.

Chemicals in the environment can provide remarkably detailed information. A mother bat, for example, uses her sense of smell to identify the unique combination of chemicals coming from her baby among the million or so other bats that may roost in the same cave. Male silkworm moths can detect bombykol, a chemical produced by female silkworm moths as a way of attracting mates, at concentrations so low as to be equivalent to humans being able to smell a barbecue 20 miles away. We may not be able to smell things this far away, but unlike moths, we can appreciate an enormous range of odors and flavors. Just look at the selection of flavors available in an ice cream store or the choices of pizza on a menu.

Chemoreceptors are found on parts of the body most likely to encounter chemicals

Chemoreceptors may be modified neurons or specialized sensory cells associated closely with neurons. Although diverse in form, all chemoreceptors trigger action potentials in response to the presence of chemicals. Different receptor proteins in the plasma membranes of chemoreceptor cells bind to different chemicals, much as the receptor proteins of target cells bind specifically to certain hormones. When a chemical binds to its matching receptor protein, the chemoreceptor cell depolarizes, setting in motion an action potential that will be passed on to the brain.

Chemoreceptor cells are concentrated in those parts of animals most likely to encounter chemicals. In humans, for example, smell receptors crowd the lining of the nose, through which the air we breathe moves in and out, and taste receptors fill some thousands of taste buds arranged on the surface of the tongue, over which all our food passes. Male silkworm moths have about 50,000 receptors specific to bombykol concentrated on paired feathery antennae that sift the air (Figure 26.1a), and female silkworm moths have taste sensors on their legs that allow them to taste plant leaves as they search for the right place to lay their eggs. For many fish, chemoreception is a whole-body experience, thanks to dense arrays

(a)

(b)

Figure 26.1 Chemoreception Can Be a Heady or a Whole-Body Experience

(a) Male silkworm moths can detect as little as one molecule of bombykol with their feathery antennae. (b) Catfish have chemoreceptors not only inside their mouths and nostrils, but also scattered over the body surface, and especially on the whiskerlike barbels around the mouth. For fish, as for all aquatic animals, there is no distinction between taste and smell.

of chemoreceptors in their paired nostrils, in their mouths, and spread over the surface of their skin (Figure 26.1*b*).

Smell allows animals to recognize and locate distant chemicals

When we smell, receptor cells lining the nasal cavity respond to chemicals that have been released by some substance in our environment (Figure 26.2). For example, even with our eyes closed, we know when there's a barbecue somewhere nearby. Chemicals released from

the grilling food move through the air and into our noses as we breathe in. The nasal receptor cells have hairlike projections extending into the mucus that lines the nasal passage, waiting for chemicals to come along. Molecules of various chemicals get trapped in the mucus and bind to receptor proteins in the hairs. The binding triggers the chemoreceptor cells to produce action potentials, which their axons carry through an interior bone of the skull to the **olfactory bulbs** in the brain, where they synapse with sensory neurons. Here some sorting of smells takes place before additional action potentials travel by other sensory neurons to other parts of the brain for further processing.

We are able to distinguish different odors because each kind of molecule from the environment binds with a chemoreceptor specialized to respond to that chemical. The overall pattern of action potentials from the chemoreceptors in the nose tells us that what we smell is grilling food. The almost countless action potentials processed by the brain allow us to identify the food as fish, as meat slathered in barbecue sauce, as grilled vegetables, or as some morsel burnt beyond recognition.

We can also use the chemicals released from the grilling food to locate the source of the smell, even if we cannot see the source before we begin our search. We can locate the smell by following our nose to stronger and stronger aromas. The concentration of molecules coming from the barbeque increases as we get closer to the source. Increasing the stimulus strength—in this case,

4 Neurons in the olfactory bulb send action potentials to other parts of the brain for more processing.

3 Receptor axons send action potentials through the bone separating the nasal cavity from the brain to the olfactory bulb.

2 Hairlike structures on the ends of chemoreceptor cells detect specific molecules.

1 A layer of mucus traps odor molecules entering the nasal cavity.

Olfactory bulb

Olfactory bulb in the brain

Nasal cavity

Bone

Receptor cells

Epithelium

Figure 26.2 Smell in Humans

the concentrations of certain chemicals in the air—causes the same chemoreceptors to produce action potentials more often.

Taste allows animals to sort safe from unsafe foods

Humans have about 4,000 oval **taste buds** recessed into the surface of the tongue. Each taste bud is lined with chemoreceptor cells associated with sensory neurons that synapse onto them. When particular molecules bind to the chemoreceptor cells, the associated sensory neurons are triggered to send action potentials back to the brain (Figure 26.3).

Taste buds are specialized for detecting four "flavors": "salty" taste buds respond to table salt (NaCl) and other salts, "sweet" buds respond to glucose and other carbohydrates, "bitter" buds respond to plant alkaloids, and "sour" buds to acids. In insects, the taste receptors for these same four classes are responsive only to molecules in those classes, but in vertebrates these categories are less rigid, as each kind of taste bud also responds weakly to some of the other kinds of chemicals. Recently, researchers have suggested a fifth category, called umami [oo-*MAH*-mee], which is a perception of "meatiness" in response to glutamate, but there is disagreement on whether this is a distinct class of taste buds or not. These different taste bud classes probably evolved to help animals determine which foods are safe to eat and which are not. For example, "good" foods generally contain carbohydrates or salts, while many poisons are alkaloids.

Some chemicals "fool" the chemoreceptors, leading us to taste something that is not what we think it is. The sweetener saccharin, for example, binds to sugar-specific receptor proteins to make us sense a sweet taste. It is several hundred times sweeter than sugar, but chemically, it is not related to sugar or other carbohydrates. Toothpaste and artichokes, on the other hand, contain chemicals that block the functioning of some taste receptors, leading to distorted tastes—try drinking some orange juice right after brushing your teeth.

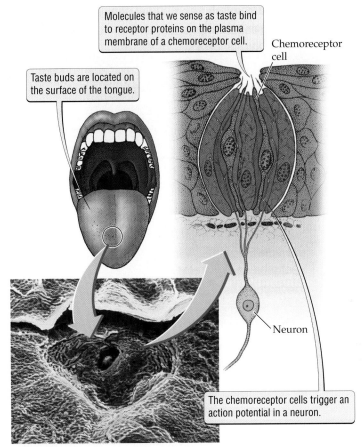

Molecules that we sense as taste bind to receptor proteins on the plasma membrane of a chemoreceptor cell.

Taste buds are located on the surface of the tongue.

Chemoreceptor cell

The chemoreceptor cells trigger an action potential in a neuron.

Neuron

A taste bud

Figure 26.3 Taste in Humans
Taste buds consist of groups of chemoreceptor cells and supporting cells. The receptor cells respond to chemicals that lodge in the tongue during eating.

26.3 Mechanoreceptors: Touch, Hearing, and Other Senses

The sense of touch informs us about our immediate environment: Is a surface smooth or rough? Does someone have a firm handshake? Is something crawling on your skin, or hurting you? Our sense of hot and cold is delivered by mechanoreceptors. Mechanoreceptors also inform us about the positioning of our bodies: Are you standing or sitting? In what position is your arm? Which way is up? How fast are you moving?

Touch tells us about our immediate internal and external environment

Many different kinds of "touch" mechanoreceptors respond to different mechanical stimuli (Figure 26.4). Like chemoreceptors, mechanoreceptors include both sensory neurons and specialized cells associated with them. For example, some receptor cells that are sensitive to deep pressure depolarize when their plasma membrane is stretched. Others respond to slight pressure on the surface of the skin. Neurons specialized to detect slight pressures are often associated with body hairs.

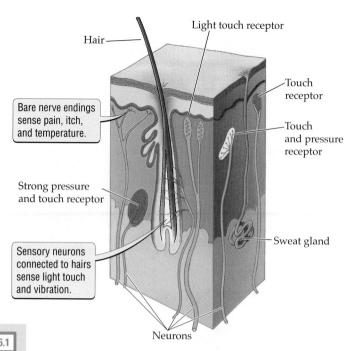

Figure 26.4 **Some Mechanoreceptors in Human Skin**
Human skin contains many different types of neurons and sensory cells that provide detailed information about mechanical stimuli.

Hair

Light touch receptor

Bare nerve endings sense pain, itch, and temperature.

Touch receptor

Touch and pressure receptor

Strong pressure and touch receptor

Sensory neurons connected to hairs sense light touch and vibration.

Sweat gland

Neurons

Learn more about receptors in the skin.

26.1

When a hair moves, it sends a mechanical message to a sensory neuron, and the mechanical energy in the movement is converted into an action potential. We feel a breeze largely due to the movement of hairs on our bodies.

The number of mechanoreceptors varies over the surface of an animal. Human fingers have more mechanoreceptors than most other parts of the body surface. This makes sense, since we use our fingers to manipulate objects in our environment, often with great delicacy. The large number of mechanoreceptors in human hands matches the relatively large portion of the human brain devoted to handling information from the hands, as described in the preceding chapter.

Hearing is stimulated by rapid pressure changes in the environment

The sounds we hear are the result of vibrations and collisions of molecules in our external environment. We call these disturbances sound waves. Sound waves can travel through any kind of matter—solid (such as the soil under your feet or the bones in your head), liquid (water), or gas (air). Our discussion here will focus primarily on the perception of sounds traveling through air. Sound waves spread out in all directions from the sound source, much as ripples on the surface of a still pond spread out from a fallen raindrop (Figure 26.5a). Water waves, however move along a surface, in two dimensions, while sound

waves move outward in all three dimensions. You can hear a drum whether you are over it, under it, behind it, or in front of it.

Hearing involves detecting waves of pressure change in air. For example, when your hands collide in a single hand clap, you hear a "smacking" sound. Before you can perceive the sound, however, several events occur. First, the air near the collision is compressed, producing a wave of compressed air that travels outward in all directions at the speed of sound (about 756 miles an hour, or 1,209 kilometers an hour). The pressure is greater within the wave of compressed air generated by the clap than it is in the air just in front of it and just behind it. When the wave of higher-pressure air enters your ears, it is transmitted to mechanoreceptors that transform it into a pattern of action potentials conveyed by neurons to your brain; then, after some processing in the brain, you "hear" the clap. How about continuous sound? When we listen to a guitar being played, we are receiving a series of pressure waves, thousands per second, which are created as the vibrating strings compress air (Figure 26.5b). Sound, then, consists of pressure changes that can occur singly (as with a clap) or many times a second (as with the strings of a guitar).

Humans can distinguish two aspects of sound: pitch and loudness. In the case of a continuing source of sound,

(a)

(b)

Waves of air pressure

Higher pressure

Lower pressure

Air pressure

Normal atmospheric pressure

Time ⟶

Figure 26.5 **Sound Is Composed of Pressure Waves**
(a) Waves on a pond's surface from a raindrop radiate outward much like sound waves. (b) Sound caused by a vibrating object consists of rapidly changing waves of higher and lower air pressure.

the number of pressure changes per second is known as the sound's frequency, which we perceive as pitch. The higher the frequency, the higher the pitch we hear. But our ears are limited: the lowest frequencies we can hear are about 20 pressure changes per second, and the highest about 20,000. Some animals, such as elephants, can hear sounds at lower frequencies than we can, especially those that travel many miles through the ground.

The loudness of a sound reflects the intensity of the pressure changes in the sound waves. Clapping softly doesn't create as great a pressure change as does clapping loudly. The ability to detect sound depends on the fact that the mechanoreceptors in the ear are arranged so that they can respond to rapid and regular pressure changes as these changes arrive in waves. Sound waves diminish in intensity as they travel, which is why distant sounds are softer than the same sounds originating nearby.

A related characteristic of sounds is that they travel differently through different kinds of matter. It may be hard to believe, but sound waves travel best through solids, less well through liquids, and least well through gases such as air. That is why you can put an ear to the ground to pick up distant sounds that might be hard to hear through the air. That is also why your voice sounds different to you than to other people—which you probably noticed with surprise the first time you listened to your recorded voice. You hear your own voice from two different sources—sound waves traveling from your mouth to your ears through the air, plus vibrations traveling through your skull bones—while others hear your voice traveling through only one medium, the air. The sound waves traveling through your skull dissipate too quickly to make it to their ears.

Hearing is a highly refined form of mechanoreception

Animals use various ear designs to concentrate and convert waves of changing air pressure into meaningful patterns of action potentials. The human ear illustrates the essential features of many vertebrate hearing systems, in which sound waves are concentrated by an outer ear and funneled to thin, delicate membranes and mechanoreceptors (Figure 26.6a). Let's look more closely at how this works.

The **outer ear** consists of the **pinna**—the curled, cartilaginous part that we identify as "the ear"—and the **auditory canal**. The characteristic funnel shape of our outer ear collects sound over an area about 20 times that of the **eardrum**, the delicate membrane at the end of the auditory canal. The pinna thus concentrate sound energy

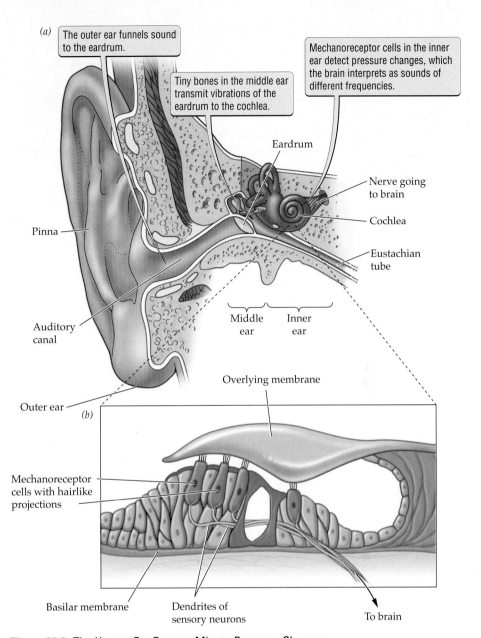

Figure 26.6 **The Human Ear Detects Minute Pressure Changes**

(a) The anatomy of the human ear allows it to convert tiny, rapid changes in air pressure into movements that mechanoreceptors can detect. The outer ear acts as a funnel to collect sound over a large area and concentrate it onto the small area of the eardrum. In the middle ear, the eardrum converts changes in air pressure into physical movements, which three tiny bones transmit to the cochlea, in the inner ear. (b) In the cochlea, patches of mechanoreceptor cells in the organ of Corti convert these movements into action potentials, which are sent to the brain for processing.

Learn more about the human ear.

on the eardrum, allowing us to detect fainter sounds than would otherwise be possible. By vibrating in response to the rapid changes in air pressure it is exposed to, the eardrum converts sound energy into physical movements. This mechanical energy passes across the eardrum and into the middle ear.

For our hearing apparatus to work properly, the air pressures between the middle and outer ears must be equal. The **eustachian tube** [you-*STAY*-shun . . .], which runs from the middle ear to the throat, allows us to "pop" our ears to equalize these pressures. The need to do this is particularly noticeable during airplane flights, which can cause us to experience pressure imbalances when the plane is ascending or descending.

In the **middle ear**, the three tiniest bones in the body receive the vibrations from the eardrum and transmit them to a second membrane, which channels them to the **cochlea** [*COKE*-lee-uh] (Latin for "snail shell"). The cochlea is a coiled, fluid-filled tube in the **inner ear**. Within the cochlea lies the **organ of Corti** [*KOR*-tee], which converts mechanical sound (sensory input) into a pattern of action potentials.

In the organ of Corti, mechanoreceptor cells with hair-like projections lie in close association with the **basilar membrane** (Figure 26.6*b*). The basilar membrane vibrates in response to vibrations passed to it from the middle ear. These vibrations push the mechanoreceptor cells up against an overlying membrane, which bends their hairlike projections. This bending triggers action potentials in individual sensory neurons that synapse directly with the mechanoreceptor cells.

We can distinguish loudness because louder sounds produce greater movement of the basilar membrane. And we can perceive pitch because different regions of the basilar membrane vibrate in response to different sound frequencies. Not all animals that can hear can distinguish sound frequencies because many lack the equivalent of the organ of Corti found in vertebrates.

Specialized mechanoreceptors tell us about body position and balance

An often-overlooked aspect of mechanical stimuli is the role they play in telling us what our own bodies are doing at any given moment. We are constantly informed of the position of our bodies in space by vast numbers of **proprioceptors** [*PROH*-pre-oh-*SEP*-ters], mechanoreceptors that respond to stretching of, or pressure on, muscles, tendons, and joints. Proprioceptors tell you that your arm is extended or bent at the elbow, that your left foot has just hit the ground, or that you are smiling.

This positional awareness is largely unconscious, yet without it, we would be completely helpless to get around independently. Proprioceptor cells in the soles of our feet, for example, play an important role in maintaining our physical balance when we are upright (standing, walking, or running). When the body shifts position, these cells detect pressure changes in the soles. As people age, however, these pressure-sensitive cells become less sen-

sitive, depriving us of some information about our posture. This is one factor contributing to the increased chances of falling among the elderly.

In addition to proprioceptors, a structure in our ears, called the **vestibular apparatus** [ves-*TIB*-you-ler . . .], helps us maintain our sense of balance (Figure 26.7*a*). The mechanoreceptors in the vestibular apparatus fall into two classes: those involved in telling up from down, and those involved in detecting head movements. One part of the vestibular apparatus—the **vestibule**—contains chambers with patches of mechanoreceptor cells. These cells have hairlike projections extending into a gelatinous layer topped by a layer of small, dense crystals of calcium carbonate called **otoliths** (from *oto*, "ear"; *lith*, "stone"). Because they are relatively dense, the otoliths tend to settle in the direction of the pull of gravity. As they settle, they push on and bend down the hairs in the direction in which gravity is pulling them (Figure 26.7*b*). The pattern of stimulation of the mechanoreceptor cells allows us to distinguish up from down even when our eyes are closed.

The vestibular apparatus also contains three fluid-filled **semicircular canals** (see Figure 26.7). The canals are oriented at right angles to one another, forming a three-dimensional grid. When you rotate or nod your head, the fluid inside the semicircular canals moves relative to the walls of the canals. Mechanoreceptors with hairlike projections inside these canals detect the direction of flow of the fluid. For example, when you shake your head from side to side, the fluid in the canal most nearly parallel to the plane of movement flows to the left and right relative to the hairs lining the wall of that canal. The characteristic pattern of action potentials from the mechanoreceptor cells lets you know that you are shaking your head.

Pain and temperature are detected by bare nerve endings

We have discussed five senses, but there are two sensations that are familiar to all of us that we haven't yet mentioned: temperature and pain (see Figure 26.4). Heat and cold are detected by a class of mechanoreceptors called **thermoreceptors**, which are located mostly in the skin. These neurons are sensitive to changes in temperature.

Pain receptors are neurons that respond to extreme heat and cold and to injuries. Pain may not be a pleasant sensation, but as a way of letting us know when we are sick or injured, it was (and still is) critical to our survival. You do not leave your hand in a hot candle flame, and keeping your feet in a bucket of ice water soon becomes

Helpful to know

Note that the second half of *proprioceptor*, unlike the names of other receptor types, is *ceptor*, not *receptor*. This variation in spelling has no significance, though it may make the word a little easier to pronounce.

painful. Our perception of "itch" comes from receptors similar to pain receptors.

26.4 Photoreceptors: Vision

At the simplest level, light detection requires only a light-sensitive pigment. If the pigment molecule changes shape in response to light, and that change in turn triggers either a signaling molecule or an action potential, then the organism can respond to light or dark. For certain organisms, including some protists and simple animals, this rudimentary form of vision is adequate. The ability to distinguish light from dark provides them with information about the time of day and how they are positioned in their local environment. If an animal had just two light-sensitive cells, one on each side of its head, it could detect the changes in light patterns as something moved in front of it. So even simple light-detecting structures are capable of providing a great deal of information.

Forming images requires a way to focus light

Most animals go well beyond merely responding to light and dark: their eyes can form images. All of the great variety of image-forming eyes found among animals have two features in common:

1. The eye gathers light and focuses it onto an array of photoreceptor cells in such a way that all of the light from a single point in the view the animal is looking at goes to a single point in the array of photoreceptors. Only in this way can the eyes produce a clear image of the view.
2. The photoreceptor cells convert light energy into action potentials that are sent by sensory neurons to the brain for processing.

Animals have developed these two features in several different ways, one of which is the **single-lens eye**. In a typical single-lens eye, a **lens** focuses incoming light point for point, and all the photoreceptors are clustered at the back of the eye in a sheet called the **retina**, which converts the light stimuli into action potentials. Vertebrates, including humans, have single-lens eyes (Figure 26.8). Let's take a look at vision in a single-lens eye, using the human eye as an example.

The formation of a sharp image on the retina requires a lens. The shape of the lens allows it to bend light so that the light coming from each particular point on an object

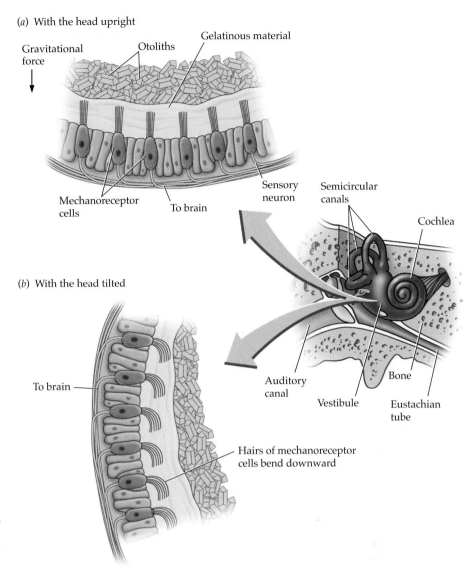

Figure 26.7 Our Ears Help Us Keep Our Balance
The vestibule of the inner ear contains otoliths, or "ear stones." (*a*) When the head is upright, even distribution of the otoliths over the gelatinous material produces no deformation of the hairs of the mechanoreceptor cells. (*b*) When the head is tilted, the otoliths tumble in response to gravity, pulling some of the gelatinous material downward. This in turn bends the hairs of the mechanoreceptor cells, triggering action potentials in sensory neurons associated with them.

is focused on a different particular point on the retina, stimulating the photoreceptors there. The result is a pattern of stimulated photoreceptors that re-creates the object. (The object appears upside down on the retina; further processing in the brain turns the image "right side up" again.) Eyes don't have to have lenses, but without them, images are weak and fuzzy (Figure 26.9*a*).

One of the most delicate tasks of a single-lens eye is adjusting the focus of the lens to accommodate objects at different distances from the eye. In order to keep an image focused on the retina, the lens must bend light

Figure 26.8 Major Features of the Human Eye

A pupil adjusts the amount of light entering through the lens. The lens focuses images on the retina, where photoreceptor cells generate action potentials in sensory neurons that are closely associated with them.

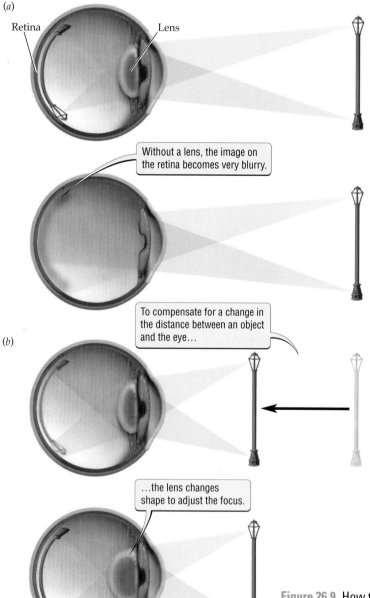

Without a lens, the image on the retina becomes very blurry.

To compensate for a change in the distance between an object and the eye...

...the lens changes shape to adjust the focus.

from nearby objects more sharply than light from distant objects. In many vertebrates, including humans, the focus is adjusted by changing the shape of the lens (Figure 26.9*b*). Accomplishing this requires intricate neuronal connections between the retina and the brain that allow the nervous system to continually monitor the quality of the image and send instructions to the muscles that control the shape of the lens.

Vertebrates, especially those living on land, encounter a wide range of light intensities. The amount of light reaching the retina needs to be regulated: too much light hitting the retina can overload the photoreceptors, while too little light makes for poor images. An adjustable opening called the **pupil** controls the amount of light entering the eye. In bright external light, the pupil becomes smaller, while in dim light, it opens wider to let more light in.

In the human retina, individual neurons carry information from one or a few photoreceptor cells to the brain. We have two classes of photoreceptors, called rods and cones, which get their names from their shapes (Figure 26.10). **Rods**, which are more numerous than cones, cannot detect color, but they do allow us to see images in shades of gray in dim light. **Cones**, on the other hand, transmit information in color, but they function best in bright light. Both rods and cones contain light-sensitive pigments. At night, we can detect shapes only in shades of black and gray because only the rods receive enough light to function.

In addition to our lenses, several other features contribute to the ability of our eyes to produce sharp, crisp

Figure 26.9 How the Human Eye Forms Sharp Images

(*a*) The lens brings the image into focus on a layer of photoreceptor cells in the retina.
(*b*) As the distance to an object changes, the eye must adjust to keep the image focused on the retina. It does this primarily by changing the shape of the lens.

Figure 26.10 Human Photoreceptors

Shown here are diagrams and a scanning electron micrograph of the photoreceptors found in the human retina. Rods (green in micrograph) detect dim light, while cones (blue) detect color in relatively bright light.

images. One is having about 100 million photoreceptors in each retina. Just as the number of pixels affects how sharp a computer image will be and the number of pigment molecules on film affects how sharp a photograph will be, the number of photoreceptors in an eye helps to determine how detailed the image can be. Furthermore, due to efficient processing by other cells in the retina, the information gathered by this huge number of photoreceptors is condensed so that "only" about a million axons are needed to carry signals from rods and cones back to the brain for further processing and integration (Figure 26.11).

Our vision depends not only on how many photoreceptors we have, but also on where they are located on the retina. The more densely packed the photoreceptors are, the crisper the image. The center of the human retina has a region called the **fovea** (see Figure 26.8), where rods and cones are most densely packed. This higher density of photoreceptors allows us to form especially sharp, colorful images of whatever is in the center of our view. For example, as you read this page, you are automatically moving your eyes so that each word you read is projected onto the fovea. To see the importance of the fovea, look at a particular word and then, without moving your eyes, try to make out surrounding words. You can read the word in the center of your view—the one projected onto your fovea—but not the words off to the sides.

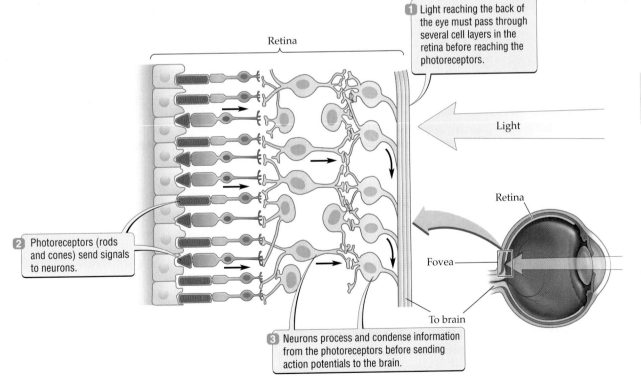

1 Light reaching the back of the eye must pass through several cell layers in the retina before reaching the photoreceptors.

2 Photoreceptors (rods and cones) send signals to neurons.

3 Neurons process and condense information from the photoreceptors before sending action potentials to the brain.

Retina

Light

Retina

Fovea

To brain

Learn more about the human eye.

26.3

Figure 26.11
The Human Retina

Rods and cones are connected to neurons in the retina that carry out some processing of visual information before sending action potentials to the brain.

Some vertebrates perceive their world three-dimensionally

Most animals capable of forming images have at least two eyes positioned on their heads. In vertebrates, how the eyes are positioned on the head often depends on how the animal "makes a living." Because most herbivores are also prey animals, their eyes tend to be situated far apart on either side of the head. This positioning gives them the widest possible panoramic view, the better to watch out for predators. In contrast, most predators have eyes situated close enough together on the front of the head to allow both eyes to focus on the same point at the same time. This positioning gives them **depth perception**, the ability to determine the distances of objects accurately in space—that is, to have a three-dimensional view of their world. Depth perception is a great advantage for chasing and successfully catching prey.

Primates (including humans) have the best depth perception of all animals. Most likely this is because our primate ancestors, like most primates today, lived in trees, where the need for depth perception was (and still is) essential to avoid falling while moving—often leaping—from branch to branch. This evolutionary perspective helps to explain why even primates that are not predators, such as gorillas, have depth perception.

Let's see how depth perception works by examining our own eyes. Each of our eyes sees a slightly different view of the world because our eyes are positioned slightly apart. You can observe how this works for yourself: without moving your head or eyes, cover first one eye, then the other, with your hand. Objects in your view will appear to shift from side to side, and objects that are nearby will seem to shift more than distant ones. In the brain, these two different views are compared and assembled into a three-dimensional image. If we lose vision in one eye, the brain can no longer compare two views of the world, so our ability to perceive three dimensions is diminished.

Some invertebrates have well-developed visual systems

Visual systems have also developed among the invertebrates. The capacities of these systems range from simply distinguishing light and dark to an ability to form images and detect motion. Some insects can detect motion far better than we can. Dragonflies can see and snatch mosquitoes (their prey) when both are flying in midair, and you probably know from experience how difficult it is to swat a fly.

One simple visual system found in many arthropods consists of **simple eyes** that can distinguish light and dark, and sometimes distance and color. Many spiders, for example, have several (usually three or four) pairs of simple eyes on their heads (Figure 26.12a). Jumping spiders, which hunt visually, can form sharp color images with the largest of their eight eyes.

Other arthropods, such as most insects, also have a pair of **compound eyes** that can form images. Honeybees, for example, use their compound eyes to see patterns of color on flowers. A compound eye consists of many individual light-receiving units, each with its own lens and small cluster of photoreceptors. These units, sometimes thousands of them, are "compounded" together into a single functioning unit (Figure 26.12b). In the brain, the combined information from all of these units forms an image. In dragonflies, the light-receiving units are densely

(a) Simple eyes on a spider

(b) Compound eyes on a fly

Figure 26.12 Arthropod Eyes
(a) Many spiders have several pairs of eyes; this zebra jumping spider (*Salticus scenicus*) has eight altogether, the rounded simple eyes on the front and side of its head that allow it to stalk its prey and pounce at the last minute. (b) Most flies have separate simple and compound eyes; this fly has small simple eyes at the top and middle of its head, and large compound eyes on the side.

clustered on certain portions of the eye, which, like the human fovea, can form especially sharp images.

One group of invertebrates—the cephalopods, including octopi and squid—has eyes that resemble the vertebrate eye in spite of having evolved completely independently. These paired eyes have retinas capable of forming detailed images and lenses to focus light. In contrast to vertebrate eyes, in which the lens focuses by changing shape, the typical cephalopod eye focuses by moving the lens away from or toward the retina (much as a camera lens focuses).

26.5 The Brain's Role in Integrating Sensory Information

Our different sensory systems do not work in isolation. Instead, sensory input from chemoreceptors, mechanoreceptors, and photoreceptors arrives at different parts of the brain, which processes and integrates this information, allowing us to make sense of the world around us (unless something interferes with our ability to integrate this information properly; see the box on page 512).

Our perception of food results from combining several senses

What does it mean to say that a particular food or dish "tastes good"? Our perception of flavor comes from integrating in the brain the sensory input from several senses, not just from the taste buds. Smell plays a huge role in our perception of food and its flavors. We are continuously smelling the food as it sits on the plate and is brought close to the nose with each bite. The act of eating itself increases the influence of smell on our perception of flavor: chewing causes some molecules from the food to become airborne, entering the nasal passages from the back of the throat. You probably know from experience that you cannot sense much flavor when you have a stuffed-up nose from a cold. Elderly people sometimes lose interest in eating, in part because the sense of smell often diminishes with age. Professional tasters (they do exist!) focus on a combination of smell and taste when evaluating a food product.

Although taste and smell are the primary senses that give food flavor, two other senses also play a role. Mechanoreceptors in the tongue and mouth send information to the brain regarding the texture of what we are chewing and swallowing. If you have ever eaten anything soon after getting dental work done and before the anesthetic has worn off, you know that not being able to feel the texture of the food you're eating makes it "taste funny."

Vision also contributes to our experience of food. For many people, eating from a plate on which the foods show a variety of colors is preferable to eating from a plate of foods that are all the same, or close to the same, color. This preference for a colorful combination of foods may have survival value, since the different colors of fruits and vegetables indicate the presence of different essential vitamins and minerals.

Information from different sensory systems sometimes conflicts

When the information coming from all of our sensory systems agrees, the body functions well, but when it does not, trouble can ensue. For example, the brain integrates information from our mechanoreceptors and from our eyes to keep track of body position. The image we see, the signals from our proprioceptors, and the information from our vestibular apparatus all inform us about how the body is oriented. If the signals from our eyes tell us something different from the information from our vestibular system, for example, we can suffer from motion sickness. If you have ever sat close to a large movie screen watching scenes that take you on a ride, as though you were in an airplane swooping into and out of valleys, turning left and right, you have experienced conflicting messages. Your eyes tell you that you are turning left or right, or swooping down into a valley, but your vestibular apparatus tells you (correctly in this case) that you are not moving at all. In some people, this apparent conflict in the signals reaching the brain leads to feelings of dizziness or nausea. Much the same happens to roughly half of all astronauts while they are weightless in space. The lack of a gravity signal from the vestibular apparatus conflicts with what their eyes tell them, leading to motion sickness.

In a slightly different way, we coordinate sensory information to determine how acceptable food is. We can see and smell pizza, and if this combination of senses says everything seems OK, then we'll eat it. But we wouldn't eat a pizza that smelled like perfume, or use perfume or cologne that smelled like pizza. Mashed potatoes colored green with food dye might taste just like white mashed potatoes, but the uncharacteristic color would make them visually unappealing. Most people prefer white mashed potatoes to green mashed potatoes, even if they know that the potatoes taste the same.

Biology Matters

Driving and Drinking: Be Aware and Beware

WHAT IS A DRINK?

A standard drink is:

- One 12-ounce bottle of beer* or wine cooler
- One 5-ounce glass of wine
- 1.5 ounces of 80-proof distilled spirits

*Different beers have different alcohol content. Malt liquor has a higher alcohol content than most other brewed beverages.

From the point of view of perceiving and processing sensory information, driving a car is no simple matter. A driver needs to react rapidly and appropriately to sensory input that is constantly changing, usually at a fast pace.

Alcohol impairs every aspect of a driver's ability to use critical senses. Eye movements cannot be well coordinated with the patterns of impulses being delivered to the brain by rods and cones. The pupils of the eyes close less rapidly in response to intense light, making the glare from oncoming headlights more blinding. The pupils can then take up to 8 seconds to reopen sufficiently; during that time, a night driver is virtually blind. Alcohol also affects depth and distance perception, so it becomes more difficult to judge how far away a tree or a pedestrian is. Reactions to sound slow down. By the time an impaired driver responds to the warning sound from a blaring horn, it may be too late to prevent a collision.

Drinking and driving is a dangerous business. Each year over 2 million college students drive while under the influence of alcohol, and over a thousand are killed in alcohol-related accidents. The danger increases dramatically with increasing amounts of alcohol in the blood.

Levels of alcohol in the blood are expressed as blood alcohol concentrations (BAC), which are simply percentages of alcohol in the blood. For example, if a man weighing 160 pounds consumes two standard drinks on an empty stomach, he will have a BAC of about 0.04 percent after 1 hour. The risk of a single-vehicle fatal accident at this level is almost 50 percent greater than at a BAC of 0 percent. At about three times that level, at a BAC of 0.12 percent, the risk increases dramatically. The risk is not just 3 times greater; it is 48 times greater. Just one more drink and the risk leaps to 380 times greater! Women are at even greater risk because females metabolize alcohol differently from males. Drinking the same amounts can lead to even higher BACs in a woman.

Knowing more about how your senses work, and what makes them work poorly, can help you beware and be aware of the consequences of drinking and driving.

The Evolution of the Eye

The human eye is so complex that it is hard to imagine how it ever evolved. A close look at the array of light-detecting structures in the animal kingdom, however, reveals that the eye did not emerge all at once in this complex form. Instead, we can find among animals a whole range of light-sensing structures that reveal the origins of many of the features that make the human eye such a remarkable structure.

The evolution of complex eyes could have begun with a primitive photoreceptor, such as a light-sensitive pigment associated with a sensory cell. Within most animal groups, including creatures such as jellyfish, lizards, and sea snakes, we find evidence of a dermal light sense, a diffuse and poorly defined sensitivity of the body surface to light. Having a dermal light sense by itself would have allowed primitive animals to tell day from night and detect whether they were shaded or exposed.

In jellyfish and many worms we find a logical next step, photoreceptors collected into small patches, or eyespots (Figure 26.13). Collecting light-sensitive cells into aggregates increases sensitivity to light because having many photoreceptors lying next to one another and associated with a single nerve increases the chance that even low light levels will trigger action potentials in the nerve. Thus eyespots allow animals to detect light in low-light environments. Such a patch of photoreceptors provides the raw material for a retina.

In a further increase in the sophistication of the eyespot, many animals have the eyespot placed in a pocket. Animals gain two advantages by having their photoreceptors in such a structure: the photoreceptors are in a more protected position, and they can better detect the direction from which light is coming. The more deeply the photoreceptors lie within a pocket, the more restricted the direction from which light must come to reach them.

Lenses almost certainly evolved initially as a way of increasing sensitivity to light, rather than as way of forming images. If the transparent coating that protected photoreceptors took on more of a lens shape, it could collect and focus light on an eyespot. In this way, more light energy would hit the photoreceptors, further lowering the light intensity needed to trigger action potentials. For example, large numbers of eyespots are spread over the upper surfaces of sea stars. Each eyespot has a small lens that focuses light on the photoreceptors,

Figure 26.13
One Feature of Complex Eyes Is Evident in Jellyfish
Jellyfish have photoreceptor cells gathered together into eyespots.

even though there are not enough photoreceptors to form an image. We find a similar structure in many other simple animals.

Thus the two necessary building blocks of a true image-forming single-lens eye—a lens and a retina—are evident in the rudimentary light-sensing structures of many animals. Strong evidence exists for an evolutionary progression from simple light-detecting structures to eyes as complex as ours.

Chapter Review

Summary

26.1 Sensory Structures: Making Sense of the Environment

• All human senses rely on input from just three types of sensory receptors. Chemoreceptors respond to chemical stimuli and are used in smell and taste. Mechanoreceptors respond to physical changes in the environment and are found in our senses of touch, pain, temperature, hearing, body position, and balance. Photoreceptors respond to light and are used only in vision.

- Some animals can detect magnetic and electrical fields.
- All receptors convert environmental stimuli into action potentials carried by sensory neurons to the CNS.
- The more receptor cells an animal has that are sensitive to a particular stimulus, the more details about that feature of the environment the animal can detect.
- Paired sensory organs allow animals to locate the sources of distant stimuli.

26.2 Chemoreceptors: Smell and Taste

- Animals use chemical stimuli to identify food, detect predators, and find mates.
- In terrestrial animals, smell and taste are distinct senses: smell involves detecting airborne chemicals at a distance, while taste involves detecting dissolved chemicals in contact with the body. In aquatic animals, there is no distinction between these two senses because all chemicals come to them dissolved in water.
- Chemoreceptors are concentrated in those parts of an organism most likely to encounter chemicals. Chemicals in the environment bind with receptor proteins embedded in the plasma membranes of chemoreceptor cells.
- We detect smells when chemicals bind with proteins in the hairlike projections of chemoreceptors embedded in our nasal passages. The stimulated receptors send action potentials to sensory neurons in the olfactory bulbs, where some processing occurs, before other sensory neurons send additional action potentials to other parts of the brain for further processing.
- We can identify specific smells because our chemoreceptors respond to specific chemicals. The intensity of an odor is signaled by how frequently these receptors produce action potentials.
- Our taste buds have chemoreceptors that detect salty, sweet (carbohydrate), bitter (alkaloid), and sour (acid) chemicals. This specialization probably evolved to help animals distinguish safe from unsafe foods.

26.3 Mechanoreceptors: Touch, Hearing, and Other Senses

- Mechanoreceptors tell us about our immediate internal and external environment. Human skin contains various kinds of mechanoreceptors.
- Hearing involves detecting waves of rapid pressure changes. The frequency of these pressure changes determines a sound's pitch, and their intensity determines its loudness. Sound waves travel differently through solids, liquids, and gases.
- In many vertebrates (including humans), sound stimuli are first funneled through the outer ear (pinna and auditory canal) to the eardrum. Vibrations in the eardrum are passed along to tiny bones in the middle ear. The eustachian tube keeps pressures equal on both sides of the eardrum.
- From the middle ear, vibrations are passed to the cochlea in the inner ear. The organ of Corti in the cochlea contains mechanoreceptor cells with hairlike projections that respond to vibrations of the basilar membrane.

- Proprioceptors in joints, muscles, and other tissues provide ongoing information about the position of the body in space.
- The vestibular apparatus, part of the inner ear, helps with our sense of balance. Mechanoreceptor cells responding to the movements of otoliths in the vestibule tell us whether we are upright or not, and mechanoreceptor cells responding to fluid movements in the semicircular canals inform us about head movements.
- Pain and temperature are detected by bare nerve endings. Thermoreceptors respond to moderate changes in temperature, while pain receptors respond to injuries and to extreme heat and cold.

26.4 Photoreceptors: Vision

- Most animals can distinguish light from dark. Many animals can form images.
- Most image-forming eyes have two features: (1) a way of focusing light onto photoreceptors so that the pattern of stimulation of the photoreceptors re-creates the object or scene being viewed, and (2) a way to convert light energy into action potentials, which are conveyed by sensory neurons to the brain for processing.
- In a single-lens eye, a lens focuses an image on the retina, an array of photoreceptor cells at the back of the eye. An opening called the pupil adjusts the amount of light entering the eye.
- There are two types of photoreceptors, rods and cones. Rods detect images in dim light, but not in color. Cones respond to color, but require brighter light to function.
- A region of the retina called the fovea, which is densely packed with photoreceptor cells, provides the sharpest vision.
- Depth perception is possible when the eyes are close enough together on the head to see the same view at the same time, but from slightly different viewpoints. Prey animals tend not to have depth perception; their eyes are set far apart, giving them a panoramic view that helps them to detect predators. Predators tend to have depth perception, which helps them catch their prey. Primates (including humans) have the best depth perception.
- Invertebrates have well-developed visual systems. Some have paired simple eyes. Others have a combination of compound and simple eyes.
- A compound eye consists of thousands of small units, each with its own lens and photoreceptors. The information from all these units is processed in the brain to form an image.

26.5 The Brain's Role in Integrating Sensory Information

- Different sensory systems do not work in isolation. Information from chemoreceptors, mechanoreceptors, and photoreceptors arrives in the brain, where it is processed and integrated to make sense of the world.
- Flavor in food is primarily a combination of taste and smell, but touch (texture) and vision (eye appeal) contribute to giving us a full experience of our food.
- Information from different sensory systems can conflict, causing disorientation and even nausea.

Review and Application of Key Concepts

1. Describe the means by which a stimulus from the environment is converted to our perception of that stimulus.

2. If you smelled something burning in your home, how would your sense of smell allow you to know that something was burning and then help you find your way to the source?

3. Describe how taste and smell work together to provide us with such a "flavorful" world. What other senses play a role?

4. Suppose you are listening to a guitar being played. Describe how the sense of touch in the player's fingers and your sense of hearing are related to the performance. What mechanoreceptors are involved in the player and in the listener?

5. The pigments in the rods and cones of our eyes briefly change chemically when light hits them. The chemical change creates an action potential that travels to the brain. Do you think that these receptors should be called chemoreceptors instead of photoreceptors? Why or why not?

Key Terms

auditory canal (p. 505)	otolith (p. 506)
basilar membrane (p. 506)	outer ear (p. 505)
chemoreceptor (p. 499)	pain receptor (p. 506)
cochlea (p. 506)	photoreceptor (p. 499)
compound eye (p. 510)	pinna (p. 505)
cone (p. 508)	proprioceptor (p. 506)
depth perception (p. 510)	pupil (p. 508)
eardrum (p. 505)	retina (p. 507)
eustachian tube (p. 506)	rod (p. 508)
fovea (p. 509)	semicircular canal (p. 506)
inner ear (p. 506)	simple eye (p. 510)
lens (p. 507)	single-lens eye (p. 507)
mechanoreceptor (p. 499)	taste bud (p. 503)
middle ear (p. 506)	thermoreceptor (p. 506)
olfactory bulb (p. 502)	vestibular apparatus (p. 506)
organ of Corti (p. 506)	vestibule (p. 506)

Self-Quiz

1. Mechanoreceptors respond to
   a. electrical fields.   c. physical changes.
   b. chemicals.   d. light.

2. All sensory receptors convert stimuli from the environment into
   a. action potentials.
   b. chemical stimuli.
   c. magnetic fields.
   d. patterns of vibration.

3. Our overall perception of food is a combination of
   a. taste and smell.
   b. taste, smell, and vision.
   c. vision and taste.
   d. taste, smell, touch, and vision.

4. Which of the following would tell you the position of your arm?
   a. organ of Corti   c. proprioceptors
   b. rods   d. cones

5. Sound is a sophisticated form of
   a. mechanoreception.
   b. chemoreception.
   c. photoreception
   d. none of the above

6. Which of the following would detect the position of your head?
   a. the auditory canal
   b. the vestibular apparatus
   c. the organ of Corti
   d. the pinna

7. The mechanoreceptors that convert sound into action potentials are located in
   a. the eardrum.
   b. the semicircular canals.
   c. the cochlea.
   d. the vestibular apparatus.

8. In our eyes, light is focused by the
   a. rods.   c. retina.
   b. cones.   d. lens.

9. In our eyes, rods and cones are found in the
   a. lens.   c. cochlea.
   b. retina.   d. pupil.

10. We can perceive three dimensions because
    a. our lenses can change shape.
    b. rods are in one eye and cones are in the other.
    c. our eyes are positioned slightly apart.
    d. our retinas are curved.

Experts Warn of iPod Hearing Loss

By *MACWORLD* STAFF

Hearing experts are concerned that the prevalence of digital music players may cause hearing problems in the future.

A report [in] *The Scotsman* warns music player owners, particularly iPod users, to turn their music down. . . .

"Audiologists believe tens of thousands of young people are causing serious damage to themselves, and are likely to suffer tinnitus and loss of hearing in later life. The experts say MP3 players should be designed to prevent people playing music above 90 dB, about two-thirds of the maximum volume of a typical device," the report warns.

Although we do not usually think of sound as damaging, it can be. Of the 28 million people in the United States who suffer from hearing loss, about one-third have suffered noise-related damage to the inner ear. When we hear a loud noise, such as an explosion, a lot of energy hits the delicate structures of the ear. If the noise is loud enough, it may represent more energy than the sensitive mechanoreceptor cells in the inner ear can handle. Even sounds that we probably would not consider loud, such as traffic noise, can cause hearing loss if we are exposed to them for long enough. One effect of prolonged exposure to loud sounds is tinnitus [tin-*EYE*-tuss], a condition that produces constant background noises—tinkling, buzzing, ringing, even roaring or shrieks—that only the person with the condition can hear. Worse effects include loss of hearing. Once the delicate hairs on the mechanoreceptor cells are lost, they are gone for good.

Sound intensities are measured in decibels (dB). An increase of 10 dB represents a tenfold increase in loudness. Common background noises that we can easily ignore, such as the humming of a refrigerator, measure 40 dB or lower. A normal conversation measures about 55 dB. City traffic measures about 80 dB, and a ride in a subway can expose you to about 90 dB. Loud sounds, such as motorcycles or jets taking off, can measure 120 dB. Hearing experts are concerned that most people play their portable music devices at 90 to 100 dB, a potentially harmful range.

What determines the risk of hearing loss is not just how loud sounds are, but also how long the ears are exposed to them. Several factors encourage people to use their portable players at high volumes for longer periods of time than in the past. For one thing, today's digital devices have much longer battery lives and are much lighter than earlier designs. In addition, their hard drives have considerably more memory; because they can store a greater number of songs and pieces of music, the potential boredom of repetition entailed by a smaller song collection is eliminated. In addition, because music is more enjoyable than plain noise, people are willing to listen to loud music for longer time periods in order to drown out the increasing noise in their environment.

To avoid damage, many experts recommend listening to a portable player for no more than an hour at a time and not above 90 dB—which is considerably below the maximum volume of today's devices.

Evaluating the news

1. How realistic do you think the experts' recommendations for "safe" listening are?
2. Most governments have clear guidelines on sound pollution in the workplace designed to protect workers' hearing. These guidelines set out how loud the workplace can be for a given length of exposure time. In most cases, North American and European legislation requires

hearing protection for workers spending an 8-hour workday exposed to sound louder than 90 decibels (sound as loud as a lawn mower or motorcycle). Make an educated guess as to the decibel levels of your non-work environment (including listening to music through earphones); does it meet these guidelines? Are there any non-work activities that pose a particular risk to your hearing? If so, what could you do to protect your hearing?

3. Many communities in recent years have enacted restrictions on noise. Does your city or town have such restrictions? What do they allow and restrict? Are the restrictions enforced? Do you think they are fair?

SOURCE: *Macworld Daily News*, Monday, May 9, 2005.

CHAPTER 27 Muscles, Skeletons, and Movement

Key Concepts

- The human body relies mainly on bone for support, but cartilage and hydrostats play a role as well.

- Skeletons and muscles provide the framework and power to create movement in animals. Skeletons can be internal or external.

- Joints are combinations of stiff and flexible materials that allow motion in animals.

- In muscles, two kinds of microscopic filaments work together and use ATP to power muscle contractions.

- Muscles and skeletons work together to control the strength and speed of movement.

- Animals move by running, flying, or swimming. Each mode of locomotion has different energetic costs, and animals have evolved different strategies for reducing those costs.

Leaping Cats and Humans

Urban life can be tough. Humans living in cities must survive polluted air, violent crime, and speeding taxis. As our companions, domestic cats face their own hazards in a landscape of tall buildings and hard pavement. For example, the legendary curiosity of cats, combined with the promise of better hunting grounds, lures many cats through open windows or over balcony railings. Unfortunately, these adventures can turn into a quick trip ending abruptly on streets and sidewalks several stories down.

For people, such falls usually end in disaster. The death rate increases rapidly as people fall from greater heights—up to the seventh story, at which point the chances of dying level off at almost 100 percent. In sharp contrast, information on 115 cats admitted to animal hospitals in New York City reveals two amazing facts: although the cats fell an average distance of about five and a half stories, with one falling from the thirty-second story (this high-diving cat survived with minor injuries), only 11 of the cats died from their injuries. Even more incredibly, cats stood a better chance of surviving a fall from *above* the fifth story than from *below* the fifth story. How did these cats avoid catastrophe?

How Do Cats Survive Falls?
Does the cat's legendary ability to land on its feet contribute to its surprising ability to survive falls from great heights?

Animals can move themselves through their environment. The ability of animals to walk, fly, or swim provides an almost endless list of potential advantages. Think how different our own lives would be if we could not move: we could not explore our surroundings, we would not have dance and sports, we could not seek mates, and we could not even drive a car.

Animals can move so well because they have a unique combination of structural support tissues, or **skeletons**, and **muscle** tissues that can contract and relax connected to those skeletons. We begin this chapter with a brief introduction to the skeletons of humans and other animals, concentrating on the features that give an animal a strong framework that also allows motion. We then consider how muscles work in concert with skeletons to move an animal. Finally, with this basic understanding in hand, we consider some of the physical forces that determine how animals move through different environments by walking, swimming, or flying.

27.1 Basic Features of the Human Skeleton

We will use the human skeleton as a starting point in our overview of animal skeletons because it is most familiar to us. Understanding some basic features of the human skeleton will give us a point of reference for exploring some general patterns in the way skeletons work in other animals.

The human skeleton provides a framework for protecting organs and allowing complex movements

The human skeleton is the internal framework that supports the body, gives it shape, and allows us to move. The skeleton has two major components (Figure 27.1a). The **axial skeleton** supports and protects the long axis of the body. As its name indicates, it extends from head to "tail" end, consisting of the skull, a long bony spinal column, and the ribs. While the axial skeleton plays a role in movement—we can, for instance, nod and turn our heads based on the way the spinal column joins up with the skull—its primary purpose is to protect vital organs. Ribs form a strong cage around the heart, liver, pancreas, and lungs. The skull is a concrete-hard assembly of bones "welded" together to protect the delicate

brain from injury. The hard bones of the spinal column surround the nerves in the spinal cord that send and receive vital information.

The bones of the arms and legs (our appendages) and the pelvis make up the **appendicular skeleton** [ap-en-DIK-you-ler . . .]. These bones have more to do with motion than protection. In sports, arms and legs, with their attached muscles, do most of the throwing and running, while the axial skeleton does most of the protecting. But both components of the skeleton work together to create the graceful, fluid motions that we see in a running back or a ballet dancer. A wide receiver's head fake is mostly axial. A dancer's long leap is mostly appendicular.

The human skeleton is composed of bone tissue

Although much of a bone is made up of nonliving material, bone is a living tissue that has a blood and nerve supply. Specialized bone cells, called **osteocytes**, surround themselves with a hard, nonliving mineral matrix composed largely of calcium and phosphate compounds that we accumulate from our diet. Bones are built for strength. Figure 27.1b shows the structure of a typical bone, the humerus in the upper arm, which contains two major types of bone tissue. **Compact bone** forms the hard white outer region. **Spongy bone** lies inside the compact bone; the holes in spongy bone allow the remaining bone tissue to act like thousands of tiny buttresses to resist stress. Spongy bone contains **marrow**, a tissue that produces new red and white blood cells.

Our bones are constantly changing in response to how we live our lives. Physical activity builds stronger bones. The bones in a pitcher's throwing arm, for example, are stronger and have larger ridges to which muscles can attach than the same bones in the other arm. It is for this reason that doctors encourage older women to exercise to help offset the effects of osteoporosis, a condition common in postmenopausal women that leads to a gradual weakening of the bones. In contrast, inactivity leads to weaker bones, as osteocytes eliminate what appears to be excess bone tissue that is not needed while, for example, you are sitting on a couch and watching television.

Bone is stiff, yet somewhat flexible, meaning that it does not change shape easily or snap under stress, which suits it well to providing a framework for the body. If you snap your fingers, you can feel the force you are exerting, yet your finger and thumb maintain their overall shapes. Without bones for support, your thumb and forefinger would simply collapse against one another as if made of jelly.

Many medical and anatomical terms relating to bone contain the prefix *osteo* ("bone"). For example, an *osteocyte* (from *cyte*, "cell") is a bone cell. *Osteoarthritis* (from *arthr*, "joint") is a form of arthritis marked by degeneration of the cartilage and bone in joints. And *osteoporosis* (literally "porous bone") is a condition in which the bones become increasingly brittle due to loss of calcium.

Figure 27.1 The Human Skeleton

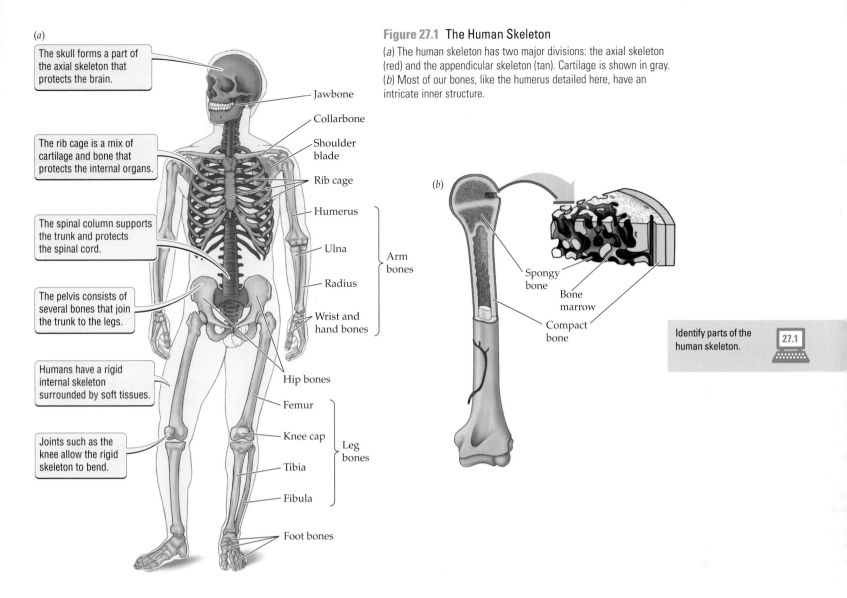

(a)

The skull forms a part of the axial skeleton that protects the brain.

The rib cage is a mix of cartilage and bone that protects the internal organs.

The spinal column supports the trunk and protects the spinal cord.

The pelvis consists of several bones that join the trunk to the legs.

Humans have a rigid internal skeleton surrounded by soft tissues.

Joints such as the knee allow the rigid skeleton to bend.

Jawbone
Collarbone
Shoulder blade
Rib cage
Humerus
Ulna
Radius
Wrist and hand bones

Arm bones

Hip bones
Femur
Knee cap
Tibia
Fibula
Foot bones

Leg bones

Figure 27.1 The Human Skeleton
(a) The human skeleton has two major divisions: the axial skeleton (red) and the appendicular skeleton (tan). Cartilage is shown in gray. (b) Most of our bones, like the humerus detailed here, have an intricate inner structure.

(b)

Spongy bone
Bone marrow
Compact bone

Identify parts of the human skeleton.

27.1

Cartilage and hydrostats provide additional support

Cartilage, which consists of cells that produce a matrix of the protein **collagen** [*KOLL*-uh-jin], supplements the support offered by bone. In the human skeleton, cartilage gives form to the nose, the pinnae of our ears, and part of the rib cage (see Figure 27.1a), among other things. Compared with bone, cartilage is more pliable, as you can observe by gently bending your ear or the tip of your nose.

Another component of the human skeleton is found in the tongue and in the female's clitoris or the male's penis. These parts of our bodies are examples of **hydrostats**, structures that stiffen when a fluid under pressure pushes against an elastic collagen matrix or muscle tissue (Figure 27.2). In hydrostats, the water under pressure acts in much the same way that air under pressure pushes against an elastic balloon to stiffen it.

Flexible ligaments connect inflexible bones to form joints

The stiff bones that give us shape would make movement impossible were it not for our joints. **Joints** are junctions in the skeletal system that let the skeleton move in specific ways. Walking, for example, requires movement at the hips and knees as well as many other joints. Joints in

Figure 27.2 Our Tongues Are Hydrostats

our shoulders, elbows, wrists, and fingers allow us to throw a ball, swing a bat, or hold a pen (Figure 27.3). The lower jaw, which connects to the rest of the skull at a joint, can move relative to the rest of the skull so that we can chew and talk. Specialized, flexible sheets of tissue called **ligaments** join bone to bone to help hold a joint together.

27.2 Comparing Skeletons among Animals

Other animals have skeletons that may differ from ours in appearance but share two basic functional features with ours: stiff structures that maintain shape, and joints that let the stiff parts of the skeleton move relative to one another.

Many skeletons depend on tissues that produce a stiff matrix

Although animals have evolved a diversity of skeletons, all animals, including humans, rely on specialized tissues to support the rest of their bodies. The support tissues consist of specialized cells that produce a strong, stiff matrix around themselves. This matrix can be made up of minerals (as in human bones) or long chains of amino acids (as in cartilage) or sugars (as in the exoskeletons of insects). The minerals that form hard matrices typically contain calcium.

Among animals as a whole, the most widespread non-mineral substances in the matrices of support tissues are collagen and chitin. Collagen is the protein found in cartilage and many other animal support tissues, and

chitin [*KITE*-in] is a complex sugar that forms the exoskeletons of arthropods.

Terrestrial animals need stiff skeletons to counter the pull of gravity and to support their weight. However, the support tissues themselves add weight to the body. The skeletons of larger animals make up a greater proportion of the animal's body weight than do those of smaller animals. For example, a shrew, a tiny animal that weighs about 6 grams (about as much as a couple of pennies), has a skeleton that makes up only 4.6 percent of its body weight, whereas the skeleton of a 6,600-kilogram elephant makes up 24 percent of its body weight. Elephants need heavier skeletons in part to support their heavier skeletons. However, larger animals give up some skeletal strength in exchange for their increasing size. Considering the properties of bone, we can estimate that for an elephant's skeleton to be as strong relative to the animal's body weight as that of a shrew, the elephant's skeleton would have to weigh almost five times its actual weight. Because it weighs so much less than its "ideal" weight, the elephant's skeleton is less strong relative to the animal's body weight than a shrew's (Figure 27.4).

Soft-bodied animals depend on hydrostats to provide support

Many familiar animals, including earthworms, caterpillars, and octopi, have bodies that are like the human tongue in that they lack an obvious means of support. These animals depend mainly on hydrostats, which provide support through an interaction between elastic tissues and a fluid under pressure (Figure 27.5). Water is a fluid within the temperature range from 0°C to 100°C, and air is a fluid at all temperatures at which animals live. Most animals use watery blood, not air, as the fluid in their hydrostats. Nevertheless, filling a balloon with air illustrates how hydrostats work. The more air we blow into a balloon, the greater the pressure of the air inside the balloon. Balloons illustrate two important features of hydrostats: first, as pressure increases, stiffness increases; and second, it takes energy to create the pressure on which the hydrostat depends. Blowing up a balloon takes a lot of effort.

The hydrostats in animals work much like balloons. Penises and clitorises become stiffer when a fluid—blood—is pumped in under pressure. Other animal hydrostats accomplish the same thing by causing an elastic membrane to shrink around a fixed volume of water. Imagine partially inflating a balloon to a point at which it is still rather limp. Squeezing off the base of the balloon with your hand forces the air in the balloon into a smaller

Figure 27.3 Our Jointed Skeletons Allow Many Different Movements
The joints in our body enable the movements required in baseball, from the players pitching and hitting the ball, to the spectator chewing a hot dog in the stands.

(a)

Shrew

Elephant

(b) With increasing size, the proportion of a mammal's body weight devoted to the skeleton also increases.

	Body weight	Actual skeleton weight	"Ideal" skeleton weight
Shrew	6 g	4.6%	4.6%
Human	80 kg	13%	54%
Elephant	6600 kg	24%	495%

To maintain an "ideal" skeleton, as strong relative to body weight as that of a shrew, humans and elephants would have to have bodies containing impossible amounts of bone.

Figure 27.4

Animals Devote Different Proportions of Their Body Weight to Their Skeletons

(a) Relative to its body weight, a shrew's skeleton is lighter than an elephant's. (b) For an elephant to have a skeleton as strong as a shrew's (an "ideal" skeletal weight), its skeleton would have to be almost five times larger than it is. Thus large animals sacrifice some skeletal strength in exchange for their size.

volume, making that small volume of balloon stiff. Similarly, when the muscles in your tongue contract, their internal volume decreases. This causes the muscle to stretch around the fixed volume of water that it contains, so that the muscle becomes stiffer. Because generating stiffness in a hydrostat requires muscle contraction, stiffening a hydrostat requires energy.

Skeletons can be internal or external

Over evolutionary time, two distinct ways of arranging stiff support tissues have arisen in the animal kingdom. Some animals, including humans and most other vertebrates, have an **endoskeleton** (*endo*, "inner, inside"), in which the support tissues lie inside the body and the soft tissues they support, such as skin and muscles, surround them (Figure 27.6a). Other animals—such as lobsters, insects, and many other invertebrate groups—have an **exoskeleton**, an external skeleton that surrounds and encloses the soft tissues it supports (Figure 27.6b). Joints are common to both types of skeletons.

In addition to support, exoskeletons provide an advantage that endoskeletons cannot offer: a protective outer armor. A disadvantage of exoskeletons, however, is that they weigh more than endoskeletons offering similar support, so terrestrial animals with exoskeletons (such as insects and scorpions) face even more severe size constraints than those with endoskeletons. For example, one of the largest land animals with an exoskeleton is the giant weta, a cricket from New Zealand, which weighs up to 71 grams (Figure 27.6c). Yet this is only about 3 times the weight of a house mouse, while the largest land animal with an endoskeleton (the elephant) weighs about 93,000 times as much as the weta.

Figure 27.5 Nematodes Use Hydrostats for Support

When muscles in the outer skin of tiny nematodes (such as this *Strongyloides filariform*) contract, they decrease the internal volume of the nematode. The "shrunken" skin must then stretch around the fluid inside the nematode's body, which does not change in volume. This stretching generates a pressure that stiffens the nematode.

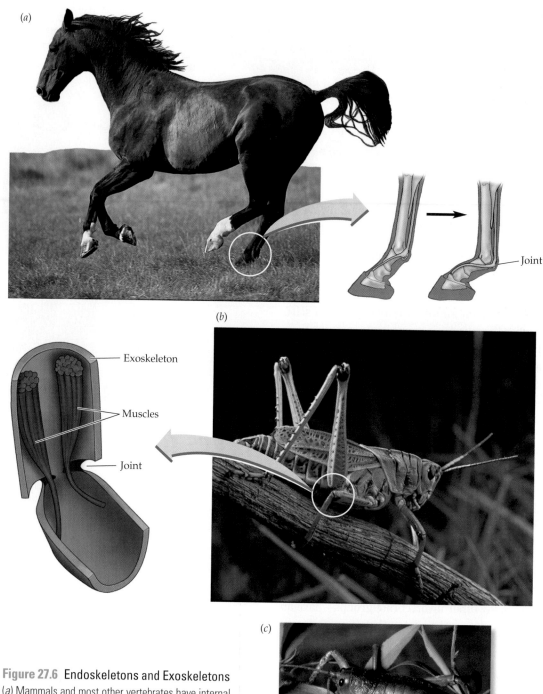

(a)

Joint

Exoskeleton

Muscles

Joint

(b)

(c)

Figure 27.6 Endoskeletons and Exoskeletons
(a) Mammals and most other vertebrates have internal skeletons with joints. (b) Grasshoppers and most other arthropods have external skeletons with joints. (c) Weta crickets are large for an animal with an exoskeleton, but not compared with most animals with endoskeletons.

27.3 The Human Knee

Although our knees differ in detail from other joints in our bodies and from the joints in other animals, they make a good general model for how flexible and stiff materials are combined in a joint to allow controlled motion (Figure 27.7).

The bones that meet in the human knee joint support body weight and define how the leg bends. The lower end of the femur (the thigh bone) rides in a pair of grooves at the end of the tibia, which is the larger of the two lower leg bones. This arrangement allows the lower leg to swing forward and back like a hinge, but not side to side, thereby providing both motion and stability. The knee, however, must allow for the slight twisting of the tibia relative to the femur that takes place when we walk.

Ligaments connect bone to bone in the knee, as they do in other joints. Two pairs of ligaments, one on the front and back of the knee and a second on the sides of the knee, connect the femur to the lower leg bones. These ligaments prevent sliding along the joint surface when the knee bends, keeping the femur and the lower leg bones in their proper place. The collagen of the ligaments can stretch slightly, allowing some bending and twisting motion in the knee. **Tendons**, which are also made of collagen, connect muscle to bone; tendons help hold the knee together by connecting the upper leg muscles to the bones of the lower leg.

Wherever two moving parts rub against each other, as in a joint, wear can destroy bone and friction can waste energy. Layers of cartilage in the knee cushion the points where the femur meets the tibia. Surrounding the cartilage are **synovial sacs** filled with lubricating fluid, which reduces friction, allowing femur and tibia to slide past each other more easily than a skate slides over ice.

When all these components—bone, cartilage, ligaments, tendons, and synovial sacs—work together, the result is a joint that can control and withstand motion over many decades. When one of these components fails, however, potentially serious medical problems can result (Table 27.1).

Tears in the cartilage and ligaments of the knee are common and sometimes disabling sports injuries (Figure 27.8). The knee joint can handle vertical forces up to about ten times body weight, which is quite remarkable, but it is less able to resist horizontal forces, such as blocks and tackles in football or being hit in the knee by a hockey

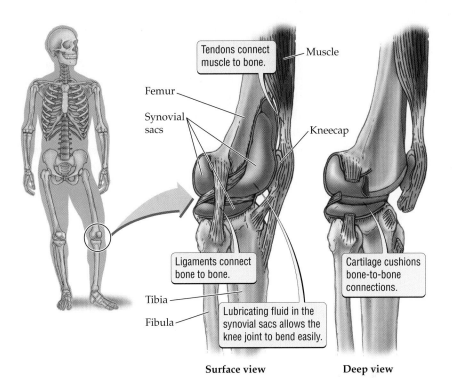

Surface view   **Deep view**

Figure 27.7 The Human Knee
The human knee joint is a good example of how rigidity and flexibility combine to allow movement. Fluid-containing synovial sacs are shown in purple and cartilage in blue.

Explore the function of the human knee.

27.2

puck. Injuries to the knee can also happen when a person rapidly speeds up or slows down while turning. Torn cartilage can lead to increased wear on the knee joint. Unless the tear is repaired surgically, the remaining cartilage and underlying bone may become permanently damaged, leading to arthritis. In various kinds of arthritis, damage

Figure 27.8 Knee Injuries Can Be Disabling
Damage to the knee joint can cause severe pain and sometimes ends careers.

Table 27.1

Common Ailments of the Knee Joint

Condition	Affects			
	Bone	**Cartilage**	**Ligaments**	**Synovial sacs**
Osteoarthritis	X	X		
Rheumatoid arthritis	X	X		X
Torn cartilage		X		
Torn ligament			X	
Bursitis				X

to the cartilage and bone in the joint make bending difficult and painful.

Ligaments tear when stretched to more than one and a half times their normal length, either by a severe blow to the front or sides of the knee or by severe twisting of the lower leg relative to the femur. Torn ligaments cannot hold the femur in place relative to the tibia and fibula, leading to movement in the knee joint that may make walking impossible. Like torn cartilage, torn ligaments require surgery to repair.

27.4 How Muscles Work

Muscle tissue is unique to animals. The skeleton and its joints are the framework for motion, but muscles provide the power necessary for movement. Muscle tissue performs one crucial function: it contracts.

Muscle tissue is built for contraction

We are all familiar with muscles in a general way. Most of the features that relate to the role of muscles as a source of power for movement, however, remain hidden to the unaided eye. Each muscle—for example, the biceps muscle of the upper arm (Figure 27.9a)—consists of many bundles of **muscle fibers**. Each fiber, in turn, is packed with smaller **myofibrils**, which contain units called **sarcomeres** (from *sarc*, "flesh"; *mere*, "part"), which do the work of contraction.

Sarcomeres are visible as bands when seen through a microscope (Figure 27.9b). At extremely high magnification, additional details of their structure become evident. Each sarcomere is based on a very specific arrangement of actin and myosin filaments, which allows muscles to contract. Each **actin filament** consists of two molecules of the protein actin, and each **myosin filament** is made up of many molecules of the protein myosin. Zigzagging dark lines called **Z discs** mark the ends of each sarcomere and provide anchor points for the actin filaments. Between the ends of the actin filaments, extending toward the middle of the sarcomere, lie myosin filaments.

When a muscle contracts, actin and myosin filaments slide past each other within its sarcomeres. If you could watch your biceps muscle at high magnification while lifting a weight, you would see the two Z discs that define each sarcomere pull closer together (see Figure 27.9b). An even higher magnification would reveal that the protruding "head" of each myosin molecule can bind to specific sites on adjacent actin filaments. In slow motion,

here's how a muscle contraction works: The myosin heads at each end of the myosin filaments attach to binding sites on the surrounding actin filaments. This binding changes the shape of the myosin heads, causing them to pull the myosin filament closer to the two Z discs. Then the myosin heads release from their binding sites before attaching to new binding sites closer to the Z discs (see Figure 27.9b). In this way, the myosin filaments "walk" their way from binding site to binding site along the actin filaments, pulling the myosin filaments and Z discs closer together.

Each "walking step" of each myosin head requires the presence of calcium ions and the energy stored in one molecule of ATP. The result of countless myosin and actin filaments interacting in this way is a muscle contraction. Contracting muscles bulge because all the actin and myosin filaments must squeeze into a shorter length of muscle.

Muscles must work in pairs

Because the myosin heads can only pull the Z discs together, a contracted muscle cannot relax itself. Thus most muscles are arranged in opposing pairs so that the contraction of one causes the other to "relax," or stretch back to its condition before contraction. For example, the triceps muscle (on the lower side of your upper arm) contracts to stretch the biceps muscle (on the upper side of your upper arm), and vice versa (Figure 27.10). This pattern of paired muscles is repeated in other parts of the body.

27.5 Converting Muscle Contraction into Motion

Within the animal kingdom, mechanisms have evolved that convert simple muscle contractions into the wide range of motions of which most animals are capable.

Animals increase strength by increasing the amount of muscle tissue

All animals have essentially identical sarcomeres that contain identical actin and myosin molecules arranged in much the same pattern. Because of this similarity, a muscle fiber from an ant can lift about the same weight as a muscle fiber from an elephant.

Differences in strength between species and between muscles within an individual depend almost entirely on the cross-sectional area of the muscles. The greater the cross-sectional area, the more muscle fibers can lie side

(a)

Both ends of a muscle are anchored to support structures by tendons.

Tendon

Muscle

Bundle of muscle fibers

Each muscle consists of many muscle fibers.

Single muscle fiber

Each muscle fiber, in turn, consists of many myofibrils.

Z disc   Single sarcomere   Z disc

Single myofibril

Figure 27.9
The Microscopic Structure of Muscle
(*a*) Muscles contain bundles of muscle fibers, which contain myofibrils, made up of sarcomeres. (*b*) Muscle contraction depends on the movement of actin and myosin filaments. The electron micrograph shows one sarcomere, bounded by Z discs.

(b)

Sarcomeres, which represent the basic functional unit of muscles, extend from one Z disc to another.

Single sarcomere

The sliding of myosin filaments along actin filaments causes muscles to contract.

Contraction

Z disc   Actin filament   Myosin filament

The heads of myosin molecules bind reversibly to special sites on actin filaments.

Z disc

Z disc

See the contraction of the biceps muscle.

27.3

by side and contract simultaneously. In other words, bigger muscles are stronger than smaller muscles. For example, in absolute terms, elephants are stronger than ants because elephants have muscles with bigger cross-sectional areas. But pound for pound, ants are stronger because they can lift objects many times their own body weight, while elephants cannot.

Exercise can change muscle strength as well. With weight training, muscles initially increase in strength without increasing in cross-sectional area. The number of muscle fibers in the muscles remains the same, but a greater proportion of them contract at any one time. Thus the initial increase in strength comes about because muscle fibers are being used more efficiently. With continued weight training, however, the number of muscle fibers in the muscles increases, leading to the greater muscle bulk for which bodybuilders are known.

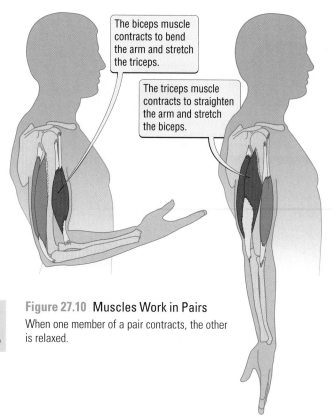

The biceps muscle contracts to bend the arm and stretch the triceps.

The triceps muscle contracts to straighten the arm and stretch the biceps.

Figure 27.10 Muscles Work in Pairs
When one member of a pair contracts, the other is relaxed.

Watch how muscles compete.

27.4

Match characteristics of fast and slow muscle fiber types.

27.5

The speed of muscle contraction depends on the length and type of muscle

Animals have different types of muscles that contract at different speeds. As with strength, these differences arise because muscle tissues differ not in their fundamental units, but in how those units are arranged. In particular,

differences in contraction speed depend on a muscle's length and its response to action potentials.

In general, longer muscles can contract faster than shorter ones. Here's why: When fully contracted, a muscle shortens to about 60 percent of its resting length. Thus a 10-centimeter-long muscle in the arm shortens by 4 centimeters (to 6 centimeters) when fully contracted. In the same time, a 30-centimeter-long muscle in the thigh shortens by 12 centimeters (to 18 centimeters) when fully contracted. The longer leg muscle contracts three times the distance in the same amount of time as the shorter arm muscle.

Many groups of animals also have two types of muscle fibers that occur in different proportions within a muscle depending on its function. What distinguishes these fiber types is how they respond to the action potentials from the motor neurons that trigger contractions. The synapses between motor neurons and muscle fibers are called **neuromuscular junctions**. Muscle fibers that contract quickly, like those that generate the explosive power of a runner sprinting from the starting blocks, are called **fast muscle fibers**. Fast muscle fibers contract completely in an all-or-nothing response to action potentials. In contrast, **slow muscle fibers** respond to action potentials by contracting in small steps. Slow muscle fibers are most often used by the body for activities sustained over long periods of time, as during the slower pace of a marathon. Table 27.2 summarizes the characteristics of these two types of muscle fibers.

Whereas fast muscle fibers generally sacrifice power for speed, slow muscle fibers sacrifice speed for power. Fast muscle fibers contract so rapidly that they quickly use up the available oxygen and have to rely on fermentation, which takes place when oxygen runs out, to supply them with ATP (see Chapter 8). Slow muscle fibers, on the other hand, rely on aerobic respiration. To ensure a steady oxygen supply, slow muscle fibers contain a pigment similar to hemoglobin, called **myoglobin**, that binds and holds oxygen until it is needed by the muscle.

We can usually distinguish slow muscle tissue by its darker color, which reflects its myoglobin content. Thus, whereas the white meat of a turkey consists mostly of fast muscle fibers used to power the wings, the dark meat of the legs consists mostly of slow muscle fibers that must contract for long periods with enough force to hold up the body of the bird.

Muscles and skeletons work together to control the strength and speed of movement

The speed or strength of the movement that results from a muscle contraction also depends on the arrangement

Table 27.2

Fast and Slow Muscle Fibers

Characteristic	Type of muscle fiber	
	Fast	**Slow**
Speed of contraction	Fast	Slow
Force of contraction	Weak	Powerful
Length of contraction	Brief	Sustained
Response of sarcomeres	Either no contraction or complete contraction	Partial contraction possible
Source of ATP	Fermentation	Aerobic respiration
Human example	The quadriceps muscle (in the thigh)	The gluteus maximus muscle (the largest muscle in the buttocks)

of muscles and skeletons to form **lever systems**. We have all used mechanical levers in the course of daily life. All lever systems consist of a **fulcrum** that serves as a pivot point for two stiff lever arms (Figure 27.11a). We apply a force to the **power arm**, which moves the **load arm**, which does the work we want done. In biological lever systems, the power and load arms consist of stiff components of the skeleton, such as bones, and the joint forms the fulcrum.

Mechanical lever systems tend to be designed to increase the force generated by a given muscle contraction. For example, we use the claw of a hammer to pull nails that are too firmly stuck in wood to extract with our bare hands (Figure 27.11b). In such cases, the load arm is

shorter than the power arm. Thus the tip of the load arm moves a much shorter distance than the power arm, and the force applied to the power arm is "concentrated" over the short distance that the load arm moves. In this case, speed in the power arm is sacrificed for power in the load arm. If the load arm is longer than the power arm, the force is reduced and the speed increased at the tip of the load arm relative to that of the power arm.

In animals, the power arm is almost always shorter than the load arm. Thus the force generated at the tip of the load arm is usually less than that generated by the muscle, but the load arm can be moved a greater distance more rapidly. The leg arrangements of mammals with different lifestyles show how natural selection can act on

Figure 27.11 Mechanical and Biological Lever Systems
(a) In a mechanical lever system, force applied to the power arm (red) causes the lever to pivot about the fulcrum (triangle) and do work at the load arm (blue). The system at the top shows how most mechanical lever systems are designed, while the system at the bottom represents how lever systems are arranged in animal bodies. (b) A human arm working with a hammer to extract a tightly held nail illustrates how a biological lever system and a mechanical lever system can work together; it also reveals a common difference between the two. In the arm, the elbow joint forms the fulcrum of the biological lever system. The biceps muscle (see Figure 27.10) attaches to bones in the forearm to define the power arm.

lever systems to maximize either strength and endurance or speed (Figure 27.12).

In humans, only the upper and lower leg bones contribute to leg length. In contrast, some of the bones equivalent to those found in the human foot add to leg length in dogs, and all the foot bones contribute to making a longer leg in speedy runners such as horses and deer. Increasing the leg length increases the length of the load arm, which means that a given muscle contraction moves the foot a greater distance in the same amount of time. The human leg, with its relatively short load arm, works best for steady walking over long distances; dogs are suited to running over moderate distances, and deer with long legs can run very rapidly over short distances.

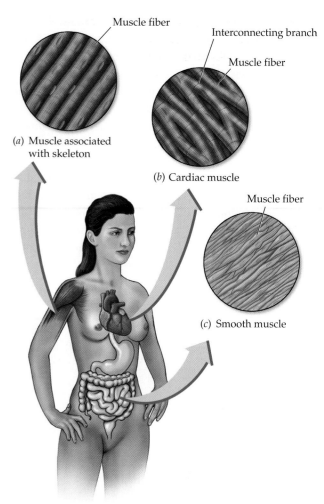

(a) Muscle associated with skeleton

(b) Cardiac muscle

(c) Smooth muscle

Figure 27.13 Some Specialized Muscle Types Create Internal Movements
(a) Muscles associated with the skeleton have a banded appearance.
(b) Cardiac muscle is also banded, and its muscle fibers are branched.
(c) Smooth muscle does not have bands.

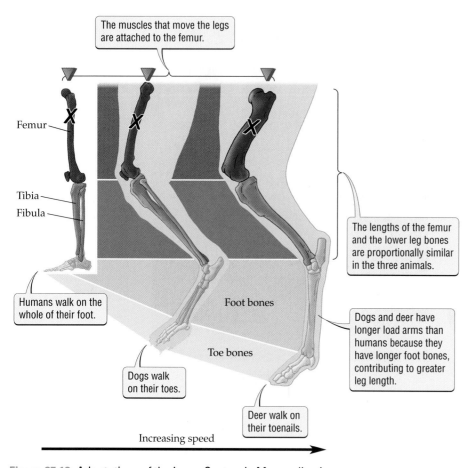

The muscles that move the legs are attached to the femur.

Femur

Tibia

Fibula

The lengths of the femur and the lower leg bones are proportionally similar in the three animals.

Humans walk on the whole of their foot.

Foot bones

Dogs and deer have longer load arms than humans because they have longer foot bones, contributing to greater leg length.

Toe bones

Dogs walk on their toes.

Deer walk on their toenails.

Increasing speed

Figure 27.12 Adaptations of the Lever System in Mammalian Legs
Humans cannot walk as fast as most other animals because our legs have a short load arm (leg length) relative to their power arm (the distance from the fulcrum at the tip of the femur, marked by a triangle, to the point on the femur where the muscle attaches, marked by an X). All of our foot and toe bones touch the ground when we walk, meaning that only the upper (red) and lower (orange) leg bones contribute to leg length. Speedier animals have changed these proportions to give them more favorable leg lengths. By increasing the length of the load arm (the full length of the leg) relative to the power arm, dogs and deer increase the speed at which they can run.

Specialized muscles create other kinds of movements

Even when an animal is very still, there is movement going on within its body. A heart is pumping blood, and food is being moved along a digestive tract by muscular contractions. These muscles do their work without our having to consciously think about moving them.

The vertebrate heart is a muscular organ and the only organ that contains **cardiac muscle**. Cardiac muscle has bands like the muscles associated with the skeleton, and its muscle fibers are joined by interconnecting branches that help them to produce the coordinated contractions we know as heartbeats (Figure 27.13b).

For digestion to proceed, food must move along the digestive tract. A specialized muscle type called **smooth**

muscle (Figure 27.13*c*) causes this movement as it contracts in coordinated waves along the intestine. Smooth muscle is also found in the walls of blood vessels. It is called smooth muscle because it lacks the bands associated with other muscle types.

27.6 How Animals Move in Their Environment

Animals have three main modes of locomotion: running, swimming, and flying. Animals expend energy to generate thrust with their muscles. Running and walking animals, such as humans, expend this energy to push off against the ground with their legs. Flying and swimming animals move through fluid environments that require a different way of generating thrust. Flying and swimming animals have evolved wings and fins, which provide broad surfaces with which to push off against air and water, respectively (Figure 27.14).

Natural selection favors bodies that minimize drag

As animals move through their environment, they convert the contractions of their muscles into **thrust**, which propels them forward. However, an animal's motion is always resisted by drag, the same force that impedes blood moving through vessels (see Chapter 23). Actually, drag is a broad concept that includes two somewhat different forces—pressure drag and frictional drag—both of which resist motion.

Imagine you are riding a bicycle on a calm day: As you set off, there seems to be wind blowing in your face, but it is really just air colliding with your face and body as you move. And this air makes you work harder. High pressure in front of you combined with low pressure created behind you in the void left by your moving body creates **pressure drag**. The combination of high pressure in front and low pressure behind appears to push you backward while you do your best to drive yourself forward. The faster you go, the stronger the wind, because the resistance caused by pressure drag increases very quickly as speed increases. Pressure drag also increases as the surface area exposed to it—the area presented by the animal when viewed from the direction in which it is moving—increases.

Frictional drag, on the other hand, affects other parts of the moving body. The air that piles up in front of you

(a)

(b) (c)

Figure 27.14 Modes of Locomotion in Animals
(*a*) Running animals, such as this horse, generate thrust by pushing against the ground with their legs. (*b*) Swimming organisms, such as this orange-finned anemone fish, push against the surrounding water with paddle-like flippers or fins. (*c*) Flying animals, such as this ruby-throated hummingbird, use the broad surfaces of their wings to push against air.

as you ride your bike has to flow past you sooner or later, but it tends to stick to your sides briefly before doing so. Thus, as you pedal down the road, you move not only yourself, but extra "baggage" in the form of the air molecules that rub against you. This tendency of molecules touching a moving object to briefly stick to the object's surface is frictional drag.

Frictional drag can become a serious burden, especially in water, which is a much denser fluid than air. If you drive your bicycle through a deep puddle, you can feel the much greater frictional drag offered by water. Frictional drag increases most significantly as the total surface area of the organism increases and as the viscosity (thickness) of the fluid it is moving through increases. Frictional drag has its greatest effect on animals with a large surface area and on those that live in water. Small swimming animals must expend relatively more energy overcoming frictional drag because they have a higher surface area-to-volume ratio.

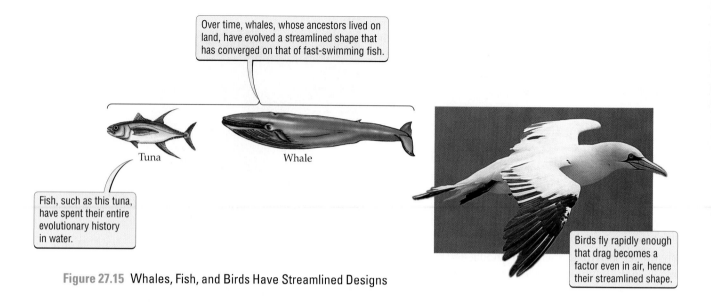

Over time, whales, whose ancestors lived on land, have evolved a streamlined shape that has converged on that of fast-swimming fish.

Tuna

Whale

Fish, such as this tuna, have spent their entire evolutionary history in water.

Birds fly rapidly enough that drag becomes a factor even in air, hence their streamlined shape.

Figure 27.15 Whales, Fish, and Birds Have Streamlined Designs

Because organisms expend energy to generate thrust, drag is costly in a biological sense. Evolution favors adaptations that reduce drag because they free up energy for other uses. Reducing the two kinds of drag requires different strategies. The most effective strategy to minimize pressure drag seems to be a streamlined shape (Figure 27.15). In spite of their very different evolutionary histories, fast-moving members of bird, aquatic mammal, and fish groups have all converged on a similar body shape. This shape makes it easy for water or air to flow smoothly around the animal. Animals reduce frictional drag by having "nonstick" surfaces. The smooth, slimy surface that makes many aquatic animals so unappealing to touch effectively reduces frictional drag.

Different modes of locomotion have different energetic costs

How much energy do animals expend to move from one place to another? Scientists have calculated energy expenditures for animals of different sizes and using different means of locomotion (Figure 27.16).

Not surprisingly, heavy animals use more energy to move a given distance than lighter animals. Pound for pound, however, heavy animals expend *less* energy in relation to their body weight. A cat, for example, uses less than ten times as much energy to run a given distance as does a squirrel weighing one-tenth as much. Why do heavier animals move more efficiently than lighter animals? In part, larger animals have relatively less surface area per unit of body weight than smaller ones, and thus are subjected to relatively less drag.

Notice in Figure 27.16 that the line representing the energy cost for running (and walking) animals lies above the line representing the energy cost for fliers; in other words, runners expend more energy moving a given distance than do fliers of the same size. (The box on page 533 looks in more detail at the energetics of walking and running in humans.) Fliers use less energy than runners, and swimmers use the least. Why this is so has to do with how bodies move in different fluid environments. Half the energy spent on running is "wasted" on moving

Energy cost per kilogram of body weight to move 1 kilometer

Fliers

Runners and walkers

Heavier animals use relatively less energy than lighter ones to move the same distance.

Swimmers

Running human

Relative to their weight, swimmers use less energy than fliers, and fliers use less than runners.

Bicyclist

Body weight (in kilograms)
0.00001 0.0001 0.001 0.01 0.1 1 10 100 1000 10,000

Figure 27.16 The Relative Costs of Locomotion
The energy cost of locomotion is determined by two factors: the fluid medium—in air, over land, or through water—in which the locomotion takes place, and the weight of the animal. Note that if the line for swimmers were extended farther to the right, human bicyclists would fall between swimmers and fliers. Within a given medium, the size of the animal affects how much energy per unit of body weight it requires to move.

Science Toolkit

The Mathematics of Biology

Many of you may be taking biology because you thought it would involve less mathematics than other science courses. Although this textbook hides much of it, math is a critical tool for biologists, who use it to quantify behavior and to reveal tendencies among living organisms.

Take, for example, the study of how animals move. At low speeds, humans walk, planting the heel of one leg on the ground in front of us, then vaulting the body over this planted leg bone as a pole vaulter uses a pole. When we run, our legs change from relatively rigid "poles" into bouncy "springs" that emphasize the elastic tendons and ligaments that connect our bones. Biologists studying human movement have found that by converting our motion into numbers, they can use statistics to understand why we walk sometimes and run at other times. By measuring the speed at which a large sample of people walked and ran, biologists learned that most people switched from walking to running at a speed of between 2.3 and 2.5 meters per second, even though it was possible to walk at much higher speeds and run at lower speeds.

Mathematical observations can also provide insights into the physics of walking and running. When we look at energy transformation during walking, we find that energy is stored as potential energy as the body is lifted over the "pole" formed by the planted leg, and converted into kinetic energy (the energy a body has because it is in motion) as the body falls to plant the other leg (see part *a* of the accompanying figure). When running, the pattern is different: as the leading foot hits the ground, the kinetic energy of downward and forward motion is stored in the stretching of ligaments and tendons, then converted into potential energy as the body bounces upward and forward at the beginning of the next stride. By calculating these energy exchanges, biologists learned that the transition from walking to running at 2.3 to 2.5 meters per second corresponds to the speed at which running becomes energetically less costly than walking (see part *b* of the figure).

Mathematics can also help biologists identify general patterns in complex behaviors across different groups of organisms. A striking example involves the application of Froude's number—originally developed

Math Helps Us Understand the Energetics of Running and Walking in Humans

(*a*) The pattern of conversion of energy between kinetic, potential, and elastic energy differs greatly between walking and running. (*b*) The speed at which humans make the transition from walking to running corresponds to the speed at which it costs less energy to run than to walk.

(continued)

to explore ways of designing faster ships—to our investigation of walking and running. Froude's number relates the forces that result from the pull of gravity to the forces needed to change the direction or speed of a moving object. It turns out that for all sorts of mammals, including humans, the transition from a walk to a run takes place at a Froude's number of 0.6. This means that small mammals with shorter legs make the transition from walking to running at a slower speed than larger, long-legged mammals. The generality of Froude's number suggests that in spite of their outward differences, common factors influence walking and running in all mammals. In these different ways, through statistics and numerical models, mathematical tools illuminate and inform the biological sciences.

the body up and down and on changing the momentum of the legs as they alternate between moving forward and moving backward. Bicycling, which involves fewer changes in momentum, is more efficient than running, as the graph shows. Animals that fly use less energy to change momentum than runners, but must expend energy to keep themselves aloft. Despite the movement of their fins and tails, swimmers lose little energy to changes in momentum; moreover, they receive support from the surrounding water.

The Ninth Life of Cats

Skydiving domestic cats, as mentioned at the outset of this chapter, actually survive falls from greater heights better than falls from intermediate heights (Figure 27.17). The secret to their survival lies in their skeletons, their muscles, and the physics of their movement through the air. In partic-ular, four factors combine to increase their chances of surviving falls from heights that would be fatal to humans.

First, cats have an exceptional ability to soften their landing. Perhaps because falls are a normal part of life for tree-climbing animals such as cats, natural selection strongly favors cats that instinctively rotate their bodies as they fall so that they land feet first. Humans, in contrast, may land with arms or feet first. By landing as they do, cats spread the force of impact over four limbs, whereas we concentrate a larger force over only two limbs. In addition, cats' limbs are stronger, pound for pound, than ours.

In addition to flexing to land on four legs, cats prepare for their landing in other ways. They relax some muscles a bit during the fall and bend their legs as they land so that the ligaments and tendons associated with the joints can act as springs. This strategy further reduces the force of the impact by spreading it over a slightly greater period of time. Although the difference between slowing down from a fall to a dead stop in 0.002 second instead of 0.001 second may seem trivial, doubling the deceleration time cuts the force of impact in half. For similar reasons, a padded dashboard in a car or

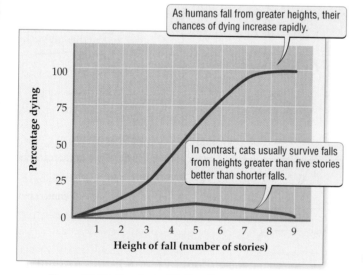

As humans fall from greater heights, their chances of dying increase rapidly.

In contrast, cats usually survive falls from heights greater than five stories better than shorter falls.

Percentage dying — 100, 75, 50, 25, 0

Height of fall (number of stories) — 1 2 3 4 5 6 7 8 9

Figure 27.17 Survival of Falling Cats and Humans
Statistics collected by doctors and veterinarians show that domestic cats survive falls from high places much better than humans do. Not only are cats much less likely to die from a fall than are humans, the mortality rate for cats actually decreases for falls from heights greater than the fifth story.

a bicycle helmet can dramatically reduce the injuries suffered in a collision. The drop in the death rate when cats fall from heights greater than the fifth story (see Figure 27.17) may reflect the increased time that they have to prepare properly for impact.

A third factor is that, after it has fallen a certain distance, the speed of a falling animal no longer increases. A cat reaches this maximum speed when the pull of gravity, which accelerates the rate of falling, is balanced by the increasing resistance from drag as speed increases. Cats attain this balance after falling about eight stories.

Finally, there is the matter of skeletons. Both cats and humans have a strong endoskeleton made of collagen-reinforced bone. But the cat skeleton, relative to body weight, is stronger than the human skeleton. Remember that as animals become larger, the weight of the skeleton increases more rapidly than the weight it can support. The disadvantages of a heavier skeleton outweigh the advantages of a stronger skeleton, resulting in an evolutionary compromise: humans, like all large terrestrial animals, have sacrificed skeletal strength to reduce weight. Whereas we must pay for the benefit of a tolerably light skeleton by looking carefully before we leap, smaller and more strongly constructed cats can get away with less caution.

Because cats climb, and because they are so curious, it is tempting to conclude that their amazing ability to survive long falls represents a marvelous adaptation to their way of being. In part, their ability to survive falls seems to be an adaptation to life in the trees. Equally important, however, may be the relative strength of cat leg bones, a side effect of their size rather than an adaptation to their lifestyle.

Chapter Review

Summary

27.1 Basic Features of the Human Skeleton

- The axial skeleton supports and protects vital organs along the long axis of the body. The appendicular skeleton is composed of the bones in the arms, legs, and pelvis, and is primarily involved with movement.
- Compact bone consists of dense deposits of hard calcium and phosphate compounds. Spongy bone lies inside compact bone and contains bone marrow, which makes red and white blood cells, in spaces between the hard materials.
- Cartilage, which is more pliable than bone, supplements the support provided by bone.
- Hydrostats are structures that stiffen when a fluid under pressure pushes against an elastic collagen matrix or muscle tissue.
- Ligaments hold together the bones at a joint.

27.2 Comparing Skeletons among Animals

- All animals have tissues that support their bodies, consisting of specialized cells that produce a nonliving matrix. In humans and most other vertebrates, the support tissue is bone (made of minerals) and cartilage (made of protein). In arthropods, the support tissue is made up of chitin, a complex sugar.
- The larger an animal is, the greater the proportion of its body weight that is in its skeleton, and the less strong its skeleton is.
- Soft-bodied animals depend on hydrostats to provide support. In a hydrostat, a fluid (usually water) exerts pressure on elastic tissues to produce movement or structure.
- Most vertebrates have endoskeletons, while many invertebrates (especially insects and other arthropods) have exoskeletons. Exoskeletons are heavier, limiting animals that have them to smaller sizes than animals with endoskeletons.

27.3 The Human Knee

- In our knees, as well as in other joints, ligaments connect bone to bone. Tendons help hold the knee together by connecting upper leg muscles to the bones of the lower leg. Tendons and ligaments are made of collagen.
- Cartilage in the knee cushions the points where femur and tibia meet. Fluid-filled synovial sacs lubricate the cartilage, reducing friction.

- Tears in the cartilage and ligaments, which are common sports injuries, require surgery to repair.

27.4 How Muscles Work

- Muscles provide the power necessary for movement.
- Muscles consist of muscle fibers, each of which is packed with myofibrils; myofibrils contain the sarcomeres, which do the work of contraction.
- Sarcomeres are visible as bands when seen through a microscope. They extend between two Z discs, which are the attachment points of actin filaments. The thicker myosin filaments lie in the center of the sarcomere.
- During contraction of a muscle, the myosin heads on each end of the myosin filaments repeatedly attach to binding sites on nearby actin filaments. This binding pulls the ends of the myosin filaments closer to the Z discs at the ends of each sarcomere, pulling the Z discs closer together. The energy stored in ATP powers these movements.
- Muscles work in pairs. When one muscle in a pair contracts, the other is relaxed.

27.5 Converting Muscle Contraction into Motion

- All animals have essentially identical sarcomeres with the same molecules arranged in much the same way.
- Animals increase the strength of their muscles by increasing the cross-sectional area of muscle tissue.
- A longer muscle contracts proportionally more than a short muscle, so a long muscle contracts faster.
- Action potentials arriving at neuromuscular junctions cause muscle fibers to contract.
- Fast muscle fibers contract all at once, are used in shorter-term but faster activities, and rely on fermentation. Slow muscle fibers contract in small steps, are used in sustained activities, and rely on aerobic respiration; myoglobin provides them with a steady supply of oxygen.
- The speed or strength of movement depends on the arrangement of muscles and skeletons to form lever systems, which consist of a fulcrum, a load arm, and a power arm. In animals, the power arm is usually shorter than the load arm, so the force generated at the tip of the load arm is usually less than that generated by the muscle, but the load arm can move a greater distance more rapidly.
- Faster-running animals have longer legs (load arms) relative to the length of the muscles that power them (power arms).
- Internal movements are created by specialized muscle types. Cardiac muscle is found in the heart, where its contractions pump blood. Smooth muscle, which is found in the digestive tract and blood vessels, contracts in waves.

27.6 How Animals Move in Their Environment

- Among animals, there are three main modes of locomotion: running, swimming, and flying.

- Drag resists the thrust of forward motion and is more severe in water than in air. A streamlined body design and, in water, a slippery body surface reduce pressure drag and frictional drag, respectively.
- Within each mode of locomotion, heavier animals expend less energy per pound of body weight than lighter animals. Runners expend the most energy per pound per distance, and swimmers the least.

◉ Review and Application of Key Concepts

1. Describe how bone and cartilage provide a framework of support for humans.

2. Knee joints are designed for one type of "back and forth" motion. Design a joint that would allow circular motion, such as swinging an arm to throw an underhand softball pitch. Where would you place cartilage?

3. Turn your hand palm up and bring it to your shoulder. Draw the arrangement of bones, joint, and muscles that you used to do this simple motion as a lever system. Include tendons and ligaments.

4. Sarcomeres are the basic unit of muscle contraction. Describe how a sarcomere works.

5. Describe how drag would work as a bird flaps its wings and moves forward. How could drag actually assist a bird's motion as a wing flaps?

Key Terms

actin filament (p. 526)
appendicular skeleton (p. 520)
axial skeleton (p. 520)
cardiac muscle (p. 530)
cartilage (p. 521)
chitin (p. 522)
collagen (p. 521)
compact bone (p. 520)
endoskeleton (p. 523)
exoskeleton (p. 523)
fast muscle fiber (p. 528)
frictional drag (p. 531)
fulcrum (p. 529)
hydrostat (p. 521)
joint (p. 521)
lever system (p. 529)
ligament (p. 522)
load arm (p. 529)
marrow (p. 520)

muscle (p. 520)
muscle fiber (p. 526)
myofibril (p. 526)
myoglobin (p. 528)
myosin filament (p. 526)
neuromuscular junction (p. 528)
osteocyte (p. 520)
power arm (p. 529)
pressure drag (p. 531)
sarcomere (p. 526)
skeleton (p. 520)
slow muscle fiber (p. 528)
smooth muscle (p. 530)
spongy bone (p. 520)
synovial sac (p. 525)
tendon (p. 525)
thrust (p. 531)
Z disc (p. 526)

Self-Quiz

1. Which of the following would you find in the appendicular skeleton?
   a. legs
   b. arms
   c. toes
   d. all of the above

2. Which of the following is a hydrostat?
   a. compact bone
   b. spongy bone
   c. cartilage
   d. tongue

3. Ligaments
   a. connect muscles to other muscles.
   b. connect muscles to bones.
   c. connect bones to other bones.
   d. connect tendons to muscles.

4. Which of the following would have the strongest skeleton relative to its body weight?
   a. a giraffe
   b. a human
   c. a squirrel
   d. a buffalo

5. Which of the following would *not* be found in the human knee joint?
   a. synovial sac
   b. chitin
   c. cartilage
   d. ligaments

6. Sarcomeres contain
   a. muscle fibers.
   b. myofibrils.
   c. actin.
   d. cartilage.

7. Which of the following is *not* part of a sarcomere?
   a. actin
   b. myosin
   c. Z disc
   d. collagen

8. Which of the following athletes would have the most muscles containing large amounts of myoglobin?
   a. a marathon runner
   b. a 100-meter sprinter
   c. a high jumper
   d. a gymnast

9. A joint in the skeleton acts as which part of a lever system?
   a. power arm
   b. load arm
   c. fulcrum
   d. none of the above

10. Which of the following, similarly sized animals would have the biggest problem with frictional drag?
    a. a slow-moving bird in flight
    b. a fast-moving bird in flight
    c. a slow-moving fish
    d. a fast-moving fish

Biology in the News

Powell the Fastest Man on Earth

By Duncan Mackay

Asafa Powell set a world record for the 100 metres last night in the Athens stadium where he flopped in the Olympic Games 10 months ago.

The Jamaican, who has been competing for less than five years, ran 9.77 sec in the Tsiklitiria 2005 meeting to take one-hundredth of a second off the previous record with a performance that, in the end, was as predictable as it was spectacular.

Powell, the son of a preacher from Spanish Town, becomes the ninth man to hold the world record since Jim Hines first dipped below 10 seconds in 1968.

Every year, athletes set new world records. On the face of it, this makes perfect sense: setting world records is what world-class athletes do. Powell had even predicted that he would set a world record in 2005.

The trend in times recorded in the men's 100-meter sprint over the past 40 years suggests that we can expect more world records in the years to come. In 1968, Jim Hines became the first sprinter to break the 10-second barrier by running 100 meters in 9.95 seconds. Before that run, no one had ever covered 100 meters in under 10 seconds. Since then, sprinters have done it well over 250 times. Nowadays, a less-than-10-second sprint is still good, but by no means exceptional.

Does this mean that humans will continue to improve their times indefinitely in the 100-meter sprint and other races? Some of the improvement in times over the past several decades results from improvements in technology. For example, the relatively basic running shoes of years gone by have evolved into intricately engineered shoes that combine light weight with stiffness and suppleness in just the right places. Similarly, the shorts and muscle-Ts that served sprinters well in the 1960s have morphed into the aerodynamic bodysuits of today.

Other advances have come from an improved understanding of the physics and psychology of running. Today sprinters have their strides digitized and analyzed by computer so that they can be adjusted to convert muscle action into forward motion as efficiently as possible. Sports psychologists help sprinters overcome mental obstacles that might add a thousandth of a second to their time.

Even though the record time for the 100-meter sprint continues to fall, there is some evidence that overall performance may have peaked. In 1998, the average of the ten fastest times was 9.86 seconds. In 2003, this average increased to 9.97 seconds, over a tenth of a second longer than 5 years earlier. Perhaps the technology of running has begun to approach the limits of what it can do.

Evaluating the news

1. In many sports, including sprinting, the women's world records are being broken more frequently than are those in men's sports. Can you think of reasons why this might be so?

2. When a world record is broken, it is always big news. Do you think that when the limit of human performance has been reached in a particular sport, and world records in that sport are being broken only very rarely, that the sport will become less exciting to watch? Do you think that when world records stop being broken, that fans of track and field will begin to lose interest?

3. If improved training methods and improved technologies have played an important part in the ever faster times recorded by runners, how is this different from the use of ever more effective drugs to enhance performance? How should we view the roles of different ways of enhancing performance in athletics?

Source: *The Guardian*, Wednesday, June 15, 2005.

28 Defenses Against Disease

Key Concepts

◉ Our first line of defense against pathogens consists of physical and chemical barriers that keep them from crossing from the outside to the inside of the body.

◉ When pathogens do succeed in getting inside, the body mounts several internal defenses. Some of these defenses are immediate, short-term, and general, fighting anything considered "foreign"; others are slower, longer-term, and specific to the invading pathogen.

◉ All animals can distinguish their own cells from "foreign" cells and proteins, a key requirement in fighting pathogens.

◉ When tissues are injured, the body acts as if pathogens are present and immediately sets in motion physiological changes that make combating pathogens easier.

◉ The human immune system has several types of defensive cells and proteins that target specific pathogens. The immune system "remembers" prior infections and responds more quickly to subsequent invasions of a particular pathogen.

◉ Many common diseases are the result of the immune system turning against the body's own tissues.

The Continuing Spread of HIV

The problem first came to the world's attention in the early 1980s when doctors in the United States began to notice that young gay men were falling victim to a variety of previously rare diseases, including a skin cancer called Kaposi sarcoma (which normally affects only elderly people) and an unusual kind of pneumonia. By the mid-1980s, it became clear that these rare diseases were attacking people who had experienced a failure of the immune system—the human body's main internal defense against disease. Researchers also learned that this failure, now called AIDS (acquired immunodeficiency syndrome), was caused by HIV, the human immunodeficiency virus. Those who developed AIDS invariably died.

The number of reported cases of this frightening new disease rapidly increased in North America and Europe until it claimed tens of thousands of people each year. Initially, most cases were among gay men, intravenous drug users, and people requiring frequent blood transfusions. The common factor among these three groups was intimate contact with the body fluids of others: gay men during sex, drug users by exchanging dirty needles, and those needing blood transfusions by receiving contaminated blood. In 2004, 21,000 new cases appeared in Western and Central Europe, and an additional 44,000 new cases were reported in North America.

As sobering as the impact of AIDS in North America and Europe may be, less than 1 percent of the population has been infected. In contrast, the impact of AIDS on many African countries almost defies comprehension. Since we published the previous edition of this book, the number of Africans infected with HIV has tripled to 30 million. In two of the hardest-hit nations, Zimbabwe and Botswana, a staggering one of every three adults tests positive for HIV. A child born in Botswana in 1998, when AIDS was just

The Human Toll of AIDS
Leopards Hill Cemetery in Lusaka, Zambia, is a free city cemetery taken up mostly with AIDS victims.

541

beginning to tighten its grip on the country, could expect to live 62 years. A child born in the AIDS-ravaged Botswana of today can expect to live only 37 years. In Africa, AIDS has attacked people in all walks of life, all classes of society, including men and women, adults and children. The social fabric of the hardest-hit nations has begun to come apart as one in ten children have been orphaned by the disease and the workforce has been decimated.

The spread of AIDS in Africa probably stems from two main causes. First, lack of funds forces hospitals to reuse often improperly sterilized needles for blood transfusions and vaccinations. This fact may be responsible for over half of the AIDS transmission among adults in Africa. Second, the social upheaval that followed the withdrawal of the

European colonial powers from Africa during the 1960s created a dangerous mix of conditions: frequent use of prostitutes by men in labor camps promoted the spread of many sexually transmitted diseases; this often resulted in open sores, which allows for the blood of male and female partners to come into contact during sex. In addition, regular migrations of individuals and whole ethnic groups enabled the rapid spread of disease.

Once established in Africa, AIDS spread quickly and widely, and its eventual toll is likely to be staggering. The most optimistic estimates suggest that, even if effective technology and financial aid for fighting the disease were provided immediately, another 8 million Africans would become infected with HIV by 2010. What makes HIV such a deadly virus?

Long ago, scientists began referring to agents that cause diseases as "germs." The air we breathe is full of germs, and almost any surface we touch is crawling with germs. Even hospitals are not exempt. Germs that can cause serious infections have been found on doctors' stethoscopes and hospital uniforms. In general, any agent that causes disease is called a **pathogen**, and an infected organism or cell is called a **host**. Another term for a pathogen is *infectious agent*,

meaning an organism or a virus that can cause disease if it infects the host's body. **Infectious diseases** are caused by bacteria, protists, and viruses as well as by some fungi and multicellular animals, such as parasitic worms.

Though pathogens abound in nature, our bodies have a remarkable collection of weapons to keep most of them from hurting us. In addition, we have invented ways to expand that arsenal with antibiotics and vaccines (although overuse of antibiotics may be playing a role in the emergence of resistant bacterial strains, as we saw in Chapter 17). Even without those extra weapons, however, we generally do pretty well. A cold doesn't last forever because our bodies have ways of fighting and eventually eliminating the infectious agent that caused it. If we get the measles as children, we don't need to worry about getting that disease again because our bodies have a means of "remembering" the measles virus and pulling out the right weapon to fight it. This defense against a second attack by the measles virus works so quickly and powerfully that we probably won't even notice any symptoms.

You might ask, If our bodies are so good at "remembering" measles, why can't they "remember" a cold? Why do we get colds again and again? The symptoms of all colds are about the same, so we tend to think of all colds as having the same cause. But a huge number of different viruses cause the common cold. You won't ever get the specific cold you once had again, because your body will "remember" that one. But the next one will be new to your system (Figure 28.1).

(a)

(b)

Figure 28.1 There Are Many Kinds of Cold Viruses

There are approximately 200 different viruses that produce the symptoms we lump together as "the common cold." Shown here are transmission electron micrographs of single particles of (a) adenovirus and (b) coronavirus.

In this chapter we examine the body's injury- and disease-fighting arsenal. We begin by exploring our first lines of defense against invasion by pathogens, and how invading pathogens can sometimes get past those first defenses to enter the body. Then we look at the body's second and third lines of defense, which comprise the remarkable internal defense systems that humans and other vertebrates have evolved to fight invaders. We close with a look at what happens when these defense systems turn against the body they are intended to protect.

pathogens. The tubes of the respiratory system, as we saw in Chapter 22, trap invading pathogens in mucus, which is then removed, with its contents, from the respiratory system.

The digestive system is less vulnerable to invasion than the skin or the respiratory system because the acids and digestive enzymes in the stomach destroy most of the pathogens we swallow. The few pathogens that can survive the harsh environment of the stomach are largely kept from entering the body by the lining of the intestine.

28.1 Our First Lines of Defense: Physical and Chemical Barriers

Our first lines of defense against invading pathogens consist of barriers that work to keep them from entering the tissues of the body. Some of these barriers are physical, just like the walls, closed windows, and locked doors of a home. Some are sticky traps that bind up invaders so that they can be disposed of. Others barriers are chemical, and destroy the invaders.

Linings in the lungs, digestive system, and skin keep most pathogens at bay

Pathogens can infect us only if they can find a way into our bodies. Linings that separate the "outside" from the "inside" of the body make up the first line of defense against invaders. These linings—the skin and the linings of the lungs and digestive system—act as a barrier to keep out most potentially dangerous pathogens. Our skin may not seem like a strong defense to us, but unless it is injured, its outer layer of dead cells and the interlocking live cells just beneath it make the skin virtually impenetrable to bacteria and other pathogens.

For areas not protected by a tough outer layer of skin, we have other outer defenses. For instance, even when we are not crying, our eyes are bathed by the same fluid that forms tears. Bacteria that land on the outer surfaces of our eyes are destroyed by antibacterial chemicals in this fluid. Similarly, many bacteria entering our bodies in food or drink are killed by chemicals in our saliva. Hairs in our noses filter out large airborne particles, such as dust, that may carry

Pathogens have ways of crossing the body's physical barriers

Although the body's physical barriers do a good job of keeping out most pathogens, they are still vulnerable to penetration in different ways. Consider the skin. Wounds, in the form of cuts, abrasions, and punctures, are common. Many pathogens take advantage of breaks in the skin to gain entry to their hosts. For example, the tetanus-causing bacterium, *Clostridium tetani*, occurs commonly in soil and on other surfaces worldwide. Although we constantly come in contact with this bacterium, it becomes a threat only if it enters the body through a poorly aerated wound, such as a deep puncture wound. In the absence of oxygen, *C. tetani* can multiply rapidly, producing a **toxin** (a poisonous protein) that causes severe muscle spasms. In most cases, our body's defenses alone cannot act quickly enough to prevent fatal amounts of this toxin from being produced; for those who do not receive regular tetanus shots and boosters, death is the usual outcome. In countries where infants are born under poor sanitary conditions and are not vaccinated, *C. tetani* kills more than half a million babies annually by entering through the cut made at birth to remove the umbilical cord.

Many important infectious diseases and parasites of humans and other animals rely on blood-feeding insects to penetrate the host's skin. For example, the saliva of certain mosquito species carries *Plasmodium*, the protist that causes malaria, an infectious disease that kills 1 to 2 million people annually. When biting through skin to take a blood meal, these mosquitoes inject *Plasmodium* into the bloodstream. Flea bites were responsible for spreading bubonic plague from person to person in Europe in past centuries, and lice have done the same during typhus epidemics (Figure 28.2). Other well-known insect-borne diseases include West Nile virus, yellow fever, and sleeping sickness.

Figure 28.2 Getting Past Our First Line of Defense
Blood-feeding insects can inject pathogens directly into an animal's body, bypassing the skin's defenses. Shown here are (a) an *Anopheles funestus* mosquito sucking blood from a human, (b) a flea biting a dog, and (c) a louse biting a pig.

(a)

(b)

(c)

28.2 If the Invasion Is On, How Do We Fight It?

Once invading pathogens get through the physical and chemical barriers of the body, defense becomes a much more complicated matter. Humans, like all animals, face two distinct challenges: first, we must have a way of detecting the presence of pathogens, and second, we must mount defenses against them within our own bodies.

Detecting pathogens requires distinguishing "self" from "other"

Before we can mount an internal defense that kills, disables, or isolates invading pathogens, our body must have a way of recognizing that an invader is present. Although we are not consciously aware of it, our body can distinguish foreign invaders from our own cells. If our internal defenders could not tell "self" from "other," they would attack our own cells—and indeed, this sometimes happens, as we shall see later in this chapter.

We make the same sort of distinction in team sports. We know which team is which because each team wears a different uniform. We know who's on our home team and who's on the opposing team by the uniforms the players wear. Similarly, the cells in our bodies wear our "home team uniform," and each kind of invading pathogen wears the uniform of its own specific team. The "uniforms" for both the home team and the opposing teams consist of proteins or complex carbohydrates that are associated with the plasma membranes of cells or the outer coats of viruses.

Our internal defenses are both nonspecific and specific

When invaders have breached our first lines of defense—our physical and chemical barriers—we have additional ways of defending ourselves. Our second line of defense is immediate and consists of **nonspecific responses**. As soon as the body detects cells or molecules that don't belong, certain specialized cells take measures to eliminate the "nonself" pathogens and substances. Using the team sports analogy again, in a nonspecific response, the defending team is prepared to play against any team not wearing the home team uniform.

In addition, some animals—most notably vertebrates like us—have a third line of defense, which operates over the long term and consists of **specific responses**: the body goes beyond simply recognizing something as "nonself" to identify particular invaders, such as *C. tetani* or the measles virus. To do this, the body has separate defending teams of cells, each specialized to recognize and play against only one particular opposing team. These specialized teams of cells constitute our immune system. In addition to their specificity, these teams of cells have long memories: each team can "remember" being attacked by a particular kind of pathogen in the past, allowing the body to mount an exceptionally effective defense if the same opposing team shows up again. The immune system can "hold a grudge" for decades. It is because of this memory of our own specific response to a pathogen that, for example, we become immune to attack by the virus that causes measles after we suffer the disease a first time.

The players for the "home teams" include several types of white blood cells and a battery of proteins and other molecules. **White blood cells** are a kind of defensive cells found in the interstitial fluid between the cells in the most vulnerable parts of the body and circulating in the blood, where they constantly intercept invading pathogens. As we shall see, some types participate only in nonspecific responses, and others only in specific responses, while a few are involved in both kinds of responses. Most of these players have several roles—although they specialize, they also back up other players.

With this brief overview of our internal defenses, let's examine the nonspecific and specific responses in greater detail.

28.3 Our Second Line of Defense: Nonspecific Responses

When a pathogen has breached our physical and chemical barriers, our bodies have several nonspecific means of defending against the intruder. Some of these defenses are local, occurring at the point of entry, while others involve the whole body. Most of these nonspecific responses occur not only in humans, but in most other vertebrates as well.

Macrophages engulf or encapsulate stray pathogens

In addition to the mechanical and chemical protection our outer barriers provide (our first line of defense), certain cells patrol these borders from within. In the skin, the alveoli of our lungs, and many other places in the body, white blood cells called **macrophages** (*macro*, "big"; *phage*, "eater") are constantly on the prowl. They destroy invading cellular organisms the same way many single-celled eukaryotes, such as amoebas, eat their food, engulfing the pathogens by phagocytosis (Figure 28.3). Any pathogens the macrophages find indigestible they can isolate from other cells by **encapsulation**, in which they form a protective capsule around the invaders.

Individual macrophages die once they are full of pathogens. They are removed from the body by various means. Dead macrophages from the lining of the alveoli, for example, are swept out of the lungs, along with whatever they have ingested, in the mucus that leaves the respiratory system.

Although many invaders enter the body by crossing the portion of the small intestine devoted to absorbing nutrients, this part of the body wall does not contain large numbers of macrophages. Pathogens that pass directly into the circulatory system along with absorbed nutrients are delivered to the liver, where they are quickly destroyed by macrophages.

The inflammatory response prepares an injured site for defense against pathogens

When cells have been irritated, damaged, or killed, the result is tissue damage. The body responds to tissue damage by mounting an immediate and coordinated sequence of events called the **inflammatory response**. These nonspecific events occur whether or not any pathogens are present in the wound—the body "assumes" that invaders will get in any time the body wall is breached. Although the inflammatory response can occur anywhere inside the body, let's consider what happens following a deep cut or puncture wound to the skin. Though we've laid out these events sequentially, many of them actually happen at the same time.

The inflammatory response begins as damaged body cells or pathogens in the wound release chemicals. Injured **mast cells**, a type of white blood cells that live in connective tissue, release **histamine**. Histamine has two effects. One is to cause arterioles to dilate, bringing more blood to the area around the wound. Histamines do this by stimulating damaged cells to release chemical messages called **prostaglandins**, which stimulate increased blood flow to the area. We experience this increased flow as warmth and reddening in the injured area, but it serves an important purpose: more oxygen and nutrients, which are needed for repair, can reach the area more quickly.

Histamine also causes capillaries in the area to become more porous. This allows large molecules and white blood cells, which normally cannot cross capillary walls because of their large size, to enter the tissues in the area and begin their work. In addition, blood plasma leaks into the interstitial fluid in the area. We experience this leakage as swelling, but again, it has important functions, including diluting any toxins that invaders may be producing and making the region hypotonic to the blood plasma, which draws even more cells and proteins out of the blood and into the area.

When the skin barrier is broken, our bodies reduce the risk posed by an open wound by sealing off the damage. Tissue damage triggers events that cause proteins normally dissolved in the blood to combine with sticky cell fragments, called **platelets**, that also normally

Figure 28.3

Macrophages Eat Our Microscopic Enemies

Macrophages render pathogens harmless either by ingesting them (by phagocytosis, as in this transmission electron micrograph of a *Staphylococcus* bacterium being engulfed) or by walling them off from other cells (encapsulation).

circulate in the blood, forming a gel-like mesh that entraps blood cells (Figure 28.4). The result is a blood clot, which helps prevent any pathogens that may be present in the wound from spreading to surrounding tissues. Clotting can begin as quickly as 15 seconds after tissue damage occurs. The subsequent growth of new tissues then repairs the wound more permanently.

Macrophage precursor cells are permanently stationed just below the surface of the skin. The proteins involved in clotting stimulate them to develop into large, active macrophages. In addition to their other functions, macrophages release a chemical that causes us to become drowsy and our body's average temperature to rise. The resulting fever may be uncomfortable, but as long as they are moderate, fevers are beneficial: they limit bacterial growth, enhance phagocytosis, and speed repair of damaged tissues. The drowsiness, along with pain signals initiated by prostaglandins, signal us to slow our pace of activity and attend to the wound while we heal.

White blood cells and defensive proteins mount nonspecific attacks on any pathogens present

Circulating in the blood plasma are about 20 **complement proteins** (so called because they complement, or work together with, white blood cells). When they arrive at the wound, they bind to and kill invading cells by destroying the invaders' plasma membranes. In addition, by remaining bound only to the invaders, complement proteins help white blood cells identify and destroy foreign cells. For example, complement proteins marking the surfaces of invading *C. tetani* cells help macrophages bind to these cells in preparation for phagocytosis. The few activated macrophages initially present at the wound then begin engulfing *C. tetani* bacteria and damaged cells.

Complement proteins also signal another type of white blood cell, the **neutrophils** (Figure 28.5), which

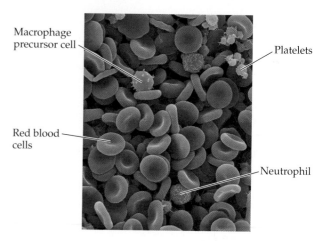

Figure 28.5
Some Players on the Nonspecific Defense Team

normally circulates in the blood, to come to the wound. Individual neutrophils are smaller than macrophages, so they destroy fewer invaders and cannot surround invading cells bigger than bacteria, but they compensate for their smaller size with numbers. Within a few hours of injury, the neutrophil concentration at the wound may soar from 4,000 to 25,000 per milliliter of blood. During the next several days, additional macrophages make their way to the wound from a population circulating in the blood in precursor form. Although macrophages accumulate much more slowly than neutrophils, they eventually become the dominant white blood cell at the site. As the infection progresses, macrophages shift from attacking pathogens to cleaning up dead cells and debris at the site.

When the pathogen is a virus rather than a bacterium or protist, the body's response is somewhat different. Defensive cells respond not to the viral particles directly, but to the cells infected by them. Bits of the viral protein coat, abandoned on the plasma membrane when the virus penetrates a cell, mark infected human cells. Macrophages and neutrophils devour these marked cells. In addition, cells infected by viruses release **interferon**. This chemical attaches to the plasma membranes of nearby cells, interfering (hence its name) with the ability of the virus to enter and infect those cells. Interferon also summons another type of white blood cell, **natural killer (NK) cells**, to the site. NK cells destroy any cell whose plasma membrane is marked with foreign proteins, including viral proteins. NK cells kill either by causing so many holes to form in the other cell's plasma membrane that it disintegrates, or by inducing the infected cell to destroy itself.

Figure 28.4
Blood Clot Sealing the Broken Skin Barrier
An injury activates platelets and fibrin (yellow) to form a mesh and trap blood cells (red), creating a barrier that stops bleeding and prevents pathogens from spreading around the wound.

28.4 Our Third Line of Defense: The Immune System

In a puncture wound like the one described earlier, the chances are high that some live *C. tetani* bacteria have been thrust deep into the wound, thus bypassing our first line of defense. They then find themselves in a favorable environment for proliferating and producing their deadly toxin. In addition to our second line of defense—the various nonspecific responses, including the inflammatory response—the body automatically puts into play a third line of defense, the immune system. The **immune system** is a coordinated combination of specialized cells and molecules that give us *immunity*—the ability to resist specific pathogens. Our description of the human immune system applies to almost all vertebrates, which are the only animals to have evolved this system of specific responses.

The immune system depends on specialized cells, proteins, and the lymphatic system

The human immune system consists of about 7 trillion white blood cells of several types and an assortment of defensive proteins. Many of these cells and proteins circulate in the body in an inactive form. When activated, they move out of the blood into the interstitial fluid to fight an infection. From there, these defensive proteins and white blood cells are collected, along with the interstitial fluid, in the **lymphatic ducts**, a network of tubes that returns the interstitial fluid to the circulatory system. At various points along the lymphatic ducts lie **lymph nodes**, pockets of tissue containing huge numbers of white blood cells that trap bacteria, viruses, and foreign proteins. The lymphatic ducts and the lymph nodes, along with several other organs, make up the **lymphatic system** (Figure 28.6a).

Lymphocytes (Figure 28.6b) are a class of white blood cells that are found primarily in the lymphatic system and play a critical role in the immune system. They allow us to mount specific responses to particular species of invading pathogens. Lymphocytes that bind to *C. tetani*, for example, do not bind to other pathogens, and thus distinguish *C. tetani* from all other invaders.

Lymphocytes develop from stem cells in bone marrow. Immature lymphocytes differentiate into two basic types of mature lymphocytes. **B cells** (B for *bone*) mature in the bone marrow, and **T cells** (T for *thymus*) migrate from the bone marrow to mature in the thymus, a gland found in the chest (Figure 28.7). Once mature, both types

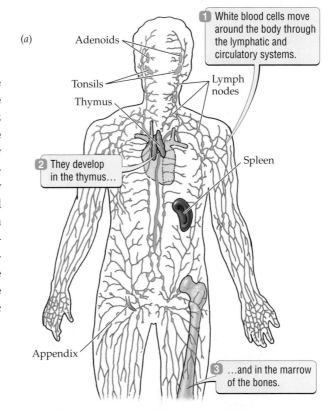

(a)

Adenoids

Tonsils

Thymus

Lymph nodes

① White blood cells move around the body through the lymphatic and circulatory systems.

② They develop in the thymus...

Spleen

③ ...and in the marrow of the bones.

Appendix

(b)   (c)

          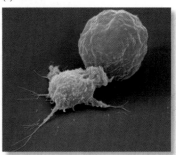

Figure 28.6 The Human Immune System

(a) The human lymphatic system consists of a series of lymphatic ducts and lymph nodes (both in green) and associated organs (as labeled). Lymphocytes circulate in the lymphatic and circulatory systems, and they accumulate in lymph nodes and other organs, such as the spleen and the tonsils, that defend points at which invaders can most easily enter the body. Several types of lymphocytes are involved in mounting a defense against specific pathogens, including (b) helper T cells (pink) and B cells (green), and (c) killer T cells (yellow), here shown attacking a cancer cell.

Test your knowledge of the vertebrate immune system.

 28.1

of lymphocytes move into the lymphatic system, congregating at places where foreign cells and foreign proteins tend to accumulate (especially the lymph nodes and organs associated with the lymphatic system).

Both types of lymphocytes can recognize specific pathogens. Each kind of pathogen that invades the body carries one or more unique molecules on its surface.

Figure 28.7
Lymphocyte Origins
Lymphocytes originate from stem cells in bone marrow. B cells mature in the bone marrow, while T cells mature in the thymus.

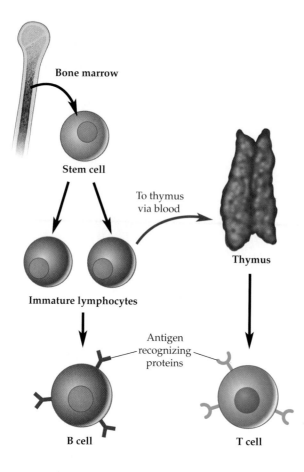

Bone marrow

Stem cell

To thymus via blood

Thymus

Immature lymphocytes

Antigen recognizing proteins

B cell

T cell

bers of the immune system defense team (Figure 28.9*a*). This type of immunity is called **humoral immunity** because the antibodies circulate in the "humors" of the body (the blood and interstitial fluid). Another term for humoral immunity is *antigen-mediated immunity*. T cells, on the other hand, keep their antigen-binding proteins in their plasma membranes, so when these membrane proteins bind to antigens, the whole T cell remains bound to the invader (Figure 28.9*b*). Since the result is a cell-to-cell bond, immunity involving the action of T cells is called **cell-mediated immunity**.

With this overview of the immune system in hand, let's see how its players work together. We pick up the story of a puncture wound to follow two different scenarios:

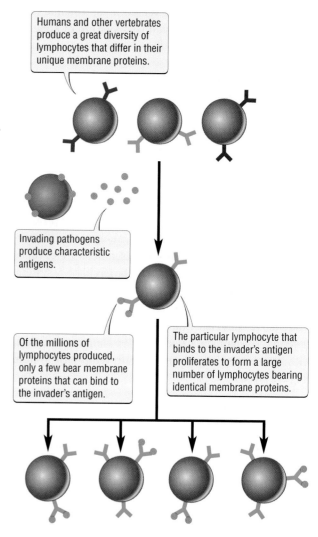

Humans and other vertebrates produce a great diversity of lymphocytes that differ in their unique membrane proteins.

Invading pathogens produce characteristic antigens.

Of the millions of lymphocytes produced, only a few bear membrane proteins that can bind to the invader's antigen.

The particular lymphocyte that binds to the invader's antigen proliferates to form a large number of lymphocytes bearing identical membrane proteins.

Figure 28.8
How the Human Body Recognizes Specific Invaders
Those lymphocytes that bind to an antigen on an invading pathogen multiply rapidly, giving rise to identical lymphocytes, all of which can bind specifically to the same antigen.

These molecules, called **antigens**, characterize pathogens in much the same way that unique combinations of letters and colors on their uniforms distinguish different opposing teams in our team sports analogy. Different lymphocytes have unique proteins embedded in their plasma membranes that can bind with different antigen molecules. Thus each lymphocyte, like a key that fits a particular lock, is individually specialized to recognize and bind with only its corresponding antigen. There are a huge number of antigen-binding proteins, but each "team" of lymphocytes carries only one specific kind in its plasma membrane (Figure 28.8). For example, only one group of B cells and one group of T cells have copies of the membrane protein that recognizes and binds with antigen molecules on the surface of *C. tetani*. Other B cells and T cells may recognize a specific cold virus, or the bacterium that causes strep throat, or even proteins on the surface of a particular kind of pollen grain.

While both types of lymphocytes have antigen-binding membrane proteins, B cells also make and release many free-standing copies of their membrane proteins, which circulate in the blood and lymphatic system as **antibodies**. By binding with the antigens of invading pathogens, antibodies tag these cells for destruction by other mem-

See how the human body recognizes and attacks invaders.

28.2

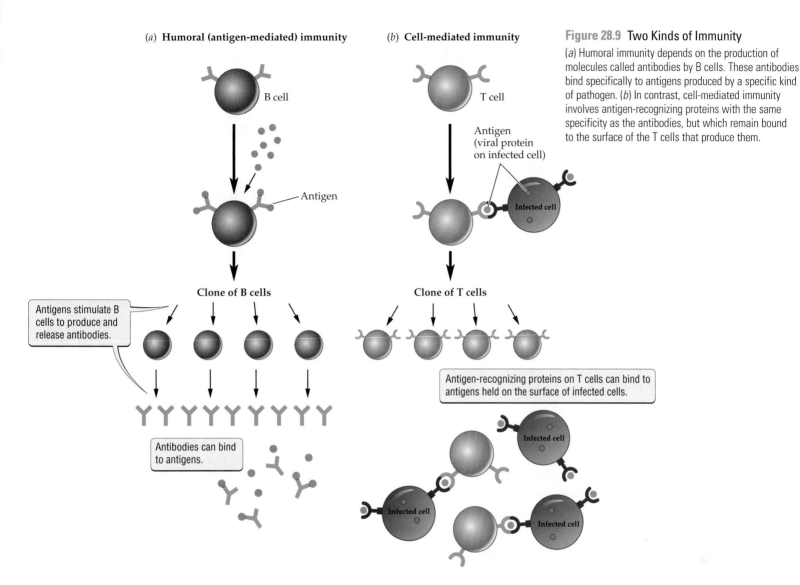

(a) Humoral (antigen-mediated) immunity

B cell

Antigen

Clone of B cells

Antigens stimulate B cells to produce and release antibodies.

Antibodies can bind to antigens.

(b) Cell-mediated immunity

T cell

Antigen (viral protein on infected cell)

Infected cell

Clone of T cells

Antigen-recognizing proteins on T cells can bind to antigens held on the surface of infected cells.

Infected cell

Infected cell

Infected cell

Infected cell

Figure 28.9 Two Kinds of Immunity

(*a*) Humoral immunity depends on the production of molecules called antibodies by B cells. These antibodies bind specifically to antigens produced by a specific kind of pathogen. (*b*) In contrast, cell-mediated immunity involves antigen-recognizing proteins with the same specificity as the antibodies, but which remain bound to the surface of the T cells that produce them.

(1) a puncture wound in which *C. tetani* enters the body for the first time, and (2) a puncture wound that leads to a subsequent exposure to *C. tetani*, either because the person survived an earlier infection or because the person received a tetanus vaccination prior to initial exposure.

Our first infection with a pathogen stimulates a slow, mild immune response

Our first exposure to a particular antigen brings on a **primary immune response**. It takes time—about two weeks—to get this primary response up to full steam. Because the primary immune response takes place relatively slowly, and because *C. tetani* multiplies so rapidly, people infected with *C. tetani* often lose the race, become ill, and die. Similarly, any pathogen that is new to humans is particularly dangerous (see the box on page 550). When combating most other pathogens, the primary response mobilizes B cells and T cells into a defense that usually prevails.

Once a B cell or T cell has bound to its specific antigen, it reproduces rapidly. This process, called **clonal selection**, yields large numbers of identical lymphocytes that carry exactly the same antigen-recognizing proteins as their parent cells (see Figure 28.9). Some of these cloned lymphocytes become **effector cells**, which then set about fighting the pathogen. It takes anywhere from a week to two and a half weeks for the system to get up to full power, but these hordes of identical effector cells reach a point at which they are capable of producing thousands of copies of precisely the antigen-recognizing proteins needed to attack the invading pathogen. Though we are never aware of it, this proliferation literally saves our lives many times.

Once activated by an antigen, B cells and T cells work in different ways. The antibodies released by effector B cells specifically target the antigen on the invading pathogen—in this case, *C. tetani*. When antibody molecules bind to an invading cell, they enhance the action of complement proteins; they also increase the

Biology Matters

Avian Flu: A World at Deadly Risk?

Each year we face the threat of a flu epidemic, and each year between 35,000 and 40,000 people in the United States die from these outbreaks of flu. Flu epidemics are caused by viruses that have been circulating for some time, so some measure of immunity exists in the human population. Flu pandemics, on the other hand, arise when a type of flu virus infecting pigs or poultry, but which has not previously infected humans, develops the capacity to do so and to spread easily from person to person. Since these viruses are totally "new" to humans, they are far more serious and deadly.

Flu pandemics are catastrophic. In 1918 a flu pandemic killed 500,000 people in the United States and an estimated 20 million to 40 million worldwide. The threat of a new pandemic, perhaps even more deadly, is on the horizon. An avian flu virus that has been killing poultry by the millions in parts of Asia has now infected a small number of people. In 1997 eighteen people contracted avian flu from infected poultry, and six of them died. The virus has been transmitted from person to person, but only very rarely.

Experts fear that if the avian flu virus mutates into a form that can be easily transmitted from human to human, the global death toll could be more than double the cost in lives of the 1918 pandemic. Dr. Julie Gerberding, director of the U.S. Centers for Disease Control in Atlanta, Georgia, has described this virus as the world's number one health threat. She warns that the world may see a new strain of this virus to which we have no immunity.

A vaccine to prevent infection by the avian flu virus has been developed, but needs further testing to assess its safety and determine the proper dosage. Even when a fully tested vaccine does become available, however, vaccine producers have difficulty getting enough chicken eggs, in which the virus for vaccines is grown, so it will be some time before supplies will be adequate to vaccinate the general population. Thus, as teams of experts monitor the situation in Asia, governments around the world are preparing emergency plans to treat large numbers of infected people and to implement quarantines to stem the rate of infection.

---

ease with which macrophages and neutrophils can bind to and destroy the invaders. In addition, antibodies can bind to and neutralize small foreign molecules, such as toxins (like those produced by *C. tetani*) and viruses (Figure 28.10), which then cannot penetrate a host's cells. Viruses can proliferate only by using a cell's own biochemical apparatus; if the virus cannot get into a cell, it cannot reproduce.

T cells come in two types. **Helper T cells** are so called because they help other cells in the immune system. When helper T cells become activated by binding to their specific antigen, they not only proliferate rapidly by clonal selec-

tion, but also stimulate the reproduction of B cells and killer T cells that bind to the same antigen (Figure 28.11). Activated helper T cells also help macrophages bind more rapidly to invaders. **Killer T cells** (also called *cytotoxic T cells*) are so named because they destroy the body's own cells that are damaged or have been infected by viruses. Because viruses need working host cells to reproduce, every infected host cell that gets destroyed lessens the chance that the viral particles inside it will be able to complete their reproductive cycle (Figure 28.12). After about three weeks, the antibodies produced in the primary immune response decline to very low levels.

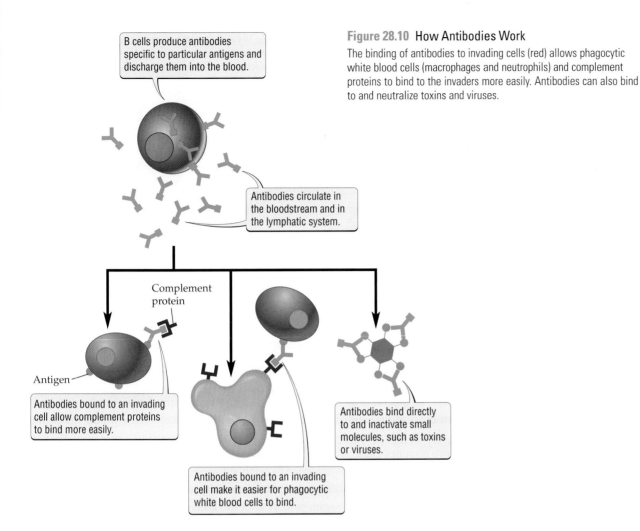

B cells produce antibodies specific to particular antigens and discharge them into the blood.

Antibodies circulate in the bloodstream and in the lymphatic system.

Complement protein

Antigen

Antibodies bound to an invading cell allow complement proteins to bind more easily.

Antibodies bound to an invading cell make it easier for phagocytic white blood cells to bind.

Antibodies bind directly to and inactivate small molecules, such as toxins or viruses.

Figure 28.10 How Antibodies Work
The binding of antibodies to invading cells (red) allows phagocytic white blood cells (macrophages and neutrophils) and complement proteins to bind to the invaders more easily. Antibodies can also bind to and neutralize toxins and viruses.

Subsequent exposures to the same pathogen provoke an even more effective immune response

The immune system produces a faster, more dramatic response to pathogens when it encounters them a second time (Figure 28.13). This subsequent response is called a **secondary immune response**. For example, after our first exposure to the toxin produced by *C. tetani*, we can survive an exposure more than 100,000 times as severe as the one that would have killed us the first time around; thus we are essentially immune to tetanus.

The key to the body's ability to learn from its experience with a particular invader lies with the lymphocytes that recognize particular antigens. Some of the B cells that recognize *C. tetani* and are cloned during the first infection, for example, become **memory cells**, designed to remain in the body as a long-term record of the primary immune response to the bacterium and its toxin. Memory cells are the members of the home team that

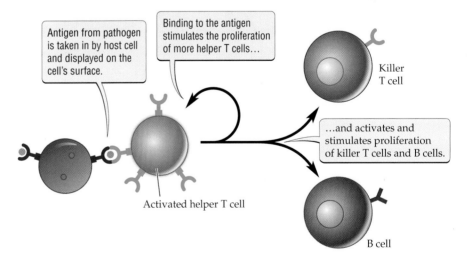

Antigen from pathogen is taken in by host cell and displayed on the cell's surface.

Binding to the antigen stimulates the proliferation of more helper T cells...

...and activates and stimulates proliferation of killer T cells and B cells.

Killer T cell

Activated helper T cell

B cell

Figure 28.11 Helper T Cells
When they become activated by binding to an antigen, helper T cells proliferate rapidly. They also activate B cells and killer T cells, which then generate clones of cells with the same antigen-binding specificity.

Figure 28.12 Killer T Cells Destroy Infected or Damaged Host Cells

One type of protein released by killer T cells forms a hole in the plasma membrane of a host cell that has been infected by a virus, allowing other proteins to enter the cell and destroy it.

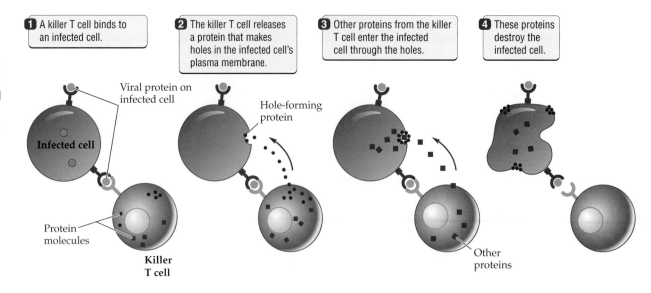

1 A killer T cell binds to an infected cell.

2 The killer T cell releases a protein that makes holes in the infected cell's plasma membrane.

3 Other proteins from the killer T cell enter the infected cell through the holes.

4 These proteins destroy the infected cell.

Viral protein on infected cell

Infected cell

Protein molecules

Killer T cell

Hole-forming protein

Other proteins

hold a grudge. During a second exposure to *C. tetani*, the memory cells specific for *C. tetani* and its toxin multiply rapidly, forming a vast fleet of lymphocytes. Having so many *C. tetani*–specific lymphocytes this time around dramatically speeds up all steps of the immune response and increases the strength of the response as well. Later exposures to this pathogen produce yet more memory cells.

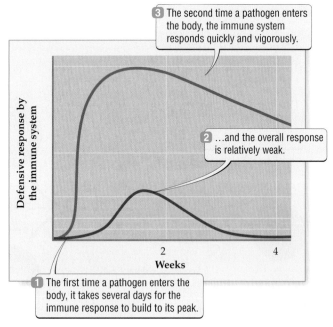

3 The second time a pathogen enters the body, the immune system responds quickly and vigorously.

2 ...and the overall response is relatively weak.

1 The first time a pathogen enters the body, it takes several days for the immune response to build to its peak.

Defensive response by the immune system

2   4
Weeks

Figure 28.13

Primary versus Secondary Immune Responses

Lymphocytes in the human immune system defend the body against a particular pathogen more quickly and more vigorously in second and later attacks than during the original encounter.

Immunity can be gained actively or passively

We can acquire immunity in two ways: either actively or passively.

We acquire **active immunity** to a particular pathogen when the antibodies to its antigens are produced by our own bodies, not received from an outside source. We can acquire active immunity naturally, as when we are exposed to certain diseases, such as measles. We can also acquire active immunity to certain diseases through vaccination. A **vaccine** consists of one or more injections containing a killed or harmless form of a particular bacterium or virus. Even though the pathogen has been rendered into a form that prevents it from making us ill, its antigens are still present. (Tetanus shots, for example, contain an inactive form of the *C. tetani* toxin, but not the live bacterium or the active form of the toxin.) Our bodies respond to these antigens as if they were the real thing— by producing antibodies. This response primes the immune system so that it is ready to attack immediately if we are exposed to that particular pathogen later on. Put another way, vaccination causes a primary immune response so that all our natural exposures to a pathogen will produce secondary responses.

Preventive medicine relies heavily on vaccines. Most vaccines are given to us as children, when the immune system is strong enough to mount a response, but before we are likely to encounter the pathogen naturally. Some vaccines protect against several pathogens at once (see the box on page 553).

We acquire **passive immunity** by receiving antibodies that were not made by our own bodies. A human fetus receives all the nutrients and oxygen it requires from exchanges between its blood and its mother's blood. In

Science Toolkit

Responding to Risk: Individuals, Populations, and Vaccination

Most human activities have a small risk associated with them. Our decision to do something depends on how we balance our perception of its risk (as opposed to its actual risk) against our perception of its benefits. Information about the actual risk usually is expressed from the perspective of large populations: how many individuals out of, say, a million are likely to have something bad happen to them? In contrast, our perception of risk depends on what the "something bad" is and on how much control we think we have over the risk. The difference between real and perceived risk often leads us to act in ways that, statistically, make little sense.

The conflict between real and perceived risk becomes clear in vaccination programs. From a population perspective, vaccination makes a lot of sense. In exchange for a small risk of harmful side effects, vaccines provide people with protection against diseases that carry a significant risk of death or disability. Nevertheless, more and more people are choosing to not have their children vaccinated.

For example, the MMR vaccine protects against three important childhood diseases: measles, mumps, and rubella. Before the 1960s, when vaccines for these diseases were first introduced, measles, mumps, and rubella were viewed as a normal part of childhood. Parents expected their children to catch these illnesses, miss a week or two of school, and recover. For a small proportion of the infected children, however, the consequences were much more severe—and they were even more so for people who caught these diseases as adults. In the 1950s, 2 of every 1,000 measles patients died of measles-related complications. Mumps killed between 1 and 2 of every 1,000 people who caught the disease, and it could cause sterility when it infected teenage or adult males. Rubella is a relatively mild disease of childhood, but a women who gets infected during the first trimester of pregnancy stands a greatly increased chance of miscarriage and stillbirth. If the baby survives to birth, it stands a 1 in 4 chance of suffering from heart defects, deafness, and blindness.

From a public health perspective, the individual vaccines for these diseases and, more recently, the combined MMR shot, have been a great success. Only one in a million children receiving the MMR vaccine suffered dangerous side effects. At the same time, the vaccine has greatly reduced measles, mumps, and rubella. For example, in the United States, 212,000 cases of mumps were recorded in 1964, prior to release of the mumps vaccine, but in 2001 only 201 cases were reported.

Despite the success of the MMR vaccine, many parents choose not to have their children vaccinated. In England, the proportion of children receiving the MMR vaccine has dropped from about 90 percent in the 1990s to about 82 percent. Some parents decide not to give their child the MMR vaccine for religious reasons, but most parents make the decision based on their perception of the benefits and risks.

The benefits of vaccination may not seem that great, in part because parents today are less familiar with the risks associated with these diseases. Many parents did not experience measles, mumps, or rubella themselves, so they think of these diseases as mild conditions. Moreover, because of ongoing vaccination programs, the chances of their child catching the diseases today are small.

At the same time, the perceived risks of the MMR vaccine loom larger. The thought of deliberately injecting even a weakened form of a pathogen has always been a source of concern for the public. In choosing to have their child vaccinated, parents are actively deciding to affect their child's health. Should their child suffer a reaction to the vaccine, parents are likely to perceive it as being their fault, whereas catching the disease is considered simply an unavoidable event with no associated blame.

The perceived risks grew in 1998, when a paper appeared in the medical literature hinting at a link between autism and the MMR vaccine. Children with autism often seem incapable of normal social interaction, exhibit odd repetitive behaviors, and have learning disabilities. This paper did not claim to show a causal connection between the MMR vaccine and autism; therefore, contrary to what was widely reported in the media, the paper made no predictions about the chances that the MMR vaccine would lead to autism. Several large studies have since failed to find any connection between the MMR vaccine and autism, but the perceived risk is now firmly embedded in many parents' minds. The result is an increase in the actual risks that their children face.

MMR Vaccine

the same way, the fetus acquires antibodies from its mother's blood. Her immune system has made many antibodies in her lifetime, which are perhaps the earliest gift she passes on to her child. Another form of passive immunity comes from injections of snakebite antivenin, which contains antibodies to antigens in a particular snake's venom. Since passive immunity produces no memory cells, it wears off as the received antibodies degrade, usually within a few weeks or months.

28.5 Autoimmune Diseases: When Our Defenses Turn Against Us

Sometimes the immune system mistakenly responds to the body's own cells as if they were foreign. The result is an **autoimmune disease**, in which the body destroys some of its own tissues in an attempt to protect itself against a misidentified "invader."

The list of autoimmune diseases is long, and it includes some familiar afflictions. In multiple sclerosis (MS), for example, the immune system attacks neurons in the brain, causing them to lose some of the myelin sheath that is essential to their proper function. The result is a loss of

muscular coordination that eventually confines those with MS to a wheelchair. In rheumatoid arthritis, another autoimmune disease, the immune system attacks the cartilage that lines the joints, leading to crippling pain and loss of joint function (Figure 28.14). The organs and tissues of people with lupus are chronically inflamed because they are targeted by the person's own immune system. Scleroderma is an autoimmune disease that causes hardening of the skin.

Figure 28.14
Rheumatoid Arthritis Is a Common Autoimmune Disease
The disfigurement and pain caused by rheumatoid arthritis can be severe.

Explore why AIDS is so deadly.

28.3

Why Is AIDS So Deadly?

HIV, the virus that causes AIDS, kills its host not directly, but by weakening the host's immune system. HIV attacks two components of the human immune system: the macrophages, which devour infected cells, and the helper T cells, which stimulate the activity of antibody-producing B cells and killer T cells.

Evidence suggests that immediately after infection, the human immune system does a good job of keeping HIV at bay. When activated, killer T cells recognize and destroy helper T cells infected by the virus. By doing so, the killer T cells greatly slow the rate at which HIV reproduces and spreads, and they allow replacement of helper T cells lost to the virus. Most people infected with

HIV go through a period of up to 15 years—between the time of infection and the onset of AIDS symptoms—during which the immune system holds its ground.

There is increasing evidence, however, that HIV mutates quickly, evolving rapidly inside an infected person. The antigens on the protein coat of HIV thus change rapidly over time. As the infection progresses, the body's lymphocytes seem to have an increasingly difficult time identifying HIV, which leads to an increasingly less effective immune response. Thus lymphocytes that fit the HIV antigen types dominating at the time of infection may not recognize the most common HIV antigen types found in the body a few years later.

Eventually HIV destroys helper T cells more quickly than they can be replaced. The downfall of the helper T cells leads to a decrease in the effectiveness of the B cells and the killer T cells. Ultimately the victim's immune system collapses, and opportunistic infection by pathogens that a healthy immune system could easily defeat leads to death. Over 3 million people die each year worldwide because their immune systems have been severely compromised by HIV.

But research is progressing, and new drugs are being developed. One class of drugs interrupts the early stages of the virus's reproduction, and another interrupts later stages. Yet a third class of drugs acts by blocking the ability of the virus to merge with plasma membranes, which it must do to enter a cell. While they do not offer a cure, these drugs, used in combination, have improved the health of many AIDS victims. And some people infected with HIV live for many years without AIDS symptoms. Studying these long-term survivors is a fast-growing area of research that may yield valuable new approaches to attacking the disease.

Chapter Review

Summary

28.1 Our First Lines of Defense: Physical and Chemical Barriers

- Human and other animal bodies have ways of keeping pathogens from gaining entry.
- Our skin and the linings of our respiratory and digestive systems form barriers against pathogens.
- Many pathogens take advantage of breaks in the skin to gain entry to their hosts. Some pathogens rely on other organisms, such as blood-feeding insects, to penetrate the skin.

28.2 If the Invasion Is On, How Do We Fight It?

- Our bodies can distinguish "self" (the body's own cells) from "nonself" (cells and molecules that are not part of the body).
- Once pathogens are detected, the body mounts a series of immediate, short-term, nonspecific responses (in which defensive cells can only distinguish self from nonself) and various slower, longer-term, but specific responses (in which specific defensive cells can recognize specific pathogens).
- Several types of white blood cells and many defensive proteins are involved in specific responses.

28.3 Our Second Line of Defense: Nonspecific Responses

- Large white blood cells called macrophages engulf or encapsulate stray pathogens that get past the physical barriers of the skin and lungs. Pathogens absorbed across the small intestine circulate to the liver, where they are engulfed by macrophages.
- The inflammatory response prepares an injured site for battle with pathogens.

- Histamine and prostaglandins cause blood vessels to dilate, bringing more blood, and hence more nutrients and oxygen, to the injured area. Histamine also makes capillaries more porous so that large molecules and white blood cells can more easily pass from the blood into the interstitial fluid. The leakiness of the capillaries also allows blood plasma to move into the area. We experience these effects as warmth, redness, and swelling.
- Tissue damage stimulates platelets and proteins in the blood to form a blood clot, which helps prevent any pathogens present in the wound from spreading.
- A chemical released by macrophages causes fever and drowsiness, and prostaglandins initiate pain signals. These effects encourage us to slow our ordinary activities while we heal.
- A battery of white blood cells and defensive proteins mounts nonspecific defenses against any pathogens present.
- Complement proteins circulate to the wound, where they kill invading cells by destroying their plasma membranes and tag pathogens for destruction by white blood cells.
- Together, activated macrophages and the smaller neutrophils destroy and isolate cellular pathogens and virus-infected body cells.
- Virus-infected body cells release interferon, which prevents infection of nearby cells by viruses. Interferon also summons NK cells, which destroy any cells (including body cells infected by viruses) with foreign proteins on their surfaces.

28.4 Our Third Line of Defense: The Immune System

- The immune system is a coordinated combination of specialized cells and molecules that provides long-term defenses against specific pathogens and parasites. While all animals have internal defenses, only vertebrates have immune systems.

- The lymphatic system provides the primary sites for immune system activity. White blood cells called lymphocytes confer immunity. Immature lymphocytes differentiate into B cells in the bone marrow and T cells in the thymus. Mature lymphocytes are found primarily in the lymphatic system.

- Each lymphocyte has a unique membrane protein that binds only with its corresponding antigen, a unique protein that characterizes the surface of a specific pathogen.

- B cells provide humoral immunity: they make and release antibodies, additional free-standing copies of their antigen-binding membrane proteins. Antibodies circulate in the circulatory and lymphatic systems, where they bind to antigens on invading pathogens, "tagging" them for destruction by other immune system cells.

- T cells confer cell-mediated immunity: the whole T cell binds to the pathogen or infected body cell that bears its specific antigen.

- T cells come in two types: helper T cells, which help other defensive cells in various ways, and killer T cells, which destroy body cells damaged or infected by viruses.

- First exposure to a pathogen brings on a primary immune response to that pathogen. The B cell or T cell that binds to the pathogen's antigen undergoes clonal selection, making many copies of itself. Some of these copies become effector cells, which then combat the new pathogen. A primary immune response is relatively slow and mild.

- In a secondary immune response, the immune system produces a faster, stronger response to a pathogen that it has already encountered one or more times. Some of the B cells that are cloned to recognize the pathogen during the first exposure become memory cells, which stay in the body and respond rapidly to the pathogen when it is encountered again.

- Active immunity to a pathogen results when our own bodies produce antibodies to its antigens. Active immunity can be acquired through natural exposure to a pathogen or through a vaccine. Vaccines "prime" the immune system by provoking a mild primary immune response so that during subsequent natural exposures to the pathogen, the immune system can respond immediately and forcefully.

- Passive immunity comes from receiving antibodies that were not made by our own bodies, as when a fetus acquires antibodies from its mother.

28.5 Autoimmune Diseases: When Our Defenses Turn Against Us

- When the immune system identifies the body's own cells as foreign, the result is an autoimmune disease, in which the immune system destroys some of its own tissues in an attempt to protect us against a misidentified "invader."

◉ Review and Application of Key Concepts

1. Describe the first line of defense that humans and other vertebrates have to prevent the entry of pathogens.

2. How do disease-fighting cells in the body tell the difference between foreign invaders and our own cells?

3. Suppose you were an intelligent bacterium that had penetrated a breach in a person's skin. How would you go about defeating the forces arrayed against you in the second and third lines of defense (nonspecific and specific responses)?

4. Autoimmune diseases can cause immense suffering. Describe how an autoimmune disease develops.

Key Terms

active immunity (p. 552)
antibody (p. 548)
antigen (p. 548)
autoimmune disease (p. 554)
B cell (p. 547)
cell-mediated immunity (p. 548)
clonal selection (p. 549)
complement protein (p. 546)
effector cell (p. 549)
encapsulation (p. 545)
helper T cell (p. 550)
histamine (p. 545)
host (p. 542)
humoral immunity (p. 548)
immune system (p. 547)
infectious disease (p. 542)
inflammatory response (p. 545)
interferon (p. 546)
killer T cell (p. 550)
lymphatic duct (p. 547)
lymphatic system (p. 547)

lymph node (p. 547)
lymphocyte (p. 547)
macrophage (p. 545)
mast cell (p. 545)
memory cell (p. 551)
natural killer (NK) cell (p. 546)
neutrophil (p. 546)
nonspecific response (p. 544)
passive immunity (p. 552)
pathogen (p. 542)
platelet (p. 545)
primary immune response
    (p. 549)
prostaglandin (p. 545)
secondary immune response
    (p. 551)
specific response (p. 544)
T cell (p. 547)
toxin (p. 543)
vaccine (p. 552)
white blood cell (p. 544)

Self-Quiz

1. Which of the following would be the most likely location for invasion by a pathogen?
   a. stomach
   b. lung
   c. small intestine
   d. thymus gland

2. Which of the following would engulf and digest bacteria in a wound?
   a. a killer T cell
   b. a B cell
   c. a mast cell
   d. a macrophage

3. The inflammatory response
   a. occurs only in lymph nodes.
   b. includes the release of histamine.
   c. requires helper T cells.
   d. is an immune response.

4. Macrophages
   a. are part of our immune response system.
   b. produce antibodies.
   c. engulf and digest microscopic invaders.
   d. produce antigens.

5. Which of the following are a part of the immune response?
   a. neutrophils
   b. mast cells
   c. macrophages
   d. none of the above

6. Which of the following statements about antigens is true?
   a. They are produced by helper T cells.
   b. They are present on the surfaces of pathogens.
   c. They are part of a nonspecific response.
   d. They are produced by neutrophils.

7. Which of the following statements about antibodies is true?
   a. They are produced by helper T cells.
   b. They are produced by killer T cells.
   c. They are produced by B cells.
   d. all of the above

8. Which of the following would be able to recognize a specific kind of invading bacterium?
   a. B cells
   b. helper T cells
   c. killer T cells
   d. all of the above

9. Cell-mediated immunity involves
   a. neutrophils.
   b. B cells.
   c. T cells.
   d. histamine.

10. Memory in B cells
    a. prevents autoimmune diseases.
    b. increases the speed and strength of the immune response.
    c. produces specific interferons.
    d. is stored in stem cells.

Biology in the News

Canada Can Learn from Dutch Hospital Standards

AMSTERDAM—Dutch hospitals are winning the fight against antibiotic-resistant infections, thanks to strict patient isolation policies and a "seek-and-destroy" approach to infection control that some experts say should be imported to Canada.

A *CBC News* investigation this week suggested a lack of infection control resources and regulations to enforce cleanliness standards has contributed to the spread of "superbugs" that could be killing as many as 8,000 Canadians a year.

Across Canada, about 250,000 people a year come down with a hospital-acquired infection.

Methods used in the Netherlands could dramatically cut those numbers, if Canada is willing to pay for them, Dutch medical staff say.

Hospitals are supposed to be places where we go to get well. More and more often, however, hospitals in North America have become places where the sick get sicker. In Canada, an estimated 250,000 people each year get bacterial infections acquired while they were in the hospital for other reasons. In the United States, with its much larger population, an estimated 2 million patients get a bacterial infection while in the hospital each year. In contrast, in Dutch hospitals, it is rare for a patient to be infected by bacteria.

As if the high number of infections acquired in hospitals were not bad enough, many bacteria that cause these infections are particularly dangerous. A patient has a 2 percent to 5 percent chance of dying once infected by one of these "superbugs." They are so much deadlier than other bacteria because they are resistant to most of the commonly used antibiotics.

Why are people so prone to infection in North American hospitals? Patients are at risk of infection because they often arrive in the hospital with an immune system already weakened by the illness or injury that put them there to begin with. As a result, their bodies are ill-prepared to defend themselves against pathogens. In the hospital, moreover, patients find themselves closely surrounded by other sick people that may harbor the superbugs. Hospital ventilation systems and staff aid in moving bacteria among patients.

Dutch hospitals have combated the spread of superbugs where they can: by reducing the ability of the bacteria to move between patients. Patients who have previously been treated at hospitals outside the Dutch system are immediately isolated in private rooms away from uninfected patients until it is clear they do not carry a superbug infection. Infected patients are carefully tracked, as are the staff who take care of them. The doctors and nurses take seriously the need to scrub their hands carefully between patients and to properly dispose of gloves and gowns. At the first sign that a patient has been infected by a superbug while in the hospital, two steps are immediately taken: First, the room in which the patient stayed is shut down and thoroughly disinfected. Second, the staff who came into contact with the patient are tested and perhaps even restricted from working until given a clean bill of health.

By contrast, North American hospitals take a much less active role in controlling within-hospital disease transmission. Cleaning staff are often the first to be cut when hospital budgets are tight. Hospitals have less power to enforce standards of cleanliness. And a culture of observing strict hygiene for staff who move between patients has not yet developed in North American hospitals.

SOURCE: *CBC News*, March 23, 2005.

Evaluating the news

1. Why do you think that superbugs survive in hospitals, where antibiotics are widely used?

2. What challenges do you think a hospital faces in trying to prevent the spread of superbugs? Do you think that North American hospitals could accomplish what Dutch hospitals seem to have done?

3. What would you do to protect yourself from a superbug infection if you found yourself hospitalized?

CHAPTER 29

Reproduction and Development

Key Concepts

- In sexual reproduction, gametes from two parents of different sexes join to create a genetically unique individual offspring. Asexual reproduction requires only one parent, and all the offspring are genetically identical to that parent.

- Sexually reproducing animals use various strategies for bringing their gametes close together and releasing them in a favorable environment. The female is often choosier about her mating partner than the male is because she has more energy and resources invested in her gametes.

- The development of a human fetus follows a set pattern from zygote to birth. The early embryo undergoes more rapid cell division than at any other time in its life. After implantation, three basic cell layers differentiate. Cell types become more defined as development continues.

- A cell's fate is determined by which genes are activated in the cell and at what time during development. Specific groups of proteins control the activation of developmental genes.

- Small genetic changes can have large developmental effects, which in turn can affect evolution.

- Humans, like most animals, continue developing after birth. Animals have various ways of replacing lost tissue and dead cells throughout the life span. Unlike most animals, humans today live well past their reproductive years.

Don't Drink While Reproducing

A child developing in its mother's womb has no power to make choices that will affect its health or its future chances for success in the world it is about to enter. But its mother does. If a woman drinks alcohol while she is pregnant, the consequences for the child growing in her womb can be devastating. Alcohol can wreak havoc with an unborn child's rapidly dividing cells—cells destined to form vital organs and intricate nerve connections that will give the child a fair chance of coping well with the challenging outside world it will soon face.

Consider the universe of a child growing in its mother's womb. That child relies for its survival, and its future, on exchanging vital materials with the mother through an intimate connection between the mother's blood and its own. The child receives oxygen and nutrients from the mother's blood and discharges carbon dioxide and wastes into it. Alcohol moves easily from a mother's bloodstream into the blood of her child.

Alcohol can damage the growing child at every stage of development in the womb, producing a range of effects, the most severe of which are collectively called fetal alcohol syndrome. Alcohol kills significant numbers of brain cells, resulting in mental retardation, stunted growth, malformations of the skull and face, and behavioral problems. The child with fetal alcohol syndrome may never be able to function on its own as an adult. In the worst case, the child may die before it is born.

The risk of causing these effects is greatest during the early days of pregnancy, especially the period—which can be almost a month—between conception and the moment a women realizes she is pregnant. Consuming alcohol during that critical time can do enormous damage to

The Effect of Alcohol on the Brain
At 6 weeks after birth, the brain of this normal baby is twice as big as the brain of one with an extreme case of fetal alcohol syndrome.

an unborn child, who has no way of protecting itself. The advice from experts is clear: "Don't drink alcohol, if you are even expecting to be expecting."

Human development is a delicate process, and alcohol isn't the only environmental factor that can disrupt it. What other substances are harmful, and what can be done to avoid their effects?

Reproduction—giving rise to offspring—is a basic feature of all life. Although reproduction always involves making new individuals, for most animals, including humans, it also involves a combining and reshuffling of genes from two parents. These genetic rearrangements play an important role in maintaining the genetic variation needed for evolution by natural selection.

Because an animal's evolutionary success hinges on its reproductive success, the animal devotes much of its time and energy to producing offspring and ensuring that those offspring will themselves survive to reproduce. In a sense, the function of all the structures and processes described so far in Unit 5 is to allow animals to accumulate and manage the resources they need to reproduce.

In this chapter we explore the different ways in which animals reproduce and develop. We begin with a basic overview of reproduction and development in our own species. Then we explore sexual and asexual reproduction in animals and look at the advantages and disadvantages of each. Next we look at how animals deal with two important challenges associated with reproduction: getting eggs and sperm together to produce a fertilized egg, and choosing a mating partner of high genetic quality. We then discuss how the fertilized egg, or zygote, develops into a complex individual consisting of many different cell types, and we look at what controls the differentiation of those cell types. We finish with a brief look at human development following birth.

Test your knowledge of the human reproductive system.

29.1

2 ...and are transported through the vas deferens...

4 Eggs are produced in the ovaries...

5 ...from which they pass into the oviduct.

3 ...to the penis, through which they are ejaculated.

6 The penis deposits sperm into the vagina.

1 Sperm are produced in the testes...

3 ...into the oviducts, where one sperm fertilizes the egg.

Figure 29.1 Sexual Reproduction in Humans

A complex series of events in males and females allows eggs and sperm to meet in the female body. The result is a zygote that can develop in the sheltered environment of the uterus.

Key

Path of sperm (male) →

Path of egg (female) →

4 The zygote moves down the oviduct to the uterus, where it implants to continue development.

2 ...through the cervix and uterus...

1 Sperm swim from the vagina...

29.1 Human Reproduction and Development: A Brief Overview

Reproduction and development in humans provides a good starting point for our exploration of how animals produce offspring. You will recall from Chapter 9 that in sexually reproducing species, males and females produce specialized reproductive cells called gametes. Male gametes (**sperm**) are produced in reproductive organs called **testes** (singular: testis), while female gametes (eggs, also called **ova**; singular: ovum) are produced in organs called **ovaries**. Human sperm can swim through fluid, whereas eggs cannot actively propel themselves.

Fertilization is the first step in development

Reproduction in humans, as in most animals, begins with a sequence of activities that increase the chances that a sperm will encounter an egg and that their fusion will result in a zygote capable of developing into a new individual, who in turn can grow up to reproduce. During courtship, we spend a great deal of energy, time, and effort on finding and impressing a potential mate. Once we manage to win someone's heart, we consummate the relationship by having sexual intercourse, during which the male releases sperm directly into the female's body.

Figure 29.1 shows the general anatomy of the human male and female reproductive systems and the series of events that lead to a fertilized egg. As we saw in Chapter 24, a woman releases from her ovaries, or ovulates, about one egg every 28 days. The egg moves down a tube called the **oviduct** (or fallopian tube). If the woman has sexual intercourse, the man's **penis** ejaculates almost 300 million sperm into her **vagina**. Huge numbers of sperm swim from the vagina through the opening of the uterus, called the cervix, into the uterus, and then up the oviduct in response to a chemical signal released by the ovary. The woman aids the progress of the swimming sperm by a combination of muscular contractions of the reproductive tract and the release of chemical signals that stimulate sperm to swim. Of the many millions of sperm ejaculated into the vagina, only a few hundred manage to reach the egg in the oviduct. Only one of those sperm fertilizes the egg to create a zygote. Fertilization is often called "the moment of conception."

Development in the uterus involves differentiation

Fertilization, though a critical step in human reproduction, marks only the beginning of a 9-month-long period of development within the mother's uterus. During development, the single-celled zygote is transformed into a complex organism consisting of many different types of cells, arranged to form tissues, organs, and organ systems specialized to carry out specific functions. The process by which unspecialized cells acquire specific identities is called **differentiation**, and it follows an orderly, predictable pattern. During this time, the mother's blood provides the developing baby with oxygen and nutrients and carries away the wastes it produces.

At birth, a baby becomes physically independent of its mother. It no longer obtains its oxygen and nutrients directly from her blood, and it must eat, breathe, and maintain homeostasis on its own. Human parents, however, continue to invest a great deal of time and effort in caring for their offspring, protecting them from danger, providing them with food, and teaching them how to function on their own over a childhood that lasts for many years.

29.2 Sexual and Asexual Reproduction in Animals

Animals have two basic means of reproducing. In **sexual reproduction**, the mode of reproduction used by humans, haploid gametes from a male and a female combine to form a diploid zygote, which develops into an individual that is genetically unique and different from either parent. In **asexual reproduction**, cells from only one parent are involved in producing the offspring, and all of the offspring's genes come from that parent.

Sexual reproduction requires meiosis

Recall from Chapter 9 that our gametes differ in chromosome number from other cells in our bodies: they are haploid (denoted n), containing just half the chromosomes found in our other cells, which are diploid ($2n$). To produce mature gametes thus requires meiosis (see Figure 9.7 for a review of this process). The fusion of a haploid egg and a haploid sperm during fertilization produces a diploid zygote.

With some exceptions, animals that reproduce sexually have two distinct sexes—male and female—and the two sexes produce different kinds of gametes. Let's follow

Learn more about the production of male and female gametes.

29.2

gamete production in humans, since it is similar to the pattern found in many other animals.

The production of sperm, or **spermatogenesis** (Figure 29.2a), occurs in tubules that permeate the male's testes. Spermatogenesis involves several steps. In the tubules, diploid precursor cells are constantly undergoing mitosis to produce **primary spermatocytes**. Primary spermatocytes, in turn, undergo meiosis to become mature sperm. Each primary spermatocyte divides to complete the first stage of meiosis (meiosis I), resulting in two haploid **secondary spermatocytes**. The secondary spermatocytes divide again during the second stage of meiosis (meiosis II) to produce four haploid cells that mature into sperm capable of fertilizing an egg. In this way, millions of sperm are produced each day.

Eggs also develop through a series of divisions. **Oogenesis** [*OH*-uh-*JEN*-uh-siss], the production of mature eggs capable of being fertilized, begins before a woman's birth, when undifferentiated precursor cells multiply and develop into diploid cells called **primary oocytes** that can develop into mature eggs. Still before birth, these primary oocytes begin, but do not complete, the first meiotic division on the way to producing haploid eggs. At birth, a human female has in her ovaries between 1 and 2 million primary oocytes whose development has been suspended in the middle of meiosis. By the time she reaches puberty, at about 10 to 12 years of age, only about 400,000 viable primary oocytes remain.

When a female reaches puberty, the primary oocytes resume their development into mature eggs. On average, during her reproductive years (from puberty to menopause, at about 50), one primary oocyte a month successfully develops into a mature egg (Figure 29.2b). The release of follicle-stimulating hormone stimulates several of the small primary oocytes to resume growing. Of these, only the most rapidly growing one completes its meiosis, and the others degenerate. Meiosis in the fastest-growing primary oocyte produces two haploid cells of very different sizes: a large **secondary oocyte** and a smaller **polar body**, which plays no further role. The haploid secondary oocyte then begins a second meiotic division. In the middle of meiosis, the secondary oocyte is released into the oviduct. The second meiotic division is completed only if the secondary oocyte is fertilized by a sperm.

There are several important differences between the production of ova and of sperm. The supply of a female's primary oocytes is limited, and once a primary oocyte develops into a mature ovum, it is deleted from the supply. In contrast, the supply of sperm precursor cells in males constantly replenishes itself because when these cells divide, only some of the resulting cells are primary spermatocytes; the others are new precursor cells. Also noteworthy is that in a normal 28-day menstrual cycle, a human female produces only one mature egg, while a male produces about 200 million sperm every day. These two observations tell us that, in evolutionary terms, a single egg is far more precious than a single sperm.

In addition to these differences, ova are typically much larger than sperm (Figure 29.3a). For example, whereas the human egg is visible (just barely) to the naked eye, seeing individual sperm requires a microscope. Sperm contain little substance beyond chromosomes and the cellular machinery needed to move them to an egg (Figure 29.3b).

For fertilization to take place (Figure 29.3c), a sperm must first work its way through the follicle cells surrounding the outer surface of the egg and then penetrate the egg's outer, gel-like covering. The **acrosome**—the front tip of a sperm—contains enzymes that digest this outer covering. Once inside the egg, the sperm's nucleus separates from the rest of the sperm (which degenerates) and fuses with the egg's nucleus to form a genetically unique diploid zygote. Although the chromosomes of the zygote are contributed equally by the two parents, the organelles and other cellular machinery of the zygote come almost entirely from the female. For example, all of your mitochondria, which generate life-sustaining energy for your body, came from your mother.

Sexually reproducing animals may be male, female, or both

Sex in animals is more variable than our human perspective might lead us to expect. As already noted, humans may be either sperm-producing males or egg-producing females. The majority of animal groups, however, include species in which some individuals produce both functional testes and functional ovaries and are therefore both female and male. We call such individuals **hermaphrodites** [her-*MAF*-roh-dites].

Hermaphrodites live in a wide variety of habitats. They are most common among invertebrates—for example, the common earthworm and most flatworm species are hermaphrodites—but they also exist among vertebrates. Many fish species change sex (and develop different reproductive organs) depending on their size, their stage of life, or environmental conditions. A common misconception about hermaphrodites is that they fertilize their own eggs. Most hermaphrodites, even those, like the earthworm, that have functional testes and ovaries in the same individual at the same time, must still mate with another individual. Nonetheless, unlike animals that are only male or only female throughout their lives, hermaphrodites can mate with any individual they encounter.

Asexual reproduction is also common in animals

In asexual reproduction, no gametes or fertilization are involved, and the offspring are genetically identical to their single parent. Animals reproduce asexually in several different ways. Some, like the hydra, reproduce asexually by budding, in which a fully formed offspring grows on the parent animal (Figure 29.4a). Sponges (Figure 29.4b) can reproduce asexually by fragmentation: a piece of the sponge's body simply breaks off and develops into a whole new sponge. Asexually reproducing animals have lifestyles as various as those of marine jellyfish and desert-dwelling lizards.

Although some animals rely exclusively on asexual reproduction, most asexually reproducing species switch between sexual and asexual reproduction depending on environmental conditions. Aphids, the small insects that sometimes infest our greenhouses or garden plants (see Figure 34.1), usually reproduce asexually. The crowds of aphids that we see sucking sap from a plant in the summer are all genetically identical copies of a female that originally colonized that plant. Aphids rely on sexual reproduction, however, when the temperature drops and the days grow shorter as winter approaches, driving them from their summer food plant.

The two modes of reproduction have different advantages. Sexual reproduction produces offspring with a new mix of genes—their genotypes differ not only from those of their parents, but also from those of their siblings. The resulting genetic variation increases the

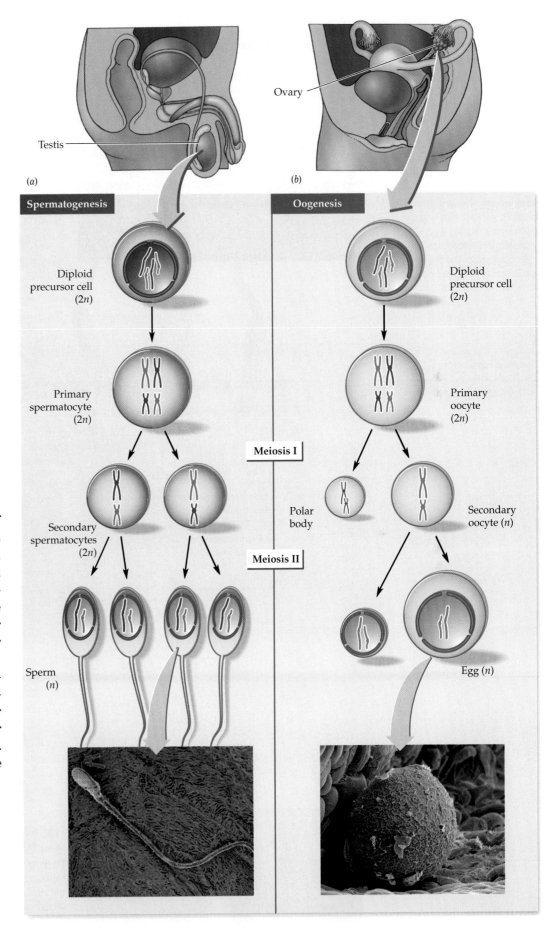

Figure 29.2
The Production of Male and Female Gametes
Sexual reproduction requires the production of haploid gametes by meiosis. (a) Spermatogenesis produces haploid sperm. (b) Oogenesis produces haploid eggs (ova).

Figure 29.3 Fertilization Occurs When a Single Sperm Enters the Egg

(a) Sperm are many times smaller than eggs, as can be seen in this scanning electron micrograph of a swarm of sperm surrounding a single human egg. (b) The tail of a sperm consists of a whiplike flagellum, which helps it wriggle up through the vagina and into the oviduct to meet the egg. The acrosome contains enzymes that help the sperm penetrate the egg's outer covering. (c) Although many sperm are present on the surface of the egg, only one can penetrate to fertilize the egg.

Learn more about the structure of human sperm.

(a)

(b)

Tail (flagellum)

Head

Midpiece containing mitochondria

Nucleus containing chromosomes

Acrosome containing enzymes

(c)

Nucleus

Acrosome

Follicle cells surrounding egg

1 A sperm moves past the follicle cells that surround the egg.

Enzymes from acrosome

Outer covering of egg

2 Enzymes from the acrosome digest a hole in the covering of the egg.

Egg

3 The sperm enters the egg.

4 The haploid nuclei of the egg and sperm fuse to form a diploid zygote nucleus.

Sperm nucleus (n)

Egg nucleus (n)

Plasma membrane

Diploid zygote nucleus (2n)

(a)

(b)

Figure 29.4
Asexual Reproduction Doesn't Require Mates or Gametes

(a) Budding is one type of asexual reproduction, as can be seen in this common freshwater Cnidarian hydra, with the larger parent form on the left and the offspring emerging on the right. (b) Some animals reproduce by fragmenting, such as this Caribbean sponge that had pieces break off and grow next to it.

chances that one of the parents' offspring will survive in a changing environment. Some of the offspring produced by two sexually reproducing parents, for example, may be better able to survive a colder-than-normal winter than their parents and siblings are. The offspring of asexually reproducing animals do not have this genetic variation. They all arise from the same set of genes, and if their environment changes severely enough, all, not just some, of the offspring are threatened.

In contrast, asexual reproduction has the advantage of producing offspring more rapidly and at a lower energy cost—the parent, for example, doesn't have to expend energy to find a mate. For this reason, asexually reproducing animals tend to be found in more stable and steady environments where genetic variation is not needed from one generation to the next.

29.3 Finding and Selecting Mates

For sexually reproducing animals, fertilization is only the last step in a complex sequence of events. A major task in animal reproduction is for members of the same species and opposite sexes to find each other, and thus bring their sperm and eggs close enough together to give fertilization a good chance of occurring.

Successful fertilization requires strategies to bring sperm and egg closer together

Mating with a member of the wrong species cannot produce viable offspring, as we saw in Chapter 18, so animals have developed ways of recognizing and reacting to members of their own species. Animals can find other individuals of their species by using signals of various kinds (Figure 29.5). The most common are sounds and chemical signals, both of which can travel long distances. Cricket chirps, for example, are generally heard within the ranges of many potential mates, and chemical signals released by female moths can draw in males from distances of many kilometers, as we saw in Chapter 26. The mating calls and songs produced by male birds, crickets, and frogs all guide females to their source. Some animals, such as fireflies and deep-sea fish, produce flashes of light to signal mates in their dark environments.

Finding a potential mate may seem less challenging to humans than the next component of mating behavior—choosing a high-quality mate—because we often live in concentrated groups, primarily in towns and cities. So we need not travel far to find a member of the opposite sex. For many animals, however, encountering a potential mate can be difficult. For slowly moving animals and animals of species whose members are solitary and widely scattered, finding a mate requires a lot of time and energy.

Once members of the same species have identified each other as potential mates, they must get close enough together to ensure that sperm and egg have a reasonable chance of encountering one another. The major biological significance of human mating, for example, is to get sperm close enough to an egg to fertilize it.

Aquatic animals that cannot move as adults, such as corals, nonetheless find ways of getting sperm and egg together. These animals commonly release their sperm and eggs into water, relying on currents to direct their gametes to their destinations (Figure 29.6). Nonmobile animals typically live in dense concentrations to counteract the tendency of currents to disperse small particles (such as eggs and sperm) over long distances.

A male song sparrow singing in his territory.

Fireflies signal mates after dark using flashes of light.

Female moths attract males by releasing a chemical signal from brushlike structures.

Figure 29.5 Finding a Mate
Animals use diverse signals to find potential mates, only a few of which are shown here.

This spawning hard coral has just released sperm (the milky cloud) into the ocean.

Gametes are usually released into a favorable environment

Gametes need a moist, relatively constant environment. In open air, these tiny cells, with their large surface-to-volume ratio, dry out and die quickly. In aquatic environments, moisture is not a problem. Many males of aquatic animal species simply release their sperm into the water surrounding a female, who releases her eggs into the water at the same time (Figure 29.7a).

In terrestrial animals, eggs are released from ovaries into internal oviducts that are supportive to their well-being, and sperm are delivered to essentially the same favorable surroundings. In mammals, for example, **copulation** (Figure 29.7b) allows a male to release his sperm through a penis directly into a female's reproductive system (as in human sexual intercourse).

Many animals are choosy about their mates

For many animals, simply finding a mate to fertilize one's gametes is not good enough. Humans, for example, try to choose their mates carefully. Only after considerable thought and familiarity do most of us decide with whom we want to produce children. There are several reasons for this, all of which apply to most animals.

Careful mate selection is important for successful reproduction. Some animals, including humans, select mates for their ability to provide parental care and sustenance for the offspring for a period of time after birth.

Of broader concern, however, is the quality of the potential mate's genes. An animal that selects a poor genetic candidate with which to mate risks combining its own genes with inferior ones to produce offspring with a reduced chance of surviving and reproducing.

Because males and females produce different kinds of gametes, they tend to choose their mates differently. Whereas each sperm costs a male relatively little in the way of resources, each egg represents a much greater investment of the female's resources. As we have seen, a human male produces many millions of sperm each month, whereas a human female produces only one mature egg. Furthermore, in mammals, the female must expend considerable resources sustaining the fertilized zygote within her body. Females of many animal species spend additional time and energy nurturing, protecting,

(a)

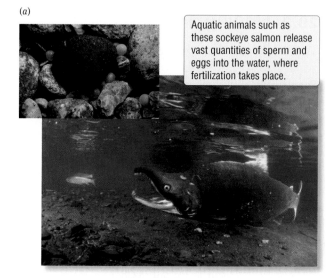

Aquatic animals such as these sockeye salmon release vast quantities of sperm and eggs into the water, where fertilization takes place.

(b)

On land, males typically transfer sperm directly into the female's body to avoid exposing it to the environment.

Figure 29.7 Sperm Transfer in Animals
(a) Many aquatic animals release their eggs and sperm directly into the water, often timing their releases to coincide so as to increase the chances of fertilization. (b) Terrestrial animals usually copulate in order to bring gametes together.

and training offspring after birth or hatching. Although both parents contribute time and resources to caring for offspring in some species, among most animals, this task predominately falls to the female parent. Thus a female stands to lose more if one of her eggs results in an off-spring with a poor chance of survival than does the male that fertilized that egg. As a result, in many animal species, females tend to choose mates carefully to ensure that their offspring have the greatest possible chance of success, while males tend to mate with many females in an effort to fertilize as many eggs as possible. In some species, such as humans and most species of western bluebirds in the United States, individuals tend to choose one mate for life.

29.4 Development from Zygote to Birth

The fertilization of an egg by a sperm to form a single-celled diploid zygote represents only the first step in the development of a new offspring. Although the earliest stages of development are similar in all animals, we concentrate on vertebrates, and particularly mammals, in this section, referring often to the human case to illustrate important steps in the process. For all their outward differences in size, shape, and complexity, most vertebrate animals go through a remarkably similar sequence of steps as the zygote develops into a functional organism.

Rapid cell division characterizes early development

In the first stage of animal development, the single-celled zygote divides again and again to form a multicellular **embryo**. This term applies to the early period of development extending from the zygote to the very early development of organs and organ systems. For humans, this amounts to the first 8 of the roughly 40 weeks required for a baby's normal development.

Never again will an animal's cells divide as rapidly as they do immediately after fertilization. Unlike later cell division, this early cell division involves mainly subdivision of the cells rather than cell growth in size. At this point, cells increase in number while decreasing in size. Even after many cell divisions, a frog embryo containing tens of thousands of cells is still about the same size as the original zygote (Figure 29.8). At this early stage, the embryo, which has the shape of a hollow sphere, is called a **blastula**, and

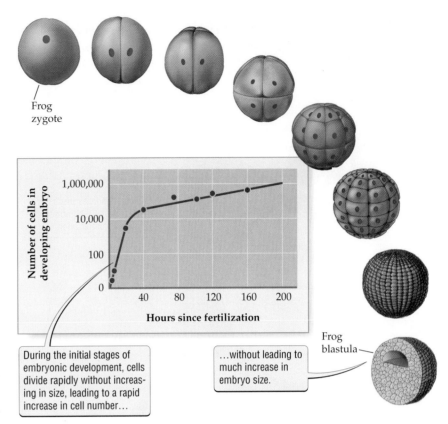

During the initial stages of embryonic development, cells divide rapidly without increasing in size, leading to a rapid increase in cell number…

…without leading to much increase in embryo size.

Frog zygote

Frog blastula

Figure 29.8 Early Development of Vertebrates
Immediately after fertilization, most vertebrate zygotes, like that of a frog shown here, develop rapidly as cells divide without increasing in size. The rapidly dividing cells in a frog embryo form a blastula, a hollow sphere consisting of thousands of cells.

usually consists of cells that all look very much alike. In mammals, the blastula is referred to as the **blastocyst**.

Mammals, including humans, develop in a protected environment inside their mother's body, in an organ called the **uterus**. Although mammalian cells divide without growing in size early in development, they divide more slowly than in other animals. It takes about a week for a human egg to travel from the site of fertilization in the oviduct to the uterus. At the end of that week the embryo consists of only about a hundred cells. Compare that with the frog embryo, which after 1 week (168 hours) is closer to a million cells (see Figure 29.8).

A second difference is that the mammalian blastocyst is not just a simple, hollow sphere, like the blastulas of most other animals, but one that contains two distinct cell types. The outer layer of the blastocyst, called the **trophoblast**, will become the developing embryo's portion of the **placenta**. This uniquely mammalian structure consists of tissues from both the mother's uterus and the embryo and permits the transfer of nutrients and wastes between the blood of the mother and the blood of the

Explore a woman's reproductive years. 29.4

embryo. Inside the trophoblast, clustered at one end, are cells of a second type—the **inner cell mass**—which will develop into the actual embryo (Figure 29.9).

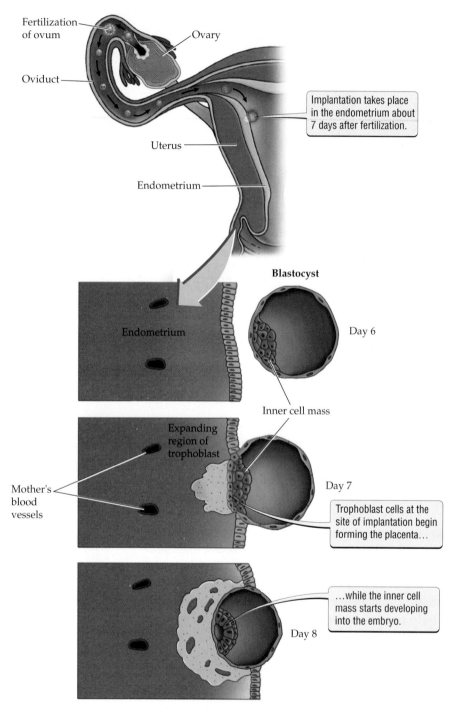

Figure 29.9 Implantation in Mammalian Development
Unlike the blastulas of other vertebrates, the mammalian blastocyst has two cell types, whose fates become evident after the blastocyst is implanted in the endometrium. The trophoblast expands into the endometrium to begin forming a placenta, and the inner cell mass begins to differentiate on its path to becoming the actual embryo.

The blastocyst implants in the wall of the uterus, and a placenta forms

On its journey down the oviduct to the uterus, the human zygote has been developing into a blastocyst. When it encounters the **endometrium** (the layer of cells lining the wall of the uterus), cells in the trophoblast secrete enzymes that allow the blastocyst to implant itself in the endometrium (see Figure 29.9). The trophoblast expands into the endometrium and eventually contributes to the formation of the placenta, which includes a network of blood vessels that are closely associated with the mother's blood. The two-way exchanges that occur between the mother's circulatory system and that of the embryo provide the embryo with the nutrients and oxygen it needs and a means of getting rid of carbon dioxide and other metabolic wastes (Figure 29.10). The **umbilical cord** contains the veins and arteries that convey materials to and from the placenta.

The human embryo lives in a sac called the **amnion**, which is filled with a watery fluid that protects the developing child. Just before birth this sac usually ruptures and the amniotic fluid flows out of the mother's vagina. Notice also that the developing human has a **yolk sac** (see Figure 29.10). Unlike the yolk in a chicken's egg, the human yolk sac contains no yolk material. Instead, it produces blood cells and the cells that will eventually produce gametes in the adult human.

Further development of the embryo produces three cell layers

As animal development continues, two critical events take place. First, depending on their position within the blastula, the cells in the embryo form three layers. These layers follow three distinct developmental paths, each giving rise to different tissues in the mature animal. The **endoderm** gives rise to the epithelial tissues of the gut and lungs, the liver, and the glands of the endocrine system, among other structures. The **mesoderm** gives rise to the muscles, heart, kidneys, reproductive system, and skeletal structures. The **ectoderm** gives rise to the nervous system and the outer layer of the skin (Table 29.1). These three tissue types are found in most animals and represent the first visible evidence of a trend that continues throughout the embryo's development: that the identity, or role, of any particular cell becomes increasingly defined as development progresses.

The second important change results from the movement of cells in these layers. Initially, all three layers lie on the outside of the embryo; there are no infoldings. However, if the endoderm is to give rise to the lining of

Learn more about the placenta.

Figure 29.10 The Placenta
The placenta provides the life-giving connection between mother and developing child, allowing the child to receive oxygen and nutrients from the mother's bloodstream and to dispose of wastes.

the gut and the mesoderm to internal organs, these cell layers must move to new positions inside the embryo. This remarkable rearrangement takes place during **gastrulation**, a complex sequence of events in which the three layers move relative to one another. The resulting embryonic stage is called the **gastrula**. The endoderm now lies inside the embryo and is surrounded by mesoderm; as a result, the ectoderm forms the entire outer layer of the embryo. In humans, the formation of the three cell layers and their rearrangement during gastrulation (Figure 29.11) take place during the 2 weeks following implantation.

Table 29.1

The Fates of the Three Cell Layers in the Developing Embryo

Cell layer	Corresponding adult structures
Endoderm	Liver, pancreas, thyroid, and linings of the gut and lungs
Mesoderm	Skeleton, muscles, reproductive structures, kidneys, circulatory and lymphatic systems, blood, and the inner layer of skin
Ectoderm	Skull, nerves and brain, the outer layer of skin, and teeth

Cell fate becomes increasingly defined as development proceeds

The single cell of the zygote has the potential to give rise to any cell type in the mature animal. Once cells have differentiated into endoderm, mesoderm, and ectoderm, however, each cell faces a more restricted future. As the animal continues to develop, the range of potential fates for each cell in its body becomes increasingly limited, until by the time the animal is born, its cell types have become fixed.

This growing limitation creates a problem: for multicellular animals to reproduce, they must retain some cells that can produce all the structures needed in their offspring. Animals overcome this problem by setting aside a few cells early in development to form gamete precursor cells. These unique cells, which originate in the yolk

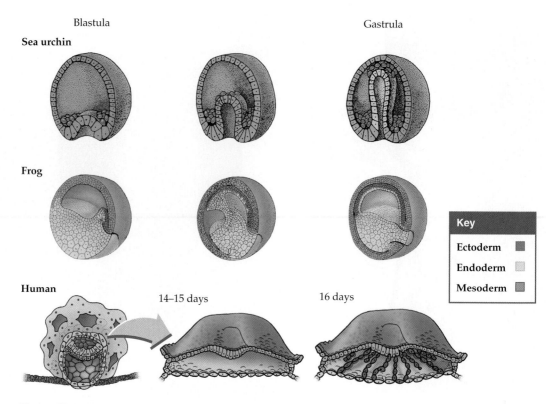

Blastula   Gastrula

Sea urchin

Frog

Key

Ectoderm ■

Endoderm ▫

Mesoderm ▪

Human

14–15 days   16 days

Explore the gastrulation phase. 29.6

Figure 29.11 The Three Cell Layers Have Developmentally Distinct Fates
In spite of the difference in their patterns of early cell division, all animals have recognizable layers of endoderm, mesoderm, and ectoderm cells once the initial stage of rapid cell division ends. During gastrulation, these cell layers move to their proper positions in the embryo.

sac but eventually move to the egg-producing ovaries or sperm-producing testes, remain unspecialized and separate from the other cells.

Development of a human fetus is divided into trimesters

The 40 weeks of human development in the uterus are divided into three stages called **trimesters**, each about 3 months long (Figure 29.12). During the first trimester, the zygote develops from a single cell into an organism that has most of the structures and organs that will be present at birth. After gastrulation, all animals undergo a period of rapid differentiation and rearrangement of the cells in their three cell layers into the various organs needed to make up a functioning individual. For example, just 3 weeks into our own 40-week-long development (see Figure 29.12), the human embryo has an identifiable heart. By the end of the eighth week of development, the developing human is 3.5 centimeters long—little more than an inch—but the head is clearly identifiable, the endoderm has differentiated into an identifiable liver, the mesoderm has differentiated into red blood cells

and kidneys, and the beginnings of all the organs found in an adult are present. From this point on, the developing human is called a **fetus**. Near the end of the first trimester, the fetus is 12 centimeters long, the fingers and toes have nails, the external genitalia are recognizably male or female, and the gut has developed from the endoderm to the point that it can absorb sugars.

Human development is particularly vulnerable to disruption during the first trimester. Anything that alters the normal course of events during this critical time of rapid cell differentiation and organ formation can lead to serious problems. In humans, most miscarriages take place during the first trimester of pregnancy because any genetic problems in the developing embryo (such as those described in Chapter 11) are most likely to be expressed at this time. Environmental factors can also disrupt development. We saw, for example, the disastrous effects of placing alcohol in the environment and blood of a developing embryo at the beginning of this chapter. Nutritional deficiencies can also disrupt development (see the box on page 574).

In the second and third trimesters, the emphasis switches from rapid differentiation to growth and

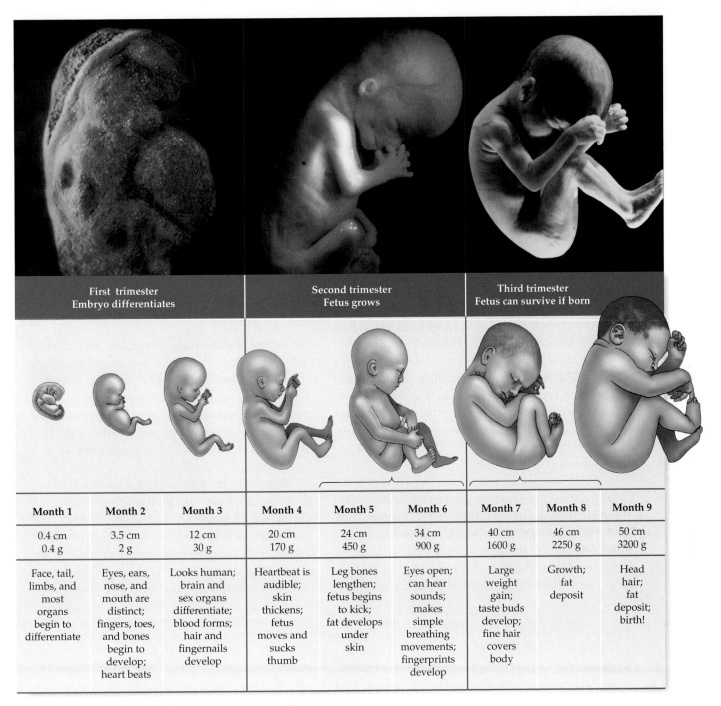

	First trimester Embryo differentiates		Second trimester Fetus grows			Third trimester Fetus can survive if born		
Month 1	**Month 2**	**Month 3**	**Month 4**	**Month 5**	**Month 6**	**Month 7**	**Month 8**	**Month 9**
0.4 cm 0.4 g	3.5 cm 2 g	12 cm 30 g	20 cm 170 g	24 cm 450 g	34 cm 900 g	40 cm 1600 g	46 cm 2250 g	50 cm 3200 g
Face, tail, limbs, and most organs begin to differentiate	Eyes, ears, nose, and mouth are distinct; fingers, toes, and bones begin to develop; heart beats	Looks human; brain and sex organs differentiate; blood forms; hair and fingernails develop	Heartbeat is audible; skin thickens; fetus moves and sucks thumb	Leg bones lengthen; fetus begins to kick; fat develops under skin	Eyes open; can hear sounds; makes simple breathing movements; fingerprints develop	Large weight gain; taste buds develop; fine hair covers body	Growth; fat deposit	Head hair; fat deposit; birth!

Figure 29.12 Development of the Human Embryo and Fetus
The development and growth of a fetus occurs over 9 months. For convenience, this period is divided into three trimesters, each 3 months long.

Investigate the process of human fetal development.

29.7

further development of already existing organs (see Figure 29.12). During this time, the fetus grows from the size of a large mouse to the typical birth weight of over 3 kilograms. During the second trimester, weight increases faster than length. By the end of that trimester, the developing baby has gained about half of the length and about 25 percent of the weight it will have at birth.

The baby becomes active in the uterus and can kick. At the end of the third trimester, the baby is ready for its sudden transition from the uterus to the outside world. It gains a good deal of weight during the third trimester, and its circulatory and respiratory systems prepare for living in a gaseous atmosphere rather than the watery world of the amniotic fluid.

Biology Matters

Folic Acid: Motherhood Can Begin Long Before Conception

A mother's body begins nurturing her child at the moment it is conceived. But some aspects of nurturing occur long before conception. Good nutrition, even before a woman becomes pregnant, plays a big role, and folic acid (folate), a B vitamin, provides a good example of how important a woman's diet can be.

Experts agree that folate plays a major role in preventing spina bifida, a serious and common birth defect of the skeletal and nervous systems. A child with spina bifida cystica, the most serious and common form, is shown in the accompanying photograph. The protruding cyst results when the spinal column fails to close around the spinal cord, as it should during the first few weeks of development. Infections occur continually because the wall of the cyst is thin and ruptures easily. Children born with spina bifida are often partly paralyzed, and later may be confined to wheelchairs or must use crutches or wear braces for a lifetime. Some victims suffer from varying degrees of learning disabilities. Babies born with this condition require surgeries and extensive medical care.

Spina bifida occurs in about 1 out of every 1,000 births and is more common than muscular dystrophy, cystic fibrosis, and multiple sclerosis combined. However, the condition is largely preventable. Studies have shown that the occurrence of spina bifida can be reduced by as much as 70 percent with adequate amounts of folate in a woman's diet. And thinking ahead

is important, because taking in adequate amounts must begin long before conception. Scientists and physicians who specialize in this area recommend that all women of childbearing age include adequate amounts folic acid in their diets long before they become pregnant. Taking the vitamin only when pregnancy begins is too late. Nutritionists recommend 0.4 milligrams per day as a part of a woman's regular diet.

Citrus fruits, leafy green vegetables, whole grains, egg yolks, and beans are good sources of folic acid, and many packaged foods, such as breads, pasta, and cereals, are now fortified with folate. Folate is also a component of many vitamin supplements. Given the power of this vitamin to prevent spina bifida, responsible motherhood can begin well before pregnancy.

Childbirth occurs in stages

In the last few weeks of pregnancy, the level of the hormone estrogen in the mother's blood increases. This increase causes the muscles of the uterus to become sensitive to **oxytocin**, a hormone secreted by the fetus and, later in the birth process, by the mother's pituitary gland. Oxytocin stimulates the uterine muscles and causes the placenta to secrete prostaglandins, which stimulate those muscles further. Labor begins when the muscles of the uterus begin to contract in response to these hormones (Figure 29.13). In a positive feedback loop, more contractions cause the production of more oxytocin, and the strength of the contractions increases as more oxytocin is produced. The cervix begins to open, and the increasingly strong contractions eventually push the fetus out

of the mother's body. At this point the positive feedback ends, and the contractions subside as oxytocin levels decrease. The placenta, often referred to as the "afterbirth," is expelled during the last stage of childbirth.

29.5 How Development Is Controlled

Our descriptions of animal development so far provide only an overview of how diversity in form and function develops in the cells, tissues, and organs of a multicellular organism. But these descriptions do lead us to observe that the development of a single-celled zygote into a mature organism requires precise control over the fates of cells as they divide. What exerts this control?

Cell fates are determined by which genes are activated during development

As we saw in Chapter 14, cell types arranged into tissues in an organism differ from one another not because they contain different genetic information—all cells in an individual's body have the same DNA molecules—but because cells express that information differently at different times. A major advance in biology in the past few decades has been the rapid growth in our understanding of how the expression of the genetic information in cells is controlled during development. A few small, simple animals, such as nematode worms and fruit flies, have played key roles in these discoveries (see the box on page 576).

In the embryo, the potential fates of a cell depend on the particular genes in its DNA that can be activated. You will recall from Chapter 13 that the conversion of information in the DNA blueprint into specialized activity in each cell proceeds in two stages: First, during transcription, the information encoded in DNA is converted in the nucleus into an RNA transcript, and that transcript

is processed to produce messenger RNA (mRNA). Then, during translation, the mRNA leaves the nucleus and crosses into the cytoplasm, where it serves as a template for protein production. Preventing or promoting the transcription of certain genes, or altering their processing into mRNA, can cause the identical genetic information in each cell's DNA to be expressed differently in different cells or at different times in the same cells.

Humans and other mammals, for example, produce different kinds of oxygen-binding hemoglobin pigments at different stages of their development: embryonic hemoglobin at 0 to 8 weeks, fetal hemoglobin at 8 to 12 weeks, and then progressively greater proportions of adult hemoglobin at 12 weeks and beyond. The embryonic and fetal hemoglobins have a greater tendency to bind to oxygen than adult hemoglobin does, allowing the developing baby to efficiently pull oxygen from its mother's blood by way of the placenta. The genes that encode the embryonic, fetal, and adult hemoglobin proteins are associated with other DNA sequences that, when bound by regulatory proteins, allow transcription of those genes (see Chapter 13). At certain stages during development, transcription of the appropriate hemoglobin genes is activated in red blood cells, while the other hemoglobin genes remain inactive.

In other cases, different cells transcribe a particular gene, but the processing of the resulting RNA differs

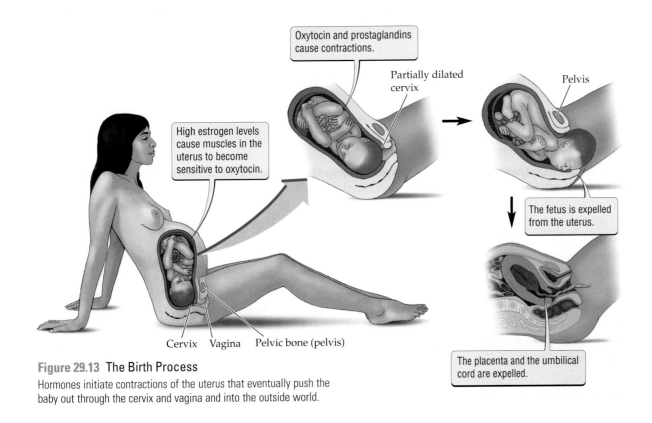

Oxytocin and prostaglandins cause contractions.

Partially dilated cervix

Pelvis

High estrogen levels cause muscles in the uterus to become sensitive to oxytocin.

The fetus is expelled from the uterus.

Cervix   Vagina   Pelvic bone (pelvis)

The placenta and the umbilical cord are expelled.

Figure 29.13 The Birth Process
Hormones initiate contractions of the uterus that eventually push the baby out through the cervix and vagina and into the outside world.

Science Toolkit

Why Worms?

Scientific research often depends on animals that are of no particular interest to most people. Fruit flies, for example, play a key role in genetic research, as mice do in medical research. Many studies of animal development rely on a particularly unimpressive experimental subject, *Caenorhabditis elegans*. This transparent nematode worm is tiny—much shorter than its name—and lives inconspicuously in the soil, where it feeds on bacteria. It causes no diseases, does not affect any crops, and has no economic use. Why, then, has *C. elegans* become a star in the realm of developmental biology?

One reason is that *C. elegans*, like most successful laboratory animals, lends itself to easy laboratory culture: place these worms on a petri dish containing a good bacterial culture (or a commercially available nematode food), and they thrive. The short life span of *C. elegans* (only 3 weeks) means that experiments can be completed quickly, allowing researchers to publish their results more quickly.

Caenorhabditis elegans, Developmental Superstar
Since its genome was sequenced in 1998, this nematode, already helpful as a research tool, has became even more valuable.

What has placed *C. elegans* firmly on the path to biological stardom, however, is that its development has been thoroughly mapped. An adult worm has just under a thousand cells. Developmental biologists have documented exactly how a single-celled *C. elegans* zygote divides multiple times to give rise to each of its adult cells. In other words, researchers know the complete family tree for every cell in the adult worm. Moreover, the position of each cell at each stage of development has also been documented.

In 1998, scientists finished sequencing the entire genome of *C. elegans*, which consists of just under 20,000 genes. Given the importance of the activation and inactivation of genes in regulating development, this latest advance will only increase the stature of *C. elegans*. Researchers have already begun to determine which genes are active and which are inactive at various stages in the worm's development. Even though the development of *C. elegans* itself might not particularly interest us, the general developmental principles that we can uncover using this little worm certainly do.

Sea urchin

Helpful to know

Morphogens literally means "form producers"—an appropriate name for proteins that turn on and off the genes whose expression determines the form an animal will have.

in cells with different fates. An example is the gene in sea urchins that controls the production of actin, an important protein molecule in muscle tissue. Early in its development, actin is produced specifically in the animal's ectoderm. Although ectoderm, mesoderm, and endoderm cells all transcribe the actin gene into RNA, only in the ectoderm cells is the transcribed RNA converted into a piece of mRNA that can move out of the nucleus into the cytoplasm to produce actin.

Another way animals can regulate gene expression during development is to cut up the transcribed RNA in different ways to make different mRNA products. For example, male and female fruit flies form different mRNAs from the transcription of a single gene that they both express. The resulting proteins, which determine sex in the developing flies, play an important role in controlling the sexual development of these animals.

Specific molecules activate genes during development

Genes are generally switched on or off by the presence of certain proteins called **morphogens**. One method by which morphogens can work is to spread outward from a cell that produces them, decreasing in concentration with increasing distance from the source. Cells with the proper receptor proteins in their plasma membranes can respond to the binding of a morphogen by activating or inactivating certain genes. In this way, genes can be expressed at specific locations in the developing animal.

The role of morphogens in controlling development has been studied extensively in fruit flies of the genus *Drosophila*. In the embryonic fruit fly, patterns in the concentrations of several morphogens provide developing cells with information about their position relative to the head and tail (Figure 29.14*a*). By experimentally introducing these morphogens into a developing embryo, researchers can disrupt these morphogen patterns and, as a result, the developmental pattern of the embryo. If a morphogen that identifies the head end is injected into the tail end of the embryo, for example, a two-headed embryo will result (Figure 29.14b).

The bodies of insects are divided into segments, and studies of the development of segments in fruit flies have revealed how morphogens work in those insects: they act on a family of homeotic (master-switch) genes that distinguish the different segments in the body of the fly. These genes appear to regulate other genes that control the form of a particular segment. Genes and morphogens that resemble those in fruit flies play a major role in directing head-to-tail differences in the development of tissues in all animals, including humans (see Chapter 14 for a more detailed discussion of homeotic genes).

In other cases, morphogens simply act on adjacent tissues to provide their cells with information on the identity of the surrounding cells. In birds, for example, the interaction between mesoderm-derived and ectoderm-derived tissues in the skin determines the form of the skin, as scientists have shown by transplant experiments (Figure 29.15). Chicken ectoderm from the wing develops into wing feathers when placed next to mesoderm from the wing, but it develops into scales and claws when placed next to mesoderm from the foot. Chemical interactions between the mesoderm and ectoderm tissue determine which genes in the ectoderm tissue are activated.

Hormones also help control animal development. Sex hormones, for example, influence the development of reproductive structures to determine whether individuals will become males or females, as described in Chapter 24.

The environment can influence development

It's important to remember that genes and their protein products do not work in isolation during development. The developing animal also responds to various environmental influences. These influences include cues from the internal environment, such as the amount of stored resources available, as well as external cues such as temperature, day length, or even the chemicals produced by potential predators and competitors. We offer just a few examples here.

Sex in many turtles and in alligators depends on the temperature of the nest in which the eggs develop: turtle eggs from cool nests tend to develop into males, while alligator eggs from cool nests tend to develop into females. Many aquatic invertebrates develop spines or bristles when exposed to chemicals released by nearby

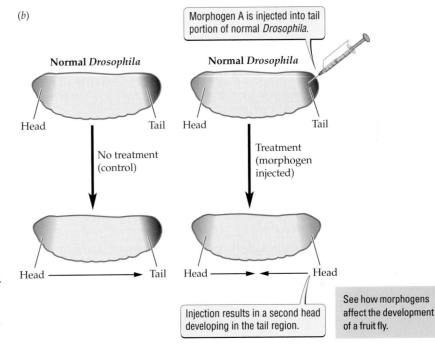

See how morphogens affect the development of a fruit fly. **29.8**

Figure 29.14 Morphogens Often Control Development

(a) Patterns of increasing or decreasing morphogen concentrations provide cells with information about their position relative to the head and tail ends of a developing fruit fly embryo. (b) Experimental manipulation of morphogen concentration gradients can dramatically disrupt development. By injecting morphogen A (which identifies the head end) into the tail end of an embryo, cells near the tail can be "fooled" into developing into head tissues.

predators in their environment. The spines and bristles make it harder for the predators to eat their prey. Aphids show several environmental influences on their development: scarcity of food and high densities of other aphids nearby stimulate the development of winged forms, and temperature and day length determine whether the newly formed individuals will reproduce sexually or asexually.

Figure 29.15
The Identity of Surrounding Cells Influences Development
Cells in the ectoderm-derived layers of chicken skin develop differently depending on the source (wing or foot) of the mesoderm-derived layers of chicken skin they are adjacent to. When placed next to ectoderm from the wing, mesoderm from the wing stimulates the production of feathers typical of wings. When placed next to the same wing ectoderm, mesoderm from the foot stimulates the production of scales and claws typical of feet.

Chimpanzee

29.6 The Effect of Development on Evolution

Changes in development that may seem small can cause dramatic changes in the form and function of mature animals. Because of this, simple changes in genes involved with development can lead to complex changes in its outcome. As a result, developmental changes have played

an important role in the evolution of life. Consider the following two examples.

Chicken feet have four distinct toes. Chick embryos, however, have webbed feet more like those of an adult duck than those of an adult chicken. The disappearance of the webbing between the chick embryo's toes seems to depend on a single gene. This particular gene, when active in a cell, causes the cell to die. During development, this gene is activated specifically in the cells of the webbing, but not in the toes (Figure 29.16). This process of programmed cell death, called **apoptosis** [*APE*-uh-*TOH*-siss], occurs in the development of all animals. For example, the webs between human fingers and toes, like those between the toes in chicken claws, disappear during development. A chick embryo with a mutant form of this gene—one that failed to be activated in cells in the webbed areas of the embryonic feet—might well grow up to have feet resembling those of a duck rather than normal, chick-enlike feet. In this way, a relatively minor genetic change could lead to dramatic morphological change—such as the development of a completely different kind of foot.

As a second example, recall from Chapter 2 that chimpanzees are our closest relatives. Many of the outward differences that distinguish humans from chimpanzees may have arisen from a relatively few developmental changes. Outwardly, adult humans and adult chimpanzees don't look alike: adult chimpanzees have jaws that protrude more than human jaws, and their skulls are much less domed than ours. The heads of baby chimpanzees, on the other hand, are shaped much like those of human babies, and newborn chimpanzees share with adult humans a relatively flat face and domed skull. During

Figure 29.16 Why Chickens Don't Have Duck Feet
In chicken embryos, the webbing between the toes disappears when a gene that is activated in the cells of the webbing triggers the death of those cells. This activation gives adult chickens separate toes. In duck embryos, this gene is not activated in the foot region, resulting in adult ducks with webbed feet.

The feet of adult chickens lack the webbing...

...that stretches between the toes of adult ducks, allowing them to swim.

In developing chicken feet, the apoptosis gene is active in the webbing between the embryonic toes, as indicated by the dark areas.

human development, the shape of the skull changes little, but during chimpanzee development, skull shape changes greatly (Figure 29.17).

The difference in skull shape between adult humans and adult chimpanzees may simply reflect differences in the timing of the development of their skulls relative to the rate of sexual maturation that defines adulthood. Evidence suggests that, compared with the chimpanzee skull, the rate at which the human skull develops has slowed relative to the rate at which the reproductive organs develop. In other words, humans gain sexual maturity while still in a young animal's body. Here again, the effects of developmental changes are pronounced and complex, though the causes may be relatively simple.

29.7 Development after Birth

Development doesn't end when a human child—or a juvenile of any animal species—is born or hatched. We humans, in particular, spend about a quarter of our lives reaching our full adult size. Most of this growth occurs during childhood, before we become sexually mature.

In many ways, the development of the human baby represents a continuation of its fetal development. During the first 2 years of life, our bodies continue to increase in size at a rate comparable to that of a fetus, rather than at the slower rate typical of infant development in other mammals. Our brains also continue to increase in weight at the fetal rate for the first 2 years of life and do not reach their mature weight for another 8 years.

This increase in body size requires cell proliferation, but interestingly, development also requires programmed cell death. Apoptosis continues to remove cells from particular parts of the baby's body in response to a combination of environmental cues and genetic programming. In the brain, for example, apoptosis continues to prune neurons that are not stimulated by environmental signals.

The developmental changes that take place after birth follow a sequence as predictable as the sequence of events before birth. We highlight just two aspects of post-birth development here: the role of stem cells and events following the reproductive years.

Parts of the body must regenerate during the lifetime of the animal

Cells have a limited life span and must be replaced as they wear out, become less effective at performing their

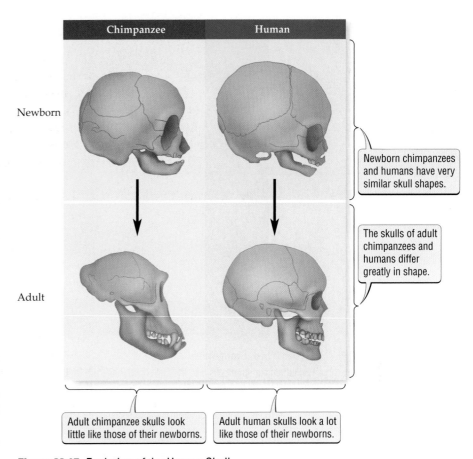

Figure 29.17 Evolution of the Human Skull
The skulls of adult chimpanzees differ in shape from the skulls of newborns much more than is the case in humans.

duties, or die. Human skin, for example, must regenerate itself as layer after layer of skin cells is worn away, and the food-absorbing cells lining the small intestine live for less than a week and so must be continuously replaced with new cells.

Depending on the tissue involved, cells are regenerated in one of two ways. Some tissues, such as the epithelial cells that form the inner walls of blood vessels, regenerate by simple cell division (mitosis). Other tissues, such as the skin and the lining of the small intestine, depend on **stem cells** (Figure 29.18). Stem cells are populations of unspecialized cells that, in response to external signals from other cells, differentiate into particular cell types, depending on where they are in the body. Stem cells have the ability to keep dividing throughout the lifetime of the animal, making them a continuous source of new cells.

However, it is important to note that stem cells are unspecialized to various degrees, depending on the tissue and the life stage of the animal. At the blastocyst

Figure 29.18 Stem Cells

Populations of stem cells, like the cell shown here, are found in the developing embryo and in select parts of the body after birth. When stimulated by chemical signals from surrounding cells, stem cells divide to differentiate into specific cell types.

red blood cells and lymphocytes, but they cannot differentiate into skin cells or gut cells. Similarly, the sperm precursor cells in the testes are stem cells that can differentiate into spermatocytes or more sperm precursor cells; however, they cannot differentiate into blood cells. Throughout the body are located similar populations of stem cells, able to differentiate into a few cell types, usually those needed for regenerating surrounding or nearby tissues.

As we saw in Chapter 15, researchers are finding ways to use stem cells—from humans and other animals—to improve some medical conditions. Many diseases and conditions for which there are yet no cures—including Alzheimer's and Parkinson's diseases—may benefit from future therapies involving stem cells. However, stem cell research, particularly research using human embryonic stem cells, remains controversial. Of all stem cell populations, embryonic stem cells have the greatest potential for creating successful therapies because these cells are so undifferentiated. Yet using them involves destroying young embryos, which people who believe human life begins at conception find morally unacceptable. Work on adult stem cells is less controversial, and biologists are also beginning to have success in tricking adult stem cells into losing their semi-specialized nature and behaving more like embryonic stem cells.

stage of the early embryo, all cells are almost completely undifferentiated, lacking the specific characteristics of particular tissues. However, at later stages of fetal development and in adult animals, the range of cell types into which cells can differentiate is restricted. For example, bone marrow contains stem cells that divide to produce all the different cell types found in the blood, including

Humans today live past their reproductive years

In humans, both males and females reproduce less effectively as they age. Females show a clear drop in their ability to produce normal eggs and bear children when they pass 40 years of age (Table 29.2). If eggs from younger women are implanted into women over 40, the pregnancy rate equals that for the age group of the women who donated the eggs. This finding suggests that as women age, it is their eggs that decline in quality, a conclusion supported by the increased frequency of birth defects in children born to older mothers. For example, Down syndrome (trisomy 21) occurs much more commonly in children born to mothers over the age of 40 than in those whose mothers are younger. Human females reach menopause, the end of their reproductive lives, around the age of 50 (see Chapter 24).

Similar patterns have been found for sperm produced by older and younger men. Children fathered by older men tend to live shorter lives than children fathered by the same men at a younger age, suggesting that the genetic quality of sperm declines during

Table 29.2

The Effect of Age on the Reproductive Success of Females

Age (years)	Cycles that result in a pregnancy (%)	Cycles that result in an embryo that fails to complete development (%)
Under 30	29.0	14.9
30–35	19.8	16.5
35–40	17.1	22.4
40–45	12.8	33.2

the father's lifetime. In addition, men produce fewer sperm as they age, decreasing their chances of fertilizing an egg. Males do not undergo the clearly identifiable menopause characteristic of females. Instead, their ability to produce sperm and their sex drive slowly decrease as they age.

What could be the reason for these patterns? Because egg and sperm precursor cells are present in our bodies from birth, they accumulate mutations over time from exposure to various environmental factors. The older these cells are, the more likely they are to accumulate mutations that are harmful or even lethal to a developing fetus.

Most species rarely live beyond their reproductive years. Even among the other apes, which are so similar to us, no species has a significant period of life after reproduction ends. Archaeological evidence suggests that prehistoric humans rarely lived longer than 50 years. Even into modern times, people in some parts of the world have had similarly short average life spans. However, today this is no longer generally true. The average life expectancy for men and women in the developed countries is 77 to 81 years. Women tend to live longer than men. This means that people, especially women, are living on average two to four decades past the end of their reproductive years.

Alcohol Is Not the Only Threat to Normal Development

The first trimester is the most dangerous time to expose a fetus to alcohol. But experts point out that drinking alcohol at any stage of pregnancy should be considered dangerous. No one claims to know how much is enough to cause harm. And alcohol is not the only substance that can harm a developing fetus.

Tobacco use also has negative effects. The bloodstream of a developing fetus concentrates the nicotine it receives from its mother at levels higher than those in the mother's blood. Nicotine also finds its way into amniotic fluid and breast milk. One ingredient in tobacco smoke is the highly toxic gas carbon monoxide. Nicotine and carbon monoxide both interfere with oxygen uptake by the fetus. The effects of smoking can cause premature births, reduced birth weights, and, especially in the first trimester, spontaneous abortions. The risks increase in proportion to how much a mother smokes. Studies show that infants born to mothers who smoke exhibit signs of nicotine withdrawal.

Heroin is another addictive drug whose use can dramatically affect a developing fetus. Mothers who use heroin are at greater risk of having a miscarriage. Their babies have a greater chance of suffer-

ing sudden infant death syndrome (SIDS). Pregnant mothers using heroin give birth to babies who show signs of heroin dependence. These children can be treated safely in the nursery, but their early exposure to heroin may affect their future in countless ways.

Other drugs carry risks for unborn children. Marijuana can cause premature births and reduce the fertility of couples who want to conceive. Methamphetamine (speed), a highly popular illegal "street" drug, raises the blood pressure of the mother, which can cause premature birth. Methamphetamine also constricts blood vessels in the placenta, reducing the supply of oxygen and nutrients to the fetus. Cocaine also has many effects, among them increasing the likelihood that the placenta will separate from the wall of the uterus during pregnancy. This can cause serious bleeding and even the death of both mother and child.

Educational programs and efforts to inform the public about the risks of these substances can help prevent some of the harm they do to developing children. But, in the final analysis, it's up to parents, before they actually become parents, to start caring for their baby even before it is conceived.

Chapter Review

Summary

29.1 Human Reproduction and Development: A Brief Overview

- In humans, male and female gametes are called sperm and eggs (ova), respectively. Males produce sperm in testes, and females produce eggs in ovaries.
- An egg released from a woman's ovary moves into the oviduct, where it can be fertilized.
- During sexual intercourse, the man's penis releases nearly 300 million sperm into the woman's vagina, of which only one can fertilize the egg.
- During development inside the mother's uterus, the single-celled zygote is transformed, through a series of precisely ordered events, into a complex individual. Through the process of differentiation, its unspecialized cells acquire specific identities.

29.2 Sexual and Asexual Reproduction in Animals

- Spermatogenesis and oogenesis—the production of male and female gametes—requires meiosis to reduce diploid ($2n$) precursor cells to mature haploid (n) gametes (sperm and eggs). Fertilization fuses the haploid sperm and egg to produce a diploid ($2n$) zygote.
- During fertilization, one sperm, aided by enzymes in its acrosome, enters the ovum. The sperm's nucleus then fuses with the egg's nucleus to create a diploid zygote.
- Some animals are hermaphrodites, producing both eggs and sperm. They can mate with others of their species, but not with themselves. Some hermaphrodites can switch from male to female, or vice versa.
- Asexual reproduction is common among animals. Some animals can reproduce asexually or sexually, depending on environmental conditions.
- Sexual reproduction tends to be found in variable environments, where the genetic variation of the offspring provides a hedge against unpredictable changes.
- Asexual reproduction requires less time and energy than sexual reproduction. It tends to be found in stable environments where genetic variation is not required to ensure reproductive success.

29.3 Finding and Selecting Mates

- Successful fertilization requires that sperm be deposited on or near eggs. Chemical, sound, and visual signals are used to attract and locate individuals of the same species and the opposite sex.

- Mobile animals actively seek out potential mates. Stationary aquatic animals tend to live in dense groups of individuals and rely on water currents to bring eggs and sperm together.
- Sperm and eggs need a moist, relatively constant environment between release and fertilization. Most aquatic animals release their gametes directly into the water near potential mates. On land, males release sperm directly into the female's reproductive system through copulation, and females release their own eggs into the same protected environment.
- Choosing a mate of high genetic quality helps to ensure that one's offspring will also be of high quality, increasing their chances of survival and reproduction. Female animals tend to be more choosy about mates than males are because eggs require a greater investment of resources than sperm and are in more limited supply. In addition, females tend to invest more energy and resources in caring for offspring.

29.4 Development from Zygote to Birth

- In the first stage of animal development, the single-celled zygote divides several times to form a multicellular blastula, consisting of cells that all look alike. This early development involves rapid cell division with little cell growth or differentiation.
- Mammals, including humans, develop in the protected environment of the mother's uterus.
- The mammalian blastula, called a blastocyst, has two cell types: the inner cell mass, which will develop into the actual embryo, and the trophoblast, which will become part of the placenta.
- When the blastocyst implants in the endometrium, trophoblast cells expand into the endometrium, eventually forming part of the placenta. Found only in mammals, the placenta is a two-way exchange surface for receiving nutrients and oxygen from the mother's blood and giving up carbon dioxide and other metabolic wastes.
- The embryo floats within the amnion, a sac of watery fluid. The human embryo has a yolk sac that produces blood cells and gamete precursor cells.
- The blastocyst forms three cell layers: endoderm, mesoderm, and ectoderm.
- During gastrulation, the three cell layers move relative to one another to produce the gastrula, in which endoderm (inside) and mesoderm (middle) are surrounded by ectoderm (outside). After gastrulation, the three cell layers differentiate into specialized organs.
- To reproduce successfully, the future adult must retain some cells that can differentiate into all the structures that will be needed in

its offspring. These cells are kept apart from the other cells in the developing embryo, which become increasingly restricted in the range of cell types into which they can differentiate. The gamete precursor cells eventually come to reside in the ovaries or the testes.

- Human development in the uterus (which takes about 40 weeks) is divided into trimesters. During the first trimester, the embryonic cells differentiate rapidly into the various organs and structures present at birth. From the ninth week of development on, the developing human is called a fetus.
- During the second and third trimesters the fetus grows rapidly.
- Childbirth occurs in stages. The hormone oxytocin, secreted by both mother and fetus, signals uterine muscles to contract. The contractions become increasingly stronger as positive feedback increases the amounts of oxytocin produced. The mother's cervix opens, and the fetus is eventually expelled from the uterus. The placenta soon follows.

29.5 How Development Is Controlled

- Development of a zygote into a mature animal requires precise control over cell fates. Gene expression is regulated by which genes are transcribed and when, and by how RNA transcripts are processed to make mRNAs.
- Genes are switched on or off by morphogens. Patterns of morphogen concentrations in tissues inform cells of their position within the developing embryo and influence the expression of homeotic genes. Homeotic genes, in turn, influence the expression of many genes that affect structure and function.
- A similar set of genes and morphogens seems to be involved in directing "head" to "tail" differences in the embryos of all animals, including humans.
- Hormones also influence patterns of development.
- During development, animals respond to internal and external environmental cues.

29.6 The Effect of Development on Evolution

- Small changes in how genes influencing development are expressed can have profound effects on the resulting animal, which in turn can influence evolution.
- Changes in the pattern of apoptosis during development can influence structure and function, as can changes in the rate of growth of one structure in relation to another.

29.7 Development after Birth

- Development for most animals continues after birth. Human babies grow at rates similar to that of a mammalian fetus, not a young mammal.
- Although cellular proliferation is needed for growth in size, apoptosis continues to prune cells from select parts of the body in response to use and environmental cues.
- Some tissues replace dead cells through simple mitosis. Others depend on stem cells: unspecialized, continuously dividing cells that differentiate into particular types after receiving the right signals from surrounding tissues. The earlier the stage of development, the more undifferentiated they are, and the wider the range of cell types into which they can differentiate.
- Stem cells have great potential for treating or curing a number of difficult diseases; however, stem cell research, especially on embryonic stem cells, remains controversial.
- Unlike most other animals, modern humans live part of their lives past their reproductive years. Both eggs and sperm can accumulate harmful mutations, making successful reproduction less likely as people age.

⊙ Review and Application of Key Concepts

1. Describe the gametes produced by animals in terms of their structure and function.

2. Compare sexual and asexual reproduction. What are some advantages and disadvantages of each?

3. Describe the fertilization of a human egg by a human sperm.

4. The development of a human fetus follows a pattern. Describe that pattern from implantation of the blastocyst to birth. Which trimester is most susceptible to disruption? Why?

5. During development, a cell's fate becomes more defined. Explain what this means and how it happens.

6. Terrestrial vertebrates are thought to have evolved from fish. Describe a way in which apoptosis could have led to an adaptation that allowed better mobility on land.

7. If intelligent species inhabit other planets, do you think they would live beyond their reproductive years? What would some of the advantages be?

Key Terms

acrosome (p. 564)
amnion (p. 570)
apoptosis (p. 578)
asexual reproduction (p. 563)
blastocyst (p. 569)
blastula (p. 569)
copulation (p. 568)
differentiation (p. 563)
ectoderm (p. 570)
embryo (p. 569)

endoderm (p. 570)
endometrium (p. 570)
fetus (p. 572)
gastrula (p. 571)
gastrulation (p. 571)
hermaphrodite (p. 564)
inner cell mass (p. 570)
mesoderm (p. 570)
morphogen (p. 576)
oogenesis (p. 564)

ovary (p. 563)
oviduct (p. 563)
ovum (p. 563)
oxytocin (p. 574)
penis (p. 563)
placenta (p. 569)
polar body (p. 564)
primary oocyte (p. 564)
primary spermatocyte (p. 564)
secondary oocyte (p. 564)
secondary spermatocyte
    (p. 564)

sexual reproduction (p. 563)
sperm (p. 563)
spermatogenesis (p. 564)
stem cells (p. 579)
testis (p. 563)
trimester (p. 572)
trophoblast (p. 569)
umbilical cord (p. 570)
uterus (p. 569)
vagina (p. 563)
yolk sac (p. 570)

Self-Quiz

1. Which of the following is a gamete?
   a. an ovary   c. a sperm
   b. a testis   d. a stem cell

2. A zygote is formed
   a. just prior to fertilization.
   b. when more than one sperm fertilizes the same egg.
   c. within the ectoderm of a blastula.
   d. when sperm and egg nuclei fuse.

3. Hermaphrodites
   a. always reproduce asexually.
   b. have both testes and ovaries.
   c. cannot produce zygotes.
   d. result from unfertilized eggs.

4. Human fertilization normally takes place in the
   a. oviduct.   c. ovary.
   b. vagina.   d. uterus.

5. Which of the following stages comes first in the development of a human?
   a. gastrula   c. fetus
   b. zygote   d. blastocyst

6. During implantation, the blastocyst implants itself in
   a. the endometrium.   c. an ovum.
   b. the oviduct.   d. the trophoblast.

7. A human baby's skin develops from
   a. ectoderm.   c. endoderm.
   b. mesoderm.   d. trophoderm.

8. As development of an embryo continues,
   a. stem cells develop from the blastula.
   b. gastrulation produces a blastula.
   c. mesoderm gives rise to heart muscle.
   d. endoderm gives rise to ectoderm.

9. Morphogens play a role in directing head-to-tail differences in
   a. fruit flies only.   c. all animals.
   b. humans only.   d. birds only.

10. Apoptosis describes a process of
    a. cell differentiation.   c. cell specialization.
    b. cell proliferation.   d. cell death.

Biology in the News

Vatican Flexes Muscle in Italy

The Result of the Two-Day Nationwide Consultation Was a Paradox

By David Willey

Only just over a quarter of Italian voters turned out to cast their ballots on the reform of one of Europe's most restrictive fertility treatment laws.

So the referendum will have no legal effect. Yet an overwhelming majority of 80–90% of those who did vote chose to answer "Yes" to a change in the law.

The referendum was skewed by the massive intervention of the Roman Catholic Church. Italian bishops urged a boycott of the poll and parish priests thundered to the faithful from their pulpits: "Life cannot be put to a vote. Do not go out and vote!"

The new Pope, Benedict XVI, gave the Italian bishops his personal support.

Italy recently held a referendum in which citizens could decide whether to keep or relax what are some of Europe's most restrictive laws on reproductive technology. In a backlash against what used to be some of the laxest regulation, Italy passed laws that stopped sperm and egg donation, banned the screening of embryos for genetic disorders, and limited to three the number of embryos that could be created during a round of fertility treatments. As a result of these strict laws, many Italians went outside the country to seek treatment when they had difficulty conceiving naturally. If Italians had voted in favor of changing the current laws, then they would have had access to these treatments within their own country once again.

The Catholic Church supports the existing laws and, therefore, opposed the referendum. The Church holds that conception and birth is something with which people should not interfere. In previous attempts to influence referendums related to reproduction, the side taken by the Catholic Church lost overwhelmingly. In this referendum, however, the Church changed its tactics from urging its members to vote in a particular way to urging them not to vote at all. Under Italian law, if too small a percentage of the voting population turns out, then the referendum does not count.

Possibly because doing nothing is less contentious than taking a stand, the Church's new tactics proved effective, since only a quarter of the voters showed up at the polls. As a result, a referendum in which a huge majority of the voters favored repealing the current laws on reproductive technology will have no effect.

Evaluating the news

1. Why do you think that human interference in reproduction (and in death, for that matter) causes such strong feelings?
2. In spite of the results of this referendum, Italians seeking some of the restricted reproductive technologies can still have access to them elsewhere in Europe. In what way, then, does the retention of the current laws affect access to reproductive technology? Now imagine if the situation were closer to home, and the United States had restrictive reproductive laws similar to the one in Italy while Canada and Mexico did not.
3. Are you confident you have thought enough about the ethics and science of reproductive technology to make an informed decision about what kinds of reproductive technologies should be allowed where you live and what kinds should be restricted there? If not, which questions would you like to have answered, and where would you go to get answers?

Source: *BBC News*, June 13, 2005.

Animal Behavior

Key Concepts

⊙ Behavior is a coordinated response to stimuli.

⊙ Behavior can be inborn or learned. Most animals have some inborn behaviors, which are usually triggered by simple external stimuli, while learned behaviors allow individual animals to respond to their environment flexibly.

⊙ All behaviors have a genetic basis. Differences in the genetic makeup of individuals can be observed as differences in their behaviors or abilities. Natural selection can work on behaviors as well as on morphological traits.

⊙ Communication is an important type of behavior. Communication ranges from simple chemical signals to elaborate displays and songs to the intricate languages of humans.

⊙ Group living and social behavior offer many advantages, including extra protection from predators and greater access to resources. Social behavior involves cooperation among members of a group, usually of the same species.

Main Message: Animal behavior consists of
coordinated movements by animals in response
to stimuli in their environment.

Bright Butterflies, Bad Meals

In many fields and forests of the southeastern United States you are likely to encounter an iridescent blue butterfly called the pipevine swallowtail. If you happen to be watching a female butterfly that has fertilized eggs to deposit, you may see her laying clusters of rust-colored eggs under the leaves of a plant. If you watch carefully, you may notice that she chooses only one kind of plant, which belongs to a group of vines called pipevines. Laying eggs under the leaf instead of on top provides them with added protection from predators, and choosing the correct plant is critical because the caterpillars that hatch from these eggs will not feed on other plants. So if the female butterfly makes a mistake—and some do—her young will not survive.

But this is more than just a case of avoiding the wrong leaves: if the eggs end up on the right pipevine leaves—and they usually do—they won't have much competition. The leaves of the pipevine plant contain toxins that are poisonous to most animals, but not to pipevine swallowtail caterpillars. As they feed and grow, the toxins build up in their bodies. Thus any bird that eats a pipevine swallowtail caterpillar will get very sick, and will avoid that particular menu item in the future. The same toxins remain concentrated in the adult butterflies, offering them the same protection.

Adults and caterpillars of the pipevine swallowtail may both have the same chemical protection, but in form and behavior they share virtually nothing else. As we saw in Chapter 24, hormones cause dramatic physical changes as caterpillars metamorphose into adult butterflies. This is true of many other insects as well. Dramatic behavioral changes accompany this type of metamorphosis. One is diet: while caterpillars of the pipevine swallowtail eat only pipevine leaves, the adults consume some something completely different:

The Female Pipevine Swallowtail Butterfly Must Be Choosy
Pipevine swallowtail caterpillars will feed on leaves from only a few species of pipevines, so selecting the right leaves on which to lay her eggs is critical for the female butterfly.

the nectar in flowers of other plants, such as honey-suckle, thistle, and milkweed. Their behavior is shaped by responses to the color of particular flowers and substances in their nectar, rather than to the color or substances in the leaves of the plant they preferred in their "caterpillar days."

These behaviors raise some intriguing questions. How does a female butterfly recognize a pipevine leaf? How does a caterpillar or a butterfly recognize its preferred food? How does the feeding behavior that makes its tissues toxic to predators protect that individual? After all, an attack by a predatory bird will probably be lethal to the prey, before the predator realizes the consequences of eating it and before the pipevine swallowtail can pass on its genes to the next generation.

A behavior is a coordinated movement or series of movements made by one animal in response to another animal or the physical environment. This broad definition encompasses behaviors that we recognize, such as courtship in humans and plant-searching activities by butterflies such as the pipevine swallowtail, as well as many more subtle behaviors. Behavior allows animals to respond quickly and flexibly to an ever-changing environment. Moreover, it allows animals to coordinate their own activities with those of other animals. The nervous systems and muscles of animals suit them particularly well to coordinating sensory information with movement to generate behavior. Neurons can relay signals more rapidly than hormones can, and their collections of sensory neurons give animals a unique ability to process sensory information. Muscles and nervous systems together give animals the ability to produce the coordinated movements we see as behavior.

We begin this chapter by looking at the behavioral interaction between a human mother and her nursing baby as a way of emphasizing the relationship between behavior and the endocrine and nervous systems. We then introduce two different kinds of behaviors: innate behaviors that are expressed from birth, and behaviors that are learned as the animal gains experience with its surroundings. Next we look at the genetic bases of both innate and learned behaviors. Then we consider communication, which allows two or more animals to coordinate their behavior. We end with a look at forms of social behavior.

1 Pressure on a nipple triggers nerve signals to the mother's brain …

2 …which stimulate the pituitary gland to release the hormone oxytocin…

3 …which travels through the circulatory system to the breast, where it stimulates the contraction of cells that deliver milk to the nipple.

Figure 30.1 The Milk-Letdown Reflex
Nursing by an infant stimulates the contraction of cells in the breast to release milk in a reflex coordinated by both the nervous system and hormones.

30.1 Breast-Feeding in Humans as an Example of Behavior

Behavior is a crucial part of human life. To understand the function and importance of behavior, let's look at one of the first behavioral interactions in human lives: that between newborns and their mothers during breast-feeding (Figure 30.1).

When a mother brings her breast to the mouth of her child for the first time, the child searches for the nipple with its mouth in an attempt to get milk. This behavior, called the **rooting reflex**, is stimulated by a touch against the infant's cheek or mouth. The touch triggers sensory signals that are carried to the infant's brain by its nervous system. The infant's brain, in turn, coordinates nursing behavior by stimulating contractions in its mouth muscles.

The baby's nursing behavior stimulates a less obvious behavioral response in the mother: the **milk-letdown reflex**. For the first 30 to 60 seconds after the baby begins nursing, no milk is released. The reason for the delay is that the milk storage cells in the mother's breasts release milk only in response to nursing. Pressure-sensitive cells around the nipple send signals to the mother's brain, which trigger the release of a hormone called oxytocin from the pituitary gland (see Figure 30.1). Oxytocin circulates in the blood to target cells in both breasts. These cells respond by contracting, which forces milk out of the storage cells and into tubes leading to the nipples.

The rooting and milk-letdown reflexes ensure that a baby tries to feed only when a breast is available, and that a mother releases milk only when her hungry baby is in close contact with her. The baby's mouth and the mother's breast are structures that allow babies to be fed, and the behavioral interaction between mother and child allows for efficient use of these structures.

Over weeks, as her breast-feeding experience increases, a mother's milk-letdown reflex changes in an intriguing way: instead of releasing milk in response to her baby's nursing, she releases milk in response to hearing her baby's cries, which signal to her that her baby is hungry. The switch from the touch of the baby's lips to the baby's cry as the stimulus triggering the milk-letdown reflex emphasizes the flexibility of behavioral responses. The mother's response can now be brought about by two different kinds of sensory stimuli.

30.2 Fixed and Learned Behaviors in Animals

Behavioral patterns can either be innate (inborn) features of an animal's behavior or be developed through the animal's experiences. For instance, males of most bird species sing a distinctive song to attract females. Males of some species know their courtship songs from birth, but others must learn them. Male cowbird chicks that are raised in captivity, and thus never hear an adult male cowbird sing, still give a flawless rendition of their species' courtship song when they mature. In contrast, male white-crowned sparrows raised in captivity fail miserably in their attempts to sing a song attractive to females. To learn their courtship song, male white-crowned sparrow chicks must hear other males of their species sing.

Animals display fixed behaviors in response to single stimuli

A **fixed behavior** is an innate behavior in which a single stimulus leads predictably to a single response. Many examples of such behaviors are familiar to us. For instance, newborn babies display the rooting reflex, and most first-time mothers display the milk-letdown reflex, without having learned how. Kittens reflexively cover their droppings when first introduced to a litter box, even those kittens removed from their mothers too young to have ever seen their mothers demonstrate this behavior. Butterfly caterpillars know which food plants to eat and which to avoid from the moment they hatch.

Fixed behaviors allow animals to behave appropriately when they have no chance to learn from experience or when the risks associated with the wrong behavior are great. A baby who had to learn to nurse, for example, might not learn the behavior in time to keep from starving. Female cowbirds lay their eggs in the nests of other bird species. As a result, male cowbird chicks grow up hearing only the courtship songs of their foster parents, and have no opportunity to learn their own species' song from their fathers.

Cowbird

The stimuli that trigger fixed behaviors are often simple. An example is found in herring gull chicks, which beg eagerly for food when their parents return to the nest. In a famous series of experiments, Nobel prize–winning behavioral scientist Niko Tinbergen showed that the chicks "aim" their begging behavior at a conspicuous red dot on their parents' bills. Oddly colored or shaped models of herring gull heads trigger begging behavior just as effectively as lifelike models do, as long as they feature the red dot (Figure 30.2a). A simple stimulus that causes a fixed behavior, such as the red dot in herring gulls, is called a **releaser**.

Similarly, the sight of the red patches on the wings of a male red-winged blackbird stimulates neighboring male red-winged blackbirds to defend their nesting territories vigorously. It is the color of the wing patch, rather than the overall appearance of the bird, that stimulates this behavior (Figure 30.2b). A person can cause a frenzy of territorial defense by putting on a red shirt and marching into a marsh full of nesting red-winged blackbirds during the spring breeding season. (But do not do this, or you will disrupt the birds' nesting!)

The stimuli that trigger fixed behaviors often do so only under certain circumstances. An individual's physical condition may affect whether a stimulus triggers a fixed behavior. The rooting reflex of newborn babies is a fixed behavior, but the baby exhibits it only when it is hungry. Similarly, an animal's response to a stimulus may

White-crowned sparrow

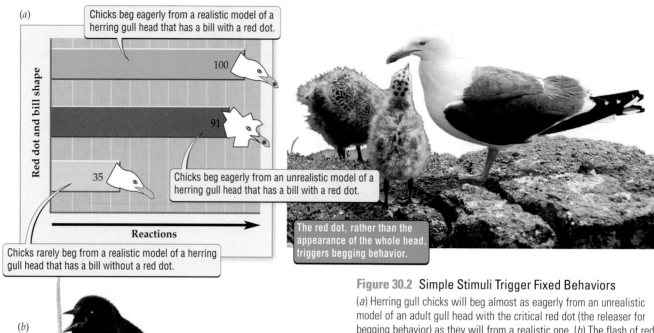

(a) Chicks beg eagerly from a realistic model of a herring gull head that has a bill with a red dot.

100

91

35

Chicks beg eagerly from an unrealistic model of a herring gull head that has a bill with a red dot.

Chicks rarely beg from a realistic model of a herring gull head that has a bill without a red dot.

Red dot and bill shape

Reactions

The red dot, rather than the appearance of the whole head, triggers begging behavior.

(b)

The red patches on the wings of male red-winged blackbirds play a key role in territorial behavior.

Figure 30.2 Simple Stimuli Trigger Fixed Behaviors
(*a*) Herring gull chicks will beg almost as eagerly from an unrealistic model of an adult gull head with the critical red dot (the releaser for begging behavior) as they will from a realistic one. (*b*) The flash of red wing patches stimulates territorial behavior in male red-winged blackbirds.

Explore how simple stimuli trigger fixed behaviors.

30.1

depend on the timing of the stimulus. For example, the aggressive response of male red-winged blackbirds to the color red occurs only during the breeding season, when levels of the male sex hormone testosterone are at their highest.

Learned behaviors add flexibility to animal responses

In **learned behaviors**, the response to a stimulus depends on an animal's past experience. For instance, we can train a dog to sit in response to a whistle or to the command "Sit." A dog trained to the oral command sits in response to "Sit"; a dog trained to the whistle also sits, but in response to a different stimulus. The two dogs learn to respond appropriately to their different learning experiences.

Much of our own behavior is learned rather than fixed. Mothers learn to release milk in response to a baby's cry. We learn appropriate behaviors for many different social situations. We learn rules of games and how to play musical instruments. We refine learning by practice. Students learn to answer questions correctly on biology exams in response to information gained from lectures and textbooks. Most other animals also modify their behavior in light of their previous experiences.

The feeding behavior of rats illustrates some advantages of learned behavior. Rats in laboratory cages can learn to press a lever that releases food pellets. When first placed in the cage, a rat will sometimes accidentally bump into the lever and discover the food pellet. The rat soon learns to press the lever deliberately to get more pellets. This capacity to learn from experience may be related to one aspect of the biology and behavior of rats that has made them so successful: they thrive on the ever-changing variety of garbage that humans discard. However, that garbage is a mix of edible and inedible substances. Because of the diversity of their diet, rats cannot have a set of fixed rules about what to eat and what to avoid. Instead, they combine a specific behavior with the capacity to learn.

When rats encounter a food for the first time, they sample only an amount small enough to avoid serious consequences should the material prove to be inedible.

(a)

(b)

Figure 30.3
Some Animals Learn Who Their Parents Are by Imprinting

(*a*) Imprinting usually takes place only during a brief "window" of sensitivity. A study of ducklings found that their peak sensitivity to imprinting occurred at about 15 hours after hatching. (*b*) To these young geese, from a famous experiment, Konrad Lorenz represents "mother."

If the food sample sickens them, they avoid that food in the future. If the food causes no harm, they add it to their learned list of good things to eat. By learning what to avoid based on their experiences, rats can cope with new and unexpected kinds of food, something they could not do with inflexible fixed behaviors.

Imprinting helps offspring learn quickly who their parents are

An interesting type of learning, called **imprinting**, occurs in animal species in which parental involvement in rearing the young is important. You may have seen a mother goose with her goslings trailing along behind her. Within a short time after hatching, the goslings develop a "vision" of who their mother is and what a goose is supposed to look like. This sense of "mother goose" is critical to the survival of young animals that must spend time with parents in order to learn how to survive on their own. Imprinting usually takes place only during a specific period of time in the early life of offspring. In ducklings, for example, imprinting occurs during (but not before or after) the period about 7 to 23 hours after hatching, with a peak sensitivity to imprinting at about 15 hours (Figure 30.3*a*).

Generally this form of learning works well, because after the young hatch, the mother is close by and is the animal the young are most likely to encounter. However, in a series of experiments conducted in the mid-twentieth century, behavioral scientist Konrad Lorenz found that when goslings saw him and heard "goose" sounds only from him—and not from their actual mother goose—in their early hours after hatching, they imprinted on him. These goslings treated him as "mother goose" and followed him, not their mother, about (Figure 30.3*b*).

30.3 The Evolution of Animal Behavior

Although behaviors are not morphological traits, and although behaviors—learned behaviors in particular—depend strongly on environmental factors, behaviors do have a genetic basis. For this reason, natural selection acts on behaviors just as it does on morphological traits—by selecting those behaviors that contribute to survival.

Genes control fixed behaviors

We can easily imagine the genetic control of fixed behaviors. One of the clearest examples of the direct genetic control of a fixed behavior involves the nest-cleaning behavior of honeybees. Hives of honeybees sometimes fall victim to a contagious bacterial disease that kills larvae (immature bees) in the hive. The hives of some genotypes of honeybees, however, rarely suffer from the disease. Hives of these "resistant" genotypes do become infected, but they reduce the spread of the bacteria by quickly removing infected larvae from the hive. This nest cleaning involves two fixed behaviors: cutting open cells containing larvae killed by the bacteria, and removing the dead bodies from the hive. Genetic crosses have revealed that each behavior is under the control of a different gene. In some "susceptible" genotypes, the honeybees neither cut open cells nor dispose of dead larvae. In other susceptible genotypes, the bees either fail to open cells of infected larvae or fail to remove dead larvae from opened cells (Figure 30.4).

More typically, genes influence behavior, but interactions between genes and the environment ultimately determine the behavioral patterns shown by an individual. Careful studies of identical and fraternal twins in humans, for example, have revealed that schizophrenia—a mental illness in which a person hallucinates, feels persecuted, and behaves strangely—has a strong genetic component.

Identical twins arise from a single fertilized egg that splits into two genetically identical individuals during the blastula stage, early in its development. Fraternal twins result when the ovaries release two eggs simultaneously, and these eggs are then fertilized by two different sperm, so fraternal twins are no more similar genetically than

are typical brothers and sisters. If genes control schizophrenia, then we would predict that either both members of a pair of identical twins should have schizophrenia or both should be healthy. In addition, we would expect both members of a pair of fraternal twins to have schizophrenia less frequently because they differ genetically.

Researchers have found that it is indeed much more common for both members of a pair of identical twins to have schizophrenia than for both members of a pair of fraternal twins to have the disorder. On the other hand, if one member of a pair of identical twins has schizophrenia, the other may still be healthy, even though he or she carries all the same genes. This observation indicates that environmental influences play a role in determining whether a person with the genes for schizophrenia actually develops the disease (Figure 30.5).

Genes also influence learned behaviors

Learned behaviors, as we have seen, are strongly influenced by the environment: what animals learn depends on what they experience and pay attention to during their lifetime. Nonetheless, differences in what, how much, and when animals can learn reflect genetic differences in their learning patterns and capacities.

Most laboratory rats, for example, can learn to negotiate mazes, but some rats learn faster than others. By mating fast learners with other fast learners, researchers can produce offspring that also learn to negotiate mazes quickly. Mating slow learners with other slow learners produces rats that have a difficult time learning to find their way through a maze.

Some species of birds, such as Clark's nutcrackers, excel at remembering where they have stored food. These birds have brains that differ structurally, and thus presumably genetically, from the brains of closely related species with only an average ability to find stored food. Researchers have even identified genes that seem to control the ability of birds to learn the songs that are characteristic of their species.

Clark's nutcracker burying a seed

A *UuRr* × *ur* cross yields the following phenotypes:

These bees are **susceptible** because they carry both the *U* and the *R* alleles.

UuRr

These bees are **susceptible** because they carry the *U* allele, but they will remove dead larvae if cells are uncapped.

Uurr

These bees are **susceptible** because they carry the *R* allele. They will uncap cells of dead larvae, but won't remove them.

uuRr

These bees are **resistant** because they carry neither the *U* nor the *R* alleles.

uurr

Figure 30.4

Genes Control the Nest-Cleaning Behavior of Honeybees

The results of a cross between a *UuRr* female bee and a *ur* male bee reveal that nest-cleaning behavior has a genetic basis. (Male honeybees are haploid, and hence have just one allele for each gene instead of two alleles, like diploid females.) Bees that carry either the *U* allele or the *R* allele are susceptible to a bacterial disease because they do not perform behaviors that keep their hive disease-free. Bees with the *U* allele do not open cells containing larvae killed by the disease, while bees with the *R* allele do not remove the dead larvae from opened cells.

30.4 Communication Allows Behavioral Interactions Between Individuals

Communication is a type of behavior that allows one individual to exchange information with another, thereby making it possible for an animal to coordinate its activities with those of other individuals. By communicating with one another, animals in groups can do things that no individual could do on its own, such as fend off large predators, capture large prey, or build bridges (Figure 30.6).

One individual produces signals that stimulate a response in one or more others

The communication behavior of animals varies widely in its complexity and includes just about any type of signal that other animals can sense: sounds, visual signals, odors, electrical pulses, touch, and tastes.

At the simple end of the spectrum, the release of a chemical signal, called a **pheromone**, by one individual informs others of the same species about its identity, its location, its physical condition, or a situation in the environment. As we saw in Chapter 26, female silkworm moths release bombykol, a sex pheromone, to attract males. The pheromone not only reveals to a male the identity of the individual that released it as a female of his species, and that she is interested in finding a mate; it also tells him where she is. If he responds to her signal by moving up the concentration gradient of the pheromone, which increases the closer he gets to the female, the male can locate her.

There is evidence that we humans also produce pheromones that influence sexual behavior. But pheromones communicate other conditions besides sexual readiness. For example, alarm pheromones released by ants, honeybees, and other social insects when a colony or hive is disturbed inform other individuals in the community about the situation. They interpret the disturbance as an attack and respond accordingly.

An example of a more complex signal is the intricate dance used by honeybees to communicate the distance and location of food to other members of their hive (Figure 30.7). And at the extreme end of the complexity spectrum of communication lies human **language**. Most human languages consist of thousands of words representing everything from objects to actions to abstract ideas.

Why do animals communicate?

Animals most often communicate to identify themselves, to avoid conflict, or to coordinate their activities. Self-identification is probably the most common function of communication. An animal uses a variety of signals to inform other members of its species that it is a member of the same species, as well as to indicate its sex, its physical condition, and its location. When a dog sprays a fire

Figure 30.5 The Genetics of Schizophrenia

Schizophrenia is a genetically based mental disorder in humans. However, the environment clearly affects the expression of the genes causing schizophrenia, because even among genetically identical twins there are many cases in which only one twin develops the condition.

hydrant or a post with its urine, the scent communicates the dog's sex, breeding condition, health, and status to other dogs. The mating songs and behaviors of male birds, frogs, and crickets tell females their species, location, and potential quality as mates (Figure 30.8a).

A second important function of communication in animals is to avoid potentially harmful conflicts. Physical conflict over food or mates can lead to serious injury for both the winner and the loser. To reduce the risk of injury

Figure 30.6 Communication Is Vital to Human Behavior

The design and construction of bridges and other structures is just one of many areas of human behavior in which communication is essential. The Millau Viaduct in France is the highest bridge in the world at over 1,000 feet, taller than the Eiffel Tower.

Figure 30.7
Honeybees Communicate by Dancing in Their Hive

Bees use a dance language to communicate complex information about the location of distant food sources to other worker bees in their hive.

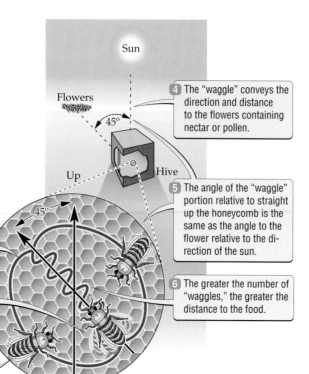

4 The "waggle" conveys the direction and distance to the flowers containing nectar or pollen.

5 The angle of the "waggle" portion relative to straight up the honeycomb is the same as the angle to the flower relative to the direction of the sun.

6 The greater the number of "waggles," the greater the distance to the food.

1 Honeybee workers perform a dance on the honeycomb in the hive that moves in a figure eight pattern…

2 …centered on a "waggle" portion in which the worker vibrates her body from side to side.

3 Other workers watch the dance closely to learn where to find food.

Pattern of waggle dance

See bees communicate by dancing. 30.2

in such encounters, many species communicate their fighting ability through ritual displays. When male red-winged blackbirds flash red wing patches at one another, for example, their brightness and size signal a male's quality as a fighter to other males. This ritual allows the poorer fighter to back down without engaging in a potentially dangerous fight.

Sheep, antelope, and their relatives bear horns that vary greatly in shape—from stout, straight weapons capable of doing great damage to impressive but ineffective weapons. Those species with the most lethal horns usually engage in displays that communicate their suitability as mates rather than in actual fights, whereas those species with less lethal horns more often engage in physical combat (Figure 30.8b).

Animals that live in groups communicate with one another to coordinate their behaviors to accomplish a shared goal. Humans do this when they work as a team (Figure 30.8c). Wolf packs and lion prides communicate when they hunt for food. And as we saw in Figure 30.7, the dances performed by honeybees allow the members of a hive to coordinate their foraging activities.

(a)

(b)

(c)

Figure 30.8 Why Animals Communicate

(a) A singing male frog advertises to listening females its species, its location, and its quality as a mate. (b) Male bighorn sheep communicate their genetic "quality" to potential mates by engaging in head-butting contests. Their backward-curved horns allow them to engage in these contests with a relatively low risk of injury. (c) Humans communicate to coordinate their activities and act as a team, which can be vital in a frenetic atmosphere like that of a hospital emergency room.

Language may be a uniquely human trait

Humans rely heavily on spoken, written, and grammatically based sign languages for communication. Much of our human identity depends on our ability to express complex and abstract ideas through language. One of our innate behavioral characteristics is that we are problem solvers. Language allows us to exchange ideas and solve problems, even in very stressful or dangerous situations (see the box on page 597).

There is no strong evidence of language among other animals. Many birds and mammals can produce a variety of sounds, each of which conveys a particular message, but do not assemble the sounds into ideas. Research on the ability of chimpanzees to communicate indicates that they can string together symbols provided by humans to express abstract ideas, but there is no indication that chimpanzees use such sophisticated communication in nature.

Recent research on humans suggests that language goes beyond communicating simple information, such as "I'm hungry," or "I'm angry." Language can affect our ability to form abstract concepts, such as beauty in art and numbers. For example, one study has indicated that the concept of counting beyond two may require words for those numbers. The language of the Pirahã, a group of indigenous people living in the Amazon Basin of Brazil, has only a few words to represent numbers of things. In particular, the language can express only quantities of "one," "two," or "many." Experiments involving the Pirahã suggest that their limited counting vocabulary affects their ability to work with abstractions based on counting. For example, some Pirahã speakers were asked to line up a number of objects that matched the number of objects set out by the researchers. When the number of objects exceeded two (the limit of the ability of the language to communicate a number precisely), they started to have problems with the task. Quantities of three or more objects are all included in the category "many." These results suggest that the lack of appropriate words for numbers higher than two limits the ability to form a concept of numbers greater than two.

30.5 Social Behavior in Animals

Many animals besides humans live in closely interacting groups. **Social behavior**—behavioral interactions among members of a group, usually of the same species—offers its members advantages not available to solitary animals, but requires communication to be successful (see the box on page 598 to see how social structure can break down).

Group living offers many benefits

The many benefits of group living can more than compensate for the increased competition for resources that results from living in close proximity to other animals of the same species. Groups can find some foods better than individuals can. Groups can stir up and capture prey more effectively than individuals can, and unsuccessful hunters or hunters-to-be can watch and learn while others in their group demonstrate successful hunting behaviors. By working together, members of a group can acquire foods that are not available to individuals. The hunting of large mammals by groups of cooperating wolves or humans illustrates such cooperative behavior well (Figure 30.9a).

Living in groups also gives animals two effective defenses against predators that are not available to single individuals. First, the larger the group, the more eyes, ears, and noses it contains, making early detection of approaching predators more likely (see Figure 35.8). Second, the individuals on the edge of a group are more likely to be attacked than those on the inside, offering members of the group in the middle additional warning and the advantage of being in less vulnerable positions. Gazelles in the middle of a herd, for example, benefit because a cheetah is most likely to kill an individual at the edge of the herd. Moreover, all gazelles in the herd benefit because at least one is likely to notice and warn of approaching cheetahs before the cheetahs are close enough to attack (Figure 30.9b).

In some cases, living in a group gives animals access to a scarce resource. Groups of Florida scrub jays include some members that do not breed, but instead help care for the offspring of a breeding pair, feeding the nestlings and helping to defend the breeding territory against intruders. These nonbreeding birds appear to benefit because belonging to the group puts them on a "waiting list" to inherit the territory from the breeding pair. The chance that a solitary scrub jay can find and hold control over a high-quality breeding territory is so small that it pays to put off breeding for several years and help breeding pairs while waiting to inherit a territory (Figure 30.9c).

Individuals in groups may act to benefit other members more than themselves

Animals that live in groups are often **altruistic**; that is, they do things that help other members of their group survive or reproduce while decreasing their own chances of doing so. For many years, altruism seemed to contradict Darwin's idea that only traits that improve the *individual's* reproductive success can spread through a population.

(a)

(b)

Figure 30.9 The Benefits of Group Life
(a) Wolves can hunt larger prey as a group than they can as individuals. (b) A predator circles a herd of gazelles, and even if it picks an individual on the edge of the group, the rest will have a greater chance of survival. (c) Family groups of scrub jays work together to feed young, defend nests, and hold on to valuable breeding territories.

(c)

The scrub jays mentioned earlier provide a good example of altruism: nonbreeding members of a group help the breeding pair raise their young while forgoing reproduction themselves. This behavior improves the reproductive success of the breeding pair, while presumably lessening the reproductive success of nonbreeding individuals.

In general, social groups consist of closely related individuals. A pride of lions, for example, centers around a group of closely related females. Similarly, among scrub jays, the helpers most often turn out to be older offspring of the breeding pair, so the young birds they are helping to raise are their younger siblings. In these cases, the individuals benefiting from altruistic behavior carry many of the same genes as the altruistic individuals, so altruistic behavior often helps to spread many of the same genes that the altruist carries.

Do social insects represent the ultimate in group living?

Social insects such as ants, bees, and termites include some of the most successful species on Earth. Although ants, bees, and termites evolved social behavior independently of one another, these groups share some remarkable traits.

Each of these groups lives in large colonies containing individuals belonging to distinct classes that serve distinct functions within the colony. The workers work together to build complex nests (Figure 30.10), forage widely for food, and defend the colony effectively against predators in ways that individuals acting by themselves never could. The queen spends her life producing massive numbers of eggs. The workers in a hive are the female offspring of the queen and are therefore closely related to her and to one

another. The workers are sterile; they do not reproduce, but contribute to reproduction by carefully tending to the queen and her larvae. Their altruistic behaviors, including defending to the death the queen and the colony or hive, do not reduce the reproductive capacity of the colony, but rather increase its chances of survival.

Figure 30.10 Cooperation Creates Termite Mounds
This giant termite mound was built through the cooperative effort of millions of closely related, sterile termite workers. Such workers make up most of the population of each colony. Some individual termite mounds have existed for hundreds of years.

Biology on the Job

Using Communication to Move Hostage Takers from Violence to Problem Solving

Jim Maguire is an experienced member of the Hostage Negotiating Team for the Chattanooga, Tennessee, Police Department, where he has worked for 20 years. As a hostage negotiator, he must walk a fine line between patiently negotiating with a hostage taker and recommending decisive action by a Special Weapons and Tactics (SWAT) team.

Jim, when you and other team members arrive at a scene where hostages have been taken, what happens first? A lot has already happened before we arrive. We always have a plan before we open negotiations. We have assembled a SWAT team and a hostage negotiation team. We have designated an overall on-scene commander who is in charge of both the negotiation team and the SWAT team. We also know something about the situation. For example, hostages may have been taken as a result of a domestic problem, a home intrusion, a robbery, or an act of terror. We usually know at least what general type of situation we will be facing, and our plans can be tailored to meet it. We gather every detail possible about the house or building and the surroundings. If the situation is taking place in a school or other large building, we obtain architectural drawings. Even the smallest details, such as whether a particular door opens in or out, can become important if a SWAT team has to gain entry.

What do you do when you get to the scene? For the sake of public safety we establish a wide perimeter and move everyone out of that area. Then we determine how we will attempt to contact the hostage taker. Is there a phone that the hostage taker could use? What's the number? Telephone dialogue provides us with a lot more information about the attitude and potential behavior of a hostage taker than shouting through a bullhorn and listening to shouted responses. In any case, the most important first step is starting a dialogue. Once a hostage taker begins talking, we have made progress, because we have begun moving away from what we call an action mode to a problem-solving mode.

What do you mean by "action mode" and "problem-solving mode"? The initial phase in these situations is called the action mode, because hostage takers are almost always highly agitated when negotiations begin. Rage, fear, desperation, or a combination of these emotions is very often shaping a person's behavior at this point, and he or she is more likely to take

Scene from the Movie *The Negotiator*
Successfully dealing with hostage takers to minimize loss of life requires communication among the negotiation team members as well as skillful communication under stressful circumstances with hostage takers.

violent action. If I can start a dialogue, I can begin to develop an understanding of how this problem situation arose from the point of view of the hostage taker. Then I can begin to play a role that helps the person to think about the situation as a problem to be solved. That's real progress, because one innate consistency in human behavior is that we are all problem solvers. No matter what the situation, we feel a sense of reward if we are solving a problem. That's the feeling I want to build in the hostage taker. A negotiator's approach is to become a partner in the problem-solving process by asking guiding questions and using active listening.

So you don't try to solve the problem for them. No, the goal is to help the person become engaged in solving the problem. As negotiations continue with this emphasis, the risk to everyone involved diminishes. Once problem solving begins, chances for a safe and injury-free resolution increase with every exchange of words.

It's surprising to find that human behavior has such predictable components in these seemingly unpredictable situations. Are there other predictable behaviors that develop in hostage situations? One dynamic that plays out along predictable lines is something called the Stockholm syndrome. The name arose from a 1973 bank robbery in Stockholm, Sweden, in which hostages were held for 6 days. A strong bond developed between the hostages and their captors. When hostages are held for long periods, such as in the Stockholm robbery, these bonds consistently develop. Hostages even begin to think of the police and negotiators as the enemy.

That must make your job much more difficult. Actually it helps. It becomes less likely that the captors will harm hostages, and it means that we have more time to negotiate. And, in negotiating, time is our ally. More time means that we increase the likelihood that the situation will be resolved by reason rather than force. Our goal is to prevent injury, or worse, to any of the people involved, including the hostage takers.

(continued)

Negotiation is one thing, but aren't there times when force is the only solution? Use of force is always a last resort, and making the decision to use force is a very tough call. However, if hostages are killed or tortured, we have no choice. The negotiations are declared over, and the SWAT team goes in.

But there must be times when the call isn't that clear-cut. That's true. In one instance a man had been holding his own 74-year-old mother hostage. She escaped, and he shot at her as she ran from the house where she had been held. She escaped, but her son was still inside, armed, and dangerous. We heard two more shots right after the mother escaped, then silence. After 5 hours of trying to make contact without success, the SWAT commander asked me for a judgment. My experience told me that it was very likely that he had either committed suicide or had somehow gotten away. The SWAT commander decided to go in with his team. I heard about 40 shots exchanged. One officer was wounded and the man was subdued. He had been hiding motionless and silent for 5 hours at the bottom of the cellar stairs. In this instance the only option was to send in the SWAT team. The incident demonstrates that many aspects of human behavior are simply not predictable. That's what makes hostage negotiations a combination of expertise and guesswork.

The Job in Context

Jim Maguire deals regularly with the complexities of human behavior in volatile and potentially deadly situations. He must make judgments based on his experience with similar behaviors that he has either studied or seen firsthand. A hostage negotiator must be an expert on human behavior in high-stress situations, and must be able to use that knowledge when lives are on the line. In negotiations, a lot rides on every word and gesture. Constant training and practice for responding to different situations are part of Jim Maguire's life.

Biology Matters

Drinking and Dark Sides of Human Behavior

Human behavior changes dramatically under the influence of alcohol, as does our ability to regulate that behavior. Among college students, the consequences of abusing alcohol go far beyond the fatalities and injuries caused by drinking and driving. Each year in the United States approximately 600,000 physical assaults and 70,000 cases of sexual assault involving college students are alcohol-related. In addition, 400,000 students have unprotected sex, and 100,000 students report that they were too intoxicated to know whether they had consented to sexual intercourse. About 1 in 100 students attempt suicide because of alcohol or other drugs. On the academic side, students who drink miss more classes, get lower grades, and are more likely to fall behind in their work.

And the consequences of drinking spill over into the lives of those who are not even doing the drinking. Among students who live on campus or in sororities or fraternities and don't drink, or drink moderately, 60 percent have had sleep interrupted, almost half have had to take care of a drunken student, and 15 percent have had property damaged. One in three has been insulted or humiliated by or has had a serious quarrel with a drunken student, and almost 1 in 10 has been pushed or hit. One study found that alcohol is one of the most significant contributing factors to sexual aggression by male college students. One in 100

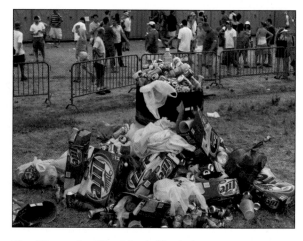

The Aftermath of Too Much Alcohol

college-age women have been victims of alcohol-related sexual assault or acquaintance rape. The damage to the psychological well-being and academic performance of these victims is difficult to measure.

Alcohol, when consumed in excess, very often causes drastic and dangerous changes in human behavior. An awareness of the costs and personal consequences associated with those changes can build a better appreciation for the advice we often see in the media: "If you drink, drink responsibly."

Pipevine Swallowtails Lead Two Behavioral Lives

It seems simple enough to say that a female pipevine swallowtail lays her eggs on only one type of plant. But before she can do so, she must know what to look for. Her brain must constantly process information that leads her to the right leaf. Her brain must translate that information and use it to control the muscles that power and direct her wings, so that she gets to where she needs to go. Only when she has done so does the coordinated behavior of depositing eggs on a relatively safe part of the leaf begin.

Once the eggs hatch, the feeding behavior of a caterpillar is the result of a coordinated effort among sensory organs, the brain, and the muscles that steer its crawling movements and manipulate its mouthparts. A caterpillar responds to sensory input unique to pipevine leaves. Without the appropriate stimuli, it will not eat. Adults rely on the same type of coordination, but they respond to an entirely different set of stimuli, such as flower color and substances in nectar. The functions of the two behaviors are the same—to direct the animal to the right food—but the actual behaviors are governed by stimuli as different as the bodies of caterpillars and adults.

Behavior and anatomy are intimately related. Consider the differences in feeding apparatus between adults and caterpillars of the pipevine swallowtail. A caterpillar's mouthparts are made for chewing leaves, and the muscles that move them are directed by a brain specialized for coordinating the "caterpillar" eating behavior of chewing and grinding. An adult's feeding apparatus is completely different. It is an extensible proboscis [pro-*BAH*-skiss] that works like a straw, and is made for sipping nectar from flowers, not for chewing leaves. Adult feeding behavior consists of extending the pro-

boscis and drawing in liquid. Thus an adult's brain coordinates different muscles that move a different feeding apparatus.

As we have seen, the stimuli that elicit feeding on pipevine leaves by caterpillars end up protecting adults as well. The tastes that caterpillars prefer end up protecting them all of their lives. How does that help an individual if it is devoured or fatally injured—the usual result of an attack by a predator? And how does it help the pipevine swallowtail as a species to survive? Consider how this works with respect to a predator's behavior.

Most caterpillars and butterflies share a clearly recognizable, yet generalized, appearance that signals "potential prey" to certain predators, such as birds. In addition, butterflies and caterpillars have specific characteristics, such as colors and spines on caterpillars and distinct color patterns on the wings of adults, which predators can learn to distinguish. A bird that devours a pipevine swallowtail caterpillar or an adult pipevine swallowtail based on its general appearance will regret it. That individual pipevine swallowtail is killed, and is thus unable to pass its genes on directly to the next generation. However, the bird that ate it is left with such a vividly unpleasant memory that it will refrain from attacking other pipevine swallowtails. So the sacrifice of a few protects the many, and the species as a whole.

This behavior works so well that other butterfly species have acquired some protection from predators over evolutionary time by mimicking the color pattern and form of pipevine swallowtails. The learning that affects the behavior of predators toward pipevine swallowtails protects these look-alike species as well.

Chapter Review

Summary

30.1 Breast-Feeding in Humans as an Example of Behavior

- A behavior is a coordinated movement or series of movements made by one animal in response to another animal or to the physical environment.
- Breast-feeding in humans involves the rooting reflex in infants and the milk-letdown reflex in mothers.
- The rooting reflex is initiated by a gentle touch of the baby's mouth or cheek and results in the baby searching with its mouth for the mother's nipple to get milk.
- In the milk-letdown reflex, pressure-sensitive cells in the mother's nipples stimulate the release of oxytocin into the bloodstream, which in turn causes the breasts to release milk.
- The behavioral interaction between infant and mother during breast-feeding coordinates their responses so that the infant tries to feed only when milk is available and the mother releases milk only when her hungry baby is nearby.

30.2 Fixed and Learned Behaviors in Animals

- Behaviors can be fixed (innate, requiring no learning) or learned (requiring response to changes in the environment and memory of those experiences).
- Fixed behaviors have a simple pattern: a single (usually simple) stimulus (called a releaser) leads to a single response. Fixed behaviors allow animals to behave appropriately when they have no chance to learn by experience or when the risks associated with the wrong behavior are great. Fixed behaviors may be stimulated only under certain conditions.
- Learned behaviors are responses based on past experience of the individual animal. The same response can be brought about by different stimuli, depending on an animal's past learning experience. Much of human behavior is learned.
- Imprinting is a type of behavior in which offspring learn to identify their parents and other members of their species according to who or what is nearby during a certain period in their development.

30.3 The Evolution of Animal Behavior

- Both fixed and learned behaviors have a genetic basis, so all behaviors can evolve.
- Behaviors involve interactions between genes and the environment. Fixed behaviors are more strictly governed by inheritance than learned behaviors, but the capacity to learn may also be under genetic control.

30.4 Communication Allows Behavioral Interactions Between Individuals

- Communication is behavior in which one individual exchanges information with another. It allows individuals to coordinate their activities with those of other individuals. Information is exchanged through the production of signals by one animal that stimulate a response in another.
- Communication can range from simple to highly complex. Pheromones are a simple form of communication used to communicate identity, location, physical condition, or a situation in the environment. Language is the most complex form of animal communication.
- Animals communicate to identify themselves (as individuals and as members of a species), to avoid conflict, to coordinate their activities in order to perform shared tasks, to indicate their sex and sexual readiness, and to convey information about their physical condition and location.
- Language may be a uniquely human trait because it allows humans to form abstract concepts such as numbers.

30.5 Social Behavior in Animals

- Social behavior—the interactions among members of a group of the same species—serves several different purposes. For prey animals, it provides added protection against predation. Social behavior also allows animals to gain access to resources, including food and breeding territories, more effectively, in some cases allowing animals to obtain foods not available to individuals working alone.
- Individuals living in groups often act in ways that benefit other group members more than themselves. Such altruistic behavior can evolve within groups of closely related animals. Some individuals sacrifice their lives or reproductive success to increase the overall survival and success of the group.
- Social insects (including ants, bees, and termites) are among the most successful social animal species. Sterile female workers forgo reproduction to promote the success of the colony or hive as a whole. They carry out specific tasks such as foraging for food, protecting the colony or hive, and tending to the needs of the queen, who lays large numbers of eggs.

⊙ Review and Application of Key Concepts

1. Describe the behaviors that occur in human nursing. Include learned behaviors associated with nursing.

2. Do you think humans are born with an innate fear of spiders? Explain and defend your answer.

3. Some people are born with athletic or musical talent. But even the most talented must learn and practice in order to excel. Pick a sport or a musical instrument and describe fully the dimensions of learning that must take place in order to fully develop a talent. Keep in mind that learning includes more than practice.

4. Language is a major component of the human ability to communicate. But we also communicate without words. What nonverbal forms of human communication have you responded to or used?

5. What kinds of fixed behaviors might be advantageous to animals that live in a group, and why?

Key Terms

altruism (p. 595)
behavior (p. 588)
communication (p. 592)
fixed behavior (p. 589)
imprinting (p. 591)
language (p. 593)

learned behavior (p. 590)
milk-letdown reflex (p. 589)
pheromone (p. 593)
releaser (p. 589)
rooting reflex (p. 588)
social behavior (p. 595)

Self-Quiz

1. The rooting reflex
   a. occurs in response to milk production.
   b. is an example of a fixed behavior.
   c. occurs in response to a visual stimulus.
   d. is an example of a learned behavior.

2. Behavior
   a. depends primarily on innate responses.
   b. always involves communication between two animals.
   c. always requires learning.
   d. allows animals to respond quickly to changes in their environment.

3. Fixed behaviors
   a. derive from experience.
   b. are always imprinted.

c. are genetically inherited.
   d. require language.

4. Learning
   a. occurs only in humans.
   b. overrides all genetic control of behavior.
   c. depends on an animal's past experience.
   d. requires pheromones.

5. Genes
   a. have no influence on behavior.
   b. influence only social behavior.
   c. affect only fixed behaviors.
   d. can affect fixed and learned behaviors.

6. The dance of a honeybee worker communicates
   a. the location of a food source.
   b. the location of a mate.
   c. the sex of the dancer.
   d. the location of other hives.

7. Which of the following animals would rely most on learned behaviors?
   a. termite   c. rat
   b. frog   d. spider

8. Which of the following is the most complex form of communication?
   a. mating displays   c. pheromones
   b. the Pirahã language   d. bird songs

9. Which of the following is *not* an example of altruistic behavior?
   a. A butterfly lays an egg.
   b. A scrub jay chases a snake away from its parents' nest.
   c. A sterile worker ant defends the nursery of its anthill from invading ants.
   d. A lioness regurgitates food for her sister's cubs.

10. Animals that live in groups
    a. have no fixed behaviors.
    b. usually rely on language for communication.
    c. can reap benefits that compensate for increased competition among group members.
    d. have no genetically controlled behaviors.

Biology in the News

Teenage Girl Charged in Pit-Bull Attack

OTTAWA—A 17-year-old girl has been charged with criminal negligence after three pit bulls attacked a toddler and his father in Ottawa last week.

The teenager, who can't be named under the Youth Criminal Justice Act, was charged with three counts of criminal negligence causing bodily harm on Tuesday—the same day Ontario approved a ban on pit bulls and similar dogs.

Police also said they expect to lay criminal charges against a 21-year-old this week in connection with the attack but they wouldn't identify the man.

The dogs jumped the fence and attacked a two-year-old boy and his father as they walked through a park on Feb. 24. One pit bull clamped down on the boy's head before the man and his neighbour managed to beat the animals away. The adults were badly bitten.

Stories about attacks by pit bulls on people seem to appear in the news every week. Pit bull owners are quick to point out that most pit bulls are not aggressive toward people and make wonderful pets. On the other hand, statistics show an increase in the number of people bitten by dogs over the past decade. The American Humane Society has called dog bites the most important public health problem for children, who make up almost half of dog-bite victims. Statistics also show that most serious maulings can be attributed to a relatively short list of dog breeds that includes the pit bull. Every year an average of 20 people, mostly children, are killed by dogs. The problem of injury or death from dog bites has reached the point that many communities have enacted or are considering legislation banning dog breeds considered especially dangerous.

Pit bulls were originally bred as fighting dogs. They have a compact and muscular body, powerful jaws, and a tendency to be aggressive. Although dogs of all breeds will occasionally attack people, the consequences of a pit bull attack are generally more severe than the consequences of an attack by a Chihuahua or toy poodle. Between them, pit bulls and rottweilers (another large, powerful breed) account for 50 percent of all fatalities from dog bites, even though these breeds are not the ones involved in the most biting incidents (a distinction held by German shepherds and chow chows).

The old saying, "There are no bad dogs, only bad owners," also has some truth to it. The two factors most likely to contribute to a dog attack are poor training and failure to sterilize the dog. Well-socialized dogs are unlikely to attack people. Unfortunately, many people buy dogs with aggressive reputations as guard dogs and specifically train them to respond aggressively to strangers. Dogs that have not been sterilized retain strong territorial behavior and are three times more likely to bite than sterilized dogs.

Evaluating the news

1. Do you think that a ban on pit bulls and other dog breeds considered dangerous is an effective way of preventing dog-bite injuries and fatalities? Why or why not?
2. If you owned a Great Dane, a large breed of dog involved in relatively few attacks but which has caused some deaths, what would you do to make others feel comfortable around your pet?
3. How would you go about increasing the chances that dog owners will raise and maintain their dogs responsibly?

SOURCE: *CBC News*, Wednesday, March 2, 2005.

Smoking: Beyond Lung Cancer

A Brief History of Tobacco

Tobacco originated in the Americas. Archaeological evidence indicates that from the beginnings of agriculture in South America, roughly 8,000 years ago, native Americans have cultivated two species of tobacco, *Nicotiana rustica* and *N. tabacum* (Figure E.1a). These species, like tomatoes and potatoes, belong to the nightshade family. They grow as weeds in disturbed soils, such as those produced by farming. As agriculture spread through the Americas, so did tobacco. Among American crops, only corn was more widely grown.

Tobacco was probably first used by chewing it or by infusing it in hot water as a tea. Tubular pipes probably used to inhale tobacco smoke have been discovered in burial sites dating back to 1,500 to 3,000 years ago. The earliest indisputable evidence of tobacco smoking—Mayan pottery bearing pictures of a person smoking tobacco—dates back to just over 1,000 years ago.

Tobacco spread widely throughout the Americas because of its role in rituals connecting the natural world with the supernatural. When taken in large amounts, the potent leaves of tobacco—especially of *N. rustica*—can cause hallucinations. In more moderate amounts, tobacco acts as a stimulant and can improve night vision. Tobacco proved to be the mind-altering drug of choice of native farming societies throughout the Americas. In this role, tobacco use was largely limited to priests and shamans. Tobacco also had a medicinal role, proving to be a powerful insecticide and an effective way of ridding the intestines of parasitic worms.

When Europeans arrived in the New World in the late fifteenth century, they also found tobacco being used much as it is today—as a stimulant used by the general population (Figure E.1b). It was this role that tobacco took on when it arrived in Europe and, from there, spread around the world.

(a) *N. tabacum*

(b) Example of tobacco use by Native Americans

Figure E.1 Native Americans Were the First to Cultivate Tobacco

(a) Tobacco comes from two closely related plant species, *Nicotiana rustica* and *N. tabacum* (pictured). (b) Native Americans used tobacco as a stimulant, for medicinal purposes, and in rituals, as illustrated in this Theodor de Bry engraving.

Main Message: Tobacco smoke, with its addictive properties, causes a host of health problems and is one of the leading causes of preventable death worldwide.

European herbalists (the doctors and pharmacists of their day) also found a number of medical uses for tobacco, including treatment for toothaches, worms, bad breath, and—ironically—cancer. The popularity of tobacco throughout the world gradually increased over time, finally exploding with the widespread acceptance of inexpensive cigarettes during World War I. During the two world wars, cigarettes were included in the ration packages distributed to all soldiers by many governments. Currently, over a billion people worldwide smoke tobacco.

Today we recognize that using tobacco, and especially smoking tobacco, does great harm to our health, so it may come as a surprise that the health effects of tobacco have been known for centuries. Sir Francis Bacon reported in 1610 that tobacco was addictive. In 1761, the British doctor John Hill linked the taking of snuff (tobacco in powdered form, taken up through the nose) to nasal cancer. In 1856, the British medical journal *The Lancet* published a debate on the health effects of tobacco. By 1912, evidence that smoking causes lung cancer was beginning to accumulate. Today, even the tobacco companies accept that tobacco harms their customers' health.

Even though we understand perfectly well that tobacco is bad for us, people worldwide continue to smoke in increasing numbers. Since the early 1980s, smoking has contributed to 1 out of every 10 deaths worldwide. Tobacco smoke is the leading cause of preventable deaths worldwide.

So we see that tobacco is not new to human culture, and that we have known about its potentially harmful health effects for a long time. In this essay we investigate why tobacco continues to kill people. We begin with an overview of how tobacco affects our health, then examine why tobacco affects our bodies as it does. We conclude by considering how, despite the highly addictive nature of tobacco, smokers can kick the habit.

The Adverse Effects of Smoking

Tobacco kills. All of us have seen or heard this message; it has appeared on every package of cigarettes sold in the United States since 1965 (Figure E.2). Most of us know about the now well-established link between smoking and lung cancer. However, most of us don't appreciate the many different ways in which smoking can affect our health. Tobacco kills most of its victims in one of three interrelated ways: by causing cancer, by causing diseases of the circulatory system, and by causing lung disease. Let's look at these adverse health effects in their order of severity: of the three, diseases of the heart and circulatory system take the greatest toll, followed by various cancers, and then by lung disease (Table E.1).

Smoking damages the circulatory system

Smoking-related diseases of the circulatory system include atherosclerosis and congestive heart failure. As we saw in Chapter 23, in atherosclerosis, plaques (fatty deposits) build up in blood vessels, eventually blocking them, and in congestive heart failure, fluid builds up in the lungs and other tissues because the heart can no longer adequately move blood through the circulatory system. How does smoking relate to these two conditions?

Smoking greatly increases the chances of atherosclerosis by damaging the epithelial cells on the inner walls of major blood vessels and by decreasing the proportion of high-density lipoproteins (HDL, the "good" cholesterol) in the blood. Both factors greatly increase the chances that plaques will form on blood vessel walls, gradually narrowing the diameter of the blood vessels. Narrowing of a blood vessel increases the chance that not enough

Table E.1

How Tobacco Smoke Kills

Illness	Annual deaths worldwide due to tobacco smoke
Diseases of the circulatory system	1, 690,000
Cancers	1,469,000
Respiratory diseases	1,399,000

blood will get through, starving the tissues supplied by that blood vessel.

If the narrowed blood vessel supplies the heart, it causes coronary artery disease (also called ischemic [iss-*KEE*-mik] heart disease), in which heart tissue is starved of blood. If some of the plaque in an artery supplying a part of the heart breaks off, it can lead to the formation of a blood clot that completely cuts off the supply of blood, leading to a heart attack (Figure E.3). Besides causing atherosclerosis, smoking contributes to the chances of a heart attack by causing blood to clot more readily and by decreasing the amount of oxygen that blood can carry.

The narrowing or blockage of blood vessels elsewhere in the body as a result of atherosclerosis can have similarly bad effects. Atherosclerosis of the arteries supplying the legs and arms causes peripheral artery disease, a condition found almost exclusively among smokers. In severe cases, peripheral artery disease can lead to gangrene (death of soft tissues due to loss of blood supply), requiring amputation of the limb. Atherosclerosis of the

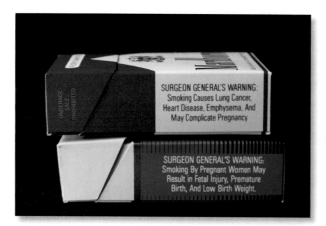

Figure E.2 Smoking Tobacco Is Hazardous to Your Health
Surgeon General's warnings printed on packages of cigarettes.

Figure E.3 Smoking Greatly Increases the Chances of Heart Disease and Heart Attack

arteries supplying the brain can lead to a stroke, in which part of the brain tissue dies. Severe strokes can kill, while nonfatal strokes can lead to paralysis, memory loss, and changes in behavior, depending on the part of the brain affected.

Congestive heart failure results when an overworked or damaged heart loses its ability to effectively pump blood through the circulatory system. The narrowing of blood vessels increases the blood pressure needed to push blood through the circulatory system, which forces the heart to work harder. Prolonged hypertension, such as that caused by atherosclerosis, can cause the heart to enlarge, which interferes with its ability to pump blood. Similarly, a heart damaged by a heart attack may not be able to pump blood as well as a healthy heart. If the heart can no longer pump out blood quickly enough, the blood returning to the heart in the veins backs up, leading to congestion as the blood plasma begins to diffuse back out of the blood vessels and into the surrounding tissues. This movement of plasma has several effects: it can cause **edema** (swelling of the tissues); it can cause fluid to collect in the lungs, which interferes with breathing; and it can reduce the kidneys' ability to eliminate sodium and water from the body. Over time, congestive heart failure leads to potentially fatal lung and kidney failure.

Smoking causes cancer

Cancer, as we saw in Interlude B, is a group of diseases caused by the out-of-control proliferation of abnormal cells and the spread of those cells to other parts of the body. As a result, cancers can cause their damage either in the tissue in which they originate or in tissues to which they spread. Cancerous cells form when damage to DNA in normal cells causes those cells to lose their control over growth and division. Tobacco smoke contains **carcinogens**, compounds that can cause cancer by damaging DNA. Because the chemical compounds in tobacco smoke enter the body through the lungs, their damage is concentrated there. Once dissolved in the blood, however, they can spread to and affect cells in other organs.

In the United States, more people die each year from lung cancer—almost 170,000 people—than from any other kind of cancer (Figure E.4). Of these deaths, 75 percent (in women) to 90 percent (in men) result from smoking. If you contract lung cancer, your chances of surviving for 5 years (the length of time doctors use to pronounce a cancer patient "cured" after treatment) are less than 1 in 6.

Although lung cancer is the most conspicuous type of cancer we associate with tobacco smoking, it is by no

Figure E.4 Most Cases of Lung Cancer Result from Exposure to Tobacco Smoke
The arrows on this X-ray point to malignant tumors, indicating that these lungs are riddled with cancer.

means the only kind of cancer caused by smoking. Other smoking-related cancers combine to account for about one-third as many deaths as lung cancer. Some of these cancers affect other parts of the respiratory system, such as the mouth, nasal sinuses, larynx (or voice box), and throat. Others target less obvious organs. Smoking is a major cause of cancers of the urinary tract, including the kidneys and bladder. It also causes cancers in organs linked to the digestive system, such as the stomach, pancreas, and liver. Smoking is also implicated in one kind of leukemia (a cancer of the white blood cells) and one kind of cervical cancer in women.

It is worth pointing out that even forms of tobacco that do not involve smoking (for example, snuff and chewing tobacco), as well as smoking without inhaling, pose significant cancer risks. Using snuff or chewing tobacco greatly increases your risk of cancers of the mouth and throat. Smoking a cigar without inhaling still exposes you to secondhand smoke, which carries with it all of the same risks as inhaling.

Smoking damages the respiratory system

The third important way in which smoking affects our health is by damaging the respiratory system, sometimes irreparably. The chemicals in tobacco smoke can damage the tissues of our lungs and airways.

Regular exposure to tobacco smoke irritates the tissue lining the bronchi that carry air to the lungs. This constant irritation leads to inflammation and thickening of the lining of the bronchi, a condition called chronic bronchitis. As the lining thickens, the diameter of the airway decreases, making it harder to draw air into and out of the lungs. The irritated bronchial lining leads to the production of a lot of mucus, which further decreases the airway diameter, provides a breeding ground for bacteria, and results in a persistent "smoker's cough." Although not directly fatal, chronic bronchitis can reduce a person's ability to be active and is associated with an increased number of lung infections.

Tobacco smoke also interferes with the function of the microscopic cilia that line the bronchi. Normally these cilia help to keep small particles and bacteria from reaching the lungs by sweeping them out of the bronchi. Tobacco smoke paralyzes these cilia, allowing the particles and bacteria to reach the alveoli in the lungs. These irritants lead to inflammation that, over time, destroys the delicate walls of the alveoli. As these walls break down, the alveoli combine into fewer, larger, less elastic air sacs that provide less surface area for gas exchange. As this condition, known as emphysema, worsens, it becomes harder and harder to breathe (Figure E.5).

Both chronic bronchitis and emphysema belong to a family of diseases known as chronic obstructive pulmonary diseases, which interfere with the ability to exchange gases. The body is starved of oxygen, leading to shortness of breath, chronic fatigue, and a restricted ability to be active. As they progress, chronic obstructive pulmonary diseases become life-threatening. The vast majority of cases of chronic bronchitis and emphysema stem from exposure to tobacco smoke.

What else does smoking do to us?

Smoking can affect people's health in a number of other ways, both large and small. For example, smoking during pregnancy harms the health of the developing child, as described in Chapter 29. Other side effects of smoking include damage to oral health and accelerated aging. Smoking stains teeth and dulls the sense of taste. In addition, it increases the chances of gingivitis (chronic gum disease), which can lead to loss of teeth.

People who smoke get more wrinkles, and at an earlier age, than those who do not. But the effect of smoking on apparent aging may be more than skin-deep: recent research has shown that smokers have telomeres that are shorter than those of nonsmokers. **Telomeres** are bits of DNA that cap the ends of, and apparently protect, chromosomes (Figure E.6). Telomeres typically shorten as a person ages, but the rate of shortening is faster in smokers than in nonsmokers.

Figure E.5 Emphysema

The breakdown of the alveoli by smoking-induced emphysema, as in the more transparent areas of the left lung of this computed tomography (CT) scan, lead to fewer but larger alveoli with reduced overall surface area. The infected lung becomes permanently enlarged and it can become all but impossible to breathe.

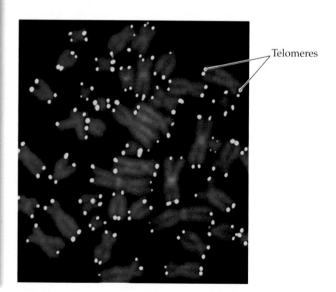

Telomeres

Figure E.6

Smoking Speeds the Aging of Our Chromosomes

The telomeres at the ends of chromosomes, shown here in yellow, become shorter with age. Smokers have shorter telomeres than nonsmokers of the same age.

Smoking exacts a tremendous personal and societal cost

Tobacco kills just under half a million Americans each year. Worldwide, it kills almost 5 million people annually. About 80 percent of these people are men. The life expectancy of smokers is 10 to 15 years less than that of nonsmokers. These statistics, disturbing enough by themselves, cannot convey the negative effects smoking has on the quality of the life that precedes a premature death due to failing lungs or heart.

Even people who don't smoke can fall victim to the ill effects of tobacco smoking. **Secondhand smoke** (tobacco smoke inhaled by people near the smoker) is recognized as a Group 1 carcinogen (the most strongly carcinogenic class of pollutants) by the U.S. Environmental Protection Agency, a distinction shared with such notable carcinogens as asbestos and benzene. Annually in the United States, almost 30,000 premature deaths result from exposure to secondhand smoke. Those most exposed to secondhand smoke are the smoker's spouse and children. A nonsmoking spouse of a smoker has a 20 to 30 percent greater chance of getting lung cancer than a nonsmoker does in a nonsmoking household. And about 1,000 babies die each year in the United States because of the side effects of their mothers' smoking. To put this in perspective, this is more than the number of babies (400 to 800) killed annually by child abuse in the United States.

Given the health problems caused by tobacco smoke, it should come as no surprise that the economic cost of smoking is enormous. In the United States, the annual cost of health care related to cigarette smoking is about $1,600 per smoker, and the economic loss in terms of productivity lost to smoking-caused health problems is about $1,700 per smoker. Added to these indirect costs of smoking are its direct costs: a one-pack-a-day cigarette habit costs the smoker as much as $2,500 (based on the $7 cost for a pack in New York City in 2005). Thus the total annual smoking-related cost per smoker is $5,800. An estimated 46.5 million Americans smoke, meaning that the total annual cost of cigarette smoking in the United States is about $250 billion.

Perhaps worst of all, the addictive properties of cigarette smoking make it a hard habit to kick.

The Problem with Tobacco Smoke

Why has tobacco, a plant originally used for ceremonial purposes in the Americas, become the leading cause of

preventable deaths worldwide? To address this question, we must consider why tobacco does so much damage, as well as why smokers do not stop smoking.

Tobacco smoke contains harmful chemical compounds and particulates

The smoke produced by burning tobacco consists of a complex mixture of over 4,000 different chemical compounds. The major compounds responsible for the adverse health effects of tobacco are listed in Table E.2, many by their group names.

Table E.2

The Major Components of Cigarette Smoke with Adverse Health Effects

Component	Known effects
Smoke particulates	Increase blood clotting, irritate respiratory system
Carbon monoxide	Reduces oxygen-carrying capacity of blood
Polycyclic aromatic hydrocarbons	Carcinogenic
Aza-arenes [AZZ-uh-AH-reens]	Carcinogenic
N-Nitrosamines [EN-nigh-TROZ-uh-meens]	Carcinogenic
Aromatic amines	Carcinogenic
Heterocyclic aromatic amines	Carcinogenic
Aldehydes	Carcinogenic
Inorganic compounds (such as arsenic)	Carcinogenic
Miscellaneous tar constituents	Stain skin and teeth; may have other effects
Cyanide	Toxic
Formaldehyde	Toxic
Ammonia	Toxic
Nicotine	Toxic; addictive drug

Most of these components are similar to those produced by the burning of wood or any other organic material, though a few are unique to tobacco smoke. The microscopic dust particles, or **particulates**, found in any kind of smoke can embed themselves in lung tissues, where they cause inflammation, or they can pass into the bloodstream, where they increase the risk of blood clotting, which plays a role in heart attacks and strokes.

Smoke also contains the gas carbon monoxide (CO). Like the oxygen (O_2) we breathe, CO binds with hemoglobin, but unlike O_2, CO is not easily released once bound to hemoglobin. Thus CO reduces the ability of red blood cells to carry oxygen. When inhaled in large quantities, as when car exhaust fills a closed garage, CO quickly occupies enough of the O_2-carrying sites on hemoglobin molecules to cause death. When inhaled in small quantities, as with smoking, CO reduces the amount of O_2 obtained from each breath.

At least 55 chemical compounds found in tobacco smoke are known carcinogens. The polycyclic aromatic hydrocarbon group includes benzo[*a*]pyrene, which has been linked directly to the DNA damage associated with cancer. Tobacco smoke also contains inorganic carcinogens such as arsenic, chromium, and cadmium. The *N*-nitrosamines are of particular interest because they include a chemical known as NNK (the acronym for "nicotine-derived nitrosamino ketone"), which is found only in tobacco smoke and is known to produce lung tumors in many different kinds of laboratory animals.

In addition to releasing particulates and chemicals, burning tobacco leaves produce **tar**, a sticky brown material that condenses as the smoke cools. Most tar condenses in the lungs, although it also stains the fingers and teeth of smokers. Tar includes many of the chemicals mentioned earlier, with their known adverse effects, but it also contains other compounds whose effects have not yet been determined.

A number of other noxious and unpleasant chemicals are found in tobacco smoke. These chemicals include cyanide, which is a poisonous gas; formaldehyde, a preservative; and ammonia, which occurs in household cleaners.

Of special importance is **nicotine**, a chemical compound that is characteristic of tobacco (Figure E.7). Nicotine is a highly toxic chemical that was once widely used as an insecticide, but because of its toxicity has fallen out of favor. At high doses, nicotine is a nerve poison that fatally interferes with breathing. At the lower doses derived from tobacco smoke, nicotine acts as a stimulant by triggering the release of the neurotransmitter **dopamine**, which is associated with feelings of happiness. Nicotine also stimulates the release of the hormone epinephrine from the adrenal glands. Epinephrine, in turn, increases blood glucose levels, heart rate, and blood pressure. These low-dose effects of nicotine are what help to make it a highly addictive drug.

Tobacco smoking is highly addictive

As anyone who smokes will recognize, smokers generally understand that smoking is not good for them. For this reason, most smokers try to stop smoking. However, most attempts to quit smoking end in failure. Recent data show that 70 percent of smokers want to quit, but only 20 percent of those who try succeed in not smoking for an entire year. Smokers who successfully quit usually do so after several tries and with the aid of outside support.

Research has shown that tobacco is highly addictive and that the chemical responsible for its addictive properties is nicotine. As with any addictive drug, once a person has become addicted to nicotine, this substance generates pleasurable feelings when taken into the body and physical discomfort when not taken. Nicotine's addictive properties are evident in people's patterns of usage: 95 percent of smokers smoke at least one cigarette per day and find it difficult to imagine going a day without smoking. Almost half of smokers have their first cigarette within 30 minutes of waking up. Smokers smoke an average of 14 cigarettes per day.

What makes nicotine addictive? Like two other addictive drugs, cocaine and heroin, nicotine stimulates the release of dopamine in the **nucleus accumbens** (Figure E.8), a part of the brain shown experimentally to be associated with addictive behavior (in rats as well as humans). With increased exposure to nicotine, smokers' brains become less sensitive to nicotine and other triggers that release dopamine. As a result, they need to smoke more to get the same effect.

When addicted smokers stop smoking, they no longer get the pleasure reward that comes with the release of dopamine stimulated by nicotine. In the absence of nicotine,

Figure E.7 The Chemical Structure of Nicotine
Nicotine is toxic at high doses, but an addictive stimulant at the low doses found in cigarettes.

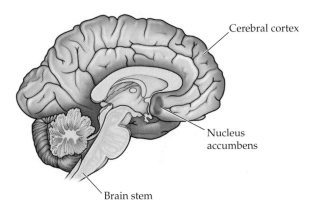

Figure E.8 Addictive Drugs Stimulate Cells in the Nucleus Accumbens

The nucleus accumbens is associated with a part of the brain involved in the release of dopamine, the neurotransmitter involved with feelings of pleasure that plays a role in many addictions.

other triggers for dopamine release in the desensitized brain no longer elicit much of a pleasure response. Without nicotine, smokers no longer get the feelings of alertness and energy that stimulation of the adrenal gland provides. As a result, smokers trying to quit often become irritable and anxious, have difficulty sleeping and concentrating, and tend to eat more.

Interestingly, the craving for nicotine, which often persists for months after successfully quitting smoking, is a phenomenon distinct from nicotine withdrawal itself. Former smokers given nicotine in the form of gum or patches may still crave the act of smoking a cigarette. A possible explanation for this is that smoking provides pleasurable associations not related directly to nicotine—such as the smell and feel of a cigarette—which in turn promote dopamine release. Recent research also suggests that there may be a second chemical in tobacco smoke that slows the rate at which dopamine is broken down by enzymes in the brain, thus prolonging dopamine's effects on brain cells (Figure E.9). Thus, in smokers, dopamine and its effects may persist longer than in non-smokers getting nicotine from another source.

Kicking the Habit

Once you start smoking, quitting is hard to do. As statistics indicate, most people who try to quit on their own fail within a week. People who successfully quit smoking usually do so on a second or third try. Once you succeed, however, the health benefits can be significant.

Most successful efforts at quitting involve outside help of some sort

Public health agencies agree on a number of factors that improve the chances of succeeding at the difficult process of breaking a nicotine addiction:

Set the stage for a successful attempt to quit. Two key elements in preparing to quit are to aim for a clear target date and to remove all temptation (such as cigarettes and ashtrays) in the workplace and at home.

Have a support system in place. Friends and family who are willing to help to reduce the temptation to smoke and who understand just how hard it is to quit can greatly increase the chances of success. Trying to deal with a craving for a cigarette when surrounded by smokers can often derail your plan to quit smoking. Physicians or counselors can also help keep your plan on track.

Establish alternative behaviors to keep the craving to smoke at bay. The craving for a cigarette can be incredibly strong when you are quitting. It helps to come up with behaviors to distract you from these cravings. For example, go for a walk when the cravings develop.

Use alternative sources of nicotine to ease withdrawal. Nicotine patches or nicotine gum, when properly used, can help blunt the withdrawal symptoms and reduce the craving to smoke (Figure E.10). Although many people want to quit without using alternative sources of nicotine, such aids, when properly used, can double your

Figure E.9

Tobacco May Contain Chemicals That Prolong Pleasure

Our brains produce MAO B, an enzyme that breaks down dopamine, to ensure that it does not remain in the brain for too long after its release. Tobacco smoke appears to contain a chemical that interferes with the production of MAO B, causing dopamine to persist longer after its production. In these two series of PET scans, notice that the brain of the nonsmoker shows higher concentrations of MAO B (with the highest amount being in red and the lowest in blue) than does the brain of the smoker.

Figure E.10 Aids to Quitting Smoking
Gradually reducing doses of nicotine through the use of aids such as a nicotine patch (shown here) or nicotine gum can double a smoker's chances of successfully breaking a nicotine addiction.

chances of success. Health care providers can give advice on the proper use of these medications.

Recognize that it is hard to quit smoking and that you may not succeed on your first try. Try again until you succeed.

Because it is so easy to give in to a nicotine addiction, it is important to avoid giving in to the urge to smoke. It helps to avoid situations likely to lead to smoking and to understand that there are other ways to deal with stress or depression than smoking. And if you don't succeed in quitting on your first try, don't give up!

Quitting improves your health and that of your family

If you stop smoking early enough in life, you can expect to live as long as a nonsmoker. A long-term study of male British doctors who smoked from their early twenties indicated that those who stopped smoking before they reached the age of 35 had the same chance of surviving to a given age as doctors who had never smoked (Figure E.11). Doctors who quit smoking later in life, between the ages of 45 and 55, had a better chance of living to an old age than those who continued to smoke, though not as good a chance as the lifelong nonsmokers.

Some of the adverse health effects of smoking clear up relatively quickly after quitting. For example, the risk to unborn children during pregnancy drops quickly when mothers stop smoking. The risk of a heart attack drops by half within a year of quitting. The risk of cancer of the lung or larynx begins to drop within a year or two of quitting, and at 10 years after quitting, the risk falls to 30 to 50 percent of that in continuing smokers.

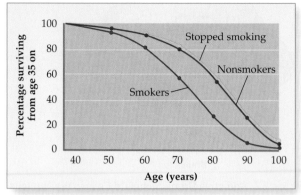

(a) Stopping smoking at age 25–34

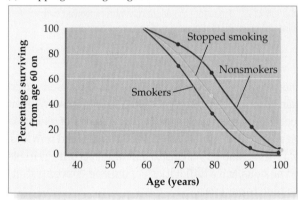

(b) Stopping smoking at age 55–64

Figure E.11 If You Stop Smoking, Your Life Expectancy Increases
These two graphs compare the percentage of male British doctors who survived to a particular age if they were lifelong cigarette smokers (red), lifelong nonsmokers (blue), or quit smoking at a particular age (yellow). (a) Those who quit smoking before age 35 had essentially the same chances of survival as nonsmokers. (b) Those who quit smoking later in life saw some improvement in their longevity, but had lower chances of survival than the nonsmokers.

On the other hand, recovery from some smoking-related damage can prove more difficult. For someone who stops smoking after a diagnosis of emphysema, the emphysema will no longer get worse. However, the lung damage already caused by the disease cannot easily be undone.

For many smokers, the decision to quit smoking is made not so much to benefit their own health as to safeguard the health of their families. Many women quit smoking when they learn they are pregnant so as to give their developing child the best chance possible of success. As awareness of the dangers of secondhand smoke increases, many smokers are choosing to quit to protect the health of their children and their partners. Thus, even though tobacco is highly addictive, smokers can take back control over their own health, and the health of those around them, by kicking the habit.

⊙ Review and Discussion

1. Smoking costs society a lot of money. Even if you remain a nonsmoker all your life, the health care costs associated with smoking increase the health insurance premiums you must pay. Given this, do you think smokers have the right to smoke, or should society make smoking illegal? Pick one position or the other, and with some of your classmates, put together a convincing argument to present to the class as a whole.

2. How does secondhand smoke affect your health?

3. What makes tobacco so addictive?

4. In spite of the huge effort that has been put into educating people of all ages about the risks associated with smoking, many people continue to smoke. Worldwide, the number of smokers is increasing. How would you explain this?

5. Interview some people who have tried to quit smoking. Find out if they succeeded, and, if so, if they had a hard time. For those who tried and failed, what interfered with their success?

6. How do the health consequences of other forms of tobacco use, such as chewing tobacco, snuff, or bidis (a kind of Asian cigarette wrapped in a leaf rather than paper), resemble or differ from those of smoking cigarettes?

Key Terms

carcinogen (p. E4)
dopamine (p. E7)
edema (p. E4)
nicotine (p. E7)
nucleus accumbens (p. E7)

particulate (p. E6)
secondhand smoke (p. E6)
tar (p. E7)
telomere (p. E5)

CHAPTER 31

Plant Structure, Nutrition, and Transport

Key Concepts

- The plant body has a belowground root system and an aboveground shoot system. It has three basic organs: roots in the root system, and stems and leaves in the shoot system. Each organ consists of three tissue types: dermal, ground, and vascular tissues.

- Dermal tissues protect the plant and control the flow of materials in and out of the plant. Ground tissues support the plant, heal wounds, and perform photosynthesis. Vascular tissues transport food and water throughout the plant.

- Roots absorb water and nutrients, anchor the plant, and store food. Stems support the plant, while leaves are the primary site of photosynthesis. Roots, stems, and leaves can be highly modified to perform specialized tasks.

- Plants get essential mineral nutrients from the soil. However, in terms of their dry weight (water excluded), most of the plant body comes from carbon dioxide absorbed from the air and converted to sugars by photosynthesis.

- Some plants that live in nutrient-poor environments get extra nutrients by eating animals.

- The transport of food in the phloem requires energy and relies on water pressure. The transport of water and mineral nutrients in the xylem requires no energy and is driven by the physical forces resulting when water evaporates from the leaves.

Plants Rule, Animals Drool

As anyone who has watched a cow eat, fed a baby, or been licked by a Saint Bernard knows, many animals do indeed drool. But do plants rule? They do, in two important ways.

First, life as we know it depends on plants. When plants capture energy from the sun during photosynthesis, they release oxygen as a by-product—thus plants help to make the very air we breathe. In addition, by using solar energy to make carbohydrates, plants provide food for much of life on Earth. We depend on plants for our food, as do zebras that eat grass, and lions that eat zebras that ate grass. Moreover, plants provide humanity with many essential services. These services include erosion control (by holding soil in place with their roots), flood prevention (a single tree can soak up hundreds of gallons of water per day), and pollution control (plant roots can absorb pollutants such as heavy metals from the soil and water, while plant leaves can absorb pollutants such as sulfur dioxide from the air). Finally, we get enormous economic value from products made from plants, such as food and beverages, paper, lumber, cotton, pyrethrin [pie-*REE*-thrin] (an insecticide derived from marigolds), and many medicinal drugs.

Plants also rule in a second, very different sense: they solve tough environmental challenges in novel and interesting ways. To appreciate what we mean, consider the challenges a moose calf might face over the course of its first year. It may starve or freeze to death during a cold winter, suffer from lack of sodium (an essential nutrient), or be killed by a wolf or human predator. In many ways, the moose has it rough. But plants have it *really* rough. Rooted in the ground, a plant has nowhere to run, nowhere to hide. If the nutrients it needs are scarce, it must make do or die. Unlike the moose, it cannot run from its enemies or travel to new locations to find food. Also unlike the

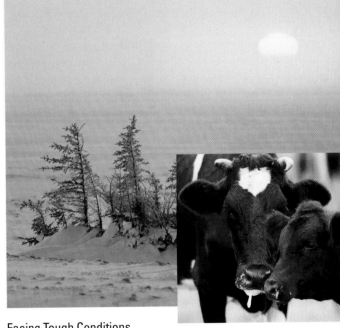

Facing Tough Conditions
Unlike cows and other animals, plants cannot move, and hence must have ways to cope with harsh environmental conditions, such as this snow-covered taiga in Canada.

moose, the plant does not have fur to keep it warm, nor can it seek shelter from wind or freezing rain. Moose can live in places such as Minnesota or Canada, where temperatures in winter can plummet well below −35°C (−31°F), cold enough to damage human skin within minutes. But some plants are exposed to even colder temperatures. Why don't plant cells freeze and burst open when left exposed to such temperatures, sometimes for days on end?

Or consider the opposite extremes of temperature—life in a desert. Here a plant must cope with blistering heat and relentless bombardment by ultraviolet (UV) light. Temperatures in the desert can be so hot that rocks can burn human skin on contact, and UV light damages DNA, causing cancer. Animals in the desert seek shade during the hot parts of the day, but the plants that provide those animals with shade don't have that option. How do plants tolerate the heat? Similarly, although they are constantly exposed to high levels of UV light, plants don't get cancer. Why not?

In this chapter and the next, we describe how plants meet these and other tough environmental challenges. We introduce you to how plants are built, how they get the nutrients they need, how they prevent their cells from bursting open in freezing cold, and how they survive the punishing desert heat. And we describe other novel features of plants, such as how they poison their enemies, reproduce without moving, and—in some species—switch sex from male to female and back again. You'll find that plants are not the simple, defenseless life-forms you might have imagined. Indeed, you may join us in thinking that plants do, in fact, rule.

Earth has a rich diversity of plants—over 280,000 species—ranging from low-lying mosses to ferns to giant redwood trees to spectacular flowering plants (Figure 31.1). We can make sense of this overwhelming diversity by dividing plants into four major groups:

1. **Bryophytes** (represented by the mosses) lack a true vascular system for transporting materials.
2. Seedless vascular plants (represented by the ferns) have a vascular system, but lack seeds.
3. **Gymnosperms** (represented by conifers such as pine or spruce trees) have a vascular system and seeds, but lack flowers and fruits.
4. **Angiosperms**, or **flowering plants**, have a vascular system, seeds, flowers, and fruits.

In terms of sheer numbers, flowering plants dominate the planet (see Figure 19.3 for the time periods when other plant groups dominated life on Earth). Today there are over 250,000 species of flowering plants, compared with 18,000 bryophytes, 13,000 seedless vascular plants, and 720 gymnosperms. Flowering plants literally cover Earth: they include the most common plant species found in tundra, tropical forests, temperate forests (many of which are dominated by trees that lose their leaves each year), grasslands, deserts, and other plant communities. Flowering plants are also of overwhelming importance in terms of what we eat: worldwide, people get over 80 percent of their calories from flowering plants such as grasses (wheat, rice, corn), legumes (peas, beans, peanuts), potatoes, manioc, and sweet potatoes.

For all of these reasons, we focus on flowering plants in this unit. In this chapter we describe how plants are built, how they obtain nutrients, and how they transport materials throughout their bodies.

31.1 An Overview of the Plant Body

Compared with animals, plants have a simple structure. First, the plant body can be broken down into two basic systems, the belowground **root system** and the aboveground **shoot system** (Figure 31.2). These two systems are specialized for life in two very different environments: roots in soil, shoots in air.

Compared with animals, plants also have fewer types of cells, tissues, and organs. (Recall that a tissue is a group

(a)

(b)

(c)

(d)

Figure 31.1 A Panorama of Plants

(a) This spring moss is a type of bryophyte; (b) maidenhair ferns (*Adiantum pedatum*) are a type of seedless vascular plant; (c) the coast redwood (*Sequoia sempervirens*) is a type of gymnosperm; (d) purple saxifrage (*Saxifraga oppositifolia*) wildflowers are flowering plants, or angiosperms.

of cells with a particular function, and an organ is a group of different tissues that work together to perform major functions in the body.) Plants have only three different types of organs—roots, stems, and leaves—each of which is made of the same three tissue types (dermal, ground, and vascular tissues; see Figure 31.2).

Working from the bottom up, the root system contains a single type of plant organ, the roots. **Roots** anchor the plant, absorb water and nutrients from the soil, transport food and water, and (often) store food. The shoot system contains two types of plant organs: stems and leaves. **Stems** provide the plant with structural support

and also serve to transport food and water. **Leaves** are where photosynthesis occurs; thus their main function is to produce food.

You may think we've forgotten to mention a fourth plant organ: flowers, which carry out the all-important function of reproduction. However, recent research has shown that flowers can be produced from leaves by turning a small number of genes on or off. Thus, although they look very different, flowers are actually modified leaves, and so do not represent a fourth organ type. Other unique-looking plant structures are also modified from one of the three basic plant organs; examples include carrots (modified roots), thorns (modified stems), and the bulb of the onion—the part that we eat (modified leaves). We'll see other examples of modified stems and leaves throughout this chapter.

Plants also differ from animals in the way their bodies are constructed. Unlike most animals, plants grow throughout their entire lives. Furthermore, plants grow in a unique way: Aboveground, they are built like Tinkertoys, in that the same basic unit, the bud-stem-leaf unit shown in Figure 31.2, is added over and over again. As we'll see in Chapter 32, this modular approach to building an organism gives plants great flexibility to respond to changing environmental conditions.

Learn more about how plants are built.

31.1

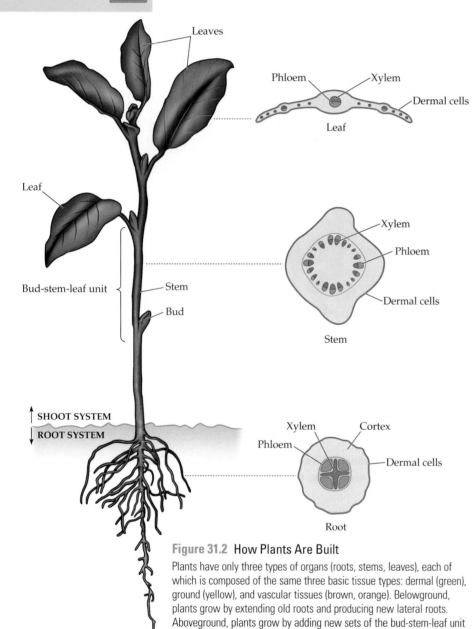

Figure 31.2 How Plants Are Built
Plants have only three types of organs (roots, stems, leaves), each of which is composed of the same three basic tissue types: dermal (green), ground (yellow), and vascular tissues (brown, orange). Belowground, plants grow by extending old roots and producing new lateral roots. Aboveground, plants grow by adding new sets of the bud-stem-leaf unit shown in the diagram (the "Tinkertoy" approach to growth described in the text).

31.2 The Three Plant Tissue Types

Each of the three basic tissue types in plants—dermal, ground, and vascular—is composed of cells that work together to perform particular functions. **Dermal tissues**, which form the outermost layer of the plant, protect the plant from the outside environment and control the flow of materials into and out of the plant. **Ground tissues**, which form the intermediate layer, make up the bulk of the plant body and perform a wide range of functions, including support, wound repair, and photosynthesis. Finally, in or near the center of the plant body are the **vascular tissues**, which function to transport food and water throughout the plant.

Dermal tissues interact with the environment

Dermal tissues play a key role in the ways in which plants meet their many tough environmental challenges. They help protect the plant from enemies such as herbivores (plant-eating organisms) or pathogens. They also increase the uptake of water and nutrients, control gas exchange, and limit water loss (Figure 31.3).

Most dermal tissue consists of a single layer of rectangular protective cells that are not specialized to perform any particular task. In aboveground plant parts, these unspecialized cells are covered with a waxy **cuticle**, which helps to prevent water loss and keep enemies such as fungi from invading the plant.

Certain dermal tissue cells have special shapes for performing particular functions. For example, the leaves and stems of many plants are covered with dermal-cell outgrowths called **trichomes** [*TRY*-komes], which serve such functions as impaling insect herbivores (see Figure 31.3a) and secreting and storing salt (in plants that can tolerate high-salt environments). **Root hairs**, trichomes found in root dermal cells, play an essential role in the uptake of nutrients. **Guard cells** are specialized cells on leaves and stems that can open and close, thereby controlling the rates at which CO_2 is brought into the plant and O_2 and water are lost from the plant (see Figure 31.3b).

Ground tissues have many essential roles

Have you ever wondered why nonwoody plants—which do not have an obvious support system like the skeleton of an animal or the woody tissue of a tree—don't just flop over? One reason is that they contain ground tissue cells with thick walls that provide support (Figure 31.4). In

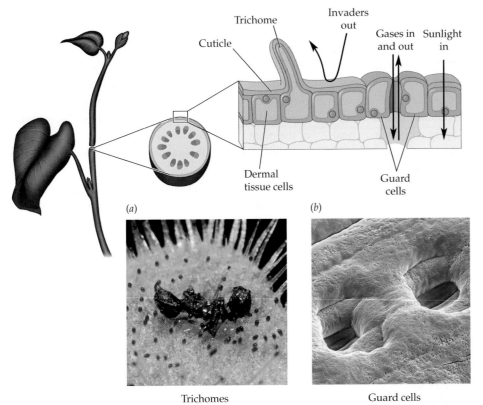

Learn how plants regulate water loss.

Trichome

Invaders out

Cuticle

Gases in and out

Sunlight in

Dermal tissue cells

Guard cells

(a)

(b)

Trichomes

Guard cells

Figure 31.3 Dermal Tissue Is Where Plants Meet the Environment

Dermal tissues protect the plant from attack and control the flow of materials into and out of the plant. Some dermal tissue cells perform specialized tasks. (a) Trichomes serve a wide range of functions, such as protecting the plant from insects like this unfortunate ant. (b) Guard cells open and close to control the loss of water and the exchange of gases such as CO_2 and O_2; most guard cells are found on the leaf, but some are found on the stem. Water loss is also limited by the waxy cuticle that covers dermal tissues in the stem and leaf.

Photosynthetic ground tissue

Supportive ground tissue

Figure 31.4 Ground Tissue Makes Up the Bulk of the Plant Body

Ground tissues play a number of essential roles in plants, including support, wound repair, and photosynthesis. The majority of ground tissue cells that perform photosynthesis are found in the leaves, but some are in the stems, as shown here.

some cases, these ground tissue cells provide enough support that the plant does not wilt even when it dries out. In other plants, pressure from water within the cells is needed to support the plant, much as air within a balloon is needed to prevent the balloon from collapsing.

Ground tissue cells also provide food to the plant, since they are the primary sites where photosynthesis occurs. Finally, some ground tissue cells play a key role in healing wounds, allowing the plant to regrow damaged parts.

Vascular tissues transport food and water

Like any multicellular organism, plants must transport materials throughout their bodies. We'll discuss how they do it later in this chapter; here we describe how the vascular tissues that make it possible are built.

Identify the parts of the vascular system. 31.3

(a) Phloem   (b) Xylem

Vascular bundle

Tracheid

Vessel element

Sieve element

Companion cell

Figure 31.5 Vascular Tissue Transports Food and Water
Vascular tissues form a continuous set of pipes that run through the plant. (a) Food is transported in the phloem, which is built from two types of living cells: sieve elements and their associated companion cells. (b) Water and mineral nutrients are transported in the xylem. There are two types of xylem cells, tracheids and vessel elements. The cells that make up both types of xylem are dead; hence they are hollow, which allows water and mineral nutrients to move relatively easily from one cell to the other.

Plants have two types of vascular tissues: **phloem** [FLOME], which transports food (the carbohydrate products of photosynthesis), and **xylem** [ZY-lum], which transports water and mineral nutrients (Figure 31.5). The vascular tissues form a continuous set of pipes that run through the entire plant. Food must be shipped from the leaves, where it is produced, to living cells in every part of the plant. Water and minerals absorbed from the soil must move upward from the roots and outward from the central stem to the leaves.

Phloem tissue is made up of living cells, connected to one another by pores (plasmodesmata) through which food can move. Unlike phloem cells, functional xylem cells are dead, consisting of empty cylinders arranged end to end. Connections between the xylem cells allow a continuous column of water to form between the roots and the leaves. As we shall see on page 619, this thin column of water is critical to the movement of water and mineral nutrients in xylem tissues.

31.3 The Three Plant Organs

Let's take a closer look at the structure of each of the three essential plant organs: roots, stems, and leaves.

Roots absorb water and nutrients, anchor the plant, and store food

Plants invest a great deal of resources in making roots, which testifies to the importance of their various functions: absorption of water and nutrients, physical anchoring of the plant, and food storage. In terms of structure, roots have a region of active growth, protected by a layer of cells known as the **root cap** (Figure 31.6). The root hairs, which greatly increase the surface area over which water and nutrients can be absorbed from the soil, are produced above the region of active growth. Mature roots can also produce many lateral (side) roots, forming a tight mass that anchors the plant. Roots are made up of an outer layer of dermal cells, then the cortex (which consists of ground tissue cells that can store starch), and at the center, bundles of vascular tissues—the phloem and xylem.

Plants have one of two kinds of root systems. In a **taproot system**, a dominant taproot grows directly downward, giving rise to branch roots that extend to the sides before also descending downward (Figure 31.7a). In a **fibrous root system**, no one root predominates (Figure 31.7b). Fibrous roots can form dense mats that hold soil firmly in place. Taproots, on the other hand, usually hold the soil less firmly, but penetrate more deeply than fibrous roots. Some taproots descend great distances, reaching a record depth of 53.3 meters (nearly 175 feet) in mesquite, a desert shrub. Perhaps even more astonishing is the vast amounts of roots that can be produced, even by small plants. For example, a 4-month-old rye plant grown in 52 liters (about 14 gallons) of soil made enough roots to cover 639 square meters; about 400 square meters of this area resulted from its 14 billion root hairs, which, if placed end-to-end, would have extended over 10,000 kilometers.

Stems support the plant

Stems provide support to the plant, allowing it to grow vertically against the force of gravity. Viewed in cross section, stems show many similarities to roots in their structure. As in other parts of the plant body, the arrangement of tissues in the stem consists of dermal tissue at the external surface and ground tissue in the interior, with bundles of vascular tissue embedded in the ground tissue (Figure 31.8). The arrangement of the vascular tissue in stems can vary from a single ring of vascular bundles to

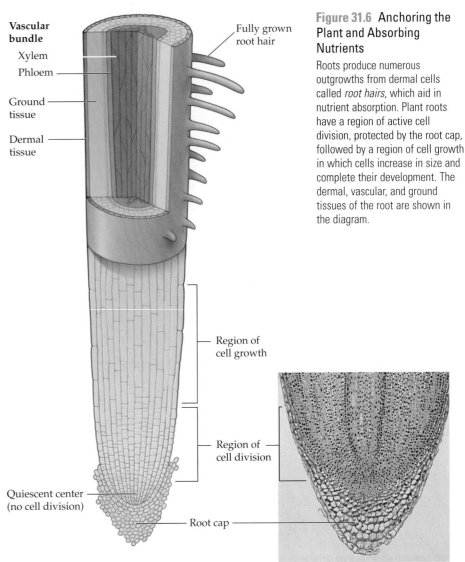

Figure 31.6 Anchoring the Plant and Absorbing Nutrients
Roots produce numerous outgrowths from dermal cells called *root hairs*, which aid in nutrient absorption. Plant roots have a region of active cell division, protected by the root cap, followed by a region of cell growth in which cells increase in size and complete their development. The dermal, vascular, and ground tissues of the root are shown in the diagram.

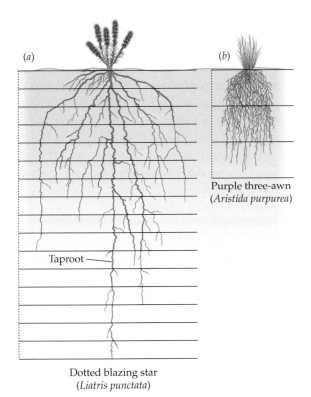

Figure 31.7 Two Kinds of Root Systems
(a) The dotted blazing star, a wildflower found on the prairie, has a taproot system. (b) Purple three-awn, a prairie grass, has a fibrous root system. The scale bars show the relative size of each plant.

a series of vascular bundles scattered throughout the ground tissue. Stems can be highly modified to perform specialized tasks, including protection (hawthorn thorns are stems), climbing (grape tendrils are stems), and belowground food storage (potatoes, believe it or not, are stems).

Leaves feed the plant

Although cells in the stems of many plants can perform photosynthesis, most of the plant's food is produced by the leaves. In cross section, leaves look different from roots or stems (Figure 31.9), but the basic structural arrangement of tissues is the same. The outer layer—the top and bottom of the leaf—is made up of dermal tissues, including the specialized guard cells that control gas exchange (see Figure 31.3b). The ground tissues—the primary site of photosynthesis—are sandwiched between

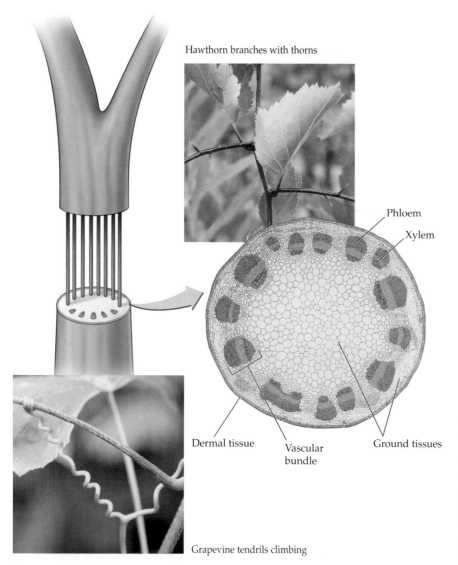

Hawthorn branches with thorns

Phloem

Xylem

Dermal tissue

Vascular bundle

Ground tissues

Grapevine tendrils climbing

Figure 31.8 Stems Support the Plant
Ground tissues in stems support for the plant against gravity. The vascular bundles resemble a set of pipes running through the stem. In some species, the vascular bundles are arranged in a roughly circular shape, as shown here; in other species, they are scattered throughout the ground tissue. Some plants use modified stems for functions such as climbing (grape tendrils), protection (thorns), and storage (the potatoes we eat).

the two dermal tissue layers. And the vascular tissues (phloem and xylem, which are contained in veins) are located roughly in the middle of the leaf.

Many kinds of leaves are covered with what feels like "fuzz" to us; this fuzz is actually a thick mat of trichomes, which can be deadly to insects (see Figure 31.3*a*). Leaves also contain chemicals that may protect the plant from UV light, as well as from harmful byproducts of photosynthesis. Finally, the structure of some leaves is modified greatly to perform functions other than photosynthesis. Examples of modified leaves include the nutrient-collecting "pots" of the flower pot

plant (see Figure 31.9), the tendrils of peas and beans (used for climbing), and the protective spines of cacti.

31.4 How Plants Obtain Nutrients

Because of the importance of plants in our diets, people have long been interested in how plants function, including how they get nutrients. Aristotle thought plants got everything they needed for growth from the soil. Almost two thousand years later, however, the Belgian physician Jan Baptista van Helmont showed that soil alone was not enough. He did this by planting a small willow tree in a pot and adding only water to the pot. After 5 years of growth, the willow had increased in weight by 74.4 kilograms, but the soil had decreased in weight by only 0.06 kilograms. He concluded that soil had nothing to do with plant growth, and that plants needed only water to grow!

Roughly a hundred years later, a variety of scientists in the 1700s suggested that plants got most of what they needed to grow from air, not soil or water. It turns out that Aristotle, van Helmont, and these scientists each were partly right (Figure 31.10). Plants get essential mineral nutrients (inorganic molecules) from the soil, as Aristotle thought. But nutrients from the soil contribute only a small amount to the weight of the plant, as van Helmont showed in his experiment. Instead, much of the weight of a living plant comes from water, which is held in the central vacuoles of its cells (see Figure 5.2). Finally, if you weigh a plant after removing its water (using a drying oven), most of the material in the plant body comes from air, as suggested by the scientists of the 1700s. In particular, most of the plant's dry weight comes from CO_2 absorbed from the air by the leaves and used during photosynthesis to produce sugars.

As you'll recall, photosynthesis is a series of chemical reactions that use sunlight, CO_2 from the atmosphere, and water from the soil to produce sugars, releasing O_2 as a by-product. We described the process of photosynthesis in detail in Chapter 8; thus, for the remainder of this section, we'll focus on the mineral nutrients needed by plants.

Plants need seventeen mineral nutrients to grow

The label on a typical package of plant fertilizer shows that it provides plants with three mineral macronutrients—nitrogen (N), phosphorus (P), and potassium (K)—and an assortment of mineral micronutrients (Figure 31.11). Plants

Figure 31.9 Leaves Feed the Plant

Plant feed themselves through their leaves, the primary site of photosynthesis. CO_2 enters, and O_2 and water leave, from the guard cells, which are more commonly found on the bottom surface of the leaf. Some plants use modified leaves for functions such as nutrient capture (such as the pouches in the flower pot plant that contain its roots and collect debris), defense (spines in cacti), and climbing (tendrils in pea plants, which look similar to the grape tendrils in Figure 31.8, although those tendrils are modified stems, not modified leaves).

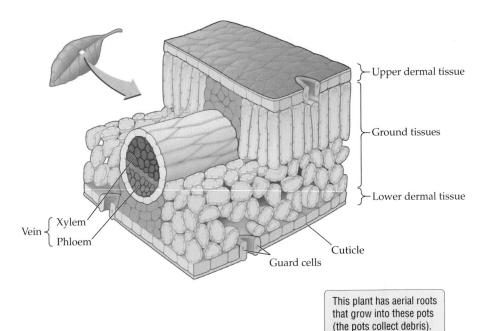

Upper dermal tissue

Ground tissues

Lower dermal tissue

Vein { Xylem Phloem

Cuticle

Guard cells

Learn more about leaves.

31.4

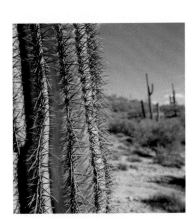

Spines from a saguaro cactus

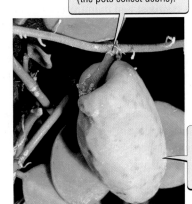

This plant has aerial roots that grow into these pots (the pots collect debris).

A modified leaf or "pot" of the flower pot plant, a species that grows on other plants, not in soil.

The "pot" of the flower pot plant
(*Dischidia rafflesiana*)

require **macronutrients** in relatively large amounts (at least 1,000 milligrams of the nutrient per kilogram of plant dry weight), but require much smaller amounts of **micronutrients** (less than 100 milligrams per kilogram of plant dry weight). In total, plants need nine macronutrients (carbon, oxygen, hydrogen, nitrogen, sulfur, phosphorus, potassium, calcium, and magnesium) and at least eight micronutrients (including iron, zinc, and copper).

Soils typically contain water and some (but not all) of these nutrients in amounts sufficient for plant growth. Thus plant fertilizers do not provide all seventeen of the mineral nutrients plants need. Instead, they provide only those nutrients that are usually found in soil in low amounts (for example, N, P, and K). To "feed" plants with fertilizer, we mix it into the soil, which allows the nutrients in the fertilizer to dissolve in the water found in soil. Plant roots can then absorb the dissolved nutrients. Note that as long as the right nutrients are present, it is also possible to grow plants without soil, as is done in many commercial operations (see the box on page 616).

Figure 31.10
How Plants Get Raw
Materials for Growth

To grow, plants need CO_2, mineral nutrients, and water. CO_2 enters the plant through the leaves (and to a lesser degree, the stem), while mineral nutrients and water enter the plant through the roots. As a by-product of photosynthesis, plants release oxygen to the air, some of which diffuses from the air into the soil, where it can be used by root cells to perform cellular respiration.

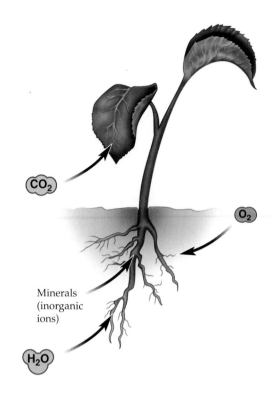

CO₂

O₂

Minerals
(inorganic
ions)

H₂O

Helpful to know

Don't let the name *endodermis* (from Greek words meaning "inner skin") fool you; the endodermis is not a dermal tissue, but rather a *ground* tissue.

The label in Figure 31.11 comes from a fertilizer that contains no organic (carbon) compounds. Other commonly used fertilizers, such as compost and manure, contain many organic compounds, yet provide relatively small amounts of the nutrients plants need. Nevertheless, such organic fertilizers improve plant nutrition in two ways. First, the spongelike texture of rotting organic material soaks up and holds water into which mineral nutrients can dissolve. Second, particles in organic fertilizers act as "magnets" that hold mineral nutrients in the soil, keeping them from being washed out of reach of the plant roots.

Finally, note that plant fertilizers do not provide sources of energy. Plants don't need chemical energy sources because photosynthesis allows them to use energy from the sun to convert carbon dioxide into sugars. Humans and other animals, in contrast, must eat foods that contain energy stored in chemical forms—in carbohydrates, fats, and proteins (see Figure 21.4). Overall, our nutritional requirements differ greatly from those of plants, as a comparison of Figures 21.4 and 31.11 makes clear.

Roots absorb nutrients from the soil through cells and along cell walls

To understand the two different ways in which roots absorb nutrients (dissolved in water) from the soil, consider a simplified diagram of a plant cell (Figure 31.12*a*).

Plant cells have three main compartments: the cell wall, the cytosol, and the central vacuole. The cytosol is surrounded by the plasma membrane, which lies just inside the cell wall; the central vacuole is also surrounded by a membrane. Throughout the plant, adjacent plant cells are connected to one another by membrane-lined openings in the cell wall called **plasmodesmata** (singular: plasmodesma) (see Figure 6.7*a*).

Because of the plasmodesmata, materials that enter the cytosol of one root cell can pass to other root cells without having to repeatedly cross cell walls and plasma membranes. Of the two routes by which roots absorb nutrients, this "cell-interior route" is the most common: mineral nutrients enter the root through root hairs and then travel from one cell's cytosol to another's via the plasmodesmata. However, roots can also absorb nutrients by a route that bypasses cell interiors; in this "cell-wall route," nutrients dissolved in water travel along cell walls (Figure 31.12*b*).

However, nutrients traveling by either route must cross at least one plasma membrane, which allows the root to block entry of harmful materials. Nutrients traveling by the cell-interior route must pass through a plasma membrane (that of the root hair) right away. Nutrients that follow the cell-wall route, however, aren't forced to encounter a plasma membrane until they reach the endodermis, a layer of ground tissue cells that surrounds vascular tissue in the root. The endodermis has a waxy layer (the Casparian strip) joining the walls of adjacent cells that blocks passage into the xylem, forcing water and

Plants get their
energy from the sun,
so the nutrients they
need to obtain are
mainly minerals.

GUARANTEED MINIMUM ANALYSIS
Total Nitrogen (N).........................15%
Available Phosphoric Acid (P₂O₅)....15%
Soluble Potash (K₂O)......................18%

Micronutrients:
Iron (Fe)....................................0.10%
Manganese (Mn)......................0.05%
Zinc (Zn)..................................0.05%
Copper (Cu)............................0.05%

Mineral micronutrients
like these are essential
to plants in small amounts.

Figure 31.11 What Plants Eat

This label, from a fertilizer for house plants, shows that it contains three essential plant macronutrients—nitrogen, phosphorus (as phosphoric acid), and potassium (as potash)—and four essential micronutrients (iron, manganese, zinc, and copper). Compare this fertilizer label with the label from a cereal box shown in Figure 21.4, which lists important sources of energy and nutrients for humans.

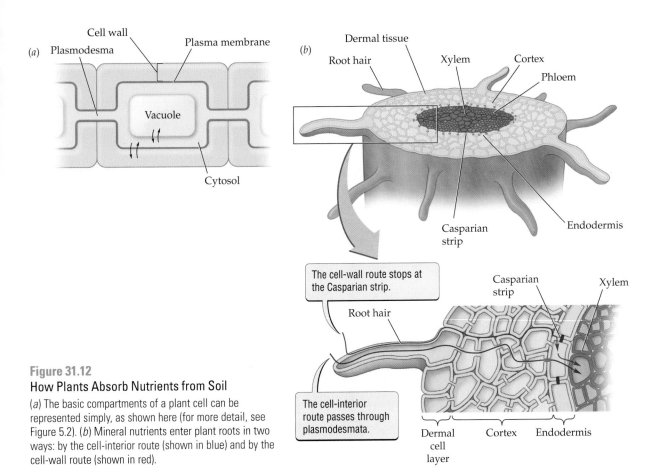

Figure 31.12

How Plants Absorb Nutrients from Soil

(*a*) The basic compartments of a plant cell can be represented simply, as shown here (for more detail, see Figure 5.2). (*b*) Mineral nutrients enter plant roots in two ways: by the cell-interior route (shown in blue) and by the cell-wall route (shown in red).

nutrients to pass through the plasma membrane of an endodermal cell if they are to proceed any farther.

Nutrient concentrations in roots can be 10 to 10,000 times greater than those in soil. As we saw in Chapter 6, molecules can cross a plasma membrane up a concentration gradient only by active transport, which requires energy. This means that plants must expend energy to absorb nutrients into their roots. If no energy is available, nutrient uptake stops. For example, when scientists reduced the energy stores of corn plants (by placing them in the dark for 4 days), the amount of phosphate ions absorbed by their roots dropped to just 5 percent of normal levels; absorption levels returned to normal after 2 days of sunlight. If plants did not expend energy to absorb nutrients, eventually the nutrients would leak (by diffusion) from the plant into the soil—just the opposite of what the plant needs.

Adding complexity to this story of nutrient absorption from the soil is the fact that most plants receive help from a mutually beneficial relationship with fungi. Plant roots literally unite with fungal hyphae (see page 56) to form structures called **mycorrhizae** (singular: mycorrhiza). The fungus benefits from this arrangement because the plant provides it with carbohydrates produced by photosynthe-

sis, and the plant benefits because mycorrhizae increase the surface area over which the plant can absorb nutrients. Benefits for the plant can be substantial—up to 3 *meters* of hyphae can extend from only 1 centimeter of plant roots!

31.5 Plants That Eat Animals

We can think of plants as living "between the devil and the deep blue sea." The "devil" they face is the vast array of animals that eat them, including over 300,000 species of herbivorous insects. Plants must also cope with a "deep blue sea" of soil in which essential nutrients (such as nitrogen and phosphorus) are scarce. About 600 plant species that live in very nutrient-poor environments have solved the problem of the deep blue sea while turning the tables on the devil—they eat animals.

Plants use a variety of strategies to eat animals. The Venus flytrap, for example, has modified leaves that look like fanged jaws, yet attract insects with a sweet-smelling nectar (Figure 31.13*a*). If an insect is lured within these jaws and trips their touch-sensitive hairs, the leaf snaps shut in about a tenth of a second. Once the insect is captured,

Biology on the Job

Growing Plants Without Soil

Ruben Houweling is the propagation manager for a company that grows plants using hydroponics, which is the growing of plants in nutrient solutions. Worldwide, greenhouse vegetables are a multibillion-dollar industry. Nearly all vegetables grown in greenhouses are produced hydroponically.

What do you do during a typical day at work? I manage how our company raises hydroponic vegetable seedlings for greenhouse vegetable producers. Each crop we grow is custom-tailored to the specific needs of each individual greenhouse and the crop they have ordered. A typical day consists of managing administrative, logistical, sales, and growing activities. The growing, for example, refers to the activities such as irrigation, climate, and pest control. As we figure out how best to grow a new crop, I spend a lot of time talking with the customer and with the people in our company that grow the seedlings.

What got you started in this field? Our company began as a family operation started by my grandfather, Cornelius Houweling, an immigrant from the Netherlands. My grandfather passed on the growing knowledge he learned in Europe to his sons, and they developed the company to where it is today. I started working in the greenhouses when I was 7 and what started as weekend and vacation employment eventually became a career.

What training do you need to grow plants using hydroponics? Formal education and/or on-the-job training is required for specialized positions such as growers. People interested in these positions often seek a college degree in horticulture. Experienced growers constantly upgrade their knowledge through trade publications, networking, trade associations, courses, and consultants.

How does the root system of a plant grown hydroponically compare to the roots that form if the same species is grown in soil? Diseases, pathogens, and pests are often present in soil and, without sterilization, these can quickly harm the plant. Also, depending on the design of the container in which the roots grow, we can encourage the root system to be evenly distributed throughout the container, thus maximizing the uptake of nutrients. And finally, we can place the containers in specific positions in the greenhouse to help provide the plants with optimal rooting temperatures.

Growing Plants Hydroponically

How has hydroponics changed in recent years? There have been many technological advances in recent years, such as irrigation systems that distribute nutrients in equal amounts to many plants. These dripper lines, as they are commonly called, can maintain equal pressure over larger distances and are manufactured with a high degree of accuracy so that nutrient flow rates remain consistent. The nutrients in fertilizers are also being developed in purer and more consistent percentages. Finally, computerized monitoring of the nutrient levels in the irrigation solution allows us to control precisely what the plant eats.

Are hydroponic systems good for the environment? Natural substrates are being introduced that will allow growers to compost their growing materials after use and thereby avoid landfills. Technology continues to improve in the sterilization and filtration of excess irrigation solution so that it may be reused in the crops. Consumer demand for high-quality produce year-round will probably see a greater volume and diversity of food crops grown using hydroponic systems. Greenhouse advances in the next 5 years will likely include increased knowledge in the use of artificial lighting, advances in cogeneration of energy (e.g., using waste heat given off when electricity is generated), and increased use of geothermal heating and cooling.

What do you enjoy most about your job? One of the most enjoyable aspects of my job is watching a single seed develop into a plant capable of producing hundreds of pounds of food. The work is demanding, but being part of the process in producing quality and healthy food is very rewarding. Not all companies may be family operations anymore, but the agricultural community is like a family and great to be part of.

The Job in Context

The specific nutrients and growing conditions each crop needs really matter to companies like the one in which Ruben Houweling works—they would go bankrupt if they did not get it right. For example, as a result of their modular ("Tinkertoy") growth form, plants grow differently in different environments. Hydroponic growers can take advantage of this flexibility by controlling the growth environment of their crops very carefully, thus ensuring that the plant grows in a way that best suits the needs of their customers. The simple nutrient requirements of plants also make it possible for hydroponic growers to reuse leftover nutrient solutions given to their crops. This is not possible in a farm field, where fertilizer is added to the soil and excess nutrients from the fertilizer are washed by rainfall into rivers and streams (where they can cause serious pollution problems; see page 623).

As we shall see in Interlude F, agriculture faces many challenges as humans strive to feed our growing population while minimizing our impact on the environment. Hydroponically grown crops have great potential in this regard. Because the plants are grown in a simple, artificial environment, we have considerable control over the raw materials used to grow them, the use of energy, and the wastes generated by the growing process, not to mention where the plants are grown. As hydroponics continue to improve, we may be able to produce increasingly large amounts of high-quality food while reducing the environmental impact of our crops.

the trap tightens further, forming an airtight seal around its prey. The leaf then secretes enzymes that digest its prey over the course of 5 to 12 days, absorbing the nitrogen and other nutrients from its body. Once the insect is digested, the trap opens to wait for another insect. After three to five meals, the trap no longer functions to capture insects, but the leaf continues to supply the plant with food by photosynthesis.

Like the Venus flytrap, other plants have moving parts that help them capture, then kill and eat animals. Sundew leaves, for example, have sticky, club-shaped hairs that attract insects. When an insect becomes stuck on these hairs, they bend inward, causing the leaf to curl around its hapless prey, which is then slowly killed and digested. A group of aquatic flowering plants, members of the bladderwort genus *Utricularia* [yoo-trick-yuh-*LAIR*-ee-uh], are even more aggressive: they literally suck their prey into a digestive chamber (Figure 31.13*b*). This chamber has a trapdoor entry with trigger hairs on the outside.

If a small animal, such as a mosquito larva, bumps into these hairs, the door swings open and then—in 1/30 of a second—water rushes into the chamber, sucking the prey into the trap. The door then closes, and the plant digests its animal prey.

Other plants lack moving parts, yet still manage to eat animals. Pitcher plants use nectar to lure insects into a pitcher-shaped trap. The inside of the pitcher may have downward-facing hairs, which make it easy to crawl in, but hard to crawl out. And in many pitcher plants, the surface of the pitcher about halfway down is covered with flaky wax. Any insect that steps onto this wax is doomed: the wax sticks to its feet, causing it to lose its grip and tumble into a vat of deadly digestive juices. Although pitcher plants usually eat small animals such as insects, they hold the record for the biggest animals eaten by plants: some species on occasion eat vertebrate animals, including frogs, lizards, birds, and even members of the class of animals to which we belong, the mammals (Figure 31.13*c*).

(b)

(a)

Note the shadow of a fly (at left) caught in the jaws of this Venus flytrap (*Dionaea*).

(c)

The bladderwort (*Utricularia neglecta*) sucks underwater animals into its bladders.

Perfect mouse skeletons have been discovered inside tropical pitcher plants, such as this *Nepenthes rafflesiana*.

Figure 31.13 Death Awaits
These plants may look pretty and innocent, but each of them eats animals. Plants usually eat small animals such as insects, but tropical pitcher plants have been known to eat vertebrates, including frogs, lizards, birds, and mice.

31.6 How Plants Transport Food and Water

Although the human circulatory system is impressive in terms of the pressure it can generate, it pales in comparison to what trees can do. Against the pull of gravity, trees must lift water and dissolved nutrients from their roots to their leaves, which often are 20 meters (65 feet) or more above the ground. To manage this, a 20-meter-tall tree must generate 4,500 mm Hg of pressure, which is more than 25 times the 175 mm Hg of pressure generated by the human heart during exercise. Now consider the world's tallest living tree, a 114-meter-tall coast redwood (*Sequoia sempervirens* [suh-*KWOY*-uh sem-per-*VEER*-uns]) growing in California. To supply its uppermost leaves with water and nutrients, this tree must generate an astonishing 25,500 mm Hg of pressure, more than enough to burst a human heart.

Plants do not have a muscular pump like a heart to transport water from their roots to their leaves—indeed, such a pump would have to be unrealistically large to produce the pressures needed in trees. Nor do plants use pumps to transport food from the leaves to the rest of the body. Instead of relying on pumps to move food and water, plants use clever alternative approaches that develop huge pressures at the cost of little or no energy.

The movement of food relies on water pressure

As described earlier, plants transport food in the phloem. To do this, specialized cells in the leaves expend energy to pump the sugars produced by photosynthesis into the phloem against a concentration gradient (Figure 31.14).

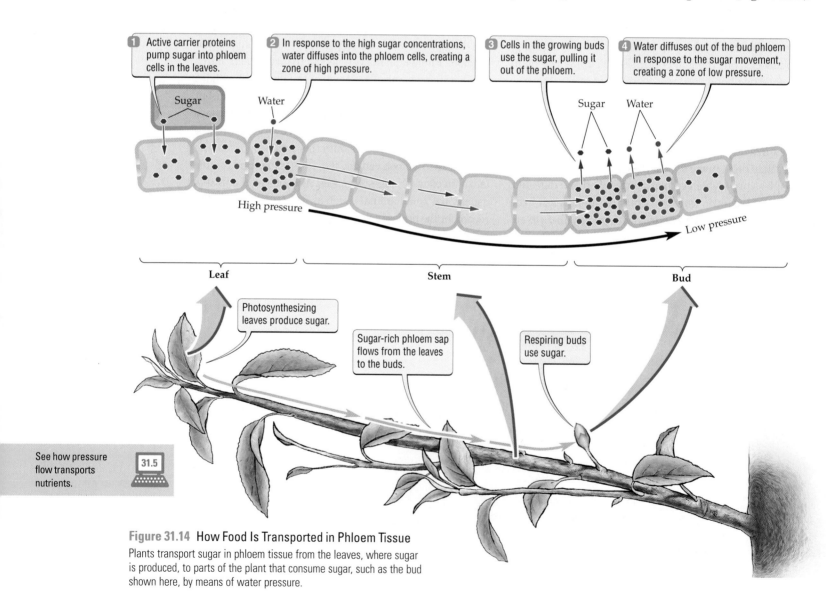

1 Active carrier proteins pump sugar into phloem cells in the leaves.

2 In response to the high sugar concentrations, water diffuses into the phloem cells, creating a zone of high pressure.

3 Cells in the growing buds use the sugar, pulling it out of the phloem.

4 Water diffuses out of the bud phloem in response to the sugar movement, creating a zone of low pressure.

Sugar

Water

Sugar   Water

High pressure

Low pressure

Leaf

Stem

Bud

Photosynthesizing leaves produce sugar.

Sugar-rich phloem sap flows from the leaves to the buds.

Respiring buds use sugar.

See how pressure flow transports nutrients.

31.5

Figure 31.14 How Food Is Transported in Phloem Tissue
Plants transport sugar in phloem tissue from the leaves, where sugar is produced, to parts of the plant that consume sugar, such as the bud shown here, by means of water pressure.

This process can build up sugar concentrations of 10 to 30 percent in the phloem. When the sugar concentration rises this high, the water concentration drops, which causes water to diffuse (by osmosis) from the surrounding leaf tissue into the sugar-rich phloem cells. As a result, the phloem cells fill with water and hence have high pressures compared with surrounding tissues.

In nonphotosynthesizing plant parts, which may be distant from the leaves, cells take up sugars from nearby phloem tissue. Once the sugars have been removed from the phloem, water diffuses from the now sugar-poor phloem cells into the sugar-rich surrounding cells. The diffusion of water out of the phloem cells leads to a region of relatively low pressure there (see Figure 31.14).

The difference in pressure between sugar-rich and sugar-poor phloem cells can easily reach 700 mm Hg, much greater than the pressure a human heart generates during exercise. This difference in pressure pushes food in the phloem from the leaves (where the pressure is high) to all plant tissues that use sugars for energy (where the pressure is low).

Transport through the xylem relies on water evaporating from the leaves

The method plants use to move water and minerals through the xylem is remarkable: it generates enormous pressure, yet at no energetic cost. The process requires no energy because it is driven by physical forces that automatically result when water evaporates from a leaf (Figure 31.15). Here's how it works:

1. Water brought to the leaf by the xylem sticks to and coats the surfaces of ground tissue cells within the leaf, forming a film that is continuous with the water in the xylem. The cells are in direct contact with the outside air because the leaf is filled with small air spaces. As water evaporates from the cells into these interior air spaces, and from these air spaces into the outside air, the remaining water that coats the cells is drawn into microscopic cracks between cell walls. This causes the water surface to curve, which in turn creates a pull on the water column in the xylem. The strength of this pull, or tension, can be measured in the same units as pressure. It turns out that the tensions generated by the curving film of water can be very large—more than –67,000 mm Hg in some species. (When a force pulls rather than pushes, the pressure is reported as a negative number—hence the minus sign—but the effect is the same as a push.)

2. The tension generated in the leaf pulls a column of water through the xylem, all the way up from the roots. (In contrast, fluid is pushed by water pressure in the phloem and by the muscular contractions of hearts in animals.) To see that water is pulled up the xylem by the leaves, rather than pushed up by the roots, consider cut flowers in a vase: they have no roots, yet evaporation of water from their leaves continues to draw water out of the vase.

3. For plants to pull a column of water from the roots to the leaves, the water in the xylem must act like a thin cable that is strong enough to withstand very high tensions, such as the 25,500 mm Hg in a coast redwood. Most of us would have a hard time thinking of water as a cable. By some calculations, however, the molecules in a column of water hold together so tightly that they can withstand greater tensions than a steel cable of equal diameter. Measured values in plants fall below this, but are still more than twice the strength needed to transport water in the tallest redwood.

Overall, as water evaporates from a leaf, a "cable" of water is pulled up from the roots, transporting dissolved nutrients and replacing the water lost to evaporation.

Explore how plants transport water via xylem tissue.

31.6

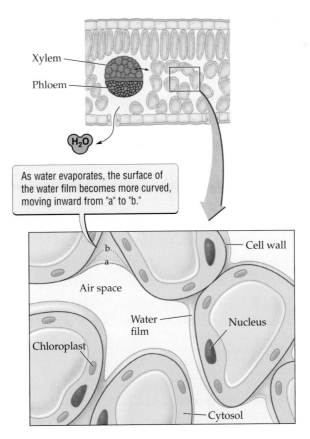

As water evaporates, the surface of the water film becomes more curved, moving inward from "a" to "b."

Figure 31.15 How Plants Move Water in Xylem Tissue

Plants transport water and dissolved mineral nutrients in xylem tissue. Water coats the surfaces of cells within a leaf, creating a film of water surrounding air spaces within the leaf. When water evaporates from this film into one of these air spaces, the film of water is drawn into cracks between plant cell walls. This creates a curve in the water surface, which in turn exerts a pull, or tension, on the water column in the xylem to which the film is connected. Enormous pressures can be generated in this way, requiring no input of energy from the plant.

Anything that breaks the continuous column of water, such as an air bubble, breaks the cable and prevents the lifting of water in that particular water column. Tall plants such as trees have billions to trillions of water columns, and hence have multiple backup transport routes available when a particular column is blocked. Both drought and freezing temperatures promote air bubble formation, thus blocking columns; if enough columns are blocked, xylem transport may become impossible during dry or cold seasons.

Ruling in Extreme Cold

One of the ways plants "rule," as we saw at the outset of this chapter, is by tolerating extreme climatic conditions. Here we examine one of these conditions: extreme cold. To appreciate the magnitude of the challenge plants face, imagine you are stranded outside in your pajamas for a few minutes during a winter night in central Canada. If there is no wind and the air temperature is −40°C (−40°F), how long do you think it will take for frostbite to set in?

The answer, according to U.S. Army data, is not very long: just 11 minutes in some people, and well under an hour in even the most resistant individuals. Now think what a plant must endure. In central Canada, plants regularly face temperatures below −40°C for hours on end. Some plants can even tolerate temperatures as low as −50°C to −100°C (−58°F to −148°F). How do they do it?

Plants have several ways to cope with cold. First, when days get shorter (indicating the approach of winter) and temperatures drop close to 0°C (32°F), plants that tolerate cold prepare to enter **dormancy**, a state of arrested growth. Their preparations involve physical and chemical changes, including covering their buds with thick protective scales, producing sugars and amino acids that act as antifreeze, and decreasing the water content of their cells.

The danger to plants from cold comes from ice. Some plants cannot tolerate temperatures at or below the freezing point of water (0°C); in such

species, sustained temperatures below 0°C cause ice to form within their cells, ripping the cells apart and killing the plant. Other species cope with temperatures below 0°C by using sugars and amino acids as a type of antifreeze that limits ice formation. This approach can be very effective: as a result of the "antifreeze" they use, water does not freeze inside the cells of oak, maple, and fir trees until temperatures reach −40°C. At temperatures below −40°C, however, ice forms spontaneously, no matter how much "antifreeze" is used. Thus, to tolerate extreme cold (temperatures of −50°C to −100°C), a plant must be able to limit the damage caused by ice.

Here's how species that tolerate extreme cold do it. Once temperatures drop below −40°C, ice crystals begin to form within their cell walls. As this happens, the plants gradually transport water from the inside of the cell through the cell wall. Once outside the cell wall, the water freezes to become part of the growing ice crystals, but the contents of the cell remain unfrozen. Plants that tolerate extreme cold have enough space between their cells to make room for the growing ice crystals, and they can tolerate the extreme lack of water that results as they remove the water from the inside of their cells. This fact probably explains why some trees that are very resistant to cold are also resistant to drought.

We'll return to how plants respond to changing environmental conditions in the next chapter, in which we explore plant growth and reproduction.

Chapter Review

Summary

31.1 An Overview of the Plant Body

- The plant body contains two basic systems, the belowground root system and the aboveground shoot system.
- The root system is composed of one of the three types of plant organs, the roots, while the shoot system contains the other two types of plant organs, stems and leaves.
- Plants are built differently from animals: aboveground, plants grow in size by the repeated addition of the same basic bud-stem-leaf unit. Unlike most animals, plants grow throughout their lives.

31.2 The Three Plant Tissue Types

- Plant bodies are made from three basic tissue types: dermal, ground, and vascular tissues.
- Dermal tissues interact with the environment, controlling the flow of materials into and out of the plant and protecting the plant from attack.
- Most cells in dermal tissues are unspecialized. Some dermal cells have specialized shapes and functions: trichomes help defend the plant against enemies, root hairs absorb nutrients from soil, and guard cells control gas exchange and restrict the rate of water loss.
- Ground tissues make up the bulk of the plant body and play many essential roles, including support, wound repair, and photosynthesis.
- Phloem and xylem, the two types of vascular tissue, form a continuous set of pipes running through the entire plant body.
- Phloem transports food from the leaves throughout the plant. Phloem is made up of living cells connected to one another by plasmodesmata, through which the food moves.
- Xylem transports water and mineral nutrients from the roots to the leaves. Xylem is made up of hollow, dead cells arranged end to end.

31.3 The Three Plant Organs

- Roots absorb water and mineral nutrients, anchor the plant, and store food.
- There are two types of root systems. In a taproot system, a dominant root grows directly downward. In a fibrous root system, no one root dominates.
- Root hairs greatly increase the surface area through which plants can absorb water and mineral nutrients.
- Stems support the plant against gravity. In many species, stems also contain ground tissue cells that perform a limited amount of photosynthesis.

- Stems can be highly modified to perform specialized tasks such as protection, climbing, and food storage.
- Leaves feed the plant: ground tissue cells in the leaf perform photosynthesis and produce most of the plant's food.
- Guard cells in leaf dermal tissues control the rates at which CO_2 is absorbed from the air and O_2 and water are lost to the air.
- Leaves can be highly modified to perform specialized tasks such as protection, climbing, food storage, and collection of mineral nutrients.

31.4 How Plants Obtain Nutrients

- To grow, plants need CO_2, water, and mineral nutrients. Most of the dry weight of the plant comes from CO_2.
- Plants need seventeen mineral nutrients to grow, including nine macronutrients and seven micronutrients.
- Plant roots absorb water and mineral nutrients from the soil by two routes: through their cells and outside their cells along cell walls.
- Because mineral nutrient concentrations are much higher in roots than in the soil, plants must expend energy to absorb essential nutrients.
- Most plants form structures called mycorrhizae in which plant roots are united with fungal hyphae. Mycorrhizae greatly increase the ability of plants to absorb nutrients from soil.

31.5 Plants That Eat Animals

- Some plants that live in nutrient-poor environments get mineral nutrients by eating animals.
- Plant strategies for eating animals include modified leaves that close rapidly upon their prey, sticky hairs that attract and trap insects, digestive chambers that suck prey inside, and pitchers filled with digestive juices that lure animals inside with nectar.

31.6 How Plants Transport Food and Water

- Plants require much higher pressures to transport food and water than animals do. Plants do not have a muscular pump similar to an animal heart because such a heart would have to be unrealistically large to generate the pressures needed in plants.
- Plants use water pressure to transport food through the phloem. Cells in the leaves pump sugar into phloem cells, lowering the concentration of water in those cells. Water then flows into the sugar-rich phloem cells, creating a zone of high pressure that pushes food from the leaf to parts of the plant that need sugar (where the pressure is low).

- The transport of water and mineral nutrients through the xylem relies on physical forces generated by the evaporation of water from the leaves. As water evaporates from a film that surrounds cells within the leaf, the remaining water is drawn into cracks between leaf cell walls. This creates a curve in the water surface, which in turn creates tension on the water column in the xylem to which the film is connected.

◉ Review and Application of Key Concepts

1. List the three organs and the three tissue types found in plants. For each of the tissue types, describe at least two functions. Similarly, for each organ, list typical and specialized tasks performed by that organ.

2. Describe the two routes by which mineral nutrients are absorbed by plant roots.

3. If a plant is placed in a drying oven to remove its water, do nutrients absorbed by the roots account for most of the (dry) weight of the plant? Explain your answer.

4. Give examples of how plants manage to eat animals, including (in some cases) vertebrates.

5. How do tall trees generate enough pressure to move water from their roots to their uppermost leaves? Similarly, how do plants generate enough pressure to move food produced by their leaves to other parts of their bodies?

Key Terms

angiosperm (p. 606)	mycorrhizae (p. 615)
bryophyte (p. 606)	phloem (p. 610)
cuticle (p. 608)	plasmodesmata (p. 614)
dermal tissue (p. 608)	root (p. 607)
dormancy (p. 620)	root cap (p. 610)
fibrous root system (p. 610)	root hair (p. 608)
flowering plant (p. 606)	root system (p. 606)
ground tissue (p. 608)	shoot system (p. 606)
guard cell (p. 608)	stem (p. 607)
gymnosperm (p. 606)	taproot system (p. 610)
leaf (p. 607)	trichome (p. 608)
macronutrient (p. 613)	vascular tissue (p. 608)
micronutrient (p. 613)	xylem (p. 610)

Self-Quiz

1. The waxy covering on leaves and stems is called the
   a. dermal tissue.
   b. cuticle.
   c. endodermis.
   d. Casparian strip.

2. Water and mineral nutrients are transported throughout the plant body in
   a. phloem tissue.
   b. ground tissue.
   c. dermal tissue.
   d. xylem tissue.

3. The most common route by which plants absorb mineral nutrients into their roots is
   a. directly into the phloem.
   b. along cell walls.
   c. through root cells.
   d. passively, by diffusion from the soil.

4. Compared with taproot systems, fibrous root systems usually
   a. hold soil more firmly.
   b. penetrate deeper into the soil.
   c. have no ground tissues.
   d. transport only water, not mineral nutrients.

5. The small openings that connect adjacent cells throughout the plant body are called
   a. trichomes.
   b. guard cells.
   c. plasmodesmata.
   d. xylem.

6. Flowers are modified
   a. leaves.
   b. roots.
   c. stems.
   d. dermal tissue.

7. The vast majority of plants alive today are
   a. nonvascular plants.
   b. gymnosperms.
   c. conifers.
   d. angiosperms.

8. Outgrowths of dermal cells that have special functions, such as protection and nutrient absorption, are called
   a. cell walls.
   b. trichomes.
   c. guard cells.
   d. phloem.

Biology in the News

"Dead Zone" Spreads in Gulf of Mexico

Oxygen-Poor Sea May Spark Spate of Shark Bites

A huge "dead zone" of water so devoid of oxygen that sea life cannot live in it has spread across more than 5,800 square miles (15,000 square kilometers) of the Gulf of Mexico this summer, in what has become an annual occurrence caused by pollution.

The extensive area of uninhabitable water may be contributing indirectly to an unusual spate of shark bites along the Texas coast, experts said.

How can pollution lead to shark bites? Are we putting something into the water that makes sharks aggressive? Nothing so dramatic, say scientists at the Louisiana Universities Marine Consortium. Instead, look no further than the food you eat: fertilizers used to increase crop production cause the dead zone and may contribute to the increase in shark bites.

The problem begins when rainfall washes fertilizer from agricultural fields into rivers and streams. Fertilizer from an enormous area—much of the farm belt of the United States—eventually drains into the Mississippi and other rivers. From there, the fertilizer flows into the Gulf of Mexico, polluting its waters.

Once the fertilizer reaches the Gulf, it stimulates a bloom of marine algae. Algae require some of the same mineral nutrients that plants do, so the minerals in the fertilizer, particularly nitrogen and phosphorus, cause them to grow and reproduce rapidly. When the algae die, their bodies are consumed by bacterial decomposers. The bacteria require oxygen to consume the dead algae. As a result, nearly all of the oxygen in the water is used up by the bacteria, suffocating any animals trapped there. "Fish and swimming crabs escape [from the dead zone]," said Dr. Nancy Rabalais, the consortium's chief scientist for research into low-oxygen conditions. "Anything else dies."

The oxygen content of the water drops slowly enough that sharks, like fish and crabs, can escape. But in 2004 it seems that the sharks were driven close to shore by low-oxygen waters, thus causing the observed increase in the frequency of shark bites.

SOURCE: Reuters, Tuesday, August 3, 2004.

Evaluating the news

1. We apply fertilizer to farm fields to supplement the low amounts of key nutrients, such as nitrogen, typically found in soils. The extra nutrients allow crops to grow well, thereby increasing the productivity of our farms. Increased productivity allows more people to be fed at a lower cost. Do such advantages to people justify the death of untold millions of marine animals that cannot escape the low-oxygen conditions brought about by our use of fertilizers? Why or why not?

2. According to Dr. Rabalais and other scientists, virtually nothing is being done to stop the flow of nutrients into the Gulf of Mexico. Do you think our failure to take actions to stop this problem is wrong? If so, consider some possible solutions.

3. We could decrease our use of fertilizers, but this would probably reduce crop production, thereby decreasing the supply of food and increasing its cost. Alternatively, we could continue to use fertilizers, yet strive to reduce their effects by blocking the flow of fertilizers into rivers. There are a variety of ways to accomplish this (such as restoring habitat adjacent to rivers and streams), but all of these methods cost money and therefore could increase the price of food. What do you think we should do?

4. When nutrients flow into the Gulf of Mexico, it is difficult to trace the original sources of those nutrients. If, as a society, we decide to continue to use fertilizers but seek to block their flow into rivers and streams, who should pay? Should a tax be added to the cost of fertilizer production, thereby increasing the cost of food production and, ultimately, the cost of food? Or should federal or state taxes be raised, allowing food prices to remain as they are, but reducing the expendable income of taxpayers?

Plant Growth and Reproduction

Key Concepts

◉ Undifferentiated groups of cells called meristems, distributed throughout their bodies, enable plants to grow their entire lives, adding new body parts as needed.

◉ Plants increase in length when meristem cells located at the tips of their roots or branches divide. Many plants can also increase in thickness.

◉ Flowers provide plants with a way of transporting sperm to eggs. Angiosperms have a multicellular haploid stage plus two fertilization events—one that produces a zygote, the other that produces a tissue that stores food for the developing embryo.

◉ After fertilization, the embryo and its surrounding tissues mature to form a seed containing a young plant of the next generation. When conditions are favorable, the seed germinates, and a seedling begins to grow.

◉ Sexual reproduction in plants varies greatly among species. In some plant species, individuals are either male or female. In others, individuals are both male and female. In still others, individuals switch sex repeatedly.

◉ Plants have a well-developed set of hormones that control how they grow, reproduce, and respond to their environment.

◉ Plants defend themselves against herbivores with physical and chemical weapons. Plants protect themselves from pathogens with a tough outer surface and an immune system that provides both specific and nonspecific defenses.

A Forest Walks

Villains in literature have a problem with forests that walk. Consider the lead character in *Macbeth*, Shakespeare's tale of ambition, murder, and madness. Obsessed with power after he kills the king and usurps the throne, Macbeth is astonished by the prophecy that he will never be defeated until a forest walks. Later, his astonishment turns to horror when what appears to be a moving forest (actually an army carrying branches) marches on his castle. Macbeth is not alone: a moving forest also spells doom for the evil wizard Saruman in J. R. R. Tolkien's *Lord of the Rings*. In the midst of his efforts to seize great power, Saruman's plans are dashed when his fortress is surrounded by an army of what look like trees, commanded by the usually slow-moving (but now hopping mad) treelike creatures called Ents.

Of course, forests do not move in the sense of *Macbeth* and *Lord of the Rings*—rooted in the ground, a tree cannot pull up its roots and walk from one place to another. But if we look into the biology of trees more deeply, we find that the notion of a walking forest contains elements of truth.

For example, many trees make genetically identical copies of themselves, which causes them to move, albeit slowly. Trees do this by producing rhizomes, horizontal stems that grow from the base of the trunk. After growing one to a few meters across the ground, the rhizomes turn upward and sprout, giving rise to a new tree that is genetically identical to the parent tree from which it came. Two events have happened here: the parent tree has reproduced asexually (by making a genetic copy, or "clone," of itself), and movement has occurred (the clone has moved a short distance away from its parent). Some trees are masters at this slow form of travel. In trembling aspen, for example, what appears to be a forest may actually be

Aspen Clones

many genetically identical copies of a single tree. In such cases, the forest truly does move: from where its first tree was located, the forest spreads to cover a large area—up to 81 hectares (200 acres) in one aspen clone in Colorado.

Forests also move in another, very different way. Over long periods of time, trees migrate north or south, following changes in temperature as the climate warms or cools. Here, too, it is not individual rooted trees that move, but rather their offspring—seeds. As we shall see in this chapter, each tree seed contains a young tree. Most young trees move only short distances from their parent—a meter to a few tens of meters—but some travel very far indeed. Blown by wind, floating on water, or sticking to the fur or feathers of an animal, tree seeds occasionally move 40 kilometers (25 miles) or more. As a result, forests really do move from one place to another, just not as rapidly as imagined by Shakespeare or Tolkien. The fossil record shows, for example, that the northern edge of the geographic range of spruce trees has shifted over 1,000 kilometers to the north during the last 18,000 years; similar patterns have been found for many other tree species.

Many other plants besides trees move in these two ways—within their local environment by producing clones, and from one region to another by traveling inside of seeds. These observations raise further questions: How far can clones move, and how long can they live? How can a young tree lie within a seed, survive long-distance travel, and then begin to grow at just the right time of year?

Although there are over 280,000 species of plants alive today, we can group them into just three categories: annuals, biennials, and perennials. **Annuals** (from *annus*, "year") are plants that complete their entire life cycle in one year. Thus, in angiosperms (flowering plants)—the focus of this chapter—an annual has one year to grow from a seed, produce a mature plant, produce flowers, and make the seeds that will start the next generation. It can be difficult to achieve all this in a single year, so some other angiosperms have a built-in "hedge" to cover that difficulty: they are **biennials** (*bi*, "two"; *ennial*, "years"), which means that they require 2 years to complete their life cycle. In its first year, a biennial grows from a seed to produce a plant that is mature in all aspects but one: it does not reproduce. Reproduction occurs in the second year of growth. Finally, many angiosperms are **perennials** (*per*, "throughout"), which means that they live for 3 years or more—sometimes many more.

In this chapter we use these three terms to help us understand how flowering plants grow and reproduce. We begin with a closer look at the unique ways in which plants grow. Then we examine how plants reproduce, how new individuals begin their development, and the roles hormones play in the lives of plants. Finally, we discuss the powerful mechanisms that plants use to defend themselves from attack.

32.1 How Plants Grow: Modularity

To appreciate how plants grow, first consider your own body. Like all humans, you grew rapidly from a fertilized egg according to a fixed developmental plan (one head, two arms, two legs, and so on). Once you reach adulthood (if you haven't already), you will stop growing. You may gain weight, but you won't get taller, and you won't add new arms or heads or other major body parts.

Plants are completely different because they don't grow according to a fixed developmental plan. Instead, they exhibit **indeterminate growth**, which means that they grow throughout their entire lives, adding new body parts as needed. In Chapter 31 we referred to this type of growth as a Tinkertoy approach. We made this analogy because the plant body, like a structure built with Tinkertoys, is **modular**, formed by the repeated addition of the same basic units. Aboveground, plants get bigger through the repeated addition of the bud-stem-leaf unit shown in Figure 32.1. Belowground, plants get bigger through the repeated formation of new lateral roots.

The modular approach to growth helps plants adjust to varying environmental conditions, such as high or low levels of sunlight or nutrients or attack from enemies. If conditions are favorable, a plant may add many new body parts and take on one shape, but if conditions are

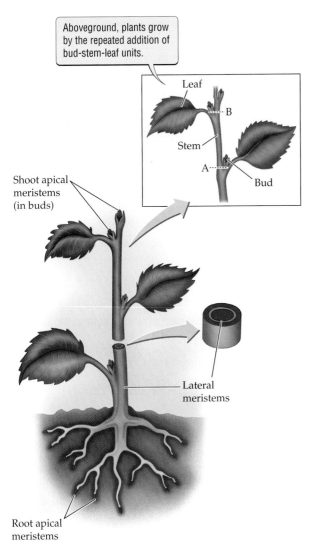

Aboveground, plants grow by the repeated addition of bud-stem-leaf units.

Leaf

B

Stem

A

Bud

Shoot apical meristems (in buds)

Lateral meristems

Root apical meristems

Figure 32.1 How Plants Grow
Plants grow throughout their lives, increasing in size and adding new parts as needed. They can do this because they have two types of undifferentiated (perpetually young) meristem tissue, apical meristem tissue (shown in red) and lateral meristem tissue (shown in blue). Apical meristems allow the plant to increase in length by repeatedly adding bud-stem-leaf units aboveground and lateral roots below ground. Lateral meristems allow many plants to increase the thickness of their stems and roots as they grow.

unfavorable, the plant may add few new body parts and take on another shape. Chicory, for example, is an herbaceous (nonwoody) flowering plant that grows 4 feet tall in good conditions, producing dozens of flowers. However, when the top of a chicory plant is repeatedly cut off (as by a cow or a lawn mower), the plant eventually "gives up" its attempts to grow to its usual height and produce many flowers. Instead, it reaches a maximum height of only a few inches, producing a total of two or three flowers. Similarly, when two trees grow very close to each

other, each may resemble half of a typical tree, and thus, from a distance, the two trees look like one tree.

How are plants able to grow and add new body parts throughout their lives? (Again, to keep the significance of this question in perspective, consider what would be required for you to continue to add new heads, arms, or other body parts as you grew.) It turns out that plants can grow as they do because they have several different types of **meristems**, which are groups of perpetually young, undifferentiated cells from which new cells arise (see Figure 32.1). Vertebrates, including us, have similar cells, called stem cells. We have a number of different types of stem cells, but the ones that can differentiate into nearly any type of human cell are active only in embryos, not in adults. In contrast, meristem cells that can differentiate into any type of plant cell remain active in adult plants, thus enabling the plant to add new body parts as needed. In fact, plants are so flexible in their growth plan that whole new plants can be cloned from single cells located throughout their bodies (see the box on page 628).

32.2 How Plants Grow: Primary and Secondary Growth

Plant meristem cells are located throughout the roots and shoot. As these cells divide and multiply, the plant can grow in two ways: in length and in thickness.

Plants increase in length by primary growth

If you were to watch a tree grow, you would find that the tree increases in length (upward and outward) only at the tips of its central stem (trunk) and branches, not throughout its whole body (Figure 32.2). Plants grow at their tips when cells in an **apical meristem** divide to produce new cells (see Figure 32.1). The increase in length that results from divisions of apical meristem cells is called **primary growth**.

When a plant germinates from a seed, it initially has one shoot apical meristem and one root apical meristem. However, as the shoot produces new branches and the original root produces new lateral roots, the meristems at the ends of these new branches and roots function like the original shoot or root apical meristem; thus we refer to any meristem at the end of a branch or root as an apical meristem.

When an apical meristem cell divides, some of the daughter cells remain as undifferentiated meristematic tissue. Other daughter cells divide repeatedly to produce

Chicory

Science Toolkit

Stirring Up Plant Clones

In 1997, headlines across the world flashed news of the birth of the lamb Dolly, the first mammal ever to be cloned from the cells of an adult. Some 40 years earlier, in a groundbreaking experiment that sounds more like an adventure in cooking than science, researchers managed to clone a new plant from a single mature plant cell.

F. C. Steward and his colleagues placed tiny pieces of phloem tissue taken from a carrot root in a large container filled with coconut milk and stirred gently. The phloem cells divided as they circulated through the coconut milk to form lumps of callus, a type of poorly organized tissue that typically forms in plants after wounding. The coconut milk provided the hormones and nutrients needed to stimulate and support the dividing cells. (What we call coconut "milk" is the unique, liquid portion of the endosperm of a coconut, while the coconut "meat," not used in this experiment, forms the more typical, solid portion of the endosperm.)

The constant movements of the callus as the coconut milk was stirred broke off individual callus cells. These cells continued to divide, but instead of forming another mass of disorganized callus tissue, they differentiated to form a recognizable root and a mass of cells that eventually formed the stems and leaves of a new carrot plant. To understand just how amazing this event was, imagine putting a few cells from your little finger into a vat of mother's milk, and then returning a few weeks later to find embryonic copies of yourself floating in the vat.

Since 1958, the year of Steward's pioneering experiment, methods of cloning have been developed and refined for many plant species. Plant cloning has many important commercial applications. One of these is the propagation of popular horticultural varieties such as orchids and house ferns, which would lose their unique characteristics if they reproduced sexually. In agriculture, plant cloning has been used to evade persistent diseases. For example, plant viruses reduce the yield of many important crops, including potatoes. Commercial varieties of potatoes can be cloned from small portions of apical meristems. These meristems are usually virus-free because the viruses, which spread from one cell to another, reach them very slowly. Because cloning allows for the consistent production of potato plants that are free of viruses, this procedure has greatly increased crop yields in potatoes.

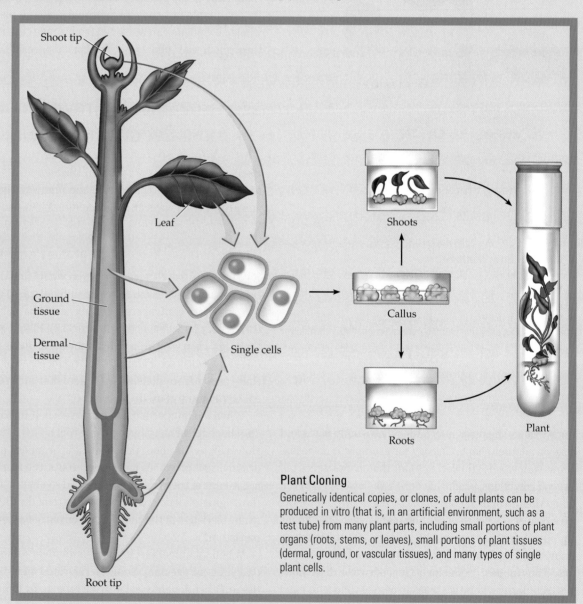

Plant Cloning

Genetically identical copies, or clones, of adult plants can be produced in vitro (that is, in an artificial environment, such as a test tube) from many plant parts, including small portions of plant organs (roots, stems, or leaves), small portions of plant tissues (dermal, ground, or vascular tissues), and many types of single plant cells.

descendant cells that begin to differentiate into new body parts. Aboveground, for example, some of these descendant cells begin a new bud, others begin a new stem, and still others begin a new leaf; this process enables the plant to add a new bud-stem-leaf unit to its body. As long as growth conditions remain favorable, plants can keep on adding new bud-stem-leaf units to the tips of their existing branches. They can also form more branches because each bud contains an apical meristem that can grow into a new branch. Once a new branch begins to grow, it (like the rest of the aboveground plant) increases in length by the repeated addition of new bud-stem-leaf units.

Woody perennials increase in thickness through secondary growth

As they grow, many plants increase in thickness as well as in height; that is, they grow out as well as up. This increase in thickness is most obvious in woody perennials such as trees, so our discussion will focus on them. And although we describe this process of growing thicker for stems, keep in mind that events are similar in roots.

Trees and other woody plants grow outward when a second type of meristem tissue—the **lateral meristem** (see Figure 32.1)—begins to divide. Trees have two types of lateral meristem tissue: **vascular cambium**, which gives rise to new vascular tissues, and **cork cambium**, which gives rise to a tough new set of dermal tissues (including the outer bark of a woody stem). Whenever the division of lateral meristem cells causes a plant to increase in thickness, **secondary growth** is said to occur.

Most of the increase in thickness in the trunk and branches of a tree comes from the first type of lateral meristem tissue, the vascular cambium. As vascular cambium cells divide, their daughter cells take on one of three identities: they may remain undifferentiated meristem cells, develop into new xylem tissue, or develop into new phloem tissue (Figure 32.3). Xylem and phloem made from vascular cambium during secondary growth are called secondary xylem and secondary phloem to distinguish them from the original, or primary, xylem and phloem that formed during primary growth (when the plant was increasing in length, not thickness).

As it grows each year, a tree adds a new layer of secondary xylem inside the vascular cambium and a new layer of secondary phloem outside the vascular cambium. Most trees produce more secondary xylem (for transporting water) than secondary phloem (for transporting food). What we call **wood** is actually a series of layers of secondary xylem. Each new layer of secondary xylem

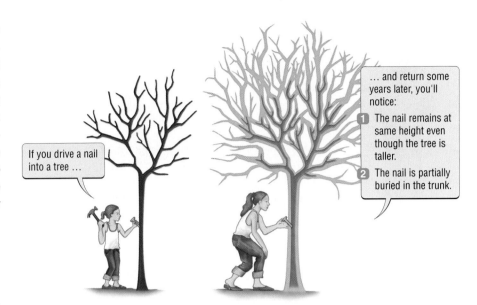

Figure 32.2 Growing Up and Out

As this simple experiment shows, as a tree grows taller, not all parts of the tree increase in length; otherwise, the nail would have moved upward. Plants get taller through primary growth, by growing at their tips. Many plants also exhibit secondary growth, increasing in thickness over time, as the buried nail demonstrates.

forms a ring around the previous year's layer, producing the "tree rings" in wood that can be used to age a tree. Over time, the older secondary xylem tissue becomes clogged and darkens. This relatively dark, centrally located xylem is called the **heartwood**; although heartwood is too clogged to transport water, it provides support to the tree. Newer secondary xylem (which can transport water) is called the **sapwood** (Figure 32.4a).

The material outside the vascular cambium, called **bark**, has two layers, the inner and outer bark (see Figure 32.4a). The outer bark contains the cork cambium, the second type of lateral meristem tissue. As the cork cambium divides, some of the daughter cells form a tough outside layer (cork). As a tree continues to increase in thickness, its bark has to make room for the added layers of wood (secondary xylem). As a result, the bark gradually cracks open and is sloughed off, to be replaced by new bark tissue produced by the cork cambium.

The inner bark is the secondary phloem. As we learned in Chapter 31, phloem is essential for transporting food. As a result, a tree can be killed if the bark on its main trunk is cut all the way around, deep enough to slice entirely through the secondary phloem (Figure 32.4b). Over time, older secondary phloem tends to get crushed and sloughed off with other parts of the bark; hence secondary phloem does not produce annual growth rings.

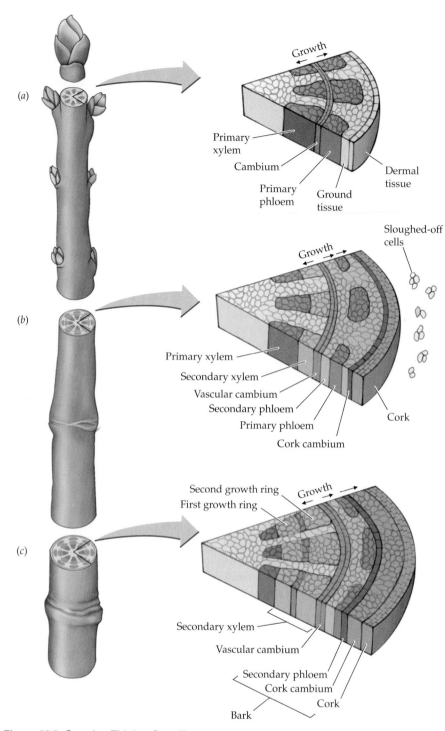

Figure 32.3 Growing Thicker Over Time

As a stem ages, it grows thicker by secondary growth. (a) Close to the apical meristem, the stem is elongating by primary growth; secondary growth has not begun to increase stem thickness. (b) Farther down the stem, secondary growth has begun, producing secondary xylem and phloem and increasing the thickness of the stem. (c) In the second year of secondary growth, you can see two layers of secondary xylem—the first two tree rings.

Nonwoody plants show little or no secondary growth

Secondary growth from lateral meristem tissue also occurs in many herbaceous (nonwoody) plants, even some annuals and biennials. Although most of the growth in these plants is up—not out—a small amount of secondary growth can occur, causing their roots or stems to thicken slightly over the course of the growing season.

The exception to this rule is the monocots, a large group of plants that includes grasses, lilies, orchids, irises, cattails, and palms. Any thickening of the stems or roots in monocots usually results from the division of cells that come from apical meristem cells, not lateral meristem cells.

32.3 Producing the Next Generation: Flower Form and Function

Although it can be difficult for animals to find or win over a mate, sex can be even harder for plants. Rooted in the ground, plants must somehow transfer the sperm of one individual to the eggs of another. The simplest plants (mosses and other bryophytes) rely on water to splash the sperm toward the eggs. But as you can imagine, this is not a very efficient way to get the job done. Angiosperms solve this problem by relying on their flowers, which are marvels of efficiency that serve just one function: reproduction.

Sexual reproduction in flowering plants differs from reproduction in animals in key ways

As we learned in the opening pages of this chapter, many flowering plants can reproduce asexually, as when a parent plant forms a genetically identical clone. Flowering plants can also reproduce sexually. They do so using their flowers, a typical example of which is shown in Figure 32.5.

Flowers are built with four whorls of modified leaves. Moving from the outermost whorl inward, these four whorls are called sepals, petals, stamens, and carpels. **Sepals** [*SEE*-puls] enclose and protect the flower before it opens, while **petals** often serve to attract animals to the flower. **Stamens** [*STAY*-mens]—the male reproductive part of the flower—consist of an anther on top of a slender filament. **Carpels**—the female reproductive part of the flower—consist of a stigma, style, and ovary. The egg is contained within the ovule, which is located inside the ovary.

To understand better how sex works in plants, let's review some terms from Chapter 9. As we learned there, *diploid* refers to a species' number of chromosomes (46 in humans), usually represented as $2n$, while *haploid* refers to half the species' number of chromosomes (23 in

(a)

Outer bark (including cork and cork cambium)

Inner bark (secondary phloem)

Heartwood

Sapwood

Bark

Vascular cambium

(b)

Figure 32.4 The Insides of a Tree

Cutting through a tree trunk or branch reveals several distinct layers. (*a*) Most of a tree's thickness is due to added layers of secondary xylem; one layer is produced each year, resulting in the rings we use to age a tree. Outer layers of secondary xylem, called sapwood, can still transport water. The inner layers of secondary xylem, called heartwood, have become clogged and can no longer transport water, but continue to serve as support for the tree. The inner bark consists of secondary phloem, and the outer bark consists of the cork cambium and the layer of cork it produces. (*b*) Girdling a tree—cutting through the layer of inner bark all the way around the main trunk—will kill it.

humans), usually represented as *n*. In humans, only the gametes (sperm or egg cells) are haploid. Sperm and eggs are produced by meiosis, the special cell division process that reduces the number of chromosomes from 2*n* to *n* (from 46 to 23 in humans). Recall that our gametes must have 23 chromosomes so that the species number of 46 chromosomes can be restored when the sperm and egg fuse during fertilization.

When you look at a plant in flower, the plant you see is a diploid individual. As in all sexually reproducing organisms, plants must perform meiosis before the sperm and egg unite in fertilization. However, unlike what happens in humans, meiosis in plants does not directly produce sperm and eggs. Instead, meiosis produces haploid cells called spores, which then divide by mitosis and grow into small, multicellular haploid stages: male **pollen grains**, formed in the anther, or female **embryo sacs**, formed in the carpel. As shown in Figure 32.6, sperm are produced within pollen grains, and eggs are produced within embryo sacs.

Flowering plants also differ from humans in the details of fertilization. A fully mature pollen grain contains three cells: two sperm cells and one tube cell. When the pollen grain lands on a stigma, the tube cell begins to grow through the style, providing a pathway for the sperm to enter the embryo sac. One of the two sperm cells fertilizes the egg, producing a diploid cell (zygote) that will grow into the next generation's flowering plant. The other

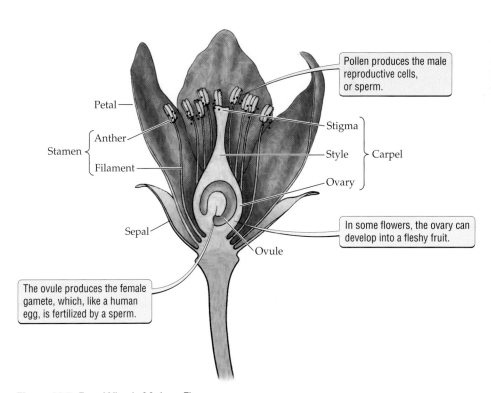

Petal

Stamen { Anther, Filament

Stigma

Style } Carpel

Ovary

Sepal

Ovule

Pollen produces the male reproductive cells, or sperm.

In some flowers, the ovary can develop into a fleshy fruit.

The ovule produces the female gamete, which, like a human egg, is fertilized by a sperm.

Figure 32.5 Four Whorls Make a Flower

Beginning at the outermost whorl and working our way inward, a typical flower consists of four whorls of modified leaves: sepals, petals, stamens, and carpels.

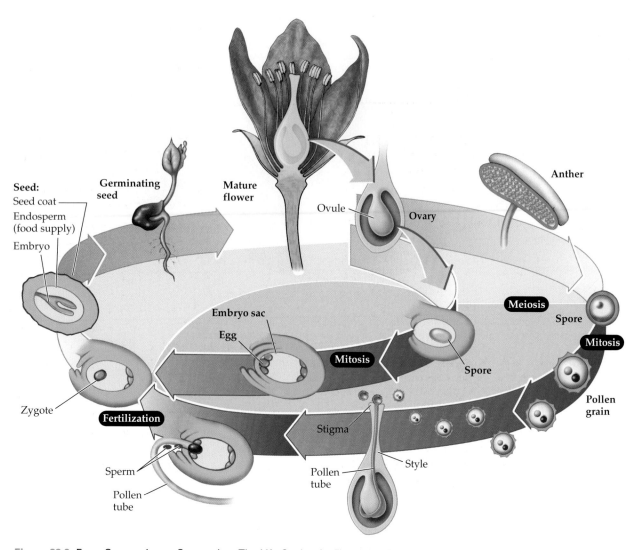

Figure 32.6 From Generation to Generation: The Life Cycle of a Flowering Plant

Within the flower of a mature plant, haploid spores are produced by meiosis. Each haploid spore divides and grows into a small, multicellular haploid stage. In male reproductive parts of the flower, the haploid stage is a pollen grain, which contains sperm. In female reproductive parts of the flower, the haploid stage is an embryo sac, which contains an egg. When fertilization occurs, a single-celled zygote is formed. The zygote grows into an embryo within a seed, and ultimately into a new flowering plant. Haploid stages of the life cycle are shown in orange.

Learn more about the life cycle of a flowering plant.
32.1

Test your knowledge of this life cycle.
32.2

sperm cell fertilizes a large central cell within the embryo sac, producing **endosperm**, a tissue that provides food for the developing embryo (see Figure 32.6). Because these two fertilization events occur when flowering plants reproduce sexually, they are said to have **double fertilization**.

Many plants use animals to bring pollen to flowers

Given that plants cannot travel to find mates, how do pollen grains containing sperm reach the female reproductive parts of a flower containing eggs? In some species,

Ragweed pollen

pollen is transported by wind. The air is filled with pollen in spring and fall, causing hay fever in millions of people who have an allergic reaction to the proteins that cover the outside of pollen grains. Wind can carry pollen long distances, but most pollen blown by wind lands in inappropriate places (such as parking lots or lakes), not on the stigma of a flower of its own species. Most flowering plants sidestep this problem by not relying on wind: instead, their sperm is transported by animals, such as insects, birds, and even mammals (such as bats).

How do angiosperms get animals to transport their sperm? The answer is bribery: they use their flowers to attract **pollinators**, which are animals that carry pollen grains from the stamens of one flower to the stigmas of other flowers of the same species (Figure 32.7). Plants bribe pollinators to visit their flowers with foods such as sugary nectar or protein-rich pollen. Pollinators are often attracted to just one or a few types of flowers, which makes the transport of sperm by pollinators much more efficient than wind. For example, if you follow a honeybee through a field, you will find that it visits only a few of the available flower types. With each visit to a flower, as the honeybee collects nectar and pollen for food, it accidentally deposits and picks up pollen that gets caught in the hairs on its body.

Some plants have taken the next step: they cheat. In such species, the plant gets the pollinator to transport pollen without providing a reward. Consider the plight of male thynnine [*THIGH*-neen] wasps. Ordinarily, males of this species locate females by smell, then carry off the females (which have no wings) to another location to mate. This process has been hijacked by the elbow orchid, a small plant whose flowers contain chemicals that mimic the smell of a female wasp that is ready to mate (Figure 32.8). On its first encounter with such a flower, the male wasp is hopelessly confused: it tries to carry off the portion of the flower that resembles a female wasp, but gets nothing in return except a ball of pollen stuck to its back. When the male repeats these actions at another flower, it may transfer the pollen to that flower, fertilizing the plant. However, the male wasps are not totally at the mercy of these devious flowers: they remember where their failed mating attempts occurred, and so tend to be fooled only once by any particular flower.

Honeybees carry the pollen that dusts their bodies from flower to flower as they search for food.

Birds play an important role as pollinators as they search for nectar in flowers.

The distinctive colors, shapes, and smells of flowers induce pollinators to selectively visit one or a few flower types.

Figure 32.7 Bribing Animals to Do the Work
Pollinators provide stationary plants with a way of transporting sperm to eggs. The spectacular colors, shapes, and odors of flowers, in combination with food rewards such as nectar, convince pollinating animals to visit several flowers of the same species, accidentally transferring pollen in the process.

(a)

(b)

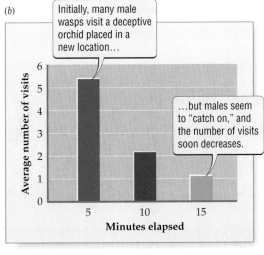

Initially, many male wasps visit a deceptive orchid placed in a new location…

…but males seem to "catch on," and the number of visits soon decreases.

Average number of visits

Minutes elapsed

Figure 32.8 Sexual Trickery
When a male thynnine wasp locates a female of his species, he carries the flightless female to another location, then mates with her. Flowers of the elbow orchid bear a slight resemblance to female wasps, but more important is their similarity in odor: the orchid flowers contain the same chemicals used by female wasps to attract males. (*a*) On his first visit to one of these flowers, a male wasp tries to carry off the flower (as he would a female wasp), but comes away with nothing but a ball of pollen stuck to his back. (*b*) Male wasps are not totally helpless, however: somehow, they remember where such flowers are located, so that they are not fooled repeatedly by any particular flower.

32.4 Producing the Next Generation: From Zygote to Seedling

When fertilization occurs, a single-celled zygote is formed, and the species' number of chromosomes ($2n$) is restored by the union of a haploid sperm (n) and a haploid egg (n). Next, the zygote begins to divide, forming a heart-shaped embryo connected to the wall of the ovule (Figure 32.9). Cell division and development continue, eventually forming a mature embryo. At this stage, we can identify the essential parts of a mature plant, including the shoot apical meristem, the root apical meristem, and the cotyledons, the first leaves of the plant (see Figure 32.9).

Let's look at these stages of development in a little more detail. To help you keep your bearings, remember that the zygote and developing embryo are located within the ovule, which is contained within the ovary. As the embryo develops inside the ovule, the parent plant transfers nutrients to the embryo and to the dividing endosperm cells, the food source for the mature embryo. At the same time that the embryo and endosperm are forming, a third event also occurs: the outer portion of the ovule begins to harden into what will become the outer layer of a seed. A **seed** is a mature ovule, consisting of a mature embryo, a food supply, and a protective outer layer, the seed coat. The ovary surrounding the developing seed develops into a **fruit**.

Once the embryo within a seed is mature, it enters **dormancy**, halting its growth for a period of time. A key event in the plant life cycle occurs during dormancy: the seeds are dispersed; that is, they move away from the parent plant. Seed dispersal is influenced by the type of fruit that the plant has. Fruit structure can vary widely, as shown in Figure 32.10. Some fruits (and the seeds they carry or contain) are dispersed by wind; others attach to the fur of animals. Some plants produce fleshy fruits that are eaten by animals. When dispersal occurs by this means, the seeds typically pass unharmed through the animal's digestive system and then are deposited on the ground along with a ready supply of fertilizer: animal feces.

Seeds may remain dormant from a few days in some species to one to many years in other species. Dormancy ends once the embryo is stimulated by favorable conditions, such as rainfall (in a desert) or warm weather after months of cold (in a region with cold winters), to resume growing. The seed is said to germinate when the embryo breaks out of the seed and a seedling begins to grow (see Figure 32.9). Many of these young plants contained in

---

Figure 32.9 From Zygote to Seedling

After fertilization occurs, the single-celled zygote divides to produce two cells, one of which goes on to produce the embryo (shown in green); the other produces a structure (yellow) that anchors the embryo to the ovule. (For simplicity, only one ovule is shown developing within this flower's ovary, as shown in Figure 32.6; however, an ovary can contain many ovules, each with its own developing embryo.) The ovule develops into a seed, which consists of the embryo, its food source (the endosperm), and a protective outer coat (the seed coat). Once the seed is fully formed, the embryo enters dormancy until it is stimulated by environmental conditions to begin growing again. At this point the seed germinates, giving rise to a seedling.

Ovary
Ovule
Zygote

Seed coat
Endosperm
Cotyledons
Embryo

Zygote within the ovule

Embryo development and seed formation

seeds are killed before they disperse—think of how many die when you eat corn on the cob—and many more die after germinating. Seedlings often die because the seed from which they grew happened to land in a place without enough soil, water, or sunlight. Other seeds land in a favorable environment that allows them to grow to maturity and produce flowers, signaling the onset of yet another generation.

32.5 Gender-Bending Plants

We usually assume that it is easy to identify whether an individual is male or female. This assumption is based in part on human experience: with few exceptions, people live their entire lives as a sperm-producing male or an egg-producing female. For many organisms, however, human notions of gender have limited relevance.

Consider the fact that the majority of animal phyla contain species whose individuals can produce both sperm and eggs and are, therefore, both male and female. We call such individuals **hermaphrodites** [her-*MAF*-roh-dites]. The earthworm is probably the most familiar hermaphrodite. Mature earthworms can produce sperm and eggs at the same time, but other hermaphrodites, including some fish species, change their sex depending on their size. Freshwater sunfish, for example, switch irreversibly from female to male once they are large enough to defend a breeding territory from other males.

And what about plants? For a minority of species (such as the ginkgo tree), human ideas of gender work well: some individuals produce only male flowers, while other individuals produce only female flowers. However, if you

Figure 32.10 A Panorama of Fruits

Dandelion fruits have a parachute-like structure and are spread by wind.

Fruits of the African plant *Harpagophytum* have "grappling hooks" that attach to the fur of large mammals.

The seeds on the surface of a strawberry fruit are dispersed when the fruit is eaten by a mammal or bird.

Shoot apical meristem

Root apical meristem

Mature embryo (seed) on the ground

Cotyledons (two)

Primary leaf

The cotyledons are out of the ground, and the primary leaves are opening.

Primary leaf

Primary leaf

Cotyledon

The new seedling is rooted, and the cotyledons are about to rise out of the ground.

Germination and growth

Jack-in-the-pulpit

reexamine Figure 32.5, you'll see that flowers typically have both male and female reproductive parts. In such species, both the individual and its flowers are hermaphroditic. In still other species, each individual has a mixture of male and female flowers: some of the flowers have male parts only, while others have female parts only. Here, the individual is hermaphroditic, but its flowers are not. Finally, some species really shake our notions of gender: from one year to the next, the plant can change from male to female and back again.

Let's look at one example, the jack-in-the-pulpit, which grows in moist forests throughout eastern North America. It is named for its unique flower, which consists of a modified leaf shaped like a church pulpit surrounding a vertical spadix ("Jack"); although it looks like a single object, the spadix is actually densely crowded with tiny flowers. A close look at jack-in-the-pulpits reveals that some are males, producing only powdery yellow pollen, and others are females, which eventually produce a cluster of bright red berries.

But all is not as simple as it seems. Among jack-in-the-pulpits, this year's male may have been last year's female, and vice versa. The sex of a jack-in-the-pulpit depends not on the presence of a Y chromosome, as in humans, but on the amount of energy the plant can store in its roots. When it can store a lot of energy, the plant has more resources for growth in the following year. Seeds require more resources to produce than pollen, so jack-in-the-pulpits switch from male to female when they have enough stored energy to produce seeds. Likewise, females that store little energy during the year turn into small male plants the next year. And, of course, any individual that continues to store about the same amount of energy from year to year will remain the same sex from year to year.

32.6 Controlling Plant Growth and Reproduction: Hormones

All multicellular organisms must coordinate their body activities. Animals do this in two ways: through chemical signaling with hormones and through electrical signaling with nerve impulses. Plants lack a nervous system, but they have hormones that control how they grow, reproduce, and respond to their environment. Plant hormones stimulate fruit ripening, regulate cell division, activate defenses against pathogens, control seed germination, turn on genes that provide resistance to drought, cause sugars to be transported from leaves to developing seeds, induce dormancy, and stimulate lateral root production, among many other effects. Here we discuss two examples of plant hormones in action: apical dominance and the onset of flowering.

Auxin causes apical dominance

All gardeners know about **apical dominance**, in which an apical meristem inhibits the growth of lateral buds (Figure 32.11). To counteract apical dominance and make a plant bushier, a gardener nips off the shoot tip, thus removing the apical meristem and causing increased growth of side branches. The plant hormone **auxin** causes apical dominance, which allows the plant to control how much primary growth occurs at any one time. Auxin has many other important effects, such as controlling the way plants bend toward sunlight and stimulating the differentiation of vascular tissue; gardeners use it to promote the formation of roots on plant cuttings.

(a)

Shoot tip (auxin source)

High concentrations of auxin inhibit lateral buds.

Auxin

Developing lateral bud

(b)

Shoot tip removed

Deprived of auxin, these lateral buds are no longer inhibited.

Figure 32.11 Removing the Tip Lets the Branches Grow

(a) The hormone auxin (shown in red) is produced in apical meristem cells and ordinarily acts to inhibit the growth of lateral buds. (b) When the concentration of auxin drops for whatever reason (here, through removal of the shoot tip), the lateral buds begin to grow, producing branches more rapidly.

One or more unknown hormones control flowering

The timing of flowering is one of the key decisions a plant must make. In some regions, for example, small herbaceous plants known as spring ephemerals grow rapidly and flower in the spring (before new tree leaves block most of the sunlight). Other species flower in the fall, when their pollinators are abundant. How does a plant "know" when to flower?

The answer has two parts. First, the leaves of many plant species contain a light-sensitive chemical that detects how long the days are. Since day length varies with the season, growing shorter in winter and longer in summer, the plant can sense the season indirectly by detecting day length. Some plants (for example, irises, tulips, spinach, and some cereal grains) do this by responding to a period of nighttime darkness that is *shorter* than a certain number of hours (Figure 32.12), others (such as poinsettias and chrysanthemums) by responding to nighttime darkness that is *longer* than a particular number of hours.

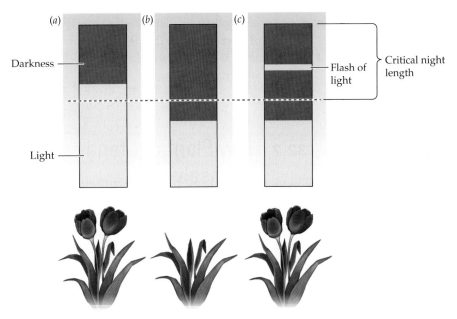

Figure 32.12 Plants Can Detect Day Length

(a) Like many plants, tulips flower when nights become *shorter* than a critical length, as happens in spring and early summer. (b) If nights are longer than the critical length, the plant does not flower. (c) In the laboratory, a flash of light during an otherwise "too long" night can cause flowering in such plants. In other plants that flower when nights become *longer* than a critical length, a flash of light during the night prevents flowering.

Several weeks later

Second, once a leaf detects that it is the appropriate time of year for flowering, it releases a hormone that causes flowering. Scientists have not yet identified this hormone, but experiments like the one in Figure 32.13 show that a chemical signal (that is, a hormone) must be involved.

The control of flowering in plants is an example of **photoperiodism** (from *photo*, "light"), which is a physiological response to day length. In addition to flowering, day length influences when seeds germinate, when buds go dormant for the winter, and when growth resumes in the spring. The consequences could be disastrous if responses like these were initiated at the wrong time of

Figure 32.13 Hormones Control Flowering

Two plants were initially raised under conditions that prevented flowering. Next, the two plants were subjected to different day lengths: the plant on the left to a day length that causes flowering, the plant on the right to a day length that does not. When the two plants were grafted together, the right-hand plant flowered, even though the day length it had experienced was not appropriate for flowering. This result indicates that a chemical signal passed from the plant on the left to the plant on the right, stimulating flowering in the plant on the right.

year. For example, in a region with cold winters, imagine the fate of a tree that tried to flower in the middle of winter, and then—if it was still alive—stopped all growth once spring arrived!

32.7 How Plants Defend Themselves

For hundreds of millions of years, life on Earth has been a struggle between species trying to eat and species trying to avoid being eaten. Plants are no exception. Over time, plants have developed a rich variety of mechanisms to deter attacks by herbivores and by pathogens such as fungi, bacteria, and viruses.

Plants use physical and chemical defenses against herbivores

If you walk through a thicket of thorny plants, you quickly experience one of the many physical defenses plants have against herbivores: you may have a foot punctured (right through the shoe), clothes ripped, or your body covered with scratches. Plants are also well protected by defenses that seem harmless to humans, such as leaf trichomes that feel like "fuzz" to us but can be fatal to insect herbivores, as we saw in Chapter 31.

In addition to their physical defenses, many plants contain chemical substances that are toxic to most herbivores. Nicotine, the addictive chemical in cigarette smoke, protects tobacco plants from most insects. In fact, nicotine is sold as a potent insecticide, and a few drops of pure nicotine will kill a human. Similarly, plants in the mustard family (such as cabbage and horseradish) are protected by chemicals so powerful that most insects, even if they are starving, will not eat their leaves. However, the same chemicals that drive most herbivores away may serve as attractants for the few species that can tolerate the plant's toxic compounds. For example, caterpillars of the cabbage white butterfly, when presented with the chemicals from cabbage that repel most insect herbivores, extend their mouthparts and act as they would when feeding.

Plants have three lines of defense against pathogens

Humans are protected against pathogens by a sophisticated defense system, including mechanisms that give us increased protection against previous attackers (see Chapter 28). Although plants do not have the ability to acquire immunity against previous attackers, they do have a highly effective immune system.

Like animals, plants have three ways to protect themselves against pathogens. First, as we have seen, plants have tough outer surfaces that keep pathogens out. But, as in animals, that line of defense can be broken. Once a pathogen enters the plant body, it encounters the other two lines of defense: specific and nonspecific immune responses.

The specific immune response is based on **gene-for-gene recognition**: plants have a number of genes that collectively have hundreds of alleles (known as R alleles) that provide resistance against specific attackers. To be resistant to a particular attacker, the plant must have a particular R allele that "recognizes" a particular allele in the pathogen. Of course, these alleles don't literally recognize one another; instead, the R allele in the plant encodes a protein that defends against pathogens with a particular genetic makeup. This system breaks down if the plant lacks the appropriate R allele.

Plants are not helpless even when they lack specific resistance to an attacker. Plants also rely on nonspecific immune responses, which involve the production of chemicals that accomplish various tasks. Some are antimicrobial chemicals, including those that attack the cell walls of bacteria. Other chemicals released in one part of the plant alert nearby cells to an impending attack. Still other chemicals cause the point of attack to be sealed off, in effect setting up a barricade to limit the invader's spread.

Finally, plants produce compounds that alert the entire plant body to the presence of an invader. One such hormone is salicylic acid, which is closely related to the active ingredient in aspirin (see Chapter 7). Long before aspirin was sold for pain relief, the people of some cultures knew that chewing the bark of willow trees lessened the pain from headaches. Willow bark contains a lot of salicylic acid, so we can think of aspirin as "medicine" that works in both plants and people, albeit with very different effects.

From Macbeth to Methuselah

As we saw in this and the previous chapter, plants, unlike most animals, cannot run from their enemies, seek shelter from heat or cold, or travel to find food or mates. Although they are challenged by their environment in many ways, there is one sense in which plants have a clear advantage over animals: longevity.

The oldest animals, such as the tubeworms pictured in Figure 32.14, have a life span of 250 years. Other long-lived animals include the Galápagos land tortoise, which can live 150 to 190 years, and a clam that can live 220 years. With respect to humans, many people in developed nations now live into their eighties or nineties, and a few live to be well over 100. Some cultures record early members as having lived even longer, such as the biblical patriarch Methuselah [muh-THOO-zuh-luh], whose life span was supposed to have been 969 years.

Although a life span of 100 years or more in humans and other animals is impressive, it pales in comparison with the life spans of certain plants. One of the world's oldest living trees is a bristlecone pine (given the appropriate nickname "Methuselah") that is over 4,700 years old. By human standards, this plant was already old when construction began (about 2650 BC) on the oldest known Egyptian pyramid!

Ancient as this tree may be, other plants are thought to live even longer. Contenders for the world's oldest living plant include "King Clone," a creosote bush in California thought to be 12,000 years old; a box-huckleberry shrub in Pennsylvania estimated to be 13,000 years old; and the granddaddy of them all, a shrub in Tasmania called King's holly, believed to be more than 43,000 years old.

How can plants live to be tens of thousands of years old? Each of the three plants just described germinated from a seed 12,000 to 43,000 years ago, then grew into a mature plant. Once grown, each plant reproduced asexually by forming clones. The asexual offspring slowly "moved" from where the original seedling was located, forming a small, circular clone in the case of "King Clone," and growing to a very large size (about 2 kilometers long) in the case of the box-huckleberry clone.

In shrubs that reach ages of 10,000 years or more, it is the asexual offspring that are still alive,

Tubeworm (*Lamellibrachia luymesi*): up to 250 years old.

Jeanne Calment: lived to be 122 years old (120 in this photo).

Figure 32.14
How Long Can Organisms Live?

Bristlecone pine (*Pinus aristata*): up to 4,700 years old.

Creosote bush clonal ring, "King Clone" (*Larrea divaricata*): 12,000 years old.

King's holly (*Lomatia tasmanica*): 43,000 years old.

not the original stem (which is long dead). So in one sense, the bristlecone pine "Methuselah" is the oldest living plant. But if you think of an individual as beginning life when sperm and egg fuse to form a new genetic individual, and ending life when the last cell that is part of that genetic individual dies, then the shrubs we described are the world's oldest plants. In principle, these shrubs are immortal: they can keep forming new, genetically identical copies of themselves forever. But eventually, something will destroy the entire clone, such as a fire or an advancing glacier. In the meantime, however, the clone will live to reach an age that we humans—with our maximum life span of roughly 100 years—can barely imagine.

Chapter Review

Summary

32.1 How Plants Grow: Modularity

- Plant growth is unlike the growth of most animals in two ways: plants have indeterminate growth, and the plant body is modular, formed by the repeated addition of the same basic units.
- Indeterminate and modular growth allow plants to respond flexibly to their environment. They tend to add many new parts when conditions are favorable and few new parts when conditions are not.
- Groups of meristem cells located throughout their bodies allow plants to add new body parts throughout their lives.

32.2 How Plants Grow: Primary and Secondary Growth

- Plants increase in length by primary growth, which results from the division of apical meristem cells located at the tips of the roots, of the main shoot, and of any side branches.
- During primary growth, the plant increases in length by the repeated addition of the same basic units: the bud-stem-leaf unit aboveground and new lateral roots belowground.
- Plants increase in thickness by secondary growth. In woody perennials, this growth results from the division of two types of lateral meristem tissues: vascular cambium and cork cambium.
- Dividing vascular cambium cells produce new vascular tissues (secondary xylem, or wood, and secondary phloem). Dividing cork cambium cells produce a tough new set of dermal tissues that form the outer bark.
- Secondary growth in trees produces wood. Wood consists of many layers of secondary xylem tissue. Each year, a new layer of secondary xylem grows on the outside of the previous year's layer, producing the "tree rings" we use to age a tree.
- A small amount of secondary growth occurs in some nonwoody plants.

32.3 Producing the Next Generation: Flower Form and Function

- Male and female reproductive parts of plants are contained in flowers, which consist of four whorls of modified leaves: sepals, petals, stamens, and carpels. Sepals protect the flower before it opens, petals often attract animals to the flower, stamens are the male reproductive part of the flower, and carpels are the female reproductive part of the flower.
- Meiosis in plants (unlike meiosis in animals) does not produce sperm and egg cells directly. Instead, it produces a spore, which undergoes mitosis and grows into a small, multicellular haploid stage—a pollen grain in males and an embryo sac in females. The pollen grains produce sperm, and the embryo sacs produce eggs.
- Flowering plants have double fertilization. A pollen grain contains two sperm, one that fertilizes the egg to produce a zygote, and another that fertilizes a large central cell in the embryo sac to produce endosperm (stored food for the developing embryo).
- Animal pollinators provide immobile plants with a way of transporting sperm-containing pollen to eggs.

32.4 Producing the Next Generation: From Zygote to Seedling

- Fertilization produces a single-celled, diploid ($2n$) zygote, which begins to divide and develop into an embryo. The embryo is located within the ovule, which is contained within the ovary.
- The parent plant provides nutrients to the developing embryo and the dividing endosperm cells. The outer portion of the ovule begins to harden into a protective seed coat. Each seed (mature ovule) contains the ingredients for growing a young plant of the next generation: a mature embryo, the endosperm, and the seed coat. The ovary surrounding the seed forms a fruit.

- Once a seed is formed, the embryo enters dormancy, and the seed is dispersed from its parent. Seeds are often dispersed by wind, or by attaching to or being eaten by animals.
- Dormancy ends when the embryo is stimulated by favorable conditions to start growing again. Then the seed germinates, and a seedling begins to grow.

32.5 Gender-Bending Plants

- Most plants are hermaphrodites: they contain both male and female reproductive parts and hence produce both sperm and eggs. In many species, all of the flowers on a plant are hermaphroditic. In other species, individuals are hermaphroditic, but their flowers are not.
- Some plants can switch from being male to female, or vice versa, repeatedly over the course of their lives, depending on environmental conditions.

32.6 Controlling Plant Growth and Reproduction: Hormones

- Hormones control how plants grow, reproduce, and respond to their environment.
- The plant hormone auxin influences apical dominance, in which an apical meristem inhibits the growth of lateral buds. Apical dominance allows the plant to control how much primary growth occurs at any one time.
- In many plants, a light-sensitive chemical in the leaves detects changes in day length. Once the appropriate day length has been reached, an unknown plant hormone initiates flowering.
- Day length also influences other important plant responses, such as when seeds germinate and when buds go dormant for the winter. Such physiological responses to day length are called photoperiodism.

32.7 How Plants Defend Themselves

- Plants defend themselves against herbivores with physical weapons (such as thorns and leaf trichomes) and a rich variety of toxic chemicals.
- Plants defend themselves against pathogens with a tough outer covering and a well-developed immune system. The plant immune system has two components, a specific response and a nonspecific response.
- The specific plant immune response is based on a system of gene-for-gene recognition, by which plants with specific resistance alleles (R alleles) can recognize and attack pathogens with particular alleles.
- Nonspecific immune responses include producing antimicrobial chemicals, signaling nearby cells of an impending attack, sealing off the point of invasion, and sending signals throughout the entire plant.

◉ Review and Application of Key Concepts

1. How can plants grow in length throughout their entire lives, adding new body parts as needed?

2. Label each flower part in the diagram below.

3. Provide a detailed description of the life cycle of a flowering plant.

4. What is wood and how does it form?

5. In plants with fleshy fruits that attract animals, the fruits often taste bitter until the seeds they contain are fully mature. Explain why this might increase the plant's chance of reproducing successfully.

6. Imagine that it became a crime to kill a unique genetic individual of any plant species. List five common foods that you would no longer be able to eat. (*Hint:* Whenever a sperm and an egg fuse to form a zygote, a new, unique genetic individual is formed.)

7. Describe how the sex of plant individuals and plant flowers varies in different plant species.

8. Consider a cultivated plant species that has two genetic mutations: one that results in a feature that people value (such as a hypothetical genetic mutation in clover that causes the plant to produce four leaves consistently, instead of the usual three), and another that impairs the ability of the plant to detect day length. If the plant could be mass-produced (say, by some method of cloning), and if people throughout the world planted it in their gardens, in what parts of the world would the mutant be likely to thrive? In what parts of the world would it be likely to grow poorly? Explain your answer.

9. Describe how plants defend themselves against herbivores and pathogens.

Key Terms

annual (p. 626)
apical dominance (p. 636)
apical meristem (p. 627)
auxin (p. 636)
bark (p. 629)
biennial (p. 626)
carpel (p. 630)
cork cambium (p. 629)
dormancy (p. 634)
double fertilization (p. 632)
embryo sac (p. 631)
endosperm (p. 632)
fruit (p. 634)
gene-for-gene recognition
   (p. 638)
heartwood (p. 629)
hermaphrodite (p. 635)

indeterminate growth (p. 626)
lateral meristem (p. 629)
meristem (p. 627)
modular (p. 626)
perennial (p. 626)
petal (p. 630)
photoperiodism (p. 637)
pollen grain (p. 631)
pollinator (p. 633)
primary growth (p. 627)
sapwood (p. 629)
secondary growth (p. 629)
seed (p. 634)
sepal (p. 630)
stamen (p. 630)
vascular cambium (p. 629)
wood (p. 629)

Self-Quiz

1. The increase in length that results from the division of apical meristem cells is called
   a. primary growth.
   b. secondary growth.
   c. apical dominance.
   d. secondary xylem.

2. Angiosperm pollen grains have two sperm cells. What is their function?
   a. One fertilizes an egg to produce a zygote, the other fertilizes a cell to produce fruit.
   b. Each one fertilizes a different egg, producing two genetically distinct zygotes.
   c. Both fertilize a single egg, producing two genetically identical zygotes.
   d. One fertilizes an egg to produce a zygote; the other fertilizes a another cell to produce stored food.

3. In terms of plant structure, what we call the wood of a tree consists of
   a. layers of secondary xylem.
   b. layers of secondary phloem.
   c. endosperm.
   d. cork cambium.

4. A mature ovule is called
   a. an ovary.
   b. an embryo.
   c. a seed.
   d. the endosperm.

5. Plants have a specific immune response that depends on
   a. salicylic acid.
   b. gene-for-gene recognition.
   c. apical dominance.
   d. Plants do not have a specific immune response.

6. Apical dominance
   a. occurs when apical meristems inhibit the growth of lateral buds.
   b. is caused by auxin.
   c. allows the plant to detect day length.
   d. a and b

7. Most of the increase in the thickness of tree trunks and branches comes from the division of cells in the
   a. apical meristems.
   b. vascular cambium.
   c. cork cambium.
   d. lateral meristems.

8. In flowering plants, which of the following is produced directly by meiosis?
   a. haploid spores
   b. pollen grains
   c. embryo sacs
   d. endosperm

Biology in the News

Smoke Chemical That Causes Seed Germination Found

Australian scientists have identified the chemical in smoke that makes plant seeds germinate after bushfires, a discovery that could reap huge benefits for the agricultural sector.

A team of Australian scientists has become the world's first research team to pinpoint the previously unknown chemical, called a butenolide, which induces germination in a range of plant species including celery, parsley and Echinacea.

Plant seeds have an uncanny ability to germinate when conditions are right. But sometimes people would like to highjack the process—for example, to cause weeds to germinate all at once (making them easier to kill), or to stimulate the germination of plants we want to nurture (such as native plants when we seek to restore natural vegetation). However, controlling germination has proved hard to do—until now. The research described in this news article may give people the edge we've been looking for—a way to stimulate plants to germinate when and where we want them to.

Following wildfires, the seeds of many plant species germinate in large numbers, which allows such species to recover rapidly from burning. Scientists have long thought that what stimulates seeds to germinate after a fire is heat or chemicals in the ash. Heat from a fire does indeed cause the cones of some pine trees to open, releasing their seeds. But 15 years ago, botanists in South Africa showed that it was smoke, not heat or ash, that stimulated seed germination in a collection of grassland plants. Since then, researchers have been hot on the trail of the chemicals in smoke that stimulate germination.

The discovery that germination is stimulated by butenolide [byoo-TEEN-oh-lide], a chemical in smoke, "represents one of the most significant advances in seed science with benefits in the natural, agricultural, conservation, and restoration sciences," says Geoff Gallup, science minister in the state of Western Australia. In agriculture alone, this discovery, with further testing, could give farmers a multimillion-dollar edge in weed control. Weeds that germinate throughout the growing season require frequent herbicide applications. Making weeds germinate all at once could lead to less frequent herbicide use, or perhaps no herbicide use at all if the weeds could be forced to germinate in winter, when the seedlings would die.

Evaluating the news

1. The research described in this article has the potential to benefit people around the world in several ways: First, it might lower the cost of food because of savings related to the control of weeds. Second, it might reduce the health and environmental risks that stem from herbicide use. In addition, if a patent were issued for butenolide, the patent holder could potentially make a lot of money. This general situation is common to many research projects: the research may provide a public good, and it may provide someone with the opportunity to make money. Who should benefit financially from discoveries with commercial applications—those who do the research? Or those who fund the research (often, members of the public whose taxes fund publicly sponsored research)?

2. If butenolide turns out to make it easier to restore natural vegetation in areas where people have caused environmental damage, would this potential remedy make people less cautious about causing environmental damage in the first place?

3. Using a chemical like butenolide, as proposed in this article, can be thought of as an example of people controlling events in nature. Do you think this is right? Should people seek to control nature?

SOURCE: Reuters, Friday, July 9, 2004.

Feeding a Hungry Planet

The Great Challenge

The contemporary age is not short of terrible and nasty happenings, but the persistence of extensive hunger in a world of unprecedented prosperity is surely one of the worst. . . . What makes this even more of a tragedy is the way we have come to accept and tolerate it, as if it was essentially unpreventable.

Amartya Sen, winner of the 1998 Nobel prize in economics, wrote these words. As he suggests, we live in a world of plenty, yet billions of people get so little to eat that their health is poor and they are at risk of starvation (Figure F.1). How can this be? What can we do about it?

At first glance, it seems that there should be a simple answer to questions like these. At present, enough food is produced to meet the basic needs of the world's 6.5 billion people. In 2002, for example, the available food could have provided each person with 2,700 calories per day, 1.5 times the 1,800-calorie minimum requirement. Since there is enough food to feed everyone, we could end world hunger right now by doing a better job of distributing existing food supplies. Unfortunately, saying so does not make it so: as we'll see, improving the distribution of food is a hard thing to do. It is also difficult to increase the amount of food we produce. And what is hard now is only getting harder. Currently, the human population

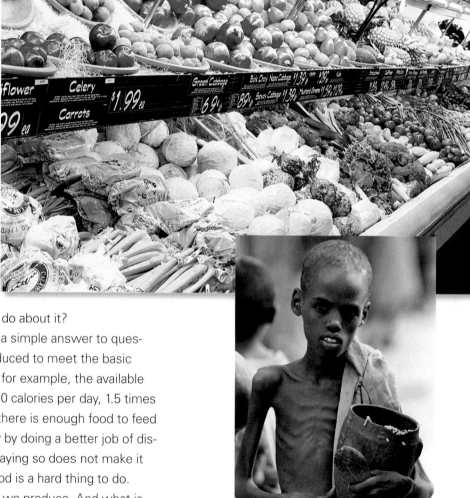

Figure F.1 Hunger in a World of Plenty

Figure F.3 A Drop in World Grain Production

On a per person basis, world production of rice, wheat, and corn peaked in 1984 and has declined since then.

Table F.1

Food Production from the Three Traditional Food Systems

In the second half of the twentieth century, world food production per person from each of the three traditional food systems grew rapidly, then declined.

Food system	Food	Years of growth	Growth per person (%)	Years of decline	Decline from peak (%)
Cropland	Grain	1950–1984	+38	1984–2000	−11
Rangeland	Beef and mutton	1950–1972	+44	1972–2000	−15
Oceanic fisheries	Seafood	1950–1988	+112	1988–1998	−17

continues to increase in size, by themselves, rangelands and oceanic fisheries will be hard pressed to meet our growing need for animal protein.

Not only are there more people to feed, but standards of living are increasing for many people. When people have more disposable income, they tend to eat more animal protein, including meat (beef, poultry, pork, and mutton, which is the meat from adult sheep), milk products, and fish. Meat production has risen rapidly over the last 50 years. This growth, more than double the growth of the human population, allowed yearly meat intake to increase from 17 kilograms per person to 38 kilograms per person. As incomes rise in countries such as China, the demand for meat and other sources of animal protein is expected to keep rising over the next 20 to 30 years.

Animal protein comes increasingly from managed systems

With demand on the rise but production falling in rangelands and oceanic fisheries, we have started to get more of our animal protein from highly managed agricultural systems, such as feedlots (confined areas for the controlled raising and feeding of animals) and aquaculture (Table F.2). While this change has the potential to allow the production of enough animal protein to keep up with demand, it also brings new challenges: managed systems can produce large amounts of pollution, and the animals often are raised under poor conditions (such as extreme crowding). If these problems can be solved, the shift to managed systems should help efforts to combat world hunger.

Table F.2

Sources of Animal Protein

"Annual rate of growth" is the average yearly increase (1990–2000) in world production of animal protein from five sources: aquaculture, three feedlot systems (poultry, pork, and beef), and oceanic fisheries. In aquaculture and the three feedlot systems, increases in production were greatest for those sources that use less grain per kilogram (kg) of animal live weight produced.

Source of animal protein	Annual rate of growth (%)	Amount of grain (in kg) needed to produce 1 kg animal live weight
Aquaculture	11.4	1.5–2
Poultry	4.9	2
Pork	2.5	4
Beef	0.5	7
Oceanic fisheries	0.1	Not applicable

Of the managed systems listed in Table F.2, **aquaculture**, which includes the cultivation of both fish and shellfish for human use, has increased so quickly that it is poised to overtake global beef production by 2010. China leads the world in aquacultural production, having produced in 1998 a combined total of 21 million tons of fish and shellfish (roughly equal amounts of each). Some aquacultural operations, such as the production of salmon (a predatory fish) and shrimp, have been criticized as having negative effects on the environment. Worldwide, however, aquaculture is dominated by industries that cause less environmental damage, such as the production of herbivorous fish including carp, catfish, and tilapia [tih-LOP-ee-uh] (Figure F.4). It can be less damaging to the environment to raise herbivorous fish than predatory fish because the herbivores eat plants, hence there is no need to harvest other fish to feed them.

The production of poultry has also grown rapidly in recent years, with an annual growth rate of nearly 5 percent per year. Both aquaculture and poultry production use less grain than cattle production (see Table F.2). Thus, if aquaculture and poultry production continue to grow rapidly, relatively less of the world's grain will be needed to feed animals, and more can be used to feed hungry people.

Another way to produce more animal protein while reducing world hunger is to feed corn stalks, rice straw, and wheat straw to cattle, sheep, and goats. This approach would allow us to squeeze more food out of a single crop: in addition to the corn, rice, or wheat that we eat, we could also produce meat or milk from animals fed on the parts of each crop that we can't eat (the stalks or straw).

Steps to End World Hunger

Today, roughly 840 million people are malnourished because they don't get enough calories, protein, or both (Figure F.5). Many others suffer from poor mineral nutrition (for example, insufficient iron) or lack of vitamins. For example, an estimated 1.2 billion people are weakened by iron-deficiency anemia, and 600 million people suffer from iodine-deficiency disorders. All told, an estimated 40,000 deaths per day are linked to malnutrition; nearly half of those who die are children.

Although sources of animal protein are important, increasing the production of croplands lies at the heart of efforts to end hunger. This is true for two reasons: we get most of our calories and protein directly from crop plants, and we also use crop plants to feed animals in managed systems.

Figure F.4 A Fish Farm in China

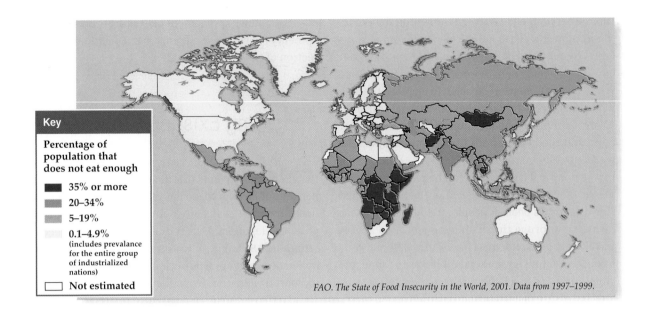

Key

Percentage of population that does not eat enough

- 35% or more
- 20–34%
- 5–19%
- 0.1–4.9% (includes prevalance for the entire group of industrialized nations)
- Not estimated

FAO. The State of Food Insecurity in the World, 2001. Data from 1997–1999.

Agricultural productivity has already increased nearly 500 percent over the last 300 years due to high-yield crops, fertilizer use, pesticide use, irrigation, and improvements in farm machinery. In particular, the last 50 years has seen a remarkable increase in productivity known as the **Green Revolution**. The Green Revolution was driven by the development of new varieties of wheat, rice, and corn that allowed the production of large amounts of grain on a given amount of farmland (Figure F.6). To give just one example, pest-resistant varieties of wheat were developed that increased yields from 0.75 tons per hectare to 8 tons per hectare.

Crop varieties developed in the Green Revolution greatly helped efforts to end hunger by enabling farmers to produce more food. But the Green Revolution could not end hunger by itself because it did not address root causes of hunger, such as poverty, overpopulation, and poor distribution of food. In addition, gains made in the Green Revolution came with environmental costs, such as increased pollution of soils, water, and air (from fertilizer, pesticide, and energy use).

The challenge we face now is to continue to improve yields while simultaneously seeking to reduce poverty, control population growth, improve the distribution of food, and limit damage to the environment. Meeting this challenge will require social as well as scientific innovations. Some people, for example, are hungry because existing food supplies cannot reach them. In essence, this problem represents a failure of government: food does not reach hungry people because of war, absence of the rule of law, or corruption. Other people are hungry because they are too poor to buy the food that is available.

Addressing social factors that cause hunger is challenging, but not impossible. Consider what happened in Thailand. In 1982, 35 percent of the Thai people suffered from mild malnutrition, 8 percent from moderate malnutrition, and 2 percent from severe malnutrition. By 1998, these numbers had improved greatly: mild malnutrition dropped to 8 percent, moderate malnutrition was

Figure F.6 High Yields from Short Plants
New varieties of wheat, rice, corn, and other crop plants developed during the Green Revolution greatly increased crop yields. In wheat and rice, for example, high-yielding dwarf varieties were developed that devote more of their energy to producing grain and less to producing straw and leaves (which people don't eat).

less than 1 percent, and severe malnutrition had virtually disappeared. These gains resulted from actions designed to improve the education and health of poor people, thus enabling them to work more and make more money. In addition, steps were taken to increase the ability of poor farmers to produce and sell agricultural products. For example, low-cost loans were made to poor farmers, and roads were built that connected isolated rural areas to city markets.

As was done in Thailand, we can help end hunger by increasing our investment in education, health care, and jobs for the world's poorest people. We must also devise new ways to improve the distribution of food and to increase the access of poor farmers to world markets. Poor countries often have abundant labor and land, yet they may be unable to sell their crops to rich nations (the United States, Europe, and Japan). One reason for this is that rich countries commonly impose high taxes on imported foods in order to protect their own farmers, effectively pricing crops grown by farmers in poor countries out of the market. In some cases, farmers in rich nations also receive government subsidies that allow them to sell their crops below cost to poor nations; this practice drives crop prices down in poor countries and hence reduces the income of farmers there.

Reducing barriers to the fair pricing of food imports and exports is a tough problem to solve—the 2003 General Agreement on Tariffs and Trade (GATT) talks broke down over just this issue—but efforts are under way to develop solutions that protect farmers in both poor and rich countries. Improving the distribution of food in the face of ongoing wars and political corruption is also a major challenge, but again, it is a challenge many people are working to meet.

Overall, four key steps must be taken to end world hunger:

Step 1: Increase investment in poor people.
Step 2: Create new ways to distribute food and provide poor farmers with fair access to world agricultural markets.
Step 3: Increase crop production.
Step 4: Reduce the human impact on the global environment.

Except for the brief description we've given here, the first two of these steps are outside the scope of this book; you can find more information about these issues in books such as *Ending Hunger in Our Lifetime* and *Agriculture in the Global Economy: Hunger 2003*. In the rest of this essay, we'll focus on steps 3 and 4 by discussing the improvement of existing crops, the development of new crops and new farming methods, and protection of the environment.

Improving Existing Crops

There are two ways to increase crop production: increase the **yield** of crop plants (the amount of the crop produced per unit of farmland), or increase the area farmed. The best farmland is already being used, and most of the potential farmland that remains lies in fragile ecosystems, such as tropical forests (Figure F.7). Thus future increases in crop production will come mostly from increased yields, not from increases in the area farmed.

In addition to increasing the amount of food we produce, we need to increase its quality. Billions of people suffer malnutrition because their diets do not include enough protein, minerals, or vitamins. Three of the most promising approaches for improving the yield and quality of existing crops are plant breeding, genetic engineering, and the preservation of crop diversity.

Figure F.7 Future Farmland?
If crop production cannot be increased on existing farmland, marginal lands like this tropical forest may have to be used. Many policymakers, scientists, and members of the public hope to avoid this outcome, for two reasons: tropical forests are ecologically important (they contain roughly half of the world's plant and animal species), and they often have poor soils that require high inputs of fertilizer and energy if they are to support agriculture for more than a few years.

Genetic engineering can potentially increase crop yields and improve human nutrition

As we saw in Chapter 15, we can now use **genetic engineering** to isolate a gene from one species and insert it into the DNA of a second, very different species. This approach is being widely used to create genetically modified (GM) crops (Figure F.8), such as Bt varieties of cotton, tomatoes, potatoes, and tobacco. **Bt crops** are crops into which a gene from the bacteria *Bacillus thuringiensis* [. . . ther-ing-*JEN*-siss] has been inserted, causing the plant to produce a toxin that provides protection from insect pests. A wide variety of GM crops have been produced, including some that are resistant to drought, salty soils, insect attack, and viral attack. Many of these crops have the potential to reduce hunger by increasing yields and improving human nutrition, but their widespread use will depend on public acceptance and on their effects on the environment.

GM crops are widely used in some countries, such as the United States, where roughly half of all corn and soybean crops are genetically engineered. In other countries, the use of GM crops is highly controversial. Recent studies have shown that the use of GM crops increased yields in cotton, corn, beets, and oilseed rape (the source of canola oil), and in some cases led to decreased herbicide use (thereby reducing pollution). For example, a 2003 study found that the use of Bt cotton in India nearly doubled cotton yields (to 1,501 kilograms per hectare for Bt cotton versus 818 kilograms per hectare for non-Bt cotton). Other recent studies have shown that GM crops pose little direct health risk to people, but may reduce the numbers of bees, butterflies, and other insects. However, few commercial or farm-scale studies have been conducted, so we still have much to learn about the benefits and costs of GM crops.

So far, relatively few GM crops have targeted improved human nutrition, but considerable potential exists for such crops. Rice, for example, is the staple food for roughly 3 billion people, but it is a poor source of several important minerals and vitamins. As a result, millions of people that depend on rice for food have diets deficient in iron, iodine, zinc, and vitamin A. In particular, about 250 million people suffer from vitamin A deficiency, which causes many children to suffer from blindness, poor growth, and increased risk of death. In 2000, a GM rice variety, known as golden rice (see Figure F.8), was produced that contains extra beta-carotene, which our bodies use to make vitamin A. A few years later, modified versions of golden rice were produced that contain enough beta-carotene to meet 20 to 40 percent of the minimum daily vitamin A requirement. As golden rice and other similar crops become approved for public use, GM crops may begin to realize their potential for reducing human suffering from malnutrition.

Careful plant breeding may be the best way to improve desired traits

People have long been creating new crop varieties using traditional **plant breeding** techniques. Plant breeders use artificial selection to improve crops, allowing only plants with particular phenotypic characteristics to produce offspring (see page 308). Over the last 70 years, plant breeders have made enormous strides, particularly through the use of hybrid plants. **Hybrids** are produced when two highly inbred varieties of plants are mated with each another. Inbred varieties, which result from self-fertilization or from mating among close relatives, are vulnerable to pests and disease and often produce relatively low yields. Hybrids, in contrast, tend to have much

(a)

(b)

(c)

Figure F.8 A Panoply of Genetically Modified Crops
(a) GoldenRice™ has the potential to improve human nutrition. (b) Bt cotton nearly doubled cotton yields in India. (c) GM oilseed rape has increased canola oil yield.

Biology Matters

The Importance of Vitamin E

Researchers continue to find links between a healthy diet and the prevention of disease, but getting people to take advantage of their recommendations is a challenge. For example, according to the American Dietetics Association, "an estimated 75 percent of adult Americans are not meeting the 15 milligrams-per-day recommended daily amount for vitamin E. In general, men get only 10 milligrams a day, while women consume as little as 7 milligrams a day—that's less than half of what the body needs."

Plants manufacture vitamin E in chloroplasts and similar organelles. Vitamin E is a powerful antioxidant that protects cell membranes and other cell compartments from damage due to substances called free radicals. Living cells produce free radicals as a natural by-product. The red blood cells that carry oxygen to all parts of our bodies can break down without the protection of vitamin E. Free radicals can also damage DNA, so vitamin E provides protection for our genetic material. Some studies have indicated that the risk of Alzheimer's disease is significantly less among people who consume more vitamin E. And there is some evidence that this vitamin might play a role in preventing heart attacks and cancer.

Nutritionists list nuts, vegetable oils, and leafy green vegetables as major sources of vitamin E. Since salad dressings contain vegetable oils, they are also good sources. With so many sources of vitamin E available it seems surprising that many people are not getting enough. Part of the problem may be that we eat many foods that are subjected to a lot of heat. Vitamin E is degraded by heat and oxidation. Deep frying and other forms of heat processing, so common in food preparation, destroys the vitamin, as does oxidation due to exposure to the air. So, consuming fresh vegetables and lightly processed foods is helpful. The accompanying table shows the vitamin E content of a few foods. For comparison, two slices of white bread contain less than 1 mg of vitamin E, whereas 24 almonds (about one handful) would provide almost 7.5 mg, about half of the recommended daily

Selected whole food sources of vitamin E	Mg
Almonds, 1 oz. (24 nuts)	7.4
Hazelnuts, 1 oz. (20 nuts)	4.3
Canola oil, 1 tbsp.	2.9
Broccoli, 1 cup cooked	2.6
Peanuts, 1 oz. (28 nuts)	2.2
Olive oil, 1 tbsp.	1.7
Wheat germ, 1 tbsp.	1.3
Red bell pepper, 1 cup	1.0
Kiwifruit, 1 medium	0.9
Olives, 5 large	0.7
Spinach, 1 cup raw	0.6
Avocado, 1 oz.	0.4
Brown rice, 1 cup, long-grain	0.4
Apple, 1 medium	0.4
Banana, 1 medium	0.3
Sesame seeds, 1 tbsp.	0.2
Romaine lettuce, 1 cup	0.2
Beef, ground, 3 oz.	0.2

amount. Reading more about good sources of this vitamin and examining the labels on food and vitamin containers could help you to become more aware of how sufficient your vitamin E intake is.

Should you, for example, take a vitamin supplement? Our bodies typically extract only a small amount—roughly 15 percent—of the vitamin in a vitamin supplement, so if you are consuming a diverse and healthy diet, you don't need to supplement it. In normal quantities, however, vitamin supplements won't cause any harm, either.

---

higher yields. For example, the use of hybrids helped to increase corn yields in the United States from 22 bushels of corn per acre in 1930 to 142 bushels per acre in 2003.

There remains considerable potential to improve yields by using plant breeding to develop new crop varieties. While newer techniques, such as genetic engineering, will also play a role, plant breeding may be the best way to improve many agriculturally important traits. This is true because traits such as yield often are controlled by many genes, not just one or a few genes. Genetic engineering is most useful when the insertion of a single gene has a large effect, but plant breeding works well regardless of the number of genes involved.

One example of the power of plant breeding involves wheat's resistance to the Hessian fly, which causes millions of dollars in crop losses each year. To improve the ability of wheat to resist the fly—hence allowing farmers to harvest more wheat—scientists crossed several different types of wheat (including bread wheat and durum wheat, which is used to make pasta), each of

Hessian fly

which contained different genes for resistance to the fly. They succeeded in breeding a wheat variety that contains three of the thirty known resistance genes. With three resistance genes in one wheat variety, the scientists estimate that it will take the fly roughly 50 years to overcome that resistance, as opposed to the 8 to 10 years observed for varieties with a single resistance gene.

Preserving crop diversity protects crops from large losses

In modern agriculture, fields are often planted with a single, high-yielding genetic variety. This practice allows for large harvests, but it also provides an abundant, genetically uniform supply of food for any pest able to attack the variety planted. A lack of genetic diversity in crop fields can have devastating economic consequences. In 1970, over 70 percent of North American cornfields were planted with just six varieties of corn. That year, a new strain of the southern corn leaf blight fungus wiped out 15 percent of the entire corn crop, at a cost of $1 billion. In some regions, over 50 percent of the corn crop was destroyed as the fungus spread rapidly through large, genetically uniform cornfields.

One way to insure against new pests having such devastating effects is to preserve many genetically different crop varieties. Worldwide, much effort is being devoted to this task. For example, by 2000, the International Rice Research Institute had stored more than 81,000 varieties of rice seeds, up from 12,000 varieties stored in 1970 (Figure F.9).

It is equally important to maintain a high diversity of wild species that are closely related to crop plants. Such wild species can provide novel genes that may improve crop yields. In tomatoes, for example, plant breeding that incorporated genes from wild species improved yield and fruit characteristics by 20 to 50 percent. Similarly, the recent discovery of a rare wild species of teosinte [tay-oh-SIN-tee] (the precursor of corn) has the potential to improve corn yields: the wild species carries previously unknown genes for resistance to seven of the nine major viruses that threaten the corn crop each year.

Developing New Crops and New Farming Methods

In addition to improving existing crops, we can also increase food production by developing new crops and new farming methods.

Figure F.9 Storing Rice for the Future
A seed storage bank at the National Institute of Agrobiological Resources in Tsukuba Science City, Japan.

There is great potential for new crops

Of the more than 250,000 species of flowering plants, only 3,000 have ever been cultivated, and only 150 species have ever been used widely. Today, as noted above, a mere six species provide more than 80 percent of our calories. Our reliance on so few crop plants suggests that there may be great potential to reduce hunger by developing new crop species. Two examples illustrate how new crops could increase both the quantity and quality of food produced.

Amaranths (also called pigweeds) once were widely cultivated in Central and South America. These plants, which produce small, nut-flavored seeds, offer several advantages over existing crops. First, the seeds are rich in protein: they are 18 percent protein by weight, compared with 14 percent or less for cereals such as wheat, rice, and corn. In addition, amaranths are high in lysine, one of the eight essential amino acids that people cannot make and hence must include in their diets. In contrast, rice, corn, and wheat have low amounts of lysine. What's more, amaranth leaves are highly nutritious, and, like spinach, can be eaten

Amaranth

cooked or raw. Finally, amaranths are very hardy. They grow in conditions not tolerated by other crops, such as salty, acidic, and alkaline soils. At present, amaranth production is limited by pest species and weeds, as well as by a lack of methods for harvesting its small seeds.

As a second example, people in arid lands from Africa to China have cultivated grass pea for thousands of years. In many respects, grass pea is a wonder plant. It is extremely drought-tolerant, and its seeds contain 28 percent protein—twice that of wheat. Grass pea can also grow in low-nitrogen soils because its roots house bacteria that perform nitrogen fixation, a process in which nitrogen from the air is converted to forms of nitrogen that plants can use. Unfortunately, grass pea has a downside: it is poisonous. If a person gets 30 percent of their calories from grass pea for several months, their nervous system will be damaged, causing spasms and eventually loss of use of the legs. Ongoing research may solve this problem. Grass pea usually contains about 1 percent (by weight) of the poison that damages human nerves, but new varieties developed in Syria have only 0.04 percent of the poison.

Figure F.10 Using Water Only Where Needed
Grapevines being watered by a drip irrigation system in California.

New farming methods can increase yields and decrease costs

Many new farming methods are under development, ranging from organic farming to precision farming. **Organic farming** is a way of farming that avoids the use of synthetic chemicals and genetically modified organisms (GMOs), while **precision farming** uses intensive data collection, computers, and in some cases, satellite data to inform farmers when to fertilize or use pesticides.

Common to all new farming methods is the goal of increasing yields, decreasing costs, or both. Organic farming methods, for example, have led to increased profits from apples; profits went up because yields were comparable to those from conventional farming, but apple prices were higher and energy costs were lower. However, for some crops, organic farming produces lower yields, a major problem when seeking to end hunger. For important crops such as corn and wheat, for which organic methods often reduce yields, precision farming may be able to maintain or increase yields while reducing fertilizer or pesticide use, thus increasing profits and benefiting the environment.

Other ways to cut costs and protect the environment include efforts to control erosion and reduce water use. New drip irrigation systems apply water directly to crop plants, instead of flooding or spraying the whole field with water (Figure F.10). This method allows less water to be used; it may also provide a secondary benefit of helping to control weeds, since the weeds have access to less water.

Protecting the Global Environment

Worldwide, the productivity of our farmland has dropped because of environmental problems such as increased soil erosion, decreased availability of water, and increased pollution. In addition, agriculture faces threats when wild species are harmed; the negative effects on agriculture range from the loss of insect pollinators (which are essential for the production of billions of dollars of crops each year) to the inability to harvest useful genes from species we drive extinct.

Farming involves a large number of activities (Figure F.11), and some of the environmental problems that agriculture faces are caused directly by those activities. For example, from late fall to early spring, farm fields often have few plants growing on them, leading to increased soil erosion from both wind and water. Farming also causes soil fertility to decline over time because nutrients are removed from the soil by crop plants, then shipped to other areas when the crops are harvested, not returned to the soil when the plants die and decompose. The decline of soil fertility makes it necessary to use fertilizer, which, along with the use of pesticides and energy, pollutes our soils, air, and water. As a final example, irrigation can cause soils to become more salty. When irrigation water evaporates, it leaves behind concentrated salts, eventually making the soil useless for farming.

In some cases, agricultural practices can be changed to lessen their impact on the environment. As we've seen, new methods are under development to reduce the use of fertilizer and pesticides and use water more efficiently. But farms themselves are by no means the only source of environmental problems that can make crops hard to grow. To ensure that farms continue to have high yields, we must transform human societies so that the sum of all of our activities does not cause long-term environmental damage.

As part of this effort, we must stabilize the human population size. Equally important, we must stop using resources more rapidly than they are renewed (see Interlude G for some ways to achieve this goal). If we do not accomplish these two goals, agriculture will be in the hopeless position of trying to feed more people at a time when farms are less able to produce food because of ever-increasing damage to the environment.

Can We End Hunger in Our Lifetime?

Recent economic analyses indicate that the answer to this question is yes. By 2050, assuming that food production continues to grow and that environmental problems are brought under control, these analyses suggest that hunger can be reduced dramatically from current levels. Results from such analyses are encouraging, but they are also sobering. Ending hunger will require that we make many changes—business as usual will not work. We will need to control human population growth, improve the distribution of food, make investments that favor poor people, alter trade policies, increase the conservation of water, increase food production, and reduce human impacts on the environment. It is within our power to make all of these changes. Will we do so?

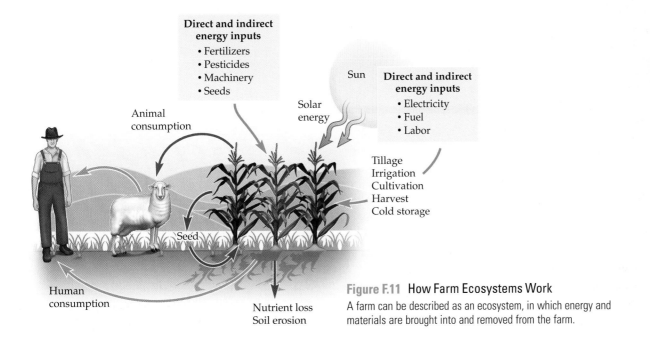

Figure F.11 How Farm Ecosystems Work

A farm can be described as an ecosystem, in which energy and materials are brought into and removed from the farm.

◉ Review and Discussion

1. The demand for animal protein is increasing even faster than the human population size. Traditional sources of animal protein, such as rangelands and oceanic fisheries, are not keeping pace with this demand. Do you support the use of managed systems as a way to increase the global production of animal protein? What are the pros and cons of your position?

2. Assume that agriculture begins to rely much more heavily on organic farming methods than it does at present. Would such a change improve or hinder efforts to end world hunger? Why do you think so?

3. Explain what the Green Revolution was, and what it did—and did not—achieve.

4. Describe the effects of pollution, threats to wild species, and other global environmental problems on efforts to end world hunger.

5. This essay discusses a number ways to increase crop production. Select the three methods you think are the most promising. Explain why you selected these methods, and describe how implementation of these methods would help to feed our growing population.

Key Terms

aquaculture (p. F5)
Bt crops (p. F8)
cropland (p. F3)
genetic engineering (p. F8)
Green Revolution (p. F6)
hybrid (p. F8)

oceanic fishery (p. F3)
organic farming (p. F11)
plant breeding (p. F8)
precision farming (p. F11)
rangeland (p. F3)
yield (p. F7)

CHAPTER 33 The Biosphere

Key Concepts

◉ Ecology is the study of interactions between organisms and their environment. All ecological interactions occur in the biosphere, which consists of all living organisms on Earth together with the environments in which they live.

◉ Climate has a major effect on the biosphere. Climate is determined by incoming solar radiation, global movements of air and water, and major features of Earth's surface.

◉ The biosphere can be divided into major terrestrial and aquatic life zones called biomes.

◉ Terrestrial biomes cover large geographic regions and are usually named for the dominant plants that live there. The locations of terrestrial biomes are determined by climate and by the actions of humans.

◉ Aquatic biomes are usually characterized by physical conditions of the environment. Aquatic biomes are heavily influenced by the surrounding terrestrial biomes, by climate, and by human activity.

◉ Because components of the biosphere depend on one another in complex ways, human actions that affect the biosphere can have unexpected side effects.

Main Message: Organisms interact with one another and their physical environment, forming a web of interconnected relationships that is heavily influenced by climate.

The Tapestry of Life

Take a walk outside, find a place where plants grow, and look around. If it is winter and cold where you live, at first glance you may not see much. But if you look carefully, you will find hidden signs of life everywhere. These signs lurk behind the bark of trees, in footprints left in snow, under rocks, locked within the seeds and buds of plants waiting to burst forth come spring. If you walk outside in spring, summer, or fall, the wild riot of life is plain to see: it floats or buzzes by you in the air, crawls around your feet, spreads its leaves over you, and scampers nearby. Such is the stuff of which the tapestry of life is made.

The rich abundance of life that surrounds you is but a small fraction of the biosphere: all organisms on Earth plus their environments (see Chapter 1). The biosphere includes grasslands, deserts, tropical rainforests, streams, and lakes, to name just a few of its many parts. It also includes the bottoms of our feet and the bacteria that grow there, as well as hydrothermal vents on the bottom of the ocean and the bacteria that grow there.

The biosphere is very complex. To put this statement in perspective, consider the challenge researchers faced in trying to understand how cells store and copy the billions of bits of genetic information contained in your DNA (see Chapter 12). Although challenging, problems such as understanding how DNA is copied pale before the difficulty of understanding the biosphere. While an organism may have billions of bases of DNA, there are many trillions of individual organisms on Earth, and they interact with one another and their environment in a vast number of ways.

The biosphere is not only complex, but also extremely important to us—our very lives depend on it. Interactions between organisms and their environment affect the quality of the air we breathe, the water we drink, the food

A Troublesome Invader

The introduction of the zebra mussel from Eurasia to North America, probably in ship ballast water, has caused a number of problems; in this picture they are being removed after clogging a power station.

we eat. And people are not just passive cogs in the wheel of life. Human actions have an enormous impact on the biosphere, sometimes producing surprising effects. Consider the seemingly harmless practice in which ships take up water in one port—to provide ballast (weight) to stabilize the ship's motion—and discharge it in another. Each time this is done, hundreds of aquatic species are moved from one part of the world to another. This is probably how the zebra mussel got from Eurasia to North America, where it now clogs waterways, fouls pipes and boat bottoms, outcompetes some native species, and causes billions of dollars of economic damage.

When people began the practice of taking up ballast water in one port and discharging it in another, no one suspected that doing so would affect native species and cause economic harm. Zebra mussels, for example, were not a major problem in Eurasia. Why does the zebra mussel cause more trouble in North America than in its original home? And how can we prevent this and other similar problems in the future? To begin to answer questions like these, we must understand how organisms interact with one another and with their environment.

A view of Earth from space highlights the beauty and fragility of the **biosphere**, which consists of all organisms on Earth together with the physical environments in which they live. Since the biosphere supplies us with food and raw materials, we depend on the biosphere for all aspects of human society. In Unit 7 we discuss ecology, the branch of science devoted to understanding how the biosphere works.

Ecology can be defined as the scientific study of interactions between organisms and their environment, where the environment of an organism includes both biotic (other organisms) and abiotic (nonliving) factors. Thus ecologists are interested in how the two parts of the biosphere—organisms and the environments in which they live—interact with and affect each other. In the chapters of this unit, our study of ecology will cover several levels of the biological hierarchy (see Figure 1.11): individual organisms, populations, communities, ecosystems, and the biosphere.

All ecological interactions, at whatever level they occur, take place in the biosphere, the focus of this chapter. We begin by discussing why ecology is important and what types of information ecologists must have to understand how the biosphere works. We go on to discuss climate and other factors that shape the biosphere. Then we take a brief look at the variety of terrestrial and aquatic biomes that those factors give rise to.

33.1 Why Is Ecology Important?

The science of ecology helps people to understand the natural world in which we live. Such an understanding is becoming increasingly important because we are changing the biosphere in ways that can be expensive—and in some cases, difficult or even impossible—to fix. Consider species such as zebra mussels that are accidentally or deliberately brought by people to new geographic regions (**introduced species**). Unfortunately, zebra mussels are not the only introduced species causing trouble. In the United States, for example, people have introduced thousands of other species, some of which have become pests that collectively cause an estimated $120 billion in economic losses each year, a huge cost similar to those due to smoking ($150 billion per year) and obesity ($90 billion per year). By studying the ecology of introduced species, we can understand how people help them to spread to new regions, why they increase dramatically in abundance, how they affect natural communities, and how they cause economic disruption—all of which can be helpful in limiting the damage these species cause.

As another example, consider chlorofluorocarbons (CFCs), which are synthetic chemicals used as refrigerants, in aerosol sprays, and in foam manufacture. These chemicals have created a hole in the ozone layer of the atmosphere over Antarctica (Figure 33.1). Ecologists seek to understand both how the ozone hole formed and what its consequences are for life on Earth. Ozone in the atmosphere prevents much of the sun's ultraviolet (UV) light from reaching Earth, thus protecting organisms from DNA mutations caused by exposure to UV light. Scientists have discovered that decreases in the ozone layer have many consequences, including more skin cancers in people, reduced yields for some crops, and reduced populations of phytoplankton, the small photosynthetic aquatic organisms on which all other aquatic organisms

Biology Matters

The Costs of Introduced Species

Most of the estimated 50,000 introduced species in the United States cause few problems, and we get enormous benefits from many of them. For example, more than 98 percent of the food produced in the country comes from introduced crops, such as rice, wheat, and corn, and introduced livestock, such as cows and poultry. But it does not take many unwelcome arrivals to cause huge economic losses. For example, the Asian tiger mosquito, introduced from Japan in the mid-1980s and now spreading in many regions, attacks more hosts than any other mosquito in the world, including many mammals, birds, and reptiles. It can thus transmit diseases such as encephalitis and yellow fever from one species to another, including humans.

Here we list 20 types of introduced species that collectively cost the U.S. economy an estimated $114 billion each year. The total cost of the damage caused by all introduced species is not known precisely, but may be as high as $150 billion each year.

Introduced species	Cost (in US $ per year)
Weeds in crops	$29,000,000,000
Diseases in crops	$23,500,000,000
Rats	$19,000,000,000
Insects in crops	$14,500,000,000
Weeds in gardens	$6,500,000,000
Organisms that cause or carry human diseases	$6,500,000,000
Plant diseases in gardens	$3,000,000,000
Zebra mussels	$3,000,000,000
Insects in gardens	$2,500,000,000
Insects in forests	$2,100,000,000
Birds	$2,100,000,000
Asiatic clam	$1,000,000,000
Fish	$1,000,000,000
Purple loosestrife and other plants	$250,000,000
Feral pigs	$200,000,000
Elm disease	$100,000,000
Mongoose	$50,000,000
Green crab	$44,000,000
Gypsy moth	$22,000,000
Fire ants	$10,000,000

(a) September 1979

(b) September 2000

Figure 33.1 The Antarctic Ozone Hole

These satellite images show average ozone levels over Antarctica for the months of (a) September 1979 and (b) September 2000. Ozone levels declined slowly in the 1970s, then dropped dramatically in the 1980s. Thus the September 1979 image represents near-normal conditions. In the September 2000 image, regions with the greatest ozone loss appear purple. Although average ozone levels have been declining, the size of the ozone hole in any year is strongly influenced by weather conditions. At the time this book went to press, the largest ozone hole ever recorded occurred on September 16, 2000, and covered an area larger than all of North America.

depend for food. Decreases in phytoplankton can result in decreased fish populations, which in turn lead to decreased catches by people and hence economic losses.

As the examples of introduced species and the ozone hole illustrate, the actions people take can affect natural systems, which in turn affect us. Ecologists study how natural systems work, and they use what they learn to document how we are affecting life on Earth and to predict the consequences of our actions. As such, ecology is an important area of applied biology because it can help us fix current environmental problems and prevent future ones.

33.2 Interactions with the Environment

Dingo

All organisms interact with their environment. These interactions go both ways: organisms affect their environment (as when a beaver builds a dam that blocks the flow of a stream and creates a pond or lake), and the environment affects organisms (as when an extended drought limits the growth of plant species on which the beaver depends for food). Because the interactions go both ways, the organisms and physical environments of the biosphere can be thought of as forming a web of interconnected relationships.

Thinking of the biosphere this way can help us gain insight into the natural world. For example, plants have simple food requirements, one of which is nitrogen. Nitrogen is often in short supply in natural communities;

hence we might think that soil high in nitrogen would necessarily be good for plants. The actual situation is more complex.

First, it matters which molecular form of nitrogen is present. Plants are able to use nitrogen in only two forms: ammonium (NH_4^+) and nitrate (NO_3^-). The extent to which these forms of nitrogen are available to plants depends on the activities of certain bacteria found in the soil. If soil conditions change so as to inhibit these bacteria, plants will grow poorly even when there are large amounts of nitrogen in the soil. In addition, large amounts of nitrogen do not benefit all plant species equally. As a result, when extra nitrogen becomes available to plants, those species that are best able to use the extra nitrogen may drive other species extinct. Thus the effect of the amount of nitrogen in the soil depends on how the form of nitrogen, soil conditions, soil bacteria, and plant species affect one another.

Thinking of the biosphere as an interconnected web can also help us to understand the consequences of human actions. Consider what happened when people used fencing, poison, and hunting to remove dingoes from a large region of rangelands in Australia. Dingoes, a type of wild dog, were removed to prevent them from eating sheep. As shown in Figure 33.2, in areas where dingoes were removed, the population of their preferred prey, red kangaroos, increased dramatically (166 times). Kangaroos decrease the food available to sheep because kangaroos and sheep like to eat the same plants. In addition, in times of drought, kangaroos resort to a behavior not found in sheep: they dig up and eat belowground plant parts. This behavior has the potential to change the numbers and

Figure 33.2
An Explosion in Numbers
Red kangaroo numbers increased 166 times when their dingo predators were removed from the rangelands located south of the world's longest fence (shown in red).

types of plant species found in rangelands, thus further increasing the effect kangaroos have on sheep.

When people removed dingoes, the subsequent effects were not what they expected or desired because red kangaroos outcompeted their sheep for food. With the advantage of hindsight, the negative side effects of removing dingoes seem predictable, because changes that affect one part of the biosphere (such as removal of dingoes) can have a ripple effect to produce changes elsewhere (such as increases in red kangaroos and decreased food available to sheep).

The two examples given in this section illustrate how natural systems can be viewed as an interconnected web. To understand such connections further, we need to learn more about the physical factors that affect the biosphere.

33.3 Climate Has a Large Effect on the Biosphere

Weather refers to the temperature, precipitation (rainfall and snowfall), wind speed and direction, humidity, cloud cover, and other physical conditions of Earth's lower atmosphere at a specific place over a short period of time. **Climate** is more inclusive in area and time frame: it refers to the prevailing weather conditions experienced in a region over relatively long periods of time (30 years or more). Weather, as we all know, changes quickly and is hard to predict, whereas climate is more predictable.

Climate has major effects on ecological interactions because organisms are more strongly influenced by climate than by any other feature of their environment. On land, for example, whether a particular region is desert, grassland, or tropical rainforest depends primarily on such features of climate as temperature and precipitation. Let's look at the factors that determine the climate of a given geographic region.

Incoming solar radiation shapes climate

Sunlight strikes Earth directly at the equator, but at a slanted angle near the North and South poles. This difference causes more solar energy to reach Earth's surface near the tropics than near the poles. As a result, tropical regions receive 2.5 times the solar radiation that reaches polar regions, making them much warmer than the poles. Tropical regions also show low seasonal fluctuations in temperature, so organisms that live there experience a relatively warm, stable climate throughout the year.

Global movements of air and water shape climate

Near the equator, intense sunlight heats moist air, causing the air to rise from the surface of Earth. Warm air rises because heat causes it to expand and therefore to be less dense, or lighter, than air that has not been heated. The warm, moist air cools as it rises. As a result, rain falls, because cool air cannot hold as much water as warm air can (Figure 33.3).

Figure 33.3 Earth Has Four Giant Convection Cells
Two giant convection cells are located in the Northern Hemisphere and two in the Southern Hemisphere. In each convection cell, relatively warm, moist air rises, cools, and then releases moisture as rain or snow. The cool, dry air then sinks to Earth and flows back toward the region where the warm air is rising. In temperate regions (30° to 60° latitude), the airflow pattern is less predictable.

Learn more about these convection cells. 33.1

Usually, cool air sinks. However, the cool air above the equator cannot sink immediately because of the warm air that is rising beneath it. Instead, the cool air moves to the north and south, tending to sink back to Earth at about 30° latitude. The cool air warms as it descends, which allows it to hold more water. As the air flows back toward the equator, it absorbs moisture from Earth's surface. By the time it reaches the equator, the air is once more warm and moist, so it rises, repeating the cycle.

Earth has four giant **convection cells** in which warm, moist air rises and cool, dry air sinks (see Figure 33.3). Two of the four convection cells are located in tropical regions and two in polar regions, where they generate relatively consistent wind patterns. In temperate regions (roughly 30° to 60° latitude), winds are more variable, and there are no stable convection cells. The variable winds form when cool, dry air from polar regions collides with warm, moist air moving north from the tropics.

The winds produced by the four giant convection cells do not move straight north or straight south. Instead, Earth's rotation causes these winds to curve as they travel near Earth's surface (Figure 33.4). Winds that travel toward the equator, for example, curve to the west. When winds curve to the west, they blow from the east; hence such winds are called easterlies ("from the east"). Similarly, winds that travel toward the poles blow from the

west and are called westerlies. Thus, at any given location, the winds usually blow from a consistent direction; these patterns of air movement are known as prevailing winds. In southern Canada and much of the United States, for example, winds blow mostly from the west; thus storms in these regions usually move from west to east.

Ocean currents also have major effects on climate. The rotation of Earth, differences in water temperatures between the poles and the tropics, and the directions of prevailing winds all contribute to the formation of ocean currents. In the Northern Hemisphere, ocean currents tend to run clockwise between the continents; in the Southern Hemisphere, they tend to run counterclockwise (Figure 33.5).

Ocean currents carry a huge amount of water and can have a great influence on regional climates. The Gulf Stream, for example, moves 25 times the amount of water carried by all the world's rivers combined. Without the warming effect of the water carried by this current, the climate in countries such as Great Britain and Norway would be subarctic to arctic instead of temperate. Overall, the Gulf Stream causes cities in Europe to be much warmer than cities at similar latitudes in North America, as illustrated by comparing Rome with Boston, Paris with Montréal, and Stockholm with Fort-Chimo (a town of 1,400 people in Québec, Canada).

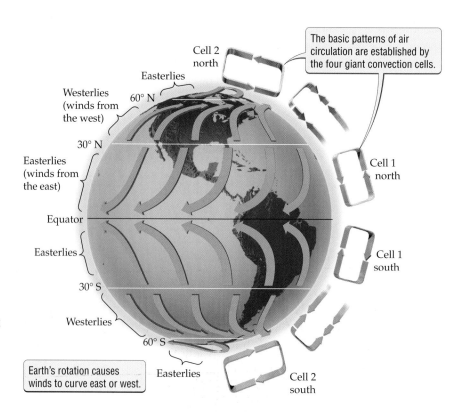

Figure 33.4 Global Patterns of Air Circulation
The four giant convection cells determine the basic pattern of air circulation on Earth. Earth's rotation causes winds to curve to the east or west. The direction in which they curve depends on their latitude, but for any given geographic region on Earth, the winds usually blow from a consistent direction. In temperate regions, in which there are no stable convection cells, the winds are more variable, but often blow in the directions shown.

In the Northern Hemisphere, most
ocean currents run clockwise.

In the Southern Hemisphere, most
ocean currents run counterclockwise.

Figure 33.5 The World's Major Ocean Currents

Ocean currents can be cold (blue) or warm (red), depending on a combination of factors, including water depth and latitude.

The major features of Earth's surface also shape climate

Heat is absorbed and released more slowly by water than by land. Therefore, because they retain heat comparatively well, oceans and large lakes moderate the climate of the surrounding lands. Mountains can also have a large effect on a region's climate. For example, mountains often cause a **rain shadow** effect, in which little precipitation falls on the side of the mountain that faces away from the prevailing winds (Figure 33.6). In the Sierra Nevada of North America, five times as much precipitation falls on the west side of the mountains (which faces toward winds that blow in from the ocean) as on the east side, where the lack of precipitation contributes to the formation of deserts. Mountain ranges in northern Mexico, South America, Asia, and Europe also create rain shadows.

33.4 Life on Land

Now that we've discussed factors that influence climate, let's explore the effects of climate on the biosphere in more detail.

The biosphere can be divided into biomes

The biosphere can be divided into several major terrestrial and aquatic zones of life, called **biomes**. Biomes on land, such as grasslands and tropical forests, cover large geographic regions and are usually named after the dominant vegetation of the region. Earth's terrestrial environments are usually divided into the following seven major biomes: **tropical forest**, **temperate forest**, **grassland**, **chaparral**, **desert**, **boreal forest** (from *borealis*, "northern"), and **tundra** (which includes the tops of some

Learn more about a rain shadow.

33.2

Figure 33.6 The Rain Shadow Effect
The side of a high mountain that faces the prevailing winds (the windward side) receives more precipitation than the side of the mountain that faces away from the prevailing winds (the leeward side). The leeward side is thus said to be in a rain shadow.

On the windward side of the mountain, air rises and cools. Because cool air holds less water than warm air, rain or snow falls.

Prevailing winds pick up moisture from bodies of water.

Mountain range

On the leeward side of the mountain, air descends and warms, producing little rain or snow.

Rain shadow area

Ocean

high mountains) (Figure 33.7). Figure 33.8 illustrates each of these biomes with a representative photograph.

Keep in mind, however, that maps and photographs of a biome can give the misleading impression that the entire biome is the same. To produce a worldwide map of terrestrial biomes like the one shown in Figure 33.7, regions that actually are very different must be lumped together into a single biome. If this were not done, the map would be so complicated that it would not be very helpful.

Consider grasslands. The grassland biome includes arid grasslands (characterized by short, drought-resistant grasses) in southwestern North America, tallgrass prairie (characterized by tall grasses and many wildflowers) in north central North America, and savanna (grasslands

Identify the major terrestrial biomes.

33.3

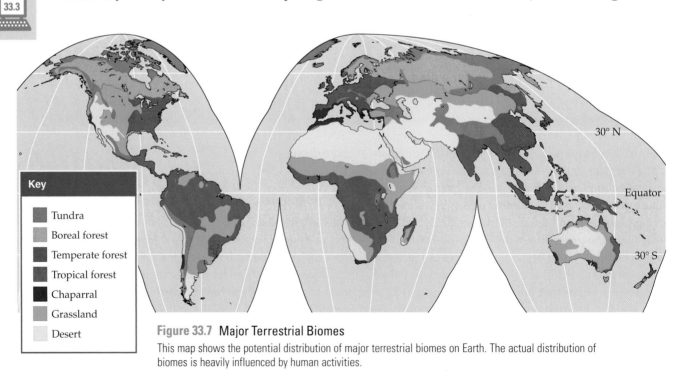

Key

Tundra

Boreal forest

Temperate forest

Tropical forest

Chaparral

Grassland

Desert

30° N

Equator

30° S

Figure 33.7 Major Terrestrial Biomes
This map shows the potential distribution of major terrestrial biomes on Earth. The actual distribution of biomes is heavily influenced by human activities.

(a) Tropical forest

(b) Temperate forest

(c) Grassland

(d) Chaparral

(e) Desert

(f) Boreal forest

(g) Tundra

Figure 33.8 Terrestrial Biomes

(a) Tropical forests form in warm, rainy regions and are dominated by a rich diversity of trees, vines, and shrubs. (b) Temperate forests are dominated by trees and shrubs that grow in regions with cold winters and moist, warm summers. (c) Grasslands are common throughout the world and are dominated by grasses and many different types of wildflowers. They often occur in relatively dry regions with cold winters and hot summers. (d) Chaparral is characterized by shrubs and small, nonwoody plants that grow in regions with mild summers and winters and low to moderate amounts of precipitation. (e) Deserts form in regions with low precipitation, usually 25 centimeters per year or less. (f) Boreal forests are dominated by coniferous trees that grow in northern or high-altitude regions with cold, dry winters and mild, humid summers. (g) Tundra is found at high latitudes and high elevations and is dominated by low-growing shrubs and nonwoody plants that can tolerate extreme cold.

dotted with occasional trees) in Africa (Figure 33.9). What is true of grasslands is true of all biomes: both the species found in the biome and the physical conditions of the biome can vary greatly from place to place.

The location of terrestrial biomes is determined by climate and human actions

Climate is the single most important factor controlling the potential (natural) location of terrestrial biomes. The climate of a region—most importantly, the temperature and the amount and timing of precipitation—allows some species to thrive and prevents other species from living there. Overall, the effects of temperature and moisture on different species cause particular biomes to be found under a consistent set of conditions (Figure 33.10). Biomes change as one moves from the equator to the poles, and also as one moves from the bottom to the top of mountains.

Climate can exclude species from a region directly or indirectly. Species that cannot tolerate the climate of a region are directly excluded from that region. Species that can tolerate the climate but are outperformed by other organisms that are better adapted to that climate are excluded indirectly.

Although climate places limits on where biomes can be found, the actual extent and distribution of biomes in the world today are very strongly influenced by human activities (see the box on page 657). We will return to the effects of humans on natural biomes when we discuss global change in Chapter 38.

33.5 Life in Water

Life evolved in water billions of years ago, and aquatic ecosystems cover about 75 percent of Earth's surface.

Arid grassland

Tallgrass prairie

 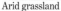
Test your knowledge of biomes and climate zones.
33.4

Figure 33.9
Diversity Within the Grassland Biome
As is true of all biomes, the species found in grassland biomes and the physical conditions of the grassland can vary greatly from place to place.

A savanna in Kenya

There are eight major aquatic biomes: **river**, **lake**, **wetland**, **estuary**, **intertidal zone**, **coral reef**, **open ocean**, and **benthic zone** (Figure 33.11). Unlike terrestrial biomes, aquatic biomes are usually characterized by physical conditions of the environment, such as salt content, water temperature, water depth, and the speed of water flow.

As with terrestrial biomes, Figure 33.11 shows only a small portion of the diversity of aquatic biomes. Lake biomes, for example, include bodies of water that range in size from small ponds to lakes covering thousands of square kilometers. In addition, very different species can be found in two parts or types of a given aquatic biome. For example, algae may thrive in a lake that contains large amounts of nitrogen and phosphorus. When the algae die, their remains are consumed by bacteria. As the bacteria grow and reproduce, they may use up so much of the oxygen dissolved in the water that fish in the lake die. In contrast, a lake containing little nitrogen and phosphorus may have few algae, few bacteria, high oxygen levels, and many species of fish. Thus, although both areas are lakes, the species that live in them differ tremendously.

Figure 33.10 The Location of Terrestrial Biomes Depends on Temperature and Precipitation

Aquatic biomes are influenced by terrestrial biomes and climate

Aquatic biomes, especially lakes, rivers, wetlands, and the coastal portions of marine biomes, are heavily influenced by the terrestrial biomes that they border or through which their water flows. High and low points of the land, for example, determine the locations of lakes and the speed and direction of water flow. In addition, when water drains from a terrestrial biome into an aquatic biome, it brings with it dissolved nutrients (such as nitrogen, phosphorus, and salts) that were part of the terrestrial biome. Because nutrients are available only in low amounts in many aquatic biomes, nutrients imported from the surrounding terrestrial biome can have a significant effect. For example, rivers and streams carry nutrients from terrestrial environments to the ocean, where they may stimulate large increases in the abundance of phytoplankton.

Aquatic biomes also are strongly influenced by climate. In temperate regions, for example, seasonal changes in temperature cause the oxygen-rich water near the top of a lake to sink in the fall and the spring, bringing oxygen to the bottom of the lake. In tropical regions, seasonal differences in temperature are not great enough to cause a similar mixing of water. This lack of mixing causes the

deep waters of tropical lakes to have low oxygen levels and relatively few forms of life.

Climate also helps to determine the temperature, level, and salt content of the world's oceans. Such physical conditions of the ocean have dramatic effects on the organisms that live there, and hence climate has a powerful effect on marine life. Consider the El Niño [ell-*NEEN*-yoh] events that are often reported in the news. These events begin when warm waters from the west deflect the cold Peru Current along the Pacific coast of South America (Figure 33.12). The results of this change are spectacular, including dramatic decreases in numbers of fish, die-offs of seabirds, reduced catches of fish by people, storms along the Pacific coast of North America that destroy underwater "forests" dominated by long strands of brown algae called kelp, crop failures in Africa and Australia, and drops in sea level in the western Pacific that kill huge numbers of coral reef animals.

Aquatic biomes are also influenced by human activity

Like terrestrial biomes, aquatic biomes are strongly influenced by the actions of humans. Portions of some aquatic

Figure 33.11
Aquatic Biomes

(*a*) Rivers are relatively narrow bodies of fresh water that move continuously in a single direction. (*b*) Lakes are standing bodies of fresh water of variable size, ranging from a few square meters to thousands of square kilometers. (*c*) Wetlands are characterized by shallow waters that flow slowly over lands that border rivers, lakes, or ocean waters. (*d*) Estuaries are tidal ecosystems where rivers flow into the ocean. (*e*) Intertidal zones are found in coastal areas where the tides rise and fall on a daily basis, periodically submerging a portion of the shore. (*f*) Coral reefs form in warm, shallow waters located in the tropics. Corals are tiny animals that build up long-lasting structures on which many of the reef's other organisms depend. (*g*) Open oceans cover the majority of Earth's surface. They include a shallow layer (100 to 200 meters deep) in which photosynthesis can occur as well as deeper ocean waters that little light can penetrate. (*h*) Benthic zones, located on the bottom surfaces of rivers, lakes, wetlands, estuaries, and oceans, are home to a wide variety of organisms.

(*a*) River

(*b*) Lake

(*c*) Wetland

(*d*) Estuary

(*g*) Open ocean

(*e*) Intertidal zone

(*h*) Benthic zone

(*f*) Coral reef

A deep-ocean-water fish

Science Toolkit

The Interplay of Science, Philosophy, and Ethics

We usually think of scientific knowledge as being gained by using the scientific method. As we learned in Chapter 1, the scientific method is a four-step process in which scientists (1) observe nature and ask a question about those observations; (2) use previous knowledge or intuition to develop one or more hypotheses or possible answers to that question; (3) use these hypotheses to make predictions that can be tested; and (4) perform experiments or gather carefully selected observations in order to evaluate competing hypotheses. Scientists then use the results of those experiments or observations to modify their hypotheses, to pose new questions, or to draw conclusions about the natural world. This process can continue indefinitely: new observations lead to new questions, which stimulate scientists to formulate and test new ideas about how nature works.

But scientists do not operate in a vacuum. Every scientist brings his or her individual view of the world to work. A scientist's personal background, philosophy of life, and ethical views can influence the questions about nature he or she finds interesting or important enough to study. And once a question is posed, viewpoint and background can also affect which of several possible answers will be most appealing to the researcher. Thus science, philosophy, and ethics can be inextricably intertwined.

This interplay between science, philosophy, and ethics can be illustrated by considering some issues surrounding the following question, which is related to the ideas presented in this chapter: What is a natural biome? To answer this question, first locate the region where you live in Figure 33.7. Does the biome shown on the map match your actual surroundings? For many of you, the answer is no. Figure 33.7 shows *potential* biomes—the vegetation types that could thrive given the climate of each particular region if humans were not present. But in reality, people have converted major portions of these potential biomes to urban (housing and industry) and agricultural areas. Vast regions of North America no longer have their original biomes, as illustrated by the almost complete conversion of the original grasslands (prairies) to urban and agricultural areas.

In some parts of the world, humans have altered the landscape so thoroughly, and over such long periods of time, that we now consider the altered landscape to be the natural one. Consider the moors of England and Scotland, made famous by Sherlock Holmes stories and travel advertisements. In the biological hierarchy, the moors would be considered a community. Although many people view the moors as a

Is This Scottish Moor a "Natural" Community?

beautiful natural landscape, regions now covered with moors once were covered with oak woodlands. In fact, the moors are completely dependent on humans for their existence: they were produced when people cut the trees down and then used the land for grazing.

Moors are just one example of a common phenomenon: in many instances, people's actions have created or are maintaining landscapes that we now find beautiful and consider natural. Are landscapes that are dependent on the actions of people natural? What exactly is "nature"? Is it a pristine state that does not include people? Or does "nature" include—at least to some degree—people and our effects on the world? And if people are part of nature, does that mean that anything we do is by definition natural?

There are many possible answers to these questions, depending on your philosophical and ethical views about such issues as whether people have an ethical obligation not to harm other species, including species that are of no practical value to us. In turn, how the questions posed in the previous paragraph are answered can influence important decisions about whether and how to regulate the human impact on the environment. So it is with many issues in ecology, as well as in other branches of biology—indeed, in all of science. What a person views as good or interesting or important depends on how he or she views the world; this simple fact applies to everyone, including scientists.

As a group, scientists are well aware that personal and cultural biases can influence how science is done. Although it is impossible to be completely unbiased, scientists can and do try to remove as much bias as they can. Individual scientists do this by taking great care to examine their assumptions and logic. The effort to remove bias also occurs at the level of the community of scientists who work in a particular field. This "community check" on bias is accomplished in part by peer review of scientific papers before they are published (a paper will be rejected for publication if it is found by other scientists to be biased) and by the esteem with which the scientists in a field hold a paper after it is published. Scientists also frequently perform experiments designed to test conclusions reached by other scientists. If those conclusions do not hold up in a series of repeated tests, those scientists may suggest that the conclusions of the original investigator may have been influenced by hidden biases. Individual scientists know that others in their field will judge their research, both before and after it is published; this knowledge provides a powerful motive to expunge bias.

Ordinarily, the cold Peru Current flows west near the equator, warming as it goes.

Pacific Ocean

Western equatorial Pacific

During an El Niño event, warm water flows from west to east and turns aside the Peru Current.

Weakened Peru Current

biomes, such as wetlands and estuaries, are often destroyed to allow for development projects. Rivers, wetlands, lakes, and coastal marine biomes are negatively affected by pollution in most parts of the world. Aquatic biomes also suffer when humans destroy or modify the terrestrial biomes in which they are situated. For example, when forests are cleared for timber or to make room for agriculture, the rate of soil erosion increases dramatically because trees are no longer there to hold the soil in place. Increased erosion can cause streams and rivers to become clogged with silt, which harms or kills invertebrates, fish, and many other species.

Figure 33.12 El Niño Events
During an El Niño event, winds from the west push warm surface water from the western Pacific to the eastern Pacific. The resulting changes in sea surface temperatures cause changes in ocean currents (shown here in blue for cold, red for warm). El Niño events cause many additional changes (not shown here), altering wind patterns, sea levels, and patterns of precipitation throughout the world.

The Human Impact on the Biosphere

The discharge of ships' ballast water opened the way for zebra mussels to colonize the waterways of North America. These mussels have harmed native species and caused serious economic problems, including the shutdown of a Ford Motor Company plant and the closure of a Michigan town's power plant.

Species like the zebra mussel that become major pests in a new environment are referred to as **invasive species**. Such species often cause much less trouble in their native environment. Zebra mussels, for example, come from Eurasia, where they are held in check by a wide range of predators and parasites. In North America, fewer predators and para-

sites attack zebra mussels, which allowed the numbers of these mussels to increase explosively after their arrival.

What is true of zebra mussels is true of other invasive species: when introduced by people to novel environments, the population size of a non-native species may increase greatly because its usual predators, parasites, and competitors are not present. When this occurs, the introduced species may harm the native species of its new home and use large amounts of resources or space. In Interlude A we saw how native species are harmed by invasive species such as the Nile perch in Lake Victoria and the brown tree snake on Guam. In cases in which an invasive species has a major and disruptive effect on its new environment, people can suffer great economic losses, as seen for the zebra mussel and for dozens of other troublesome invaders.

The effects of humans on the biosphere are not limited to introductions of invasive species. When we control the flow of rivers or eliminate (or even reduce the numbers of) a predator such as the dingo (see page 648), we also may cause outcomes we did not expect or want. Thinking of the biosphere as an interconnected web suggests that when we change one component of the biosphere, our actions may have unintended and undesirable effects on other parts of the biosphere. What's more, global patterns of air and water movement can cause actions in one area to have consequences that reach distant locations. This is especially evident when considering pollution: though emitted in one country or continent, it can cause problems in other countries or continents (for example, sulfur dioxide emitted in Great Britain can cause acid rain in Scandinavia, as we shall see in Chapter 37).

People can—indeed, to feed and house ourselves, we must—take actions that affect the natural world. Given the view of the biosphere as an interconnected web, what do you think people should do when we consider taking actions whose possible outcomes we either don't understand well or suspect may have harmful side effects? Should we let fear or lack of knowledge paralyze us into inaction? How can we use what we know about how the biosphere works to solve current environmental problems and prevent future ones? Although there are no easy answers to such questions, we'll examine how people have begun to address such issues at the close of this unit, in Interlude G.

Chapter Review

Summary

33.1 Why Is Ecology Important?

- Human actions affecting natural systems can produce unintended consequences that are expensive or impossible to remedy.
- Ecology is an important area of applied biology because an understanding of how natural systems work can help us to address current environmental problems and to predict the effects of future human actions on the biosphere.

33.2 Interactions with the Environment

- Organisms affect—and are affected by—their environment, which includes not only their physical surroundings, but also other organisms. As a result, a change that affects one organism can affect other organisms and the physical environment.
- It is helpful to think of organisms and their environment as forming a web of interconnected relationships.
- Because components of the biosphere depend on one another in complex ways, human actions that affect the biosphere can have surprising side effects.

33.3 Climate Has a Large Effect on the Biosphere

- Weather describes the physical conditions of Earth's lower atmosphere in a specific place over a short period of time. Climate describes a region's long-term weather conditions.
- Climate depends on incoming solar radiation. Tropical regions are much warmer than polar regions because sunlight strikes Earth directly at the equator, but at a slanted angle near the poles.
- Climate is strongly influenced by four giant convection cells that generate relatively consistent wind patterns over much of Earth.
- Ocean currents carry an enormous amount of water and can have a large effect on regional climates.
- Regional climates are also affected by major features of Earth's surface, as when mountains create rain shadows.

33.4 Life on Land

- There are seven major terrestrial biomes: tropical forest, temperate forest, grassland, chaparral, desert, boreal forest, and tundra.
- Climate is the most important factor controlling the potential (natural) location of terrestrial biomes. Climate can exclude a species

from a region directly (if it finds the climate intolerable) or indirectly (if it is outcompeted for resources by other species in the region).

- The actual location and extent of terrestrial biomes are heavily influenced by human activities.

33.5 Life in Water

- There are eight major aquatic biomes: river, lake, wetland, estuary, intertidal zone, coral reef, open ocean, and benthic zone.
- Aquatic biomes are usually characterized by physical conditions of the environment, such as temperature, salt content, and water movement.
- Aquatic biomes are strongly influenced by the surrounding terrestrial biomes, by climate, and by human actions.

◉ Review and Application of Key Concepts

1. This chapter suggests that the organisms and environments of the biosphere can be thought of as forming an "interconnected web." Why do ecologists think of the biosphere this way? Do you think it is a useful analogy?

2. Explain in your own words how global patterns of air and water movement can cause local events to have far-reaching ecological consequences. Give an example that shows how local ecological interactions can be altered by distant events.

3. List all 15 of the major terrestrial and aquatic biomes. How many of those biomes are located within 100 kilometers (about 60 miles) of your home?

4. Discuss the difference between potential and actual locations of biomes. What factors control the potential and actual locations of biomes?

5. Explain how climate can exclude species from a region in both terrestrial and aquatic biomes.

6. In human medicine, we do not use poorly understood drugs to treat disease, nor do we deliberately introduce pathogens into our bodies. Given what you've learned about the biosphere, consider whether you agree with the following statement, and evaluate its implications: We should treat the biosphere with the same respect that we treat our own bodies.

Key Terms

benthic zone (p. 655)
biome (p. 651)
biosphere (p. 646)
boreal forest (p. 651)
chaparral (p. 651)
climate (p. 649)
convection cell (p. 650)
coral reef (p. 655)
desert (p. 651)
ecology (p. 646)
estuary (p. 655)
grassland (p. 651)
intertidal zone (p. 655)
introduced species (p. 646)
invasive species (p. 658)
lake (p. 655)
open ocean (p. 655)
rain shadow (p. 651)
river (p. 655)
temperate forest (p. 651)
tropical forest (p. 651)
tundra (p. 651)
weather (p. 649)
wetland (p. 655)

Self-Quiz

1. Which of the following is *not* a level of the biological hierarchy commonly studied by ecologists?
   a. ecosystem
   b. individual
   c. organelle
   d. population

2. Earth has four stable regions ("cells") in which warm, moist air rises, and cool, dry air sinks back to the surface. Such cells are known as
   a. temperate cells.
   b. latitudinal cells.
   c. rain shadow cells.
   d. giant convection cells.

3. The biosphere consists of
   a. all organisms on Earth only.
   b. only the environments in which organisms live.
   c. all organisms on Earth and the environments in which they live.
   d. none of the above

4. What aspect(s) of climate most strongly influence the locations of terrestrial biomes?
   a. rain shadows
   b. temperature and precipitation
   c. only temperature
   d. only precipitation

5. Winds that blow from the west across warm waters in the Pacific Ocean become warm and moist. By analogy to what happens in a rain shadow, what do you think would happen if such warm, moist winds blew across the cold Peru Current (see Figure 33.5)?
   a. An El Niño event would occur.
   b. The winds would continue to pick up moisture from the ocean currents.
   c. The warm, moist winds would cool, causing rain to fall.
   d. The warm, moist winds would cool, but rain would not fall.

6. Which of the following is best defined as a major terrestrial or aquatic life zone?
   a. a population
   b. a community
   c. the biosphere
   d. a biome

Biology in the News

Cuban Tree Frogs Moving Toward South Carolina

Everything from fire ants and coyotes to armadillos, lionfish and beach kudzu have made new homes in South Carolina over the years. So can the Cuban tree frog be far behind?

The frogs, which are fat, toxic and voracious, have been spotted as far north as Savannah, GA. So it may not take long for the frogs to migrate across the border into South Carolina, scientists say.

This news article was based on the discovery in September 2004 of a single Cuban tree frog in the backyard pond of a home in Savannah, Georgia, where it was happily devouring the frogs and goldfish that lived in the pond. Although that particular Cuban tree frog is now dead, scientists fear that it is the harbinger of bad news: more frogs. Why such concern over one frog? In short, these tiny frogs pack a powerful ecological punch, and scientists are bracing themselves for ecological problems should the Cuban tree frog reach South Carolina.

As Steve Bennett, a biologist with the state Department of Natural Resources, puts it, "It may not be time for widespread panic, but it is certainly time to take notice." These frogs were first noticed in the United States about 80 years ago, when they were brought accidentally to the Florida Keys from Cuba, probably in vegetable crates. Since then, they have migrated north, decimating native species as they go. The invader attempts to eat anything that moves and fits into its mouth. It eats so many native frogs that it can take over their range. The frog also secretes a toxic chemical that covers its skin and causes painful stings if a person picks it up.

In South Carolina, millions of native tree frogs can be found along the coast. But an invader like the Cuban tree frog could reduce a native frog population from a million to none in a fairly short period—in only a few decades. Prior to the September 2004 observation in Georgia, the Cuban tree frog had never been seen north of Jacksonville, Florida. Scientists thought the invader was prevented from migrating north of Florida because it could not tolerate the colder winters. But little is known about how cold-tolerant the frogs actually are, and scientists now worry that the frog may be able to survive north of Florida after all. If the Cuban tree frog can tolerate the South Carolina winters, the native frogs of that state may be in for a nasty shock.

Evaluating the news

1. Some people in the United States keep Cuban tree frogs as pets. Do you think this is a good idea? Should it be illegal to import or own species, such as the Cuban tree frog, that can cause considerable ecological or economic damage?

2. Consider the following hypothetical situation: An introduced ornamental plant species escapes from cultivation and can grow in the wild. It harms native plant species—it outcompetes and eventually replaces them—but it causes little economic damage. Do you think such a species should be considered a problem species, one that we should spend money on controlling? Or should we spend money only on controlling introduced species that harm human economic interests?

3. As mentioned in this news article and discussed in the text, many introduced species cause problems, some costing society huge sums of money. Do you think governments should spend money to reduce the number of introduced species (for example, by inspecting vegetable crates more thoroughly) and to control those that are already here? If so, should this be a problem for states to handle, or should it be a federal responsibility? At either the state or federal level, would you be willing to pay for government efforts to control invasive species by having your taxes go up?

SOURCE: Associated Press, Wednesday, October 27, 2004.

CHAPTER 34

Growth of Populations

Key Concepts

- A population is a group of interacting individuals of a single species located within a particular area.

- Populations increase in size when birth and immigration rates exceed death and emigration rates, and they decrease when the reverse is true.

- A population that increases by a constant proportion from one generation to the next exhibits exponential growth.

- Eventually, the growth of all populations is limited by environmental factors such as lack of space, food shortages, predators, disease, and environmental deterioration.

- Different populations may exhibit different patterns of growth over time, including J-shaped curves (which result from exponential growth), S-shaped curves, population cycles, and irregular fluctuations.

- The world's human population is increasing exponentially. Rapid human population growth cannot continue indefinitely; either we will limit our own growth, or the environment will do it for us.

The Tragedy of Easter Island

Imagine standing at the edge of a cliff on Easter Island, looking into the long-abandoned quarry of Rano Raraku. Scattered about the grassy slopes of the quarry lie hundreds of huge, eerie statues carved from stone hundreds of years ago. The scene is beautiful, yet also ghostly and disturbing. Some of the statues stand upright but unfinished; they look almost as if the artists dropped their tools in mid-stroke. Others are complete, but lie fallen at odd angles. Hundreds more statues are scattered along the coast of Easter Island. Who carved these statues? Why were so many left unfinished? What happened to the people who made them?

The mystery deepens when we consider where Easter Island is and what it looks like today. Extremely isolated, the island is a small, barren grassland, just 166 square kilometers in area, with little water and little potential for agriculture. How could such a remote and forbidding place support a civilization capable of carving, moving, and maintaining these enormous stone statues?

The answers to these questions provide a sobering lesson for people today. Easter Island was not always a barren grassland; at one time most of the island was covered by forest. According to archaeological evidence, no humans lived on the island until about AD 400. At that time, about 50 Polynesians arrived in large canoes, bringing with them crops and animals with which to support themselves. These people developed a well-organized society capable of sophisticated technological feats, such as moving 15- to 20-ton stone statues long distances without the aid of wheels (they probably rolled the statues on logs).

Easter Island
Located 3,700 kilometers west of the coast of Chile and 4,200 kilometers east of the nearest major population center in Polynesia (Tahiti), Easter Island is small and extremely isolated.

Easter Island

By the year 1500 the population had grown to about 7,000 people. By this time, however, virtually all the trees on the island had been cut down to clear land for agriculture and (presumably) to provide the logs used to roll the statues from one place to another. The cutting of trees, along with other forms of environmental destruction, increased soil erosion and decreased crop production, leading to mass starvation.

With no large trees remaining on the island, the people could not build canoes to escape the ever-worsening conditions. The society collapsed, resorting to warfare, cannibalism, and living in caves for protection. The population also crashed; even 400 years later (in 1900) only 2,000 people remained on the island, less than one-third the number that had lived there in 1500.

What caused the events on Easter Island? In essence, the number of people and the patterns of resource use on the island increased above the level the land could support. And that's the scary part: many scientists now think that Earth's human population is using more resources than the planet can support. Does the story of Easter Island provide a preview of what will happen to the whole human race?

Recall from Chapter 33 that ecology is the study of interactions between organisms and their environment. Many ecologists are concerned with questions that relate to how many organisms live in a particular environment, and why. The answers to such questions not only provide insight into the natural world, but are also essential for the solution of real-world problems, such as the protection of rare species or the control of pest species. To set the stage for our study of the factors that influence how many individuals are in a population, we begin by defining what populations are. We then describe how populations grow over time, and we consider the limits to growth that are faced by all populations, including the human population.

34.1 What Are Populations?

A **population** is a group of interacting individuals of a single species located within a particular area. The human population of Easter Island, for example, consists of all the people who live on the island.

Ecologists usually describe the number of individuals in a population by the **population size** (the total number of individuals in the population) or by the **population density** (the number of individuals per unit of area). Population density can be calculated by dividing the population size by the total area. To illustrate, let's return to the Easter Island example: in the year 1500, the population size was 7,000 people. If we divide that number by 166 square kilometers (the size of the island), we get a population density of 42 people per square kilometer ($^{7,000}/_{166} = 42$). The density of people that lived on Easter Island in 1500 was thus higher than the 31.7 people per square kilometer that lived in the United States in 2005.

Easter Island is an easy example for determining what constitutes a population: islands have well-defined boundaries, and human individuals are relatively easy to count. But often it is more difficult to determine the size or density of a population. Suppose a farmer wants to know whether the aphid population that damages his or her crops is increasing or decreasing (Figure 34.1). Aphids are small and hard to count. More importantly, it is not obvious how the aphid population should be defined. What do we mean by "a particular area" in this case? Aphids can produce winged forms that can fly considerable distances, so should only the aphids in the farmer's field be counted? What about the aphids in the next field over?

In general, what constitutes a population often is not as clear-cut as in the Easter Island example. Overall, the area appropriate for defining a particular population depends on the questions being asked and on aspects of the biology of the organism of interest, such as how far and how rapidly the organism moves.

34.2 Changes in Population Size

All populations change in size over time—sometimes increasing, sometimes decreasing. In one year, abundant rainfall and plant growth may cause mouse populations to increase; in the next, drought and food shortages may

Figure 34.1
Populations of Aphids Can Cause Extensive Crop Damage
Aphids are small insects with sucking mouthparts. They are pests on many plant species, which they infest in such large numbers that they can be difficult to count.

Aphids insert their mouthparts into a plant and withdraw nutrients, thus damaging the plant.

Aphids have infested this rose in large numbers.

archs' overwintering sites were hit with an unusual storm that first drenched the butterflies with rain, then subjected them to freezing cold. The combination of wet and cold proved lethal: an estimated 70 to 80 percent of the butterflies—roughly 500 million of them—died overnight, the worst die-off in the last 25 years.

Fortunately for the butterflies, this huge increase in winter death rates was followed by an equally spectacular rise in birth rates during the summer of 2002. Monarch birth rates shot up that year because it turned out to be a great summer for the monarch's primary food plant,

Deer mouse

Figure 34.2 Before the Crash
Huge numbers of monarch butterflies overwinter in mountains west of Mexico City, then migrate each spring to eastern North America. In 2002, 70 to 80 percent of the overwintering butterflies died in an unusual winter storm. (Not shown are smaller populations of monarchs found west of the Rockies that overwinter along the Pacific coast of California.)

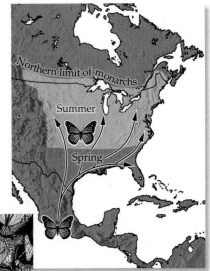

Spring migration

cause mouse populations to decrease dramatically. Such changes in the population sizes of other organisms can have important consequences for people. For example, an increase in the number of deer mice, carriers of hantavirus, is thought to have been responsible for the 1993 outbreak of this deadly disease in the southwestern United States.

Whether a population increases or decreases in size depends on the number of births and deaths in the population, as well as on the number of individuals that immigrate to (enter) or emigrate from (leave) the population. A population increases in size whenever the number of individuals entering the population (by birth and immigration) is greater than the number of individuals leaving the population (by death and emigration). We can express this in equation form as

birth + immigration > death + emigration.

The environment plays a key role in the increase or decrease of a population because birth, death, immigration, and emigration rates all depend on environmental factors. Consider how features of the environment caused monarch butterfly populations to fluctuate wildly in 2002. Each spring, monarchs make a spectacular migration that begins in the mountains west of Mexico City (where the butterflies overwinter) and ends in eastern North America (Figure 34.2). On January 13, 2002, the mon-

Overwintering monarchs

Helpful to know
Remembering the meanings of two common prefixes can help you distinguish three similar-sounding terms. *Migration* comes from the Latin meaning "to move." The prefix *in* or *im* can mean "into," while *ex* or *e* means "out of, away from." So *immigration* is movement into an area, and *emigration* is movement out of an area.

milkweed. Because of this chance good fortune, butterflies that survived the winter produced so many young in the summer of 2002 that monarch numbers rebounded to almost the historic average.

34.3 Exponential Growth

Like monarch butterflies, many organisms produce vast numbers of young. If even a small fraction of those young survive to reproduce, a population can grow extremely rapidly.

Exponential growth results in rapid population increases

An important type of rapid population growth is **exponential growth**, which occurs when a population increases by a constant proportion from one generation to the next (Figure 34.3). We can represent exponential growth from one year to the next by the equation

$$N_{\text{next year}} = \lambda \times N_{\text{this year}},$$

where N is the number of individuals in the population and λ (lambda) is the proportional increase in population

Biology Matters

How Fast Can Populations Grow?

To figure out how fast populations can grow, we turn to data that scientists have collected during periods when real populations grew exponentially. In exponential growth, the number of individuals in the population increases each year by a constant multiplier, the proportional increase (λ). But just how big is that constant multiplier?

Although the global human population is increasing rapidly in size, the multiplier (λ) for human populations is relatively small compared with that of other species. The multiplier for some bacterial populations is unimaginably large. Even the populations of relatively large multicellular animals, such as field voles, can increase rapidly in size. As a result, if exponential growth were to continue, the populations of some species could increase from two individuals to more than a billion individuals in a remarkably short period of time. What's more, physicists estimate that the universe contains a total of 10^{80} atoms. Assuming that exponential growth could continue indefinitely (even though we know it can't), how long would it take a population that began with two individuals to produce more individuals than there are atoms in the universe?

Species	Annual proportional increase (λ)	Time for population to grow from 2 to more than 1,000,000,000 individuals	Time required to produce more individuals than there are atoms in the universe
Humans (in 2003)	1.012	1,680 years	15,385 years
Wild ginger	1.1	211 years	1,926 years
Reindeer	1.2	110 years	1,007 years
Gray kangaroos	1.9	32 years	286 years
Field voles	24	7 years	58 years
Rice weevils	10^{17}	7 months	5 years
E. coli bacteria	10^{5274}	15 hours	6 days

size, a constant multiplier that determines the population size from one year to the next. For example, if $\lambda = 1.5$ and the current population size is 40, then the population size in the following year will be 60, and in successive years will be 90, 135, and so on.

In exponential growth, the proportional increase (λ) is constant, but the numerical increase—the number of individuals added to the population—becomes larger with each generation. For example, the population in Figure 34.3 doubles every generation (that is, $\lambda = 2$). With respect to its *numerical* increase, however, the population increases vary: the population increases by only 1 individual between generations 1 and 2, but by 16 individuals between generations 5 and 6. When plotted on a graph, exponential growth forms a **J-shaped curve**, as seen in Figure 34.3.

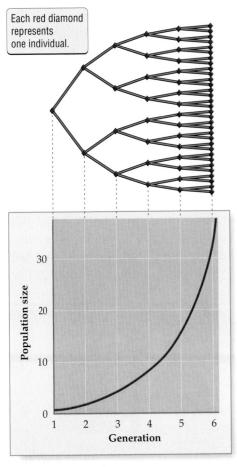

Figure 34.3 Exponential Growth
In this hypothetical population, each individual produces two offspring, so the population increases by a constant proportion with each generation (in this case, where $\lambda = 2$, the population doubles). The number of individuals added to the population increases each generation, resulting in the J-shaped curve that is characteristic of exponential growth. Exponential growth curves are always J-shaped, regardless of the value of λ, though different λ values will change the curve's steepness.

The time it takes a population to double in size—the **doubling time**—can be used as a measure of how fast the population is growing. We like it when our bank accounts double rapidly, but when populations grow exponentially in nature, problems eventually result, as we shall see in the following sections.

Exponential growth often occurs when a species moves into a new area

Populations can increase exponentially, at least initially, when they migrate to, or are introduced into, a new area. Consider the following tale of woe: In 1839, a rancher in Australia imported from South America a species of *Opuntia* (prickly pear cactus) and used it as a "living fence," since a thick wall of this cactus is nearly impossible for human or beast to cross. Unlike a real fence, however, the *Opuntia* cactus did not stay in one place: it spread rapidly throughout the landscape. As the cactus spread, whole fields were turned into "fence," crowding out cattle and destroying good rangeland.

In about 90 years, *Opuntia* cacti spread across eastern Australia, covering more than 243,000 square kilometers (over 60 million acres) and causing great economic damage. All attempts at control failed until 1925, when scientists introduced a moth species, appropriately named *Cactoblastis cactorum*, whose caterpillars feed on the growing tips of the cactus. This moth killed billions of cacti, which successfully brought the cactus population under control (Figure 34.4). The moths' success, however, also caused their own numbers to plummet due to lack of food; both the cactus and the moth still exist in eastern Australia in low numbers.

Overall, the *Opuntia* population in Australia increased exponentially at first, then declined even more rapidly after the introduction of the moth. Exponential growth has also been observed in other species introduced by people to new areas, as well as in species that have expanded naturally to new areas.

34.4 Limits to Population Growth

A giant puffball mushroom can produce up to 7 trillion offspring (Figure 34.5). If all of these offspring survived and reproduced at this same (maximal) rate, the descendants of a giant puffball would weigh more than Earth in just two generations. Humans and *Opuntia* cacti have

See a simulation of unrestrained population growth. 34.1

(a)

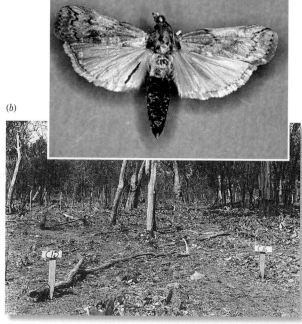

(b)

Figure 34.4 Blasting the Cactus

In Australia, the moth *Cactoblastis cactorum* was introduced in 1925 to halt the exponential growth of populations of an introduced cactus species. (*a*) Two months before release of the moth, *Opuntia* cacti were growing in a dense stand. (*b*) Three years after the introduction of the moth, the same stand had been almost completely eliminated.

much longer doubling times than giant puffballs, but given enough time, they, too, can produce an astonishing number of descendants. Obviously, however, Earth is not covered with giant puffballs, *Opuntia* cacti, or even humans. These examples illustrate an important general point: No population can increase in size indefinitely. Limits exist.

Growth is limited by essential resources and other environmental factors

The most obvious reason that populations cannot continue to increase indefinitely is simple: food and other resources become diminished. Imagine that a few bacteria are placed in a closed jar that contains a source of food. The bacteria absorb the food and then divide, and their offspring do the same. The population of bacteria grows exponentially, and in short order there are billions of bacteria in the jar. Eventually, however, the food runs out and metabolic wastes build up. All the bacteria die.

This example may seem extreme because it involves a closed system: no new food is added, and the bacteria and the metabolic wastes cannot go anywhere. In many respects, however, the real world is similar to a closed system. Space and nutrients, for example, exist in limited amounts. In the *Opuntia* example of the previous section, even if humans had not introduced the *Cactoblastis* moth, the cactus population could not have

sustained exponential growth indefinitely. Eventually, the growth of the cactus population would have been limited by an environmental factor, such as a lack of suitable **habitat** (the type of environment in which an organism lives).

The growth pattern of some populations can be represented by an **S-shaped curve**. Such populations grow close to exponentially at first, but then stabilize at the maximum population size that can be supported indefi-

Figure 34.5 Will They Overrun Earth?

A giant puffball mushroom has the potential to produce 7 trillion offspring in a single generation, based on the number of spores it produces. Large giant puffballs weigh 40 to 50 kilograms each; a medium-sized one is shown here.

 See a simulation of restrained population growth. 34.2

nitely by their environment. This level is known as the **carrying capacity** (Figure 34.6). The growth rate of the population decreases as the population size nears the carrying capacity because resources such as food and water begin to be in short supply. At the carrying capacity, the population growth rate is zero.

In the 1930s, the Russian ecologist G. F. Gause carried out experiments on *Paramecium caudatum*, a common protist. He found that laboratory populations of paramecia increased to a certain size and then remained there (see Figure 34.6). In these experiments, Gause added new nutrients to the protists' liquid medium at a steady rate and removed the old solution at a steady rate. At first, the population increased rapidly in size. But as the population continued to increase, the paramecia used nutrients so rapidly that food began to be in short supply, slowing the growth of the population. Eventually the birth and death rates of the protists equaled each other, and the population size stabilized.

In contrast to natural systems, there was no immigration or emigration in Gause's experiments. In natural systems, populations reach and remain at a constant population size when

birth + immigration = death + emigration

for extended periods of time.

Like laboratory populations of bacteria and paramecia, natural populations also experience limits (Figure 34.7). Their growth can be held in check by a number of environmental factors, including food shortages, lack of space, disease, predators, habitat deterioration, weather, and natural disturbances. When there are many individuals in a population, birth rates may drop or death rates may increase; either effect may limit the growth of the population, and sometimes both effects occur. Let's take a brief look at how this works.

Any area contains a limited amount of food and other essential resources. Thus, as the number of individuals in a population increases, fewer resources are available to each individual. As resources diminish, each individual, on average, produces fewer offspring than when resources are plentiful, causing the birth rate of the population to decrease.

In addition, when there are large numbers of individuals in a population, disease spreads more rapidly (since individuals tend to encounter one another more often), and predators may pose a greater risk (since many predators prefer to hunt abundant sources of food). Disease and predators obviously increase the death rate.

Large populations can also damage or deplete their resources. If a population exceeds the carrying capacity

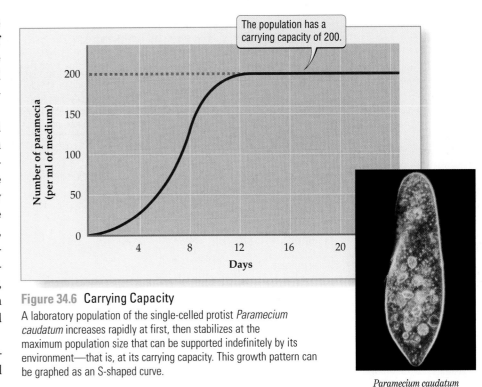

Figure 34.6 Carrying Capacity
A laboratory population of the single-celled protist *Paramecium caudatum* increases rapidly at first, then stabilizes at the maximum population size that can be supported indefinitely by its environment—that is, at its carrying capacity. This growth pattern can be graphed as an S-shaped curve.

Paramecium caudatum

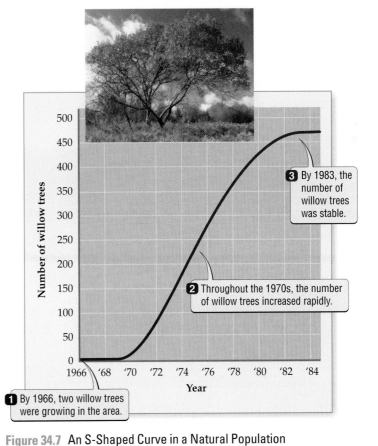

Figure 34.7 An S-Shaped Curve in a Natural Population
At a site in Australia, rabbits heavily grazed young willow trees, preventing willows from growing in the area. The rabbits were removed in 1954. Willows were present in the area by 1966. They increased rapidly in number, and the population then leveled off at about 475 trees.

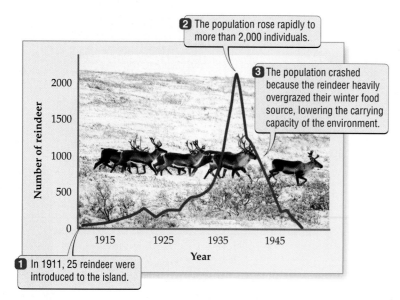

2 The population rose rapidly to more than 2,000 individuals.

3 The population crashed because the reindeer heavily overgrazed their winter food source, lowering the carrying capacity of the environment.

1 In 1911, 25 reindeer were introduced to the island.

Figure 34.8 Boom and Bust

When reindeer were introduced to Saint Paul Island, off the coast of Alaska, in 1911, their population increased rapidly at first, then crashed. By 1950, only eight reindeer remained.

of its environment, it may damage that environment so badly that the carrying capacity may be lowered for a long time. A drop in the carrying capacity means that the habitat cannot support as many individuals as it once could. Such habitat deterioration may cause the population to decrease rapidly (Figure 34.8).

Plantago major

Figure 34.9 It's Getting Crowded

The number of seeds produced per plant drops dramatically under crowded conditions in plantain, a small herbaceous plant.

Some growth-limiting factors depend on population density; others do not

Food shortages, lack of space, disease, predators, and habitat deterioration—all these factors influence a population more strongly as it grows and therefore increases in density. The birth rate, for example, may decrease or the death rate may increase when there are many individuals in the population. When birth and death rates change as the density of the population changes, such rates are said to be **density-dependent**. In natural populations, the number of offspring produced (Figure 34.9) and the death rate are often density-dependent.

In other cases, populations are held in check by factors that are not related to the density of the population; such factors cause the population to change in a **density-independent** manner. Density-independent factors can prevent populations from reaching high densities in the first place. Year-to-year variation in weather, for example, may cause conditions to be suitable for rapid population growth only occasionally. Poor weather conditions may reduce the growth of a population directly (by freezing the eggs of an insect, for example) or indirectly (by decreasing the number of plants available as food to that insect). Natural disturbances such as fires and floods also limit the growth of populations in a density-independent way. Finally, the effects of environmental pollutants such as DDT are density-independent; such pollutants can threaten natural populations with extinction.

34.5 Patterns of Population Growth

Different populations may exhibit a number of different growth patterns over time. We'll discuss four such patterns: J-shaped curves, S-shaped curves, population cycles, and irregular fluctuations.

Under favorable conditions, the population size of any species increases rapidly. An initial period of rapid population growth can be seen in growth patterns that are J-shaped (see Figure 34.3) or S-shaped (see Figure 34.7). With the J-shaped growth pattern, rapid population growth may continue until resources are depleted, causing the population size to drop dramatically (see Figure 34.8). In contrast, with the S-shaped growth pattern, the rate of population growth slows as the population size nears the carrying capacity. Predators, disease, and other

> The number of lynx rises and falls with the number of hares.

Figure 34.10 **Population Cycles**

Populations of two species occasionally increase and decrease together. The Canada lynx depends on the snowshoe hare for food, so the number of lynx is strongly influenced by the number of hares. Experiments indicate that hare populations are limited by their food supply and by their lynx predators. (Numbers of lynx and hares were estimated from the number of furs sold by trappers to the Hudson's Bay Company, Canada.)

See predator-prey interactions.

See consumer-victim interactions.

factors may then keep the population near the carrying capacity for a long time.

As we have seen, populations change in size over time, increasing at some times and decreasing at others. Even populations with an S-shaped growth pattern do not remain indefinitely at a single, stable population size; instead, they fluctuate slightly over time, yet remain close to the carrying capacity.

In some cases, the population sizes of two species change together in a tightly linked cycle (Figure 34.10). Such **population cycles** can occur when at least one of the two species involved is very strongly influenced by the other. The Canada lynx, for example, depends on the snowshoe hare for food, so lynx populations increase when hare populations increase and decrease when hare populations drop.

There are relatively few examples from nature in which the populations of two species show regular cycles like those of the hare and lynx. However, as illustrated dramatically by monarch butterflies, the populations of most species do rise and fall over time—just not as regularly as

in Figure 34.10. **Irregular fluctuations** are far more common in nature than is the smooth rise to a stable population size shown in Figure 34.7.

Finally, different populations of the same species may experience different patterns of growth. Understanding the reasons for these differences can provide critical information on how best to manage endangered or economically important species. The first step toward such an understanding is to perform population counts (see the box on page 672) to determine whether different patterns of population growth are present. If there are different growth patterns, the next step is to figure out why.

During the 1980s, forest managers needed to decide where and how much (if any) mature or old-growth forest could be cut without harming the rare spotted owl. For each owl population, researchers first gathered data on the birth rate and the amount of habitat used by each individual. The researchers then used these data to predict how the growth of spotted owl populations would be affected by the number, size, and

Spotted owl

Science Toolkit

Population Counts Help Save Endangered Species

Populations of many species are threatened by human actions, some to the point of being in danger of extinction. One of the most important tools for people trying to save endangered species is the population count. Like the censuses used by demographers studying human populations, a population count gives the number of individuals in a particular population. Results from such counts can prod organizations and legislatures into taking action, as happened in the case of the bald eagle.

Bald eagles are relatively easy to count: they have large, conspicuous nests to which they return year after year. The bald eagle was one of many bird species in the United States that were severely affected by DDT poisoning—birds with high levels of DDT in their bodies produced such fragile eggshells that they could not reproduce. By the early 1960s, population counts revealed that only 417 breeding pairs of bald eagles remained in the lower 48 states—a huge drop from the estimated 100,000 breeding pairs present in 1800.

The effect of DDT on these birds prompted a 1972 ban on its use within the United States. This ban gave populations of bald eagles the opportunity to bounce back from extinction. Today there are more than 6,400 breeding pairs of in the lower 48 states. The encouraging results from ongoing population counts of bald eagles illustrate that when people recognize a problem and take decisive measures to fix it, populations sometimes can return from the brink of extinction. Furthermore, actions taken with one species in mind can give other species a chance to recover; this was the case for the DDT ban, which allowed other birds, such as the peregrine falcon, to recover from their perilously low numbers.

Not all species are as easy to count as bald eagles. Bog alkaligrass [AL-kuh-lie-GRASS] (*Puccinellia parishii* [puh-chuh-NELL-ee-uh pah-REE-shee]) is designated as an endangered plant species under state and tribal statutes in Arizona, New Mexico, and the Navajo Nation. It is a small, inconspicuous grass that germinates in the winter and typically dies by late spring or early summer. Only 30 populations are known, which makes the species vulnerable to extinction. Population counts are chal-

Bog alkaligrass

lenging because the numbers of seeds that germinate can fluctuate wildly: one year there may be millions of plants, and the next there may be few or none (the number depends on growing conditions). Because no plants are visible in some years, it is hard to know whether all existing populations have been located. Furthermore, depending on how many plants there are, different techniques must be used in different years to estimate the size of the 30 known populations.

Although more difficult to perform, population counts are as essential for bog alkaligrass as they are for bald eagles. They document which populations are increasing and which are decreasing—information that can help scientists and policymakers to decide how best to protect the species from the threat of extinction.

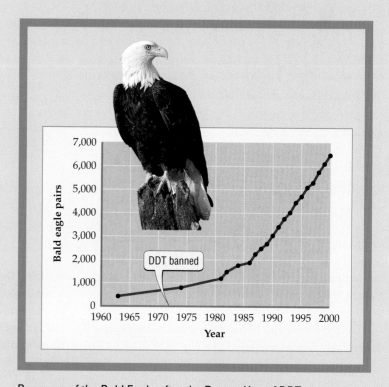

Recovery of the Bald Eagle after the Ban on Use of DDT

location of patches of the bird's preferred habitat, old-growth forest. (Patches are portions of a particular habitat that are surrounded by a different habitat or habitats.) The amount—that is, the total area—and arrangement of old-growth forest patches was found to have a large effect on owl population growth rates (Figure 34.11).

34.6 Human Population Growth: Surpassing the Limits?

The human population is growing today at a spectacular rate (Figure 34.12). It took more than 100,000 years for our population to reach a billion people, but now it increases by a billion people every 13 years. Our use of resources and our overall impact on the planet has increased even faster than our population size. For example, from 1860 to 1991, the human population increased fourfold, but our energy consumption increased 93-fold.

The global human population passed the 6.4 billion mark early in 2005. At present, the human population is growing exponentially, increasing by about 74 million people each year, or over 8,400 people per hour. These numbers are all the more sobering when considering the following facts:

- More than 1.3 billion people live in absolute poverty.
- 2 billion people lack basic health care or safe drinking water.
- More than 2 billion people have no sanitation services.
- Each year 14 million people, mostly children, die from hunger or hunger-related problems.

By the year 2025, the global human population is projected to increase to over 8 billion people. Even if our birth rate dropped immediately from the current 2.8 children per female to a level that ultimately would allow the human population to simply replace itself, but not increase (about 2.1 children per female), the human population would continue to grow for at least another 60 years. The population will continue to increase long after birth rates drop because a huge number of existing children have not yet had children of their own.

How did the human population increase so rapidly, apparently escaping the limits to population growth described in this chapter? There are several reasons.

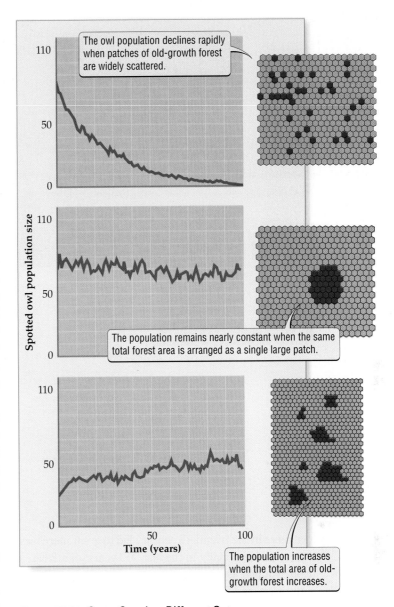

Figure 34.11 Same Species, Different Outcomes
Different populations of the endangered spotted owl are predicted to show different patterns of growth over time, depending on the arrangement and area of their preferred habitat, old-growth forest. Patches of old-growth forest are shown in blue.

First, as our ancestors emigrated from Africa (see page D7), they encountered and prospered in many kinds of new habitats. Few other species can thrive in places as different as grasslands, coastal environments, tropical forests, deserts, and arctic regions. Second, people increased the carrying capacity of the places where they lived. The development of agriculture, for example, allowed more people to be fed per unit of land area. More recently, our heavy use of fossil fuels and nitrogen

Learn more about human population growth.

Figure 34.12 Rapid Growth of the Human Population

An indication of our growing numbers on Earth, this image from space shows the planet brightly lit by its most populous cities' lights. The image was created by researchers at NASA using satellite image data.

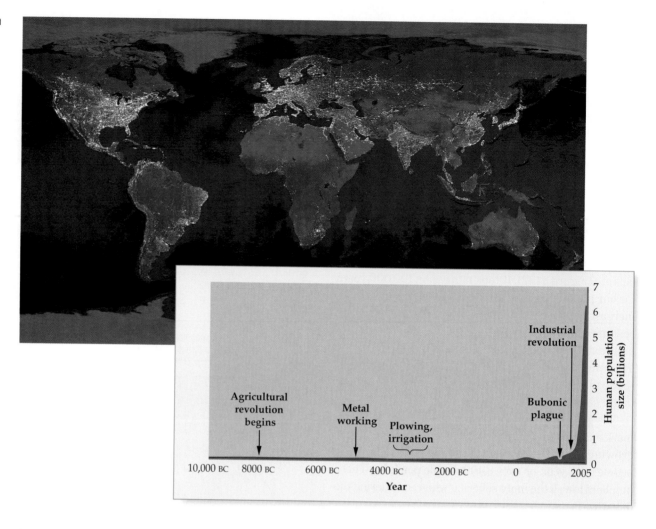

fertilizers in the twentieth century led to great increases in crop yields. Finally, in the last 300 years, death rates have dropped as a result of improvements in medicine, sanitation, and food storage and transportation. However, birth rates did not drop at the same time; hence our population has continued to grow.

Viewed broadly, human inventiveness and technology have allowed us to sidestep limits to population growth for some time. However, like all other populations, our population cannot continue to increase without limit; ultimately we will be subject to the environmental factors that limit the growth of all species.

What Does the Future Hold?

As described at the opening of this chapter, the people who colonized Easter Island initially maintained a culturally rich and densely populated society. But their society did not persist. The people of Easter Island temporarily increased the carrying capacity of the island by cutting down the forest to create farm fields. Ultimately, however, cutting

down the trees led to environmental deterioration, starvation, and the collapse of their civilization.

As on Easter Island, many of the problems facing humans today relate to population growth and environmental deterioration. More people means more environmental deterioration, which in turn makes it harder to feed the people we already have. Already much of Africa depends on imported food to prevent starvation, and cities in California persist only because of water imported from other states. In addition, many people think that our society, like that of Easter Island, is not based on the sustainable use of resources. The term **sustainable** describes an action or process that can continue indefinitely without using up resources or causing serious damage to the environment.

Many lines of evidence suggest that the current human impact on Earth is not sustainable. For example, water tables are dropping throughout the world, global fish populations have plummeted in response to overharvesting, and if current rates of logging continue (14,000,000 hectares or 35,000,000 acres per year), scientists estimate that all tropical forests will be gone in 100 to 150 years.

One measure of sustainability is the **ecological footprint**, which is the area of productive ecosystems needed throughout the year to support a population and cope with its waste materials. Early results of calculating our ecological footprint are not encouraging. Scientists recently estimated that the average person's ecological footprint is 2.3 hectares (5.75 acres), which is about 20 percent higher than the 1.9 hectares (4.75 acres) that could be sustained for each of the world's 6.4 billion people. The ecological footprint of individuals in some countries, such as the United States (9.7 hectares per person) and the United Kingdom (5.7 hectares per person), is 3 to 5 times what is sustainable. Overall, such estimates suggest that since the early 1980s, people have been using resources faster than they can be replenished (Figure 34.13), a pattern of resource use that, by definition, is not sustainable.

Will people limit the growth and impact of our global population, or will the environment do it for us? There are some hopeful signs: the growth rate of the human population has slowed in recent years, and people throughout the world are conscious of the risks of environmental degradation. But much remains to be done. To limit the growth and impact of the human population, we must address the interrelated issues of population growth rates, poverty, unequal use of resources, environmental deterioration, and sustainable development. It is especially important for people who live in North America, Japan, and Europe to address such issues because people in these regions have such large ecological footprints.

Our hope for the future—for your future and the future of your children—lies in realistically assessing the problems we face, and then committing ourselves to take bold actions to address those problems (see Interlude G). In the end, it is up to all of us to help ensure that humankind does not repeat on a grand scale the tragic lessons of Easter Island.

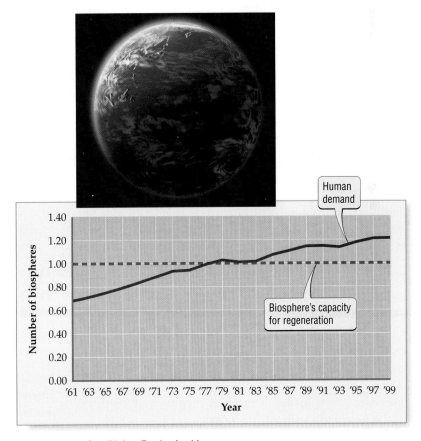

Figure 34.13 Our Rising Ecological Impact
The global ecological impact of people has increased steadily over the past 40 years. This graph compares human demands on the biosphere in each year with the capacity of the biosphere to regenerate itself. One vertical unit on the graph represents the entire capacity of the biosphere to regenerate itself in a given year. Human demand has exceeded the biosphere's entire regenerative capacity since the early 1980s.

Chapter Review

Summary

34.1 What Are Populations?

- A population is a group of interacting individuals of a single species located within a particular area.
- Two basic concepts used in studying populations are population size (the total number of individuals in the population) and population density (the number of individuals per unit of area).
- What constitutes an appropriate area for determining a population depends on the questions of interest and the biology of the organism under study.

34.2 Changes in Population Size

- All populations change in size over time.
- Populations increase when birth and immigration rates are greater than death and emigration rates, and they decrease when the reverse is true.
- Because birth, death, immigration, and emigration rates all depend on environmental factors, the environment plays a key role in changing the size of populations.

34.3 Exponential Growth

- A population grows exponentially when it increases by a constant proportion from one generation to the next. Exponential growth produces a J-shaped curve.
- The doubling time is one measure of how fast a population is growing.
- Populations may grow at an exponential rate when organisms are introduced into or migrate to a new area.

34.4 Limits to Population Growth

- Because the environment contains a limited amount of space and resources, no population can continue to increase in size indefinitely.
- Some populations increase rapidly at first, then level off and stabilize at the carrying capacity, the maximum population size that their environment can support. This growth pattern is represented by an S-shaped curve.
- Density-dependent environmental factors limit the growth of a population more strongly when the density of the population is high. These factors include food shortages, diminishing space, disease, predators, and habitat deterioration.
- Density-independent factors, such as weather and natural disturbances, limit the growth of populations without regard to their density.

34.5 Patterns of Population Growth

- Different populations (including those of the same species) can exhibit different patterns of growth over time, including J-shaped curves, S-shaped curves, population cycles, and irregular fluctuations.
- Two populations of different species can change together in tightly linked cycles when one or both species is strongly influenced by the other.
- In natural systems, a growth pattern of irregular fluctuations is much more common than an S-shaped growth pattern or tightly linked cycles.
- Understanding why different populations have different patterns of growth can provide critical information on how best to manage endangered species.

34.6 Human Population Growth: Surpassing the Limits?

- The global human population is growing exponentially, increasing by 1 billion people every 13 years.
- So far, we have been able to postpone dealing with limits to our population growth by increasing the carrying capacity of our environment. We have accomplished this through inventiveness and technology, in particular through agriculture and the use of fossil fuels.
- However, our use of resources is now increasing even faster than our rate of population growth. The same environmental factors that limit other species' growth will eventually limit ours, too, unless we take steps to limit it before that happens.

⊙ Review and Application of Key Concepts

1. Explain why it can be difficult to determine what constitutes a population.

2. Populations increase in size when birth and immigration rates are greater than death and emigration rates. Keeping this basic principle in mind, what actions might a scientist or policymaker take in order to protect a population threatened by extinction?

3. Assume that a population grows exponentially, increasing by a constant proportion of 1.5 per year. Thus, if the population initially contains 100 individuals, it will contain 150 individuals in the next year. Graph the number of individuals in the population

versus time for the next 5 years, starting with 150 individuals in the population.

4. Population growth cannot increase without limit.
   a. What environmental factors prevent limitless growth?
   b. Why is it common for populations of species that enter a new region to grow exponentially for a period of time?

5. Describe the difference between density-dependent and density-independent factors that limit population growth. Give two examples of each.

6. Different populations of a species can have different patterns of population growth. Explain how an understanding of the causes of these different patterns can help managers protect rare species or control pest species.

7. List five specific actions that you can take to limit the growth or impact of the human population.

Key Terms

carrying capacity (p. 669)
density-dependent (p. 670)
density-independent (p. 671)
doubling time (p. 667)
ecological footprint (p. 675)
exponential growth (p. 666)
habitat (p. 668)
irregular fluctuations (p. 671)

J-shaped curve (p. 667)
population (p. 664)
population cycle (p. 671)
population density (p. 664)
population size (p. 664)
S-shaped curve (p. 668)
sustainable (p. 675)

Self-Quiz

1. A group of interacting individuals of a single species located within a particular area is a(n)
   a. biosphere.
   b. ecosystem.
   c. community.
   d. population.

2. A population of plants has a density of 12 plants per square meter and covers an area of 100 square meters. What is the population size?
   a. 120
   b. 1,200
   c. 12
   d. 0.12

3. A population that is growing exponentially increases
   a. by the same number of individuals each generation.
   b. by a constant proportion each generation.
   c. in some years and decreases in other years.
   d. none of the above

4. In a population with an S-shaped growth curve, after an initial period of rapid increase, the number of individuals
   a. continues to increase.
   b. drops rapidly.
   c. remains near the carrying capacity.
   d. cycles regularly.

5. The growth of populations can be limited by
   a. natural disturbances.
   b. weather.
   c. food shortages.
   d. all of the above

6. Factors that limit the growth of populations more strongly at high densities are said to be
   a. density-dependent.
   b. density-independent.
   c. exponential factors.
   d. sustainable.

7. The maximum number of individuals in a population that can be supported indefinitely by the population's environment is called the
   a. exponential size.
   b. J-shaped curve.
   c. sustainable size.
   d. carrying capacity.

8. A population that initially has 40 individuals grows exponentially with an (annual) proportional increase (λ) of 1.6. What is the size of the population after 3 years? (*Note:* Round down to the nearest individual.)
   a. 16.
   b. 163.
   c. 192.
   d. 102,400.

Desert City a Pool of West Nile Cases

By Beth DeFalco

With triple-digit heat and nearly non-existent rainfall, Phoenix seems an unlikely spot for this year's West Nile virus epicenter. Yet, federal officials say Arizona is the only state where the mosquito-borne virus is an epidemic.

"Minnesota may be the land of a thousand lakes, but we are the land of thousands of abandoned swimming pools," says Will Humble, head of disease control of the Arizona Department of Health Services.

Deserts and mosquitoes usually do not go together, but Arizona has one species, *Culex tarsalis* [KYOO-lex tar-SAY-liss], that tolerates heat well and can increase rapidly in numbers if there is enough standing water for their young to develop into adults. (Mosquitoes lay their eggs in water, and their larvae are fully aquatic.) *Culex tarsalis*, like all mosquito species, can breed in standing water found in wheelbarrows, old tires, or even bottles—the kinds of miscellaneous small sources of standing water found in towns and cities throughout the country. But in some parts of Arizona, *C. tarsalis* mosquitoes recently hit the "double jackpot" with two additional sources of standing water.

A view from the air shows that the Phoenix area is covered with swimming pools. A well-maintained pool has too many chemicals in it for mosquitoes to breed. But out of 600,000 residential pools in the state, health department officials estimate that 10,000 of them can support mosquitoes. These abandoned pools are like miniature backyard swamps—perfect homes for mosquito larvae.

In addition, many homeowners have created lush landscapes in their yards, which in Phoenix's dry climate require large amounts of water to maintain. With irrigation canals, sprinklers, and lawn misters, there seems to be no shortage of human-related water sources for mosquitoes to breed in.

Altogether, there is so much standing water in the Phoenix area that mosquitoes have become a problem. Their populations have increased greatly in size, which in turn has contributed to the sharp increase in the number of West Nile cases reported in Arizona. "It didn't used to be this bad. You never saw a mosquito," said resident Gary Clark, who

walks every morning in an area where a high number of West Nile cases have been reported. At the time this article was published, 290 of the nation's 500 West Nile cases were in Arizona, as were 3 of the 14 deaths. Nearly all of these cases were in the state's most heavily populated county, Maricopa, which includes Phoenix.

Evaluating the news

1. In many parts of the country, mosquitoes breed in the standing water found in abandoned pools, old tires, and many other places.
   a. Should heavy fines be imposed on people who leave standing water in their yards?
   b. To reduce the chance of outbreaks of diseases that are transmitted by mosquitoes, such as West Nile virus, should additional health department officials be hired to inspect yards for standing water and to treat abandoned pools with chemicals that kill mosquito larvae? If so, who should pay for this?

2. In some desert regions, homeowners and farmers apply so much water to their yards and fields that the humidity has steadily increased over the past 20 years. One consequence is that evaporative cooling systems (called "swamp coolers") don't work very well any more, causing people to switch to more expensive cooling systems (such as standard air conditioners). Do you think this trend of modifying local climates to such an extent is good, bad, or not significant? Why?

3. People who live in Phoenix or other cities located in deserts—such as Tucson, Arizona; El Paso, Texas; Las Cruces, New Mexico; and

Las Vegas, Nevada—must water their lawns and gardens heavily if they want to create lush landscapes. Far too little rain falls in these areas to contribute significantly to the maintenance of these landscapes, so water must be taken from rivers, pumped out of underground reservoirs, or imported from other states. Such high levels of water use in a desert are not sustainable because water is used more rapidly than it is replenished.

a. Do you think it is appropriate to use large amounts of water on plants in lawns and gardens that otherwise could not live in a desert?
b. Should cities encourage people to conserve water, and if so, how should they do that?

SOURCE: Associated Press, Wednesday, August 18, 2004.

CHAPTER 35

Interactions among Organisms

Key Concepts

◉ Organisms interact with one another in many different ways, the three most important of which are mutualisms, exploitation, and competition.

◉ In mutualisms, two species interact for the benefit of both species. Mutualisms evolve when the benefits of the interaction outweigh the costs for both species.

◉ In exploitation, one species in the interaction benefits (the consumer) while the other is harmed (the species that is eaten). Species that are eaten by other organisms have evolved elaborate ways of defending themselves against their consumers.

◉ In competition, two species that share resources have a negative effect on each other. Competition can result in the evolution of greater differences between species.

◉ Mutualisms, exploitation, and competition help determine where organisms live and how abundant they are.

◉ Changes in interactions among organisms can change the community of species that live in an area and the underlying nature of the ecosystem.

Lumbering Mantises and Gruesome Parasites

Praying mantises have been seen to walk to the edge of a river, throw themselves in, and drown shortly thereafter. If they are rescued from the water, they will immediately throw themselves back in. What causes mantises to do this?

This bizarre behavior appears to be driven not by the mantises themselves, but by a parasitic worm (*Gordius*). Less than a minute after the mantis lands in the water, a worm emerges from its anus. This worm attacks and infects terrestrial insects, such as praying mantises, but it also depends on an aquatic host for part of its life cycle. The worm has performed a neat trick: it has evolved the ability to cause its insect host to jump into the river, an act that kills the insect but increases the chance that the worm will eventually reach its aquatic host.

Moving from the bizarre to the gruesome, examine the accompanying photographs. The fungus that killed the ant first grew throughout the ant's entire body, dissolving portions of its body and using them for food. Eventually, the fungus sprouted reproductive structures (indicated by arrows), which allowed it to spread and attack other ants. Fungi attack many other species, including crops such as the corn plant shown. While you may have trouble empathizing with ants or corn, another photograph shows the effects of one of the many parasites that attack people. Such attacks cause enormous health problems: hundreds of millions of people are disabled every year by more than a thousand different types of bacterial, protist, fungal, and animal parasites, each of which has unique and harmful effects on the human body.

Parasites, such as the worms that plague praying mantises and the fungi that riddle the bodies of ants, are organisms that live in or on other

(a)

(b)

(c)

Parasitic Relationships
(a) An ant has been killed by a fungus. (b) An ear of corn has been destroyed by a fungus known as corn smut. (c) A person has been infected with a protist that attacks the skin.

organisms (known as their hosts). They obtain nutrients from their hosts, often causing them harm but not immediate death. The effects of parasites on their hosts illustrate one important type of interaction among organisms: a relationship in which one species benefits and the other is harmed. How do such relationships work?

Helpful to know

Symbiosis (from *sym*, "with"; *biosis*, "life") describes some (but not all) mutualistic and parasitic interactions. Individuals of one species, for at least part of their life cycle, must live close to, in, or on individuals of another species for the interaction to be considered symbiotic.

As we have seen in the previous two chapters, ecology focuses on interactions between organisms and their environment. An organism's environment includes the other organisms that live there. Thus the subject of this chapter—interactions among organisms—is central to the definition of ecology.

Interactions among organisms have huge effects on natural communities. For example, as we saw in Chapter 34, the moth *Cactoblastis cactorum*, by feeding on the cactus *Opuntia*, caused *Opuntia* populations to crash throughout a large region of Australia. Overall, interactions among organisms have an influence at every level of the biological hierarchy at which ecology is studied.

The millions of species on Earth can interact in many different ways. In this chapter we classify interactions among organisms by whether the interaction is beneficial (+) or harmful (−) to each of the interacting species. We focus on the three most common and most important kinds of ecological interactions:

+/+ interactions, in which both species benefit (mutualisms)

+/− interactions, in which one species benefits and the other is harmed (exploitation)

−/− interactions, in which both species are harmed (competition)

Each type of interaction plays a key role in determining where organisms live and how abundant they are. We also discuss how changes in interactions among organisms can alter ecological communities.

See mutualism in action. `35.1`

Citrus mealybug

35.1 Mutualisms

Mutualisms (+/+ interactions) are associations between two species in which both species benefit. Mutualisms are common and important to life on Earth: many species receive benefits from, and provide benefits to, other species. These benefits increase the survival and reproduction of both of the interacting species.

Mutualisms can occur when two or more organisms of different species live together, an association known as **symbiosis**. Insects such as aphids and mealybugs, both of which feed on the nutrient-poor sap of plants, often have a mutualistic, symbiotic association with bacteria that live within their cells. The bacteria receive food and a home from the insects, while the insects receive nutrients that the bacteria (but not the insects) can synthesize from sugars in the plant sap. Such symbiotic associations can be amazingly complex. Scientists have recently discovered that a second species of bacteria lives within the bacteria that live inside of citrus mealybug cells; it is not yet clear whether this second species benefits or harms the bacteria in which they live.

This open question illustrates an important point: although some symbiotic associations are clearly mutualistic, benefiting both organisms (a +/+ interaction), many cases of symbiosis involve one species harming rather than helping the other species in the association (a +/− interaction). This is true of many parasites, which spend all or most of their lives within their hosts, deriving benefits from their hosts yet harming rather than benefiting their hosts. By some estimates, nearly half of all species on Earth are parasites.

There are many types of mutualisms

Nature abounds with varieties of mutualisms; here we describe only some of the most common types. In **gut inhabitant mutualisms**, organisms that live in an animal's digestive tract receive food from their host and benefit the host by digesting foods, such as wood or cellulose, that the host otherwise could not use. The interaction between a mealybug and the first bacterial species living in its gut is an example of this type of mutualism. So are the bacteria that live in the guts of termites, enabling them to digest wood, as well as some of the bacteria that live in our intestines and help us to digest and absorb nutrients.

In **seed dispersal mutualisms**, an animal, such as a bird or mammal, eats a fruit that contains plant seeds, then later defecates the seeds far from the parent plant. Such dispersal by animals is the primary way that many plant species reach new areas of favorable habitat. For example, most of the plant species that live on isolated oceanic islands (those that are farther than 1,000 kilometers from land) are thought to have arrived there by bird dispersal.

Mutualisms in which each partner has evolved to alter its behavior to benefit the other species are called **behavioral mutualisms**. The relationship between certain shrimps and fish is a good example of a behavioral mutualism (Figure 35.1). Shrimps of the genus *Alpheus* live in an environment with plenty of food, but little shelter. They dig burrows in which to hide, but they see poorly, and so are vulnerable to predators when they leave their burrows to feed. These shrimps have formed a fascinating relationship with some go by fish in the genera *Crypto-centrus* and *Vanderhorstia*. When a shrimp ventures out of its burrow to eat, it keeps an antenna on an individual goby with which it has formed a special relationship. If a predator or other disturbance causes the fish to make a sudden movement, the shrimp darts back into the burrow. Thus the goby acts as a "seeing-eye" fish for the shrimp, warning it of danger. In return, the shrimp shares its burrow with the goby, thereby providing the fish with a safe haven.

In **pollinator mutualisms**, an animal, such as a honeybee, transfers pollen (which contains male reproductive cells) from one flower to the female reproductive organs of another flower of the same species. These animals are known as **pollinators**, and without them many plants could not reproduce. To ensure that pollinators come to their flowers, plants offer a food reward, such as pollen or nectar. Thus both species benefit from the interaction. Pollinator mutualisms are important in both natural and agricultural ecosystems. For example, the oranges we buy at the supermarket are available only because honeybees have pollinated the flowers of orange trees, thus enabling the trees to produce their fruit.

Mutualists are in it for themselves

Although both species in a mutualism benefit from the relationship, what is good for one species may come at a cost to the other. For example, a species may use energy or increase its exposure to predators when it acts to benefit its mutualistic partner. From an evolutionary perspective, mutualisms evolve when the benefits of the interaction outweigh the costs for both species. However, even in mutualisms, the interests of the two species may be in conflict.

Consider the pollinator mutualism between the yucca plant and the yucca moth. A female yucca moth collects pollen from yucca flowers, flies to another group

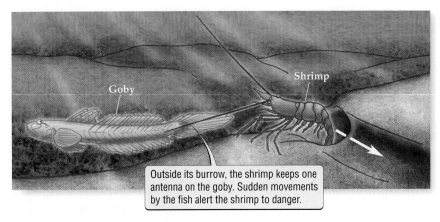

Outside its burrow, the shrimp keeps one antenna on the goby. Sudden movements by the fish alert the shrimp to danger.

Figure 35.1 A Behavioral Mutualism

Each *Alpheus* shrimp builds a burrow for shelter, which it shares with a goby fish. The fish provides an early-warning system to the nearly blind shrimp when the shrimp leaves the burrow to feed.

of flowers, and lays her eggs at the base of the carpel of a newly opened flower. After she has laid her eggs, the female moth climbs up the carpel and deliberately places the pollen she collected earlier onto the stigma of the flower, thus fertilizing the eggs of that second yucca plant (Figure 35.2; for a review of the reproductive parts of a flower, see Figure 32.6). When the moth larvae hatch, they feed on the seeds of the yucca plant.

Honeybee

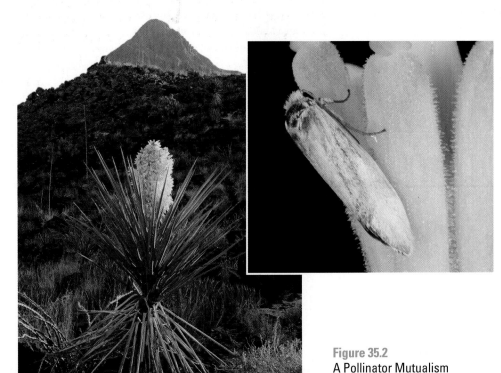

Figure 35.2
A Pollinator Mutualism
The yucca and the yucca moth are dependent on each other for survival.

In this mutualism, the plant gets pollinated (a reproductive benefit provided to the plant by the moth) and the moth eats some of its seeds (a food benefit provided to the moth by the plant). In fact, plant and pollinator each depend absolutely on the other—the yucca is the moth's only source of food, and this moth is the only species that pollinates the yucca—so this association is mutualistic, not parasitic. But there are costs for both species—let's examine these costs more closely.

In a cost-free situation for the plant, the moth would transport pollen, but would not destroy any of the plant's seeds. In a cost-free situation for the moth, the moth would produce as many larvae as possible, and they would consume many of the plant's seeds. In actuality, an evolutionary compromise has been reached: the moth usually lays only a few eggs per flower, and the plant tolerates the loss of a few of its seeds. Yucca plants have a defense mechanism that helps to keep this compromise working: if a moth lays too many eggs in one of the plant's flowers, the plant can selectively abort that flower, thereby killing the moth's eggs or larvae.

Mutualisms are everywhere

Mutualisms are very common. Most of the plant species that dominate forests, deserts, grasslands, and other biomes are mutualists. For example, as we saw in Chapter 31, most plant species have mutualistic associations with fungi, called **mycorrhizae**. The fungi help the plant roots absorb nutrients and water from the soil, and the plant provides the fungi with carbohydrates produced by photosynthesis (Figure 35.3).

Figure 35.3 Mycorrhizae
Fungal hyphae surround a plant root and penetrate some of its cells, helping the plant roots absorb mineral nutrients and water from the soil and allowing carbohydrates to be transported from the plant to the fungus.

As mentioned earlier, many animal species are pollinators involved with plants in pollinator mutualisms. Other examples of mutualisms involving animals include the spectacular reefs found in tropical oceans (Figure 35.4). These reefs are built by corals (soft-bodied animals), most of which house photosynthetic algae—their mutualistic partners—inside their bodies. The corals provide the algae with a home and several essential nutrients, such as phosphorus, and the algae provide the corals with carbohydrates produced by photosynthesis.

Mutualisms can determine the distribution and abundance of species

Mutualisms can influence the **distribution** (the geographic area over which a species is found) and abundance of organisms in two ways. First, because each species in a mutualism survives and reproduces better where its partner is found, the two species strongly influence each other's distribution and abundance. For example, because the yuccas and yucca moths described earlier depend absolutely on each other, each species is found only where the other is present.

Second, a mutualism can have indirect effects on the distribution and abundance of species that are not part of the mutualism. Coral reefs, for example, are home to many different plant and animal species. Since the corals that build the reefs depend on their mutualisms with algae, the many other species that live in coral reefs depend on those mutualisms indirectly.

35.2 Exploitation

Exploitation (+/− interactions) includes a variety of interactions in which one species (the consumer) benefits and the other (the species that is eaten, or "food organism") is harmed. The consumers in such interactions can be classified into three main groups:

1. **Herbivores** are consumers that eat plants or plant parts.
2. **Predators** are animals (or, in rare cases, plants) that kill other animals for food; the animals that are eaten are called **prey**.
3. **Parasites** are consumers that live in or on the organisms they eat (which are called **hosts**). An important group of parasites are **pathogens**, which cause disease in their hosts.

These three major types of +/− interactions are very different from one another. For example, whereas predators (such as wolves) kill their food organisms immediately, herbivores (such as cows) and parasites (such as fleas) usually do not. Although the three types of exploitation have obvious and important differences, in this section we look at some general principles that apply to all three.

Consumers and their food organisms can exert strong selection pressure on each other

The presence of consumers in the environment has caused many species to evolve elaborate strategies to avoid being consumed. Many plants, for example, produce spines and toxic chemicals as defenses against herbivores. Some plants rely on **induced defenses**, responses that are directly stimulated by an attack from herbivores. Spine production is an induced defense in some cacti: an individual cactus that has been partially eaten, or grazed, is much more likely to produce spines than is an individual that has not been grazed (Figure 35.5).

Many organisms have evolved bright colors or striking patterns that warn potential predators that they are heavily defended, usually by chemical means (Figure 35.6a). Such warning coloration can be highly effective. Blue jays, for example, quickly learn not to eat monarch butterflies, which are brightly colored and whose tissues contain cardiac glycosides; these chemicals cause nausea in birds (and people) (Figure 35.6b) and, at high doses, sudden death from heart failure. Other prey have evolved to avoid predators by being hard to find or hard to catch. In addition, animals have evolved molecular defenses (immune systems) to help them fight off the effects of microbial diseases and parasitic infections.

The many ways in which species have evolved to protect themselves against consumers indicate that consumers often apply strong selection pressure to their food organisms. Selection occurs in the other direction as well. If a plant or prey species evolves a particularly powerful defense against attack, its consumers, in turn, experience strong selection pressure to overcome that defense. There are many examples of defenses that work against all consumers except for a few species that have evolved the ability to overcome them. Consider the rough-skinned newt, whose skin contains unusually large amounts of the potent neurotoxin TTX (tetrodotoxin)—enough to kill 25,000 mice. The newt is so toxic that only one predator, the garter snake, can tolerate its poison well enough to eat the newt and survive (see Figure 16.2).

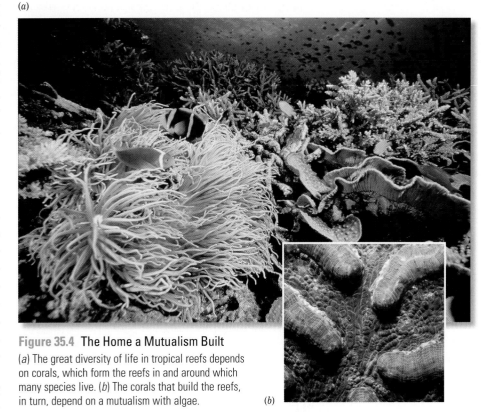

(a)

(b)

Figure 35.4 The Home a Mutualism Built
(a) The great diversity of life in tropical reefs depends on corals, which form the reefs in and around which many species live. (b) The corals that build the reefs, in turn, depend on a mutualism with algae.

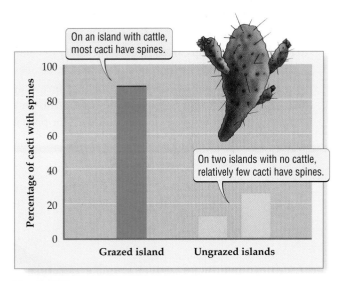

On an island with cattle, most cacti have spines.

On two islands with no cattle, relatively few cacti have spines.

Percentage of cacti with spines

Grazed island   Ungrazed islands

Figure 35.5

Spines on Some Cacti Are an Induced Defense

On three islands off the coast of Australia, the percentage of cacti with spines is higher on the island that has cattle than on the two islands that do not. Field and laboratory experiments show that grazing by cattle directly stimulates the production of spines in this species of cactus.

(a)

(b)

Figure 35.6 Warning Coloration Can Be Highly Effective
(a) The bright colors of this poison dart frog warn potential predators of the deadly chemicals contained in its tissues. (b) An inexperienced blue jay vomits after eating a brightly colored monarch butterfly.

Consumers can alter the behavior of the organisms they eat

The bizarre story of the praying mantises that jump to their deaths in rivers at the opening of this chapter provides a dramatic example of how consumers can alter the behavior of their food organisms. But exploitation can alter the behavior of food organisms in more subtle ways as well.

Predators can be a driving force that causes animals to live or feed in groups. In some cases, several prey individuals acting together may be able to thwart attacks from predators (Figure 35.7). Large groups of prey may also be able to provide better warning of a predator's attack. Because more individuals can watch for preda-

tors, a large flock of wood pigeons detects the approach of a goshawk [GOSS-hawk] (a predatory bird) much sooner than a single pigeon does. The success rate of goshawk attacks drops from nearly 80 percent when attacking single pigeons to less than 10 percent when attacking flocks of more than 50 birds (Figure 35.8).

Consumers can restrict the distribution and abundance of their food organisms

The American chestnut used to be a dominant tree species across much of eastern North America. Within its

Figure 35.7 Come and Get Us
Although a single musk ox may be vulnerable to predators such as wolves, a group that forms a circle makes a difficult target.

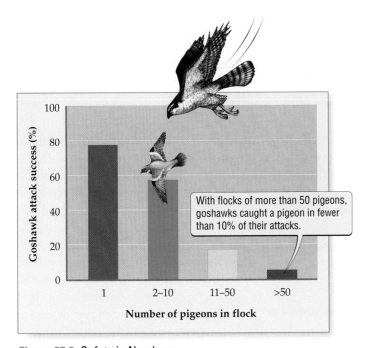

Figure 35.8 Safety in Numbers
The success of goshawk attacks on wood pigeons decreases greatly when there are many pigeons in a flock.

Graph axis labels: Goshawk attack success (%) / Number of pigeons in flock

Graph callout: With flocks of more than 50 pigeons, goshawks caught a pigeon in fewer than 10% of their attacks.

Consumers can drive their food organisms to extinction

Laboratory experiments with protists and with mites have shown that predators can drive their prey extinct (see the box on page 688 for a description of how experiments are used in ecology). Exploitation can drive food organisms to extinction in natural systems as well. The effect of chestnut blight on the American chestnut provides one clear example: although the chestnut tree is not extinct throughout its entire range, many local populations have been driven to extinction. Similarly, *Cactoblastis* moths drove many populations of the *Opuntia* cactus in Australia extinct

range, anywhere from a quarter to a half of all trees were chestnuts. They were capable of growing to large size: trunks up to 10 feet in diameter were noted by colonial settlers. In 1900, however, a fungus that causes a disease called chestnut blight was introduced into the New York City area. This fungus spread rapidly, killing most of the chestnut trees in eastern North America. Today the American chestnut survives throughout its former range only in isolated patches, primarily as sprouts that arise from the base of otherwise dead trunks. With few exceptions, the new sprouts die back from reinfection with the fungus before they can grow large enough to generate new seeds.

The effect of chestnut blight on the American chestnut shows how a consumer (the fungus) can limit the distribution and abundance of its food organism (the chestnut): in this case, a formerly dominant tree species was virtually eliminated from its entire range. The effects of consumers on the distribution and abundance of the species they eat are also shown by what can happen when a food organism is freed from its consumers. As we saw in Chapter 33, nonnative species introduced by people to new regions sometimes disrupt the ecological communities there. A series of recent studies suggests that some introduced species are able to increase rapidly in number in their new areas in part because they have many fewer parasites there than in their original homes (Figure 35.9).

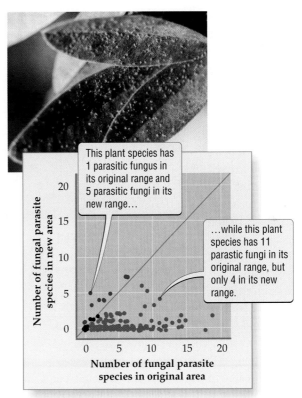

Graph axis labels: Number of fungal parasite species in new area / Number of fungal parasite species in original area

Graph callouts: This plant species has 1 parasitic fungus in its original range and 5 parasitic fungi in its new range… / …while this plant species has 11 parasitic fungi in its original range, but only 4 in its new range.

Figure 35.9 Leaving Their Parasites Behind
Most introduced plant species have fewer fungal parasites in their new homes than in their original homes. Each point on the graph represents a different plant species that has been introduced to a new area. Points below the diagonal line represent plants with fewer fungal parasites in their new home than in their old home; points above the diagonal line represent the opposite. Points falling on the diagonal line indicate plants showing no difference in the number of fungal parasite species between their new and old ranges. The photograph shows the perennial bush clover (*Lespedeza capitata*) infected with the pathogenic rust fungus (*Uromyces lespedezae-procumbentis*), each individual "bump" is an individual fungus that infected the leaf from a separate microscopic spore, and is now producing its own spores.

Science Toolkit

Answering Ecological Questions with Experiments

Like all scientists, ecologists observe nature and ask questions about their observations. They then rely on three approaches to answer these questions: experiments, additional observations, and models. Observations are helpful if the questions concern events that cover large geographic regions or occur over long periods of time—experiments designed to answer such questions may not be feasible. Models, such as those that predict changes in the global climate, can also be used to answer questions that are hard to examine experimentally. Nevertheless, experiments are one of the most important parts of the modern ecologist's scientific toolkit.

In an **ecological experiment**, an investigator alters one or more features of the environment and observes the effect of that change. Experiments in ecology range from laboratory experiments to experiments conducted outside in an artificial environment (such as an artificial pond) to field experiments conducted in a natural environment in which one or more factors are manipulated by the experimenter, but all else is left as it is in the local environment. In any ecological experiment, the purpose is to examine the effects of different treatments—such as a high amount of pesticide, a low amount of pesticide, and no pesticide—on natural processes.

The "no pesticide" treatment is an example of a **control** (applied to a **control group**), an important feature included in most ecological experiments (as well as most other kinds of experiments). The control group is subjected to the same environmental conditions as the experimental groups, except that the factor or factors being tested in the experiment are omitted. Including a control group in an experiment strengthens the argument that any effects seen in the experimental group (but not in the control group) result from the factor or factors under investigation.

When performing an experiment, an ecologist replicates each treatment (that is, performs it more than once), including the control. The advantage of replication is that as the number of **replicates** increases, it becomes less likely that the results are actually due to a variable not being measured or controlled in the study. Consider an experiment designed to test whether the presence of pesticides in a pond causes frogs to have more deformities. If the experiment were performed with only two ponds, one with detectable levels of pesticides and the other without, the results would be hard to interpret. Suppose that frog deformities were more common in the pond that contained pesticides. While pesticides may have caused this result, the two ponds might have differed in many other ways, one or more of which might have been the real cause of the result. On the other hand, if many ponds were used—that is, many replicates—it becomes much less likely that each pond with pesticides also contained something else that increased the chance of frog deformities.

Ecologists also seek to limit the effects of unmeasured variables by assigning treatments at random. Suppose an experiment was designed to test whether insects that eat plants decrease the number of seeds the plants produce. To test this idea, a natural area could be divided into a series of experimental plots. Each plot would receive one of two treatments: it would either be sprayed regularly with an insecticide, reducing the number of plant-eating insects, or be left alone. If plant-eating insects have much effect, seed production should be higher in plots that are sprayed. Usually, the decision of whether a particular plot will be sprayed (or not) would be made at random at the start of the experiment. This would be done to make it less likely that the plots receiving a particular treatment would share other features that might influence seed production, such as high or low levels of soil nutrients.

Ecological experiments allow us to test whether we understand how nature works. The results may answer the question the experiment was designed to address, and they may also stimulate a whole new set of questions. As new questions lead to new discoveries, what we know about ecology constantly changes. Thus our understanding of ecology is, and always will be, a work in progress.

Field Experiment with Fish

An ecologist studying lake fish might perform experiments in laboratory aquariums, in natural lakes, or in artificial (human-constructed) ponds as seen here.

(see Figure 34.4). If a consumer eats only one species, then if that consumer drives a population of the species it eats to extinction, the consumer must either locate a new population of food organisms or go extinct itself. This is exactly what happened to *Cactoblastis* in eastern Australia: the moth drove most populations of the cactus it eats extinct, and now both species are found in low numbers.

35.3 Competition

In **competition** (−/− interactions), each of two interacting species has a negative effect on the other. Competition is most likely when two species share an important resource, such as food or space, that is in short supply. When two species compete, each has a negative effect on the other because each uses resources (such as a source of food) that otherwise could have been used by its competitor. This is true even when one species is so superior as a competitor that it ultimately drives the other species extinct: until the inferior competitor actually becomes extinct, it continues to use some resources that could have been used by the superior competitor.

There are two main types of competition:

1. In **interference competition**, one organism directly excludes another from the use of a resource. For example, individuals from two species of birds may fight over the tree holes that they both use as nest sites.
2. In **exploitative competition**, species compete indirectly for a shared resource, each reducing the amount of the resource available to the other. For example, two plant species may compete for a resource that is in short supply, such as nitrogen in the soil.

Competition can limit the distribution and abundance of species

Competition between species often has important effects on natural populations. These effects, as shown by a great deal of field evidence, include limiting the distributions and abundances of species. Let's explore two examples.

Along the coast of Scotland, the larvae of two species of barnacles, *Balanus balanoides* [buh-LAY-nus bah-luh-NOY-deez] and *Chthamalus stellatus* [thuh-MAY-lus stell-AY-tus], both settle on rocks on high and low portions of

the shoreline. However, *Balanus* adults appear only on the lower portion of the shoreline, which is more frequently covered by water, and *Chthamalus* adults are found only on the higher portion of the shoreline, which is more frequently exposed to air (Figure 35.10).

In principle, the distributions of these two barnacles could have been caused either by competition or by environmental factors. In an experimental study, however, ecologists discovered that *Chthamalus* could thrive on low portions of the shoreline, but only when *Balanus* was removed. Hence competition with *Balanus* ordinarily prevents *Chthamalus* from living low on the shoreline. This interaction is an example of interference competition because *Balanus* individuals often crush the smaller and more delicate *Chthamalus* individuals. The distribution of *Balanus*, on the other hand, depends mainly on environmental factors: the increased heat and dryness found at higher levels of the shoreline prevent *Balanus* from surviving there.

A second case of competition affecting distribution and abundance concerns wasps of the genus *Aphytis* [aye-FITE-us]. These wasps attack scale insects, which can cause serious damage to citrus trees. Female wasps lay eggs on a scale insect, and when the wasp larvae hatch, they pierce the scale insect's outer skeleton and then consume its body parts.

Learn more about competition.

35.2

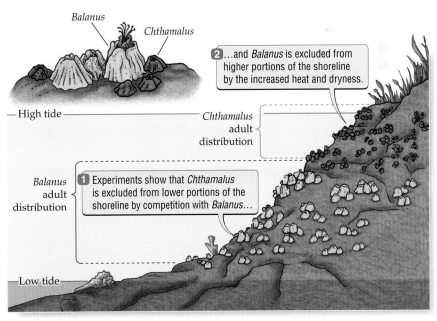

Figure 35.10 What Keeps Them Apart?
On the rocky coast of Scotland, the larvae of *Balanus* and *Chthamalus* barnacles settle on rocks on both high and low portions of the shoreline. However, adult *Balanus* barnacles are not found on high portions of the shoreline, and adult *Chthamalus* individuals are not found on low portions of the shoreline.

In 1948, the wasp *Aphytis lingnanensis* was released in southern California in order to curb the destruction of citrus trees caused by scale insects. A closely related wasp, *A. chrysomphali*, was already living in that region at the time. *A. lingnanensis* was released in the hope that it would provide better control of scale insects than *A. chrysomphali* did. *A. lingnanensis* proved to be a superior competitor (Figure 35.11), driving *A. chrysomphali* to extinction in most locations by exploitative competition. As hoped for, *A. lingnanensis* also provided better control of scale insects.

Although competition between species is very common, it is important to note that it does not always occur when two species share resources or space. This is especially true when the resources are abundant. Competition among leaf-feeding insects, for example, is relatively uncommon for this reason. A huge amount of leaf material is available for the insects to eat, and usually there are too few insects to cause their food to be in short supply. As long as their food remains abundant, little competition occurs.

Competition can increase the differences between species

As Charles Darwin realized when he formulated the theory of evolution by natural selection, competition between species can be intense when the two species are very similar in form. For example, birds whose beaks are similar in size eat seeds of similar sizes and thus compete intensely, whereas birds whose beaks differ in size eat seeds of different sizes and compete less intensely. Intense competition between similar species may result in **character displacement**, in which the forms of the competing species evolve to become more different over time. By reducing the similarity in form between species, character displacement reduces the intensity of competition. As we saw in Chapter 17, however, species can evolve in this way only if their populations vary genetically for traits (in this case, beak size) on which natural selection can act.

Some evidence for character displacement comes from observations that the forms of two species are more different when they live together than when they live in separate places. In the Galápagos Islands, for example, the beak sizes of two species of Galápagos finches, and hence the sizes of the seeds the birds eat, are more different on islands where both species live than on islands that have only one of the two species (Figure 35.12). Recent experiments with other groups, such as fish and lizards, also suggest that character displacement is important in nature.

Wasps of the genus *Aphytis* lay eggs on scale insects. When an egg hatches, the immature wasp burrows into the scale insect to feed.

Key

A. chrysomphali

A. lingnanensis

Figure 35.11 A Superior Competitor Moves In

After being introduced to southern California in 1948, the wasp *Aphytis lingnanensis* rapidly drove its competitor, *A. chrysomphali*, extinct in most locations. Both species of wasps prey on scale insects that damage citrus crops (such as lemons and oranges).

35.4 Interactions among Organisms Shape Communities and Ecosystems

Throughout this chapter we have seen how interactions among organisms help determine their distribution and

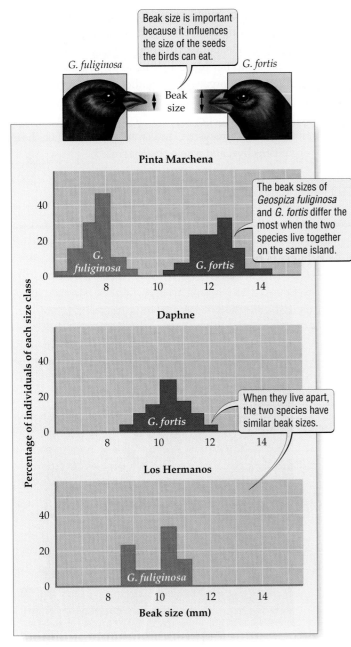

Beak size is important because it influences the size of the seeds the birds can eat.

G. fuliginosa   *G. fortis*

Beak size

Pinta Marchena

The beak sizes of *Geospiza fuliginosa* and *G. fortis* differ the most when the two species live together on the same island.

G. fuliginosa   *G. fortis*

Daphne

G. fortis

When they live apart, the two species have similar beak sizes.

Los Hermanos

G. fuliginosa

Beak size (mm)

Percentage of individuals of each size class

Figure 35.12 Character Displacement

Competition for resources may cause the competing species to become more different over time. Competition between two species of Galápagos finches, the small ground finch (*Geospiza fuliginosa*) and the medium ground finch (*G. fortis*), may be the driving force that causes the beak sizes of these birds to be more different when they live on the same island than when they live apart.

abundance. Interactions among organisms also have large effects on the communities and ecosystems in which those organisms live.

When dry grasslands are overgrazed by cattle, for example, grasses may become less abundant and desert shrubs may become more abundant. These changes in the abundances of grasses and shrubs can change the physical environment. The rate of soil erosion may increase because shrubs do not stabilize soil as well as grasses do. Ultimately, if overgrazing is severe, the ecosystem can change from a dry grassland to a desert.

Changes in interactions among organisms can have complex effects on natural communities. Recall what happened when people removed dingoes from rangelands in Australia to prevent them from eating sheep (see page 648). The removal of this predator caused other, unintended changes, including increases in the number of red kangaroos and decreases in the availability of plants that sheep like to eat.

Mutualisms can have similar large effects. For example, the fire tree was brought to Hawaii by Portuguese immigrants for use as an ornamental plant and as firewood (Figure 35.13). The fire tree forms a mutualistic association with bacteria that can convert N_2 from the air into ammonium (NH_4^+) in the soil, a form of nitrogen that

Figure 35.13 A Tree That Fertilizes Itself

The fire tree (*Myrica faya*) was brought to Hawaii by people, but has since escaped from cultivation. The fire tree forms a mutualism with bacteria that provides the tree with extra nitrogen. The mutualism gives this tree an advantage over other species, allowing it to invade ecological communities such as recent volcanic deposits and exclude other species from those sites.

plants can use. The fire tree has escaped cultivation and has invaded ecological communities such as those found on recent volcanic deposits. Because of its mutualism, the fire tree causes four times the usual amount of nitrogen to enter volcanic site ecosystems. In essence, the mutualism allows the fire tree to fertilize itself, which helps it to grow rapidly and exclude other species from colonizing volcanic sites.

In the examples discussed in this section, a change in an interaction between organisms had a ripple effect, changing the abundances of populations, the community of species living in an area, and even, in the case of the dry grasslands, converting one ecosystem (dry grassland) into another (desert shrubland). In general, interactions among organisms can affect all the levels of the biological hierarchy at which ecology can be studied: the individual organisms involved in an interaction, populations of those organisms, the communities in which those organisms live, and whole ecosystems.

Parasites Often Alter Host Behavior

Parasites affect their hosts in ways that range from merely annoying (fleas) to downright deadly (fungal parasites of ants). In addition, many parasites cause their hosts to perform unusual or even bizarre behaviors that harm the host but benefit the parasite. Recall the parasitic worms described at the opening of this chapter that cause praying mantises to throw themselves into rivers and drown. Similarly, the protist *Toxoplasma gondii* causes its rat host to become more curious and less fearful. Such changes make infected rats easier prey for cats, the other host of the protist. As these examples show, some parasites cause broad changes in their hosts' behavior, such as making the host less cautious or causing it to move from one habitat to another.

Other parasites cause much more specific changes in their hosts' behavior. A parasitic wasp called *Hymenoepimecis* [HIGH-men-oh-EP-ee-MEE-sis] attacks the spider *Plesiometa argyra* [PLEZ-ee-oh-MEET-uh are-JIRE-uh]. A female wasp stings the spider into temporary paralysis, then lays an egg on its body. The spider recovers quickly and builds normal webs for the next week or two (Figure 35.14a).

Figure 35.14 My Parasite Made Me Do It
(a) A typical web of the spider *Plesiometa argyra*. (b) A "cocoon web," produced by a *P. argyra* spider infected by a parasitic wasp that alters the spider's web-spinning behavior. The wasp's cocoon can be seen hanging down from the center of the cocoon web.

(a)

(b)

During this period, the wasp egg hatches, and the wasp larva feeds by sucking body fluids from the spider. Then, one evening, the larva injects a chemical into the spider, causing the spider to spin a unique "cocoon web" (Figure 35.14*b*). In response to the injected chemical, the spider performs many repetitions of one part of its normal web-building process, suppressing the other parts. Thus the wasp has evolved the ability to cause a very particular change in how the spider builds its web.

Why does the wasp alter spider behavior in this way? As soon as the spider finishes the web, the larva kills and consumes the spider. The larva then spins a cocoon, in which it will complete its development. The larva uses the spider's altered web as a strong support from which to hang its cocoon, thus protecting itself from being swept away by heavy rains. In effect, the wasp not only consumes the spider, but also forces the spider to build it a safe haven.

Chapter Review

Summary

35.1 Mutualisms

- Mutualisms are associations between two species in which both species benefit (+/+ interactions).
- A symbiosis is an association of two species that live together. A symbiosis may or may not be mutualistic.
- There are many types of mutualisms, including gut inhabitant, seed dispersal, behavioral, and pollinator mutualisms.
- Mutualisms evolve when the benefits of the interaction to both partners are greater than its costs for both partners.
- Mutualisms are very common in nature. Most plant species form mycorrhizae, a type of mutualism in which plant roots associate with fungi for mutual benefit.
- Mutualisms help determine the distribution and abundance of the mutualist species as well as other species that depend directly or indirectly on the mutualist species.

35.2 Exploitation

- In exploitation (+/− interactions), one species (the consumer) benefits and the other (the species that is eaten) is harmed.
- Consumers include herbivores, predators, and parasites.
- Consumers can be a strong selective force, leading their food organisms to evolve various ways to avoid being eaten. Many plants have evolved induced defenses, such as growing spines, that are directly stimulated by attacking herbivores.
- Food organisms, in turn, exert selection pressure on their consumers, which evolve ways to overcome the defenses of the species they eat.
- Consumers can restrict the distribution and abundance of the species they eat, in some cases driving their food organisms to extinction.

35.3 Competition

- In competition (−/− interactions), each of two interacting species has a negative effect on the other.
- In interference competition, one species directly excludes another species from the use of a resource.
- In exploitative competition, species compete indirectly, each reducing the amount of a resource available to the other species.
- Competition can have a strong effect on the distribution and abundance of species.
- Competition between similar species may result in character displacement, in which the forms of the competing species evolve to become more different over time.

35.4 Interactions among Organisms Shape Communities and Ecosystems

- Interactions among organisms affect individuals, populations, communities, and ecosystems.

⊙ Review and Application of Key Concepts

1. A mutualism typically has costs for both of the species involved. Why, then, are mutualisms so common?

2. Consumers affect the evolution of the organisms they eat, and vice versa. Explain how this occurs, and illustrate your reasoning with an example described in the text.

3. How can a species that is an inferior competitor nevertheless have a negative effect on a superior competitor?

4. How do ecological interactions affect the distribution and abundance of organisms?

5. Rabbits can eat many plants, but they prefer some plants over others. Assume that the rabbits in a grassland that contains many plant species prefer to eat a species of grass that happens to be a superior competitor. If the rabbits were removed from the region, which of the following do you think would be most likely to happen?
   a. The plant community would have fewer species.
   b. The plant community would have more species.
   c. The plant community would remain largely unchanged.
   Explain and justify your answer.

Key Terms

behavioral mutualism (p. 683)
character displacement (p. 690)
competition (p. 689)
control (p. 688)
control group (p. 688)
distribution (p. 684)
ecological experiment (p. 688)
exploitation (p. 684)
exploitative competition (p. 689)
gut inhabitant mutualism (p. 682)
herbivore (p. 684)
host (p. 684)
induced defense (p. 685)

interference competition (p. 689)
mutualism (p. 682)
mycorrhizae (p. 684)
parasite (p. 684)
pathogen (p. 684)
pollinator (p. 683)
pollinator mutualism (p. 683)
predator (p. 684)
prey (p. 684)
replicate (p. 688)
seed dispersal mutualism (p. 682)
symbiosis (p. 682)

Self-Quiz

1. Which of the following statements about consumers is true?
   a. Consumers cannot drive the species they eat to extinction.
   b. Consumers are not important in natural communities.
   c. Consumers can apply strong selection pressure to their food organisms.
   d. Consumers cannot alter the behavior of their food organisms.

2. In what type of interaction do species directly confront each other over the use of a shared resource?
   a. interference competition
   b. exploitative competition
   c. exploitation
   d. distribution competition

3. Interactions among species
   a. do not influence the distribution or abundance of organisms.
   b. are rarely beneficial to both species.
   c. have a strong influence on communities and ecosystems.
   d. cannot drive species to extinction.

4. The advantages received by a partner in a mutualism can include
   a. food.
   b. protection.
   c. increased reproduction.
   d. all of the above

5. The shape of a fish's jaw influences what the fish can eat. Researchers found that the jaws of two fish species were more similar when they lived in separate lakes than when they lived together in the same lake. The increased difference in jaw structure when the fish live in the same lake may be an example of
   a. warning coloration.
   b. character displacement.
   c. mutualism.
   d. exploitation.

6. Which of the following statements about symbiosis is *not* correct?
   a. Symbiosis is an association in which two or more organisms of different species live together.
   b. Symbiosis almost always involves species that benefit each other.
   c. Mutualisms can occur between the species in a symbiotic association.
   d. One species in a symbiotic association may harm the other species.

7. Experiments with the barnacle *Balanus balanoides* showed that
   a. where this species was found on the shoreline was not influenced by physical factors.
   b. competition with *Chthamalus* restricted *Balanus* to high portions of the shoreline.
   c. competition with *Chthamalus* restricted *Balanus* to low portions of the shoreline.
   d. this species restricted *Chthamalus* to high portions of the shoreline.

8. The American chestnut used to be a dominant tree species in eastern North America, but it is now virtually gone from its entire range. This species is much less common than it once was because
   a. a consumer (an introduced fungus) nearly drove it extinct.
   b. other tree species outcompeted it.
   c. insect herbivores evolved the ability to overcome the tree's defenses against herbivore attack.
   d. it could not form mycorrhizae because acid rain killed its fungal partner.

Biology in the News

At the Old Swimming Hole, a Vicious Cycle Thrives

By Donald G. McNeil, Jr.

Kwa'al, Nigeria—The pond was about the size of a school swimming pool, except it was surrounded by dry mud pocked with hundreds of hoof prints.

A herd of goats was at one edge, drinking and defecating in the same spot. The sun was going down behind a thorn tree, backlighting 50 naked boys splashing one another in the warm dusk.

Where a colonialist romantic would have seen a landscape of native innocence, I saw a horror movie: there were worms in the brown water invisibly digging right through the boys' skin.

More than a thousand different parasites can attack people. The boys in this article were at risk from a parasitic worm that can cause seizures or paralysis, and typically damages organs such as the liver, intestines, and lungs. Over 200 million people are infected by this worm, which causes the disease schistosomiasis.

The infection cycle begins when an infected person urinates or defecates into water, contaminating the water with the worm's eggs. The eggs hatch, and the parasites grow and develop inside snails. Next, the parasite leaves the snail and enters the water, where it can survive for about 48 hours. If the parasite comes into contact with the skin of a person in the water, the worm burrows through and begins to grow inside the blood vessels. Within weeks, the worms produce eggs, and the vicious cycle is ready to begin again.

In Kwa'al, Nigeria, the village in this article, a third of the children have heavy infestations of the worm. Heavy infestations have a clear warning sign—the children have blood in their urine. Children with lots of worms are stunted in their growth, do poorly in school, and suffer from ongoing medical problems.

Schistosomiasis can be treated with a drug that kills virtually all of the worms in the body. Not-for-profit organizations purchase the drug and distribute it for free to infected children. The pills are relatively inexpensive (only 7 cents per pill), but they are only made by three companies. These companies make 89 million of the pills each year, but unfortunately, nearly five times that number would be needed each year to treat the 200 million people infected with schistosomiasis.

Evaluating the news

1. This news article discusses just one parasitic disease. If we broaden our perspective to include all human parasitic infections, the annual cost in lives and human suffering is enormous. Do you think most people are aware of and care about the global effects of parasites? If not, why, and what can be done to make people better informed?

2. Do you think governments should provide subsidies to companies making relatively inexpensive drugs that can combat human parasites in developing nations? Or should market forces alone be used to address these diseases?

3. Former Costa Rican president and winner of the 1987 Nobel peace prize Oscar Arias Sanchez has estimated that universal health care could be provided to everyone in the world's developing nations for an annual cost equal to 12 percent of the combined annual military budget of those nations. He also noted that in developing nations, there are about 20 soldiers for each physician, although the chances of dying from malnutrition and preventable diseases are 33 times greater than the chances of dying from a war. Developing nations are not unique in spending much more money on military expenditures than on solutions to global health and environmental problems. Why do human societies consistently behave in this way? Should we—and could we—do things differently?

SOURCE: *The New York Times,* Tuesday, November 2, 2004.

CHAPTER 36

Communities of Organisms

Key Concepts

- A community is an association of populations of different species that live in the same area.

- Food webs describe the feeding relationships within a community.

- Keystone species play a critical role in determining the types and abundances of species in a community.

- All communities change over time. As species colonize new or disturbed habitat, they tend to replace one another in a directional and fairly predictable process called succession. Communities also change over time in response to changes in climate.

- Communities can recover rapidly from some forms of natural and human-caused disturbance, but it may take some communities thousands of years to recover from other forms of human-caused disturbance.

The Formation of a New Community

The origin of new habitat—as when an island rises out of the sea or a new lake is formed—marks the beginning of a huge and exciting natural experiment. What organisms will colonize the new habitat first? How will those organisms interact and evolve over time? Will the new habitat come to have unusual communities of organisms? Or will the communities of the new habitat be similar to other, nearby ecological communities?

The outcome of such grand experiments can be spectacular, especially when the new communities are located far from existing ones. Consider what has happened on the Hawaiian Islands, a remote chain of volcanic islands, the most recent of which (Hawaii) was formed about 600,000 years ago.

The Hawaiian Islands are so isolated (they lie 4,000 kilometers from the nearest continent) that they have been colonized by relatively few species. Over time, however, the few species that have colonized the islands have evolved to form many new species. As a result, the communities of organisms on Hawaii are very different from those anywhere else on Earth. Such a series of events is not restricted to Hawaii: unusual communities often form when new habitat is located far from existing communities.

Today, many of the unique communities on Hawaii are threatened by various human activities—including habitat destruction and the introduction of nonnative species. People brought beard grass, for example, to Hawaii as forage for cattle. By the late 1960s, beard grass had invaded the seasonally dry woodlands of Hawaii Volcanoes National Park. Before that time, fires occurred there, on average, every 5.3 years, and each fire burned an average of 0.25 hectare (about five-eighths of an acre). Since

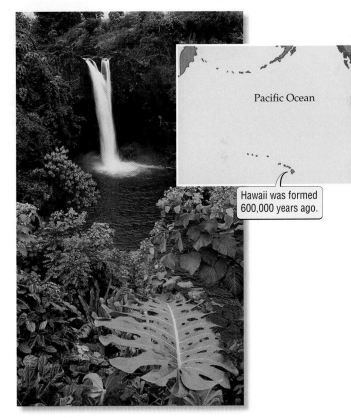

Pacific Ocean

Hawaii was formed 600,000 years ago.

A Natural Experiment
The Hawaiian Islands are part of a chain of volcanic islands that have risen from the sea over the past 70 million years. As newly formed islands were colonized by species from the mainland and as new species evolved on the islands, unique new ecological communities like this one formed.

the introduction of beard grass, fires have occurred at a rate of more than one per year, and the average burn area of each fire has increased to more than 240 hectares (about 600 acres). The fires are now so frequent and intense that the seasonally dry woodlands that once thrived in the park have disappeared.

Why did the introduction of beard grass increase the frequency and size of fires in Hawaii Volcanoes National Park? Is there something about Hawaii and other island communities that makes them particularly vulnerable to human disturbance? This example raises questions about communities of all types, not just those on islands: in the absence of dramatic disturbances such as fires or volcanic explosions, do ecological communities remain constant, or do they change over time? And finally, how well do communities bounce back from disturbances, including those caused by people?

A **community** is an association of populations of different species that live in the same area. There are many different types of communities, ranging from those found in grasslands and forests to those found in the digestive tract of a cow or deer (Figure 36.1). Most communities contain many species, and as we learned in Chapter 35, the interactions among those species can be complex.

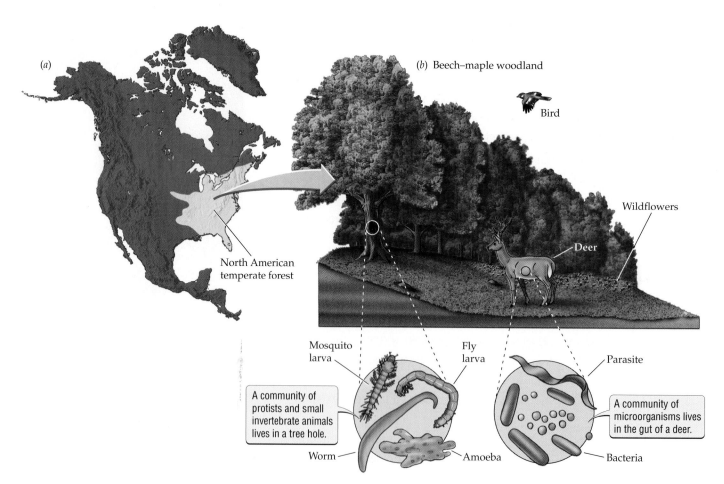

Figure 36.1 Ecological Communities

(a) Temperate forests in North America contain many types of woodland communities. (b) Smaller communities can be nested within a larger community. This beech-maple woodland community contains the community of a temporary pool of water in a tree hole and the community of a deer's gut.

Ecologists seek to understand how interactions among organisms influence natural communities, thus helping to answer questions like those we have just posed.

Ecologists also seek to understand how human actions affect communities. At present, people are having a profound effect on many different kinds of ecological communities. When we cut down tropical forests, we destroy entire communities of organisms, and when we give antibiotics to a cow, we alter the community of microorganisms that live in its digestive tract. To prevent such actions from having effects that we do not anticipate or want, we must understand how ecological communities work. In this chapter we describe the factors that influence what species are found in a community. We pay particular attention to how communities change over time and how they respond to disturbance, including disturbance caused by people.

36.1 The Effects of Species Interactions on Communities

Communities vary greatly in size and complexity, from the community of microorganisms that inhabits a small temporary pool of water to the community of plants that lives on the floor of a forest to a forest community that stretches for hundreds of kilometers. Communities can also be nested within one another, as Figure 36.1 shows. Whatever its size or type, an ecological community can be characterized by its composition, or diversity. The **diversity** of a community has two components: the number of different species that live in the community, and the relative abundances of those species (Figure 36.2).

Overall, ecological communities are influenced by the individual species that live in them, by interactions among those species, and by interactions between those species and the physical environment.

Food webs consist of multiple food chains

One important aspect of a community is which species eat what other species. These feeding relationships can be described by **food chains**, each of which describes a single sequence of who eats whom in a community. The movement of energy and nutrients through a community can be summarized by connecting the different food chains to one another to form a **food web**, which describes

Figure 36.2 **Which Community Has Greater Diversity?**
Community A and community B have the same four species of trees. However, community A is dominated by a single species, whereas all four species are equally represented in community B. Because community A is dominated by a single species, ecologists would consider it to be less diverse than community B.

the interconnected and overlapping food chains of a community (Figure 36.3).

As we saw in Chapter 1, food webs and the ecological communities they describe are based on a foundation of producers. **Producers** are organisms that use energy from an external source, such as the sun, to produce their own food without having to eat other organisms or their remains. On land, photosynthetic plants, which harvest energy from the sun, are the major producers. In aquatic biomes, a wide range of organisms serve as producers, including phytoplankton in the oceans, algae in intertidal zones and lakes, and bacteria in deep-sea hydrothermal vents.

Consumers are organisms that obtain energy by eating all or parts of other organisms or their remains. Important groups of consumers include decomposers (which we'll discuss in Chapter 37) and the herbivores, predators, and parasites (including pathogens) described in Chapter 35. **Primary consumers** are organisms, such as cows or grasshoppers, that eat producers. **Secondary consumers** are organisms, such as humans or birds, that feed on primary consumers as part or all of their diet. This sequence of organisms eating organisms that eat other organisms can continue: a bird that eats a spider that ate a beetle that ate a plant is an

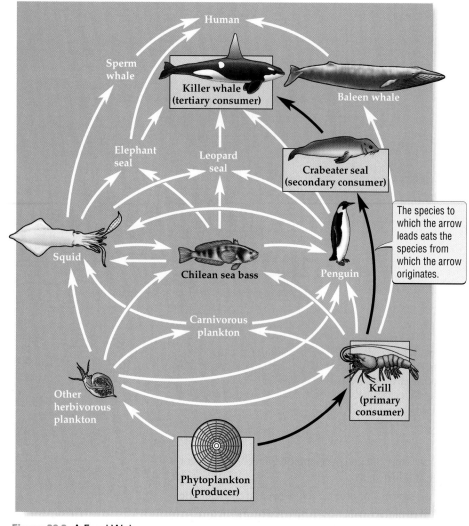

Figure 36.3 A Food Web

Food webs summarize the movement of food through a community. This figure is a simplified version of the food web in the Antarctic Ocean. Food webs are composed of many specific sequences of one species eating another known as food chains. One of the food chains found in this food web is highlighted with black arrows and yellow boxes.

Labels within figure:

Human

Sperm whale

Killer whale (tertiary consumer)

Baleen whale

Elephant seal

Leopard seal

Crabeater seal (secondary consumer)

The species to which the arrow leads eats the species from which the arrow originates.

Squid

Chilean sea bass

Penguin

Carnivorous plankton

Other herbivorous plankton

Krill (primary consumer)

Phytoplankton (producer)

Learn more about this tangled food web.   36.1

---

example of a tertiary consumer. Similarly, in the food chain highlighted in Figure 36.3, the killer whale is a tertiary consumer.

Keystone species have profound effects on communities

Interactions among organisms such as mutualism, exploitation, and competition influence the number of species found in a community, as shown by the coral reef, dingo, and barnacle examples discussed in Chapter 35. In addition, there are certain species that, relative to their own abundance, have a disproportionately large effect on the types and abundances of the other species in a community; these species are called **keystone species**.

In an experiment conducted along the rocky Pacific coast of Washington State, ecologist Robert Paine demonstrated that the sea star *Pisaster ochraceus* [pih-ZASS-ter oh-KRAY-see-us] is a keystone species in its intertidal-zone community. He removed sea stars from one site and left an adjacent, undisturbed site as a control. In the absence of sea stars, all of the original 18 species in the community except mussels disappeared (Figure 36.4). When the sea stars were present, they ate the mussels, thereby keeping the number of mussels low enough that the mussels did not crowd out the other species.

Although *Pisaster* is a predator, organisms other than predators can be keystone species. Plants such as fig trees, herbivores such as snow geese and elephants, and pathogens such as the distemper virus that kills lions have been found to be keystone species. In addition, humans often function as a keystone species: we have a large effect on interactions among other species, including species far more abundant than we are (for example, insects and small abundant plants such as grasses).

In general, the term "keystone species" can include any producer or consumer of relatively low abundance that has a large influence on its community. Although the most abundant or dominant species in a community (such as the corals in a coral reef or the mussels in Paine's intertidal zone) also have large effects on their communities, they are not considered keystone species because their abundance is not low.

Finally, it is important to note that we usually do not know in advance which species are keystone species. As a result, it is often not until after people remove a species from a community, and then observe large changes in that community, that a species is discovered to have been a keystone species. When people removed rabbits from a region in England, for example, grasslands with a variety of plant species, including many nongrasses, were converted (unintentionally) into grasslands consisting primarily of just a few species of grasses. This change occurred because the rabbits had held the grasses in check. In the absence of rabbits, the grasses crowded out other plant species.

36.2 Communities Change over Time

All communities change over time. The number of individuals of different species in a community often changes as the seasons change. For example, although they might be abundant in summer, we would not find butterflies flying in a North Dakota field in the middle of winter. Similarly, every community shows year-to-year changes in the abundances of organisms, as we saw in Chapter 34.

In addition to such seasonal and yearly changes, communities show broad, directional changes in species composition over longer periods of time.

Succession establishes new communities and replaces disturbed communities

A community may begin when new habitat is created, as when a volcanic island like Hawaii rises out of the sea. New communities may also form in regions that have been disturbed, as by a fire or hurricane. Some species arrive early in such new or disturbed habitat. These early colonists tend to be replaced later by other species, which in turn may be replaced by still other species. These later arrivals replace other species because they are better able to grow and reproduce under the changing conditions of the habitat.

The process by which species in a community are replaced over time is called **succession**. In a given location, the order in which species will replace one another is fairly predictable (Figure 36.5). Such a sequence of species replacements sometimes ends in a **climax community**, which, for a particular climate and soil type, is a community whose species are not replaced by other species. But in many—perhaps most—ecological communities, **disturbances** such as fires or windstorms occur so frequently that the community is constantly changing in response to a previous disturbance event, and a climax community never forms.

Primary succession occurs in newly created habitat, as when an island rises from the sea or when rock and soil are deposited by a retreating glacier. In such a situation, the process begins with a habitat that does not contain any species. The first species to colonize the new habitat usually have one of two advantages over other species: either they can disperse more rapidly (and hence reach the new habitat first), or they are better able to grow and reproduce under the challenging conditions of the newly formed habitat.

In some cases of primary succession, the first species to colonize the area alter the habitat in ways that allow later-arriving species to thrive. In other cases, the early colonists hinder the establishment of other species. An experimental study of primary succession was conducted on marine intertidal communities on the rocky coast near Santa Barbara, California, where researchers created

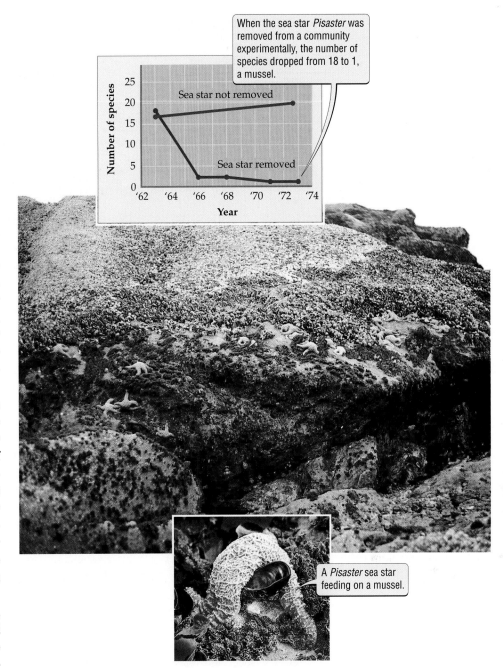

When the sea star *Pisaster* was removed from a community experimentally, the number of species dropped from 18 to 1, a mussel.

A *Pisaster* sea star feeding on a mussel.

Figure 36.4 A Keystone Species
The sea star *Pisaster ochraceus* is a predator that feeds on mussels, thereby preventing the mussels from crowding out other species in their community.

new habitats by placing concrete blocks along the shoreline. The first species of algae to colonize the concrete blocks initially inhibited other species—the ones that ultimately replaced them—from establishing themselves. In cases like this one, the early colonists eventually lose their hold on a habitat by being more susceptible than later species to some particular feature of the environment,

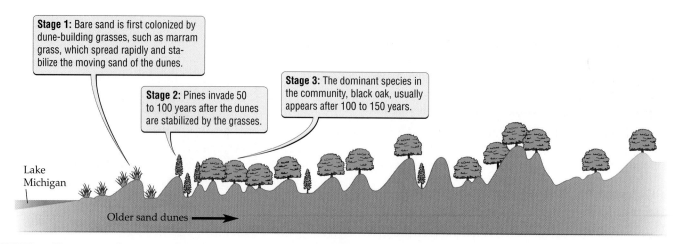

Stage 1: Bare sand is first colonized by dune-building grasses, such as marram grass, which spread rapidly and stabilize the moving sand of the dunes.

Stage 2: Pines invade 50 to 100 years after the dunes are stabilized by the grasses.

Stage 3: The dominant species in the community, black oak, usually appears after 100 to 150 years.

Lake Michigan

Older sand dunes ⟶

See this transition from shrubs to trees. 36.2

Figure 36.5 Succession

When strong winds cause sand dunes to form at the southern end of Lake Michigan, succession often leads to a community dominated by black oak. Succession on such dunes occurs in three stages and forms black oak communities that have lasted up to 12,000 years. Under different local environmental conditions, succession on Michigan sand dunes can lead to the establishment of stable communities as different as grasslands, swamps, and sugar maple forests.

such as disturbance, grazing by herbivores, or extremes of heat or cold.

Secondary succession is the process by which communities recover from disturbance, as when natural vegetation recolonizes a field that has been taken out of agriculture, or when a forest grows back after a fire (Figure 36.6). In contrast to primary succession, habitats undergoing secondary succession often have well-developed soil containing seeds from species that usually predominate late in the successional process. The presence of such seeds in the soil can considerably shorten the time required for the later stages of succession to be reached.

(a)

(b)

(c)

Figure 36.6 Secondary Succession

Forests in the United States grow back after being cut by people, blown down by windstorms, or burned by fire. These photographs, taken in different locations, show the stages of regrowth following the large fire that struck Yellowstone National Park in 1988: (a) shortly after the fire in 1988; (b) 4 years later, lodgepole pine saplings are growing in a stand of trees killed by the fire; (c) a mature lodgepole pine forest (not burned in the fire).

Communities change as climate changes

Some groups of species stay together for long periods of time. For example, an extensive plant community once stretched across the northern parts of Asia, Europe, and North America. As the climate grew colder during the past 60 million years, plants in these communities migrated south, forming communities in Southeast Asia and southeastern North America that are similar to one another—and similar in composition to the community from which they originally came.

Although groups of plants can remain together for millions of years, the community located in a particular place changes as the climate of that place changes. The climate at a given location can change over time for two reasons: global climate change and continental drift.

Consider first the climate of Earth as a whole, which changes over time. What we experience today as a "normal" climate is warmer than what has been typical for the past 400,000 years. Over even longer periods of time, the climate of North America has changed greatly (Figure 36.7), causing dramatic changes in the plant and animal species that live there. For example, fossil evidence indicates that 35 million years ago, the areas of southwestern North America that are now deserts were covered with tropical forests. Historically, changes in the global climate have been due to relatively slow natural processes, such as the advance and retreat of glaciers. However, evidence is mounting that human activities are now causing rapid changes in the global climate (see Chapter 38).

Second, as the continents move slowly over time (see Figure 19.7), their climates change. To give a dramatic example of continental drift, 1 billion years ago Queensland, Australia, which is now located at 12° south latitude, was located near the North Pole. Roughly 400 million years ago, Queensland was at the equator. The

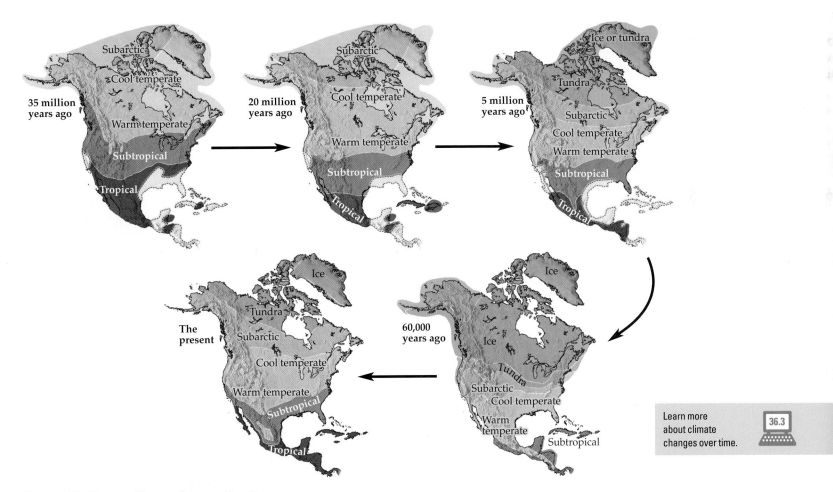

Figure 36.7 Climates Change, Communities Change

The climate of North America has changed greatly during the past 35 million years. As the climate changed, the communities found in particular places changed as well. The white regions surrounded by dotted lines are below sea level.

Learn more about climate changes over time. **36.3**

species that thrive at the equator and in the Arctic are very different; thus continental drift has resulted in large changes in the communities of Queensland over time.

36.3 Recovery from Disturbances

Ecological communities are subject to many natural forms of disturbance, such as fires, floods, and windstorms. Following a disturbance, secondary succession can reestablish the previously existing community. Thus communities can and do recover from some forms of disturbance. Depending on the community, the time required for recovery varies from years to decades or centuries.

Communities have been exposed over long periods of time to natural forms of disturbance such as windstorms. In contrast, people may introduce entirely new forms of disturbance, such as the dumping of hot wastewater into a river. People may also alter the frequency of an otherwise natural form of disturbance—for example, causing a dramatic increase or decrease in the frequency of fires or floods.

Communities can recover from some human-caused disturbances

Can communities recover from disturbances caused by people? For some forms of human-caused disturbance, the answer is yes. Throughout the eastern United States, for example, there are many places where forests were cut down and the land used for farmland; years later, the farmland was abandoned. Second-growth forests have grown on these abandoned farms, often within 40 to 60 years after farming stopped.

Second-growth forests are not identical to the forests that were originally present. The sizes and abundances of the tree species are different, and fewer plant species grow beneath the trees of a second-growth forest than beneath a virgin forest (one that has never been cut down). However, the second-growth forests of the eastern United States already have recovered partially from cutting. If current trends continue, over the next several hundred years there will be fewer and fewer differences between such forests and the original forests.

In some cases, communities can also recover from pollution. Consider Lake Washington, which is a large, clear lake in Seattle, Washington (Figure 36.8). As the city of Seattle grew, raw sewage was dumped into the lake. This practice declined after 1926 and was stopped by 1936. Beginning in 1941, treated sewage was discharged into Lake Washington from newly constructed sewage treatment plants.

A major effect of discharging sewage—treated or not—into Lake Washington was the addition of extra nutrients to the lake, in the form of phosphorus in the sewage. As a result, the numbers of algae soared, decreasing the clarity of the water. As the algae reproduced and then died, their bodies provided an abundant food source for bacteria, whose populations also increased. Bacteria use oxygen when they consume dead algae, so concentrations of dissolved oxygen in the water decreased. The lowered oxygen concentrations killed invertebrates and fish. These events illustrate how a body of water can be degraded by **eutrophication** [*YOU*-truh-fih-*KAY*-shun] (from *eu*, "well"; *troph*, "nourished"), a process in which enrichment of water by nutrients (often from sewage or runoff from fertilized agricultural fields) causes bacterial populations to increase and oxygen concentrations to decrease.

By the early 1960s Lake Washington was so degraded that it was referred to in the local press as "Lake Stinko." From 1963 to 1968, the dumping of sewage into the lake was reduced, and virtually none was dumped after 1968. Once inputs of sewage stopped, algae populations declined, oxygen concentrations increased, and Lake Washington returned to its former, clear state.

Figure 36.8 Lake Stinko
In the early 1960s, Lake Washington (in Seattle, Washington) was so polluted that it was dubbed "Lake Stinko" by the local press.

Other, even larger bodies of water, including Lake Erie in North America and the Black Sea in Eurasia, have also shown signs of recovery from disturbances caused by people. In the 1980s, high nutrient inputs caused the Black Sea to become eutrophic [you-*troh*-fik]. Conditions in the Black Sea were made even worse by the introduction (probably in ships' ballast water) of the North American comb jelly, *Mnemiopsis* [*nem*-ee-*op*-sis], a voracious predator that ate the plankton that anchovies and other fish depend on for food (Figure 36.9). When comb jellies were introduced to the Black Sea, they found themselves in an environment with abundant food and relatively few predators. As a result, comb jelly numbers exploded, which caused anchovy populations to crash and the Turkish anchovy fishing industry to collapse. A mere 10 years later, however, the situation had improved: through reductions of nutrient inputs into the Black Sea and the accidental introduction (again, probably in ballast water) of a predator that eats *Mnemiopsis*, the sea became less eutrophic, *Mnemiopsis* numbers plummeted, and anchovy populations recovered.

People can cause long-term damage to communities

It is encouraging that complex ecological communities such as those in Lake Washington and the Black Sea can recover rapidly from disturbances caused by people. However, communities do not always recover so quickly from human-caused disturbances, as a few examples will show.

Northern Michigan once was covered with a vast stretch of white pine and red pine forest. Between 1875 and 1900, nearly all of these trees were cut down, leaving only a few scattered patches of virgin forest. The loggers left behind large quantities of branches and sticks, which provided fuel for fires of great intensity. In some locations, the pine forests of northern Michigan have never recovered from the combination of logging and fire.

Throughout South America and Southeast Asia, large tracts of tropical forest have been converted into grasslands by a combination of logging and fire (Figure 36.10). Scientists estimate that it will take tropical forest communities hundreds to thousands of years to recover from such changes.

Anchovies swimming

A *Beroe* jelly eating another comb jelly

Figure 36.9 Changes in the Black Sea
In the late 1980s, the accidental introduction of the comb jelly *Mnemiopsis* wreaked havoc on the food webs of the Black Sea, causing a decline in the abundance of many species and the collapse of the anchovy fishing industry. Conditions began to improve after the introduction of *Beroe*, a predator capable of controlling *Mnemiopsis* populations.

Figure 36.10 Communities Do Not Always Recover from Disturbance

(a) This photograph of a low-elevation Indonesian rainforest was taken in 1980. One to two years after this picture was taken, the forest was partially logged. Partial logging makes forests more susceptible to fire, and in 1982, the rainforest burned. (b) Three years after the fire, as seen here, few trees remained alive. (c) A second fire occurred in 1998, killing the remaining trees and converting the former rainforest into a grassland dominated by an introduced grass species.

(a)

(b)

(c)

Finally, in some areas of the American Southwest, overgrazing by cattle has transformed dry grasslands into desert shrublands (Figure 36.11). How do cattle cause such large changes? Grazing and trampling by cattle decrease the abundance of grasses in the community. With less grass to cover the soil and hold it in place, the soil dries out and erodes more rapidly. Desert shrubs thrive under these new soil conditions, but grasses do not. These changes in soil characteristics can make it very difficult to reestablish grasses, even when the cattle are removed.

Grazing can have a dramatic effect on dry grasslands, but how do you think the effects of grazing would compare to the effects of an atomic bomb? The first outdoor explosion of an atomic bomb occurred on July 16, 1945, at the Trinity site in New Mexico (Figure 36.12). Fifty years later, dry grasslands that had been destroyed by the bomb blast (but which had never been grazed) had

(a)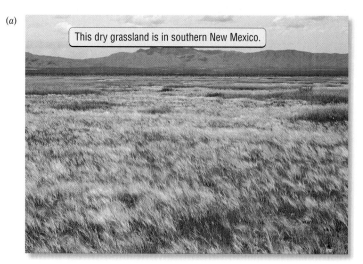

This dry grassland is in southern New Mexico.

(b)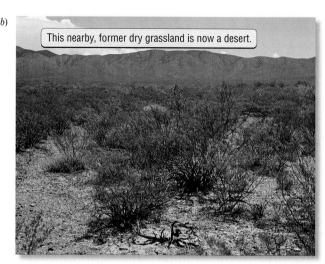

This nearby, former dry grassland is now a desert.

Figure 36.11 Overgrazing Can Convert Grasslands into Deserts

(a) More than 200 years ago, large regions of the American Southwest were covered with dry grasslands.
(b) Most of these grasslands have been converted into desert shrublands, in large part because of overgrazing by cattle.

Figure 36.12 At the Trinity Site
The world's first outdoor atomic explosion occurred in July 1945 at the Trinity site in New Mexico. Fifty years later, researchers learned that grasslands destroyed by the explosion recovered more rapidly than grasslands that had been overgrazed by cattle.

recovered. In contrast, nearby dry grasslands that had been heavily grazed (but not destroyed by the bomb blast) had not recovered, even though they had not been grazed since the time of the blast. Thus the plant community recovered more rapidly from the effects of a nuclear explosion than from the effects of grazing—a dramatic example of how strongly ecological interactions can affect natural communities.

In the three examples discussed in this section, people have altered ecological communities so greatly that it will take hundreds or thousands of years for those communities to recover. Do such changes differ in any way from changes that occur in the absence of people?

Changes in communities affect their value to humans

Human actions have the power to change communities rapidly across large geographic regions. Disturbances not caused by people can also occur rapidly and affect large geographic regions. For example, most scientists think that the impact of a large asteroid contributed to the sudden extinction of many species, including the last of the dinosaurs, 65 million years ago (see Chapter 19). However, changes in communities caused by people are different in one important sense: we can consider the consequences of our actions and use that information to decide what to do. Many people think our unique ability to control our actions brings with it the responsibility to use our power to change communities wisely.

What human values are affected by human-caused community change? First, human actions that degrade or destroy ecological communities have ethical implications. When people disrupt communities, our actions kill individual organisms, alter communities that may have persisted for thousands of years, and threaten species with extinction. According to results from surveys in North America and Europe, many people think that such effects of human actions on individuals, communities, and species are ethically unacceptable.

Second, human-caused changes can reduce the aesthetic value of a community. Tropical forests, for example, have unique aesthetic value to many people. When our actions cause tropical forests to be destroyed and replaced by introduced grasses (see Figure 36.10), we deprive current and future generations of experiencing the beauty of those forests.

Finally, in many cases, when people change a community, its economic value is reduced. Economic value is lost, for example, when human actions convert grasslands into deserts, in part because there is more plant material to support grazing in grasslands than in deserts. In general, when our actions cause long-term damage to ecological communities, we run the risk of harming our own long-term economic interests.

Overall, communities change constantly in the face of both natural and human-caused disturbances. Can human societies manage Earth's changing communities while maintaining the aesthetic and economic values they provide for us? One way to work toward this goal is to design the indoor and outdoor spaces that we use in ways that reduce their environmental impact (see the box on page 708). We'll return to efforts to reduce the human impact on the environment in Interlude G, where we focus on how to build a sustainable society.

Biology on the Job

Designing with Nature

When we plant a garden, construct a pond, build a road, or plan for the long-term growth of a city, people think about how to alter outdoor space for our use. All such activities involve changing existing ecological communities and establishing new ones. It is possible to alter outdoor space in ways that serve human needs and protect key aspects of natural communities. We discussed how this can be done with Carlyn Worstell, a landscape architectural designer who works at the Bronx Zoo.

What is landscape architecture, and what drew you to this field?
Landscape architecture is the design of outdoor spaces for use by people. I was drawn to landscape architecture because of my interest in nature and the environment. As a teen, I designed a backyard pond for fish, and while in college I worked as an intern at botanical gardens and zoos. I also had a strong interest in environmental education and in the challenge of re-creating exotic environments.

What is your educational and professional background, and how did you get your first job in landscape architecture? I was a student at Cornell University, where I received both an undergraduate and a master's degree in landscape architecture. After I graduated, I got a scholarship that enabled me to spend one and a half years studying how gorilla habitats were designed in 20 different zoos in Europe. When I returned to the States, I got a job as a landscape architectural designer here at the Bronx Zoo, where I've been working for the past 3 years.

What do you do during a typical day's work? I do many different things related to exhibit design, including sketch different possibilities for fencing, water features (such as pools or moats), or paths. I research animals and the plants that are part of their ecosystem. I also design pavilions and seating areas, and hold meetings with administrators to understand what they want the public to learn about animals, what the budget is to build an exhibit, and when the exhibit is scheduled to open.

What are the foremost concerns in designing a landscape? Well-being and safety for animals, and creating an environment that will elicit the most natural behaviors from the animals. We also have to consider the aesthetic quality, and such urban design concerns as accessibility and the flow of people through a space. An enclosure must accommodate the needs of the animals and the animal keepers—and it must appeal to children and other visitors in order to provide the best opportunities for education.

What elements are animals most sensitive to? Plants play a big role—they provide shade and they break up the space so animals can hide from one another if they need privacy. Plants also provide enrichment because the animals are stimulated by the different textures around them and have things to do. Water features such as moats serve as a more natural-looking barrier between the animals and people, but they also are a good way for animals to cool off.

Are some ecological communities more difficult to re-create than others? Why? Desert or tropical communities are difficult in New York. To create a more realistic-looking environment, we sometimes use plants that might be native to New York, but look like or are pruned to resemble plants native to the habitat we are trying to re-create. Indoor enclosures can be challenging since they must be climate controlled yet energy efficient. That way we can both protect wildlife and conserve Earth's natural resources.

What has been your most challenging or memorable assignment? I am currently working on an indoor LEED (leadership and energy efficient) design in a historic building from 1911. It's a landmark building so the outside façade cannot be changed, but they are trying to build an extensive Madagascar exhibit inside. To conserve energy they need to use low lighting levels, but the plants need high light to live. This is a tough one!

A Hippo Exhibit at the Hanover Zoo, Germany

What do you enjoy most about your job? I love learning about new animals, plants, and ecosystems. And it is a lot of fun to meet people who are out there doing hands-on conservation in different places all around the world. My job gives me a chance to blend creative work (the aesthetic concerns of design) with practical work (trying to make a design suit the needs of all the people who use it).

The Job in Context

To do her job well, Carlyn Worstell has to understand the animals she works with and the communities in which they live. When she helps build an exhibit, she mimics a natural community, building it from scratch. It takes roughly a year to design an exhibit, and then an additional 2 years to construct it. This process leaps over many of the typical steps in ecological succession—machines are used to move earth, and animals and plants are brought from other locations to form a human-made ecological community. But succession and other aspects of ecological change cannot be completely circumvented. The plants in the exhibit must be given time to establish themselves before the animals are introduced—otherwise, the plants will not be able to tolerate the tendency of animals to dig them up, scratch against them, and sometimes eat them. Zoo staff must also prevent the colonization of unwanted species (such as a plant species that might outcompete those in the exhibit), they must monitor the quality of the water, and animals or plants that are not healthy must be treated or replaced. Like any ecological community, a zoo exhibit is a dynamic place—it is a challenging and rewarding job to help ensure that it continues to serve the needs of the animals that live there.

Introduced Species: Taking Island Communities by Stealth

The Hawaiian Islands are the most isolated chain of islands on Earth. Because the islands are so remote, entire groups of organisms that live in most communities never reached them. For example, there are no native ants or snakes in Hawaii, and there is only one native mammal (a bat, which was able to fly to the islands).

The few species that did reach the Hawaiian Islands found themselves in an environment that lacked most of the species from their previous communities. The sparsely occupied habitat and the lack of competitor species resulted in the evolution of many new species. For example, genetic evidence indicates that the many different species of Hawaiian silversword [SIL-ver-SORD] plants found on the islands today (Figure 36.13) all arose from a single ancestor. The new silversword species evolved very different forms, enabling them to live

Figure 36.13 Great Diversity from a Single Ancestor
Hawaiian silverswords are found only on the Hawaiian Islands. Genetic evidence indicates that this diverse group of plant species evolved from a single ancestor. Although the three silversword species shown here are closely related, they live in very different habitats and differ greatly from one another in form.

in a number of different habitats. Because other groups of plants and animals on the islands also evolved many new species, the Hawaiian Islands have many unique natural communities.

Since the arrival of people on the islands, about 1,500 to 2,000 years ago, Hawaii's unique communities have been threatened by habitat destruction, overhunting, and introduced species. Of these threats, the effects of introduced species can be the easiest to overlook because such species often wreak their havoc quietly, behind the scenes. Introduced Argentine ants, for example, may drive native insects to extinction, but it can take years before even a trained biologist realizes what the introduced ants have done.

Island communities are particularly vulnerable to the effects of introduced species. Relatively few species colonize newly formed islands, and those species then evolve in isolation. For this reason, species on islands may be ill-equipped to cope with new predators or competitors that are brought by people from the mainland. In addition, introduced species often arrive without the predator and competitor species that held their populations in check on the mainland. Thus, on islands, the potential exists for populations of introduced species to increase dramatically and become invasive.

In some cases, introduced species can destroy entire communities. For example, if most of the native plants are not adapted to fire, an introduced species that alters the frequency or intensity of fire can have devastating effects. Recall the description at the opening of this chapter of what happened after beard grass was introduced to Hawaii Volcanoes National Park: the frequency of fires in the park increased by more than five times, and the burn area of an average fire increased by more than 960 times.

Why did the introduction of beard grass have these effects? As beard grass grows, it deposits a large amount of dry plant matter on the ground. This material catches fire easily, and the fires burn much hotter than they would in the absence of beard grass. Beard grass recovers well from large, hot fires, but the native trees and shrubs of the seasonally dry woodland do not. As a result, former woodlands have now been converted into open meadows filled with beard grass and other, even more fire-prone, introduced grasses.

The Hawaiian dry woodland community has been destroyed, probably forever. Because there is no hope of restoring the native community, ecologists are now trying to construct a new community that is tolerant of fire yet contains native trees and shrubs. This is a difficult challenge, and it is uncertain whether the effort will succeed. If not, what was once woodland is likely to remain indefinitely as open meadows filled with introduced grasses.

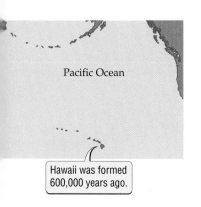

Pacific Ocean

Hawaii was formed 600,000 years ago.

Chapter Review

Summary

36.1 The Effects of Species Interactions on Communities

- Communities can be described by food webs, which summarize the interconnected food chains describing who eats whom in a community.
- Producers produce their own energy from an external source, such as the sun. Consumers get their energy by eating all or parts of other organisms. Primary consumers eat producers, while secondary consumers feed on primary consumers.

- A keystone species has a large effect on the composition of a community relative to its abundance.
- Keystone species alter the interactions among organisms in a community, thus changing the types or abundances of species in the community.

36.2 Communities Change over Time

- All communities change over time.

- Directional changes that occur over relatively long periods of time have two main causes: succession and climate change.
- Primary succession occurs in newly created habitat. Secondary succession occurs in communities recovering from disturbance.
- The climate at a given location can change because of global climate change or continental drift.

36.3 Recovery from Disturbances

- Communities can recover from some forms of natural and human-caused disturbance. The time required for recovery varies from years to decades or centuries.
- Degradation of water bodies resulting from eutrophication can be reversed if the sources of nutrient enrichment are removed.
- It can take hundreds to thousands of years for communities to recover from some forms of human-caused disturbance.
- Community change caused by people is unique in that we can consider the consequences of our actions and use that information to decide what to do. We can choose to act in ways that do not reduce the aesthetic or economic value of ecological communities.

◉ Review and Application of Key Concepts

1. Describe how each of the following factors influences ecological communities: (a) species interactions, (b) disturbance, (c) climate change, (d) continental drift.

2. What is the difference between primary and secondary succession?

3. Provide an example of how the presence or absence of a species in a community can alter a feature of the environment, such as the frequency of fire.

4. Consider two forms of human disturbance to a forest:
   a. All trees are removed, but the soils and low-lying vegetation are left intact.
   b. The trees are not removed, but a pollutant in rainfall alters the soil chemistry so greatly that the existing trees can no longer thrive.
   Which form of disturbance do you think would require the longest recovery time before a healthy forest community could once again be found at the site? Explain the assumptions you used to answer this question, and justify the conclusion that you reached.

5. Do you think it is ethically acceptable for people to change natural communities so greatly that it takes thousands of years for the communities to recover? Why or why not?

6. Explain the difference between a keystone species and one of the most abundant or dominant species of a community.

Key Terms

climax community (p. 701)
community (p. 698)
consumer (p. 699)
disturbance (p. 701)
diversity (p. 699)
eutrophication (p. 704)
food chain (p. 699)
food web (p. 699)

keystone species (p. 700)
primary consumer (p. 699)
primary succession (p. 701)
producer (p. 699)
secondary consumer (p. 699)
secondary succession (p. 702)
succession (p. 701)

Self-Quiz

1. A species that has a large effect on a community relative to its abundance is called a
   a. predator.   c. keystone species.
   b. herbivore.   d. dominant species.

2. Organisms that can produce their own food from an external source of energy without having to eat other organisms are called
   a. suppliers.   c. producers.
   b. consumers.   d. keystone species.

3. Ecological communities
   a. cannot recover from disturbance.
   b. can recover from natural but not human-caused disturbance.
   c. can recover from all forms of disturbance.
   d. can recover from some, but not all, forms of natural and human-caused disturbance.

4. Which of the following was *not* caused by the introduction of beard grass to Hawaii?
   a. an increase in the growth of native trees and shrubs
   b. an increase in the frequency and intensity of fire
   c. the decline of native trees and shrubs
   d. the conversion of dry woodlands into grasslands

5. A directional process of species replacement over time in a community is called
   a. global climate change.   c. competition.
   b. succession.   d. community change.

6. A community whose species are not replaced by other species is known as a ———— community.
   a. primary succession
   b. climax
   c. competitive
   d. disturbance-based

7. A single sequence of feeding relationships describing who eats whom in a community is a
   a. life history.
   c. food web.
   b. keystone relationship.
   d. food chain.

8. The process in which the enrichment of water by nutrients causes bacterial populations to increase and oxygen concentrations to decrease is called
   a. eutrophication.
   c. fertilization.
   b. disturbance.
   d. nutrient loading.

Biology in the News

Linking Rivers, With Happy Results

BY HENRY FOUNTAIN

The engineering marvel that is the Panama Canal created a link between the Atlantic and Pacific Oceans. But as a byproduct it also connected rivers that had been separated by the Continental Divide.

That river link, between the Rio Chagres on the Caribbean side and the Rio Grande on the Pacific, allowed freshwater fish species from both sides to mingle. Now, a researcher has studied how those fish have mixed in the 90 years since the canal was built, with an eye to answering some basic ecological questions about the nature of communities.

Human actions sometimes drastically alter the ecological communities of an area, causing a shift from one community type to another. In less extreme cases, the community remains similar to what it once was, but some species are driven to extinction while others are catapulted to new dominance. Even in these less extreme cases, what often happens is that more species are lost than are gained, so the number of species in the community drops. But in the example reported here, human actions appear to have led to an increase—not a decrease—in the number of species.

Working with several other scientists, Dr. Scott A. Smith, of McGill University and the Smithsonian Tropical Research Institute, made use of a detailed biological survey that had been conducted just before the Panama Canal was built. "We knew which species were where," Dr. Smith said. "Then we were able to sample [the same two] rivers and examine how community composition had changed over time."

They found that, at least for the fish in the Rio Chagres and the Rio Grande, the effect of building the canal has been positive: no species in either river went extinct, and some species from the Rio Chagres successfully colonized the Rio Grande, and vice versa.

These results may surprise some ecologists, especially those who think of communities as being full, or "saturated." If a community is saturated, all of the possible food sources and microenvironments are being used by species in the community. According to this view, colonization events should occur only if the newcomers replace one or more of the existing species, thereby driving local species extinct.

But the results published in the August issue of the prestigious *Proceedings of the Royal Society* showed that the number of species in the two rivers in Panama actually rose rather than fell. As such, Dr. Smith said that his results support the idea that communities are "neutral" groups of species, in which species are not so precisely adapted and the community is not so tightly knit, thus allowing new species to thrive without driving existing species extinct.

Evaluating the news

1. Given the results found by Dr. Smith and his colleagues, do you think we need to worry about human actions that may introduce new species to ecological communities? Justify your answer.
2. Many members of the public think that each species in nature has a specific, unique, and irreplaceable role to play. People who hold this view may conclude that it is unethical to drive other species extinct (because humans have no right to disrupt the tightly knit relationships found in ecological communities). Many ecologists, on the other hand, do not hold this view. Instead, they view communities as being composed of more loosely connected groups of interacting species. In this view, it may be possible to add or remove some species without greatly affecting other species in the community.

SOURCE: *The New York Times,* Tuesday, September 21, 2004.

Ecologists, however, recognize that all events in nature are interconnected, and that actions that affect one species may affect other species. What are the ethical implications—if any—of the view that events in nature are interconnected?
3. Imagine that you were a policymaker who had to decide what to do about a complex global environmental problem, such as global warming or declining fish populations in the ocean. Do you think it would be best to take decisive action now to prevent possible future harmful effects of human actions? Bear in mind that taking decisive action now would risk spending money needlessly if the problem should turn out to be less serious than we feared. Or do you think it would be better to proceed cautiously, spending limited sums to gather more information, but waiting until that information becomes available before taking decisive action? The risk with this second course of action is that the problem will get worse in the interim and may cost much more to fix than it would have cost if decisive action had been taken earlier.

Key Concepts

◉ An ecosystem consists of a community of organisms together with the physical environment in which those organisms live. Energy, materials, and organisms can move from one ecosystem to another.

◉ Energy enters an ecosystem when producers capture it from an external source, such as the sun. A portion of the energy captured by producers is lost as heat at each step as it moves through a food chain. As a result, energy cannot be recycled.

◉ Earth has a fixed amount of nutrients. If nutrients were not recycled between organisms and the physical environment, life on Earth would cease. Human activities affect some nutrient cycles.

◉ Ecosystems provide humans with essential services, such as nutrient recycling, at no cost. Our civilization depends on these and many other ecosystem services.

Is There a Free Lunch?

Next time you drink a glass of water, think for a moment about where your water comes from. If you are like many of us, you may not know. Does it come from surface waters, such as rivers, lakes, or reservoirs? Or does it come from deep underground? Whatever its source, the delivery of safe drinking water can make for an interesting story.

Consider New York City. The 8 million people who live in that city get about 90 percent of their water from the Catskill Watershed, with the remainder coming from the Croton Watershed. Together, these watersheds store 580 billion gallons of water in nineteen reservoirs and three controlled lakes. Over 1.3 billion gallons of this water is delivered to New York each day. The water flows by gravity to the city in a vast set of pipes, some of them large enough to drive a bus through.

For years, New Yorkers drank high-quality water, essentially for free: their water was kept pure by the root systems, soil micro-organisms, and natural filtration processes of forests in the Catskill and Croton watersheds. By the late 1980s, however, pollutants such as sewage, fertilizers, pesticides, and oil began to overwhelm these purification processes, causing the quality of the water to decline. New York had a problem, one that could be very expensive to fix.

The standard way of dealing with surface water contamination is to build a water treatment plant. Faced with violations of Environmental Protection Agency (EPA) water quality regulations, New York readily agreed to build such a plant for the Croton

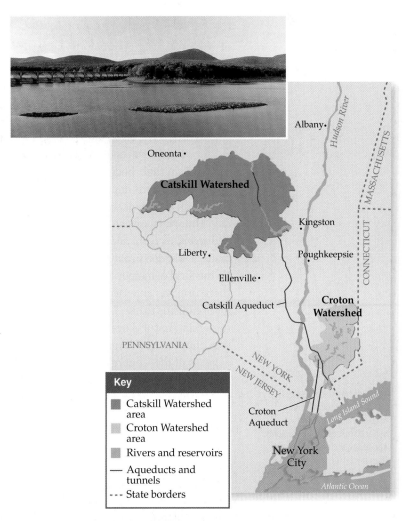

New York City's Water Supply System
The Ashokan Reservoir, located in the Catskill Forest Preserve near Woodstock, New York, is one of several reservoirs that supplies water to New York City.

Key
- Catskill Watershed area
- Croton Watershed area
- Rivers and reservoirs
- Aqueducts and tunnels
- State borders

Watershed, for an estimated cost of $300 million. But the city balked at the price tag for treating the much larger supply from the Catskill Watershed. Early estimates put the cost of that treatment plant at $6 billion to $8 billion, plus another $300 million per year for its operation. Could a less expensive solution be found?

The answer turned out to be yes. For a projected cost of $1.5 billion, the city, in the early 1990s, embarked on an ambitious but simple plan: protect the environment of the watershed so that natural systems can resume supplying the city with clean water. The city is buying land that borders rivers and streams in the Catskills, protecting the land from development in order to minimize inputs of fertilizers, pesticides, and other pollutants into the water. The city is also building new storm sewers and septic systems, upgrading existing sewage treatment plants, and providing funds to encourage environmentally friendly forms of development.

New York City is not alone. Cities throughout the world are struggling to cope as increasing development causes natural water purification systems to lose their effectiveness. And in at least one case, a private company has taken actions similar to those taken in New York. The bottled water company Perrier Vittel became concerned in the late 1980s that pollutants would threaten the quality of its water. These concerns were driven home a few years later when the company had to pull its water from the shelves due to contamination with benzene, a carcinogen found in gasoline. Rather than get its water from a new source—as some other companies had done when faced with similar issues—Perrier Vittel decided to invest in the environment. To protect its water, the company spent $9 million to acquire land and to reach agreements with farmers to reduce fertilizer and pesticide use.

As the New York City and Perrier Vittel examples suggest, there really can be a free lunch: unless we overtax their ability to do so, ecological communities provide us with many free services, such as the purification of water. How do ecosystems accomplish this?

To survive, all organisms need energy to fuel their metabolism and materials to construct and maintain their bodies. For their energy needs, most organisms depend directly or indirectly on solar energy, an abundant supply of which reaches Earth each day. Materials, on the other hand, such as the carbon, hydrogen, oxygen, and other elements of which we are made, are added to our planet in relatively small amounts (in the form of meteoric matter from outer space). Earth, therefore, has an essentially fixed amount of materials for organisms to use. This simple fact means that for life to persist, natural systems must recycle materials. In this chapter our focus will be on these two essential aspects of life—energy and materials—as we discuss ecosystem ecology, the study of how energy and materials are used in natural systems.

Helpful to know

In discussing the biology of humans and other animals, *nutrients* refers to vitamins, minerals, essential amino acids, and essential fatty acids (see Chapter 21). In the ecosystem context of this chapter, however, *nutrients* has a more restricted meaning, referring only to the essential elements required by producers.

37.1 How Ecosystems Function: An Overview

An **ecosystem** consists of a community of organisms together with the physical environment in which those organisms live. Like communities, ecosystems may be small or very large: a puddle teeming with protists is an ecosystem, as is the Atlantic Ocean. In fact, global patterns of air and water circulation (see Chapter 33) may be viewed as linking all the world's organisms into one giant ecosystem, the biosphere. But whether they are large or small, ecosystems can be challenging to study because organisms, energy, and materials often move from one ecosystem to another.

Figure 37.1 gives a broad overview of how ecosystems work. First, examine the orange arrows in the figure, which show the movement of energy through the ecosystem. At each step in a food chain (see Figure 36.3 for an example of a food chain), a portion of the energy captured by producers is lost as metabolic heat, which is released as the inevitable by-product of the chemical breakdown of food (see Chapter 7). Organisms lose a lot of energy as metabolic heat, as can be seen by the fact that a small room crowded with people rapidly becomes hot. Because of this steady loss of heat, energy flows in only one direction through ecosystems: it enters Earth's ecosystems from the sun and leaves them as metabolic heat.

In contrast to energy, **nutrients**—the chemical elements required by producers (see page 612)—are recycled between living organisms and the physical environment. As shown in Figure 37.1, nutrients are absorbed from the environment by producers, cycled among consumers for varying lengths of time, and eventually returned to the environment when the ultimate consumers—decomposers—break down the dead bodies of organisms.

To understand how ecosystems work, ecologists focus on the two processes we've just discussed: the one-way flow of energy and the cycling of nutrients. As we'll describe, they study how organisms capture energy and nutrients from the environment, transfer energy and nutrients to one another, and ultimately, return nutrients to the environment. We turn now to a more detailed look at the first of these steps, energy capture in ecosystems.

37.2 Energy Capture in Ecosystems

Most life on Earth depends directly or indirectly on the capture of solar energy by producers. The energy captured by plants and other photosynthetic organisms is stored in their bodies in the form of chemical compounds, such as carbohydrates. Herbivores (which eat plants), predators (which eat herbivores and other predators), and decomposers (which consume the remains of dead organisms) all depend indirectly on the solar energy originally captured by plants.

To better understand these points, imagine a world with no plants. Although bathed in sunlight, herbivores and predators alike would starve because they could not use that energy to produce food. This line of thinking also helps us to understand why some environments can support more animals than other environments can. In a tropical forest, for example, there are many plants that can capture energy from the sun (Figure 37.2). As a result, a large amount of energy from the sun is stored in chemical forms that can be used as food by animals. In contrast, in environments with few plants (such as arctic or desert regions), relatively little energy is captured from the sun. Hence less food is available in such environments, and fewer animals can live there.

Overall, the amount of energy captured by plants is an important first step in determining how an ecosystem works: it influences the amount of plant growth and hence the amount of food available to other organisms. Each of these factors, in turn, influences the type of ecosystem found in a region and how that ecosystem functions.

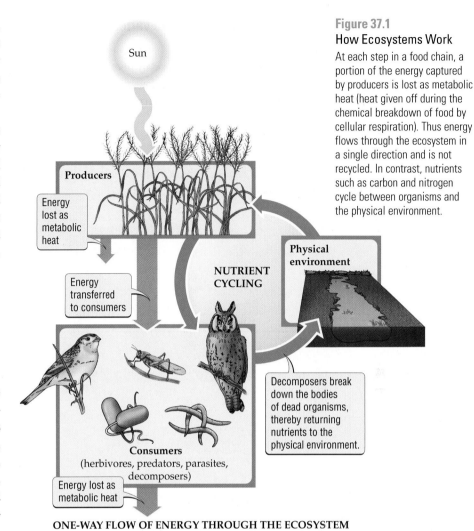

Figure 37.1
How Ecosystems Work
At each step in a food chain, a portion of the energy captured by producers is lost as metabolic heat (heat given off during the chemical breakdown of food by cellular respiration). Thus energy flows through the ecosystem in a single direction and is not recycled. In contrast, nutrients such as carbon and nitrogen cycle between organisms and the physical environment.

Sun

Producers

Energy lost as metabolic heat

Energy transferred to consumers

NUTRIENT CYCLING

Physical environment

Decomposers break down the bodies of dead organisms, thereby returning nutrients to the physical environment.

Consumers
(herbivores, predators, parasites, decomposers)

Energy lost as metabolic heat

ONE-WAY FLOW OF ENERGY THROUGH THE ECOSYSTEM

Figure 37.2 What's for Dinner?
(a) In tropical rainforests, the abundant producers (plants) store a lot of chemical energy, which in turn is available to consumers. (b) In arctic regions, little chemical energy is stored for consumers due to the sparse plant life there.

The rate of energy capture varies across the globe

The amount of energy captured by photosynthetic organisms, minus the amount they lose as metabolic heat, is called **net primary productivity** (**NPP**). Although it is defined in terms of energy, it is usually easier to estimate NPP as the amount of new **biomass** (the mass of organisms) produced by the photosynthetic organisms in a given area during a specified period of time. In a grassland ecosystem, for example, ecologists would estimate NPP by measuring the average amount of new grass and other plant matter produced in a square meter each year. Such NPP estimates based on biomass can be converted to units based on energy.

NPP is not distributed evenly across the globe. On land, NPP tends to decrease from the equator toward the poles (Figure 37.3a). This decrease occurs because the amount of solar radiation available to plants also decreases from the equator to the poles (as we saw in Chapter 33). But there are many exceptions to this general pattern. For example, there are large regions of very low NPP in northern Africa, central Asia, central Australia, and the southwestern portion of North America. Each of these regions is the site of one of the world's major deserts.

The low NPP in deserts emphasizes the fact that sunlight alone is not sufficient to produce high NPP; water is also required. In addition to water and sunlight, productivity on land can be limited by temperature and the availability of nutrients in the soil. The most productive terrestrial ecosystems are tropical rainforests and farmland, while the least productive are deserts and tundra (including some mountaintop communities).

The global pattern of NPP in marine ecosystems (Figure 37.3b) is very different from that on land. There is little tendency for NPP to decrease from the equator to the poles. Instead, the general pattern relates to distance from shore: the productivity of marine ecosystems is often high in ocean regions close to land, but relatively low in the open ocean, which is, in essence, a marine "desert." Exceptions occur where upwelling of water from the ocean bottom provides nutrients to marine organisms. Streams and rivers that drain from the land into the ocean carry nutrients that are in short supply in the ocean. Addition of these nutrients to ocean water stimulates the growth and reproduction of phytoplankton, the small producers that form the foundation of aquatic food webs. As on land, the NPP in aquatic ecosystems can be strongly limited by sunlight and temperature.

Although the world's oceans generally have very low productivity relative to terrestrial ecosystems (compare the maps in Figures 37.3a and 37.3b), coral reefs and estuaries are among Earth's most productive ecosystems (Figure 37.4), rivaling tropical forests and farmland; wetlands such as swamps and marshes can also match the productivities of tropical forests and farmland.

(a)

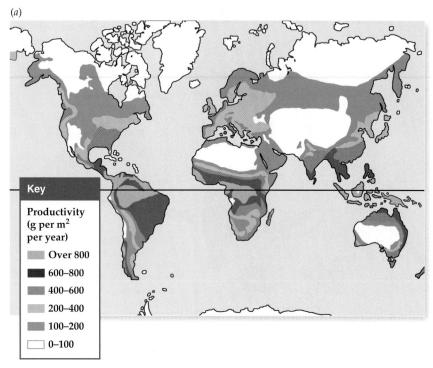

Key

Productivity (g per m² per year)

- Over 800
- 600–800
- 400–600
- 200–400
- 100–200
- 0–100

(b)

Key

Productivity (g per m² per year)

- >90
- 55–90
- 35–55
- <35

Figure 37.3 Net Primary Productivity Varies Across the Globe

Net primary productivity varies greatly among both (a) terrestrial ecosystems and (b) marine ecosystems. Net primary productivity is measured as grams of new biomass made by plants or other producers each year in a square meter of area (g per m² per year).

(a)

(b)

Figure 37.4 Most Productive Ecosystems
(a) Coral reefs like this one in the northern Bahamas and (b) wetlands like this estuary can harness energy from sunlight and oxygen and translate it into high net primary productivity (NPP) on par with terrestrial systems like tropical forests and farmlands.

Human activities can increase or decrease NPP

Human activities can change the amount of energy captured by ecosystems on local, regional, and global scales. For example, rain can cause fertilizers to wash from a farm field into a stream, which then flows into a lake. When extra nutrients are added to a lake, the lake becomes eutrophic, as we saw in Chapter 36. In a eutrophic lake, NPP usually increases because the added nutrients cause

photosynthetic algae to become more abundant, and hence they absorb more energy from the sun.

It is not necessarily a good thing when human actions cause NPP to increase. In some instances, nutrient enrichment has increased NPP to such an extent that large bodies of water have become nearly devoid of life. Each summer, large amounts of nitrogen are brought by the Mississippi and other rivers to the Gulf of Mexico (Figure 37.5). Algae

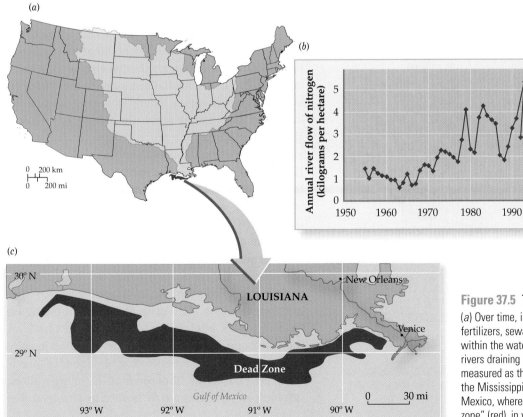

Figure 37.5 The Dead Zone
(a) Over time, increasingly large amounts of nitrogen (N) from fertilizers, sewage, and industrial by-products have entered waters within the watersheds (yellow) of the Mississippi River and other rivers draining into it. (b) The resulting nutrient addition can be measured as the amount of nitrogen flowing past particular points on the Mississippi per year. (c) This nitrogen then drains into the Gulf of Mexico, where each summer the extra nutrients create a large "dead zone" (red), in which virtually all organisms are killed.

populations (and hence NPP) increase in these nutrient-rich waters, which float on top of the colder, saltier waters of the Gulf. As the algae die, they drift into deeper waters, where bacteria decompose their bodies, using oxygen in the process. As a result, oxygen levels in these deeper waters drop so low that virtually all organisms die, creating a large "dead zone" (see Figure 37.5c). This dead zone threatens to diminish the fish and shellfish industry in the Gulf, which produces an annual catch worth about $500 million. In the summer of 2002, the dead zone reached its largest size ever—about 22,000 square kilometers (8,500 square miles)—covering an area greater than the state of Massachusetts.

Human activities can also change NPP on land. For example, NPP decreases when logging and fire convert tropical forest to grassland (see page 704). Globally, scientists estimate that human activities leading to such land conversions have decreased NPP in some regions while increasing it in other regions, but the net effect is a 5 percent decrease in NPP worldwide.

Learn more about the idealized energy pyramid.

37.1

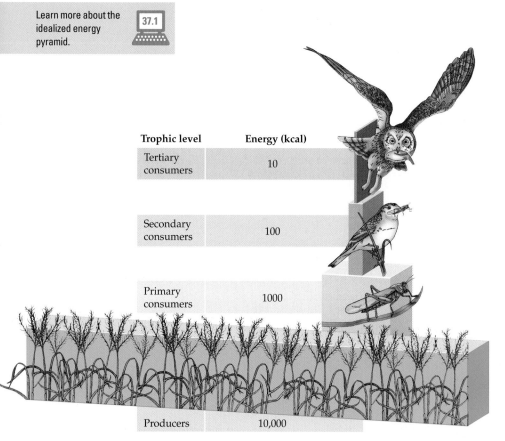

Figure 37.6 An Idealized Energy Pyramid
Of each 10,000 kilocalories (kcal; 1 kilocalorie equals 1,000 calories) of energy from the sun captured by producers, primary consumers harvest only about 10 percent. Roughly 10 percent of the energy at each trophic level is then transferred to the next trophic level.

37.3 Energy Flow Through Ecosystems

As described in Chapter 36, there are two major ways in which organisms capture energy. Producers get their energy from abiotic sources, such as the sun. The organisms at the bottom of a food web, such as plants, algae, and photosynthetic bacteria, are producers. Consumers get their energy by eating all or parts of other organisms or their remains. Consumers include decomposers, such as bacteria and fungi, that break down the dead bodies of organisms, as well as the herbivores, predators, and parasites described in Chapter 35.

An energy pyramid shows the amount of energy transferred up a food chain

Energy from the sun is stored by plants and other producers in the form of chemical compounds. This stored energy can be transferred from one organism to another up a food chain, but eventually every unit of energy captured by producers is lost as heat. Thus energy cannot be recycled.

To illustrate these points, let's follow the fate of energy from the sun after it strikes the surface of a grassland. A portion of the energy captured by grasses is transferred to the herbivores that eat the grasses, and then to the predators that eat the herbivores. The transfer of energy from grasses to herbivores to predators is not perfect, however. When a unit of energy is used by an organism to fuel its metabolism, some of that energy is lost from the ecosystem as unrecoverable heat, as we have seen. Thus energy moves through ecosystems in a single direction: as one proceeds up a food chain (for example, from grass to grasshopper to bird, as shown in Figure 37.6), portions of the energy originally captured by photosynthesis are steadily lost. Because of this steady loss of energy, more energy is available toward the bottom than toward the top of a food chain.

The amounts of energy available to organisms in an ecosystem can be represented by a pyramid. Each level of an energy pyramid corresponds to a step in a food chain and is called a **trophic level** (see Figure 37.6). The grass-grasshopper-bird example shown in the figure has four trophic levels: grass is on the first trophic level, the grasshopper is on the second, the insect-eating bird is on the third, and the bird-eating bird is on the fourth. On average, roughly 10 percent of the energy at one trophic level is transferred to the next trophic level. The energy that is not transferred to the next trophic level is not consumed (for example, when we eat an apple, we eat only a small part of the apple tree), is not taken up by the body (for example, we cannot digest the cellulose that is contained in the apple), or is lost as metabolic heat.

Secondary productivity is highest in areas of high NPP

The rate of new biomass production by consumers is called **secondary productivity**. As we've seen, because consumers depend on producers for both energy and materials, secondary productivity is highest in ecosystems with high net primary productivity. A tropical forest, for example, has a much higher NPP than tundra. For this reason, per unit of area, there are many more herbivores and other consumers in tropical forests than in tundra, and hence secondary productivity is much higher in tropical forests than in tundra.

In natural ecosystems, new biomass made by plants and other producers is consumed either by herbivores or by decomposers. In some ecosystems, 80 percent of the biomass produced by plants is used directly by decomposers. Eventually, since all organisms die, all biomass made by producers, herbivores, predators, and parasites is consumed by decomposers (Figure 37.7). In some instances, people bypass the decomposers, as when we use crops or agricultural refuse to produce fuels (see the box on page 722).

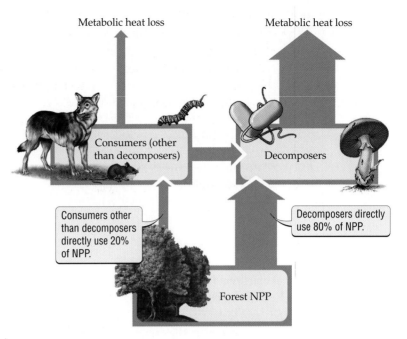

Figure 37.7 Decomposers Consume Most of NPP
Decomposers such as bacteria and fungi use more than 50 percent of net primary productivity in ecosystems of all types. In this forest, 80 percent of NPP is used directly by decomposers, and the remaining 20 percent is used by other consumers (such as herbivores and predators).

to producers, such as rocks, ocean sediments, or fossilized remains of organisms. Weathering of rocks, geologic uplift, human actions, and other forces can move nutrients back and forth between such reservoirs and **exchange pools**—sources such as soil, water, and air where nutrients are available to producers.

Helpful to know

Reservoir in this chapter has two meanings: It can refer to a body of water held aside for use by people, as in the New York City water supply. More generally, a *reservoir* is any place or substance that holds something (not necessarily water) in storage, as is meant with regard to nutrient cycling.

37.4 Nutrient Cycles

Nutrients—chemical elements such as carbon, hydrogen, oxygen, and nitrogen—are used by producers and other organisms to construct their bodies. Producers obtain these and other essential nutrients from the soil, water, or air in the form of ions such as nitrate (NO_3^-) or inorganic molecules such as carbon dioxide (CO_2). Consumers obtain them by eating producers or other consumers. Because these nutrients are essential for life, their availability and movement through ecosystems influences many aspects of ecosystem function.

Nutrients are transferred between organisms and the physical environment in a cyclical pattern called a **nutrient cycle**. Figure 37.8 provides a general description of how nutrient cycles work. Nutrients can be stored for long periods of time in reservoirs that are inaccessible

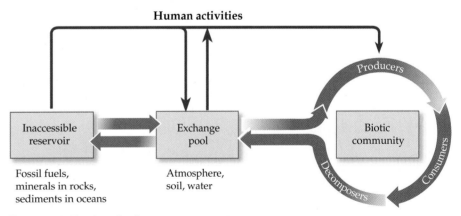

Figure 37.8 Nutrient Cycling
Nutrients cycle between reservoirs that are inaccessible to producers, exchange pools that are available to producers, and living organisms. Human activities, such as fossil fuel use and fertilizer synthesis, can move nutrients between inaccessible reservoirs and exchange pools, altering their availability to producers and changing nutrient cycles.

Biology on the Job

From Corn to Cars

With gasoline prices soaring and dependence on oil produced in other countries a concern, many nations are looking for alternative sources of automotive fuel. Some countries have turned to ethanol, a form of alcohol that can be made from corn or other organic materials. Brazil, for example, invested heavily in ethanol fuels beginning in the 1970s; now, 40 percent of the fuel used to drive cars in that country is ethanol, compared with 3 percent in the United States. Ben P. Sever is a lawyer at a company in the United States that makes ethanol for use as automotive fuel.

What are your day-to-day tasks? What does your job entail? I draft and revise contracts and advise the company on contract negotiations. Sometimes I perform negotiations that lead to contracts with shippers, corn suppliers, enzyme suppliers, and people buying the ethanol. The ethanol we sell is blended with gasoline for motor vehicle fuel.

What do you like most about your job? I believe in the product that we produce, and think it is environmentally sound. Ethanol is good for farmers and it decreases the nation's dependence on foreign oil. Also, I get to look at a variety of legal issues, so every day brings something new.

What are the legal issues surrounding ethanol and ethanol production? Ethanol is an alcohol, so its production is regulated by the federal government. To operate an ethanol company, you have to get a variety of permits from the Tax and Trade Bureau, which used to be part of the Bureau of Alcohol, Tobacco, and Firearms. The ethanol has to be altered where it is made so that it cannot be used for human consumption.

What are some of the environmental impacts of ethanol production? In terms of the environmental implications of operating an ethanol production plant, you don't have to be very concerned about wastes or air emissions. This is true because part of the corn or biomass that you are using is turned into ethanol, and leftover wastes are sold to feed companies. The carbon dioxide that is produced when you make ethanol can be sold to companies that collect it.

What future do you see for ethanol-based alternative fuels? Traditionally, ethanol fuels are 10 to 15 percent ethanol blended with gasoline. Emerging E-85 fuel is 85 percent ethanol. A standard automobile can be modified to run on that, and although there aren't many gas stations that currently sell it, there will be.

Do you have a scientific background? Not at all. I have a B.S. in political science and I began my career in securities and corporate law. In a job like mine, you have to deal with permits and you have to negotiate intellectual property contracts. Certain companies have proprietary blends of enzymes (used to break down the corn), and you have to negotiate to be able to use those enzymes. It would be handy to have a scientific background, since otherwise you face a pretty steep learning curve. But I learn faster on the job than in the classroom anyway. If you are good student—not necessarily of academic things but of life—you'll do well no matter what subject your degree is in, or what you go into.

An Alternative Source of Automotive Fuel
New ethanol fuels may prove to be an important alternative source of energy.

The Job in Context

When fully combusted, ethanol is a clean-burning fuel—only carbon dioxide and water are produced. Although carbon dioxide contributes to global warming (see page 743), use of ethanol fuels can help reduce emissions of other pollutants into the atmosphere and could thus reduce smog in many large cities. Despite these potential benefits, efforts to expand the use of ethanol as an automotive fuel have been criticized by some people, largely because studies conducted in the 1970s and early 1980s showed that it took more energy to produce ethanol fuels than could be gained from those fuels. How can that be?

The answer requires keeping track of all energy inputs and outputs, just as we do when we study the flow of energy in a natural ecosystem. Although corn, for example, captures energy from the sun, and microorganisms are then used to produce ethanol from the corn, other energy inputs are required before ethanol can be pumped into a car. Energy must be expended to grow the corn (including energy used to run farm machinery and to manufacture fertilizers and pesticides), to transport the corn, to culture the microorganisms that produce the ethanol, and to transport the ethanol to the pump. Over the last 20 years, steady progress has been made, and by 2002, the U.S. Department of Agriculture calculated that 134 units of energy are produced for every 100 units of energy expended to produce ethanol. Further progress lies ahead, as innovative new approaches (including production of ethanol from waste products such as wood chips and rice hulls) begin to be used commercially.

Once captured by a producer, a nutrient can be passed from the producer to an herbivore, then to one or more predators or parasites, and eventually to a decomposer. Decomposers break down once-living tissues into simple chemical components, thereby returning nutrients to the physical environment. Without decomposers, nutrients could not be repeatedly reused, and life would cease because all essential nutrients would remain locked up in the bodies of dead organisms.

The length of time it takes for a nutrient to cycle from a producer to the physical environment and back to another producer depends on which of two main types of nutrient cycles the nutrient has: an atmospheric cycle or a sedimentary cycle.

Sulfur is one of several important nutrients with an atmospheric cycle

Nutrients that cycle between terrestrial ecosystems, aquatic ecosystems, and the atmosphere are said to have an **atmospheric cycle**. These nutrients enter the atmosphere easily because they commonly occur as gases. Once in the atmosphere, they can be transported by wind from one region of Earth to another. When nutrients are transported long distances in this way, they can affect nutrient cycles in distant ecosystems.

Carbon, hydrogen, oxygen, nitrogen, and sulfur are all essential nutrients that have atmospheric cycles. Let's look at the cycling of sulfur as an example. There are three natural ways by which sulfur enters the atmosphere from terrestrial and aquatic ecosystems (Figure 37.9): in sea spray, as a metabolic by-product (the gas hydrogen sulfide, H_2S) released by some types of bacteria, and, least important in terms of overall amount, as a result of volcanic activity. Human activities also cause sulfur to enter the atmosphere, as we shall see shortly.

Sulfur enters terrestrial ecosystems through the weathering of rocks and as sulfate (SO_4^{2-}) that is lost from the atmosphere. Sulfur enters the ocean in stream runoff from land and, again, as sulfate lost from the atmosphere. Once in the ocean, sulfur cycles within marine ecosystems before being lost in sea spray or deposited in sediments on the ocean bottom. Sulfur, like most nutrients with atmospheric cycles, cycles through terrestrial and aquatic ecosystems relatively quickly.

Learn more about the sulfur cycle.

37.2

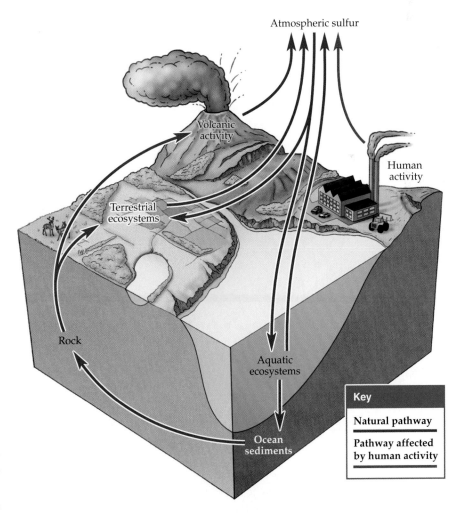

Figure 37.9 The Sulfur Cycle

Phosphorus is the only major nutrient with a sedimentary cycle

A nutrient that does not enter the atmosphere easily is said to have a **sedimentary cycle**. Such nutrients first cycle within terrestrial and aquatic ecosystems for variable periods of time (from a few years to many thousands of years); then they are deposited on the ocean bottom as sediments. Nutrients may remain in sediments, unavailable to most organisms, for hundreds of millions of years. Eventually, however, the bottom of the ocean is thrust up by geologic forces to become dry land, and once again the nutrients in the sediments may be available to organisms. Sedimentary nutrients usually cycle very slowly, so they are not replaced easily once they are lost from an ecosystem.

Among the major nutrients that cycle within ecosystems, phosphorus is the only one with a sedimentary cycle (Figure 37.10). The reason for this is that soil conditions usually do not allow bacteria to carry out the chemical reactions required for the production of a gaseous form of phosphorus (phosphine, PH_3).

Explore the phosphorous cycle.

Phosphorus is important to ecosystems because it strongly affects net primary productivity, especially in aquatic ecosystems. NPP usually increases, for example, when phosphorus is added to lakes. As we have seen, such an increase in productivity can have undesirable effects, including eutrophication, which leads to the death of aquatic plants, fish, and invertebrates (see Figure 37.5 and the story of Lake Washington on page 704).

Human activities can alter nutrient cycles

Human activities can have major effects on nutrient cycles. Ecologists have shown, for example, that the clear-cutting of a forest, followed by spraying with herbicides to prevent regrowth, causes the forest to lose large amounts of nitrate, an important source of nitrogen for plants (Figure 37.11).

On a larger geographic scale, nutrients such as nitrogen and phosphorus used to fertilize crops can be carried by streams to a lake or an ocean hundreds of kilometers away, where they can increase NPP and cause the body of water to become eutrophic. Finally, on a still larger geographic scale, many human activities, such as shipping crops and wood to distant locations, transport nutrients around the globe. Many human activities also release chemicals into the air, where global wind patterns move them over long distances.

When people alter atmospheric nutrient cycles, the effects are often felt across international borders. Consider sulfur dioxide (SO_2), which is released into the atmosphere when we burn fossil fuels such as oil and coal. The burning of fossil fuels has altered the sulfur cycle greatly: annual human inputs of sulfur into the atmosphere are more than one and a half times the inputs from all natural sources combined.

Most human inputs of sulfur into the atmosphere come from heavily industrialized areas such as northern Europe and eastern North America. Once in the atmosphere, SO_2 is dissolved in water and converted into sulfuric acid (H_2SO_4), which then returns to Earth in rainfall. Rainfall normally has a pH of 5.6, but sulfuric acid (as well as nitric acid, caused by nitrogen-containing pollutants) has caused the pH of rain to drop to values as low as 2 or 3 in the United States, Canada, Great Britain, and Scandinavia (see Chapter 4 for a review of pH). Rainfall with a low pH is called **acid rain**.

Acid rain can have devastating effects on human-made structures (such as statues) and on natural ecosystems. Acid rain has drastically reduced fish populations in thousands of Scandinavian and Canadian lakes. Much of the acid rain that falls in these lakes is caused by sulfur

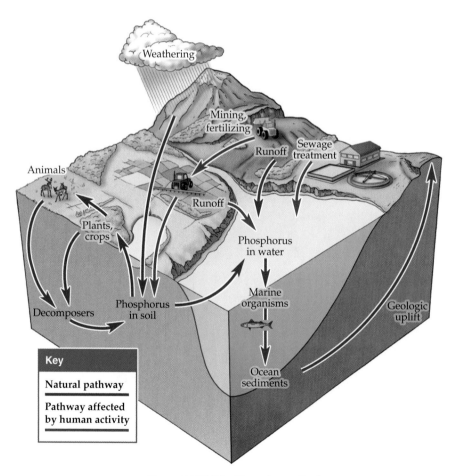

Figure 37.10 The Phosphorus Cycle

Figure 37.11

Altering Nutrient Cycles in a Forest Ecosystem

(a) This portion of the Hubbard Brook Experimental Forest in New Hampshire has been clear-cut. (b) A portion of the forest was first clear-cut, then sprayed with herbicides for 3 years to prevent regrowth. A second portion of the forest was not clear-cut or sprayed and served as a control plot. Nitrate, a form of nitrogen that is important to plants, was lost from the ecosystem in streams at a much higher rate in the clear-cut portion of the forest than in the control portion. (The loss of nitrate was measured in milligrams of nitrate in the stream per liter of water.)

(a)

(b)

dioxide pollution that originates in other countries (such as Great Britain, Germany, and the United States). Acid rain has also caused extensive damage to forests in North America and Europe (Figure 37.12).

The international nature of the acid rain problem has led nations to agree to reduce sulfur emissions. In the United States, annual sulfur emissions were cut by nearly 40 percent between 1980 and 2001 (see Figure 37.12). Such reductions are a very positive first step, but the problems resulting from acid rain will be with us for a long time: acid rain alters soil chemistry and thus has effects on ecosystems that will last for many decades after the pH of rainfall returns to normal levels.

37.5 Ecosystem Design

Throughout the world, much effort is currently being devoted to the design and construction of ecosystems. One reason for these efforts is the desire to restore or replace ecosystems that have been degraded or destroyed by human activities. In the Netherlands, for example, where the land has been so heavily modified by people for so long that few natural ecosystems remain, intensive work is under way to rebuild some of the original ecosystems. In many other countries, similar efforts are in progress to replace heavily damaged ecosystems, such as native prairie ecosystems in the United States.

People also seek to design and build ecosystems for economic reasons. There is often a potential for conflict between developers who want to build homes or industrial parks and environmental activists who seek to prevent such development in order to protect natural ecosystems. Some argue that such conflict between developers and environmentalists could be avoided by playing a type of "zero sum" game: if an ecosystem such as a wetland is destroyed or degraded by development, a comparable ecosystem should be built to replace the damaged natural ecosystem.

In principle, such a policy could work if people could build ecosystems with properties similar to those of natural ecosystems. Are we able to do this? Results to date indicate that we have much to learn. With respect to our ability to build ecosystems from scratch, consider what

Figure 37.12 Acid Rain

Acid rain can damage many ecosystems, such as this spruce forest killed by acid rain in the Jizerske Mountains of Czechoslovakia. As shown in the graph, however, the amount of sulfur dioxide, a major contributor to acid rain, emitted into the atmosphere each year in the United States has fallen by more than 6.5 million tons since 1980.

happened with Biosphere II, an experimental 3-acre enclosure containing several artificial ecosystems (Figure 37.13). Biosphere II—named in reference to Biosphere I, or Earth—was designed to house eight people and be self-sustaining for a 2-year period. Despite the expenditure of more than $200 million on this project, it had to be shut down early because of a slew of problems, including pest population outbreaks, the extinction of all pollinators, the extinction of 19 of 25 vertebrate species, a drop in oxygen levels (from 21 percent to 14 percent, the level usually found at elevations of 5,300 meters, or 17,500 feet, at which most people develop altitude sickness), and spikes in the concentration of nitrous oxide high enough to impair human brain function.

With respect to efforts to build specific ecosystems, a 2001 National Academy of Sciences report on wetland restoration concluded that we can build ponds and cattail marshes, but we cannot replace fens, bogs, and many other complex, species-rich wetland ecosystems. Similar challenges face efforts to restore terrestrial ecosystems. Restored prairies, for example, can resemble native prairies in some respects after as few as 5 to 10 years. However, even 30 to 50 years after restoration work has begun, critical differences may remain in the way nutrients are cycled in restored versus native prairie ecosystems. Most ecosystems are complex, and we often do not know enough about fundamental ecosystem processes to replicate those processes in a human-built ecosystem.

Figure 37.13 Biosphere II: A Lesson in Humility

Despite an expenditure of over $200 million, the Biosphere II experiment, which began in 1991 and was intended to run for 2 years, had to be shut down early. The artificial ecosystems in Biosphere II, such as the inset underwater view of the ocean zone, could not duplicate the services found in natural communities, and hence were not self-sustaining.

The Economic Value of Ecosystem Services

New York City had a compelling economic reason to restore the ability of forest ecosystems to purify the city's water supply: faced with spending an estimated $6 to $8 billion on a water treatment plant, planners chose instead to spend roughly $1.5 billion to purchase land, update sewer and septic systems, and promote environmentally friendly forms of development.

More recent estimates put the price tag for the treatment plant at closer to $4 billion, while costs for restoring the ecosystem's water purification services are now expected to be over $2 billion. Still, the New York City example shows that what is good for the environment can also be sound economic policy. Is this an isolated example? Or are there other cases in which ecosystems provide people with services that have economic value?

Consider the floods that struck the western United States in 1996 and 1997. Damage amounted to billions of dollars in the states of Nevada, California, Oregon, and Washington (Figure 37.14). Most news reports about the floods and associated mudslides said they were caused by unusually large amounts of rainfall and snowfall. But a huge flood is not always something that just happens, beyond our control. Some human actions prevent ecosystems from responding as they normally would to heavy rainfall, thus helping to set the stage for a flood.

How can human actions increase the chance of flooding? People often build dikes or divert the flow of rivers to protect homes or industrial areas located in what were once floodplains. By preventing rivers from overflowing into floodplains, we reduce the ability of the ecosystem to handle periods of heavy rainfall. Floodplains normally function as huge sponges: when streams and rivers overflow, floodplains absorb the excess water, thus preventing even more severe floods from occurring farther downstream. By building on floodplains and attempting to control floods, we unintentionally make it more likely that when a flood does occur, it will be a big one.

Like the forests that filter pollutants from New York City's water, floodplains provide us with a free service: they act as safety valves for major floods, preventing even larger floods. Ecological communities provide us with many such **ecosystem services** (Figure 37.15), including removal of pollutants from the air by plants, pollination of plants by

(a)

(b)

Figure 37.14 Flood Devastation in the Pacific Northwest
(a) Following heavy flooding in Oregon City during the 1990s, people used boats and float tubes to get around. (b) Mudslides can kill people, destroy homes, contaminate stream ecosystems, and have undesirable aesthetic effects.

Natural services provided by ecosystems

Removal of pollutants; clean air and clear water

Climate moderation

Replacement of groundwater

Erosion control

Overall effect: Productive ecosystem

Damage to ecosystems resulting in loss of natural services

Pollution

More severe climate

Depletion of groundwater

Severe runoff and erosion

Overall effect: Loss of ecosystem productivity

Figure 37.15 Services Provided by Ecosystems
Ecosystems provide humans with many essential services for free. These same services are also essential to maintaining the productivity of ecological communities. Only a few of the many ecological services provided by ecosystems are diagrammed here.

insects (essential for many crops), maintenance of breeding grounds for shellfish and fish in marine ecosystems, prevention of soil erosion by plants, screening of dangerous ultraviolet light by the atmospheric ozone layer, moderation of the climate by the ocean, and, as we have seen throughout this chapter, nutrient cycling. Such ecosystem services are essential for maintaining healthy ecological communities, and they also provide people with enormous economic benefits.

What is the total economic value of ecosystem services? In one sense, their value is infinite: we cannot live without the clean air and water they provide. But even in a more restricted sense, the monetary value of ecosystem services is enormous, as illustrated by the billions of dollars New York City would need to spend to replace just one of those services, water purification. Globally, the value of ecosystem services provided by lakes and rivers alone has been estimated by some researchers to be a staggering $1.7 trillion each year. Other examples of the value of ecosystem services include the world's fish catch (valued at between $50 billion and $100 billion per year) and the billions of dollars' worth of crops that could not be produced if plants were not pollinated by insects.

Human civilization depends on many essential services provided by ecosystems. At present, while we can design landscapes that incorporate aspects of natural ecosystems, we cannot duplicate with technology what ecosystems provide us for free. Although we usually have only a rough idea of the economic value of particular ecosystem services, we do know that when we destroy or degrade ecosystems, we place ourselves—and our wallets— at peril.

Biology Matters

How Much Are Lakes and Rivers Worth?

In the 1990s, a group of 13 researchers set out to estimate the dollar value of ecosystem services to the global economy. They estimated, for example, that insects and other pollinators, by fertilizing plants, contribute a service that is worth more than $19 billion each year in terms of the production of grains and fruits that we eat, or that are eaten by animals that we eat. The researchers added up similar sums for 17 ecosystem services provided by each of 16 biomes (which they defined and grouped together in a slightly different way than we did in Chapter 33). They estimated that the total economic value of ecosystem services is probably at least $33 trillion per year—nearly double the amount of the total global gross national product ($18 trillion per year).

Most of the $33 trillion value of ecosystem services is not bought or sold on world economic markets. This means that we receive trillions of dollars in benefits from these services, but we do not have to pay for them. If such costs were reflected in day-to-day transactions, these researchers say, society would pay more attention to what is lost when natural communities are damaged or destroyed.

The table below shows the value of ecosystem services provided by just one of the 16 biomes, lakes and rivers.

Biome	Global area (hectares × 10⁶)	Ecosystem services (1994 US $ per hectare, per year)					Total global value (billions per year)
		Food production	Water regulation	Water supply	Waste treatment	Recreation	
Lakes/ rivers	200	41	5,445	2,117	665	230	1,700

Chapter Review

Summary

37.1 How Ecosystems Function: An Overview

- Energy and materials can move from one ecosystem to another.
- A portion of the energy captured by producers is lost as metabolic heat at each step in a food chain. Thus energy moves through ecosystems in just one direction.
- Nutrients are recycled in ecosystems. They pass from the environment to producers to various consumers, then back to the environment when the ultimate consumers—decomposers—break down the bodies of dead organisms.

37.2 Energy Capture in Ecosystems

- Most life on Earth depends on energy that is captured from sunlight by producers and stored in their bodies as chemical compounds.
- Energy capture in an ecosystem is measured as net primary productivity (NPP); the amount of NPP in an area is an important first step in determining the type of ecosystem found there and how it functions.
- On land, NPP tends to decrease from the equator toward the poles. In marine ecosystems, NPP tends to decrease from relatively high values where the ocean borders land to low values in the open ocean

(except where upwelling provides scarce nutrients to marine organisms). Aquatic ecosystems on land (such as wetlands) can also show high NPP.

- Human activities can increase or decrease NPP on local, regional, and global scales.

37.3 Energy Flow Through Ecosystems

- The energy captured by producers can be transferred from organism to organism up a food chain.
- Once an organism uses energy to fuel its metabolism, that energy is lost from the ecosystem as heat.
- The amounts of energy available to organisms at different trophic levels in an ecosystem can be represented in the form of a pyramid.
- Secondary productivity is highest in areas of high net primary productivity.

37.4 Nutrient Cycles

- Nutrients are cycled between organisms and the physical environment.
- Decomposers return nutrients from the bodies of dead organisms to the physical environment.
- Nutrients that enter the atmosphere easily (in gaseous form) have atmospheric cycles, which occur relatively rapidly and can transfer nutrients between distant parts of the world.
- Nutrients that do not enter the atmosphere easily have sedimentary cycles, which usually take a long time to complete.
- Human activities can alter nutrient cycles on local, regional, and global scales.
- Human activities that alter atmospheric nutrient cycles can affect ecological communities located in distant parts of the world.
- Human inputs to the sulfur cycle exceed those from all natural sources combined, creating problems of international scope, such as acid rain.

37.5 Ecosystem Design

- People seek to design and restore ecosystems for aesthetic and economic reasons.
- Attempts to create self-sustaining ecosystems from scratch or restore damaged ecosystems have not fully succeeded. Humans do not yet know enough about the fundamental, complex processes of ecosystems to build ecosystems that function like those in the natural world.

◉ Review and Application of Key Concepts

1. Some people think the current U.S. Endangered Species Act should be replaced with a law designed to protect ecosystems, not species. The intent of such a law would be to focus conservation efforts on what its advocates think really matters in nature—whole ecosystems. Given how ecosystems are defined, do you think it would be easy or hard to determine the boundaries of what should and should not be protected if such a law were enacted? Give reasons for your answer.

2. What prevents energy from being recycled in ecosystems?

3. What is the essential role of decomposers in ecosystems?

4. Explain why human alteration of nutrient cycles can have international effects.

5. Describe some key ecosystem services and discuss the extent to which human economic activity depends on such services.

Key Terms

acid rain (p. 724)
atmospheric cycle (p. 723)
biomass (p. 718)
ecosystem (p. 716)
ecosystem services (p. 727)
exchange pool (p. 721)
net primary productivity (NPP) (p. 718)

nutrient (p. 716)
nutrient cycle (p. 721)
secondary productivity (p. 721)
sedimentary cycle (p. 724)
trophic level (p. 720)

Self-Quiz

1. The amount of energy captured by photosynthesis, minus the amount lost as metabolic heat, is
   a. secondary productivity.
   b. consumer efficiency.
   c. net primary productivity.
   d. photosynthetic efficiency.

2. The movement of nutrients between organisms and the physical environment is called
   a. nutrient cycling.
   b. ecosystem services.
   c. net primary productivity.
   d. a nutrient pyramid.

3. Free services provided to humans by ecosystems include
   a. prevention of severe floods.
   b. prevention of soil erosion.
   c. filtering of pollutants from water and air.
   d. all of the above

4. Each step in a food chain is called a(n)
   a. trophic level.   c. food web.
   b. exchange pool.   d. producer.

5. What type of organisms consume 50 percent or more of the NPP in all ecosystems?
   a. herbivores   c. producers
   b. decomposers   d. predators

6. Sources of nutrients that are available to producers, such as soil, water, or air, are
   a. called essential nutrients.
   b. called exchange pools.
   c. considered eutrophic.
   d. called limiting nutrients.

7. Select the most representative term for an organism that gets its energy by eating all or parts of other organisms or their remains.
   a. fungus   c. consumer
   b. predator   d. producer

8. Nutrients that cycle between terrestrial and aquatic ecosystems and are then deposited on the ocean bottom
   a. have a short cycling time.
   b. have an atmospheric cycle.
   c. are more common than nutrients with a gaseous phase.
   d. have a sedimentary cycle.

Marvel Chemicals Pop Up in Animals All Over World

Teflon and Scotchgard, found from the Arctic to Lake Michigan, are raising health concerns

BY MICHAEL HAWTHORNE

Chemicals used to make Teflon and Scotchgard have been promoted as modern marvels for their ability to keep food from sticking to pots and fast-food packaging, repel stains on carpets and furniture, and make water roll off coats and clothing.

Now scientists are finding that the chemicals also have managed to spread throughout the world.

Researchers have detected them in polar bears roaming near the Arctic Circle, dolphins swimming in the Mediterranean Sea off the coast of Italy, and gulls flying above ocean cliffs outside Tokyo.

Over the past 40 years, the widespread use of a group of non-stick chemicals known as perfluoronated compounds seems to have slowly created an environmental problem. As this article indicates, these chemicals have been found in animals throughout the globe—and they have also been detected in the blood of nearly every person ever tested for them in the United States. The levels of these chemicals found in some children have been close to those that caused developmental problems in laboratory rats. Some studies also link these chemicals to cancer and liver problems in laboratory animals, although the applicability of such results to humans remains sharply disputed.

Finding perfluoronated compounds in the environment is a concern: as explained by Timothy Kropp, senior scientist at the Environmental Working Group, "there are rocks that break down faster than these compounds." For this reason, once the chemicals get into the environment, they stay there, and they are passed from one organism to another through the food web.

That seems to be exactly what has happened. It is not known whether perfluoronated compounds got into the environment by wear and tear on the products that contain them or by unreported releases of the chemicals into the water and air. But however they got into the environment, perfluoronated compounds have now been detected in the Great Lakes, the source of drinking water for 40 million people in the United States and Canada, and in a variety of foods, including bread, apples, green beans, and beef.

The U.S. Environmental Protection Agency (EPA), after decades of little oversight, has recently begun to focus on these chemicals. The agency pressured 3M, once the world's leading manufacturer of perfluoronated compounds, to limit the use of the original formulation of Scotchgard. In response, 3M decided to stop making perfluoronated compounds and now uses another chemical that, the company says, does not accumulate in the environment. A few other companies, such as DuPont, the maker of Teflon and Stainmaster products, still make perfluoronated compounds.

Evaluating the news

1. The detection of perfluoronated compounds in humans and other organisms suggests that we cannot release chemicals into the environment and expect them to "go away." As suggested by this example

and by the material in this chapter, there is no "away"—everything goes somewhere, and chemicals that break down slowly can cycle among organisms and persist in the environment for long periods of time. Do these basic ecological principles have implications for how we regulate the use of novel chemicals, such as perfluoronated compounds, that are synthesized by people but do not occur naturally?

2. According to the EPA, DuPont failed to report evidence it found in 1981 that perfluoronated compounds killed some laboratory rats and caused others to be born with eye and face defects; that same year, two of the five babies born to DuPont workers who handled perfloronated compounds had similar birth defects. So in 2004 the EPA initiated a class action lawsuit against DuPont, claiming that for more than 20 years the company withheld evidence that perfluoronated compounds might cause birth defects in humans. DuPont denies these allegations. Do you think the government should initiate such lawsuits against companies? Why or why not?

3. Perfluoronated compounds are the basis of a multibillion-dollar industry. If studies initiated recently indicate that these compounds pose a serious threat to human health, alternative chemicals will have to be found, and human societies may face large health and environmental cleanup costs for decades to come. Do you think we should act now to limit the use of perfluoronated compounds? Or should we decide what to do only after more information is available about whether these chemicals really are a threat? Defend your answer.

SOURCE: *Chicago Tribune,* Tuesday, July 27, 2004.

CHAPTER 38 Global Change

Key Concepts

◉ The effects of human actions on the world's lands and waters are thought to be the main causes of the current high rate of extinction of species.

◉ Human activities have added natural and synthetic chemicals to the environment, which in turn has altered how natural chemicals cycle through ecosystems.

◉ Human inputs to the global nitrogen cycle now exceed those of all natural sources combined. If unchecked, these changes to the nitrogen cycle are expected to have negative effects on many ecosystems.

◉ The concentration of carbon dioxide (CO_2) gas in the atmosphere is increasing at a dramatic rate, largely because of the burning of fossil fuels. Increased CO_2 levels are expected to have large but hard-to-predict effects on ecosystems.

◉ Increased concentrations of CO_2 and certain other gases in the atmosphere are predicted to cause a rise in temperatures on Earth. Most scientists think that such global warming is occurring, but its extent and consequences remain uncertain.

◉ Because global change caused by humans is expected to have large, negative consequences for many species, including our own, most ecologists think we must learn to use Earth's ecosystems in a sustainable fashion.

Main Message: Global change caused by human actions is occurring at a rapid pace.

Devastation on the High Seas

Nearly 75 percent of Earth's surface is covered with oceans. The oceans are so deep and so vast that many scientists once thought people could never drive marine species extinct. No matter how much we overhunted a species or polluted local portions of its habitat, it was thought there would always be places where the species could thrive. Now it seems this assumption was wrong.

Consider the white abalone's tale of woe. This large marine shellfish once was common along 1,200 miles of the California coast. It lives on rocky reefs in relatively deep water (25 to 65 meters or deeper). The fact that it lives in deep water protected it for a while: white abalone is delicious to eat, but people first hunted other species of abalone that live in shallower waters and hence are easier to find. When the shallow-water species became rare, fishermen turned to the white abalone. After only 9 years of commercial fishing, the fishery collapsed. This species, which once covered the seafloor with up to 10,000 individuals per hectare, is now on the verge of extinction.

In general, when people begin fishing in a new region, the amount of fish they catch drops sharply—typically by 80 percent—within the first 15 years. In 1958, for example, having depleted fishing grounds in the western Pacific, the Japanese fleet still was able to catch many large predatory fish in portions of the Indian Ocean, southern Pacific, and Atlantic. Six years later, the regions that yielded many fish in 1958 gave much lower returns, and by 1980, many of them had been abandoned.

Overall, people have had a large negative effect on fish populations worldwide. Recent studies indicate that 66 percent of the world's marine fisheries are in trouble due to overfishing. Furthermore, over the past 45 years, the catch has included fewer large predatory fish and more invertebrates and

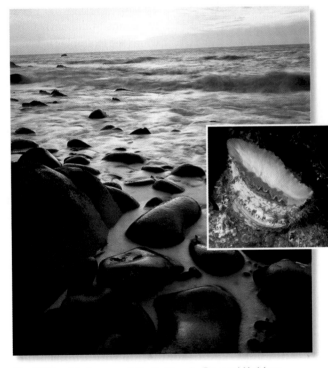

The White Abalone and Its California Coastal Habitat

small fish that feed on plankton. This has happened because large predatory fish were preferred by fishermen worldwide. As a result, populations of these fish plummeted. Thus, in addition to reducing populations of individual species, people have altered the food webs of ocean communities (by removing top predators).

The near-extinctions of species and the other effects of human actions on marine ecosystems are just a few of the changes we are making to the biosphere. Other examples include changes in the global sulfur cycle (see Chapter 37), effects on the locations of biomes (see Chapter 33), and the extinctions or declines of many species worldwide (see Interlude A). What are the consequences of such changes for life on Earth? How long can we continue to alter the biosphere to the extent that we do now?

Statements by politicians, talk show hosts, and others can give the impression that worldwide change in the environment, or **global change**, is a controversial topic. Such statements cause many in the general public to think that global change may not really be occurring, or cause them to wonder whether anything really needs to be done about it.

This impression of controversy is unfortunate because we know with certainty that global change is occurring. Invasions of nonnative species have increased worldwide (see examples in Chapters 34 and 36), large losses of biodiversity have occurred (see Interlude A), and pollution has altered ecosystems throughout the world (see Chapter 37). These are important types of global change that we know with certainty are happening today.

Although these three types of global change are caused by people, the biosphere has always changed over time. As we saw in Chapter 36, the continents move, the climate changes, and succession and natural forms of disturbance change the composition of communities. Thus, even in the absence of human actions, we know that ecological communities face—and always have faced—global change.

In this chapter we describe how people have influenced global change. We first discuss two types of global change that we know have occurred and that we know are caused by people: changes in land and water use, and changes in the cycling of nutrients through ecosystems. We finish with a look at global warming and its possible consequences for our future.

38.1 Land and Water Transformation

People make many physical and biotic changes to the land surface of Earth, which collectively are referred to as **land transformation**. Such changes include the destruction of natural habitat to allow for resource use (as when a forest is clear-cut for lumber), agriculture, or urban growth. Land transformation also includes many human activities that alter natural habitat to a lesser degree, as when we graze cattle on grasslands.

Similarly, **water transformation** refers to physical and biotic changes that people make to the waters of our planet. For example, we have drastically altered the way water cycles through ecosystems. People now use more than half of the world's accessible fresh water, and we have altered the flow of nearly 70 percent of the world's rivers. Since water is essential to all life, our heavy use of the world's waters has many and far-reaching effects, including changing where water is found and altering which species can survive at a given location.

Many of our effects on the lands and waters of Earth are local in scale, as when we cut down a single forest or pollute the waters of one river. However, such local effects can add up to have a global impact.

There is ample evidence of land and water transformation

Aerial photos, satellite data, changing urban boundaries, and local instances of destruction of natural habitats show how humanity is changing the face of Earth (Figure 38.1). Together, many such lines of evidence show that land and water transformation is occurring, is caused by human actions, and is global in scope.

To estimate the total amount of land that has been transformed by people, the effects of many different human activities must be added together for every acre of the world. This task may seem nearly impossible, but with the aid of satellites and other new technologies, we now can measure our total impact on Earth for the first time in history. Although researchers are just beginning this task, one reasonable estimate is that people have substantially altered one-third to one-half of Earth's land

(a)

(b)

Figure 38.1 Examples of Land Transformation

(a) A clear-cut forest in Washington State. (b) An open-pit copper mine in Arizona. (c) Change in the boundaries (in red) of urban regions near Baltimore, Maryland, and Washington, D.C., between 1850 and 1992 (d).

(d) 1992

(c) 1850

See a city growing over time.

38.1

surface. Thus, although the exact amount of land transformed by humans is not yet known, we know that we have altered a large percentage of it. Although we cannot yet measure our total impact on the waters of Earth (in part because we know so little about the deep ocean), global problems with water pollution and observed declines in many aquatic populations make it clear that we have transformed Earth's waters.

In modifying lands and waters for our own use, we have had dramatic effects on many ecosystems. For example, large regions of the United States were once covered by wetlands. Many of these wetlands have been drained so that the land could be used for agriculture or other purposes. During the 200-year period beginning in the 1780s, wetlands declined in every state in the United

States (Figure 38.2). They are still declining today. Other examples of human effects on ecosystems include the ongoing destruction of tropical rainforests (Figure 38.3) and the conversion of once vast grasslands in the American Midwest to croplands.

Land and water transformation have important consequences

Many ecologists think that land transformation and water transformation are—and will remain for the immediate future—the two most important components of global change. There are several reasons for this assessment.

First, as we alter the lands and waters to produce goods and services for an increasing number of people, we use a very large share of the world's resources. Estimates suggest that humans now control (directly and indirectly) roughly 30 to 35 percent of the world's total net primary productivity (NPP) on land (see Chapter 37 for a definition of NPP). By controlling such a large portion of the

Figure 38.2
Disappearing Wetlands

Wetlands in the United States declined greatly over 200 years. These maps show the percentage of the land area of each state covered by wetlands (*a*) in the 1780s and (*b*) in the 1980s.

(*a*) Wetland distribution, circa 1780s

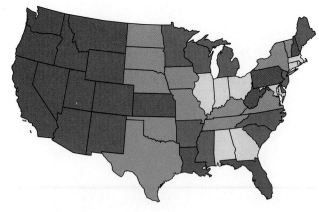

(*b*) Wetland distribution, circa 1980s

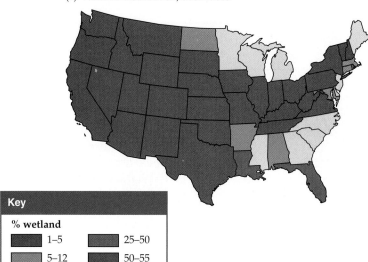

Key

% wetland

■ 1–5	■ 25–50
■ 5–12	■ 50–55
■ 12–25	

38.2 Changes in the Chemistry of Earth

Life on Earth depends on, and is heavily influenced by, the cycling of nutrients in ecosystems. NPP often depends on the amount of nitrogen and phosphorus available to producers, and the amount of sulfuric acid in rainfall has many effects on ecological communities, as we saw in Chapter 37. The nitrogen and phosphorus that stimulate NPP and the sulfur in acid rain are just two of many examples of naturally occurring chemicals that cycle through ecosystems.

Synthetic chemicals (chemicals made only by humans) also cycle through ecosystems. The pesticide DDT, whose effects on bald eagles we described on page 672, is one such synthetic chemical. Another example is **chlorofluorocarbons (CFCs)** [*KLOR*-oh-*FLOR*-oh . . .], the

6.0 miles

Figure 38.3 Transformation of Tropical Forests

In this satellite image of a Bolivian portion of the Amazon rainforest, healthy forest shows as red; in the absence of human impact, the entire image would be red. Loggers have cut long, straight paths through the forest, while ranchers have cleared large areas for cattle grazing. Four human settlements (radial arrangements of fields and farms that resemble starbursts) are visible near the middle of the image, and several irrigated crop fields (which resemble small circles) can be seen at the bottom right.

world's land area and resources, we have reduced the amount of land and resources available to other species, causing some to go extinct. Water transformation has similar effects. As we saw at the opening of this chapter, when people overfish or pollute Earth's waters, we may cause dramatic changes in the abundances and types of species found in the world's aquatic ecosystems (see also Chapters 36 and 37).

The transformation of lands and waters has other effects as well. One of these effects is change in local climates. For example, when a forest is cut down, the local temperature may increase and the humidity may decrease. Such climatic changes can make it less likely that the forest will regrow should the logging stop. In addition, as we'll see shortly, the cutting and burning of forests increases the amount of carbon dioxide in the atmosphere, an aspect of global change that may alter the climate worldwide.

chemical compounds used in aerosol spray cans, as coolants in refrigerators, and in foam manufacture. Because CFCs are synthetic, there were no CFCs in the environment until recently. CFCs are not toxic, but their use and subsequent release into the atmosphere had a large and unexpected effect: they caused a decrease in the ozone layer of the atmosphere across the globe, including the formation of the Antarctic ozone hole (see Figure 33.1). Because the ozone layer shields the planet from harmful ultraviolet light (which can cause mutations in DNA), damage to the ozone layer poses a serious threat to all life. Fortunately, the international community responded quickly to this threat, and the ozone layer has recently begun to show signs of a recovery (see the box on page 740).

By adding synthetic and naturally occurring chemicals to the environment, people have altered the ways in which many chemicals cycle through ecosystems. In some cases, some of the harm caused by changes in chemical cycles has been undone (see pages 740 and 672). In other cases, such as those of the global nitrogen cycle and the global carbon cycle, great challenges lie ahead.

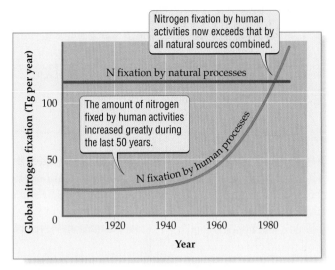

Figure 38.4 Human Effects on the Global Nitrogen Cycle
Nitrogen is fixed naturally by bacteria and by lightning at a rate of about 130 teragrams (Tg) per year (1 Tg = 1,012 grams, or 1.1 million tons). Human activities such as the production of fertilizers now fix more nitrogen than all natural sources combined.

38.3 Changes in the Global Nitrogen Cycle

There is a large amount of nitrogen in Earth's atmosphere, where N_2 gas makes up 78 percent of the air we breathe. However, plants and most other organisms cannot use N_2 directly. Instead, the nitrogen in N_2 gas must be converted to other forms, such as ammonium (NH_4^+) or nitrate (NO_3^-), before it can be used by plants and other producers. The conversion of N_2 to NH_4^+, called **nitrogen fixation**, is accomplished by several species of bacteria and, to a much lesser degree, by lightning. Once nitrogen is converted to NH_4^+, other bacteria can convert it to NO_3^-. These two forms of nitrogen then cycle among plants, animals, and microorganisms. The amount of nitrogen that cycles among organisms is much smaller than the amount found in gaseous form in the atmosphere.

Human technology is also capable of fixing nitrogen. In recent years, the amount of nitrogen fixed by human activities has exceeded the amount fixed by all natural processes combined (Figure 38.4). Much of this nitrogen fixation by humans is the result of industrial production of fertilizers. Other major sources of nitrogen fixation include car engines (in which the heat from combustion converts some of the N_2 found in the air to NO and NO_2 in exhaust) and fixation by bacteria that have a mutual-

istic relationship with certain crop plants known as legumes (such as peas, beans, and peanuts). The fact that human inputs of nitrogen to ecosystems exceed natural inputs tells us that our activities have greatly changed the global nitrogen cycle.

The potential effects of changing the nitrogen cycle are far-reaching. When nitrogen is added to terrestrial communities, NPP usually increases, but the number of species often decreases (Figure 38.5). This happens because those species that are best able to use the extra nitrogen outcompete other species. For example, the addition of nitrogen to grasslands in the Netherlands that historically were poor in nitrogen has resulted in the loss of more than 50 percent of the species from some of those communities.

Similarly, when nitrogen is added to nitrogen-poor aquatic ecosystems, such as many ocean communities, productivity increases, but species are lost (see page 704). In general, an increase in productivity caused by the addition of nitrogen is not necessarily a good thing for the ecosystem.

Learn more about the nitrogen cycle. 38.2

38.4 Changes in the Global Carbon Cycle

Vast amounts of carbon are found throughout the biosphere, and this carbon cycles readily among organisms, soils, the atmosphere, and the ocean (Figure 38.6). Here

Science Toolkit

From Science to Public Policy and Back: Repairing the Ozone Layer

Human societies often have to make tough choices regarding scientific issues such as genetic engineering, gene therapy, cloning, causes of cancer, effects of pollutants on public health, and the environmental impacts of human actions. The science behind such issues can be complex, and hence scientists must provide input to the policymakers who evaluate possible courses of action and the consequences of those actions. In any particular case, policymakers must consider questions such as the following: Should science be allowed to progress unhindered, regardless of the consequences? How can we prevent a new technology from having unintended and undesirable consequences? Who should pay to fix problems caused by existing technologies or policies?

We can illustrate how science influences public policy (and vice versa) with what is shaping up to be a major environmental success story: efforts to repair the damage humans caused to the ozone layer. In 1974, Mario Molina and F. Sherwood Rowland wrote an influential scientific article suggesting that chlorofluorocarbons (CFCs) in aerosol spray cans damaged the ozone layer; for this work, Molina and Rowland shared the 1995 Nobel prize in chemistry with Paul Crutzen, who also performed research on the ozone layer. Various scientists described the threats posed by CFCs to members of the public, and almost immediately, sales of aerosol products began to drop as people worried that ozone loss would cause skin cancer and other health problems. By 1978, CFC-containing aerosols were banned in the United States, and other countries, including Canada, Sweden, and Norway, soon followed suit.

Although some individuals and some nations took action early, large amounts of CFCs continued to be produced for use in refrigeration, air conditioning, and foam manufacture. Many policymakers and members of the business community balked at taking further action, citing high projected costs and arguing that the science was too weak to support restrictions on ozone production. Throughout the 1980s, however, the scientific evidence that CFCs were destroying the ozone layer got stronger and stronger, and in 1985 people around the world were prodded into action by the discovery of the now-familiar Antarctic ozone hole.

After 2 years of tough negotiations, the Montreal Protocol, an international treaty designed to halt the production and use of CFCs and other chemicals that harm the ozone layer, was signed in 1987, ratified

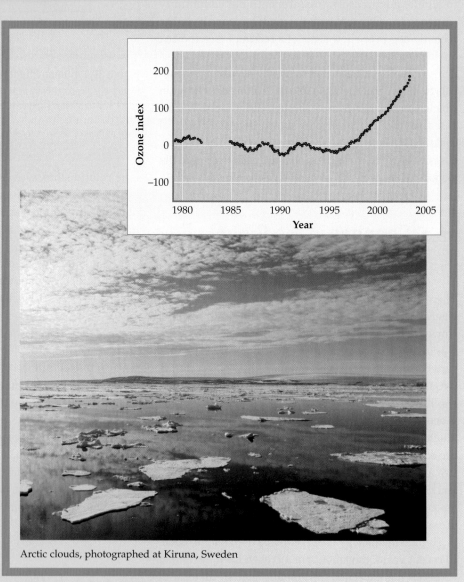

Arctic clouds, photographed at Kiruna, Sweden

The Ozone Layer Begins to Recover

In this Arctic sky and elsewhere in the world, the destruction of the ozone layer has begun to slow down. The "ozone index" in the graph would remain close to zero if the rate at which ozone was destroyed did not change from the rate observed between 1979 and 1996. Instead, by 1999, as shown by the points drawn in red, the index was significantly greater than zero, indicating that the rate of ozone destruction was decreasing. The air samples used to calculate the index were taken at latitudes 30° to 50° north, at an altitude of 35 to 45 kilometers.

in 1989, and has been strengthened several times since. New scientific evidence indicates that the Montreal Protocol is working. CFC emissions measured at Earth's surface have dropped, and satellite data show that the atmosphere is now losing ozone less rapidly than it did from 1979 to 1996. Although we are not gaining ozone yet, results like these provide an early sign that the ozone layer has begun to recover. If current trends continue, scientists expect that the ozone layer will recover completely in about 50 years.

It is hard to overstate the importance of these results. Assuming we stay on track, the policies we have formed in response to global thinning of the ozone layer show that people can tackle and solve monumental environmental problems. What's more, solving this problem has proved far less costly than originally estimated: when forced by treaties to stop using CFCs, researchers working at universities and corporations rapidly developed cost-effective alternatives.

Scientists cannot fix large-scale environmental problems—such as damage to the ozone layer—on their own. But scientists do have two important responsibilities: they must seek to understand the problem from a scientific perspective, and they must clearly explain what they learn to nonscientists. As our approach to repairing the ozone layer illustrates, the clear communication of scientific information to nonscientists can prod policymakers, business leaders, and members of the public to take actions to solve major environmental problems. Science, and the communication of science to nonscientists, play essential roles in efforts to help society solve existing environmental problems and prevent future ones.

we focus our discussion on one portion of the global carbon cycle that has been altered by human activities: the concentration of carbon dioxide (CO_2) gas in the atmosphere.

Atmospheric carbon dioxide levels have risen dramatically

Although CO_2 makes up less than 0.04 percent of Earth's atmosphere, it is far more important than its low concentration might suggest. As we have seen in earlier chapters, CO_2 is an essential raw material for photosynthesis, on which most life depends. CO_2 is also the most important of the atmospheric gases that contribute to global warming. Therefore, scientists took notice in the early 1960s when new measurements showed that the concentration of CO_2 in the atmosphere was rising rapidly.

Scientists have directly measured the concentration of CO_2 in the atmosphere since 1958. By measuring

(a)

(b)

Figure 38.5 A Nitrogen Addition Experiment
Native grasslands in Minnesota often have 20 to 30 plant species per square meter. (*a*) No nitrogen was added to this control plot, and it lost no species between 1984 and 1994. (*b*) Researchers added nitrogen to this nearby experimental plot during the same time period. Most of the native species disappeared from this plot, and an introduced species, European quackgrass, became dominant.

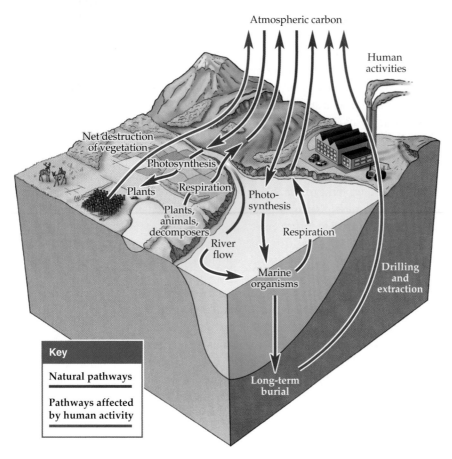

Figure 38.6 The Global Carbon Cycle

Key

Natural pathways

Pathways affected
by human activity

CO$_2$ concentrations in air bubbles that were trapped in ice for hundreds to hundreds of thousands of years, scientists have also estimated the concentration of CO$_2$ in both the recent and relatively distant past (Figure 38.7). For ice formed recently, direct measurements of CO$_2$ in the air match estimates from ice bubbles, giving us confidence that the ice bubble measurements for the past are accurate. Both types of measurements show that CO$_2$ levels have risen greatly during the past 200 years.

Overall, of the current yearly increase in atmospheric CO$_2$ levels, about 75 percent is due to the burning of fossil fuels. Logging and burning of forests is responsible for most of the remaining 25 percent.

The recent increase in CO$_2$ levels is striking for two reasons. First, the increase happened quickly: the concentration of CO$_2$ increased from 280 parts per million (ppm) to 370 ppm in roughly 200 years. Measurements from ice bubbles show that this rate of increase is greater than even the most sudden increase that occurred naturally during the past 420,000 years. Second, although the concentration of CO$_2$ in the atmosphere has ranged from about 200 ppm to 300 ppm during the past 420,000 years, CO$_2$ levels are now higher than those estimated for any time during this period. Thus global CO$_2$ levels have changed very rapidly in recent years and have reached concentrations that are unmatched in the last 420,000 years.

Increased carbon dioxide concentrations have many biological effects

An increase in the concentration of CO$_2$ in the air can have large effects on plants (Figure 38.8). Many plants increase their rate of photosynthesis and use water more efficiently, and thus grow more rapidly, when more CO$_2$ is available. When CO$_2$ levels remain high, some plant species keep growing at higher rates, but others drop their growth rates over time. As CO$_2$ concentrations in the atmosphere rise, species that maintain rapid growth at high CO$_2$ levels might outcompete other species in their current ecological communities or invade new communities.

Differences in how individual species respond to higher CO$_2$ levels may cause changes to entire communities. However, it is difficult (at best) to predict exactly how communities will change under higher CO$_2$ levels. Increased CO$_2$ levels in the atmosphere are likely to cause Earth's climate to warm, as we shall see in the following section. As both temperatures and CO$_2$ levels change, many different competitive and exploitative interactions

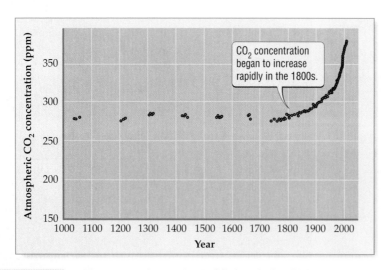

Figure 38.7 Atmospheric CO$_2$ Levels Are Rising Rapidly
Atmospheric CO$_2$ levels (measured in parts per million, or ppm) have increased greatly in the past 200 years. The red circles show results from direct measurements of the concentration of CO$_2$ in the atmosphere. The blue circles indicate CO$_2$ levels measured from bubbles of air trapped in ice.

Learn more about the carbon cycle.

38.3

Figure 38.8 High CO$_2$ Levels Can Increase Plant Size

These three *Arabidopsis thaliana* plants all have the same genotype, but were grown under different CO$_2$ concentrations: (*a*) 200 ppm, a level similar to that found roughly 20,000 years ago. (*b*) 350 ppm, the level found in 1988, and (*c*) 700 ppm, a predicted future level. Notice that plants grew larger as CO$_2$ concentrations increased.

may also change, but usually in ways that will not be known in advance. As we learned in Chapter 35, when interactions among species change, entire communities can change dramatically.

global warming, has proved controversial in both the media and the political arena.

We know that CO$_2$ concentrations in the atmosphere are increasing, but is the global climate getting warmer? Although the 10 hottest years in over 140 years of recorded measurements all occurred after 1990, year-to-year variation in the weather can make it hard to show that the climate really is getting warmer. In 1995, however, the United Nations–sponsored Intergovernmental Panel on Climate Change (IPCC) concluded for the first time that our climate is warming (Figure 38.9). The IPCC also concluded that the increase in global temperatures is most likely a result of human-caused increases in the concentration of CO$_2$ and other greenhouse gases in the atmosphere.

Since 1995, new statistical analyses have supported the IPCC's conclusion that recent rises in global temperatures represent a significant trend, not just ordinary variation in the weather. In addition, results from new scientific studies suggest that recent temperature increases have already changed ecosystems. For example, as temperatures increased in Europe during the twentieth century, dozens of bird and butterfly species shifted their geographic ranges to the north. Similarly, plants in northern latitudes have increased the length of their growing season as temperatures have warmed since 1980.

Other studies published since 1995 indicate that the warming since 1950 has been caused largely by human activities. For example, computer simulations performed on data for the second half of the twentieth century were

Explore whether temperatures are rising.

38.4

38.5 Global Warming

Some gases in Earth's atmosphere, such as carbon dioxide (CO$_2$), water vapor (H$_2$O), methane (CH$_4$), and nitrous oxide (N$_2$O), absorb heat that radiates from Earth's surface to space. These gases are called **greenhouse gases** because they function much as the walls of a greenhouse or the windows of a car do: they let in sunlight, but trap heat. As the concentration of greenhouse gases in the atmosphere goes up, more heat is trapped, thus raising temperatures on Earth.

Global temperatures appear to be rising

Carbon dioxide is the most important of the greenhouse gases because so much of it enters the atmosphere. Scientists have predicted that the ongoing increases in atmospheric CO$_2$ concentrations will cause temperatures on Earth to rise. This aspect of global change, known as

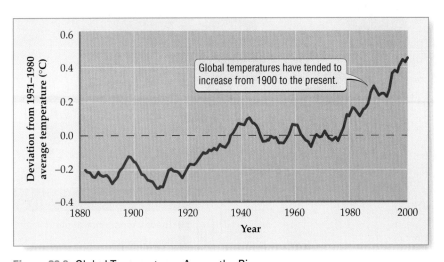

Figure 38.9 Global Temperatures Are on the Rise

Global air temperatures are plotted relative to the 1951–1980 average temperature (dashed line). Portions of the curve below and above the dashed line represent lower than average and higher than average temperatures, respectively.

able to predict the observed 0.1°C rise in temperature of the top 2,000-meter layer of the world's oceans only when human activities (such as greenhouse gas emissions) were included in the computer's calculations. Overall, an increasing amount of scientific evidence suggests that global warming is already happening, that it is affecting ecological communities, and that it is caused at least in part by human activities.

What will the future bring?

Because there is no end in sight to the rise in CO_2 levels, the current trend of increasing global temperatures seems likely to continue. How will increased temperatures affect life on Earth? Not surprisingly, the effects will depend on how much, and how fast, global warming occurs.

Scientists predict that 100 years from now, average temperatures on Earth will have risen by 1.4°C to 5.8°C. The effects of such a rise will depend on what the actual average increase turns out to be. Let's begin by examining a middle-of-the-road value within that range: a 3.5°C increase would probably have a large effect on Earth's biomes (Figure 38.10). With a global temperature increase of 3.5°C or higher, some species might go extinct simply because they were unable to migrate north fast enough to keep up with the changing climate. In addition, such an increase would be likely to have severe negative effects on the world's agricultural systems, especially since by then there will probably be 4 billion to 5 billion more people to feed.

Now let's look at the high and low ends of the predicted range of climate change: at the high end (a 5.8°C increase), humanity is likely to face a rise in the global sea level of over 5 meters. This change would submerge many cities and even entire island nations. Even at the low end of the global warming predictions (a 1.4°C increase), the effects of increased global temperatures are likely to be considerable, ranging from negative effects on agriculture in some regions to increases in the number and severity of weather events (for example, floods, hurricanes, and extremes of heat and cold).

Estimates related to the timing, extent, and consequences of global warming are filled with uncertainty. This uncertainty puts us in the difficult position of choosing to act now, perhaps unnecessarily, or choosing to wait, perhaps until it is too late to do anything about the problem. Efforts to curb global warming will have social costs, but delays may have even greater costs. Given such uncertainties, what do you think we should do?

(a) Current climate

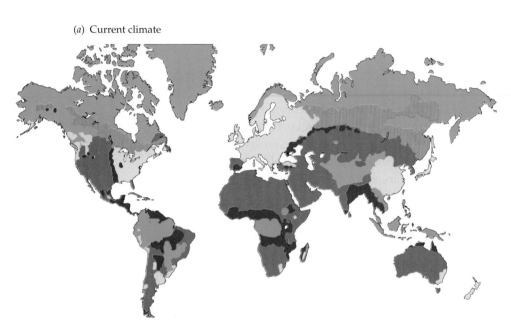

(b) Future climate (3.5°C increase in temperature)

Key

- Tundra, tundra-boreal forest
- Boreal forests
- Temperate forests
- Tropical forests
- Savannas, dry forests, woodlands
- Grasslands, shrublands, deserts

Figure 38.10 Biomes on the Move

If Earth's temperatures warm by an average of 3.5°C, the distribution of forests, grasslands, deserts, and other biomes could be altered by global climate change.

A Message of Ecology

The science of ecology has important and timely messages for humanity, such as the one we learned in Chapter 34: No population can continue to increase without limit. Although related, the message of this chapter is more complex. As one ecologist has written, "We are changing the world more rapidly than we are understanding it." In a very real sense, the world is in our hands. What we do to change it will determine our future and the future of all other species on Earth.

As we have seen in this chapter, human activities have had profound effects on life on Earth. Many scientists are convinced that people are causing global change at a rate and intensity unmatched by natural patterns of change. Depending on the actions we take, global change has the potential to have even greater effects in the future.

As scientists, we believe that the main message provided by our knowledge of how people have changed the planet is that we must reduce the rate at which humans alter Earth's ecosystems. Such a change in our behavior not only will be good for other species; it is in our own self-interest. Our entire civilization depends on the many services that ecosystems provide to us at no cost (see page 729). If we continue to ignore the effects of our actions on these natural systems, ultimately we will harm ourselves.

To reduce our effects on natural systems, we must limit the growth of the human population, and equally important, we must use Earth's resources more efficiently. Simply put, we must strive to have a sustainable impact on Earth—that is, our impact should be one that can continue indefinitely without causing serious environmental damage and without using up resources faster than they are replenished. Thus, as with money saved in a bank, we must learn to live on the "interest" provided by nature, leaving the original deposit or principal intact. For example, if we want the world's oceans to continue to provide their bounty, we must stop harvesting fish populations more rapidly than they can regenerate. Otherwise, their numbers will crash (Figure 38.11). In general, to have a sustainable impact, we must stop altering natural systems in ways and at speeds that lead to short-term gain but result in long-term damage.

To achieve the goal of having a sustainable impact on the planet, we must anticipate the effects of our actions before they have disastrous consequences. No other species is capable of such forethought. Will we use that capability? Will we be bold enough, creative enough, and intelligent enough to take responsibility for our impact on Earth? As the cases described throughout Unit 7 suggest, there is hope that the answers to these questions will be yes. For example, our response to the threat that CFCs posed to the ozone layer shows that people can face reality and solve challenging environmental problems. The conversion to a sustainable society is an even bigger challenge, but the first steps to meet that challenge have already been taken, as we shall see in Interlude G.

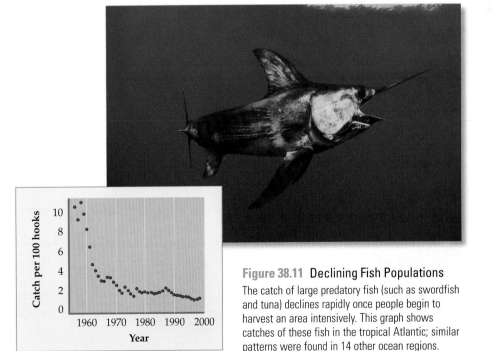

Figure 38.11 Declining Fish Populations
The catch of large predatory fish (such as swordfish and tuna) declines rapidly once people begin to harvest an area intensively. This graph shows catches of these fish in the tropical Atlantic; similar patterns were found in 14 other ocean regions.

Biology Matters

What's the Size of Your Footprint?

Each person on Earth has an ecological footprint, which is the acreage of productive land and water required to supply the resources you use and to dispose of the wastes you generate. At present, the average footprint of a person on Earth is 2.3 ha (5.7 acres), which is greater than the amount of biologically productive land available per person on a sustainable basis (1.9 ha per person, or 4.7 acres per person). The 2.3 ha per person is for an "average" person—the ecological footprint is much higher in some countries, such as the United States (9.7 ha per person, or 24 acres) and Canada (8.8 ha per person, or 21.7 acres) and much lower in others (0.5 ha per person, or 1.2 acres, in Bangladesh). As the world population grows, the amount of biologically productive land available per person will decline, increasing the speed at which the Earth's resources are consumed.

Although wealthier countries do tend to have higher average footprint sizes, there is a large variation in footprints; for example, average European footprints are substantially lower than American and Canadian footprints. As countries such as China and India develop, their average footprints will probably grow. Currently, the size of a person's footprint is most closely correlated with the size of a person's residence and the amount of traveling one does, especially by car or airplane.

What is your ecological footprint? To find out, take the footprint quiz at the Web site www.earthday.net/footprint/info.asp. If you are a typical college student, your footprint is probably close to the U.S. average of 9.7 ha (24 acres). In other words, it would take about five planet Earths to support the human population if everyone on Earth enjoyed the same standard of living.

Chapter Review

Summary

38.1 Land and Water Transformation

- Many lines of evidence show that human activities are changing lands and waters worldwide.
- Land and water transformation has caused extinctions of species and has the potential to alter local and global climate.

38.2 Changes in the Chemistry of Earth

- Human activities are changing the way many chemicals, both natural and synthetic, are cycled through ecosystems.

- An important class of synthetic chemicals is chlorofluorocarbons (CFCs). Their release into the atmosphere thinned the ozone layer over Earth, posing a serious threat to all life.

38.3 Changes in the Global Nitrogen Cycle

- Nitrogen requires fixation—conversion from N_2 gas to ammonium (NH_4^+) and nitrate (NO_3^-)—before it can be used by producers. In nature, most nitrogen fixation is performed by certain bacteria.
- Human activities (fertilizer production, car exhaust, and legume crops) fix more nitrogen than all natural sources combined.

- The extra nitrogen fixed by human activities has altered the global nitrogen cycle, leading to increases in productivity that can cause losses of species from ecosystems.

38.4 Changes in the Global Carbon Cycle

- Concentrations of atmospheric CO_2 have increased greatly in the past 200 years and are higher now than in the past 420,000 years. These CO_2 increases are caused by the burning of fossil fuels and the destruction of forests.
- Increased CO_2 concentrations can alter the growth of plants in ways that will probably cause changes in many ecological communities.

38.5 Global Warming

- Carbon dioxide and other greenhouse gases in the atmosphere trap heat that radiates from Earth's surface. As the concentration of greenhouse gases increases, average temperatures on Earth are expected to rise.
- Global warming during the twentieth century is, at least in part, a result of human activities.
- Although the amount of global warming that will occur in the twenty-first century is uncertain, if high-end predictions are correct, the social and economic costs will be very large.

◉ Review and Application of Key Concepts

1. Summarize the major types of global change caused by human activities. What consequences do such types of global change have for species other than humans?

2. Compare examples of human-caused global change with examples of global change not caused by people. What is different or unusual about human-caused global change?

3. Producers such as algae and plants require nitrogen to grow, and nitrogen is often in limited supply in both aquatic and terrestrial ecosystems. People are adding considerable amounts of nitrogen to ecosystems. Is that a good thing? Explain why or why not.

4. How do scientists determine the current atmospheric concentration of CO_2, as well as the concentrations thousands of years ago? How does the current atmospheric CO_2 concentration compare with concentrations over the past 420,000 years?

5. The future magnitude and effects of global warming remain uncertain. Do you think we should take action now to address global warming, despite those uncertainties? Or do you think we should wait until we are more certain what the ultimate effects of global warming will be? Support your answer with facts that are already known about global warming.

6. What changes to human societies would have to be made for people to have a sustainable impact on Earth?

7. Take the "Footprint Quiz" described in the box on page 746. Is your impact on Earth sustainable? If not, what changes could you make so that your impact would be sustainable?

Key Terms

chlorofluorocarbons (CFCs) (p. 738)
global change (p. 736)
global warming (p. 743)
greenhouse gas (p. 743)
land transformation (p. 736)
nitrogen fixation (p. 739)
synthetic chemical (p. 738)
water transformation (p. 736)

Self-Quiz

1. Which of the following do most ecologists think is the most important component of global change?
   a. increased CO_2 concentrations in the atmosphere
   b. global warming
   c. synthetic chemicals and overfishing
   d. land and water transformation

2. CO_2 absorbs some of the ——— that radiates from the surface of Earth to space.
   a. ozone
   b. heat
   c. ultraviolet light
   d. smog

3. The conversion of N_2 gas to a form of nitrogen that can be used by plants is called
   a. nitrogen fixation.
   b. fertilizer production.
   c. nitrogen cycling.
   d. nitrate conversion.

4. The concentration of CO_2 in the atmosphere is now about 370 ppm, a level that is roughly ——— the levels of 200 years ago.
   a. the same as
   b. 300 percent higher than
   c. 30 percent higher than
   d. 30 percent lower than

5. Human-caused changes to the nitrogen cycle are expected to result in
   a. an increase in acid rain.
   b. an increase in the loss of species from ecosystems.
   c. higher concentrations of a greenhouse gas.
   d. all of the above

6. The release of chlorofluorocarbons (CFCs) into the atmosphere caused what aspect of the global environment to change?
   a. the carbon cycle
   b. the ozone layer of the atmosphere
   c. acid rain
   d. the sulfur cycle

7. Most scientists think that three of the following four statements related to global warming are correct. Select the exception.
   a. The concentration of greenhouse gases in the atmosphere is not increasing.
   b. Dozens of species have shifted their geographic ranges to the north.
   c. Plant growing seasons are longer now than they were before 1980.
   d. Human actions, such as the burning of fossil fuels, contribute to global warming.

8. Compared with 45 years ago, fish catches worldwide now include more small fish and invertebrates and fewer large predators. What has caused this trend?
   a. People no longer want to eat large predators such as tuna and swordfish.
   b. Pollution has reduced the abundance of large predators (but not the abundance of small fish and invertebrates).
   c. Introduced species have wiped out many large predators.
   d. Global populations of large predators have been reduced by overfishing.

Biology in the News

Risk to State Dire in Climate Study

Unless checked, global warming could reduce the Sierra snowpack up to 89% by century's end, new research says

By Miguel Bustillo

Global warming could raise average temperatures as much as 10 degrees in California by the end of this century—sharply curtailing water supplies, causing a rise in heat-related deaths and reducing crop yields—if the world does not dramatically cut its dependence on fossil fuels, according to a study by 19 scientists published Monday.

The study, in the *Proceedings of the National Academy of Sciences*, contemplated the consequences of two distinct paths the industrialized world could take in response to a changing climate: maintaining its current reliance on coal, oil and gas, or massively investing in new technologies and alternative energy sources. Burning fossil fuels adds carbon dioxide to the atmosphere, which increases global temperatures by trapping more of the sun's heat.

Results from two new, detailed climate models that focus specifically on the state of California have scientists and policymakers concerned about the future. The report warns that if we take a "business as usual" approach to global warming (that is, if we postpone taking serious action to mitigate climate change and fail to invest heavily in alternative fuels and energy conservation), temperatures in California are likely to increase by 7°C to 10°C, and the state's economy will take a big hit.

"The choices we make today and in the near future will determine the outcome of this giant experiment we are undertaking with our planet," said Katharine Hayhoe, a climate consultant who was the lead author of the report. A 7°C to 10°C increase in temperature "is enough to make many coastal cities feel like inland cities do today, and enough to make inland cities feel like Death Valley," Hayhoe said.

The predicted effects of a 7°C to 10°C increase include an 89 percent reduction in snow in the Sierra Nevada. Less snowfall in this mountain range means less snowmelt in the spring, which means less water for agriculture throughout a large region of the state. If fossil fuel use does not decrease, the scientists warned, heat waves could occur six to eight times more often in Los Angeles, and heat-related deaths could increase by 500 to 700 percent.

Even the good news is not all that good. According to the model's results, if aggressive actions are taken now to reduce global warming, temperatures in California will still rise by 4°C to 6°C, again causing a reduction in the water available for agriculture. A big worry is the Napa and Sonoma wine harvest, which experts said could be hit hard by even a slight rise in temperature. At higher temperatures, grapes fall off the vine more quickly, and the quality of the grapes can go down. Thus even a 4°C to 6°C rise in temperature could harm the state's position as the leading producer of wine grapes.

SOURCE: *The Los Angeles Times,* Tuesday, August 17, 2004.

Evaluating the news

1. In June of 1991, Mount Pinatubo in the Philippines exploded in the second largest volcanic eruption of the twentieth century. In the fall of 1991, scientists used a climate model to predict that the eruption would cool the average temperature of Earth by 0.5°C (1°F) over a 15-month period, and that this cooling would start early in 1992. That is exactly what happened: Earth began to cool early in 1992, and temperatures dropped by 0.5°C over a 15-month period. Many policymakers took notice of the accuracy of this model's predictions. Since that time, such policymakers have given considerable weight to predictions like those described in this news article. Yet some pundits and some politicians dismiss the predictions of climate models as worthless and hence as nothing to worry about. What do you think?

2. As one of the authors of the California study commented in the news article, "The question is, are you going to wait 25 years to solve this, or are you going to act now on the vast preponderance of evidence that we are accumulating?" What do you think we should do?

3. Would you be willing to pay a gasoline tax whose proceeds would be used to fund aggressive actions to reduce global warming? If so, how much tax per gallon would you be willing to pay—50 cents per gallon, a dollar per gallon, two dollars per gallon? If not, why not?

Building a Sustainable Society

The State of the World

Each year, representatives of nations and corporations give speeches and produce reports that summarize what they've done in the past year and where they are headed for the upcoming year. If such an update could be provided for Earth, it would tell us how the planet's air, water, soil, and living organisms had changed in the past year. No one makes anything close to a complete version of such a report, and indeed, no one could: we do not even know how many species there are on Earth, let alone the current status of each of those species and the environments in which they live.

While we cannot give a complete "State of the World" address, we do know how some pieces of the planetary puzzle are changing over time. Some of that news is good. As we've described in Unit 7, populations of some endangered species (such as the bald eagle) are increasing in size, sulfur emissions that cause acid rain have decreased by almost 40 percent in the United States, and the ozone layer is showing early signs of recovery. Other news is bad. Nitrogen pollution continues to have negative effects on ecosystems worldwide, populations of many species are in serious decline, and global CO_2 levels continue to rise rapidly.

In addition to being able to provide particular bits of good and bad news, we also know enough about Earth to make an overall assessment (Figure G.1). Unfortunately, that assessment indicates that the current human impact on the biosphere is not sustainable. As we'll

Figure G.1 Measuring the State of the World
New tools allow us to monitor Earth's vital signs in unprecedented detail, as in this computer image made using four different types of satellite data. Fires over land are shown in red. The large plume that extends from Africa over the Atlantic (and ranges in color from red to orange to yellow to green) was caused by the burning of vegetation and by windblown dust.

Main Message: Aware that the current human impact on the global environment cannot be sustained, many individuals, corporations, and governments are taking innovative actions to help build a sustainable society.

see, people are using and damaging many of Earth's resources more rapidly than they can be renewed.

Although scientific evidence indicates that our current impact is not sustainable, there are many hopeful signs for the future. Five aspects of human society—education, individual action, research, government, and business—have already begun to contribute to the formation of a sustainable society. In this essay we first describe some of the evidence that the current human impact on the biosphere is not sustainable. With that material as background, we turn to our main focus: sources of hope for the future, and case studies that provide clues to how to build a sustainable society.

The Current Human Impact Is Not Sustainable

Many different lines of evidence suggest that the current human impact on the biosphere is not sustainable. What does this mean? An action or process is **sustainable** if it can be continued indefinitely without using up resources or causing serious damage to the environment. To begin with a simple example, modern societies depend on fossil fuels such as oil and natural gas to power our vehicles, heat our homes, and generate electricity. Although they provide abundant energy now, our use of fossil fuels is not sustainable: these fuels are not renewable, and hence supplies will run out, perhaps sooner rather than later (Figure G.2). Already, the volume of new sources of oil discovered worldwide has dropped steadily from over 200 billion barrels during the period 1960–1965 to less than 30 billion barrels during 1995–2000. Other types of data tell the same story.

Actions that cause serious damage to the environment are also considered unsustainable, in part because our economies depend on clean air, clean water, and healthy soils. But what constitutes "serious" environmental damage? One way to tell if an action causes serious damage is to see whether it disrupts important features of an ecosystem. As we have seen, many human actions have such effects. Human inputs to the global nitrogen

Figure G.2
Running Out of Oil
Many experts predict that the annual global production of oil will peak, then decline, sometime before 2020. Actual annual global oil production is shown in red; forecast production is shown in black.

and sulfur cycles, for example, now exceed all natural inputs combined (see Chapters 37 and 38). Such changes to the world's nutrient cycles can cause large disruptions to ecosystems, as manifested by such global problems as acid rain (see page 724), the Antarctic ozone hole (see page 646), rising CO_2 concentrations (see page 741), and lethal drops in oxygen levels in lakes and oceans (see pages 704 and 719).

Let's consider human use of two important resources: water and forests.

Declining water resources are a serious problem

People currently use over 50 percent of the world's annual supply of available fresh water, and demand is expected to rise over time. Many regions of the world already experience problems with either the amount of water available or its quality and safety (Figure G.3). Declining water resources are a serious issue today, and experts are worried that matters may get much worse.

To illustrate the problem, let's look at water pumped from underground sources, or **groundwater**. We use groundwater to drink, to irrigate our crops, and to run our industries. How does the rate at which people use groundwater compare with the rate at which it is replenished by rainfall? The answer is that we often use water in an unsustainable way: we pump it from **aquifers** (underground bodies of water, sometimes bounded by impermeable layers of rock) much more rapidly than it is renewed.

In Texas, for example, in 1995, more than 6 million acre-feet of water were removed from the vast Ogallala aquifer (an acre-foot is enough water to cover an acre of land with a foot of water). That amount—which would cover a football field with a wall of water more than 1,000 miles high—is more than 20 times the 0.3 million acre-feet of new water each year. Water has been pumped from the Ogallala aquifer faster than it is replenished for 100 years, causing the Texan portion of the aquifer to lose half its original volume. If that rate of use were to continue, in another 100 years the water would be gone, and many of the farms and industries that depend on it would collapse.

Texas is not alone. Pumping has caused groundwater levels to drop at rates of 1 to 3 feet per year in many regions of the world (Figure G.4a). Rapid drops in groundwater levels (about 1 meter per year) in China pose a severe threat to its recent agricultural and economic gains, and at current rates of use, large agricultural regions in India will completely run out of water in 5 to 10 years. In addition, because there is less water underground to provide support, the land surface may sink when groundwater levels drop. Such land **subsidence** [sub-*SIGH*-dence] can force people to stop pumping long before the water runs out. In Mexico City, for example, pumping has caused land within the city to sink by an average of 7.5 meters (more than 24 feet) since 1900, damaging buildings, destroying sewers, and causing floods.

Global deforestation continues at an alarming rate

For centuries, people have cut down trees at rapid rates. Much of Europe, for example, was deforested from AD 900 to AD 1500. Forest losses have approached 100 percent in some cases, as in Kenya and in São Paulo State, Brazil. Today we continue to cut some forests at unsustainable rates. Forested area in developing regions of the world declined by 9 percent from 1980 to 1995. These regions include most of the world's tropical forests, which are currently being cut at the rate of 14 million hectares (35 million acres) per year. If this rate of loss were to continue,

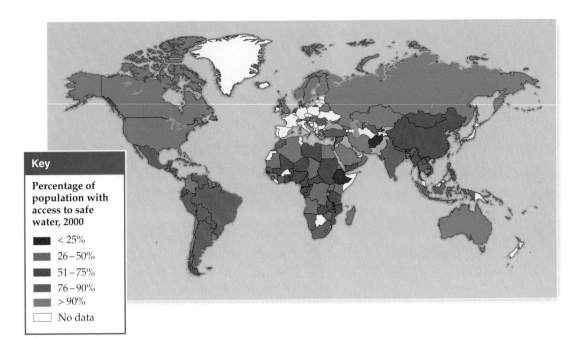

Key

Percentage of population with access to safe water, 2000

- ◼ < 25%
- ◼ 26–50%
- ◼ 51–75%
- ◼ 76–90%
- ◼ > 90%
- ☐ No data

the world's tropical forests would be gone in 100 to 150 years.

The news is not all bad for global forests. In industrial regions of the world, including Europe, the United States, Canada, and Japan, forested area increased by 3 percent from 1980 to 1995. Even so, this good news must be qualified: these increases were mainly due to growth in young forests, not to an increase in the area covered by older, more mature forests (which often harbor unique species). Few old-growth forests remain in industrial regions of the world (Figure G.5), and their regeneration takes hundreds of years. As a result, even if all harvesting of old-growth forests were to stop immediately, it would take several hundred years before the area covered by such forests began to increase.

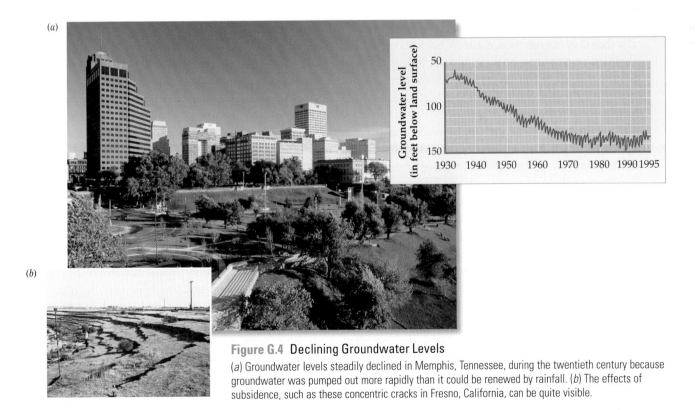

(a)

(b)

Figure G.4 Declining Groundwater Levels

(*a*) Groundwater levels steadily declined in Memphis, Tennessee, during the twentieth century because groundwater was pumped out more rapidly than it could be renewed by rainfall. (*b*) The effects of subsidence, such as these concentric cracks in Fresno, California, can be quite visible.

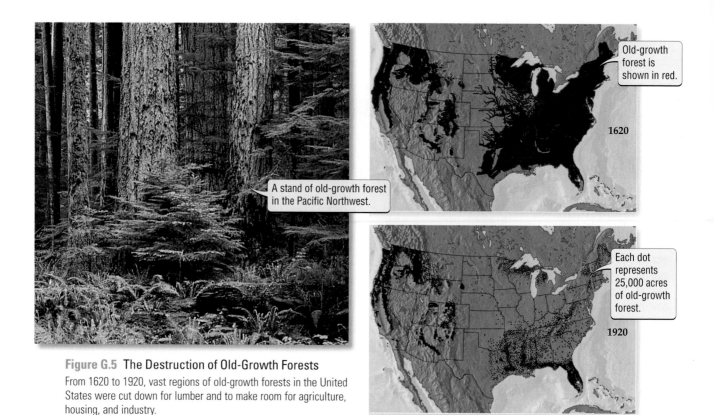

Figure G.5 **The Destruction of Old-Growth Forests**
From 1620 to 1920, vast regions of old-growth forests in the United States were cut down for lumber and to make room for agriculture, housing, and industry.

Sources of Hope for the Future

As everyone knows, if you start with a fixed amount of money in the bank and you consistently spend more than you earn, the funds in your bank account will shrink. Similarly, if people consistently use resources more rapidly than they are renewed, nature's "bank accounts" will shrink. For resources such as forests, we can think of nature's "bank account" as consisting of the total amount of forest in the world. Let's consider how the amount of forest changes over time.

Each year, total forest area can decrease due to natural disturbances (such as fire or windstorms) and human actions (including logging) that remove forests, and it can increase by natural growth and by human actions (such as planting trees) that regenerate forests. Currently, increases and decreases due to natural factors roughly balance one another, but logging is causing the global area of tropical forests to decrease each year.

For forests, or for any other resource, sustainable use requires that we not use resources faster than they are replenished. Unfortunately, data collected over the past few decades suggest that we are using many resources, including forests, more rapidly than they are being replenished. As a result, nature's "bank accounts" are shrinking. These findings are discouraging, for they suggest that future generations may have fewer resources than we do today.

But thinking in terms of nature's bank accounts also provides a source of hope: we can alter our actions to avoid using resources faster than they are replenished. This attitude—along with the realization of the dire consequences if we continue to use resources unsustainably—has prompted many to embrace the monumental challenge of helping to change our societies and economies to make our collective impact on Earth sustainable. Sources of hope are being provided in many arenas of human society—education, individual action, research, government, and business.

Awareness and understanding are the first steps to solving environmental problems

Efforts to build a sustainable society depend on education, in part because people cannot solve problems they don't know about. Many people are aware, in general terms, that environmental problems exist, but often they are not aware of the extent of those problems or the amount of scientific evidence demonstrating how serious those problems are. For example, after listening to a presentation about the status of the world's oceans, one judge for the U.S. Ninth Circuit Court said, "I thought I knew a lot about the environment. But I was staggered by what I didn't know." In addition to playing a critical role in informing people about the scope of environmental problems, education can also help people explore connections between the economy and the environment—showing, for example, that a "jobs versus the environment" trade-off is not always necessary (as we shall see later in this essay).

Education also plays a key role in shaping attitudes about nature. As various commentators have written, we will save only what we love, and we love only what we know. In many human societies, the dominant view has been that a species or community is worth saving only to the extent that it provides direct benefits to people. The appreciation for natural communities that can result from education provides a powerful alternative view: communities can be worth saving for their own sake. For example, some people who live in desert regions refer to the land that surrounds their city or town as "worthless." Such views can change when a person spends time in the desert, learns more about it, and comes to appreciate its beauty and unique value (Figure G.6).

Because it is fundamental to how we view the world and to our knowledge of particular problems, education is central to all efforts to build a sustainable society. Education has already produced great changes. People are far more aware of environmental issues today than they were 30 or 40 years ago, and they place more value on solving environmental problems than they once did.

One national poll in the United States asked people whether they agreed with the statement, "Protecting the environment is so important that requirements and standards cannot be too high, and continuing environmental improvements must be made regardless of cost." The percentage of people who agreed with this statement rose from 45 percent in 1981 to 74 percent in 1990. Polls conducted in other nations have obtained similar results, indicating that environmental issues are of worldwide concern.

Individual actions can have a ripple effect

Each day we make many choices that affect the environment. For example, we can choose to purchase energy-efficient cars and appliances—or not. Similarly, we can refuse to buy throwaway products, such as disposable cameras—or not. Many people do make choices with the environment in mind, and as a result, "green" products that minimize impacts to the environment are becoming increasingly common. As one example of this trend, sales of organic food in the United States skyrocketed from $180 million in 1986 to $11 billion in 2002. Similarly, consumers in many parts of the industrial world can now opt to receive electricity generated from renewable sources, such as wind or solar power (although consumers still must pay extra for renewable energy).

In addition to using their wallets as leverage, some individuals take other actions to help support the conversion

Figure G.6 "Worthless" Desert?
Desert regions, such as the Sonoran desert of Arizona pictured here, are a vital ecosystem that is home to the likes of jackrabbits, cacti, wrens, hawks, prairie dogs, Gila monsters, rattlesnakes, and coyotes—all of which can be seen as valuable in their own right.

to a sustainable society. For example, individuals, corporations, and local governments have established greenroof projects in Europe, North America, and Japan. A **greenroof** is a 2- to 4-inch-thick "living" rooftop: it has a layer of soil or other material in which plants grow, under which there are one or more layers that absorb water and prevent roots and water from damaging the underlying roof structure (Figure G.7). Greenroofs are well established in countries such as Germany, where more than 13 million square meters of rooftops had greenroof systems by 2002.

Individuals who build greenroofs enjoy a number of benefits, including reduced stormwater flow, decreased heating and cooling bills (because the roofs insulate the building), and reduced levels of dust and pollutants (because they are absorbed by the plants). As part of a series of steps designed to convert a 600-acre, $2 billion assembly plant to sustainable forms of manufacturing, managers at the Ford Motor Company decided to install greenroofs on 454,000 square feet of roofing (see Figure G.7). These roofs, which are covered with the groundcover sedum, can absorb an inch of rainfall with no runoff. Other benefits include reduced energy costs and increased absorption (by the plants) of CO_2, the most important greenhouse gas. Some greenroof projects, such as one initiated by a husband-and-wife team in Nashville, Tennessee, offer additional benefits by providing a place to grow rare or endangered species (Figure G.8).

Efforts by individuals to build a sustainable society can start small—sometimes in a person's backyard—and then grow into something larger. Such was the case with greenroofs in the city of Portland, Oregon: a greenroof was started in 1997 on a garage by one person, and that effort has now snowballed into several large greenroof projects sponsored by the city. Efforts by one person can even blossom into national movements. For example, in 1977, Wangari Maathai (winner of the 2004 Nobel peace prize) founded the Green Belt Movement (GBM) in Kenya (Figure G.9). The GBM, initially a small organization with no staff or funds, began by planting seven trees in a small park in Nairobi. By 2003, the GBM had gone on to plant over 20 million trees in Kenya. At present, the movement has over 3,000 tree nurseries and has provided jobs for thousands of people.

New research tools are being used to measure human impacts

Documenting the human impact on the planet is a monumental task: we must understand our impact, acre by acre, across the entire globe. Now, for the first time in history, we have the research tools needed to accomplish this seemingly impossible task. Satellite images,

Chicago City Hall

A residential rooftop garden in Tokyo, Japan

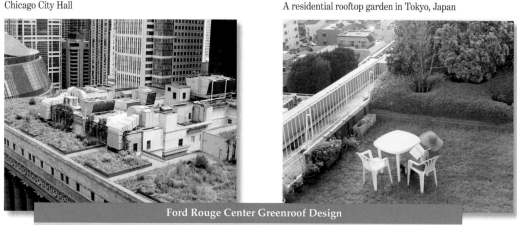

Ford Rouge Center Greenroof Design

Plant layer — Traps dust, absorbs carbon dioxide, creates habitat

Matrix — A mixture of shale, sand, peat, and compost, into which plant roots grow

Absorbent layer — A felt-like mat that holds water

Drainage layer

Protective membane — Protects the underlying roof from water and root damage

A school in Unterensingen, Germany, combining solar panels with a greenroof

Figure G.7
Greenroofs Around the World
Greenroofs in Germany, the United States, Japan, and other countries differ in appearance. But they tend to be similar in their basic design, which includes a plant layer, a soil or soil-like matrix, an absorbent layer, a drainage layer, and a protective membrane, as illustrated here for the greenroof design used at the Ford Rouge Center, located near Dearborn, Michigan.

Figure G.8 Putting Endangered Species on Top
As part of their plan to redevelop an abandoned meat packing plant in Nashville, Tennessee, a husband-and-wife team installed a greenroof on one of the plant's buildings. Their greenroof uses species found in Tennessee's endangered Cedar Glade community, shown here. Cedar Glade is home to many rare species, including the Tennessee coneflower, the first plant species listed under the Endangered Species Act.

Figure G.9 One Tree at a Time
In 1977, Wangari Maathai founded the Green Belt Movement, initially a small organization with no staff or funds. The GBM has grown and has gone on to plant millions of trees in Kenya. Ms. Maathai is pictured here after she received the 2004 Nobel peace prize for her efforts toward sustainable development, democracy, and peace.

for example, allow us to monitor Earth in unprecedented detail (Figure G.10). The preliminary results from such monitoring efforts show that we are causing serious damage to the world's ecosystems. But our recently acquired ability to recognize how much we are changing the planet provides a powerful source of hope: we can use that information to motivate change and to guide our efforts to build a sustainable society.

Consider the new research capabilities unveiled by NASA in April of 2003: a satellite system able to generate the world's first consistent and continuous measurements of net primary productivity (NPP; see page 718) on a global scale (Figure G.11). Such measurements show how much new plant growth is occurring throughout the entire world, thus providing a snapshot of Earth's "metabolism." Early results from this satellite system are fascinating, showing, for example, the speed with which land plants respond to changing weather conditions. (NPP rises and falls considerably in a matter of days.) Researchers can now use such data to monitor how changes in climate affect plant growth, to measure the

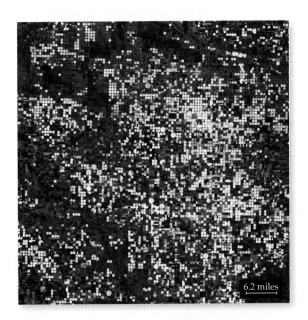

Figure G.10 A Polka-Dotted Landscape
Satellite images provide a unique perspective on our world, as seen in this image of irrigated farmland near Garden City, Kansas. The red and white circles are farmland watered by circular or "center pivot" irrigation systems; red colors indicate healthy vegetation.

6.2 miles

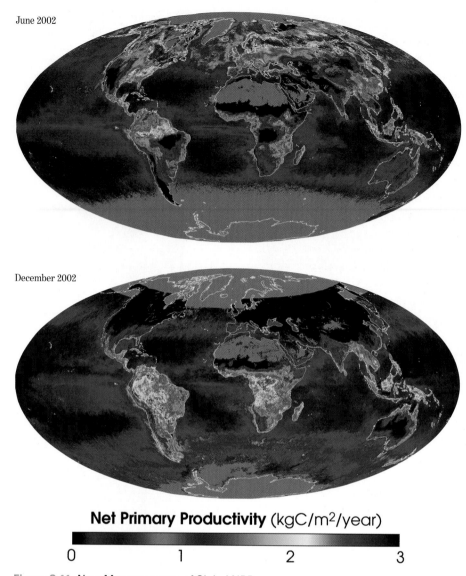

June 2002

December 2002

Net Primary Productivity (kgC/m²/year)

0   1   2   3

Figure G.11 New Measurements of Global NPP

Data used to construct these images were collected using NASA's new MODIS satellite system. NPP is expressed as kilograms of new growth produced per square meter per year.

rate at which deserts are expanding globally, and to analyze the effects of droughts on ecosystems. By monitoring plant growth in real time, such data could also be used to improve crop production forecasts and to help ranchers decide when to move cattle from one pasture to another.

Finally, research in fields other than the sciences can help build a sustainable society. For example, traditional measures of a nation's economic output, such as **gross domestic product (GDP)**, keep track of goods produced, but do not consider the social and environmental costs that result from producing those goods. Such costs include billions of tax dollars spent to clean up polluted

waters and soils, billions of dollars in medical bills for lung conditions caused by air pollution, and long-term problems caused by unsustainable practices that increase GDP in the short run but harm our environment and economy in the long run. To account for such costs, economists are developing measures of economic output, such as the **index of sustainable economic welfare**, that include a wider range of the benefits and costs of economic activities than does the traditional GDP. The goal is to create a new system of accounting that provides a realistic measure of the true long-term health of our economy, thus enabling us to recognize and correct problems before a crisis occurs.

Government has an important role to play

Efforts to build sustainable societies require government action. Treaties among governments have reduced emissions of pollutants that cross national boundaries, as illustrated by international agreements to curb the production of CFCs that damage the ozone layer (see page 740). Considerable potential remains for governments to take other actions that would promote a sustainable society, such as initiating major efforts to develop new sources of energy or enacting legal reform so that tax laws would penalize polluters and provide incentives for environmentally friendly practices. Some governments provide such incentives; in Germany, for example, tax breaks are given to families who use thatch roofs, which are made of renewable plant materials rather than asphalt-derived materials.

Governments can also play an essential role by placing limits on resource use. While many fisheries are in trouble, as we saw in Chapter 38, that is not the case for the lobster fishery in Maine, where the catch of lobsters was stable for many years and then increased in recent years (Figure G.12). The lobster fishery in Maine is regulated by strict laws that forbid the catch of several categories: (1) females bearing eggs, (2) small individuals (which probably have not yet reproduced), and (3) large individuals (each of which can produce many offspring). These and other regulations, which help prevent lobsters from being caught more rapidly than they can reproduce, allowed the catch of lobsters to remain roughly stable from 1950 to 1990. In the past 10 years, the lobster catch has increased dramatically, suggesting an increase in the lobster population. Declining populations of fish species, such as flounder and cod, that feed on lobster eggs and immature lobsters may have contributed to this dramatic increase.

As a final example, consider how government regulations reduced sulfur dioxide (SO_2) emissions from power plants by nearly 40 percent in the United States (see Figure 37.12). Reductions in SO_2 emissions are required by the Acid Rain Program, established by Congress in 1990 as an amendment to the Clean Air Act. Reduced SO_2 emissions have already decreased acid rain, and additional reductions mandated by the law are expected to save more than $50 billion per year in medical costs by 2010. These impressive results were achieved using an innovative approach to pollution control, the cap-and-trade system.

Here's how a **cap-and-trade system** works (Figure G.13). First, the government sets a nationwide limit, or cap, on the amount of a pollutant that can be added to the atmosphere each year. In the case of SO_2, the cap for

Figure G.12   A Sustainable Fishery
Statewide, the catch of lobsters was fairly stable from 1950 to 1990, indicating that lobster harvests in Maine were sustainable during that time period. In the early 1990s, the catch increased dramatically, in part because of declining populations of fish that eat lobster eggs and young lobsters.

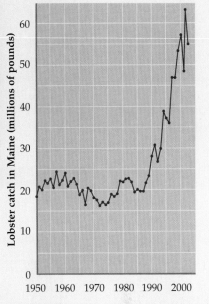

2010 is 8.95 million tons, a decrease of nearly 50 percent from 1980 levels. Second, each factory is given a certain number of "emission allowances" in each year, which allow it to emit a specified number of tons of SO_2. At the end of the year, the factory must provide government officials with enough allowances to cover its emissions for the year. Unused allowances can be sold, traded, or saved (banked) for future use. Factories that cannot declare enough in allowances to cover their yearly emissions are fined and must use future allowances to cover the shortfall, just as we would use future earnings to repay present debts.

The cap-and-trade system reduced SO_2 emissions at a fraction of the estimated costs. When the system was proposed in the 1980s, it was estimated that reaching the 2010 caps would cost the industry over $12 billion. By 1998, however, updated estimates placed the total cost of meeting the 2010 caps at $0.87 billion, significantly less than originally feared.

Why has the cap-and-trade system worked so well? First, the cap restricts total emissions. Thus, even if the industry grows, total emissions must go down, resulting in both health and environmental benefits. Second, although the government limits total emissions, it does not specify how the reductions are to be achieved. This frees companies to seek the most cost-effective way to meet the caps. In addition, because the total number of allowances is limited, emission allowances are scarce,

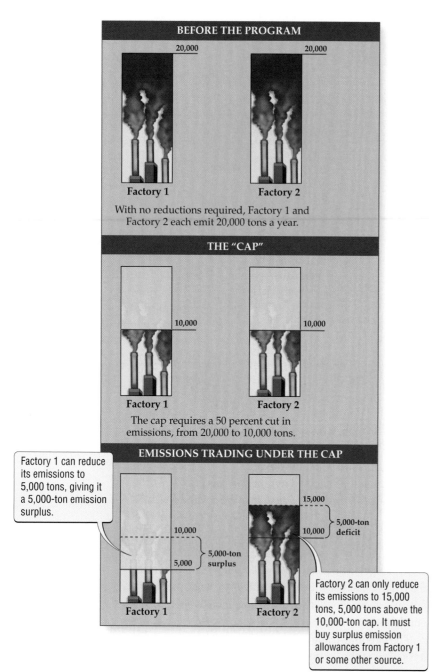

BEFORE THE PROGRAM

20,000   20,000

Factory 1   **Factory 2**

With no reductions required, Factory 1 and
Factory 2 each emit 20,000 tons a year.

THE "CAP"

10,000   10,000

Factory 1   **Factory 2**

The cap requires a 50 percent cut in
emissions, from 20,000 to 10,000 tons.

EMISSIONS TRADING UNDER THE CAP

Factory 1 can reduce
its emissions to
5,000 tons, giving it
a 5,000-ton emission
surplus.

10,000

15,000

5,000-ton
deficit

10,000

5,000

5,000-ton
surplus

Factory 1   **Factory 2**

Factory 2 can only reduce
its emissions to 15,000
tons, 5,000 tons above the
10,000-ton cap. It must
buy surplus emission
allowances from Factory 1
or some other source.

Figure G.13 How a Cap-and-Trade System Works

and hence valuable. Simple economics then takes over: because allowances are worth money and can be bought and sold, the profit motive stimulates the development of new ways to reduce emissions. Such market forces may soon provide a new source of hope for reducing CO_2 emissions: the European Union recently established the world's first international trading scheme for carbon dioxide emissions, which began as scheduled in 2005.

Business participation is crucial

Successful conversion to a sustainable economy will be an ongoing and complex process. For this process to work, business must play a central role. There are two reasons for this. First, corporations use large amounts of resources and emit large quantities of pollutants. Hence there is enormous potential for corporations to reduce the environmental impact of their activities. Second, as we've seen with the cap-and-trade system for SO_2, the profit motive can help drive the rapid development of innovative new technologies that reduce the human impact on the environment.

Based on current patterns of investment, a growing number of corporations anticipate huge profits if they are well positioned to help societies convert to sustainable forms of development. For example, the energy giant British Petroleum (BP) is investing $1 billion in alternative sources of energy, such as fuel cells, solar power, and wind power. Similarly, ABB, a Swiss company with revenues of $24 billion per year, sold its large electric power plants in 1999 as part of a move to become an industry leader in small-scale and renewable sources of energy. ("Small-scale" here refers to power sources that serve a large building or a cluster of homes.)

What other steps are businesses taking to help build a sustainable society? Consider global warming. Most scientists think global warming is real and is caused at least in part by such human actions as fossil fuel use, which cause CO_2 levels to rise. Although some politicians and members of the public argue against the scientific consensus, many corporations view climate change as a reality that threatens both society and their business interests. Such companies include economic powerhouses such as the energy firms Shell and BP, the chemical company DuPont, and the aluminum manufacturer Alcoa.

To lessen the impact of projected global warming, these and other large corporations have set targets that are more ambitious than actions required by the **Kyoto Protocol**, an international agreement designed to reduce global warming by reducing CO_2 emissions. The Kyoto Protocol, which was signed in 1997 and has been ratified by 141 countries (but not by the United States, the world's largest producer of CO_2 emissions), calls for industrial nations to reduce CO_2 emissions by an average of 7 percent from 1990 levels. BP has already beat that goal: by 2001, after just 4 years of effort, it had reduced CO_2 emissions by 10 percent from 1990 levels at no net cost to the company. Other companies have reduced emissions of CO_2 and other greenhouse gases by even greater levels, as shown in Figure G.14.

Finally, firms such as The Home Depot, McDonald's, and Alcoa seek to document the total environmental impact of their products, from manufacture to disposal. This approach, called **life cycle engineering**, gives corporations the information they need to develop new business practices that cause less environmental damage, often saving money at the same time.

Life cycle engineering, the cost-free reductions in CO_2 emissions achieved by BP, the success of cap-and-trade systems, and savings from greenroofs all suggest that what is good for the environment can also be good for the bottom line. As a result, some economists think we can protect the environment without causing economic harm or large job losses. Even if these economists are only partly right, in some cases we will be able to avoid the tough choices required by a "jobs versus the environment" trade-off. Even when such trade-offs are necessary, it is clear that people can draw on a range of motivating factors—including the profit motive, concern for the environment, and concern for human welfare—in efforts to build a sustainable society.

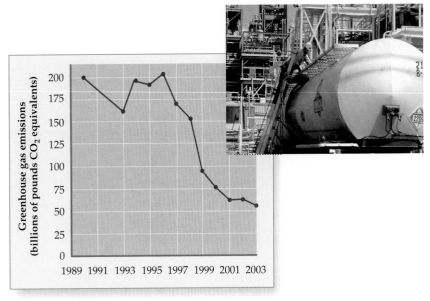

Figure G.14 Declining Greenhouse Gas Emissions
Greenhouse gas emissions at the chemical company DuPont declined by 72 percent from 1990 to 2003. In the graph, DuPont's total greenhouse gas emissions—including CO_2, nitrous oxide, and several types of CFCs—are expressed as the number of pounds of CO_2 emissions that would have the same effect on global warming had they been released.

◉ Review and Discussion

1. Some critics object to cap-and-trade systems because they do not think it is right for a company that does not meet the cap to be able to purchase emission allowances and thereby continue to pollute at a high level. Other people counter that because the cap-and-trade system reduces *total* emissions, it does not matter if some companies continue to pollute as usual. What do you think?

2. Describe in detail two lines of evidence from this chapter (or elsewhere in Unit 7) indicating that the impact of people on the environment is not sustainable.

3. Many new technologies and products that reduce human impact on the environment (such as renewable energy sources; lumber certified not to be from old-growth forests; fish raised using low-impact aquaculture) cost more than standard technologies, at least initially. Would you be willing to pay more for use of such technologies or products? Why or why not?

4. Do you think governments should switch from traditional measures of economic productivity, such as the gross domestic product (GDP), to new measures that incorporate the environmental and social costs of producing goods (such as the index of sustainable economic welfare)?

5. Many corporations are investing large sums of money in products and technologies that reduce the human impact on the environment. Why are they doing this? What unique features of corporations enable them to rapidly develop innovative new products?

Key Terms

aquifer (p. G3)
cap-and-trade system (p. G10)
greenroof (p. G7)
gross domestic product (GDP) (p. G9)
groundwater (p. G3)

index of sustainable economic welfare (p. G9)
Kyoto Protocol (p. G11)
life cycle engineering (p. G12)
subsidence (p. G3)
sustainable (p. G2)

Appendix: The Hardy-Weinberg Equilibrium

In this appendix we describe the conditions under which populations do not evolve. Specifically, we discuss the conditions for the Hardy-Weinberg equation, a formula that allows us to predict genotype frequencies in a hypothetical nonevolving population. As we described in Chapter 17 (see the box on page 324), this equation provides a baseline with which real populations can be compared in order to figure out whether evolution is occurring.

A population can evolve as a result of mutation, gene flow, genetic drift, or natural selection. Put another way, a population *does not* evolve when the following four conditions are met:

1. There is no net change in allele frequencies due to mutation.
2. There is no gene flow. This condition is met when new alleles do not enter the population via immigrating individuals, seeds, or gametes.
3. Genetic drift does not change allele frequencies. This condition is met when the population is very large.
4. Natural selection does not occur.

The Hardy-Weinberg equation is derived by assuming that all four of these conditions are met. In reality, these four conditions are rarely met completely in natural populations. However, many populations meet these conditions well enough that the Hardy-Weinberg equation is approximately correct, at least for some of the genes within the population.

To derive the Hardy-Weinberg equation, consider a hypothetical population of 1,000 moths. The dominant allele for orange wing color (W) has a frequency of 0.4, and the recessive allele for white wing color (w) has a frequency of 0.6. What we seek to do now is predict the frequencies of the WW, Ww, and ww genotypes in the next generation for a population that is not evolving.

If mating among the individuals in the population is random (that is, if all individuals have an equal chance of mating with any member of the opposite sex), and if the four conditions described above are also met, we can use the approach described in the accompanying figure to predict the genetic makeup of the next generation. This approach is similar to mixing all the possible gametes in a bag and then randomly drawing one egg and one sperm to determine the genotype of each offspring. With such random drawing, the allele and genotype frequencies in our moth population do not change from one generation to the next, as the figure shows.

The Hardy-Weinberg Equation

When mating is random and certain other conditions are met, allele and genotype frequencies in a population do not change. p = frequency of the W allele, q = frequency of the w allele.

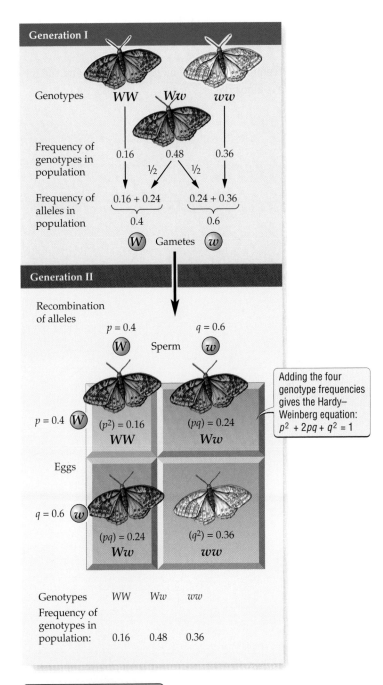

Because the *WW*, *Ww*, and *ww* genotypes are the only three types of zygotes that can be formed, the sum of their frequencies must equal 1. As shown in the figure, when we sum the frequencies of the three genotypes, we get the Hardy-Weinberg equation:

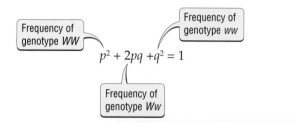

Frequency of genotype *WW*

Frequency of genotype *ww*

$$p^2 + 2pq + q^2 = 1$$

Frequency of genotype *Ww*

In this equation, the frequency of the *W* allele is labeled *p* and the frequency of the *w* allele is labeled *q*.

In general, once the genotype frequencies of a population equal the Hardy-Weinberg frequencies of p^2, $2pq$, and q^2, they remain constant over time if the four conditions listed above continue to be met. A population in which the observed genotype frequencies match the Hardy-Weinberg predicted frequencies is said to be in Hardy-Weinberg equilibrium.

Suggested Readings

Unit 1   The Diversity of Life

For Further Exploration

Futuyma, D. J. 1995. *Science on Trial: The Case for Evolution.* Sinauer Associates, Sunderland, MA. A leading evolutionary biologist summarizes the evidence for evolution and refutes the arguments made by creationists.

May, R. M. 1988. "How Many Species Are There on Earth?" *Science* 241: 1441–1449. Robert May, a distinguished English biologist, evaluates Terry Erwin's estimate of the number of species on Earth, which was based on fogging of canopies.

Mayr, E. 1997. *This is Biology: The Science of the Living World.* Harvard University Press, Cambridge, MA. A description of what biology is and how the science works by one of the most famous and respected biologists.

The New York Times Book of Science Literacy. Volume 2: *The Environment from Your Backyard to the Ocean Floor.* 1994. Times Books, New York. Collected articles from *The New York Times* on biodiversity and related environmental issues.

Pool, R. 1990. "Pushing the Envelope of Life." *Science* 247: 158–247. An exploration of what is known about Archaebacteria and their extreme lifestyles.

Takacs, D. 1997. *The Idea of Biodiversity.* Johns Hopkins University Press, Baltimore, MD. A discussion of the history of biodiversity and its meaning for biologists, written by a biologist–historian.

Wilson, E. O. 1999. *The Diversity of Life.* W. W. Norton and Co., New York. A lyrical and informative overview of the diversity of life on Earth by a leading evolutionary biologist at Harvard University.

For Your Enjoyment

Allmon, W. 2001. *Rock of Ages / Sands of Time.* The University of Chicago Press, Chicago. An intriguing book of paintings of fossils showing one page for every million years of the history of life.

Attenborough, D. 1995. *The Private Life of Plants.* Princeton University Press, Princeton, NJ. A popular natural history about plant "behavior."

Brodo, I. M., Sharnoff, S. D. and Sharnoff, S. 2001. *Lichens of North America.* Yale University Press, New Haven, CT. A beautifully photographed book describing scientific and folkloric knowledge of these species, each of which is a combination of two major groups or kingdoms, Protists and Fungi.

Gould, S. J. 1980. *The Panda's Thumb.* W. W. Norton and Co., New York. Essays by a leading evolutionary biologist on evolution and evolutionary relationships.

Lowman, M. D. 2000. *Life in the Treetops: The Adventures of a Woman in Field Biology.* Yale University Press, New Haven, CT. A fascinating first-person account by one of the first women to study life in the canopies of tropical trees.

Raup, D. 1991. *Extinction: Bad Genes or Bad Luck?* W. W. Norton and Co., New York. In this book for lay audiences, renowned paleontologist David Raup examines why some organisms survive mass extinctions and others do not.

Watson, J. D. 1968. *The Double Helix: A Personal Account of the Discovery of the Structure of DNA.* Atheneum, New York. A lively account of the Nobel prize–winning discovery of the structure of DNA, written by one of its discoverers.

Unit 2   Cells: The Basic Units of Life

For Further Exploration

Atkins, P. W. 1987. *Molecules.* Scientific American Library, New York. Accurate and revealing renderings of important molecules.

Cooper, G. M. 1998. *The Cell: A Molecular Approach.* Sinauer Associates, Sunderland, MA. A concise introductory text on the molecular biology of cells.

Hall, D. O., and Rao, K. K. 1999. *Photosynthesis (Studies in Biology).* Cambridge University Press, Cambridge. An easily understandable and methodical treatment of photosynthesis.

Harold, F. M. 2001. *The Way of the Cell: Molecules, Organisms and the Order of Life.* Oxford University Press, New York. Thoughtful and witty discussion of what constitutes life, with a particular emphasis on single-cell microorganisms.

Needham, J. (ed.) 1970. *The Chemistry of Life.* Cambridge University Press, Cambridge. A selection of thoughtful essays on the history of biochemistry.

Salway, J. G., and Salway, J. D. 1994. *Metabolism at a Glance.* Blackwell Science Inc., Cambridge, MA. A concise and complete overview of metabolism with clear diagrams.

Stossel, T. 1994. "The Machinery of Cell Crawling." *Scientific American,* September. Insightful discussion of how different cytoskeletal proteins contribute to cell movement.

Stryer, L. 1995. *Biochemistry,* 4th Ed. W. H. Freeman and Co., New York. Exceptionally well written and organized introductory text on biochemistry.

For Your Enjoyment

Angier, N. 1999. *Natural Obsessions: Striving to Unlock the Deepest Secrets of the Cancer Cell.* Houghton Mifflin Co., New York. Engaging account of what it is like to do biological research. Reveals the process of scientific inquiry as it unfolded for several important breakthroughs in molecular and cell biology.

Bloch, K. 1997. *Blondes in Venetian Paintings, the Nine-Banded Armadillo, and Other Essays in Biochemistry.* Yale University Press, New Haven. Colorful anecdotes by a Nobel prize–winning scientist on the chemical basis for diverse human and biological phenomena.

Emsley, J. 1998. *Molecules at an Exhibition: Portraits of Intriguing Materials in Everyday Life.* Oxford University Press, New York. Fascinating and entertaining anecdotes about the many chemical compounds encountered in daily life.

Galston, A. W. 1992. "Photosynthesis as a Basis for Life Support in Space." *Bioscience,* July/August. Provocative discussion of how photosynthetic organisms may be used to support artificial ecosystems in space.

Goodsell, D. S. 1998. *The Machinery of Life.* Springer Verlag, New York. Beautifully illustrated tour of the molecular structure of cells.

Loewenstein, W. R. 1999. *The Touchstone of Life: Molecular Information, Cell Communication, and the Foundations of Life.* Oxford University Press, New York. A lucid discussion of how the flow of information, molecular biology, and cell structure work together in living processes.

Rasmussen, H. 1991. *Cell Communication in Health and Disease: Readings from* Scientific American *Magazine.* W. H. Freeman and Co., New York. A collection of articles from *Scientific American* in which researchers examine the complexity of cell communication and its role in various human diseases such as atherosclerosis and diabetes.

Weinberg, R. A. 1998. *Racing to the Beginning of the Road: The Search for the Origin of Cancer.* W. H. Freeman and Co., New York. The fascinating history of cancer's origins, written by one of the most prominent researchers in cancer biology. Contains revealing renderings of important molecules.

Unit 3   Genetics

For Further Exploration

Carroll, S. B., Grenier, J. K. and Weatherbee, S. D. 2001. *From DNA to Diversity: Molecular Genetics and the Evolution of Animal Design.* Blackwell Science, Inc., Malden, MA. An exploration of how genes influence the shape and design of animals, including people.

Griffiths, A. J. F. et al. 2000. *An Introduction to Genetic Analysis,* 7th Ed. W. H. Freeman and Co., New York. A general genetics textbook.

"Making Gene Therapy Work." *Scientific American,* June 1997. A collection of articles outlining current progress in human gene therapy.

Mange, E. J. and A. P. Mange. 1999. *Basic Human Genetics,* 2nd Ed. Sinauer Associates, Sunderland, MA. An introduction to human genetics.

"Microchip Arrays Put DNA on the Spot." *Science,* October 16, 1998. A collection of news articles discussing the current and likely future impact of DNA chips.

Orel, V. 1996. *Gregor Mendel, the First Geneticist.* Oxford University Press, New York. A biography of the founder of the science of genetics.

Subramanian, S. 1995. "The Story of Our Genes." *Time* magazine, January 16. An overview piece on genetic screening and what our genes can tell us.

For Your Enjoyment

"GM foods: are they safe?" *Scientific American*, April 2001. A special report that explores the benefits and risks associated with our widespread use of genetically modified organisms in the food we eat.

Judson, H. F. 1996. *The Eighth Day of Creation: The Makers of the Revolution in Biology.* Cold Spring Harbor Laboratory Press, Cold Spring Harbor, NY. An engaging history of molecular genetics.

Keller, E. F. 1983. *A Feeling for the Organism: The Life and Work of Barbara McClintock.* W. H. Freeman and Co., San Francisco. The story of the life and work of a Nobel Prize–winning female scientist working in an all-male world of science.

Kitcher, P. 1996. "Junior Comes Out Perfect." *New York Times Magazine,* September 29. An article on the potential for producing the perfect baby through genetic screening.

Ridley, M. 2000. *Genome: The Autobiography of a Species in 23 Chapters.* HarperCollins, New York. A well-written and fascinating account of a single important gene from each of our chromosomes. The discussion includes a balanced narrative of the human genome projects.

Watson, J. D. 1968. *The Double Helix.* Atheneum, New York. A personal and controversial account of the Nobel Prize–winning discovery of the structure of DNA.

Unit 4   Evolution

For Further Exploration

Axelrod, R. 1984. *The Evolution of Cooperation.* Basic Books, New York. A far-reaching book on cooperation that shows how phenomena ranging from mutualism to trench warfare can be partially explained by an evolutionary perspective.

Cowen, R. 2000. *History of Life,* 3rd Ed. Blackwell Science, Inc., Malden, MA. An interesting and highly accessible survey of the history of life on Earth.

Darwin, C. 1859. *On the Origin of Species.* J. Murray, London (reprint edition: 1964, Harvard University Press, Cambridge, MA.) The single most important book on evolution ever published. A landmark scientific work that revolutionized biology and had a large impact on many other areas of human society. With careful reading, this text is very clear to a nonscientific audience.

Freeman, S. and Herron, J. C. 2000. *Evolutionary Analysis,* 2nd Ed. Prentice Hall, Englewood Cliffs, NJ. A general textbook on evolution.

Futuyma, D. J. 1995. *Science on Trial: The Case for Evolution.* Sinauer Associates, Sunderland, MA. A leading evolutionary biologist summarizes the evidence for evolution and refutes the arguments made by creationists.

Slatkin, M. (ed.) 1995. *Exploring Evolutionary Biology: Readings from* American Scientist. Sinauer Associates, Sunderland, MA. An interesting collection of articles on evolution, originally published in the journal *American Scientist* and intended for both scientists and the general public.

For Your Enjoyment

Diamond, J. 1992. *The Third Chimpanzee: The Evolution and Future of the Human Animal.* HarperCollins, New York. A fascinating account of human evolution and the future of our species by one of the leading writers on science for the general public.

Johanson, D. and Shreeve, J. 1989. *Lucy's Child.* Avon Books, New York. The story of the discovery of the *Australopithecus afarensis* skeleton that became known in the popular press as Lucy.

Nesse, R. and Williams, G. 1995. *Why We Get Sick: The New Science of Darwinian Medicine.* Vintage Books, New York. A compelling argument for why both doctors and patients need to understand basic evolutionary principles.

Palumbi, S. R. 2001. *The Evolution Explosion: How Humans Cause Rapid Evolutionary Change.* W. W. Norton & Co., New York. A fascinating description of the many evolutionary changes caused by people and the effects such changes have on human society.

Weiner, J. 1994. *The Beak of the Finch: A Story of Evolution in our Time.* Knopf, New York. A Pulitzer Prize–winning account of evolution in the Galápagos finches.

Wilson, E. O. 1992. *The Diversity of Life.* W. W. Norton and Co., New York. A thought-provoking description of how the diversity of life evolved and how it is being threatened by human actions.

Units 5 and 6 Animal and Plant Form and Function

For Further Exploration

Bonner, J. T. 1980. *The Evolution of Culture in Animals.* Princeton University Press, Princeton, NJ. An interesting treatment of the evolution of cultural behavior.

Eisenberg, A., Murkoff, H. E. and Hathaway, S. E. 1996. *What to Expect When You're Expecting.* Workman Publishing, New York. This and other books on human pregnancy give a good overview of the steps in and timing of human reproduction and development.

Gordon, J. E. 1978. *Structures.* Pelican Books, London. Draws many parallels between the structure of plant and animal support systems and the way in which engineers assemble materials into bridges and buildings.

Lovejoy, C. O. 1988. "Evolution of human walking." *Scientific American* 259(5): 118–125. The evolution of walking and its relationship to the size of the birth canal in human females.

Olshansky, S. J., Carnes, B. A., and Butler, R. N. 2001. "If humans were built to last." *Scientific American* 284(3): 50–55. An interesting redesign of the human body to better accommodate the changes that come with aging.

Owen, J. 1980. *Feeding Strategy.* University of Chicago Press, Chicago. A broad and elementary review of nutrition among living things.

Ricklefs, R. E., and Finch, C. E. 1995. *Aging: A Natural History.* Scientific American Library, New York. An overview of the ecology and evolution of aging.

Schmidt-Nielsen, K. 1972. *How Animals Work.* Cambridge University Press, London. This book presents diverse and entertaining examples of how and why animals function the way that they do.

Schmidt-Nielsen, K. 1984. *Scaling: Why Is Animal Size So Important?* Cambridge University Press, New York. A readable but thorough overview of the effect of size on animal biology.

Vogel, S. 1988. *Life's Devices.* Princeton University Press, Princeton, NJ. A thorough overview of the mechanics of how organisms deal with their environment.

For Your Enjoyment

Andrews, M. 1976. *The Life that Lives on Man.* Taplinger Publishing Company, New York. A novel look at how various microbes, fungi, and animals make their living on the surfaces of our own bodies.

Bakker, R. T. 1986. *The Dinosaur Heresies.* Zebra Books, New York. A thorough presentation of the idea that dinosaurs were endotherms.

Bonner, J. T. 1980. *The Evolution of Culture in Animals.* Princeton University Press, Princeton, NJ. An interesting treatment of cultural evolution.

Gordon, J. E. 1978. *Structures.* Pelican Books, London. Draws many parallels between the structure of plant and animal support systems and the way in which engineers assemble materials into bridges and buildings.

Unit 7 Interactions with the Environment

For Further Exploration

Gates, D. M. 1993. *Climate Change and its Biological Consequences.* Sinauer Associates, Sunderland, MA. A clear description of past and present examples of climate change and how those changes affect life on Earth.

Kareiva, P. (ed.) 1998. *Exploring Ecology and its Applications: Readings from American Scientist.* Sinauer Associates, Sunderland, MA. An interesting collection of articles on ecology, originally published in the journal *American Scientist* and intended for both scientists and the general public.

Levin, S. A. 1999. *Fragile Dominion: Complexity and the Commons.* Perseus Books, Reading, MA. An assessment by a leading ecologist of natural ecosystems and how they are affected by people.

Newton, L. H. and Dillingham, C. K. 2002. *Watersheds 3: Ten Cases in Environmental Ethics.* Wadsworth Publishing Company, Belmont, CA. An excellent collection of environmental case studies, each of which provides readers with a clear description of the important biological and ethical considerations that relate to the issue at hand.

Primack, R. B. 1998. *Essentials of Conservation Biology,* 2nd Ed. Sinauer Associates, Sunderland, MA. A good introduction to the fast-developing field of conservation biology, a branch of science focused on the conservation of the diversity of life on Earth.

Ricklefs, R. E. and Miller, G. L. 1999. *Ecology,* 4th Ed. W. H. Freeman and Co., New York. A general ecology textbook.

For Your Enjoyment

Botkin, D. B. 1996. *Our Natural History: The Lessons of Lewis and Clark.* Perigee, New York. A provocative and interesting book that uses the journeys of Lewis and Clark as a springboard from which to discuss how ecological systems are always in a state of change and how endangered ecosystems can be saved.

Carson, R. 1962. *Silent Spring.* Houghton Mifflin Co., New York. An award-winning book that galvanized public interest in ecology and in the environmental movement.

Diamond, J. 1999. *Guns, Germs, and Steel: The Fates of Human Societies.* W. W. Norton & Co., New York. A leading scientist's riveting analysis of how the rise of human civilizations—and the dominance of some groups over others—was shaped by geography and by features of the natural environment.

Leopold, A. 1949. *A Sand County Almanac and Sketches Here and There.* Oxford University Press, New York. A collection of essays from one of the founders of the conservation movement.

McKibben, B. 1995. *Hope, Human and Wild: True Stories of Living Lightly on the Earth.* Little, Brown and Co., Boston. Beautifully written and penetrating essays that focus on successful community efforts to preserve wilderness and reverse environmental damage.

Orr, D. W. 1994. *Earth in Mind: On Education, Environment, and the Human Prospect.* Island Press, Washington, D.C. A thought-provoking collection of essays about causes of and solutions to global environmental problems.

Table of Metric–English Conversion

Common conversions

Length		To convert	Multiply by	To yield
nanometer (nm)	$0.000000001\ (10^{-9})$ m	inches	2.54	centimeters
micrometer (μm)	$0.000001\ (10^{-6})$ m	yards	0.91	meters
millimeter (mm)	$0.001\ (10^{-3})$ m	miles	1.61	kilometers
centimeter (cm)	$0.01\ (10^{-2})$ m			
meter (m)	—	centimeters	0.39	inches
kilometer (km)	$1000\ (10^{3})$ m	meters	1.09	yards
		kilometers	0.62	miles

Weight (mass)				
nanogram (ng)	$0.000000001\ (10^{-9})$ g	ounces	28.35	grams
microgram (μg)	$0.000001\ (10^{-6})$ g	pounds	0.45	kilograms
milligram (mg)	$0.001\ (10^{-3})$ g			
gram (g)	—	grams	0.035	ounces
kilogram (kg)	$1000\ (10^{3})$ g	kilograms	2.20	pounds
metric ton (t)	$1{,}000{,}000\ (10^{6})$ g (=10^3 kg)			

Volume				
microliter (μl)	$0.000001\ (10^{-6})$ l	fluid ounces	29.57	milliliters
milliliter (ml)	$0.001\ (10^{-3})$ l	quarts	0.95	liters
liter (l)	—			
kiloliter (kl)	$1000\ (10^{3})$ l	milliliters	0.034	fluid ounces
		liters	1.06	quarts

Temperature

degree Celcius (°C)	—

To convert Fahrenheit (°F) to Centigrade (°C):
$$°C = \tfrac{5}{9}(°F - 32°)$$

To convert Centigrade (°C) to Fahrenheit (°F):
$$°F = \tfrac{9}{5}°C + 32°$$

Answers to Self-Quiz Questions

Chapter 1
1. *b*
2. *a*
3. *c*
4. *d*
5. *c*
6. *b*
7. *a*
8. *d*

Chapter 2
1. *c*
2. *c*
3. *a*
4. *d*
5. *c*
6. *c*

Chapter 3
1. *d*
2. *c*
3. *b*
4. *a*
5. *b*
6. *a*
7. *a*
8. *c*

Chapter 4
1. *a*
2. *c*
3. *d*
4. *a*
5. *c*
6. *b*
7. *c*
8. *b*
9. *d*

Chapter 5
1. *b*
2. *a*
3. *c*
4. *d*
5. *b*
6. *d*
7. *b*
8. *a*
9. *b*
10. *c*

Chapter 6
1. *d*
2. *b*
3. *a*
4. *d*
5. *b*
6. *a*
7. *d*
8. *d*
9. *b*
10. *c*

Chapter 7
1. *c*
2. *a*
3. *b*
4. *c*
5. *a*
6. *a*
7. *c*
8. *d*
9. *d*
10. *c*

Chapter 8
1. *d*
2. *d*
3. *d*
4. *a*
5. *b*
6. *b*
7. *c*
8. *a*
9. *b*
10. *c*

Chapter 9
1. *b*
2. *a*
3. *d*
4. *b*
5. *d*
6. *c*
7. *c*
8. *c*
9. *d*
10. *a*

Chapter 10
1. *a*
2. *c*
3. *b*
4. *d*
5. *d*
6. *a*
7. *d*
8. *d*

Chapter 11
1. *d*
2. *b*
3. *c*
4. *c*
5. *a*
6. *c*
7. *a*
8. *d*

Chapter 12
1. *c*
2. *c*
3. *b*
4. *d*
5. *a*
6. *c*
7. *d*
8. *d*

Chapter 13
1. *b*
2. *c*
3. *b*
4. *d*
5. *a*
6. *d*
7. *b*
8. *d*

Chapter 14
1. *b*
2. *c*
3. *b*
4. *a*
5. *d*
6. *d*
7. *b*
8. *b*
9. *d*

Chapter 15
1. *c*
2. *a*
3. *b*
4. *d*
5. *a*
6. *c*
7. *a*
8. *d*

Chapter 16
1. *d*
2. *d*
3. *c*
4. *a*
5. *b*
6. *c*
7. *d*
8. *a*
9. *b*

Chapter 17
1. *b*
2. *a*
3. *b*
4. *d*
5. *c*
6. *a*
7. *c*
8. *d*

Chapter 18
1. *b*
2. *b*
3. *c*
4. *a*
5. *d*
6. *d*
7. *b*
8. *c*

Chapter 19
1. *d*
2. *d*
3. *b*
4. *a*
5. *c*
6. *c*
7. *a*
8. *d*

Chapter 20

1. c
2. c
3. d
4. c
5. a
6. c
7. b
8. a
9. d
10. d

Chapter 21

1. c
2. d
3. c
4. b
5. c
6. d
7. d
8. b
9. c
10. d

Chapter 22

1. c
2. c
3. a
4. d
5. d
6. b
7. b
8. a
9. b
10. c

Chapter 23

1. c
2. d
3. b
4. c
5. c
6. c
7. a
8. a
9. d
10. d

Chapter 24

1. d
2. c
3. d
4. b
5. c
6. a
7. a
8. c
9. b
10. d

Chapter 25

1. c
2. d
3. a
4. b
5. c
6. b
7. d
8. a
9. a
10. c

Chapter 26

1. c
2. a
3. d
4. c
5. a
6. b
7. c
8. d
9. b
10. c

Chapter 27

1. d
2. d
3. c
4. c
5. b
6. c
7. d
8. a
9. c
10. d

Chapter 28

1. b
2. d
3. b
4. c
5. d
6. b
7. c
8. d
9. c
10. b

Chapter 29

1. c
2. d
3. b
4. a
5. b
6. a
7. a
8. c
9. c
10. d

Chapter 30

1. b
2. d
3. c
4. c
5. d
6. a
7. c
8. b
9. a
10. c

Chapter 31

1. b
2. d
3. c
4. a
5. c
6. a
7. d
8. b

Chapter 32

1. a
2. d
3. a
4. c
5. b
6. d
7. b
8. a

Chapter 33

1. c
2. d
3. c
4. b
5. c
6. d

Chapter 34

1. d
2. b
3. b
4. c
5. d
6. a
7. d
8. b

Chapter 35

1. c
2. a
3. c
4. d
5. b
6. b
7. d
8. a

Chapter 36

1. c
2. c
3. d
4. a
5. b
6. b
7. d
8. a

Chapter 37

1. c
2. a
3. d
4. a
5. b
6. b
7. c
8. d

Chapter 38

1. d
2. b
3. a
4. c
5. d
6. b
7. a
8. d

Answers to Review Questions

Chapter 1

1. **Observation:** Huge numbers of fish were being found dead; their bodies, covered with bleeding sores, were found floating by the millions in the estuaries of North Carolina.

   Hypothesis: Based on a previous experience in which laboratory fish died suddenly after exposure to local river water, Dr. Burkholder hypothesized that *Pfiesteria*, a protist found in high numbers in the tanks containing the dead lab fish, was also responsible for the fish die-offs in local estuaries.

   Experiment: Dr. Burkholder isolated samples of *Pfiesteria* and exposed healthy fish to the samples. The fish were quickly killed, thus upholding the prediction Dr. Burkholder had formulated based on her hypothesis.

2. **Hypothesis:** Perhaps *Pfiesteria* kills fish by depriving them of something that is essential to their existence, such as oxygen in the water.

   Prediction: If the protist is responsible for removing oxygen from the water or making it difficult for the fish to use, then this process would necessarily affect other organisms living in the local estuaries that also rely on a supply of oxygen in the water.

   Experiment: Expose other inhabitants of the local estuaries that rely on oxygen in the water to the protist to determine whether they are affected in any way by its presence.

3. All living organisms share all of the following characteristics: they are built of cells; they reproduce themselves using DNA; they develop; they capture energy from their environment; they sense their environment and respond to it; they show a high level of organization; they evolve. Prions do not share many of these characteristics and therefore cannot be considered living organisms. Prions are not built of cells; they are protein molecules. Prions do not reproduce themselves using DNA; rather, they reproduce by causing other molecules to mimic their shape. Because prions lack these essential characteristics of life, they are not alive.

4. Levels of the biological hierarchy from smallest to largest (with examples): molecules (DNA); cells (bacteria); tissues (muscle tissues); organs (heart); organ system (stomach, liver, intestines in digestive system); individual organism (human); population (field mice in one field); community (different species of insects living in a forest); ecosystem (river ecosystem); biome (the arctic tundra, coral reefs); biosphere (Earth).

5. Energy flows from the sun to photosynthesizing organisms such as grasses. The grasses, which are producers, use the sun's energy to produce chemical energy in the form of sugars and starches. Antelope, which are consumers, feed on the grasses to produce energy for their own use. Lions, also consumers, then eat the antelope. Ticks, consumers as well, feed on both the antelope and the lions.

6. a. The grasses are producers because they capture sunlight and convert it to energy. The antelope, lions, and ticks are consumers because they eat either plants or other organisms that derive energy from plants.

   b.

   Sunlight ⟶ Grasses ⟶ Antelope ⟨ Lions / Ticks ⟩

Chapter 2

1. By analyzing the similarities and differences in both structural and behavioral features between an unknown organism and known organisms, biologists can determine how closely or distantly related the unknown organism is to an organism already on the tree of life and can thus place the new organism on the tree. The most useful features for this purpose are those known as shared derived features.

2. Both family trees and evolutionary trees demonstrate how different members of a group are biologically related to one another. Common ancestors shared by individuals can be traced using both family and evolutionary trees.

3. Unique structural or behavioral features that have evolved in a group's most recent common ancestor and are then shared by the descendant species of that ancestor are known as *shared derived features*. The panda's thumb and the human thumb are not shared derived features because this trait (an opposable thumb) did not evolve in these different species from their most recent common ancestor. Rather, pandas and humans each evolved opposable thumbs independently; such features are referred to as *convergent features*.

4. Biologists view an evolutionary tree as a hypothesis because it is the best approximation of the relationships of organisms to one another given what we know today. They continue to study and reevaluate these relationships as new information becomes available.

5. Using an evolutionary tree like the one in Figure 2.3, which shows the relationship of crocodilians to birds, biologists are able to speculate about the behavior of dinosaurs. Both crocodilians and birds are dutiful parents: they build nests and defend their young. Parental behavior is therefore presumed to be a shared derived feature of crocodilians and birds inherited from, and exhibited in, their most recent common ancestor. Crocodilians and birds share their most recent common ancestor with dinosaurs, so we can hypothesize that dinosaurs, too, exhibited parental behavior.

6. Figure 2.3 shows that birds and crocodilians share a common ancestor with dinosaurs. If both birds and crocodilians are known to either sing or make chirping vocalizations, it can be assumed, given these groups' relationship to each other, that this behavior is a shared derivative feature inherited from their common ancestor. Because these two groups share their most recent common ancestor with dinosaurs, one could conclude that singing (or chirping) dinosaurs existed.

7. From the smallest to the largest, the groupings of the Linnaean hierarchy are species, genus, family, order, class, phylum, kingdom.

8. DNA allows us to see connections or separations between species that are not expressed in structural or behavioral features, thus changing our understanding of certain organisms' placement on the tree of life and their relationships to one another. For example, organisms that exhibit similar traits may be genetically very distantly related, while organisms that appear to be very different on the outside can have very similar DNA. Through the study of DNA, biologists have begun to hypothesize that at its base, the tree of life is structured more like an interconnected web than a tree.

DNA studies of different organisms have resulted in the discovery of bacterial DNA in archaeans and in eukaryotes. This finding raises the question of how these three distinct lineages—Bacteria, Archaea, and Eukarya—came to share similar DNA. Dr. W. Ford Doolittle suggests that throughout the early history of life, organisms within the three lineages were freely exchanging genes, thus passing genetic information horizontally—in a *horizontal gene transfer*—as well as vertically to their descendants. Therefore, we can attribute this new understanding of the shape of the tree of life to the study of DNA.

Chapter 3

1. The three major systems used to categorize living organisms are the evolutionary tree of life, the Linnaean hierarchy, and the system of life's three domains.

2. The two kingdoms that make up the prokaryotes are Bacteria and Archaea. Some factors that contributed to the success of prokaryotes are their simple and efficient structure, their uncomplicated reproduction, their ability to reproduce rapidly and prolifically, the diverse ways in which they obtain nutrition, and their ability to persist in extreme environments.

3. *Giardia lamblia*, unlike many eukaryotes, has two nuclei, no chloroplasts, and no mitochondria. Biologists theorize that this unique cell structure is a result of one experiment in engulfment that has gone on over time, a conclusion that supports the idea that eukaryotic cells first formed from prokaryotic cells engulfing other prokaryotic cells.

Slime molds provide biologists with insight into the early evolution of multicellularity. Because these protists live their lives in two phases—as independent, single-celled organisms, and then as individual members of a multicellular society—they are studied by biologists who hope to understand the evolutionary transition from single-celled organisms to multicellular organisms.

4. When cells began to work together in a coordinated fashion, some became specialized in order to aid the new multicellular organisms in survival. Sponges, which are among the simplest of animals, are loose collections of specialized cells. One of the earliest animal groups to develop true tissues—specialized, coordinated collections of cells—was the Cnidaria. Cnidarians evolved stinging cells used to stun their prey, as well as nervous tissues, musclelike tissues, and digestive tissues. With cells working in conjunction with one another as tissues, animals were able to evolve organs: body parts composed of different tissues organized to carry out specialized functions. Flatworms were one of the earliest groups to evolve true organs.

5. Viruses are not classified in any kingdom or domain because they occupy a gray area between living organisms and nonliving matter. Viruses lack many of the qualities present in living organisms. They cannot reproduce outside of a host organism, and they lack a clear evolutionary relationship to any one group.

6. When plants colonized land, they were forced to evolve in ways that allowed them to persist in an environment where they were no longer surrounded and supported by water. To deal with the problem of obtaining and retaining water, plants evolved root systems, which allow them to absorb water and nutrients from soil, and the waxy covering over their stems and leaves known as the cuticle, which prevents their tissues from drying out when exposed to sun and air.

No longer floating in water, plants had to meet the challenge of gravity. Plants evolved rigid cell walls composed of cellulose, which allowed them enough rigidity to grow up and into the air. In conjunction with strong cell walls, plants evolved vascular systems—networks of specialized tissues that can transport fluids—extending from their roots throughout their bodies. When the plant is properly hydrated, these systems act like a water-filled balloon and allow the plant to remain and grow upright.

7. Plant cells contain an organelle lacking in animal cells: the chloroplast. Chloroplasts allow plants to produce their own food from carbon dioxide and sunlight. As photosynthesizing organisms and producers, plants provide an important source of food for nearly all organisms on land and are the basis of terrestrial food webs.

8. Plants have developed many characteristics that utilize the mobility of animals. Animals are attracted to various features that plants have evolved in order to reproduce or spread their seeds. Some plants—specifically, angiosperms—evolved flowers, specialized reproductive structures that produce nectar, a sugary liquid used as food by some animals. Animals attracted by nectar may transport pollen between very distant plants, or from flower to flower.

Angiosperms have evolved another food that attracts animals for seed distribution. Fruits develop from the ovary surrounding an angiosperm embryo. Animals eat fruits and later excrete the seeds, usually in a location far from the parent plant where the embryo will not be in competition with its parent for food, water, or sunlight. The animal's feces may also provide a nutrient-rich environment for the germination of the new plant.

9. It is difficult to accommodate viruses in either the Linnaean hierarchy or the domain system, as they lack clear evolutionary relationships to any one group. Furthermore, if we consider that viruses are not living organisms, they cannot be placed anywhere in either classification system, as these systems are used only for living organisms.

Interlude A

1. Terry Erwin fogged a single rainforest tree with insecticide and found more than 1,100 species of beetles living in its canopy. He estimated that 160 of these species were likely to be specialists on a single tree species. There are an estimated 50,000 or so tropical tree species. Thus if the tree species Erwin was studying were typical, beetle species in tropical trees should number 8,000,000 ($50,000 \times 160$). Beetle species are thought to make up about 40 percent of all arthropod species. If that is the case, then the total number of arthropod species in tropical tree canopies should be 20 million.

Many scientists believe that the total number of arthropod species in the canopy is double the number found in other parts of tropical forests, suggesting that there are another 10 million arthropod species in other parts of the tropical forest environment. Thus Erwin estimated that the total number of arthropod species in the tropics at 30 million. Such estimates are difficult to do because they involve making a lot of assumptions, any of which could be wrong.

2. We should take seriously the claims about the extinctions of species around the world because, even though we do not know the exact number of species or their rate of extinction, reasonable estimates of these numbers are invariably quite high. The lack of precise data does not lessen the actual problem of species extinction.

3. Biodiversity has fluctuated ever since life began, declining rapidly during mass extinctions. Five previous mass extinctions (about 440, 350, 250, 206, and 65 million years ago) are thought to have been caused by natural forces, including climate changes, volcanic eruptions, changes in sea level, and atmospheric dust from the collision of Earth with an asteroid. After each of these events, biodiversity recovered slowly, over millions of years. Some scientists believe that the current rapid expansion of human populations could lead to another mass extinction.

4. As human populations grow, cities, suburbs, and commercial areas expand and natural habitats are destroyed and degraded, driving out other species. Human population growth has increased pollution, further threatening biodiversity. Expansion of human populations has also increased the introduction of nonnative species, which constitute another threat to biodiversity.

5. Human beings require places to live, just as other species do. If we have a right to exist, then other species must as well. Where humans' rights end and those of other species begin will depend on how we, as a society, decide to live and to what degree we agree to either protect or encroach on the habitats and lives of other species.

6. Compromise between development and habitat protection for endangered species can be a better outcome than prolonged conflict between staunch environmentalists and determined developers that results in one party winning and the other losing. In such compromises, parties can often come to a speedier, less costly, and more practical solution that provides benefits for humans and other species as well.

Chapter 4

1. Polymers are complex molecules composed of monomers, often with attached functional groups. Combinations of these functional groups give polymers chemical properties not present in the smaller "building block" molecules. These added properties allow polymers to carry out an enormous number of different functions.

2. The pH of pure water should be 7. Units on the pH scale represent the concentration of free hydrogen ions in water. In the presence of a base, the pH of a solution will be above 7, indicating that there are more hydroxide ions than hydrogen ions; thus the solution is basic. In the presence of an acid, the pH will be below 7, indicating more free hydrogen ions than hydroxide ions; thus the solution is acidic. Pure water has equal amounts of hydrogen and hydroxide ions and is thus neutral.

3. Water molecules are polar; that is, they have regions that are charged. The region around the oxygen atom is slightly negative, and the regions around the two hydrogen atoms are slightly positive. This property provides for the formation of hydrogen bonds between water molecules, as well as between water and other polar molecules. Hydrogen bonds are critical components of the structures of biological polymers such as DNA and proteins.

4. Each carbon atom can form strong covalent bonds with up to four other atoms, including other carbons, creating large molecules containing hundreds, even thousands, of atoms. These molecules play many different roles critical to life.

5. Polymers of amino acids are proteins. Some proteins serve as structural components of organisms. Some proteins in muscles allow us to move. Others are enzymes that speed up life's chemical reactions.

6. Cells use carbohydrates as a readily available energy source. Nucleic acids such as DNA and RNA, which carry genetic information, are polymers of nucleotides. Some nucleotides act as energy carriers. Proteins make up the physical structures of organisms as well as the enzymes that catalyze biochemical reactions. Fats are the most common means of long-term energy storage.

Chapter 5

1. The plasma membrane, which is a feature of all cells, provides a necessary boundary between a cell and its surrounding environment. The plasma membrane is selectively permeable so that it controls what gets in and what flows out. Both prokaryotic and eukaryotic cells also contain DNA, cytosol, and ribosomes. DNA contains the information for producing the proteins needed by each cell. Cytosol is the watery medium in which biochemical reactions take place. Ribosomes are the workbenches for producing proteins.

2. The major components of a plasma membrane are a phospholipid bilayer and an assortment of proteins. The phospholipid molecules are oriented so that their hydrophilic heads are exposed to the watery environments both inside and outside the cell. Their hydrophobic fatty acids tails are grouped together inside the membrane away from the watery surroundings. Some membrane proteins extend all the way through the phospholipid bilayer and act as gateways for the passage of selected ions and molecules into and out of the cell. Other membrane proteins are used by the cell to detect changes in and signals from the environment outside the cell. Those proteins that are not anchored to structures within the cell are free to move sideways within the phospholipid bilayer. This freedom of movement supports what is known as the fluid mosaic model, which describes the plasma membrane as a highly mobile mixture of phospholipids and proteins. This mobility is essential for many cellular functions, including movement of the cell as a whole and the ability to detect external signals.

3. Prokaryotic cells lack the specialized internal compartments of eukaryotic cells. These compartments, called organelles, contain, isolate, and concentrate the molecules necessary for biochemical processes. This internal specialization may have allowed eukaryotic cells themselves to become specialized. Multicellular organisms are communities of coordinated specialized cells. Prokaryotic cells are relatively unspecialized internally and have not specialized in the same way as eukaryotic cells. If mitochondria and chloroplasts were originally prokaryotic organisms, the only route for prokaryotes to multicellularity seems to have been to hitchhike as part of the innards of evolving eukaryotic cells.

4. The nucleus houses the cell's DNA; it directs the activities of the cell and determines what proteins will be produced by the cell in response to messages received from both inside and outside the cell. The endoplasmic reticulum is the site of manufacture of proteins and lipids. The Golgi apparatus directs proteins and lipids produced in the ER to their final destinations. Lysosomes, which are found in animal cells, are compartments that contain enzymes used to break down macromolecules for cell use. Vacuoles, found in plant and fungal cells, contain enzymes that break down substances and can store nutrients for later use by the cell. Mitochondria use chemical reactions to transform the energy from many different molecules into ATP, the universal cellular fuel. Chloroplasts, found in plant cells, capture energy from sunlight and convert it into chemical energy to be used by the cell.

5. A fibroblast cell uses microfilaments—twisted polymers of the protein actin—to move. Microfilaments can change rapidly in length, and a fibroblast cell is able to "crawl" as different arrangements of microfilaments alter the structure of the leading and trailing parts of the cell. In the leading parts of the fibroblast cell, the pseudopodia, microfilaments are aligned pointing outward in a forward direction. As the filaments lengthen, they push against the plasma membrane, thereby extending the pseudopodia in the direction the cell is moving. Microfilaments at the trailing end of the cell tend not to be well organized. While the well-aligned filaments in the pseudopodia expand, the filaments in the trailing end of the fibroblast shrink, so that the cell appears to be pulling up its rear end as it moves forward.

   A bacterium with a flagellum moves quite differently from a fibroblast. The corkscrew-like flagellum is turned by a unique rotary "motor." The rotary action of the motor depends on specialized proteins that the bacterium uses to pump hydrogen ions out of the cell. When these ions diffuse back into the cell, they pass through the motor, rotating the flagellum and propelling the cell.

6. Both mitochondria and chloroplasts exhibit characteristics that have led scientists to hypothesize that they are descendants of primitive prokaryotic cells. They both have their own DNA, are able to make some proteins, and reproduce independently of the cell by dividing in two.

Chapter 6

1. The phospholipid bilayer that constitutes the plasma membrane blocks large and hydrophilic molecules from entering or leaving the cell. Channel proteins and passive carrier proteins in the plasma membrane allow certain molecules to cross the membrane by moving down their concentration gradients. Active carrier proteins can transport molecules or ions against their concentration gradients.

2. If a cell's external environment is hypertonic, water will tend to flow out of the cell toward the area of higher solute concentration, causing the cell to shrink. In a hypotonic environment, water will diffuse into the cell and can cause it to swell or burst. In an isotonic environment, the solute concentrations inside and outside of the cell are equal, and there will be no net flow of water into or out of the cell.

3. Exocytosis is the release of substances from a cell when vesicles join with the plasma membrane and open to the outside. Endocytosis is the opposite process, in which substances are brought into a cell by membrane vesicles that bud inward. The cell can import specific substances through receptor-mediated endocytosis, in which specialized receptor proteins on the outer surface of the plasma membrane recognize surface characteristics of the material to be brought into the cell.

4. The two organizing principles of multicellular organisms are cell specialization and cell communication. If all the cells in an organism were specialized but could not communicate with one another, their activities would not be coordinated, and the organism as a whole could not function. Similarly, multicellular life would not be possible if none of the cells were specialized for certain tasks—for instance, structural support or movement—even if a system of communication existed. Without both cell specialization and cell communication, cells would not be able to work together in a multicellular organism, and life would consist entirely of unicellular organisms.

5. In animals, leak-proof barriers are provided by tight junctions, which bind cells together with protein strands. Anchoring junctions consist of protein hooks that hold neighboring cells together and allow passage of materials between them. Gap junctions are direct channel protein connections between plasma membranes. In plants, plasmodesmata provide cytoplasmic connections through cell walls.

6. Slow-acting cell signals affect target cells far away from their source, and usually travel in the bloodstream (in animals) or in the sap (in plants). Fast-acting signals are short-lived and travel only between neighboring cells.

Chapter 7

1. The second law of thermodynamics holds that systems tend toward disorder. In a living system such as a cell, the order maintained by chemical reactions is counterbalanced by the release of heat energy (disorder) into the surroundings.

2. The reactions of photosynthesis obtain carbon from the air as CO_2, using light energy to combine that CO_2 with water to synthesize sugars. Organisms ultimately break down these sugars for energy and release the stored carbon back into the atmosphere as CO_2.

3. Concentrating enzymes and their substrates in a compartment, as with components of the citric acid cycle in the mitochondrial matrix, can increase catalytic efficiency by making collisions between enzymes and their substrates more likely. Another method for increasing efficiency is the sequential arrangement of enzymes for a series of reactions, as found in the arrangement of enzymes for ATP synthesis on the inner membrane of the mitochondrion.

4.

Chapter 8

1. The transfer of electrons down an ETC produces a proton gradient in both chloroplasts and mitochondria. The protons move down that gradient through a membrane channel protein known as ATP synthase. The

movement of the protons releases energy, which is used by ATP synthase to phosphorylate ADP to form ATP.

2. In the light reactions of photosynthesis, light energy is first captured by chlorophyll. The energy excites electrons, which travel down the electron transport chain (ETC) in the thylakoid membrane of the chloroplast. This process produces a proton gradient that drives ATP synthesis. Finally, $NADP^+$ is converted to NADPH. In the dark reactions, the ATP and NADPH help power the incorporation of CO_2 into sugars (carbon fixation). Electrons that move out of photosystem II are replaced by the stripping away of electrons from hydrogen atoms in water molecules that are split by photolysis.

3. When protons pass through ATP synthase, their energy is used to convert ADP into ATP. Drugs that allow protons to bypass ATP synthase will deter the production of ATP.

4. Electrons are removed from water in photosystem II, then transferred to electron-accepting proteins in the thylakoid membrane. The electrons pass from the electron transport chain to photosystem I. From there, the electrons pass to another electron transport chain and finally to $NADP^+$, which is reduced to NADPH.

Chapter 9

1. A horse cell undergoing mitosis would have 64 chromosomes. A horse cell in meiosis II would contain 32 chromosomes.

2. Interphase consists of the G_1, S, and G_2 phases. During the gap phases (G_1 and G_2), the cell has time to grow and make the proteins it will need for the next phase. In between the two gap phases is the S phase, during which the DNA of the cell is replicated. During mitosis, two sets of daughter chromosomes are formed, and during cytokinesis, two daughter cells are formed. Cells in the G_0 stage do not divide.

3. In mitosis, spindle microtubules from each centrosome attach to the centromere of a chromosome at the kinetochores. During anaphase, these microtubules pull the sister chromatids of the chromosome to opposite poles of the cell. Each sister chromatid is a new chromosome and occupies one of the two daughter cells that arise after cytokinesis. In meiosis I, microtubules from only one pole attach to each homologous chromosome. During anaphase I, the homologous chromosomes (sister chromatids still joined) are pulled to opposite poles. This process halves the number of chromosomes in each daughter cell.

   Both processes are the same in that genetic material is redistributed. The processes differ in that the cells that result from mitosis are diploid, but the cells that result from meiosis are haploid.

4. A human male has a karyotype of 44 non-sex chromosomes (called autosomes, as we will see in Unit 3) in homologous pairs, plus one X and one Y sex chromosome. A female has 44 autosomes plus two X chromosomes. Thus the haploid gamete of the female would contain 22 autosomes plus an X chromosome, while a male haploid gamete would contain 22 autosomes plus either an X or a Y sex chromosome.

5. If gametes of sexually reproducing organisms were produced by mitosis, the offspring of each succeeding generation would have double the number of chromosomes of the parent generation.

Interlude B

1. When a proto-oncogene mutates to form an oncogene, some cellular processes may become uncontrolled and lead to cancer.

2. Cell signals must activate proto-oncogenes in order to promote cell division, and must also inactivate tumor suppressor genes in order for cell division to occur.

3. Colon cancer begins with a benign growth of cells called a polyp, in which mutations have inactivated both copies of a tumor suppressor gene and transformed a proto-oncogene into an oncogene. Loss of a part of chromosome 18 deletes two other tumor suppressor genes, allowing for more aggressive and rapid cell division in the tumor. Finally, the loss of the tumor suppressor gene *p53* removes all remaining cell division controls and allows the cancer cells of the now malignant tumor to metastasize (spread to other parts of the body).

4. The possibility of carcinogens in many commonly consumed products should persuade people to lessen their risk of cancer by taking more care in choosing what products to use and what lifestyle choices to make. Studies should be done to identify unknown carcinogens and determine what levels constitute a health risk to consumers.

6. It is important to provide sufficient information about cancer risks so that consumers can make informed decisions about their product purchases. Food labels could be simplified to provide more concise information about possible cancer risks, and public health programs could raise public awareness of cancer risks present in certain foods and other products.

Chapter 10

1. Genes are the basic units of inheritance; as such, they carry genetic information for specific traits. Genes are composed of DNA and are located on chromosomes. Most genes contain instructions for building proteins.

   Mendel's theory of inheritance can be summarized as follows: (1) Alternative versions of genes, called alleles, cause variation in inherited traits. (2) Offspring inherit one copy of a gene from each parent. (3) An allele is dominant if it determines the phenotype of an organism even when paired with a different allele. (4) The two copies of a gene separate during meiosis and end up in different gametes. (5) Gametes fuse without regard to which alleles they carry.

2. A sexually reproducing organism contains two copies of each gene because it gets one copy from each parent. If an individual is homozygous for a particular gene—say, it has genotype *gg*—each parent of that individual must have had at least one copy of the *g* allele. Thus the mother could have had genotype *Gg* or genotype *gg*; the same is true of the father.

3. New alleles arise when mutations in genes occur. A mutation is any change in the DNA that makes up a gene. When a mutation occurs, the new allele that results may contain instructions for a protein with a form different from that of the protein specified by the original allele. By specifying different versions of proteins, the different alleles of a gene cause hereditary differences among organisms.

4. (1) (2)

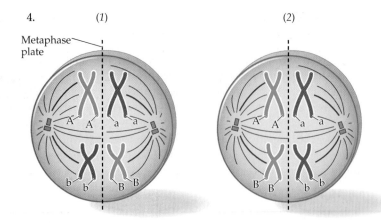

Metaphase plate

The diagrams shown here illustrate metaphase I of meiosis for an individual of genotype *AaBb*; the DNA has already replicated, so each of the four chromosomes consists of two identical chromatids (see Chapter 9). In these diagrams, the maternal chromosomes are shown in blue and the paternal chromosomes are in red. In diagram 1, by chance one maternal chromosome lined up to the left of the metaphase plate, while the other maternal chromosome lined up to the right of the plate. When the chromosomes line up in this way, the gametes have genotype *Ab* or *aB*; if you are not clear why this is so, see Figure 9.8 for more information. Alternately, and also by chance, the chromosomes could have lined up as shown in diagram 2, causing the gametes to have genotype *AB* or *ab*. Overall, genes on different chromosomes are inherited independently of each other (in this case, producing gametes of genotype *AB, Ab, aB,* or *ab* because the chromosomes on which they are found line up at random on the spindle microtubules during meiosis.

5. In order to determine the genotype of the purple-flowered plant (*PP* or *Pp*), you could cross it with a white-flowered plant (*pp*). When either genotype is crossed with the homozygous recessive (the white-flowered plant), the resulting phenotypes of the offspring will indicate whether the purple-flowered parent plant is heterozygous or homozygous. A plant with the genotype *Pp*, when crossed with a plant possessing the *pp* phenotype, will produce white-flowered offspring 50 percent of the time and purple-flowered offspring 50 percent of the time. A dominant homozygous plant (*PP*), when crossed with a recessive homozygous plant (*pp*), will always produce purple-flowered offspring.

	P	p
p	Pp	pp
p	Pp	pp

	P	P
p	Pp	PP
p	Pp	PP

6. Although identical twins are genetically identical, their phenotypes can differ because environmental factors can alter the effects of genes. A twin who is well nourished in childhood, for example, may grow up to be tall, while one who is malnourished in childhood may grow up to be short. Similarly, exposure to different amounts of sunlight will cause their skin color to differ. The phenotypes of twins can differ radically if one is exposed to environmental factors that trigger the onset of a genetic disorder that both are predisposed to get, but the other is not.

7. Dominant alleles for lethal human diseases may be uncommon because a person carrying such an allele would be likely to develop the disease and perish before producing any offspring, so the frequency of such an allele in

the human population would be low. Conversely, individuals possessing a recessive allele for a lethal genetic disorder can live as carriers, unaffected by the gene, and pass the allele on to their children, who, depending on the genetic makeup of their other parent, might develop the disease, or might become carriers in turn, remaining unharmed by the allele and passing it on to the next generation.

Sample Genetics Problems

1. a. *A* and *a*
   b. *BC, Bc, bC,* and *bc*
   c. *Ac*
   d. *ABC, ABc, Abc, AbC, aBC, aBc, abC,* and *abc*
   e. *aBC* and *aBc*

2. a. genotype ratio: 1:1   phenotype ratio: 1:1

	A	a
a	Aa	aa

   b. genotype ratio: 1:0   phenotype ratio: 1:0

	B
b	Bb

   c. genotype ratio: 1:1   phenotype ratio: 1:1

	AB	Ab
ab	AaBb	Aabb

   d. genotype ratio: 1*BBCC*:1*BBCc*:2*BbCC*:2*BbCc*:1*bbCC*:1*bbCc*
      phenotype ratio: 6:2, reduced to 3:1

	BC	Bc	bC	bc
BC	BBCC	BBCc	BbCC	BbCc
bC	BbCC	BbCc	bbCC	bbCc

   e. genotype ratio:
      1*AABbCC*:2*AABbCc*:1*AABbcc*:1*AAbbCC*:2*AAbbCc*:1*AAbbcc*:1*AaBbCC*:
      2*AaBbCc*:1*AaBbcc*:1*AabbCC*:2*AabbCc*:1*Aabbcc*
      phenotype ratio: 6:2:6:2, reduced to 3:1:3:1

	ABC	ABc	AbC	Abc	aBC	aBc	abC	abc
AbC	AABbCC	AABbCc	AAbbCC	AAbbCc	AaBbCC	AaBbCc	AabbCC	AabbCc
Abc	AABbCc	AABbcc	AAbbCc	AAbbcc	AaBbCc	AaBbcc	AabbCc	Aabbcc

3.

	S	s
S	SS	Ss
s	Ss	ss

   genotype ratio: 1*SS*:2*Ss*:1*ss*
   phenotype ratio: 3 healthy:1 sickle-cell anemia
   Each time two *Ss* individuals have a child there is a 25% chance that the child will have sickle-cell anemia.

4. 100% of the offspring should be chocolate labs.

5. a. *NN* and *Nn* individuals are normal; *nn* individuals are diseased.
   b.

	N	n
N	NN	Nn
n	Nn	nn

   genotype ratio: 1*NN*:2*Nn*:1*nn*
   phenotype ratio: 3 normal:1 diseased

c.

	N	n
N	NN	Nn

genotype ratio: 1:1   phenotype ratio: 2 healthy:0 diseased

6. a. *DD* and *Dd* individuals are diseased; *dd* individuals are normal.

b.

	D	d
D	DD	Dd
d	Dd	dd

genotype ratio: 1*DD*:2*Dd*:1*dd*   phenotype ratio: 3 diseased:1 normal

c.

	D	d
D	DD	Dd

genotype ratio: 1:1   phenotype ratio: 2 diseased:0 healthy

7. The parents are most likely *BB* and *bb*. The white parent must be *bb*. The blue parent could potentially be *BB* or *Bb*, but if it were *Bb*, we would expect about half of the offspring to be white. Therefore, if the cross yields many offspring and all are blue, it is extremely likely that the blue parent's genotype is *BB*.

8. The allele for green seed pods is dominant. Since each parent breeds true, that means each parent is homozygous. When a homozygous recessive parent is bred to a homozygous dominant parent, the F_1 generation will exhibit only the dominant phenotype. Therefore, the phenotype of the F_1 generation, green seed pods, is produced by the dominant allele.

Chapter 11

1. A gene is a small region of the DNA molecule in a chromosome. Genes are located on chromosomes.

2. Human females have two X chromosomes, while human males have one X and one Y chromosome; thus human males have only one copy of each gene that is unique to either the X or the Y chromosome. As a result, patterns of inheritance for genes located on the X chromosome differ between males and females. A mother can pass an X-linked allele, such as one for a genetic disorder, to her male or female offspring. However, a male can pass an X-linked allele only to his female offspring (since his male offspring receive his Y chromosome, not his X chromosome).

3. Genes located on different chromosomes separate into gametes independently of one another during meiosis; hence such genes are not linked. If the genes for the traits shown in Figure 11.3 were inherited independently of each other, Morgan would have obtained approximately equal numbers of flies for each of the four genotypes shown in the figure. Since the numbers of the two parental genotypes outnumbered the other two genotypes by a wide margin, Morgan concluded that the genes must be located on the same chromosome. Because they are physically connected to each other, they are inherited together, or linked.

4. Crossing-over occurs when genes are physically exchanged between homologous chromosomes during meiosis. Part of the chromosome inherited from one parent is exchanged with the corresponding region from the other parent. Two genes that are far apart from each other on a chromosome are more likely to be separated from each other than two genes that are close

to each other. Understanding this, one can assume that genes *A* and *C* are more likely to be separated during crossing over than are genes *A* and *B*.

5. Nonparental genotypes are formed during crossing-over. The exchange of genes that takes place during crossing-over makes possible the formation of gametes with combinations of alleles that differ from those found in either parent.

6. Relatively few human genetic disorders are caused by inherited chromosomal abnormalities, probably because most large changes in the chromosomes kill the developing embryo. Genetic disorders caused by single-gene mutations appear to be more common because the survival rate of embryos with single-gene mutations is higher.

Sample Genetics Problems

1. a. Males inherit their X chromosome from their mothers, since their Y chromosome must come from their fathers. Their mothers do not have a Y chromosome to give them, and they must have one in order to be male.

   b. No, she does not have the disorder. If she has only one copy of the recessive allele, her other X chromosome must then have a copy of the dominant allele. She is a carrier, but she does not have the disorder herself.

   c. Yes, he does have the disorder. The trait is X-linked, he has only one X chromosome, and that X chromosome carries the recessive disorder-causing allele. His Y chromosome does not carry an allele for this gene, so cannot contribute to the male's phenotype relative to this trait.

   d. If the female is a carrier of an X-linked recessive disorder, her genotype is $X^D X^d$, where D = the dominant allele and d = the recessive, disorder-causing allele. This means she can produce two types of gametes relative to this trait: X^D and X^d. Only the X^d gamete carries the disease-causing allele.

   e. None of their children will have the disorder, since the mother will always contribute a dominant non-disorder-causing allele to each child. However, all of the female children will be carriers, since their second X chromosome comes from their father, who only has one X chromosme to contribute, and it carries the disorder-causing allele.

2. a. 50% chance of *aa* cystic fibrosis genotype

	A	a
a	Aa	aa
a	Aa	aa

   b. 0% chance of *aa* cystic fibrosis genotype

	A	A
A	AA	AA
a	Aa	Aa

   c. 25% chance of *aa* cystic fibrosis genotype

	A	a
A	AA	Aa
a	Aa	aa

   d. 0% chance of *aa* cystic fibrosis genotype

	A	A
a	Aa	Aa
a	Aa	Aa

3. a. 50% chance of Huntington's disease genotype, Aa

	A	a
a	Aa	aa
a	Aa	aa

b. 100% chance of Huntington's disease genotype, AA or Aa

	A	A
A	AA	AA
a	Aa	Aa

c. 75% chance of Huntington's disease genotype, AA or Aa

	A	a
A	AA	Aa
a	Aa	aa

d. 100% chance of Huntington's disease genotype, Aa

	A	A
a	Aa	Aa
a	Aa	Aa

4. a. 0% chance of a child with hemophilia

	X^a	Y
X^A	$X^A X^a$	$X^A Y$
X^A	$X^A X^a$	$X^A Y$

b. 50% chance of a child with hemophilia

	X^a	Y
X^A	$X^A X^a$	$X^A Y$
X^a	$X^a X^a$	$X^a Y$

c. 25% chance of a child with hemophilia

	X^A	Y
X^A	$X^A X^A$	$X^A Y$
X^a	$X^A X^a$	$X^a Y$

d. 50% chance of a child with hemophilia

	X^A	Y
X^a	$X^A X^a$	$X^a Y$
X^a	$X^A X^a$	$X^a Y$

e. No, male and female children do not have the same chance of getting the disease. Male children are more likely to have hemophilia since they do not possess a second allele for this trait to mask a recessive allele that they may inherit.

5. The terms "homozygous" and "heterozygous" refer to pairs of alleles for a given gene. Since a male has only one copy of any X-linked gene, it does not make sense to use these pair-related terms.

6. The disease-causing allele (d) is recessive and is carried by both parents, because although neither the mother nor the father expresses the trait in question, some of their children do. The disease-causing allele is located on an autosome. If it were on the X chromosome, the father would express the gene, since we have already determined that he must carry one recessive copy of the gene, and he would not have another copy of the gene to mask this recessive allele. Both individuals 1 and 2 of generation I have the genotype Dd.

7. Designate the dominant, X-linked allele D and the recessive normal allele d. Based on the Punnett squares shown in a and b (below), males are not more likely than females to inherit a dominant x-linked genetic disorder.

a. There are two possible Punnett squares, depending on whether the affected female has genotype $X^D X^d$ or genotype $X^D X^D$:

	X^d	Y
X^D	$X^D X^d$	$X^D Y$
X^d	$X^d X^d$	$X^d Y$

or

	X^d	Y
X^D	$X^D X^d$	$X^D Y$

b. This cross is $X^d X^d \times X^D Y$, which gives the Punnett square:

	X^D	Y
X^d	$X^D X^d$	$X^d Y$
X^d	$X^D X^d$	$X^d Y$

8. The disorder allele is a recessive allele, located on the X chromosome. We know the allele is recessive because individual 2 in generation II carries the allele but does not have the condition. If the allele were located on an autosome, the parents in generation I would be of genotype AA (the male) and aa (the female). In this case, none of the individuals in generation II could have the condition—yet two of them do have the condition, which implies that the allele is on a sex chromosome. Finally, we know that the allele is on the X chromosome because, otherwise, only males could get the condition.

9. a. If the two genes are completely linked:

	AB	ab
aB	$AaBB$	$aaBb$

b. If the two genes are on different chromosomes:

	AB	Ab	aB	ab
aB	$AaBB$	$AaBb$	$aaBB$	$aaBb$

Chapter 12

1. Frederick Griffith's experiment with two strains of bacteria and mice showed that harmless strain R bacteria could be transformed into deadly strain S bacteria when exposed to heat-killed strain S bacteria. This finding suggested that genetic material from the heat-killed strain S bacteria had somehow changed living strain R bacteria into strain S bacteria.

   Oswald Avery, Colin MacLeod, and Maclyn McCarty isolated and tested different compounds from the bacteria in Griffith's experiments and found that only DNA from heat-killed strain S bacteria was able to transform harmless strain R bacteria into deadly strain S bacteria. This finding led to the conclusion that DNA, not protein, is the genetic material.

   Alfred Hershey and Martha Chase studied a virus that consists only of a DNA molecule surrounded by a coat of proteins. By using radioisotopes to selectively label either the DNA or the protein portion of the virus, Hershey and Chase showed that the DNA, but not the proteins, entered a bacterium to take over the bacterial cell and produce the next generation of viruses.

2. The three main components of a nucleotide from a DNA molecule are the sugar deoxyribose, a phosphate group, and a nitrogen base. The base is what makes one type of nucleotide different from others. The four bases are adenine (A), cytosine (C), guanine (G), and thymine (T). When covalent bonds link nucleotides to one another, one strand of the double helix that is a DNA molecule is formed.

3. The two strands that make up the double helix of the DNA molecule are held together by hydrogen bonds between the nucleotides' nitrogen bases.

4. The genetic information of the alleles is contained in the sequence of the nitrogen bases—adenine (A), cytosine (C), guanine (G), and thymine (T)—found within the segment of DNA that constitutes each allele. At any genetic locus, different alleles differ in the sequence of bases they contain. Thus the DNA segments of the two codominant alleles A^1 and A^2 differ in the sequence of bases found in each allele.

5. The double-helix structure of DNA and the base-pairing rules theorized by Watson and Crick suggested a simple way that genetic material could be copied. Because A pairs only with T and C pairs only with G, each strand of DNA contains the information needed to duplicate the complementary strand. DNA replication using the double-helix model results in two identical copies of the original DNA molecule.

6. The sequence of bases in DNA is the basis of inherited variation. A change in the sequence of bases in DNA, whether because of an error during replication or because of exposure to a mutagen, is a mutation. Such a change could result in a new allele that encoded a new version of the protein encoded by the gene in which the mutation occurred. If the new allele produced a protein that did not function properly (or at all), serious damage could be done to the cell, and consequently to the organism; such an allele could cause a genetic disorder.

7. DNA is repaired by enzymes and other proteins. When DNA is being replicated, enzymes check for and immediately correct mistakes in pair bond formation. Mistakes that escape this process, called mismatch errors, are caught and corrected by repair proteins.

   DNA repair is essential for cells to function normally because DNA is constantly being damaged by chemical, physical, and biological agents. If none of this damage were repaired, genes that encoded proteins that were critical to life would eventually cease to function, thus disabling the production of those proteins and killing the cell, and ultimately the organism.

Chapter 13

1. A gene is a DNA sequence that contains information for the synthesis of one of several types of RNA molecules used to make proteins. A gene stores information in its sequence of nitrogen bases.

2. Genes control the production of a variety of RNA products (mRNA, rRNA, tRNA). mRNA encodes the amino acid sequence of proteins, rRNA is an essential component of ribosomes (the site of protein synthesis), and tRNA carries amino acids to the ribosomes during protein construction. Thus each of the RNA products specified by genes functions in the synthesis of proteins. Proteins are essential for many functions that support life. In cells and organisms, proteins provide structural support, transport materials through the body, and defend against disease-causing organisms. Enzymes are a class of proteins that control chemical reactions.

3. Genes contain instructions for the synthesis of proteins. Each gene is composed of a segment of DNA on a chromosome and consists of a sequence of the four bases: adenine (A), cytosine (C), guanine (G), and thymine (T). The sequence of bases specifies the amino acid sequence of the gene's protein product. Through the processes of transcription and translation, proteins are produced using the information stored in genes. In transcription, mRNA is synthesized directly from the sequence of bases in one DNA strand inside the nucleus of a cell. Translation occurs in the cytoplasm and converts the sequence of bases in an mRNA molecule into the sequence of amino acids in a protein. Proteins, by their many and various functions, influence the phenotype of an individual.

4. For a protein to be made, the information in a gene must be sent from the gene, which is located in the nucleus, to the site of protein synthesis, on a ribosome. This transfer of information requires an intermediary molecule because DNA does not leave the nucleus, whereas ribosomes are located in the cytoplasm. In eukaryotes, a newly formed mRNA molecule usually must be modified before it can be used to make a protein. The reason for this is that most eukaryotic genes contain internal sequences of bases (introns) that do not specify part of the protein encoded by the gene. DNA sequences copied from introns must be removed from the initial mRNA product if the protein encoded by the gene is to function properly.

5. mRNA is the product of transcription, and is a version of the genetic information stored in a gene. The mRNA moves from the nucleus to the cytoplasm, where it binds with a ribosome to guide the construction of a protein.

   rRNA is a major component of ribosomes. Translation occurs at ribosomes, which are molecular machines that make the covalent bonds linking amino acids together into a particular protein.

   tRNA molecules carry the amino acids specified by the mRNA to the ribosome. At the ribosome, a three-base sequence (anticodon) on the tRNA binds by means of complementary base pairing with the appropriate codon on the mRNA. Each tRNA molecule carries the amino acid specified by the mRNA codon to which its tRNA anticodon can bind.

6. If a tRNA molecule does not function properly because of a mutation, each protein that it helps to build will be altered in some way. By failing to bind properly with the mRNA codons of many different genes, a mutant tRNA may significantly affect the structure of many different protein products. Because their structure is altered, the function of these protein products may be impaired. Since proteins are key components of many metabolic reactions, changing the function of many different proteins can result in a series of metabolic disorders.

Chapter 14

1. Two features allow cells to pack an enormous amount of DNA into a very small space: the thinness of the DNA molecule and a highly organized, complex packing system. Each portion of a chromosome, which contains one DNA molecule, consists of many tightly packed loops. Each loop is composed of a fiber made up of many histone spools, which are made of proteins called histones. A segment of DNA winds around each spool, and if that DNA were unwound, it would reveal its double helix structure.

2. Prokaryotes have less DNA than eukaryotes. All DNA in a prokaryote is located on one chromosome, in contrast to a eukaryotic cell, in which the DNA is distributed among several chromosomes. Eukaryotes have more genes than prokaryotes do, and genes constitute only a small portion of the eukaryotic genome. Most prokaryotic DNA encodes proteins, and very little of it is noncoding DNA and transposons. Functionally related genes in prokaryotes are grouped together on the chromosome, while eukaryotic genes with related functions often are not located near one another.

3. Crossing-over occurs more often between genes that are located farther apart. Thus the presence of long segments of noncoding DNA will increase the frequency of crossing-over between genes.

4. The bacterium would begin expressing the gene that encodes the enzyme that breaks down arabinose. Organisms can turn genes on and off in response to short-term changes in food availability or other features of the environment.

5. In multicellular organisms, different cell types express different genes by controlling transcription, along with other methods. By switching specific genes on or off, cells can vary their structure and perform specialized metabolic tasks, even though each has exactly the same genes (and alleles).

6. From gene to protein, the steps at which gene expression can be controlled are as follows: (1) Tightly packed DNA is not expressed, in part because the proteins necessary for transcription cannot reach them. (2) Transcription can be regulated by regulatory proteins that bind to regulatory DNA, effectively switching a gene on or off. (3) The breakdown of mRNA molecules can be regulated such that mRNA is destroyed hours or weeks after it is made. (4) Translation can be inhibited when proteins bind to mRNA molecules to prevent their translation. (5) Proteins can be regulated after translation, either when the cell modifies or transports them or when they are rendered inactive by repressor molecules. (6) Synthesized proteins can be destroyed.

7. Regulatory DNA sequences such as the tryptophan operator switch genes on and off. The tryptophan operator controls whether or not the gene that encodes tryptophan is transcribed. If tryptophan is present, it binds to a repressor protein, which is then able to bind to the operator and prevent transcription, since the presence of tryptophan indicates that the cell does not need to waste energy by making more. If tryptophan is absent, the repressor is unable to bind to the operator, thus allowing the gene to be transcribed so that the tryptophan needed by the cell is produced.

8. A DNA chip consists of thousands of samples of DNA placed on a small glass surface. Often the DNA at each position on a chip corresponds to the DNA of one gene. When a gene is expressed, an mRNA copy of the information in that gene is produced. To study many genes at once, mRNA is isolated from the organism or cells being tested, labeled (as with a dye that glows red or green), and then washed over the DNA chip. The labeled mRNA can bind to the DNA representing the gene from which it was originally produced. Since the gene that corresponds to each location on the chip is known, results from this procedure can tell us which of the organism's genes produced mRNA—and hence which of the organism's many genes were expressed (and which were not).

Chapter 15

1. In order to produce domesticated species, humans have manipulated the reproduction of other organisms, selecting for desirable qualities that, over time, became standard in domesticated species. Although such selection practices do lead to changes in the DNA of organisms (that is, they lead to an increase in the frequencies of alleles that control the inheritance of the traits we select for), we can make much greater changes today. We can now manipulate the DNA of organisms directly, and we can transfer genes from one species to another. Transfers of DNA from, say, a human to a bacterium (as done in the production of human insulin) far exceed the scope of DNA transfers that occur in nature. We can also selectively change specific DNA sequences—something we could never do before. Overall, we can now manipulate DNA with greater power and precision than we could when we domesticated species such as corn and cows.

2. By using restriction enzymes and gel electrophoresis together, geneticists can examine differences in DNA sequences. The man and woman could be tested for the sickle-cell allele with the use of the restriction enzyme *Dde*I, which cuts the normal hemoglobin allele into two pieces, but cannot cut the sickle-cell allele. Their doctor might also want to use a DNA probe to test for the sickle-cell allele in these would-be parents. A DNA probe is a short, single-stranded segment of DNA with a known sequence, usually tens to hundreds of bases long. A probe can pair with another single-stranded segment of DNA if the sequence of bases in the probe is complementary to the sequence of bases in the other segment.

3. A gene is said to be cloned when geneticists isolate it and produce many copies of it. Once a gene is cloned, automated sequencing machines can quickly determine its DNA sequence. Two of the most common methods of cloning are constructing a DNA library and using the polymerase chain reaction. To build a DNA library, a vector such as a plasmid is used to transfer DNA fragments from the organism whose gene is to be cloned to a host organism, such as a bacterium. To clone a gene by PCR, primers are synthesized, allowing DNA polymerase to produce billions of copies of the gene in a few hours.

4. The advantage of gene cloning is that it is easier to study a gene once you have many copies of it. The cloned gene can then be sequenced, transferred to other organisms, or used in various experiments.

5. Bacterial colonies acting as hosts in the DNA library are screened by DNA hybridization to see if their DNA can pair with a probe for the gene of interest. Colonies whose DNA can pair with the probe contain all or part of the desired gene. It is often necessary to screen the colonies on many petri dishes before such a colony is found.

6. Genetic engineering is used to alter the phenotype (especially the performance or productivity) of the genetically modified organism (GMO) or to produce many copies of a DNA sequence or a gene's protein product. In genetic engineering, a DNA sequence (often a gene) is isolated, modified, and inserted back into the same species or into a different species. Genetic engineering works because, with rare exceptions, all organisms share an identical genetic code.

Interlude C

1. A human and a tomato plant would probably share sets of genes that control DNA transcription, cell division, glycolysis, and cell signaling cascades. Humans would not share with plants sets of genes for proteins involved with vision, various sensory receptors, or nerve function.
2. Single nucleotide polymorphisms (SNPs) are single-base-pair differences in the DNA of unrelated individuals. SNPs or groups of SNPs can be linked to the propensity to develop certain diseases, so SNP testing can reveal whether an individual is at risk for developing one of those diseases.
3. Knowing the genome of the mosquito that carries malaria may assist in the development of pesticides that could specifically disrupt certain biological processes in that species and be more effective at killing it.
4. Genetic screening of patients could help doctors diagnose diseases and assign proper treatments or preventative measures. On the other hand, if a genetic screening of an individual shows the likelihood of a serious illness whose treatment would be costly, insurance companies may deny coverage. Genetic screening could raise concerns of privacy violation or increase the likelihood that unborn children inheriting undesirable genetic conditions would be aborted.
5. Widespread genetic screening might preclude the prescription of certain drugs to some patients if the nature of possible diseases could be more accurately predicted. Pharmaceutical trade organizations might be afraid of lower revenues from those drugs that would no longer be prescribed by doctors relying on traditional methods of diagnosis.
6. Some considerations for this question: Because of privacy concerns, patients should have the option of whether or not to undergo genetic screening, but it would often be in the best interest of patients to know what preventive steps to take. Doctors should inform their patients that possible genetic dispositions toward certain medical conditions do not always lead to the illnesses.

also share less puzzling characteristics because of convergent evolution resulting from similar selective pressures.
4. Overwhelming evidence indicates that evolution occurred and continues to occur. Support for evolution comes from five lines of evidence:
   a. The fossil record provides clear evidence of the evolution of species over time and documents the evolution of major groups of organisms from previously existing organisms.
   b. Organisms contain evidence of their evolutionary history. For example, scientists find that studies of proteins and DNA support the evolutionary relationships determined by anatomical data; that is, the proteins and DNA of closely related organisms are more similar than those of organisms that do not share a recent common ancestor. In this and many similar examples, the extent to which organisms share characteristics other than those used to determine evolutionary relationships is consistent with scientists' understanding of evolution.
   c. Scientists' understanding of evolution and continental drift has allowed them to predict the geographic distributions of certain fossils depending on whether the organisms evolved before or after the breakup of Pangaea.
   d. Scientists have gathered direct evidence of small evolutionary changes in thousands of studies by documenting genetic changes in populations over time.
   e. Scientists have observed the evolution of new species from previously existing species.
5. In any area of science, new pieces of information are continually being added to our knowledge. The debate among scientists as to which mechanisms of evolution are most important means only that evolution is not fully understood, not that it does not occur.
6. Genetic drift has a greater effect on smaller populations. If the plant population were larger, the likelihood that all plants of a certain genotype would die in a windstorm would be smaller, and the dramatic shift in the frequencies of the A and a alleles would therefore be less likely.

Chapter 16

1. Evolution is change in the genetic characteristics of a population over time, which can occur through mechanisms such as genetic mutation or natural selection. Since the genotypes of individuals do not change, a population can evolve, but an individual cannot.
2. In the new habitat, larger lizards will have an advantage over smaller ones. The larger lizards of the species will therefore be more likely to pass on the trait of large size to their offspring, and the average size of lizards in the population will increase over time because of this selective advantage.
3. Each of these aspects of life on Earth can be explained by evolution. (a) Adaptations, which improve the performance of an organism in its environment, result from natural selection. (b) The diversity of life results from speciation, which occurs when one species splits to form two or more species. (c) Organisms can share puzzling characteristics because of common descent. Consider the wing of a bird, the flipper of a whale, and the arm of a human. Even though these appendages are used for very different purposes—and hence we would not expect them to be structurally similar—they are composed of the same set of bones. This occurs because birds, whales and humans share a common ancestor that had these bones. Organisms can

Chapter 17

1. **Mutation:** A nonlethal mutation of a particular allele can be inherited by offspring, thereby increasing the frequency of the mutant allele in a population over time.
   Gene flow: The exchange of alleles between populations can change the frequencies at which alleles are found in the populations by introducing new alleles. Populations affected by gene flow tend to become more genetically similar to one another.
   Genetic drift: Random events (such as chance events that influence the survival or reproduction of individuals) can cause one allele to become dominant in a small population. By chance alone, drift can lead to the fixation of alleles in small populations; if these alleles are harmful, the population may decrease in size, perhaps to the point of extinction.
   Natural selection: If an inherited trait provides a selective advantage for individuals in a certain population, individuals with that trait will be more likely to reproduce, and the frequency of the allele for that trait will increase in succeeding generations. Likewise, an allele for a disadvantageous trait will be selected against, and will be found with decreasing frequency over time.

2. Genetic variation in a population provides the "raw material" on which evolution can work. In the absence of genetic variation, evolution cannot occur. Natural selection, for example, causes individuals of some genotypes to leave more offspring than individuals of other genotypes, but this sorting process cannot occur if all individuals have the same genotype.

3. Recombination (a term that refers collectively to fertilization, crossing-over, and the independent assortment of chromosomes) causes offspring to have combinations of alleles that differ from one another and from those found in their parents. Thus recombination greatly increases the genetic variation on which evolution can act.

4. Gene flow is the exchange of alleles between populations. Gene flow makes populations more similar to one another in their genetic makeup. Genetic drift is a process by which alleles are sampled at random over time. Genetic drift can have a variety of causes, such as chance events that cause some individuals to reproduce and prevent others from reproducing. Natural selection is a process in which individuals with particular inherited characteristics survive and reproduce at a higher rate than other individuals. Sexual selection is a form of natural selection in which individuals with certain traits have an advantage in attracting mates, and consequently in passing on those traits to offspring.

5. The potential benefits include making the population larger, and therefore less susceptible to genetic drift, and providing an input of new alleles on which natural selection can operate. The potential drawbacks include the introduction of individuals with genotypes that are not well matched to the local environmental conditions of the smaller population. Throughout time, some species have gone extinct locally or worldwide. Extinction is a natural process. However, humans have greatly increased the rate at which populations and species have become extinct. If numerous other populations of a species exist, it may not be worth introducing new members to the smaller population. If the smaller population is one of the few populations of that species left, however, it may be important to introduce new individuals in an attempt to allow the population to recover and survive.

6. The genotype frequencies for the original population are

$$AA: \frac{280}{280 \ + \ 80 \ + \ 60} = 0.67$$

$$Aa: \frac{80}{280 \ + \ 80 \ + \ 60} = 0.19$$

$$aa: \frac{60}{280 \ + \ 80 \ + \ 60} = 0.14$$

The allele frequencies for this population are

Frequency of A allele =

$$p \ = \ \frac{2(280)+80}{2(280+80+60)} \ = \ 0.76$$

Frequency of a allele =

$$q \ = \ \frac{2(60)+80}{2(280+80+60)} \ = \ 0.24$$

The Hardy–Weinberg equation predicts that the frequency of genotype AA should be

$$p^2 = (0.76)(0.76) = 0.58,$$

that the frequency of genotype Aa should be

$$2pq = (2)(0.76)(0.24) = 0.36,$$

and that the frequency of genotype aa should be

$$q^2 = (0.24)(0.24) = 0.06.$$

Note the sum of the genotype frequencies,

$$p^2 + 2pq + q^2 = 1.0.$$

These calculated genotype frequencies do not match those of the original population. This difference could be due to mutation, nonrandom mating, gene flow, a small population size, and/or natural selection.

7. Using an antibiotic drug to kill large numbers of bacteria tends to give a considerable reproductive advantage to those bacteria that possess resistance to the drug. Since bacteria reproduce extremely rapidly, the entire population of bacteria will soon be resistant. Reducing human exposure to the bacteria would not allow resistant strains to have as great a degree of reproductive advantage over normal bacterial strains. Slowing the growth of bacteria would likewise help limit the reproductive advantage of resistant strains.

Chapter 18

2. Individuals with inherited traits that allow them to survive and reproduce better than other individuals replace those with less favorable traits. This process, by which natural selection improves the match between organisms and their environment over time, is called adaptive evolution. In our efforts to kill or control bacteria that cause infectious diseases, we are creating a new environment in which bacteria that cannot withstand antibiotics are eliminated. Often there are some bacteria in the population that are not killed by antibiotics; these bacteria reproduce, thus increasing the frequency of resistant bacteria in the population. These evolutionary changes in disease-causing bacteria are harmful to us, as more and more of the diseases we encounter will be resistant to medical treatment.

3. Species that hybridize in nature may still be distinct species, due to a host of alleles that do not affect their ability to interbreed but may cause them to look different or to differ from each other ecologically. For this reason, many people would argue that the rare species is separate from the common species and should remain classified as rare and endangered.

4. Defining species by their inability to reproduce sexually with other species is convenient, but there are many alleles that do not affect reproductive isolation yet could cause the two oaks to be different enough that they can be classified as separate species, even though they can produce hybrids.

5. This storm-blown population will be connected by little or no gene flow to other populations of its species. As a result, genetic changes due to mutation, genetic drift, and natural selection will accumulate over time. Natural selection is likely to cause genetic change in the population because the new environment is different from the parent population's environment; genetic drift will probably also be important because the island population is small (making drift more likely). As a by-product of genetic changes due to selection, drift, and mutation, the island population may become reproductively isolated from the parent population. Thus, if the island population remains isolated for a long enough time, enough genetic changes may accumulate for it to evolve into a new species.

6. Some of the cichlid populations of Lake Victoria may have had so little contact with one another that they can be said to have evolved into separate species in geographic isolation, despite the fact that they live in the same lake. Other populations may have evolved into new species in the absence of geographic isolation.

7. New plant species can form in the absence of geographic isolation as a result of polyploidy, a condition in which an individual has more than two sets of chromosomes. There is strong evidence that sympatric speciation occurred in the cichlids that live in Lake Bermin and Lake Barombi, and evidence is accumulating that apple and hawthorn populations of the apple maggot fly are diverging into two species despite living in the same area. In apple maggot flies and cichlids, sympatric speciation is promoted by ecological factors (such as selection for specialization on different food items) and sexual selection.

There is a greater potential for gene flow between populations whose geographic ranges overlap than between populations that are geographically isolated from one another. Gene flow tends to cause populations to remain (or become) similar to one another. Thus, in the absence of geographic isolation, it can be difficult for genetic differences great enough to cause reproductive isolation to accumulate over time. As a result, sympatric speciation occurs less readily than allopatric speciation.

Chapter 19

1. One example of the evolution of one group of organisms from another is the emergence of mammals from reptiles. The emergence of the mammalian jaw and teeth can be used to illustrate the steps in this process. The first step of mammalian evolution in this respect was the development of an opening in the jaw behind the eye. Then more powerful jaw muscles and specialized teeth appeared with the therapsids. Finally, a subgroup of these reptiles, the cynodonts, emerged with more specialized teeth and a more forward hinge of the jaws, completing this aspect of mammalian evolution.

2. The emergence of photosynthesis in ancient organisms gradually led to the buildup of O_2 in the atmosphere, which killed many organisms to which oxygen was toxic. However, the oxygen supplied by photosynthetic organisms made possible the evolution of eukaryotes and later multicellular life forms.

3. The Cambrian explosion was a large increase in the diversity of life forms over a relatively short time about 530 million years ago. Larger organisms of most phyla emerged during this period, setting the stage for the colonization of land.

4. The colonization of land led to another great increase in the diversity of life forms. Life on land required different means of mobility and reproduction, adaptations to obtain and retain water, and ways to breathe in air rather than water. Early terrestrial organisms had the opportunity to expand into new types of largely unoccupied habitat, which provided ample resources for those organisms able to survive the challenges of life on land.

5. A mass extinction event may be associated with rapid environmental changes that have no relation to the conditions that favored a particular adaptation. Thus organisms with wonderful adaptations can (and have) become extinct during mass extinction events.

6. Although speciation can happen within a single year, it often takes hundreds of thousands to millions of years to occur. Thus it is not surprising that it usually takes 10 million years for the number of species found in a region to rebound after a mass extinction event. The time required to recover from mass extinction events provides a powerful incentive for humans to halt the current, human-caused losses of species; otherwise, it will take millions of years for biodiversity to recover.

7. Microevolution refers to the changes in allele or genotype frequencies within a population of organisms. This concept is fundamentally different from macroevolution, which deals with the rise and fall of entire groups of organisms, the large-scale extinction events that cause some of these changes, and the evolutionary radiations that follow extinction events. Macroevolutionary changes cannot be predicted solely from an understanding of the evolution of populations. Of the evolutionary processes we have studied in Unit 4, speciation occupies a "middle ground" between macroevolution and microevolution.

Interlude D

2. If other human species shared the world with us today, there would undoubtedly be instances of social or cultural friction between the other species and modern humans, just as there exist tensions between "different" ethnic groups today. In order for there to be peaceful cohabitation, societies would need to recognize the humanity of other human species and discuss ways of cooperation in sharing living space and resources.

3. It is generally recognized that people have an ethical responsibility to care for the living things in the environment. The loss of so many species would have negative consequences not only for the environment, but for people as well. Human actions that could potentially lead to a mass extinction should therefore be seriously examined and measures taken to avert such an occurrence.

4. An ethical case for preventing people who carry harmful genes from reproducing would be very difficult to make. The designation of harmful alleles that would be detrimental to the societal good would be largely subjective, and mandatory testing would be a serious violation of human rights and privacy. Living under such a system would probably be very restrictive to personal freedom.

Chapter 20

1. Most animals must have the right amount of water in their cells (typically 40–65 percent of their body weight) to survive and remain active, although some can remain dormant with almost no water. Too little water impedes the movement of solutes and the action of enzymes in cells, while too much water can interfere with biochemical reactions, alter protein structure, and cause cells to burst. Animal cells must also stay within a certain temperature range. The freezing of water at 0°C can kill cells by bursting plasma membranes. Above about 40°C many enzymes become denatured and unable to catalyze biochemical reactions.

2. Gains must equal losses; regulation requires feedback; homeostasis is more easily maintained in larger animals; more energy is required to maintain homeostasis as the difference between internal and external conditions increases.

3. In a cold environment such as that of the North Pole, homeostasis might be maintained by wearing several layers of insulating clothing to minimize heat loss, drinking sufficient fluids, and increasing caloric intake. Technology, such as indoor heating and air conditioning, can assist humans in their maintenance of homeostasis by minimizing differences between body conditions and the external environment. Special fabrics used in clothing can help dissipate excess heat or prevent heat loss.

4. When your body senses that it needs food, signals are sent that increase your appetite. When enough food is eaten, your body senses that it is satiated, and signals are sent that diminish your appetite.

5. In a nephron, plasma from blood that enters the capillaries of the glomerulus is filtered through pores in the capillary walls. The filtrate enters and flows down a tubule. Along the way, valuable solutes, water, and nutrients are reabsorbed into the bloodstream, and the filtrate becomes more concentrated. The tubules join to form larger tubes, and the end product of the filtration and reabsorption process, called urine, flows down these tubes and out of the body.

6. In conduction, heat flows between an animal's body and a substance with which it is in direct contact. In radiation, an animal's body absorbs radiant (infrared) energy from, or emits radiant energy to, its surroundings. In evaporation, the body loses heat by the conversion of water to water vapor on a wet body surface.

7. The loss of water and the high body temperature that result from heatstroke would disrupt metabolic reactions in a cell and its surroundings as enzymes become denatured and solute movement is impeded.

Chapter 21

1. The elements animals need to obtain in relatively large amounts are carbon, sulfur, phosphorus, oxygen, nitrogen, hydrogen, magnesium, calcium, potassium, sodium, and chlorine. Carbon, sulfur, phosphorus, oxygen, nitrogen, and hydrogen are found in molecules such as nucleotides, amino acids, fatty acids, and carbohydrates.

2. In the mouth, teeth bite and chew pieces of the apple. The apple pieces are mixed with saliva by the tongue, then swallowed. They travel down the esophagus to the stomach, where acid and pepsin begin breaking them down as the stomach muscles churn the mixture. From here the mixture is released into the small intestine, where digestive enzymes released by the pancreas, gallbladder, and wall of the small intestine break down the apple's carbohydrates, proteins, and fats. Cells in the lower small intestine absorb the nutrients and water into the bloodstream. The remaining fiber from the apple is consolidated in the colon and passed out of the body as feces. Indigestible material in the apple would probably include the skin and fiber in the fruit.

3. The digestive systems of most animals have hard surfaces that mechanically grind food into smaller pieces, enzymes that break down complex carbohydrates, proteins, and fats, and mechanisms for absorption of nutrients. These features are widespread because all animals must convert the large, complex molecules present in food into small molecules that can be absorbed by the cells of the body.

4. Herbivores have broad, flat teeth that are suitable for extensive grinding of plant material. Microorganisms in the multi-chambered stomach of an herbivore break down cellulose from plant material. Finally, the relatively large surface area of an herbivore's small intestine maximizes the absorption of the scarce protein in its plant diet.

Chapter 22

1. Air arrives at the pharynx after passing through either the mouth or the nasal cavity. From the pharynx, air passes down the trachea and into the bronchi leading to the lungs. In the lungs, each bronchus branches into many smaller bronchioles, which end at the balloonlike alveoli, where gas exchange occurs.

2. Oxygen from the air diffuses through the thin wall of the alveolus and into the capillaries surrounding it. In the oxygen-deficient blood, it is picked up by hemoglobin molecules in the red blood cells. The red blood cells are carried through the bloodstream to respiring cells, where oxygen diffuses to the cells and carbon dioxide from the cells diffuses into the blood plasma. When the blood returns to the lungs, carbon dioxide diffuses out of the blood plasma into the air in the alveoli.

3. Because of their one-way flow and their direct exposure to the oxygenated medium, gills allow for the passage of the large amount of water necessary for oxygen extraction in an aquatic environment, where oxygen is in relatively short supply, and keep energy expenditure low for the process. Gills would be disadvantageous in a terrestrial environment because their exposed moist surface would cause a significant water loss. On land, lungs provide an internal surface for gas exchange that helps retain moisture. Lungs would not work well in an aquatic environment because oxygen would not be able to diffuse into the body quickly enough, and it would be impossible to pump a sufficient amount of water in and out.

4. A molecule of hemoglobin can bind up to four molecules of oxygen, so hemoglobin has a much higher oxygen-carrying capacity than does blood plasma. This binding is reversible, so hemoglobin not only picks up oxygen easily from oxygen-rich surroundings (as in the alveoli of the lungs), but releases it easily to surroundings with relatively low oxygen concentrations (near respiring cells).

5. Aquatic endotherms that breathe air can more easily obtain the amount of oxygen needed to sustain their high body temperatures. The oxygen concentration of water is much lower than that of air, so it would be difficult, and would require a much greater expenditure of energy, for a water-breathing endotherm to obtain enough oxygen.

Chapter 23

1. Diffusion of gases (O_2 and CO_2) can occur effectively only over short distances, but these gases must be transported long distances to and from tissues in the body. In the human circulatory system, gases diffuse over short distances in the alveoli of the lungs and at the capillaries near respiring cells, but the gases are transported long distances to their destinations by bulk movement of red blood cells through the blood vessels.

2. From the left atrium, the drop of blood enters the left ventricle, and from there it is pumped through the pulmonary artery to the lungs. The blood releases CO_2 and picks up O_2 in the alveoli of the lung, then flows back to the heart through the pulmonary vein, completing the pulmonary circuit. It

enters the right atrium and then the right ventricle. Contraction of the right ventricle pumps the oxygenated blood into the aorta, beginning the systemic circuit. From there the blood travels through arteries and successively smaller arterioles, eventually reaching the capillaries surrounding respiring cells. Here the red blood cells deliver O_2 and pick up CO_2, then flow through successively larger venules and veins as the blood is carried away from the tissues. The blood then flows through the vena cava and back to the left atrium of the heart.

3. As a red blood cell is pumped from the heart, the cell (along with the oxygen it carries) is transported long distances to tissues around the body. When the cell reaches a capillary near a respiring cell, the oxygen itself is transported a short distance from the red blood cell to the respiring tissues by diffusion.

4. As the diameter of a blood vessel decreases, the ratio of the surface area to the volume of the vessel increases. Blood flowing over a larger surface area encounters more frictional drag, decreasing its velocity and reducing the amount of blood that can flow through the vessel in a certain amount of time.

5. In a human, an open circulatory system could not facilitate the transport and exchange of gases efficiently enough to sustain life. The large size and high metabolic demand of humans requires a rapid and directional transport of blood. Gas exchange in most animals with open circulatory systems is not coupled with the transport of blood.

6. In nephrons, large volumes of blood are filtered through the tiny, thin capillaries of the glomerulus. Blood passing through these capillaries under high pressure could cause tears in the capillaries, resulting in leakage and less capacity for filtration. This failure of kidney function could result in the buildup of toxic wastes in the blood.

Chapter 24

1. Islet cells in the pancreas produce insulin and release it directly into the bloodstream, where the insulin molecules circulate until reaching target cells in the liver. In response to the insulin signal, the liver cells absorb higher amounts of glucose from the blood and assemble glucose molecules into chains of glycogen for storage.

2. In human males, testosterone controls the development of the male sex organs before birth, interacts with brain cells to bring about behavioral changes, and stimulates sperm production in the testes, the growth of facial hair, and growth in cells throughout the body.

3. When you are confronted with a threatening dog, nerve signals from your brain travel to the muscles of your arms and legs, causing you to begin running away. At the same time, the brain signals the adrenal glands to secrete epinephrine, which stimulates glycogen breakdown in your liver and speeds up your heartbeat, providing the quick energy boost you need for running.

4. In an animal that breeds only once a year, a rise in the levels of estrogens could be triggered by an annual environmental signal, such as warm temperatures or long days in the spring. The high estrogen levels in the female animal would trigger an increase in levels of gonadotropins, hormones that would induce ovulation and prepare the uterus for a pregnancy. The production of these hormones would be regulated at much lower levels during the rest of the year.

5. If too much growth hormone is produced in the years before adulthood, gigantism results, while if too little is produced, dwarfism results. An excess of growth hormone after a person reaches adulthood results in acromegaly, a condition characterized by enlargement of the hands, feet, and bones of the head and face.

Chapter 25

1. When a stimulus of sufficient strength is received, channels in the neuron's plasma membrane open, causing an influx of Na^+ ions and depolarization, or loss in the electrical charge across the cell membrane of the axon, at that point. Depolarization causes sodium channels to open in adjacent portions of the axon's cell membrane, which leads to depolarization of the "downstream" cell membrane. In this way the action potential travels down the length of the neuron. When the action potential reaches an axon terminal, neurotransmitters are released from vesicles by exocytosis into the synaptic cleft between the neuron and its target cell. The neurotransmitter molecules diffuse across the cleft and bind to receptor proteins on the plasma membrane of the target cell, where they can either trigger or inhibit action potentials.

2. In this nervous system, two sensory neurons, one responding to each of the two different stimuli, would send their signals to an interneuron that relays those signals to a motor neuron, which would convey the "move" message to the appropriate muscles. The interneuron could also process the signals so that no signal (the "stay still" message) is sent to the motor neuron.

3. The nervous system is designed for effecting rapid responses to various stimuli, while the endocrine system generally works more slowly by means of circulating hormones in the bloodstream. The hypothalamus integrates the two systems: its specialized neurons detect hormone levels in the body and release hormones of their own to signal for the release of other hormones by the glands of the endocrine system, particularly the pituitary gland.

4. It is not known why the size of the human brain has increased so rapidly over the past 2 million years. Most of the increase is due to enlargement of the neocortex, which has enabled humans to develop language, the ability to think complex thoughts and perform mathematical calculations, and invent new tools.

Chapter 26

1. Sensory receptors must first receive the stimulus from the environment. The stimulus must be converted to action potentials, which travel along nerves to the brain. The brain processes those signals and integrates them to create a sensory perception.

2. The particles from the burning object would bind to specific chemoreceptors in your nose, and action potentials would be sent to your brain. Your brain would process the signals and recognize the smell as "burning." The signals would intensify closer to the source as a greater amount of particles were bound, allowing you to locate the source of the odor.

3. To perceive flavor, the brain processes and integrates sensory information from the taste buds and the olfactory receptors in the nose. In addition,

vision (seeing the food) and touch (feeling the food's texture) contribute to the perception of flavor.

4. The guitar player relies in part on the feel of the fingers on the strings to create the music in the performance. The listener perceives the sounds produced by the vibrating guitar strings as music. The player uses mechanoreceptors near the surface of the fingertips that sense slight changes in pressure, while the listener depends on hair cells in the basilar membrane of the organ of Corti.

5. The pigments in the eye might be more properly called photoreceptors because even though a chemical change occurs in response to a stimulus, the stimuli received are actually photons of light, not chemicals.

Chapter 27

1. The stiffness and internal makeup of bones allow body parts to maintain their shape and withstand stress. Bones also provide attachments for muscles and ligaments and support for muscle contraction. Cartilage supports the structure of the nose and ear pinnae while allowing for greater flexibility than bone, and is also found in joints.

2. A joint allowing for circular motion would be similar to a ball-and-socket joint, such as the one found in the hips. In such a joint cartilage would line the entire socket to facilitate rotation of the connected bone.

3.
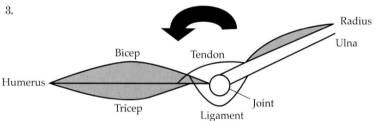

4. A sarcomere consists of actin and myosin filaments. The actin filaments are anchored to Z discs at the ends of the sarcomere. During contraction, the heads of the myosin molecules "walk" down the actin filaments by attaching to and detaching from binding sites. This action pulls the Z discs closer together, and the sarcomere contracts.

5. As a bird flaps its wings, it experiences drag when the pressure from air buildup in front of the wings pushes backward as the bird moves its wings forward. Alternately, the drag from air pushing against the wings can help move the bird forward as it pulls its wings backward.

Chapter 28

1. The first line of defense against pathogens consists of physical and chemical barriers. Physical barriers include the skin and the linings of the lungs and intestines, which most pathogens are unable to penetrate. Chemicals in tears and saliva kill pathogens that come into contact with them, and mucus in the respiratory system traps invading pathogens.

2. Disease-fighting cells can recognize distinctive proteins and carbohydrates on the plasma membranes of other cells in the body. These cells recognize the proteins and carbohydrates on the surfaces of pathogens and viruses as being "nonself," and initiate a response.

3. An intelligent bacterium might produce large amounts of toxins that would overwhelm the nonspecific responses. The bacterium could also produce proteins that would mimic the antigens on the host body's cells, and so avoid destruction by specific immune responses.

4. An autoimmune disease develops when cells of the immune system fail to recognize some tissues as self and instead attack them as if they were foreign.

Chapter 29

1. Sperm, the male gametes, are small and mobile. They deliver the male parent's genetic material. The female gamete, the ovum or egg, is large and does not move on its own. It contributes the female parent's genetic material and nearly all of the cellular organelles to the offspring.

2. Sexual reproduction requires meiosis and results in genetically unique individuals. Whereas sexual reproduction involves the combining of genetic material from two individuals, offspring produced by asexual reproduction are genetically identical to their parent. The genetic variation of sexually produced offspring makes it more likely that some individuals in a population will be able to survive environmental changes, while a population of offspring produced asexually is more vulnerable to such changes. Sexual reproduction requires time and energy, whereas asexual reproduction takes less energy and can produce offspring more rapidly.

3. Sperm swim into the oviduct to reach the egg, following a chemical signal released by the ovary. A sperm makes its way through the follicle cells surrounding the egg to the outer covering of the egg, where it releases enzymes from its acrosome. The enzymes digest the egg's outer covering and allow the sperm to enter the egg, where its nucleus detaches and fuses with the nucleus of the egg, forming a diploid zygote.

4. The blastocyst first implants itself in the wall of the uterus. The inner cell mass grows and divides, forming the three main cell layers. These layers differentiate into the various organs of the body. The heart is identifiable at 3 weeks, and all the major organs have begun to form by the end of the eighth week. By the end of the first trimester external structures such as nails and genitalia have formed. In the second and third trimesters, the existing organs grow and develop further, and the fetus gains weight. Before birth, the circulatory and respiratory systems prepare for the transition to a gaseous atmosphere. The developing fetus is most susceptible to disruption in the first trimester, when cells are differentiating and organs are developing rapidly.

5. At the beginning of development, cells are undifferentiated and have the ability to give rise to cells of any type. As development progresses, the range of cell types that a cell can give rise to becomes increasingly limited. This process occurs by activation and expression of specific genes. Gene expression can be controlled by morphogens, hormones, and environmental factors.

6. The apoptosis of extra skin cells on the fins of early terrestrial animals could have improved mobility by aiding the development of limbs.

7. Intelligent species on other planets might be able to live beyond their reproductive years if they had, for example, medical technology or few natural health threats. Advantages of living longer could include societal benefits from the ability of older generations to pass down skills and knowledge to younger generations.

Chapter 30

1. There are a number of behaviors associated with nursing. The child searches for its mother's nipple with its mouth in an attempt to get milk, a behavior called the rooting reflex. The touch of the nipple triggers sensory signals to the child's brain, which, in turn, stimulates contractions in its mouth muscles. The baby's nursing behavior triggers the milk-letdown reflex in the mother; pressure-sensitive cells around the nipple trigger a release of hormones that result in the release of milk. As the child grows, the milk-letdown reflex changes to let the mother release milk when the baby cries, which is a learned, rather than a fixed, behavior.

2. Humans would develop a fear of spiders only after learning that spider bites are painful. For organisms to learn, however, they must have a nervous system and brain that are sufficiently well developed to allow them to interpret and remember what they have experienced, and to respond flexibly to this experience. The development of these structures is under genetic control.

3. Learning to play an instrument or a sport requires both learning the techniques and strategies needed to be effective (how to hold the instrument or how to kick a soccer ball) and learning which forms or positions are most effective (which sounds made by the instrument sound nice to the ears or which poses in dance look the most graceful).

5. Fixed behaviors that might be important to group members might involve responses to stimuli about which an organism might never have a chance to learn. Group-living animals might have fixed behaviors that allow them to recognize members of their own group (for example, recognizing an odor unique to their group). As an another example, a group-living animal might have a fixed behavior that leads it to avoid snakelike objects or to hide at the sight of a silhouette that resembles that of a predatory bird. There might not be a chance to learn the appropriate behaviors from others, because one incorrect response could easily be fatal. Behaviors associated with surviving during the first few days following birth or hatching, before the creature has a chance to learn, are likely to be fixed behaviors.

Interlude E

1. Some arguments for and against allowing smoking: **For:** costs of enforcing a ban on smoking might be high; a ban would restrict personal choice; addiction would make it very difficult for smokers to stop smoking. **Against:** the health care and economic costs arising from smoking are a serious burden on society; smoking causes many preventable deaths, including those of many unborn babies; tobacco use lowers quality of life and can affect the health of others through secondhand smoke.

2. Long-term exposure to secondhand smoke carries the same health risks as smoking cigarettes: chronic bronchitis, emphysema, lung cancer, and other health problems, including increased risk of heart attack and stroke.

3. The chemical compound nicotine is responsible for the addictive effects of tobacco. Nicotine acts as a stimulant by causing release of the neurotransmitter dopamine, which is associated with pleasurable feelings, and of epinephrine, which increases blood pressure, heart rate, and blood glucose levels. The pleasurable effects of nicotine stimulation associated with tobacco use contribute to addiction.

4. Most smokers are aware of the health risks of smoking, but it is very difficult to quit. In the absence of nicotine, smokers no longer get the pleasurable effects of dopamine or feel the alertness caused by epinephrine release, and they experience strong cravings for nicotine. Those who start smoking, especially young people, may feel that they will be able to quit easily when they want to and that health damage will not appear for a long while, but the addiction increases over time, and much of the long-term health damage is irreversible.

6. Using chewing tobacco or snuff greatly increases the risk of throat and mouth cancer and, like smoking, can cause gingivitis.

Chapter 31

1. The three plant organs are (1) roots, which anchor the plant, absorb and transport water and nutrients, and can be specialized to store food; (2) stems, which give plants support, transport water and nutrients, and can be specialized to form structures such as thorns; and (3) leaves, which provide food by means of photosynthesis and can be specialized to form reproductive organs (flowers) or spines. The three plant tissue types are (1) dermal tissues, which protect the plant and control the exchange of gases and water; (2) ground tissues, which support the plant, heal wounds, and perform photosynthesis; and (3) vascular tissues, which transport food and water throughout the plant.

2. Mineral nutrients dissolved in water can be absorbed by plant roots through cells (the "cell-interior" route) or along cell walls (the "cell-wall" route).

3. Most of the dry weight of a plant comes not from nutrients absorbed by the roots, but from sugars synthesized by photosynthesis using CO_2 obtained from the air.

4. Some carnivorous plants capture insects with traplike modified leaves (Venus flytrap) or water-filled chambers (bladderworts) or attract prey with sticky hairs (sundews). Some pitcher plants can trap large prey, including vertebrates, in their slippery fluid-filled pitchers.

5. Water evaporating from leaves causes the water in the leaf tissue to curve, creating tension that pulls on the continuous column of water in the xylem that extends to the roots of the tree. The water acts like a cable, leaving the upper part of the plant through evaporation as more is drawn up from the roots. Plants generate the pressure required to transport food from the leaves to other areas by pumping sugars into the phloem. The movement of sugar into the phloem cells decreases the concentration of water in those cells, causing water to diffuse into the phloem cells. This creates a pressure difference that effectively pushes the food through the phloem to other areas of the plant.

Chapter 32

1. Groups of undifferentiated cells called meristems, distributed at various locations in the plant body, retain the ability to differentiate into any type of plant cell. Meristems can produce new bud-stem-leaf modules, adding to the length of the plant body.

2.

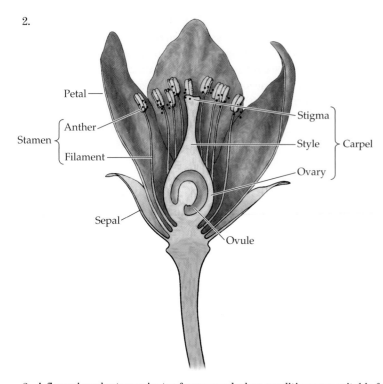

Petal —
Stamen { Anther
Filament
Sepal
Stigma
Style
Ovary
Ovule
Carpel

3. A flowering plant germinates from a seed when conditions are suitable for growth. The plant grows and produces flowers, which contain the reproductive structures of the plant. Small, haploid pollen grains (in which sperm are produced) are formed in the anther of a flower; small, haploid embryo sacs (in which eggs are produced) are formed in the carpel. When a pollen grain from a male anther lands on the stamen of a flower of the same species, a pollen tube containing two sperm cells grows downward to the ovule. Double fertilization occurs in the ovule: one sperm fertilizes the egg, producing an embryo, and the other fertilizes a large, central cell in the female embryo sac, producing endosperm (a tissue that will provide food for the embryo). The embryo becomes encased in a protective seed coat, and the surrounding ovary forms a fruit. The fruit and its enclosed seeds are dispersed by means of wind, water, or animals. When a seed germinates, the cycle begins again.

4. Wood consists of layers of secondary xylem, which are formed as new rings of cells are put down each year; these cells are produced by the vascular cambium during secondary growth.

5. Animals will be more likely to eat the fruit, and consequently disperse the enclosed seeds, when it is ripe and no longer bitter. Thus the animals are likely to disperse only mature seeds, which increases the chance that the plant will reproduce successfully.

6. Wheat bread, corn on the cob, peanuts, beans, oatmeal, pancakes, etc.

7. Some plant species possess both all-male and all-female flowers on the same individual plant. In many other species, all the flowers of an individual plant possess both male and female parts. In a few species, such as the gingko tree, an entire individual only possesses either male or female flowers. Some species, such as the jack-in-the-pulpit, alternate sex between male and female in response to growing conditions.

8. Plants with an impaired ability to detect day length would probably survive best in tropical areas of the world, where the climate is favorable for plant growth year-round. The plants would probably do poorly at more northerly or southerly latitudes or in other places where there is more seasonal vari-

ation in environmental conditions. In these areas, the plant's inability to respond to seasonal changes, such as the onset of cold winters, would interfere with its reproducing at the proper time.

9. Plants may protect themselves from herbivores with thick bark, spines or thorns, or toxic compounds in their leaves. Plants also possess specific and nonspecific immune responses to pathogens. The R alleles in plant genes can recognize foreign pathogens, and plants produce chemicals that attack bacterial cell walls. Other chemical signals cause the sealing off of pathogen entry points.

Interlude F

1. Pros and cons of managed systems: **Pros:** Managed systems can increase the supply of animal protein, often using less grain for feed and leaving more grain for human consumption; some kinds of managed systems have low potential for environmental damage. **Cons:** potential production of large amounts of pollution, animal overcrowding, possible negative environmental effects when farming some species.

2. While organic farming of crops can help preserve the environment by reducing use of chemicals, in some crop systems it leads to reduced yields. Precision farming may be able to help maintain yields for some crops, but for other crops, relying on organic methods would be detrimental for ending world hunger.

3. The Green Revolution refers to the large increase in crop productivity in the last 50 years. Many new disease- and pest-resistant varieties of crop plants were developed, and farmers were able to grow more food on the same amount of land. The Green Revolution did not end world hunger because it did not address the root causes of hunger, and it did not improve the environment because it increased pollution.

Chapter 33

1. The description of the biosphere as an "interconnected web" is apt because all organisms within it are connected by their interactions with one another, as shown by examples from various food chains to symbiotic relationships between species. This analogy is useful because it reminds us that any change we make in one part of the biosphere has effects—sometimes quite unexpected ones—on other parts.

2. Giant convection cells in the atmosphere and ocean currents carry the results of local events (such as a volcanic eruption or an oil spill) to distant areas around the globe. For example, oil spilled into an ocean current next to one continent's shore may be carried by that current and end up coating the shores of other continents. If shorebirds on those continents are killed when they become coated with oil from the spill, they may no longer keep populations of their food organisms under control, and they are no longer available as a food source for their predators.

3. Earth's terrestrial biomes are tropical forest, temperate forest, grassland, chaparral, desert, boreal forest, and tundra. The major aquatic biomes are river, lake, wetland, estuary, intertidal zone, coral reef, open ocean, and benthic zone.

4. The potential or natural location of a biome is where conditions are such that the biome could, in principle, be found, while the actual location is where the biome currently is found. The potential locations of terrestrial biomes are most strongly influenced by climate, particularly by temperature and the amount and timing of precipitation, while the potential locations of aquatic biomes are strongly influenced by neighboring terrestrial biomes as well as by climate. The actual locations of both terrestrial and aquatic biomes are strongly influenced by human actions.

5. Climate can exclude species from a region directly, as when the lack of rainfall excludes many plant and animal species from desert biomes. Climate can also exclude species from a region indirectly, as when a species that can tolerate the region's climate is excluded by competitors that are better adapted to that climate.

6. Humans, like all species, depend on the biosphere to survive, and it follows that human actions affecting the biosphere have consequences not only for the environment, but for people as well. Just as people do not intentionally harm themselves by introducing pathogens into their bodies, they should not introduce invasive species or harmful chemicals that could disrupt the biosphere, since these actions will come back to affect them, economically or otherwise.

Chapter 34

1. A population may be difficult to define if the boundaries of its range are unclear, if its members move around frequently, or if its members are small and hard to count.

2. If a population is threatened with extinction, possible options for saving it might be protecting it from human disturbances, treating diseases, reducing the number of predators, or moving the population to an area with greater food supply (limiting death or emigration), introducing individuals from other populations of the species (increasing immigration), or instituting captive breeding programs (increasing the birth rate).

3. Coordinates for graph, in the notation (*x* coordinate, *y* coordinate): (1, 150) (2, 225) (3, 337.5) (4, 506.3) (5, 759.4)

4. a. Some factors that limit population growth include available habitat, available food and water, disease, weather, natural disturbances, and predators.
   b. Species new to an area often do not have established predators. In addition, they have not yet reached the carrying capacity of their habitat.

5. A density-dependent factor is one whose intensity increases as the density of the population increases. An example of a density-dependent factor is an infectious disease that spreads more rapidly in densely populated areas. A density-independent factor is not affected by the density of the population. Temperature is a density-independent factor. If the temperature drops below what a certain plant species can tolerate, it does not matter how dense the plant population is; the plants will still die.

6. If the pattern of population growth is understood, managers may be able to manipulate the factors that most directly affect the population's growth rate. If a population of organisms is more successful because, for example, it has adequate access to water, a manager would be sure that nearby rivers are not drained off for agricultural needs.

7. Limit reproduction to no more than one child per parent; reduce the consumption of unnecessary goods; reuse and recycle items to promote sustainable use of resources; work to develop and follow environmentally friendly policies and activities (for instance, using energy-efficient cars and light bulbs); purchase goods that have a lower impact on the environment, including organically grown clothing and food items.

Chapter 35

1. Mutualisms are common because their costs are outweighed by the benefits they provide. Yucca plants, for example, may lose a few seeds to the offspring of their moth pollinators, but they still end up with more seeds than if the moths had not pollinated them to begin with.

2. Organisms eaten by consumers are under selection pressure to develop defenses against those consumers. Likewise, consumers experience selection pressure to overcome the defenses of their food organisms. Adaptations that improve the survival of individuals in either group will therefore be likely to spread throughout the population. An example would be the poison of the rough-skinned newt, which can kill nearly all predators; garter snakes, however, have evolved the capacity to tolerate the toxin and eat the newt.

3. The inferior competitor is still using resources that the superior competitor needs, thus possibly limiting its distribution or abundance.

4. A superior competitor can reduce the abundance of an inferior competitor in a particular habitat, and even restrict the other species' distribution by competitively excluding it from a habitat. A consumer can have a similar effect on its food organisms. In contrast, each partner in a mutualism is likely to be found only where the other partner is present.

5. The plant community would have fewer species (a). When the rabbits were removed, the grass they prefer would no longer be eaten, so it would assert its dominance as the superior competitor and would probably drive some of the other grass species in the area to extinction.

Chapter 36

1. a. Food webs influence the movement of energy and nutrients through a community. Some species, called keystone species, have a disproportionately large effect, relative to their abundance, on the types and abundances of other species in the community.
   b. Disturbances such as fire occur so often in many ecological communities that the communities are constantly changing and hence may never establish climax communities (relatively stable endpoints of ecological succession). Depending on the type and severity of the disturbance, a given community may or may not be able to recover.
   c. Climate is a key factor in determining what organisms can live in a given area, so if the climate changes, the community changes.
   d. As continents move to different latitudes, their climates, and thus their communities, change.

2. Primary succession occurs in a newly created habitat that contains no species. Secondary succession is the process by which communities recover from disturbance.

3. The introduction of beard grass to Hawaii has increased the frequency and size of fires on the island. This change is due to the large amount of dry matter the grass produces, which burns more easily and hotter than the native vegetation. In this way, the presence of one species in the community has profoundly altered its disturbance pattern, leading to other large changes in the community.

4. The disturbance described in (b) would probably require a longer recovery time, assuming that no other disturbances, such as fire, were to occur. In disturbance (a), the soil and ground cover would be left intact, so new trees would be able to sprout according to natural succession, eventually growing to replace the trees that were removed. In (b), however, the pollutant would have damaged the soil, which would hinder the ability of the trees and ground vegetation to grow. The soil chemistry would need to return to normal before the forest vegetation would be able to grow back and thrive again.

5. Change is a part of all ecological communities. However, human-caused change is unique in that we can consider the impact of our actions, and we can decide whether or not to take actions that cause community change. Whether or not a particular change is viewed as ethically acceptable will depend on the type of change, the reason for it, and the perspective of the person evaluating the change. For example, a person might find it ethically acceptable to alter a region so as to produce a long-term source of food for the growing human population, yet not ethically acceptable to take actions that result in short-term economic benefit but cause long-term economic loss and ecological damage.

6. A keystone species is one that has a disproportionate effect, relative to its own abundance, on other species in a community. As such, keystone species often control the numbers of other species that otherwise would be dominant and far more abundant in a community.

Chapter 37

1. An ecosystem consists of a community of organisms together with the physical environment in which those organisms live. The organisms in an ecosystem interact with one another in various ways; organisms can also move from one ecosystem to another. For this reason, determining the boundaries of protection for a particular ecosystem would probably be difficult. Such a plan would require an understanding of the roles certain organisms play in the overall function of an ecosystem.

2. Energy captured by producers from an external source, such as the sun, is stored in the bodies of producers in chemical forms, such as carbohydrates. At each step in a food chain, a portion of the energy captured by producers is lost from the ecosystem as metabolic heat. This steady loss prevents energy from being recycled.

3. Decomposers break down the tissues of dead organisms into simple chemical components, thereby returning nutrients to the physical environment so that they can be used again by other organisms.

4. Nutrients can cycle on a global level. When sulfur dioxide pollution, for example, enters the atmosphere in one area of the world, winds can move that pollution around the world, where it can affect other ecosystems.

5. Human economic activity is interwoven with several key ecosystem services. Pollination is essential for the production of crop (and other) plants. Floodplains act as safety valves for major floods if we do not build on them or

separate them from the bodies of water they help control. Forests act as water filtration systems. We rely on nutrient cycling to keep us alive. When ecosystem services such as these are damaged, human economic interests are damaged as well.

Chapter 38

1. Major types of global change caused by humans include global warming, land and water transformation, and changes in the chemistry of Earth (for example, changes to nutrient cycles). By altering the conditions under which species live, all of these changes could result in increased dominance of certain types of species and the disappearance of others from various ecosystems.

2. Human-caused global changes often happen at a much more rapid rate than changes due to natural causes. The speed of continental drift or natural climate change is much slower than that of the measurable increases humans have caused in atmospheric carbon dioxide levels and nitrogen fixation. In addition, humans have a choice about the global changes we cause.

3. The addition of large amounts of nitrogen to the environment by human activities is not necessarily a good thing. When more nitrogen becomes available than would exist naturally, a few species may outcompete others for the extra nitrogen, causing the other species to disappear from their communities.

4. Scientific instruments can directly measure the amount of carbon dioxide in the atmosphere. By measuring CO_2 levels in bubbles of air trapped in ancient ice, scientists can estimate the amount of CO_2 that was present in the atmosphere up to hundreds of thousands of years ago. The present levels of atmospheric CO_2 are higher than any seen in the previous 420,000 years.

5. It would be prudent to take action on global warming sooner rather than later, despite present uncertainties as to its extent. There is already evidence of climate changes that are consistent with the predicted effects of global warming. In addition, the correlation of rising CO_2 levels and worldwide temperature increases suggests that these increases will continue in the future if carbon dioxide emissions are not reduced. If action is delayed for too long, it may be too late to undo many of the effects of global warming.

6. For people to have a sustainable impact on Earth, we must reduce the rate of growth of the human population, and we must reduce the rate of resource use per person. To achieve these goals, many aspects of human society would have to change. For example, our view of nature would have to change from one of a limitless source of goods and materials that can be exploited for short-term economic gain to one that accepts limits and seeks always to take only those actions that can be sustained for long periods of time. Many specific actions would follow from such a change in our view of the world, such as an increase in recycling; the development and use of renewable sources of energy; a decrease in urban sprawl; an increase in the use of technologies with low environmental impact (such as organic farming); and a concerted effort to halt the ongoing extinction of species.

7. Some examples of actions you could take to make your impact on Earth more sustainable are reducing the quantity of nonfood items purchased; reusing items until they are no longer usable; buying used items rather than getting everything new; recycling paper, plastic, glass, and metal; bringing reusable cloth bags when shopping; rarely using paper cups, plates, or

towels; planting trees and other native plants, especially those that help feed native wildlife; reducing water use by not leaving water running when brushing teeth, by adjusting the water level of washing machines to match the size of the load, and by using water-saving fixtures; reducing fossil fuel use by choosing a fuel-efficient car and by using household heating and air conditioning only as needed; using compact fluorescent light bulbs, and turning off lights that are not in use; supporting organic farmers by purchasing organically grown food.

Interlude G

2. **Deforestation:** The destruction of forests can have many negative consequences for both the environment and for humans. At the current rate of forest destruction, nearly all the world's tropical rainforests will be gone in 100 to 150 years. The rate of new forest growth is not keeping up with the rate of forest loss. In addition, old-growth forests take centuries to recover from logging, and contain many unique species that are not found in second-growth forests. Some areas of the world have already lost virtually all of their original forested area.

Water use: Humans use more than half of the world's fresh water, and population growth will add to the demand. In many parts of the world, more water is drawn out of underground aquifers than can be naturally replenished. The rate of water use in some agricultural areas of China and India is not sustainable and will result in water shortages and potential economic damage in only a few years.

4. In contrast to the GDP, methods of measurement such as the Index of Sustainable Economic Welfare give a more realistic picture of the output and value of an economy by taking into account the environmental costs of producing goods and services.

5. Corporations have the motive of making a profit, and if there are environmental restrictions in place for certain products or activities, businesses will have an incentive to develop innovative and marketable solutions. In a market economy, competition and availability of capital help corporations develop new products rapidly.

Glossary

acid A chemical compound that can give up a hydrogen ion. Compare *base* and *buffer*.

acid rain Rainfall with a low pH. Acid rain is a consequence of the release of sulfur dioxide and other pollutants into the atmosphere, where they are converted to acids that then fall back to Earth in rain or snow.

acrosome The front tip of a sperm, which contains enzymes that digest the outer covering of the unfertilized ovum.

actin A protein found in the cytoskeleton, in muscle tissue, and in bacterial flagella.

actin filament One of the two types of filaments that make up muscle, consisting of two molecules of the protein actin. Compare *myosin filament*.

action potential An electrical signal generated by the flow of ions across the plasma membrane of a neuron. Action potentials are self-sustaining and can travel down a neuron in only one direction.

activation energy The small input of energy required for a chemical reaction to proceed.

active carrier protein A protein in the plasma membrane of a cell that, using energy from an energy storage molecule such as ATP, changes its shape to transfer a molecule across the plasma membrane. Compare *passive carrier protein*.

active immunity The immunity of an organism to a pathogen that depends on the production of antibodies specific to that pathogen by the organism's own body. Compare *passive immunity*.

active site The specific region on the surface of an enzyme where substrate molecules bind.

active transport Movement of molecules that requires an input of energy. Compare *passive transport*.

adaptation A characteristic of an organism that improves that organism's performance in its environment.

adaptive evolution The process by which natural selection improves the match between organisms and their environment over time.

adaptive radiation An evolutionary expansion in which a group of organisms takes on new ecological roles and forms new species and higher taxonomic groups.

adenosine triphosphate See *ATP*.

adrenal gland One of a pair of endocrine glands that sit atop the mammalian kidney and which release the hormones epinephrine and norepinephrine.

adrenaline See *epinephrine*.

aerobic Of or referring to a metabolic process or organism that requires oxygen gas. Compare *anaerobic*.

aerobic respiration A general term used to describe a series of oxidation reactions that use oxygen to produce ATP.

allele One of several alternative versions of a gene. Each allele has a DNA sequence different from that of all other alleles of the same gene.

allele frequency The proportion (percentage) of a particular allele in a population.

allopatric speciation The formation of new species from populations that are geographically isolated from one another. Compare *sympatric speciation*.

altruism A behavior that benefits another individual but has a cost to the individual performing the behavior.

alveolus (pl. alveoli) A small sac in the mammalian lung where gas exchange takes place.

amino acid An organic compound that has an amino group, a carboxyl group, and a variable R group attached to a single carbon atom. Proteins are polymers of amino acids.

amniocentesis A procedure in which a needle is inserted through the abdomen into the uterus to extract a small amount of amniotic fluid from the pregnancy sac that surrounds the fetus; this fluid contains fetal cells that can be used to test for genetic disorders.

amnion A sac filled with a watery fluid that surrounds and protects the developing vertebrate embryo.

amplification The chain of events set in motion when the binding of a single hormone molecule to a receptor activates thousands of proteins in the target cell.

anabolic See *biosynthetic*.

anaerobic Of or referring to a metabolic process or organism that does not require oxygen gas. Compare *aerobic*.

analogous Of or referring to a characteristic shared by two groups of organisms because of convergent evolution, not common descent. Compare *homologous*.

anaphase The stage of mitosis during which sister chromatids separate and move to opposite poles of the cell.

anchoring junction A protein structure that acts as a "hook" between two animal cells or between a cell and the extracellular matrix.

androgen One of a class of steroid hormones that stimulate cells to develop the characteristics of maleness. The primary androgen is testosterone. Compare *estrogen*.

angiosperms The flowering plants, a group that includes most plants on Earth today; named for the protective tissues covering the plant's embryo in the seed. Compare *gymnosperms*.

Animalia The kingdom made up of animals, multicellular eukaryotes that have evolved specialized tissues, organs and organ systems, body plans, and behaviors.

annual A plant that completes its entire life cycle in one year.

antagonistic Of or referring to the opposing effects of two hormones that together serve to regulate a process.

antenna complex An arrangement of chlorophyll molecules in the thylakoid membrane of a chloroplast that harvests energy from sunlight.

antibody A protein that is produced by a B cell and binds specifically to a particular antigen.

anticodon A sequence of three nitrogen bases on a transfer RNA molecule that can bind to a particular codon on an mRNA molecule. Compare *codon*.

antigen A characteristic membrane protein or molecule produced by an invading pathogen that is recognized by particular lymphocytes and the antibodies produced by those lymphocytes.

anus The opening of the digestive system through which undigested food and other solid wastes leave the body.

apical dominance Inhibition of the growth of lateral buds by the apical meristem.

apical meristem A region of rapidly dividing cells in the tips of plant branches and roots that gives rise to new stem and root tissues.

apoptosis Programmed or intentional cell death.

appendicular skeleton A collective term for the bones of the arms, legs, and pelvis. Compare *axial skeleton*.

aquaculture Managed agricultural systems in which fish and shellfish are cultivated for human consumption.

aquifer An underground body of water that is sometimes bounded by impermeable layers of rock.

Archaea A domain of microscopic, single-celled prokaryotes that arose after the Bacteria.

artery A blood vessel that carries blood from the heart. Compare *vein*.

arthropods A group of animals characterized by a hard exoskeleton; includes millipedes, crustaceans, insects, and spiders.

artificial selection A process in which only individuals that possess certain characteristics are allowed to breed; used to guide the evolution of crop plants and domestic animals in ways that are advantageous for people.

asexual reproduction The production of genetically identical offspring without the exchange of genetic material with another individual. Compare *sexual reproduction*.

atmospheric cycle A type of nutrient cycle in which the nutrient enters the atmosphere easily. Compare *sedimentary cycle*.

atom The smallest unit of a chemical element that still has the properties of that element.

atomic mass number The sum of the number of protons and neutrons found in the nucleus of an atom of a particular chemical element.

atomic number The number of protons found in the nucleus of an atom of a particular chemical element.

ATP Adenosine triphosphate, a molecule that is commonly used by cells to store energy and to transfer energy from one chemical reaction to another.

atrioventricular (AV) node A signaling center in the heart that relays the signal for cardiac muscle to contract from the sinoatrial node to the ventricles.

atrium (pl. atria) A chamber in a heart that receives blood from the body and pumps it into the ventricle. Compare *ventricle*.

auditory canal The tubular opening leading from the ear pinna to the eardrum.

autoimmune disease A disease caused when the body destroys some of its own tissues that the immune system misidentifies as foreign tissues.

autosome Any chromosome that is not a sex chromosome. Compare *sex chromosome*.

auxin A plant hormone that causes apical dominance and has other important effects, such as controlling the way plants bend toward sunlight and stimulating the differentiation of vascular tissue.

axial skeleton A collective term for the skull, the spinal column, and the ribs. Compare *appendicular skeleton*.

axon An extension of a neuron that carries action potentials through an animal's body.

B cell A lymphocyte that produces antibodies.

Bacteria A domain of microscopic, single-celled prokaryotes that were the first organisms to arise.

bacterial flagellum (pl. flagella) A rotary motor unique to bacteria that, in response to the flow of hydrogen ions across the bacterial plasma membrane, propels a bacterium. Compare *eukaryotic flagellum*.

bark The protective outer covering of stems in woody plants that contains the cork cambium and the secondary phloem.

base (1) A chemical compound that can accept a hydrogen ion. Compare *acid* and *buffer*. (2) A nitrogen-containing molecule that is part of a nucleotide. See *nitrogen base*.

base pairing A process in which complementary nitrogen bases form hydrogen bonds with each other. In DNA, A pairs with T, and C pairs with G; in RNA, U replaces T.

basilar membrane A thin membrane lying in the inner ear that vibrates in response to vibrations passed to it from the middle ear.

behavior A coordinated response to a stimulus; particularly, a response that involves movement.

behavioral mutualism A mutualism in which each of two interacting species alters its behavior to benefit the other.

benign Of or referring to a relatively harmless cancerous growth that is confined to a single tumor and does not spread to other tissues in the body. Compare *malignant*.

benthic zone An aquatic biome, the communities of which are home to a wide variety of organisms that live on the bottom surfaces of rivers, lakes, wetlands, estuaries, and oceans.

biennial A plant that completes its life cycle in two years.

bile A substance produced by the liver that helps break up large fat globules in the small intestine.

biodiversity The variety of organisms on Earth or in a particular location, ranging from the genetic variation and behavioral diversity of individual organisms or species through the diversity of ecosystems.

biological hierarchy The hierarchy in which all living things are organized, from molecules at the lowest level to the entire biosphere at the highest level.

biology The study of life.

biomass The mass of organisms per unit of area.

biome A major terrestrial or aquatic life zone, defined either by its vegetation (terrestrial biomes) or by the physical characteristics of the environment (aquatic biomes). There are seven terrestrial biomes (tundra, boreal forest, temperate forest, chaparral, grassland, desert, and tropical forest) and eight aquatic biomes (lake, river, wetland, estuary, intertidal zone, coral reef, ocean, and benthic zone).

biosphere All living organisms on Earth, together with the environments in which they live.

biosynthetic Of or referring to chemical reactions that manufacture complex molecules in living cells. Compare *catabolic*.

bipedal Of or referring to an organism that walks upright on two legs.

bivalent A pair of homologous chromosomes. Bivalents form during prophase I of meiosis.

blastocyst The name for the blastula in mammalian development.

blastula The early stage of animal development in which the embryo forms a hollow ball of undifferentiated cells.

blood A fluid tissue consisting of liquid plasma and the blood cells that it carries. Blood plays a central role in carrying nutrients, oxygen, and wastes through the circulatory system.

blood plasma The fluid portion of animal blood.

blood pressure The pressure that pushes the blood from the heart to the respiring tissues. In humans, it is typically measured as the pressure in the arteries leading from the left ventricle.

blood vessel One of the interconnected tubes that make up the circulatory system.

boreal forest A terrestrial biome, the communities of which are dominated by coniferous trees that grow in northern or high-altitude regions with cold, dry winters and mild, humid summers.

brain A complex collection of neurons located at the head end of an animal that serves as the major clearinghouse for information in the nervous system.

brain stem A collective term for the midbrain and hindbrain, which together control many of the functions of the body that are not under conscious control.

breathing The process of drawing air into and out of the lungs to extract oxygen and expel carbon dioxide.

bronchiole One of the tiny tubes inside the lung that carry air from the bronchi to the alveoli.

bronchus (pl. bronchi) One of two tubes that carry air from the trachea to the two lobes of the vertebrate lung.

brush border A series of microscopic, brushlike extensions of the cells lining the small intestine that help to provide a greatly increased surface area for nutrient absorption.

bryophytes Plants, such as mosses, that lack a true vascular system for transporting materials (nonvascular plants).

Bt crop A crop plant into which a gene from the bacteria *Bacillus thuringiensis* has been inserted, causing the plant to produce a toxin that provides protection from insect pests.

buffer A chemical compound that can both give up and accept hydrogen ions. Buffers can maintain the pH of water within specific limits. Compare *acid* and *base*.

calcitonin A hormone produced by the mammalian parathyroid gland that promotes the storage of calcium in bones. Compare *parathyroid hormone*.

Cambrian explosion A major increase in the diversity of life on Earth that occurred about 530 million years ago, during the Cambrian period. The Cambrian explosion lasted 5 to 10 million years; during this time large and complex forms of most living animal phyla appeared suddenly in the fossil record.

cancer A group of diseases caused by rapid and inappropriate cell division.

canopy The habitat in the branches of forest trees.

cap-and-trade system An approach to pollution control in which a government sets a nationwide limit, or cap, on the amount of a pollutant that can be added to the environment each year. Each factory that emits the pollutant is given a certain number of "emission allowances" in each year. Unused allowances can be sold, traded, or saved (banked) for future use.

capillary A tiny blood vessel in a closed circulatory system, across the walls of which all exchange with the surrounding tissues takes place.

carbohydrate Any of a class of organic compounds that includes sugars and their polymers, in which each carbon atom is linked to two hydrogen atoms and an oxygen atom. See also *sugar*.

carbon fixation The process by which carbon atoms from carbon dioxide gas are incorporated into sugars; occurs in the chloroplasts of plants.

carcinogen A physical, chemical, or biological agent that causes cancer.

cardiac cycle The series of events involved in the pumping of blood by the heart. The cardiac cycle consists of two phases: diastole and systole.

cardiac muscle The specialized muscle of the vertebrate heart. Cardiac muscle has bands like the muscles associated with the skeleton, and its muscle fibers are joined by interconnecting branches that help them to produce the coordinated contractions known as heartbeats.

cardiovascular system The closed circulatory system of vertebrates, which consists of a muscular, chambered heart and loops of blood vessels.

carnivore A consumer that relies on living animal tissues for nutrients. Compare *herbivore*.

carpel The female reproductive part of a flower, consisting of a stigma, style, and ovary. Carpels are the innermost of the four whorls of modified leaves that make up a flower. Compare *sepal*, *petal*, *stamen*.

carrier An individual that carries a disease-causing allele but does not get the disease.

carrying capacity The maximum population size that can be supported indefinitely by the environment in which a population is found.

cartilage A support structure found in animals made up of cells that secrete an extracellular matrix rich in the protein collagen.

catabolic Of or referring to a chemical reaction that breaks down complex molecules to release energy for use by the cell. Compare *biosynthetic*.

catalysis A process in which a chemical substance (a catalyst) participates in a reaction in such a way as to lower the amount of activation energy required, greatly increasing the rate at which the reaction proceeds.

catalyst A molecule that speeds up a specific chemical reaction without being permanently altered in the process. Enzymes are protein catalysts.

cell The smallest self-contained unit of life, enclosed by a membrane.

cell communication The process by which one cell can affect the activities of another via signaling molecules.

cell cycle A series of distinct stages in the life cycle of a cell that culminate in cell division.

cell junction A structure connecting two cells that holds them together and allows them to communicate with one another.

cell-mediated immunity A form of immunity that involves the action of T cells. Compare *humoral immunity*.

cell specialization An organizational principle stating that different types of cells in a multicellular organism differ in their structure and function.

cell wall A support layer that lies outside the plasma membrane of the cells of many prokaryotes, fungi, and plants.

cellulose A carbohydrate produced by plants and some other organisms that makes up much of their cell walls.

central nervous system (CNS) The portion of the nervous system devoted to the exchange and integration of information among neurons. The central nervous system consists of the brain and spinal cord. Compare *peripheral nervous system*.

central vacuole A large vacuole that usually occupies more than a third of a plant cell's total volume.

centromere A physical constriction that holds sister chromatids together.

centrosome A protein structure in the cytosol that helps organize the mitotic spindle and defines the two poles of a dividing cell.

cerebral cortex The highly folded outer portion of the human brain that is specialized for coordinating information from sensory cells with the activity of muscles.

cerebrum The largest part of the human forebrain; includes the cerebral cortex.

channel protein A protein in the plasma membrane of a cell that forms an opening through which certain molecules can pass.

chaparral A terrestrial biome, the communities of which are characterized by shrubs and small nonwoody plants that grow in regions with mild summers and winters and low to moderate amounts of precipitation.

character displacement A process by which intense competition between species causes the forms of the competing species to evolve to become more different over time.

chemical compound An association of atoms of different chemical elements linked by covalent bonds.

chemical reaction A process that rearranges atoms in chemical compounds.

chemoreceptor A sensory cell that initiates an action potential in response to a chemical stimulus.

chitin A carbohydrate that serves as an important support material in the cell walls of fungi and in animal exoskeletons.

chlorofluorocarbons (CFCs) Synthetic chemical compounds whose release into the atmosphere can damage the ozone layer.

chlorophyll A green pigment that is used to capture energy from light in photosynthesis.

chloroplast An organelle found in plants and algae that is the primary site of photosynthesis.

cholesterol A steroid constructed from saturated fats that is important in cell membranes and forms the molecular backbone of steroid hormones.

chorionic villus sampling (CVS) A procedure in which a flexible tube is inserted through a woman's vagina and into her uterus. The tip of this tube is placed next to a cluster of cells that attaches the pregnancy sac to the wall of the uterus; cells are then removed from this cluster by gentle suction so they can be tested for genetic disorders.

chromatid Either of two identical, side-by-side copies of a chromosome that are linked at the centromere.

chromatin The combination of DNA and proteins that makes up chromosomes.

chromosome Any of several elongated structures found in the nucleus of a cell, each composed of DNA packaged with proteins. Chromosomes become visible under the microscope during mitosis and meiosis.

chromosome theory of inheritance A theory, supported by much experimental evidence, stating that genes are located on chromosomes.

cilium (pl. cilia) A hairlike structure found in some eukaryotes that uses a rowing motion to propel the organism or to move fluid past the organism. Compare *eukaryotic flagellum*.

circulatory system A system of interconnected tubular vessels through which blood is pumped by the heart; the main internal transport system in animals.

citric acid cycle A series of oxidation reactions that produce high-energy electrons stored in NADH and release CO_2 as a waste product. In eukaryotic cells, the citric acid cycle takes place in mitochondria.

class The level in the Linnaean hierarchy that is above orders and therefore comprises orders, but that is below phyla and so composes phyla.

climate The prevailing weather conditions experienced in an area over relatively long periods of time (30 years or more). Compare *weather*.

climax community A community, typical of a given climate and soil type, whose species are not replaced by other species. A climax community is the end point of succession for a particular location; in many cases, however, ongoing disturbances such as fire or windstorms prevent the formation of a stable climax community.

clonal selection The process by which a lymphocyte that has bound to its specific antigen reproduces rapidly, yielding large numbers of genetically identical lymphocytes with exactly the same antigen-recognizing proteins as the parent cell.

clone (of a gene) A copy of a gene or other DNA sequence.

clone (of an organism) A genetically identical copy of an individual organism, as produced by reproductive cloning.

closed circulatory system An internal transport system found in animals in which blood vessels carry blood throughout the body and back to the heart. Compare *open circulatory system*.

cochlea A coiled, fluid-filled tube in the inner ear.

coding strand Of the two strands in a DNA molecule, the strand whose DNA sequence is exactly duplicated by the sequence of bases in an mRNA molecule (except that U substitutes for T). Compare *template strand*.

codominance A situation in which the phenotype of a heterozygote is determined equally by each allele.

codon A sequence of three nitrogen bases in an mRNA molecule. Each codon specifies either a particular amino acid or a signal to start or stop the translation of a protein. Compare *anticodon*.

collagen A protein produced by the connective tissues of animals.

colon The portion of the digestive system in which undigested material is prepared for release from the body. Also called *large intestine*.

communication A type of behavior that allows one individual to exchange information with another.

community An association of populations of different species that live in the same area.

compact bone The hard, white outer region of vertebrate bone. Compare *spongy bone*.

comparative genomics A field of scientific study that analyzes and compares the genomes of multiple species.

competition An interaction between two species in which each has a negative effect on the other.

complement protein A type of protein that circulates in the blood plasma and concentrates at sites of tissue damage, stimulating macrophage and neutrophil production and activity.

complementary strand A strand of DNA whose sequence of bases can pair (according to base-pairing rules) with the sequence of bases found in a focal DNA strand.

concentration gradient A change in the concentration of molecules from one location to another.

conduction The transfer of heat by direct contact between two materials.

cone A type of photoreceptor cell in the vertebrate eye that detects color and functions best in bright light. Compare *rod*.

connective tissue A type of animal tissue that connects and supports other cells in the body. Many connective tissues secrete an extracellular matrix of nonliving material.

consumer An organism that obtains its energy by eating other organisms or their remains. Consumers include herbivores, carnivores, and decomposers. Compare *producer*.

continental drift The movement of Earth's continents over time.

control A treatment in an experiment in which the factor or factors being tested in the experiment are omitted, but other conditions are the same as in the experimental treatments.

convection The physical movement of heat in air or water.

convection cell A large and consistent atmospheric circulation pattern in which warm, moist air rises and cool, dry air sinks. Earth has four stable giant convection cells (two in tropical regions and two in polar regions) and two less stable cells (located in temperate regions).

convergent evolution Evolutionary change that occurs when natural selection causes distantly related organisms to evolve similar structures in response to similar environmental challenges.

convergent feature A feature shared by two groups of organisms not because it was inherited from a common ancestor, but because it arose independently in the two groups.

copulation A behavior in which a male animal places his sperm directly into the body of a female; the typical mode of sperm transfer in land animals.

coral reef An aquatic biome, the communities of which form in warm, shallow waters located in the tropics. Corals are tiny animals that build up long-lasting structures, the reefs on which many of the other organisms in the community depend.

cork cambium A lateral meristem tissue that contributes to secondary growth by dividing to produce tough dermal tissues (including the outer surface of bark). Compare *vascular cambium*.

corpus luteum (pl. corpora lutea) The cells of the ruptured follicle that remain behind in the ovary after ovulation to produce the hormone progesterone.

covalent bond A strong chemical linkage between two atoms based on the sharing of electrons. Compare *hydrogen bond* and *ionic bond*.

Cretaceous extinction A mass extinction that occurred 65 million years ago, wiping out many marine invertebrates and terrestrial plants and animals, including the last of the dinosaurs.

crista (pl. cristae) A fold in the inner mitochondrial membrane.

crossing-over A physical exchange of genes between homologous chromosomes in which part of the genetic material inherited from one parent is replaced with the corresponding genetic material inherited from the other parent.

cuticle A waxy layer that covers aboveground plant parts, helping to prevent water loss and to keep enemies, such as fungi, from invading the plant.

cynodont A member of a group of mammal-like reptiles from which the earliest mammals arose, roughly 220 million years ago.

cytokinesis The stage following mitosis, during which the cell physically divides into two daughter cells.

cytoplasm The contents of a cell enclosed by the plasma membrane, but, in eukaryotes, excluding the nucleus. Compare *cytosol*.

cytoskeleton A complex network of protein filaments found in the cytosol of eukaryotic cells. The cytoskeleton maintains cell shape and is necessary for the physical processes of cell division and movement.

cytosol The contents of a cell enclosed by the plasma membrane, but, in eukaryotes, excluding all organelles. Compare *cytoplasm*.

dark reactions A series of chemical reactions that directly use carbon dioxide to synthesize sugars. The dark reactions do not require light and take place in the stroma of chloroplasts. Compare *light reactions*.

decomposer An organism that breaks down dead tissues into simple chemical components, thereby returning nutrients to the physical environment.

deletion A mutation in which one or more nitrogen bases are removed from the DNA sequence of a gene or chromosome. Compare *insertion* and *substitution*.

dendrite An extension of a neuron that receives signals from adjacent cells.

density-dependent Of or referring to a factor, such as food shortage, that limits the growth of a population more strongly as the density of the population increases. Compare *density-independent*.

density-independent Of or referring to a factor, such as weather, that can limit the size of a population but does not act more strongly as the density of the population increases. Compare *density-dependent*.

deoxyribonucleic acid See *DNA*.

depolarization A decrease in the charge difference between the inside and the outside of the plasma membrane of a neuron; initiates an action potential. Compare *polarization*.

depth perception The ability to accurately determine the relative distances of objects in space.

dermal tissue The tissue type in plants that forms the outer covering of the body.

desert A terrestrial biome, the communities of which are dominated by plants that grow in regions with low precipitation, usually 25 centimeters per year or less.

deuterostome Any of a group of animals, including sea stars and vertebrates, in which the second opening to develop in the early embryo becomes the mouth. Compare *protostome*.

development The process by which an organism grows from a single cell to its adult form.

diastole The relaxation phase of the cardiac cycle. The diastolic pressure corresponds to the second and lower of the two values reported in a blood pressure reading. Compare *systole*.

differentiation The process by which a cell develops physical characteristics that make it suited for a specialized function.

diffusion The passive movement of a molecule from areas of high concentration of that molecule to areas of low concentration of that molecule.

digestion The chemical breakdown of food.

digestive system A tubular passageway through which food generally moves in a one-way flow and in which animals convert the food they eat into a form their bodies can absorb and use.

diploid Of or referring to a cell or organism that has two complete sets of homologous chromosomes (2*n*). Compare *haploid*.

directional selection A type of natural selection in which individuals with one extreme of an inherited characteristic have an advantage over other individuals in the population, as when large individuals produce more offspring than small and medium-sized individuals. Compare *disruptive selection* and *stabilizing selection*.

disruptive selection A type of natural selection in which individuals with either extreme of an inherited characteristic have an advantage over individuals with an intermediate phenotype, as when both small and large individuals produce more offspring than medium-sized individuals. Compare *directional selection* and *stabilizing selection*.

distribution The geographic area over which a species is found.

disturbance An event, such as a fire or windstorm, that kills or damages some organisms in a community, thereby creating an opportunity for other organisms to become established.

diversity The composition of an ecological community, which has two components: the number of different species that live in the community and the relative abundances of those species.

DNA Deoxyribonucleic acid, a polymer of nucleotides that stores the information needed to synthesize proteins in living organisms.

DNA chip A small surface, or "chip," roughly the size of a dime on which thousands of samples of DNA are placed in a regimented order.

DNA fingerprinting The use of DNA analysis to identify individuals and determine the relatedness of individuals.

DNA hybridization Base pairing of DNA from two different sources.

DNA library A collection of an organism's DNA fragments that is stored in a host organism, such as a bacterium.

DNA polymerase The key enzyme that cells use to copy their DNA; in DNA technology, used in the polymerase chain reaction to make many copies of a gene or other DNA sequence.

DNA primer A short segment of DNA used in PCR amplification that is designed to pair with one of the two ends of the gene being cloned by PCR.

DNA probe A short sequence of DNA (usually tens to hundreds of bases long) that can pair with a particular gene or other specific region of DNA.

DNA repair A three-step process in which damage to DNA is repaired. Damaged DNA is first recognized, then removed, and then replaced with newly synthesized DNA.

DNA replication The duplication, or copying, of a DNA molecule. DNA replication begins when the hydrogen bonds connecting the two strands of DNA are broken, causing the strands to unwind and separate. Each strand is then used as a template for the construction of a new strand of DNA.

DNA segregation The process by which the DNA of a dividing cell is divided equally between two daughter cells.

DNA sequence The sequence or order in which the nitrogen bases adenine (A), cytosine (C), guanine (G), and thymine (T) are arranged throughout all or part of an organism's DNA.

DNA sequencing A procedure, usually automated, used to determine the sequence of bases in a DNA fragment.

DNA technology The set of techniques that scientists use to manipulate DNA.

domain A level of biological classification above the kingdom. The three domains are the Bacteria, the Archaea, and the Eukarya.

dominant Of or referring to an allele that determines the phenotype of an organism when paired with a different (recessive) allele. Compare *recessive*.

dopamine A neurotransmitter released in the brain that is associated with feelings of happiness.

dormancy A state of arrested growth in a plant. Seed dormancy occurs when a mature plant embryo (inside of a seed) halts its growth, and ends when the embryo is stimulated by favorable conditions to begin growing again. In some plant species, bud dormancy occurs during winter and ends when the plant resumes growth in the spring.

double fertilization The two fertilization events that occur when a flowering plant reproduces sexually. In one of these fertilization events, a sperm cell fertilizes the egg, producing a diploid cell (zygote) that will grow into the next generation's flowering plant; in the other, a second sperm cell fertilizes a large central cell within the embryo sac, producing the endosperm that will nourish the developing embryo.

double helix The structure of DNA, in which two long strands of covalently bonded nucleotides are held together by hydrogen bonds and twisted into a spiral coil.

doubling time The time it takes a population to double in size. Doubling time can be used as a measure of how fast a population is growing.

drag A force that resists the motion of any moving object.

duplication A mutation in which a fragment from one chromosome fuses to the homologous chromosome, increasing the length of the chromosome that receives the fragment.

eardrum A delicate membrane at the end of the auditory canal that converts pressure changes in the environment into physical movements.

ecological experiment A procedure in which an investigator alters one or more features of the environment and observes the effect of that change.

ecological footprint The area of productive ecosystems needed throughout a year to support a population and cope with its waste materials.

ecology The scientific study of interactions between organisms and their environment.

ecosystem A community of organisms, together with the physical environment in which the organisms live. Global patterns of air and water circulation link all the world's organisms into one giant ecosystem, the biosphere.

ecosystem service An action or function of an ecosystem that provides a benefit to humans, such as pollination by insects or water filtration by wetlands.

ectoderm The cell layer in animal development that forms the exterior of the gastrula and gives rise to the epidermis and nervous system. Compare *endoderm, mesoderm*.

ectotherm An organism that relies on environmental heat for most of its heat input. Compare *endotherm*.

edema Swelling of the tissues.

effector cell A lymphocyte that is actively involved in attacking pathogens.

egg The large, immobile, haploid gamete that is produced by sexually reproducing female eukaryotes. Compare *sperm*.

electrocardiogram (EKG) A recording of the signals from the sinoatrial and atrioventricular nodes of the heart.

electron A negatively charged particle found in atoms. Each atom contains a characteristic number of electrons. Compare *proton*.

electron transport chain (ETC) A group of membrane-associated proteins that can both accept and donate electrons. The transfer of electrons from one ETC protein to another releases energy that is used to manufacture ATP in both chloroplasts and mitochondria.

element A substance made up of only one type of atom. The physical world is made up of 92 natural elements.

embryo The early stage of animal development extending from the zygote to the early development of organs and organ systems. Compare *fetus*.

embryo sac The small, multicellular haploid stage of the flowering plant life cycle in which eggs are produced. Compare *pollen grain*.

encapsulation The formation of a protective capsule around especially large pathogens by macrophages.

endangered In danger of extinction.

endocrine gland A structure that produces hormones and releases them directly into the circulatory system.

endocrine system The collection of glands and other tissues in animals that produce signaling molecules called hormones and release them directly into the circulatory system without using ducts.

endocytosis A process by which a section of a cell's plasma membrane bulges inward as it envelops a substance outside of the cell, eventually breaking free to become a closed vesicle within the cell.

endoderm The cell layer in animal development that forms the interior of the gastrula and gives rise to the gut and associated organs. Compare *ectoderm, mesoderm*.

endometrium The layer of cells lining the wall of the mammalian uterus.

endoplasmic reticulum (ER) An organelle composed of many interconnected membrane sacs and tubes; the major site of protein and lipid synthesis in eukaryotic cells.

endoskeleton A skeleton that is surrounded by the soft tissues it supports. Compare *exoskeleton*.

endosperm The nutritive tissue produced within a plant seed.

endotherm An organism that relies on metabolic heat for most of its heat input. Compare *ectotherm*.

energy carrier A molecule that can store energy and donate it to another molecule or a chemical reaction. ATP is the most commonly used energy carrier in living organisms.

enzyme A protein that acts as a catalyst, speeding the progress of chemical reactions. All chemical reactions in living organisms are catalyzed by enzymes.

epinephrine A hormone produced by the adrenal glands that regulates the availability of glucose in conjunction with norepinephrine.

epistasis A gene interaction in which the phenotypic effect of the alleles of one gene depends on which alleles are present for another, independently inherited gene.

epithelial tissue A type of animal tissue that is specialized to regulate how the animal's internal environment exchanges water, solutes, and heat directly with the external environment.

ER See *endoplasmic reticulum*.

esophagus The portion of the digestive system that uses waves of muscular contraction to push food from the mouth to the stomach.

essential amino acid An amino acid that a consumer cannot synthesize and must therefore obtain from its food.

estrogen One of a class of steroid hormones that stimulate cells to develop the characteristics of femaleness. Compare *androgen*.

estuary An aquatic biome found in tidal regions where rivers flow into the ocean.

ETC See *electron transport chain*.

eugenics movement An effort to breed better humans by encouraging the reproduction of people with certain genetic characteristics and discouraging the reproduction of people with other genetic characteristics.

Eukarya The domain that encompasses the eukaryotes.

eukaryote A single-celled or multicellular organism in which each cell has a distinct nucleus and cytoplasm. All organisms other than the Bacteria and the Archaea are eukaryotes. Compare *prokaryote*.

eukaryotic flagellum (pl. flagella) A hairlike structure found in eukaryotes that propels the organism by means of waves passing from its base to its tip. Compare *bacterial flagellum* and *cilium*.

eustachian tube A thin tube that connects the middle ear to the throat, allowing mammals to equalize the pressure in these two structures.

eutrophication A process in which enrichment of water by nutrients (often from sewage or runoff from fertilized agricultural fields) causes bacterial populations to increase and oxygen concentrations to decrease.

evaporation The conversion of a substance from a liquid to a gas.

evolution Change over time in a lineage of organisms. See also *macroevolution* and *microevolution*.

evolutionary innovation A key adaptation of a group that originated in that group.

evolutionary tree A diagrammatic representation showing the order in which different lineages arose, with the lowest branches having arisen first.

exchange pool A source such as the soil, water, or air where nutrients are available to producers.

exocytosis A process by which a vesicle approaches and fuses with the plasma membrane of a cell, releasing the substance it contains into the cell's surroundings.

exon A DNA sequence within a gene that encodes part of a protein.

exoskeleton A skeleton that surrounds the soft tissues of the animal it supports. Compare *endoskeleton*.

experiment A controlled manipulation of nature designed to test a hypothesis.

exploitation An interaction between two species in which one species benefits (the consumer) and the other species is harmed (the food organism). Exploitation includes the killing of prey by predators, the eating of plants by herbivores, and the harming or killing of a host by a parasite or pathogen.

exploitative competition A type of competition in which species compete indirectly for shared resources, with each reducing the amount of a resource available to the other. Compare *interference competition*.

exponential growth A type of rapid population growth in which a population increases by a constant proportion from one generation to the next.

extracellular matrix (pl. matrices) A coating of nonliving material, released by the cells of multicellular animals, that often holds those cells together.

extremophile An organism, such as many Archaea, that lives in extreme environments, such as boiling hot geysers or on salted meat.

F_1 generation The first generation of offspring in a genetic cross. Compare *P generation*.

F_2 generation The second generation of offspring in a genetic cross.

family The level in the Linnaean hierarchy that is above genera and therefore comprises genera, but that is below orders and so composes orders.

fast muscle fiber A type of muscle fiber that contracts quickly, but can sustain a contraction for only a short time. Compare *slow muscle fiber*.

fat An organic compound that consists of glycerol linked to three fatty acids. Fats are solid at room temperature and can be used by living organisms to store energy.

fatty acid An organic compound composed primarily of a long hydrocarbon chain; found in lipids and fats.

feces The solid waste product of animal digestion, which consists mostly of indigestible food and bacteria.

female An individual of a sexually reproducing species that produces eggs. Compare *male*.

fermentation A series of catabolic reactions that produce small amounts of ATP without the use of oxygen. Fermentation is similar to glycolysis, except that pyruvate is converted to other products, such as ethanol or lactic acid.

fertilization The fusion of haploid gametes (egg and sperm) to produce a diploid zygote (the fertilized egg).

fetus The stage of animal development from the time when most of the major organs and tissues are identifiable to birth. Compare *embryo*.

fibrous root system A root system in which no one root predominates and which can form dense mats that hold soil firmly in place. Compare *taproot system*.

fight-or-flight response A response to a potentially dangerous situation that includes increased heart rate, blood pressure, blood flow to the brain, and energy availability—all of which contribute to increased activity throughout the body.

filtrate The liquid left over after filtration.

filtration The process of removing certain substances from a liquid by passing it through a filter, such as a semipermeable membrane.

first law of thermodynamics The law stating that energy can be neither created nor destroyed, only transformed or transferred from one molecule to another.

fixation The removal from a population of all alleles at a genetic locus except one; the allele that remains has a frequency of 100 percent.

fixed behavior A predictable response to a particular, often simple, stimulus that does not involve learning. Compare *learned behavior*.

flower A specialized reproductive structure that is characteristic of the plant group known as the angiosperms, or flowering plants.

flowering plants See *angiosperms*.

fluid mosaic model A model that describes the plasma membrane of a cell as a mobile phospholipid bilayer with embedded proteins that can move laterally in the plane of the membrane.

follicle An organized group of cells that surrounds the developing ovum in the ovary.

follicle-stimulating hormone (FSH) A hormone produced by the pituitary gland that works with luteinizing hormone to regulate sperm development in males and to regulate the menstrual cycle in females.

food chain A single sequence of feeding relationships describing who eats whom in a community. Compare *food web*.

food web A summary of the movement of energy through a community. A food web is formed by connecting all of the food chains in the community to one another. Compare *food chain*.

forebrain The part of the brain that consists of the thalamus, the cerebrum, and the hypothalamus and is involved in many of the conscious functions of the brain.

fossil Preserved remains of or an impression of a formerly living organism. Fossils document the history of life on Earth, showing that past organisms were unlike living forms, that many organisms have gone extinct, and that life has evolved through time.

founder effect A genetic bottleneck that results when a small group of individuals from a larger source population establishes a new population far from the original population.

fovea A region of especially densely packed photoreceptors in the retina of the eye that allows for the formation of sharp images.

frameshift A change in how the information in a DNA sequence is translated by the cell that results when a deletion or insertion is not a multiple of three base pairs (a codon).

friction drag Drag that results from the interaction of a fluid with the surface of a moving object. Compare *pressure drag*.

fruit A mature ovary surrounding a seed.

fulcrum The point around which a lever pivots.

functional group A specific arrangement of atoms that helps define the properties of a chemical compound.

Fungi The kingdom of mushroom-producing species, yeasts, and molds, most of which make their living as decomposers.

G₀ phase The stage of the cell cycle during which the cell pauses between mitosis and S phase. No preparations for S phase are made during this period.

G₁ phase The stage of the cell cycle following mitosis and before S phase. The cell makes preparations for DNA synthesis during G₁ phase.

G₂ phase The stage of the cell cycle following S phase and before mitosis.

gallbladder A small sac where bile secreted by the liver is stored before its release into the small intestine.

gamete A haploid sex cell that fuses with another sex cell during fertilization. Eggs and sperm are gametes.

ganglion (pl. ganglia) A structure consisting of densely packed neuronal cell bodies that allows the integration of signals.

gap junction A direct channel protein connection between the plasma membranes of two animal cells that allows the passage of ions and small molecules between them.

gastrula The stage of vertebrate development that results from the rearrangement of the three cell layers during gastrulation.

gastrulation The movement of the three cell layers (endoderm, mesoderm, and ectoderm) during animal development to the positions appropriate for the tissues to which these layers give rise.

gel electrophoresis A process in which DNA fragments are placed in a gelatin-like substance (a gel) and subjected to an electrical charge, which causes the fragments to move through the gel. Small DNA fragments move farther than large DNA fragments, thus causing the fragments to separate by size.

gene A DNA sequence that contains information for the synthesis of a protein or an RNA molecule. Genes are located on chromosomes.

gene cascade A process in which the protein products of different genes interact with one another and with signals from the environment, thereby turning on other sets of genes in some cells, but not in other cells. Organisms use gene cascades to control how genes are expressed during development.

gene expression The synthesis of a gene's protein or RNA product. Gene expression is the means by which a gene influences the cell or organism in which it is found.

gene flow The exchange of alleles between populations.

gene therapy A treatment approach that seeks to correct genetic disorders by repairing the genes that cause them.

gene-for-gene recognition In plants, a specific immune response in which many genes, which collectively have hundreds of alleles (known as R alleles), provide resistance against specific attackers.

genetic bottleneck A drop in the size of a population that results in low genetic variation or causes harmful alleles to reach a frequency of 100 percent in the population.

genetic code The code according to which each set of three nitrogen bases in mRNA specifies either an amino acid or a signal to start or stop the construction of a protein. The genetic code allows the cell to use the information in a gene to build the protein called for by that gene.

genetic cross A controlled mating experiment, usually performed to examine the inheritance of a particular characteristic.

genetic drift A process in which alleles are sampled at random over time, as when chance events cause certain alleles to increase or decrease in a population. The genetic makeup of a population undergoing genetic drift changes at random over time, rather than being shaped in a nonrandom way by natural selection.

genetic engineering A three-step process in which a DNA sequence (often a gene) is isolated, modified, and inserted back into an individual of the same or a different species. Genetic engineering is commonly used to change the performance of the genetically modified organism, as when a crop plant is engineered to resist attack from an insect pest.

genetic linkage The situation in which different genes that are located close to one another on the same chromosome do not follow Mendel's law of independent assortment.

genetic screening The examination of an individual's genes to assess current or future health risks and status.

genetic variation The genetic differences among the individuals of a population.

genetically modified organism (GMO) An individual into which a modified gene or other DNA sequence has been inserted, typically with the intent of improving some aspect of the recipient organism's performance.

genetics The scientific study of genes.

genome All the DNA of an organism, including its genes; in eukaryotes, the term "genome" refers to a haploid set of chromosomes, such as that found in a sperm or egg.

genomics The study of the structure and expression of entire genomes and how they change during evolution.

genotype The genetic makeup of an organism. Compare *phenotype*.

genotype frequency The proportion (percentage) of a particular genotype in a population.

genus The level in the Linnaean hierarchy that is above species and therefore comprises species, but that is below families and so composes families.

geographic isolation The physical separation of populations from one another by a barrier such as a mountain chain or a river. Geographic isolation often causes the formation of new species, as when populations of a single species become physically separated from one another and then accumulate so many genetic differences that they become reproductively isolated from one another. Compare *reproductive isolation*.

gill The gas exchange surface of an aquatic animal, typically consisting of finely folded sheets of thin epithelial tissue. Compare *lung*.

gizzard A portion of the digestive system that breaks down food by grinding it against rocks or sand collected from the environment.

global change Worldwide change in the environment. There are many causes of global change, including climate change caused by the movement of continents and changes in land and water use by humans.

global warming A worldwide increase in temperature. Earth appears to be entering a period of global warming caused by human activities; specifically, by the release of large quantities of greenhouse gases such as carbon dioxide into the atmosphere.

glomerulus The tiny capillary bed through which blood is filtered from the circulatory system into the tubules of the vertebrate kidney.

glucagon A hormone produced by the pancreas that acts as an antagonist to insulin in regulating blood sugar levels. Compare *insulin*.

glucose A monosaccharide that is the primary metabolic fuel in most cells.

glycogen A storage carbohydrate found primarily in the liver.

glycolysis A series of catabolic reactions that split glucose to produce pyruvate, which is then used in either oxidative phosphorylation or fermentation.

GMO See *genetically modified organism*.

Golgi apparatus An organelle composed of flattened membrane sacs that routes proteins and lipids to various parts of the eukaryotic cell.

gonad An animal sex gland that produces gametes and sex hormones. See also *testis, ovary*.

gonadotropin Either of a pair of hormones, luteinizing hormone (LH) and follicle-stimulating hormone (FSH), that are produced by the pituitary gland and regulate the development and function of the reproductive organs.

granum (pl. grana) A structure made up of a stack of membrane sacs, called thylakoids, that is part of the interconnected internal membrane system within a chloroplast.

grassland A terrestrial biome, the communities of which are dominated by grasses and many different types of wildflowers. Grasslands often occur in relatively dry regions with cold winters and hot summers.

Green Revolution A large increase in agricultural productivity that occurred in the last 50 years due to the development of new varieties of wheat, rice, and corn that allowed the production of large amounts of grain on a given amount of farmland.

greenhouse gas Any of several gases in Earth's atmosphere that let in sunlight, but trap heat.

greenroof On a building, a 2- to 4-inch-thick "living" rooftop that has a layer of soil or other material in which plants grow, under which there are one or more layers that absorb water and prevent roots and water from damaging the underlying roof structure.

gross domestic product (GDP) A traditional measure of a nation's economic output that records the value of goods produced, but does not consider the social and environmental costs that result from producing those goods. Compare *index of sustainable economic welfare*.

ground tissue The tissue type in plants that forms most of the photosynthetic tissue in leaves and surrounds the vascular system.

groundwater Water from an underground source, such as an aquifer or belowground river.

growth hormone (GH) A hormone produced by the pituitary gland that plays a key role in regulating the growth of bones and muscles.

guard cell A specialized cell type found on leaves and stems that can open and close, thereby controlling the rates at which CO_2 is brought into the plant and O_2 and water are lost from the plant.

gut inhabitant mutualism A mutualism involving organisms that live in the digestive tract of a host, receiving food from the host and digesting foods that the host otherwise could not use.

gymnosperms A group of plants that includes pine trees and other conifers, ginkgos, and cycads. Gymnosperms were the first plants to evolve seeds. Compare *angiosperms*.

habitat A characteristic place or type of environment in which an organism lives.

haploid Of or referring to a cell or organism that has only one complete set of homologous chromosomes (n). Compare *diploid*.

Hardy–Weinberg equation An equation ($p^2 + 2pq + q^2 = 1$) that predicts the genotype frequencies in a population that is not evolving.

heart The muscular pump that pushes blood through the animal circulatory system.

heart rate The number of heartbeats per minute.

heartbeat The series of events corresponding to one cardiac cycle.

heartwood Older, centrally located secondary xylem tissue that provides support to a tree but has become clogged over time and hence no longer functions in the transport of materials. Compare *sapwood*.

helper T cell A type of lymphocyte that stimulates killer T cells and B cells.

hemoglobin An oxygen-binding pigment that is used to carry oxygen from the gas exchange surfaces to the tissues.

herbivore A consumer that relies on living plant tissues for nutrients. Compare *carnivore*.

hermaphrodite An individual that can produce both eggs and sperm.

heterozygote An individual that carries one copy of each of two different alleles (for example, an *Aa* individual). Compare *homozygote*.

high-density lipoprotein (HDL) A molecule that carries excess cholesterol through the circulatory system to the liver. Compare *low-density lipoprotein*.

hindbrain The portion of the brain that includes the medulla oblongata, the pons, and the cerebellum and coordinates breathing rhythms and balance.

histamine A protein released by mast cells that causes blood vessels to dilate and capillaries to become more porous.

histone spool A group of proteins around which the DNA of a eukaryotic chromosome is wound.

homeostasis The process of maintaining appropriate and constant conditions inside cells.

homeotic gene A master-switch gene that plays a key role in the control of gene expression during development. Each homeotic gene controls the expression of a series of other genes whose protein products direct the development of an organism.

hominid Any of a group of primates that encompasses humans and our now extinct humanlike ancestors.

homologous Of or referring to a characteristic shared by two groups of organisms because of their descent from a common ancestor. Compare *analogous*.

homologous chromosomes The two members of a specific chromosome pair found in diploid cells, one of which comes from the individual's mother and the other from its father.

homologue One of a pair of homologous chromosomes.

homozygote An individual that carries two copies of the same allele (for example, an *AA* or an *aa* individual). Compare *heterozygote*.

horizontal gene transfer The movement of genes from one organism or group of organisms to another, not vertically within a lineage via reproduction, but horizontally to another lineage altogether by some other means.

hormone A signaling molecule released into the circulatory system of an animal or the vascular system of a plant in very small amounts that affects the functioning of target tissues.

host An organism in which a parasite or pathogen lives.

housekeeping gene A gene that has an essential role in the maintenance of cellular activities and is expressed by most cells in the body.

Human Genome Project (HGP) A publicly funded effort on the part of an international consortium created by the U.S. National Institutes for Health and the U.S. Department of Energy to determine the sequence of the human genome.

humoral immunity A form of immunity that relies on the marking of pathogens by the antibodies produced by B cells. Compare *cell-mediated immunity*.

hybrid An offspring that results when two different species mate.

hybridize To cause hybrid offspring to be produced.

hydrogen bond A chemical linkage between a hydrogen atom, which has a slight positive charge, and another atom with a slight negative charge. Compare *covalent bond* and *ionic bond*.

hydrophilic Of or referring to molecules or parts of molecules that interact freely with water. Hydrophilic molecules dissolve easily in water, but not in fats or oils. Compare *hydrophobic*.

hydrophobic Of or referring to molecules or parts of molecules that do not interact freely with water. Hydrophobic molecules dissolve easily in fats and oils, but not in water. Compare *hydrophilic*.

hydrostat A support structure that depends on the interaction between a fluid under pressure and an elastic membrane for stiffness.

hypertension Abnormally high blood pressure.

hypertonic solution A solution that has a higher solute concentration than the cytosol of a cell, causing more water to flow out of the cell than into it. Compare *hypotonic solution* and *isotonic solution*.

hypha (pl. hyphae) In fungi, a threadlike absorptive structure that grows through a food source. Mats of hyphae form mycelia, the main bodies of fungi.

hypothalamus (pl. hypothalami) A structure at the base of the vertebrate brain that controls the release of hormones by the pituitary gland. Along with the pituitary gland, the hypothalamus helps regulate interactions between the nervous and endocrine systems.

hypothesis (pl. hypotheses) A possible explanation of how a natural phenomenon works. A hypothesis must have logical consequences that can be proved true or false.

hypotonic solution A solution that has a lower solute concentration than the cytosol of a cell, causing more water to flow into the cell than out of it. Compare *hypertonic solution* and *isotonic solution*.

immune system A system of defensive proteins and white blood cells in vertebrate animals that destroys invading pathogens.

imprinting A type of learning in which an offspring forms an association or bond with its parent early in its development.

incomplete dominance The situation in which heterozygotes (*Aa* individuals) are intermediate in phenotype between the two homozygotes (*AA* and *aa* individuals) for a particular gene.

independent assortment of chromosomes The random distribution of maternal and paternal chromosomes into gametes during meiosis.

indeterminate growth A feature of some organisms that grow throughout their entire lives, adding new body parts as needed.

index of sustainable economic welfare A measure of economic output that includes a wider range of the benefits and costs of economic activities than does the traditional gross domestic product (GDP). Compare *gross domestic product*.

individual A single organism, usually physically separate and genetically distinct from other individuals.

induced defense A plant defensive response that is directly stimulated by attacking herbivores.

infectious disease A disease that is transmitted from one host to another by the transfer of a pathogen.

inflammatory response The characteristic swelling and reddening of an area of tissue damage; part of the nonspecific response to invading pathogens.

ingestion The act of taking food into the mouth, or eating.

inner cell mass The cluster of cells inside the trophoblast that eventually develops into the embryo.

inner ear The portion of the ear where physical movements in the middle ear are converted into the action potentials that vertebrates detect as sound.

insect Any of a group of six-legged arthropods that includes grasshoppers, beetles, ants, and butterflies; the most species-rich group of animals on Earth.

insertion A mutation in which one or more nitrogen bases are inserted into the DNA sequence of a gene. Compare *deletion* and *substitution*.

insulin A hormone produced by the pancreas that regulates blood sugar levels in conjunction with its antagonist, glucagon. Compare *glucagon*.

interference competition A type of competition in which one organism directly excludes another from the use of resources. Compare *exploitation competition*.

interferon A chemical released by virus-infected cells that attaches to the plasma membranes of nearby cells, interfering with the ability of the virus to enter and infect those cells.

intermediate filament One of a diverse class of ropelike protein filaments that serve as structural reinforcements in the cytoskeleton.

intermembrane space The space between the inner and outer membranes of a chloroplast or a mitochondrion.

interneuron A neuron in the central nervous system that processes signals from sensory neurons, then passes those signals on to other neurons. Compare *sensory neuron*, *motor neuron*.

interphase The period of time between two successive mitotic divisions, during which most of the preparations for cell division occur.

interstitial fluid The body fluid that surrounds and bathes the cells.

intertidal zone An aquatic biome found in coastal areas where the tides rise and fall on a daily basis, periodically submerging a portion of the shore.

introduced species A species that does not naturally live in an area but has been brought there either accidentally or on purpose by humans.

intron A sequence of nitrogen bases within a gene that does not specify part of the gene's final protein or RNA product. Enzymes in the nucleus must remove introns from mRNA, tRNA, and rRNA molecules for these molecules to function properly.

invasive species An introduced species that proliferates rapidly and becomes a major pest in its new environment.

inversion A mutation in which a fragment of a chromosome breaks off and returns to the correct place on the original chromosome, but with the genetic loci in reverse order.

ion An atom or group of atoms that has either gained or lost electrons and therefore has a negative or positive charge.

ionic bond A chemical linkage between two atoms based on the electrical attraction between positive and negative charges. Compare *covalent bond* and *hydrogen bond*.

irregular fluctuations A pattern of population growth in which the number of individuals in the population changes over time in an irregular manner.

islet cells Specialized endocrine cells within the pancreas that produce and release the hormones insulin and glucagon.

isotonic solution A solution that has the same solute concentration as the cytosol of a cell, resulting in an equal amount of water flowing into the cell and out of it.

isotope A variant form of a chemical element that differs in its number of neutrons, and thus in its atomic mass number, from the most common form of that element.

J-shaped curve A pattern of population growth in which the number of individuals in the population rises rapidly over time, as in exponential growth.

joint A flexible connection between the rigid elements that make up an animal skeleton.

karyotype The specific number and shapes of chromosomes found in the cells of a particular species.

keystone species A species that, relative to its own abundance, has a large effect on the presence and abundance of other species in a community.

kidney An organ that regulates the body's supply of water and solutes and helps dispose of metabolic wastes.

killer T cell A lymphocyte that binds to cells that have been infected by a virus and helps destroy them.

kinetochore A plaque of protein on the centromere of a chromosome where spindle microtubules attach during mitosis and meiosis.

kingdom The largest taxonomic category in the Linnaean hierarchy. Generally six kingdoms are recognized: Bacteria, Archaea, Protista, Plantae, Fungi, and Animalia.

Kyoto Protocol An international agreement designed to reduce global warming by reducing CO_2 emissions.

lake An aquatic biome, the communities of which live in standing bodies of fresh water of variable size, ranging from a few square meters to thousands of square kilometers.

land transformation Changes made by humans to the land surface of Earth that alter the physical or biological characteristics of the affected regions. Compare *water transformation*.

large intestine See *colon*.

lateral meristem Perpetually young, undifferentiated cells that cause a plant to increase in thickness when they divide. There are two types of lateral meristem tissue: vascular cambium and cork cambium.

law of independent assortment Mendel's second law, which states that when gametes form, the separation of alleles for one gene is independent of the separation of alleles for other genes. We now know that this law does not apply to genes that are linked.

law of segregation Mendel's first law, which states that the two copies of a gene separate during meiosis and end up in different gametes.

leaf A plant organ whose function is to produce food by photosynthesis.

learned behavior A behavior acquired by trial and error or by watching others. Compare *fixed behavior*.

lens A structure that concentrates light on photoreceptors.

lever system A system consisting of a rigid structure that pivots around a fulcrum; it can either increase the force generated at the expense of speed or increase the speed generated at the expense of force.

lichen A symbiosis of an alga (kingdom Protista) and a fungus (kingdom Fungi).

life cycle engineering An approach in which a business seeks to document (and reduce, as necessary) the total environmental impact of its products, from manufacture to disposal.

ligament A collagen-rich connective structure that attaches bone to bone in vertebrate skeletons. Compare *tendon*.

ligase An enzyme that can connect two DNA fragments to each other; used in DNA technology when a gene from one species is inserted into the DNA of another species.

light reactions A series of chemical reactions that harvest energy from sunlight and use it to produce energy-rich compounds such as ATP and NADPH. The light reactions occur at the thylakoid membranes of chloroplasts and produce O_2 as a waste product. Compare *dark reactions*.

lineage A group of closely related individuals, species, genera, or the like, depicted as a branch on an evolutionary tree.

linked genes See *genetic linkage*.

Linnaean hierarchy The classification scheme used by biologists to organize and name organisms. Its seven levels are species, genus, family, order, class, phylum, and kingdom.

lipid A hydrophobic molecule that contains fatty acids; a key component of cell membranes. See also *phospholipid bilayer*.

liver An organ associated with the upper small intestine and with the circulatory system. The liver produces bile and plays an important role in carbohydrate and lipid metabolism.

load arm The section of a lever that extends from the fulcrum to the point at which work is done. Compare *power arm*.

locus (pl. loci) The physical location of a gene on a chromosome.

low-density lipoprotein (LDL) A molecule that carries cholesterol through the circulatory system to cells that need it. Compare *high-density lipoprotein*.

lower respiratory system A collective term for the trachea, bronchi, and lungs. Compare *upper respiratory system*.

lumen The space enclosed by the membrane of an organelle.

lung The gas exchange structure of a terrestrial animal, formed by an infolding of epidermal tissue. Compare *gill*.

luteinizing hormone (LH) A hormone produced by the pituitary gland that works with follicle-stimulating hormone to regulate sperm development in males and to regulate the menstrual cycle in females.

lymph node A pocket of tissue lying along a lymphatic duct that contains large numbers of white blood cells and traps pathogens.

lymphatic duct One of a network of tubes that return interstitial fluid to the circulatory system.

lymphatic system The network of vessels that returns interstitial fluid from the body to the circulatory system.

lymphocyte Any of several types of white blood cells that bind to specific antigens and then contribute in various ways to the destruction of the pathogens that bear those antigens.

lysosome A specialized vesicle with an acidic lumen, containing enzymes that break down macromolecules.

macroevolution The rise and fall of major taxonomic groups due to evolutionary radiations that bring new groups to prominence and mass extinctions in which groups are lost; the history of large-scale evolutionary changes over time. Compare *microevolution*.

macromolecule A large organic molecule formed by the bonding together of small organic molecules.

macronutrient A chemical element required by an organism in relatively large amounts. Compare *micronutrient*.

macrophage A type of white blood cell that engulfs or encapsulates invading pathogens.

magnetic resonance imaging (MRI) An imaging technology that can detect which parts of the brain are active at any particular time.

male An individual of a sexually reproducing species that produces sperm. Compare *female*.

malignant Of or referring to a cancerous growth that begins as a single tumor and then spreads to other tissues in the body with life-threatening consequences. Compare *benign*.

marrow A tissue lying within vertebrate bone in which red and white blood cells are produced.

mass extinction A event during which large numbers of species become extinct throughout most of Earth.

mast cell A type of white blood cell that lives in connective tissue and releases histamine.

mechanoreceptor A sensory cell that initiates an action potential in response to a mechanical stimulus.

meiosis A specialized process of cell division in eukaryotes during which diploid cells divide to produce haploid cells. Meiosis has two division cycles and occurs exclusively in cells that produce gametes. Compare *mitosis*.

meiosis I The first cycle of cell division in meiosis. Meiosis I produces haploid daughter cells, each with half the chromosome number of the diploid parent cell.

meiosis II The second cycle of cell division in meiosis. Meiosis II is essentially mitosis, but in a haploid cell.

memory cell A type of B cell that remains in the body as a long-term record of the primary immune response to a particular pathogen. During a second exposure to that pathogen, the memory cells specific to that pathogen multiply rapidly to produce large numbers of lymphocytes that can attack that pathogen.

menopause The end of a human female's reproductive life.

menstrual cycle A series of hormonally controlled, cyclical changes that take place in the reproductive system of human females, which include the shedding of the lining of the uterus (menstruation) about every 28 days.

meristem A group of perpetually young, undifferentiated plant cells from which new cells arise.

mesoderm The cell layer in animal development that lies between the endoderm and ectoderm in the gastrula and gives rise to muscle tissue, connective tissue, and the kidney. Compare *ectoderm, endoderm*.

messenger RNA (mRNA) A type of RNA that specifies the order of amino acids in a protein.

metabolic heat Heat generated as a product of respiration.

metabolic pathway A series of enzyme-controlled chemical reactions in a cell in which the product of one reaction becomes the substrate for the next.

metabolism All the chemical reactions that occur in a cell.

metamorphosis A dramatic developmental transformation from a reproductively immature to a reproductively mature form, involving great change in the form and function of the animal.

metaphase The stage of mitosis during which chromosomes become aligned at the equator of the cell.

microevolution Changes in allele or genotype frequencies in a population over time; the smallest scale at which evolution occurs. Compare *macroevolution*.

microfilament A protein fiber composed of actin monomers. Microfilaments are part of a cell's cytoskeleton and are important in cell movements.

micronutrient A chemical element required by an organism in tiny amounts. Compare *macronutrient*.

microtubule A protein fiber composed of tubulin monomers. Microtubules are part of the cell's cytoskeleton.

midbrain A portion of the brain stem that helps to maintain muscle tone and sends some sensory data to higher brain centers.

middle ear The portion of the vertebrate ear that contains three tiny bones that receive vibrations from the eardrum and transmit them to the inner ear.

milk-letdown reflex The reflexive release of milk by a human mother in response to the suckling behavior of an infant.

mineral A small inorganic molecule. Many minerals are needed by organisms in small amounts for proper nutrition.

mismatch error The insertion of an incorrect nitrogen base during DNA replication that is not detected and corrected.

mitochondrion (pl. mitochondria) An organelle with a double membrane that is the site of oxidative phosphorylation. Mitochondria break down simple sugars to produce most of the ATP needed by eukaryotic cells.

mitosis The process of cell division in eukaryotes that produces two daughter nuclei, each with the same chromosome number as the parent nucleus. Compare *meiosis*.

mitotic spindle An arrangement of microtubules that guides the movement of chromosomes during mitosis.

modular Of or referring to the body of an organism, such as a plant, that grows by the repeated addition of a basic structural unit.

molecule An arrangement of two or more atoms linked by chemical bonds.

monoglyceride Along with fatty acids, one of the two components of lipids.

monomer A molecule that can be linked with other related molecules to form a larger polymer.

monosaccharide A simple sugar that can be linked to other sugars, forming a polysaccharide. Glucose is the most common monosaccharide in living organisms.

morphogen A chemical signal that influences the developmental fate of a cell.

morphology The form and structure of an organism.

most recent common ancestor The ancestral organism from which a group of descendants arose.

motor neuron A neuron that carries signals from the central nervous system to the body, thereby causing changes in muscles, organs, or endocrine glands. Compare *sensory neuron, interneuron*.

motor protein A protein that uses the energy of ATP to move organelles or proteins along microtubules.

mRNA See *messenger RNA*.

multicellular Made up of more than one cell.

multiregional hypothesis A hypothesis stating that anatomically modern humans evolved from *Homo erectus* populations scattered throughout the world. According to this idea, worldwide gene flow caused different human populations to evolve modern characteristics simultaneously and to remain a single species. Compare *out-of-Africa hypothesis*.

muscle A tissue unique to animals that can contract, or shorten, to produce movement.

muscle fiber The basic unit of muscle tissue, consisting of a collection of myofibrils.

mutagen A substance or energy source that alters DNA.

mutation A change in the sequence of an organism's DNA. New alleles arise only by mutation, so mutations are the original source of all genetic variation.

mutualism An interaction between two species in which both species benefit.

mutualist An organism that interacts with another organism to the mutual benefit of both.

mycelium (pl. mycelia) The main body of a fungus, composed of hyphae.

mycorrhiza (pl. mycorrhizae) A mutualism between a fungus and a plant, in which the fungus provides the plant with mineral nutrients while receiving organic nutrients from the plant.

myelin A fatty material that forms an insulating sheath around the axon of a neuron and greatly speeds the rate at which action potentials move along the axon.

myofibril A unit of a muscle fiber consisting of many sarcomeres attached end-to-end.

myoglobin An oxygen-binding pigment that helps store oxygen in muscle tissue.

myosin filament One of the two types of filaments that make up muscle, consisting of many molecules of the protein myosin. Compare *actin filament*.

NADH An energy carrier molecule that acts as a reducing agent in the catabolic reactions that produce ATP from the breakdown of sugars into water and carbon dioxide.

NADPH An energy carrier molecule that acts as a reducing agent in photosynthesis.

natural killer (NK) cell A type of white blood cell that destroys any cell whose plasma membrane is marked with foreign proteins, including viral proteins.

natural selection An evolutionary mechanism in which those individuals in a population that possess particular inherited characteristics survive and reproduce at a higher rate than other individuals in the population because of those characteristics. Natural selection is the only evolutionary mechanism that consistently improves the survival and reproduction of the organism in its environment.

negative feedback A mechanism for slowing down or damping the fluctuations in a process. In response to a change in conditions, negative feedback acts to bring conditions back to a set point. Compare *positive feedback*.

neocortex A part of the cerebral cortex found only in mammals, where the ability to think complex thoughts and use language resides.

nephron The basic unit of the vertebrate kidney involved in urine production, consisting of a glomerulus, across which blood is filtered under pressure into a U-shaped tubule, which facilitates reabsorption of water and valuable solutes from the filtrate.

nerve A major communication pathway in the animal body made up of the axons of many individual neurons bundled together.

nerve cell See *neuron*.

nervous system A high-speed internal communication system in animals that is made up of specialized cells called neurons.

net primary productivity (NPP) The amount of energy that producers capture by photosynthesis, minus the amount lost as metabolic heat. NPP is usually measured as the amount of new biomass produced by photosynthetic organisms per unit of area during a specified period of time. Compare *secondary productivity*.

neuromuscular junction A synapse between a motor neuron and a muscle fiber.

neuron A type of animal cell that is highly specialized for transmitting action potentials from one part of the body to another.

neurotransmitter Any of various signaling molecules that transmit signals across the gaps, or synapses, that separate neurons from their target cells.

neutron A particle found in the nucleus of an atom that has no electrical charge.

neutrophil A type of white blood cell that destroys invading bacterial cells by phagocytosis.

nicotine The chemical in tobacco that is responsible for addiction.

nitrogen base Any of the five nitrogen-rich compounds found in nucleotides. The four nitrogen bases found in DNA are adenine (A), cytosine (C), guanine (G), and thymine (T); in RNA, uracil (U) replaces thymine.

nitrogen fixation The process by which nitrogen gas (N_2), which is readily available in the atmosphere but cannot be used by plants, is converted to ammonium (NH_4^+), a form of nitrogen that can be used by plants. Nitrogen fixation is accomplished naturally by bacteria and by lightning, and by humans in industrial processes such as the production of fertilizer.

node of Ranvier An unmyelinated portion of the axon of a vertebrate neuron. Action potentials jump from node to node, greatly speeding signal transmission.

noncoding DNA A segment of DNA that does not encode proteins or RNA. *Introns* and *spacer DNA* are two common types of noncoding DNA.

noncovalent bond Any chemical linkage between two atoms that does not involve the sharing of electrons. Hydrogen bonds and ionic bonds are examples of noncovalent bonds.

nonpolar Of or referring to a molecule or a portion of a molecule that has an equal distribution of electrical charge across all its constituent atoms. Nonpolar molecules do not form hydrogen bonds and therefore tend not to dissolve in water. Compare *polar*.

nonspecific response A defensive response found in most animals that leads to the destruction of cells not recognized as belonging to the organism. Compare *specific response*.

norepinephrine A hormone produced by the adrenal glands that regulates the availability of glucose in conjunction with epinephrine.

NPP See *net primary productivity*.

nuclear envelope The double membrane that encloses the nucleus of a eukaryotic cell.

nuclear pore A channel in the nuclear envelope that allows selected molecules to move into and out of the nucleus.

nucleic acid A polymer made up of nucleotides. There are two kinds of nucleic acids: DNA and RNA.

nucleotide Any of a class of organic compounds that serve as energy carriers and as the chemical building blocks of nucleic acids such as DNA and RNA. A nucleotide is made up of a phosphate group, a five-carbon sugar, and one of four nitrogen-containing molecules called bases (see *nitrogen base*). Nucleotides are linked together to form a single strand of DNA or RNA.

nucleus (pl. nuclei) The organelle in a eukaryotic cell that contains the genetic blueprint in the form of DNA.

nucleus accumbens A part of the human brain that is associated with addictive behavior.

nutrient In an ecosystem context, an essential element required by a producer. See also *macronutrient* and *micronutrient*.

nutrient cycle The cyclical movement of a nutrient between organisms and the physical environment. There are two main types of nutrient cycles: atmospheric and sedimentary. See also *atmospheric cycle* and *sedimentary cycle*.

observations With respect to the scientific method, facts learned by a scientist by observing that are used to formulate hypotheses.

olfactory bulb A part of the vertebrate brain that receives signals from the chemoreceptors lining the nasal passages and where some initial sorting of smells takes place prior to further processing elsewhere in the brain.

omnivore An organism that uses both plant and animal tissues as a source of nutrition.

oncogene A mutated gene that promotes excessive cell division, leading to cancer.

oogenesis The production of a mature ovum capable of being fertilized.

open circulatory system An internal transport system in which blood flows through the body cavity that surrounds the organs. Compare *closed circulatory system*.

open ocean An aquatic biome that covers the majority of Earth's surface and includes communities found in a shallow layer (100 to 200 meters deep) in which photosynthesis can occur, as well as in deeper waters where little light can penetrate.

operator In prokaryotes, a regulatory DNA sequence that controls the transcription of a gene or group of genes.

opposable In primates, of or referring to a thumb (or big toe) that moves freely and can be placed opposite other fingers (or toes).

oral cavity Another term for the mouth, where food first enters an animal's digestive system.

order The level in the Linnaean hierarchy that is above families and therefore comprises families, but that is below classes and so composes classes.

organ A self-contained collection of tissues, usually of a characteristic size and shape, that is organized for a particular function.

organ of Corti A structure in the cochlea of the inner ear that converts mechanical energy into a pattern of action potentials that vertebrates detect as sound.

organ system A group of organs that work together to carry out a particular function.

organelle A distinct, membrane-enclosed structure in a eukaryotic cell that has a specific function.

organic compound A carbon-containing compound of biological origin.

organic farming A method of farming that avoids the use of synthetic chemicals and genetically modified organisms.

osmoregulation The process of maintaining an internal water concentration that supports biological processes.

osmosis The passive movement of water across a selectively permeable membrane.

osteocyte A cell that either produces or reabsorbs the matrix of calcium and phosphorus compounds that makes up vertebrate bone.

otolith A dense crystal of calcium carbonate in the vestibular apparatus that indicates the direction of gravitational pull.

outer ear The outer portion of the mammalian ear, which consists of the pinna and the auditory canal and which funnels pressure changes from the environment to the eardrum.

out-of-Africa hypothesis A hypothesis stating that anatomically modern humans evolved in Africa within the past 200,000 years, then spread throughout the rest of the world. According to this idea, as they spread from Africa, modern humans completely replaced older forms of *Homo sapiens*, including advanced forms such as the Neandertals. Compare *multiregional hypothesis*.

ovary The reproductive structure in female animals that produces ova.

oviduct The tube down which ova produced by female animals pass after leaving the ovary.

ovulation The release of an ovum from the ovary.

ovum (pl. ova) The female gamete.

oxidation The loss of electrons by one atom or molecule to another. Compare *reduction*.

oxidative phosphorylation The shuttling of electrons down an electron transport chain in mitochondria that results in the production of ATP.

oxygen-binding pigment Any of various complex molecules used by animals to increase the oxygen capacity of their tissues.

oxytocin A hormone released by the pituitary gland that triggers, among other things, milk production in mammals.

P generation The parent generation of a genetic cross. Compare F_1 *generation* and F_2 *generation*.

pacemaker See *sinoatrial node*.

pain receptor A sensory cell that initiates an action potential in response to extreme heat or cold or to injury.

pancreas An organ associated with the digestive system that produces a number of important digestive enzymes as well as hormones.

Pangaea An ancient supercontinent that contained of all of the world's landmasses. Pangaea formed 250 million years ago and began to break apart 200 million years ago, ultimately yielding the continents we know today.

parasite An organism that lives in or on another organism (its host) and obtains nutrients from that organism. Parasites harm and may eventually kill their hosts, but do not kill them immediately.

parathyroid gland A gland lying next to the thyroid gland that produces parathyroid hormone.

parathyroid hormone (PTH) A hormone that stimulates the release of calcium from bones and increases the reabsorption of calcium in the kidneys. Compare *calcitonin*.

particulates Microscopic particles in smoke that can penetrate the lung wall and enter the blood.

passive carrier protein A protein in the plasma membrane of a cell that, without the input of energy, changes its shape to transport a molecule across the membrane from the side of higher concentration to the side of lower concentration. Compare *active carrier protein*.

passive immunity The immunity of an organism to a pathogen that depends on antibodies specific to that pathogen that come from a source outside the organism's body. Compare *active immunity*.

passive transport Movement of molecules from areas of lower concentration to areas of higher concentration without the expenditure of energy. Compare *active transport*.

pathogen An organism or virus that infects a host and causes disease, harming and in some cases killing the host.

PCR See *polymerase chain reaction*.

pedigree A chart that shows genetic relationships among family members over two or more generations of a family's history.

penis A reproductive structure used by male animals to introduce sperm directly into a female's reproductive tract.

peptide bond A covalent bond between the amino group of one amino acid and the carboxyl group of another that links amino acids together.

perennial A plant that lives for three or more years.

peripheral nervous system (PNS) The portion of the nervous system that carries signals to and from the central nervous system. The peripheral nervous system consists of the sensory organs and all the nerves that are not part of the central nervous system. Compare *central nervous system*.

Permian extinction The largest mass extinction in the history of life on Earth; it occurred 250 million years ago, driving up to 95 percent of the species in some groups to extinction.

petal A part of a flower that often serves to attract pollinators to the flower. Petals are the second (from the outside) of the four whorls of modified leaves that make up a flower. Compare *sepal, stamen, carpel*.

pH The concentration of hydrogen ions in a solution. The pH scale runs from 1 to 14. A pH of 7 is neutral; values below 7 indicate acids, and values above 7 indicate bases.

phagocytosis A form of endocytosis by which a cell engulfs a large particle, such as another cell; "cell eating."

pharynx The portion of the respiratory system where the mouth and the nasal cavity join to form a single passageway.

phenotype The observable physical characteristics of an organism. Compare *genotype*.

pheromone A chemical signal produced by one individual to communicate its identity and location to another individual.

phloem A tissue composed of living cells through which a plant transports the products of photosynthesis. Compare *xylem*.

phospholipid A lipid molecule with an attached phosphate group. Phospholipids are the major components of all biological membranes.

phospholipid bilayer A double layer of phospholipid molecules arranged so that their hydrophobic "tails" lie sandwiched between their hydrophilic "heads." A phospholipid bilayer forms the basic structure of all biological membranes.

phosphorylation The addition of a phosphate group to an organic molecule.

photoperiodism A physiological response by a plant to day length; examples include the timing of flowering, seed germination, and bud dormancy (during winter).

photoreceptor A sensory cell that initiates an action potential in response to light.

photosynthesis A process by which organisms capture energy from sunlight and use it to synthesize sugars from carbon dioxide and water.

photosystem A large complex of proteins and chlorophyll that captures energy from sunlight. Two distinct photosystems (I and II) are present in the thylakoid membranes of chloroplasts.

photosystem I The photosystem that is primarily responsible for the production of NADPH.

photosystem II The photosystem in which light energy is used to initiate an electron flow along the electron transport chain that leads to the production of ATP.

phylum The level in the Linnaean hierarchy that is above classes and therefore comprises classes, but that is below kingdoms and so composes kingdoms.

pinna (pl. pinnae) The curled, cartilaginous part of the ear that is visible externally. The pinna funnels pressure changes from the environment into the auditory canal.

pinocytosis A form of nonspecific endocytosis by which cells take in fluid; "cell drinking."

pituitary gland A gland associated with the vertebrate brain that releases hormones that control the release of hormones by other glands. Along with the hypothalamus, the pituitary helps regulate interactions between the nervous and endocrine systems.

placenta A structure found in mammals that transfers nutrients and gases from the blood of the mother to the blood of the developing fetus in her uterus.

plant breeding A traditional breeding technique used to create new crop varieties by allowing only plants with particular phenotypic characteristics to produce offspring.

Plantae The kingdom that encompasses plants.

plaque A fatty growth that develops on the walls of arteries as a result of excess cholesterol in the human diet.

plasma membrane The phospholipid bilayer that surrounds the cell.

plasmid A small circular segment of DNA found naturally in bacteria. Plasmids are involved in natural gene transfers among bacteria and can be used as vectors in genetic engineering.

plasmodesma (pl. plasmodesmata) A tunnel-like channel between two plant cells that provides a cytoplasmic connection allowing the flow of small molecules and water between them.

platelet A type of sticky cell fragment that circulates in the blood and that helps to form a blood clot by combining with proteins that also circulate in the blood to form a gel-like mesh that traps blood cells.

polar Of or referring to a molecule or a portion of a molecule that has an uneven distribution of electrical charge. Polar molecules can easily interact with water molecules and are therefore soluble. Compare *nonpolar*.

polar body The smaller of the two haploid cells that result from the first meiotic division of the primary oocyte. The polar body plays no role in development.

polarization The state of an axon in which there is a difference in electrical charge between the inside and the outside of the plasma membrane. Compare *depolarization*.

pollen grain The small, mobile, multicellular haploid stage of a seed plant's life cycle in which sperm are produced. Compare *embryo sac*.

pollinator An animal that carries pollen grains from the stamens of one flower to the stigmas of other flowers of the same species.

pollinator mutualism A mutualism in which an animal transfers pollen grains from one flower to the female reproductive organs of another flower of the same species and receives food as a reward for this service.

polygenic Of or referring to inherited traits that are determined by the action of more than one gene.

polymer A large organic molecule composed of many monomers linked together.

polymerase chain reaction (PCR) A method of DNA technology that uses the DNA polymerase enzyme to make multiple copies of a targeted sequence of DNA.

polyploidy A condition in which an organism has three or more complete sets of chromosomes (rather than the usual two complete

sets). Polyploidy can cause new species to form rapidly without geographic isolation.

polysaccharide A polymer composed of many linked monosaccharides. Starch and cellulose are examples of polysaccharides.

population A group of interacting individuals of a single species located within a particular area.

population cycle A pattern in which the population sizes of two species increase and decrease together in a tightly linked cycle; this pattern can occur when at least one of the two species involved is very strongly influenced by the other.

population density The number of individuals in a population, divided by the area covered by the population.

population size The total number of individuals in a population.

positive feedback A mechanism for speeding up or amplifying a process. Compare *negative feedback*.

positron emission tomography (PET) An imaging technology that can detect which parts of the brain are active at any particular time by following the location of radioactively labeled chemicals.

power arm The section of a lever that extends from the fulcrum to the point at which force is applied. Compare *load arm*.

precision farming A method of farming that uses intensive data collection, computers, and in some cases, satellite data to inform farmers when to fertilize or use pesticides.

predator An organism that kills other organisms for food.

predictions With respect to the scientific method, statements about logical consequences that should be observed if a hypothesis is correct.

preimplantation genetic diagnosis (PGD) A procedure used in in vitro fertilization in which one or two cells are removed from a developing embryo and tested for genetic disorders; embryos that are free of genetic disorders are then implanted into the mother's uterus.

pressure drag Drag that results from a difference in the pressures in front of and behind a moving object. Compare *friction drag*.

prey Animals that predators kill and eat.

primary consumer An organism that eats a producer. Compare *secondary consumer*.

primary growth An increase in the length (or height) of a plant that results from divisions of apical meristem cells.

primary immune response The relatively slow mobilization of B cells and T cells following the first exposure to a pathogen. Compare *secondary immune response*.

primary oocyte A diploid animal cell that can undergo meiosis to form mature ova.

primary spermatocyte A diploid animal cell that can undergo meiosis to form mature sperm.

primary structure The sequence of amino acids in a protein.

primary succession Ecological succession that occurs in newly created habitat, as when an island rises from the sea or a glacier retreats, exposing newly available bare ground. Compare *secondary succession*.

primate An order of mammals whose living members include lemurs, tarsiers, monkeys, humans, and other apes. Primates share characteristics such as flexible shoulder and elbow joints, opposable thumbs or big toes, forward-facing eyes, and brains that are large relative to body size.

producer An organism that uses energy from an external source, such as sunlight, to produce its own food without having to eat other organisms or their remains. Compare *consumer*.

productivity The mass of plant matter that can be produced in a given area from the available nutrients and sunlight.

progesterone A steroid hormone that helps maintain pregnancy in female mammals.

prokaryote A single-celled organism that does not have a nucleus. All prokaryotes are members of the domains Bacteria or Archaea. Compare *eukaryote*.

prometaphase The stage of mitosis during which chromosomes become attached to the mitotic spindle.

promoter The DNA sequence in a gene to which RNA polymerase binds to begin transcription.

prophase The stage of mitosis during which chromosomes first become visible under the microscope.

proprioceptor A pressure-sensitive sensory receptor that provides information on the position of an animal's body relative to its surroundings and of the animal's body parts relative to one another.

prostaglandin One of a class of chemical messages released by damaged cells in response to histamines to stimulate increased blood flow to the area.

protein A linear polymer of amino acids linked together in a specific sequence. Most proteins are folded into complex three-dimensional shapes.

Protista The oldest eukaryotic kingdom, consisting of a diverse collection of mostly single-celled but some multicellular organisms.

proton A positively charged particle found in atoms. Each atom contains a characteristic number of protons. Compare *electron*.

proton gradient An imbalance in the concentration of protons across a membrane.

proto-oncogene A gene that promotes cell division in response to normal growth signals.

protostome Any of a group of animals, including insects, worms, and snails, in which the first opening to develop in the early embryo becomes the mouth. Compare *deuterostome*.

pseudopodium (pl. pseudopodia) A dynamic protrusion of the plasma membrane that enables some cells to move. The extension of pseudopodia depends on actin filaments inside the cell.

puberty In humans, the transition from childhood to reproductive maturity.

pulmonary circuit The loop of the circulatory system through which blood travels from the heart to the lungs and back.

Punnett square A diagram in which the possible types of male and female gametes are listed on two sides of a square, providing a graphic way to predict the genotypes of the offspring produced in a genetic cross.

pupil The opening through which light enters the eye.

pyruvate A three-carbon molecule produced by glycolysis that is processed in the mitochondria to generate ATP.

quaternary structure The three-dimensional arrangement of two or more separate chains of amino acids into an aggregate protein.

radiation A means of exchanging heat in the form of light or infrared (heat) waves.

radioisotope An unstable, radioactive form of an element that decays to more stable forms at a constant rate over time.

rain shadow An area on the side of a mountain facing away from moist prevailing winds where little rain or snow falls.

rainforest A forest that receives high rainfall.

rangeland An open area where domesticated animals such as cattle or sheep can roam.

reabsorption Removal of valuable solutes and water from the filtrate in the tubules of the kidney before they can leave the kidney in the urine.

reaction center A formation within an antenna complex where electrons become excited and are passed to an electron transport chain.

receptor A protein that facilitates the transmission of a signal after binding to a specific signaling molecule. Receptors may be found inside the cell or embedded in the plasma membrane.

receptor-mediated endocytosis A form of endocytosis in which receptor proteins embedded in the plasma membrane of a cell recognize certain surface characteristics of materials to be brought into the cell by endocytosis.

recessive Of or referring to an allele that does not have a phenotypic effect when paired with a dominant allele. Compare *dominant*.

recombination A collective term for the processes of fertilization, crossing-over, and independent assortment of chromosomes, all of which result in new combinations of alleles.

red blood cell A type of cell that circulates freely with the blood and that contains an oxygen-binding pigment, hemoglobin, that greatly increases the amount of oxygen that the blood can carry.

redox reaction A chemical reaction in which electrons are transferred from one molecule or atom to another.

reduction The gain of electrons by one atom or molecule from another. Compare *oxidation*.

reflex arc A neuronal connection in the spinal cord between a sensory cell and a muscle cell that allows a simple stimulus to be translated rapidly into a movement.

regulatory DNA A DNA sequence that can turn the expression of a particular gene or group of genes on or off. Regulatory DNA sequences interact with regulatory proteins to control gene expression.

regulatory protein A protein that signals whether or not a particular gene or group of genes should be expressed. Regulatory proteins interact with regulatory DNA to control gene expression.

releaser A simple stimulus that triggers a fixed behavior.

replicate An independent run or performance of an experiment.

repressor protein A protein that prevents the expression of a particular gene or group of genes.

reproductive cloning A technology used to produce an offspring that is an exact genetic copy (a "clone") of another individual. The first two steps in reproductive cloning are the same as those in therapeutic cloning, but stem cells are not removed from the embryo. Instead, the embryo is transferred to the uterus of a surro-

gate mother, where, if all goes well, the birth of a healthy offspring ultimately results; this offspring is genetically identical to the individual who provided the donor nucleus. Compare *therapeutic cloning*.

reproductive isolation A condition in which barriers to reproduction prevent or strongly limit two or more populations from reproducing with one another. Many different kinds of reproductive barriers can result in reproductive isolation, but it always has the same effect: no or few genes are exchanged between the reproductively isolated populations.

respiration See *aerobic respiration*.

resting potential The electrical state of an axon when it is not transmitting an action potential.

restriction enzyme Any of a number of enzymes that cut DNA molecules at a specific target sequence; a key tool of DNA technology.

retina A field of photoreceptors in the eye that allows organisms to form an image.

reversible binding The capacity of a molecule to bind and to release another molecule under different conditions.

RFLP analysis A method of DNA technology in which restriction enzymes are used to cut an organism's genome into small pieces, which are sorted by size using gel electrophoresis. Next, a DNA probe is used to form a profile, whose pattern depends on the number and size of the fragments that can bind to the probe.

ribonucleic acid See *RNA*.

ribosomal RNA (rRNA) A type of RNA that is an important component of ribosomes.

ribosome A particle composed of proteins and RNA at which new proteins are synthesized. Ribosomes can be either attached to the endoplasmic reticulum or free in the cytosol.

ring species A species whose populations loop around a geographic barrier (such as a mountain chain) and in which the populations at the two ends of the loop are in contact with one another, yet cannot interbreed.

river An aquatic biome, the communities of which live in relatively narrow bodies of fresh water that move continuously in a single direction.

RNA Ribonucleic acid; a polymer of nucleotides that is necessary for the synthesis of proteins in living organisms.

RNA polymerase The key enzyme in DNA transcription, which links together the nucleotides of the RNA molecule specified by a gene.

rod A type of photoreceptor cell in the vertebrate eye that detects light and dark and functions well in dim light. Compare *cone*.

root A plant organ that serves to anchor the plant, absorb water and nutrients from the soil, transport food and water, and (often) store food.

root cap A layer of cells at the base of the root that protects the root apical meristem.

root hair An absorptive structure formed by root dermal cells.

root system A collection of fingerlike growths into the soil that absorb water and nutrients; one of the two basic systems of the plant body.

rooting reflex The reflexive search for a nipple by a human infant when its cheek or lips are touched lightly.

rough ER A region of the endoplasmic reticulum that has attached ribosomes. Compare *smooth ER*.

rRNA See *ribosomal RNA*.

rubisco The enzyme that catalyzes the first reaction of carbon fixation in photosynthesis.

S phase The stage of the cell cycle during which the cell's DNA is replicated.

S-shaped curve A pattern of population growth in which the number of individuals in a population at first increases at a rate similar to exponential growth; however, as the number of individuals increases, the growth rate gradually decreases and the population stabilizes at the size that can be supported indefinitely by the environment.

saliva A fluid that contains enzymes that help to digest carbohydrates and which is released into the mouth by the salivary glands.

salivary gland A gland in the mouth that produces saliva.

sapwood Relatively recently formed secondary xylem cells that function to transport water. Compare *heartwood*.

sarcomere A unit of a myofibril consisting of an arrangement of actin and myosin filaments that extends between two Z discs.

saturated Of or referring to a fatty acid that has no double bonds between its carbon atoms. Compare *unsaturated*.

saturated fat A fat containing saturated fatty acids. Saturated fats are harmful to human health if eaten in excess because the body converts them into cholesterol.

science A method of inquiry that provides a rational way to discover truths about the natural world.

scientific method A series of steps in which a scientist develops a hypothesis, tests its predictions by performing experiments, and then changes or discards the hypothesis if its predictions are not supported by the results of the experiments.

scientific name The unique two-part name given to each species that consists of, first, a Latin name designating the genus and, second, a Latin name designating that species.

second law of thermodynamics The law stating that all systems, such as a cell or the universe, tend to become more disordered, and that the creation and maintenance of order in a system requires the transfer of disorder to the environment.

secondary consumer An organism that eats a primary consumer. Compare *primary consumer*.

secondary growth An increase in plant thickness caused by the division of lateral meristem cells.

secondary immune response The rapid defensive response to a second exposure to a pathogen that is mediated by memory B cells produced during the first exposure to that pathogen. Compare *primary immune response*.

secondary oocyte The larger of the two cells resulting from the first meiotic division of the primary oocyte. The secondary oocyte gives rise to the mature ovum through the second meiotic division.

secondary productivity The rate of new biomass production by consumers per unit of area. Compare *net primary productivity*.

secondary spermatocyte The haploid cells that result from the first meiotic division of the primary spermatocyte, and which give rise to mature sperm through the second meiotic division.

secondary structure The folding of regions in a protein into spirals or sheets.

secondary succession Ecological succession that occurs as communities recover from disturbance, as when a forest grows back when a field ceases to be used for agriculture. Compare *primary succession*.

secondhand smoke Tobacco smoke inhaled by people near the smoker.

secretion Active transport of materials out of an organ or an organism.

sedimentary cycle A type of nutrient cycle in which the nutrient does not enter the atmosphere easily. Compare *atmospheric cycle*.

seed A structure produced by a plant in which a plant embryo is encased in a protective covering.

seed dispersal mutualism A mutualism in which an animal eats a fruit provided by a plant, then later deposits the seeds contained within the fruit far from the parent plant.

selectively permeable Of or referring to a structure, such as the plasma membrane of a cell, that controls which materials can pass through it.

semicircular canal One of the three fluid-filled tubes in the vestibular apparatus of the ear that are oriented at right angles to one another, forming a three-dimensional grid. Movement of the fluid within the tubes informs the animal about head movement and orientation.

sensory input An action potential that conveys information about stimuli.

sensory neuron A neuron that carries sensory information to the central nervous system. Compare *interneuron, motor neuron*.

sepal A part of a flower that encloses and protects the flower before it opens. Sepals are the outermost of the four whorls of modified leaves that make up a flower. Compare *petal, stamen, carpel*.

sex chromosome Either of a pair of chromosomes that determines the sex of an individual. Compare *autosome*.

sex hormone One of a class of steroid hormones that regulate the development of sex and sexual behavior in animals.

sex-linked Of or referring to genes located on a sex chromosome. Genes located on the X chromosome are called *X-linked*; genes located on the Y chromosome are called *Y-linked*.

sexual reproduction Reproduction in which genes from two individuals are recombined. Compare *asexual reproduction*.

sexual selection A type of natural selection in which individuals that differ in inherited characteristics differ, as a result of those characteristics, in their ability to get mates.

shared derived feature A feature unique to a common ancestor that is passed down to all of its descendants, clearly defining them as a group.

shoot system One of the two basic systems of the plant body, adapted for life in air. Compare *root system*.

signaling molecule A molecule produced and released by one cell that affects the activities of another cell (referred to as a target cell). Signaling molecules enable the cells of a multicellular organism to communicate with one another and coordinate their activities.

simple eye A non-image-forming eye that can distinguish light from dark.

single-lens eye An eye in which a lens focuses light on an image-forming surface called a retina.

single nucleotide polymorphism (SNP) A single-base-pair difference among the genomes of individuals.

sinoatrial (SA) node A signaling center in the heart that regulates the cardiac cycle.

skeleton The structural support tissue of animals.

slow muscle fiber A type of muscle fiber that is specialized for slow, sustained contraction. Compare *fast muscle fiber*.

small intestine The portion of the digestive system where digestion of food and most nutrient absorption takes place.

smooth ER A region of the endoplasmic reticulum that does not have attached ribosomes. Compare *rough ER*.

smooth muscle The specialized muscle found in the walls of the digestive system and blood vessels, so called because it lacks the bands associated with other muscle types when viewed under a microscope.

social behavior Behavior that involves cooperation among members of a group of animals, usually of the same species.

soluble Of or referring to a chemical compound that will dissolve in water.

solute A substance dissolved in water.

solution Any combination of a solute and a solvent.

solvent A liquid (in biological systems, usually water) into which a solute has dissolved.

spacer DNA A region of noncoding DNA that separates two genes. Spacer DNA is common in eukaryotes, but not in prokaryotes.

specialist A species that requires very specific conditions to survive, such as an insect that can eat only one kind of plant, as opposed to being able to eat and survive on many different kinds of plants.

speciation The process by which one species splits to form two or more species that are reproductively isolated from one another.

species A group of interbreeding natural populations that is reproductively isolated from other such groups.

specific response An immune response in which an organism recognizes and responds much more rapidly to a particular parasite or pathogen to which it has had previous exposure. Compare *nonspecific response*.

sperm The small, mobile, haploid gamete that is produced by sexually reproducing male eukaryotes. Compare *egg*.

spermatogenesis The production of sperm in animals.

spinal cord A dense collection of neuronal cell bodies and axons that carries information between the brain and the tail end of a vertebrate. The spinal cord of vertebrates forms an intermediate site of information processing that acts as a filter between the brain and the muscles and sensory neurons.

spinal reflex See *reflex arc*.

spiracle An opening to the system of tracheoles that supplies oxygen to insect tissues.

spongy bone The porous bone that lies within the compact bone of vertebrates. Compare *compact bone*.

spore (1) The reproductive cell of a fungus, which is typically encased in a protective coating that shields it from drying or rotting. (2) A structure produced by plants from which a haploid stage develops.

stabilizing selection A type of natural selection in which individuals with intermediate values of an inherited characteristic have an advantage over other individuals in the population, as when medium-sized individuals produce offspring at a higher rate than small or large individuals. Compare *directional selection* and *disruptive selection*.

stamen The male reproductive part of a flower, consisting of an anther on top of a slender filament. Stamens are the third (from the outside) of the four whorls of modified leaves that make up a flower. Compare *sepal, petal, carpel*.

start codon A three-nucleotide sequence on an mRNA molecule (usually the codon AUG) that signals where translation should begin.

stem A plant organ that provides the plant with structural support and also serves to transport food and water.

stem cell An undifferentiated cell that can divide to produce one or more specialized cell types.

steroid A class of lipids formed from cholesterol.

steroid hormone Any of a class of hydrophobic signaling molecules that can pass through the plasma membrane of a target cell.

stimulus Anything that can cause a sensory cell to generate a signal.

stomach The portion of the digestive system that is specialized for mixing and storing food.

stop codon A three-nucleotide sequence on an mRNA molecule that signals where translation should end.

stroma The space enclosed by the inner membrane of the chloroplast, in which the thylakoid membranes are situated.

subsidence Sinking of the land surface when groundwater levels have dropped and hence there is less water underground to support the land above.

substitution A mutation in which one nitrogen base is replaced by another at a single position in the DNA sequence of a gene. Compare *deletion* and *insertion*.

substrate The specific molecule on which an enzyme acts. Only the substrate will bind to the active site of the enzyme.

succession A process by which species in a community are replaced over time. For a given location, the order in which species will be replaced over time is fairly predictable.

sugar An organic compound that has the general chemical formula $(CH_2O)n$.

sustainable Of or referring to an action or process that can continue indefinitely without using up resources or causing serious damage to ecosystems.

symbiosis A relationship in which two or more organisms of different species live together in close association.

sympatric speciation The formation of new species from populations that are not geographically isolated from one another. Compare *allopatric speciation*.

synapse The narrow gap that separates a neuron from its target cell.

synaptic cleft The minute gap between a neuron and its target cell across which neurotransmitters, rather than action potentials, carry the signal.

synovial sac A fluid-filled sac that forms a cushion in vertebrate joints.

synthetic chemical A chemical made only by humans.

systematist A scientist who studies evolutionary relationships among organisms and builds evolutionary trees.

systemic circuit The loop of the circulatory system through which blood travels from the heart to respiring tissues and back to the heart.

systole The pumping phase of the cardiac cycle. The systolic pressure corresponds to the first and higher of the two values reported in a blood pressure reading. Compare *diastole*.

T cell Any of a specialized class of lymphocytes that form part of the vertebrate immune system. See *helper T cell, killer T cell*.

taproot system A root system in which a dominant taproot grows directly downward, giving rise to branch roots that extend to the sides before also descending downward. Compare *fibrous root system*.

tar A sticky brown material that condenses as tobacco smoke cools.

target cell A cell that receives and responds to a signaling molecule.

taste bud A contact chemoreceptor located on the vertebrate tongue.

taxon A group defined within the Linnaean hierarchy; for example, a species or a kingdom.

telomere A bit of DNA that caps the end of, and apparently protects, a chromosome.

telophase The stage of mitosis during which chromosomes arrive at the opposite poles of the cell and new nuclear envelopes begin to form around each set of chromosomes.

temperate forest A terrestrial biome, the forest communities of which are dominated by trees and shrubs that grow in regions with cold winters and moist, warm summers.

template strand Of the two strands in a DNA molecule, the strand that contains the promoter and hence serves as the template from which an mRNA molecule is synthesized. Compare *coding strand*.

tendon A collagen-rich connective structure that attaches muscle to bone in vertebrate skeletons. Compare *ligament*.

terminator A DNA sequence that, when reached by RNA polymerase, causes transcription to end and the newly formed mRNA molecule to separate from its DNA template.

tertiary structure The overall folding of a protein into a three-dimensional form.

testis (pl. testes) The reproductive structure in male animals that produces sperm.

testosterone One of a group of steroid hormones, called androgens, that maintain male sexual and behavioral characteristics.

thalamus (pl. thalami) The portion of the vertebrate brain that relays sensory information to the cerebral cortex.

therapeutic cloning A technology used to produce stem cells in which the (haploid) nucleus of an unfertilized egg cell is replaced with the (diploid) nucleus of a nonreproductive donor cell, such as a skin cell. Next, chemicals are used to stimulate the egg to divide so that it begins to form an embryo. Finally, stem cells are removed from the developing embryo and stimulated to grow into a wide range of human cell types. Compare *reproductive cloning*.

thermoreceptor A sensory cell that initiates an action potential in response to heat.

thrust The force generated by an organism that propels it forward.

thylakoid One of a series of flattened, interconnected membrane sacs that lie one on top of another within a chloroplast in stacks called grana.

thylakoid membrane The membrane that encloses the thylakoid space inside a chloroplast. The thylakoid membrane houses both photosystems and their associated electron transport chains.

thylakoid space The space enclosed by the thylakoid membrane inside a chloroplast; the innermost compartment of the chloroplast.

thyroid gland An endocrine gland in the neck that releases calcitonin.

thyroid-stimulating hormone (TSH) A hormone released by the pituitary gland in response to thyrotropin-releasing hormone. Thyroid-stimulating hormone acts on the thyroid gland to trigger the release of thyroid hormones.

thyrotropin-releasing hormone (TRH) A hormone released by the hypothalamus that acts on the pituitary gland to trigger the release of thyroid-stimulating hormone.

tight junction A structure made up of strands of protein arranged in a belt beneath the plasma membrane of each of two animal cells that holds those cells together and prevents the passage of ions and small molecules between them.

tissue A collection of coordinated and specialized cells that together fulfill a particular function for the organism.

toxin A poisonous protein.

trachea (pl. tracheae) In vertebrates, the wide, rigid tube, also called the windpipe, that connects the pharynx with the bronchi. In insects, one of the tubes that carries air between the environment and the tracheoles.

tracheole In insects, a tube that carries air between the tracheae and respiring cells.

trait A feature of an organism, such as its height, flower color, or the chemical structure of one of its proteins.

transcription Synthesis of an RNA molecule from a DNA template. Transcription is the first of the two major steps in the process by which genes specify proteins; it produces mRNA, tRNA, and rRNA molecules, all of which are essential in the production of proteins. Compare *translation*.

transfer RNA (tRNA) A type of RNA that transfers the amino acid specified by mRNA to the ribosome during protein synthesis.

transformation A change in the genotype of a cell as a result of the incorporation of external DNA by that cell.

translation The conversion of a sequence of nitrogen bases in an mRNA molecule to a sequence of amino acids in a protein. Translation occurs at the ribosomes and is the second of the two major steps in the process by which genes specify proteins. Compare *transcription*.

translocation A mutation in which a segment of a chromosome breaks off and is then attached to a different, nonhomologous chromosome.

transposon A DNA sequence that can move from one position on a chromosome to another, or from one chromosome to another; known informally as a "jumping gene."

trichome An outgrowth from a dermal cell in a leaf, stem, or root. Such outgrowths serve a variety of functions, including the uptake of nutrients, defense against herbivores, and the secretion and storage of salt (in plants that can tolerate high-salt environments).

trimester One of the three stages of human pregnancy, each lasting about 3 months.

trisomy In diploid organisms, the condition of having three copies of a chromosome (instead of the usual two).

tRNA See *transfer RNA*.

trophic level A level or step in a food chain. Trophic levels begin with producers and end with predators that eat other organisms but are not fed on by other predators.

trophoblast The outer layer of the mammalian blastocyst, which will become the developing embryo's portion of the placenta.

tropical forest A terrestrial biome, the forest communities of which are dominated by a rich diversity of trees, vines, and shrubs that grow in warm, rainy regions.

tubule The portion of the nephron in the vertebrate kidney where valuable water and solutes are reabsorbed before the filtrate entering the nephron is released as urine.

tubulin The protein monomer that makes up microtubules.

tumor suppressor A gene that inhibits cell division under normal conditions.

tundra A terrestrial biome, the communities of which are dominated by low-growing shrubs and nonwoody plants that can tolerate extreme cold.

umbilical cord The structure that connects a developing mammalian fetus to the placenta, and which contains the veins and arteries that convey materials to and from the placenta.

unsaturated Of or referring to a fatty acid that has one or more double bonds between its carbon atoms. Compare *saturated*.

upper respiratory system A collective term for the nose, mouth, and pharynx. Compare *lower respiratory system*.

urine The filtrate from the kidneys that animals expel from their bodies.

uterus (pl. uteri) The organ in female mammals in which the zygote implants and develops.

vaccine A preparation of killed or weakened pathogens that is used to stimulate the vertebrate immune system as a protection against future attack by that pathogen.

vacuole A large water-filled vesicle found in plant cells. Vacuoles help maintain the shape of plant cells and can also be used to store food molecules.

vagina The reproductive structure in female animals into which the penis of a male deposits sperm.

vascular cambium A type of lateral meristem tissue that contributes to secondary growth by dividing to form new vascular tissues. Compare *cork cambium*.

vascular system The tissue system in plants that is devoted to internal transport.

vector In DNA technology, a piece of DNA that is used to transfer a gene or other DNA fragment from one organism to another.

vein A blood vessel that carries blood to the heart. Compare *artery*.

ventricle A chamber in a heart that receives blood from the atrium and contracts to propel blood away from the heart. Compare *atrium*.

vertebrates A group of animals with backbones. Vertebrates include fish, amphibians, mammals, birds, and reptiles.

vesicle A small, membrane-enclosed sac found in the cytosol of eukaryotic cells.

vestibular apparatus A collection of mechanoreceptors and associated structures in the ear that plays a central role in maintaining the sense of balance in vertebrates.

vestibule The portion of the vestibular apparatus that contains the otoliths and mechanoreceptors that make it possible to detect gravitational pull.

vestigial organ A structure or body part that served a purpose in an ancestral species, but is currently of little or no use to the organism that has it.

villus (pl. villi) One of numerous tiny projections that increase the absorptive surface of the small intestine.

virus An infectious particle consisting of nucleic acids and proteins. A virus cannot reproduce on its own, and must instead use the cellular machinery of its host to reproduce.

vitamin An organic compound required by consumers in small amounts.

water transformation Changes made by humans to the waters of Earth that alter their physical or biological characteristics. Compare *land transformation*.

weather Temperature, precipitation, wind speed, humidity, cloud cover, and other physical conditions of the lower atmosphere at a specific place over a short period of time. Compare *climate*.

wetland An aquatic biome, the communities of which live in shallow waters that flow slowly over lands that border rivers, lakes, or ocean waters.

white blood cell Any of several types of cells that are part of an animal's nonspecific and specific internal defenses against pathogens.

wood A series of layers of secondary xylem. Each new layer forms a ring around the previous year's layer, producing the "tree rings" that can be used to age a tree.

X-linked See *sex-linked*.

xylem A tissue composed of dead cells through which a plant transports mineral nutrients and water from the soil to the leaves. Compare *phloem*.

Y-linked See *sex-linked*.

yield The amount of a crop produced per unit of farmland.

yolk sac The structure in most developing animals that becomes the yolk of the egg, but that in the developing mammal gives rise to blood cells and the cells that produce gametes.

Z disc The disc that separates the sarcomeres in a myofibril.

zygote The diploid cell formed by the fusion of two haploid gametes; a fertilized egg.

Credits

Richardson/Animals Animals. *30.3c:* Thomas D. McAvoy/Time & Life Pictures/Getty Images. *30.6:* Chris Hellier/Corbis. *30.8a:* © Christopher Crowley/Tom Stack & Associates. *30.8b:* © Stan Olinsky/Dembinsky Photo Associates. *30.8c:* Ed Kashi/Corbis. *Biology on the Job:* © 1998 Monarchy Ent. B. V. & Regency Ent. USA/Photofest. *Biology Matters:* John Brecher/Corbis. *30.9a:* © Thomas Kitchin/Tom Stack & Associates. *30.9b:* Kevin Schafer/ Corbis. *30.9c:* © Glen Woolfenden. *30.10:* © Gunter Ziesler/Peter Arnold, Inc.

Interlude E *E.1a:* © David Sieren/Visuals Unlimited. *E.1b:* Mary Evans Picture Library. *E.2:* James Leynse/Corbis. *E.3:* Pete Saloutos/Corbis. *E.4:* James Cavallini/Photo Researchers, Inc. *E.5:* Du Cane Medical Imaging Ltd./Photo Researchers, Inc. *E.6:* © Peter Lansdorp/Visuals Unlimited. *E.9:* Courtesy Brookhaven National Lab Center for Imaging and Neurosciences. *E.10:* Francoise Sauze/Photo Researchers, Inc.

Chapter 31 *Plant:* Stephen Krasemann/Photo Researchers, Inc. *Cow:* © Reuters/Corbis. *31.1a:* Gary W. Carter/Corbis. *31.1b:* Malcolm Richards/SPL/Photo Researchers, Inc. *31.1c:* Jack Coulthard/SPL/Photo Researchers, Inc. *31.1d:* Ralph Lee Hopkins/Photo Researchers, Inc. *31.3a:* © Jerome Wexler/Visuals Unlimited. *31.3b:* Andrew Syred/Photo Researchers, Inc. *31.4:* SPL/Photo Researchers, Inc. *31.5:* Andrew Syred/Photo Researchers, Inc. *31.6:* M. I. Walker/Photo Researchers, Inc. *31.8 stem:* Ed Reschke. *31.8 hawthorn:* © David Sieren/Visuals Unlimited. *31.8 grape:* Charles D. Winters/Photo Researchers, Inc. *31.9 cactus:* Royalty-Free/Corbis. *31.9 flower pot plant:* © Barry Rice/Visuals Unlimited. *Biology on the Job:* Richard T. Nowitz/Corbis. *31.13a:* W. Perry Conway/Corbis. *31.13b:* Biophoto Associates/Photo Researchers, Inc. *31.13c:* © Brian Rogers/ Visuals Unlimited.

Chapter 32 *Aspens:* © Inga Spence/Visuals Unlimited. *32.4b:* Layne Kennedy/Corbis. *32.7 honeybee:* Peter Arnold. *32.7 hummingbird:* Anthony Marceica/National Audubon Society Collection/Photo Researchers, Inc. *32.8a:* © John Alcock/Visuals Unlimited. *32.9:* Sheila Terry/SPL/Photo Researchers, Inc. *32.10a:* Lynwood M. Chace/Photo Researchers, Inc. *32.10b:* Martin Harvey/Peter Arnold, Inc. *32.10c:* Michael P. Gadomski/ Photo Researchers, Inc. *32.14 woman:* © Parrot Pascal/Corbis Sygma. *32.14 tubeworm:* Ian R. MacDonald, Texas A&M University-Corpus Christi. *32.14 pine:* Tony Craddock/SPL/Photo Researchers, Inc. *32.14 creosote:* Jerry L. Ferrara/Photo Researchers, Inc. *32.14 holly:* A. M. Gray, Tasmania.

Interlude F *F.1 supermarket:* © Mark E. Gibson/Corbis. *F.1 boy:* Reuters/ Corbis. *F.2a:* Royalty-Free/Corbis. *F.2b:* Melanie Acevedo/PictureArts/Corbis. *F.2c:* George H. H. Huey/Corbis. *F.2d:* Sheila Terry/SPL/Photo Researchers, Inc. *F.2e:* Wolfgang Kaehler/Corbis. *F.2f:* TH FOTO-WERBUNG/SPL/ Photo Researchers, Inc. *F.3:* Paul A. Souders/Corbis. *F.4:* Joseph Sohm; ChromoSohm, Inc./Corbis. *F.6:* Ted Streshinsky/Corbis. *F.7:* B. Bird/zefa/ Corbis. *F.8a:* Keren Su/Corbis. *F.8b:* B. Peebles/Stock Photos/zefa/Corbis. *F.8c:* Duncan Shaw/SPL/Photo Researchers, Inc. *F.9:* Marcello Bertinetti/ Photo Researchers, Inc. *F.10:* Carol Cohen/Corbis.

Chapter 33 *Power station:* Peter Yates/Photo Researchers, Inc. *Zebra mussel:* Ted Kinsman/Photo Researchers, Inc. *Biology Matters:* James Gathany, Public Health Image Library, Center for Disease Control. *33.1a,b:* NASA. *Dingo:* © Martin Harvey/Corbis. *33.2:* © Frans Lanting/Corbis. *33.8a:* © Gerry Ellis/ Minden Pictures. *33.8b:* © Michael P. Gadomsky/Earth Scenes. *33.8c:* Willard Clay/Dembinsky Photo Associates. *33.8d:* © Ken Lucas/Visuals Unlimited. *33.8e:* Willard Clay/Dembinsky Photo Associates. *33.8f:* © Frances Lepine/Earth Scenes. *33.8g:* Scott T. Smith/Dembinsky Photo Associates. *33.9 grassland:* Mark De Fraeye/SPL/Photo Researchers, Inc. *33.9 prairie:* © Tom Bean/Corbis. *33.9 savanna:* © Paul A. Souders/Corbis. *33.11a:* Terry Donnelly/Tom Stack & Associates. *33.11b:* Carr Clifton/Minden Pictures. *33.11c:* Scott T. Smith/ Dembinsky Photo Associates. *33.11d:* Kevin Magee/Tom Stack & Associates. *33.11e:* Jim Zipp/National Audubon Society Collection/Photo Researchers, Inc. *33.11f:* Mark J. Thomas/Dembinsky Photo Associates. *33.11g:* Chet Tussey/ National Audubon Society Collection/Photo Researchers, Inc. *33.11h:* Norbert Wu/Peter Arnold, Inc. *33.11h inset:* Gregory Ochocki/National Audubon Society Collection/Photo Researchers, Inc. *Science Toolkit:* Dembinsky Photo Associates.

Chapter 34 *Easter Island:* © James L. Amos/Peter Arnold, Inc. *34.1 top:* © Jeremy Burgess/National Audubon Society Collection/Photo Researchers, Inc. *34.1 bottom:* © Volker Staeger/SPL/Science Source/Photo Researchers, Inc. *34.2:* © Frans Lanting/Minden Pictures. *34.4a,b:* Reproduced with permission of the Department of Natural Resources, Queensland, Australia. *34.5:* Scott Camazine/National Audubon Society Collection/Photo Researchers. *34.6:* Laguna Design/Photo Researchers, Inc. *34.7:* Ecoscene/ Corbis. *34.8:* Mark Newman/Tom Stack & Associates. *34.9:* G. A. Matthews/ SPL/Photo Researchers, Inc. *34.10:* Alan G. Nelson/Dembinsky Photo Associates. *Science Toolkit:* © Frans Lanting/Minden Pictures. *34.12:* Craig Mayhew and Robert Simmon, NASA GSFC. *34.13:* © G. Schuster/zefa/Corbis.

Chapter 35 *Ant:* © Gregory G. Dimijian/National Audubon Society Collection/Photo Researchers, Inc. *Corn:* © Inga Spence/Tom Stack & Associates. *Person:* © Science VU/Visuals Unlimited. *35.2 yucca plant:* Willard Clay/Dembinsky Photo Associates. *35.2 yucca moth:* © Ken Wagner/Visuals Unlimited. *35.3:* Courtesy of Mycorrhizal Applications, Inc. www.mycorrhizae.com. *35.4a:* © F. Bravendam/Peter Arnold, Inc. *35.4b:* © Jeffrey L. Rotman/Corbis. *35.6a:* Mike Bacon/Tom Stack & Associates. *35.6b:* © Dr. Lincoln Brower. *35.7:* © Fred Bruemmer/Peter Arnold, Inc. *35.9:* Courtesy of Dr. Charles Mitchell, University of North Carolina at Chapel Hill. *Science Toolkit:* © Mauro Fermariello/Photo Researchers, Inc. *35.11:* J. K. Clark/University of California, Davis. *35.13:* Forest and Kim Starr/USGS. *35.14a,b:* Reprinted by permission from *Nature,* vol. 406, pp. 255–256. © 2000 MacMillan Magazines, Ltd.

Chapter 36 *Waterfall:* © Greg Vaughn/Tom Stack & Associates. *36.4:* Courtesy Dr. Robert T. Paine. *36.6a:* Stan Osolinski/Dembinsky Photo Associates. *36.6b,c:* Howard Garrett/Dembinsky Photo Associates. *36.8:* Charles E. Rotkin/ Corbis. *36.9 anchovies:* © Jeffrey L. Rotman/Corbis. *36.9 jelly:* © David Wrobel/ Visuals Unlimited. *36.10:* Courtesy Dr. Johann Goldammer. *36.11:* Courtesy Robert Gibbens, Jornada Experimental Range, USDA. *36.12:* © Corbis. *Biology on the Job:* Courtesy of Carlyn Worstell. *36.13:* © Dr. Gerald D. Carr.

Chapter 37 *Ashokan Reservoir:* © Joseph Sohm; ChromoSohm Inc./Corbis. *37.2a:* © Martin Harvey/Corbis. *37.2b:* © Simon Fraser/Photo Researchers, Inc. *37.4a:* Michael Patrick O'Neill/Photo Researchers, Inc. *37.4b:* Kevin Magee/Tom Stack & Associates. *Biology on the Job:* © Salem Krieger/ZUMA/ Corbis. *37.11a:* © Dr. John D. Cunningham/Visuals Unlimited. *37.12:* © Richard Packwood/Oxford Scientific Films. *37.13:* © Adam Hart-Davis/Photo Researchers, Inc. *37.13 inset:* © Mauro Fermariello/Photo Researchers, Inc. *37.14a:* © Bob Galbraith/AP Photo. *37.14b:* © Steven Holt.

Chapter 38 *California coast:* © Carr Clifton/Minden Pictures. *White abalone:* © Ron McPeak. *38.1a:* © D. Dancer/Peter Arnold, Inc. *38.1b:* Doug Sokell/Tom Stack & Associates. *38.1c:* U.S. Geological Survey. *38.3:* Courtesy of NASA Landsat Project Science Office and USGS National Center for EROS. *Science Toolkit:* Ralph Lee Hopkins/Photo Researchers, Inc. *38.5a,b:* Courtesy of David Tilman, University of Minnesota. *38.8:* Courtesy of Joy Ward and Anne Hartley. *38.11:* © Jeffrey L. Rotman/Corbis. *Biology Matters:* © Brownie Harris/Corbis.

Interlude G *G.1:* Courtesy Rudolf B. Husar, Washington University at St. Louis. *G.2:* © Royalty-Free/Corbis. *G.4a:* Raymond Gehman/Corbis. *G.4b:* U.S. Geological Survey. *G.5:* © Gerry Ellis/Minden Pictures. *G.6:* Jeffrey Lepore/Photo Researchers, Inc. *G.7 Chicago:* Courtesy of Mark Farina/City of Chicago. *G.7 Tokyo:* Reuters/Corbis. *G.7 Germany:* Copyright ZinCo; further information available at www.ZinCo.de. *G.8:* Courtesy of Eric Shriner. *G.9 top:* Radu Sigheti/Reuters/Corbis. *G.9 bottom:* William Campbell/Corbis. *G.10:* Courtesy of NASA Landsat Project Science Office and USGS National Center for EROS. *G.11:* NASA. *G.12:* Owaki-Kulla/Corbis. *G.14:* Courtesy of DuPont.

"Biology in the News" Credits

Chapter 1 Reprinted with permission from the Akron Beacon Journal/Knight Ridder Digital.

Chapter 2 USA TODAY, June 6, 2001. Reprinted with permission.

Chapter 3 From "News in Science" first broadcast Monday, 31 March 2003, on ABC Radio National and reproduced by permission of the Australian Broadcasting Corporation and ABC Online. © 2003 ABC. All rights reserved.

Chapter 4 newstarget.com. We have made diligent efforts to contact the copyright holder to obtain permission to reprint this selection. If you have information that would help us, please write to W. W. Norton & Company, Inc., 500 Fifth Avenue, New York, NY 10110.

Chapter 5 Annals of Internal Medicine, reprinted by permission.

Chapter 6 Reprinted by permission of Dan Shaw.

Chapter 7 Reprinted by permission of the American Medical Association.

Chapter 8 Wired News.

Chapter 9 The Muse. We have made diligent efforts to contact the copyright holder to obtain permission to reprint this selection. If you have information that would help us, please write to W. W. Norton & Company, Inc., 500 Fifth Avenue, New York, NY 10110.

Chapter 10 © 2004, Los Angeles Times. Reprinted with permission.

Chapter 11 Used with permission of The Associated Press. Copyright © 2005. All rights reserved.

Chapter 12 From BBC News at bbcnews.com, © bbc.co.uk.

Chapter 13 © 2004, Los Angeles Times. Reprinted with permission.

Chapter 14 Reprinted by permission from Macmillan Publishers Ltd., *Nature* © 2004.

Chapter 15 © The Sacramento Bee, 2005.

Chapter 16 © The New York Times. Reprinted by permission.

Chapter 17 © 2003, The Washington Post. Reprinted with permission.

Chapter 18 © 2004, Los Angeles Times. Reprinted with permission.

Chapter 19 © The New York Times. Reprinted by permission.

Chapter 20 Used with permission of The Associated Press. Copyright © 2005. All rights reserved.

Chapter 21 © 2004, Los Angeles Times. Reprinted with permission.

Chapters 22–25 From BBC News at bbcnews.com, © bbc.co.uk.

Chapter 26 MacWorld UK.

Chapter 27 The Guardian.

Chapter 28 From cbc.ca.

Chapter 29 From BBC News at bbcnews.com, © bbc.co.uk.

Chapter 30 From cbc.ca.

Chapters 31–32 © Reuters.

Chapters 33–34 Used with permission of The Associated Press. Copyright © 2005. All rights reserved.

Chapters 35–37 © The New York Times. Reprinted by permission.

Chapter 38 © 2004, Los Angeles Times. Reprinted with permission.

Index

Page numbers in *italics* refer to illustrations. Page numbers preceded by a letter refer to pages from the Interludes.

A

ABB, G11
ABO blood typing system, 190, 197, *197,* 202
abortion, 21
   prenatal screening and, 217
absorption, of nutrients, 411, 414
Acanthamoeba polyphaga, 66
acetylcholine, *480,* 481
acetyl CoA, 163
acid rain, 129, 132, 169, 658, 724–25, *726,* 731, 738, G1, G3
Acid Rain Program, G10
acids, 77, 92
acromegaly, 466, *467,* 471–72
acrosome, 564, 582
actin, 109, 111, 113, 576
actin filiment, 526, 536
action potentials, 474, 478–84, *479,* 496, 499, 501, 502, 514, 536
   features of, 480, 492
activation energy, 141–42, *141,* 148
activation genes, 575–77
active carrier proteins, 120–21, *120,* 130
active immunity, 552, 556
active transport, 119, *119,* 120–21, 130, 390
Acyrthosiphon pisum (pea aphid), C8, *C8*
adaptations, 18, 309, 314, 340–44
   barriers to, 343–44, *344*
   behavioral, 340
   characteristics of, 340–41, *341*
   defined, 13, 298, 303, 338, 340
   developmental limitations and, 343–44, *344,* 352
   ecological trade-offs and, 344
   environment and, 340–43, *341, 343*
   evolution and, 298, 303, 314
   genes and, 343–44, *344*
   kinds of, 340
   lack of genetic variation and, 343, 352
   in populations, 341–43

adaptive evolution, 338, 340, 352
adaptive radiation, 356
   definition of, 368
   diversity and, 368, 374
   evolution and, 368–70, 374
   of mammals, 368–70
adducts, B10
adenine, 81, *83,* 211, 212, 216, 220, 221, 226, 246, 255, 258
adenosine, 490–91
adenosine deaminase (ADA) deficiency, 292–93, *292*
adenosine triphosphate (ATP), 83–84, *84,* 92, 101, 107, 110, 136, 140–41, *141,* 144, 148, 152, 154, 155, 157–58, 159, 160, 161, 163, 167, 168, 459, 518, 536
   biosynthesis of DNA, 144–45
   in muscle function, 526, 528
   oxidative phosphorylation and, 163–64, *165*
   as universal cellular fuel, 106, 109
adenovirus, B1–B2, *B2,* B12, *542*
Adiantum pedatum (maidenhair fern), *607*
ADP, 161, 163
adrenal glands, 460, *462,* 471, 491, E7, E8
adrenaline, *see* epinephrine
aerobic organisms, 42, 163
aerobic respiration, 152, 163–64, 361
Africa, 317, 654, 655, 673, 674–75
African seed crackers, 329, *330*
afterbirth, 574
agriculture, 673–74, 736, 737
   control of germination and, 643
   crop improvement and diversity in, F2–F3, F6, F7–F8, F10
   environment and, F2, F6, F11–F12
   genetic engineering and, F7, F8, F9
   Green Revolution and, F6, *F6*
   hydroponics and, 617
   increasing productivity in, F5–F6
   managed systems of, F4–F5
   new methods of, F10–F11
   organic farming and, F11
   plant breeding and, F8–F10
   precision farming and, F11
Agriculture Department, U.S. (USDA), 12, A3, 185, 291, 468, 723

Agriculture in the Global Economy: Hunger 2003, F7
AIDS (acquired immunodeficiency syndrome), 28, *47,* 56, 210, 322, 350, D11, 541–42, *541,* 554–55
   vaccine for, 289
air pollution, 436
Alba (rabbit), 279, *279*
Alcoa, G11–G12
alcohol, 161, 168, 512
   behavior and, 598, *598*
   fetal development and, 561–62, *561,* 572, 581
alcoholism, 209, C12
Alexandra, Tsarina of Russia, 187, 200
algae, 16, 64, 169, 361, 362, 623, 655, 699, 701, 704
   corals and, 684
   eutrophication and, 704, 711, 719–20, *719,* 724
   fungi and, 56
   known species of, A4
alkaptonuria, 246
alleles, 186
   codominance in, 297
   DNA sequences in, 232
   dominant and recessive, 189, 192
   frequency of, 318, 320–22, 325, 326, 331–32
   incomplete dominance in, 295–97, *295*
   independent assortment of, *194*
   mutation of, 190–91, 201, 213, 239–40
allergies, 436
alligators, 577
allopatric speciation, 348–49, *349,* 352
alpha carbon, 84
alpha helixes, 86
Alpheus (shrimp), 60, 683, *683*
Alpine androsace, *A9*
altitude sickness, 421
altruistic behavior, 595–96, 600
AluI enzyme, 281, *281*
alveoli, 425, 431, 434, 441, 545
Alzheimer's disease, C9–C10, 488, 580, F9
amaranths (pigweeds), *F10,* F10–F11
Ama'uma'u fern, *50*
Ambulocetus, 360

American Board of Genetic Counselors, 235
American Board of Medical Genetics, 234
American chestnut tree, 686–87
American Diabetes Association, 395
American Dietetics Association, F9
American Heart Association, 115
American Humane Society, 602
American Medical Association (AMA), 302
American Museum of Natural History (AMNH), 45
American Society for Microbiology, 302
amino acids, 8, 20, 78, 84–85, 91, 92, 143, 248, 321, 406, 407, 411, 412, 413, 460, 522, 620, F10
   diversity of, *85*
   essential, 408, *409*
   gene mutations and, 244
   in protein synthesis, 244, 250–51, 258
   R group component of, 84–85
   structure of, *84*
Amish, 419
ammonia, 77, 388, 398
ammonium, 648, 691–92, 739
amniocentesis, 217, *217,* 222
amnion, amniotic fluid, 570, 573, 581, 582
amoebas, 46, 66, 545
amphibians, *59,* 61, 64, A1–A4, *A2,* A9, 306, 307, 358, 391, 398, 447
   in fossil record, 363–64, *364*
amplification, by hormones, 459
amylase, 413
amyotrophic lateral sclerosis (Lou Gehrig's disease), 33, *33*
Anableps anableps, 340–41, *341*
anabolic steroid, 90
anaerobes, 42
anaerobic reactions, 161
anaphase, 176–77
Anastasia, Princess, 187, *187,* 200–201
anchoring junctions, 124, 130
anchovies, 705, *705*
Anderson, Anna, 187–88, 200–201
Andes Mountains, 39, 421
androgens, 463, 471
anemia, 431, F5

mutations (continued)
single-gene, 216–18, 222, 239–40
speciation and, 346–48, 352
mutualisms, 53, 680, *685,* 691–92, 693, 700
types of, 682–83, 693
mycelium, 55, *55,* 64
mycorrhizae, 615, 621
mycorrhizal fungi, 56, 64, 684, *684,* 693
myelin, 476, 492, 554
myelinated axon, 478–80, *479,* 492
myelinated neurons, 476
myofibrils, 526, 536
myoglobin, 528, 536
myosin filament, 526, 536
Myrica faya (fire tree), 691–92, *691*
myxomas (heart tumor), 205

N

NADH, 152, 161, 163–64, *165,* 167, 168
NADPH, 152, 155, 157–58
nanobes, 3–4, *3,* 17–18
National Academy of Sciences, 726
National Aeronautics and Space
Administration (NASA), 70, G8
National Cancer Institute, B1
National Center for Health Statistics, 12
National Center for Science Education
(NCSE), 310
National Heart, Lung, and Blood Institute,
451, 473
National Institute for Mental Health, 261
National Institutes of Health (NIH), C3,
C5, C8, 469
National Sickle Cell Anemia Foundation,
256
National Society of Genetic Counselors, 235
natural gas, G2
natural killer (NK) cells, 546, 555
natural selection, 298, 301–3, 305, *305,*
307, 308, 309–11, 314, 318, 320,
321–24, 326–29, 331, 335, 371, 534,
562
as adaptive evolution, 338, 352
behavior and, 586, 590
competition and, 690
defined, 301, 326
evolution and, 301–3, 305, *305,* 307,
308, 309–11, 318, 320, 321–24,
326–29, 331–32, 335
in future human evolution, D8–D9
locomotion and, 531
mammalian leg arrangement and,
529–30, *530*
medium ground finch example of,
312–13, *313*
speciation and, 346–48, 352
types of, 328–29, 335
see also adaptations; evolution
Nazis, D12

Neandertals, D1, *D1,* D6, D7–D8
nectar, 51
negative feedback, 382, 391, 397, 460, 462,
490
nematode worms (*Caenorhabditis elegans*),
145–47, 265, *265,* C3, C5–C7, *523,*
575, 576, *576*
Nemoria arizonaria, 341
Nenet people, 382
Neoceratodus fosteri (lungfish), 307–8, *309*
neocortex, 486–87, *487,* 489, 492–93, 499
Nepal, 421
Nepenthes rafflesiana (pitcher plant), 617,
617
nephron, 389–90, 398, 445
nerves, 474, 476, 492
nervous system, 458, 459, 460, 471,
474–94, 508, 588
basic features of, 476–77, 492
brain and, *see* brain
central (CNS), 481–82, 492, 499, 501, 511
endocrine system and, 474, 489–90,
489, 493
hormones and, 474, 489–90, *489,* 493
lead poisoning and, 495
organization of, *483*
peripheral (PNS), 474, 482, *483,* 492,
499
transport of electrical signals in,
478–81
Netherlands, 558, 725, 739
net primary productivity (NPP), 718–20,
718, 729–30, 737–38, 739, G8
human activity and, 719–21
measuring of, G9
secondary productivity and, 721
neuromuscular junctions, 528
neurons, 474, 476–81, 492, 588
definition of, 476
diffusion process in, 480–81
function of, 476, *477*
main classes of, 482–83
motor, 483–84, 492, 499, 528
synapses and, 480, 492
transport of electrical signals by,
478–81
neurotransmitters, 480–81, 482, 491
functions of, *481*
neutrons, 71, 91
neutrophils, 546, 549–50, 555
Nevada, 727
New England Journal of Medicine, 416
New Guinea, A8, 348, D8
New Mexico, 706
New York, 715–16, *715,* 727–28
New Zealand, A6
Nicholas II, Tsar of Russia, 187, 200
Nicotiana:
N. rustica, E1, *E1*
N. tabacum, E1, *E1*

nicotine, B10, 581, *E7,* 638
addiction to, E7–E8, *E8*
Nile perch, A8, 340, 351, 658
nitrate, 46, 63, 648, 721, 724, *725,* 739, 746
nitrogen, 46, 68, 73, 75, 77, 78, 406, 612,
613, 615, 623, 648, 655, 673–74,
691–92, 719–20, *719,* 721, 723, 724,
738, *741,* G1
nitrogen base, 81
nitrogen cycle, 730, G2–G3
nitrogen fixation, F11, 739, 746
nitrous oxide, 500, 743
Nixon, Richard, B1
n-nitrosamines, E7
NNK, E7
nodes of Ranvier, 478–80, 492
noncoding DNA, 266, 275, 286, C3
noncovalent bonds, 68, 73, 92
nonspecific responses, 544
norepinephrine, 460, 471, 490–91
North America, A6, 364, 646, 675, 707,
725
changing climate of, 703, *703*
northern elephant seal, 326, *326*
Norway, 650, 740
nuclear envelope, 103
nuclear pores, 103
nucleic acids, 68, 78, 81, 92
nucleoid region, 103
nucleotides, 78, 81–84, 92, 143, 148, 246,
406
biosynthesis of, 144–45, *145*
as building blocks of DNA, 226, 231
components of, 81, *83*
functions of, 83–84
nucleus, of atom, 71, 91
nucleus, of cell, 101, *103,* 109, 245, 248
nucleus accumbens, E7, *E8*
nutrient cycle, 714, 716–17, 721–25, *721, 725,*
728, 729, 730
atmospheric, 723–24, 730
exchange pools and, 721
human activity and, 724–25, 731
phosphorus, 724, *724*
sedimentary, 723, 724
sulphur, 723, *723,* 731
nutrients, nutrition, 12, 44, 414, 440
animals' need for, 405–8, *406, 407*
chemical elements and, 402, 406, *406,*
417
energy and, 406–8, *407*
in plants, 604, 612–15, *614, 615,* 621
see also digestive systems

O

oak trees, 620
obesity, 404–5, 646
observation, 2, 5, 18
occipital lobe, 485

ocean currents, 650, *651,* 659
ocean fisheries, F3, F4
octopi, *58,* 511, 522
Odysseus, 263
Ogallala aquifer, G3
oil, G2
oilseed rape, F8, *F8*
Old Faithful, 8, 40
Old Order Amish, 419
olfactory bulbs, 502, 514
omnivores, 406, 417
oncogenes, B5, B7, B9
O'Neil, Steve, 80–81
onions, 608
oogenesis, 564, 582
open ocean biome, 655, *656,* 660
opium poppy, *A13*
Opportunity (Mars rover), 70
Opuntia (prickly pear cactus), 667–68,
668, 682, 687, 689
oral cavity, 409, 417
oral contraception, 466
orangutans, 33, D2
orb-weaving spiders, 412, *413*
orchids, 39, *39, 50,* 51, 628
Orcinus, 360
orders, biological, 22, 29, 35, 36
Oregon, 727
organelles, 48, 55, 64, 96, 101–2, 113, 144,
564
evolution of, 111–12
organic compounds, 78
organic farming, 94, F11
organic foods, 468–69, G6
organisms:
in biological hierarchy, 2, 14–16
communities of, *see* communities
consumer, 16–17, *17,* 19, 699–700, 710,
721
convergent features of, 26
decomposer, 17, *17,* 19, 699, 717, 721, *721,*
723, 730
DNA and evolutionary relationships
of, 26–27
DNA and reproduction of, 9–10
energy capture by, 10, *10, 11,* 12, 18, 19
interactions among, *see* organisms,
interactions among
producer, 16, *17,* 19, 699, 710, 723, 731
scientific names of, 29
sensing of environment by, 10–11, *11,* 18
organisms, interactions among, 680–95
competition in, 680, 689–690, *689,*
690, 693
ecosystems and, 690–692, *691,* 693
exploitation in, 680, 684–89, 693
mutualisms in, 680, 682–84, *685,*
691–92, 693
parasitic, 681–82, 692–93, *692*
symbiosis in, 682, 693